The Periodic Table of Elements (long form). Atomic weights are shown above the symbols; atomic numbers, below.

NONMETALS

METALS

TRANSITION METALS

INNER TRANSITION METALS

PERIODS	IA	IIA	IIIB	IVB	VB	VIB	VIIB	VIIIB	VIIIB	VIIIB	IB	IIB	IIIA	IVA	VA	VIA	VIIA	VIIIA
1	1.0079 H 1																1.0079 H 1	4.00260 He 2
2	6.941 Li 3	9.01218 Be 4											10.81 B 5	12.011 C 6	14.0067 N 7	15.9994 O 8	18.998403 F 9	20.179 Ne 10
3	22.98977 Na 11	24.305 Mg 12											26.98154 Al 13	28.0855 Si 14	30.97376 P 15	32.06 S 16	35.453 Cl 17	39.948 Ar 18
4	39.0983 K 19	40.08 Ca 20	44.9559 Sc 21	47.90 Ti 22	50.9415 V 23	51.996 Cr 24	54.9380 Mn 25	55.847 Fe 26	58.9332 Co 27	58.70 Ni 28	63.546 Cu 29	65.38 Zn 30	69.72 Ga 31	72.59 Ge 32	74.9216 As 33	78.96 Se 34	79.904 Br 35	83.80 Kr 36
5	85.4678 Rb 37	87.62 Sr 38	88.9059 Y 39	91.22 Zr 40	92.9064 Nb 41	95.94 Mo 42	(98) Tc 43	101.07 Ru 44	102.9055 Rh 45	106.4 Pd 46	107.868 Ag 47	112.41 Cd 48	114.82 In 49	118.69 Sn 50	121.75 Sb 51	127.60 Te 52	126.9045 I 53	131.30 Xe 54
6	132.9054 Cs 55	137.33 Ba 56	* [57-71]	178.49 Hf 72	180.9479 Ta 73	183.85 W 74	186.207 Re 75	190.2 Os 76	192.22 Ir 77	195.09 Pt 78	196.9665 Au 79	200.59 Hg 80	204.37 Tl 81	207.2 Pb 82	208.9804 Bi 83	(209) Po 84	(210) At 85	(222) Rn 86
7	(223) Fr 87	226.0254 Ra 88	† [89-103]	(261) Rf 104	(262) Ha 105	(263) — 106	(262) — 107			(266) — 109								

*LANTHANIDE SERIES

138.9055 La 57	140.12 Ce 58	140.9077 Pr 59	144.24 Nd 60	(145) Pm 61	150.4 Sm 62	151.96 Eu 63	157.25 Gd 64	158.9254 Tb 65	162.50 Dy 66	164.9304 Ho 67	167.26 Er 68	168.9342 Tm 69	173.04 Yb 70

†ACTINIDE SERIES

227.0278 Ac 89	232.0381 Th 90	231.0359 Pa 91	238.029 U 92	237.0482 Np 93	(244) Pu 94	(243) Am 95	(247) Cm 96	(247) Bk 97	(251) Cf 98	(252) Es 99	(257) Fm 100	(258) Md 101	(259) No 102

COLLEGE CHEMISTRY

WITH QUALITATIVE ANALYSIS

COLLEGE CHEMISTRY

WITH QUALITATIVE ANALYSIS

SEVENTH EDITION

HENRY F. HOLTZCLAW, JR.
University of Nebraska—Lincoln

WILLIAM R. ROBINSON
Purdue University

WILLIAM H. NEBERGALL

D. C. HEATH AND COMPANY
Lexington, Massachusetts   Toronto

Copyright © 1984 by D. C. Heath and Company.

Previous editions copyright © 1980, 1976, 1972, 1968, 1963, and 1957 by D. C. Heath and Company.

Published simultaneously in Canada.

Printed in the United States of America.

International Standard Book Number: 0-669-06333-9

Library of Congress Catalog Card Number: 83-80991

PREFACE

From its new cover to its new and revised coverage, this textbook differs significantly from its previous editions, having continued to evolve and to improve through this seventh edition. Over two years of work have gone into a careful revision based on the experiences of our own teaching and the suggestions and ideas of our many users, colleagues, and collaborators. During the revision we have always tried to make the text more interesting and readable but we retained the same clear discussion and the same concern for the student that has characterized the previous editions and made them valuable to over one million general chemistry students. We are excited about this new edition of our book. We hope that students and faculty share our excitement.

For several years there has been an emphasis on principles of chemistry (theory) in the general chemistry curriculum. Our book contains a complete coverage of these aspects of chemistry. However, over the past few years, there has also been an increasing awareness of the importance of descriptive inorganic chemistry (reaction chemistry) in the general chemistry curriculum. As might be expected, there is not yet a consensus as to the one best method by which this should be presented; therefore, we provide the instructor with several options for teaching descriptive chemistry. For example, an introduction to the general aspects of chemical reactivity is available in Chapters 8, 9, 12, and 14. These chapters introduce the periodic variation of chemical reactivity, provide examples of this behavior in a selected set of compounds, describe the general chemical behavior of acids and bases, introduce some general aspects of oxidation-reduction, and describe chemistry involving water. A more extensive introduction to chemical reactivity is available in the descriptive chemistry chapters toward the end of the text. Thus in developing an appropriate plan of study, the instructor may choose to cover as much or as little descriptive chemistry as desired.

College Chemistry, Seventh Edition, is actually one of a pair of textbooks we have written. It presents the chemistry of the metals according to their location in the qualitative analysis scheme. The book also contains a qualitative analysis laboratory scheme. Compared to other books, this scheme has proven its dependability, having been continually tested by thousands of students and improved over a period of many years. Our other text, *General Chemistry,* Seventh Edition, is identical in level and coverage. It differs only in that it presents the chemistry of the metals according to the groups of the Periodic Table and it does not contain a qualitative analysis laboratory scheme.

A great deal of help is available to the student and the instructor in this book. We have included new For Review sections containing an end-of-chapter Summary and Key Terms and Concepts. The Summary provides a quick review of the salient features of the chapter. The references in Key Terms and Concepts identify the section where a concept is introduced. As in previous editions, we have included many cross references. A student wishing to review a previously introduced idea will often find a direct reference identified within the sentence. Alternatively, students can use the extensive index at the end of the book. We worked hard to make this index complete and accurate. It has over 6,000 entries; no general chemistry textbook has a better one.

We have provided a variety of new Examples throughout the text, ranging from very simple introductory examples to more involved ones. In several examples we have deliberately selected simple chemistry where the complexity of a chemical system might obscure the principle that an example is intended to illustrate. The examples proceed from simple straightforward problems to more complex ones. The relevance of chemistry to many other disciplines is amply illustrated in the textual Examples and in the end-of-chapter Exercises.

Exercises at the end of each chapter are grouped by topic, for easy location of problems of a particular type, so that students can use the text more effectively. In this edition, we have included many simple drill problems, with no frills to distract the student, to help reinforce the introduction to a topic. We have also included plenty of problems that will challenge the student, including problems that apply chemistry to a wide variety of other fields. Note, for example, the range of stoichiometry problems on pages 52–53 (problems 50–59) and the range of interests addressed on pages 24, 72, 73, 76, 277, 501, and 563.

Some students get frustrated with chemistry problems, chemistry texts, and even chemistry instructors, in part because they often look for a universal approach they can "plug in" to solve problems. Unfortunately there is no single way to solve problems, but we do try to help the student resolve this issue. For example, we first emphasize the steps common to solving almost all stoichiometry problems and then introduce the idea that solving these problems is somewhat like working jigsaw puzzles; one must juggle the arrangement of the pieces so that they fit together. We help the student as much as possible to recognize how the pieces of the puzzle can fit together, and we provide many exercises for practice. Answers to all numerical problems are provided so that the student can tell immediately whether or not the problem has been worked correctly.

Throughout the text we have tried to relate theoretical principles to descriptive (reaction) chemistry. In Chapter 8, for example, we have added a new section describing the chemistry of several significant industrial inorganic compounds. Section 8.7 gives some idea of the enormous scale of the chemical industry. It describes the preparation and uses of seventeen or eighteen important products and illustrates the chemical behavior of these compounds. This section includes covalent and ionic compounds, acids and bases, oxidation-reduction processes, elements, and salts. Students can then use their knowledge of these compounds as a guide to the behavior of other compounds.

A number of supplemental materials have been designed for use with this text. These include:

- *Problems and Solutions for College Chemistry and General Chemistry, Seventh Editions,* by John H. Meiser, F. Keith Ault, Henry F. Holtzclaw, Jr., and William R. Robinson
- *Study Guide for College Chemistry and General Chemistry, Seventh Editions,* by Norman E. Griswold
- *Instructor's Guide for College Chemistry and General Chemistry, Seventh Editions,* by Norman E. Griswold
- *Basic Laboratory Studies in College Chemistry With Semimicro Qualitative Analysis,* by Grace R. Hered
- *Software.* A set of fourteen disks for the Apple® IIe and II+ computer, *Introduction to General Chemistry,* available by special arrangement with COM-Press, Inc. These programs, by Stanley G. Smith, Ruth Chabay, and Elizabeth Kean, feature color graphics and animation in chemical tutorials, drills, laboratory simulations, and games. They have been directly keyed to this text at the "menu" level, thus linking the software to the text in a unified learning program. A special demonstration disk is available through your D. C. Heath representative or by writing to D. C. Heath directly.

No authors can take the entire credit for the success of a text; many others play an important role in its development. We are particularly indebted to those who read all or part of the manuscript and made many thoughtful suggestions. In addition to John Meiser, F. Keith Ault, Norman Griswold, and Grace Hered, they include the reviewers of this new edition and the many present users of this text, who submitted helpful ideas based on their own teaching experience.

ACKNOWLEDGMENTS

Lawrence Albright, Missouri Southern University; David Becker, Oakland Community College, Orchard Ridge; Jon Bellama, University of Maryland; Marshall Bishop, Southwestern Michigan College; George Bodner, Purdue University; Alfred Boyd, University of Maryland; Ingetraud Brieger, Oakland Community College, Orchard Ridge; David Brooks, University of Nebraska—Lincoln; Lynn Buffington, Carleton College; Thomas Bydalek, University of Minnesota, Duluth; William Child, Carleton College; Ronald Clark, Florida State University; Henry Connor, Kentucky Wesleyan University; Duane Converse, Joliet Junior College; Nathan Coward, University of Wisconsin, Superior; Derek Davenport, Purdue University; Anne Deckerd, Grand Rapids Junior College; John DeKorte, Northern Arizona University; Harry Eick, Michigan State University; Lawrence Epstein, University of Pittsburgh; Dennis Flentge, Cedarville College; Sharon Gardulund, Macomb Community College, Warren Campus; Patrick Garvey, Des Moines Area Community College; Roy Garvey, North Dakota State University; Martin Gouterman, University of Washington; Edward Garvin, Highland Park Community College; Robert Hammer, Michigan State University; Henry Heikkinen, University of Maryland; Paul Hunter, Michigan State University; Peggy Hurst, Bowling Green State University; William Husa, Middle Georgia College; Robert Jensen, Elgin Community College; Joseph Kagy, University of Toledo College & Technical College; David Katz, Community College of Philadelphia; Philip Kensey, University of Evansville; Stanley Kirschner, Wayne State University; Richard Kolaczkowski, Indiana University of Pennsylvania; John Konitzer, McHenry County College; Norman Kulevsky, University of North Carolina; Knox Landers, Clemson University; Kenneth Larson, Gogebic Community College; Norbert Leach, Macomb Community College, Mt. Clemens Campus; Maurice Lindauer, Valdosta State College; Edward Lingafelter, University of Washington; George Linington, Macomb Community College, Warren Campus; Garry McGlaun, Gainesville Junior College; Ronald Marks, Indiana University of Pennsylvania; Clinton A. Medbery, The Citadel, The Military College of South Carolina; Steven Miller, Oakland University; Steven Monts, Kankakee Community College; William Movius, Kent State University; R. Thomas Myers, Kent State University; Robert Nakon, West Virginia University; Joyce Neiburger, Purdue University; Sigurd Nelson, Grand Rapids Junior College; James O'Neal, Kerkhof College; Joseph Oravec, Wayne State University; Conrad Pillar, Benedictine College; Melvin Potts, Pittsburgh State University; Ronald Ragsdale, University of Utah; Roland Roskos, University of Wisconsin, La Crosse; Neil Sands, North Hennepin Community College; George Schenk, Wayne State University; Robert Taylor, Olivet Nazarene College; Larry Thompson, University of Minnesota, Duluth; Louis Trudell, Macomb Community College, Warren Campus; Janice Turner, Augusta College; Andrew Watson, Schoolcraft College; Alan Wilson, Kaskaskia College; Ralph Yingst, Youngstown State University; William Yockey, Morningside College.

We wish to thank Harvey Pantzis and Peggy J. Flanagan of the editorial staff at D. C. Heath, and Libby Koponen, our developmental editor; their thoughtful work on our manuscript has made it a better textbook. We sincerely appreciate the efforts of Diana Huffer and Becky Fagan, who typed the manuscript, and Nancy Jensen, who checked all Examples in the chapters and the answers to all problems in the Exercises.

Henry F. Holtzclaw, Jr.
University of Nebraska—Lincoln

William R. Robinson
Purdue University

CONTENTS

1
SOME FUNDAMENTAL CONCEPTS 1

Introduction • *1.1 Chemistry* • *1.2 Matter and Energy* • *1.3 Law of Conservation of Matter and Energy* • *1.4 Chemical and Physical Properties* • *1.5 The Classification of Matter* • *1.6 Atoms and Molecules* • *1.7 The Scientific Method*

Measurement in Chemistry • *1.8 Units of Measurement* • *1.9 Conversion of Units; Dimensional Analysis* • *1.10 Uncertainty in Measurements and Significant Figures* • *1.11 Length and Volume* • *1.12 Mass and Weight* • *1.13 Density and Specific Gravity* • *1.14 Temperature and Its Measurement* • *1.15 Heat and Its Measurement* • For Review • Summary • Key Terms and Concepts • Exercises

2
SYMBOLS, FORMULAS, AND EQUATIONS; ELEMENTARY STOICHIOMETRY 29

2.1 Symbols and Formulas • *2.2 Atomic Weights and Moles of Atoms* • *2.3 Avogadro's Number* • *2.4 Molecular Weights, Empirical Weights, and*

6

CHEMICAL BONDING, PART 2— MOLECULAR ORBITALS 149

6.1 The Molecular Orbital Theory • 6.2 Molecular Orbital Energy Diagrams • 6.3 The Hydrogen Molecule, H_2 • 6.4 Two Diatomic Helium Species, He_2^+ and He_2 • 6.5 The Lithium Molecule, Li_2 • 6.6 The Instability of the Beryllium Molecule, Be_2 • 6.7 The Boron Molecule, B_2 • 6.8 The Carbon Molecule, C_2 • 6.9 The Nitrogen Molecule, N_2 • 6.10 The Oxygen Molecule, O_2 • 6.11 The Fluorine Molecule, F_2 • 6.12 The Instability of the Neon Molecule, Ne_2 • 6.13 Diatomic Systems with Two Different Elements • 6.14 Bond Order • For Review • Summary • Key Terms and Concepts • Exercises

7

MOLECULAR STRUCTURE AND HYBRIDIZATION 167

Valence Shell Electron-Pair Repulsion *• 7.1 Prediction of Molecular Structures • 7.2 Rules for Predicting Molecular Structures*

Hybridization of Atomic Orbitals *• 7.3 Methane, CH_4, and Ethane, C_2H_6 (sp^3 Hybridization) • 7.4 Beryllium Chloride, $BeCl_2$ (sp Hybridization) • 7.5 Boron Trifluoride, BF_3 (sp^2 Hybridization) • 7.6 Other Types of Hybridization • 7.7 Prediction of Hybrid Orbitals • 7.8 Ethylene, C_2H_4, and Acetylene, C_2H_2 (π Bonding) • For Review • Summary • Key Terms and Concepts • Exercises*

8

CHEMICAL REACTIONS AND THE PERIODIC TABLE 185

8.1 Classification of Chemical Compounds • 8.2 Classification of Chemical Reactions • 8.3 Metals, Nonmetals, and Metalloids • 8.4 Variation in Metallic and Nonmetallic Behavior of the Representative Elements • 8.5 Periodic Variation of Oxidation Numbers • 8.6 Prediction of Reaction Products • 8.7 Chemical Properties of Some Important Industrial Chemicals • For Review • Summary • Key Terms and Concepts • Exercises

9

OXYGEN, OZONE, AND HYDROGEN 215

Oxygen *• 9.1 Chemical Properties of Oxygen • 9.2 Occurrence and Preparation of Oxygen • 9.3 Physical Properties of Oxygen • 9.4 Chemical Reactivity of*

10

THE GASEOUS STATE AND THE KINETIC-MOLECULAR THEORY 241

11

THE LIQUID AND SOLID STATES 279

12

WATER AND HYDROGEN PEROXIDE   313

13

SOLUTIONS; COLLOIDS   339

14

ACIDS AND BASES 383

The Brönsted-Lowry Concept of Acids and Bases • *14.1 The Protonic Concept of Acids and Bases* • *14.2 Amphiprotic Species* • *14.3 The Strengths of Acids and Bases* • *14.4 The Relative Strengths of Strong Acids and Bases*

The Lewis Concept of Acids and Bases • *14.5 Definitions and Examples*

Acids and Bases in Aqueous Solution • *14.6 Brönsted Acids in Aqueous Solution* • *14.7 Formation of Brönsted Acids* • *14.8 Polyprotic Acids* • *14.9 Brönsted Bases in Aqueous Solution* • *14.10 Formation of Hydroxide Bases* • *14.11 Acid-Base Neutralization* • *14.12 Salts* • *14.13 Quantitative Reactions of Acids and Bases* • *14.14 Equivalents of Acids and Bases* • For Review • Summary • Key Terms and Concepts • Exercises

15

CHEMICAL KINETICS AND CHEMICAL EQUILIBRIUM 407

Chemical Kinetics • *15.1 The Rate of Reaction* • *15.2 Rate of Reaction and the Nature of the Reactants* • *15.3 Rate of Reaction and the State of Subdivision of the Reactants* • *15.4 Rate of Reaction and Temperature* • *15.5 Rate of Reaction and Concentration; Rate Equations* • *15.6 Order of a Reaction* • *15.7 Half-Life of a Reaction* • *15.8 Collision Theory of the Reaction Rate* • *15.9 Activation Energy and the Arrhenius Equation* • *15.10 Elementary Reactions* • *15.11 Unimolecular Reactions* • *15.12 Bimolecular Reactions* • *15.13 Termolecular Reactions* • *15.14 Reaction Mechanisms* • *15.15 Catalysts*

Chemical Equilibrium • *15.16 The State of Equilibrium* • *15.17 Law of Mass Action* • *15.18 Determination of Equilibrium Constants* • *15.19 Effect of Change of Concentration on Equilibrium* • *15.20 Effect of Change in Pressure on Equilibrium* • *15.21 Effect of Change in Temperature on Equilibrium* • *15.22 Effect of a Catalyst on Equilibrium* • *15.23 Homogeneous and Heterogeneous Equilibria* • For Review • Summary • Key Terms and Concepts • Exercises

16

IONIC EQUILIBRIA OF WEAK ELECTROLYTES 457

16.1 The Ionization of Water • *16.2 pH and pOH*

Ionic Equilibria of Weak Acids and Weak Bases • *16.3 Ion Concentrations in Solutions of Strong Electrolytes* • *16.4 The Ionization of Weak Acids* • *16.5 The Ionization of Weak Bases* • *16.6 The Salt Effect* • *16.7 The Common Ion Effect* • *16.8 Buffer Solutions* • *16.9 The Ionization of Weak Diprotic Acids* • *16.10 The Ionization of Weak Triprotic Acids* • *16.11 Hydrolysis* • *16.12 Hydrolysis of a Salt of a Strong Base and a Weak Acid* • *16.13 Hydrolysis of a Salt of a Weak Base and a Strong Acid* • *16.14 Hydrolysis*

23
NITROGEN AND ITS COMPOUNDS   665

24
PHOSPHORUS AND ITS COMPOUNDS   683

25
CARBON AND ITS COMPOUNDS   697

28
NUCLEAR CHEMISTRY   783

29
COORDINATION COMPOUNDS   807

30
METALS OF ANALYTICAL GROUP I   833

31
METALS OF ANALYTICAL GROUP II   849

32

METALS OF ANALYTICAL GROUP III 871

33
METALS OF ANALYTICAL GROUP IV   901

34
METALS OF ANALYTICAL GROUP V   909

SEMIMICRO QUALITATIVE ANALYSIS   925

35
GENERAL LABORATORY DIRECTIONS   927

36
THE ANALYSIS OF GROUP I   935

42

THE ANALYSIS OF SOLID MATERIALS   993

APPENDIXES   A-1

1

SOME FUNDAMENTAL CONCEPTS

Our society depends on science and technology. First-class medical care, sufficient and varied food supplies, comfortable housing, convenient transportation, rapid communication, and overall personal comfort are usually taken for granted. But these benefits are a direct result of scientific and technological developments in the past century.

Chemistry has played an important role in these developments. For example, engineering materials are prepared, analyzed, and fabricated using processes based on chemical technology. The manufacture of L-dopa, a drug used for the treatment of Parkinson's disease, takes advantage of chemical properties of hydrogen discovered by chemists in several academic laboratories. Research on recombinant DNA led to the bacterial synthesis of human insulin for treatment of diabetes. Enzyme-mapping techniques, developed by chemists and biologists in the late 1960s, were essential to the design and synthesis of diphenyl ether herbicides. Study of certain compounds of phosphorus led to the discovery of organophosphorus pesticides, which, along with other pesticides, reduced crop losses enough to feed an estimated 500 million more people. Even everyday items such as rubber bands, toothpaste, and the dye that gives blue jeans their color reflect the input of chemists in the development of these products.

However, within the last two decades, science and technology have acquired an ambivalent reputation because some of their many contributions to our lives have had undesirable side effects. For example, even though energy is the key to our technological society, we deplore the destruction of natural beauty caused by strip mining for coal, the pollution of natural waters by oil spills, the possibility of radioactive contamination from accidents in nuclear plants, and the alarmingly rapid depletion of natural sources of energy. The fertilizers and insecticides that have increased food production have polluted our air and water. Automobiles and jet airplanes have revolutionized transportation, but they also have contaminated the air. Some medicines that have saved countless lives have toxic side effects.

We have sometimes been slow to recognize and have always been slow to act on the vexing and often unexpected problems that result from the interaction and interdependence of various aspects of our society. Still, most of the problems created by the development of new products and processes can also be solved by scientists through cooperative effort with others in many different occupations.

Chemists can take well-deserved credit for many scientific and technological advances, but they must also assume the responsibility for solving the problems created by some of these advances. The responsibility cannot be assigned to chemists alone, however. It is the *joint* responsibility of physicists, sociologists, biologists, theologians, political scientists, humanists, *and* chemists. Together we must work to solve the problems of a world that sometimes seems to be destroying itself with its own technological progress.

No single discipline offers the total perspective or the expertise to accomplish this task. Each discipline has specialized knowledge that, together with that of other disciplines, can contribute to bringing about a better life for everyone. All of us must draw on all of our knowledge if we are to achieve that goal and still preserve ethical and moral standards. Chemists must know their own field extremely well and other fields well enough to ensure the wise use of their discoveries. People in other fields must know enough chemistry and other sciences to apply their own specialized knowledge to problems involving science. The questions of why a person should study chemistry or any other science, regardless of the intended main field of endeavor, and why a scientist should study other fields in addition to science have never had a more demanding or obvious answer.

INTRODUCTION

1.1 Chemistry

In our study of chemistry we shall be concerned with the composition and structure of matter and with the changes that matter undergoes. Such forms of matter as oil and nylon, gasoline and water, salt and sugar, and iron and gold differ strikingly from each other in many ways. Their differences are due to differences in composition and structure. Some forms of matter can be produced from or converted into other forms. For example, when gasoline is burned, water and other substances are formed. Oil can be converted into nylon by a series of chemical reactions. Our very existence depends on changes that occur in matter. Plants convert simple varieties of matter into more complex forms, which serve as food. The chemical changes that take place when this food is digested and assimilated by the body are essential to life processes.

As we proceed in our study of chemistry, we will examine some of the changes in the composition and structure of matter, the causes that produce these changes, the changes in energy that accompany them, and the principles and laws involved in these changes. In brief, then, the science of **chemistry may be defined as the study of the composition, structure, and properties of matter and of the reactions by which one form of matter may be produced from or converted into other forms.**

Knowledge about chemistry has increased so much during the last century that chemists usually specialize in one of the field's several principal branches. **Analytical chemistry** is concerned with the identification, separation, and quantitative determination of the composition of different substances. **Physical chemistry** is primarily concerned with the structure of matter, energy changes, and the laws, principles, and theories that explain the transformation of one form of matter into another. **Organic chemistry** is the branch dealing with reactions of the compounds of carbon. **Inorganic chemistry** is concerned with the chemistry of elements other than carbon and of their compounds. **Biochemistry** is the chemistry of the substances comprising living organisms. The boundaries between the branches of chemistry are arbitrarily defined, however, and much work in chemistry cuts across them.

1.2 Matter and Energy

Matter, by definition, is anything that occupies space and has mass. All the objects in the universe, since they occupy space and have mass, are matter. The property of occupying space is often easily perceived by our senses of sight and touch. The **mass** of an object pertains to the quantity of matter that the object contains. The force required to change the speed at which an object moves, or to change the direction in which it moves, is a measure of its mass.

Matter can exist in three different states, which are designated solid, liquid, and gas (Fig. 1-1). Each state can be distinguished by certain characteristics. A **solid** is rigid, possesses a definite shape, and has a volume that is very nearly

Figure 1-1. The three states of matter as illustrated by water.

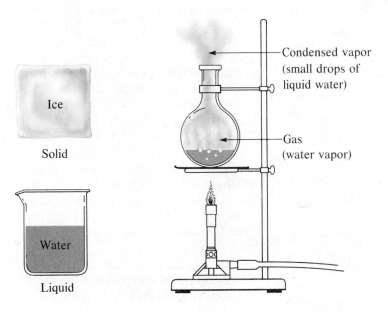

independent of changes in temperature and pressure. A **liquid** flows and thus takes the shape of its container, except that it assumes a horizontal upper surface. Liquids, like solids, are only slightly compressible and so, for practical purposes, have definite volumes. A **gas** takes both the shape and the volume of its container. Gases are readily compressible and capable of infinite expansion.

Energy can be defined as the capacity for doing work, where **work** is simply the process of causing matter to move against an opposing force. For example, when we pump up a bicycle tire, we are doing work—we are moving matter (the air in the pump) against the opposing force of the air already in the tire.

Like matter, energy exists in different forms. The energy for running the pump to carry out the work of pumping up a tire could be chemical energy released in our muscles as we use a hand pump, electrical energy used to drive an electric motor on a compressor, or heat used in a steam engine running a compressor. Light is also a form of energy. Solar cells can convert light into electrical energy, which could run our compressor or do other work.

Energy can be further classified as either potential energy or kinetic energy. A piece of matter is said to possess **potential energy** by virtue of its position, condition, or composition. Water *at the top* of a waterfall possesses potential energy because of its position; if it falls in a hydroelectric plant, it does work that leads to the production of electricity. A compressed spring, because of its condition, possesses potential energy and can do work such as making a clock run. Natural gas possesses potential energy. Because of its composition, it will burn, producing heat, another kind of energy. When a piece of matter is in motion, it also has energy, the capacity for doing work. The energy that a piece of matter possesses because of its motion is called **kinetic energy.** As water falls from the top of a waterfall, its potential energy becomes kinetic energy.

1.3 Law of Conservation of Matter and Energy

When a piece of calcium metal is exposed to dry air, it unites with oxygen in the air. If the product (calcium oxide) is collected and weighed, it is found to have a mass greater than that of the original piece of metal. If, however, the mass of the oxygen in the air that combined with the metal is taken into consideration, the final mass is found to be equal to the sum of the masses of the calcium metal and the oxygen. This behavior of matter is in accord with the **Law of Conservation of Matter: During an ordinary chemical change, there is no detectable increase or decrease in the quantity of matter.**

The conversion of one type of matter into another (chemical change) is always accompanied by the conversion of one form of energy into another. Usually heat is evolved or absorbed, but sometimes the conversion involves light or electrical energy instead of, or in addition to, heat. When calcium metal combines with oxygen, potential energy is converted into heat. Under certain conditions the calcium may actually burn, producing light as well. Many transformations of energy, of course, do not involve chemical changes. Electrical energy can be changed into light, heat, or potential energy without any chemical change. Potential and kinetic energy can each be converted into the other. Many other conversions are possible, but all of the energy involved in any change always exists in some form after the change is completed. This fact is expressed in the **Law of Conservation of Energy: During an ordinary chemical change, energy can be neither created nor destroyed, although it can be changed in form.**

In precise terms matter and energy are not distinct, but are two forms of a single entity. In nuclear reactions both conversion of matter into energy and conversion of energy into matter can occur. So, although the laws of conservation of matter and of energy both hold in ordinary chemical reactions, to be more precise we should combine these two laws into one statement: **The total quantity of matter and energy available in the universe is fixed.**

Since energy can be neither created nor destroyed in a chemical reaction, why is there concern about a shortage of energy? The answer to this question is that we are not running out of energy itself, but we are running out of useful forms of energy. The energy contained in a fossil fuel such as oil can be converted, in part, into work when the fuel is burned; the rest is generally converted into waste heat, a form of energy that cannot be used. Furthermore, millions of years are required for nature to produce oil reserves to replace those used. If we are not to run out of *available* energy, we must be willing to devote money and time to research to develop more efficient methods of using the energy we have and to find new methods of using other energy present in the universe.

1.4 Chemical and Physical Properties

The characteristics that enable us to distinguish one substance from another are known as properties. **Chemical properties** refer to how one kind of matter transforms into another kind. Iron exhibits a chemical property when it combines with oxygen and water to form the reddish-brown iron oxide we call rust. A chemical property of aluminum is that it does not rust. The **physical properties** of a particular kind of matter are those characteristics that do not involve a change in the chemical identity of the matter. Familiar physical properties include color, hardness, physical state, melting temperature, boiling temperature, and electrical conductivity. Iron, for example, melts at a temperature of 1535°C; aluminum melts at a different temperature, 660°C. As they melt, the metals change from solids to liquids but are still iron and aluminum, since no change in chemical identity has occurred.

The properties of various kinds of matter enable us to distinguish one kind from another. We can distinguish between iron wire and aluminum wire by their physical properties (their colors, for example) or by their chemical properties (for example, iron rusts, but aluminum does not).

In order to identify a chemical property, we look for a chemical change. A **chemical change** always produces one or more different kinds of matter from those that were present before the change occurred. Iron rust contains oxygen as well as iron, and it is therefore a different kind of matter from the iron, air, and water from which it formed. Thus the formation of iron rust is a chemical change. When milk sours, the sugar (lactose) in the milk is converted into lactic acid by a chemical change, and the composition and the properties of the acid differ from those of the sugar.

The absence of a chemical change is also a chemical property. Since aluminum does not rust, one of its chemical properties is that it does not undergo this chemical change.

A **physical change** is one that does not involve a change of one kind of matter into another. The melting of iron, the freezing of water, the conversion of liquid water to steam, and the condensation of steam to liquid water are all examples of physical changes. In each of these there is a change in one or more properties, but there is no alteration of the chemical composition of the substance involved. Water, whether in the solid, liquid, or gaseous state, maintains the same chemical composition. Iron, whether molten or solid, is the same kind of matter. Some physical properties can be observed only when matter undergoes a physical change. The freezing temperature of water, for example, is determined by measuring the temperature as water changes from a liquid to a solid. Other physical properties can be measured without a physical change. For example, the density of a sample can be determined without an accompanying physical change.

1.5 The Classification of Matter

All samples of matter can be classified as either pure substances or mixtures (Fig. 1-2). A **mixture** is composed of two or more kinds of matter that can be separated by physical means. Moreover, the composition of a mixture can be varied continuously. Blood consists of varying amounts of proteins, sugar, salt, oxygen, carbon dioxide, and other components mixed in water. It can be recognized as a mixture because it can be separated into solids and plasma by centrifugation, a physical

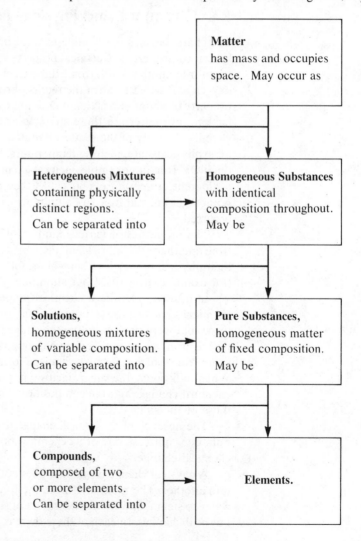

Figure 1-2. The classification of matter.

Matter
has mass and occupies space. May occur as

Heterogeneous Mixtures
containing physically distinct regions.
Can be separated into

Homogeneous Substances
with identical composition throughout.
May be

Solutions,
homogeneous mixtures of variable composition.
Can be separated into

Pure Substances,
homogeneous matter of fixed composition.
May be

Compounds,
composed of two or more elements.
Can be separated into

Elements.

In precise terms matter and energy are not distinct, but are two forms of a single entity. In nuclear reactions both conversion of matter into energy and conversion of energy into matter can occur. So, although the laws of conservation of matter and of energy both hold in ordinary chemical reactions, to be more precise we should combine these two laws into one statement: **The total quantity of matter and energy available in the universe is fixed.**

Since energy can be neither created nor destroyed in a chemical reaction, why is there concern about a shortage of energy? The answer to this question is that we are not running out of energy itself, but we are running out of useful forms of energy. The energy contained in a fossil fuel such as oil can be converted, in part, into work when the fuel is burned; the rest is generally converted into waste heat, a form of energy that cannot be used. Furthermore, millions of years are required for nature to produce oil reserves to replace those used. If we are not to run out of *available* energy, we must be willing to devote money and time to research to develop more efficient methods of using the energy we have and to find new methods of using other energy present in the universe.

1.4 Chemical and Physical Properties

The characteristics that enable us to distinguish one substance from another are known as properties. **Chemical properties** refer to how one kind of matter transforms into another kind. Iron exhibits a chemical property when it combines with oxygen and water to form the reddish-brown iron oxide we call rust. A chemical property of aluminum is that it does not rust. The **physical properties** of a particular kind of matter are those characteristics that do not involve a change in the chemical identity of the matter. Familiar physical properties include color, hardness, physical state, melting temperature, boiling temperature, and electrical conductivity. Iron, for example, melts at a temperature of 1535°C; aluminum melts at a different temperature, 660°C. As they melt, the metals change from solids to liquids but are still iron and aluminum, since no change in chemical identity has occurred.

The properties of various kinds of matter enable us to distinguish one kind from another. We can distinguish between iron wire and aluminum wire by their physical properties (their colors, for example) or by their chemical properties (for example, iron rusts, but aluminum does not).

In order to identify a chemical property, we look for a chemical change. A **chemical change** always produces one or more different kinds of matter from those that were present before the change occurred. Iron rust contains oxygen as well as iron, and it is therefore a different kind of matter from the iron, air, and water from which it formed. Thus the formation of iron rust is a chemical change. When milk sours, the sugar (lactose) in the milk is converted into lactic acid by a chemical change, and the composition and the properties of the acid differ from those of the sugar.

The absence of a chemical change is also a chemical property. Since aluminum does not rust, one of its chemical properties is that it does not undergo this chemical change.

A **physical change** is one that does not involve a change of one kind of matter into another. The melting of iron, the freezing of water, the conversion of liquid water to steam, and the condensation of steam to liquid water are all examples of physical changes. In each of these there is a change in one or more properties, but there is no alteration of the chemical composition of the substance involved. Water, whether in the solid, liquid, or gaseous state, maintains the same chemical composition. Iron, whether molten or solid, is the same kind of matter. Some physical properties can be observed only when matter undergoes a physical change. The freezing temperature of water, for example, is determined by measuring the temperature as water changes from a liquid to a solid. Other physical properties can be measured without a physical change. For example, the density of a sample can be determined without an accompanying physical change.

1.5 The Classification of Matter

All samples of matter can be classified as either pure substances or mixtures (Fig. 1-2). A **mixture** is composed of two or more kinds of matter that can be separated by physical means. Moreover, the composition of a mixture can be varied continuously. Blood consists of varying amounts of proteins, sugar, salt, oxygen, carbon dioxide, and other components mixed in water. It can be recognized as a mixture because it can be separated into solids and plasma by centrifugation, a physical

Figure 1-2. The classification of matter.

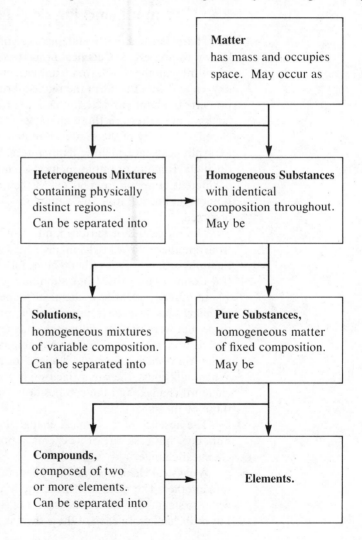

method of separation of these two components. Other mixtures include granite, which is a mixture of quartz, feldspar, and mica; sugar syrup, which is a mixture of sugar and water; and air, which is a mixture of oxygen, nitrogen, carbon dioxide, water vapor, and small amounts of other gases.

A solution of sugar in water is an example of a **homogeneous mixture;** that is, a mixture whose composition is uniform throughout. At any point in a container holding a solution of sugar and water, the amount of sugar and water is the same. Homogeneous mixtures are often called **solutions.** Other examples of homogeneous mixtures include air, soft drinks, gasoline, and a solution of salt in water. A mixture whose composition differs from point to point is a **heterogeneous mixture.** Chocolate chip ice cream is an example of a heterogeneous mixture. At some points it consists of chocolate chips; at other points, vanilla ice cream. Other heterogeneous mixtures include granite (you can often see the separate bits of mica, quartz, and feldspar), sand in water, blood, and concrete.

Pure substances are similar to homogeneous mixtures in that both are of uniform composition throughout. However, while the composition of a homogeneous mixture can vary from one sample to another, a specific pure substance always has the same composition. Thus a **pure substance** is defined as a homogeneous sample of matter, all specimens of which have identical compositions as well as identical chemical and physical properties.

A carbonated soft drink is a homogeneous mixture, typically of water, sugar, coloring and flavoring agents, and carbon dioxide. It is not a pure substance by our definition since its composition can vary. The amount of carbon dioxide in the drink decreases after the bottle is opened and the drink begins to lose its fizz. A sample of carbon dioxide alone, however, is a pure substance. It is always composed of 12 parts of carbon and 32 parts of oxygen, by weight, and has the same melting temperature, boiling temperature, and other properties, no matter what brand of beverage or other source it is isolated from.

Chemists divide pure substances into two classes: elements and compounds. **Elements are pure substances that cannot be decomposed by a chemical change.** Familiar examples are iron, silver, gold, aluminum, sulfur, oxygen, and carbon. One hundred and eight elements are known at the present time; a list of these is printed on the inside front cover of this book. Eighty-eight elements occur naturally on the earth, and the other twenty have been created in laboratories, the most recent during 1982. Eleven of the eighty-eight naturally occurring elements make up about 99% of the earth's crust and the atmosphere (Table 1-1).

Table 1-1   Percentages of Elements in the Atmosphere and the Earth's Crust by Mass

Oxygen	49.20%	Chlorine	0.19%
Silicon	25.67	Phosphorus	0.11
Aluminum	7.50	Manganese	0.09
Iron	4.71	Carbon	0.08
Calcium	3.39	Sulfur	0.06
Sodium	2.63	Barium	0.04
Potassium	2.40	Nitrogen	0.03
Magnesium	1.93	Fluorine	0.03
Hydrogen	0.87	Strontium	0.02
Titanium	0.58	All others	0.47

Oxygen constitutes nearly 50% and silicon about 25% of the total quantity of the elements in the atmosphere and the earth's crust. Only about one-fourth of the elements ever occur on the earth in the free state; the others are found only in chemical combination with other elements.

Compounds are pure substances that are composed of two or more different elements and that can be decomposed by chemical changes. The properties of elements in combination are different from those of elements in their free, or uncombined, states. However, the term *element* is used to designate an elemental substance whether free or in combination. For example, white crystalline sugar is a compound consisting of the element carbon, which is usually a black solid when free, and the two elements hydrogen and oxygen, which are colorless gases when uncombined. If heated sufficiently in the absence of air, sugar decomposes to carbon and water. Water, a compound, can be decomposed by an electric current into its two constituent elements, hydrogen and oxygen. Sodium chloride, a compound that is the principal constituent of table salt, can be broken down by an electric current into the elements sodium and chlorine.

There are only 108 known elements but several million chemical compounds result from different combinations of these elements. Each compound possesses definite chemical and physical properties by which chemists can distinguish it from all other compounds.

1.6 Atoms and Molecules

An atom is the smallest particle of an element that can enter into a chemical combination. If we could take a sample of the element gold and divide it into smaller and smaller pieces, eventually we would have a single atom of gold. This atom would no longer be gold if it were divided. Such a single atom of gold is incredibly small, about 0.0000000113 inch in diameter. It would require about 88 million gold atoms lying side by side to make a line 1 inch long.

The first suggestion that matter is composed of atoms is attributed to a few of the early Greek philosophers, notably Democritus. These philosophers argued that matter is composed of small indivisible particles. Indeed, the word *atom* comes from the Greek word *atomos,* which means indivisible. However, it was not until the early nineteenth century that John Dalton (1766–1844), an English chemist and physicist, presented an atomic theory *and* supported this theory with quantitative measurements. Since that time repeated experiments have verified his theory, which is still used with only minor revisions.

Only a few elements, such as the gases helium, neon, and argon, consist of a collection of individual atoms that move about independently of one another (Fig. 1-3). Other elements, such as the gases nitrogen, oxygen, and chlorine, consist of pairs of atoms, each pair moving as a single unit. The element phosphorus consists of units composed of four phosphorus atoms; sulfur, of units composed of eight sulfur atoms. These units are called molecules.

A molecule is the smallest particle of an element or compound that can have a stable, independent existence. A molecule may consist of a single atom, as in helium, of two or more identical atoms, as in nitrogen and sulfur, or of two or more different atoms, as in water. Water has a definite composition and a set of chemical properties that enable us to recognize it as a distinct substance. One might ask, "To what extent can a drop of water be subdivided and still be water?"

Figure 1-3. Representations of molecules of the elements helium, oxygen, and sulfur and of the compound water.

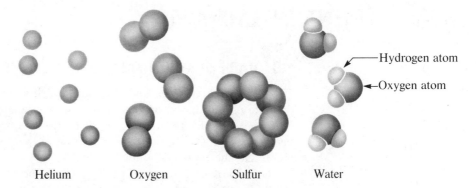

—Hydrogen atom

←Oxygen atom

Helium   Oxygen   Sulfur   Water

The limit to which such a subdivision can be carried is the water molecule. Each water molecule is a unit that contains two hydrogen atoms and one oxygen atom (Fig. 1-3). Subdivision of a water molecule results in the formation of the gases hydrogen and oxygen, each of which has properties quite different from those of water and those of each other. A molecule of the sugar glucose contains 6 carbon atoms, 12 hydrogen atoms, and 6 oxygen atoms.

Atoms and molecules are too small to be seen even with very powerful optical microscopes, and only relatively large molecules can be seen even with electron microscopes. An appreciation of the minute size of molecules can be gained from the fact that if a glass of water were somehow enlarged to the size of the earth, the water molecules would be about the size of golf balls.

1.7   The Scientific Method

Your study of chemistry will be concerned with the observations, theories, and laws that give this science its foundation and that form a framework into which information fits to make an integrated area of knowledge. This framework develops from the pursuit of answers to many questions, each of which can be subjected to experimental investigation by an approach often called the **scientific method.**

Scientists identify problems or questions through their own observations and experiments or from the observations, experiments, and conclusions of others. (Problems are often identified when experiments do not go as expected.) The first step in applying the scientific method to a problem involves carefully planning experiments to gather facts and obtain information about all phases of the problem. The results are examined for general relationships that will unify the observations. Sometimes a wide variety of observations can be summarized in a general verbal statement or mathematical equation known as a **law.** One example is the Law of Conservation of Matter (Section 1.3), which summarizes the results of thousands of experimental observations. More often, however, a tentative explanation of physical or chemical phenomena is suggested. Such a proposal is called a **hypothesis.** For instance, Dalton attempted to explain why mass is conserved in a chemical reaction when he first presented his ideas, which were really hypotheses, of the atomic nature of matter. A hypothesis is tested by further experiments, and, if it is capable of explaining the large body of experimental data, it is dignified by the name **theory.** Dalton's ideas have been so extensively tested that we now refer to the atomic theory. Theories themselves can prompt new questions or suggest new directions in which additional information can be sought.

MEASUREMENT IN CHEMISTRY

1.8 Units of Measurement

The hypotheses, theories, and laws that describe the behavior of matter and energy are usually based on quantitative measurements. When properly reported these measurements convey three ideas: the size or magnitude of the property being measured (a number), an indication of the possible error in the measurement, and the units of the measurement.

Units provide a standard of comparison for a measurement. A well-known sandwich is prepared from a quarter pound of hamburger meat (0.250 pound). The mass of the meat has a magnitude of 0.250 times the mass of the arbitrary standard of 1 pound. If the unit of reference is changed, the property being measured (in this case, the mass) does not change. The same mass of hamburger meat can be reported as being a quarter pound, 4.00 ounces, or 0.000125 ton (a 0.000125-tonner?); the actual amount of meat is the same no matter what the unit of measurement.

The results of scientific measurements are usually reported in the metric system originally devised in France in 1799. The original metric units for length, mass, and time were the meter, the gram, and the second. Various combinations of these units are still in use. One such system that has been used frequently by chemists is the **centimeter-gram-second, or cgs, system.** More recently scientists have begun to use an updated version of the metric system of units in which the units for length, mass, and time are the meter, the kilogram, and the second. This system of units is the **International System of Units (SI units).** These units have been used by the U.S. National Bureau of Standards since 1964. The SI system consists of the following seven base units, from which other units of weight and measure can be derived.

Physical Property	Name of Unit	Symbol
Length	Meter	m
Mass	Kilogram	kg
Time	Second	s
Electric current	Ampere	A
Temperature	Kelvin	K
Luminous intensity	Candela	cd
Amount of substance	Mole	mol

In both the SI and other metric systems, measurements may be reported as fractions or multiples of ten times a base unit by using the appropriate prefix with the name of the base unit. For example, a length can be reported in units of meters, kilometers (10^3 meters), millimeters (10^{-3} meter), or picometers (10^{-12} meter). The prefixes and their symbols denoting the powers to which 10 is raised are given in Table 1-2.

Since both SI and cgs units are currently in use by scientists, it is necessary to be familiar with both systems. Consequently, both systems of units will be used in this text.

Table 1-2   Common Prefixes Used in the Metric System

Prefix	Symbol	Factor	Example
pico	p	10^{-12}	1 picometer (pm) = 1×10^{-12} m (0.000000000001 m)
nano	n	10^{-9}	1 nanogram (ng) = 1×10^{-9} g (0.000000001 g)
micro	μ*	10^{-6}	1 microliter (μL) = 1×10^{-6} L (0.000001 L)
milli	m	10^{-3}	2 milliseconds (ms) = 2×10^{-3} s (0.002 s)
centi	c	10^{-2}	5 centimeters (cm) = 5×10^{-2} cm (0.05 m)
deci	d	10^{-1}	1 deciliter (dL) = 1×10^{-1} L (0.1 L)
kilo	k	10^{3}	1 kilometer (km) = 1×10^{3} m (1000 m)
mega	M	10^{6}	3 megagrams (Mg) = 3×10^{6} g (3,000,000 g)
giga	G	10^{9}	5 gigameters (Gm) = 5×10^{9} m (5,000,000,000 m)
tera	T	10^{12}	1 teraliter (TL) = 1×10^{12} L (1,000,000,000,000 L)

*Greek letter mu.

1.9 Conversion of Units; Dimensional Analysis

In addition to the two systems of units in use within the scientific community, many other units are employed by other disciplines. For example, units of mass range from grams and kilograms in science, to pounds and tons in engineering, to grains and drams in medicine. The relationships between some common units are given in Table 1-3. A more complete table is given in Appendix C.

Table 1-3   Common Conversion Factors (to Four Significant Figures)

Length

1 meter = 1.094 yards   1 inch = 2.54 centimeters*

Volume

1 liter = 1.057 quarts   1 cubic foot = 28.316 liters

Mass

1 kilogram = 2.205 pounds   1 ounce = 28.35 grams

*Exact conversion.

Using a technique called dimensional analysis simplifies conversions between these various units. If units are treated like algebraic quantities, they can be carried through calculations and multiplied together, divided into each other, or canceled. The final units resulting from such a treatment will provide valuable assistance in determining whether or not a calculation has been set up correctly.

Consider the conversion of a length from meters to yards. As given in Table 1-3, 1 meter = 1.094 yards. This may be read as indicating that there is exactly 1 meter per 1.094 yards or, alternatively, 1.094 yards per 1 meter. Very often a slash (/) or a division line is used instead of the word *per*. This gives the following unit conversion factors:

$$1 \text{ meter}/1.094 \text{ yards}, \quad \text{or} \quad \frac{1 \text{ meter}}{1.094 \text{ yards}}$$

$$1.094 \text{ yards}/1 \text{ meter}, \quad \text{or} \quad \frac{1.094 \text{ yards}}{1 \text{ meter}}$$

A **unit conversion factor** is used to convert a quantity in one system of units to the corresponding quantity in another system of units. For example, the second factor above can be used to convert a length in meters to the corresponding length in yards. We simply multiply the quantity in meters by the conversion factor.

EXAMPLE 1.1   How long in yards is a run of 100.0 meters?

From the relationship 1 meter = 1.094 yards, we can get the unit conversion factor 1.094 yards/1 meter. Multiplication gives

$$100.0 \text{ meters} \times \frac{1.094 \text{ yards}}{1 \text{ meter}} = 109.4 \text{ yards}$$

Keep in mind that the distance covered by the runner does not change when we switch from units of meters to units of yards. Multiplying by a unit conversion factor is equivalent to multiplying by 1. In this case, since 1 meter and 1.094 yards are identical distances, 1.094 yards/1 meter = 1.

Note that the unit of meters in 100.0 meters and in the conversion factor cancel, as indicated by the cancellation lines in color. Since the answer is in the correct unit, it is highly likely that the conversion has been set up correctly. If an incorrect unit conversion factor had been used, an incorrect unit would have been obtained. Consider the following conversion:

$$100.0 \text{ meters} \times \frac{1 \text{ meter}}{1.094 \text{ yards}} = 91.41 \frac{\text{meters}^2}{\text{yard}}$$

The final unit, meters^2/yard, indicates that the conversion does not convert meters into yards, since the units do not cancel to give yards.

1.10 Uncertainty in Measurements and Significant Figures

Suppose you weigh an object on an inexpensive balance and find, as best you can determine, that its mass is closer to 13.4 grams than to either 13.3 or 13.5 grams. You would report the mass as 13.4 grams. The uncertainty in this measurement is almost 0.1 gram, or 1 part in 134 (about 0.75%), since the true mass of the object could lie anywhere between 13.35 and 13.45 grams. If the same object were weighed on an analytical balance in your laboratory, your best efforts might show that its mass is 13.384 grams; that is, the mass is closer to 13.384 grams than to either 13.383 or 13.385 grams. The uncertainty in this case would be about 0.001 gram (1 part in 13,384, or about 0.0075%), since the true mass could lie anywhere between 13.3835 and 13.3845 grams. Any experimental measurement, no matter how carefully made, may contain this type of uncertainty. **All measurements may**

be taken to have an uncertainty of at least one unit in the last digit of the measured quantity.

The mass 13.4 grams has an uncertainty of 0.1 gram; 13.384 grams, an uncertainty of 0.001 gram; a measured volume of 25 milliliters, an uncertainty of 1 milliliter; and a measured length of 0.001378 meter, an uncertainty of 0.000001 meter. All of the measured digits in the determination are called **significant figures.**

Results calculated from measurements are as uncertain as the measurement itself. Suppose a sample weighs 78.7 grams and exactly one third of it is iron. In order to calculate how much iron is in the sample, we need only multiply 78.7 grams by one third. On an electronic calculator, this comes out to 26.233333 grams. This is not the correct answer, however. Since the mass of the sample is uncertain to 1 part in 787, or about 0.1%, the mass of iron in the sample is subject to the same uncertainty. To report the mass of iron as 26.233333 grams indicates an uncertainty of 0.000001 gram, or about 0.000004% (1 part in 26,233,333). The mass of iron should be reported as 26.2 grams (three significant figures). In calculations involving experimental quantities, the significant figures in these quantities must be taken into account in order not to overestimate or underestimate the uncertainty in the calculated result.

The following rules may be used to determine the number of significant figures in a measured quantity:

1. *Find the first nonzero digit on the left and count to the right, including this first digit and all remaining digits. Unless the last digit is a zero lying immediately to the left of the decimal point, this is the number of significant figures in the number.*

first nonzero digit on the left

0.00120530

six significant figures

first nonzero digit on the left

126.7309

seven significant figures

first nonzero digit on the left

97

two significant figures

The first three zeros on the left in the first example above, 0.00120530, are not significant. They merely tell us where the decimal point is located. We could equally well express this number in exponential notation as 1.20530×10^{-3}. In this case the number 1.20530 contains all six significant figures, and 10^{-3} serves to locate the decimal.

2. *When a number ends in zeros that are to the left of a decimal point, the trailing zeros may or may not be significant.* A mass given as 1300 grams may indicate a mass closer to 1300 grams than to 1200 or 1400 grams. In this case the zeros serve only to locate the decimal point and are not significant. If the mass is closer to 1300 grams than to 1290 or 1310 grams, then the zero to the right of the 3 is

significant. If the mass is closer to 1300 than to 1299 or 1301, both zeros are significant. The ambiguity can be avoided by using exponential notation: 1.3×10^3 (two significant figures), 1.30×10^3 (three significant figures), 1.300×10^3 (four significant figures). For convenience in this text, we will assume that when a number ends in zeros that are to the left of a decimal point, these trailing zeros are not significant unless otherwise specified.

The two rules given above apply to measured quantities. If we count items rather than measure them, the result is exact, with no uncertainty. If we count eggs in a carton, we know, without any uncertainty, exactly how many eggs the carton contains. Defined quantities are also exact. By definition 1 foot is exactly 12 inches, 1 kilogram is exactly 1000 grams, and 1 inch equals exactly 2.54 centimeters. These exact numbers, numbers with no uncertainty, can be considered to have an infinite number of significant figures.

The correct use of significant figures in a calculation carries the uncertainty of measured quantities into the result. The following rules govern the number of significant figures that should be in an answer:

1. *When adding or subtracting, report the results with the same number of decimal places as that of the number with the least number of decimal places.*

EXAMPLE 1.2   (a) Add 4.383 g and 1.0023 g.
(b) Subtract 421 g from 486.39 g.

$$
\begin{array}{rl}
\text{(a)} & 4.383 \ \text{g} \\
& +1.0023 \ \text{g} \\
\hline
& 5.385 \ \text{g}
\end{array}
\qquad
\begin{array}{rl}
\text{(b)} & 486.39 \ \text{g} \\
& -\ 421 \ \text{g} \\
\hline
& 65 \ \text{g}
\end{array}
$$

---

2. *When multiplying or dividing, report the product or quotient with no more digits than the least number of significant figures in the numbers involved in the computation.*

EXAMPLE 1.3   Multiply 0.6238 by 6.6.

$$0.6238 \times 6.6 = 4.1$$

In rounding numbers off at a certain point, simply drop the digits that follow if the first of them is less than 5; 8.7235 rounds off to 8.7. If the first digit to be dropped is greater than 5 or if it is 5 followed by other nonzero digits, increase the preceding digit by 1; 3.8679 rounds off to 3.87; 23.3501, to 23.4. If the digit to be dropped is 5 or 5 followed only by zeros, a common practice is to increase the preceding digit by 1 if it is odd and to leave it unchanged if it is even. Dropping a 5 always leaves an even number. Thus 3.425 rounds off to 3.42, while 7.53500 rounds off to 7.54.

1.11 Length and Volume

The standard unit of length in both metric and SI units is the **meter (m).** In 1899 the meter was defined by international agreement to be the length of a certain platinum-iridium alloy bar kept in France. In 1960 the meter was redefined as the length equal to exactly 1,650,763.73 wavelengths, in vacuum, of the orange line in the emission spectrum of ^{86}Kr, a particular isotope of krypton (spectra and isotopes will be discussed in subsequent chapters). Within the error of the measurement, the length of the meter was not changed; only the standard used to define a meter was changed.

In more familiar units of reference, a meter is about 3 inches longer than a yard; 1 meter is approximately 1.094 yards; 1 meter equals 100 centimeters or 1000 millimeters (Fig. 1-4).

Figure 1-4. Relative lengths of (a) 1 meter, (b) 1 yard, (c) 1 centimeter, and (d) 1 inch (not actual size) and a comparison of 1 inch and 2.54 centimeters (actual size). A meter is a little longer than 1 yard and exactly 100 times longer than 1 centimeter.

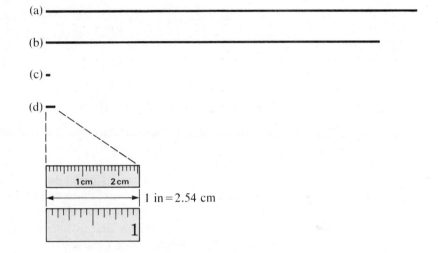

Dime   1 cubic centimeter = 1 milliliter

Figure 1-5. Comparison between a cubic-centimeter block and a dime. A cubic centimeter is equal to 1 milliliter; 1 cubic centimeter of water has a mass of 1 gram at 4°C. (*Copyright © 1983 by William S. Nawrocki.*)

Volumes are defined in terms of the standard length. The standard SI volume, a **cubic meter (m^3),** is the volume of a cube that is exactly 1 meter on an edge. A more convenient unit is the volume of a cube with an edge length of exactly 10 centimeters (cm), or 1 decimeter (10 cm = 1 dm). This volume of 1000 cm^3, or 1 dm^3, is often called a **liter (L)** and is 0.001 times the volume of a cubic meter; 1 liter is about 1.06 quarts. A **milliliter (mL)** is the volume of a cube with an edge length of exactly 1 centimeter (Fig. 1-5) and thus is equal in volume to 1 cubic centimeter (cm^3); 1 liter contains 1000 milliliters.

EXAMPLE 1.4   What is the volume in liters of 1.000 oz, given that 1 L = 1.06 qt and that 32 oz = 1 qt?

From the two relationships it is possible to determine the exact unit conversion factor 1 qt/32 oz and the inexact unit conversion factor 1 L/1.06 qt. First convert ounces to quarts.

$$1.000 \text{ oz} \times \frac{1 \text{ qt}}{32 \text{ oz}} = 0.03125 \text{ qt}$$

Then convert quarts to liters.

$$0.03125 \text{ qt} \times \frac{1 \text{ L}}{1.06 \text{ qt}} = 2.95 \times 10^{-2} \text{ L}$$

Thus

$$1.000 \text{ oz} = 2.95 \times 10^{-2} \text{ L}$$

These two steps could have been combined into one operation as follows:

$$1.000 \text{ oz} \times \frac{1 \text{ qt}}{32 \text{ oz}} \times \frac{1 \text{ L}}{1.06 \text{ qt}} = 2.95 \times 10^{-2} \text{ L}$$

The number of significant figures in the result is determined by the inexact experimental relationship that exactly 1 liter is equal to 1.06 quarts.

1.12 Mass and Weight

We said in Section 1.2 that the mass of a body is the quantity of matter that it contains. The mass of a body of matter is an invariable quantity. On the other hand, the **weight** of a body is the force that gravity exerts on the body and is variable, since the attraction is dependent on the distance of the body from a planet's center. If this book were taken up in an airplane, it would weigh less than it does at sea level. If it were taken far out into space, its weight would become negligible. It is well known that astronauts experience weightlessness while in outer space. However, neither the mass of the book nor that of the astronauts changes as the distance from the center of the earth changes. Scientists measure quantities of matter in terms of mass rather than weight because the mass of a body remains constant whereas the weight of a body is an accident of its environment. It should be noted, however, that the term *weight* is often used loosely for the mass of a substance.

The mass of an object is found by comparing it to other objects of known mass. The instrument used to make this comparison is called a **balance.** Figure 1-6 shows a two-pan balance. The object to be weighed is placed in one pan, and weights of known mass are added to the other until the two pans balance and the pointer comes to the center of the scale. At this point both pans contain equal masses, and the mass of the object equals the total mass of the weights added. Weighing on a balance makes use of the fact that the gravitational attraction for objects of equal mass is the same; i.e., their weights are equal. A modern analytical balance is shown in Fig. 1-7.

The standard object to which all SI and other metric units of mass are referred is a cylinder of platinum-iridium alloy, which is also kept in France. The unit of mass, the **kilogram (kg),** is defined as the mass of this cylinder; 1 kilogram is about 2.2 pounds. The **gram (g)** is equal to exactly 0.001 times the mass of the standard kilogram (1×10^{-3} kg) and is very nearly equal to the weight of 1 cubic centimeter of water at 4°C, the temperature at which water has its maximum density. A dime has a mass of about 2.3 grams.

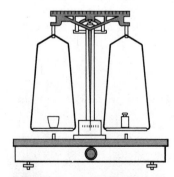

Figure 1-6. A traditional two-pan balance. Weights of known mass are added to the right pan until they balance the object in the left pan. At this point both pans contain equal weights.

1.13 Density and Specific Gravity

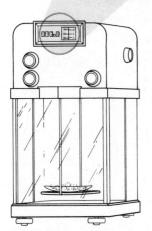

Figure 1-7. A modern analytical balance. Inside the upper part of the balance is a beam with a counterweight at the back and movable weights at the front (the end of the beam with the balance pan). When an object is placed in the pan, the front end of the beam becomes heavier than the back. The dials are used to *remove* weights until balance is re-established. The mass of the object is equal to the mass removed to restore the balance.

One of the physical properties of a solid, a liquid, or a gas is its density. **Density is defined as mass per unit volume.** This may be expressed mathematically as

$$\text{Density} = \frac{\text{mass}}{\text{volume}} \quad \text{or} \quad D = \frac{M}{V}$$

Substances can often be distinguished by measuring their densities since any two substances usually have different densities (Table 1-4).

Table 1-4 Densities of Common Materials at 25°C and 1 atm

Air	1.29 g/L
Helium gas	0.179 g/L
Water	0.997 g/cm³
Glycerin	1.26 g/cm³
Mercury	13.6 g/cm³
Salt	2.17 g/cm³
Iron	7.86 g/cm³
Silver	10.5 g/cm³

The term **specific gravity** denotes the ratio of the density of a substance to the density of a reference substance. The reference substance for solids and liquids is usually water. Common reference substances used in specifying the specific gravities for gases are air and hydrogen.

$$\text{Specific gravity of substance} = \frac{\text{density of substance}}{\text{density of reference substance}}$$

Note that specific gravity, being the ratio of two densities, has no units. The units in the numerator and denominator cancel each other.

When measured in the metric system of units, the density of any substance has practically the same numerical value as its specific gravity referred to water as the reference substance. For example, the density of glycerin is 1.26 g/cm³, whereas its specific gravity referred to water (0.997 g/cm³) as the reference substance is 1.26 (no units).

EXAMPLE 1.5   Calculate the density and specific gravity of a body that has a mass of 321 g and a volume of 45.0 cm³ at 25°C.

STEP 1 $$\text{Density} = \frac{\text{mass}}{\text{volume}} = \frac{321 \text{ g}}{45.0 \text{ cm}^3} = 7.13 \text{ g/cm}^3$$

STEP 2 $$\text{Specific gravity} = \frac{\text{density of sample}}{\text{density of water}}$$

The density of water at 25°C is 0.99707 g/cm³.

$$\text{Specific gravity} = \frac{7.13 \text{ g/cm}^3}{0.99707 \text{ g/cm}^3} = 7.15$$

EXAMPLE 1.6   What is the mass in kilograms of 10.5 gal (39.7 L) of gasoline with a specific gravity of 0.82?

Because gasoline has a specific gravity of 0.82, the density of 1 mL of gasoline is 0.82 times that of 1 mL of water.

$$\text{Density} = \text{specific gravity} \times \text{density of water}$$
$$= 0.82 \times 0.997 \text{ g/cm}^3$$
$$= 0.82 \text{ g/cm}^3 \text{ (or 0.82 g/mL)}$$

Conversion of 39.7 L to mL followed by multiplication by the mass of 1 mL of gasoline gives the total mass.

$$\text{Volume} = 39.7 \, \cancel{L} \times \frac{1000 \text{ mL}}{1 \, \cancel{L}}$$
$$= 39,700 \text{ mL}$$

$$\text{Mass} = \text{density} \times \text{volume}$$
$$= 0.82 \text{ g/}\cancel{mL} \times 39,700 \, \cancel{mL}$$
$$= 32,000 \text{ g} = 32 \text{ kg}$$

1.14 Temperature and Its Measurement

The word **temperature** refers to the hotness or coldness of a body of matter. In order to measure a change in temperature, some physical property of a substance that varies with temperature must be used. Practically all substances expand when temperature increases and contract when it decreases. This property is the basis for the common glass thermometer. The mercury or alcohol in the tube rises when the temperature increases because the volume of the mercury or alcohol expands more than that of the glass container.

To agree on a set of temperature values, we need **standard reference temperatures** that can be readily determined. Two such temperatures are the freezing and boiling points of water at a pressure of 1 atmosphere. On the **Celsius** scale the freezing point of water is taken as 0° and the boiling point as 100°. The heights of the mercury column at these two reference temperatures determine the 0° and 100° points on a thermometer. The space between these two points is divided into 100 equal intervals, or degrees. On the **Fahrenheit** scale the freezing point of water is taken as 32° and the boiling point as 212°. The space between these two points on a thermometer is divided into 180 equal parts, or degrees. Thus, a degree Fahrenheit is exactly $\frac{100}{180}$, or $\frac{5}{9}$, of a degree Celsius (Fig. 1-8). The relationships are shown by the equations

$$\frac{°F - 32}{180} = \frac{°C}{100}, \qquad °C = \frac{5}{9}(°F - 32), \qquad °F = \frac{9}{5}°C + 32$$

The readings below 0° on either scale are treated as negative.

The SI unit of temperature is called the **kelvin (K),** named after Lord Kelvin, a British physicist. The zero point on the Kelvin scale corresponds to $-273.15°C$. The size of the degree on the Kelvin scale is the same as that on the Celsius scale.

Figure 1-8. The relationships among the Kelvin, Celsius, and Fahrenheit temperature scales.

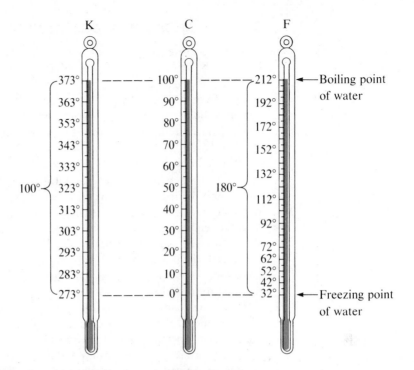

The freezing temperature of water on the Kelvin scale is 273.15 K (0°C), and the boiling temperature is 373.15 K (100°C). A temperature on the Celsius scale is converted to its equivalent on the Kelvin scale by adding 273.15 to the Celsius reading; a temperature on the Kelvin scale is converted to its equivalent on the Celsius scale by subtracting 273.15 from the Kelvin reading. (Note that temperatures on the Kelvin scale are, by convention, reported without the degree sign.)

$$K = °C + 273.15, \qquad °C = K - 273.15$$

Figure 1-8 shows the relationships among the three temperature scales. Temperatures in this book are in Celsius unless otherwise specified. The following examples demonstrate conversions between the three scales.

EXAMPLE 1.7   The boiling temperature of grain alcohol is 78.5°C at 1 atmosphere of pressure. What is its boiling point on the Fahrenheit scale and on the Kelvin scale?

$$°F = \frac{9}{5}°C + 32 = \left(\frac{9}{5} \times 78.5\right) + 32 = 141 + 32 = 173°F$$

$$K = °C + 273.15 = 78.5 + 273.15 = 351.6 \text{ K}$$

EXAMPLE 1.8   Convert 50°F to the Celsius scale and the Kelvin scale.

$$°C = \frac{5}{9}(°F - 32) = \frac{5}{9}(50 - 32) = \frac{5}{9} \times 18 = 10°C$$

$$K = °C + 273.15 = 10 + 273.15 = 283 \text{ K}$$

1.15   Heat and Its Measurement

The temperature of an object tells us how hot or cold the object is. We can increase the temperature of the object by adding heat, or some other form of energy, to it or decrease the temperature by removing heat, or some other form of energy, from it. For example, when hot coffee is poured into a mug, the coffee becomes cooler and the mug, warmer. Heat moves from the coffee, resulting in a decrease in its temperature, into the mug, thus increasing its temperature.

The SI unit of heat is the joule. A **joule (J)** is equal to the kinetic energy ($\frac{1}{2} Mv^2$) of an object with mass M of exactly 2 kilograms moving with a velocity v of exactly 1 meter per second. The **kilojoule (kJ)** is equal to 10^3 joules. The older metric unit for heat, as for other forms of energy, is the **calorie (cal).** For many years the calorie was defined as the amount of heat, or other energy, necessary to raise the temperature of exactly 1 gram of water from 14.5°C to 15.5°C. With the adoption of SI units, the calorie was redefined to be exactly 4.184 joules.

Specific heat is a physical property of a substance that may be used to measure quantities of heat. The **specific heat** of a substance is the quantity of heat required to raise the temperature of 1 gram of the substance 1 degree Celsius (or kelvin). Every substance has its own specific heat. The specific heat of water, which is one of the largest specific heats, is 4.184 joules per gram per degree (4.184 J/g °C). The specific heat of copper is 0.38 joule per gram per degree, that of aluminum is 0.88 joule per gram per degree, and that of zinc is 0.39 joule per gram per degree. Thus if a 1-gram sample of aluminum picks up enough heat to raise its temperature by 1 degree, it must have absorbed 0.88 joule.

The specific heat of a substance describes a property of 1 gram of the substance. The heat capacity of a given quantity of matter describes a property of the entire quantity of matter. The **heat capacity** of a body of matter is the quantity of heat required to increase its temperature by 1 degree Celsius (or kelvin).

$$\text{Heat capacity} = \text{specific heat (J/g °C)} \times \text{mass (g)}$$

For example, 10 grams of aluminum (specific heat = 0.88 J/g °C) has a heat capacity of 8.8 J/°C; 25 grams of aluminum, a heat capacity of 22 J/°C. The greater the mass of a substance, the greater its heat capacity.

The following examples deal with the measurement of heat.

EXAMPLE 1.9   Calculate the quantity of heat in joules required to raise the temperature of 135 g of water from 11°C to 41°C.

$$\text{Heat required} = \text{mass} \times \text{specific heat} \times \text{temperature change}$$

$$= 135 \text{ g} \times 4.184 \frac{\text{J}}{\text{g °C}} \times (41 - 11)°C$$

$$= 1.7 \times 10^4 \text{ J}$$

EXAMPLE 1.10   Calculate the quantity of heat in calories required to raise the temperature of 3.00×10^2 g of mercury from 15°C to 95°C. The specific heat of mercury is 0.033 cal/g °C.

$$\text{Heat required} = \text{mass} \times \text{specific heat} \times \text{temperature change}$$

$$= 3.00 \times 10^2 \text{ g} \times 0.033 \frac{\text{cal}}{\text{g} \cdot ^\circ\text{C}} \times (95 - 15)^\circ\text{C}$$

$$= 790 \text{ cal}$$

The calculated answer is 792 cal. However, only two significant figures are justified, and hence the answer is properly expressed in two significant figures as 790 cal. It could be expressed exponentially as 7.9×10^2 cal.

EXAMPLE 1.11 Calculate the temperature rise that results when 1625 J are supplied to 75 g of ethyl alcohol, the specific heat of which is 2.4 J/g °C.

$$\text{Heat required} = \text{mass} \times \text{specific heat} \times \text{temperature change}$$

$$\text{Temperature change} = \frac{\text{heat required}}{\text{mass} \times \text{specific heat}}$$

$$= \frac{1625 \text{ J}}{75 \text{ g} \times 2.4 \dfrac{\text{J}}{\text{g} \cdot ^\circ\text{C}}} = 9.0\,^\circ\text{C}$$

FOR REVIEW

SUMMARY

Since chemistry deals with the composition, structure, and properties of matter and with the reactions by which various forms of matter may be interconverted, it occupies a central place in the study of science and technology.

Matter is anything that occupies space and has **mass.** The basic building block of matter is the **atom,** the smallest unit of an element that can combine with other elements. In many substances, atoms are combined into **molecules.** Matter exists in three states: **solids,** of fixed shape and essentially incompressible; **liquids,** of variable shape but essentially incompressible; and **gases,** of variable shape and compressible or expansible. Most matter is a **mixture** of substances. These mixtures can be **heterogeneous** or **homogeneous.** Homogeneous mixtures are called **solutions.** Pure substances are also homogeneous. A **pure substance** may be either an **element** composed of only one type of atom or a **compound** consisting of two or more types of atoms. A pure substance exhibits characteristic **chemical and physical properties** by which it can be identified.

Chemistry is a science based on laws and theories that are derived from observations by the **scientific method.** Quantitative measurements utilize the metric system, usually with **SI units.** Common units in use include the **gram** and **kilogram** for mass, the **centimeter** and **meter** for length, the **liter** (cubic decimeter) and **milliliter** (cubic centimeter) for volume, **kelvin** or degrees **Celsius** for temperature, and **joules** and **calories** for heat. Experimental measurements have an associated error that can be expressed by attention to significant figures. When using quantities in calculations, you must keep track of the units involved.

KEY TERMS AND CONCEPTS

Note: Section numbers are given for each term.

Atom (1.6)
Calorie (1.15)
Chemical change (1.4)
Chemical properties (1.4)
Compound (1.5)
Density (1.13)
Element (1.5)
Energy (1.2)
Gas (1.2)
Gram (1.12)
Heat capacity (1.15)
International System of Units (SI Units) (1.8)
Joule (1.15)
Kelvin (1.14)
Kilogram (1.12)
Liquid (1.2)
Liter (1.11)
Mass (1.2)
Matter (1.2)
Meter (1.11)
Mixture (1.5)
Molecule (1.6)
Physical change (1.4)
Physical properties (1.4)
Scientific method (1.7)
Significant figures (1.10)
Solid (1.2)
Solutions (1.5)
Specific gravity (1.13)
Specific heat (1.15)
Temperature (1.14)
Unit conversion factor (1.9)

EXERCISES

Note: In all exercises the symbol [S] before the number indicates that the solution to that problem is worked out in the manual prepared by Meiser, Ault, Holtzclaw, and Robinson titled *Problems and Solutions for College Chemistry and General Chemistry, Seventh Editions, by Holtzclaw, Robinson, and Nebergall.*

You should make a special effort to develop the good habit of expressing answers to problems to the correct number of significant figures. To this end we have paid careful attention to significant figures in the answers provided. Our answers were obtained by carrying all figures in the calculator, and rounding off the final answer. If you choose to do the problems in steps, rounding off after each step, you may get answers that differ from ours by one or two units in the least significant figure. These differences do not indicate an error but simply reflect two equally acceptable, but different, ways of rounding off.

Introduction

1. With what is the science of chemistry concerned?
2. What properties distinguish each of the three states of matter?
3. Describe a chemical change that illustrates the Law of Conservation of Matter.
4. Why is a consideration of energy, which is not matter, so important in chemistry, which is the study of matter?
5. How does the example of hot coffee poured into a cold mug (Section 1.15) illustrate the concept of conservation of energy?
6. Trace the progression from the formulation of a problem, or question, to the establishment of a theory.

Classification of Matter

7. Explain how you would distinguish between an element, a compound, and a mixture.
8. How does a heterogeneous mixture differ from a homogeneous mixture? In what ways are they alike?
9. Classify each of the following as a heterogeneous mixture or a homogeneous mixture:
   (a) blood
   (b) ocean water
   (c) air
   (d) blueberry pancakes
   (e) pancake syrup
   (f) gasoline
   (g) a milkshake
   (h) concrete
   (i) a bowl of vegetable soup
10. How does a pure substance differ from a homogeneous mixture? In what ways are they alike?

11. Classify each of the following homogeneous materials as a solution, an element, or a compound:
(a) iron (e) oxygen
(b) water (f) sulfur
(c) shampoo (g) glucose
(d) carbon dioxide (h) blood plasma

12. Classify each of the following changes as physical or chemical:
(a) condensation of steam
(b) burning of gasoline
(c) souring of milk
(d) dissolving of sugar in water
(e) melting of gold
(f) "turning" of leaves to color
(g) explosion of a firecracker
(h) magnetizing of a screwdriver
(i) melting of ice

13. How does an atom differ from a molecule? In what ways are they alike?

14. A helium atom and a helium molecule are identical, while a sulfur atom and a sulfur molecule are not. What is the difference? Figure 1-3 may prove helpful.

15. How do molecules of elements and compounds differ?

Significant Figures

16. The addition of two numbers, each of which contains three significant figures, can give an answer that has three or four significant figures. Explain how both cases can arise, and provide an example of each.

17. Indicate whether each of the following can be determined exactly or must be measured with some degree of uncertainty:
(a) the number of eggs in a basket
(b) the mass of a dozen eggs
(c) the number of gallons of gasoline necessary to fill an automobile gas tank
(d) the number of cm in exactly 2 m
(e) the mass of this textbook
(f) the number of seconds in 1 day
(g) the time required to drive from San Francisco to Kansas City at an average speed of 53 mi/h

18. In a 1950 handbook 1 inch (in) was given as being equal to 2.540005 cm. Since that time 1 in has been defined as being exactly 2.54 cm. Are there more, less, or the same number of significant figures in the newer value than in the older one?

19. Tell how many significant figures are contained in each of the following numbers:
(a) 38.7 (e) 2×10^{18}
(b) 3,486,002 (f) 9.74150×10^{-4}
(c) 0.0613 (g) 17.0
(d) 0.01400

Ans. 3; 7; 3; 4; 1; 6; 3

20. Round off each of the following numbers to two significant figures:
(a) 0.436 (d) 9.000
(b) 27.2 (e) 135
(c) 1.497×10^{-3} (f) 0.445

Ans. 0.44; 27; 1.5×10^{-3}; 9.0; 140; 0.44

21. Express each of the following numbers in exponential notation:
(a) 711.0 (e) 0.05499
(b) 0.239 (f) 10000.0
(c) 90743 (g) 0.000000738592
(d) 134.2

Ans. 7.110×10^2; 2.39×10^{-1}; 9.0743×10^4; 1.342×10^2; 5.499×10^{-2}; 1.00000×10^4; 7.38592×10^{-7}

22. Perform the following calculations, and give each answer with the correct number of significant figures.
(a) 628×342 *Ans. 215,000 (or 2.15×10^5)*
[S](b) $(5.63 \times 10^2) \times (7.4 \times 10^3)$ *Ans. 4.2×10^6*
[S](c) $\frac{2734}{28.0}$ *Ans. 97.6*
(d) 8119×0.000023 *Ans. 0.19*
(e) $14.98 + 27.340 + 84.7593$ *Ans. 127.08*
[S](f) $\frac{42.7 + 0.259}{28.4445}$ *Ans. 1.51*

23. Assume it is necessary to determine the density of a solution to three significant figures. The volume of solution can be measured to the nearest 0.1 cm³.
(a) What is the minimum volume of sample that can be used for the measurement?
Ans. 10.0 cm³
(b) Assuming the minimum-volume sample determined in (a), how accurately must the sample be weighed (to the nearest 0.1 g, 0.01 g, . . .) if the density of the solution is greater than 1.00 g/cm³?
Ans. Nearest 0.1 g

Metric System and SI Units

24. Much of Chapter 1 deals with measurement. Why is measurement so important in chemistry?
25. Distinguish between the terms *mass* and *weight*.
26. Why is it unnecessary to consider the distance from the earth's center when the mass of an object is determined by weighing it on an analytical balance as described in Section 1.12?
27. Indicate the SI base units that appropriately express the following measurements:
    (a) the length of a marathon race
    (b) the mass of an automobile
    (c) the volume of a swimming pool
    (d) the speed of an airplane
    (e) the density of gold
    (f) the area of a football field
    (g) the maximum temperature at the South Pole on April 1, 1913
28. Name the prefixes used with SI units to indicate multiplication by the following exact quantities: 10^3; 10^{-2}; 0.1; 10^{-3}; 1,000,000; 0.000001.
29. Use exponential notation to express the following quantities in terms of SI base units (see Section 1.8):

    Ⓢ(a) 0.13 g   *Ans. 1.3×10^{-4} kg*
    Ⓢ(b) 232 Gg   *Ans. 2.32×10^8 kg*
    (c) 5.23 pm   *Ans. 5.23×10^{-12} m*
    (d) 86.3 mg   *Ans. 8.63×10^{-5} kg*
    (e) 37.6 cm   *Ans. 3.76×10^{-1} m*
    (f) 54 Mm   *Ans. 5.4×10^7 m*
    (g) 1 Ts   *Ans. 1×10^{12} s*
    (h) 27 ps   *Ans. 2.7×10^{-11} s*
    (i) 0.15 mK   *Ans. 1.5×10^{-4} K*
30. Many medical laboratory tests are run using 5.0 μL of blood serum. What is this volume in liters and in milliliters?   *Ans. 5.0×10^{-6} L; 5.0×10^{-3} mL*
31. Complete the following conversions.
    (a) 612 g = _____ kg   *Ans. 0.612 kg*
    (b) 8.160 m = _____ mm   *Ans. 8.160×10^3 mm*
    (c) 3779 mg = _____ g   *Ans. 3.779 g*
    (d) 781 mL = _____ L   *Ans. 0.781 L*
    (e) 4.18 kg = _____ mg   *Ans. 4.18×10^6 mg*
    (f) 27.8 m³ = _____ L   *Ans. 2.78×10^4 L*
    (g) 0.13 mL = _____ cm³   *Ans. 0.13 cm³*
    (h) 17.38 km = _____ cm   *Ans. 1.738×10^6 cm*

Conversion of Units

A table of useful conversion factors can be found in Appendix C.

32. The mass of a competition-type Frisbee is 125 g. What is the mass in ounces?   *Ans. 4.41 oz*
Ⓢ33. Calculate the length of a football field (100.0 yd) in meters and kilometers.   *Ans. 91.41 m; 9.141×10^{-2} km*
34. Soccer is played with a round ball having a circumference between 27 and 28 in and a weight between 14 and 16 oz. What are these specifications in the metric units of centimeters and grams?   *Ans. 69–71 cm; 400–450 g*
Ⓢ35. How many milliliters of a soft drink are contained in a 12.0-oz can?   *Ans. 355 mL*
36. A barrel of oil is exactly 42 gal. How many liters of oil are in a barrel?   *Ans. 159.0 L*
37. A track and field athlete is 160 cm tall and weighs 45.8 kg. Is she more likely to be a distance runner or a shot-putter?   *Ans. At 5 ft 3 in and 101 lb, a runner*
38. The diameter of a red blood cell is about 3×10^{-4} in. What is its diameter in centimeters?   *Ans. 8×10^{-4} cm*
39. Is a 197-lb weight lifter light enough to compete in a class limited to those weighing 90 kg or less?   *Ans. Yes, weight = 89.4 kg*
40. A very good 197-lb weight lifter lifted 152 kg in the snatch and 192 kg in the clean and jerk. What were the masses of the weights lifted in pounds?   *Ans. 335 lb; 423 lb*
Ⓢ41. Gasoline is now often sold by the liter. How many liters are required to fill a 12.0-gal gas tank?   *Ans. 45.4 L*
Ⓢ42. In a recent Belgian Grand Prix, the leader turned a lap with an average speed of 182.83 km/h. What was his speed in miles per hour, meters per second, and feet per second?   *Ans. 113.61 mi/h; 50.786 m/s; 166.62 ft/s*
43. In 1980, a quantity of 490,000,000 lb of rayon was manufactured in the United States for use in carpets, automobile tires, and fabrics. What is this mass in kilograms?   *Ans. 2.2×10^8 kg*
44. If the western Antarctic ice sheet and the Arctic Sea ice melted due to a climatic change, the level of the oceans could rise about 6 m. How much would sea level rise in feet?   *Ans. 20 ft (one significant figure)*

45. Make the conversion indicated in each of the following:
(a) the record long jump, 29 ft 2.5 in, to meters *Ans. 8.903 m*
[S](b) the height of Mt. Kilimanjaro, at 19,565 ft the highest mountain in Africa, to kilometers *Ans. 5.9634 km*
(c) the greatest depth of the ocean, about 6.5 mi, to meters *Ans.* 1.0×10^4 *m*
[S](d) the area of the state of Oregon, 96,981 mi^2, to square kilometers *Ans.* 2.5118×10^5 km^2
(e) the volume of 1 gill, exactly 4 oz, to milliliters *Ans. 118.3 mL*
[S](f) the displacement of an automobile engine, 161 in^3, to liters *Ans. 2.64 L*
(g) the estimated volume of the oceans, 330,000,000 mi^3, to cubic kilometers. *Ans.* 1.4×10^9 km^3
(h) the volume of 1 gal of milk to liters *Ans. 3.785 L*
(i) the mass of a bushel of rye, 32.0 lb, to kilograms *Ans. 14.5 kg*
(j) the mass of a 2.3-oz egg to grams *Ans. 65 g*
(k) the mass of a 5.00-grain aspirin to milligrams (1 grain = 0.00229 oz) *Ans. 325 mg*
(l) the mass of a 3525-lb car to kilograms *Ans. 1599 kg*

46. Solutions are often prepared in chemistry laboratories in 250.0-mL flasks. What is the volume of such a flask in fluid ounces and in pints? *Ans. 8.454 oz; 0.5284 pt*

47. The distance between the centers of the two oxygen atoms in an oxygen molecule is 1.21 Å (angstrom units). What is this distance in centimeters and in inches? (One Å = exactly 1×10^{-8} cm.) *Ans.* 1.21×10^{-8} *cm;* 4.76×10^{-9} *in*

[S]48. If a line of 1.0×10^8 water molecules is 1.00 in long, what is the average diameter of a water molecule in centimeters, angstrom units, and nanometers? *Ans.* 2.5×10^{-8} *cm; 2.5 Å; 0.25 nm*

[S]49. The density of liquid bromine is 2.928 g/cm^3. What is the mass of 25 mL of bromine? What is the volume of 100.0 g of bromine? *Ans. 73 g; 34.15* cm^3

50. Calculate the density of aluminum if 27.6 cm^3 has a mass of 74.6 g. *Ans. 2.70 g/*cm^3

[S]51. Osmium is the densest element known. What is its density if 2.72 g has a volume of 0.121 cm^3? *Ans. 22.5 g/*cm^3

52. What is the mass of each of the following?
(a) 6.00 cm^3 of mercury, density = 13.5939 g/cm^3 *Ans. 81.6 g*
(b) 25.0 mL of octane, density = 0.702 g/cm^3 *Ans. 17.6 g*
(c) 4.00 cm^3 of sodium, density = 0.97 g/cm^3 *Ans. 3.9 g*
(d) 125 mL of gaseous chlorine, density = 3.16 g/L *Ans. 0.395 g*

53. What is the volume of each of the following?
(a) 25 g of iodine, density = 4.93 g/cm^3 *Ans. 5.1 mL*
[S](b) 3.28 g of gaseous hydrogen, density = 0.089 g/L *Ans. 37 L*
(c) 11.3 g of graphite, density = 2.25 g/cm^3 *Ans. 5.02 mL*

[S]54. What is the specific gravity of uranium at 25 °C relative to water if 37.4 g of uranium has a volume of 2.00 cm^3? *Ans. 18.8*

55. What is the specific gravity of natural gas (density = 0.893 g/L) with respect to air (density = 1.2929 g/L)? with respect to hydrogen (density = 0.08987 g/L)? *Ans. 0.691; 9.94*

Heat and Temperature

56. A burning match and a bonfire may have the same temperature, yet you would not sit around a burning match on a fall evening in order to stay warm. Why not?

57. Explain the difference between the heat capacity and the specific heat of a substance.

58. Normal body temperature is 98.6 °F. What is this temperature on the Celsius and Kelvin scales? *Ans. 37.0 °C, 310.2 K*

[S]59. The *Voyager 1* flyby of Saturn revealed that the surface temperature of the moon Titan is 93 K. What is the surface temperature in degrees Celsius and degrees Fahrenheit? *Ans. −180 °C; −292 °F*

60. The temperature 0 K is equal to −273.15 °C. What is 0 K equal to on the Fahrenheit scale? *Ans. −459.67 °F*

61. Make the conversion indicated in each of the following:
[S](a) temperature of the interior of a rare roast, 125 °F, to degrees Celsius *Ans. 52 °C*

(b) temperature of scalding water, 54°C, to degrees Fahrenheit and kelvin *Ans. 129°F; 327 K*

(c) coldest area in a freezer, −10°F, to degrees Celsius and kelvin *Ans. −23°C; 250 K*

[S](d) temperature of dry ice, −77°C, to degrees Fahrenheit and kelvin *Ans. −107°F; 196 K*

(e) the melting temperature of gold, 1336 K, to degrees Celsius and degrees Fahrenheit *Ans. 1063°C; 1945°F*

(f) the boiling temperature of gold, 2966°C, to degrees Fahrenheit and kelvin *Ans. 5371°F; 3239 K*

(g) the boiling temperature of liquid ammonia, −28.1°F, to degrees Celsius and kelvin *Ans. −33.4°C; 239.8 K*

[S]62. If 1.506 kJ of heat are added to 30.0 g of water at 26.5°C, what is the resulting temperature of the water? *Ans. 38.5°C*

63. How much heat, in joules and in calories, must be added to a 75.0-g iron block to increase its temperature from 25°C to its melting temperature? Its melting temperature is 1535°C, and its specific heat is 0.451 J/g °C. *Ans. 5.11×10^4 J; 1.22×10^4 cal*

[S]64. The specific heat of ice is 1.95 J/g °C. How much heat in joules and calories is required to heat a 35.5-g ice cube from −23.0°C to −1.0°C? *Ans. 1.52×10^3 J; 364 cal*

65. Calculate the heat capacity of each of the following:

[S](a) 22.6 g of water (specific heat = 4.184 J/g °C) *Ans. 94.6 J/°C*

(b) 2.37 kg of lead (specific heat = 0.129 J/g °C) *Ans. 306 J/°C*

(c) 45.8 g of nitrogen gas (specific heat = 1.04 J/g °C) *Ans. 47.6 J/°C*

66. Calculate the specific heat of water in the unit cal/lb °F. The specific heat of water is 4.184 J/g °C. *Ans. 252.0 cal/lb °F*

[S]67. How many milliliters of water at 25°C must be mixed with 150 mL (about 5 oz) of coffee at 100°C so that the resulting combination will have a temperature of 70°C? Assume that coffee and water have the same density and specific heat. *Ans. 100 mL*

68. How much will the temperature of 225 mL of coffee at 95°C be reduced when a 45-g silver spoon (specific heat 0.24 J/g °C) at 25°C is placed in the coffee and the two are allowed to reach the same temperature? Assume that coffee and water have the same density and specific heat. *Ans. 1°C*

69. A 10.0-kg piece of metal at 50.0°C is placed in 1.00 kg of water at 10.0°C. The metal and water come to the same temperature of 31.4°C. What is the specific heat of the metal? *Ans. 0.481 J/g °C*

Additional Exercises

70. How do the densities of most substances change as a result of an increase in temperature?

71. Explain how you could determine whether the outside temperature is higher or lower than 0°C without using a thermometer.

72. In 1980 a quantity of 18.4 billion lb of phosphoric acid was manufactured in the United States for use in fertilizer. The average cost of this acid was $318 per ton (1 ton = 2000 lb). What was the total value of the phosphoric acid produced? What was its cost per kilogram? *Ans. $\$2.93 \times 10^9$; $0.351/kg*

73. Given that 1 bushel (bu) is defined as 32 dry qt (1 dry qt = 1.1012 L), what is the mass in kilograms of a liter of each of the following types of grain?

(a) wheat, 60.0 lb/bu *Ans. 0.772 kg/L*

(b) oats, 32.0 lb/bu *Ans. 0.412 kg/L*

[S](c) corn, 56.0 lb/bu *Ans. 0.721 kg/L*

(d) barley, 48 lb/bu *Ans. 0.62 kg/L*

74. In the United States there are about 2.5 million farms. Annually, these farms consume some 2.4×10^{10} L of gasoline and diesel fuel, 4.90×10^9 m^3 of natural gas, 5.7×10^9 L of liquefied petroleum gas (LPG), and 1.16×10^{14} kJ of electricity. How many gallons of gasoline and diesel fuel, cubic feet of natural gas, gallons of LPG, and kilowatt-hours (1 kWh = 3.6×10^6 J) of electricity were consumed? *Ans. 6.3×10^9 gal; 1.73×10^{11} ft^3; 1.5×10^9 gal; 3.22×10^{10} kWh*

[S]75. Moon rock is estimated to contain about 0.1% water. What mass of moon rock (in pounds) would a moon base have to process to recover 1 L of water (density = 1.0 g/cm^3)? *Ans. 2000 lb*

[S]76. A solid will float in a liquid that has a density greater than its own. A pound of butter forms a block about $6.5 \times 6.5 \times 12.0$ cm. Will this butter float in water (density = 1.0 g/cm³)?
    Ans. Yes, its density is 0.89 g/cm³

77. A 5.0-pt bottle of sulfuric acid contains 9.0 lb of sulfuric acid. What is the density of sulfuric acid in grams per cubic centimeter?
    Ans. 1.7 g/cm³

78. Commercial solvents are commonly sold in 55-gal drums. What is the weight in pounds of the carbon tetrachloride (density = 1.5942 g/cm³) contained in a 55-gal drum?   *Ans. 730 lb*

79. A European recipe calls for the use of 1.50 kg of milk. An American who wants to follow the recipe has no balance and only volume-measuring containers marked in ounces. Assuming the density of milk is 1.03 g/cm³, what volume of milk in ounces is needed?   *Ans. 49.2 oz*

80. The density of aluminum is 2.70 g/cm³. Tell which of the following masses of aluminum occupies the greatest volume: 1.0 lb, 0.50 kg, 0.050 L.   *Ans. 0.50 kg*

81. What is the density of sugar (sucrose) in grams per cubic centimeter if 100.0 lb occupies 1.008 ft³?   *Ans. 1.589 g/cm³*

[S]82. What mass, in kilograms, of concentrated hydrochloric acid is contained in a standard 5.0-pt container? The specific gravity of concentrated hydrochloric acid is 1.21.   *Ans. 2.9 kg*

83. Assume that a 450-lb automobile engine is made entirely of iron with a specific heat of 0.45 J/g °C. How much heat, in calories, is required to raise the engine from a starting temperature of 25°C to its operating temperature of 185°C?   *Ans. 3.5×10^6 cal*

2

SYMBOLS, FORMULAS, AND EQUATIONS; ELEMENTARY STOICHIOMETRY

When speaking and writing about matter and the changes that it undergoes, chemists use symbols, formulas, and equations to indicate the elements present, the relative amounts of each element, and how the combinations of elements vary during a chemical change. In addition, they can use this information along with the masses of the substances involved to determine how much product will form during the course of a reaction or how much reactant will be required.

Measurement of changes in mass and the interpretation of the meaning of these changes have played an important role in the development of the science. For example, there was little real progress in the development of chemistry before the French chemist Lavoisier (1743–1794) put this science on a quantitative basis. His simple observation that mass is gained when metals are burned (oxidized) had a profound effect on the chemical theories of the time and provided the basis for the later work of John Dalton and his contemporaries.

2.1 Symbols and Formulas

Symbols are abbreviations the chemist uses to indicate elements. Several common elements and their symbols are listed in Table 2-1. Some of these symbols are abbreviations of the Latin names of the elements. To avoid confusion with other notations, the second letter of a two-letter symbol is never capitalized: Co is the symbol for the element cobalt; CO is the notation for the compound carbon

Table 2-1 Some Common Elements
and Their Symbols

Aluminum	Al
Bromine	Br
Calcium	Ca
Carbon	C
Chlorine	Cl
Chromium	Cr
Cobalt	Co
Copper	Cu (from *cuprum*)
Fluorine	F
Gold	Au (from *aurum*)
Helium	He
Hydrogen	H
Iodine	I
Iron	Fe (from *ferrum*)
Lead	Pb (from *plumbum*)
Magnesium	Mg
Mercury	Hg (from *hydrargyrum*)
Nitrogen	N
Oxygen	O
Potassium	K (from *kalium*)
Silicon	Si
Silver	Ag (from *argentum*)
Sodium	Na (from *natrium*)
Sulfur	S
Tin	Sn (from *stannum*)
Zinc	Zn

monoxide containing the elements carbon (C) and oxygen (O). The symbols for all known elements are given on the inside front cover of this book.

A **formula** represents the composition of a substance, using symbols for the elements present in the substance. Thus NaCl is the formula for sodium chloride (common table salt) and identifies the elements sodium (Na) and chlorine (Cl) as its constituents. Subscripts are used to indicate the relative numbers of atoms of each type in the compound, but only if more than one atom of a given element is present. For example, the formula for water, H_2O, indicates that each molecule contains two atoms of hydrogen and one atom of oxygen. The formula of sodium chloride, NaCl, indicates the presence of equal numbers of atoms of the elements sodium and chlorine. The formula for aluminum sulfate, $Al_2(SO_4)_3$, specifies two atoms of aluminum, Al, for each three sulfate, SO_4, groups. Each sulfate group contains one atom of sulfur and four atoms of oxygen. Hence the formula shows a total of two atoms of aluminum, three atoms of sulfur, and twelve atoms of oxygen. The ratios obtained from such formulas are exact to any number of significant figures.

A **molecular formula** gives the actual numbers of atoms of each element in a molecule. An **empirical formula** (sometimes called a simplest formula) gives the simplest whole-number ratio of atoms. For example, the molecular formula H_2O indicates that water has exactly two atoms of hydrogen and one atom of oxygen in one molecule (2 atoms H/1 atom O). The simplest ratio of atoms is 2:1, so the empirical formula is also H_2O. (A subscript of 1 is not written.) The molecular formula C_6H_6 indicates that the benzene molecule is composed of exactly six carbon atoms and six hydrogen atoms. The simplest ratio of these atoms is 1:1. Thus the empirical formula of benzene is CH.

Some molecules contain a single type of atom; their formulas contain only a single symbol. A molecule of the most common form of sulfur consists of eight atoms (see Fig. 1-3 in Chapter 1); the molecular formula is S_8. The molecular formulas for elemental hydrogen and oxygen (diatomic molecules) are H_2 and O_2, while the formula for neon (a monatomic molecule) is Ne.

Note that H_2 and 2H do not mean the same thing. The formula H_2 represents a molecule of hydrogen consisting of two atoms of the element, chemically combined. The expression 2H, on the other hand, indicates that the two hydrogen atoms are not in combination as a unit but are separate particles (Fig. 2-1).

Figure 2-1. Illustration of the difference in meaning between the symbols H, 2H, and H_2.

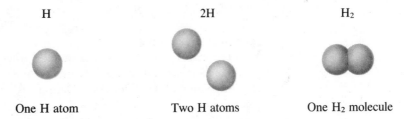

H	2H	H_2
One H atom	Two H atoms	One H_2 molecule

Many compounds do not contain discrete molecules but instead are composed of particles called **ions.** Ions are atoms or groups of atoms that are electrically charged. (See Section 5.1.) Thus the composition of compounds such as NaCl, $Al_2(SO_4)_3$, and $CuSO_4$ cannot be identified in terms of the number of atoms of each type in a molecule, but only in terms of a *formula unit,* since such compounds do not contain physically distinct and electrically neutral molecules. The formula units of such compounds are empirical formulas because they give only the simplest whole-number ratios of elements in the compounds. The formula for copper sulfate, $CuSO_4$, indicates that there is one atom of copper for each atom of sulfur and each four atoms of oxygen in a sample of the compound, but it is an empirical formula since copper sulfate does not contain distinct molecules composed of a copper atom, a sulfur atom, and four oxygen atoms. The total numbers of atoms of the three elements in any sample of copper sulfate will be proportional to these numbers.

EXAMPLE 2.1   A molecule of blood sugar, glucose, contains 6 carbon atoms, 12 hydrogen atoms, and 6 oxygen atoms. What are the empirical and molecular formulas of glucose?

The molecular formula is $C_6H_{12}O_6$ since one molecule actually contains 6C, 12H, and 6O atoms. The simplest whole-number ratio of C to H to O atoms in glucose is 1:2:1, so the empirical formula is CH_2O.

EXAMPLE 2.2   How many oxygen atoms are present in an aspirin tablet, which contains 1.08×10^{21} aspirin molecules with the formula $C_8H_8O_4$?

The molecular formula of aspirin, $C_8H_8O_4$, indicates that there are exactly 4 atoms of oxygen per $C_8H_8O_4$ molecule (4 atoms O/1 molecule $C_8H_8O_4$). This ratio can be used as a unit conversion factor.

$$1.08 \times 10^{21} \text{ molecules } C_8H_8O_4 \times \frac{4 \text{ atoms O}}{1 \text{ molecule } C_8H_8O_4} = 4.32 \times 10^{21} \text{ atoms O}$$

2.2 Atomic Weights and Moles of Atoms

Although a chemical formula tells *how many* atoms of each type must be assembled to make a particular compound, it is not necessary to count the actual number of atoms involved in the chemical reaction used to produce that compound. It is sufficient to get the correct ratios of the numbers of atoms. To make calcium chloride, $CaCl_2$, simply requires that a given number of calcium atoms be combined with twice as many chlorine atoms. The correct ratio of Ca to Cl can be obtained by weighing.

The mass of an atom is very small. Rather than describe such a small mass in grams, chemists use another arbitrary unit called either the **atomic mass unit (amu)** or the **dalton** (after John Dalton). By international agreement, an atomic mass unit (or dalton) is taken to be exactly $\frac{1}{12}$ of the mass of a ^{12}C atom, a particular carbon isotope. **Isotopes** are atoms that exhibit identical chemical properties but differ in mass (see Section 4.6). On the average the atomic mass of a hydrogen atom is 1.0079 amu (or daltons) on this arbitrary scale. The average atomic mass of a fluorine atom is 18.998 amu, and that of a phosphorus atom, 30.974 amu. Average atomic masses are given here because many elements consist of mixtures of different isotopes. For example, 98.89% of naturally occurring carbon atoms are ^{12}C atoms with a mass of exactly 12 amu (by definition); the remainder have different masses, with 1.11% possessing a mass of 13.00335 amu and less than $10^{-8}\%$ having a mass of 14.0032 amu. The average atomic mass of all of the carbon atoms in the naturally occurring mixture is 12.011 amu.

The **atomic weight** of an element is a number, without units, equal to the average atomic mass of an atom of the element (in amu). Thus the atomic weight of hydrogen is 1.0079, that of carbon is 12.011, that of fluorine is 18.998, and that of phosphorus is 30.974.

Let us now look at the relationship between the mass of a sample of an element and the number of atoms in it. A mass of 18.998 amu of fluorine contains one atom of fluorine; a mass of 30.974 amu of phosphorus contains one atom of phosphorus. A mass of 75.992 amu (4×18.998) of fluorine contains four atoms of fluorine. Similarly, a mass of 123.90 amu (4×30.974) of phosphorus contains four atoms of phosphorus. The ratio of the mass of the sample of fluorine to the mass of the sample of phosphorus (75.992 amu to 123.90 amu) is the same as the ratio of their atomic weights (18.998 to 30.974), and the samples contain the same number of atoms—four. This is true of any pair of elements. **If the masses of samples of two elements have the same ratio as the ratio of their atomic weights, the samples contain identical numbers of atoms.**

Balances in chemical laboratories measure in units of grams rather than atomic mass units, but this is of little consequence in weighing out relative numbers of atoms. Just as 75.992 amu of fluorine and 123.90 amu of phosphorus contain the same number of atoms, 75.992 grams of fluorine and 123.90 grams of phosphorus also contain identical numbers of atoms, because the ratio of the masses is still 18.998 to 30.974.

For use in the laboratory, a convenient quantity of an element is a mole of atoms of the element. **A mole of atoms,** sometimes called a gram-atomic weight, or gram-atom, **contains the same number of atoms as is contained in exactly 12 grams of ^{12}C.** The mass of a mole of atoms is numerically equal to the atomic weight of the atom, and a mole of atoms of any element can be measured out by weighing out a mass of the element, in grams, equal to its atomic weight. The quantity 30.974 grams of phosphorus contains 1 mole of phosphorus atoms (1 gram-atom, or gram-atomic weight, of phosphorus). Similarly, 18.998 grams of fluorine contains 1 mole of fluorine atoms, and 12.011 grams of carbon contains 1 mole of carbon atoms. All of these quantities contain the same number of atoms. **A mole of atoms of any element contains the same number of atoms as are found in a mole of atoms of any other element.**

It is easy to count out relative numbers of atoms in the laboratory. Suppose, for example, that you have 15.999 grams of oxygen atoms and that you want an equal number of zinc atoms. From the atomic weight of oxygen, 15.999, you can recognize that 15.999 grams of oxygen atoms equals 1 mole of oxygen atoms. To get 1 mole of zinc atoms, you simply find the atomic weight of zinc (65.38) and then weigh out 1 mole of zinc atoms, or 65.38 grams. Finding fractional parts of moles simply requires an extra arithmetical step.

EXAMPLE 2.3   How many moles of nitrogen atoms are contained in 9.34 g of nitrogen?

The atomic weights on the inside front cover indicate that the atomic weight of nitrogen is 14.0067; hence 1 mol of nitrogen atoms contains 14.0067 g, or there is 1 mol of nitrogen atoms per 14.0067 g (1 mol N atoms/14.0067 g). The quantity of nitrogen in 9.34 g can be converted to moles of nitrogen atoms as follows:

$$9.34 \text{ g} \times \frac{1 \text{ mol N atoms}}{14.0067 \text{ g}} = 0.667 \text{ mol N atoms}$$

Notice that the gram units cancel, indicating that the correct conversion factor has been used. Three significant figures in the answer are justified by the mass of nitrogen given (9.34 g).

EXAMPLE 2.4   How much sodium contains the same number of atoms as are in 18.29 g of chlorine?

From the table of atomic weights, we find that 1 mol of sodium atoms equals 22.990 g (22.990 g/1 mol Na atoms) and 1 mol of chlorine atoms equals 35.453 g (1 mol Cl atoms/35.453 g). First find the number of moles of chlorine atoms present.

$$18.29 \text{ g} \times \frac{1 \text{ mol Cl atoms}}{35.453 \text{ g}} = 0.5159 \text{ mol Cl atoms}$$

A quantity of 0.5159 mol of Cl atoms contains the same number of atoms as are in 0.5159 mol of Na atoms. The next step, then, is to find the mass of 0.5159 mol of Na atoms.

$$0.5159 \text{ mol Na atoms} \times \frac{22.990 \text{ g}}{1 \text{ mol Na atoms}} = 11.86 \text{ g}$$

2.3   Avogadro's Number

It can be shown experimentally that **any sample containing one mole of atoms, whatever its mass, contains 6.022×10^{23} atoms.** This number is called Avogadro's number in honor of the Italian professor of physics Amadeo Avogadro (1776–1856). **Avogadro's number is the number of atoms in one mole of atoms of any element;** for example, Avogadro's number of atoms (6.022×10^{23}) is contained in 15.999 grams of oxygen, in 18.998 grams of fluorine, and in 30.974 grams of phosphorus.

EXAMPLE 2.5   How many chlorine atoms are contained in 18.29 g of chlorine?

In Example 2.4 it was shown that 18.29 g of chlorine contains 0.5159 mol of chlorine atoms. Since 1 mol contains 6.022×10^{23} atoms, the conversion factor is 6.022×10^{23} atoms/1 mol.

$$0.5159 \text{ mol} \times \frac{6.022 \times 10^{23} \text{ atoms}}{1 \text{ mol}} = 3.107 \times 10^{23} \text{ atoms}$$

The number of atoms in a mole is so large that it is difficult to appreciate exactly how big it is. To gain some appreciation of the size of the number, consider that if the entire population of the United States (approximately 225 million people) each spent 12 hours a day, 365 days a year, counting atoms at the rate of 1 atom per second, it would require about 170 million years to count the atoms in 1 mole of hydrogen atoms.

The mass of a single atom can be calculated from the mass of a mole of atoms and the number of atoms in a mole of atoms (Avogadro's number).

$$\text{Mass of atom} = \frac{\text{mass of mole}}{\text{Avogadro's number}}$$

The mass of a mole of oxygen atoms is 15.999 grams. The mass of a single atom is

$$\frac{15.999 \text{ g}}{1 \text{ mol}} \times \frac{1 \text{ mol}}{6.022 \times 10^{23} \text{ atoms}} = 2.657 \times 10^{-23} \text{ g/atom}$$

2.4   Molecular Weights, Empirical Weights, and Moles of Molecules

The **molecular weight** of a molecule is the sum of the atomic weights of all the atoms in the molecule. For example, the molecular weight of chloroform, $CHCl_3$, equals the sum of the atomic weights of one carbon atom, one hydrogen atom, and three chlorine atoms; the molecular weight of vitamin C, $C_6H_8O_6$, equals the sum of the atomic weights of six carbon atoms, eight hydrogen atoms, and six oxygen atoms.

For $CHCl_3$:

1C = 1 × 12.011	=	12.011
1H = 1 × 1.0079	=	1.0079
3Cl = 3 × 35.453	=	106.359
	mol. wt =	119.378

For $C_6H_8O_6$:

6C = 6 × 12.011	=	72.066
8H = 8 × 1.0079	=	8.0632
6O = 6 × 15.9994	=	95.9964
	mol. wt =	176.126

Since the average mass of an atom in amu is equal to its atomic weight, the average mass of a molecule in amu is equal to its molecular weight.

As we pointed out earlier, ionic compounds such as $NaCl$, $Al_2(SO_4)_3$, and $CuSO_4$ do not contain discrete molecules and cannot be characterized properly by a molecular weight. But for such compounds we can calculate the **formula weight,** or **empirical weight,** which is the sum of the atomic weights of the atoms found in one unit as indicated by the empirical formula.

$$\text{For } Al_2(SO_4)_3:$$
$$2Al = 2 \times 26.981 = 53.962$$
$$3S = 3 \times 32.06 = 96.18$$
$$\underline{12O = 12 \times 15.999 = 191.988}$$
$$\text{formula wt} = 342.13$$

Although many chemists refer loosely to the formula weight of an ionic compound as its molecular weight, there is a difference.

A mole of molecules contains the same number of molecules as there are atoms in exactly 12 grams of ^{12}C. Thus a mole of molecules contains 6.022×10^{23} molecules. **The mass of a mole of molecules, in grams, is numerically equal to the molecular weight of the molecule.** This mass is sometimes called the **molar mass** or the **gram-molecular weight** of the substance. Thus 119.378 grams of chloroform contains 1 mole of $CHCl_3$ molecules, and this mass is the molar mass of chloroform; 176.126 grams of vitamin C contains 1 mole of vitamin C molecules, and this mass is the molar mass of vitamin C. A mole of molecules also contains Avogadro's number of molecules; hence 119.378 grams of $CHCl_3$ and 176.126 grams of $C_6H_8O_6$ each contains 6.022×10^{23} molecules. It is also possible to speak of a mole of an ionic compound such as $NaCl$ or KOH. A mole of an ionic compound contains 6.022×10^{23} of the units described by the formula. A mole of $CuSO_4$, therefore, contains Avogadro's number of $CuSO_4$ units, each unit being composed of one copper atom, one sulfur atom, and four oxygen atoms. The mass of a mole of an ionic compound, in grams, is equal numerically to the formula weight and is referred to as the **molar mass** or the **gram-formula weight.**

We said earlier that a chemical formula indicates the number of atoms in one molecular or empirical unit of that compound (Section 2.1). **A chemical formula also indicates the number of moles of atoms in one mole of the compound.** Ethanol (C_2H_5OH, often called ethyl alcohol) contains two carbon atoms in one molecule. It also contains two moles of carbon atoms in one mole of ethanol molecules ($2 \times 6.022 \times 10^{23}$ C atoms in 6.022×10^{23} C_2H_5OH molecules).

The following examples illustrate the use of the concepts of chemical formulas, atomic weights, molecular weights, and moles of atoms and molecules.

EXAMPLE 2.6   How many moles of sulfur atoms are contained in 80.3 g of sulfur?

The atomic weight of sulfur is 32.06, so 1 mol S atoms = 32.06 g (1 mol S atoms/32.06 g).

$$80.3 \text{ g} \times \frac{1 \text{ mol S atoms}}{32.06 \text{ g}} = 2.50 \text{ mol S atoms}$$

EXAMPLE 2.7   How many moles of sulfur molecules are contained in 80.3 g of sulfur if the molecular formula is S_8?

The molecular weight of S_8 is $8 \times 32.06 = 256.5$; hence 1 mol of $S_8 = 256.5$ g (1 mol S_8/256.5 g).

$$80.3 \text{ g} \times \frac{1 \text{ mol } S_8}{256.5 \text{ g}} = 0.313 \text{ mol } S_8$$

EXAMPLE 2.8   The recommended minimum daily dietary allowance of vitamin C, $C_6H_8O_6$, for a young woman of average weight is 4.6×10^{-4} mol. What is this allowance in grams?

The molecular weight of $C_6H_8O_6$ is 176.1, to one decimal place. Hence 1 mol $C_6H_8O_6 = 176.1$ g (176.1 g $C_6H_8O_6$/1 mol $C_6H_8O_6$).

$$4.6 \times 10^{-4} \text{ mol } C_6H_8O_6 \times \frac{176.1 \text{ g } C_6H_8O_6}{1 \text{ mol } C_6H_8O_6} = 0.081 \text{ g } C_6H_8O_6$$

EXAMPLE 2.9   How many moles of hydrogen atoms are contained in 0.500 g (500 mg) of vitamin C?

First calculate the number of moles of vitamin C, $C_6H_8O_6$; then use the information from the chemical formula, which indicates exactly 8 mol of H atoms per 1 mol of $C_6H_8O_6$.

$$0.500 \text{ g } C_6H_8O_6 \times \frac{1 \text{ mol } C_6H_8O_6}{176.1 \text{ g } C_6H_8O_6} = 2.84 \times 10^{-3} \text{ mol } C_6H_8O_6$$

$$2.84 \times 10^{-3} \text{ mol } C_6H_8O_6 \times \frac{8 \text{ mol H atoms}}{1 \text{ mol } C_6H_8O_6} = 2.27 \times 10^{-2} \text{ mol H atoms}$$

2.5 Percent Composition from Formulas

All samples of a pure compound contain the same elements in the same proportion by mass. All samples of pure water, for example, contain 11 parts of hydrogen to 89 parts of oxygen. This **Law of Definite Proportion,** or **Law of Definite Composition,** helped convince Dalton of the atomic nature of matter and led him to outline his atomic theory.

Since all samples of a pure compound contain the same relative amounts of elements by mass, the proportion of each element present in a compound can be used to identify the compound. The proportion commonly used is the **percent composition,** that is, the percent by mass of the element in the sample, or the fraction of the total mass of the sample due to the element, multiplied by 100.

EXAMPLE 2.10   Calculate the percent by mass of hydrogen and of oxygen in the compound water, H_2O.

A convenient quantity to use is 1 mol of water.

$$\text{Percent hydrogen by mass in } H_2O = \frac{2 \times \text{atomic weight of H}}{\text{molecular weight of } H_2O} \times 100$$

$$= \frac{2 \times 1.01}{(2 \times 1.01) + (1 \times 16.0)} \times 100 = 11.2\%$$

$$\text{Percent oxygen by mass in } H_2O = \frac{1 \times \text{atomic weight of O}}{\text{molecular weight of } H_2O} \times 100$$

$$= \frac{1 \times 16.0}{(2 \times 1.01) + (1 \times 16.0)} \times 100 = 88.8\%$$

2.6 Derivation of Formulas

An empirical formula shows the relative numbers of moles of atoms in a compound; thus, in order to write the empirical formula of a compound, we must know the relative numbers of moles of atoms in a sample of it.

EXAMPLE 2.11   A sample of the principal component of natural gas contains 0.36 mol of hydrogen and 0.090 mol of carbon. What is the empirical formula of this compound?

For every 0.090 mol of C in the compound, there is 0.36 mol of H; thus the formula might be written as $C_{0.090}H_{0.36}$. But chemical formulas are customarily written in terms of whole-number ratios, so this formula must be reduced. The ratio of C to H is 1:4, so the *empirical* formula of the compound must be CH_4. To reduce two nonintegers to integers, divide each by the smaller noninteger. For this example,

$$\frac{0.36}{0.090} = 4, \qquad \frac{0.090}{0.090} = 1$$

Information about the composition of a compound is usually given in terms of either the masses of the elements in a sample or the percents by mass of the elements. If the masses of the elements are given, convert these to moles and reduce the mole ratio to the simplest whole numbers to find the empirical formula. If percents by mass are given, find the mass of each element present in a specific mass of sample (100 g is a convenient mass), convert to moles, and reduce the mole ratios to the simplest whole numbers.

EXAMPLE 2.12   A sample of the gaseous compound formed during bacterial fermentation of grain is found to consist of 27.29% carbon and 72.71% oxygen. What is the empirical formula of the compound?

Percent composition of an element in a compound is understood to indicate percent by mass. The mass of an element in a 100.0-g sample of a compound is equal in grams to the percent of that element in the sample; hence 100.0 g of this sample would contain 27.29 g of carbon and 72.71 g of oxygen.

$$100.0 \text{ g sample} \times \frac{27.29 \text{ g C}}{100 \text{ g sample}} = 27.29 \text{ g C}$$

$$100.0 \text{ g sample} \times \frac{72.71 \text{ g O}}{100 \text{ g sample}} = 72.71 \text{ g O}$$

The relative number of moles of carbon and oxygen atoms in the compound can be obtained by converting grams to moles, as shown in the following table.

Element	Mass of Element in 100.0 g of Sample	Relative Number of Moles	Divide by the Smaller Number	Smallest Integral Number of Moles
Carbon	27.29 g	$27.29 \text{ g C} \times \frac{1 \text{ mol C}}{12.01 \text{ g C}} = 2.272 \text{ mol}$	$\frac{2.272}{2.272} = 1.000$	1
Oxygen	72.71 g	$72.71 \text{ g O} \times \frac{1 \text{ mol O}}{16.00 \text{ g O}} = 4.544 \text{ mol}$	$\frac{4.544}{2.272} = 2.000$	2

For every 2.272 mol of carbon, there are 4.544 mol of oxygen. Reduced to the smallest whole numbers, the number of moles of oxygen is twice the number of moles of carbon. Hence, the number of oxygen *atoms* in the compound is twice the number of carbon *atoms*. The simplest formula for the gaseous compound produced during fermentation must therefore be CO_2.

The empirical formula CO_2 indicates a formula weight of 44. It can be shown by experiment that the gas actually has a molecular weight of 44. Thus the empirical formula and the molecular formula of CO_2 are the same. If the molecular weight for the gas were 88, then the molecular formula would be C_2O_4, indicating twice as many atoms as in the simplest formula, CO_2.

EXAMPLE 2.13   A sample of the mineral hematite, an oxide of iron found in iron ores, contains 34.97 g of iron and 15.03 g of oxygen. What is the empirical formula of hematite?

The steps in the solution of this problem are outlined in the following table.

Element	Mass of Element	Relative Number of Moles	Divide by the Smaller Number	Smallest Integral Number of Moles
Iron	34.97 g	$34.97 \text{ g Fe} \times \frac{1 \text{ mol Fe}}{55.85 \text{ g Fe}}$ $= 0.6261 \text{ mol}$	$\frac{0.6261}{0.6261} = 1.00$	$2 \times 1.00 = 2$
Oxygen	15.03 g	$15.03 \text{ g O} \times \frac{1 \text{ mol O}}{16.00 \text{ g O}}$ $= 0.9394 \text{ mol}$	$\frac{0.9394}{0.6261} = 1.50$	$2 \times 1.50 = 3$

Here, division by the smaller number of moles does not give two integers, so a third step is necessary. We must multiply by the smallest whole number that will give whole numbers for the relative numbers of moles, and hence of atoms, of each element: $2 \times 1.00 = 2$ and $2 \times 1.50 = 3$. The simplest whole-number mole ratio (and also whole-number atom ratio) is $2:3$, and the empirical formula of hematite is Fe_2O_3.

EXAMPLE 2.14   Pure oxygen is sometimes prepared in a general chemistry laboratory by heating a compound containing potassium, chlorine, and oxygen. What is the empirical formula of this compound if a 3.22-g sample decomposes to give gaseous oxygen and 1.96 g of KCl?

The mass of oxygen produced, and therefore the mass of oxygen atoms in the 3.22-g sample, is $3.22\ g - 1.96\ g = 1.26\ g$. We can determine the number of moles of oxygen atoms by converting 1.26 g of oxygen atoms to moles of oxygen atoms. The moles of potassium atoms and of chlorine atoms can be determined from the number of moles of KCl since there is 1 mol K and 1 mol Cl per 1 mol KCl. The steps in the solution of this problem are outlined in the following table:

Atom	Number of Moles	Divide by Smaller Number	Smallest Integral Number of Moles
O	$1.26\ g\ O \times \dfrac{1\ mol\ O}{16.0\ g\ O} = 0.0788\ mol\ O$	$\dfrac{0.0788}{0.0263} = 3.00$	3
K	$1.96\ g\ KCl \times \dfrac{1\ mol\ KCl}{74.6\ g\ KCl}$ $\times \dfrac{1\ mol\ K}{1\ mol\ KCl} = 0.0263\ mol\ K$	$\dfrac{0.0263}{0.0263} = 1.00$	1
Cl	$1.96\ g\ KCl \times \dfrac{1\ mol\ KCl}{74.6\ g\ KCl}$ $\times \dfrac{1\ mol\ Cl}{1\ mol\ KCl} = 0.0263\ mol\ Cl$	$\dfrac{0.0263}{0.0263} = 1.00$	1

Since the ratio of K to Cl to O is $1:1:3$, the simplest formula of the compound is $KClO_3$.

2.7 Chemical Equations

When atoms, molecules, or ions regroup to form other substances, chemists use a shorthand type of expression called a **chemical equation** to describe the chemical change. Formulas indicate the composition of each substance involved. For example, the equation for the reaction of methane with oxygen to form carbon dioxide and water (Fig. 2-2) is

$$CH_4 + 2O_2 \longrightarrow CO_2 + 2H_2O$$

Figure 2-2. The chemical equation for the reaction of one molecule of methane with two molecules of oxygen to give one molecule of carbon dioxide plus two molecules of water. The drawing shows how the atoms are redistributed during the chemical change.

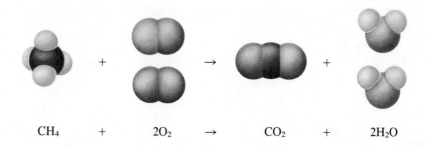

$$CH_4 \quad + \quad 2O_2 \quad \rightarrow \quad CO_2 \quad + \quad 2H_2O$$

The formulas for the reactants are written to the left of the arrow, and the formulas for the products are written to the right. The arrow is read as "gives," "produces," "yields," or "forms." A plus sign on the left side of an equation means "reacts with," and one on the right side, "and." Neither use of the plus sign indicates mathematical addition. Since matter cannot be created or destroyed in a chemical reaction, a chemical equation must have the same number of atoms in the products as there are in the reactants. In Fig. 2-2 we can see that one carbon atom, four hydrogen atoms, and four oxygen atoms are present in both the reactants and the products.

The statement, "when water is decomposed by an electric current, hydrogen and oxygen are formed," can be expressed by the chemical equation

$$H_2O \longrightarrow H_2 + O_2 \tag{1}$$

However, this equation is incomplete because it does not contain the same number of atoms on both sides of the arrow. One molecule of water, which contains only one atom of oxygen, cannot be rearranged into a product that contains two atoms of oxygen. Two molecules of water, however, can provide two atoms of oxygen. We can correct the equation by balancing it—placing the proper coefficients on each side of the arrow.

$$2H_2O \longrightarrow 2H_2 + O_2 \tag{2}$$

Of course, the subscripts in a formula cannot be changed to make an equation balance: subscripts indicate a compound's atomic composition.

Most simple chemical equations can be balanced by examination. This involves looking at the equation and adjusting the coefficients so that equal numbers of atoms of each type are present on both sides of the arrow. Let's consider the reaction of aluminum, Al, with hydrogen chloride, HCl, producing aluminum chloride, $AlCl_3$, and hydrogen, H_2. First we write the unbalanced equation:

$$Al + HCl \longrightarrow AlCl_3 + H_2$$

Now we determine the number of atoms of each type on both sides. In this equation there are one Al, one H, and one Cl atom on the left and one Al, two H, and three Cl atoms on the right. The equation is not balanced because the numbers of atoms are not the same on both sides. To find the proper coefficients, it is usually best to consider the molecule with the most atoms first, which in this case is $AlCl_3$. This molecule contains three Cl atoms, all of which must come from HCl. Therefore, we place a coefficient of 3 in front of HCl. At this stage we have

$$Al + 3HCl \longrightarrow AlCl_3 + H_2$$

Although the Cl and Al atoms are now balanced, the H atoms are not. Three HCl molecules, which collectively contain three hydrogen atoms, would give $1\frac{1}{2}$ H$_2$ molecules. Thus we have

$$Al + 3HCl \longrightarrow AlCl_3 + 1\tfrac{1}{2} H_2$$

This equation is not in its most conventional form because it indicates that half of a molecule of hydrogen is formed, and half-molecules do not exist. Therefore, we multiply the equation by 2 and obtain a balanced equation:

$$2Al + 6HCl \longrightarrow 2AlCl_3 + 3H_2$$

There are two Al, six H, and six Cl atoms on both sides.

Balancing equations by inspection is a trial and error process; you may have to try several sets of coefficients before you find the correct ones.

In addition to identifying reactants and products, chemical equations often give other information, such as the state of the reactants and products, whether heat is evolved or absorbed during the reaction, the temperature of the reaction, or other special conditions. For example, the formula for a gaseous reactant or product is followed by (g); a liquid by (l); a solid by (s); and a reactant or product that is dissolved in water by (aq). The reaction of solid sodium, Na, with liquid water to give hydrogen gas, H$_2$, and a solution of sodium hydroxide, NaOH, in water would be indicated by this equation:

$$2Na(s) + 2H_2O(l) \longrightarrow H_2(g) + 2NaOH(aq)$$

Sometimes it is useful to identify the amount of heat absorbed or produced during a reaction. A reaction that occurs with absorption of heat is called an **endothermic reaction;** a reaction that occurs with production of heat, an **exothermic reaction.** The quantity of heat may be written as if it were a reactant or product, as in the following equations:

$$H_2(g) + Cl_2(g) \longrightarrow 2HCl(g) + 184,610 \text{ joules}$$
$$131,300 \text{ joules} + H_2O(g) + C(s) \longrightarrow CO(g) + H_2(g)$$

However, it is conventional to write this information to the right of the equation using the symbol ΔH to indicate a heat change. A positive value for ΔH indicates that heat is absorbed (an endothermic reaction), while a negative value for ΔH indicates that heat is produced (an exothermic reaction).

$$H_2(g) + Cl_2(g) \longrightarrow 2HCl(g) \qquad\qquad \Delta H = -184,610 \text{ J}$$
$$\text{or} \quad \Delta H = -44,124 \text{ cal}$$

$$H_2O(g) + C(s) \longrightarrow CO(g) + H_2(g) \qquad\qquad \Delta H = 131,300 \text{ J}$$
$$\text{or} \quad \Delta H = 31,380 \text{ cal}$$

Recall that calories (cal) and joules (J) are simply different units to express quantities of heat (Section 1.15).

Special conditions such as temperature, the presence of a catalyst, or any other special circumstances that characterize the reaction may be written above or below the arrow. The electrolysis of water described in Equation (2) is sometimes written as follows:

$$2H_2O(l) \xrightarrow{\text{Elect.}} 2H_2(g) + O_2(g)$$

where the abbreviation *Elect*. over the arrow indicates that the reaction occurs by means of an electric current. A reaction carried out by heating may be indicated with a triangle over the arrow.

$$CaCO_3(s) \xrightarrow{\triangle} CaO(s) + CO_2(g)$$

2.8 Mole Relationships Based on Equations

The chemical reaction

$$2H_2O \longrightarrow 2H_2 + O_2$$

indicates that exactly two molecules of H_2O decompose to give two molecules of H_2 and one molecule of O_2. But the relative proportions of molecules, atoms, or ions is identical to the relative proportions of *moles* of molecules, atoms, or ions. Thus the equation also indicates relative numbers of moles.

$2H_2O$	$\longrightarrow$	$2H_2$	$+$	O_2
2 molecules		2 molecules		1 molecule
$2 \times 6.022 \times 10^{23}$ molecules		$2 \times 6.022 \times 10^{23}$ molecules		6.022×10^{23} molecules
2 moles		2 moles		1 mole

The coefficients in a chemical equation indicate the relative numbers of moles of atoms or moles of molecules that react or are formed. The equation indicates that 2 moles of H_2 are formed for each 2 moles of H_2O decomposed (2 mol H_2/2 mol H_2O), that 1 mole of O_2 is formed for each 2 moles of H_2O decomposed (1 mol O_2/2 mol H_2O), and that 1 mole of O_2 forms for each 2 moles of H_2 formed (1 mol O_2/2 mol H_2). These unit conversion factors are exact to any number of significant figures.

It is therefore quite straightforward to determine, from the information provided by a chemical equation, the numbers of atoms and molecules, or the numbers of moles of these atoms and molecules, that react or are produced in a chemical process.

EXAMPLE 2.15 How many moles of Al_2I_6 are produced by the reaction of 4.0 mol of aluminum according to the following equation?

$$2Al(s) + 3I_2(s) \longrightarrow Al_2I_6(s)$$

The balanced equation indicates that exactly 1 mol of Al_2I_6 is produced per 2 mol of Al consumed (1 mol Al_2I_6/2 mol Al). This relationship can be used to convert the number of moles of Al reacting to the number of moles of Al_2I_6 produced.

$$4.0 \text{ mol Al} \times \frac{1 \text{ mol } Al_2I_6}{2 \text{ mol Al}} = 2.0 \text{ mol } Al_2I_6$$

The unwanted units cancel, so the correct factor was used. The number of significant figures in the answer was determined by the amount of Al, since conversion factors determined from equations are exact.

EXAMPLE 2.16   How many moles of I_2 are required to react exactly with 0.429 mol of aluminum?

The balanced equation indicates that exactly 3 mol of I_2 reacts with 2 mol of Al. This gives us the conversion factor: 3 mol I_2/2 mol Al.

$$0.429 \text{ mol Al} \times \frac{3 \text{ mol } I_2}{2 \text{ mol Al}} = 0.644 \text{ mol } I_2$$

2.9 Calculations Based on Equations

In this section we will introduce a method of chemical calculation that relates the quantities of substances involved in chemical reactions. All such calculations proceed from the facts that (1) equations give the ratios of moles of reactants and products involved in chemical reactions, and (2) the masses of the reactants and products can be determined from the numbers of moles involved, using the atomic, molecular, or formula weights of the substances participating in the reaction. The following examples illustrate the method.

EXAMPLE 2.17   Calculate the moles of oxygen produced during the thermal decomposition of 100.0 g of potassium chlorate to form potassium chloride and oxygen according to the following reaction:

$$2KClO_3(s) \xrightarrow{\triangle} 2KCl(s) + 3O_2(g)$$

We are asked to calculate the number of moles of O_2 produced from 100.0 g of $KClO_3$ and we are given the equation for the reaction by which O_2 forms from $KClO_3$. If we can determine the number of moles of $KClO_3$ that react, we can also determine the number of moles of O_2 produced, since the equation indicates that exactly 3 mol of O_2 is produced for each 2 mol of $KClO_3$ that reacts (3 mol O_2/2 mol $KClO_3$). The number of moles of $KClO_3$ used can be found from the mass of $KClO_3$ (100.0 g), the formula weight of $KClO_3$ [39.0983 + 35.453 + (3 × 15.9994) = 122.5, to one decimal place], and the relationship 1 mol $KClO_3$ = 122.5 g $KClO_3$ (Section 2.4).

$$100.0 \text{ g } KClO_3 \times \frac{1 \text{ mol } KClO_3}{122.5 \text{ g } KClO_3} = 0.8163 \text{ mol } KClO_3$$

$$0.8163 \text{ mol } KClO_3 \times \frac{3 \text{ mol } O_2}{2 \text{ mol } KClO_3} = 1.224 \text{ mol } O_2$$

Thus 100.0 g of $KClO_3$ will produce 1.224 mol of O_2. In this case, the chain of calculations involves the following unit conversions:

$$\boxed{\begin{array}{c} \text{Mass of} \\ KClO_3 \end{array}} \longrightarrow \boxed{\begin{array}{c} \text{Moles of} \\ KClO_3 \end{array}} \longrightarrow \boxed{\begin{array}{c} \text{Moles of} \\ O_2 \end{array}}$$

Note how dimensional analysis with the resulting cancellation of units is helpful in checking to see if the correct conversion factors have been used.

It is not necessary to write each step of the calculations separately. Although the logic of the problem is perhaps best approached in two steps, the calculations could have been written in one step as follows:

$$100.0 \text{ g KClO}_3 \times \frac{1 \text{ mol KClO}_3}{122.5 \text{ g KClO}_3} \times \frac{3 \text{ mol O}_2}{2 \text{ mol KClO}_3} = 1.224 \text{ mol O}_2$$

EXAMPLE 2.18   Calculate the mass of oxygen gas, O_2, required for the combustion of 702 g of octane, C_8H_{18}.

$$2C_8H_{18} + 25O_2 \longrightarrow 16CO_2 + 18H_2O$$

In this example we are asked to calculate the mass of O_2 that reacts with 702 g of octane, C_8H_{18}, in a reaction indicated by the balanced equation. We can calculate the number of grams of O_2 required (by using the molecular weight of O_2) if we know the number of moles of O_2 required. The equation indicates the number of moles of O_2 and the number of moles of C_8H_{18} that react with each other, so the number of moles of O_2 can be determined if we can determine the number of moles of C_8H_{18} present. We can determine the number of moles of C_8H_{18} from the mass of C_8H_{18} and its molecular weight. Thus, the chain of calculations requires the following conversions:

$$\boxed{\begin{array}{c}\text{Mass of} \\ C_8H_{18}\end{array}} \longrightarrow \boxed{\begin{array}{c}\text{Moles of} \\ C_8H_{18}\end{array}} \longrightarrow \boxed{\begin{array}{c}\text{Moles of} \\ O_2\end{array}} \longrightarrow \boxed{\begin{array}{c}\text{Mass of} \\ O_2\end{array}}$$

From the molecular weight of C_8H_{18}, 1 mol $C_8H_{18} = 114.2$ g C_8H_{18} [$(8 \times 12.011) + (18 \times 1.0079)$, rounded to one decimal place].

$$702 \text{ g C}_8\text{H}_{18} \times \frac{1 \text{ mol C}_8\text{H}_{18}}{114.2 \text{ g C}_8\text{H}_{18}} = 6.147 \text{ mol C}_8\text{H}_{18}$$

From the balanced chemical equation, 2 mol of C_8H_{18} reacts for each 25 mol of O_2 (25 mol O_2/2 mol C_8H_{18}), so

$$6.147 \text{ mol C}_8\text{H}_{18} \times \frac{25 \text{ mol O}_2}{2 \text{ mol C}_8\text{H}_{18}} = 76.84 \text{ mol O}_2$$

From the molecular weight of O_2, 1 mol $O_2 = 32.00$ g O_2. Hence

$$76.84 \text{ mol O}_2 \times \frac{32.00 \text{ g O}_2}{1 \text{ mol O}_2} = 2459 \text{ g O}_2 = 2.46 \times 10^3 \text{ g O}_2$$

In this example one more digit than is justified by the data was carried through the calculations, and the final answer was rounded off to three significant figures. Although the logic of the problem involves three steps, the calculations could have been handled in one step as follows:

$$702 \text{ g C}_8\text{H}_{18} \times \frac{1 \text{ mol C}_8\text{H}_{18}}{114.2 \text{ g C}_8\text{H}_{18}} \times \frac{25 \text{ mol O}_2}{2 \text{ mol C}_8\text{H}_{18}} \times \frac{32.00 \text{ g O}_2}{1 \text{ mol O}_2} = 2.46 \times 10^3 \text{ g O}_2$$

EXAMPLE 2.19   What mass of sodium hydroxide, NaOH, would be required to produce 16 g of the antacid milk of magnesia [magnesium hydroxide, $Mg(OH)_2$] by the reaction of magnesium chloride, $MgCl_2$, with NaOH?

$$MgCl_2(aq) + 2NaOH(aq) \longrightarrow Mg(OH)_2(s) + 2NaCl(aq)$$

In order to determine the mass of NaOH required, we need to go through the following steps:

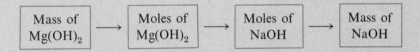

From the formula weight of $Mg(OH)_2$, 58.3 g $Mg(OH)_2$ = 1 mol $Mg(OH)_2$. Therefore

$$16 \text{ g } Mg(OH)_2 \times \frac{1 \text{ mol } Mg(OH)_2}{58.3 \text{ g } Mg(OH)_2} = 0.274 \text{ mol } Mg(OH)_2$$

From the chemical equation, 2 mol of NaOH reacts to give 1 mol of $Mg(OH)_2$ [2 mol NaOH/1 mol $Mg(OH)_2$]. Hence

$$0.274 \text{ mol } Mg(OH)_2 \times \frac{2 \text{ mol NaOH}}{1 \text{ mol } Mg(OH)_2} = 0.548 \text{ mol NaOH}$$

From the formula weight of NaOH, 40.0 g NaOH = 1 mol NaOH. Thus

$$0.548 \text{ mol NaOH} \times \frac{40.0 \text{ g NaOH}}{1 \text{ mol NaOH}} = 22 \text{ g NaOH}$$

The data, 16 g $Mg(OH)_2$, requires two significant figures for the answer.

The calculations here could have been handled in one step, as demonstrated in previous examples.

2.10 Limiting Reagents

Under certain conditions, all of the reactants in a chemical reaction may not be completely consumed. Consider, for example, the reaction of lithium with oxygen. The equation

$$4Li + O_2 \longrightarrow 2Li_2O$$

indicates that exactly 4 moles of Li react per mole of O_2. If the ratio of reactants actually used in this reaction differs from 4:1, then one of the reactants will be present in excess and not all of it will be consumed. If, for example, we have 5 moles of Li and 1 mole of O_2, 1 mole of Li must remain unreacted at the end of the reaction because 1 mole of O_2 can react with only 4 moles of Li. The reactant that will be completely consumed is called the **limiting reagent.** This reactant limits the amount of product that can be formed and determines the theoretical yield of the reaction. Other reactants are said to be present in excess.

To determine which reagent is the limiting reagent, we calculate the amount of product expected from each reactant. The reactant that gives the smallest theoretical yield is the one that limits the amount of product actually formed.

EXAMPLE 2.20   A mixture of 5.0 g of $H_2(g)$ and 10.0 g of $O_2(g)$ is ignited. Water forms according to the following equation:

$$2H_2(g) + O_2(g) \longrightarrow 2H_2O(g)$$

Which reactant is limiting? How much water will be produced by the reaction?

First, assume that all of the H_2 will react and calculate the theoretical yield of water (to the two significant figures justified), using the following steps:

$$\boxed{\begin{array}{c}\text{Mass of}\\ H_2\end{array}} \longrightarrow \boxed{\begin{array}{c}\text{Moles of}\\ H_2\end{array}} \longrightarrow \boxed{\begin{array}{c}\text{Moles of}\\ H_2O\end{array}}$$

$$5.0 \text{ g } H_2 \times \frac{1 \text{ mol } H_2}{2.02 \text{ g } H_2} = 2.48 \text{ mol } H_2$$

$$2.48 \text{ mol } H_2 \times \frac{2 \text{ mol } H_2O}{2 \text{ mol } H_2} = 2.5 \text{ mol } H_2O$$

Now, assume that all of the O_2 will react and calculate the theoretical yield of water (to the three significant numbers justified), using the following steps:

$$\boxed{\begin{array}{c}\text{Mass of}\\ O_2\end{array}} \longrightarrow \boxed{\begin{array}{c}\text{Moles of}\\ O_2\end{array}} \longrightarrow \boxed{\begin{array}{c}\text{Moles of}\\ H_2O\end{array}}$$

$$10.0 \text{ g } O_2 \times \frac{1 \text{ mol } O_2}{32.00 \text{ g } O_2} = 0.3125 \text{ mol } O_2$$

$$0.3125 \text{ mol } O_2 \times \frac{2 \text{ mol } H_2O}{1 \text{ mol } O_2} = 0.625 \text{ mol } H_2O$$

At this point it should be apparent that O_2 is the limiting reactant since 10.0 g of O_2 gives 0.625 mol of H_2O, whereas 5.0 g of H_2 would give 2.5 mol of H_2O if the hydrogen were all converted to water. The following expression gives the mass of water equal to 0.625 mol of water:

$$0.625 \text{ mol } H_2O \times \frac{18.02 \text{ g } H_2O}{1 \text{ mol } H_2O} = 11.3 \text{ g } H_2O$$

2.11   Theoretical Yield, Actual Yield, and Percent Yield

In the preceding sections we have shown how to calculate the mass of product produced by a chemical reaction. These calculations are based on the assumptions that the reaction is the only one involved, that all of the reactant is converted into product, and that all of the product can be collected. The calculated amount of product based on these assumptions is called the **theoretical yield.** We rarely find these conditions satisfied either in the laboratory or in industrial production; the actual mass of product isolated (the **actual yield**) from a reaction is usually less than the theoretical yield and is used in calculating the **percent yield** of a reaction from the following relationship:

$$\text{Percent yield} = \frac{\text{actual yield}}{\text{theoretical yield}} \times 100$$

EXAMPLE 2.21   A general chemistry student, preparing copper metal by the reaction of 1.274 g of copper sulfate with zinc metal, isolated a yield of 0.392 g of copper. What was the percent yield?

$$CuSO_4(aq) + Zn(s) \longrightarrow Cu(s) + ZnSO_4(aq)$$

To calculate percent yield you need both the theoretical yield of copper and the actual yield. The actual yield is 0.392 g; the theoretical yield is the amount of copper that would have been obtained if all of the $CuSO_4$ had been converted into Cu and recovered. The theoretical yield can be calculated by the following steps:

$$\boxed{\begin{array}{c}\text{Mass of}\\ CuSO_4\end{array}} \longrightarrow \boxed{\begin{array}{c}\text{Moles of}\\ CuSO_4\end{array}} \longrightarrow \boxed{\begin{array}{c}\text{Moles of}\\ Cu\end{array}} \longrightarrow \boxed{\begin{array}{c}\text{Mass of}\\ Cu\end{array}}$$

$$1.274 \text{ g } CuSO_4 \times \frac{1 \text{ mol } CuSO_4}{159.6 \text{ g } CuSO_4} = 7.982 \times 10^{-3} \text{ mol } CuSO_4$$

$$7.982 \times 10^{-3} \text{ mol } CuSO_4 \times \frac{1 \text{ mol } Cu}{1 \text{ mol } CuSO_4} = 7.982 \times 10^{-3} \text{ mol } Cu$$

$$7.982 \times 10^{-3} \text{ mol } Cu \times \frac{63.546 \text{ g } Cu}{1 \text{ mol } Cu} = 0.5072 \text{ g } Cu$$

The theoretical yield of Cu is 0.5072 g. The percent yield may now be calculated.

$$\text{Percent yield} = \frac{\text{actual yield}}{\text{theoretical yield}} \times 100$$

$$= \frac{0.392 \text{ g}}{0.5072 \text{ g}} \times 100 = 77.3\%$$

FOR REVIEW

SUMMARY

A chemical **formula** identifies the elements in a substance using **symbols,** one or two letter abbreviations for the elements. Subscripts indicate the relative numbers of the different atoms. A **molecular formula** indicates the exact number of each type of atom in a molecule. An **empirical formula** gives the simplest whole number ratio of atoms in a substance.

The atomic mass of an atom is usually expressed in **atomic mass units, amu.** An amu is defined as exactly $\frac{1}{12}$ the mass of a ^{12}C atom. The **atomic weight** of an element is a number without units and is numerically equal to the atomic mass of the element (in amu). A **mole of atoms** of any given substance contains the same number of atoms as are found in exactly 12 grams of ^{12}C. A mole of any type of atoms contains **Avogadro's number** of atoms, 6.022×10^{23} atoms. The mass, in grams, of a mole of atoms is numerically equal to the atomic weight of the atom.

The mass of a molecule, often expressed in amu, is equal to the sum of the masses of its constituent atoms. The **molecular weight** of a compound is equal to the sum of the atomic weights of its constituent atoms. One **mole of molecules** contains 6.022×10^{23} molecules, and the mass of a mole of molecules, in grams, is numerically equal to the molecular weight. The subscripts in a molecular formula indicate the number of moles of each type of atom in a mole of molecules. Compounds that do not consist of individual molecules are characterized by a **formula weight,** which is equal to the sum of the atomic weights of the atoms in the empirical formula. A mole of such a compound is 6.022×10^{23} of the units described by the empirical formula.

The amount of a substance, in moles, may be determined from its mass, or the mass may be determined from the number of moles. Thus a given quantity of a substance, in moles, may be obtained by weighing. The empirical formula of a substance can be determined from the relative numbers of moles of the various types of atoms that make up the substance. If the **percent composition** is known, then the relative numbers of moles can readily be calculated for a sample of any specific weight of the compound.

Chemical equations indicate the regrouping of atoms that accompanies chemical reactions. Because matter can be neither created nor destroyed in a chemical reaction, the same number of atoms of each type must appear in both the reactants and the products of a reaction. The coefficients in a balanced chemical equation indicate the relative numbers of moles of reactants and products involved in a reaction. Balanced equations may be used to identify how many moles of product should result from a given quantity of reactant.

The amount of product produced by complete conversion of a reactant in a reaction is called the **theoretical yield** of the reaction. Calculation of the theoretical yield of a reaction requires a knowledge of which reagent acts as the **limiting reagent.** The amount of product that is actually isolated from a reaction, either in the laboratory or in an industrial process, is called the **actual yield** of the reaction. The **percent yield** can be calculated from the actual yield and the theoretical yield.

KEY TERMS AND CONCEPTS

Actual yield (2.11)
Atomic mass unit, amu (2.2)
Atomic weight (2.2)
Avogadro's number (2.3)
Chemical equation (2.7)

Empirical formula (2.1)
Empirical weight (2.4)
Formula (2.1)
Formula weight (2.4)
Gram-atom (2.2)

Gram-atomic weight (2.2)
Gram-formula weight (2.4)
Gram-molecular weight (2.4)
Law of Definite Proportion (2.5)
Limiting reagent (2.10)

Molar mass (2.4)
Mole of atoms (2.2)
Mole of molecules (2.4)
Molecular formula (2.1)

Molecular weight (2.4)
Percent composition (2.5)
Percent yield (2.11)

Symbols (2.1)
Theoretical yield (2.11)
Yield (2.11)

EXERCISES

Note: Your answers may differ in the last decimal place from those given. This generally reflects different ways of rounding off results of intermediate steps in the calculations. The answers given have been determined by rounding off to the correct number of significant figures at the end of the calculations. Molecular weights were calculated using all significant figures in the atomic weights of the constituent atoms, then rounding off as described in Section 1.10. It should be pointed out, however, that the calculations can be simplified, without loss of accuracy, by rounding off the result at each stage to one digit beyond that justified by the data. In such a case the rounding off may create a small difference in the answer in the last decimal place.

Symbols, Formulas, and Chemical Equations

1. The elements found in the human body include C, H, O, N, P, S, I, Na, Cl, K, Ca, Fe, Br, Cu, Zn, Co, and Mg. What are the names of these elements?
2. Explain why the symbol for the element sulfur and the formula for a molecule of sulfur differ.
3. Determine the empirical formulas of the following compounds:
   (a) hydrogen peroxide, H_2O_2
   (b) hydrazine, N_2H_4
   (c) sulfuric acid, H_2SO_4
   (d) cyclohexane, C_6H_{12}
   (e) fructose, $C_{12}H_{22}O_{11}$
   (f) caffeine, $C_8H_{10}N_4O_2$
   (g) codeine, $C_{18}N_{21}NO_3$
4. Write a balanced equation that describes each of the following chemical reactions:
   (a) Methane gas, CH_4, burns in air forming gaseous carbon dioxide, CO_2, and water.
   (b) When heated, pure iron reacts with air to give Fe_2O_3.
   (c) Limestone, $CaCO_3$, decomposes when heated giving lime, CaO, and gaseous carbon dioxide, CO_2.
   (d) Pure aluminum reacts with sulfuric acid, $H_2SO_4(aq)$, to form aluminum sulfate, $Al_2(SO_4)_3$, and gaseous hydrogen, H_2.
   (e) During photosynthesis in plants carbon dioxide and water are converted into glucose, $C_6H_{12}O_6$, and oxygen, O_2.
   (f) Acid rain, a solution of sulfurous acid, $H_2SO_3(aq)$, forms when sulfur dioxide, SO_2, reacts with water.
   (g) Water vapor reacts with sodium metal to produce gaseous hydrogen, H_2, and solid sodium hydroxide, NaOH.
   (h) Gaseous water and hot carbon react to form gaseous hydrogen, H_2, and gaseous carbon monoxide, CO.
5. Balance the following equations:
   (a) $PtCl_4 \longrightarrow Pt + Cl_2$
   (b) $H_2 + Br_2 \longrightarrow HBr$
   (c) $Sb + O_2 \longrightarrow Sb_4O_6$
   (d) $P_4 + O_2 \longrightarrow P_4O_{10}$
   (e) $Pb + H_2O + O_2 \longrightarrow Pb(OH)_2$
   (f) $Ag + H_2S + O_2 \longrightarrow Ag_2S + H_2O$
   (g) $Cu + HNO_3 \longrightarrow Cu(NO_3)_2 + H_2O + NO$
   (h) $PCl_5 + H_2O \longrightarrow POCl_3 + HCl$
6. Balance the following equations:
   (a) $PCl_3 + Cl_2 \longrightarrow PCl_5$
   (b) $TiCl_4 + H_2O \longrightarrow TiO_2 + HCl$
   (c) $Fe + H_2O \longrightarrow Fe_3O_4 + H_2$
   (d) $Sc_2O_3 + SO_3 \longrightarrow Sc_2(SO_4)_3$
   (e) $Al + H_2SO_4 \longrightarrow Al_2(SO_4)_3 + H_2$
   (f) $KClO_3 \longrightarrow KCl + KClO_4$
   (g) $Ca_3(PO_4)_2 + C \xrightarrow{\Delta} Ca_3P_2 + CO$
   (h) $Ca_3(PO_4)_2 + H_3PO_4 \longrightarrow Ca(H_2PO_4)_2$
7. Determine both the formula and the number of moles of the missing substances in the following equations:
   (a) $Fe + CuSO_4 \longrightarrow \underline{\quad} + FeSO_4$
   (b) $N_2 + 3H_2 \longrightarrow \underline{\quad}$
   (c) $\underline{\quad} + 2H_2O \longrightarrow 2NaOH + H_2$
   (d) $C_3H_8 + \underline{\quad} \longrightarrow 3CO_2 + 4H_2O$
   (e) $2MnO_2 + 4KOH + O_2 \longrightarrow 2H_2O + \underline{\quad}$
   (f) $2CsCl + \underline{\quad} + H_2O \longrightarrow 2CsOH + 2AgCl$

Atomic and Molecular Weights; Moles

8. If the mass of a ^{12}C atom were redefined to be exactly 18 nmu (new mass units), how would the

masses of other atoms change when converted to nmu? How would the atomic weights change?

[S] 9. A sample of TlCl contains 0.85217 g of Tl and 0.14783 g of Cl. From this information calculate the atomic weight of Tl using 35.453 as the atomic weight of Cl. *Ans. 204.37*

10. Cadmium bromide, $CdBr_2$, contains 41.29% Cd and 58.71% Br. If the atomic weight of bromine is 79.904, calculate the atomic weight of cadmium from this data. *Ans. 112.4*

11. Calculate the molecular weight of each of the following compounds:
    (a) hydrogen fluoride, HF   *Ans. 20.0063*
    (b) ammonia, NH_3 *Ans. 17.0304*
    (c) nitric acid, HNO_3 *Ans. 63.0128*
    [S](d) silver sulfate, Ag_2SO_4 *Ans. 311.79*
    [S](e) acetic acid, CH_3CO_2H *Ans. 60.052*
    (f) boric acid, $B(OH)_3$ *Ans. 61.83*
    (g) caffeine, $C_8H_{10}N_4O_2$ *Ans. 194.193*

12. Determine the mass, in grams, of exactly 1 mol of each of the following compounds:
    (a) hydrogen bromide, HBr   *Ans. 80.912 g*
    (b) methane, CH_4 *Ans. 16.043 g*
    (c) sulfuric acid, H_2SO_4 *Ans. 98.07 g*
    (d) sodium hydroxide, NaOH   *Ans. 39.9971 g*
    (e) hydrogen peroxide, H_2O_2 *Ans. 34.0146 g*
    (f) propane, C_3H_8 *Ans. 44.096 g*
    (g) aspirin, $C_6H_4(CO_2H)(CO_2CH_3)$
                                        Ans. 180.160 g

13. Determine the number of moles in each of the following:
    [S](a) 2.12 g of potassium bromide, KBr
                               Ans. 1.78 × 10⁻² mol
    [S](b) 235 g of calcium carbonate, $CaCO_3$
                                        Ans. 2.35 mol
    (c) 0.1488 g of phosphoric acid, H_3PO_4
                               Ans. 1.518 ×10⁻³ mol
    (d) 27.9 g of acetylene, C_2H_2 *Ans. 1.07 mol*
    (e) 78.452 g of aluminum sulfate, $Al_2(SO_4)_3$
                                        Ans. 0.22930
    (f) 0.1250 g of caffeine, $C_8H_{10}N_4O_2$
                               Ans. 6.437 × 10⁻⁴ mol

14. The approximate minimum daily dietary requirement of the amino acid leucine, $C_6H_{13}NO_2$, is 1.1 g. What is this requirement in millimoles?
                                        Ans. 8.4 mmol

[S]15. Up to 0.0100% by mass of copper(I) iodide may be added to table salt as a dietary source of iodine. How many moles of CuI would be con-

tained in 1 lb (454 g) of such supplemented table salt?   *Ans. 2.38 ×10⁻⁴ mol*

16. The herbicide Treflan, $C_{13}H_{16}N_2O_4F_3$, is applied at the rate of 450 g (two significant figures) per acre to control weeds in corn, cantaloupes, cotton, and other crops. What is this application in moles per acre?   *Ans. 1.4 mol per acre*

17. Determine the mass, in grams, of each of the following:
    [S](a) 0.250 mol of ammonia, NH_3 *Ans. 4.26 g*
    (b) 1.743 mol of potassium cyanide, KCN
                                           Ans. 113.5 g
    (c) 1.07 mol of ethylene, C_2H_4 *Ans. 30.0 g*
    (d) 2.60 × 10⁻⁴ mol of the amino acid glycine, $CH_2(NH_2)CO_2H$ *Ans. 1.95 × 10⁻² g*
    (e) 0.0241 mol of the insecticide Paris green, $Cu_3(AsO_3)_2 \cdot Cu(CH_3CO_2)_2$ *Ans. 14.9 g*

18. Determine the mass in grams of each of the following:
    (a) 0.600 mol of oxygen atoms   *Ans. 9.60 g*
    (b) 0.600 mol of oxygen molecules, O_2
                                         Ans. 19.2 g
    (c) 0.600 mol of ozone molecules, O_3
                                         Ans. 28.8 g

19. The approximate minimum daily dietary requirement of the amino acid lysine, $C_6H_{14}N_2O_2$, is 5.5 mmol. What is this requirement in grams?
                                         Ans. 0.80 g

20. An average 55-kg woman has 7.5 × 10⁻³ mol of hemoglobin (molecular weight = 64,456) in her blood. What is this quantity in grams? in pounds?
                                    Ans. 480 g; 1.1 lb.

21. Determine the moles of copper atoms, the moles of aluminum atoms, the moles of water molecules, and the moles of oxygen atoms in 0.500 mol of the mineral turquoise, $CuAl_6(PO_4)_4(OH)_8 \cdot 4H_2O$.   *Ans. 0.500 mol Cu; 3.00 mol Al; 2.00 mol H_2O; 14.0 mol O*

[S]22. Determine the mass in grams of zirconium and of oxygen in 0.3384 mol of the mineral zircon, $ZrSiO_4$.   *Ans. 30.87 g Zr; 21.66 g O*

23. Determine each of the following:
    (a) grams of sulfur in 0.36 mol SO_2
                                           Ans. 12 g S
    [S](b) grams of magnesium in 7.52 mol MgO
                                           Ans. 183 g
    (c) grams of carbon in 0.01008 mol of Novocain, $C_{13}H_{21}N_2O_2Cl$
                                         Ans. 1.574 g

(d) moles of nitric acid in 785.4 g of nitric acid, HNO_3 *Ans. 12.46 mol*

(e) moles of nitrogen atoms in 51.7 g of ammonium nitrate, NH_4NO_3 *Ans. 1.29 mol*

(f) moles of carbon atoms in 0.6163 g of niacin, $C_6H_5NO_2$ *Ans. 3.004×10^{-2} mol*

S 24. Determine which of the following contains the greatest mass of hydrogen: 1 mol of CH_4, 0.6 mol of C_6H_6, 0.4 mol of C_3H_8.   *Ans. 1 mol CH_4*

25. Determine which of the following contains the greatest mass of aluminum: 27 g of Al, 122 g of $AlPO_4$, 266 g of Al_2Cl_6, 225 g of Al_2S_3.   *Ans. 225 g of Al_2S_3*

26. Determine which of the following contains the largest mass of carbon atoms: 0.10 mol of glucose, $C_6H_{12}O_6$; 3.0 g of ethane, C_2H_6; a 1.0-g diamond (diamond is pure carbon).   *Ans. 0.10 mol $C_6H_{12}O_6$*

27. (a) Determine which of the following has the greatest mass in 1.0 mol of atoms: nitrogen, oxygen, phosphorus, argon.   *Ans. Ar*

   (b) Determine which of the following has the greatest mass in 1.0 mol of molecules: N_2, O_2, P_4, Ar.   *Ans. P_4*

S 28. Determine which of the following contains the greatest total number of atoms: 10.0 g of potassium metal; 1.02×10^{23} molecules of nitrogen, N_2; 0.10 mol of chloroform, $CHCl_3$; 5 g of carbon dioxide, CO_2.   *Ans. 0.10 mol $CHCl_3$*

Empirical Formulas and Percent Composition

29. Calculate the percent composition of each of the following compounds to three significant figures:

   S (a) potassium bromide, KBr
          Ans. 32.9% K; 67.1% Br

   (b) nitrogen dioxide, NO_2 *Ans. 30.4% N; 69.6% O*

   (c) aluminum oxide, Al_2O_3 *Ans. 52.9% Al; 47.1% O*

   (d) formaldehyde, CH_2O *Ans. 40.0% C; 6.7% H; 53.3% O*

   S (e) cryolite, Na_3AlF_6 *Ans. 32.9% Na; 12.8% Al; 54.3% F*

   (f) magnesium diphosphate, $Mg_2P_2O_7$ *Ans. 21.8% Mg; 27.8% P; 50.3% O*

30. Determine the percent composition of each of the following to three significant figures:

(a) acetic acid, CH_3CO_2H *Ans. 40.0% C; 53.3% O; 6.71% H*

S (b) vitamin C, ascorbic acid, $C_6H_8O_6$ *Ans. 40.9% C; 4.58% H; 54.5% O*

(c) trinitrotoluene, TNT, $C_6H_2(CH_3)(NO_2)_3$ *Ans. 37.0% C; 2.22% H; 42.3% O; 18.5% N*

(d) codeine, $C_{18}H_{21}NO_3$ *Ans. 72.2% C; 7.07% H; 4.68% N; 16.0% O*

(e) aspirin, $C_6H_4(CO_2H)(CO_2CH_3)$ *Ans. 60.0% C; 4.48% H; 35.5% O*

31. Determine which of the following fertilizers contains the highest percentage of nitrogen; urea, N_2H_4CO; ammonium nitrate, NH_4NO_3; ammonium sulfate, $(NH_4)_2SO_4$.   *Ans. Urea*

32. Determine each of the following to three significant figures:

   (a) The percent of water in Epsom salts, $MgSO_4 \cdot 2H_2O$ *Ans. 23.0%*

   (b) The percent of cyanide ion, CN^-, in $KAg(CN)_2$ *Ans. 26.1%*

   (c) The percent of phosphate ion, PO_4^{3-}, in $Ca_3(PO_4)_2$ *Ans. 61.2%*

33. The most common flavoring agent added to food contains 39.3% Na and 60.7% Cl. Determine the empirical formula of this compound.   *Ans. NaCl*

S 34. The light-emitting diode used in some electronic calculator displays contains a compound composed of 69.24% Ga and 30.76% P. What is the empirical formula of this compound?   *Ans. GaP*

35. A gaseous compound containing 50% S and 50% O is used to sterilize grapes used in winemaking. What is the empirical formula of this gas?   *Ans. SO_2*

36. Determine the empirical formula of sodium nitrite, a compound that contains 33.32% Na, 20.30% N, and 46.38% O and is used to preserve bacon and other meats.   *Ans. $NaNO_2$*

37. The spines of some sea urchins are composed of a mineral that contains 40.0% Ca, 12.0% C, and 48.0% O. What is this mineral?   *Ans. $CaCO_3$*

38. Most polymers are very large molecules composed of simple units repeated many times. Thus they often have relatively simple empirical formulas. Determine the empirical formula of each of the following polymers:

(a) Teflon, 24.0% C, 76.0% F   *Ans. CF_2*

(b) polystyrene, 92.3% C, 7.7% H   *Ans. CH*

(c) saran, 24.8% C, 2.0% H, 73.1% Cl

Ans. CHCl

(d) Orlon, 67.9% C, 5.70% H, 26.4% N

Ans. C_3H_3N

[S](e) Lucite (Plexiglas), 71.40% C, 9.59% H, 19.02% O *Ans. C_5H_8O*

(f) natural rubber, 88.2% C, 11.8% H

Ans. C_5H_8

[S] 39. Determine the empirical and molecular formulas of a compound with a molecular weight of 32 that contains 87.5% N and 12.5% H.

Ans. NH_2; N_2H_4

40. Dichloroethane, a compound containing carbon, hydrogen, and chlorine, is often used for dry cleaning. The molecular weight of dichloroethane is 99. If a sample of dichloroethane is found to contain 24.3% carbon and 71.6% chlorine, what are its empirical and molecular formulas?

Ans. CH_2Cl; $C_2H_4Cl_2$

41. Determine the empirical and molecular formulas of a compound with a molecular weight of 84 that contains 85.6% carbon and 14.4% hydrogen.

Ans. CH_2; C_6H_{12}

[S] 42. Plaster of paris contains 6.2% water and 93.8% calcium sulfate, $CaSO_4$. What is the empirical formula of plaster of paris?

Ans. $(CaSO_4)_2 \cdot H_2O$ or $Ca_2(SO_4)_2 \cdot H_2O$

43. Nicotine contains 74.9% carbon, 8.7% hydrogen, and 17.3% nitrogen. It is known that this compound contains two nitrogen atoms per molecule. What are the empirical and molecular formulas of nicotine? *Ans. C_5H_7N; $C_{10}H_{14}N_2$*

44. A 27.5-g sample of a compound consisting of carbon and hydrogen contains 6.87 g of hydrogen. What is the empirical formula of this compound?

Ans. CH_4

[S] 45. Determine the empirical formula of gold chloride if a 126-mg sample contains 81.8 mg of gold.

Ans. $AuCl_3$

46. A 13.83-g sample of magnesium carbonate decomposes when heated, giving gaseous CO_2 and a residue of 6.57 g of MgO. Determine the empirical formula of this compound. *Ans. $MgCO_3$*

47. Hemoglobin (molecular weight = 64,456) contains 0.35% iron by mass. How many iron atoms are present in a hemoglobin molecule?

Ans. 4

[S] 48. A 0.709-g sample of iron reacts with HCl(g) to give hydrogen gas, H_2, and 1.610 g of a compound containing iron and chlorine. Identify the

product and write the balanced equation for its formation. *Ans. $Fe + 2HCl \longrightarrow FeCl_2 + H_2$*

49. The 1.610-g sample of iron(II) chloride obtained in Exercise 48 combines with 0.450 g of chlorine to give another form of iron chloride. Determine the empirical formula of this compound.

Ans. $FeCl_3$

Chemical Calculations Involving Equations

50. If 28.1 g of Si reacts with N_2 giving 0.333 mol of Si_3N_4 according to the equation

$$3Si + 2N_2 \longrightarrow Si_3N_4$$

which of the following can be determined using only the information given in this exercise: the moles of Si reacting, the moles of N_2 reacting, the atomic weight of N, the atomic weight of Si, the mass of Si_3N_4 produced?

51. (a) How many moles of aluminum bromide are formed by the reaction of 1.5 mol of HBr according to the following equation?

$$2Al + 6HBr \longrightarrow 2AlBr_3 + 3H_2$$

Ans. 0.50 mol

(b) What mass of $AlBr_3$, in grams, is produced?

Ans. 130 g

52. (a) How many moles of H_2 are produced by the reaction of 0.256 mol of H_3PO_4 according to the following equation?

$$2Cr + 2H_3PO_4 \longrightarrow 3H_2 + 2CrPO_4$$

Ans. 0.384 mol

(b) What mass of H_2, in grams, is produced?

Ans. 0.774 g

[S] 53. (a) How many molecules of I_2 are produced by the reaction of 0.3600 mol of $CuCl_2$ according to the following equation?

$$2CuCl_2 + 4KI \longrightarrow 2CuI + 4KCl + I_2$$

Ans. 1.084×10^{23}

(b) What mass of I_2, in grams, is produced?

Ans. 45.69 g

54. Thin films of silicon, Si, used for fabrication of electronic components may be prepared by the decomposition of silane, SiH_4.

$$SiH_4(g) \xrightarrow{\triangle} Si(s) + 2H_2(g)$$

What mass of SiH_4 is required to prepare 0.2173 g of Si by this technique?   *Ans. 0.2485 g*

55. Silicon carbide, SiC, a very hard material used as an abrasive on sandpaper and in other applications, is prepared by the reaction of pure sand, SiO_2, with carbon at a high temperature. Carbon monoxide, CO, is the other product of this reaction. Write the balanced equation for the reaction and calculate how much SiO_2 is required to produce 738 g of SiC.   *Ans. 1.11 × 10³ g*

56. Calcium chloride 6-hydrate, $CaCl_2 \cdot 6H_2O$, is a solid used to melt ice and snow. What mass of $CaCO_3$ (limestone), in kilograms, is required to make 100 lb (45.4 kg) of this compound?

$$CaCO_3(s) + 2HCl(aq) + 5H_2O \longrightarrow CaCl_2 \cdot 6H_2O + CO_2$$

Ans. 20.7 kg

S 57. Tooth enamel consists of hydroxyapatite, $Ca_5(PO_4)_3(OH)$. This substance is converted to the more decay-resistant fluoroapatite, $Ca_5(PO_4)_3F$, by treatment with tin(II) fluoride, SnF_2 (commonly referred to as stannous fluoride). Products of this reaction are SnO and water. What mass of hydroxyapatite can be converted to fluoroapatite by reaction with 0.100 g of SnF_2?   *Ans. 0.641 g*

58. Silver is often extracted from ores as $KAg(CN)_2$ and then recovered by the reaction

$$2KAg(CN)_2(aq) + Zn(s) \longrightarrow 2Ag(s) + Zn(CN)_2(aq) + 2KCN(aq)$$

What mass of zinc, in grams, is required to produce 1 oz (28.35 g) of silver?   *Ans. 8.592 g*

59. Silver sulfadiazine burn-treating cream creates a barrier against bacterial invasion and releases antimicrobial agents directly into the wound. What mass of silver oxide is required to prepare 225 g of silver sulfadiazine, $AgC_{10}H_9N_4SO_2$, from sulfadiazine, $C_{10}H_{10}N_4SO_2$?

$$Ag_2O + 2C_{10}H_{10}N_4SO_2 \longrightarrow 2AgC_{10}H_9N_4SO_2 + H_2O$$

Ans. 73.0 g

Limiting Reagents

60. Which is the limiting reagent when 0.5 mol of P_4 and 0.5 mol of S_8 react according to the following equation?

$$4P_4 + 5S_8 \longrightarrow 4P_4S_{10}$$

Ans. S_8

61. Which is the limiting reagent when 0.25 mol of Cr and 0.50 mol of H_3PO_4 react according to the following equation?

$$2Cr + 2H_3PO_4 \longrightarrow 2CrPO_4 + 3H_2$$

Ans. Cr

62. Which is the limiting reagent when 1.00 g of Si and 1.00 g of C combine as in the following equation? What mass of SiC forms?

$$Si + C \xrightarrow{\triangle} SiC$$

Ans. Si; 1.43 g

S 63. Which is the limiting reagent when 42.0 g of propane, C_3H_8, is burned with 115 g of oxygen?

$$C_3H_8(g) + 5O_2(g) \longrightarrow 3CO_2(g) + 4H_2O(g)$$

Ans. O_2

64. A solution containing 2.5 × 10⁻³ mol of $H_2Cr_2O_7$ is added to a solution containing 7.5 × 10⁻³ mol of H_2FeCl_4. The two react according to the following equation:

$$H_2Cr_2O_7 + 6H_2FeCl_4 + 5H_2O \longrightarrow 2[Cr(H_2O)_6]Cl_3 + 6FeCl_3$$

Which reagent is the limiting reagent? What mass of $[Cr(H_2O)_6]Cl_3$ is produced?   *Ans. H_2FeCl_4; 0.67 g*

Percent Yield

S 65. A student spilled some of an ethanol preparation and consequently isolated only 15 g rather than the 47 g that was theoretically possible. What was the percent yield?   *Ans. 32%*

66. A sample of lime, CaO, weighing 69 g was prepared by heating 131 g of limestone. What was the percent yield of the reaction?

$$CaCO_3(s) \xrightarrow{\triangle} CaO(s) + CO_2(g)$$

Ans. 94%

S 67. Toluene, $C_6H_5CH_3$, is oxidized by air under carefully controlled conditions to benzoic acid, $C_6H_5CO_2H$, which is used to prepare the food preservative sodium benzoate, $C_6H_5CO_2Na$. What is the yield of a reaction in which 1.000 kg of toluene is converted to 1.21 kg of benzoic acid?

$$2C_6H_5CH_3 + 3O_2 \longrightarrow 2C_6H_5CO_2H + 2H_2O$$

Ans. 91.3%

68. Freon-12, CCl_2F_2, is prepared from CCl_4 by reaction with HF. HCl is the other product of this reaction. What is the percent yield of a reaction in which 12.5 g of CCl_2F_2 is produced from 32.9 g of CCl_4? *Ans. 48.3%*

69. Diethyl ether, $(C_2H_5)_2O$, is prepared for anesthetic use by the reaction of ethanol with sulfuric acid.

$$2C_2H_5OH + H_2SO_4 \longrightarrow (C_2H_5)_2O + H_2SO_4 \cdot H_2O.$$

What is the percent yield if 1.00 kg of C_2H_5OH yields 0.766 kg of $(C_2H_5)_2O$? *Ans. 95.2%*

[S] 70. Citric acid, $C_6H_8O_7$, a component of jams, jellies, and "fruity" soft drinks, is prepared industrially by fermentation of sucrose by the mold *Aspergillus niger*. The overall reaction is

$$C_{12}H_{22}O_{11} + H_2O + 3O_2 \longrightarrow 2C_6H_8O_7 + 4H_2O$$

What is the amount, in kilograms, of citric acid produced from 1 metric ton (1.000×10^3 kg) of sucrose if the percent yield is 85.4%? *Ans. 959 kg*

71. If 0.20 mol of $CrPO_4$ is recovered from the reaction in Exercise 61, what is the percent yield? *Ans. 80%*

[S] 72. The reaction of 3.0 mol of H_2 with 2.0 mol of I_2 produces 1.0 mol of HI.

$$H_2 + I_2 \longrightarrow 2HI$$

Determine the limiting reagent, the theoretical yield in moles, and the percent yield for this reaction. *Ans. I_2; 4.0 mol; 25%*

73. If 1.4 g of SiC is recovered from the reaction in Exercise 62, what is the percent yield? *Ans. 98%*

74. Addition of 0.403 g of sodium oxalate, $Na_2C_2O_4$, to a solution containing 1.48 g of uranyl nitrate, $UO_2(NO_3)_2$, yields 1.073 g of solid $UO_2(C_2O_4) \cdot 3H_2O$.

$$UO_2(NO_3)_2(aq) + Na_2C_2O_4(aq) + 3H_2O \longrightarrow UO_2(C_2O_4) \cdot 3H_2O(s) + 2NaNO_3(aq)$$

Determine the limiting reagent and percent yield of this reaction. *Ans. $Na_2C_2O_4$; 86.6%*

[S] 75. What is the percent yield of $NaClO_2$ if 106 g is isolated from the reaction of 202.3 g of ClO_2 with a solution containing 3.229 mol of NaOH?

$$2ClO_2(g) + 2NaOH(aq) \longrightarrow NaClO_2(aq) + NaClO_3(aq) + H_2O(l)$$

Ans. 78.2%

Additional Exercises

76. A sample of indium bromide weighing 0.100 g reacts with silver nitrate, $AgNO_3$, giving indium nitrate and 0.159 g of AgBr. Determine the empirical formula of indium bromide. *Ans. $InBr_3$*

[S] 77. Naphthalene, a compound with a molecular weight of about 130 and containing only carbon and hydrogen, is the principal component of mothballs. A 3.000-mg sample of naphthalene burns to give 10.3 mg of CO_2. Determine its empirical and molecular formula. *Ans. C_5H_4; $C_{10}H_8$*

78. The equation for the preparation of phosphorus, P_4, in an electric furnace is

$$2Ca_3(PO_4)_2 + 6SiO_2 + 10C \longrightarrow 6CaSiO_3 + 10CO + P_4$$

What mass of SiO_2 is required to prepare 0.100 kg of P_4? *Ans. 291 g*

79. The empirical formula for an ion-exchange resin used in water softeners is $C_8H_7SO_3Na$. The resin softens water by removing calcium ion, Ca^{2+}, by the reaction

$$Ca^{2+} + 2C_8H_7SO_3Na \longrightarrow (C_8H_7SO_3)_2Ca + 2Na^+$$

What mass of ion-exchange resin, in grams, is required to soften 1.00×10^3 gal of water with a Ca^{2+} content of 2.50 ppm (2.50 g Ca^{2+} per 1,000,000 g water)? A gallon of water has a mass of 8.0 lb. *Ans. 93 g*

80. Sulfur can be removed from coal by washing powdered coal with NaOH(*aq*). The reaction may be written

$$R_2S(s) + 2NaOH(aq) \longrightarrow R_2O(s) + Na_2S(aq) + H_2O(l)$$

where R_2S represents the organic sulfur present in the coal. What mass of NaOH(*s*), in kilograms, is required to react with the sulfur in 1.00 metric ton (1.00×10^3 kg) of coal containing 1.6% S by mass? *Ans. 40 kg*

81. A sample of concentrated sulfuric acid with a density of 1.84 g/cm^3 contains 97.5% H_2SO_4 by mass. What mass of phosphate fertilizer, $Ca(H_2PO_4)_2$, in kilograms, can be prepared by the reaction of 1.00 × 10^3 L of this acid according to the equation

$$2Ca_5(PO_4)_3F + 7H_2SO_4 + 14H_2O \longrightarrow 3Ca(H_2PO_4)_2 + 7(CaSO_4 \cdot 2H_2O) + 2HF$$

Ans. 1835 kg

[S] 82. Sodium bicarbonate, or baking soda, $NaHCO_3$, can be purified by dissolving in hot water (60°C), filtering to remove insoluble impurities, cooling to 0°C to precipitate solid $NaHCO_3$, and then filtering to remove the solid $NaHCO_3$ leaving soluble impurities in solution. Any $NaHCO_3$ that remains in solution is not recovered. The solubility of $NaHCO_3$ in hot water at 60°C is 164 g/L. Its solubility in cold water at 0°C is 69 g/L. What is the percent yield of $NaHCO_3$ when purified by this method? *Ans. 58%*

83. A volume of 2.27 mL of $SiCl_4$ (density = 1.483 g/mL) reacts with an excess of $H_2S(g)$ giving $HSSiCl_3$ according to the reaction

$$SiCl_4(l) + H_2S(g) \longrightarrow HSSiCl_3(l) + HCl(g)$$

If the HCl produced reacts with 8.267 × 10^{-3} mol of NaOH according to the reaction

$$HCl + NaOH \longrightarrow NaCl + H_2O$$

what is the percent yield of $HSSiCl_3$?

Ans. 41.7%

84. If a mixture of 6.0 mol of hydrogen gas, H_2, and 15 g of oxygen gas, O_2, burns, how much water is produced? Which is the limiting reagent?

$$2H_2 + O_2 \longrightarrow 2H_2O$$

Ans. 17 g; O_2

85. What mass of potassium chloride, KCl, can be prepared by the reaction of a mixture of 39.1 g of potassium, K, and 30.1 g of chlorine, Cl_2? Which is the limiting reagent?

$$2K + Cl_2 \longrightarrow 2KCl$$

Ans. 63.3 g; Cl_2

86. Zinc reacts with nitric acid according to the equation

$$Zn + 4HNO_3 \longrightarrow Zn(NO_3)_2 + 2H_2O + 2NO_2$$

If 1.05 g of zinc reacts with 2.50 mL of nitric acid (density = 1.503 g/mL), which reagent is the limiting reagent? What mass of $Zn(NO_3)_2$ is produced? *Ans. HNO_3; 2.82 g*

3

APPLICATIONS OF CHEMICAL STOICHIOMETRY

Chemistry is a quantitative science. Both the study and the use of the principles of chemistry involve a great many calculations. However, the purpose of these calculations is not merely to get numbers, but rather to provide quantitative predictions or descriptions of how matter behaves. Calculations are tools that chemists use in trying to understand the behavior of matter.

One category of chemical calculations involves relationships between quantities of reactants and of products based on the balanced equation for their reaction. This type of calculation was introduced in Chapter 2. Determinations of the masses of reactants required for a reaction, of the amount of reactant necessary to give a desired amount of product, of the concentration of a solution, of the percent composition of a product as a check for purity, and of the empirical and molecular formulas of an unknown product all fall into this category. Calculations such as these are generally referred to as chemical stoichiometry, and in this chapter we will be primarily concerned with a detailed treatment of this category.

3.1 Solutions

In Chapter 2 we used chemical equations to relate quantities of reactants and products measured by weighing. In this section and the next, we will describe how chemical equations can be used to relate quantities of materials present in solution.

When sugar is stirred into enough water, the sugar dissolves and a clear mixture, or **solution,** of sugar in water is formed. This solution consists of the **solute** (the dissolved sugar) and the **solvent** (the water). A sugar solution that contains only a small amount of sugar in comparison with the amount of water is said to be **dilute;** the addition of more sugar makes the solution more **concentrated.** The concentration limits of a solution depend on the nature of the substance dissolved (the solute) and of the medium in which the solute is dissolved (the solvent). The solute or the solvent can be a gas, a liquid, or a solid, but in this chapter we will limit our attention to solutes dissolved in liquids. Solutions will be discussed in greater detail in Chapter 13.

The relative amounts of solute and solvent present in a solution, that is, its **concentration,** can be expressed as the molarity of the solution. **The molarity, M, of a solution is the number of moles of solute per exactly 1 liter (or dm^3) of solution.** To find the molarity, divide the number of moles of solute in a given volume of solution by the volume in liters.

$$\text{Molarity} = \frac{\text{moles of solute}}{\text{liters of solution}} \tag{1}$$

EXAMPLE 3.1   The acid secreted by the cells of the stomach lining is a hydrochloric acid solution that typically contains 0.00774 mol of HCl per 0.0500 L of solution. What is the concentration of this acid?

The concentration of the acid can be calculated using Equation (1). Divide the moles of HCl by the volume in liters.

$$\text{Molarity} = \frac{\text{moles of HCl}}{\text{liters of solution}}$$

$$= \frac{0.00774 \text{ mol}}{0.0500 \text{ L}}$$

$$= 0.155 \ M, \text{ or } 0.155 \text{ mol/L, or } 0.155 \text{ mol/dm}^3$$

For calculations it is convenient to write this molarity as the unit conversion factor 0.155 mol HCl/1 L.

This example presented the volume in liters; however, many experimental measurements are in milliliters and must be converted to liters before being used to find molarities.

EXAMPLE 3.2   How many moles of sulfuric acid, H_2SO_4, are contained in 0.80 L of a 0.050 M solution of sulfuric acid?

Equation (1) can be rearranged to solve for moles of solute:

$$\text{Moles of solute} = \text{molarity} \times \text{liters of solution}$$

$$\frac{0.050 \text{ mol } H_2SO_4}{1 \ \cancel{L}} \times 0.80 \ \cancel{L} = 0.040 \text{ mol } H_2SO_4$$

EXAMPLE 3.3   What mass of iodine is required to make 5.00×10^2 mL of a 0.250 M solution of I_2 with chloroform, $CHCl_3$, as the solvent?

The moles of I_2 necessary to make a 0.250 M solution can be found by rearranging Equation (1) as shown in Example 3.2. The weight of I_2 required can then be calculated. The following steps are involved in the calculation:

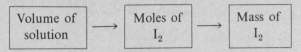

$$5.00 \times 10^2 \text{ mL} \times \frac{1 \text{ L}}{1000 \text{ mL}} = 0.500 \text{ L}$$

$$\frac{0.250 \text{ mol } I_2}{1 \text{ L}} \times 0.500 \text{ L} = 0.125 \text{ mol } I_2$$

$$0.125 \text{ mol } I_2 \times \frac{253.8 \text{ g } I_2}{1 \text{ mol } I_2} = 31.7 \text{ g } I_2$$

3.2 Titration

A known volume of a solution is usually measured with a pipet or a buret. A **pipet** is a tube that will deliver a known fixed volume of a liquid when filled to the reference line (Fig. 3-1). A **buret** is a cylinder (Fig. 3-2) that is graduated in fractions of a milliliter so that the volume of a liquid delivered from it may be accurately measured.

Figure 3-1. A 5.00-mL pipet.

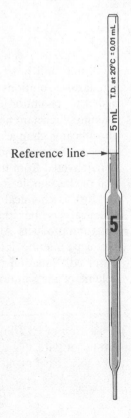

Figure 3-2. A buret is used for accurately measuring the volume of a solution (see insert).

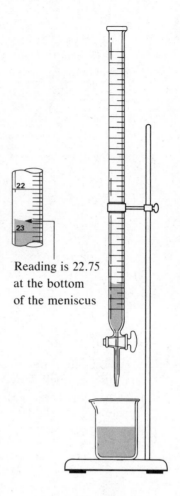

Reading is 22.75 at the bottom of the meniscus

Burets and pipets are used to determine the concentrations of solutions (to standardize solutions) by **titration.** During a titration a solution of a reactant is added to a solution of a sample, drop by drop, from a buret. When the exact amount of reactant necessary to react completely with the sample has been added, an indicator changes color. At this **end point** the titration is stopped, and the amount of reactant solution added is determined by reading the volume of solution delivered from the buret. This volume is used in determining the concentration of the sample solution.

Just as chemical equations relate the masses of reactants and products of chemical reactions, they also relate concentrations or volumes of solutions of reactants and products. A chemical equation gives the ratios of moles of reactants and products entering into a chemical reaction. If these reactants or products are dissolved, Equation (1) in Section 3.1 can be used to relate the molarity and volume of the solution to the number of moles present.

EXAMPLE 3.4   A solution of HCl in water is standardized by titrating a solution of a 0.5015-g sample of pure, dry sodium carbonate, Na_2CO_3. What is the concentration of the HCl solution if 48.47 mL is added from a buret to reach the end point?

$$Na_2CO_3 + 2HCl \longrightarrow 2NaCl + CO_2 + H_2O$$

To determine the concentration of the HCl solution, we need to know the number of moles of HCl in some volume of the solution. At the end of the titration, we know that we have added 48.47 mL of the solution and that the HCl in it has reacted exactly with 0.5015 g of Na_2CO_3. The balanced equation tells us that 2 mol of HCl has been added for every 1 mol of Na_2CO_3 present. Thus the number of moles of HCl present in 48.47 mL of solution can be determined from the number of moles of Na_2CO_3 titrated. The concentration of the solution can be determined by the following steps:

$$\boxed{\begin{array}{c} \text{Mass of} \\ Na_2CO_3 \end{array}} \longrightarrow \boxed{\begin{array}{c} \text{Moles of} \\ Na_2CO_3 \end{array}} \longrightarrow \boxed{\begin{array}{c} \text{Moles of} \\ \text{HCl} \end{array}} \longrightarrow \boxed{\begin{array}{c} \text{Molarity} \\ \text{of solution} \end{array}}$$

$$0.5015 \text{ g } Na_2CO_3 \times \frac{1 \text{ mol } Na_2CO_3}{105.99 \text{ g } Na_2CO_3} = 4.7316 \times 10^{-3} \text{ mol } Na_2CO_3$$

$$4.7316 \times 10^{-3} \text{ mol } Na_2CO_3 \times \frac{2 \text{ mol HCl}}{1 \text{ mol } Na_2CO_3} = 9.4632 \times 10^{-3} \text{ mol HCl}$$

$$\frac{9.4632 \times 10^{-3} \text{ mol HCl}}{0.04847 \text{ L}} = 0.1952 \text{ } M$$

The concentration of the HCl solution is 0.1952 M.

EXAMPLE 3.5   An acidic solution of oxalic acid, $H_2C_2O_4$, was added to a flask using a 20.00-mL pipet, and the sample was titrated with a 0.09113 M solution of potassium permanganate, $KMnO_4$. A volume of 23.24 mL was required to reach the end point. What was the concentration of the $H_2C_2O_4$ solution?

$$5H_2C_2O_4 + 2KMnO_4 + 3H_2SO_4 \longrightarrow 2MnSO_4 + K_2SO_4 + 10CO_2 + 8H_2O$$

To determine the concentration of $H_2C_2O_4$ in the solution, we need to know the number of moles of $H_2C_2O_4$ in 0.02000 L (20.00 mL) of the solution, the amount delivered by the pipet. The equation tells us that for every 5 mol of $H_2C_2O_4$ present, 2 mol of $KMnO_4$ reacts. The moles of $KMnO_4$ reacting can be determined from the volume and molarity of the $KMnO_4$ solution added during the titration.

$$\boxed{\begin{array}{c} \text{Volume of} \\ KMnO_4 \\ \text{Solution} \end{array}} \longrightarrow \boxed{\begin{array}{c} \text{Moles of} \\ KMnO_4 \end{array}} \longrightarrow \boxed{\begin{array}{c} \text{Moles of} \\ H_2C_2O_4 \end{array}} \longrightarrow \boxed{\begin{array}{c} \text{Molarity} \\ \text{of } H_2C_2O_4 \end{array}}$$

$$0.02324 \text{ L } KMnO_4 \times \frac{0.09113 \text{ mol } KMnO_4}{1 \text{ L } KMnO_4} = 2.1179 \times 10^{-3} \text{ mol } KMnO_4$$

$$2.1179 \times 10^{-3} \text{ mol } KMnO_4 \times \frac{5 \text{ mol } H_2C_2O_4}{2 \text{ mol } KMnO_4} = 5.2948 \times 10^{-3} \text{ mol } H_2C_2O_4$$

$$\frac{5.2948 \times 10^{-3} \text{ mol } H_2C_2O_4}{0.02000 \text{ L}} = 0.2647 \text{ } M$$

3.3 Chemical Calculations Involving Stoichiometry

Stoichiometry is the calculation of both material balances and energy balances in a chemical system. We will discuss material balances in this chapter and energy balances in future chapters. Most stoichiometry problems involve the basic transformations, or conversions, illustrated below. The arrows indicate that one quantity is converted to the next.

$$\boxed{\text{Quantity A}} \longrightarrow \boxed{\text{Moles A}} \longrightarrow \boxed{\text{Moles B}} \longrightarrow \boxed{\text{Quantity B}}$$

The overall conversion from Quantity A to Quantity B requires three steps (each of which may require one or more unit conversion factors of the types we have previously discussed).

1. conversion of a quantity measured in units like grams or liters into a quantity in moles
2. determination of how many moles of a second substance are equivalent to the number of moles of the first
3. conversion of the number of moles of the second substance into a mass of that substance or a volume of a solution of a certain molarity

EXAMPLE 3.6   What volume of a 0.750 *M* solution of hydrochloric acid can be prepared from the HCl produced by the reaction of 25.0 g of NaCl with an excess of sulfuric acid?

$$NaCl(s) + H_2SO_4(l) \longrightarrow HCl(g) + NaHSO_4(s)$$

In order to determine the volume of 0.750 *M* solution that we can prepare, we need the number of moles of HCl(*g*) that are available to make the solution. The chemical equation tells us the relationship between the number of moles of gaseous HCl produced and the number of moles of NaCl reacting. The number of moles of NaCl can be determined from the mass of NaCl used and its formula weight. Solution of the problem requires the following steps, each of which uses a single unit conversion factor:

$$\boxed{\text{Mass of NaCl}} \longrightarrow \boxed{\text{Moles of NaCl}} \longrightarrow \boxed{\text{Moles of HCl}} \longrightarrow \boxed{\text{Volume of HCl solution}}$$

Convert the quantity of NaCl in grams to moles of NaCl. As described in Section 2.4, this conversion requires calculating the formula weight of NaCl, 58.44; thus 1 mol NaCl = 58.44 g NaCl. The conversion is then

$$25.0 \text{ g } \cancel{\text{NaCl}} \times \frac{1 \text{ mol NaCl}}{58.44 \text{ g } \cancel{\text{NaCl}}} = 0.4278 \text{ mol NaCl}$$

Convert moles of NaCl to moles of HCl. The chemical equation indicates that 1 mol of HCl is produced for every 1 mol of NaCl that reacts. The unit conversion factor is 1 mol HCl/1 mol NaCl (Section 2.8).

$$0.4278 \text{ } \cancel{\text{mol NaCl}} \times \frac{1 \text{ mol HCl}}{1 \text{ } \cancel{\text{mol NaCl}}} = 0.4278 \text{ mol HCl}$$

Convert moles of HCl to the volume of a 0.750 M solution of HCl. As shown in Section 3.1,

$$\text{Molarity} = \frac{\text{moles of solute}}{\text{liters of solution}}$$

This can be rearranged to solve for the volume of solution:

$$\text{Liters of solution} = \text{moles of solute} \times \frac{1}{\text{molarity}}$$

Since a 0.750 M solution contains 0.750 mol HCl per exactly 1 L (0.750 mol HCl/1 L), the reciprocal of the molarity ($1/M$) is expressed as follows:

$$\frac{1}{\text{molarity}} = \frac{1\text{ L}}{0.750\text{ mol HCl}}$$

Thus

$$0.4278 \text{ mol HCl} \times \frac{1\text{ L}}{0.750 \text{ mol HCl}} = 0.570 \text{ L}$$

This example and those that follow illustrate a basic concept of stoichiometry. **The relative numbers of moles of the various species participating in a reaction are provided by the chemical equation that describes the system** (Section 2.8).

Although Example 3.6 illustrates the basic flow of almost any stoichiometry problem, such calculations often involve several additional steps. These may result from the nature of the material (a mixture rather than a pure compound, for example) used in a reaction or from the various ways of measuring it (measuring a volume rather than a mass, for example). In such cases the additional steps usually involve relating moles to measured quantities.

3.4 Common Operations in Stoichiometry Calculations

Conversion of a measured quantity of a substance to moles of the substance, or the reverse conversion, may involve one or more of the unit conversion factors discussed below. Careful attention to units and their cancellation will be a great help in correctly setting up the calculations.

1. INTERCONVERSION BETWEEN MASS AND MOLES. The number of moles of a substance, A, in a given mass of pure A can be determined by use of a conversion factor, as described in Section 2.4.

$$\text{Mol A} = \text{mass A} \times \frac{1\text{ mol A}}{\text{molar mass A}}$$

Rearrangement of this expression gives an equation that can be used to find the mass of A that corresponds to a given amount of A in moles.

$$\text{Mass A} = \text{mol A} \times \frac{\text{molar mass A}}{1\text{ mol A}}$$

2. INTERCONVERSION BETWEEN MASS OF A MIXTURE AND MASS OF A COMPONENT. The percent by mass of a component, A, in a sample of a mixture or a

compound is given by the expression

$$\text{Percent A} = \frac{\text{mass A}}{\text{mass sample}} \times 100$$

where 100 is an exact multiplier. When using percent by mass in problems, it is often convenient to use the grams of component A per 100 g of sample (g A/100 g sample), which is numerically equal to the percent of component A by mass. For example, a sample of coal that contains 0.23% sulfur by mass contains 0.23 g of sulfur per 100 g of coal.

The mass of A and the mass of the sample can be determined from the following expressions:

$$\text{Mass A} = \text{mass sample} \times \frac{\text{g A}}{100 \text{ g sample}}$$

$$\text{Mass sample} = \text{mass A} \times \frac{100 \text{ g sample}}{\text{g A}}$$

3. INTERCONVERSION BETWEEN VOLUME AND MASS USING DENSITY.

In Section 1.13 density was defined as the mass of a sample divided by its volume (density = mass/volume). The equation may be rearranged to find the mass of a given volume of a substance of known density:

$$\text{Mass} = \text{density} \times \text{volume}$$

or to find the volume of a given mass of a substance of known density:

$$\text{Volume} = \frac{\text{mass}}{\text{density}}$$

EXAMPLE 3.7   The reaction of chlorobenzene, C_6H_5Cl, with a solution of ammonia containing 28.2% NH_3 by mass is used in the preparation of aniline, $C_6H_5NH_2$, a key ingredient in the preparation of dyes for fabrics. What mass of NH_3 is required to prepare 125 mL of the ammonia solution with a density of 0.899 g/cm³?

Two steps are required to solve this problem: a conversion from volume and density to mass of a mixture (the solution), followed by a conversion from the mass of a mixture to the mass of a pure substance.

$$\boxed{\begin{array}{c}\text{Volume of}\\\text{solution}\end{array}} \longrightarrow \boxed{\begin{array}{c}\text{Mass of}\\\text{solution}\end{array}} \longrightarrow \boxed{\begin{array}{c}\text{Mass of}\\NH_3\end{array}}$$

Find the mass of 125 mL of NH_3 solution (interconversion between volume and mass using density). Since 1 mL is the same as 1 cm³, we express the volume in cm³.

$$125 \text{ cm}^3 \text{ solution} \times \frac{0.899 \text{ g solution}}{1 \text{ cm}^3 \text{ solution}} = 112.4 \text{ g solution}$$

Now we find the mass of NH_3 in the solution (interconversion between mass of a mixture and mass of a component). A 28.2% NH_3 solution by mass contains 28.2 g

NH_3 per 100 g solution.

$$112.4 \text{ g solution} \times \frac{28.2 \text{ g } NH_3}{100 \text{ g solution}} = 31.7 \text{ g } NH_3$$

4. INTERCONVERSION BETWEEN VOLUME OF SOLUTION AND MOLES OF SOLUTE USING MOLARITY.   The number of moles of a solute, A, in a given volume of a solution of known concentration; the concentration of a solution; and the volume of a solution containing a given number of moles of a solute, A, can each be determined as described in Sections 3.1 and 3.2.

$$\text{Molarity} = \frac{\text{moles A}}{\text{volume of solution}}$$

$$\text{Moles A} = \text{molarity} \times \text{volume of solution}$$

3.5 Applications of Chemical Stoichiometry

Many of the stoichiometry problems that a student sees at the beginning of a chemistry course appear formidable. However, the majority of these problems involve nothing more complicated than applying the unit conversions given above. In most of these problems the first step or steps use unit conversions to transform a quantity of a substance to moles of the substance. The numbers of moles of additional substances that are equivalent to the first are then usually found, and, finally, the amount of one of these additional substances in some unit other than moles is determined.

When students have difficulty with stoichiometry problems, it is usually in finding the order in which to do the conversions. Unfortunately, no single technique can be used to string together these conversions for all types of problems. However, there are some useful guidelines. Many problems require that you determine the quantity of B that is equivalent to a given amount of A. In most of these cases you will need a mole-to-mole conversion (to convert moles of A to moles of B). First you identify the substances (A and B) to be interconverted. Then you find the information that tells you how much of substance A is present and the information that tells you how to report the amount of substance B. When you have the problem broken down into these large steps, you work on the conversions necessary to carry them out. For example, the conversion of a quantity of A to moles of A may require only one conversion, or it may require several.

One final hint: Be certain that you understand how to do each of the various types of single conversions before you combine them in complicated problems.

EXAMPLE 3.8   What volume of a 0.2089 M solution of potassium iodide, KI, will contain enough KI to react exactly with the copper(II) nitrate, $Cu(NO_3)_2$, in 43.88 mL of a 0.3842 M $Cu(NO_3)_2$ solution according to the following equation?

$$2Cu(NO_3)_2(aq) + 4KI(aq) \longrightarrow 2CuI(s) + I_2(s) + 4KNO_3(aq)$$

This problem asks us to use the volume of a $Cu(NO_3)_2$ solution of known concentration to calculate the volume of a KI solution of known concentration. The overall conversion is

$$\boxed{\begin{array}{c}\text{Volume of}\\ Cu(NO_3)_2 \text{ solution}\end{array}} \longrightarrow \boxed{\begin{array}{c}\text{Volume of}\\ \text{KI solution}\end{array}}$$

To determine the necessary volume of the KI solution, we need to know the moles of KI required to react. This can be determined from the moles of $Cu(NO_3)_2$ that react [the moles of $Cu(NO_3)_2$ present in 43.88 mL of the $Cu(NO_3)_2$ solution] which we can find by using the conversion factor obtained from the equation. The calculation involves the following steps:

$$\boxed{\begin{array}{c}\text{Volume of}\\ Cu(NO_3)_2 \text{ solution}\end{array}} \xrightarrow{4} \boxed{\begin{array}{c}\text{Moles of}\\ Cu(NO_3)_2\end{array}} \longrightarrow \boxed{\begin{array}{c}\text{Moles of}\\ \text{KI}\end{array}} \xrightarrow{4} \boxed{\begin{array}{c}\text{Volume of}\\ \text{KI solution}\end{array}}$$

Note that the volumes must be expressed in liters. (The number above an arrow indicates the paragraph in Section 3.4 that relates to the conversion.) The calculations for each step follow.

$$\frac{0.3842 \text{ mol } Cu(NO_3)_2}{1 \text{ L}} \times 0.04388 \text{ L} = 1.686 \times 10^{-2} \text{ mol } Cu(NO_3)_2$$

$$1.686 \times 10^{-2} \text{ mol } Cu(NO_3)_2 \times \frac{4 \text{ mol KI}}{2 \text{ mol } Cu(NO_3)_2} = 3.372 \times 10^{-2} \text{ mol KI}$$

$$3.372 \times 10^{-2} \text{ mol KI} \times \frac{1 \text{ L}}{0.2089 \text{ mol KI}} = 0.1614 \text{ L}$$

The reaction requires 0.1614 L, or 161.4 mL, of KI solution.

EXAMPLE 3.9   The toxic pigment called white lead, $Pb_3(OH)_2(CO_3)_2$, has been replaced by rutile, TiO_2, in white paints. How much rutile can be prepared from 379 g of an ore containing ilmenite, $FeTiO_3$, if the ore is 88.3% ilmenite by mass? TiO_2 is prepared by the reaction

$$2FeTiO_3 + 4HCl + Cl_2 \longrightarrow 2FeCl_3 + 2TiO_2 + 2H_2O$$

This problem asks how much rutile, TiO_2, can be prepared from an ore containing ilmenite, $FeTiO_3$. The overall conversion is

$$\boxed{\begin{array}{c}\text{Mass of}\\ \text{ore}\end{array}} \longrightarrow \boxed{\begin{array}{c}\text{Mass of}\\ TiO_2\end{array}}$$

Since most conversions of this type involve mole-to-mole conversions, the chain of conversions for this problem probably includes one, too. The problem states that the $FeTiO_3$ in an ore is converted into TiO_2, so the chain of conversions should include a conversion of moles of $FeTiO_3$ to moles of TiO_2.

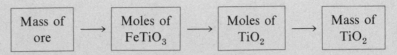

Now it only remains to convert the mass of ore to the mass of $FeTiO_3$ in the ore. The final calculation involves the following unit conversions:

| Mass of ore | $\xrightarrow{2}$ | Mass of $FeTiO_3$ | $\xrightarrow{1}$ | Moles of $FeTiO_3$ | $\longrightarrow$ | Moles of TiO_2 | $\xrightarrow{1}$ | Mass of TiO_2 |

(The number above an arrow indicates the paragraph in Section 3.4 that relates to the conversion.) The calculations for each step follow.

$$379 \text{ g ore} \times \frac{88.3 \text{ g FeTiO}_3}{100 \text{ g ore}} = 344.7 \text{ g FeTiO}_3$$

$$334.7 \text{ g FeTiO}_3 \times \frac{1 \text{ mol FeTiO}_3}{151.8 \text{ g FeTiO}_3} = 2.205 \text{ mol FeTiO}_3$$

$$2.205 \text{ mol FeTiO}_3 \times \frac{2 \text{ mol TiO}_2}{2 \text{ mol FeTiO}_3} = 2.205 \text{ mol TiO}_2$$

$$2.205 \text{ mol TiO}_2 \times \frac{79.90 \text{ g TiO}_2}{1 \text{ mol TiO}_2} = 176 \text{ g TiO}_2$$

Thus 176 g of TiO_2 can be prepared from the 379-g sample of ore.

EXAMPLE 3.10   Calcium acetate, $Ca(CH_3CO_2)_2$, is used as a mordant during the dyeing of fabric. The preparation of calcium acetate proceeds according to the equation

$$Ca(OH)_2 + 2CH_3CO_2H \longrightarrow Ca(CH_3CO_2)_2 + 2H_2O$$

How many grams of a sample containing 75.0% calcium hydroxide, $Ca(OH)_2$, by mass is required to react with the acetic acid, CH_3CO_2H, in 25.0 mL of a solution having a density of 1.065 g/mL and containing 58.0% acetic acid by mass?

Overall, this problem asks us to find the mass of a $Ca(OH)_2$ sample that will react with a given volume of CH_3CO_2H solution.

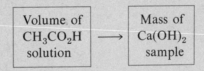

Since the solution contains CH_3CO_2H that reacts with the $Ca(OH)_2$ in the sample, we need a mole-to-mole conversion.

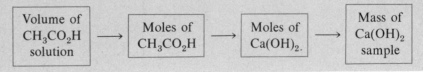

The density of the solution and the percent by mass of acetic acid in the solution are given, so we can calculate the mass of the solution and from that the mass of acetic acid in the solution. This will give us the steps necessary to convert from volume of solution to moles of acetic acid. To convert from moles of $Ca(OH)_2$ to mass of sample, we need to use the percent by mass of $Ca(OH)_2$ in the sample. The overall string of conversions is

$$\boxed{\begin{array}{c}\text{Volume of}\\CH_3CO_2H\\\text{solution}\end{array}} \xrightarrow{3} \boxed{\begin{array}{c}\text{Mass of}\\CH_3CO_2H\\\text{solution}\end{array}} \xrightarrow{2} \boxed{\begin{array}{c}\text{Mass of}\\CH_3CO_2H\\\text{solute}\end{array}} \xrightarrow{1} \boxed{\begin{array}{c}\text{Moles of}\\CH_3CO_2H\end{array}}$$

$$\longrightarrow \boxed{\begin{array}{c}\text{Moles of}\\Ca(OH)_2\end{array}} \xrightarrow{1} \boxed{\begin{array}{c}\text{Mass of}\\Ca(OH)_2\end{array}} \xrightarrow{2} \boxed{\begin{array}{c}\text{Mass of}\\\text{sample}\end{array}}$$

(The number above an arrow indicates the paragraph in Section 3.4 that relates to these conversions.) The calculations for each step follow.

$$25.0 \text{ mL solution} \times \frac{1.065 \text{ g solution}}{1 \text{ mL solution}} = 26.62 \text{ g solution}$$

$$26.62 \text{ g solution} \times \frac{58.0 \text{ g } CH_3CO_2H}{100 \text{ g solution}} = 15.44 \text{ g } CH_3CO_2H$$

$$15.44 \text{ g } CH_3CO_2H \times \frac{1 \text{ mol } CH_3CO_2H}{60.05 \text{ g } CH_3CO_2H} = 0.2571 \text{ mol } CH_3CO_2H$$

$$0.2571 \text{ mol } CH_3CO_2H \times \frac{1 \text{ mol } Ca(OH)_2}{2 \text{ mol } CH_3CO_2H} = 0.1286 \text{ mol } Ca(OH)_2$$

$$0.1286 \text{ mol } Ca(OH)_2 \times \frac{74.09 \text{ g } Ca(OH)_2}{1 \text{ mol } Ca(OH)_2} = 9.528 \text{ g } Ca(OH)_2$$

$$9.528 \text{ g } Ca(OH)_2 \times \frac{100 \text{ g sample}}{75.0 \text{ g } Ca(OH)_2} = 12.7 \text{ g sample}$$

EXAMPLE 3.11   What is the molar concentration of acetic acid in the solution described in Example 3.10?

This problem asks that we find the molarity of the acetic acid solution described in the preceding problem. Thus we must proceed from the volume of that solution to its molarity.

$$\boxed{\begin{array}{c}\text{Volume of}\\CH_3CO_2H\\\text{solution}\end{array}} \longrightarrow \boxed{\begin{array}{c}\text{Molarity of}\\CH_3CO_2H\\\text{solution}\end{array}}$$

A reaction is not involved, so we do not need to do a mole-to-mole conversion. We should remember, however, that if we know the moles of acetic acid in a given volume of solution we can calculate the molarity of the solution. The problem gives the volume of acetic acid solution (25.0 mL), so moles of acetic acid enter the chain of calculations.

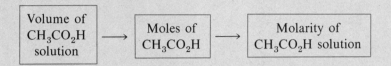

The first steps in this series of conversions have been determined in Example 3.10. Insertion of these steps gives

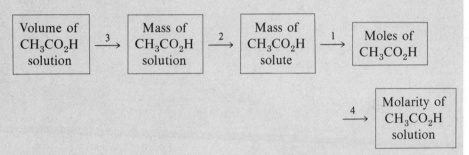

(The number above an arrow indicates the paragraph in Section 3.4 that relates to these conversions.) Steps 1 through 3 have been worked in Example 3.10, which shows that the sample contains 0.2571 mol of CH_3CO_2H.

$$Molarity = \frac{moles\ of\ CH_3CO_2H}{liters\ of\ solution}$$

$$M = \frac{0.2571\ mol}{0.0250\ L} = 10.3\ mol/L$$

Note that molarity is expressed in moles per liter; hence the volume of the solution must be expressed in liters.

EXAMPLE 3.12   A small amount of industrial sulfuric acid, H_2SO_4, is prepared by the oxidation of metal sulfides. The following sequence of reactions illustrates the manufacture of sulfuric acid from pyrites, FeS_2:

$$4FeS_2(s) + 11O_2(g) \longrightarrow 2Fe_2O_3(s) + 8SO_2(g) \tag{1}$$

$$2SO_2(g) + O_2(g) \longrightarrow 2SO_3(g) \tag{2}$$

$$SO_3(g) + H_2O(l) \longrightarrow H_2SO_4(l) \tag{3}$$

What mass of FeS_2 is required to prepare 1.00 L of H_2SO_4 (density = 1.85 g/mL)?

The problem is solved by converting the volume of H_2SO_4 into the mass of FeS_2 required to produce it. The following conversions are required:

$$\boxed{\begin{array}{c}\text{Volume of}\\ H_2SO_4\end{array}} \xrightarrow{3} \boxed{\begin{array}{c}\text{Mass of}\\ H_2SO_4\end{array}} \xrightarrow{1} \boxed{\begin{array}{c}\text{Moles of}\\ H_2SO_4\end{array}} \longrightarrow \boxed{\begin{array}{c}\text{Moles of}\\ SO_3\end{array}}$$

$$\longrightarrow \boxed{\begin{array}{c}\text{Moles of}\\ SO_2\end{array}} \longrightarrow \boxed{\begin{array}{c}\text{Moles of}\\ FeS_2\end{array}} \xrightarrow{1} \boxed{\begin{array}{c}\text{Mass of}\\ FeS_2\end{array}}$$

(The number above an arrow indicates the paragraph in Section 3.4 that relates to these conversions.)

$$1.00 \text{ L } H_2SO_4 \times \frac{1000 \text{ mL}}{1 \text{ L}} \times \frac{1.85 \text{ g}}{1 \text{ mL}} = 1850 \text{ g } H_2SO_4$$

$$1850 \text{ g } H_2SO_4 \times \frac{1 \text{ mol } H_2SO_4}{98.07 \text{ g } H_2SO_4} = 18.86 \text{ mol } H_2SO_4$$

From Equation (3):

$$18.86 \text{ mol } H_2SO_4 \times \frac{1 \text{ mol } SO_3}{1 \text{ mol } H_2SO_4} = 18.86 \text{ mol } SO_3$$

From Equation (2):

$$18.86 \text{ mol } SO_3 \times \frac{2 \text{ mol } SO_2}{2 \text{ mol } SO_3} = 18.86 \text{ mol } SO_2$$

From Equation (1):

$$18.86 \text{ mol } SO_2 \times \frac{4 \text{ mol } FeS_2}{8 \text{ mol } SO_2} = 9.430 \text{ mol } FeS_2$$

$$9.430 \text{ mol } FeS_2 \times \frac{120.0 \text{ g } FeS_2}{1 \text{ mol } FeS_2} = 1.13 \times 10^3 \text{ g } FeS_2$$

FOR REVIEW

SUMMARY

If a substance is dissolved, that is, if it is in **solution,** a given quantity of the substance, in moles, can be obtained by measuring out a known volume of the solution, provided the molarity of the solution is also known. The **molarity** of a solution is equal to the number of moles of solute dissolved in exactly 1 liter of the solution.

Stoichiometry involves calculation of the amounts of materials in chemical reactions. Such calculations reflect the fact that the relative numbers of moles of reactants and products involved in a chemical reaction are given by the balanced chemical equation describing the reaction. Thus a balanced equation can be used to determine unit conversion factors relating moles of reactants and/or products.

The quantity of a substance can be given as the mass of a pure sample of the substance, as the mass of a mixture containing a certain percent by mass of the substance, as the volume of a substance with its density, or as the volume of a solution with a certain molarity for the substance.

Volumes are often measured by use of a **pipet,** which delivers a fixed volume of a liquid or by a buret, which delivers a variable volume that can be measured. Burets are used in **titrations.**

KEY TERMS AND CONCEPTS

Buret (3.2)
Concentration (3.1)
End point (3.2)

Molarity (3.1)
Pipet (3.2)
Solution (3.1)

Stoichiometry (3.3)
Titration (3.2)

EXERCISES

Molarity and Solutions

1. Calculate the molarity of each of the following solutions:
   (a) 98.1 g of sulfuric acid, H_2SO_4, in 1.00 L of solution   *Ans. 1.00 M*
   (b) 24.5 g of sodium cyanide, NaCN, in 2.000 L of solution   *Ans. 0.250 M*
   (c) 90.0 g of acetic acid, CH_3CO_2H, in 0.750 L of solution   *Ans. 2.00 M*
   (d) 2.12 g of potassium bromide, KBr, in 458 mL of solution   *Ans. 3.89×10^{-2} M*
   (e) 0.1374 g of copper sulfate, $CuSO_4$, in 13 mL of solution   *Ans. 6.6×10^{-2} M*

2. Determine the moles of solute present in each of the following solutions:
   (a) 0.450 L of 1.00 M NaCl solution
       Ans. 0.450 mol NaCl
   (b) 2.0 L of 0.480 M $MgCl_2$ solution
       Ans. 0.96 mol $MgCl_2$
   (c) 455 mL of 3.75 M HCl solution
       Ans. 1.71 mol HCl
   (d) 2.50 mL of 0.1812 M $KMnO_4$ solution
       Ans. 4.53×10^{-4} mol $KMnO_4$
   (e) 25.38 mL of 9.721×10^{-2} M KOH solution
       Ans. 2.467×10^{-3} mol KOH

3. Determine the mass of each of the following solutes that is required to make the indicated amount of solution:
   (a) 1.00 L of 1.00 M $LiNO_3$ solution
       Ans. 68.9 g $LiNO_3$

   (b) 4.2 L of 2.45 M C_2H_5OH solution
       Ans. 4.7×10^2 g C_2H_5OH
   (c) 275 mL of 0.5151 M $KClO_4$ solution
       Ans. 19.6 g $KClO_4$
   (d) 25 mL of 0.1881 M H_2O_2 solution
       Ans. 0.16 g H_2O_2
   (e) 1856 mL of 0.1475 M H_3PO_4 solution
       Ans. 26.83 g H_3PO_4

4. The lowest limit of $MgSO_4$ that can be detected by taste in drinking water is about 0.400 g/L. What is this molar concentration of $MgSO_4$?
   Ans. 3.32×10^{-3} M

5. Cow's milk contains an average of 4.5 g of the sugar lactose, $C_{12}H_{22}O_{11}$, per 0.100 L. What is the molarity of lactose in this milk?
   Ans. 0.13 M

6. For reasons of taste, the maximum chloride ion, Cl^-, content of domestic water supplies has been set at 0.25 g/L. What is the maximum content of Cl^- in moles per liter?   *Ans. 7.0×10^{-3} M*

7. An iron content of 0.1 mg/L in drinking water can usually be detected by taste. Will the iron in a water sample with an iron(II) carbonate, $FeCO_3$, concentration of 5.25×10^{-7} M be detectable by taste?   *Ans. No, the solution contains 2.93×10^{-2} mg Fe/L*

8. How many moles of sodium hydroxide, NaOH, are required to react with the hydrochloric acid in 225 mL of a 0.100 M HCl solution?

$$NaOH(aq) + HCl(aq) \longrightarrow NaCl(aq) + H_2O(l)$$

Ans. 2.25×10^{-2} mol

S9. How many moles of silver nitrate, $AgNO_3$, are required to react with the calcium chloride in 14.96 mL of a 2.244 M solution of $CaCl_2$?

$$2AgNO_3(aq) + CaCl_2(aq) \longrightarrow$$
$$2AgCl(s) + Ca(NO_3)_2(aq)$$

Ans. 6.714×10^{-2} mol

10. What mass of calcium carbonate, $CaCO_3$, is required to react with the sulfuric acid in 375.4 mL of a 0.9734 M solution of H_2SO_4?

$$CaCO_3(s) + H_2SO_4(aq) \longrightarrow$$
$$CaSO_4(s) + H_2O + CO_2$$

Ans. 36.57 g

11. What mass of lead chromate, $PbCrO_4$, the pigment "chrome yellow," often used by artists, can be produced by addition of excess sodium chromate, Na_2CrO_4, to 25 mL of a 0.493 M solution of lead(II) nitrate, $Pb(NO_3)_2$?

$$Na_2CrO_4(aq) + Pb(NO_3)_2(aq) \longrightarrow$$
$$PbCrO_4(s) + 2NaNO_3(aq)$$

Ans. 4.0 g

12. What mass of the active metal magnesium is required to react exactly with the hydrochloric acid in a 125.0-mL sample of a 0.2110 M solution of HCl?

$$Mg(s) + 2HCl(aq) \longrightarrow MgCl_2(aq) + H_2(g)$$

Ans. 0.3205 g

13. An excess of silver nitrate, $AgNO_3$, reacts with 25.00 mL of a solution of calcium chloride, $CaCl_2$, producing calcium nitrate, $Ca(NO_3)_2$, and 4.498 g of silver chloride, AgCl. What is the molarity of the $CaCl_2$ solution?   *Ans. 0.6277 M*

S14. A 25.00-mL sample of sulfuric acid solution from an automobile battery reacts exactly with 87.42 mL of a 1.95 M solution of sodium hydroxide, NaOH. What is the molar concentration of the battery acid?

$$H_2SO_4(aq) + 2NaOH(aq) \longrightarrow$$
$$Na_2SO_4(aq) + 2H_2O(l)$$

Ans. 3.41 M

15. What is the molar concentration of barium hydroxide, $Ba(OH)_2$, in a solution formed by the reaction of 0.3400 g of barium with enough water to give 200.0 mL of solution?

$$Ba(s) + 2H_2O(l) \longrightarrow Ba(OH)_2(aq) + H_2(g)$$

Ans. 1.238×10^{-2} M

16. The lead nitrate, $Pb(NO_3)_2$, in 25.49 mL of a 0.1338 M solution reacts with all of the aluminum sulfate, $Al_2(SO_4)_3$, in 25.00 mL of solution. What is the molar concentration of the $Al_2(SO_4)_3$ in the original $Al_2(SO_4)_3$ solution?

$$3Pb(NO_3)_2(aq) + Al_2(SO_4)_3(aq) \longrightarrow$$
$$3PbSO_4(s) + 2Al(NO_3)_3(aq)$$

Ans. 4.547×10^{-2} M

Titration

S17. What is the molar concentration of $AgNO_3$ in a solution if titration of 25.00 mL of the solution with 0.300 M NaCl requires 37.05 mL of the NaCl solution to reach the end point?

$$AgNO_3(aq) + NaCl(aq) \longrightarrow$$
$$AgCl(s) + NaNO_3(aq)$$

Ans. 0.445 M

18. Titration of 10.00 mL of a 0.444 M HCl solution with a solution of LiSH required 23.2 mL of the LiSH solution to reach the end point. What is the molar concentration of LiSH?

$$HCl(aq) + LiSH(aq) \longrightarrow LiCl(aq) + H_2S(aq)$$

Ans. 0.191 M

S19. What is the molar concentration of H_2SO_4 in a solution if 31.91 mL of it is required to titrate 2.474 g of Na_2CO_3?

$$H_2SO_4(aq) + Na_2CO_3(s) \longrightarrow$$
$$Na_2SO_4(aq) + H_2O(l) + CO_2(g)$$

Ans. 0.7315 M

20. Potatoes are often peeled commercially by soaking them in a 3 M to 6 M solution of sodium hydroxide, then removing the loosened skins by spraying with water. Does a sodium hydroxide solution have a suitable concentration if titration of 10.00 mL of the solution requires 25.3 mL of 1.87 M HCl to reach the end point?

$$NaOH + HCl \longrightarrow NaCl + H_2O$$

Ans. Yes, the concentration is 4.73 M

21. In a common medical laboratory determination of the concentration of free chloride ion, Cl^-, in blood serum, a serum sample is titrated with a mercury(II) nitrate, $Hg(NO_3)_2$, solution.

$$2Cl^- + Hg(NO_3)_2 \longrightarrow HgCl_2 + 2NO_3^-$$

What is the chloride ion concentration in a 0.15-mL sample of normal serum that requires 1.62 mL of $4.96 \times 10^{-3}\ M\ Hg(NO_3)_2$ to reach the end point?   *Ans. 0.11 M*

22. Titration of a 20.0-mL sample of a particularly acidic rain collected in Scotland in 1974 required 1.7 mL of 0.0811 M NaOH to reach the end point. Assuming that the acidity was due to the presence of sulfuric acid, H_2SO_4, in the rain water, what was the molarity of this sulfuric acid solution?

$$H_2SO_4(aq) + 2NaOH(aq) \longrightarrow$$
$$Na_2SO_4(aq) + 2H_2O(l)$$

Ans. 3.4 × 10⁻³ M

23. What volume of 0.0250 M HNO_3 solution is required to titrate 125 mL of a 0.0100 M $Ca(OH)_2$ solution?

$$Ca(OH)_2 + 2HNO_3 \longrightarrow Ca(NO_3)_2 + 2H_2O$$

Ans. 1.00 × 10² mL

S 24. Crystalline potassium hydrogen phthalate, $KHC_8H_4O_4$, is often used as a "standard" acid for standardizing basic solutions because it is easy to purify and to weigh. If 1.5428 g of this salt is titrated with a solution of $Ca(OH)_2$, the reaction is complete when 42.37 mL of the solution has been added. What is the concentration of the $Ca(OH)_2$ solution?

$$2KHC_8H_4O_4 + Ca(OH)_2 \longrightarrow$$
$$Ca(KC_8H_4O_4)_2 + 2H_2O$$

Ans. 8.915 × 10⁻² M

25. A common oven cleaner contains a very caustic metal hydroxide, which can be represented as MOH. Titration of 0.7134 g of MOH with 1.000 M HNO_3 requires 17.85 mL to reach the end point.

$$MOH + HNO_3 \longrightarrow MNO_3 + H_2O$$

What is the element M and what is the formula of this hydroxide?   *Ans. Na; NaOH*

26. (a) A 5.00-mL sample of vinegar, a solution of acetic acid (CH_3CO_2H), was titrated with a

0.1198 M NaOH solution. If 33.93 mL was required to reach the end point, what was the molar concentration of acetic acid in the vinegar?

$$CH_3CO_2H + NaOH \longrightarrow$$
$$CH_3CO_2Na + H_2O$$

Ans. 0.813 M

(b) If the density of the vinegar is 1.005 g/cm^3, what is the percent by mass of acetic acid in the vinegar?   *Ans. 4.86%*

Conversions Between Quantities

27. Begin with the expression for each of the following conversions, and derive the expression for the reverse conversion:

S (a) Moles A to grams A

(b) Volume of a liquid of known density to mass

S (c) Moles of solute to molar concentration (given volume of solution)

(d) Percent A and mass of A to mass of a sample containing A

28. Derive a one-step conversion factor for each of the following:

(a) moles A to mass of a mixture containing a known percent by mass of A

(b) volume of pure A and its density to volume of a mixture of known density containing a known percent by mass of A

(c) volume of a solution of known density with a known molarity of A to percent by mass A in the solution

S 29. Determine the mass and the moles of calcium phosphate, $Ca_3(PO_4)_2$, in a 374-g sample of phosphate rock containing 75.9% $Ca_3(PO_4)_2$.
          Ans. 284 g; 0.916 mol

30. What mass and how many moles of phosphorus trichloride, PCl_3, are contained in 10.00 mL of liquid PCl_3 having a density of 1.574 g/cm^3?
          Ans. 15.74 g; 0.1146 mol

S 31. What is the molar concentration of isotonic saline solution for injection, if it has a density of 1.007 g/mL and contains 0.95% NaCl by mass?
          Ans. 0.16 M

32. Several insects are able to withstand exposures to low temperatures because their body fluids contain large amounts of glycerol, $C_3H_5(OH)_3$, which reduces the freezing point of these fluids.

What is the molar concentration of glycerol in the body fluid (density = 1.068 g/cm^3) of a wasp if it contains 28.0% glycerol by mass?   *Ans. 3.25 M*

S 33. Sulfuric acid for laboratory use is supplied in 2.5-L bottles, which contain 98.0% H_2SO_4 by mass with a density of 1.92 g/mL. What mass of pure H_2SO_4 in kilograms is contained in such a bottle?   *Ans. 4.7 kg*

34. What volume of 0.500 *M* potassium hydroxide solution can be prepared from 348 g of KOH that contains 12.8% water by mass?   *Ans. 10.8 L*

S 35. What volume of 2.00 *M* acetic acid solution can be prepared from 1 pint (0.473 L) of 99.7% (by mass) acetic acid, CH_3CO_2H, with a density of 0.960 g/cm^3?   *Ans. 3.77 L*

36. What is the concentration (molarity) of HCl in a commercial hydrochloric acid solution that has a density of 1.198 g/cm^3 and contains 36.50% HCl?   *Ans. 11.99 M*

37. Concentrated sulfuric acid produced industrially often contains 98.0% H_2SO_4 by mass and has a density of 1.92 g/mL. What is the concentration of H_2SO_4 in this acid solution?   *Ans. 19.2 M*

38. Ammonia for laboratory use is supplied as a solution containing 28.2% NH_3 (by mass) and having a density of 0.8990 g/mL. What volume of this solution is required to prepare 1.00 L of 1.00 *M* $NH_3(aq)$ for a general chemistry laboratory?   *Ans. 67.2 mL*

S 39. What volume of concentrated nitric acid, a 69.5% solution of HNO_3 with a density of 1.42 g/cm^3, is required to prepare 275 mL of a 0.150 *M* solution of HNO_3?   *Ans. 2.63 mL*

40. What mass of sodium hydroxide, NaOH, containing 19.34% water (by mass), is required to prepare exactly 100 L of a 0.2733 *M* NaOH solution?   *Ans. 1.355 kg*

S 41. What volume of gaseous carbon dioxide (density = 1.964 g/L) is required to carbonate 12 oz (355 mL) of a beverage that has a density of 1.067 g/cm^3 and contains 0.66% CO_2 by mass?   *Ans. 1.3 L*

42. The limit for occupational exposure to ammonia, NH_3, in the air is set at 2.8×10^{-3}% by mass. What mass of ammonia, in grams, per cubic meter of air (density = 1.20 g/L) is required to reach this limit?   *Ans. 0.034 g*

43. A 5.0-mL sample of human plasma with a density of 1.0299 g/mL was found to contain 0.478 g of dissolved protein and 0.077 g of dissolved nonprotein material. What are the percents of pro-

tein, of nonprotein materials, and of all dissolved solids in the plasma?   *Ans. 9.3%; 1.5%; 10.8%*

S 44. How many grams of a solution of insect repellent containing 17.0% by mass of N,N-dimethyl-*meta*toluamide, $C_{10}H_{13}NO$, can be prepared from 1.50 mol of $C_{10}H_{13}NO$?   *Ans. 1440 g*

Chemical Stoichiometry

S 45. How many grams of CaO are required for reaction with the HCl in 275 mL of a 0.523 *M* HCl solution? The equation for the reaction is

$$CaO + 2HCl \longrightarrow CaCl_2 + H_2O$$

Ans. 4.03 g

46. Limestone is almost pure calcium carbonate, $CaCO_3$. How much quicklime, CaO, can be prepared by roasting (heating) a metric ton, 1.000×10^3 kg, of limestone that contains 94.6% $CaCO_3$? $CaCO_3$ decomposes into CaO and CO_2 during the roasting process.   *Ans. 5.30×10^2 kg*

47. Aspirin, $C_6H_4(CO_2H)(CO_2CH_3)$, can be prepared in a chemistry laboratory by the reaction of salicylic acid, $C_6H_4(CO_2H)(OH)$, with acetic anhydride, $(CH_3CO)_2O$, a corrosive liquid.

$$2C_6H_4(CO_2H)(OH) + (CH_3CO)_2O \longrightarrow$$
$$2C_6H_4(CO_2H)(CO_2CH_3) + H_2O$$

What mass of aspirin can be prepared by the reaction of 5.00 mL of acetic anhydride (density = 1.0820 g/cm^3), assuming a 100% yield?

Ans. 19.1 g

48. Elemental phosphorus, P_4, is prepared by the following reaction:

$$4Ca_5(PO_4)_3F + 18SiO_2 + 30C \longrightarrow$$
$$3P_4 + 2CaF_2 + 18CaSiO_3 + 30CO$$

What mass of phosphorus can be prepared from a 1.000-kg sample of an ore that contains 86.72% $Ca_5(PO_4)_3F$, assuming a 100% yield of P_4?   *Ans. 159.8 g*

S 49. Automotive "air bags" inflate when a sample of sodium azide, NaN_3, is very rapidly decomposed.

$$2NaN_3(s) \longrightarrow 2Na(s) + 3N_2(g)$$

What mass of sodium azide is required to produce 13.0 ft^3 (368 L) of nitrogen gas with a density of 1.25 g/L?   *Ans. 712 g*

50. What volume in liters of air (density = 1.20 g/L, and percent by mass of O_2 = 21.0%) is required

to burn 1.00 L of gasoline (density $= 0.780$ g/cm^3), which can be represented as C_8H_{16}? Assume a 100% yield and complete consumption of the oxygen.

$$C_8H_{16}(l) + 12O_2(g) \longrightarrow 8CO_2(g) + 8H_2O(g)$$

Ans. 10,600 L

S 51. Reaction of rhenium metal with rhenium heptaoxide, Re_2O_7, gives a solid of metallic appearance that conducts electricity almost as well as copper. A 0.788-g sample of this material, which contains only rhenium and oxygen, was oxidized in an acidic solution of hydrogen peroxide. Addition of an excess of KOH gave 0.973 g of $KReO_4$. Determine the empirical formula of the metallic solid and write the equation for its formation.   *Ans. $Re + 3Re_2O_7 \rightarrow 7ReO_3$*

52. Bronzes used in bearings are often alloys (solid solutions) of copper and aluminum. A 2.053-g sample of one such bronze was analyzed for its copper content by the sequence of reactions given below. The aluminum also dissolves, but we do not need to consider it because aluminum sulfate, $Al_2(SO_4)_3$, does not react with potassium iodide, KI.

$$Cu + 2H_2SO_4 \longrightarrow CuSO_4 + 2H_2O + SO_2$$

$$2CuSO_4 + 4KI \longrightarrow 2CuI + I_2 + 2K_2SO_4$$

$$I_2 + 2Na_2S_2O_3 \longrightarrow Na_2S_4O_6 + 2NaI$$

If 35.11 mL of 0.8375 M $Na_2S_2O_3$ is required to react with the I_2 formed, what is the percent Cu in the sample?   *Ans. 91.02%*

53. Cementite, which contains iron and carbon, is a compound found in cast iron. A 1.724-g sample of cementite was dissolved in a dilute solution of sulfuric acid, giving a solution of iron(II) sulfate, $FeSO_4$, and various carbon-containing compounds, which may be neglected because they do not react in the subsequent analysis. A volume of 34.88 mL of a 0.2753 M solution of potassium chromate, K_2CrO_4, was required to react with all of the $FeSO_4$ according to the equation

$$6FeSO_4 + 2K_2CrO_4 + 8H_2SO_4 \longrightarrow$$
$$3Fe_2(SO_4)_3 + 2K_2SO_4 + Cr_2(SO_4)_3 + 8H_2O$$

What is the empirical formula of cementite?

Ans. Fe_3C

S 54. Glauber's salt, $Na_2SO_4 \cdot 10H_2O$, is an important industrial chemical that is isolated from naturally occurring brines in New Mexico. A 0.3440-g sam-

ple of this material was allowed to react with an excess of barium nitrate, $Ba(NO_3)_2$; 0.2398 g of barium sulfate, $BaSO_4$, was isolated. What is the percent of $Na_2SO_4 \cdot 10H_2O$ in the sample?

$$Na_2SO_4 \cdot 10H_2O(aq) + Ba(NO_3)_2(aq) \longrightarrow$$
$$BaSO_4(s) + 2NaNO_3(aq) + 10H_2O$$

Ans. 96.23%

55. Hydrogen peroxide may be prepared by the following reactions:

$$2NH_4HSO_4 \xrightarrow{\text{Elect.}} H_2 + (NH_4)_2S_2O_8$$
$$(NH_4)_2S_2O_8 + 2H_2O \longrightarrow 2NH_4HSO_4 + H_2O_2$$

What mass of ammonium hydrogen sulfate, NH_4HSO_4, is initially required to prepare 1.00 mol of H_2O_2? What mass of H_2O is required?

Ans. 2.30×10^2 g; 36.0 g

S 56. Potassium perchlorate, $KClO_4$, may be prepared from KOH and Cl_2 by the following series of reactions:

$$2KOH + Cl_2 \longrightarrow KCl + KClO + H_2O$$

$$3KClO \longrightarrow 2KCl + KClO_3$$

$$4KClO_3 \longrightarrow 3KClO_4 + KCl$$

What mass of $KClO_4$ can be prepared from 25.0 g of KOH?   *Ans. 7.72 g*

57. Sodium thiosulfate, $Na_2S_2O_3$, a photographic fixing agent known as hypo, can be prepared from Glauber's salt, $Na_2SO_4 \cdot 10H_2O$, by the following series of reactions:

$$Na_2SO_4 \cdot 10H_2O(s) \xrightarrow{\triangle} Na_2SO_4(s) + 10H_2O(g)$$

$$Na_2SO_4(s) + 4C(s) \xrightarrow{\triangle} Na_2S(s) + 4CO(g)$$

$$2Na_2S(aq) + Na_2CO_3(aq) + 4SO_2(aq) \longrightarrow$$
$$3Na_2S_2O_3(aq) + CO_2(g)$$

What mass of $Na_2SO_4 \cdot 10H_2O$ is required to prepare 50.0 g of $Na_2S_2O_3$?   *Ans. 67.9 g*

Additional Exercises

58. Tab, a sugar-free soft drink, contains 0.11 g of the artificial sweetener saccharin, $C_7H_5SNO_3$, per 12-oz can. What is the molar concentration of saccharin in this drink?   *Ans. 1.7×10^{-3} M*

59. The average volume of plasma in an adult is 39 mL per kilogram of weight. The average concentration of sodium ion, Na^+, in plasma is 0.142 M. What is the mass of sodium ion present

in the serum of a 75-kg (165-lb) adult? What mass of NaCl, which is composed of Na^+ and Cl^- ions, would be required to provide this much sodium?   *Ans. 9.5 g Na⁺; 24 g NaCl*

S 60. The concentrations of sodium ion, potassium ion, chloride ion, and dihydrogen phosphate ion in Gatorade are Na^+, 21.0 mM; K^+, 2.5 mM; Cl^-, 17.0 mM; and $H_2PO_4^-$, 6.8 mM. These, incidentally, match the relative concentrations of these ions lost through perspiration. What masses of Na^+, K^+, Cl^-, and $H_2PO_4^-$ are contained in an 8.00-oz glass of Gatorade?

Ans. 0.114 g Na⁺; 0.023 g K⁺;
0.143 g Cl⁻; 0.16 g H₂PO₄⁻

61. A copper ion, Cu^{2+}, concentration of 1.0 mg/L causes a bitter taste in drinking water. What mass of copper sulfate, $CuSO_4$, will give a copper concentration of 1.0 mg/L, and what is the $CuSO_4$ concentration in mol/L?

Ans. 2.5 mg; 1.6 × 10⁻⁵ M

62. A mixture of morphine, $C_{17}H_{19}NO_3$, and an inert solid is analyzed by combustion with O_2. The *unbalanced* equation for the reaction of morphine with O_2 is

$$C_{17}H_{19}NO_3 + O_2 \longrightarrow CO_2 + H_2O + NO_2$$

The inert solid does not react with O_2. If 4.000 g of the mixture yields 8.72 g of CO_2, calculate the percent by mass of morphine in the mixture.

Ans. 83.1%

S 63. What mass of a sample that is 98.0% sulfur would be required in the production of 75.0 kg of H_2SO_4 by the following reaction sequence?

$$S_8 + 8O_2 \longrightarrow 8SO_2$$
$$2SO_2 + O_2 \longrightarrow 2SO_3$$
$$SO_3 + H_2O \longrightarrow H_2SO_4$$

Ans. 25.0 kg

S 64. For many years, the standard medical laboratory technique for determination of the concentration of calcium (as Ca^{2+}) in blood serum used the following reaction sequence involving ammonium oxalate, $(NH_4)_2C_2O_4$, and potassium permanganate, $KMnO_4$:

$$Ca^{2+}(aq) + (NH_4)_2C_2O_4(s) \longrightarrow$$
$$CaC_2O_4(s) + 2NH_4^+(aq)$$
$$CaC_2O_4(s) + H_2SO_4(aq) \longrightarrow$$
$$H_2C_2O_4(aq) + CaSO_4(aq)$$

$$2KMnO_4(aq) + 5H_2C_2O_4(aq)$$
$$+ 3H_2SO_4(aq) \longrightarrow K_2SO_4(aq)$$
$$+ 2MnSO_4(aq) + 10CO_2(g) + 8H_2O(l)$$

What is the molar concentration of calcium in a 2.0 mL serum sample if 0.68 mL of 2.44 × 10^{-3} M $KMnO_4$ is required for the final reaction? How many milligrams of calcium is contained in 1.00 × 10^2 mL of the serum?

Ans. 0.0021 M; 8.4 mg

65. The amount of active ingredient per tablet of several antacids and equations for how these ingredients react with stomach acid (HCl) are given below. Assuming that the stomach acid is undiluted (0.155 M), calculate what volume of acid in milliliters will react with the given mass of the active ingredient in each tablet.

(a) Phillip's Tablets, 0.311 g $Mg(OH)_2$

$$Mg(OH)_2 + 2HCl \longrightarrow MgCl_2 + 2H_2O$$

Ans. 68.8 mL

(b) Tums, 0.500 g $CaCO_3$

$$CaCO_3 + 2HCl \longrightarrow CaCl_2 + H_2O + CO_2$$

Ans. 64.5 mL

(c) Rolaids, 0.334 g $NaAl(CO_3)(OH)_2$

$$NaAl(CO_3)(OH)_2 + 4HCl \longrightarrow$$
$$NaCl + AlCl_3 + 3H_2O + CO_2$$

Ans. 59.9 mL

(d) Gelusil, 0.500 g $Mg_2Si_3O_8$, 0.075 g $Mg(OH)_2$

$$Mg_2Si_3O_8 + 4HCl \longrightarrow$$
$$2MgCl_2 + 3SiO_2 + 2H_2O$$
$$Mg(OH)_2 + 2HCl \longrightarrow MgCl_2 + 2H_2O$$

Ans. 66.1 mL

66. Hach Chemical Company of Loveland, Colorado, sells a test kit for determination of chloride ion in domestic water supplies. The kit contains a silver nitrate, $AgNO_3$, solution, which is added drop by drop to a 23.0-mL water sample to which an indicator has been added. When sufficient silver nitrate has been added to convert the chloride ion completely to silver chloride, $AgCl$, the solid produced turns orange. The concentration of the silver nitrate solution is such that each drop used to reach the color change corresponds to 12.5 mg of Cl^- per liter of water tested.

(a) What mass of chloride ion (mg/L) is contained in a water sample that requires 12 drops of test solution to reach the color change? *Ans.* 1.5×10^2 *mg/L*

(b) What is the molar concentration of Cl^- in the sample tested in (a)? *Ans.* 4.2×10^{-3} *M*

(c) If the sample size used is 5.75 mL instead of 23.0 mL, to what chloride ion concentration (mg/L) does 1 drop of the test solution correspond? *Ans. 50.0 mg/L*

(d) This test kit can also be used for determination of bromide ion or iodide ion. If the reading obtained is multiplied by 2.25, the correct concentration of bromide ion (mg/L) is obtained. What factor should be used to obtain the correct concentration of iodide ion (mg/L)? *Ans. 3.58*

(e) If 20 drops of the silver nitrate test solution equals 1.00 mL, what is its molar concentration of $AgNO_3$? *Ans. 0.162 M*

[S] 67. Calculate the mass of sodium nitrate required to produce 5.00 L of O_2 (density = 1.43 g/L) according to the reaction

$$2NaNO_3 \xrightarrow{\Delta} 2NaNO_2 + O_2$$

if the percent yield of the reaction is 78.4%.

Ans. 48.4 g

68. On the average, protein contains 15.5% nitrogen by mass. This nitrogen can be converted to gaseous ammonia, $NH_3(g)$, and analyzed by the Kjeldahl technique. The ammonia is allowed to react with an excess of boric acid, H_3BO_3,

$$NH_3(g) + H_3BO_3(aq) \longrightarrow (NH_4)H_2BO_3(aq)$$

and the product is titrated with standardized hydrochloric acid.

$$(NH_4)H_2BO_3 + HCl \longrightarrow NH_4Cl + H_3BO_3$$

What is the percent by mass of protein in a sample of lobster meat if titration of the $(NH_4)H_2BO_3$ formed in a Kjeldahl analysis of 4.95 g requires 34.05 mL of a 0.2011 *M* HCl solution?

Ans. 12.5%

69. (a) Champagne can be prepared by fermentation of grape juices in the bottle. This fermentation converts sugar into ethanol and carbon dioxide.

$$C_6H_{12}O_6 \longrightarrow 2C_2H_5OH + 2CO_2$$

The carbon dioxide cannot escape and so produces a carbonated beverage. What is the percent by mass of CO_2 in 750 mL of champagne that has a density of 0.98 g/cm^3 and contains 12.0% ethanol by mass. *Ans. 11%*

(b) If exactly one-half of the CO_2 escapes from solution as bubbles of $CO_2(g)$ when the bottle is opened, what will the total volume of the bubbles be, assuming $CO_2(g)$ has a density of 1.96 g/L? *Ans. 21 L*

[S] 70. Uranium may be isolated from the mineral pitchblende, which contains U_3O_8. The pitchblende in a 4.835-g sample of an ore containing uranium was subjected to the following sequence of reactions:

$$2U_3O_8 + O_2 + 12HNO_3 \longrightarrow 6(UO_2)(NO_3)_2 + 6H_2O$$

$$(UO_2)(NO_3)_2 + 4H_2O + H_3PO_4 \longrightarrow (UO_2)HPO_4 \cdot 4H_2O + 2HNO_3$$

$$2[(UO_2)HPO_4 \cdot 4H_2O] \xrightarrow{\Delta} (UO_2)_2P_2O_7 + 9H_2O$$

What is the percent by mass of U_3O_8 in the ore if 1.432 g of $(UO_2)_2P_2O_7$ is isolated?

Ans. 23.29%

71. A 25-mL volume of a solution of triethylaluminum, $Al(C_2H_5)_3$, a substance used in the production of the plastic polyethylene, was allowed to react with 25.0 mL of a 0.103 *M* HCl solution.

$$Al(C_2H_5)_3 + 3HCl \longrightarrow AlCl_3 + 3C_2H_6$$

The $Al(C_2H_5)_3$ was the limiting reagent; hence some HCl remained unreacted. The unreacted HCl was titrated with a 0.142 *M* solution of NH_3; 16.75 mL was required.

$$HCl + NH_3 \longrightarrow NH_4Cl$$

What was the concentration of $Al(C_2H_5)_3$ in the solution? *Ans.* 2.6×10^{-3} *M*

[S] 72. What is the limiting reagent when 5.0×10^{-2} mol of nitric acid, HNO_3, reacts with 225 mL of 0.10 *M* calcium hydroxide, $Ca(OH)_2$, solution?

$$Ca(OH)_2(aq) + 2HNO_3(aq) \longrightarrow Ca(NO_3)_2(aq) + 2H_2O(l)$$

Ans. $Ca(OH)_2$

73. What is the limiting reagent when 25.00 mL of a 0.2338 *M* lead nitrate, $Pb(NO_3)_2$, solution is added to 25.00 mL of a 0.0971 *M* aluminum sulfate, $Al_2(SO_4)_3$, solution?

$$3Pb(NO_3)_2(aq) + Al_2(SO_4)_3(aq) \longrightarrow 3PbSO_4(s) + 2Al(NO_3)_3(aq)$$

Ans. $Pb(NO_3)_2$

4

STRUCTURE OF THE ATOM
AND THE PERIODIC LAW

As John Dalton examined the experimental observations that led to his theory that matter was composed of atoms, he found no evidence that atoms were composed of smaller particles. Thus Dalton assumed atoms were simple indivisible bodies. However, a series of discoveries beginning in the later part of the nineteenth century has shown that atoms are complex systems composed of a number of smaller particles.

Many different experiments have contributed to our ideas about the structure of atoms, yet a few have been particularly important in the development of these ideas. In this chapter we will examine some of these experiments and how they have contributed. Then we will discuss the details of atomic structure itself and describe how several kinds of chemical behavior follow from this structure. In fact, observed regularities in chemical behavior, as shown by the Periodic Table, provide additional evidence for our view of the structure of the atom.

As will be seen in later chapters, the chemical behavior of atoms and molecules is determined primarily by the arrangement of their electrons. Thus this study of the structure of atoms is the first step in developing a model of matter that will help you to understand why substances exhibit particular kinds of chemical properties and to recall these properties systematically.

THE STRUCTURE OF THE ATOM

4.1 Electrons

Much of our knowledge about the structure of atoms resulted from experiments involving the passage of electricity through different gases at low pressures. The apparatus used for experiments of this type, called a **discharge tube,** or **Crookes tube,** was developed by Sir William Crookes (1832–1919). Crookes passed an electric current through a gas-filled glass tube with electrodes sealed into both ends [Fig. 4-1(a)]. As gas was pumped out of the tube, a pressure was reached at which the remaining gas glowed. With more pumping, the gas ceased to glow. However, close observation revealed that the glass at one end of the tube was glowing. Crookes suggested that this glow was produced by negative particles, which he called **cathode rays,** passing from the negative electrode **(cathode)** and moving toward the positive electrode **(anode).** Those particles that missed the anode hit the glass and made it glow.

Cathode rays were shown to have the following characteristics:

1. They travel from the cathode to the anode in straight lines, as indicated by the fact that an object placed in their path casts a sharp shadow on the end of the tube [Fig. 4-1(b)].
2. They cause a piece of metal foil to become hot by striking it for a period of time.
3. They are deflected by magnetic and electrical fields [Fig. 4-1(a)], and the direction of the deflection is the same as that shown by negatively charged particles passing through such fields (a charged object will attract an object of opposite charge, whereas objects having charges of the same sign repel each other).

These characteristics are best explained by assuming that cathode rays are streams of small negatively charged particles. These particles are called **electrons.**

In 1897 the English physicist Sir J. J. Thomson measured the deflection of cathode-ray particles in both a magnetic field and an electric field. From the strengths of the fields, he determined the ratio of the charge, e, to the mass, m, of the cathode-ray particles. He found e/m to be identical for all of these particles,

Figure 4-1. Two types of discharge tubes. (a) In the absence of a charge on the plates, electrons flow from the cathode to the anode (black path). When the plates are charged, the electron stream curves toward the positive plate (colored path). (b) A shadow being cast by an object in a stream of electrons. The tube glows where there is no shadow.

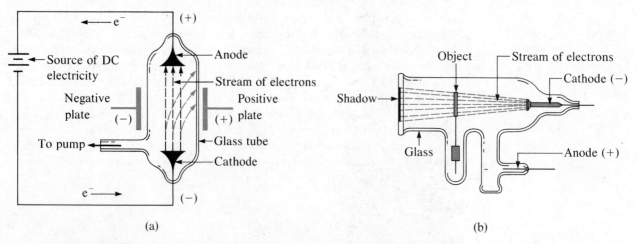

(a)

(b)

irrespective of the metal the electrodes were made of or the kind of gas in the tube. The charge was measured by the American physicist R. A. Millikan in 1909 (Fig. 4-2). From the values of e/m and e, the mass of the electron was calculated to be 1/1837 of that of the hydrogen atom, or 0.00055 amu. This quantitative observation that all electrons are identical, regardless of their source or the method of liberating them from matter, along with the observation that they can be liberated from any kind of atom, proves quite conclusively that electrons are parts of all types of matter.

Figure 4-2. The apparatus used by Millikan. A fine drop of oil drifts from the top of the apparatus through the hole into the region between the two plates. The X rays cause the drop to pick up a negative charge, and the drop is attracted toward the positive plate. At one particular voltage the electrical and gravitational forces are balanced, and the drop remains stationary. From the charge on the plates and the mass of the drop, the charge on the drop can be calculated.

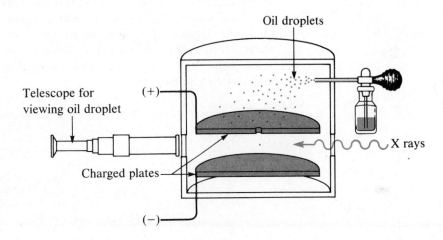

Electrons are emitted by certain metals when the metals are heated to high temperatures. This process is called **thermal emission** and is the source of electrons in X-ray tubes and cathode-ray tubes such as television picture tubes. Electrons are also emitted from metals, most readily from active metals such as cesium, sodium, and potassium, when they are exposed to light (Fig. 4-3). This type of emission is known as the **photoelectric effect** and is the basis of the electric eye, or photoelectric cell, used in some automatic door openers.

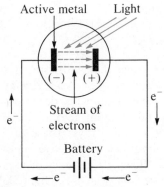

Figure 4-3. A photoelectric cell in an electric circuit. Light striking the active metal electrode causes electrons to be emitted and a current to flow.

4.2 Protons

In 1886 Eugen Goldstein, a German physicist, using a Crookes tube with holes in the cathode, observed that another kind of ray was emitted from the anode and passed through the holes in the cathode. Three years later Wilhelm Wien showed these rays to be positively charged particles. Their charge-to-mass ratio was found to be much smaller than that for electrons, and it varied with the kind of gas in the tube. Since the ratio varied, it follows that either the charge varied or the mass varied or both. Actually, both vary. Measurements showed the charge on each particle to be a positive charge, equal to that on the electron, or to some whole-number multiple of that charge, but opposite in sign. The mass of the positive particle was found to be least when hydrogen was used as the gas in the discharge tube. From the values of e and e/m for the positive particles when hydrogen was used, the value of m was calculated as 1.0073 amu. These particles are called **protons** and are one of the basic units of structure of all atoms.

The formation of positively charged particles, called **ions,** from the molecules of gas in a Crookes tube is caused by the loss of electrons from the molecules when they are struck by high-speed cathode rays. Since any neutral atom or molecule can be made to form positive ions by the loss of one or more electrons, it follows that every atom or molecule must contain one or more positive units (or protons). The simplest atom is that of hydrogen—it contains one proton and one electron.

4.3 Neutrons

The existence of the **neutron,** a particle with a mass close to that of the proton but with no charge, had been postulated for more than ten years before the English physicist James Chadwick detected this third basic unit of the atom in 1932. He showed that uncharged particles (neutrons) are emitted when atoms of beryllium (and other elements) are bombarded with high-velocity helium atoms with all electrons removed (alpha particles). Neutrons were determined to have a mass of 1.0087 amu. They are unstable outside of an atom and slowly disintegrate to form a proton and an electron.

The basic particles present in an atom are summarized in Table 4-1.

Table 4-1   Basic Particles in an Atom

Particle	Mass, amu	Charge
Electron	0.00055	−1
Proton	1.0073	+1
Neutron	1.0087	0

4.4 Radioactivity and Atomic Structure

In 1896 additional evidence that atoms are complex rather than indivisible (as stated in Dalton's atomic theory) came with the discovery of radioactivity by Antoine Becquerel, a French physicist. **Radioactivity** is the spontaneous decomposition of atoms of certain elements, such as radium and uranium, into atoms of other elements with the simultaneous production of so-called rays. These rays were characterized as being of three types by placing samples of radioactive material in a narrow hole bored in a block of lead and allowing the beam of rays coming out of the hole to pass through a strong electric field (Fig. 4-4). One type of ray curves toward the negative part of the electric field and so must consist of positively charged particles. These are called **alpha (α) particles.** An alpha particle has a mass of approximately 4 amu and a charge of +2. It is a helium ion—i.e., a helium atom that has lost two electrons. A second type of ray curves toward the

Figure 4-4. The effect of an electric field on rays from the radioactive substance radium. Beta (β) rays are electrons. Alpha (α) rays are nuclei of helium atoms. Gamma (γ) rays are similar to X rays.

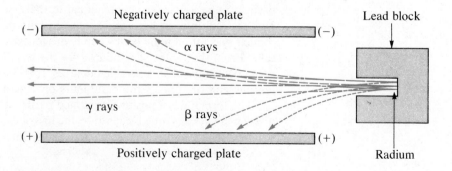

positive part of the field, with a deflection much greater than that of the alpha particles. This behavior indicates that these particles, called **beta (β) particles,** are negatively charged and have a much smaller mass than alpha particles. Beta particles have been shown to be electrons. The third type of ray is not deflected at all when it passes through the electric field, indicating that the particles are neither negatively nor positively charged. These rays are similar to X rays but have higher energies than X rays (and higher energies than α particles and β particles as well). These high-energy particles are called **gamma (γ) rays.** Several other types of rays (radiation) are also known and will be discussed in Chapter 28.

Early experimenters used particles from radioactive materials to investigate the structure of atoms. For example, convincing proof that atoms contain protons was obtained in England by Ernest Rutherford in 1919. Using high-velocity α particles from radium as projectiles, he bombarded atoms such as nitrogen and aluminum and found that protons were ejected from the atoms as a result of these collisions. These experiments indicated quite definitely that the proton is a unit of the atom.

4.5 The Nuclear Atom

Although the experiments described above indicated that atoms can be characterized as assemblies of electrons, protons, and neutrons, they gave no clue as to how these particles are arranged. The first insight into the architecture of atoms resulted from α-particle scattering experiments by Ernest Rutherford.

When Rutherford projected a beam of α particles from a radioactive source onto very thin gold foil, he found that most of the particles passed through the solid foil without deflection, a few of them were diverted from their paths, and a very few of them were deflected back toward their source. From the results of a series of such experiments, Rutherford concluded that (1) the volume occupied by an atom must be largely empty space because most of the α particles pass through the foil undeflected, and (2) each atom must contain a heavy, positively charged body (the nucleus) because of the abrupt change in path by a few α particles. A relatively heavy and positively charged α particle can change its path only when it hits or closely approaches another body (such as a nucleus) with a highly concentrated, positive charge. The atom, then, was presumed to consist of a very small, positively charged **nucleus,** in which most of the mass of the atom is concentrated, surrounded by the number of electrons necessary to produce an electrically neutral atom (Fig. 4-5). Rutherford's nuclear theory of the atom, proposed in 1911, is the model we still use.

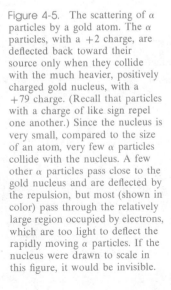

Figure 4-5. The scattering of α particles by a gold atom. The α particles, with a +2 charge, are deflected back toward their source only when they collide with the much heavier, positively charged gold nucleus, with a +79 charge. (Recall that particles with a charge of like sign repel one another.) Since the nucleus is very small, compared to the size of an atom, very few α particles collide with the nucleus. A few other α particles pass close to the gold nucleus and are deflected by the repulsion, but most (shown in color) pass through the relatively large region occupied by electrons, which are too light to deflect the rapidly moving α particles. If the nucleus were drawn to scale in this figure, it would be invisible.

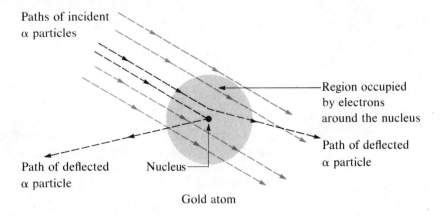

Paths of incident α particles

Region occupied by electrons around the nucleus

Path of deflected α particle

Path of deflected α particle

Nucleus

Gold atom

Rutherford determined the diameter of the nucleus to be at least 100,000 times smaller than the diameter of the atom (the diameter of an atom is of the order of 10^{-8} cm). Thus a nucleus is almost unbelievably small. If the nucleus of an atom were as large as the period at the end of this sentence, the atom would have a diameter of about 40 yards.

From this same series of experiments, Rutherford found that, for many of the lighter elements, the number of positive charges in the nucleus (and thus the number of protons in the nucleus) is approximately equal to half of the atomic weight of the element. In 1914 another English physicist, Henry Moseley, reported a method for determining the number of protons in the nucleus of any atom, and hence the number of positive charges in the nucleus. The number of protons in the nucleus is called the **atomic number** of the element. Since a neutral atom contains the same number of protons as electrons, the atomic number represents the number of electrons *for an atom* as well.

Moseley measured the X rays produced by the elements. A modern X-ray tube (Fig. 4-6) is a modified cathode-ray tube in which electrons are produced by thermal emission from a filament heated by an electric current. When a solid target is placed in the beam of electrons, very penetrating rays, called X rays, are produced by the elements in the target. Using a series of different elements as targets, Moseley showed that the energy of the X rays produced by an element depends on its atomic number. This fact was used to determine the atomic numbers of the heavier elements.

Figure 4-6. In the X-ray tube electrons striking a target such as tungsten cause the production of X rays. In Moseley's experiment various elements were used as targets.

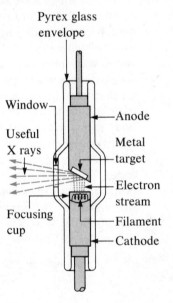

Pyrex glass envelope

Window

Useful X rays

Anode

Metal target

Electron stream

Focusing cup

Filament

Cathode

4.6 Isotopes

At about the same time that Moseley was carrying out his experiments, A. J. Dempster and F. W. Aston, working separately, showed that some elements consist of atoms with different masses. Magnesium, for example, was shown to consist of three different types of atoms. Approximately 79% of magnesium atoms have a mass of about 24 amu; 10%, about 25 amu; and 11%, about 26 amu.

Both Dempster and Aston used early versions of a **mass spectrometer** for their work. In the mass spectrometer (Fig. 4-7) gaseous atoms or molecules are bombarded by high-energy electrons, which knock off electrons and produce positively charged ions. These ions are then directed through a magnetic or electric field, which deflects their paths to an extent dependent on their mass-to-charge ratios (m/e). Ions of any mass-to-charge ratio can be made to strike the detector by varying the field. Since most of the ions are singly charged ($e = 1$), the separation is primarily on the basis of mass. Taking into account the magnetic field and other characteristics of the instrument, the mass of each fragment can be determined.

Figure 4-7. A diagram of a mass spectrometer. Ions leave the ionization chamber and move into the magnetic field. Light ions experience a large deflection and collide with the upper wall of the tube. Heavy ions experience less deflection and collide with the lower wall. Only those ions with a specific value of m/e pass through the field into the detector.

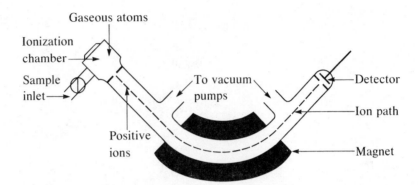

A mass spectrum is determined by varying the magnetic or electric field so that ions with progressively higher masses strike the detector and plotting the relative intensities (proportional to the numbers of ions of particular masses), as indicated by the detector, versus their mass-to-charge ratios. A mass spectrum showing the three types of magnesium atoms is shown in Fig. 4-8.

These mass spectrometric observations explained why the atomic weights of many elements are not integral numbers and suggested that atoms contain uncharged particles with a mass of 1 amu. Magnesium atoms, for example, were known to have an atomic number of 12 and thus to contain 12 protons and 12 electrons with a combined mass of about 12 amu. The additional mass in these atoms (having total masses of about 24, 25, and 26 amu) must then be due to a particle with a mass of about 1 amu but with no charge, for if it had a charge, the number of positive charges and negative charges within the atoms would not be the same and the atoms would not be electrically neutral. The three types of magnesium atoms, therefore, must contain 12, 13, and 14 of these neutral particles, respectively. The existence of such a neutral particle with a mass of 1 amu (the neutron) was subsequently confirmed by Chadwick (Section 4.3).

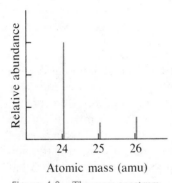

Figure 4-8. The mass spectrum of magnesium, showing three peaks due to the three naturally occurring isotopes of magnesium.

We now know that the nuclei of atoms contain both protons and neutrons (except for the nucleus of the common hydrogen atom, which contains only a proton). For a given element the number of protons does not vary, but the number of neutrons may vary within a limited range. Atoms with the same atomic number and different numbers of neutrons are called **isotopes.** Isotopes are identified with the atomic number as a subscript and the **mass number** (the sum of the number of protons and neutrons) as a superscript to the left of the symbol for the element. The naturally-occurring isotopes of magnesium would be indicated as $^{24}_{12}Mg$, $^{25}_{12}Mg$, and $^{26}_{12}Mg$. The compositions of the nuclei of the naturally-occurring elements with atomic numbers 1 through 10 are given in Table 4-2.

Table 4-2 Nuclear Compositions of Atoms of the Very Light Elements

	Symbol	Atomic Number	Number of Protons	Number of Neutrons	Mass, amu	% Natural Abundance
Hydrogen	$^{1}_{1}H$	1	1	0	1.0078	99.985
	$^{2}_{1}D$	1	1	1	2.0141	0.015
Helium	$^{3}_{2}He$	2	2	1	3.0160	0.00013
	$^{4}_{2}He$	2	2	2	4.0026	100
Lithium	$^{6}_{3}Li$	3	3	3	6.0151	7.42
	$^{7}_{3}Li$	3	3	4	7.0160	92.58
Beryllium	$^{9}_{4}Be$	4	4	5	9.0122	100
Boron	$^{10}_{5}B$	5	5	5	10.0129	19.6
	$^{11}_{5}B$	5	5	6	11.0093	80.4
Carbon	$^{12}_{6}C$	6	6	6	12.0000	98.89
	$^{13}_{6}C$	6	6	7	13.0033	1.11
	$^{14}_{6}C$	6	6	8	14.0032	$<10^{-8}$
Nitrogen	$^{14}_{7}N$	7	7	7	14.0031	99.63
	$^{15}_{7}N$	7	7	8	15.0001	0.37
Oxygen	$^{16}_{8}O$	8	8	8	15.9949	99.759
	$^{17}_{8}O$	8	8	9	16.9991	0.037
	$^{18}_{8}O$	8	8	10	17.9992	0.204
Fluorine	$^{19}_{9}F$	9	9	10	18.9984	100
Neon	$^{20}_{10}Ne$	10	10	10	19.9924	90.92
	$^{21}_{10}Ne$	10	10	11	20.9940	0.257
	$^{22}_{10}Ne$	10	10	12	21.9914	8.82

Since each proton and each neutron contributes approximately one unit to the atomic mass of an atom and each electron contributes far less, the mass of an atom is approximately equal to the mass number of the atom (a whole number). However, many atomic weights (the average mass of a great many atoms of an element) are not whole numbers because most elements exist as mixtures of two or more isotopes. The atomic weights are the weighted averages of the masses of each isotope present in a sample of the element; that is, they are equal to the sum of the masses of each isotope, each mass being multiplied by the fraction of that isotope present in the sample of the element.

EXAMPLE 4.1   Dempster found magnesium to contain 78.70% $^{24}_{12}Mg$ atoms (mass = 23.98 amu), 10.13% $^{25}_{12}Mg$ atoms (mass = 24.99 amu), and 11.17% $^{26}_{12}Mg$ atoms (mass = 25.98 amu). Calculate the weighted average mass of a Mg atom.

$$\text{Weighted average mass} = \frac{78.70 \; ^{24}_{12}\text{Mg atoms}}{100 \; \text{Mg atoms}} \times \frac{23.98 \; \text{amu}}{1 \; ^{24}_{12}\text{Mg atom}}$$

$$+ \frac{10.13 \; ^{25}_{12}\text{Mg atoms}}{100 \; \text{Mg atoms}} \times \frac{24.99 \; \text{amu}}{1 \; ^{25}_{12}\text{Mg atom}}$$

$$+ \frac{11.17 \; {}^{26}_{12}\text{Mg atoms}}{100 \; \text{Mg atoms}} \times \frac{25.98 \; \text{amu}}{1 \; {}^{26}_{12}\text{Mg atom}}$$

$$= 24.31 \; \text{amu/Mg atom}$$

More precise measurements have shown this value to be 24.305 amu.

4.7 The Bohr Model of the Atom

Rutherford's nuclear model of the atom, with its very small, massive nucleus surrounded by a diffuse sea of lightweight electrons, accounted nicely for the properties of the atom as revealed by the discharge-tube experiments of Crookes and others, by the α-particle scattering experiments, and by the varying weights of the elements due to the presence of isotopes. However, one basic problem remained. According to the physical principles known at the time, such an atom should not be stable. Since there is an attractive force between the positively charged nucleus and the negatively charged electrons, the electrons would be expected to fall into the nucleus. If it were assumed that the electrons moved around the nucleus in circular orbits, centrifugal force acting on the electrons could counterbalance the force of attraction, and the electrons would stay in their orbits. However, according to classical physics, an electron moving in such a circular orbit would radiate energy continuously; and, since this would mean that it was continuously losing energy, it should move in smaller and smaller orbits and finally fall into the nucleus.

In 1913 Niels Bohr, a Danish scientist, proposed a solution to the problem by suggesting a new theory for the behavior of matter. He proposed that the energy of an electron in an atom cannot vary continuously, but is **quantized;** that is, it is restricted to discrete, or individual, values. The success of this theory in explaining the spectra of the hydrogen atom (Section 4.8) and of hydrogenlike ions containing only one electron moving about a nucleus led to its general acceptance.

The Bohr model of a hydrogen atom or a hydrogenlike ion assumed that the electron moved about the nucleus in a circular orbit and that the centrifugal force due to this motion counterbalanced the electrostatic attraction between the nucleus and the electron. The energy of the electron was *assumed* to be restricted to certain values, each of which corresponded to an orbit with a different radius. Each of these orbits could be characterized by an integer, n. Bohr showed that the energy of an electron in one of these orbits is given by the equation

$$E = \frac{-kZ^2}{n^2}$$

where k is a constant, Z is the atomic number (the number of units of positive charge on the nucleus), and n is the integer characteristic of the orbit ($n = 1, 2, 3, 4, \ldots$). For a hydrogen atom $Z = 1$ and $k = 2.179 \times 10^{-18}$ J, which gives the energy in joules. Other units of energy commonly used include electron-volts ($1 \; \text{eV} = 1.602 \times 10^{-19}$ J) and ergs ($1 \; \text{erg} = 10^{-7}$ J).

EXAMPLE 4.2   A spark promotes the electron in a hydrogen atom into an orbit with $n = 2$. What is the energy, in joules, of an electron with $n = 2$?

The energy of the electron is given by the equation

$$E = \frac{-kZ^2}{n^2}$$

The atomic number, Z, of hydrogen is 1; $k = 2.179 \times 10^{-18}$ J; and the electron is characterized by an n value of 2. Thus

$$E = -2.179 \times 10^{-18} \text{ J} \times \frac{1^2}{2^2}$$

$$= -5.448 \times 10^{-19} \text{ J}$$

The distance of the electron from the nucleus was also related to the value of n as follows:

$$\text{Radius of orbit} = \frac{n^2 a_0}{Z}$$

where a_0 is the radius of the orbit in the hydrogen atom for which $n = 1$ (0.529 Å, where 1 Å $= 10^{-10}$ m $= 100$ pm). According to this model the closest an electron in a hydrogen atom can get to the nucleus is in an orbit with $n = 1$ and with a radius of 0.529 Å; thus the electron cannot fall into the nucleus. As the electron moves away from the nucleus, its energy becomes higher (Table 4-3), reaching zero at an infinite distance ($n = \infty$). As it gets closer to the nucleus, its energy becomes lower (more negative).

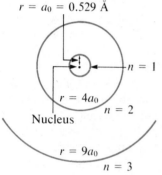

$r = a_0 = 0.529$ Å

$n = 1$

$r = 4a_0$

Nucleus

$n = 2$

$r = 9a_0$

$n = 3$

$r = 16a_0$

$n = 4$

$r = 25a_0$

$n = 5$

$r = 36a_0$

$n = 6$

Figure 4-9. A sketch of the hypothetical circular orbits of the Bohr model of the hydrogen atom drawn to scale. If the nucleus were drawn to scale, it would be invisible.

Table 4-3 Calculated Energies and Radii for a Hydrogen Atom

n	Shell	Energy, eV	Distance from Nucleus, Å	Distance from Nucleus, pm
1	K	− 13.595	0.529	52.9
2	L	− 3.399	2.116	211.6
3	M	− 1.511	4.761	476.1
4	N	− 0.850	8.464	846.4
5	O	− 0.544	13.225	1322.5
∞	—	0	∞	∞

Thus the Bohr model of the hydrogen atom (Fig. 4-9) postulates a single electron that moves in circular orbits about the nucleus. The electron usually moves in the orbit for which $n = 1$, the orbit in which it has the lowest energy. When the electron is in this lowest-energy orbit, the atom is said to be in its **ground state.** If the atom picks up energy from an outside source, it changes to an **excited state** as the electron moves to one of the higher-energy orbits.

The motion of the electron in a circular orbit is not the most important concept of the Bohr model, however. The most significant new idea is that the electron is restricted to discrete energies that are identified by n.

The integer n is called a **quantum number.** The properties of an electron in a Bohr atom are often identified by giving its quantum number. If we know the quantum number of the electron in a hydrogen atom, we can easily evaluate the energy of the electron or the size of the orbit that it occupies. For example, an electron with the quantum number $n = 1$ resides in the first orbit with an energy of -13.595 eV and a radius of 0.529 Å.

Instead of quantum numbers, letters are sometimes used to distinguish electrons (Table 4-3). When we speak of a K electron, we mean an electron with $n = 1$. An L electron is an electron with $n = 2$; an M electron, with $n = 3$; an N electron, with $n = 4$; etc. In Section 4.10, we will see that there are in fact three additional quantum numbers that also describe various properties of an electron.

4.8 Atomic Spectra and Atomic Structure

The Bohr model of the hydrogen atom would be merely an interesting intellectual curiosity if it could not be checked against experimental data. The fact is, it explains the spectrum of hydrogen atoms very well, and this observation tends to substantiate the new assumption in the model. Before we look at the agreement of the model with experiment, however, a brief introduction to light and to spectra is required.

Light is one form of **electromagnetic radiation** (radio waves, ultraviolet light, X rays, and γ rays are other examples). All forms of electromagnetic radiation exhibit wavelike behavior. They all can be characterized by a wavelength, λ (Fig. 4-10), and a frequency, ν. The frequency is the rate at which equivalent points on a wave pass a given spot. The wavelength and frequency are inversely proportional ($\nu \propto 1/\lambda$). The product $\lambda\nu$ is equal to the speed, c, with which all forms of electromagnetic radiation move (2.998×10^8 m/s in a vacuum). Different forms of electromagnetic radiation have different wavelengths and frequencies but they all move with the same speed in a vacuum. Electromagnetic radiation also has properties associated with particles called **photons.** The energy of a photon may be determined from the expression

$$E = h\nu = \frac{hc}{\lambda}$$

where h is Planck's constant (6.626×10^{-34} J s), c is the velocity, λ is the wavelength, and ν is the frequency of the radiation.

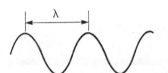

Figure 4-10. Characteristics of a wave. The distance between identical points, such as consecutive peaks, on the wave is the wavelength, λ (lambda). The number of times per second that identical points on the wave pass a given point is the frequency, ν (nu).

EXAMPLE 4.3   An experimental iodine laser emits light with a wavelength of 1.315 μm. What is the frequency of this light, and what is the energy of 1 photon?

As noted above, the product of frequency, ν, and wavelength, λ, is equal to c, the speed of light in a vacuum (2.998×10^8 m/s).

$$\lambda\nu = c$$

Rearrangement of this equation gives

$$\nu = \frac{c}{\lambda}$$

Since 1 μm $= 10^{-6}$ m, the wavelength given, 1.315 μm, equals 1.315×10^{-6} m.

Hence,

$$\nu = \frac{2.998 \times 10^8 \, \text{m/1 s}}{1.315 \times 10^{-6} \, \text{m}}$$

$$= \frac{2.280 \times 10^{14}}{1 \, \text{s}}$$

$$= 2.280 \times 10^{14} \, \text{s}^{-1}$$

The relationship of the energy of a photon to its wavelength is

$$E = h \frac{c}{\lambda}$$

$$= \frac{6.626 \times 10^{-34} \, \text{J s} \times (2.998 \times 10^8 \, \text{m/1 s})}{1.315 \times 10^{-6} \, \text{m}} = 1.511 \times 10^{-19} \, \text{J}$$

Thus the frequency of the laser light is $2.280 \times 10^{14} \, \text{s}^{-1}$ (read "per second") with an energy of $1.511 \times 10^{-19} \, \text{J}$ per photon.

Sir Isaac Newton showed that sunlight is a mixture of different kinds of light. He separated sunlight into its component colors by passing it through a glass prism. Sunlight contains all wavelengths (and thus all energies) of visible light and gives a continuous **spectrum** (Fig. 4-11), such as we see in the rainbow. Sunlight also contains ultraviolet (very short wavelengths) and infrared (very long wavelengths) light, which can be detected and recorded with instruments but which are invisible to the human eye. Incandescent solids, liquids, and gases under great pressure also give continuous spectra. However, when an electric current is passed through a gas at low pressure, the gas gives off light that shows a spectrum made up of a number of bright lines (a **line spectrum**) when passed through a prism. Each of these lines corresponds to a single wavelength of light, and so the light emitted by such a gas consists of light of discrete energies.

If a tube containing hydrogen at low pressure is subjected to an electric discharge, light of a blue-pink hue is produced. Passage of this light through a prism produces a line spectrum, indicating that it is composed of light of several energies. J. R. Rydberg measured the frequencies of the lines in the visible spectrum and found that they could be related by an equation that gives the energies, in

Figure 4-11. Sunlight passing through a prism is separated into its component colors, resulting in a continuous spectrum.

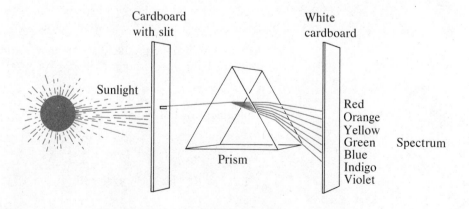

Cardboard with slit

White cardboard

Sunlight

Red
Orange
Yellow
Green
Blue
Indigo
Violet

Spectrum

Prism

joules, of the photons in each line as follows:

$$E = h\nu = 2.179 \times 10^{-18} \left(\frac{1}{n_1^{2}} - \frac{1}{n_2^{2}}\right) \text{J}$$

where n_1 and n_2 are integers with n_1 smaller than n_2. This equation is empirical; that is, it is derived from observation rather than from theory.

Bohr suggested that a hydrogen atom radiates energy as light only when its electron suddenly changes from a higher-energy orbit to one that has a lower energy (Fig. 4-12). If a hydrogen atom is excited with an electric discharge, its electron gains energy and moves to an orbit of higher energy (and higher n). As the atom relaxes, the electron moves from this higher-energy orbit to one in which its energy is lower and emits a photon with an energy equal to the difference in energy between the two orbits. According to Bohr's theory (Section 4.7), the electron in a hydrogen atom can have only those *discrete* energies, E_n, permitted by the equation

$$E_n = \frac{-kZ^2}{n^2} = \left(\frac{-2.179 \times 10^{-18}}{n^2}\right) \text{J}$$

When an electron falls from a higher-energy outer orbit characterized by n_2 to a lower-energy inner orbit characterized by n_1 (n_1 is less than n_2), the difference in energy is the absolute value of the difference in energy between the two orbits, $E_{n_1} - E_{n_2}$. This is the energy, E, emitted as a photon, or

$$E = |E_{n_1} - E_{n_2}| = 2.179 \times 10^{-18}\left(\frac{1}{n_1^{2}} - \frac{1}{n_2^{2}}\right) \text{J}$$

This theoretical expression is exactly the same as the one Rydberg found. The agreement of the two equations—one experimental, one theoretical—provides powerful and indispensable evidence for the validity of the Bohr concept of atomic structure involving discrete energy levels for electrons.

Figure 4-12. Relative energies of some of the circular orbits in the Bohr model of the hydrogen atom and electronic transitions that give rise to its atomic spectrum. Note the decreasing space between levels as n increases.

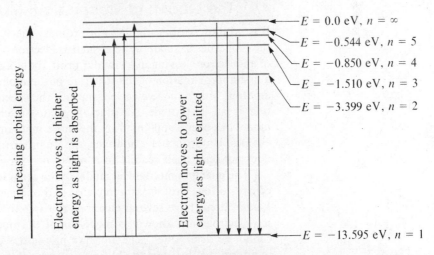

$E = 0.0$ eV, $n = \infty$

$E = -0.544$ eV, $n = 5$

$E = -0.850$ eV, $n = 4$

$E = -1.510$ eV, $n = 3$

$E = -3.399$ eV, $n = 2$

$E = -13.595$ eV, $n = 1$

Increasing orbital energy

Electron moves to higher energy as light is absorbed

Electron moves to lower energy as light is emitted

EXAMPLE 4.4 What is the energy (in joules) and the wavelength (in meters) of the photon radiated when an electron, previously excited by an input of energy, moves from the orbit with $n = 4$ to the orbit with $n = 2$ in a hydrogen atom?

In this case the electron starts out with $n = 4$, so $n_2 = 4$. It finishes with $n = 2$, so $n_1 = 2$. The difference in energy between these two states is given by the expression

$$E = |E_{n_1} - E_{n_2}| = 2.179 \times 10^{-18}\left(\frac{1}{n_1{}^2} - \frac{1}{n_2{}^2}\right) J$$

$$= 2.179 \times 10^{-18}\left(\frac{1}{2^2} - \frac{1}{4^2}\right) J$$

$$= 2.179 \times 10^{-18}\left(\frac{1}{4} - \frac{1}{16}\right) J$$

$$= 4.086 \times 10^{-19} J$$

The energy of the photon emitted is equal to the difference in energy between the two orbits, 4.086×10^{-19} J. The wavelength of a photon with this energy is found from the expression $E = hc/\lambda$. Rearrangement gives

$$\lambda = \frac{hc}{E}$$

$$= \frac{6.626 \times 10^{-34} \text{ J s} \times 2.998 \times 10^8 \text{ m s}^{-1}}{4.086 \times 10^{-19} \text{ J}}$$

$$= 4.862 \times 10^{-7} \text{ m (or 486.2 nm)}$$

Even though the idea of discrete energy levels applies to all atoms, Bohr's equation works only for simple atoms with one electron. Attempts to explain the spectra and other physical properties of more complicated atoms require the use of a more complex model, the quantum mechanical model of the atom.

4.9 The Quantum Mechanical Model of the Atom

It is now known that the laws of ordinary mechanics do not adequately explain the properties of small particles such as electrons. As a consequence, Bohr's theory of an electron moving in a circular orbit about a nucleus must be considered only a first approximation of the structure of the atom. It can explain the spectrum of an atom with a single electron, but not the spectra of atoms with more. In addition, it fails to provide a satisfactory picture of chemical bonding. Thus the Bohr model has been supplanted by the more mathematical theories of quantum mechanics. Unfortunately, however, the quantum mechanical concept of the atom does not lend itself readily to a physical representation that can be visualized.

The **quantum mechanical model of the atom** indicates that the hydrogen atom consists of one proton (as a nucleus) and one electron moving about the proton. Larger atoms have several electrons moving about a positive nucleus. Although the electrons are known to move about the nucleus, the exact path they take cannot be determined. The German physicist Werner Heisenberg expressed this in a form that has come to be known as the **Heisenberg Uncertainty Principle: It is impossible to determine accurately both the momentum and the position of a particle simultaneously.** (The momentum is the mass of the particle multiplied by its velocity.) The more accurately we measure the momentum of a moving electron (or any other particle), the less accurately we can determine its position (and conversely). An experiment designed to measure the exact position of an electron will alter its momentum, and an experiment designed to measure the exact momentum of an electron will unavoidably change its position. If position and momentum are measured at the same time, the values are inexact for one or the other or both. Although we cannot pinpoint an electron's position or path, we *can* calculate the *probability* of finding an electron at a given location within an atom.

In spite of the limitations described by the Heisenberg Uncertainty Principle, the behavior of electrons can be determined in a useful way using the mathematical tools of quantum mechanics. Both the energy of an electron in an atom and the region of space in which it may be found can be determined.

The results of a quantum mechanical treatment of the problem developed by another German physicist, Erwin Schrödinger, show that the electron may be visualized as being in rapid motion within one of several regions of space located around the nucleus. Each of these regions is called an **orbital,** or **atomic orbital.** Although the electron may be located anywhere within an orbital at any instant in time, it spends most of its time in certain high-probability regions. For example, in an isolated hydrogen atom in its ground state, the single electron effectively occupies all the space within about 1 Å of the nucleus. Within this spherical orbital the electron has the greatest probability of being at a distance of 0.529 Å from the nucleus. This gives the hydrogen atom a spherical shape. Note that a Bohr orbit and a quantum mechanical orbital are very different. An orbit is a circular path; an orbital is a three-dimensional region of space.

Some chemists describe the occupancy of an orbital in terms of **electron density.** The electron density is high in those regions of the orbital where the probability of finding an electron is relatively high, and low in those regions of the orbital where the probability is low.

Each electron in an atom can be described using quantum mechanical techniques by a mathematical expression called a **wave function,** which is given the symbol ψ. The shape of an orbital that the electron occupies, the energy of the electron in the orbital (sometimes called the energy of the orbital), and the probability of finding the electron in some region within the orbital can be determined from ψ. For example, the square of the wave function, ψ^2, is a measure of the probability of finding the electron at a given point at a distance r from the nucleus; and $4\pi r^2\psi^2$ (the radial probability density) is a measure of the probability of finding the electron within the volume of a thin spherical shell (somewhat like a layer of an onion) of radius r and thickness dr (where dr is a very small fraction of r).

Figure 4-13 illustrates the electron density in the occupied orbital of a hydrogen atom in the ground state. The electron occupies the orbital in which it will have the lowest possible energy (the lowest-energy orbital). The probability of finding the electron in a thin shell very close to the nucleus is practically zero, but increases rapidly just beyond the nucleus and becomes highest in a thin shell at a distance of 0.529 Å from the nucleus. The probability then decreases rapidly as the distance of the thin shell from the nucleus increases and becomes exceedingly small at a distance greater than about 1 Å. Most of the time, but not always, the electron will be located within a sphere with a radius of 1 Å; that is, the probability of finding the electron inside this sphere is high. Figure 4-14 shows a plot of the probability of finding the electron in a thin shell (the radial probability density, $4\pi r^2\psi^2$) as a function of r, the distance of the shell from the nucleus for three of the lowest-energy orbitals in a hydrogen atom.

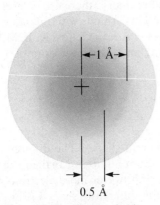

Figure 4-13. Electron density for an electron of lowest possible energy in a hydrogen atom. The electron occupies the 1s orbital.

Figure 4-14. The probability of finding the electron at a given distance r from the nucleus for three of the lowest-energy orbitals in a hydrogen atom. The energy of an electron in these orbitals increases as n increases. Note the overlap of orbitals.

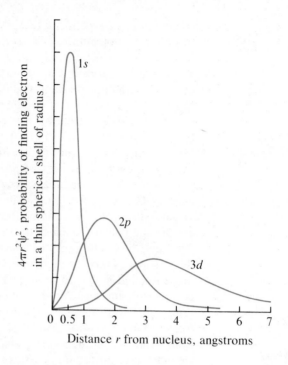

Distance r from nucleus, angstroms

4.10 Results of the Quantum Mechanical Model of the Atom

The general results of the quantum mechanical model of the atom may be summarized as follows:

1. The location of an electron cannot be determined exactly. All that can be identified is the orbital occupied by the electron—that is, the region, or volume, of space where there is a relatively high probability of finding the electron.

2. Orbitals are characterized by n, the principal quantum number, which may take on any integral value: $n = 1, 2, 3, 4, 5, \ldots$. As n increases, the orbitals extend farther from the nucleus, and the average position of an electron in these orbitals is farther from the nucleus (Fig. 4-14). The energies of the orbitals increase as the value of n increases. For values of n greater than 1, there are several different orbitals with the same value of n. A **shell** contains orbitals with the same value of n. Shells may be identified by either the values of n or the letters K, L, M, N, O, $\ldots$, respectively (an $n = 2$ shell may be called an L shell, for example).

Shell:	K	L	M	N	O	P	Q
n:	1	2	3	4	5	6	7

3. Orbitals with the same value of n may have different shapes. The different shapes of orbitals are distinguished by a second quantum number, ℓ, called the **azimuthal,** or **subsidiary, quantum number.** In atoms with two or more electrons, electrons with different ℓ values (electrons that occupy differently shaped orbitals) will have different energies. An electron with $\ell = 0$ occupies a spherical orbital [Fig. 4-15(a)]. Such an electron is called an s electron, and its orbital is called an s orbital. An electron with $\ell = 1$ occupies a dumbbell-shaped region of space [Fig. 4-15(b)] and is called a p electron; a d electron, with $\ell = 2$, occupies a

(a)

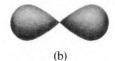

(b)

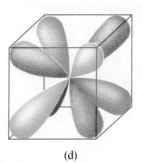

(c)

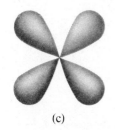

(d)

Figure 4-15. Shapes of the orbitals for (a) an *s* subshell (b) a *p* subshell (c) a *d* subshell (d) an *f* subshell. To improve the perspective the *f* orbital is shown within a cube with one lobe of each pair in color.

volume of space usually drawn with four lobes [Fig. 4-15(c)]; and an *f* electron, with $\ell = 3$, occupies a rather complex-looking volume of space usually shown with eight lobes [Fig. 4-15(d)]. Letters corresponding to the values of ℓ continue alphabetically following *f*. Although ℓ values higher than 3 are possible, electrons with ℓ greater than 3 are not found unless they have been excited by absorption of energy by an atom.

ℓ:	0	1	2	3	4	5	6
Letter Designation:	*s*	*p*	*d*	*f*	*g*	*h*	*i*

In an atom with two or more electrons, the energies of electrons in the orbitals within a given shell increase in the following order:

$$s \text{ electrons} < p \text{ electrons} < d \text{ electrons} < f \text{ electrons}$$

The mathematics of quantum mechanics limits the value that ℓ can have; an electron for which the principal quantum number has a value of *n* may have an integral ℓ value ranging from 0 to $(n - 1)$.

Shell	*n* Value	Possible ℓ Values	Types of Orbitals
K	1	0,	1*s*
L	2	0, 1	2*s*, 2*p*
M	3	0, 1, 2	3*s*, 3*p*, 3*d*
N	4	0, 1, 2, 3	4*s*, 4*p*, 4*d*, 4*f*
O	5	0, 1, 2, 3, 4	5*s*, 5*p*, 5*d*, 5*f*, 5*g*

Thus an electron in a K shell, for which $n = 1$, can only have an ℓ value of zero. An electron in a shell for which $n = 3$ can be an *s* electron ($\ell = 0$), a *p* electron ($\ell = 1$), or a *d* electron ($\ell = 2$), but cannot have an ℓ value higher than 2.

4. For ℓ values larger than zero, orbitals with the same value of ℓ in any given shell have the same shape and the same energy but differ in their orientation. There are $(2\ell + 1)$ orbitals with the same ℓ quantum number, which differ only in their orientation about the nucleus. Each orientation is characterized by a third quantum number *m*, the **magnetic quantum number,** which may have any integral value from $-\ell$ through zero to $+\ell$. A sphere may have only one orientation in space, and thus there is only one type of spherical, or *s*, orbital. This orbital is characterized by an ℓ value of zero; hence *m* here can only be zero. There are three possible values of m ($-1, 0, +1$) for an orbital with $\ell = 1$ (a *p* orbital) and thus three possible orientations. Figure 4-16 shows how the dumbbell shape that

Figure 4-16. The different orientations of the three equivalent atomic *p* orbitals.

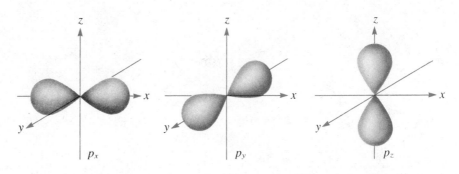

characterizes p orbitals can be oriented in three ways along the x, y, and z axes of an xyz coordinate system. The three p orbitals are designated, as shown in the figure, p_x, p_y, and p_z to indicate their directional character. Figure 4-17 shows five orientations of the d orbitals ($\ell = 2$; $m = -2, -1, 0, +1, +2$). Figure 4-18 shows seven orientations of the f orbitals ($\ell = 3$; $m = -3, -2, -1, 0, +1, +2, +3$). There is no general agreement on how best to represent the seven f orbitals.

Figure 4-17. The different orientations of the five equivalent atomic d orbitals. The drawing is oriented so that the lobes of the orbitals designated d_{z^2} and $d_{x^2-y^2}$ lie along the axes, whereas the lobes of those labeled d_{xz}, d_{yz}, and d_{xy} lie in between the axes.

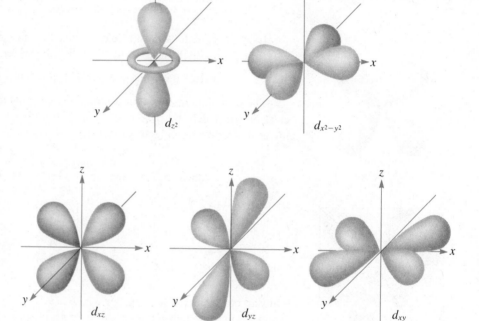

Figure 4-18. The different orientations of the seven equivalent atomic f orbitals. To improve the perspective each of the seven orbitals is shown within a cube with one lobe of each pair in color.

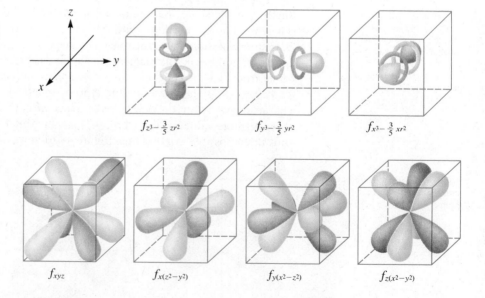

Figure 4-18 shows one way, with each orbital in a cube to help you see the three-dimensional aspect of this representation.

In a free atom the *p* orbitals in the same shell have the same energy and are referred to as **degenerate orbitals.** Degenerate orbitals are orbitals with the same energy. Likewise, the five *d* orbitals in a given shell are degenerate, as are the seven *f* orbitals. Each set of degenerate orbitals, that is, orbitals with the same ℓ value in a given shell, is referred to as a **subshell.**

EXAMPLE 4.5   Indicate the number of subshells, the number of orbitals in each subshell, and the values of *n*, ℓ, and *m* for the orbitals in the N shell of an atom.

The N shell is the shell with $n = 4$, so it contains subshells with $\ell = 0, 1, 2,$ and 3. Hence there are four subshells.

For the subshell with $\ell = 0$ (the *s* subshell), *m* can only be equal to zero. Thus there is only one 4*s* orbital. For $\ell = 1$ (the *p* subshell), $m = -1, 0,$ or $+1$ so we find three 4*p* orbitals; for $\ell = 2$ (the *d* subshell), $m = -2, -1, 0, +1,$ or $+2$, giving five 4*d* orbitals; and for $\ell = 3$ (the *f* subshell), $m = -3, -2, -1, 0, +1, +2,$ or $+3$, giving seven 4*f* orbitals.

Thus there are 16 orbitals, with the values of *n*, ℓ, and *m* indicated above, in the 4 subshells of the N shell of an atom.

---

5. Electrons have a fourth quantum number, the **spin quantum number, *s*.** Electrons behave as if each were spinning about its own axis. The spin quantum number specifies the direction of spin of an electron about *its own axis* as the electron moves around the nucleus. The spin can be either counterclockwise or clockwise and is designated arbitrarily by either $s = +\frac{1}{2}$ or $s = -\frac{1}{2}$. Electrons with the same spin quantum number are said to have **parallel spins.**

For every possible combination of *n*, ℓ, and *m*, there can be two electrons differing only in the direction of spin about their own axes. Each orbital can contain a maximum of two electrons that have identical *n*, ℓ, and *m* values but differ in their *s* values.

6. The energy of an electron in an atom is limited to discrete values. The mathematics is far more complicated than in the Bohr model, but the energy can be determined from the wave function that describes the behavior of the electron.

7. The maximum number of electrons that may be found in a shell with a principal quantum number of *n* is $2n^2$. Since each atomic orbital can hold a maximum of two electrons, the number of atomic orbitals in a shell is given by n^2. Table 4-4 summarizes all of the allowed combinations of quantum numbers that may describe an electron in the first five shells of an atom.

4.11 Orbital Energies and Atomic Structure

The energies of atomic orbitals increase as the principal quantum number, *n*, increases. If an atom contains two or more electrons, the energies of the orbitals increase within a shell as the azimuthal quantum number, ℓ, increases. Although the relative energies of certain orbitals vary from atom to atom, the increasing

Table 4-4 · Summary of Allowed Values for Each Quantum Number

Shell	n	ℓ	m	s	Maximum Number of Electrons in Subshell	Maximum Number of Electrons in Shell
K	1	0 (s)	0	$+\frac{1}{2}$	$\left.\begin{array}{c}1\\1\end{array}\right\}2$	2
	1	0	0	$-\frac{1}{2}$		
L	2	0 (s)	0	$+\frac{1}{2},\ -\frac{1}{2}$	2	
	2	1 (p)	-1	$+\frac{1}{2},\ -\frac{1}{2}$	$\left.\begin{array}{c}2\\2\\2\end{array}\right\}6$	8
	2	1	0	$+\frac{1}{2},\ -\frac{1}{2}$		
	2	1	$+1$	$+\frac{1}{2},\ -\frac{1}{2}$		
M	3	0 (s)	0	$+\frac{1}{2},\ -\frac{1}{2}$	2	
	3	1 (p)	$-1,\ 0,\ +1$	$\pm\frac{1}{2}$ for each value of m	6	18
	3	2 (d)	$-2,\ -1,\ 0,\ +1,\ +2$	$\pm\frac{1}{2}$ for each value of m	10	
N	4	0 (s)	0	$\pm\frac{1}{2}$ for each value of m	2	
	4	1 (p)	$-1,\ 0,\ +1$	$\pm\frac{1}{2}$ for each value of m	6	
	4	2 (d)	$-2,\ -1,\ 0,\ +1,\ +2$	$\pm\frac{1}{2}$ for each value of m	10	32
	4	3 (f)	$-3,\ -2,\ -1,\ 0,\ +1,\ +2,\ +3$	$\pm\frac{1}{2}$ for each value of m	14	
O	5	0 (s)	0	$\pm\frac{1}{2}$ for each value of m	2	
	5	1 (p)	$-1,\ 0,\ +1$	$\pm\frac{1}{2}$ for each value of m	6	
	5	2 (d)	$-2,\ -1,\ 0,\ +1,\ +2$	$\pm\frac{1}{2}$ for each value of m	10	50[a]
	5	3 (f)	$-3,\ -2,\ -1,\ 0,\ +1,\ +2,\ +3$	$\pm\frac{1}{2}$ for each value of m	14	
	5	4 (g)	$-4,\ -3,\ -2,\ -1,\ 0,\ +1,\ +2,\ +3,\ +4$	$\pm\frac{1}{2}$ for each value of m	18[a]	

[a]The total number of 50 electrons for the O shell, including the 18 electrons for the g subshell, is the number of electrons theoretically possible. No element presently known contains more than 32 electrons in the O shell.

Figure 4-19. Generalized energy-level diagram for atomic orbitals in an atom with two or more electrons (not to scale). Orbitals of about the same energy are indicated by braces.

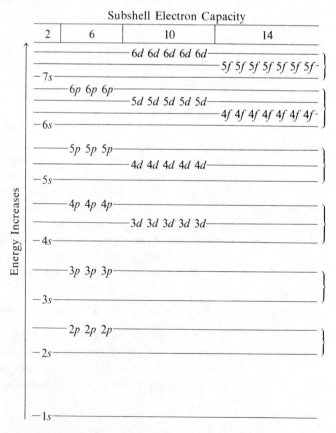

order of energy is roughly that shown in Fig. 4-19. The energy of an electron in an orbital is indicated by the vertical coordinate in the figure, the orbital with electrons of lowest energy being the $1s$ orbital at the bottom of the diagram. Orbitals that have about the same energy are indicated by the braces at the right of the figure. However, for atomic numbers greater than 20, the relative energies of the orbitals differ slightly from the order shown. The actual energies of the electrons in these orbitals vary from atom to atom. For example, the energies of the electrons in $1s$ orbitals become lower and lower as the atomic numbers of the atoms increase.

The arrangement of electrons in the orbitals of an atom (that is, the electron configuration of the atom) is described by a number in front of the subshell letter that designates the number of the principal, or major, shell and a superscript that designates the number of electrons in that particular subshell. For example, the notation $2p^4$ (read "two-p-four") indicates four electrons in the p subshell of the shell for which $n = 2$; the notation $3d^8$ ("three-d-eight") indicates eight electrons in the d subshell of the shell for which $n = 3$.

In general, electrons fill orbitals in the order shown, from the bottom to the top of Fig. 4-19. Each set of orbitals is filled before electrons occupy the next set immediately above. Figure 4-20 gives this order in a format that is easy to remember.

The distribution of electrons in the atomic orbitals of an atom in its ground state can be predicted using the following guidelines:

1. Electrons in an atom occupy the lowest possible energy levels, or orbitals. The first electron that is placed in a set of atomic orbitals goes into the $1s$ orbital. When the $1s$ orbital is filled, the next electron goes into the $2s$ orbital, and so on, up through the $2p$ orbital, the $3s$ orbital, etc. The order of orbital energies is given in Figs. 4-19 and 4-20.

2. The maximum number of electrons in an orbital is limited to two, according to the **Pauli Exclusion Principle: No two electrons in the same atom can have the same set of four quantum numbers.** Thus a $1s$ orbital is filled when it contains two electrons: one with $n = 1$, $\ell = 0$, $m = 0$, and $s = +\frac{1}{2}$; the other with $n = 1$, $\ell = 0$, $m = 0$, and $s = -\frac{1}{2}$. Since no other combination of quantum numbers is

Figure 4-20. Order of occupancy of atomic orbitals. The orbitals fill in the order indicated by the connecting lines.

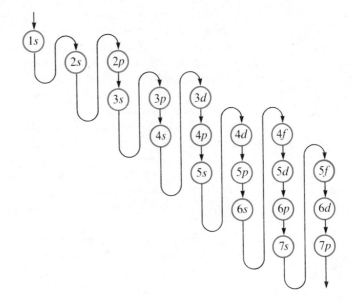

possible for an electron in the $1s$ orbital (Table 4-4), a third electron must occupy the $2s$ orbital. The $2s$ orbital, in turn, is filled when it contains two electrons, so that the fifth electron is forced into one of the three $2p$ orbitals, and so forth.

3. Subshells containing more than one orbital are filled as described by **Hund's Rule: Every orbital in a subshell is singly occupied (filled) with one electron before any one orbital is doubly occupied, and all electrons in singly occupied orbitals have the same spin;** that is, their spin quantum numbers are the same. For example, a $3p^2$ electron configuration has one unpaired electron in each of two different $3p$ orbitals. Both electrons have the same spin and the same spin quantum number. A d^6 electron configuration has a pair of electrons having different spins in one d orbital and four unpaired electrons having the same spin in the remaining four d orbitals.

4.12   The Aufbau Process

In order to illustrate the systematic variations in the electronic structures of the various elements, chemists "build" them in atomic order. Beginning with hydrogen, they add one proton to the nucleus and one electron to the proper subshell at a time until they have described the **electron configurations** of all the elements. This process is called the **Aufbau Process** from the German word *aufbau* (building up). Each added electron occupies the subshell of lowest energy available, according to the rules given in Section 4.11. Electrons enter higher-energy subshells only after lower-energy subshells have been filled to capacity.

A single hydrogen atom consists of one proton and one electron. The single electron is found in the $1s$ orbital around the proton. The electron configuration of a hydrogen atom is customarily represented as $1s^1$ (read "one-s-one").

After hydrogen the next simplest atom is that of the noble gas helium, with an atomic number of 2 and an atomic weight of approximately 4. The helium atom contains two protons and two neutrons in its nucleus and two electrons in the $1s$ orbital, which is completely filled by these two electrons. The electron configuration of helium is, therefore, $1s^2$ (read "one-s-two"). Note that in this atom the shell of lowest energy, the K shell, is completely filled.

The atom next in complexity is lithium, with an atomic number of 3, an atomic weight of approximately 7, and three protons, four neutrons, and three electrons. Two of these electrons fill the $1s$ orbital, so the remaining electron must occupy the orbital of next lowest energy, the $2s$ orbital. Thus the electron configuration of lithium is $1s^2 2s^1$.

An atom of the metal beryllium, with an atomic number of 4 and an atomic weight of approximately 9, contains four protons and five neutrons in the nucleus, and four electrons. The fourth electron fills the $2s$ orbital, making the electron configuration $1s^2 2s^2$.

An atom of boron (atomic number $= 5$) contains five electrons: two in the K shell ($n = 1$), which is filled, and three in the L shell ($n = 2$). Since the s subshell in the L shell can contain only two electrons, the fifth electron must occupy the higher-energy p subshell. The electron configuration of boron, therefore, is $1s^2 2s^2 2p^1$.

Carbon (atomic number $= 6$) has six electrons. Four of them fill the $1s$ and $2s$ subshells. The remaining two electrons occupy the $2p$ subshell, giving a $1s^2 2s^2 2p^2$

configuration. To describe the distribution of the electrons in the $2p$ subshell, we use Hund's Rule. One of these electrons occupies each of two of the three $2p$ orbitals. These electrons have the same spin quantum number and hence the same spin.

Nitrogen (atomic number = 7) has a $1s^2 2s^2 2p^3$ configuration, with one electron in each of the three $2p$ orbitals, in accordance with Hund's Rule. These three electrons have parallel spins. Oxygen atoms (atomic number = 8) have the configuration $1s^2 2s^2 2p^4$, with a pair of electrons in one of the $2p$ orbitals and a single electron in each of the other two. The two unpaired electrons in separate $2p$ orbitals have opposite spin. Fluorine atoms (atomic number = 9) have the configuration $1s^2 2s^2 2p^5$, with only one $2p$ orbital containing an unpaired electron. All of the electrons in the neon atom (atomic number = 10) are paired, with a configuration $1s^2 2s^2 2p^6$. Both the K and L shells of the noble gas neon are filled (two electrons in the K shell and eight in the L shell).

The sodium atom (atomic number = 11) has eleven electrons, one more than the neon atom. This electron must go into the lowest-energy subshell available, the $3s$ orbital, giving a $1s^2 2s^2 2p^6 3s^1$ configuration. We can abbreviate the electron configuration of sodium as $[Ne]3s^1$. The symbol $[Ne]$ represents the configuration of the two filled shells in Ne (neon), $1s^2 2s^2 2p^6$, which are identical to the two filled inner shells in a sodium atom. Similarly, the configuration of lithium may be represented as $[He]2s^1$, where $[He]$ represents the configuration of the helium atom, which is identical to that of the filled inner shell of lithium. Writing the configurations in this way emphasizes the similarity of the configurations of sodium and lithium. Both atoms have only one electron in an s subshell outside a filled set of inner shells.

The magnesium atom (atomic number = 12), with its 12 electrons in a $[Ne]3s^2$ configuration, is analogous to beryllium, $[He]2s^2$. Both atoms have a filled s subshell outside their filled inner shells. Aluminum (atomic number = 13), with 13 electrons and the electron configuration $[Ne]3s^2 3p^1$, is analogous to boron, $[He]2s^2 2p^1$. The electron configurations of silicon (14 electrons), phosphorus (15 electrons), sulfur (16 electrons), chlorine (17 electrons), and argon (18 electrons) are analogous in the compositions of their outer shells to carbon, nitrogen, oxygen, fluorine, and neon, respectively. See Table 4-5 for the configurations of these elements. The table lists the lowest-energy, or ground-state, electron configurations for atoms of each of the known elements.

Potassium (atomic number = 19) and calcium (atomic number = 20) have one and two electrons, respectively, in the N shell ($n = 4$). Hence potassium corresponds to lithium and sodium in outer shell configuration, whereas calcium corresponds to beryllium and magnesium.

Beginning with scandium (atomic number = 21; see Table 4-5), additional electrons are added successively to the $3d$ subshell after two electrons have already occupied the $4s$ subshell. After the $3d$ subshell is filled to its capacity with 10 electrons, the $4p$ subshell fills. Note that for three series of elements, scandium (Sc) through copper (Cu), yttrium (Y) through silver (Ag), and lutetium (Lu) through gold (Au), a total of 10 d electrons are successively added to the $n - 1$ shell next to the outer n shell to bring that $n - 1$ shell from 8 to 18 electrons. For two series, lanthanum (La) through lutetium (Lu) and actinium (Ac) through lawrencium (Lr), 14 f electrons are successively added to the third shell from the outside (the $n - 2$ shell) to bring that shell from 18 to 32 electrons.

Table 4-5 Electron Configurations of the Elements

Atomic Number	Symbol	Electron Configuration	Atomic Number	Symbol	Electron Configuration	Atomic Number	Symbol	Electron Configuration
1	H	$1s^1$				73	Ta	$[Xe]6s^24f^{14}5d^3$
2	He	$1s^2$	37	Rb	$[Kr]5s^1$	74	W	$[Xe]6s^24f^{14}5d^4$
			38	Sr	$[Kr]5s^2$	75	Re	$[Xe]6s^24f^{14}5d^5$
3	Li	$1s^22s^1 = [He]2s^1$	39	Y	$[Kr]5s^24d^1$	76	Os	$[Xe]6s^24f^{14}5d^6$
4	Be	$[He]2s^2$	40	Zr	$[Kr]5s^24d^2$	77	Ir	$[Xe]6s^24f^{14}5d^7$
5	B	$[He]2s^22p^1$	41	Nb	$[Kr]5s^14d^4$	78	Pt	$[Xe]6s^14f^{14}5d^9$
6	C	$[He]2s^22p^2$	42	Mo	$[Kr]5s^14d^5$	79	Au	$[Xe]6s^14f^{14}5d^{10}$
7	N	$[He]2s^22p^3$	43	Tc	$[Kr]5s^14d^6$	80	Hg	$[Xe]6s^24f^{14}5d^{10}$
8	O	$[He]2s^22p^4$	44	Ru	$[Kr]5s^14d^7$	81	Tl	$[Xe]6s^24f^{14}5d^{10}6p^1$
9	F	$[He]2s^22p^5$	45	Rh	$[Kr]5s^14d^8$	82	Pb	$[Xe]6s^24f^{14}5d^{10}6p^2$
10	Ne	$[He]2s^22p^6$	46	Pd	$[Kr]4d^{10}$	83	Bi	$[Xe]6s^24f^{14}5d^{10}6p^3$
11	Na	$[Ne]3s^1$	47	Ag	$[Kr]5s^14d^{10}$	84	Po	$[Xe]6s^24f^{14}5d^{10}6p^4$
12	Mg	$[Ne]3s^2$	48	Cd	$[Kr]5s^24d^{10}$	85	At	$[Xe]6s^24f^{14}5d^{10}6p^5$
13	Al	$[Ne]3s^23p^1$	49	In	$[Kr]5s^24d^{10}5p^1$	86	Rn	$[Xe]6s^24f^{14}5d^{10}6p^6$
14	Si	$[Ne]3s^23p^2$	50	Sn	$[Kr]5s^24d^{10}5p^2$			
15	P	$[Ne]3s^23p^3$	51	Sb	$[Kr]5s^24d^{10}5p^3$	87	Fr	$[Rn]7s^1$
16	S	$[Ne]3s^23p^4$	52	Te	$[Kr]5s^24d^{10}5p^4$	88	Ra	$[Rn]7s^2$
17	Cl	$[Ne]3s^23p^5$	53	I	$[Kr]5s^24d^{10}5p^5$	89	Ac	$[Rn]7s^26d^1$
18	Ar	$[Ne]3s^23p^6$	54	Xe	$[Kr]5s^24d^{10}5p^6$	90	Th	$[Rn]7s^26d^2$
19	K	$[Ar]4s^1$	55	Cs	$[Xe]6s^1$	91	Pa	$[Rn]7s^25f^26d^1$
20	Ca	$[Ar]4s^2$	56	Ba	$[Xe]6s^2$	92	U	$[Rn]7s^25f^36d^1$
21	Sc	$[Ar]4s^23d^1$	57	La	$[Xe]6s^25d^1$	93	Np	$[Rn]7s^25f^46d^1$
22	Ti	$[Ar]4s^23d^2$	58	Ce	$[Xe]6s^24f^2$	94	Pu	$[Rn]7s^25f^6$
23	V	$[Ar]4s^23d^3$	59	Pr	$[Xe]6s^24f^3$	95	Am	$[Rn]7s^25f^7$
24	Cr	$[Ar]4s^13d^5$	60	Nd	$[Xe]6s^24f^4$	96	Cm	$[Rn]7s^25f^76d^1$
25	Mn	$[Ar]4s^23d^5$	61	Pm	$[Xe]6s^24f^5$	97	Bk	$[Rn]7s^25f^86d^1$
26	Fe	$[Ar]4s^23d^6$	62	Sm	$[Xe]6s^24f^6$	98	Cf	$[Rn]7s^25f^{10}$
27	Co	$[Ar]4s^23d^7$	63	Eu	$[Xe]6s^24f^7$	99	Es	$[Rn]7s^25f^{11}$
28	Ni	$[Ar]4s^23d^8$	64	Gd	$[Xe]6s^24f^75d^1$	100	Fm	$[Rn]7s^25f^{12}$
29	Cu	$[Ar]4s^13d^{10}$	65	Tb	$[Xe]6s^24f^9$	101	Md	$[Rn]7s^25f^{13}$
30	Zn	$[Ar]4s^23d^{10}$	66	Dy	$[Xe]6s^24f^{10}$	102	No	$[Rn]7s^25f^{14}$
31	Ga	$[Ar]4s^23d^{10}4p^1$	67	Ho	$[Xe]6s^24f^{11}$	103	Lr	$[Rn]7s^25f^{14}6d^1$
32	Ge	$[Ar]4s^23d^{10}4p^2$	68	Er	$[Xe]6s^24f^{12}$	104	Rf	$[Rn]7s^25f^{14}6d^2$
33	As	$[Ar]4s^23d^{10}4p^3$	69	Tm	$[Xe]6s^24f^{13}$	105	Ha	$[Rn]7s^25f^{14}6d^3$
34	Se	$[Ar]4s^23d^{10}4p^4$	70	Yb	$[Xe]6s^24f^{14}$	106	—	$[Rn]7s^25f^{14}6d^4$
35	Br	$[Ar]4s^23d^{10}4p^5$	71	Lu	$[Xe]6s^24f^{14}5d^1$	107	—	$[Rn]7s^25f^{14}6d^5$
36	Kr	$[Ar]4s^23d^{10}4p^6$	72	Hf	$[Xe]6s^24f^{14}5d^2$	109	—	$[Rn]7s^25f^{14}6d^7$

EXAMPLE 4.6 Gaseous cesium (Cs) atoms have been used in an experimental laser. These atoms emit light when their outermost electron is excited to a $7p$ orbital and then falls back into the $6s$ orbital. Without reference to Table 4-5, write the electron configurations of cesium (a) in its ground state and (b) when the outermost electron is excited to a p orbital.

From the order of filling presented in Fig. 4-20, when Cs is in its ground state, we can fill the orbitals in the following order with 55 electrons:

$$1s^22s^22p^63s^23p^64s^23d^{10}4p^65s^24d^{10}5p^66s^1$$

However, note that many chemists write the orbitals in increasing order of the quantum numbers:

$$1s^2 2s^2 2p^6 3s^2 3p^6 3d^{10} 4s^2 4p^6 4d^{10} 5s^2 5p^6 6s^1$$

Either notation is acceptable. Since Cs contains one more electron than Xe, these configurations can both be abbreviated as $[Xe]6s^1$.

The outermost Cs electron is the electron in the $6s$ orbital. When this electron is excited to a higher-energy p orbital, the configuration changes to $[Xe]7p^1$. It should be noted that this configuration is not the ground-state configuration, the stable configuration. It cannot be predicted from the rules presented in Section 4.11. It occurs only when sufficient energy is put into the atomic system to move the electron from the ground-state $6s$ orbital out to the higher-energy $7p$ orbital.

Students are sometimes troubled by exceptions to the expected order for filling orbitals, which is shown in Fig. 4-20. For example, the electron configurations of chromium (atomic number = 24), copper (atomic number = 29), and lanthanum (atomic number = 57) do not conform with the behavior expected according to Fig. 4-20 (see Table 4-5). These exceptions involve the filling of subshells with very similar energies. Certain combinations of repulsions between electrons in particular electron configurations lead to minor exceptions in the expected order of filling of these subshells because the energies due to these repulsions are greater than the small differences in energy between the subshells.

Half-filled and completely filled subshells represent conditions of particular stability. The stability of half-filled or filled subshells is such that an electron shifts from the $4s$ into a $3d$ orbital in order to gain the extra stability of a half-filled $3d$ subshell in chromium and a filled $3d$ subshell in copper.

From Fig. 4-20 one would expect the electron configuration of lanthanum (atomic number = 57) to be $[Xe]6s^2 4f^1$. Instead, as shown in Table 4-5, it has no $4f$ electron but does have one $5d$ electron: $[Xe]6s^2 5d^1$. This exception in the case of lanthanum demonstrates that the $4f$ and $5d$ subshells have very similar energies (see Fig. 4-19) and that electrons change from one such subshell to the other easily. A close inspection of Table 4-5 will disclose a significant number of such exceptions. You should realize that these exceptions result from the similar energies of various subshells. Beyond this a complete understanding of the general principles of electron configuration and an ability to write the generalized configuration for each element, based on the expected order of addition of electrons, is sufficient for most purposes.

THE PERIODIC LAW

4.13 THE PERIODIC TABLE

It became evident early in the development of chemistry that certain elements could be grouped together by reason of their similar properties. The members of one such grouping are lithium, sodium, and potassium (a portion of the group called the alkali metals). These elements all look like metals, all conduct electricity well, all react with chlorine, forming white water-soluble compounds with one chlorine atom per metal atom, and all react with water, giving hydrogen gas and metal hydroxides with one hydroxide (OH) group per metal. If M is used to represent a metal atom, the formulas of the chlorides and hydroxides would be MCl and MOH, respectively. A second grouping includes calcium, strontium, and barium (a portion of the group called the alkaline earth metals), which also look like metals and conduct electricity well. However, when they react with chlorine, they give white solids that contain two chlorine atoms per alkaline earth metal atom, or MCl_2. When they react with water, they give hydrogen gas and a hydroxide that contains two hydroxide groups per metal atom, or $M(OH)_2$. Fluorine, chlorine, bromine, and iodine (in the group called the halogens) also exhibit similar properties to each other. They do not conduct electricity and are nonmetallic. They react with hydrogen to form compounds that contain one hydrogen atom and one halogen atom: HF, HCl, HBr, and HI. When they react with sodium, they form compounds that contain one sodium atom per halogen atom: NaF, NaCl, NaBr, and NaI.

The German chemist Johann Dobereiner first suggested the existence of a relationship between the atomic weights and the properties of the elements in 1829. He showed that the atomic weight of strontium lies about midway between the atomic weights of calcium and barium and that the properties of strontium are intermediate between those of calcium and barium. He later found other **triads** of similar elements. Among these are the triads of chlorine, bromine, and iodine and of lithium, sodium, and potassium. By 1854 other chemists had shown that oxygen, sulfur, selenium, and tellurium could be classed as a **family of elements** and that nitrogen, phosphorus, arsenic, antimony, and bismuth make up another family.

In 1865 in England John Newlands recognized a correlation between the magnitudes of the atomic weights and the properties of the elements. By 1870 Dmitri Mendeleev in Russia and Lothar Meyer in Germany, working independently and apparently unaware of the work of Newlands, outlined the nature of this relationship between properties and atomic weights in some detail. Their conclusions led to the statement that the properties of the elements are periodic functions of their atomic weights. As an understanding of the structure of the atom and of electronic configurations developed, however, it became apparent that the properties of atoms were actually periodic functions of their atomic *numbers* rather than their atomic weights. The modern statement of this periodic relationship is the **Periodic Law: The properties of the elements are periodic functions of their atomic numbers.**

The Periodic Table is an arrangement of the atoms in increasing order of their atomic numbers that collects atoms with similar properties in vertical columns. Figure 4-21 gives one common form of the Periodic Table. A copy of the Periodic Table is also presented inside the front cover of this text.

As mentioned earlier, lithium (Li), sodium (Na), and potassium (K) exhibit similar chemical properties. These elements are found in the leftmost vertical column of elements in the table, along with rubidium (Rb), cesium (Cs), and francium (Fr). These six elements comprise a group known as the **alkali metals,** all of which have similar chemical properties. [Hydrogen (H) is listed in both column IA and column VIIA because it possesses a small number of similarities to elements in those columns. However, it is considered to be neither an alkali metal nor a halogen.] Calcium (Ca), strontium (Sr), and barium (Ba) also exhibit similar

Figure 4-21. The Periodic Table of Elements (long form). Atomic weights are shown above the symbols; atomic numbers, below.

NONMETALS

METALS

TRANSITION METALS

PERIODS	IA	IIA	IIIB	IVB	VB	VIB	VIIB	VIIIB			IB	IIB	IIIA	IVA	VA	VIA	VIIA	VIIIA
1	1.0079 H 1																	4.00260 He 2
2	6.941 Li 3	9.01218 Be 4											10.81 B 5	12.011 C 6	14.0067 N 7	15.9994 O 8	18.998403 F 9	20.179 Ne 10
3	22.98977 Na 11	24.305 Mg 12											26.98154 Al 13	28.0855 Si 14	30.97376 P 15	32.06 S 16	35.453 Cl 17	39.948 Ar 18
4	39.0983 K 19	40.08 Ca 20	44.9559 Sc 21	47.90 Ti 22	50.9415 V 23	51.996 Cr 24	54.9380 Mn 25	55.847 Fe 26	58.9332 Co 27	58.70 Ni 28	63.546 Cu 29	65.38 Zn 30	69.72 Ga 31	72.59 Ge 32	74.9216 As 33	78.96 Se 34	79.904 Br 35	83.80 Kr 36
5	85.4678 Rb 37	87.62 Sr 38	88.9059 Y 39	91.22 Zr 40	92.9064 Nb 41	95.94 Mo 42	(98) Tc 43	101.07 Ru 44	102.9055 Rh 45	106.4 Pd 46	107.868 Ag 47	112.41 Cd 48	114.82 In 49	118.69 Sn 50	121.75 Sb 51	127.60 Te 52	126.9045 I 53	131.30 Xe 54
6	132.9054 Cs 55	137.33 Ba 56	* [57-71]	178.49 Hf 72	180.9479 Ta 73	183.85 W 74	186.207 Re 75	190.2 Os 76	192.22 Ir 77	195.09 Pt 78	196.9665 Au 79	200.59 Hg 80	204.37 Tl 81	207.2 Pb 82	208.9804 Bi 83	(209) Po 84	(210) At 85	(222) Rn 86
7	(223) Fr 87	226.0254 Ra 88	† [89-103]	(261) Rf 104	(262) Ha 105	(263) — 106	(262) — 107	(266) — 109										

INNER TRANSITION METALS

*LANTHANIDE SERIES

138.9055 La 57	140.12 Ce 58	140.9077 Pr 59	144.24 Nd 60	(145) Pm 61	150.4 Sm 62	151.96 Eu 63	157.25 Gd 64	158.9254 Tb 65	162.50 Dy 66	164.9304 Ho 67	167.26 Er 68	168.9342 Tm 69	173.04 Yb 70	174.967 Lu 71

†ACTINIDE SERIES

227.0278 Ac 89	232.0381 Th 90	231.0359 Pa 91	238.029 U 92	237.0482 Np 93	(244) Pu 94	(243) Am 95	(247) Cm 96	(247) Bk 97	(251) Cf 98	(252) Es 99	(257) Fm 100	(258) Md 101	(259) No 102	(260) Lr 103

properties, which differ from those of Li, Na, and K. The second vertical column of elements from the left in the Periodic Table includes calcium, strontium, and barium, along with beryllium (Be), magnesium (Mg), and radium (Ra), and thus it should be expected (and it is observed) that their chemical behavior is similar. These six elements are referred to as the **alkaline earth metals.** The group of elements including fluorine (F), chlorine (Cl), bromine (Br), iodine (I), and astatine (At), is located in the second vertical column from the right in the table and is known as the **halogens.**

There are changes or gradations in the properties of the elements in a vertical column, so the chemical behavior of these elements, while similar, is not identical. Adjacent elements in each horizontal row differ decidedly in both chemical and physical properties; however, these properties change in a regular way across each row. The horizontal and vertical variations in behavior will be discussed more fully in the following sections and in Chapter 8.

The modern Periodic Table, as presented inside the front cover and in Fig. 4-21, consists of 7 horizontal rows of elements (often referred to as **periods** or **series**) and 32 vertical columns of elements (referred to as **families** or **groups**). In order to fit the table on a single page, some of the elements have been written below the main body of the table. The **lanthanide series** fits between elements 56 (barium, Ba) and 72 (hafnium, Hf). The **actinide series** fits between elements 88 (radium, Ra) and 104 (rutherfordium, Rf).

The **first short period** of the table contains the elements hydrogen (H) and helium (He). The **second short period** contains eight elements, beginning with lithium (Li) and ending with neon (Ne). The **third short period** also contains eight elements, beginning with sodium (Na) and ending with argon (Ar).

The **fourth period** is the first of two long periods, each of which contains 18 elements. The fourth period includes the elements from potassium (K) through krypton (Kr). Within this period are the elements from scandium (Sc) through copper (Cu), which are known as the **first transition series.** The **fifth period** begins with rubidium (Rb) and ends with xenon (Xe). Within this period are the elements yttrium (Y) through silver (Ag), which comprise the **second transition series.**

The **sixth period,** beginning with cesium (Cs) and ending with radon (Rn), contains 32 elements. The **third transition series,** made up of lanthanum (La) and the elements hafnium (Hf) through gold (Au), is included in the sixth period. Note that the third transition series is split. Between lanthanum and hafnium is a series of 14 elements, cerium (Ce) through lutetium (Lu). These 14 elements constitute the **first inner transition series,** also referred to as the **lanthanide series** or the **rare earth elements.** Lanthanum behaves very much like the elements cerium through lutetium. Hence lanthanum is often included in the lanthanide series, even though in terms of electronic structure it is more properly considered to be the first element of the third transition series.

The **seventh period** extends from francium (Fr) through element 118; however, neither element 108 nor any element after element 109 has been characterized. The known elements in this period include a portion of the **fourth transition series** (actinium and the elements rutherfordium through 107 and 109). The elements 108, 110, and 111 would also be members of this series. The elements between actinium (Ac) and rutherfordium (Rf), namely the elements thorium (Th) through lawrencium (Lr), make up the **second inner transition series,** also referred to as the **actinide series** or the **actinide elements.** Actinium is sometimes included with the actinide series because of its similarity in properties to those elements.

The discoverer of a new element names the element. However, until the name is recognized by the International Union of Pure and Applied Chemistry (IUPAC), the recommended name of the new element will be based on the Latin words for its atomic number. These names will also be used for elements, such as element 108, which have not yet been prepared or discovered. An international dispute about credit for the discovery of element 104 led to this new set of systematic names for the heavier elements. In 1964, Soviet scientists reported element 104, which they named kurchatovium, after the leader of their nuclear research program. Five years later, scientists at the University of California suggested that the Soviets were wrong and that, in fact, *they* had prepared element 104. They named it rutherfordium, after the British scientist. Neither name was formally adopted, although each is used in its country of origin. In 1979 IUPAC recommended using the name unnilquadium (abbreviated Unq) for element 104 until the question of the original discoverer could be settled. In addition, they recommended unnilpentium (Unp) for element 105, unnilhexium (Unh) for element 106, unnilseptium (Uns) for element 107, and unniloctium (Uno) for element 108. These names are based on the following Latin words for the atomic numbers: *nil* = 0, *un* = 1, *bi* = 2, *tri* = 3, *quad* = 4, *pent* = 5, *hex* = 6, *sept* = 7, *oct* = 8, and *enn* = 9.

Many of the groups (or families) in the Periodic Table are identified by Roman numerals and letters at the top of the vertical columns. The assignment of the letters A or B to the groups is arbitrary, and some Periodic Tables, particularly European ones, have a different system from that in this text. Some of the groups also have common names. The members of Group IA (other than hydrogen) are called the alkali metals; the elements of Group IIA, the alkaline earth metals; and the elements of Group VIIA, the halogens. The elements of Group VIA are called the **chalcogens;** the elements of Group VIIIA, the **noble gases.** Within a group, or family, there are striking similarities in the chemical behavior of the elements.

4.14 Electronic Structure and the Periodic Law

When arranged in order of increasing atomic number, the elements with similar chemical properties recur at definite intervals, i.e., periodically. This behavior reflects the periodic recurrence of similar electron configurations in the outer shells of these elements. Figure 4-22 shows the electron configurations of the elements with atomic numbers 1 through 18, inclusive, arranged as these atoms are arranged in the Periodic Table. Note the periodicity with respect to the number of **valence electrons;** that is, electrons in the outermost shell.

Figure 4-22. The electron configurations of the first 18 elements arranged in the format of the Periodic Table.

IA							VIIIA
H $1s^1$	IIA	IIIA	IVA	VA	VIA	VIIA	**He** $1s^2$
Li $[He]2s^1$	**Be** $[He]2s^2$	**B** $[He]2s^22p^1$	**C** $[He]2s^22p^2$	**N** $[He]2s^22p^3$	**O** $[He]2s^22p^4$	**F** $[He]2s^22p^5$	**Ne** $[He]2s^22p^6$
Na $[Ne]3s^1$	**Mg** $[Ne]3s^2$	**Al** $[Ne]3s^23p^1$	**Si** $[Ne]3s^23p^2$	**P** $[Ne]3s^23p^3$	**S** $[Ne]3s^23p^4$	**Cl** $[Ne]3s^23p^5$	**Ar** $[Ne]3s^23p^6$

Elements in any one group have the same number of electrons in their outermost shell: lithium and sodium have one; beryllium and magnesium have two; fluorine and chlorine have seven. The outermost shell (or shells in certain heavier atoms) contains the valence electrons and is called the **valence shell.** The similarity in chemical properties among elements of the same group occurs because the number of electrons in the valence shell of an atom determines its chemical properties. **It is the loss, gain, or sharing of valence electrons that determines how elements react.** Thus the Periodic Table is simply an arrangement that puts elements with the same number of valence electrons in the same group. This arrangement is more apparent in Fig. 4-23, which shows in Periodic Table form the electron configuration of the last subshell to be filled by the Aufbau Process for each element.

It is convenient to classify the elements in the Periodic Table into four categories acording to their atomic structures.

1. *Noble gases.* Elements in which the outer shell is complete. The noble gases are the members of Group VIIIA.
2. *Representative elements.* Elements in which the last electron added enters the outermost shell but in which the outermost shell is incomplete. The outermost shell for these elements is the valence shell. The representative elements are those in Groups IA, IIA, IIB, IIIA, IVA, VA, VIA, and VIIA.

Figure 4-23. The order of occupancy of atomic orbitals in the Periodic Table. The electron configuration of the last subshell to be occupied as the atoms are built up by the Aufbau Process is shown.

	s	*s*		*d*	*d*	*d*	*d*	*d*	*d*	*d*	*d*	*d*	*d*	*p*	*p*	*p*	*p*	*s* / *p*	*s* / *p*
																		H $1s^1$	He $1s^2$
1	H $1s^1$																		
2	Li $2s^1$	Be $2s^2$												B $2p^1$	C $2p^2$	N $2p^3$	O $2p^4$	F $2p^5$	Ne $2p^6$
3	Na $3s^1$	Mg $3s^2$												Al $3p^1$	Si $3p^2$	P $3p^3$	S $3p^4$	Cl $3p^5$	Ar $3p^6$
4	K $4s^1$	Ca $4s^2$	Sc $3d^1$	Ti $3d^2$	V $3d^3$	Cr $3d^5$	Mn $3d^5$	Fe $3d^6$	Co $3d^7$	Ni $3d^8$	Cu $3d^{10}$	Zn $3d^{10}$		Ga $4p^1$	Ge $4p^2$	As $4p^3$	Se $4p^4$	Br $4p^5$	Kr $4p^6$
5	Rb $5s^1$	Sr $5s^2$	Y $4d^1$	Zr $4d^2$	Nb $4d^4$	Mo $4d^5$	Tc $4d^6$	Ru $4d^7$	Rh $4d^8$	Pd $4d^{10}$	Ag $4d^{10}$	Cd $4d^{10}$		In $5p^1$	Sn $5p^2$	Sb $5p^3$	Te $5p^4$	I $5p^5$	Xe $5p^6$
6	Cs $6s^1$	Ba $6s^2$	*	Hf $5d^2$	Ta $5d^3$	W $5d^4$	Re $5d^5$	Os $5d^6$	Ir $5d^7$	Pt $5d^9$	Au $5d^{10}$	Hg $5d^{10}$		Tl $6p^1$	Pb $6p^2$	Bi $6p^3$	Po $6p^4$	At $6p^5$	Rn $6p^6$
7	Fr $7s^1$	Ra $7s^2$	†	Rf $6d^2$	Ha $6d^3$	106 $6d^4$	107 $6d^5$		109 $6d^7$										

f

*	La $5d^1$	Ce $4f^2$	Pr $4f^3$	Nd $4f^4$	Pm $4f^5$	Sm $4f^6$	Eu $4f^7$	Gd $4f^7$	Tb $4f^9$	Dy $4f^{10}$	Ho $4f^{11}$	Er $4f^{12}$	Tm $4f^{13}$	Yb $4f^{14}$	Lu $4f^{14}$
†	Ac $6d^1$	Th $6d^2$	Pa $5f^2$	U $5f^3$	Np $5f^4$	Pu $5f^6$	Am $5f^7$	Cm $5f^7$	Bk $5f^8$	Cf $5f^{10}$	Es $5f^{11}$	Fm $5f^{12}$	Md $5f^{13}$	No $5f^{14}$	Lr $5f^{14}$

3. *Transition elements.* Elements in which the second shell counting in from the outside is building from 8 to 18 electrons. The outermost *s* subshell and the *d* subshell of the second shell from the outside contain the valence electrons in these elements. Thus the $(n-1)d$ and ns subshells are regarded as the valence shells in the transition elements. There are four transition series.

    a. First transition series: scandium (Sc) through copper (Cu); 3*d* subshell filling.

    b. Second transition series: yttrium (Y) through silver (Ag); 4*d* subshell filling.

    c. Third transition series: lanthanum (La), plus hafnium (Hf) through gold (Au); 5*d* subshell filling.

    d. Fourth transition series (incomplete): actinium (Ac), plus rutherfordium (Rf), hahnium (Ha), and elements 106, 107, and 109; 6*d* subshell filling. (Presumably, elements 108, 110, and 111 will ultimately be discovered and will complete the series.)

4. *Inner transition elements.* Elements in which the third shell counting in from the outside is building from 18 to 32 electrons. The valence shells of the inner transition elements consist of the $(n-2)f$, $(n-1)d$, and ns subshells. There are two inner transition series.

    a. First inner transition series: cerium (Ce) through lutetium (Lu); 4*f* subshell filling.

    b. Second inner transition series: thorium (Th) through lawrencium (Lr); 5*f* subshell filling.

(Lanthanum and actinium, because of their similarities to the other members of the series, are sometimes included as the first elements of the first and second inner transition series, respectively.)

4.15 Variation of Properties Within Periods and Groups

Elements within a group have identical numbers and generally identical distributions of electrons in their valence shells. Thus we expect them to exhibit very similar chemical behavior. Across a period we might expect a smoothly varying change in chemical behavior, since each element differs from the preceding element by one electron. However, sometimes the similarities or differences within a group or across a period are not as regular as we might expect. This is because the loss, gain, or sharing of valence electrons depends on several factors, including (1) the number of valence electrons, (2) the magnitude of the nuclear charge and the total number of electrons surrounding the nucleus, (3) the number of filled shells lying between the nucleus and the valence shell, and (4) the distances of the electrons in the various shells from each other and from the nucleus.

Examples of these effects on the periodic variation of some physical properties will be considered in the following paragraphs.

1. VARIATION IN COVALENT RADII. (See table inside the back cover.) There are several ways to define the radii of atoms and thus to determine their relative sizes. We will use the **covalent radius**, which is defined as half the distance between the nuclei of two identical atoms when they are joined by a single covalent bond. A covalent bond will be discussed in Chapter 5; for now, we will simply use the covalent radius as one measure of the size of an atom. In general, from left to right

across a period of the Periodic Table, each element has a smaller covalent radius than that of the one preceding it (Table 4-6). This change in size can be attributed to the increasing nuclear charge across the period, with the added electrons going into partially occupied shells. Each element in the Periodic Table has one more electron and a nuclear charge that is one higher than the preceding element. Within a period, however, the number of shells is constant. In general, within a given period, the larger nuclear charge results in a larger force of electrostatic attraction between the nucleus and the electrons because the additional electrons are in the same shell. This causes the decrease in covalent radii across the period.

Table 4-6

Atom	Covalent Radius, Å	Nuclear Charge	Electron Configuration
Na	1.86	+11	$1s^2 2s^2 2p^6 3s^1$
Mg	1.60	+12	$3s^2$
Al	1.43	+13	$3s^2 3p^1$
Si	1.17	+14	$3s^2 3p^2$
P	1.10	+15	$3s^2 3p^3$
S	1.04	+16	$3s^2 3p^4$
Cl	0.99	+17	$3s^2 3p^5$

Proceeding down a group of the Periodic Table, succeeding elements have larger covalent radii as a result of greater numbers of electron shells (Table 4-7), a factor that more than offsets the effect of the larger nuclear charge of each succeeding element. To be more specific, the increasing nuclear charge might lead us to expect that electrons would be held more tightly and pulled closer to the nucleus. However, the total number of shells increases down a group, and shells with larger principal quantum numbers have larger radii. The larger size of the shells coupled with repulsions between the increasing numbers of electrons overcome the increased nuclear attraction, so that the atoms increase in size down a group.

Table 4-7

Atom	Covalent Radius, Å	Nuclear Charge	Number of Electrons in Each Shell
F	0.64	+ 9	2, 7
Cl	0.99	+17	2, 8, 7
Br	1.14	+35	2, 8, 18, 7
I	1.33	+53	2, 8, 18, 18, 7
At	1.4	+85	2, 8, 18, 32, 18, 7

Values of covalent radii are based on interatomic distances between atoms held closely together by strong chemical bonds. Noble gases, however, do not bond this way. Thus covalent radii are not available for the noble gases.

2. VARIATION IN IONIC RADII.   As shown in the table inside the back cover, the radius of a positive ion is less than the covalent radius of its parent atom. A positive ion forms when one or more than one electron is removed from an atom.

The representative elements usually form positive ions by loss of all of their valence electrons. The loss of all electrons from the outermost shell results in a smaller radius because the remaining electrons occupy shells with smaller principal quantum numbers (and smaller radii). Even the radii of the remaining filled shells decrease (relative to their size in the neutral atom) because of an increase in *effective* nuclear charge as the valence electrons are removed, giving rise to a greater average attraction of the nucleus per remaining electron. Thus the covalent radius of a sodium atom ($1s^22s^22p^63s^1$) is 1.86 Å, whereas the ionic radius of a sodium ion ($1s^22s^22p^6$) is 0.95 Å. Proceeding down the groups of the Periodic Table, positive ions of succeeding elements have larger radii, corresponding to greater numbers of shells.

A simple negative ion is formed by the addition of one or more than one electron to the valence shell of an atom. This results in a greater force of repulsion among the electrons and a decrease in the *effective* nuclear charge per electron. Both effects cause the radius of a negative ion to be greater than that of the parent atom. For example, a chlorine atom ($[Ne]3s^23p^5$) has a covalent radius of 0.99 Å, whereas the ionic radius of a chloride ion ($[Ne]3s^23p^6$) is 1.81 Å. For succeeding elements proceeding down the groups, negative ions have more electron shells and larger radii. (These effects are apparent in the table inside the back cover.)

Ions and atoms that have the same electron configuration, such as those in the series N^{3-}, O^{2-}, F^-, Ne, Na^+, Mg^{2+}, and Al^{3+} and those in the series P^{3-}, S^{2-}, Cl^-, Ar, K^+, Ca^{2+}, and Sc^{3+}, are termed **isoelectronic.** The greater the nuclear charge, the smaller the ionic radius in a series of isoelectronic ions and atoms. This trend is illustrated in Table 4-8 for the two series just mentioned.

Table 4-8

Species:	N^{3-}	O^{2-}	F^-	Ne	Na^+	Mg^{2+}	Al^{3+}
Radius, Å:	1.71	1.40	1.36	1.12	0.95	0.65	0.50
Electron configuration: $1s^22s^22p^6$							
Species:	P^{3-}	S^{2-}	Cl^-	Ar	K^+	Ca^{2+}	Sc^{3+}
Radius, Å:	2.12	1.84	1.81	1.54	1.33	0.99	0.81
Electron configuration: $1s^22s^22p^63s^23p^6$							

As we shall see later, many of the properties of ions can best be explained in terms of their sizes and charges.

3. VARIATION IN IONIZATION ENERGIES.   The amount of energy required to remove *the most loosely bound* electron from a gaseous atom is called its **first ionization energy.** This change may be represented for any element X by the equation

$$X(g) + energy \longrightarrow X^+(g) + e^-$$

The energy required to remove the second most loosely bound electron is called the **second ionization energy;** to remove the third, the **third ionization energy;** and so forth. First ionization energies increase in an irregular way from left to right across a period (Table 4-9). This overall increase may be attributed to the fact that the electrons lost come from the same shell, while the nuclear charge increases.

Table 4-9   First Ionization Energies of Some of the Elements (Energy in electron-volts per atom; 1 eV per atom corresponds to 96.49 kJ per mole of atoms.)

1	2	3	4	5	6	7	8	9	10	11	12	13	14	15	16	17	18
1 H 13.6																	2 He 24.6
3 Li 5.4	4 Be 9.3											5 B 8.3	6 C 11.3	7 N 14.5	8 O 13.6	9 F 17.4	10 Ne 21.6
11 Na 5.1	12 Mg 7.6											13 Al 6.0	14 Si 8.1	15 P 11.0	16 S 10.4	17 Cl 13.0	18 Ar 15.8
19 K 4.3	20 Ca 6.1	21 Sc 6.5	22 Ti 6.8	23 V 6.7	24 Cr 6.8	25 Mn 7.4	26 Fe 7.9	27 Co 7.9	28 Ni 7.6	29 Cu 7.7	30 Zn 9.4	31 Ga 6.0	32 Ge 8.1	33 As 10	34 Se 9.8	35 Br 11.8	36 Kr 14.0
37 Rb 4.2	38 Sr 5.7	39 Y 6.4	40 Zr 6.8	41 Nb 6.9	42 Mo 7.1	43 Tc 7.3	44 Ru 7.4	45 Rh 7.5	46 Pd 8.3	47 Ag 7.6	48 Cd 9.0	49 In 5.8	50 Sn 7.3	51 Sb 8.6	52 Te 9.0	53 I 10.5	54 Xe 12.1
55 Cs 3.9	56 Ba 5.2	[57-71] *	72 Hf 7	73 Ta 7.9	74 W 8.0	75 Re 7.9	76 Os 8.7	77 Ir 9.2	78 Pt 9.0	79 Au 9.2	80 Hg 10.4	81 Tl 6.1	82 Pb 7.4	83 Bi 8	84 Po 8.4	85 At ...	86 Rn 10.7
87 Fr ...	88 Ra 5.3	[89-103] †	104 Rf ...	105 Ha ...	106 —	107 —		109 —									

	57 La 5.6	58 Ce 6.9	59 Pr 5.8	60 Nd 6.3	61 Pm ...	62 Sm 5.6	63 Eu 5.7	64 Gd 6.2	65 Tb 6.7	66 Dy 6.8	67 Ho ...	68 Er ...	69 Tm ...	70 Yb 6.2	71 Lu 5.0
*LANTHANIDE SERIES															
†ACTINIDE SERIES	89 Ac 6.9	90 Th ...	91 Pa ...	92 U 4	93 Np ...	94 Pu ...	95 Am ...	96 Cm ...	97 Bk ...	98 Cf ...	99 Es ...	100 Fm ...	101 Md ...	102 No ...	103 Lr ...

Figure 4-24 shows the relationship between the first ionization energies and atomic numbers of several elements. The values of the first ionization energies are provided in Table 4-9. Note that the ionization energy of boron is less than that of beryllium. This is explained by differences in the attraction of the positive nucleus for electrons in different subshells. On the average an s electron is attracted to the nucleus more than a p electron of the same principal shell; a p electron more than a d electron; and so on. This means that an s electron will be harder to remove from an atom than a p electron in the same shell of the atom; a p electron harder to remove than a d electron; and a d electron harder to remove than an f electron. The electron removed during the ionization of beryllium ($[He]2s^2$) is an s electron, whereas the electron removed during the ionization of boron ($[He]2s^22p^1$) is a p electron; this results in a smaller first ionization energy for boron even though its nuclear charge is greater by one unit. The two additional nuclear charges in carbon are sufficient to make its first ionization energy larger than that of beryllium. The first ionization energy for oxygen is slightly less than that for nitrogen because of the repulsion between the two electrons occupying the same $2p$ orbital in the oxygen atom. Since these two electrons occupy the same region of space, their repulsion overcomes the additional nuclear charge of the oxygen nucleus. Analogous changes occur in succeeding periods.

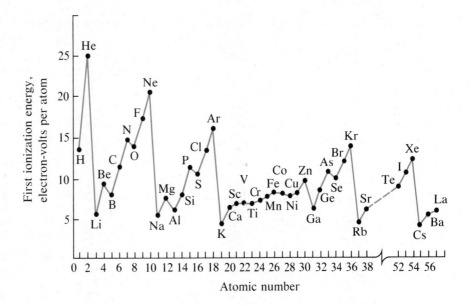

Figure 4-24. A graphic illustration of the periodic relationships between first ionization energy and atomic number for some of the elements.

The attractive force exerted on the valence electrons by the positively charged nucleus is partially counterbalanced by the repulsive forces between electrons in inner shells and the valence electrons. An electron being removed from an atom is thus shielded from the nucleus by these inner shells. This shielding and the increasing distance of the outer electron from the nucleus provide an explanation of the fact that, proceeding down the groups, succeeding elements generally have smaller first ionization energies. As a rule the elements in a group have the same outer electron configuration and the same number of valence electrons.

4. VARIATION IN ELECTRON AFFINITIES.   The **electron affinity** is a measure of the energy involved when an electron is added to a gaseous atom to form a negative ion. A positive value for electron affinity indicates that energy is produced when an electron is added to an atom. The change is expressed for any element X by the equation

$$X(g) + e^- \longrightarrow X^-(g) + \text{energy}$$

A negative value for electron affinity indicates that energy must be added to force the electron onto the atom, as represented in the equation

$$\text{Energy} + X + e^- \longrightarrow X^-$$

Elements to the left within a period have little tendency to form negative ions by gaining electrons; thus their electron affinities tend to be small or negative (Table 4-10). Elements to the right within a period tend to have larger electron affinities.

Although succeeding elements across the periods of the Periodic Table generally have progressively greater electron affinities, exceptions are found among the elements of Group IIA, Group VA, and Group VIIIA. These groups have filled *ns* subshells, half-filled *np* subshells, and all subshells filled, respectively. In each case the completely filled or half-filled subshells represent relatively stable configurations.

Table 4-10   Electron Affinities of Some Elements.
(Electron affinities in electron-volts per atom;
1 eV per atom = 96.49 kJ per mole.)

IA							VIIIA
H 0.75	**IIA**	**IIIA**	**IVA**	**VA**	**VIA**	**VIIA**	**He** −0.2[a]
Li 0.62	**Be** −2.5[a]	**B** 0.24	**C** 1.27	**N** 0.0	**O** 1.46	**F** 3.34	**Ne** −0.3
Na 0.55	**Mg** −2.4[a]	**Al** 0.46	**Si** 1.24	**P** 0.77	**S** 2.08	**Cl** 3.61	**Ar** −0.36[a]
K 0.50	**Ca** −1.6[a]	**Ga** 0.4[a]	**Ge** 1.20	**As** 0.80	**Se** 2.02	**Br** 3.36	**Kr** −0.4[a]
Rb 0.48	**Sr** −1.7[a]	**In** 0.4[a]	**Sn** 1.25	**Sb** 1.05	**Te** 1.97	**I** 3.06	**Xe** −0.4[a]
Cs 0.47	**Ba** −0.5[a]	**Tl** 0.5	**Pb** 1.05	**Bi** 1.05	**Po** 1.8[a]	**At** 2.8[a]	**Rn** −0.4[a]

[a]Calculated value.

FOR REVIEW

SUMMARY

A number of experiments have shown that atoms are composed of protons, neutrons, and electrons. **Protons** are relatively heavy particles with a charge of +1 and a mass of 1.0073 amu. **Neutrons** are relatively heavy particles with no charge and with a mass of 1.0087 amu, very close to that of protons. **Electrons** are light particles with a mass of 0.00055 amu (1/1837 of that of a hydrogen atom) and a charge of −1. Rutherford's α-particle scattering experiments showed that the atom contains a small, massive, positively charged **nucleus** surrounded by electrons. The size of the nucleus is at least 100,000 times smaller than the size of the atom. It was subsequently shown that the nucleus consists of protons and neutrons and that the number of protons in the nucleus, the **atomic number,** determines the type of atom, i.e., the element. Isotopes are atoms with the same atomic number but different **mass numbers.** An element is usually composed of several isotopes, and its atomic weight is the weighted average of the masses of the isotopes involved.

Bohr described the hydrogen atom in terms of an electron moving in a circular orbit about a nucleus. In order to account for the stability of the hydrogen atom, he postulated that the electron was restricted to certain discrete energies characterized by a **quantum number, n,** which could have only integer values (n = 1, 2, 3, . . .). Thus the energy of the electron was assumed to be **quantized.** From this model we can calculate the energy, E, of an electron:

$$E = \frac{-kZ^2}{n^2}$$

where k is a constant and Z is the atomic number. The radius of the orbit can also be calculated as follows:

$$r = \frac{n^2 a_0}{Z}$$

where a_0 is the radius of the orbit in the hydrogen atom for which $n = 1$. Bohr suggested that an atom emits energy, in the form of a photon, only when an electron falls from a higher-energy orbit to a lower-energy orbit. Because the Bohr model accounted for the line spectra produced by samples of hydrogen gas in a discharge tube, his postulation of quantized energies for the electrons in atoms was widely accepted.

Light and other forms of electromagnetic radiation move through a vacuum with a speed, c, of 2.998×10^8 m/s. This radiation shows wavelike behavior, which can be characterized by a frequency, v, and a wavelength, λ, such that $c = \lambda v$. Electromagnetic radiation also has the properties of particles called **photons.** The energy of a photon is related to the frequency of the radiation: $E = hv$, where h is Planck's constant. The line spectrum of hydrogen is obtained by passing the light from a discharge tube through a prism. From a measurement of the wavelength of each line, the energy of each can be determined. These energies correspond to the values predicted when an electron in a hydrogen atom falls from a higher-energy to a lower-energy orbit.

Atoms with two or more electrons cannot be described satisfactorily by the Bohr model, and it has been replaced by the **quantum mechanical model.** This model identifies the behavior of an electron in terms of a **wave function, ψ,** which identifies the **orbital,** or region of space, occupied by the electron. It is possible to calculate the energy of the electron from ψ or to determine the probability of finding the electron in a given location in the orbital. The distribution of an electron within an orbital may be described in terms of **electron density.**

An orbital is characterized by three quantum numbers. The **principal quantum number, n,** may be any positive integer. The energy of the orbital and its average distance from the nucleus are related to n. The **azimuthal quantum number, ℓ,** can have any integral value from 0 to $(n - 1)$. The shape of the orbital is indicated by ℓ, and within a multielectron atom the energies of orbitals with the same value of n increase as ℓ increases. The **magnetic quantum number, m,** with values ranging from -1 to $+1$, describes the orientation of an orbital. A fourth quantum number, the **spin quantum number, s** ($s = \pm\frac{1}{2}$), describes the spin of an electron about its own axis. Orbitals with the same value of n occupy the same **shell.** Orbitals with the same values of n and ℓ occupy the same **subshell.**

n	Shell	Subshells
1	K	$1s$ ($\ell = 0$)
2	L	$2s$ ($\ell = 0$), $2p$ ($\ell = 1$)
3	M	$3s$ ($\ell = 0$), $3p$ ($\ell = 1$), $3d$ ($\ell = 2$)
4	N	$4s$ ($\ell = 0$), $4p$ ($\ell = 1$), $4d$ ($\ell = 2$), $4f$ ($\ell = 3$)

There are $(2\ell + 1)$ orbitals in the ℓ subshell of any shell.

By adding electrons one by one to atomic orbitals in the order $1s$, $2s$, $2p$, $3s$, $3p$, $4s$, $3d$, $4p$, etc., and following both the **Pauli Exclusion Principle** (no two electrons can have the same set of four quantum numbers) and **Hund's Rule** (whenever possible, electrons remain unpaired in degenerate orbitals), we can predict the **electron configurations** of most atoms. The periodic recurrence of similar electron configurations led to the statement of the **Periodic Law** and the construction of a **Periodic Table** in which elements with similar electron configurations (and, consequently, similar chemical behavior) are found in the same **group,** or **family. Valence electrons** in the outermost s and p orbitals are responsible for the chemical behavior of the **representative elements:** Group IA (the alkali metals, s^1), Group IIA (the alkaline earth metals, s^2), Group IIIA (s^2p^1), Group IVA (s^2p^2), Group VA (s^2p^3), Group VIA (s^2p^4), Group VIIA (the halogens, s^2p^5), and Group VIIIA (the noble gases, s^2p^6). **Transition metals** may have valence electrons in the outer shell and first inner shell, the ns and $(n-1)d$ orbitals, respectively. The **inner transition metals** may have valence electrons in three shells: the ns, $(n-1)d$, and $(n-2)f$ orbitals. The **atomic radii** and **ionic radii** of elements, their **ionization energies,** and their **electron affinities** vary in a periodic way with the nature of the valence orbitals involved.

KEY TERMS AND CONCEPTS

Alpha particles (4.4)
Atomic number (4.5)
Aufbau Process (4.12)
Azimuthal quantum number, ℓ (4.10)
Beta particles (4.4)
Bohr model of the atom (4.7)
Cathode rays (4.1)
Covalent radius (4.15)
Degenerate orbitals (4.10)
Electron affinity (4.15)
Electron configuration (4.12)
Electron density (4.9)
Electrons (4.1)
Excited state (4.7)
Family of elements (4.13)
First ionization energy (4.15)
Gamma rays (4.4)

Ground state (4.7)
Heisenberg Uncertainty Principle (4.9)
Hund's Rule (4.11)
Ionic radius (4.15)
Ions (4.2)
Isoelectronic species (4.15)
Isotopes (4.6)
Magnetic quantum number, m (4.10)
Mass number (4.6)
Neutrons (4.3)
Noble gases (4.13)
Nucleus (4.5)
Orbital (4.9)
Pauli Exclusion Principle (4.11)
Periodic Law (4.13)
Periodic Table (4.13)

Periods of elements (4.13)
Photoelectric effect (4.1)
Photons (4.8)
Principal quantum number, n (4.10)
Protons (4.2)
Quantized (4.7)
Quantum mechanical model of the atom (4.9)
Radioactivity (4.4)
Shell (4.10)
Spectrum (4.8)
Spin quantum number, s (4.10)
Subshell (4.10)
Wave function (4.9)
Valence electrons (4.14)
Valence shell (4.14)

EXERCISES

Atomic Structure

1. What is the source of the cathode rays in a discharge tube? What is the evidence that these rays are negatively charged particles?
2. Compare the properties of cathode rays and β particles.
3. How are the positive rays in a Crookes tube and α particles similar? How do they differ?
4. Describe and interpret the results of the experiment that shows that the positive particles in an atom are located in a small nucleus of relatively large mass.
5. Give the restrictions on the number of protons,

neutrons, and electrons in each of the following:
(a) different isotopes of the same element
(b) atoms of the same isotope of an element
(c) atoms of different elements with the same mass number

6. In what way is the lightest hydrogen nucleus unique?
7. Using the data in Table 4-2, sketch the mass spectrum expected for a sample of boron atoms.
[S] 8. Using the data in Table 4-2, calculate the atomic weight of naturally occurring boron. *Ans. 10.8*
9. Explain why atomic weights of elements often differ markedly from whole numbers, even though protons and neutrons have masses very nearly equal to 1 amu.
10. Determine the number of protons and neutrons in the nuclei of each of the following elements: $^{6}_{3}Li$, $^{28}_{14}Si$, $^{37}_{17}Cl$, $^{99}_{43}Tc$, $^{227}_{89}Ac$.
11. Identify the following elements: $^{9}_{4}X$, $^{32}_{16}X$, $^{112}_{48}X$, $^{210}_{82}X$, $^{238}_{92}X$.

The Bohr Model of the Atom

12. How does the Bohr model of the atom differ from the Rutherford model of the atom?
13. What does it mean to say that the energies of the electrons in a hydrogen atom are quantized?
[S] 14. Figure 4-12 gives the energies of an electron in a hydrogen atom for values of n from 1 to 5. What is the energy, in electron-volts, of an electron with $n = 8$ in a hydrogen atom? *Ans. −0.2125 eV*
15. What is the lowest possible energy, in electron-volts, for an electron in a He^+ ion? *Ans. −54.40 eV*
[S] 16. Recently, hydrogen atoms with electrons in very high-energy orbitals have been isolated. What is the radius (in centimeters) of a hydrogen atom with an electron characterized by an n value of 110? How many times larger is this than the radius of the normal hydrogen atom? *Ans. 6.40 × 10^{-5} cm; 12,100 times larger*
17. Calculate the radius of a He^+ ion using the Bohr model of the electronic structure of this ion. *Ans. 0.264 Å*
[S] 18. According to the Bohr model, which is larger, a hydrogen atom with an electron in an orbit with $n = 4$ or a He^+ ion with an electron in an orbit with $n = 5$? *Ans. H atom, with $n = 4$*

Orbital Energies and Spectra

[S] 19. From the data in Fig. 4-12, calculate the energy, in electron-volts, of the photon emitted when an electron in an excited hydrogen atom moves from an orbit with $n = 4$ to one with $n = 2$. *Ans. 2.549 eV*
20. From the data in Fig. 4-12, calculate the energy, in electron-volts, required to ionize a hydrogen atom. Calculate the energy in joules. *Ans. 13.595 eV; 2.1782 × 10^{-18} J*
[S] 21. When lithium metal is heated, lithium atoms emit photons of red light with a wavelength of 6708 Å. What is the energy in joules and the frequency of this light? *Ans. 2.961 × 10^{-19} J; 4.469 × 10^{14} s^{-1}*
22. An FM radio station broadcasts at a frequency of 95.0×10^6 s^{-1} (95.0 megahertz). What is the wavelength of these radio waves in meters? *Ans. 3.16 m*
23. Using the Bohr model of the atom, calculate the energy, in joules, of the light emitted by the transition of an electron from the $n = 2$ (L) to the $n = 1$ (K) shell in Li^{2+}. *Ans. 1.471 × 10^{-17} J*
[S] 24. Consider a collection of hydrogen atoms with electrons randomly distributed in the $n = 1, 2, 3, 4$, and 5 shells. How many different wavelengths of light will be emitted by these atoms as the electrons fall into lower-energy orbitals? *Ans. 10*
25. Calculate the lowest and highest energies, in electron-volts, of the light produced by the transitions described in Exercise 24. *Ans. From $n = 5$: 0.3060 eV; 13.06 eV*
26. Calculate the frequencies and wavelengths of the light produced by the transitions described in Exercise 25. *Ans. $\nu = 7.399 \times 10^{13}$ s^{-1}; 3.158×10^{15} s^{-1}; $\lambda = 4.052 \times 10^{-6}$ m; 9.493×10^{-8} m*
27. Light that looks green has a frequency of 5×10^{14} s^{-1}. What is the wavelength of this green light in meters and in angstroms? What is the energy of a photon of this green light in joules? *Ans. 6 × 10^{-7} m; 6 × 10^{3} Å; 3 × 10^{-19} J*
[S] 28. Does a photon of the green light described in Exercise 27 have enough energy to excite the electron in a hydrogen atom from the $n = 1$ to the $n = 2$ shell? *Ans. No; 1.634 × 10^{-18} J is required*

S 29. X rays are produced when the electron stream in an X-ray tube knocks an electron out of a low-energy shell of an atom in the target, and an electron in a higher shell falls into the low-energy shell. The X ray is the photon given off as the electron falls into the low-energy shell. The most intense X rays produced by an X-ray tube with a copper target have wavelengths of 1.542 Å and 1.392 Å. These X rays are produced by an electron from the L or M shell falling into the K shell of a copper atom. Calculate the energy separation in electron-volts of the K, L, and M shells in copper.   *Ans. M, K: 8.907×10^3 eV; M, L: 8.67×10^2 eV; L, K: 8.040×10^3 eV*

The Quantum Mechanical Model of the Atom

30. How does the Bohr model of the atom differ from the quantum mechanical model? How are the two similar?
31. Distinguish between electron shells, subshells, and orbitals.
32. Discuss the following: probable electron density, atomic orbital, radial electron density.
33. What can be determined about the location of an electron using the quantum mechanical model of the atom?
34. Identify the four quantum numbers used in the quantum mechanical model of the atom and indicate what each describes.
35. Tell which of the following orbitals are degenerate in an atom with two or more electrons: $3d_{xy}$, $4s$, $4p_z$, $4d_{xy}$, $4d_{x^2-y^2}$.
S 36. Consider three atomic orbitals, (a), (b), and (c), whose outlines are shown below.

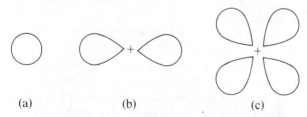

(a)   (b)   (c)

(a) What is the maximum number of electrons that can be contained in orbital (c)?
                                                    Ans. 2
(b) How many orbitals with the same value of l as orbital (a) can be found in the shell with $n = 4$? How many with the same value as

orbital (b)? How many with the same value as orbital (c)?   *Ans. 1; 3; 5*
(c) What is the smallest n value possible for an electron in an orbital of type (c)? of type (b)? of type (a)?   *Ans. 3; 2; 1*
(d) What are the l values that characterize each of these three orbitals?
                                    Ans. a, 0; b, 1; c, 2
(e) Arrange these orbitals in order of increasing energy in the M shell. Is this order different in other shells?   *Ans. a < b < c; no*

S 37. Identify the subshell in which electrons with the following quantum numbers are found:
(a) $n = 1$, $l = 0$ *Ans. 1s*
(b) $n = 3$, $l = 1$ *Ans. 3p*
(c) $n = 3$, $l = 2$ *Ans. 3d*
(d) $n = 7$, $l = 4$ *Ans. 7g*
(e) $n = 4$, $l = 3$ *Ans. 4f*

38. What type of orbital is occupied by an electron with the quantum numbers $n = 4$, $l = 1$? How many degenerate orbitals of this type are found in a multielectron atom?   *Ans. 4p; 3*

S 39. Write the quantum numbers for each electron found in an oxygen atom. For example, the quantum numbers for one of the 2s electrons are $n = 2$, $l = 0$, $m = 0$, $s = +\frac{1}{2}$.

S 40. The quantum numbers that describe the electron in the lowest energy level of a hydrogen atom (the ground state) are $n = 1$, $l = 0$, $m = 0$, $s = +\frac{1}{2}$. Excitation of the electron can promote it to energy levels described by other sets of quantum numbers. Which of the following sets of quantum numbers cannot exist in an excited hydrogen atom (or any other atom)?
(a) $n = 2$, $l = 1$, $m = -1$, $s = +\frac{3}{2}$
(b) $n = 3$, $l = 2$, $m = 0$, $s = +\frac{1}{2}$
(c) $n = 3$, $l = 3$, $m = -2$, $s = -\frac{1}{2}$
(d) $n = 4$, $l = 1$, $m = -2$, $s = +\frac{1}{2}$
(e) $n = 27$, $l = 14$, $m = -8$, $s = -\frac{1}{2}$
                                    Ans. a, c, d

Electron Configurations

41. Using complete subshell notation ($1s^2 2s^2 2p^6$, etc.), predict the electron configuration of each of the following atoms:
(a) $^{16}_{8}O$ (e) $^{40}_{20}Ca$ (i) $^{119}_{50}Sn$
(b) $^{27}_{13}Al$ (f) $^{48}_{22}Ti$ (j) $^{175}_{71}Lu$
(c) $^{32}_{16}S$ (g) $^{55}_{25}Mn$ (k) $^{222}_{86}Rn$
(d) $^{40}_{18}Ar$ (h) $^{75}_{33}As$

42. Using complete subshell notation ($1s^22s^22p^6$, etc.), predict the electron configuration of each of the following ions:
    (a) N^{3-} (e) Sc^{3+}
    (b) Mg^{2+} (f) Pb^{2+}
    (c) Al^{3+} (g) Ce^{4+}
    (d) Cl^-

43. Identify the atoms whose electron configurations are as follows:
    (a) $1s^22s^22p^3$ (d) $[Ar]4s^23d^{10}4p^6$
    (b) $1s^22s^22p^63s^23p^64s^23d^6$ (e) $[Kr]5s^24d^{10}5p^5$
    (c) $[Ar]4s^23d^{10}$ (f) $[Xe]6s^24f^6$

44. The $4s$ orbitals fill before the $3d$ orbitals when building up atomic structures by the Aufbau Process. However, electrons are lost from the $4s$ orbital before they are lost from the $3d$ orbitals when transition elements are ionized. The ionization of copper ($[Ar]3d^{10}4s^1$) gives $Cu^+([Ar]3d^{10})$, for example. The electronic structures of a number of transition metal ions (M) are given below; identify the transition metals.
    (a) $M^{3+}, [Ar]3d^1$ (d) $M^{2+}, [Ar]3d^2$
    (b) $M^{2+}, [Ar]3d^7$ (e) $M^{2+}, [Kr]4d^6$
    (c) $M^{3+}, [Ar]3d^6$ (f) $M^{2+}, [Kr]4d^8$

45. Using subshell notation ($1s^22s^22p^6$, etc.), predict the electron configurations of the following ions:
    S (a) Ti^{3+} (c) Cr^{3+} (e) Ag^+
    (b) Ni^{2+} S (d) Ru^{3+}
    (Read Exercise 44 before attempting this exercise.)

46. Which of the following electron configurations describe an atom in its ground state and which describe an atom in an excited state?
    (a) $[He]2s^12p^4$ (d) $[Ar]4s^23d^93p^3$
    (b) $[Ar]5s^1$ (e) $[Kr]6s^25d^3$
    (c) $[Ar]4s^23d^{10}$ (f) $[Xe]6s^24f^{14}5d^{10}6p^2$

S 47. Which of the following sets of quantum numbers describes the most easily removed electron in an aluminum atom in its ground state? Which of the electrons described is most difficult to remove?
    (a) $n = 1, \ell = 0, m = 0, s = -\frac{1}{2}$
    (b) $n = 2, \ell = 1, m = 0, s = -\frac{1}{2}$
    (c) $n = 3, \ell = 0, m = 0, s = \frac{1}{2}$
    (d) $n = 3, \ell = 1, m = 1, s = -\frac{1}{2}$
    (e) $n = 4, \ell = 1, m = 1, s = \frac{1}{2}$ *Ans. d; a*

48. $N^+(g)$ can be produced from $N(g)$ by removing one electron from any of the occupied orbitals of the nitrogen atom. Several of these processes are
    (a) $1s^22s^22p^3 \longrightarrow 1s^12s^22p^3$
    (b) $1s^22s^22p^3 \longrightarrow 1s^22s^12p^3$
    (c) $1s^22s^22p^3 \longrightarrow 1s^22s^22p^2$

The first ionization energy of N is the energy of which of these processes? Which process will require the most energy?   *Ans. c; a*

The Periodic Table and Periodic Properties

49. Define noble gases, representative elements, inner transition elements, and transition elements in terms of s, p, d, and f subshells.

50. State the Periodic Law. Why does the Periodic Law correlate with electronic structure?

51. In terms of electronic structure, explain why there are 2 elements in the first period of the Periodic Table, 8 elements in the second and third periods, 18 elements in the fourth and fifth periods, and 32 elements in the sixth period.

S 52. Identify the groups that have the following electron configurations in their valence shells (n represents the principal quantum number):
    (a) ns^1 (d) $ns^2(n-1)d^1$
    (b) ns^2np^3 (e) $ns^2(n-1)d^6$
    (c) ns^2np^5

53. Hydrogen is sometimes included in both Group IA and Group VIIA. In terms of the electron configuration of the H atom, explain why hydrogen can appear in both groups.

54. The formula for beryllium chloride is $BeCl_2$. What is the general chemical formula for the chlorides of the other members of Group IIA?

55. Most representative elements that form positive ions in chemical compounds do so by the loss of all of the electrons in their valence shells. Which group in the Periodic Table forms dipositive ions, M^{2+}?

56. Those representative elements that form negative ions in chemical compounds do so by filling their valence shells with electrons. Which group in the Periodic Table forms trinegative ions, X^{3-}?

S 57. Why is the radius of a positive ion smaller than the radius of its parent atom?

58. Why do negative ions have larger radii than their parent atoms?

59. Account for the increase in radius of the ions Mg^{2+}, Ca^{2+}, Sr^{2+}.

60. Account for the increase in radius of the isoelectronic ions F^-, O^{2-}, and N^{3-}.

61. Account for the decrease in radius for the atoms Li, Be, B, C, N, O, and F.

62. Which of the following would have the higher

ionization energy:

(a) Na or Mg?   (d) O or F?

(b) Rb or Cs?   (e) S or F?

(c) Mg or Al?

Explain why.

63. Basing your answers on the locations of the atoms and ions in the Periodic Table, arrange each of the following groups in order of increasing size:

(a) Cs, K, Li, Na, Rb

(b) B, Be, C, F, Li

(c) Br^-, Ca^{2+}, K^+, Se^{2-}

(d) Al^{3+}, F^-, Na^+, Mg^{2+}, O^{2-}

(e) As^{3-}, Ca^{2+}, Cl^-, K^+

Additional Exercises

S 64. The isotope of uranium used in nuclear fission is $^{235}_{92}U$. (a) How many protons, neutrons, and electrons are contained in this atom? (b) How many protons, neutrons, and electrons are contained in a $^{235}_{92}U$ ion with a charge of +3?

65. Atoms of the isotope $^{16}_{8}O$ readily form ions with a charge of −2. How many protons, neutrons, and electrons are contained in such an ion?

66. How would the results of the Rutherford α-particle scattering experiment differ if the atom were only twice as large as its nucleus?

67. (a) Are the charges on the nuclei of atoms quantized?

(b) Are the masses of individual atoms quantized?

(c) Is the atomic weight of an element that is a mixture of isotopes quantized?

68. Using the Bohr model of the atom, what is the energy, in electron-volts, of an electron with $n = 3$ in the ion Be^{3+}? What is the radius of the circular orbit?   *Ans. −24.18 eV; 1.19 Å*

69. The bright yellow light emitted by a sodium vapor lamp or by heated compounds containing sodium has wavelengths of 5896 Å and 5890 Å. These lines result from the relaxation of electrons from one or the other of two closely spaced energy levels to a common lower level. What is the energy separation, in electron-volts, of the two closely spaced levels?   *Ans. 0.002 eV*

70. In Chapter 1 the meter was defined as being 1,650,763.73 times the wavelength of a certain krypton line. What is the frequency of this radiation?   *Ans. 4.949×10^{14} s^{-1}*

71. Draw an energy-level diagram analogous to that in Fig. 4-19 showing the energies of the atomic orbitals in a hydrogen atom.

72. Write the electron configuration of each of the following elements using complete subshell notation: C, Ar, K, Co, Br, Y, Bi.

73. Sketch the shape of an orbital with $n = 1$, $l = 0$ and of an orbital with $n = 3$, $l = 1$.

74. Using [NG] to represent the electron configuration of a noble gas, write the general electron configurations for the alkali metals, the alkaline earths, the halogens, and the elements of Groups VIA, IIIB, and VIIB. The general electron configuration for the elements of Group IIIA, for example, would be $[NG]ns^2np^1$.

75. The energy required to remove an electron from a Si^{3+} ion is 45.08 eV, and that required to remove an electron from an Al^{3+} ion is 120.2 eV. Explain the large difference.

76. Assume that the values of the quantum number m are limited to zero or positive integers up to the value of l for a particular subshell; that is, for $l = 2$, m could be 0, 1, or 2. If, except for this restriction, all other quantum numbers behave as described in this chapter, describe the electron configurations of nitrogen (atomic number = 7) and of fluorine (atomic number = 9). How many unpaired electrons would these atoms contain?   *Ans. N: $1s^2 2s^2 2p^3$; F: $1s^2 2s^2 2p^4 3s^1$; one unpaired electron each*

77. Write a periodic table for the first 14 elements, assuming that the quantum numbers of these elements behave as described in Exercise 76.

78. A gaseous ion, X^+, in its most stable state has three unpaired electrons. If X is a representative element, to which group of the Periodic Table does it belong?

79. The gaseous ion X^{2+} has no unpaired electrons in its most stable state. If X is a representative element, it could be a member of one of two possible groups. Which ones are they?

80. The ion X^- has three unpaired electrons. If X is a representative element, to which group does it belong?

5

CHEMICAL BONDING,
PART 1—GENERAL CONCEPTS

Although it is important for a chemist to understand the electronic structures of isolated atoms, most chemists do not deal with isolated atoms. They usually study groups of two or more atoms and the attractive forces, called chemical bonds, that hold them together. The formation of chemical bonds between atoms creates compounds. When the atoms separate, the bonds are destroyed and the compounds no longer exist.

In this chapter we will examine two types of chemical bonds and, from a consideration of the electronic structures of the atoms involved, determine how to predict their formation. You will see that the forces that hold atoms together are electrical in nature and that formation of chemical bonds between atoms involves changes in their electronic structures.

5.1 Ionic Bonds; Chemical Bonding by Electron Transfer

When an element that loses electrons easily reacts with an element that gains electrons easily, one or more electrons are completely transferred, and **ions** are produced. The compound formed by this transfer is stabilized by **ionic bonds,** strong electrostatic attractions between the ions of opposite charge present in the compound.

An atom that loses one or more valence electrons gains a positive charge and becomes a **positive ion,** or **cation.** For example, when a sodium atom, which loses electrons easily, combines with an element that gains electrons easily, the sodium atom gives up its one valence electron and gains a positive charge of 1. When a calcium atom loses its two valence electrons, it gains a positive charge of 2.

$$\text{Na} \cdot \longrightarrow \text{Na}^+ + e^- \qquad \text{Ca} : \longrightarrow \text{Ca}^{2+} + 2e^-$$

| Sodium | Sodium | Calcium | Calcium |
| atom | cation | atom | cation |

The dot notation in these equations is often used to describe changes in the electron configurations of representative elements. Each of these electron-dot formulas, or **Lewis symbols,** consists of the symbol for the element with one dot added for each valence electron present in the atom or ion. When two dots are written adjacent to each other, as in Ca:, they represent two electrons paired (with opposite spins) in the same orbital. The Lewis symbols for the atoms of the third period are given in Table 5-1.

Table 5-1   Lewis Symbols for the
Atoms of the Third Period

Atom	Electron Configuration	Lewis Symbol
Sodium	$[\text{Ne}]3s^1$	Na ·
Magnesium	$[\text{Ne}]3s^2$	Mg :
Aluminum	$[\text{Ne}]3s^23p^1$	Àl :
Silicon	$[\text{Ne}]3s^23p^2$	· S̈i ·
Phosphorus	$[\text{Ne}]3s^23p^3$	· P̈ ·
Sulfur	$[\text{Ne}]3s^23p^4$	: S̈ ·
Chlorine	$[\text{Ne}]3s^23p^5$	: C̈l ·
Argon	$[\text{Ne}]3s^23p^6$	: Är :

For clarity, different symbols (· , ×, ∘) are sometimes used as a convenience to distinguish electrons according to their sources. You should remember, however, that all electrons are identical, regardless of origin.

Atoms that easily form cations have relatively low first ionization energies. As was noted in Section 4.15, atoms with low first ionization energies tend to be those lying to the left in a period or those lying toward the bottom of a group in the Periodic Table. These elements are metals.

Atoms that readily pick up electrons have relatively high electron affinities and lie to the right in the Periodic Table (Section 4.15). These elements, the nonmetals, are only a few electrons shy of having a filled valence shell and can pick up the electrons lost by metals, thereby filling the valence shell. A neutral atom, as it fills its valence shell by gaining one or more electrons, becomes a **negative ion,** or **anion.** Nonmetals such as F, Cl, Br, I, O, and S form anions.

$$: \ddot{\text{C}}\text{l} \cdot + e^- \longrightarrow : \ddot{\text{C}}\text{l} :^- \qquad : \ddot{\text{S}} \cdot + 2e^- \longrightarrow : \ddot{\text{S}} :^{2-}$$

| Chlorine | Chloride | Sulfur | Sulfide |
| atom | anion | atom | anion |

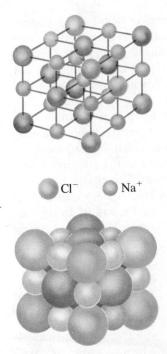

Thus, when metals combine with nonmetals, electrons are transferred from the metals to the nonmetals and **ionic compounds** (compounds that contain ions) are formed. The use of Lewis symbols to show the transfer of electrons during formation of ionic compounds is illustrated by the following examples:

Metal		Nonmetal		Ionic Compound	
$Na^{\times}$	+	$:\overset{..}{\underset{..}{Cl}}\cdot$	$\longrightarrow$	$Na^+[:\overset{..}{\underset{..}{Cl}}{\times}^-]$	(1)
Sodium atom		Chlorine atom		Sodium chloride (sodium ion and chloride ion)	
$Mg^{\times}_{\times}$	+	$:\overset{..}{O}\cdot$	$\longrightarrow$	$Mg^{2+}[:\overset{..}{\underset{\times}{O}}{\times}^{2-}]$	(2)
Magnesium atom		Oxygen atom		Magnesium oxide (magnesium ion and oxide ion)	
$Ca^{\times}_{\times}$	+	$2:\overset{..}{\underset{..}{F}}\cdot$	$\longrightarrow$	$Ca^{2+}[:\overset{..}{\underset{..}{F}}{\times}^-]_2$	(3)
Calcium atom		Fluorine atoms		Calcium fluoride (calcium ion and two fluoride ions)	

The actual reactions are not exactly as written here since the nonmetals involved exist as the diatomic molecules Cl_2, O_2, and F_2. However, the transfer of electrons from the metals to the nonmetals is an important feature of the reaction, and the products have the electronic structures shown.

A crystal of sodium chloride consists of sodium ions and chloride ions (Fig. 5-1). Their regular three-dimensional arrangement is due to each anion and cation pulling the maximum number of oppositely charged ions around itself. Each sodium ion is surrounded by six chloride ions, and each chloride ion is surrounded by six sodium ions. The force that holds the ions together in the crystal is the electrostatic attraction between ions of opposite charge. This electrostatic force of attraction is very strong in ionic compounds. For example, it requires 769 kilojoules of energy to separate the Na^+ and Cl^- ions in a mole of NaCl.

Any given ion in a crystal of sodium chloride exerts the same electrostatic force on each of its six immediate neighbors of opposite charge; therefore, it is impossible to identify any single pair of sodium and chloride ions as being a molecule of sodium chloride. In truly ionic compounds no molecules are present; a crystal of an ionic compound is an aggregation of ions. The formula of an ionic compound represents the relative numbers of ions necessary to give an algebraic balance of ionic charges. Since a crystal of sodium chloride is electrically neutral, it must contain the same numbers of Na^+ and Cl^- ions. Thus its empirical formula is NaCl. A crystal of sodium oxide contains twice as many Na^+ as O^{2-} ions. Its empirical formula is Na_2O. It follows that the term *molecular weight* has no significance in connection with ionic compounds. The formula NaCl represents one *formula weight* of sodium chloride; it cannot be said to represent a molecular weight, for there are no molecules of sodium chloride.

Figure 5-1. The arrangement of sodium and chloride ions in a crystal of sodium chloride (common salt). The smaller spheres represent sodium ions; the larger ones, chloride ions. The upper structure is an "expanded" view that shows the geometry more clearly.

Cl^- Na^+

5.2 The Electronic Structures of Ions

With a few exceptions found among the heavy atoms at the bottom of Groups IIB, IIIA, IVA, and VA, those *representative elements* (Section 4.14) that form cations do so by the loss of all of the electrons in their valence shell. The positive ions thus produced have the stable **noble gas electron configuration** (the same electronic structure as a noble gas), in which the outermost shell has the electron configuration ns^2np^6 (or $1s^2$ for cations of elements in the second period), or they have a **pseudo noble gas electron configuration,** in which the outermost shell has the electron configuration $ns^2np^6nd^{10}$. An example of the latter is the zinc ion, Zn^{2+}, with the electronic structure $1s^22s^22p^63s^23p^63d^{10}$, which forms from a zinc atom (electronic structure $1s^22s^22p^63s^23p^63d^{10}4s^2$).

The tendency of representative elements to lose all valence electrons when forming cations makes it easy to remember the charges on the various positive ions. The charge is equal to the number of the group in which the element forming the ion is found, since the group number is the same as the number of electrons in the outermost shell of a representative element. The charge is $+1$ for members of Group IA, $+2$ for members of Group IIA, $+2$ for members of Group IIB, etc. The exceptions to this rule are Hg, Tl, Sn, Pb, and Bi. In addition to the expected ions Hg^{2+}, Tl^{3+}, Sn^{4+}, Pb^{4+}, and Bi^{5+}, these elements also form $Hg_2{}^{2+}$, Tl^+, Sn^{2+}, Pb^{2+}, and Bi^{3+}.

EXAMPLE 5.1   Write the electron configurations and indicate the charges of the ions formed from the elements K, Ca, Ga, and Cd.

As members of Groups IA, IIA, IIIA, and IIB, respectively, these elements form the ions K^+, Ca^{2+}, Ga^{3+}, and Cd^{2+}. These ions are formed by the loss of all of the valence electrons from the atoms, giving noble gas electron configurations for K^+ and Ca^{2+} and pseudo noble gas electron configurations for Ga^{3+} and Cd^{2+}, as shown in the following table:

Atom		Ion	
K	$1s^22s^22p^63s^23p^64s^1$	K^+	$1s^22s^22p^63s^23p^6$
Ca	$1s^22s^22p^63s^23p^64s^2$	Ca^{2+}	$1s^22s^22p^63s^23p^6$
Ga	$1s^22s^22p^63s^23p^63d^{10}4s^24p^1$	Ga^{3+}	$1s^22s^22p^63s^23p^63d^{10}$
Cd	$1s^22s^22p^63s^23p^63d^{10}4s^24p^64d^{10}5s^2$	Cd^{2+}	$1s^22s^22p^63s^23p^63d^{10}4s^24p^64d^{10}$

Transition and inner transition elements behave differently from representative elements. Most transition metal cations have charges of $+2$ or $+3$, resulting from the loss of their outermost s electron (or electrons), sometimes followed by the loss of one or two d electrons from the next-to-outermost shell. For example, copper ($1s^22s^22p^63s^23p^63d^{10}4s^1$) forms the ions Cu^+ ($1s^22s^22p^63s^23p^63d^{10}$) and Cu^{2+} ($1s^22s^22p^63s^23p^63d^9$). Even though the d orbitals of the transition elements are the last to be filled when electronic structures are built up by the Aufbau Process, the outermost s electrons are the first to be lost when these atoms ionize.

The formation of monatomic anions by representative elements always involves formation of ions with noble gas electron configurations. Thus negative ions such as N^{3-}, S^{2-}, I^-, and O^{2-} form when a neutral atom picks up enough electrons to fill its valence shell completely. Keeping this in mind makes it simple to determine the charge on any monatomic negative ion; the charge is equal to the number of electrons necessary to fill the valence orbitals of the parent atom. Oxygen, for example, has the electron configuration $1s^22s^22p^63s^23p^4$, whereas the oxygen anion (oxide ion) has the noble gas electron configuration $1s^22s^22p^63s^23p^6$.

Since two additional electrons are required to fill the valence shell of the oxygen atom, the oxide ion has a charge of -2.

EXAMPLE 5.2   Write the electron configurations of a phosphorus atom and its negative ion, and give the charge on the anion.

Phosphorus is a member of Group VA and has the electron configuration $1s^2 2s^2 2p^6 3s^2 3p^3$, or [Ne]$3s^2 3p^3$. A phosphorus atom requires three additional electrons to fill its valence shell, so the charge of the phosphorus anion is -3. The electron configuration of P^{3-} is [Ne]$3s^2 3p^6$.

The electronic differences between an atom and its ion result in their having distinctly different physical and chemical properties. Individual sodium *atoms* combine to give sodium metal, a soft, silvery white metal that burns vigorously in air and reacts rapidly with water. Chlorine *atoms* combine to give chlorine gas, Cl_2, a greenish-yellow gas that is extremely corrosive to most metals and very poisonous to animals and plants. The vigorous reaction between sodium and chlorine forms the compound sodium chloride, common table salt, which contains *ions* of sodium and chlorine. The ions exhibit properties entirely different from those of the elements sodium and chlorine. Chlorine is poisonous, while sodium chloride is essential to life; sodium atoms react vigorously with water, while sodium chloride is stable in water.

5.3   Covalent Bonds; Chemical Bonding by Electron Sharing

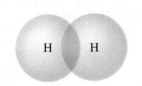

$H\cdot$ + $\cdot H$ $H \colon H$
Hydrogen atoms ⟶ Hydrogen
molecule

Figure 5-2. Combination of two hydrogen atoms to form a hydrogen molecule, H_2, through covalent bonding.

Many compounds do not contain ions, but instead consist of atoms bonded tightly together in molecules. The bonds holding such atoms together form when pairs of electrons are shared between atoms, that is, when the electrons occupy orbitals of both atoms involved in the bond. Such a bond is called a **covalent bond.** Electron pairs are shared and covalent bonds are formed when two atoms have about the same tendency to give up or to pick up electrons.

The simplest substance in which atoms are covalently bonded is the hydrogen molecule, H_2. A hydrogen atom has one electron in its $1s$ shell. The two electrons (an electron pair) from the two hydrogen atoms in a hydrogen molecule are shared by the two nuclei (Fig. 5-2). These shared electrons spend most of their time in the region between the two nuclei, and the electrostatic attraction between the two positively charged nuclei and the two negatively charged electrons holds the molecule together. The bond resulting from this attraction is very strong, as is evidenced by the large amount of energy required to break the covalent bonds in 1 mole of hydrogen molecules—436 kilojoules. Conversely, this same quantity of energy is evolved when a mole of hydrogen molecules is formed from hydrogen atoms.

In forming ionic compounds many metals and nonmetals tend to assume the electronic structures of noble gases through the transfer of electrons. Atoms forming covalent molecules by electron sharing have the same tendency. Each electron of the pair in the covalent bond in a hydrogen molecule spends an equal amount of time near each nucleus, so the $1s$ orbital of each hydrogen atom in the molecule is occupied by both electrons. In effect, each atom has the electronic structure of a helium atom.

The sharing of one pair of electrons by two atoms in a molecule of chlorine gives each atom the stable electronic structure of an atom of the noble gas argon, since a free atom of chlorine has seven electrons in its outer shell, one less than an atom of argon.

$$:\overset{..}{\underset{..}{Cl}}\cdot\;+\;\overset{\times\times}{\underset{\times\times}{\overset{\times}{Cl}}}\;\longrightarrow\;:\overset{..}{\underset{..}{Cl}}{}^{\times}_{\times}\overset{\times\times}{\underset{\times\times}{Cl}}$$

Chlorine   Chlorine
atoms   molecule

The **Lewis structure** shown above for Cl_2 indicates that each Cl atom has three pairs of electrons that are not used in bonding (called **lone pairs**) and one shared pair of electrons (written between the atoms). A dash is sometimes used to indicate a shared pair of electrons.

$$H{-}H\qquad:\overset{..}{\underset{..}{Cl}}{-}\overset{..}{\underset{..}{Cl}}:$$

A single shared pair of electrons is referred to as a **single bond.**

The bonding in the molecules of the other elements in the same family as chlorine (that is, the molecules F_2, Br_2, I_2, and At_2) is like that in the chlorine molecule, with one single bond between atoms and three lone pairs per atom. However, the electronic structures of the atoms in these molecules are like those of the noble gases Ne, Kr, Xe, and Rn, respectively.

The number of covalent bonds (shared electron pairs) that an atom can form can often be predicted from the number of electrons needed to fill its valence shell. Each atom of a Group IVA element has four electrons in its valence shell and can accept four more electrons to achieve a noble gas electron configuration. These four electrons can be gained by forming four covalent bonds, as illustrated for carbon in CCl_4 (carbon tetrachloride) and silicon in SiH_4 (silane).

$$\overset{..}{:\underset{..}{Cl}:}$$
$$:\overset{..}{\underset{..}{Cl}}:\overset{}{C}:\overset{..}{\underset{..}{Cl}}:\quad\text{or}\quad:\overset{..}{\underset{..}{Cl}}{-}\underset{\underset{..}{:\underset{..}{Cl}:}}{\overset{\overset{..}{:\underset{}{Cl}:}}{C}}{-}\overset{..}{\underset{..}{Cl}}:\qquad\underset{H}{\overset{H}{H:Si:H}}\quad\text{or}\quad\underset{H}{\overset{H}{H{-}Si{-}H}}$$

Carbon tetrachloride   Silane

Note that the Lewis structure of a molecule in general does not indicate its three-dimensional shape. As will be discussed in Chapter 7, carbon tetrachloride and silane are not flat molecules.

Nitrogen and the other elements of Group VA need only three electrons to achieve a noble gas configuration. These three electrons can be gained by formation of three single covalent bonds, as in NH_3 (ammonia). Oxygen and the other atoms in Group VIA need only two electrons to fill their valence shell; thus they can form two single covalent bonds. The elements in Group VIIA, such as fluorine, need to form only one single covalent bond in order to fill their valence shell.

$$\underset{\overset{|}{H}}{\overset{\overset{H}{|}}{:N{-}H}}\qquad\overset{\overset{H}{|}}{:\overset{..}{\underset{..}{O}}{-}H}\qquad H{-}\overset{..}{\underset{..}{F}}:$$

A pair of atoms may share more than one pair of electrons. Two or three pairs of electrons may be shared in order to give a noble gas configuration. For example, two pairs of electrons are shared between the carbon and oxygen atoms in CH_2O (formaldehyde) and between the two carbon atoms in C_2H_4 (ethylene), giving rise in both cases to a **double bond.**

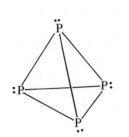

Figure 5-3. The Lewis structure of P_4.

When three electron pairs are shared by two atoms, as in CO (carbon monoxide) and N_2 (nitrogen molecule), a **triple bond** is formed.

Under normal conditions the ability to share two or three pairs of electrons and thereby to form double or triple bonds is limited almost exclusively to bonds between carbon, nitrogen, and oxygen atoms. For example, the element nitrogen forms the N_2 molecule, which contains a triple bond, whereas the element phosphorus (also in Group VA) forms the P_4 molecule, which contains only single bonds (Fig. 5-3). Sulfur and selenium sometimes form double bonds with carbon, nitrogen, and oxygen.

The molecules of CO and N_2 are said to be **isoelectronic;** they have the same arrangement of lone pairs and bonding pairs of electrons. Other examples of pairs of isoelectronic species are HCl and OH^-, and NH_3 and H_3O^+.

Note that both the ions OH^- and H_3O^+ are made up of more than one atom. Ions composed of several atoms are called **polyatomic ions** (see Section 5.11). All polyatomic ions are held together by covalent bonds. Compounds containing polyatomic ions are stabilized by both covalent bonds and ionic bonds. For example, potassium nitrate, KNO_3, contains the K^+ cation and the polyatomic NO_3^- anion. It has an ionic bond resulting from the electrostatic attraction between the ions K^+ and NO_3^- and covalent bonds between nitrogen and oxygen in NO_3^-.

5.4   Covalently Bonded Atoms Without a Noble Gas Configuration

Elements in the second row of the Periodic Table never form compounds in which they have more than eight electrons in their valence shell. However, larger atoms, in which the outermost electron shell is one where $n = 3$ or greater, can share more than four pairs of electrons with other atoms. For example, the phosphorus atom in PCl_5 shares five pairs of electrons (a total of 10 electrons). Sulfur shares six electron pairs (12 electrons) in the SF_6 molecule; and iodine shares seven electron pairs (14 electrons) in IF_7. In some molecules, such as IF_5 and XeF_4, the number of electrons in the outer shell of the central atom exceeds eight, even though some of its electron pairs are not shared. In each of these compounds with more than eight electrons in the valence shell of the central atom, the valence shell contains empty d orbitals. Although these d orbitals are empty in the free atom, they can be used to form covalent bonds.

$$\underset{\underset{\displaystyle Cl}{|}}{\overset{\overset{\displaystyle Cl}{|}}{Cl-P}}\diagdown_{Cl}^{Cl} \qquad \underset{\underset{\displaystyle F}{|}\,F}{\overset{F\,\overset{\displaystyle F}{|}}{F-S-F}} \qquad \underset{F}{\overset{F\,\,F}{F-I-F}} \qquad F-\underset{F}{\overset{\,\,\,F}{Xe}}-F$$

(The three lone pairs usually found on each Cl and F atom in these Lewis structures have been omitted for clarity.)

A very few molecules contain atoms that do not have the noble gas configuration since they have fewer than eight electrons in their valence shell. For example, boron, with three valence electrons, shares electron pairs with three chlorine atoms in the stable molecule BCl_3 (boron trichloride).

$$\overset{\cdot\cdot}{B}\,\cdot\ +\ 3:\overset{\cdot\cdot}{\underset{\cdot\cdot}{Cl}}\,\cdot\ \longrightarrow\ :\overset{\cdot\cdot}{\underset{\cdot\cdot}{Cl}}-B\diagup^{\overset{\cdot\cdot}{\underset{\cdot\cdot}{Cl}:}}_{\diagdown\overset{\cdot\cdot}{\underset{\cdot\cdot}{Cl}:}}$$

In this molecule each chlorine atom has an argon configuration. However, boron with only six electrons in its outer shell does not have a noble gas configuration, and one of its valence shell orbitals must be vacant.

An atom like the boron atom in BCl_3, which does not have a filled valence shell, is very reactive. It readily combines with a molecule containing an atom with an unshared pair of electrons. The lone pair is shared by both atoms. For example, NH_3 reacts with BCl_3 to form the compound BCl_3NH_3 because the lone pair of nitrogen is shared with the boron atom.

$$:\overset{\cdot\cdot}{\underset{\cdot\cdot}{Cl}}-B\diagup^{\overset{\cdot\cdot}{\underset{\cdot\cdot}{Cl}:}}_{\diagdown\overset{\cdot\cdot}{\underset{\cdot\cdot}{Cl}:}}\ +\ :\underset{\underset{\displaystyle H}{|}}{\overset{\overset{\displaystyle H}{|}}{N}}-H\ \longrightarrow\ :\overset{\cdot\cdot}{\underset{\cdot\cdot}{Cl}}-\underset{\underset{\displaystyle :Cl:}{|}}{\overset{\overset{\displaystyle :Cl:\ H}{|}}{B}}-\underset{\underset{\displaystyle H}{|}}{\overset{\overset{\displaystyle H}{|}}{N}}-H$$

A bond like the one between the boron and nitrogen atoms, formed when one of the atoms provides both bonding electrons, is called a **coordinate covalent,** or **dative, bond.** Coordinate covalent bonds can also occur within ions. This type of bond is formed when a water molecule combines with a hydrogen ion to form a hydronium ion and when an ammonia molecule combines with a hydrogen ion to form an ammonium ion.

$$H-\underset{\underset{\displaystyle H}{|}}{\overset{\cdot\cdot}{O}}:\ +\ H^+\ \longrightarrow\ \left[H-\underset{\underset{\displaystyle H}{|}}{\overset{\cdot\cdot}{O}}-H\right]^+ \qquad\qquad H-\underset{\underset{\displaystyle H}{|}}{\overset{\overset{\displaystyle H}{|}}{N}}:\ +\ H^+\ \longrightarrow\ \left[H-\underset{\underset{\displaystyle H}{|}}{\overset{\overset{\displaystyle H}{|}}{N}}-H\right]^+$$

| Water molecule | Hydronium ion, H_3O^+ | Ammonia molecule | Ammonium ion, NH_4^+ |

The formation of a coordinate covalent bond is only possible between an atom or ion with an unshared pair of electrons in its valence shell and an atom or ion that needs a pair of electrons to acquire a more stable electron configuration. The difference between the coordinate covalent bond and the normal covalent bond is in the mode of formation—that is, whether each atom contributes one electron or one atom contributes both. Once established, covalent bonds are indistinguishable since electrons are identical regardless of their source. All of the O—H bonds in H_3O^+ are the same, as are the N—H bonds in NH_4^+.

5.5 Electronegativity of the Elements

Electrons in a covalent bond between identical atoms are shared equally. However, the electron pair in a covalent bond between unlike atoms may not be shared equally; the electrons may spend more time closer to one atom than to the other. For example, a hydrogen atom is covalently bonded to a chlorine atom in a molecule of hydrogen chloride,

but the electrons forming the bond are not shared equally by the atoms, as they are in H_2 and Cl_2. A chlorine atom attracts electrons more strongly than a hydrogen atom, so the electrons of the shared pair are more closely associated with the chlorine atom. Thus the hydrogen atom develops a small positive charge (often referred to as a partial positive charge), and the chlorine atom develops a partial negative charge. This does not mean that the hydrogen atom has lost its electron; it means that the electrons of the shared pair spend more time *on the average* in the vicinity of the chlorine atom than they do near the hydrogen atom.

The extent of attraction of an atom for a shared pair of electrons is known as its electronegativity. **Electronegativity** is a measure of the attraction of an atom for electrons *in a covalent bond*. The larger the electronegativity of an element, the more strongly an atom of that element attracts electrons. The electronegativities of most of the elements are given in Table 5-2. These values are not absolute

Table 5-2 Approximate Electronegativities of Some of the Elements, Shown in a Periodic Table Arrangement

METALS												NONMETALS					
H 2.1																H 2.1	He ...
Li 1.0	Be 1.5	TRANSITION METALS										B 2.0	C 2.5	N 3.0	O 3.5	F 4.1	Ne ...
Na 1.0	Mg 1.2											Al 1.5	Si 1.7	P 2.1	S 2.4	Cl 2.8	Ar ...
K 0.9	Ca 1.0	Sc 1.2	Ti 1.3	V 1.4	Cr 1.6	Mn 1.6	Fe 1.6	Co 1.7	Ni 1.8	Cu 1.8	Zn 1.7	Ga 1.8	2.0	As 2.2	Se 2.5	Br 2.7	Kr ...
Rb 0.9	Sr 1.0	Y 1.1	Zr 1.2	Nb 1.2	Mo 1.3	Tc 1.4	Ru 1.4	Rh 1.4	Pd 1.4	Ag 1.4	Cd 1.5	In 1.5	Sn 1.7	Sb 1.8	Te 2.0	I 2.2	Xe ...
Cs 0.9	Ba 1.0	La-Lu 1.1-1.2	Hf 1.2	Ta 1.3	W 1.4	Re 1.5	Os 1.5	Ir 1.6	Pt 1.4	Au 1.4	Hg 1.4	Tl 1.4	Pb 1.6	Bi 1.7	Po 1.8	At 2.0	Rn ...
Fr 0.9	Ra 1.0	Ac-Lr 1.1-															

measures of electronegativity, but they do provide a measure of differences in electronegativity. For example, the difference in electronegativity between boron (2.0) and nitrogen (3.0) is the same as that between carbon (2.5) and oxygen (3.5). In addition, the electronegativity of a given element varies a little depending on the compound in which the element is found. Thus the values in Table 5-2 are average values.

In general, for the representative elements, electronegativity increases going across a period from left to right and going up a group. Thus the nonmetals, which lie toward the right in the Periodic Table, tend to have higher electronegativities than the metals, although there are some irregularities in the center of the table. This is another example of the periodic variation of properties among the elements.

Fluorine, the most chemically active nonmetal, has the highest electronegativity (4.1), and cesium and francium, the most chemically active metals, share the lowest electronegativity (0.9). Because the metals have relatively low electronegativities and tend to assume positive charges in compounds, they are often spoken of as being **electropositive;** conversely, nonmetals are said to be **electronegative.**

5.6 Writing Lewis Structures

When we write the Lewis structure for a molecule or ion that we can consider to be formed from individual atoms, we form single or multiple bonds by pairing up the unpaired electrons on the reacting atoms.

$$H\cdot + :\overset{\cdot\cdot}{\underset{\cdot\cdot}{Br}}\cdot \longrightarrow H:\overset{\cdot\cdot}{\underset{\cdot\cdot}{Br}}:$$

$$2H\cdot + :\overset{\cdot\cdot}{\underset{\cdot}{S}}\cdot \longrightarrow H:\overset{\cdot\cdot}{\underset{\overset{\textstyle |}{H}}{S}}:$$

$$\cdot\overset{\cdot\cdot}{\underset{\cdot}{N}}\cdot + \cdot\overset{\cdot\cdot}{N}\cdot \longrightarrow :N:::N:$$

Sometimes, however, a molecule or ion is more complicated, and then it is helpful to follow the general procedure outlined below. Let us determine the Lewis structures of $PO_2F_2^-$ and CO as examples in following this procedure.

STEP 1. Draw a skeleton structure of the molecule or ion showing the arrangement of atoms and connect each atom to another with a single (one electron pair) bond.

$$\begin{array}{c} F \\ | \\ F-P-O^- \\ | \\ O \end{array} \qquad C-O$$

When several arrangements of atoms are possible, as for $PO_2F_2^-$, we must use experimental evidence to choose the correct one. However, as a rule, the less electronegative element is often the central atom (as in $PO_2F_2^-$, SF_6, ClO_4^-, and PCl_5), with the exception that hydrogen is not found as a central atom.

STEP 2. Determine the total number of valence electrons in the molecule or ion.

A. For a *molecule* this is equal to the sum of the number of valence electrons on each atom.
B. For a *positive ion* this is equal to the sum of the number of valence electrons minus the number of positive charges on the ion (one electron is lost for each single positive charge).
C. For a *negative ion* this is equal to the sum of the number of valence electrons plus the number of negative charges on the ion.

$PO_2F_2^-$: Number of valence electrons =
$$5(P \text{ atom}) + 6(O \text{ atom}) + 6(O \text{ atom}) + 7(F \text{ atom})$$
$$+ 7(F \text{ atom}) + 1(\text{negative charge}) = 32$$

CO: Number of valence electrons = $4(C \text{ atom}) + 6(O \text{ atom}) = 10$

STEP 3. Deduct the two valence electrons that are used in each of the bonds written in Step 1. Distribute the remaining electrons as lone pairs, so that each atom (except hydrogen) has eight electrons if possible. If there are too few electrons to give each atom eight electrons, convert single bonds to multiple bonds, where possible. Remember, the ability to form multiple bonds is limited almost exclusively to bonds between carbon, nitrogen, and oxygen, although sulfur and selenium will sometimes form double bonds with carbon, nitrogen, and oxygen. Thus some elements may have fewer than eight electrons in their valence shells when they function as central atoms. For example, see the Lewis formula for BCl_3 in Section 5.4.

In the ion $PO_2F_2^-$ the total of 32 valence electrons can be distributed as 4 electron pairs (eight electrons) in bonds and 3 lone pairs around each of the fluorine and oxygen atoms. The 10 valence electrons in CO could be distributed as one bonding pair and 4 lone pairs, but this would leave at least one atom with fewer than 8 electrons. Filled valence shells about the carbon and oxygen atoms can only be obtained if each atom has 3 shared pairs in a triple bond and 1 lone pair.

STEP 4. In those molecules in which there are too many electrons to have only eight electrons around each atom, the central atom may have more than eight electrons in its valence shell (see, for example, the Lewis formulas of PCl_5, SF_6, and IF_5 in Section 5.4). However, note that the outer atoms contain a maximum of eight electrons in their valence shells.

Although heavier atoms can contain more than eight electrons, a Lewis formula with more than eight electrons for an atom of the first or second period is almost certainly incorrect and should be reexamined.

EXAMPLE 5.3 Jupiter's atmosphere contains small amounts of the compounds methane, CH_4, acetylene, HCCH, and phosphine, PH_3. What are the Lewis structures of these molecules?

Since hydrogen cannot be a central atom, the skeleton structure (Step 1) of CH_4 must be

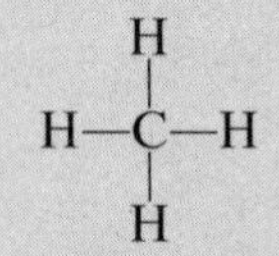

$$\begin{array}{c} H \\ | \\ H-C-H \\ | \\ H \end{array}$$

CH_4 contains eight valence electrons, four from the carbon atom and one from each of four hydrogen atoms (Step 2). All of the valence electrons are used in writing single bonds in the skeleton structure. Since none remain to be distributed, the Lewis structure is the skeleton structure.

The molecule HCCH has the skeleton structure

$$H-C-C-H$$

The molecule contains ten valence electrons, six of which are used in the single bonds in the skeleton structure leaving four electrons to be distributed. The attempts to distribute these electrons illustrated below do not result in formulas in which both carbon atoms have eight electrons each.

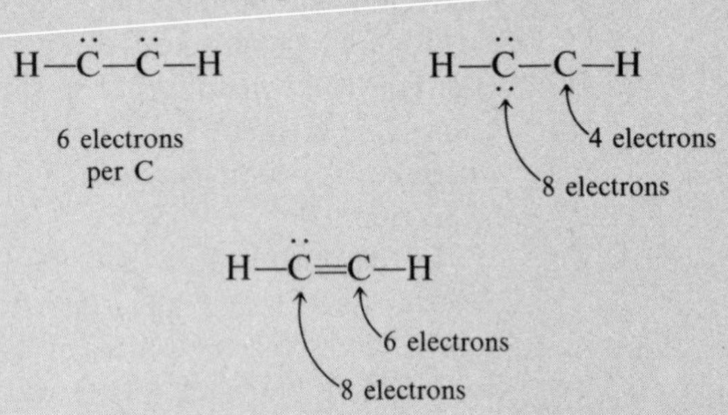

$$H-\overset{..}{\underset{}{C}}-\overset{..}{\underset{}{C}}-H \qquad\qquad H-\overset{..}{\underset{..}{C}}-C-H$$

6 electrons   4 electrons
per C   8 electrons

$$H-\overset{..}{C}=C-H$$

6 electrons
8 electrons

Only if the formula is written with a triple bond will both carbon atoms have eight electrons.

$$H-C\equiv C-H$$

The skeleton structure of PH_3 is

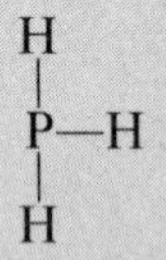

$$\begin{array}{c} H \\ | \\ P-H \\ | \\ H \end{array}$$

Six of the eight valence electrons in the molecule (five from the phosphorus atom, three from the three hydrogen atoms) are used to form single P—H bonds in the skeleton structure. The two remaining electrons must appear as a lone pair on the phosphorus atom in order to give it eight electrons.

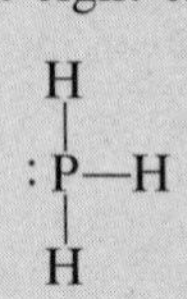

$$\begin{array}{c} H \\ | \\ :P-H \\ | \\ H \end{array}$$

EXAMPLE 5.4   Write the Lewis structure of the compound ClF_3.

Chlorine is the central atom in this molecule since it is the less electronegative element. The skeleton structure (Step 1) is

$$\begin{array}{c} F \\ | \\ Cl-F \\ | \\ F \end{array}$$

ClF_3 contains 28 valence electrons, 7 from the chlorine atom and 21 from the three fluorine atoms. Of these, 6 are used in the single bonds, leaving 22 to be placed as lone pairs. Three lone pairs can be placed on each fluorine atom leaving two pairs of electrons. Since the valence shell of chlorine has an n value greater than 2 ($n = 3$), it can hold the two remaining electron pairs. The Lewis structure is

$$\begin{array}{c} F \\ | \\ :\ddot{Cl}-F \\ | \\ F \end{array}$$

5.7 Resonance

Sometimes two or more Lewis structures can have the same arrangement of atoms but different arrangements of electrons. For example, both Lewis structures of SO_2, sulfur dioxide, have the atoms in the same positions, but some electrons in different ones.

A double bond between two given atoms is shorter than a single bond between the same two atoms. However, experiments show that both sulfur-oxygen bonds in SO_2 have the same length and are identical in all other properties. Since it is not possible to write a single Lewis structure for sulfur dioxide in which both bonds are equivalent, we must apply the concept of **resonance.** If two or more Lewis structures with the same arrangement of atoms can be written for a molecule or ion, the actual distribution of electrons is an *average* of that shown by the various Lewis structures. The actual distribution of electrons in each of the sulfur-oxygen bonds in SO_2 is an average of that shown in the two Lewis structures, that is, an average of a double bond and a single bond. The individual Lewis structures are called **resonance forms.** The actual electronic structure of the molecule (the average of the resonance forms) is called a **resonance hybrid** of the individual resonance forms. A double-headed arrow is used between Lewis structures to indicate that they are resonance forms and that *the true distribution of electrons is an average of the individual resonance forms.*

The distribution of electrons in N_2O, nitrous oxide, may be represented by two resonance forms as follows:

$$:N{=}N{=}\ddot{O}: \longleftrightarrow :N{\equiv}N{-}\ddot{O}:$$

Since the electron distribution in N_2O is an average of these two resonance forms, the distribution of electrons in the bond between the two nitrogen atoms may be seen to be greater than that in an ordinary double bond but less than that in an ordinary triple bond. The distribution of electrons in the bond between nitrogen and oxygen is between that in a single bond and that in a double bond.

It must be emphasized that a molecule described as a resonance hybrid *never* possesses an electronic structure described by a single resonance form. The actual electronic structure is *always* an average of that shown by all resonance forms. A simple analogy to a resonance hybrid is the mule, which is the hybrid offspring of a male donkey and a female horse. Just as the characteristics of a mule are fixed, so the properties of a resonance hybrid are fixed. The mule is not a donkey part of the time and a horse part of the time. It is always a mule. Correspondingly, a molecule with an electronic structure described by a resonance hybrid does not exhibit one electronic structure part of the time and another electronic structure the rest of the time. Instead, the material is always in the form of the intermediate resonance hybrid, which cannot be written with a single Lewis structure.

5.8   Polar Covalent Bonds and Polar Molecules

A bond between a pair of atoms is called a **polar covalent bond** if its center of positive charge does not coincide with its center of negative charge. Thus a polar bond has a positively charged end and a negatively charged end. When a covalent bond is formed between atoms of different electronegativities, the shared pair of electrons stays closer to the more electronegative atom, and the resulting bond is polar. As mentioned previously, the chlorine atom in the hydrogen chloride molecule attracts the pair of electrons of the covalent bond more strongly than does the hydrogen atom (Section 5.5). The hydrogen-chlorine bond is polar, with the chlorine atom becoming somewhat negative and the hydrogen atom becoming somewhat positive as the bond is formed.

The greater the difference between the electronegativities of the two atoms involved in a bond, the greater is the polarity of the bond. Thus, the polarity of the bond in the hydrogen halides increases in the order $HI < HBr < HCl < HF$, which corresponds to the increase in electronegativity of the halogens: I (2.2) $<$ Br (2.7) $<$ Cl (2.8) $<$ F (4.1). If the difference in electronegativity between two atoms is sufficiently large, the shared electron pair will spend all of its time on the more electronegative atom, resulting in ionic bonding, rather than covalent bonding. This is the case with salts such as LiF, NaCl, K_2O, Li_3N, and CsBr, where the electronegativity difference is greater than about 2. The other extreme is achieved when identical atoms share a pair of electrons, as in H—H, where the bond is covalent with no polarity. Bonds between atoms with small differences in electronegativities are primarily covalent in character. This is the case with CO_2, CCl_4, I_2O_5, NI_3, ICl, and NO. It follows that bonds that are primarily ionic are formed when elements at the extreme left of the Periodic Table react with elements at the extreme right, whereas bonds that are primarily covalent are formed when elements close together in the table react with one another.

There is no sharp dividing line between compounds in which the bonding is covalent and those in which the bonding is ionic. In intermediate cases the molecules have bonds that possess some of the nature of both covalent and ionic bonds, or **covalent bonds with partial ionic character.**

Polar molecules randomly oriented in the absence of an electric field

Positive plate   Negative plate

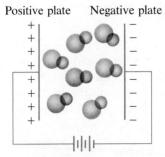

Polar molecules tending to line up in an electric field

Figure 5-4. Polar molecules, such as hydrogen chloride, tend to align in an electric field, with the positive ends oriented toward the negative plate and the negative ends toward the positive plate.

If the centers of positive and negative charge in a molecule do not coincide, the molecule has a positive end and a negative end. Such a molecule is said to be **polar** and to possess a **dipole.** Polar molecules contain polar covalent bonds. In a molecule of hydrogen chloride, for example, the centers of positive and negative charge do not coincide and the molecule has a positively charged end, the hydrogen atom, and a negatively charged end, the chlorine atom. Polar molecules tend to align when placed in an electric field, with the positive end of the molecule oriented toward the negative plate and the negative end toward the positive plate (Fig. 5-4).

Polar covalent bonds can exist in nonpolar molecules. If a molecule contains several polar covalent bonds directed in such a way as to give a symmetrical molecule, the molecule is nonpolar. This is illustrated by $HgCl_2$, in which each of the covalent bonds is polar but the molecule as a whole is nonpolar. Each chlorine atom is negative with respect to the positive mercury atom, so each mercury-chlorine bond is polar. However, these bond polarities cancel each other because they point in opposite directions, giving an electrically symmetrical molecule.

$$Cl^-—^+Hg^+—^-Cl$$

5.9 Oxidation Numbers

The oxidation number of an atom, sometimes called the oxidation state, is a useful bookkeeping concept. Oxidation numbers are of value in writing chemical formulas, naming compounds, and keeping track of the redistribution of electrons during chemical reactions. The **oxidation number** can be determined by application of the following rules:

1. The oxidation number of an atom of any element in its elemental form is considered to be zero, no matter what the complexity of the molecule in which the element is found. The atoms of Na, N_2, P_4, and S_8 all have oxidation numbers of zero.

2. The oxidation number of a monatomic ion is considered to be equal to the charge on the ion. The ions Li^+, Co^{2+}, Cl^-, and N^{3-} have oxidation numbers of $+1$, $+2$, -1, and -3, respectively.

3. The oxidation number of fluorine in a compound is always -1.

4. The elements of Group IA (except hydrogen) in compounds are assigned an oxidation number of $+1$.

5. The elements of Group IIA in compounds are assigned an oxidation number of $+2$.

6. The elements of Group VIIA are assigned an oxidation number of -1 when they are combined in compounds with less electronegative elements. Chlorine has an oxidation number of -1 in NaCl, CCl_4, PCl_3, and HCl, for example.

7. Oxygen is assigned an oxidation number of -2 with three exceptions:

   a. In compounds with fluorine it has a positive oxidation number (fluorine is more electronegative than oxygen). For example, oxygen has an oxidation number of $+2$ in oxygen difluoride, OF_2.
   b. In peroxides (compounds that contain an O—O single bond) it has an oxidation number of -1. An example is hydrogen peroxide, H_2O_2.
   c. In superoxides (compounds that contain O_2^-) it has an oxidation number of $-\frac{1}{2}$. An example is potassium superoxide, KO_2.

8. Hydrogen is assigned an oxidation number of -1 in compounds with less electronegative elements and an oxidation number of $+1$ in compounds with more electronegative elements.

9. The sum of the oxidation numbers of all atoms in a neutral compound is zero. The sum of the oxidation numbers of all atoms in an ion is equal to the charge on the ion.

If the oxidation numbers are known for all but one kind of atom in a compound, its oxidation number can be calculated.

EXAMPLE 5.5   Calculate the oxidation number of sulfur in the compounds Na_2SO_4, Na_2SO_3, and H_2S.

The oxidation number for sulfur in Na_2SO_4 can be calculated from the oxidation numbers for sodium and oxygen. The two sodium atoms, each with an oxidation number of $+1$ (Group IA, rule 4), total $+2$; the four oxygen atoms, each with an oxidation number of -2 (rule 7), total -8. For the sum of the oxidation numbers to be zero sulfur must have an oxidation number of $+6$. A similar calculation shows the oxidation number of sulfur in Na_2SO_3 to be $+4$.

In H_2S hydrogen is bonded to a more electronegative element and so has an oxidation number of $+1$ (rule 8). For the sum of the oxidation numbers of all three atoms to be zero, sulfur must have the oxidation number -2.

As demonstrated in this example, an element may have more than one oxidation number, depending on the compound in which it is found. Other examples include iron, with an oxidation number of $+2$ in $FeCl_2$ and $+3$ in $FeCl_3$, and tin, with oxidation numbers of $+2$ and $+4$ in $SnCl_2$ and $SnCl_4$, respectively. The oxidation number of chlorine in each of these examples is -1. Chlorine, however, exhibits an oxidation number of $+1$ in $NaClO$, $+3$ in $NaClO_2$, $+5$ in $NaClO_3$, and $+7$ in $NaClO_4$.

It should be emphasized that, although oxidation numbers are convenient when writing formulas and balancing oxidation-reduction equations, they are quite arbitrary. For some compounds, calculating the oxidation number of a member atom gives a fraction.

EXAMPLE 5.6   Calculate the oxidation number of Fe in the mineral lodestone, a magnetic oxide of iron with the formula Fe_3O_4.

In Fe_3O_4 oxygen is combined with a less electronegative element and is assigned an oxidation number of -2. The four oxygen atoms contribute a total of -8. For the sum of the oxidation numbers in the compound to be zero, the oxidation numbers of the three iron atoms must also total $+8$. Hence the oxidation number for iron in lodestone is $+\frac{8}{3}$ or $+2\frac{2}{3}$. (The oxidation number always refers to one atom of an element.)

5.10   Applying Oxidation Numbers to Writing Formulas

If we know the oxidation numbers of the elements in a compound, we can write the formula of the compound. The principal oxidation number of an element is, in general, predictable from the position of the element in the Periodic Table and from a knowledge of the electronic structure of the atom. The formula of a compound may be written using the oxidation numbers of the atoms involved because the sum of the oxidation numbers of the atoms in the compound must equal zero.

EXAMPLE 5.7   Write the formula for magnesium chloride using the oxidation numbers of its constituent elements.

Magnesium is a member of Group IIA and so has an oxidation number of +2. Chlorine, a member of Group VIIA, is combined with a less electronegative element and so has an oxidation number of −1.

The formula of magnesium chloride cannot be MgCl, because +2 and −1 do not add up to 0. For the total of the oxidation numbers for the compound to be zero, the atoms must be in a ratio of one magnesium ion to two chloride ions, or $MgCl_2$.

EXAMPLE 5.8   Write the formula for aluminum oxide.

Aluminum, the less electronegative element, is a member of Group IIIA and will lose three electrons, giving Al^{3+} with an oxidation number of +3. Oxygen, a member of Group VIA, needs two electrons to fill its valence shell, giving the ion O^{2-} with an oxidation number of −2. Since +3 and −2 do not add up to 0, AlO is not the correct formula for aluminum oxide. By inspection it is readily seen that two atoms of aluminum will give a total of six positive units of oxidation number, that three atoms of oxygen will give six negative units of oxidation number, and that the algebraic sum of the oxidation numbers will thus be zero. The correct simplest formula for aluminum oxide is, therefore, Al_2O_3.

5.11   Polyatomic Ions

Polyatomic ions contain more than one atom (Section 5.3). Several examples, selected from many such ions, are given in Table 5-3.

In polyatomic ions the sum of the oxidation numbers of the constituent atoms equals the charge on the ion. Hence for the OH^- ion the oxidation numbers of −2 for oxygen and +1 for hydrogen add up to the overall charge of −1.

When writing formulas of compounds that include more than one polyatomic ion, we enclose the formula of the ion in parentheses and indicate the number of such ions with a subscript. Examples include $(NH_4)_2CO_3$ and $Al_2(SO_4)_3$. In $(NH_4)_2CO_3$, two ammonium ions, each with a +1 ionic charge, balance the −2 ionic charge of the carbonate ion. In $Al_2(SO_4)_3$, two aluminum ions, each with a charge of +3, and three sulfate ions, each with a charge of −2, are present and

Table 5-3   Some Common Polyatomic Ions

Ammonium	NH_4^+	Carbonate	CO_3^{2-}	Phosphate	PO_4^{3-}
Acetate	$CH_3CO_2^-$	Sulfate	SO_4^{2-}	Diphosphate	$P_2O_7^{4-}$
Nitrate	NO_3^-	Sulfite	SO_3^{2-}	Arsenate	AsO_4^{3-}
Nitrite	NO_2^-	Thiosulfate	$S_2O_3^{2-}$	Arsenite	AsO_3^{3-}
Hydroxide	OH^-	Peroxide	O_2^{2-}		
Hypochlorite	ClO^-	Chromate	CrO_4^{2-}		
Chlorite	ClO_2^-	Dichromate	$Cr_2O_7^{2-}$		
Chlorate	ClO_3^-	Silicate	SiO_3^{2-}		
Perchlorate	ClO_4^-				
Permanganate	MnO_4^-				

the charges are balanced. It should be noted that the sum of the oxidation numbers of the various atoms, as well as the sum of the total charges on the ions, equals zero for each compound.

The following examples illustrate the process of writing formulas for compounds containing polyatomic ions using ionic charges rather than oxidation numbers.

EXAMPLE 5.9   Write the formula for iron perchlorate, given that the oxidation number for iron in the compound is +3.

The oxidation number of a monoatomic ion is equal to the charge on the ion. Thus the iron ion here is Fe^{3+}. With charges of +3 for the iron ion and −1 for the perchlorate ion (see Table 5-3), we can see that the formula $FeClO_4$ is incorrect. By combining three perchlorate ions and one iron ion, we can make the sum of the positive and negative charges be zero; the correct formula is $Fe(ClO_4)_3$.

EXAMPLE 5.10   Write the formula for calcium phosphate, which is one component of a phosphate rock used in fertilizer production.

Since the charge for the calcium ion is +2 (Ca is a member of Group IIA) and the charge on the phosphate ion is −3, the formula cannot be $CaPO_4$. The sum of the charges on the ions must be zero for the compound. By including three calcium ions and two phosphate ions, the sum becomes zero. Thus $Ca_3(PO_4)_2$ is the correct formula for calcium phosphate.

5.12   The Names of Compounds

1. BINARY COMPOUNDS.   **Binary compounds** contain two different elements. The name of a binary compound consists of the name of the more electropositive element followed by the name of the more electronegative element with its ending replaced by the suffix -ide. Some examples are as follows:

NaCl, sodium chloride

KBr, potassium bromide

CaI_2, calcium iodide

AgF, silver fluoride

HCl, hydrogen chloride

LiH, lithium hydride

Na_2O, sodium oxide

CdS, cadmium sulfide

Mg_3N_2, magnesium nitride

Ca_3P_2, calcium phosphide

Al_4C_3, aluminum carbide

Mg_2Si, magnesium silicide

A few polyatomic ions have special names and are treated as if they were single atoms in naming their compounds. Thus NaOH is called sodium hydroxide; HCN, hydrogen cyanide; and NH_4Cl, ammonium chloride.

If a binary hydrogen compound is an acid (Section 8.1) when it is dissolved in water, the prefix *hydro-* is used instead of the word *hydrogen* and the suffix *-ic* replaces the suffix *-ide* when we are referring to the aqueous solution.

HCl(*g*), hydrogen chloride

HBr(*g*), hydrogen bromide

H_2S(*g*), hydrogen sulfide

HCN(*g*), hydrogen cyanide

HCl(*aq*), hydrochloric acid

HBr (*aq*), hydrobromic acid

H_2S(*aq*), hydrosulfuric acid

HCN(*aq*), hydrocyanic acid

If a nonmetallic element with variable oxidation number can unite with another nonmetallic element to form more than one compound, the compounds can be distinguished by the Greek prefixes *mono-* (meaning one), *di-* (two), *tri-* (three), *tetra-* (four), *penta-* (five), *hexa-* (six), *hepta-* (seven), and *octa-* (eight). The prefixes precede the name of the constituent to which they refer.

CO, carbon monoxide

CO_2, carbon dioxide

NO_2, nitrogen dioxide

N_2O_4, dinitrogen tetraoxide

N_2O_5, dinitrogen pentaoxide

PbO, lead monoxide

PbO_2, lead dioxide

SO_2, sulfur dioxide

SO_3, sulfur trioxide

BCl_3, boron trichloride

A second method of naming different binary compounds that contain the same elements uses Roman numerals placed in parentheses after the name of the more electropositive element to indicate its oxidation number. This method of naming binary compounds is usually applied to those in which the electropositive element is a metal, but is occasionally applied to other compounds as well.

$FeCl_2$, iron(II) chloride

$FeCl_3$, iron(III) chloride

Hg_2O, mercury(I) oxide

HgO, mercury(II) oxide

SO_2, sulfur(IV) oxide

SO_3, sulfur(VI) oxide

NO, nitrogen(II) oxide

NO_2, nitrogen(IV) oxide

The system of nomenclature used in this book was formulated by a committee of the International Union of Pure and Applied Chemistry. However, you should also become familiar with another, older system, since you may encounter it frequently elsewhere. According to the older system, when two elements form more than one compound with each other and when both elements are nonmetals, the distinction is made by indicating only the number of atoms of the more electronegative element by a Greek prefix. NO_2, N_2O_3, N_2O_4, and N_2O_5 are nitrogen dioxide, nitrogen trioxide, nitrogen tetraoxide, and nitrogen pentaoxide, respectively. When the more electropositive element is a metal, the lower oxidation number of the metal is indicated by using the suffix -ous on the name of the metal. The higher oxidation number is designated by the suffix -ic. Sometimes the stem of the Latin name for the metal is used. Thus $FeCl_2$ is ferrous chloride, and $FeCl_3$ is ferric chloride; Hg_2O is mercurous oxide, and HgO is mercuric oxide.

2. TERNARY COMPOUNDS.

Ternary compounds contain three different elements. A few ternary compounds, such as NH_4Cl, KOH, and HCN, are named as if they were binary compounds. Chlorine, nitrogen, sulfur, phosphorus, and several other elements form oxyacids (ternary compounds with hydrogen and oxygen) that usually differ from each other in oxygen content. The most common acid of a series usually bears the name of the acid-forming element, ending with the suffix -ic, as in chloric acid ($HClO_3$), sulfuric acid (H_2SO_4), nitric acid (HNO_3), and phosphoric acid (H_3PO_4). If the "central" element (Cl, S, etc.) of a related acid has a higher oxidation number than it does in the most common form, the suffix -ic is retained and the prefix per- is added. The name perchloric acid for $HClO_4$ illustrates this rule. If the "central" element has a lower oxidation number than it does in the most common acid, the suffix -ic is replaced with the suffix -ous. Examples are chlorous acid ($HClO_2$), sulfurous acid (H_2SO_3), nitrous acid (HNO_2), and phosphorous acid (H_3PO_3). If the same central element in two acids has lower oxidation numbers than it does in its -ic acid, the acid having the lower of the two oxidation numbers is named by adding the prefix hypo- and retaining the ending -ous. Thus $HClO$ is hypochlorous acid, and H_3PO_2 is hypophosphorous acid.

Metal salts of the oxyacids (compounds in which a metal replaces the hydrogen of the acid) are named by identifying the metal and then the negative acid ion. For these salts the ending -ic of the oxyacid name is changed to -ate, and the ending -ous is changed to -ite. The salts of perchloric acid are perchlorates, those of sulfuric acid are sulfates, those of nitrous acid are nitrites, and those of hypophosphorous acid are hypophosphites. This system of naming applies to all inorganic oxyacids and their salts. The names of the oxyacids for chlorine and their corresponding sodium salts are given in Table 5-4.

Table 5-4   Names of Oxyacids of Chlorine and the Corresponding Sodium Salts

Cl Oxidation Number	Acids	Salts
+1	HClO, hypochlorous acid	NaClO, sodium hypochlorite
+3	$HClO_2$, chlorous acid	$NaClO_2$, sodium chlorite
+5	$HClO_3$, chloric acid	$NaClO_3$, sodium chlorate
+7	$HClO_4$, perchloric acid	$NaClO_4$, sodium perchlorate

Oxyacids that contain more than one replaceable hydrogen atom can form anions that contain hydrogen; H_2SO_4 gives HSO_4^- (hydrogen sulfate), and H_3PO_4 gives $H_2PO_4^-$ (dihydrogen phosphate) and HPO_4^{2-} (hydrogen phosphate). Metal salts of these ions are named like other salts of oxyacids; KH_2PO_4, for example, is potassium dihydrogen phosphate.

The system of nomenclature for a class of compounds known as coordination compounds will be described in Chapter 29.

FOR REVIEW

SUMMARY

When atoms react, their electronic structures change, giving chemical bonds. An **ionic bond** results when electrons are transferred from one atom to another. The resulting compound consists of regular three-dimensional arrangements of ions held together by the strong electrostatic attractions between ions of opposite charge. The charges on ions of the representative metals may be determined readily since, with few exceptions, these ions have a **noble gas electron configuration** or a **pseudo noble gas electron configuration.**

Covalent bonds result when electrons are shared as electron pairs between atoms, with the electrons attracted by the nuclei of both atoms. When only one atom furnishes both electrons of an electron-pair bond, the bond is called a **coordinate covalent bond.** A **single bond** results when one pair of electrons is shared between two atoms. The sharing of two or three pairs of electrons between two atoms constitutes a **double bond** or **triple bond,** respectively. The distribution of electrons in bonds and **lone pairs** in a molecule can be indicated with a **Lewis structure.** Most Lewis structures can be written by inspection or by starting with a skeleton structure consisting of the constituent atoms connected by single bonds and then distributing the remaining electrons as lone pairs or in multiple bonds to give all atoms filled valence shells, if possible. In some instances a central atom may have an empty valence orbital, or if it is an atom for which n is greater than or equal to 3 for the valence shell, it may have more than eight electrons in its valence shell.

If two or more Lewis structures with an identical arrangement of atoms but a different distribution of electrons can be written, the actual distribution of electrons in the molecule cannot be described by a single Lewis structure but is an average of the distributions indicated by the individual Lewis structures. The actual electron distribution is called a **resonance hybrid** of the individual Lewis structures (called **resonance forms**).

Even though pairs of electrons are often shared between different kinds of atoms, they may not be shared equally. The electrons of a shared pair may spend more time near one atom than near the other, giving a **polar covalent bond.** The measure of the ability of an atom to attract a pair of electrons is called its **electronegativity.** The greater the difference in the electronegativities of two atoms in a covalent bond, the more polar the bond. Some molecules that contain polar bonds also possess a **dipole;** that is, these molecules have a positive end and a negative end.

Oxidation numbers are assigned to atoms in molecules using a set of rules, as described in this chapter. The electronegativities of the elements figure prominently in these rules. The chemical formula of a compound may be written using the oxidation numbers of the atoms in the compound. A knowledge of the chemical formula is necessary to name a compound. Compounds and ions may be named systematically, using accepted nomenclature.

KEY TERMS AND CONCEPTS

Anion (5.1)
Binary compounds (5.12)
Cation (5.1)
Coordinate covalent bond (5.4)
Covalent bond (5.3)
Dipole (5.8)
Double bond (5.3)
Electronegativity (5.5)

Ionic bonds (5.1)
Ionic compounds (5.1)
Ions (5.1)
Isoelectronic (5.3)
Lewis structure (5.3)
Lewis symbols (5.1)
Lone pairs (5.3)
Oxidation number (5.9)

Polar covalent bond (5.8)
Polyatomic ions (5.3)
Resonance forms (5.7)
Resonance hybrid (5.7)
Single bond (5.3)
Ternary compounds (5.12)
Triple bond (5.3)

EXERCISES

Ions and Ionic Bonding

[S] 1. Why does a cation have a positive charge?
[S] 2. Why does an anion have a negative charge?
3. From the following list of atoms, select the ones you would expect to form positive ions and the ones you would expect to form negative ions: Al, Ba, Cl, Co, K, La, N, Ni, O, P, Se.
[S] 4. Predict the charge on the monoatomic ions formed from the following elements in ionic compounds:

   (a) Ag   (e) K   (h) O
   (b) Al   (f) Zn   (i) P
   (c) Ba   (g) N   (j) Se
   (d) Br

5. Identify the electronic structure of each of the following ions as (1) a noble gas configuration, (2) a pseudo noble gas configuration, or (3) neither:

   (a) As^{3-} (d) Cu^+ (g) Pb^{2+}
   (b) Ba^{2+} (e) Hg^{2+} (h) S^{2-}
   (c) Co^{2+} (f) Mn^{3+} (i) Sm^{3+}

6. The elements that form the greatest concentration of monoatomic ions in seawater are Cl, Na, Mg, Ca, K, Br, Sr, and F. Write the Lewis symbols for the ions formed from these elements.
7. Write the electron configuration and the Lewis symbol for each of the following atoms and for

the monoatomic ion found in ionic compounds containing the element:

   (a) Ba   (c) Ga   (e) N
   (b) F   (d) Li   (f) Se

[S] 8. Write the formulas of each of the following ionic compounds using Lewis symbols:

   (a) LiF   (d) AlF_3
   (b) Na_2S (e) Ga_2O_3
   (c) $CaBr_2$

9. In the Lewis structures listed below, M and X represent elements in the third period of the Periodic Table. Write the formula of each compound using the chemical symbols of the elements.

   (a) $(M^{2+})(:\ddot{X}:^-)_2$ (c) $(M^{3+})_2(:\ddot{X}:^{2-})_3$

   (b) $(M^+)_2(:\ddot{X}:^-)$ (d) $(M^+)_3(:\ddot{X}:^{3-})$

10. Why is it incorrect to speak of NaCl molecules in a sample of solid sodium chloride?

Covalent Bonding

11. What are the characteristics of a covalent bond that make it different from an ionic bond?
[S] 12. What are the characteristics of two atoms that will form a covalent bond?

13. When is formation of a covalent bond likely? When is a compound unlikely to contain a covalent bond?

[S] 14. Predict which of the following compounds are ionic and which are covalent, using the locations of their constituent elements in the Periodic Table:

(a) Li_2O	(e) CH_4	(i) $CaBr_2$
(b) CCl_4	(f) SCl_2	(j) K_2S
(c) CuO	(g) CS_2	(k) MnS
(d) H_2CO	(h) BrCl	(l) Al_2O_3

[S] 15. How do single, double, and triple bonds differ? How are they similar?

16. Write Lewis structures for the following:

(a) BrCl	(d) NH_4^+	(f) BF_4^-
(b) PF_3	(e) SiF_4	(g) H_3O^+
(c) SCl_2		

[S] 17. Write Lewis structures for the following:

(a) O_2	(c) Cl_2CO	(e) H_2CO
(b) CS_2	(d) ClNO	

18. Write Lewis structures for the following:

(a) CN^-	(c) HCCH
(b) ClCN	(d) NO^+

19. A molecule of sulfur consists of a ring of eight sulfur atoms, each of which is bonded only to the sulfur atoms on either side of it. Write the Lewis structure for a sulfur molecule.

20. Methanol, H_3COH, is used as a fuel in many race cars. What is the Lewis structure of methanol?

21. The skeleton structures of a number of biologically active molecules are given below. Complete the Lewis structures of these molecules.

(a) the amino acid glycine

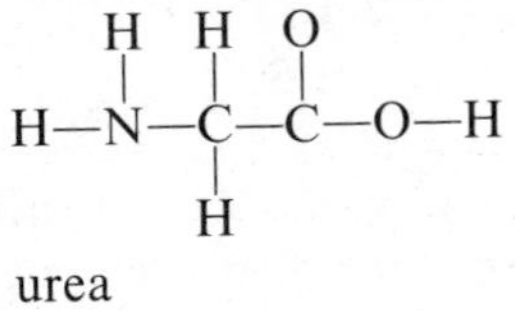

(b) urea

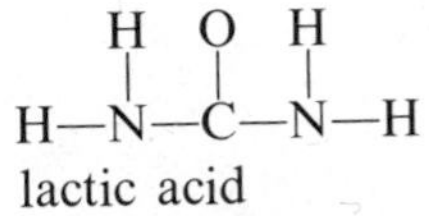

(c) lactic acid

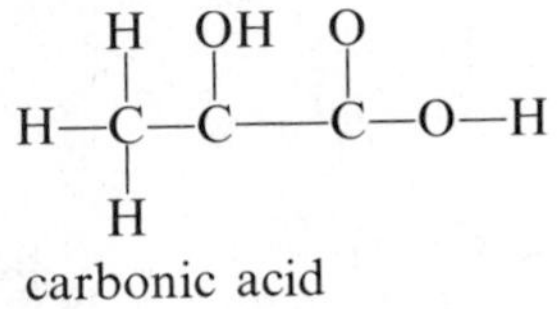

(d) carbonic acid

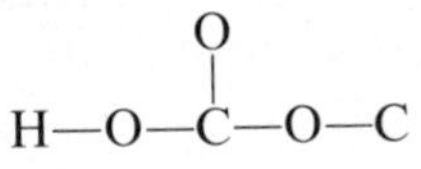

(e) histidine

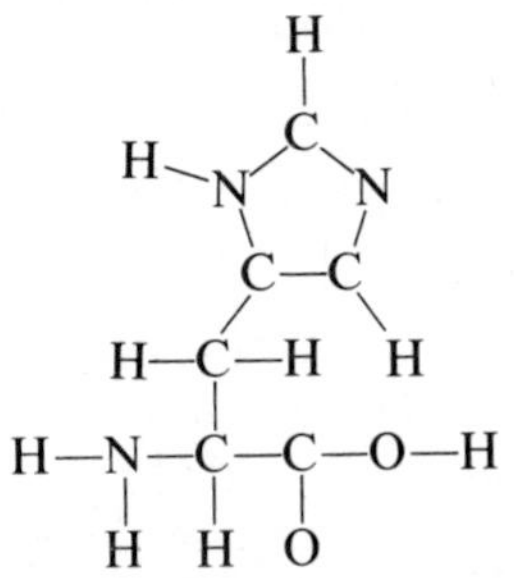

22. Write the Lewis structure for sulfuric acid, H_2SO_4, which has two oxygen atoms and two hydroxide groups bonded to sulfur.

23. The atmosphere of Titan, the largest moon of Saturn, contains methane, CH_4, and traces of ethylene C_2H_4, ethane, C_2H_6, hydrogen cyanide, HCN, propyne, H_3CCCH, and diacetylene, HCCCCH. Write the Lewis structures of these molecules.

24. The following species do not have a noble gas electron configuration. Write the Lewis structure for each.

(a) BH_3
(b) ClF_3
(c) PF_5
(d) ICl_4^-
(e) XeF_2
(f) S_2F_{10} (contains a sulfur-sulfur bond)

Resonance

25. Which of the following pairs represent resonance forms of the same species?

(a) 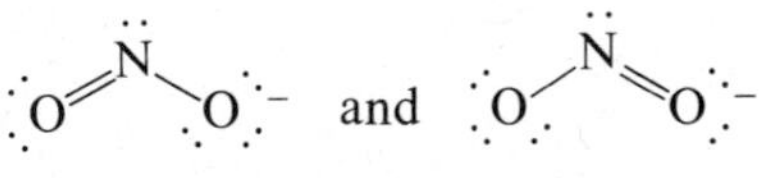

(b) 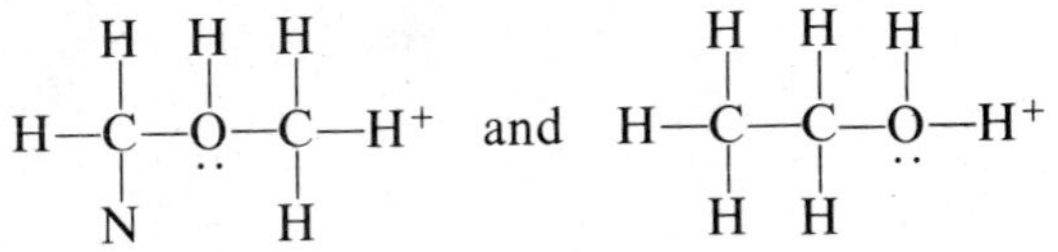

(c) 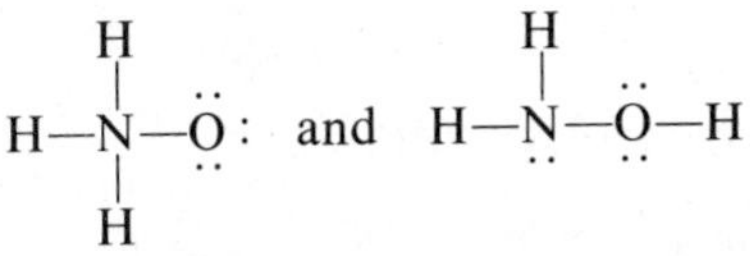

(d) 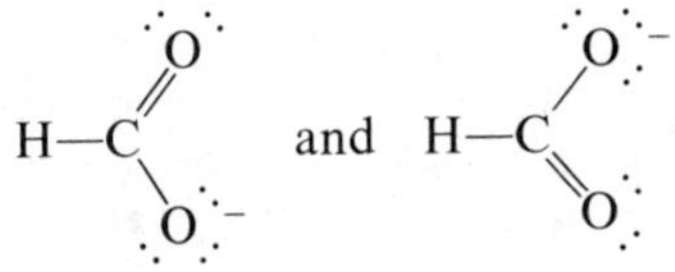

26. Draw the resonance forms for the following:
(a) selenium dioxide, OSeO
(b) the nitrite ion, NO_2^-
[S](c) the nitrate ion, NO_3^-
[S](d) the acetate ion, $H_3CCO_2^-$.

27. Ozone, O_3, the component of the atmospheric layer that protects the earth from ultraviolet radiation, is a molecule whose electronic structure is described as a resonance hybrid. Write resonance forms that describe the resonance hybrid.

28. Explain why nitric acid, $HONO_2$, contains two distinct types of nitrogen-oxygen bonds instead of three types. (*Hint*: Consider the resonance hybrid of nitric acid.)

29. Write the resonance forms for the carbonate ion, CO_3^{2-}, and show that this ion is isoelectronic with the nitrate ion, NO_3^-.

30. Sodium oxalate, $Na_2C_2O_4$, contains the oxalate anion. Write the resonance forms of this anion, which contains a carbon-carbon bond (O_2CCO_2).

31. The skeleton structure of benzene, C_6H_6, is given below. Write the resonance forms for benzene.

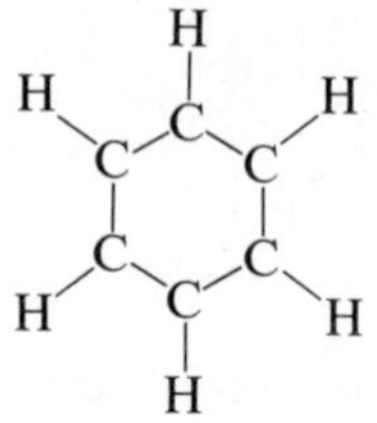

32. Bromstyrol, C_8H_7Br, has a strong hyacinth-like odor and is used in lilac and hyacinth scents. Write the resonance forms of bromstyrol, the skeleton structure of which is as follows:

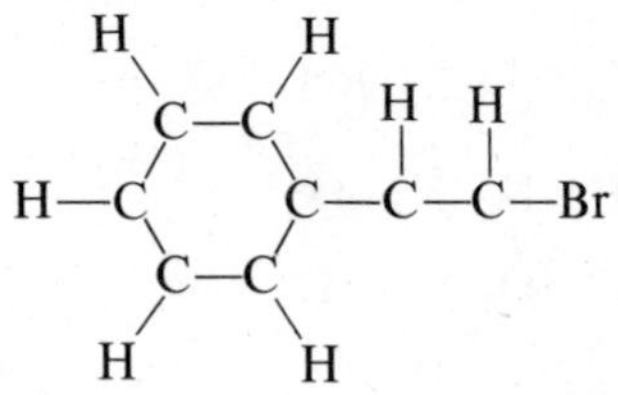

Electronegativity and Polar Bonds

33. What is meant by the electronegativity of an element?

34. How is the polarity of a covalent bond related to the electronegativity difference of the atoms in the bond?

35. How is the polarity of a covalent bond related to the polarity of a molecule containing the bond?

[S]36. Many molecules that contain polar bonds are nonpolar. How can this happen?

37. Give examples of polar covalent molecules and nonpolar covalent molecules that contain polar covalent bonds.

[S]38. Which of the following molecules and ions contain polar bonds? F_2, P_4, SO_2, NO_3^-, N_2O, H_2S, NH_4^+, HCCH

39. Which of the following molecules and ions are polar? NO_2^- (a bent ion), CN^-, NNN^- (a linear ion), BrCN (linear), H_2S (a bent molecule), Br_2, BrCl

40. One of the most important polar molecules is water, H_2O, a bent molecule with an angle of 105° between the two O—H bonds. Which end of an O—H bond in water is positive, and which is negative? Sketch the H_2O molecule and indicate the positive and negative ends.

Oxidation Numbers

41. What are the oxidation numbers of the following elements when found in compounds?
(a) F (c) Cs (e) K
(b) Ca (d) Li (f) Sr

42. Chlorine forms a binary compound with each of the following elements: Ca, Cr, F, N, Ni, O, S, and Zn. In which of these compounds will chlorine have a positive oxidation number?

43. What are the oxidation numbers of hydrogen in the binary compounds it forms with
(a) Al (d) N (f) O
(b) Cl (e) Na (g) S
(c) Mg

44. Determine the oxidation number of each of the following elements in the given compound:
[S](a) B in Na_3BO_3
[S](b) C in CBr_4
(c) Cl in $KClO_4$
(d) P in Li_3PO_4
[S](e) Co in $K_3[CoF_6]$
(f) N in HNO_3 ($HONO_2$)
(g) S in SO_2Cl_2
[S](h) P in H_2PO_3F [$(HO)_2POF$]

[S]45. Determine the oxidation number of N in each of the following compounds: Na_3N, NH_4Cl, N_2O, N_2H_4, KNO_2, $Ca(NO_3)_2$.

46. Determine the oxidation number of Mn in each of the following compounds: $LiMnO_2$, K_2MnCl_4, $ZnMn_2O_4$, Mn_2O_7, $K_5Mn(CN)_6$ (CN is an ion with a charge of -1).

47. Sodium thiosulfate, $Na_2S_2O_3$, is used as a fixing agent in photography. What is the oxidation number of each of the elements in $Na_2S_2O_3$?

48. Ammonium perchlorate, $(NH_4^+)(ClO_4^-)$, and aluminum powder are the primary components of the 2 million pounds of solid fuel in the booster rockets used to launch the space shuttle. These substances react giving $AlCl_3$, Al_2O_3, H_2O, and N_2 as the principal products. What are the oxidation numbers of each element in the components of the fuel and in the products of the reaction?

Names of Compounds

49. Do the suffixes -ous and -ic correspond to any particular oxidation state? How are they used?

50. Name the following binary compounds:
    Ⓢ(a) CaF_2 Ⓢ(e) KH Ⓢ(i) Na_2O
    (b) CuS (f) AgCl Ⓢ(j) K_3P
    Ⓢ(c) Al_2O_3 (g) MgI_2 Ⓢ(k) Mg_3N_2
    (d) HBr (h) LiCl

51. Name the following compounds, each of which contains a polyatomic ion:
    (a) $Ca(OH)_2$ (d) NaOH (g) NH_4CN
    (b) NaCN (e) $Al(CN)_3$
    (c) NH_4I (f) NH_4F

52. Name the following compounds, each of which contains a polyatomic ion:
    (a) $CaCO_3$ (e) K_2CrO_4
    (b) $KClO_4$ (f) $KMnO_4$
    (c) Na_2SO_4 (g) $Mg_3(PO_4)_2$
    (d) $AgNO_3$ (h) $Al(CH_3CO_2)_3$

Ⓢ53. Name the following compounds (since the metal atom in each can exhibit two or more oxidation states, it will be necessary to identify the oxidation number of the metal as you name the compound):
    (a) $FeBr_2$ (d) $Pb(OH)_2$ (g) SnI_4
    (b) Cu_2O (e) $TlClO_4$ (h) $CuSO_4$
    (c) MnF_3 (f) $Cr(NO_3)_3$

54. Name the following compounds, using Greek prefixes for the elements as needed:
    (a) NO_2 (c) N_2O_4 (e) Cl_2O_7
    (b) N_2O_5 (d) ClO_2 (f) SO_3

Formulas of Compounds

55. Why must a sample of calcium chloride, $CaCl_2$, contain two chloride ions for each calcium ion?

56. Using oxidation numbers as a guide, write the formula of each of the following compounds:
    (a) lithium fluoride
    (b) magnesium bromide
    (c) sodium iodide
    (d) calcium chloride
    (e) hydrogen sulfide
    (f) aluminum oxide

57. Using ionic charges as a guide, write the formula of each of the following compounds:
    (a) sodium perchlorate
    (b) magnesium nitrate
    (c) potassium sulfate
    (d) calcium hydroxide
    (e) aluminum phosphate
    (f) ammonium nitrate
    (g) sodium cyanide

Ⓢ58. Using the indicated oxidation state as a guide, write the formula of each of the following compounds:
    (a) cobalt(II) chloride
    (b) lead(IV) oxide
    (c) mercury(II) iodide
    (d) iron(III) nitrate
    (e) copper(I) sulfide
    (f) chromium(III) sulfate
    (g) tin(II) perchlorate

59. Write the formula of each of the following compounds:
    (a) sulfur trioxide
    (b) carbon tetrachloride
    (c) dinitrogen tetraoxide
    (d) sulfur hexafluoride
    (e) phosphorus pentachloride
    (f) lead dioxide

60. The following compounds have been approved by the FDA as food additives and may be used in foods as neutralizing agents, sequestering agents, sources of minerals, etc. What is the chemical formula of each compound?
    (a) potassium iodide
    (b) sulfur dioxide
    (c) calcium oxide
    (d) calcium chloride
    (e) hydrogen chloride
    (f) sodium hydroxide

(g)　magnesium carbonate
(h)　ammonium carbonate
(i)　carbon dioxide
(j)　copper(I) iodide
(k)　sulfuric acid
(l)　potassium sulfate
(m)　sodium hydrogen carbonate
(n)　ammonium aluminum sulfate
(o)　sodium dihydrogen phosphate
(p)　aluminum sulfate
(q)　iron(III) phosphate
(r)　potassium hydrogen phosphate
(s)　iron(II) sulfate
(t)　potassium hydrogen sulfite
(u)　sodium sulfite

Additional Exercises

61. Complete and balance the following equations:
(a) $Na + I_2 \longrightarrow$
(b) $Cs + S_8 \longrightarrow$
(c) $Mg + P_4 \longrightarrow$
(d) $H^+ + NH_3 \longrightarrow$
(e) $Al + O_2 \longrightarrow$
(f) $Zn + Cl_2 \longrightarrow$

62. How does the electronic structure of an isolated nitrogen atom differ from that of a nitrogen atom in a molecule of nitrogen, N_2?

63. X may indicate a different representative element in each of the following Lewis formulas. To which group does X belong in each case?

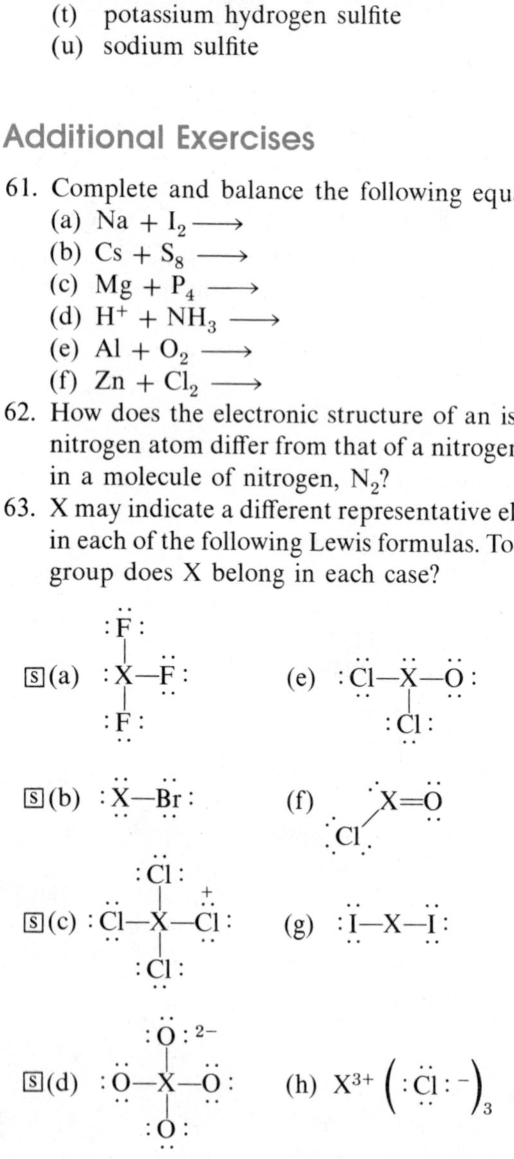

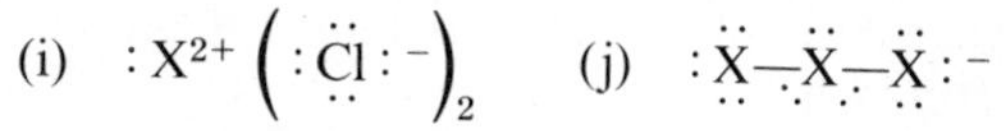

(i) $:X^{2+} \left(:\overset{..}{\underset{..}{Cl}}:^- \right)_2$　　(j) $:\overset{..}{\underset{..}{X}}-\overset{..}{\underset{..}{X}}-\overset{..}{\underset{..}{X}}:^-$

64. The number of covalent bonds between two atoms is called bond order. Bond order may be an integer, as with $:N{\equiv}N:$ (bond order = 3), or a fraction, as with SO_2 (bond order = $1\frac{1}{2}$ due to the resonance $O{=}S{-}O \longleftrightarrow O{-}S{=}O$). Arrange the molecules and ions in each of the following groups by increasing bond order for the bond indicated. Lone pairs have been omitted for clarity.

(a) C—O bond order: $C{\equiv}O$; $O{=}C{=}O$; $CH_3{-}O{-}O{-}CH_3$; HCO_2^-

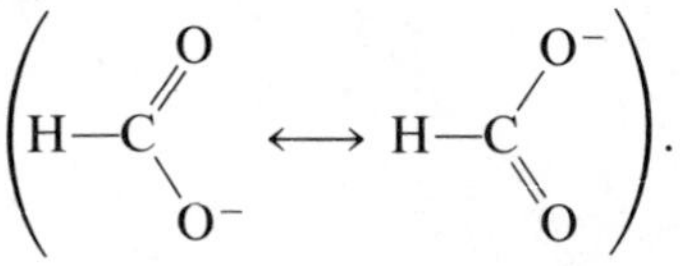

(b) N—N bond order: $CH_3{-}N{=}N{-}CH_3$; N_2; $H_2N{-}NH_2$; N_2O ($N{=}N{=}O \longleftrightarrow N{\equiv}N{-}O$).

(c) C—N bond order: $H_3C{-}NH_2$; CH_3CONH^-

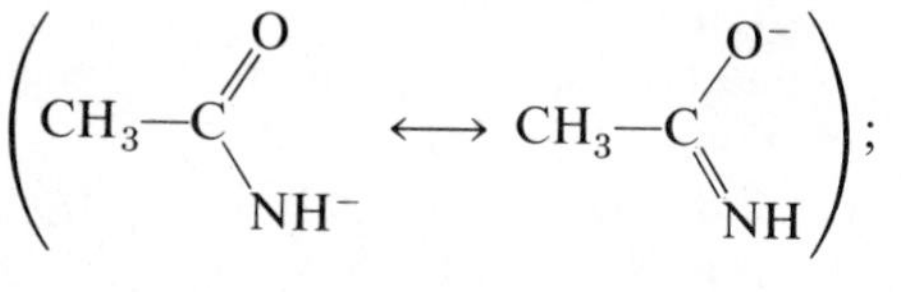

$H_2C{=}NH$; $H_3C{-}C{\equiv}N$.

(d) C—C bond order: $H_3C{-}CH_3$; $HC{\equiv}CH$; $H_2C{=}CH_2$; C_6H_6

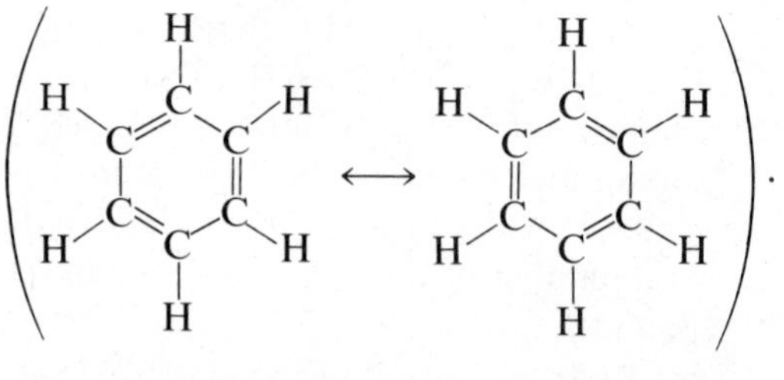

65. Arrange the following ion and molecules in increasing order of their C—O bond order (see Exercise 64): CO; CO_3^{2-} (C is the central atom since it is less electronegative); H_2CO (C is the central atom); CH_3OH (with a CO bond, three CH bonds, and one OH bond).

63 (Lewis formulas):

$\boxed{S}$(a) $:\overset{\displaystyle :\overset{..}{F}:}{\underset{\displaystyle :\overset{..}{\underset{..}{F}}:}{X}}{-}\overset{..}{\underset{..}{F}}:$　　(e) $:\overset{..}{\underset{..}{Cl}}{-}\overset{\displaystyle ..}{\underset{\displaystyle :\overset{..}{\underset{..}{Cl}}:}{X}}{-}\overset{..}{\underset{..}{O}}:$

$\boxed{S}$(b) $:\overset{.}{\underset{..}{X}}{-}\overset{..}{\underset{..}{Br}}:$　　(f) $\overset{.}{\underset{:\overset{..}{\underset{.}{Cl}}.}{X}}{=}\overset{..}{\underset{..}{O}}$

$\boxed{S}$(c) $:Cl{-}\overset{\displaystyle :\overset{..}{Cl}:}{\underset{\displaystyle :\overset{..}{\underset{..}{Cl}}:}{\overset{+}{X}}}{-}\overset{..}{\underset{..}{Cl}}:$　　(g) $:\overset{..}{\underset{..}{I}}{-}X{-}\overset{..}{\underset{..}{I}}:$

$\boxed{S}$(d) $:\overset{\displaystyle :\overset{..}{O}:^{2-}}{\underset{\displaystyle :\overset{..}{\underset{..}{O}}:}{\overset{..}{\underset{..}{O}}{-}X{-}\overset{..}{\underset{..}{O}}:}}$　　(h) $X^{3+} \left(:\overset{..}{\underset{..}{Cl}}:^- \right)_3$

66. Why is the bond order (Exercise 64) of a bond between H and another atom never greater than 1?

67. As the bond order (Exercise 64) between two atoms of the same type increases, the distance between the two atoms, the bond distance, decreases. How would you expect the bond distance in NO_2^- to compare to that in NO_3^-?

68. Separate the following species into three different groups of isoelectronic molecules and ions: NCN^{2-}, CH_3^-, NNO, H_3O^+, NO_2^- (ONO^-), O_3 (OOO), NH_3, CO_2 (OCO), ClNO.

69. A compound with a molecular weight of about 45 contains 52.2% C, 13.1% H, and 34.7% O. Write two possible Lewis structures for this compound.

CHEMICAL BONDING, PART 2—MOLECULAR ORBITALS

In Chapter 5 we considered bonding from the viewpoint that the participating electrons are principally located in the atomic orbitals of the bonded atoms. The distribution of valence electrons in a covalent molecule was treated simply in terms of lone pairs of electrons and shared pairs of electrons. Except for noting that shared pairs of electrons spend most of their time between the atoms they bond, we said little about the distribution of the electrons in a molecule and nothing about the relative energies of electrons in molecules. The Molecular Orbital Theory provides a model for describing the distribution of electrons throughout a molecule and the energies of those electrons.

6.1 The Molecular Orbital Theory

The Molecular Orbital Theory treats the distribution of electrons in molecules in much the same way as the distribution of electrons in atomic orbitals is treated for atoms. The behavior of an electron in a molecule is described, using the techniques of quantum mechanics, by a wave function, ψ, which may be used to determine the energy of the electron and the shape of the region of space within which it moves. As in an atom, the region of space in which the electron is likely to be found is called an orbital. However, an electron in a molecule may be found near the nucleus of any of the atoms in the molecule, thus the orbital extends over all of the atoms and is called a **molecular orbital.** Like an atomic orbital, a molecular orbital is full when it contains two electrons with opposite spins.

Several different types of molecular orbitals are found in molecules. Since these orbitals vary with the shapes and compositions of the molecules, let us first consider only molecules composed of two identical atoms (H_2 and Cl_2, for example). Such molecules are called **homonuclear diatomic molecules.** The types of molecular orbitals commonly observed in these diatomic molecules, **pi (π) orbitals** and **sigma (σ) orbitals,** are shown in Fig. 6-1. In three of these orbitals, called σ_s, σ_p, and π_p orbitals (read as "sigma-s," "sigma-p," and "pi-p"), the electrons are more or less between the nuclei. Electrons in these orbitals are attracted by both nuclei at the same time and are of lower energy than they would be in the isolated atoms. Thus electrons in these orbitals stabilize the molecule; the orbitals are called **bonding orbitals.** Electrons in the orbitals called σ_s^*, σ_p^*, and π_p^* (read as "sigma-s-star," "sigma-p-star," and "pi-p-star") are located well away from the region between the two nuclei. Electrons in these latter orbitals are of higher energy than they would be in the isolated atoms, so they destabilize the molecule. The orbitals are called **antibonding orbitals.**

The exact wave functions that describe the behavior of an electron in molecular orbitals like those in Fig. 6-1 are very difficult to determine. Thus approximate wave functions are usually used instead. One approximation involves using the sum or the difference of the wave functions of the constituent atoms to describe the behavior of the electron in the molecule. The wave functions of any number of atomic orbitals may be added together, or subtracted from one another, to give the wave function of a molecular orbital. At this stage, however, we will use only one atomic orbital on each of the two atoms of a diatomic molecule to approximate a molecular orbital in the molecule.

In a diatomic molecule two molecular orbitals result from each two atomic orbitals that combine. The two molecular orbitals can be thought of as being described by the addition or the subtraction of the parts of the two atomic orbitals that **overlap,** that is, share the same region of space. The molecular orbital resulting from the addition of the overlap of two atomic orbitals has a relatively large density between the two nuclei, indicating that electrons occupying such an orbital spend a large amount of time between the nuclei (Fig. 6-2, bottom right). This is a bonding situation, and the molecular orbital is a bonding orbital. A

Figure 6-1. Molecular orbitals found in homonuclear diatomic molecules. The plus signs (+) indicate the locations of nuclei.

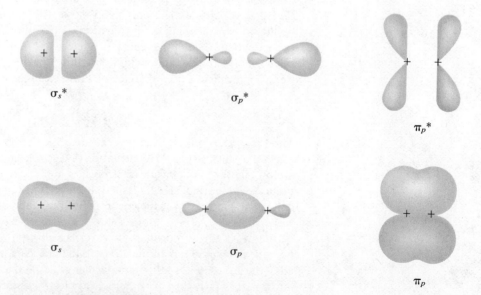

σ_s^* σ_p^* π_p^*

σ_s σ_p π_p

Figure 6-2. A representation of the formation of sigma (σ) molecular orbitals by the combination of two s atomic orbitals. The plus signs ($+$) indicate the locations of nuclei.

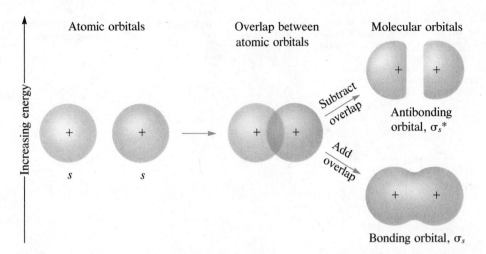

second molecular orbital, arrived at by subtraction of the overlapping parts of the atomic orbitals, has a lower density between the nuclei, indicating that the electrons spend little time between them (Fig. 6-2, top right). Such a situation is unstable, and the molecular orbital is an antibonding orbital.

Of the two molecular orbitals formed by the combination of two atomic orbitals, the bonding orbital is of lower energy than the antibonding orbital. Therefore, electrons fill the lower-energy bonding molecular orbital before the higher-energy antibonding molecular orbital, just as electrons fill atomic orbitals of lower energy before filling those of higher energy.

Figure 6-2 illustrates how a σ_s orbital is formed by the addition of two s orbitals on adjacent atoms and how a $\sigma_s{}^*$ orbital is formed by the subtraction of two s orbitals. (The subscript letters designate the type of atomic orbitals added or subtracted to give the molecular orbital.) The bonding molecular orbital produced by adding the overlap is of lower energy than either of the two atomic orbitals that combine to produce it. The antibonding orbital produced by subtracting the overlap is of higher energy than either of the two atomic orbitals. Electrons favor the lower-energy bonding molecular orbital since a given electron is attracted by both nuclei in a bonding molecular orbital but largely by only one nucleus in an antibonding molecular orbital. Note in Fig. 6-2 that an electron in an antibonding orbital has a low probability of being found in the space between the nuclei.

Four different kinds of molecular orbitals can be formed from combinations of p atomic orbitals: σ_p, $\sigma_p{}^*$, π_p, and $\pi_p{}^*$. Two p atomic orbitals, each of which possesses two lobes (Section 4.10), can be combined in either of two ways, end-to-end or side-by-side, to produce molecular orbitals (Fig. 6-3). When p atomic orbitals of two atoms are combined end-to-end (Fig. 6-3, bottom), the resulting two molecular orbitals are σ orbitals (as is the case with s atomic orbitals). However, when two p orbitals are combined side-by-side (Fig. 6-3, top), the resulting molecular orbitals are π orbitals.

Each atom contains three p atomic orbitals: p_x, p_y, and p_z (Section 4.10). One of these, say the p_x orbital, combines end-to-end with the corresponding atomic orbital of another atom to form two molecular orbitals: σ_{p_x} and $\sigma_{p_x}{}^*$. The other two p atomic orbitals, p_y and p_z, each combine side-by-side with the corresponding p atomic orbital of the other atom, giving rise to a set of two π bonding molecular orbitals and a set of two π^* antibonding molecular orbitals. The two π

Figure 6-3. A representation of the formation of sigma (σ) and pi (π) molecular orbitals by the combination of two p orbitals. The plus signs ($+$) indicate the locations of nuclei.

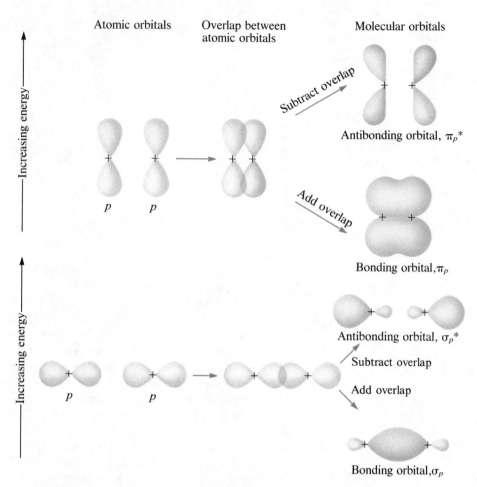

bonding orbitals are oriented at right angles to each other, as are the two π^* antibonding orbitals. It is as if one molecular orbital of a set (either π or π^*) were oriented along the y axis of a set of coordinates and the other orbital of the set were oriented along the z axis. The notations π_{p_y} and π_{p_z} (read as "pi-p-y" and "pi-p-z") are thus applied to the bonding orbitals, and $\pi_{p_y}^*$ and $\pi_{p_z}^*$ ("pi-p-y star" and "pi-p-z star") to the antibonding orbitals. Except for their orientations, the π_{p_y} and π_{p_z} orbitals are identical and have the same energy; that is, they are **degenerate orbitals.** The $\pi_{p_y}^*$ and $\pi_{p_z}^*$ antibonding orbitals are also identical except for their orientations and thus are degenerate.

A total of six molecular orbitals results from the combinations of the six p atomic orbitals in two atoms. If the σ_p and σ_p^* orbitals are oriented along the x axis and the two sets of π_p and π_p^* orbitals are oriented along the y and z axes, the six molecular orbitals are identified as σ_{p_x} and $\sigma_{p_x}^*$, π_{p_y} and $\pi_{p_y}^*$, and π_{p_z} and $\pi_{p_z}^*$.

6.2 Molecular Orbital Energy Diagrams

The relative energy levels of the lower-energy atomic and molecular orbitals of a homonuclear diatomic molecule are typically as shown in Fig. 6-4. Each colored disk represents one atomic or molecular orbital that can hold either one or two electrons. Bonding and antibonding molecular orbitals are joined by dashed lines to the atomic orbitals that combine to form them.

Figure 6-4. Molecular orbital energy diagram for a diatomic molecule containing identical atoms. A colored disk represents an atomic or molecular orbital that can hold one or two electrons.

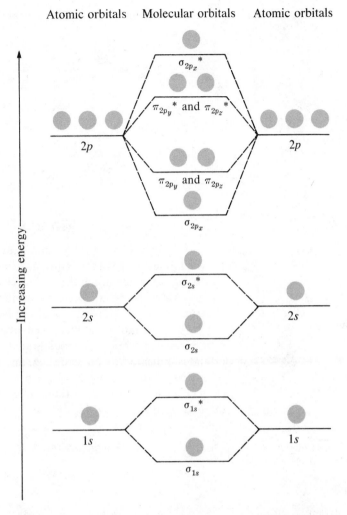

Atomic orbitals   Molecular orbitals   Atomic orbitals

The distribution of electrons in the molecular orbitals of a homonuclear diatomic molecule can be predicted by filling these molecular orbitals in the same way that atomic orbitals are filled, by the Aufbau Process (Section 4.12). The number of electrons in each orbital is indicated with a superscript. Thus a molecule containing seven electrons would have the molecular electron configuration $(\sigma_{1s})^2(\sigma_{1s}*)^2(\sigma_{2s})^2(\sigma_{2s}*)^1$.

In some cases the energy levels of the σ_p and the two π_p bonding orbitals are reversed; the π_{p_y} and π_{p_z} bonding orbitals are slightly lower in energy than the σ_{p_x} orbital. There is a logical explanation for this. We have assumed in our discussion thus far that s orbitals interact only with s orbitals to form σ_s molecular orbitals and that p orbitals interact only with p orbitals to form σ_p or π_p molecular orbitals. The main interactions do indeed occur between orbitals that are identical in energy, but an s orbital can interact with a p orbital, as shown in Fig. 6-5, if the s and p orbitals are similar in energy. This results in the σ_s molecular orbital no longer having pure s character and the σ_p molecular orbital no longer having pure p character. The mixture of s and p character shifts the energies of the molecular orbitals and in some cases even changes the relative positions of their energy levels. Hence the π_{p_y} and π_{p_z} molecular orbitals, which normally have a higher energy than the σ_{p_x} orbital, sometimes have a slightly lower energy than the σ_{p_x}

Figure 6-5. A representation of the two molecular orbitals formed by the combination of an *s* atomic orbital with a *p* atomic orbital.

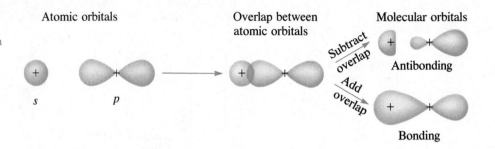

orbital. Such energy shifts are greatest when the energy difference between the *s* and *p* orbitals is small so that they may interact easily.

6.3 The Hydrogen Molecule, H_2

The hydrogen molecule (H_2) is made up of two hydrogen atoms, each of which possesses one electron in a $1s$ atomic orbital. When the atomic orbitals for the two hydrogen atoms combine, the two electrons seek the molecular orbital of lowest energy, which is the σ_{1s} bonding orbital. Each molecular orbital can hold two electrons. Hence both electrons of the hydrogen molecule are in the σ_{1s} bonding orbital, and the molecular electron configuration is $(\sigma_{1s})^2$. This can be represented by a molecular orbital energy diagram (Fig. 6-6) in which each electron within an orbital is indicated by an arrow, ⬆, and two electrons of opposite spin within an orbital are indicated by paired arrows, ⬍. The σ_{1s} orbital that contains both electrons is lower in energy than either of the two combining $1s$ orbitals. Hence the hydrogen molecule, H_2, is readily formed from two hydrogen atoms.

Figure 6-6. Molecular orbital energy diagram for the hydrogen molecule.

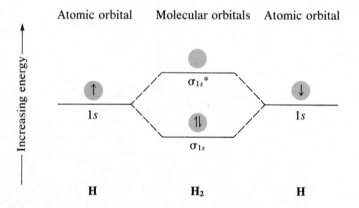

In general, the difference in energy between the atomic orbitals and the bonding molecular orbital will depend on the extent of the overlap of the two atomic orbitals. The greater the amount of overlap, the greater the energy difference and the stronger the resulting bond.

6.4 Two Diatomic Helium Species, He_2^+ and He_2

The occupation of an antibonding orbital is illustrated by the combination of a helium atom (He) and a helium ion (He⁺) to form the dihelium ion (He_2^+). The helium atom, as we have seen, has two electrons, both of which are in its $1s$ orbital

(Table 4-5). The helium ion (He$^+$) results from the loss of one of these electrons. When a helium atom and a helium ion combine to give a dihelium ion, two of the electrons go to the σ_{1s} bonding orbital, which is lower in energy than the $1s$ orbitals. The third electron must go into the molecular orbital that is next lowest in terms of energy, the σ_{1s}* antibonding orbital; hence, He$_2^+$ has the electron configuration $(\sigma_{1s})^2(\sigma_{1s}*)^1$. The locations of the electrons of the dihelium ion are illustrated in the molecular orbital energy diagram in Fig. 6-7.

Figure 6-7. Molecular orbital energy diagram for the He₂$^+$ ion.

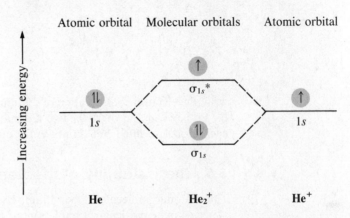

The net energy of the dihelium ion is slightly lower than that of the helium atom or ion because there are two electrons in the lower-energy molecular orbital and only one in the higher-energy molecular orbital. The lower net energy for the dihelium ion means that a helium atom and a helium ion are likely to combine to form it.

Two helium atoms do not combine to form a dihelium molecule, He$_2$, with four electrons because the two electrons in the lower-energy bonding orbital would be balanced by the two electrons in the higher-energy antibonding orbital [the configuration would be $(\sigma_{1s})^2(\sigma_{1s}*)^2$]. Since the net energy change would be zero, there is no driving force causing helium atoms to form the diatomic molecule. In fact, the element helium exists as discrete atoms rather than as diatomic molecules.

6.5 The Lithium Molecule, Li₂

The combination of two lithium atoms to form a lithium molecule, Li$_2$, is analogous to the formation of H$_2$, except that the atomic orbitals principally involved are the $2s$ orbitals. Each of the two lithium atoms, with an electron configuration of $1s^22s^1$, has one valence electron. Hence two valence electrons are available to go into the σ_{2s} bonding molecular orbital. (The lower-energy electrons in the $1s$ orbital of each atom are not appreciably involved in the bonding.) Since both valence electrons occupy the lower-energy σ_{2s} bonding orbital (Fig. 6-8), it should be possible to form Li$_2$. The molecule is, in fact, present in appreciable concentration in lithium vapor at temperatures near the boiling point of the element.

In general, electrons in inner shells of atoms are not involved in forming molecular orbitals. Because of the relatively high effective nuclear charge experienced by these electrons, the inner shells have small radii. Consequently, inner shells on adjacent atoms do not overlap. To indicate that the filled inner shells do not form molecular orbitals, we include the letters that stand for them when we

Figure 6-8. Molecular orbital energy diagram for Li_2. Lower-energy electrons in the $1s$ orbital of each atom are not shown.

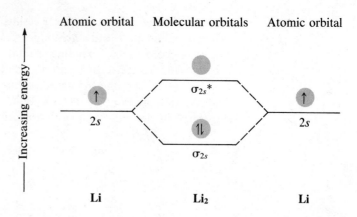

write the electron configuration of diatomic molecules. The electron configuration for Li_2 is $KK(\sigma_{2s})^2$, where the two K's indicate that the $1s$ orbitals (the K shells) on each atom are filled but do not enter into the bonding.

6.6 The Instability of the Beryllium Molecule, Be_2

The diatomic molecule of beryllium, Be_2, unlike that of lithium, would not be expected to be particularly stable, since two of the four valence electrons would go to the σ_{2s} bonding orbital and the other two to the σ_{2s}^* antibonding orbital (Fig. 6-9). The electron configuration would be $KK(\sigma_{2s})^2(\sigma_{2s}^*)^2$. The net energy change, just as for the formation of diatomic helium, is zero, indicating no tendency for beryllium atoms to combine to form the diatomic beryllium molecule. This is in accord with the experimental finding that no stable Be_2 molecule is known.

Figure 6-9. Molecular orbital energy diagram for Be_2 (not stable). Lower-energy electrons in the $1s$ orbital of each atom are not shown.

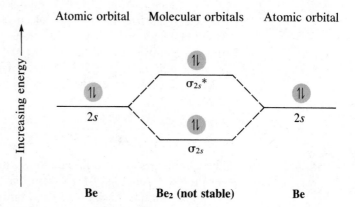

6.7 The Boron Molecule, B_2

The element boron has the electronic structure $1s^2 2s^2 2p^1$. Hence the p orbitals make an important contribution to the bonding of boron (and to that of the remaining elements in the period). We have noted that combinations of p atomic orbitals give rise to both σ and π molecular orbitals and that the two π_p bonding orbitals are of equal energy (degenerate). **Whenever two or more molecular orbitals have the same energy, electrons fill such orbitals singly with parallel spins before any pairing of electrons takes place within them.** Recall that an exactly analogous principle pertains to atomic orbitals (Section 4.11).

We have noted that the energy level of the σ_p orbital is usually slightly lower than the energy level of the two π_p orbitals but that sometimes the reverse is true. Experimental magnetic data for the B$_2$ molecule tell us that the molecule contains two unpaired electrons. This indicates that *in this particular case* the two π_p orbitals are lower in energy than the σ_p orbital and are filled first, with one electron going singly into each. If the σ_p orbital were lower in energy than the two π_p orbitals, the two electrons would be expected to pair within that orbital. Figure 6-10 is the molecular orbital energy diagram for B$_2$ and shows the two unpaired electrons. The electron configuration of B$_2$ is $KK(\sigma_{2s})^2(\sigma_{2s}{}^*)^2(\pi_{2p_y}, \pi_{2p_z})^2$. The four electrons in the σ_{2s} and $\sigma_{2s}{}^*$ orbitals balance each other in terms of energy and do not make a significant contribution to the stability of the B$_2$ molecule. However, the two π_p electrons in bonding orbitals stabilize the diboron molecule.

Figure 6-10. Molecular orbital energy diagram for B$_2$. Lower-energy electrons in the $1s$ orbital of each atom are not shown.

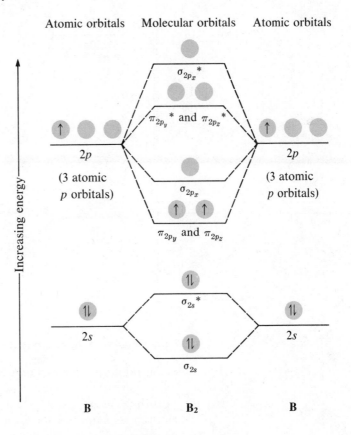

6.8 The Carbon Molecule, C$_2$

The carbon molecule, C$_2$, is similar to B$_2$. As in B$_2$, the π_{2p} orbitals of C$_2$ have a lower energy than that of the σ_{2p} orbital. The two additional electrons (one from each atom) fill the π_{2p_y} and π_{2p_z} molecular orbitals, giving the electronic structure $KK(\sigma_{2s})^2(\sigma_{2s}{}^*)^2(\pi_{2p_y}, \pi_{2p_z})^4$.

6.9 The Nitrogen Molecule, N$_2$

Each nitrogen atom has three $2p$ electrons. The nitrogen molecule, N$_2$, is well known and very stable. Experiment shows that all electrons are paired in the molecular orbitals and that the π_{2p} orbitals are lower in energy than the σ_{2p} orbital. The molecular orbital energy diagram is given in Fig. 6-11; the electron configuration is $KK(\sigma_{2s})^2(\sigma_{2s}{}^*)^2(\pi_{2p_y}, \pi_{2p_z})^4(\sigma_{2p_x})^2$. The fact that the molecule is stable is in accord with the assumption that the six electrons arising from the $2p$ atomic orbitals are all in bonding molecular orbitals.

Figure 6-11. Molecular orbital energy diagram for N_2. Lower-energy electrons in the $1s$ orbital of each atom are not shown.

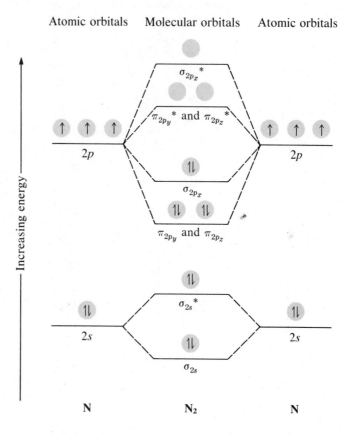

Atomic orbitals   Molecular orbitals   Atomic orbitals

6.10   The Oxygen Molecule, O_2

The molecular orbital energy diagram for the oxygen molecule, O_2, is shown in Fig. 6-12. The order of the σ_{2p} and π_{2p} orbitals changes from that for B_2, C_2, and N_2; the σ_{2p} orbital in oxygen is at a lower energy level than the π_{2p} orbitals.

The electron configuration for O_2 is $KK(\sigma_{2s})^2(\sigma_{2s}{}^*)^2(\sigma_{2p_x})^2(\pi_{2p_y}, \pi_{2p_z})^4(\pi_{2p_y}{}^*, \pi_{2p_z}{}^*)^2$, which is in accord with the known experimental fact that the oxygen molecule has two unpaired electrons. This proved to be difficult to explain on the basis of valence electron formulas, but the Molecular Orbital Theory explains it quite simply. In fact, the unpaired electrons of the oxygen molecule provide one of the strong pieces of support for the Molecular Orbital Theory.

With two p electrons in antibonding orbitals, the bond in the oxygen molecule might be expected to be weaker than that in the nitrogen molecule, in which no p electrons are in antibonding orbitals. Experimental observation shows that this is the case and that it takes much less energy to break the bond in an oxygen molecule than to break that in a nitrogen molecule (see Section 6.14).

Figure 6-12. Molecular orbital energy diagram for O_2. Lower-energy electrons in the $1s$ orbital of each atom are not shown.

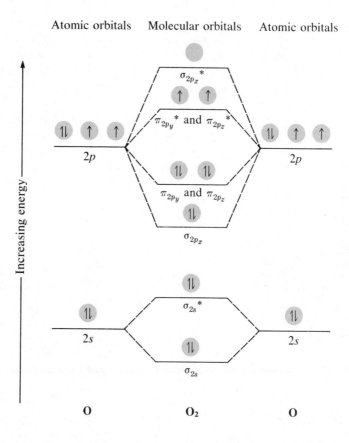

Atomic orbitals   Molecular orbitals   Atomic orbitals

6.11 The Fluorine Molecule, F_2

The fluorine molecule has no unpaired electrons; six of the p electrons are in bonding orbitals and four are in antibonding orbitals; the electronic structure is $KK(\sigma_{2s})^2(\sigma_{2s}{}^*)^2(\sigma_{2p_x})^2(\pi_{2p_y}, \pi_{2p_z})^4(\pi_{2p_y}{}^*, \pi_{2p_z}{}^*)^4$. The fluorine molecule has a weaker bond than does the oxygen molecule. In fact, the F_2 bond is one of the weakest of the covalent bonds.

6.12 The Instability of the Neon Molecule, Ne_2

The Ne_2 molecule has no appreciable tendency to form because it would have the same number of electrons in antibonding orbitals as in bonding orbitals. As with helium, the atoms remain discrete entities. If the Ne_2 molecule were to form, it would have the electron configuration $KK(\sigma_{2s})^2(\sigma_{2s}{}^*)^2(\sigma_{2p_x})^2(\pi_{2p_y}, \pi_{2p_z})^4(\pi_{2p_y}{}^*, \pi_{2p_z}{}^*)^4(\sigma_{2p_x}{}^*)^2$.

6.13 Diatomic Systems with Two Different Elements

Thus far we have considered only molecular orbitals in molecules and ions containing two identical atoms, but the Molecular Orbital Theory can be extended to **heteronuclear diatomic species,** that is, diatomic molecules and ions with bonds between atoms of two different elements.

Molecular orbitals of heteronuclear diatomic species are not fundamentally different from those of homonuclear diatomic molecules. Electrons still move over the whole molecule in molecular orbitals that can be approximated by combinations of atomic orbitals from each atom. However, in heteronuclear molecules the equivalent atomic orbitals for each contributing atom do not have the same energy. Orbitals for the more electronegative element are at a lower energy. As may be seen in Fig. 6-13, which shows a molecular orbital energy diagram for nitric oxide, NO, the energies of the atomic orbitals of oxygen are lower than those of nitrogen due to the greater charge of the oxygen nucleus. Consequently, some molecular orbitals of NO are closer in energy to the energy of the oxygen atomic orbitals than to that of the nitrogen atomic orbitals. Other molecular orbitals are closer in energy to the energy of the nitrogen atomic orbitals than to that of the oxygen atomic orbitals.

Figure 6-13. Molecular orbital energy diagram for NO. Lower-energy electrons in the $1s$ orbital of each atom are not shown.

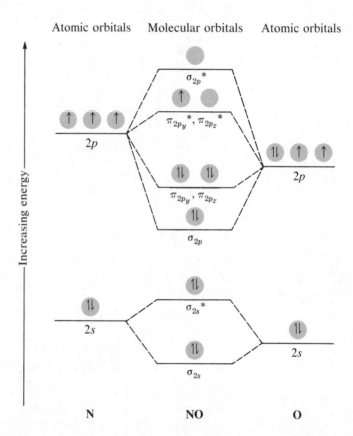

Unlike the molecular orbitals in a diatomic molecule composed of identical atoms, the molecular orbitals in a heteronuclear diatomic molecule are not symmetrical. The electron density is greater nearer the atom whose atomic orbital is closest to the energy of the molecular orbital. Figure 6-14 illustrates the distribution of electron density in the σ_{2s} and σ_{2s}^* orbitals of N_2 and NO. In the N_2 molecule the σ_{2s} and σ_{2s}^* orbitals are formed by combination of two $2s$ orbitals that have identical energies, so these molecular orbitals are symmetrical. In the NO molecule the σ_{2s} and σ_{2s}^* orbitals are formed by combination of the $2s$ orbital of oxygen with the higher-energy $2s$ orbital of nitrogen, so they are not symmetrical. The σ_{2s}^* orbital is closer in energy to the nitrogen $2s$ atomic orbital (Fig.

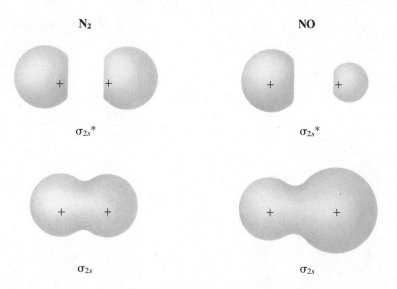

Figure 6-14. Sigma (σ) molecular orbitals found in N_2 and NO. The plus signs ($+$) indicate the locations of nuclei.

6-13); hence this antibonding molecular orbital has the greatest electron density near the nitrogen atom. The electron density of the bonding σ_{2s} orbital is greatest near the oxygen atom because the energy of this molecular orbital is closer to that of the oxygen $2s$ atomic orbital. The pair of electrons in the σ_{2s} molecular orbital spends more time, on the average, closer to the oxygen atom than to the nitrogen atom. The electrons in the σ_{2s}^{*} molecular orbital spend more time closer to the nitrogen atom.

In spite of the asymmetry of the orbitals in heteronuclear diatomic molecules, they are filled in the same way as are orbitals of homonuclear diatomic molecules. Each molecular orbital still holds only two electrons, and the lowest-energy orbitals are filled first. The electron configuration of NO is $KK(\sigma_{2s})^2(\sigma_{2s}^{*})^2(\sigma_{2p_x})^2(\pi_{2p_y}, \pi_{2p_z})^4(\pi_{2p_y}^{*}, \pi_{2p_z}^{*})^1$. Since there are more electrons in bonding molecular orbitals with greater electron density near the oxygen atom than in antibonding orbitals with greater electron density near the nitrogen atom, the bond is polar (Section 5.8).

Molecular orbital energy diagrams like the one for NO can be used to describe the electron distribution in many other heteronuclear diatomic molecules and ions, such as CO, NO^+, CN^-, and NO^-.

To explain the bonding in a molecule like HF, it is necessary to introduce the idea that only atomic orbitals that are similar in energy can be combined to give molecular orbitals. If the energies of two atomic orbitals are very different, molecular orbitals cannot be formed from them. In HF the $1s$ orbital of hydrogen has about the same energy as each of the $2p$ orbitals of fluorine (Fig. 6-15). Thus the molecular orbitals form from the hydrogen $1s$ orbital and the fluorine $2p_x$ orbital (Fig. 6-16). The energy difference, however, between the hydrogen $1s$ orbital and the fluorine $2s$ orbital is too great for these atomic orbitals to combine.

When an s orbital combines with a p orbital, it does so in the head-to-tail fashion shown in Fig. 6-16. In HF this gives the bonding and antibonding orbitals shown in Fig. 6-15. The $2s$, $2p_y$, and $2p_z$ orbitals of the fluorine atom in the HF molecule are shown at the same energies as those of the original fluorine atomic orbitals in the molecular orbital energy diagram of HF (Fig. 6-15). To a first approximation the energies of the $2s$, $2p_y$, and $2p_z$ orbitals in the molecule do not differ from their energies in a free fluorine atom since they do not combine with

Figure 6-15. Molecular orbital energy diagram for HF. The filled $1s$ orbital of F is not shown.

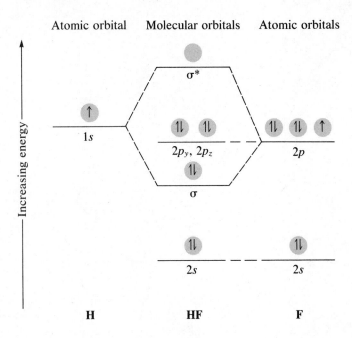

Figure 6-16. Sigma (σ) molecular orbitals found in HF. The plus signs ($+$) indicate the locations of nuclei.

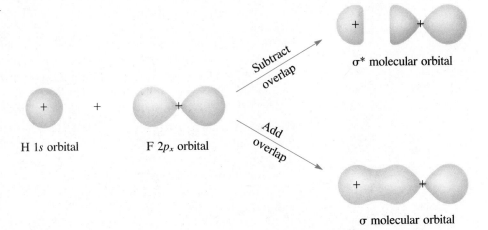

orbitals of the hydrogen atom. As may be seen in the molecular orbital energy diagram of HF (Fig. 6-15), one bonding and one antibonding orbital are formed. The eight valence electrons are located in the lowest-energy orbitals in the molecule—the $2s$, $2p_y$, and $2p_z$ atomic orbitals on fluorine and the σ bonding orbital. The molecule is stable because the two electrons in the σ orbital have a lower energy than they have in the isolated atoms.

6.14 Bond Order

When using Lewis structures to describe the distribution of electrons in molecules, **the order of a bond (bond order) between two atoms can be defined as the number**

of bonding pairs of electrons between the atoms (see Exercise 64, Chapter 5). Bond order is defined differently when the molecular orbital description of the distribution of electrons is used; however, the resulting bond order is the same.

We have seen that the bond between the atoms is weaker in O_2 than in N_2 and weaker in F_2 than in O_2 (Sections 6.9, 6.10, and 6.11). This order of bond energies is predictable from the increase in the ratio of the number of antibonding orbital electrons to the number of bonding orbital electrons, in the order N_2, O_2, and F_2. Values for the bond energies of the three molecules support the predictions. The following data show the energy necessary to break the bonds in these molecules.

Molecule	Bond Energy
N_2	955 kJ (226 kcal) per mole
O_2	498 kJ (119 kcal) per mole
F_2	159 kJ (38 kcal) per mole

The nitrogen molecule, with three pairs of bonding electrons in σ_p and π_p molecular orbitals, has a triple bond (Fig. 6-11 and Section 5.3) and a *bond order of 3*. In the oxygen molecule the two additional electrons are in the π_p*antibonding orbitals (Fig. 6-12), which weakens the bond. The effect of six bonding electrons minus the partially offsetting effect of two antibonding electrons results in a net effect that is approximately equivalent to four bonding electrons (or two pairs of bonding electrons). Hence oxygen has a double bond, which is on the average weaker than a triple bond, and a *bond order of 2*. The fluorine molecule, with six bonding and four antibonding electrons in the p orbitals has a net effect equivalent to two bonding electrons (or one pair of bonding electrons). The fluorine molecule, therefore, has a single bond and a *bond order of 1*. A single bond is, in general, weaker than either a double bond or a triple bond.

The bond order for a given bond may be determined by dividing the number of bonding electrons in the outer shell minus the number of antibonding electrons in the outer shell by 2.

$$\text{Bond order} = \frac{\text{number of bonding electrons minus number of antibonding electrons}}{2}$$

$$\text{Bond order in } N_2 = \frac{6-0}{2} = 3$$

$$\text{Bond order in } O_2 = \frac{6-2}{2} = 2$$

$$\text{Bond order in } F_2 = \frac{6-4}{2} = 1$$

These bond orders are the same as calculated from Lewis structures for N≡N, O=O, and F—F.

FOR REVIEW

SUMMARY

The **Molecular Orbital Theory** describes the behavior of an electron in a molecule by a wave function, which may be used to determine the energy of the electron and the region of space around the molecule where it is most likely to be found. The electron occupies an orbital that is called a **molecular orbital** since it may extend over all of the atoms in the molecule. An electron in a **bonding molecular orbital** stabilizes a molecule. An electron in an **antibonding molecular orbital** makes a molecule less stable. A σ_s molecular orbital is a bonding orbital created by the addition of the overlap of two s orbitals of adjacent atoms. A σ_p molecular orbital is a bonding orbital created from the addition of the end-to-end overlap of p orbitals of adjacent atoms. A π_p molecular orbital is a bonding orbital resulting from the addition of side-to-side overlap of p orbitals of adjacent atoms. The σ_s^*, σ_p^*, and π_p^* molecular orbitals are antibonding orbitals resulting from the subtraction of the overlap of atomic orbitals.

One order of increasing energies of the molecular orbitals in a **homonuclear diatomic molecule** is given by the electron configuration (σ_{1s}) $(\sigma_{1s}^*)(\sigma_{2s})(\sigma_{2s}^*)(\sigma_{2p_x})(\pi_{2p_y}, \pi_{2p_z})(\pi_{2p_y}^*, \pi_{2p_z}^*)(\sigma_{2p_x}^*)$. In some molecules the order of the σ_{2p} and π_{2p} levels is reversed. Electrons fill the orbitals of lowest energy first, with both π_{2p} orbitals (or π_{2p}^* orbitals) being occupied by one electron before either will accept two electrons. According to their molecular orbital electron configurations, H_2 and He_2^+ are stable, while He_2, which would contain the same number of bonding and antibonding electrons, is not. Similarly, Li_2, B_2, C_2, N_2, O_2, and F_2 are expected to be stable, while Be_2 and Ne_2 are not. Experimental evidence confirms these expectations. **Heteronuclear diatomic species** are similar to homonuclear molecules in the formation and filling of orbitals, but the molecular orbitals are not symmetrically distributed between the two atoms due to the differences in energy of the two types of atomic orbitals. Polar bonds result.

The **bond order** of a bond between two atoms is determined by the numbers of bonding and antibonding electrons it contains.

KEY TERMS AND CONCEPTS

Antibonding orbitals (6.1)
Bonding orbitals (6.1)
Bond order (6.14)
Degenerate orbitals (6.1)

Heteronuclear diatomic species (6.13)
Homonuclear diatomic molecules (6.1)

Molecular orbital (6.1)
Overlap (6.1)
Pi (π) orbital (6.1)
Sigma (σ) orbital (6.1)

EXERCISES

Molecular Orbitals

1. Why does the bonding molecular orbital formed from two atomic orbitals have a lower energy than that of either atomic orbital?

2. In Section 4.11 Hund's Rule was discussed in

connection with atomic orbitals. Formulate a similar rule for the filling of molecular orbitals and give specific examples of its application.

3. How do the following differ and how are they similar?
   (a) Molecular orbitals and atomic orbitals
   (b) Bonding orbitals and antibonding orbitals
   (c) σ orbitals and π orbitals

4. Describe the similarities and differences of σ orbitals formed from two s atomic orbitals, from two p atomic orbitals, and from an s and a p atomic orbital.

5. Draw diagrams showing the molecular orbitals that can be formed by combining two s orbitals, by combining two p orbitals, and by combining an s and a p orbital.

6. Consider the molecular orbitals represented by the following outlines [each plus sign ($+$) indicates the location of a nucleus]:

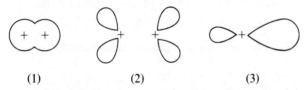

(1)   (2)   (3)

Indicate how each of these is formed either by addition or by subtraction of the overlap between s and/or p orbitals.

Homonuclear Diatomic Molecules

7. Using molecular orbital energy diagrams and the electron occupancy in the molecular orbitals, compare the stability of H_2, He_2^+, and He_2.

8. Consider the diatomic *ions* X_2^{2+}, where X is any one of the first ten elements of the Periodic Table except hydrogen. Write the molecular orbital electron configurations of these ions.

9. The acetylide ion, C_2^{2-}, is a component of calcium acetylide. Draw the molecular orbital energy diagram of this ion and write the molecular orbital electron configuration.

10. Draw a molecular orbital energy diagram for P_2 using only the orbitals in the valence shell of phosphorus that are occupied in the free atom. What is the electron configuration of P_2?

11. Draw a molecular orbital energy diagram for the ion Si_2^+ using only the orbitals in the valence shell of silicon that are occupied in the free atom. What is the electron configuration of Si_2^+?

Heteronuclear Diatomic Molecules

12. How does the molecular orbital energy diagram for a diatomic molecule involving atoms of two different elements differ from that for a diatomic molecule made up of two atoms of the same element?

13. Under what conditions does an s atomic orbital on one atom combine with a p atomic orbital on a second atom rather than with the s atomic orbital on the second atom?

14. What does the presence of a polar bond in a molecule imply about the shapes of its molecular orbitals?

15. Draw the molecular orbital energy diagrams for CO and CN^-. Show that CO and CN^- are isoelectronic and have the same bond order. Compare these molecular orbital energy diagrams with that for NO.

16. Draw the molecular orbital energy diagram of ClF using only those valence orbitals occupied in the free atoms.

17. Draw the molecular orbital energy diagram of OH^-.

18. Will the unpaired electron in NO spend more time near the nitrogen nucleus or near the oxygen nucleus, or will it spend equal amounts of time near both? Explain your answer.

Bond Order

19. Determine the bond orders of the homonuclear diatomic molecules formed by the first ten elements of the Periodic Table.

20. Determine the bond orders of the homonuclear ions, X_2^{2+}, described in Exercise 8.

21. Arrange the following in increasing order of bond energy: F_2, F_2^+, F_2^-.

22. The Lewis structure of the nitric oxide anion ($N{=}O^-$) indicates a bond order of 2 for this ion. Beginning with the molecular orbital energy diagram for NO, work out the bond order for NO^- using Molecular Orbital Theory.

23. What is the bond order of NO^+? Is the bond in NO^+ stronger or weaker than in NO?

24. Use the Molecular Orbital Theory to determine the bond orders of O_2^+, O_2, O_2^-, and O_2^{2-}. Arrange these species in order of increasing bond energy.

Additional Exercises

25. Predict whether each of the following atoms, ions, and molecules should be paramagnetic or diamagnetic. (A paramagnetic species contains unpaired electrons; a diamagnetic species has all of its electrons paired.)

(a) He   (e) O_2 (i) CO
(b) B   (f) F   (j) NO
(c) B_2 (g) G_2 (k) H_2^+
(d) O   (h) Co   (l) F_2^+

26. Consider the molecular orbitals represented by the following outlines (each plus sign indicates the location of a nucleus):

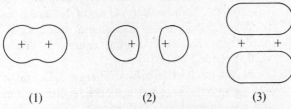

(1)   (2)   (3)

(a) What is the maximum number of electrons that can be placed in molecular orbital (1)? in orbital (2)? in orbital (3)?

(b) How many orbitals of type (1) are found in the valence shell of F_2? How many orbitals of type (3)?

(c) What homonuclear diatomic molecule formed by an element in the third period of the Periodic Table has its two highest-energy electrons in an orbital of type (1)?

(d) What homonuclear diatomic molecule formed by the elements in the third period of the Periodic Table has its four highest-energy electrons in orbitals like type (3)?

27. Identify the homonuclear diatomic molecules or ions that have the following electron configurations:

(a) X_2^+: $KK(\sigma_{2s})^1$
(b) X_2: $KK(\sigma_{2s})^2(\sigma_{2s}^*)^2(\pi_{2p_y}, \pi_{2p_z})^4$
(c) X_2^-: $KK(\sigma_{2s})^2(\sigma_{2s}^*)^2(\pi_{2p_y}, \pi_{2p_z})^3$
(d) X_2^+: $KK(\sigma_{2s})^2(\sigma_{2s}^*)^2(\sigma_{2p_x})^2(\pi_{2p_y}, \pi_{2p_z})^4$ $(\pi_{2p_y}^*, \pi_{2p_z}^*)^3$

28. The following outline indicates a bonding molecular orbital in the molecule AB. Which element, A or B, is more electronegative? Explain your answer.

A   B

7

MOLECULAR STRUCTURE AND HYBRIDIZATION

For the most part, chemists are interested in the properties of molecules or ions that contain two or more atoms. This interest includes the molecular structure of a molecule, that is, the three-dimensional arrangement of its atoms. The properties of molecules often reflect very concretely the exact spatial distribution of their constituent atoms. For example, we can digest starch from seeds such as wheat and corn but not the cellulose that forms the bodies of these plants. Starch and cellulose have the same chemical formula and are practically identical, except for certain small differences in their molecular structures. In this chapter we will examine molecular structures, how to predict them, and the models that are used to explain them.

VALENCE SHELL ELECTRON-PAIR REPULSION THEORY

7.1 Prediction of Molecular Structures

When we discuss the **molecular structure,** or **molecular geometry,** of a molecule, we mean the three-dimensional, geometrical arrangement of its atoms. Molecular

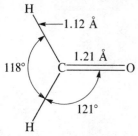

Figure 7-1. Bond distances and angles in the formaldehyde molecule, H_2CO. Note that $1 \text{ Å} = 100 \text{ pm} = 10^{-10}$ m.

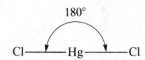

Figure 7-2. The linear structure of the $HgCl_2$ molecule. The bonds are on opposite sides of the Hg atom.

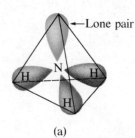

(a)

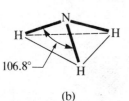

(b)

Figure 7-3. (a) Regions occupied by the lone pair (shown in color) and bonds (black) in NH_3 and (b) the resulting trigonal pyramidal molecular structure.

structures are normally described in terms of bond distances and bond angles. A **bond distance** is the distance between the nuclei of two bonded atoms along a straight line joining the nuclei. Bond distances are measured in picometers (1 pm $= 10^{-12}$ m) or in angstrom units (1 Å $= 100$ pm $= 10^{-10}$ m). A **bond angle** is the angle between any two bond distances that include a common atom. The bond distances and angles in the formaldehyde molecule, H_2CO, are illustrated in Fig. 7-1.

Approximate bond angles in a molecule can be predicted from the number of bonds and lone pairs on a central atom using the **Valence Shell Electron-Pair Repulsion Theory (VSEPR Theory).** The electrons in the valence shell of the central atom of a molecule are present either as bonding pairs, located primarily between bonded atoms, or as lone pairs occupying restricted regions of space shaped rather like those occupied by the bonding pairs. The VSEPR Theory is based on the idea that the electrostatic repulsion of these electrons is reduced to a minimum when the various regions of high electron density assume positions as far from each other as possible. For example, in a molecule such as $HgCl_2$ having only two bonds to the central atom, the bonds are as far apart as possible and the electrostatic repulsions between the electrons in these bonds are reduced to a minimum when they are on opposite sides of the central atom. The bond angle is 180° (Fig. 7-2). Another example is illustrated in Fig. 7-3, which shows the regions of space occupied by the lone pair and bonding pairs in the ammonia molecule. The repulsions among the four regions of electron density in NH_3 are minimized if they are directed toward the corners of a tetrahedron, as shown in Fig. 7-3(a). This and other geometries that reduce the repulsions among regions of high electron density (bonding pairs and/or lone pairs) to a minimum are illustrated in Table 7-1. **Linear, trigonal planar, tetrahedral, trigonal bipyramidal, and octahedral structures result for two, three, four, five, and six regions of high electron density, respectively.**

If the regions of high electron density are not identical, the bond angles may differ by several degrees from the ideal values given in Table 7-1. Nevertheless, the structures given are usually good approximations for the distribution of electron density around a central atom. In NH_3, for example, the central nitrogen atom is surrounded by three single N—H bonds and one lone pair of electrons. Since the regions of high electron density are not alike (three are bonds and one is a lone pair), the bond angle is 106.8°, as shown in Fig. 7-3(b), instead of 109.5°, as shown in Table 7-1 for symmetrical tetrahedral structures. In formaldehyde, H_2CO, the regions of high electron density consist of two single bonds and one double bond, and the bond angles differ from one another by about 3° (Fig. 7-1).

Even though the arrangement of lone pairs and bonding pairs in a molecule may correspond to one of the arrangements in Table 7-1, its molecular structure may look different. The presence of a lone pair affects the structure of a molecule, but the lone pair is invisible to the experimental techniques used to determine structure. Consequently, the molecular structure is described as if the lone pair were not there. The NH_3 molecule, for example, has a tetrahedral arrangement of regions of electron density, Fig. 7-3(a). However, one of these regions is a lone pair and is not observed when the molecular structure is determined experimentally. The molecular structure (the arrangement of atoms only) is a trigonal pyramid, Fig. 7-3(b), with the nitrogen atom at the apex and three hydrogen atoms forming the base. Table 7-2 illustrates the structures that are observed for various

Table 7-1   Arrangement of Lone Pairs and Bonds as a Result of
Electron-Pair Repulsions

Two regions of high electron density (bonding pairs and/or lone pairs	180° M	Linear 180° angle.
Three regions of high electron density	120° M	Trigonal planar. All angles 120°.
Four regions of high electron density	109.5° M	Tetrahedral. All angles 109.5°.
Five regions of high electron density	90° M 120°	Trigonal bipyramidal. Angles of 90° or 120° (an attached atom may be equatorial, in the plane of the triangle, or axial, above or below the plane of the triangle).
Six regions of high electron density	90° M	Octahedral. All angles 90°.

combinations of lone pairs and bonding pairs.

For several of the examples in Table 7-2, a different arrangement of the locations of lone pairs and bonds would give a different molecular structure. For example, the molecular structure of ClF_3 is T-shaped, as shown in Fig. 7-4(a). However, two other possible arrangements for the three bonds and two lone pairs could be written: a trigonal planar arrangement, Fig. 7-4(b), and a trigonal pyramidal arrangement, Fig. 7-4(c). The stable structure is the one that puts the lone pairs as far apart as possible and thus minimizes electron-pair repulsions. In trigonal bipyramidal arrangements (such as that of ClF_3), the positions that minimize repulsions are those in the triangular plane of the molecule because the electrons in these positions are 120° apart.

Table 7-2 Molecular Structures Based on the Valence Shell Electron-Pair Repulsion Theory

Regions of High Electron Density (Bonds and Lone Pairs)	Molecular Structures and Examples (Chemical Bonds Are Indicated in Black, Lone Pairs in Color)			
Three: trigonal planar arrangement	3 Bonds 0 Lone pairs — Trigonal planar — CO_3^{2-}, BF_3, NO_3^-	2 Bonds 1 Lone pair — Angular (120°) — NO_2^-, $ClNO$		
Four: tetrahedral arrangement	4 Bonds 0 Lone pairs — Tetrahedral — NH_4^+, CH_4	3 Bonds 1 Lone pair — Trigonal pyramidal — PCl_3, NH_3	2 Bonds 2 Lone pairs — Angular (109.5°) — NH_2^-, H_2O	
Five: trigonal bipyramidal arrangement	5 Bonds 0 Lone pairs — Trigonal bipyramidal — PF_5, $SnCl_5^-$	4 Bonds 1 Lone pair — Seesaw — SF_4, ClF_4^+	3 Bonds 2 Lone pairs — T-shaped — ICl_3, ClF_3	2 Bonds 3 Lone pairs — Linear — I_3^-, ClF_2^-
Six: octahedral arrangement	6 Bonds 0 Lone pairs — Octahedral — PCl_6^-, SF_6, IF_6^+	5 Bonds 1 Lone pair — Square pyramidal — IF_5, XeF_5^+	4 Bonds 2 Lone pairs — Square planar — ICl_4^-, XeF_4	

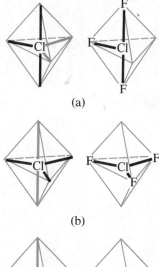

(a)

(b)

(c)

Figure 7-4. Possible arrangements of two lone pairs (shown in color) and three bonds (black) in ClF_3, and also the resulting molecular structures showing only the positions of the atoms without the lone pairs. (a) shows the stable *T*-shaped arrangement.

mize repulsions are those in the triangular plane of the molecule because the electrons in these positions are 120° apart.

The amount of repulsive force varies with the kinds of electron pairs involved. Lone pair–lone pair repulsive forces are the largest, lone pair–bond pair repulsive forces are intermediate in value, and bond pair–bond pair repulsive forces are smallest. An alternative way of thinking about these repulsions is that lone pairs require more room than bonding pairs. When several possible structures can be written, the one that minimizes as many of the strong repulsions as possible will be the most stable.

7.2 Rules for Predicting Molecular Structures

To use the VSEPR Theory to predict molecular structures, we apply the following procedure:

1. Write the Lewis structure of the molecule as described in Section 5.6.
2. Count the regions of high electron density (lone pairs and chemical bonds) around the central atom in the Lewis structure. A single, double, or triple bond counts as one region of high electron density. An unpaired electron counts as a lone pair.
3. Identify the most stable arrangement of the regions of high electron density (linear, trigonal planar, tetrahedral, trigonal bipyramidal, or octahedral) from those given in Table 7-1.
4. If more than one arrangement of lone pairs and chemical bonds is possible, choose the one that will minimize lone-pair repulsions. In trigonal bipyramidal arrangements repulsion is minimized when every lone pair is in the plane of the triangle. In an octahedral arrangement with two lone pairs, repulsion is minimized when the lone pairs are on opposite sides of the central atom.
5. Identify the molecular structure (the arrangement of atoms) from the locations of the atoms at the ends of the bonds (Table 7-2).

EXAMPLE 7.1   Predict the molecular structure of a gaseous $BeCl_2$ molecule.

The Lewis structure of $BeCl_2$ is

$$: \ddot{Cl}—Be—\ddot{Cl} :$$

The central beryllium atom has no lone pairs but is bonded to two atoms. There are therefore two regions of high electron density around it. Table 7-1 indicates that two regions of high electron density arrange themselves on opposite sides of the central atom with a bond angle of 180°. $BeCl_2$ should be a linear molecule.

EXAMPLE 7.2   Predict the molecular structure of BCl_3.

The Lewis structure of BCl_3 is

$$: \ddot{Cl} :$$
$$: \ddot{Cl}—B—\ddot{Cl} :$$

BCl$_3$ contains three bonds, and there are no lone pairs on boron. Table 7-1 indicates that the arrangement of three regions of high electron density will be trigonal planar. The bonds in BCl$_3$ should lie in a plane with 120° angles between them, a trigonal planar arrangement (Fig. 7-5).

Figure 7-5. (a) Trigonal planar arrangement of three bonds in BCl$_3$ and (b) the resulting trigonal planar molecular structure.

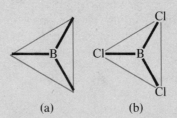

(a)   (b)

EXAMPLE 7.3 Predict the molecular structure of NH$_4$$^+$.

The Lewis structure of NH$_4$$^+$ is

$$\begin{array}{c} H \\ | \\ H-N-H^+ \\ | \\ H \end{array}$$

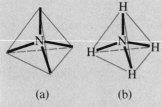

(a)   (b)

Figure 7-6. (a) Tetrahedral arrangement of four bonds in NH$_4$$^+$ and (b) the resulting tetrahedral molecular structure.

NH$_4$$^+$ contains four bonds from the nitrogen atom to hydrogen atoms. Table 7-1 indicates that four regions of high electron density arrange themselves so that they point at the corners of a tetrahedron with the central atom in the middle. The hydrogen atoms located at the ends of the bonds are located at the corners of a tetrahedron. The molecular structure of NH$_4$$^+$ (Fig. 7-6) is tetrahedral.

EXAMPLE 7.4 Predict the molecular structure of H$_2$O.

The Lewis structure of H$_2$O,

indicates that there are four regions of high electron density around the oxygen atom—two lone pairs and two chemical bonds. These four regions are arranged in a tetrahedral fashion [Fig. 7-7(a)] as indicated in Table 7-1. However, the arrangement of the atoms themselves (the molecular structure) in H$_2$O is angular with a bond angle of approximately 109.5° [Table 7-2, Fig. 7-7(b)].

Figure 7-7. (a) Tetrahedral arrangement of two lone pairs (shown in color) and two bonds (black) in H$_2$O and (b) the resulting bent molecular structure.

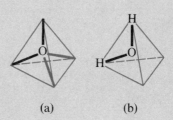

(a)   (b)

EXAMPLE 7.5   Predict the molecular structure of SF_4.

The Lewis structure of SF_4,

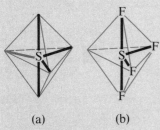

(a)   (b)

Figure 7-8. (a) Trigonal bipyramidal arrangement of one lone pair (shown in color) and four bonds (black) in SF_4 and (b) the resulting seesaw-shaped molecular structure.

indicates five regions of high electron density about the sulfur atom—one lone pair and four chemical bonds. These five regions are directed toward the corners of a trigonal bipyramid (Table 7-1). In order to minimize lone pair–bond pair repulsions, the lone pair occupies one of the locations in the plane of the triangle. The molecular structure (Fig. 7-8) is that of a seesaw (Table 7-2).

EXAMPLE 7.6   Predict the molecular structure of IF_4^-.

The Lewis structure of IF_4^-,

indicates six regions of high electron density around the iodine atom—two lone pairs and four bonds. These six regions adopt an octahedral arrangement. The two possible arrangements of lone pairs and bonds are illustrated in Figs. 7-9(a) and 7-9(b). To minimize repulsions the lone pairs should be on opposite sides of the central atom, so the structure in Fig. 7-9(a) is the more stable of the two. The five atoms are all in the same plane and have a square planar configuration.

Figure 7-9. (a),(b) Possible octahedral arrangements of two lone pairs (shown in color) and four bonds (black) in IF_4^- and (c) the stable square planar molecular structure.

  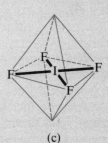

(a)   (b)   (c)

EXAMPLE 7.7   Predict the molecular structure of ClO_2F.

The least electronegative element, chlorine, is the central element, as shown in this Lewis structure:

The four regions of high electron density are arranged tetrahedrally. The atoms are arranged to give a trigonal pyramidal structure, as shown in Fig. 7-10.

Figure 7-10.   (a) Tetrahedral arrangement of one lone pair (shown in color) and three bonds in ClO_2F and (b) the resulting trigonal pyramidal molecular structure.

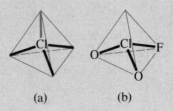

(a)   (b)

HYBRIDIZATION OF ATOMIC ORBITALS

Many of the molecules discussed in previous sections contain identical bonds joining a central atom with from two to six other atoms. These identical bonds form from various atomic orbitals on the central atom and can be described by hybridizaton, a concept introduced by Linus Pauling in 1935. Nonequivalent atomic orbitals can **hybridize** (combine) to give a set of equivalent hybrid orbitals whose orientation reflects the geometry of the molecule or ion. The theory uses atomic rather than molecular orbitals, but it is useful in explaining why the bonds in many molecules are equivalent and how the atomic orbitals on a central atom of a molecule interact with the orbitals on the other atoms.

7.3   Methane, CH_4, and Ethane, C_2H_6 (sp³ Hybridization)

A molecule of methane, CH_4, consists of a carbon atom surrounded by four hydrogen atoms at the corners of a tetrahedron (see Fig. 7-12). However, an uncombined carbon atom has only two unpaired electrons in its ground state (lowest-energy state). The electron configuration of a carbon atom is $1s^2 2s^2 2p^2$, which we will write as $1s^2 2s^2 2p^1 2p^1$ to emphasize that two of the $2p$ orbitals contain a single electron and the third is empty. With this configuration (two unpaired electrons) we might expect carbon to form only two covalent bonds, but it usually forms four, as in methane. How does the carbon atom get four equivalent electrons with which to form four equivalent covalent bonds? The answer is by promoting (exciting) one of the electrons from the $2s$ orbital to the third $2p$ orbital producing the electron configuration $1s^2 2s^1 2p^1 2p^1 2p^1$. The energy required for **promotion of electrons** is more than offset by the energy given off during bond formation. Therefore the energy state after electron promotion *and* bond information is lower than it was before these steps; the decrease in energy corresponds to a more stable state.

The electron distribution in the ground state and after promotion can be shown as follows for the carbon atom. Each orbital is represented by a circle and each electron by an arrow; arrows pointing in opposite directions designate electrons of opposite spin.

C atom
(ground state)

C atom
(one 2*s* electron promoted
to the third 2*p* orbital)

Promotion ⟶

1*s*   2*s*   2*p*   1*s*   2*s*   2*p*

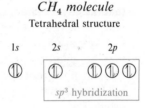

The three *p* orbitals of an uncombined carbon atom are oriented at right angles (Section 4.10); the *s* orbital is nondirectional because of its spherical shape. The four carbon-hydrogen bonds of methane, however, are equal in length and in strength and are pointed to the corners of a regular tetrahedron. When a carbon atom is bonded to four other atoms at the corners of a tetrahedron, the one 2*s* orbital and the three 2*p* orbitals of the carbon atom hybridize (combine) to form four equivalent hybrid orbitals, whose lobes point to the corners of the tetrahedron (Fig. 7-11). These are called *sp*^{3} hybrid orbitals, to designate the hybridization of one *s* orbital and three *p* orbitals; **sp³ hybridization** always results in a tetrahedral orientation of four orbitals.

When a methane molecule forms, each of the four hydrogen atoms shares its one electron (indicated by a colored arrow) with the electron in one of the four *sp*^{3} hybrid orbitals.

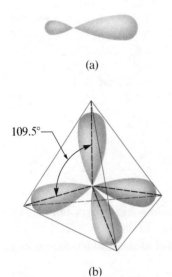

(a)

109.5°

(b)

Figure 7-11.   (a) An individual *sp*^{3} hybrid orbital. (b) The four tetrahedral *sp*^{3} hybrid orbitals of the carbon atom. For clarity only the large lobe of each hybrid orbital is shown.

CH$_4$ molecule
Tetrahedral structure

1*s*   2*s*   2*p*

sp^3 hybridization

Thus the 1*s* orbital of each of the four hydrogen atoms overlaps with one of the four *sp*^{3} orbitals of the carbon atom to form an *sp*^{3}-*s* sigma (σ) bond. (Notice the similarity between this and the overlap to produce molecular orbitals, Section 6.1.) This results in the formation of four very strong covalent bonds between one carbon atom and four hydrogen atoms to produce the methane molecule, CH$_4$ (Fig. 7-12).

Figure 7-12.   The methane molecule. (a) Diagram showing the overlap of the four tetrahedral *sp*^{3} hybrid orbitals with four *s* orbitals of four hydrogen atoms to produce the methane molecule. (b) The overall outline of the four bonding orbitals in methane.

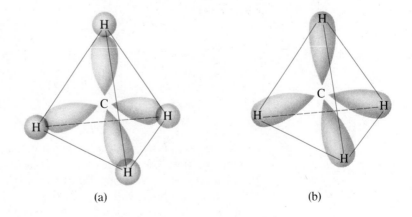

(a)   (b)

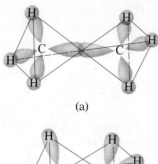

(a)

(b)

Figure 7-13. The ethane molecule (a) The overlap diagram for
the ethane molecule. (b) The
overall outline of the seven bonding orbitals in ethane.

The structure of ethane, C_2H_6, is similar to that of methane in that each carbon in ethane has four neighboring atoms arranged at the corners of a tetrahedron—three H atoms and one C atom (Fig. 7-13). In ethane an sp^3 orbital of one carbon atom overlaps end-to-end with an sp^3 orbital of a second carbon atom to form a σ bond. Each of the other three sp^3 hybrid orbitals of each carbon atom overlaps with an s orbital of a hydrogen atom to form additional σ bonds. The structure and overall outline of the bonding orbitals of ethane are shown in Fig. 7-13. Ethane is made up of two tetrahedra with one corner in common.

An sp^3 hybrid orbital can also hold a lone pair of electrons. For example, the nitrogen atom in ammonia [Fig. 7-3(a)] is surrounded by three bonding pairs of electrons and a lone pair, all directed to the corners of a tetrahedron. Thus the nitrogen atom may be regarded as being sp^3 hybridized with one hybrid orbital occupied by the lone pair (represented by the pair of black arrows).

NH₃ molecule

1s	2s	2p

sp^3 hybridization

The molecular structure of water (Fig. 7-7) is consistent with a tetrahedral arrangement of two lone pairs and two bonding pairs of electrons. Thus we say that the oxygen atom is sp^3 hybridized with two of the hybrid orbitals occupied by lone pairs and two by bonding pairs.

Any atom with a tetrahedral arrangement of lone pairs and bonding pairs may be regarded as being sp^3 hybridized. But note that hybridization can occur *only* when the atomic orbitals involved have very similar energies. It is possible to hybridize $2s$ with $2p$ orbitals, or $3s$ with $3p$ orbitals, for example, but not $2s$ with $3p$ orbitals.

7.4   Beryllium Chloride, BeCl₂ (*sp* Hybridization)

The electron configuration of beryllium, Be, is $1s^2 2s^2$, and it appears that in its ground state the atom should not form covalent bonds at all. However, in an excited state an electron from the $2s$ orbital has been promoted to a $2p$ orbital so that the configuration is $1s^2 2s^1 2p^1$, which means that two unpaired electrons are available for forming covalent bonds with atoms that can share electrons.

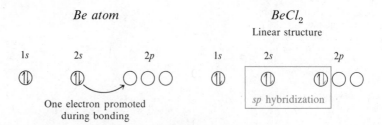

One electron promoted
during bonding

sp hybridization

In the gaseous state BeCl₂ is a linear molecule, Cl—Be—Cl; all three atoms lie in a straight line. The middle atom in a linear set of three atoms hybridizes so that the valence s orbital and one of the valence p orbitals combine to give two equivalent sp hybrid orbitals with a linear (straight-line) geometry (Fig. 7-14). In BeCl₂ these sp orbitals overlap with p orbitals of the chlorine atoms to form σ bonds.

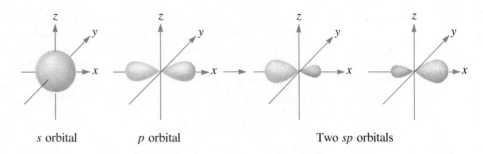

Figure 7-14. Hybridization of an s orbital and a p orbital (of the same atom) to produce two sp hybrid orbitals.

Other atoms that exhibit *sp* hybridization include the mercury atom in the linear $HgCl_2$ molecule (Fig. 7-2) and the zinc atom in $Zn(CH_3)_2$, which contains a linear C—Zn—C arrangement.

7.5 Boron Trifluoride, BF₃ (*sp*² Hybridization)

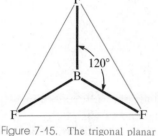

Figure 7-15. The trigonal planar structure of BF_3.

An atom surrounded by a trigonal planar arrangement of lone pairs and bonding pairs can form sp^2 hybrid orbitals directed at the corners of an equilateral triangle. One s orbital and two p orbitals combine to form these sp^2 hybrid orbitals.

The structures of BF_3 and BCl_3 suggest sp^2 hybridization for boron in these compounds. Experimental evidence shows that there are three equivalent B—F bonds in boron trifluoride. All four atoms of this molecule lie in the same plane, with the boron atom in the center and the three fluorine atoms at the corners of an equilateral triangle (a trigonal planar structure, Fig. 7-15). A similar structure is observed for BCl_3 (Fig. 7-5).

The electron configuration of boron in the ground state is $1s^2 2s^2 2p^1$. During a chemical reaction with fluorine (or chlorine), a 2s electron of boron appears to be promoted to a 2p orbital, giving the configuration $1s^2 2s^1 2p^1 2p^1$. The 2s orbital and two of the 2p orbitals hybridize to form three sp^2 hybrid orbitals (see Fig. 7-16).

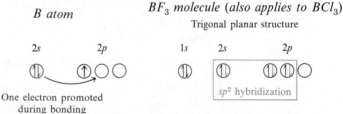

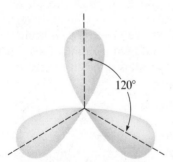

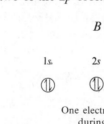

Figure 7-16. The shape and spatial orientation of trigonal planar, sp^2, hybrid orbitals.

Other atoms that exhibit sp^2 hybridization include the nitrogen atom in NO_3^- and ClNO, the sulfur atom in SO_2, and the carbon atom in CH_3^+.

7.6 Other Types of Hybridization

An atom with either a trigonal bipyramidal or an octahedral arrangement of bonding electron pairs and lone pairs cannot form hybrid orbitals using only its four s and p valence orbitals. To accommodate five pairs of electrons in a trigonal bipyramidal arrangement, an atom must hybridize five atomic orbitals—the s orbital, the three p orbitals, and one of the d orbitals in its valence shell (or shells)—giving five sp^3d hybrid orbitals. With an octahedral arrangement of six electron pairs, an atom hybridizes six atomic orbitals—the s orbital, the three p orbitals, and two of the d orbitals in its valence shell—giving six sp^3d^2 hybrid orbitals.

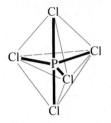

Figure 7-17. Structure of phosphorus (V) chloride, PCl_5; sp^3d hybridization.

In a molecule of phosphorus(V) chloride, PCl_5 (Fig. 7-17), there are five P—Cl bonds (and thus five pairs of valence electrons about the phosphorus atom) directed toward the corners of a trigonal bipyramid. The electron configuration of an uncombined phosphorus atom is $1s^22s^22p^63s^23p^13p^13p^13d^0$. To form a set of sp^3d hybrid orbitals on the phosphorus atom in PCl_5, one of the s valence electrons must be promoted into a d orbital and then the singly occupied orbitals must hybridize to give the five sp^3d orbitals.

P atom

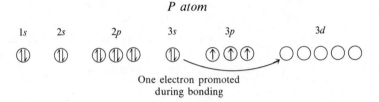

One electron promoted
during bonding

PCl_5 molecule

Trigonal bipyramidal structure

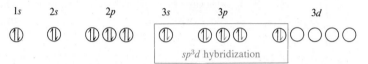

Other atoms that exhibit sp^3d hybridization include the sulfur atom in SF_4 (Fig. 7-8) and the chlorine atom in ClF_3 [Fig. 7-4(a)] and in ClF_4^+.

The sulfur atom in sulfur(VI) fluoride, SF_6, exhibits sp^3d^2 hybridization. A molecule of sulfur(VI) fluoride has six fluorine atoms surrounding a single sulfur atom (Fig. 7-18). To bond six fluorine atoms the sulfur atom must provide six bonding orbitals. The electron configuration of sulfur in the ground state is $1s^22s^22p^63s^23p^23p^13p^1$. Promotion of an s and a p valence electron to d orbitals gives the configuration $1s^22s^22p^63s^13p^13p^13p^13d^13d^1$, and hybridization of the singly occupied orbitals gives a set of six sp^3d^2 hybrid orbitals directed toward the corners of an octahedron.

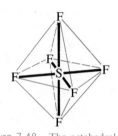

Figure 7-18. The octahedral configuration of SF_6; sp^3d^2 hybridization.

S atom

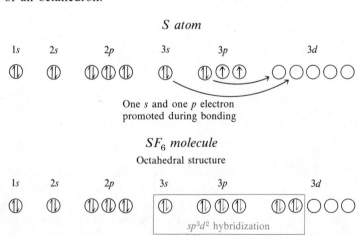

One s and one p electron
promoted during bonding

SF_6 molecule

Octahedral structure

Other atoms that exhibit sp^3d^2 hybridization include the phosphorus atom in PCl_6^-, the iodine atom in IF_6^+, IF_5, ICl_4^- (Table 7-2), and IF_4^- [Fig. 7-9(a)], and the xenon atom in XeF_4.

7.7   Prediction of Hybrid Orbitals

The purpose in formulating hybrid orbitals is to provide a model that describes how electrons can be localized in the regions between bonded atoms in a given geometry. In the preceding four sections we have described hybrid orbitals for the common distributions of bonding electrons and lone pairs. We assign an atom's hybridization after we know the geometry of the atoms around it. The idea of hybrid orbitals has little predictive value—we cannot say in advance, for example, that the oxygen atom in water uses sp^3 hybrid orbitals.

The geometrical arrangements characteristic of the various sets of hybrid orbitals are shown in Table 7-3. To find the hybridization of a central atom, we first

Table 7-3   Geometrical Arrangements Characteristic of Hybrid Orbitals

Hybridized Orbitals	Geometrical Arrangement	
sp	Linear (180° angle)	
sp^2	Trigonal planar (120° angle)	
sp^3	Tetrahedral (109.5° angle)	
sp^3d	Trigonal bipyramidal (120° and 90° angles)	
sp^3d^2	Octahedral (90° angle)	

determine the geometry of its regions of high electron density and then assign the set of hybridized orbitals from Table 7-3 that corresponds to this arrangement.

EXAMPLE 7.8   What is the hybridization of the nitrogen atom in the ammonium ion, NH_4^+?

As described in Example 7.3, the nitrogen atom in NH_4^+ has a tetrahedral arrangement of regions of high electron density. This corresponds to sp^3 hybridization of nitrogen (Table 7-3).

EXAMPLE 7.9   Urea, $NH_2C(O)NH_2$, is sometimes used as a source of nitrogen in fertilizers. What is the hybridization of the nitrogen and carbon atoms in urea?

The Lewis structure of urea is

The nitrogen atoms are surrounded by four regions of high electron density, which will arrange themselves in a tetrahedral geometry (Table 7-1). From Table 7-3 we can see that the hybridization that gives such a tetrahedral arrangement is sp^3. This is the hybridization of the nitrogen atoms in urea.

The carbon atom is surrounded by three regions of electron density, positioned in a trigonal planar arrangement (Table 7-1). Since sp^2 hybridization leads to this arrangement (Table 7-3), it is the hybridization of the carbon atom in urea.

Figure 7-19.   The ethylene molecule. (a) The overlap diagram of two sp^2 hybridized carbon atoms and four s orbitals from four hydrogen atoms. There are four σ C—H bonds, one σ C—C bond, and one π C—C bond (making the net carbon-carbon bond a double bond). The dashed lines, each connecting two lobes, indicate the side-by-side overlap of the two unhybridized p orbitals. The sp^2 hybrid orbitals are shown in grey, and the unhybridized p orbitals are shown in color. The sp^2 hybrid orbitals lie in a plane with the unhybridized p orbitals extending above and below the plane and perpendicular to it. (b) The overall outline of the bonding molecular orbitals in ethylene. The two portions of the π bonding orbital (shown in color), resulting from the side-by-side overlap of the unhybridized p orbitals, are above and below the plane.

7.8 Ethylene, C_2H_4, and Acetylene, C_2H_2 (π Bonding)

If we apply the VSEPR Theory (Section 7.1) to the Lewis structure of ethylene, C_2H_4,

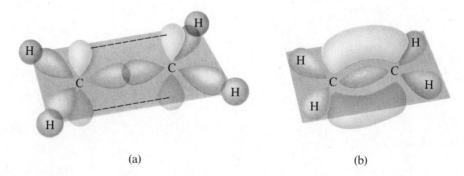

(a)   (b)

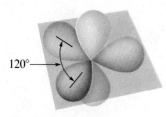

120° →

Figure 7-20. Diagram illustrating the three trigonal sp^2 hybrid orbitals of the carbon atom, which lie in the same plane, and the one unhybridized p orbital (shown in color), which is perpendicular to the plane.

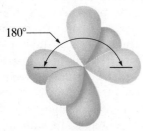

180° →

Figure 7-21. Diagram of the two linear sp hybrid orbitals of the carbon atom, which lie in a straight line, and the two unhybridized p orbitals (shown in color).

each carbon atom should be surrounded by one carbon atom and two hydrogen atoms, forming a trigonal planar array. Thus the σ bonds from each carbon atom should be formed using a set of sp^2 hybrid orbitals directed toward the apexes of a triangle. These orbitals form the C—H single bonds and one part of the C=C double bond, the σ bond (Fig. 7-19). The second part of the C=C bond, which gives it a bond order of 2, results from the $2p$ orbital that is not involved in the sp^2 hybridization.

Only two $2p$ orbitals are involved in mixing with the $2s$ orbital, so one $2p$ orbital is left unhybridized. As shown in Fig. 7-20, the unhybridized p orbital (shown in color) is perpendicular to the plane of the sp^2 orbitals. Thus when two sp^2 hybridized carbon atoms come together, one sp^2 orbital on each of them can overlap to form a σ bond while the unhybridized $2p$ orbitals can overlap in a side-by-side fashion. This overlap results in a pi (π) bond of the type discussed in Chapter 6. The overlap to form the π bond is not as efficient as the overlap to form the σ bond, and the π bond is therefore weaker than the σ bond. The two carbon atoms of ethylene are thus bound together by two kinds of bonds—one σ and one π—giving a double bond.

Note that in an ethylene molecule the four hydrogen atoms and the two carbon atoms are all in the same plane. If the two planes of sp^2 hybrid orbitals are tilted, the π bond will be weakened since the p orbitals that form it cannot overlap effectively if they are not parallel. A planar configuration for the ethylene molecule is the most stable form.

As we have seen for $BeCl_2$, if just one $2p$ orbital hybridizes with a $2s$ orbital, two sp hybrid orbitals result. This arrangement leaves two $2p$ orbitals unhybridized (Fig. 7-21). When sp hybrid orbitals of two carbon atoms combine, they overlap end-to-end to form a σ bond (Fig. 7-22). The remaining sp orbital on each carbon may be used to bond with another atom such as hydrogen, forming a linear molecule such as acetylene, H—C≡C—H. In addition to this, as indicated in Fig. 7-22, the two sets of unhybridized p orbitals are positioned so that they overlap side-by-side and hence form two π bonds. The two carbon atoms of acetylene are thus bound together by one σ bond and two π bonds, giving a triple bond.

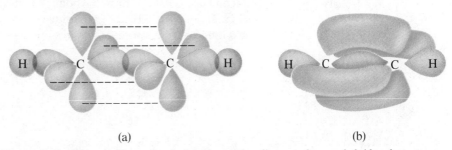

(a) (b)

Figure 7-22. The acetylene molecule. (a) The overlap diagram of two sp hybrid carbon atoms and two s orbitals from two hydrogen atoms. There are two σ C—H bonds, one σ C—C bond, and two π C—C bonds (making the net carbon-carbon bond a triple bond). The dashed lines, each connecting two lobes, indicate the side-by-side overlap of the four unhybridized p orbitals. The sp hybrid orbitals are shown in grey, and the unhybridized p orbitals are shown in color. (b) The overall outline of the bonding orbitals in acetylene. The π bonding orbitals (in color) are positioned with one above and below the line of the σ bonds and the other behind and in front of the line of the σ bonds.

FOR REVIEW

SUMMARY

The **molecular structure,** or three-dimensional arrangement of the atoms in a molecule or ion, is described in terms of **bond distances** and **bond angles.** Approximate bond angles can be predicted using the **Valence Shell Electron-Pair Repulsion (VSEPR) Theory.** According to this theory, regions of high electron density (either bonding pairs or lone pairs) located about a central atom repel each other, and the most stable arrangement is reached when they are as far away from each other as possible. For two regions of high electron density, the most stable is a **linear structure;** for three, a **trigonal planar structure;** for four, a **tetrahedral structure;** for five, a **trigonal bipyramidal structure;** and for six, an **octahedral structure.** Since a lone pair of electrons is undetectable by the techniques used to determine molecular structure, the expected arrangement of the regions of high electron density and the actual molecular structure may appear different. In ammonia three bonding pairs and one lone pair are in a tetrahedral arrangement, but the nitrogen and hydrogen atoms form a trigonal pyramid. Other geometries are illustrated in Table 7-2.

The orbitals a central atom uses to form bonds with other atoms may be described as **hybrid orbitals,** combinations of some or all of its valence atomic orbitals. These hybrid orbitals may form **sigma (σ) bonds** directed toward other atoms of the molecule or may contain lone pairs. The type of hybridization about a central atom may be determined from the geometry of the regions of high electron density about it. Two such regions imply sp hybridization; three, sp^2 hybridization; four, sp^3 hybridization; five, sp^3d; and six, sp^3d^2 hybridization.

Atomic orbitals that are not used in hybridization are available to form **pi (π) bonds.** The carbon atoms in ethylene, C_2H_4, are sp^2 hybridized, and the remaining p orbital on each is used to form a π bond, resulting in a double bond. The carbon atoms in acetylene, HCCH, are sp hybridized, and the two remaining p orbitals on each are used to form two π bonds, resulting in a triple bond.

KEY TERMS AND CONCEPTS

Bond angle (7.1)
Bond distance (7.1)
Double bond (7.7)
Hybridization (7.3)
Linear structure (7.1)
Molecular structure (7.1)
Octahedral structure (7.1)

Pi (π) bond (7.7)
Promotion of electrons (7.3)
Sigma (σ) bond (7.3)
Tetrahedral structure (7.1)
Trigonal bipyramidal structure (7.1)
Trigonal planar structure (7.1)

Triple bond (7.8)
Valence Shell Electron-Pair Repulsion (VSEPR) Theory (7.1)

EXERCISES

Molecular Structure

1. What are the bond angles in molecules having the following structures: linear, trigonal planar, tetrahedral, trigonal bipyramidal, and octahedral?

2. Predict the structure of each of the following molecules:
   (a) $AlCl_3$ (c) BeH_2 (e) $SiCl_4$
   (b) AsF_5 (d) SF_6

3. Predict the structure of each of the following ions:
   (a) CH_3^+ (c) PH_4^+ (e) $TlCl_2^+$
   (b) IF_6^+ (d) $SnCl_5^-$

4. Predict the structure of each of the following molecules:
   (a) H_2O (c) $SeCl_4$ (e) BrF_3
   (b) NH_3 (d) ClF_5 (f) XeF_4

5. Predict the structure of each of the following ions:
   (a) H_3O^+ (c) ClF_4^+ (e) SF_5^+
   (b) NH_2^- (d) ICl_4^-

6. Predict the structure of each of the following molecules:
   (a) Cl_2CO (c) HCN
   (b) $POCl_3$ (d) IOF_5

7. Predict the structure of each of the following molecules or ions:
   (a) $ClNO$ (d) $XeOF_4$ (f) SO_3^{2-}
   (b) SO_2 (e) HCO_2^- (g) XeO_2F_2
   (c) $SOCl_2$
   In each case, xenon, or the least electronegative atom other than hydrogen, is the central atom.

8. Predict the geometry around the indicated atom or atoms:
   (a) The nitrogen atom in nitric acid, HNO_3 $(HONO_2)$.
   (b) The nitrogen atom and the two carbon atoms in glycine, the simplest amino acid. The skeleton structure of glycine is

   (c) The oxygen atoms in hydrogen peroxide, H_2O_2.
   (d) The sulfur atom in sulfuric acid, H_2SO_4 $[(HO)_2SO_2]$.
   (e) The central oxygen atom in the ozone molecule, O_3.
   (f) The carbon atom in the Freon molecule, CF_2Cl_2.
   (g) The carbon atoms in benzene. The skeleton structure of benzene is

   (h) The carbon atom in the carbonate ion, CO_3^{2-}; the bicarbonate ion, HCO_3^- $(HOCO_2^-)$; and carbonic acid, H_2CO_3 $[(HO)_2CO]$.

Hybrid Orbitals

9. What is a hybrid orbital? Illustrate your answer with the hybrid orbitals that may be formed by carbon.

10. What are the angles between the following hybrid orbitals: sp, sp^2, sp^3, sp^3d, sp^3d^2?

11. Identify the hybridization of each carbon atom in each of the following molecules:
    (a) carbon tetrachloride, CCl_4
    (b) vinyl chloride, the compound used to make the plastic PVC (polyvinyl chloride), $H_2C=CHCl$
    (c) acetic acid, the compound that gives vinegar its acidic taste,

    (d) calcium cyanamide, $Ca^{2+}(NCN^{2-})$, used as a fertilizer and in preparation of the plastic Melmac

12. Identify the hybridization of the indicated atom in each of the following molecules:
    (a) Si in $SiCl_4$ (d) Sb in SbF_5
    (b) Hg in $HgCl_2$ (e) Te in TeF_6
    (c) Al in $AlCl_3$

13. Identify the hybridization of the indicated atom in each of the following molecules:
    (a) N in NH_3 (d) P in PCl_3
    (b) O in H_2O (e) I in IF_5
    (c) S in SO_2 (f) Cl in ClF_3

14. Show the distribution of valence electrons in the orbitals of the central atom in each of the following molecules or ions (1) when it is the free atom in the ground state, (2) following electron promotion, and (3) following bond formation:
    (a) $GeCl_4$ (c) BrF_3 (e) IF_6^+ (consider
    (b) $AlCl_3$ (d) BrF_5 this ion as being formed from I^+ and 6 F atoms

15. Formaldehyde, H_2CO, contains a carbon-oxygen double bond. Describe the hybrid orbitals of the carbon atom that form the σ bonds in this molecule, and the atomic orbitals of oxygen and carbon that give the π bond. Note that it is not necessary to hybridize the oxygen atom.

16. Hydrogen cyanide, HCN, contains a carbon-nitrogen triple bond. Describe the hybrid orbitals of the carbon atom that form the σ bonds in this molecule, and the atomic orbitals of nitrogen and carbon that give the π bonds. Note that it is not necessary to hybridize the nitrogen atom.

17. Write Lewis structures for the molecules and ions listed below. Utilizing both the Valence Shell Electron-Pair Repulsion Theory and the concept of hybridization and remembering that single bonds are σ bonds, double bonds are a σ plus a π bond, and triple bonds are a σ plus two π bonds, indicate the type of hybridization for each central atom in the listed molecules and ions.
    (a) BH_4^- (g) SeO_2
    (b) $AlCl_3$ (h) N_2O_4 (contains
    (c) $ClNO$ an N—N bond)
    (d) BO_3^{3-} (i) $POCl_3$
    (e) PO_4^{3-} (j) $BrCN$
    (f) $SnCl_3^-$

18. Some forms of solid phosphorus(V) chloride contain PCl_4^+ and PCl_6^- ions. Upon heating, gaseous molecules of PCl_5 are formed. Describe the hybridization of the phosphorus atom in PCl_4^+, PCl_6^-, and PCl_5.

Additional Exercises

19. Predict the geometry about the indicated atom in each of the following compounds and describe its hybridization:
    (a) The phosphorus in phosphoric acid, H_3PO_4 [$(HO)_3PO$].
    (b) The nitrogen atom in pyridine, C_5H_5N. The skeleton structure of pyridine is

    (c) The thallium atom in trimethylthallium, $Tl(CH_3)_3$.
    (d) The sulfur atoms in disulfur decafluoride, S_2F_{10} (contains an S—S bond).
    (e) The nitrogen atoms in hydrazine, H_2NNH_2 (contains an N—N bond).
    (f) The arsenic atom in arsenic trichloride difluoride, $AsCl_3F_2$.
    (g) The central carbon atom in acetone,

$$CH_3\overset{\displaystyle O}{\overset{\|}{C}}CH_3.$$

20. XF_3 has a trigonal pyramidal molecular structure. To which main group of the Periodic Table does X belong?

21. In the compound H—X≡X—H, X exhibits sp^2 hybridization. To which main group of the Periodic Table does X belong?

22. If XCl_3 exhibits a dipole moment, X could be a member of either of two groups in the Periodic Table. Which ones?

23. $B(OH)_3$, boric acid, reacts with water according to the following equation:

$$B(OH)_3 + H_2O \longrightarrow H^+(aq) + B(OH)_4^-(aq)$$

How does the hybridization of the boron atom change during this reaction?

24. Elemental sulfur consists of covalently bonded, puckered rings of eight sulfur atoms, S_8. The S—S—S angles in the ring are 107.9°. The production of sulfuric acid, $SO_2(OH)_2$, from sulfur involves oxidation of sulfur to SO_2, then to SO_3, and finally reaction with water to give sulfuric acid. Trace the changes in the hybridization of the sulfur atom during this sequence of reactions.

25. Although the location of the lone pairs in its hybrid orbitals cannot be observed, the oxygen atom in water is believed to exhibit sp^3 hybridization. Explain.

8

CHEMICAL REACTIONS AND THE PERIODIC TABLE

Chemistry is the study of the properties of matter and of the reactions by which matter is converted from one form to another. In the preceding four chapters we have considered some electronic and structural properties of atoms, molecules, and ions. In this chapter we will examine their chemical properties. In particular we will look at the kinds of chemical reactions by which common elements and compounds can be interconverted, and we will see how the products of many common types of chemical reactions can be predicted. The chemistry of the eleven elements and industrial inorganic compounds produced in greatest quantity in the United States will be used as illustration.

Much of the behavior of an element can be predicted from its position in the Periodic Table. As an example, consider the use that Mendeleev made of his periodic table in predicting the properties of unknown elements. Mendeleev's table included only 62 elements, the number known at that time. It also contained six places that were left vacant as a result of placing known elements of similar chemical properties in the same group. Mendeleev predicted the properties of the elements that would fit these gaps from the known chemical behavior of their neighbors in the Periodic Table. The elements are scandium (Sc), gallium (Ga), germanium (Ge), technetium (Tc), rhenium (Re), and polonium (Po), and they all have properties similar to those predicted by Mendeleev.

A comparison of the properties predicted by Mendeleev for germanium, which he called eka-silicon, and those determined experimentally for the element after it was isolated is given in Table 8-1. You can see that the properties are intermediate between those of silicon and tin, the neighbors of germanium in Group IVA.

Table 8-1   Predicted Properties for Eka-silicon and Observed Properties of Silicon, Tin, and Germanium

	Silicon	Tin	Predicted for Eka-silicon	Germanium
Atomic weight	28	118	72	72.59
Specific gravity	2.3	7.3	5.5	5.3
Color	Gray nonmetal	White metal	Gray metal	Gray metal
Oxidation number with oxygen	4	4	4	4
Reaction with acid	No reaction	Slow reaction	Very slow reaction	Slow reaction with conc. acid
Reaction with base	Slow reaction	Slow reaction	Slow reaction	Slow reaction with conc. base
Formula of chloride	$SiCl_4$	$SnCl_4$	$EkCl_4$	$GeCl_4$
Specific gravity of chloride	1.5	2.2	1.9	1.88
Boiling point of chloride (°C)	57.6°	114°	below 100°	83°

You can apply the same principles Mendeleev used to predict the properties of elements to correlate and recall their behavior. The Periodic Table provides a powerful framework for organizing the chemical behavior of the elements.

8.1   Classification of Chemical Compounds

Before we can begin our examination of the properties of elements, we need to consider several of the types of compounds that they form.

1. SALTS.   A **salt** is an ionic compound composed of positive ions (cations) and negative ions (anions). (See Section 5.1.) Simple salts are formed by the combination of a metal and a nonmetal—the metal forms the cation, and the nonmetal, the anion. In more complex salts the cation may be a polyatomic positive ion such as NH_4^+ or PCl_4^+, and the anion, a polyatomic negative ion such as NO_3^- or SO_4^{2-}. Ionic compounds that contain hydroxide or oxide ions are called bases rather than salts. (See Part 2 of this section.)

Compounds that are composed of ions are held together in the solid state by ionic bonds, strong electrostatic attractions between oppositely charged ions (Section 5.1). When soluble salts dissolve in water, the ions separate (Fig. 8-1) and are free to move about independently. The process for the dissolution (dissolving) of

NaCl(s)

(a)

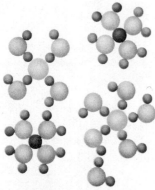

$Na^+(aq) + Cl^-(aq)$

(b)

Figure 8-1. (a) The distribution of ions in solid sodium chloride. Black spheres represent Na^+; colored spheres represent Cl^-. Each ion is in contact with six ions of opposite charge. (b) The distribution of ions in an aqueous solution of sodium chloride. Each ion is free to move independently because it is surrounded by water molecules and separated from ions of opposite charge.

the salt sodium chloride in water may be represented by the following equation, in which the abbreviation *aq* indicates that the ions are separated, surrounded by water molecules, and moving independently:

$$NaCl(s) \xrightarrow{\text{H}_2\text{O}(l)} Na^+(aq) + Cl^-(aq) \tag{1}$$

2. ACIDS AND BASES. A compound that donates a hydrogen ion, or proton (H^+), to another compound is called a **Brönsted acid,** or simply an **acid,** after Johannes Brönsted, a Danish chemist. A compound that accepts a hydrogen ion is called a **Brönsted base,** or simply a **base.** Other definitions have been developed that emphasize other aspects of the behavior of acids and bases, but for now we will concentrate on this one. The following reactions illustrate the transfer of a hydrogen ion from a Brönsted acid to a Brönsted base:

$$
\begin{array}{ccc}
\text{Acid} & \text{Base} & \\
HCl(g) & + \ NH_3(g) & \longrightarrow \ NH_4Cl(s)
\end{array}
\tag{2}
$$

$$H_2SO_4(l) + 2NaOH(s) \longrightarrow Na_2SO_4(s) + 2H_2O(l) \tag{3}$$

Salts are produced in these reactions: NH_4Cl contains the ions NH_4^+ and Cl^-, and $NaHSO_4$ contains Na^+ and HSO_4^-.

When dissolved in water, acids donate hydrogen ions to water molecules forming **hydronium ions, H_3O^+,** plus whatever anion is produced when the acid loses a hydrogen ion (Fig. 8-2).

$$HCl(g) + H_2O(l) \longrightarrow H_3O^+(aq) + Cl^-(aq) \tag{4}$$

$$H_2SO_4(l) + H_2O(l) \longrightarrow H_3O^+(aq) + HSO_4^-(aq) \tag{5}$$

$$HSO_4^-(aq) + H_2O(l) \longrightarrow H_3O^+(aq) + SO_4^{2-}(aq) \tag{6}$$

Thus an acid forms hydronium ions when dissolved in water. For convenience the hydronium ion is sometimes written as H^+ or $H^+(aq)$, but remember that H^+ and $H^+(aq)$ are abbreviations; a hydrogen ion in water is always associated with at least one water molecule. Note that in forming a hydronium ion water accepts a hydrogen ion and therefore behaves as a base.

When a base is added to an aqueous solution of an acid, protons are donated from the hydronium ions to the base. For example, a solution of hydrogen chloride in water contains hydronium ions and chloride ions. If gaseous ammonia is bubbled through the solution, the ammonia molecules pick up protons from the hydronium ions present in the solution.

$$NH_3(g) + H_3O^+(aq) + Cl^-(aq) \longrightarrow NH_4^+(aq) + Cl^-(aq) + H_2O(l) \tag{7}$$

The product of this reaction is a solution of ammonium chloride, NH_4Cl, a salt.

An acid is classified as strong or weak, depending on the extent to which it reacts with water to form hydronium ions. Upon reaction with water, a **strong acid** ionizes in dilute solution giving a 100% yield (or very nearly so) of hydronium ion and the anion of the acid. The strong acid HCl reacts with water and gives a 100% yield of H_3O^+ and Cl^-.

$$HCl(g) + H_2O(l) \longrightarrow H_3O^+(aq) + Cl^-(aq)$$

Sulfuric acid, H_2SO_4, reacts with water to give a 100% yield of hydronium ion and the hydrogen sulfate ion, HSO_4^-. The hydrogen sulfate ion is a weak acid.

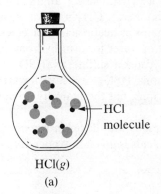

—HCl molecule

$HCl(g)$

(a)

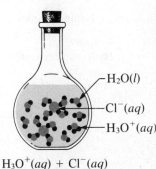

—$H_2O(l)$

—$Cl^-(aq)$

—$H_3O^+(aq)$

$H_3O^+(aq) + Cl^-(aq)$

(b)

Figure 8-2. (a) Molecules of hydrogen chloride in the gas phase. Each chlorine atom (colored sphere) is bonded to a hydrogen atom (black sphere) by a covalent bond. (b) A solution of hydrogen chloride in water (hydrochloric acid). The hydrogen ions are bonded to water molecules by a coordinate covalent bond giving a solution of H_3O^+ and Cl^- ions.

The six common strong acids and the principal anions they form in water are listed in Table 8-2.

Table 8-2   Common Strong Acids and Their Anions

Molecular Formula	Name	Anion	Name of Anion
HCl	Hydrogen chloride	Cl^-	Chloride
HBr	Hydrogen bromide	Br^-	Bromide
HI	Hydrogen iodide	I^-	Iodide
HNO_3	Nitric acid	NO_3^-	Nitrate
$HClO_4$	Perchloric acid	ClO_4^-	Perchlorate
H_2SO_4	Sulfuric acid	HSO_4^-	Hydrogen sulfate

A **weak acid** ionizes in dilute solution giving a small percent yield of hydronium ion (ordinarily 10% or less). Thus a solution of a weak acid consists primarily of covalent molecules (or ions) of the original acid with lesser amounts of hydronium ion and the anion of the acid. When acetic acid, CH_3CO_2H, a weak acid, is added to water, the majority of the acid remains as acetic acid molecules. Only about 1% of the acetic acid molecules react with water to form hydronium ions and acetate ions, $CH_3CO_2^-$. When sodium hydrogen sulfate, $NaHSO_4$, is added to water, the weakly acidic hydrogen sulfate ions, HSO_4^-, react with water to give a small yield of hydronium ions and sulfate ions, but the majority do not react and remain as HSO_4^-.

Some common weak acids and the principal anions they form in water are listed in Table 8-3. A more extensive listing of weak acids is given in Appendix G.

Table 8-3   Common Weak Acids and Their Anions

Molecular Formula	Name	Anion	Name of Anion
CH_3CO_2H	Acetic acid	$CH_3CO_2^-$	Acetate
H_2CO_3	Carbonic acid	HCO_3^-	Hydrogen carbonate
HF	Hydrogen fluoride	F^-	Fluoride
H_2S	Hydrogen sulfide	HS^-	Hydrogen sulfide
HNO_2	Nitrous acid	NO_2^-	Nitrite
H_3PO_4	Phosphoric acid	$H_2PO_4^-$	Dihydrogen phosphate
HSO_4^-	Hydrogen sulfate ion	SO_4^{2-}	Sulfate

The anion of a weak acid is a base. It will accept a proton to form the weak acid. For example, acetate ion reacts with hydrogen ion to form acetic acid. Hence, sodium acetate, the sodium salt of acetic acid, reacts with a solution of perchloric acid to form acetic acid and sodium perchlorate:

$$NaCH_3CO_2 + HClO_4 \longrightarrow CH_3CO_2H + NaClO_4 \qquad (8)$$

When the salt of a weak acid dissolves in water, the anions will accept protons from the water to a very limited extent. Sodium acetate will react with water and give a small yield of acetic acid molecules and hydroxide ions. A 0.05 M solution of sodium acetate gives only a 0.01% yield of acetic acid molecules and hydroxide ions. Even so, the one hydroxide ion produced for each acetic acid molecule formed makes the hydroxide ion concentration higher than that in pure water. Thus sodium acetate (and salts of other weak acids) increase the hydroxide ion concentration when dissolved in water, as do all compounds that act as bases. A salt of a weak acid is classified as a **weak base** since it produces a low yield of hydroxide ions when dissolved in water.

Soluble bases that contain hydroxide ion dissolve in water to give solutions containing cations and hydroxide ions. Since soluble metal hydroxides give a high concentration of hydroxide ions, they are **strong bases.** The soluble metal hydroxides include the hydroxides of the metals of Group IA and of strontium and barium.

$$NaOH(s) \xrightarrow{H_2O} Na^+(aq) + OH^-(aq) \tag{9}$$

$$Ba(OH)_2(s) \xrightarrow{H_2O} Ba^{2+}(aq) + 2OH^-(aq) \tag{10}$$

When an acid is added to an aqueous solution of a strong base, its molecules donate protons to the hydroxide ions. Adding nitric acid to a solution of barium hydroxide, for example, gives a solution of the salt barium nitrate, $Ba(NO_3)_2$, and water.

$$2HNO_3(l) + Ba^{2+}(aq) + 2OH^-(aq) \longrightarrow$$
$$Ba^{2+}(aq) + 2NO_3^-(aq) + 2H_2O(l) \tag{11}$$

A solution of ammonia in water consists primarily of solvated ammonia molecules, $NH_3(aq)$. When an acid is added, protons are transferred to the ammonia molecules producing ammonium ions, NH_4^+, and the anion of the acid.

$$HNO_3(l) + NH_3(aq) \longrightarrow NH_4^+(aq) + NO_3^-(aq) \tag{12}$$

Water itself can act as an acid with ammonia, and ammonia will accept protons from water to a very limited extent.

$$NH_3(aq) + H_2O(l) \rightleftharpoons NH_4^+(aq) + OH^-(aq) \tag{13}$$

The ammonia in a 1 M solution of NH_3 exists as 99.6% $NH_3(aq)$ and gives only a 0.4% yield of $NH_4^+(aq)$ and $OH^-(aq)$. Ammonia is regarded as a weak base since it gives only a low yield of hydroxide ion in water.

Note that water can behave as either an acid, as shown in Equation (13), or a base, as shown in Equations (4) through (6), depending on the nature of the substance dissolved in it. This dual behavior is characteristic of other compounds as well.

3. ELECTROLYTES AND NONELECTROLYTES.   Both ionic compounds such as sodium hydroxide and potassium nitrate (that dissolve in water giving solutions of ions) and covalent compounds such as sulfuric acid and ammonia (that react with water to form ions in solution) are called **electrolytes.** Ionic compounds and certain covalent compounds such as the strong acids HCl and HNO_3, which give essentially a 100% yield of ions, are called **strong electrolytes.** Compounds such as weak acids and ammonia, which give a low percentage yield of ions in water, are called **weak electrolytes. Nonelectrolytes** are compounds that do not ionize when they dissolve in water. Only covalent compounds can be nonelectrolytes. Many compounds of carbon such as methane, CH_4, benzene, C_6H_6, ethanol, C_2H_5OH, ether, $(C_2H_5)_2O$, and formaldehyde, CH_2O, are nonelectrolytes. A few inorganic

compounds such as nitrous oxide, N_2O, phosphine, PH_3, and nitrogen(III) chloride, NCl_3, are nonelectrolytes.

8.2 Classification of Chemical Reactions

The chemical behavior of the elements is characterized by the ways in which they and their compounds react. Several common types of reactions are discussed below.

1. ADDITION, OR COMBINATION, REACTIONS.   Two or more substances combine to give another substance.

$$S + O_2 \xrightarrow{\triangle} SO_2 \tag{1}$$

$$Ca + Br_2 \longrightarrow CaBr_2 \tag{2}$$

$$2K_2S + 3O_2 \xrightarrow{\triangle} 2K_2SO_3 \tag{3}$$

$$CaO + SO_3 \longrightarrow CaSO_4 \tag{4}$$

Remember that most compounds that contain both metals and nonmetals are ionic. Thus the products in Equations (2), (3), and (4) are salts.

2. DECOMPOSITION REACTIONS.   One compound breaks down (decomposes) into two or more substances.

$$2HgO \xrightarrow{\triangle} 2Hg + O_2 \tag{5}$$

$$CaCO_3 \xrightarrow{\triangle} CaO + CO_2 \tag{6}$$

$$2Cu(NO_3)_2 \xrightarrow{\triangle} 2CuO + 4NO_2 + O_2 \tag{7}$$

$$2NaHSO_4 \xrightarrow{\triangle} Na_2SO_4 + SO_3 + H_2O \tag{8}$$

3. METATHETICAL REACTIONS.   Two compounds exchange parts. We can expect a metathetical reaction to occur when an insoluble compound, a gas, a nonelectrolyte, or a weak electrolyte is formed as a product.

When solutions of the salts calcium chloride, $CaCl_2$, and silver nitrate, $AgNO_3$, are mixed, insoluble solid silver chloride, $AgCl$, forms and a solution of the salt calcium nitrate, $Ca(NO_3)_2$, remains.

$$Ca^{2+}(aq) + 2Cl^-(aq) + 2Ag^+(aq) + 2NO_3^-(aq) \longrightarrow$$
$$2AgCl(s) + Ca^{2+}(aq) + 2NO_3^-(aq) \tag{9}$$

Solid sodium chloride reacts with concentrated sulfuric acid to give gaseous hydrogen chloride and sodium hydrogen sulfate.

$$NaCl(s) + H_2SO_4(l) \longrightarrow HCl(g) + NaHSO_4(s) \tag{10}$$

A solution of $KC_2H_5CO_2$, the potassium salt of the weak acid propionic acid, reacts with a solution of HCl to give a solution containing $C_2H_5CO_2H$ molecules (propionic acid, a weak acid) and the salt KCl (potassium chloride).

$$K^+(aq) + C_2H_5CO_2^-(aq) + H_3O^+(aq) + Cl^-(aq) \longrightarrow$$
$$C_2H_5CO_2H(aq) + K^+(aq) + Cl^-(aq) + H_2O(l) \tag{11}$$

4. OXIDATION-REDUCTION REACTIONS.   Many of the reactions described above can also be classified as oxidation-reduction reactions. When an atom, either free or in a molecule or ion, loses electrons, it is **oxidized,** and its oxidation number (Section 5.9) increases. When an atom, either free or in a molecule or ion, gains electrons, it is **reduced,** and its oxidation number decreases.

Oxidation and reduction always occur simultaneously, for if one atom gains electrons and is reduced, a second atom must provide the electrons and thereby be oxidized. Reactions involving oxidation and reduction are referred to as **oxidation-reduction, or redox, reactions.** An example is the reaction between sodium and chlorine in which sodium is oxidized and chlorine is reduced.

$$\overset{0}{2Na} + \overset{0}{Cl_2} \longrightarrow \overset{+1\,-1}{2NaCl} \tag{12}$$

As sodium is oxidized, its oxidation number increases from 0 to $+1$, as indicated by the small numbers in Equation (12). As chlorine is reduced, its oxidation number decreases from 0 to -1. Other examples of oxidation-reduction reactions are given below. In each case, the species that is oxidized is written first.

$$\overset{0}{Zn} + \overset{+1}{H_2SO_4} \longrightarrow \overset{+2}{ZnSO_4} + \overset{0}{H_2} \tag{13}$$

$$\overset{+2}{SnCl_2} + \overset{+4}{PbCl_4} \longrightarrow \overset{+4}{SnCl_4} + \overset{+2}{PbCl_2} \tag{14}$$

$$\overset{+2}{2CO} + \overset{+2}{2NO} \longrightarrow \overset{+4}{2CO_2} + \overset{0}{N_2} \tag{15}$$

The **reducing agent** in an oxidation-reduction reaction is the reactant that gives up electrons to another reactant thereby causing the other reactant to be reduced. In Equations (12) through (15) the first reactant is the reducing agent. Since a reducing agent loses electrons, it is oxidized. The reactant that gains electrons thereby causing the other to be oxidized is called the **oxidizing agent.** The oxidizing agent picks up electrons during an oxidation-reduction reaction, so it is reduced.

Equation (12) may be classified as both an addition reaction and an oxidation-reduction reaction. Other reactions that fall into both categories include Equations (1), (2), and (3) in this section. Decomposition reactions may also be oxidation-reduction reactions. Equations (5) and (7) in this section represent such reactions.

5. ACID-BASE REACTIONS.   A hydrogen ion is transferred from a Brönsted acid to a Brönsted base in an **acid-base reaction.** Equations (2) through (8) and (12) and (13) in Part 2 of Section 8.1 represent acid-base reactions, as do Equations (10) and (11) in Part 3 of this section.

In general, an acid-base reaction gives a high yield of product when either a strong acid or a weak acid is added to a strong base, when a strong acid is added to a weak base, or when a strong acid is added to a salt containing the anion of a weak acid [Equation (11), Part 3 of this section]. These reactions occur when the pure substances or their solutions are mixed.

6. REVERSIBLE REACTIONS.   When acetic acid is added to water, hydronium ions and acetate ions are formed in low yield. When a solution containing hydronium ions is added to a solution of sodium acetate, the hydronium ions and acetate ions

react to give acetic acid. This reaction is the reverse of that which occurs when acetic acid is added to water. Equations describing reactions that can proceed in either direction are written with a double arrow ($\rightleftharpoons$):

$$CH_3CO_2H(aq) + H_2O(l) \rightleftharpoons H_3O^+(aq) + CH_3O_2^-(aq) \tag{16}$$

Such a reaction is called a **reversible reaction.**

The oxidation-reduction reaction of nitrogen with hydrogen is also a reversible reaction.

$$N_2(g) + 3H_2(g) \rightleftharpoons 2NH_3(g) \tag{17}$$

When a closed container of nitrogen and hydrogen at high pressure is heated at 300°C, ammonia is produced with a yield of about 50% once the reaction is complete (when the concentration of ammonia stops changing). If the container is then heated to 400°C, the reverse reaction occurs and some of the ammonia decomposes to nitrogen and hydrogen. The percentage of ammonia in the container is reduced to about 35% by the time the concentration of ammonia again stops changing. The extent to which each of the two opposite reactions takes place can be varied by changing the quantities of the reactants or by changing the temperature, pressure, or volume. The properties of reversible reactions will be considered more fully in Chapter 15. At this stage of your study, it is sufficient for you to recognize that a reversible reaction is one that can proceed in two opposing directions, and that the two opposing reactions proceed simultaneously.

8.3 Metals, Nonmetals, and Metalloids

The elements may be divided into three broad groups: metals, nonmetals, and metalloids. A pure **metal** is generally a good conductor of heat and electricity. It shows a metallic luster and is malleable and ductile (that is, it can be bent or drawn into sheets or wires without breaking). A pure **nonmetal** is generally a poor conductor. It normally shows no metallic luster and is brittle and nonductile in the solid state. Some elements, called **metalloids,** cannot be satisfactorily identified as being either metals or nonmetals, for they possess some of the properties of each. The metalloid silicon, for example, exhibits a bright metallic luster, but it is not a good conductor and it is brittle.

The distribution of metals, nonmetals, and metalloids in the Periodic Table is shown in Fig. 8-3. The nonmetals lie to the right and above the heavy line running from between beryllium (Be) and boron (B) at the upper left to between polonium (Po) and astatine (At) at the lower right. The metals lie to the left and below this line. About three-fourths of the elements are metals. Metalloids lie along the dividing line between metals and nonmetals.

If you are familiar with the general behavior patterns of metals and nonmetals, you can reasonably predict a great deal of the chemical behavior of an element simply from its position in the Periodic Table. The general types of chemical behavior characteristic of metals and nonmetals are outlined in Table 8-4. Metalloids may exhibit either metallic or nonmetallic behavior, depending on the conditions under which they react.

As an example of metallic and nonmetallic behavior, consider the behavior of sodium and chlorine and their compounds. Sodium (Na) is clearly identifiable as a metal; it lies in Group IA at the left of the Periodic Table. It loses its one valence

1 H																	2 He
3 Li	4 Be											5 B	6 C	7 N	8 O	9 F	10 Ne
11 Na	12 Mg											13 Al	14 Si	15 P	16 S	17 Cl	18 Ar
19 K	20 Ca	21 Sc	22 Ti	23 V	24 Cr	25 Mn	26 Fe	27 Co	28 Ni	29 Cu	30 Zn	31 Ga	32 Ge	33 As	34 Se	35 Br	36 Kr
37 Rb	38 Sr	39 Y	40 Zr	41 Nb	42 Mo	43 Tc	44 Ru	45 Rh	46 Pd	47 Ag	48 Cd	49 In	50 Sn	51 Sb	52 Te	53 I	54 Xe
55 Cs	56 Ba	[57-71] *	72 Hf	73 Ta	74 W	75 Re	76 Os	77 Ir	78 Pt	79 Au	80 Hg	81 Tl	82 Pb	83 Bi	84 Po	85 At	86 Rn
87 Fr	88 Ra	[89-103] †	104 Rf	105 Ha	106	107		109									

	57 La	58 Ce	59 Pr	60 Nd	61 Pm	62 Sm	63 Eu	64 Gd	65 Tb	66 Dy	67 Ho	68 Er	69 Tm	70 Yb	71 Lu
*LANTHANIDE SERIES															
†ACTINIDE SERIES	89 Ac	90 Th	91 Pa	92 U	93 Np	94 Pu	95 Am	96 Cm	97 Bk	98 Cf	99 Es	100 Fm	101 Md	102 No	103 Lr

Figure 8-3. Metals, nonmetals, and metalloids, in the Periodic Table. The nonmetals are shaded in color; the metalloids, in gray. The metals are not shaded.

Table 8-4 Chemical Behavior of Metals and Nonmetals

Metals	Nonmetals
1. Reduce elemental nonmetals (except noble gases)	1. Oxidize elemental metals and often oxidize less electronegative nonmetals
2. Form oxides that may react with water to give hydroxides	2. Form oxides that may react with water to give acids
3. Form basic hydroxides	3. Form acidic hydroxides (oxyacids)
4. React with O, F, H, and other nonmetals, usually giving ionic compounds	4. React with O, F, H, and other nonmetals, giving covalent compounds
5. Form binary metal hydrides, which, if soluble, are strong bases	5. Form binary hydrides, which may be acidic
6. React with other metals, giving metallic compounds	6. React with metals, often giving ionic compounds
7. Exhibit lower electronegativity values	7. Exhibit higher electronegativity values
8. Have one to five electrons in outermost shell; usually not more than three	8. Usually have four to eight electrons in outermost shell
9. Readily form cations by loss of electrons	9. Readily form anions by accepting electrons to fill outermost shell (except noble gases)

electron when it reduces nonmetals, giving ionic compounds containing the Na^+ ion and the anion formed by the nonmetal.

$$2Na + H_2 \longrightarrow NaH(s) \qquad (Na^+ \text{ and } H^-)$$
$$2Na(s) + F_2(g) \longrightarrow 2NaF(s) \qquad (Na^+ \text{ and } F^-)$$
$$16Na(s) + S_8(s) \longrightarrow 8Na_2S(s) \qquad (Na^+ \text{ and } S^{2-})$$
$$12Na(s) + P_4(s) \longrightarrow 4Na_3P(s) \qquad (Na^+ \text{ and } P^{3-})$$

Sodium oxide, Na_2O, is a soluble ionic oxide that reacts with water giving sodium hydroxide, a base.

$$Na_2O(s) + H_2O(l) \longrightarrow 2Na^+(aq) + 2OH^-(aq)$$

The hydride of sodium, NaH, is an ionic compound containing H^- ions. The hydride ion, H^-, is a very strong base and accepts hydrogen ion from water to form OH^- and H_2.

$$NaH(s) + H_2O(l) \longrightarrow NaOH(aq) + H_2(g)$$

Chlorine lies at the upper right in the Periodic Table in Group VIIA and thus may be readily identified as a nonmetal. Because chlorine is a nonmetal, we expect it to oxidize metals and to form ionic compounds that contain chloride ion. This ion forms when a chlorine atom picks up one electron and fills its valence shell. Ionic metal chlorides are formed when chlorine oxidizes metals lying to the left of it in the Periodic Table.

$$2Li(s) + Cl_2(g) \longrightarrow 2LiCl(s) \qquad (Li^+ \text{ and } Cl^-)$$
$$Ca(s) + Cl_2(g) \longrightarrow CaCl_2(s) \qquad (Ca^{2+} \text{ and } 2Cl^-)$$
$$Mn(s) + Cl_2(g) \longrightarrow MnCl_2(s) \qquad (Mn^{2+} \text{ and } 2Cl^-)$$

As a nonmetal, chlorine is also expected to oxidize less electronegative nonmetals (hydrogen and the nonmetals lying to the left and below chlorine in the Periodic Table) to form covalent molecules. This behavior is in fact observed; for example, the products in the following reactions are covalent:

$$H_2(g) + Cl_2(g) \longrightarrow 2HCl(g)$$
$$P_4(s) + 10Cl_2(g) \longrightarrow 4PCl_5(s)$$
$$Si + 2Cl_2(g) \longrightarrow SiCl_4(l)$$

As is expected for a compound of hydrogen and a nonmetal, HCl can act as an acid.

$$HCl(g) + H_2O(l) \longrightarrow H_3O^+(aq) + Cl^-(aq)$$
$$HCl(g) + NH_3(aq) \longrightarrow NH_4^+(aq) + Cl^-(aq)$$

The covalent oxide Cl_2O_7 reacts with water to give the acid $HClO_4$.

$$Cl_2O_7 + H_2O \longrightarrow 2HClO_4$$

Written in ionic form, this reaction is as follows:

$$Cl_2O_7(l) + 4H_2O(l) \longrightarrow 2H_3O^+(aq) + 2ClO_4^-(aq)$$

Perchloric acid, $HClO_4$, contains three oxygen atoms and one hydroxide group (OH group) covalently bonded to the chlorine:

$$\ddot{\text{O}} :$$
$$| $$
$$:\ddot{\text{O}}-\text{Cl}-\ddot{\text{O}}-\text{H}$$
$$|$$
$$:\ddot{\text{O}}:$$

Thus $HClO_4$ may be regarded as a hydroxide of the nonmetal, chlorine. Such molecules are more commonly called **oxyacids.** Other common oxyacids that contain hydroxide groups include nitric acid, which can be expressed by either the formula HNO_3 or the formula $HONO_2$; sulfuric acid, H_2SO_4 or $(HO)_2SO_2$; and phosphoric acid, H_3PO_4 or $(HO)_3PO$.

8.4 Variation in Metallic and Nonmetallic Behavior of the Representative Elements

Elements that easily lose electrons exhibit metallic behavior. The elements with only one or two electrons in the valence shell and the heavier elements lying at the bottoms of the groups of the Periodic Table generally lose electrons most readily. Atoms of nonmetals fill their valence shells either by sharing electrons with other nonmetals or by using electrons transferred from metal atoms. These are the atoms that lie in the upper right-hand portion of the table (Section 5.2). Sodium, the active metal at the left end of the third period, enters into chemical combinations by the loss of its single valence electron. An atom of chlorine, a typical nonmetal at the right end of the third period, combines by adding one electron to its valence shell of seven electrons, either by gaining an electron from a metal or by sharing an electron with a nonmetal.

Proceeding across the third period from sodium to chlorine, we encounter five other elements. In general, going across a period, valence electrons are lost with increasing difficulty. Thus the chemical behavior becomes decreasingly metallic and increasingly nonmetallic as we go from left to right. The changeover from metallic to nonmetallic behavior is gradual, but can be said to occur between aluminum and silicon in this period. Both aluminum and silicon, however, may exhibit metallic or nonmetallic properties under the appropriate conditions. To demonstrate this gradual changeover, let us examine the variation in the properties of the representative elements across the third period. Sodium, magnesium, and aluminum are shiny metals that conduct heat and electricity well. Silicon is an element with a luster characteristic of a metal, but it is a semiconductor (a poor conductor of electricity). Phosphorus, sulfur, and chlorine are dull in appearance and are nonconductors. Sodium oxide, Na_2O, reacts with water, giving the strong base sodium hydroxide, $NaOH$. Magnesium oxide, MgO, reacts slowly with water, giving the less strongly basic magnesium hydroxide, $Mg(OH)_2$. Aluminum oxide, Al_2O_3, does not react with water, but the reaction of a solution of an aluminum salt such as aluminum nitrate, $Al(NO_3)_3$, with a solution of a base produces a very weak base, aluminum hydroxide, $Al(OH)_3$. Likewise, silicon dioxide, SiO_2, will not react with water, but many covalent silicon compounds such as silicon tetrachloride, $SiCl_4$, will, giving gelatinous precipitates of silicic acid, $Si(OH)_4$, a very weak acid. The covalent oxides of phosphorus, sulfur, and chlorine (P_4O_{10}, SO_3, and Cl_2O_7) react with water, giving phosphoric acid, H_3PO_4, sulfuric acid, H_2SO_4, and perchloric acid, $HClO_4$, respectively. Of these acids, phosphoric acid is the weakest, and perchloric acid, the strongest. Thus, from left to right across any period of representative elements, the metallic character decreases and the nonmetallic character increases. *Hence the properties characteristic of metallic behavior become less pronounced and the properties characteristic of nonmetallic character become more pronounced from left to right across a period of representative elements.* For example, base strength and ionic character decrease, whereas acid strength and covalent character increase.

Because atoms of an element at the top of a group lose electrons less readily than atoms of an element at the bottom of the group, *the metallic character of the elements decreases and the nonmetallic character increases as we go up a group of representative elements.* For example, lead and tin, at the bottom of Group IVA, are metallic elements; germanium and silicon are metalloids; and carbon, at the top of the group, is a nonmetal. Elemental fluorine, at the top of Group VIIA, is a stronger oxidizing agent than is elemental iodine, near the bottom of the group. Iodine has sufficient metallic character to possess a metallic luster.

The behavior of an element also varies with its oxidation number. *The metallic behavior of an element decreases and its nonmetallic behavior increases as its oxidation number in a compound increases.* For instance, in perchloric acid, $HClO_4$ or $HOClO_3$, the oxidation number of chlorine is $+7$; in hypochlorous acid, $HOCl$, the oxidation number of chlorine is $+1$. Perchloric acid is a very strong acid (a characteristic of a compound of an element with pronounced nonmetallic behavior), whereas hypochlorous acid is a weak acid (less pronounced nonmetallic behavior). Thallium(I) chloride, $TlCl$, is an ionic compound (a characteristic of a metal chloride), whereas thallium(III) chloride, $TlCl_3$, is covalent (a characteristic of a nonmetal chloride). The chlorides of tin(II) and lead(II), $SnCl_2$ and $PbCl_2$, are solids with much more ionic character in their bonds than in the bonds of the chlorides of tin(IV) and lead(IV), $SnCl_4$ and $PbCl_4$, which are tetrahedral covalent molecules. Thus, in addition to the position of an element in the Periodic Table, we must consider its oxidation number when predicting metallic or nonmetallic behavior for one of its compounds.

8.5 Periodic Variation of Oxidation Numbers

The common oxidation numbers of representative elements when they are in compounds are tabulated in Table 8-5. This array of numbers may look formidable at first, but there are many regularities related to the positions of the elements in the Periodic Table, which should simplify your recall of these numbers. Refer to Table 8-5 as you study the regularities listed below.

1. The maximum positive oxidation number found in any group of representative elements is equal to the number of the group. Thus the maximum possible positive oxidation number increases from $+1$ for the alkali metals (Group IA) to $+7$ for all of the halogens (Group VIIA) except fluorine, which is assigned an oxidation number of -1 in its compounds. With the exceptions of thallium at the bottom of Group IIIA and mercury at the bottom of Group IIB, the maximum positive oxidation number is the only common oxidation number displayed by the metallic elements of Groups IA, IIA, IIB, and IIIA.

2. Metallic elements usually exhibit only positive oxidation numbers.

3. The most negative oxidation number of a group of representative elements is equal to the group number minus 8. Thus, for the elements in Group VA, the

Table 8-5 Commonly Observed Oxidation Numbers of the Representative Elements When in Compounds

IA	IIA	IIB	IIIA	IVA	VA	VIA	VIIA	VIIIA
H +1 −1							H +1 −1	He
Li +1	Be +2		B +3	C +4 to −4	N +5 to −3	O −1 −2	F −1	Ne
Na +1	Mg +2		Al +3	Si +4	P +5 +3 −3	S +6 +4 −2	Cl +7 +5 +3 +1 −1	Ar
K +1	Ca +2	Zn +2	Ga +3	Ge +4	As +5 +3 −3	Se +6 +4 −2	Br +7 +5 +3 +1 −1	Kr +4 +2
Rb +1	Sr +2	Cd +2	In +3	Sn +4 +2	Sb +5 +3 −3	Te +6 +4 −2	I +7 +5 +3 +1 −1	Xe +8 +6 +4 +2
Cs +1	Ba +2	Hg +2 +1	Tl +3 +1	Pb +4 +2	Bi +5 +3	Po +2	At −1	Rn
Fr +1	Ra +2							

most negative oxidation number possible is $5 - 8 = -3$. This corresponds to the number of electrons that would have to be gained to provide an outer shell of eight electrons for the elements.

4. Negative oxidation numbers are commonly limited to nonmetals and metalloids and are observed only when these elements are combined with less electronegative elements. Since fluorine is the most electronegative element known, it is never assigned a positive oxidation number.

5. Elements commonly exhibit positive oxidation numbers only when combined with more electronegative elements. Oxygen exhibits a positive oxidation number only in the few compounds it forms with the more electronegative element fluorine.

6. With the exceptions of carbon, nitrogen, oxygen, and mercury, each representative element that exhibits multiple oxidation numbers in its compounds commonly has either all even or all odd oxidation numbers. Carbon commonly

exhibits oxidation numbers from $+4$ to -4; nitrogen, from $+5$ to -3; and oxygen, -2 or -1. Mercury forms both a diatomic ion, Hg_2^{2+}, in which the oxidation number of each atom is $+1$, and a monatomic mercury(II) ion, Hg^{2+}, for which the oxidation number is $+2$.

Before we proceed, a word of explanation about the meaning of the phrase *common oxidation numbers* is necessary. The common oxidation numbers of an element are those observed in a majority of its compounds. In some cases, as for the elements in Groups IA and IIA, this majority may be all or nearly all of the known compounds of the element. Of all the thousands of known sodium compounds, for example, there is only one very reactive compound in which sodium has been shown to exhibit an oxidation number of -1. For other elements, the minority may be relatively large. Boron has an oxidation number of $+3$ in the majority of its compounds; however, in one series of compounds it has an oxidation number of $+2$, and in another it exhibits oxidation numbers that are nonintegers. Since most compounds discussed in this text contain elements having their common oxidation numbers, memorizing these numbers now will simplify studying the compounds later.

8.6   Prediction of Reaction Products

Many factors determine whether or not a chemical reaction will occur and what the products will be. These include the kinds of reactants and the conditions under which the reaction is run. Some chemists spend years developing a "feel" for the types of reactants and conditions that are likely to give a desired product. However, even they must resort to the ultimate test of their ideas; they try a proposed reaction in the laboratory to see if it works. The design and testing of reactions in order to prepare specific types of compounds or to improve the ease, yield, or simplicity with which known compounds are prepared constitute the field known as **chemical synthesis.**

In theory it should be possible to predict accurately the products of any chemical reaction using concepts of chemical bonding, kinetics, thermodynamics, solution behavior, and electrochemistry. In practice, however, most chemical systems are simply too complicated to be treated precisely by these tools. However, as we shall see in subsequent chapters, such tools do provide valuable insights into the behavior of chemical systems.

In this section we shall examine some of the general guidelines that can be helpful in answering the question "What are the likely products, if any, that will result from the reaction of two or more substances?" These guidelines are based on the metallic or nonmetallic behavior of the representative elements involved, the common oxidation numbers that they are likely to exhibit, and the similarities of the compounds involved to those types already discussed. We shall consider only those reactions that occur at room temperature in water or that occur when the pure substances are heated. Obviously, different conditions may lead to the formation of different products, but these guidelines can serve as a foundation for our considerations of chemical reactions in subsequent chapters. The following examples illustrate the approach.

EXAMPLE 8.1   What is the likely product of the reaction of gaseous ammonia, NH_3, with gaseous hydrogen iodide, HI?

From our discussion of acids and bases, we recognize that ammonia is a base and that hydrogen iodide is one of the six strong acids. Thus we can expect an acid-base reaction in which a hydrogen ion is transferred from HI, the acid, to NH_3, the base, forming the salt ammonium iodide, NH_4I.

$$NH_3 + HI \longrightarrow NH_4I$$

This reaction is analogous to the reaction of ammonia with HCl [Equation (2) in Section 8.1, Part 2]. The similarity between the predicted reaction and known behavior supports our prediction.

EXAMPLE 8.2   Predict the product of the reaction that occurs when elemental gallium (Ga) and sulfur (S_8) are warmed together.

First consider what we know about gallium and sulfur from their respective positions in the Periodic Table. Gallium is located in Group IIIA and thus should exhibit metallic behavior. As a metal, it can be expected to form compounds in which it exhibits a positive oxidation number. Since it is a member of Group IIIA, the expected oxidation number is $+3$. Sulfur is a member of Group VIA and thus should exhibit nonmetallic behavior. The common oxidation numbers of sulfur are $+6$ (its maximum positive oxidation number, equal to the group number), $+4$, and -2 (its most negative oxidation number, equal to the group number minus 8). Since sulfur in this instance is being combined with the less electronegative gallium, it is expected to exhibit the negative oxidation number of -2. As a metal, gallium can act as a reducing agent; as a nonmetal, sulfur can act as an oxidizing agent. Thus we can expect an oxidation-reduction reaction between gallium and sulfur, giving a product containing gallium with an oxidation number of $+3$ and sulfur with an oxidation number of -2. This leads us to formulate the product (Section 5.10) as Ga_2S_3.

The predicted chemical reaction therefore is as follows:

$$16Ga + 3S_8 \longrightarrow 8Ga_2S_3$$

Ga_2S_3 is, in fact, the product of the reaction of gallium with sulfur.

As a general approach to predicting the products of a reaction, you can examine the reactants to see if they are analogous to compounds or elements whose behavior you know. If, as in Example 8.1, the reactants are familiar, you can often readily predict their behavior. If analogies are not obvious, try the following method of analysis of the system.

1. Identify each element in the reactants as a metal or a nonmetal from its position in the Periodic Table and find its oxidation number (as described in Section 5.9), as well as the oxidation numbers it commonly exhibits in its compounds (Section 8.5).

2. Consider the possibility of an oxidation-reduction reaction (Section 8.2, Part 4). The following general guidelines are helpful:

    a. An elemental metal will reduce an elemental nonmetal [Equation (1)].

    b. A less electronegative metal (a more active metal) in its elemental form will reduce the ion of a more electronegative metal [Equation (2)].

    c. A more electronegative nonmetal in its elemental form will oxidize a less electronegative nonmetal [Equations (3) and (4)].

$$Mg + Cl_2 \longrightarrow MgCl_2 \tag{1}$$

$$Mg + SnCl_2 \longrightarrow MgCl_2 + Sn \tag{2}$$

$$S + O_2 \longrightarrow SO_2 \tag{3}$$

$$2SO_2 + O_2 \longrightarrow 2SO_3 \tag{4}$$

    d. The metals of Group IA and calcium, strontium, and barium of Group IIA are active enough to reduce the hydrogen in water or acids to hydrogen gas.

$$2Na + 2H_2O \longrightarrow 2NaOH + H_2 \tag{5}$$

$$Ba + 2HCl \longrightarrow BaCl_2 + H_2 \tag{6}$$

The other representative metals, with the exceptions of bismuth and lead (the least metallic representative metals), will only reduce the hydrogen in acids.

$$4Al + 6H_2SO_4 \longrightarrow 2Al_2(SO_4)_3 + 6H_2 \tag{7}$$

3. Consider the possibility of an acid-base reaction (Section 8.2, Part 5). Remember that the oxides and hydroxides of nonmetals are generally acidic and that the oxides and hydroxides of metals are generally basic. The monoatomic anions of the nonmetals of Groups IV, V, and VI are very strongly basic, and soluble compounds containing these ions accept protons from weak acids, even those as weak as water. Thus, we might find acid-base reactions involving familiar types of acids and bases:

$$2HI + Sr(OH)_2 \longrightarrow SrI_2 + 2H_2O \tag{8}$$

involving acidic oxides and bases:

$$SO_3 + Ca(OH)_2 \longrightarrow CaSO_4 + H_2O \tag{9}$$

involving acids and basic oxides:

$$6HCl + Al_2O_3 \longrightarrow 2AlCl_3 + 3H_2O \tag{10}$$

or involving acids and other types of basic anions:

$$2HCl + Na_2S \longrightarrow 2NaCl + H_2S \tag{11}$$

$$3H_2O + Li_3N \longrightarrow 3LiOH + NH_3 \tag{12}$$

4. Consider the possibility of a metathetical reaction (Section 8.2, Part 3). Metathetical reactions between ionic compounds generally occur to form at least one product that is a solid, a gas, a weak electrolyte, or a nonelectrolyte.

5. Consider whether or not the products of the reaction may react with each other or with the solvent (if any).

Some specific illustrations of the application of these ideas are provided by the following examples.

EXAMPLE 8.3   Write the balanced chemical equation for the reaction of elemental calcium (Ca) with iodine (I_2).

Let us assume that we have no specific knowledge of the chemistry of this system, and analyze it using the steps discussed above. Calcium is a member of Group IIA and is thus an active metal. Iodine is a member of Group VIIA and is thus a nonmetal. An active metal reacts with a nonmetal in an oxidation-reduction reaction. Iodine will be reduced to -1, its only available negative oxidation number as a member of Group VIIA, while calcium will be oxidized to $+2$, the oxidation number that a Group IIA element exhibits in compounds. With these oxidation numbers, the product must be CaI_2. The balanced equation is

$$Ca + I_2 \longrightarrow CaI_2$$

EXAMPLE 8.4   Write a balanced chemical equation for the reaction that occurs when tin(II) oxide, SnO, is added to a solution of perchloric acid, $HClO_4$.

$HClO_4$ is a strong acid, so we look for an acid-base reaction. Tin is a member of Group IVA. It is located near the bottom of the group and has a low oxidation number in SnO, so it should exhibit metallic properties. The oxides of metals are basic; hence we can expect an acid-base reaction between SnO and $HClO_4$ to produce a salt and water.

$$SnO + 2HClO_4 \longrightarrow Sn(ClO_4)_2 + H_2O$$

EXAMPLE 8.5   What are the two possible products of the reaction of phosphorus, P_4, with chlorine, Cl_2?

Both phosphorus (Group VA) and chlorine (Group VIIA) are nonmetals. Phosphorus has common oxidation numbers of $+5$, $+3$, and -3; chlorine, $+7$, $+5$, $+3$, $+1$, and -1. Chlorine lies to the right of phosphorus in the third period, so it is more electronegative than phosphorus. Thus we recognize that it can oxidize phosphorus. Because phosphorus is oxidized, it will exhibit a positive oxidation number—probably $+3$ or $+5$ since these are the common positive oxidation numbers exhibited by a member of Group VA. Chlorine will exhibit a negative oxidation number because it is combined with a less electronegative element. The only negative oxidation number for chlorine is -1. Thus P_4 and Cl_2 could react to give two compounds. In one phosphorus has an oxidation number of $+3$, and chlorine has an oxidation number of -1. In the other phosphorus has an oxidation number of $+5$, and chlorine has an oxidation number of -1. Thus the two compounds are PCl_3 and PCl_5, respectively.

Note that either phosphorus(III) chloride or phosphorus(V) chloride can in fact be formed, depending on the choice of reaction conditions. The reaction of 1 mol of P_4 with 6 mol of Cl_2 gives PCl_3. If the amount of Cl_2 is increased to 10 mol, then PCl_5 is produced. Hence an excess of Cl_2 favors the formation of PCl_5; conversely, an excess of P_4 favors the formation of PCl_3.

$$P_4 + 6Cl_2 \longrightarrow 4PCl_3$$
$$P_4 + 10Cl_2 \longrightarrow 4PCl_5$$

EXAMPLE 8.6   Predict the likely products of the reaction of sodium, Na, with sulfur trioxide, SO_3.

Sodium (Group IA) is an active metal. Sulfur and oxygen are both nonmetals located in Group VIA. Elemental sodium has an oxidation number of zero and can be oxidized only to $+1$. In SO_3 sulfur has an oxidation number of $+6$; oxygen, an oxidation number of -2.

In this system sodium can function only as a reducing agent. The only element that can be reduced is sulfur. Oxygen already exists with its minimum (most negative) oxidation number of -2. Thus we can expect the reaction of Na with SO_3 to give products in which the sulfur is reduced to an oxidation number of $+4$, 0, or -2. With an oxidation number of $+4$, sulfur would still be combined with oxygen, the only element in the system that exhibits a negative oxidation number. The reaction in this case would be

$$2Na + SO_3 \longrightarrow Na_2SO_3 \quad \text{(ratio of mol Na to mol } SO_3 = 2:1)$$

With additional sodium the reaction could give sulfur with an oxidation number of zero, that is, elemental sulfur, S_8.

$$48Na + 8SO_3 \longrightarrow 24Na_2O + S_8 \quad \text{(ratio of mol Na to mol } SO_3 = 6:1)$$

The sodium in this reaction must combine with oxygen to form sodium oxide since sodium can only be oxidized to an oxidation number of $+1$. Finally, even more sodium could reduce the sulfur in SO_3 to an oxidation number of -2.

$$8Na + SO_3 \longrightarrow Na_2S + 3Na_2O \quad \text{(ratio of mol Na to mol } SO_3 = 8:1)$$

Which reaction actually occurs will depend on the relative amounts of sodium and sulfur trioxide reacting, but in each case sodium functions as a reducing agent and is itself oxidized.

8.7 Chemical Properties of Some Important Industrial Chemicals

Over 1.44×10^{11} kilograms (316 billion pounds) of inorganic chemicals and elements and 6.81×10^{10} kilograms (150 billion pounds) of organic chemicals were produced in the United States during 1982. Similar amounts were also produced in other industrialized countries. The compounds and elements listed in Table 8-6 constitute almost 97% of the inorganic compounds and elements produced in the United States. The chemical properties of these compounds and elements show clearly the differences between the chemistry of metals and nonmetals and illustrate the types of compounds and reactions, as well as the oxidation numbers, described in the preceding sections. The behavior of organic materials will be described in Chapter 25.

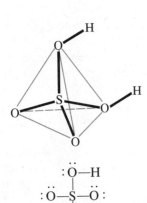

Figure 8-4. The molecular structure and Lewis structure of sulfuric acid.

1. SULFURIC ACID.   With 15% of the industrial chemical production of the United States devoted to its manufacture, sulfuric acid (Fig. 8-4) ranks as the single most significant industrial chemical. Per capita use of sulfuric acid has been taken as one index of the technical development of a nation. Perhaps surprisingly, sulfuric acid and the sulfate ion rarely appear in finished materials. Sulfuric acid is used extensively as an acid (a source of hydrogen ions) because it is the cheapest

Table 8-6 Important Industrial Inorganic Chemicals and Elements

Element or Compound	1982 U.S. Production (in billions of lb)
H_2SO_4	64.6
N_2	35.1
NH_3	31.0
O_2	29.7
CaO, $Ca(OH)_2$	28.4
NaOH	18.3
Cl_2	18.3
H_3PO_4	17.0
Na_2CO_3	15.8
HNO_3	15.2
NH_4NO_3	14.7
CO_2	7.4
HCl	5.0
$(NH_4)_2SO_4$	3.6

From *Chemical and Engineering News,* June 13, 1983, p. 29.

strong acid. It is used to manufacture fertilizer, leather, tin plate, and other chemicals, to purify petroleum, and to make and dye fabrics.

Sulfuric acid is prepared from elemental sulfur or sulfides. Sulfur is burned in air, and the nonmetal sulfur is oxidized by the more electronegative nonmetal oxygen. Covalent molecules of gaseous sulfur dioxide, with sulfur having one of its common oxidation numbers (+4), result.

$$S_8 + 8O_2 \longrightarrow 8SO_2$$

The sulfur is then oxidized to its highest oxidation number (+6) by oxygen giving sulfur trioxide in a reversible reaction.

$$2SO_2 + O_2 \overset{\triangle}{\rightleftharpoons} 2SO_3$$

Even though the yield from this reaction is highest at lower temperatures, sulfur trioxide forms slowly at these temperatures. At higher temperatures it forms more rapidly, but unfortunately the yield is lower. Consequently, the reaction is run at lower temperatures with vanadium(V) oxide, V_2O_5, as a catalyst to get as high a yield as possible as quickly as possible. **A catalyst is a substance that changes the speed of a chemical reaction without affecting the yield and without undergoing a permanent chemical change itself.**

Since sulfur trioxide is an oxide of a nonmetal in which the nonmetal has a high oxidation number, we expect it to react with water to form an acid—in this case, sulfuric acid.

$$H_2O + SO_3 \longrightarrow H_2SO_4$$

This reaction does occur, for example, when SO_3 from polluted air dissolves in rain drops to give acid rain. However, it is more efficient industrially to combine sulfur trioxide with sulfuric acid to produce pyrosulfuric acid, $H_2S_2O_7$. This acid then reacts with water to form sulfuric acid.

$$SO_3 + H_2SO_4 \longrightarrow H_2S_2O_7$$
$$H_2S_2O_7 + H_2O \longrightarrow 2H_2SO_4$$

Oxygen will also oxidize sulfur with an oxidation number of -2 in compounds. Thus sulfuric acid is sometimes prepared from the sulfur dioxide produced by burning hydrogen sulfide, an impurity separated from natural gas, or from the sulfur dioxide produced by roasting metal sulfides such as Cu_2S in air.

$$2H_2S + 3O_2 \xrightarrow{\triangle} 2SO_2 + 2H_2O$$

$$Cu_2S + 2O_2 \xrightarrow{\triangle} 2CuO + SO_2$$

Sulfuric acid is a strong acid and an oxidizing agent. Its largest use is as a source of hydrogen ion. About 65% of the sulfuric acid produced in the United States is used to manufacture fertilizers by reaction with ammonia or with calcium phosphates [typically $Ca_5(PO_4)_3F$, found principally in North Carolina and Florida as phosphate rock]. The reaction with calcium phosphates converts these extremely insoluble compounds into somewhat soluble dihydrogen phosphates that can enter a plant's roots in solution and provide it with necessary phosphorus.

$$2Ca_5(PO_4)_3F(s) + 7H_2SO_4(l) + 3H_2O(l) \longrightarrow$$
$$3Ca(H_2PO_4)_2 \cdot H_2O(s) + 2HF(g) + 7CaSO_4(s)$$

This is an acid-base reaction in which a strong acid reacts with the anions (PO_4^{3-} and F^-) of two weak acids. The solid mixture of the salts $Ca(H_2PO_4)_2 \cdot H_2O$ and $CaSO_4$ is used directly as the fertilizer. The hydrogen fluoride is recovered and used in the preparation of fluorocarbons and other fluorides. If the amount of sulfuric acid is increased, phosphoric acid is produced in a related acid-base reaction.

$$Ca_5(PO_4)_3F(s) + 5H_2SO_4(l) \longrightarrow 5CaSO_4(s) + 3H_3PO_4(l) + HF(g)$$

Both of these acid-base reactions are also metathetical reactions, which proceed because solid and gaseous products are formed.

Phosphoric acid (Fig. 8-5) is used primarily to make "triple superphosphate" fertilizers. They contain no $CaSO_4$ and thus have more phosphorus. The reaction is another acid-base reaction.

$$Ca_5(PO_4)_3F(s) + 7H_3PO_4(l) \longrightarrow 5Ca(H_2PO_4)_2(s) + HF(g)$$

The acid behavior of sulfuric acid is also evident in the acid-base reaction used to prepare the salt ammonium sulfate, a soluble solid fertilizer providing nitrogen.

$$2NH_3(g) + H_2SO_4(l) \longrightarrow (NH_4)_2SO_4(s)$$

In 1982 about 3.6 billion pounds of $(NH_4)_2SO_4$ were prepared in the United States.

Many metal oxides behave as bases and react with sulfuric acid to form ionic sulfates. For example, copper sulfate (used as a fungicide and in electroplating), aluminum sulfate (used in water treatment and in papermaking), and magnesium

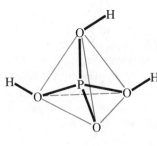

Figure 8-5. The molecular structure and Lewis structure of phosphoric acid.

sulfate (Epsom salts) are prepared this way.

$$CuO(s) + H_2SO_4(aq) \longrightarrow Cu^{2+}(aq) + SO_4^{2-}(aq) + H_2O(l)$$
$$Al_2O_3(s) + 3H_2SO_4(aq) \longrightarrow 2Al^{3+}(aq) + 3SO_4^{2-}(aq) + 3H_2O(l)$$
$$MgO(s) + H_2SO_4(aq) \longrightarrow Mg^{2+}(aq) + SO_4^{2-}(aq) + H_2O(l)$$

Since these compounds give ionic products (compounds of a metal with non-metals), they exist in solution as ions. They can be recovered as solids by evaporating the water.

Sulfuric acid also undergoes an acid-base reaction with sodium chloride, giving hydrogen chloride and sodium hydrogen sulfate. The sodium hydrogen sulfate will then react with additional sodium chloride if the mixture is heated.

$$H_2SO_4(l) + NaCl(s) \longrightarrow NaHSO_4(s) + HCl(g)$$
$$NaHSO_4(s) + NaCl(s) \longrightarrow Na_2SO_4(s) + HCl(g)$$

Even though chloride ion is a very weak base, these metathetical reactions procede because gaseous HCl is formed and escapes from the reaction.

Sulfuric acid is an oxidizing agent, but it is not often used industrially as one. In fact, its oxidizing ability prevents it from being used in two potentially useful reactions: the preparation of hydrogen bromide and of hydrogen iodide by reaction with the respective metal halides. A strong acid should react with NaBr or NaI to form gaseous molecules of HBr or HI in a manner analogous to the reaction of H_2SO_4 with NaCl. Unfortunately, sulfuric acid will oxidize the bromide ion or iodide ion, and so it cannot be used. In the reaction with bromide ion, the sulfur atom is reduced from its maximum positive oxidation number of +6 (sulfur is a member of Group VIA) to the oxidation number of +4. With the stronger reducing agent, iodide ion, oxidation numbers of +4, 0, or −2 can result depending on the relative amounts of sulfuric acid and iodide ion used in the reaction.

$$2NaBr(s) + H_2SO_4(l) \longrightarrow Na_2SO_3(s) + H_2O(l) + Br_2(l)$$
$$2NaI(s) + H_2SO_4(l) \longrightarrow Na_2SO_3(s) + H_2O(l) + I_2(s)$$
$$48NaI(s) + 32H_2SO_4(l) \longrightarrow S_8(s) + 24Na_2SO_4(s) + 32H_2O(l) + 24I_2(s)$$
$$8NaI(s) + 4H_2SO_4(l) \longrightarrow Na_2S(s) + 3Na_2SO_4(s) + 4H_2O(l) + 4I_2(s)$$

2. AMMONIA. Ammonia (Fig. 8-6) is one of the few compounds that can be prepared readily from elemental nitrogen. It is prepared by the reversible reaction of a mixture of hydrogen and nitrogen at high pressure and temperature (400–500°C) in the presence of a catalyst containing iron. The catalyst speeds up the reaction but does not increase the yield.

$$N_2(g) + 3H_2(g) \underset{\text{High pressure}}{\overset{\substack{400-500°C \\ \text{Fe catalyst}}}{\rightleftharpoons}} 2NH_3(g)$$

This reaction is simply the oxidation of a less electronegative nonmetal by a more electronegative nonmetal, giving a covalent compound.

Pure ammonia is a gas that is very soluble in water. The uses of ammonia can be attributed to two principal chemical properties of the molecule: (1) ammonia is a base, and (2) the nitrogen atom in ammonia is reactive and can readily be oxidized to higher oxidation numbers, particularly by plants, in the production of

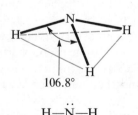

106.8°

H—N—H
|
H

Figure 8-6. The molecular structure and Lewis structure of ammonia.

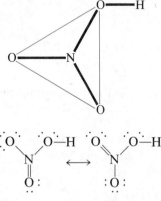

Figure 8-7. The molecular structure and Lewis structure (showing resonance forms) of nitric acid.

other nitrogen-containing compounds. Thus ammonia and its derivatives form an important source of nitrogen fertilizers.

Approximately 20 billion pounds of ammonium salts are prepared as fertilizers each year in the United States. The reactions are acid-base reactions that give salts containing the NH_4^+ ion. The principal reactions involve nitric acid, HNO_3, sulfuric acid, H_2SO_4, and phosphoric acid, H_3PO_4.

$$NH_3(g) + HNO_3(l) \longrightarrow NH_4NO_3(s)$$
$$2NH_3(g) + H_2SO_4(l) \longrightarrow (NH_4)_2SO_4(s)$$
$$NH_3(g) + H_3PO_4(l) \longrightarrow NH_4H_2PO_4(s)$$

The nitrogen in ammonia can be oxidized by oxygen. Ammonia burns in oxygen, forming nitrogen(II) oxide, NO, which can then react with additional oxygen to give nitrogen(IV) oxide, NO_2. The latter oxide reacts with water to give the oxyacid nitric acid. This reaction is more complex than most reactions of water with nonmetal oxides, because oxidation-reduction is also involved. The series of reactions is as follows:

$$4NH_3 + 5O_2 \longrightarrow 4NO + 6H_2O$$
$$2NO + O_2 \longrightarrow 2NO_2$$
$$3NO_2 + H_2O \longrightarrow 2HNO_3 + NO$$

Nitric acid (Fig. 8-7) is important industrially as a strong acid and oxidizing agent.

3. CALCIUM OXIDE AND CALCIUM HYDROXIDE.

Calcium oxide and calcium hydroxide are inexpensive bases used extensively in chemical processing, although most of the useful products prepared from them do not contain calcium.

Calcium oxide, CaO, is made by heating calcium carbonate, $CaCO_3$, which is widely and inexpensively available as limestone.

$$CaCO_3(s) \xrightarrow{\triangle} CaO(s) + CO_2(g)$$

This reaction is the reverse of the reaction between CaO and CO_2 and gives a 100% yield of CaO if the CO_2 is allowed to escape. Calcium hydroxide is prepared by the familiar acid-base reaction of a soluble metal oxide with water.

$$CaO(s) + H_2O(l) \longrightarrow Ca(OH)_2(s)$$

This reaction produces a lot of heat. In fact, farmers used to worry about the weather, for good reason, when they went to market in their wooden wagons to get calcium oxide (lime) for their fields. The reaction caused by a sudden rain on the CaO could make a wagon so hot that it would catch fire.

Since Ca^{2+} is the ion of an active metal and is therefore very difficult to reduce, and since oxygen with its oxidation number of -2 is difficult to oxidize, the oxidation-reduction chemistry of both CaO and $Ca(OH)_2$ is limited. These compounds are useful because they accept protons and thus neutralize acids.

4. OXYGEN AND NITROGEN.

Both pure oxygen and pure nitrogen can be isolated from air. Nitrogen is surprisingly unreactive for so electronegative a nonmetal. This has been attributed to the very strong nitrogen-nitrogen triple bond in the N_2 molecule (Section 6.14). The principal uses of nitrogen gas are actually based on this nonreactivity. It is used as an inert atmosphere blanket in the food industry to prevent spoilage due to oxidation. Liquid nitrogen boils at 77 K and is used to store biological materials such as blood and tissue samples.

Oxygen behaves as one would expect for a very electronegative nonmetal: it is a strong oxidizing agent. Its principal uses are in oxidations in the steel and petrochemical industries. During steel production oxygen is blown through hot liquid iron to oxidize impurities. Two common reactions are

$$C + O_2 \longrightarrow CO_2$$
$$Si + O_2 \longrightarrow SiO_2$$

Silicon dioxide, SiO_2, is less dense than liquid iron and floats to the surface where it is removed with other low-melting material called slag. Note that the two Group IVA elements carbon and silicon are each oxidized to their maximum oxidation number. During these reactions oxygen is reduced to its most negative oxidation number (-2).

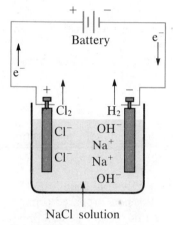

Figure 8-8. The electrolysis of an aqueous solution of NaCl. The net reaction produces hydrogen gas and hydroxide ions at the cathode and chlorine gas at the anode, resulting in a solution containing sodium hydroxide and any remaining sodium chloride for which the chloride ion has not yet been oxidized.

5. CHLORINE AND SODIUM HYDROXIDE.   Although they are very different chemically, chlorine and sodium hydroxide are paired here because they are prepared simultaneously in a very important electrochemical industrial process. The process utilizes sodium chloride, which occurs in large deposits in several parts of the country. Sodium chloride is composed of sodium ions and chloride ions, both of which are very resistant to chemical reduction or oxidation. (Recall that, although sulfuric acid does oxidize bromide and iodide ions, it does not oxidize chloride ions.) An electrochemical process, however, can be used to oxidize chloride ion to chlorine.

In the electrochemical production of chlorine, two electrodes are placed in a concentrated solution of sodium chloride (Fig. 8-8). When a direct current of electricity is passed through the solution, the chloride ions migrate to the positive electrode and are oxidized to gaseous chlorine by giving up an electron to the electrode.

$$2Cl^-(aq) \longrightarrow Cl_2(g) + 2e^- \text{ (to the electrode)}$$

The electrons are transferred through the outside electrical circuit to the negative electrode. Although the positive sodium ions migrate toward this negative electrode, metallic sodium is not produced because sodium ions are too difficult to reduce under the conditions used. (Recall that metallic sodium is active enough to react with water and hence, even if produced, would immediately react with water to produce sodium ions again.) Instead, water molecules, which are more easily reduced, pick up electrons from the electrode and are reduced, giving hydrogen gas and hydroxide ions.

$$2H_2O(l) + 2e^- \text{ (from the electrode)} \longrightarrow H_2(g) + 2OH^-(aq)$$

These changes convert the aqueous solution of NaCl into an aqueous solution of NaOH, gaseous Cl_2, and gaseous H_2.

$$2Na^+(aq) + 2Cl^-(aq) + 2H_2O(l) \xrightarrow{\text{Electrolysis}}$$
Sodium chloride

$$2Na^+(aq) + 2OH^-(aq) + Cl_2(g) + H_2(g)$$
Sodium hydroxide

The nonmetal chlorine is more electronegative than any other element except fluorine, oxygen, and nitrogen (Table 5-2). Thus we would expect elemental chlorine to oxidize all of the other elements except for these three (and the noble gases, which are quite nonreactive). Its oxidizing property, in fact, is responsible for its principal use, as an oxidizing agent. For example, phosphorus(V) chloride, an important intermediate in the preparation of insecticides, is manufactured by oxidizing the less electronegative nonmetal phosphorus with chlorine.

$$P_4 + 10Cl_2 \longrightarrow 4PCl_5$$

During World War II tin was recycled from the covering on tin cans by reaction with chlorine followed by distillation of the volatile tin(IV) chloride, $SnCl_4$, produced.

$$Sn + 2Cl_2 \longrightarrow SnCl_4$$

Since tin lies toward the right among the metals of the Periodic Table, its metallic character is not pronounced. This, coupled with its relatively high oxidation number of $+4$, results in the formation of a covalent chloride rather than an ionic chloride.

A great deal of chlorine is also used to oxidize organic or biological materials in water purification and in bleaching.

Sodium hydroxide is a strong, very water-soluble (420 g/L) base used to make concentrated solutions of base.

$$NaOH(s) \xrightarrow{H_2O} Na^+(aq) + OH^-(aq)$$

It is used in acid-base reactions during chemical processing and in the preparation of soaps, rayon, and paper.

6. NITRIC ACID.   Nitric acid (Fig. 8-7) is an important industrial chemical (15.2 billion pounds produced in the United States in 1982). It is now prepared primarily by oxidation of ammonia as discussed in Part 2 of this section. It was previously manufactured by the acid-base reaction of sulfuric acid with sodium nitrate, mined principally in Chile.

$$NaNO_3(s) + H_2SO_4(l) \longrightarrow NaHSO_4(s) + HNO_3(l)$$

The nitrate ion can be easily metabolized by most plants, making nitrates, particularly ammonium nitrate, NH_4NO_3, good fertilizers. In addition, nitric acid is both a strong acid and a strong oxidizing agent and is therefore used in both acid-base and oxidation-reduction reactions to prepare soluble metal salts. Its acid-base reactions include those with hydroxides, oxides, and carbonates.

$$KOH + HNO_3 \longrightarrow KNO_3 + H_2O$$
$$CuO + 2HNO_3 \longrightarrow Cu(NO_3)_2 + H_2O$$
$$CaCO_3 + 2HNO_3 \longrightarrow Ca(NO_3)_2 + H_2CO_3$$
$$ \longrightarrow H_2O + CO_2$$

Carbonic acid, H_2CO_3, is not very stable and decomposes into gaseous carbon dioxide and water. This decomposition reaction is reversible.

Nitrogen exhibits a wide range of oxidation numbers in its compounds. Oxidation-reduction reactions of nitric acid generally involve reduction of the nitrogen (oxidation number of $+5$) to either an oxidation number of $+4$ with formation of NO_2 or an oxidation number of $+2$ with formation of NO. Reactions involving concentrated HNO_3 usually give predominantly NO_2. Dilute HNO_3

often gives NO.

$$Zn + 4HNO_3(conc) \longrightarrow Zn(NO_3)_2 + 2NO_2 + 2H_2O$$
$$3Ag + 4HNO_3(dilute) \longrightarrow 3AgNO_3 + NO + 2H_2O$$

Since gold is not oxidized by nitric acid, the latter reaction is used to separate silver from gold. Note that, whereas most acids react with active metals such as zinc with evolution of hydrogen, nitric acid reacts by reduction of the nitrate group.

FOR REVIEW

SUMMARY

The chemical behavior of an element reflects its position in the Periodic Table. Those elements that lie to the left or toward the bottom of the table are **metals,** and all have similar characteristic behavior. **Nonmetals,** which lie in the upper right portion of the table, have characteristic nonmetallic behavior. The chemical behavior characteristic of metals becomes less pronounced from left to right across a period and from bottom to top in a group. The chemical behavior characteristic of nonmetals becomes more pronounced from left to right and from bottom to top. An element exhibits behavior increasingly more characteristic of a nonmetal as its oxidation number in its compounds increases. The elements that lie near the dividing line between metals and nonmetals, the **metalloids,** cannot be identified clearly as metals or nonmetals, and they may exhibit metallic or nonmetallic behavior depending on the conditions under which they react.

Elemental metals generally react with elemental nonmetals, resulting in **oxidation** of the metal and **reduction** of the nonmetal. The oxidation numbers of the elements in the products can often be predicted from their positions in the Periodic Table. Most representative metals have a single common positive oxidation number in their compounds, and nonmetals have a single common negative oxidation number when they form monatomic anions. Generally, representative metals and nonmetals form ionic **salts,** which dissolve in water giving solutions of **electrolytes.** The hydroxides of metals act as **bases** unless the metals exhibit unusually high oxidation numbers. These hydroxides will accept protons from acids. Soluble ionic metal oxides are also bases because the oxide ion will pick up a hydrogen ion from water to form a hydroxide ion.

In addition to oxidizing metals a nonmetal will often oxidize less electronegative nonmetals, giving covalent compounds. Fluorine, chlorine, and oxygen are particularly good oxidizing agents. Oxidation of hydrogen by nonmetals gives binary hydrides, which can act as **acids,** although HCl, HBr, and HI are the only binary **strong acids.** The other nonmetal hydrides range from **weak acids** to essentially nonacidic compounds as the nonmetallic character of the element involved decreases. Ammonia will also act as a base. Some oxides of the nonmetals react with water to produce **oxyacids** (nonmetal hydroxides), of which $HClO_4$, H_2SO_4, and HNO_3 are common strong acids. Most other oxyacids are weak.

The reactivity of unfamiliar species can be predicted by comparison with the chemistry of analogous compounds whose behavior is known or by consideration of the metallic or nonmetallic character of the elements involved, using the analysis discussed in this chapter.

The inorganic chemicals used industrially illustrate the differences between metals and nonmetals. Sulfuric acid and nitric acid are oxyacids formed from nonmetal oxides. Hydrochloric acid is a nonmetal hydride. Calcium oxide is a metal oxide that is used as a base and that reacts with water to give calcium hydroxide, another base. The hydroxide of sodium is also a base. In the reactions described involving the metal ions Na^+ and Ca^{2+}, the oxidation numbers of the metal ions do not change. The oxidation numbers of the nonmetals, however, vary widely. The nitrogen atom, for example, varies from an oxidation number of -3 in NH_3 to $+5$ in HNO_3. The stability of the most electronegative nonmetals in their reduced states is illustrated by the difficulty of preparing elemental chlorine.

KEY TERMS AND CONCEPTS

Acid (8.1)
Acid-base reaction (8.2)
Addition reactions (8.2)
Ammonia (8.7)
Base (8.1)
Brönsted acid (8.1)
Brönsted base (8.1)
Calcium hydroxide (8.7)
Calcium oxide (8.7)
Catalyst (8.7)
Chemical synthesis (8.6)
Chlorine (8.7)
Decomposition reactions (8.2)
Electrolytes (8.1)

Hydronium ion (8.1)
Metal (8.3)
Metalloid (8.3)
Metathetical reactions (8.2)
Nitric acid (8.7)
Nitrogen (8.7)
Nonelectrolytes (8.1)
Nonmetal (8.3)
Oxidation (8.2)
Oxidation number (8.5)
Oxidation-reduction reactions (8.2)
Oxidizing agent (8.2)
Oxyacids (8.3)

Oxygen (8.7)
Reducing agent (8.2)
Reduction (8.2)
Reversible reactions (8.2)
Salt (8.1)
Sodium hydroxide (8.7)
Strong acid (8.1)
Strong base (8.1)
Strong electrolytes (8.1)
Sulfuric acid (8.7)
Weak acid (8.1)
Weak base (8.1)
Weak electrolyte (8.1)

EXERCISES

Types of Compounds

1. In what ways does a salt differ from an acid? from a base? In what ways are salts and hydroxide bases similar?

2. Identify the cations and anions in the following compounds:
   (a) KI
   (b) MgF_2
   (c) K_2SO_4
   (d) NH_4Cl
   (e) LiOH
   (f) Al_2O_3
   (g) $Sr(HSO_4)_2$
   (h) $CaHPO_4$
   (i) $[PCl_4][PCl_6]$

3. When the following compounds dissolve in water, do they give solutions of acids, bases, or salts?
   (a) H_2S (e) Li_2O (i) CaO
   (b) $In(NO_3)_3$ (f) NH_4NO_3 (j) KIO_4
   (c) $Ba(OH)_2$ (g) Cl_2O_7 (k) H_3PO_4
   (d) SO_3 (h) I_2O_5 (l) NH_3

4. Hydrogen chloride is a covalent molecule, yet it dissolves in water and behaves as an electrolyte. Explain.

5. Indicate whether each of the following com-

pounds and elements dissolves in water to give a solution of a strong electrolyte, a weak electrolyte, or a nonelectrolyte:

(a) KI   (e) NH_4Cl (i) CaO
(b) N_2 (f) RbOH   (j) SO_3
(c) HCN   (g) Ne   (k) P_4O_{10}
(d) HCl   (h) H_3PO_4 (l) Li_2S

6. Sodium hydrogen carbonate, or sodium bicarbonate, $NaHCO_3$ (baking soda), is kept in chemical laboratories because it can neutralize either acid or base spills. Explain how it can act either as a base or as an acid.

7. Soil that is too acidic is often treated with slaked lime, $Ca(OH)_2$, or with calcium carbonate, $CaCO_3$. Explain why treatment with these materials neutralizes the acids in soil.

Types of Chemical Reactions

8. How do addition reactions that are also oxidation-reduction reactions differ from addition reactions that are acid-base reactions? Write balanced chemical equations illustrating each type.

9. Classify the following as addition reactions or metathetical reactions:
(a) $Li_3N + 3HCl \longrightarrow 3LiCl + NH_3$
(b) $2Li + 2HCl \longrightarrow 2LiCl + H_2$
(c) $Mg + SO_3 \longrightarrow MgSO_3$
(d) $MgO + SO_2 \longrightarrow MgSO_3$
(e) $KI + 2Cl_2 \longrightarrow K[ICl_4]$
(f) $KCl + ICl_3 \longrightarrow K[ICl_4]$

10. Classify the reactions in Exercise 9 as acid-base reactions or oxidation-reduction reactions.

11. Identify the atoms that are oxidized and reduced, the change in oxidation number for each, and the oxidizing and reducing agents in each of the following equations:
(a) $Mg + NiCl_2 \longrightarrow MgCl_2 + Ni$
(b) $PCl_3 + Cl_2 \longrightarrow PCl_5$
(c) $C_2H_4 + 3O_2 \longrightarrow 2CO_2 + 2H_2O$
(d) $Zn + H_2SO_4 \longrightarrow ZnSO_4 + H_2$
(e) $2K_2S_2O_3 + I_2 \longrightarrow K_2S_4O_6 + 2KI$
(f) $3Cu + 8HNO_3 \longrightarrow$
$\qquad 3Cu(NO_3)_2 + 2NO + 4H_2O$

Metals and Nonmetals

12. From the positions of their components in the Periodic Table, predict which of the following pairs will:

(a) conduct electricity, Ga or Br_2
(b) be ionic, CsF or HF
(c) react as a base, $P(OH)_3$ or $Al(OH)_3$
(d) oxidize Si, F_2 or Ca
(e) reduce Sn^{2+} to Sn, Al or O_2
(f) oxidize S_8, O_2 or Li
(g) contain ions, KNO_2 or $ClNO_2$
(h) react with water to give a solution of an acid, Ga_2O_3 or SeO_2

13. From the positions of the elements in the Periodic Table, predict which of the following pairs will:
(a) reduce S_8, Cl_2 or Mg
(b) neutralize H_2SO_4, LiOH or ClOH
(c) oxidize Si, Na or F_2
(d) oxidize SO_2, Al or Cl_2
(e) reduce SeO_2, Zn or O_2
(f) reduce Cu^{2+}, Fe or H_2O
(g) neutralize $NH_3(aq)$, Li_2O or SO_2

14. From the positions of the elements in the Periodic Table, predict which of the following will be:
(a) more basic, TlOH or BrOH
(b) more acidic, $Al(OH)_3$ or $B(OH)_3$
(c) more acidic, H_2CO_3 or H_2SO_3
(d) a stronger oxidizing agent, S_8 or Si
(e) more easily reduced, F_2 or I_2
(f) an ionic compound, $BaCl_2$ or ICl
(g) more easily oxidized, Sr or Se

15. Assume you have an unlabeled bottle containing a colorless gas. The gas reacts with water to give a solution of an acid but does not react with oxygen. Could this gaseous compound be Na_2SO_4? H_2S? SO_2? SO_3? Explain your answers.

16. A sample of an element that conducts electricity is broken into small pieces by cracking it with a hammer. These pieces react with chlorine to form a volatile liquid chloride. Could the element be copper? silicon? sulfur? Explain your answers.

17. An element is oxidized by fluorine but not by chlorine. Could the element be sodium? aluminum? silicon? sulfur? oxygen? Explain your answers.

Common Oxidation Numbers

18. Explain why the maximum oxidation number of an atom of a representative metal is equal to the number of electrons in its valence shell.

19. Explain why the most negative oxidation number of an atom of a representative nonmetal is equal

to the number of electrons required to fill its valence shell.

20. Which of the following elements will exhibit a positive oxidation number when combined with sulfur?
   (a) Al   (c) Ca   (e) F   (g) Si
   (b) Br   (d) Cl   (f) O

21. With which elements will phosphorus form compounds in which it has a positive oxidation number?

22. Write the formula of a binary compound (a compound containing only two elements) for each of the following elements so that the element has the indicated oxidation number:
   (a) Al, +3   (e) Br, +1
   (b) N, −3   (f) Cl, +7
   (c) P, +3   (g) N, +3
   (d) Se, −2   (h) Tl, +1

23. From their positions in the Periodic Table and without reference to Table 8-5, predict the common oxidation numbers of the following elements in their compounds:
   (a) Ga   (c) Ba   (e) P   (g) Si
   (b) I   (d) Rb   (f) Se   (h) Pb

24. Without reference to Table 8-5, identify those representative metals that may display two or more common oxidation numbers in their compounds.

Industrial Chemicals

25. The following reactions are all similar to those of the industrial chemicals described in Section 8.7. Complete and balance the equations for these reactions.
   (a) preparation of an explosive
   $$NH_3 + HClO_4 \longrightarrow$$
   (b) preparation of a soluble source of calcium and phosphorus used in animal feeds
   $$Ca_3(PO_4)_2 + H_3PO_4 \longrightarrow$$
   (c) formation of an air pollutant when burning coal that contains zinc sulfide
   $$ZnS + O_2 \longrightarrow$$
   (d) preparation of a soluble silver salt for silver plating
   $$Ag_2O + HNO_3 \longrightarrow$$
   (e) preparation of pure phosphoric acid for use in soft drinks
   $$P_4O_{10} + H_2O \longrightarrow$$
   (f) neutralization of a basic solution of nylon in order to precipitate the nylon
   $$Na^+(aq) + OH^-(aq) + H_2SO_4 \longrightarrow$$
   (g) hardening of plaster containing slaked lime
   $$Ca(OH)_2 + CO_2 \longrightarrow$$
   (h) removal of sulfur dioxide from the flue gas of power plants
   $$CaO + SO_2 \longrightarrow$$
   (i) neutralization of acid drainage from coal mines
   $$CaCO_3 + H_2SO_3 \longrightarrow$$
   (j) the reaction of baking powder that produces carbon dioxide gas and causes bread to rise
   $$NaHCO_3 + NaH_2PO_4 \longrightarrow$$
   (k) separation of silver from gold with concentrated nitric acid
   $$Ag + HNO_3 \longrightarrow$$
   (l) pickling of steel in hydrochloric acid
   $$Fe_2O_3 + HCl \longrightarrow$$
   (m) preparation of potassium hydroxide by electrolysis of a solution of potassium chloride
   $$KCl(aq) \xrightarrow{\text{Electrolysis}}$$

Prediction of Reaction Products

26. Complete and balance the equations for the following oxidation-reduction reactions. In some cases there may be more than one correct answer depending on the amounts of reactants used, as illustrated in Examples 8.5 and 8.6.
   (a) $Al + S_8 \longrightarrow$
   (b) $Mg + Cl_2 \longrightarrow$
   (c) $K + P_4 \longrightarrow$
   (d) $S_8 + F_2 \longrightarrow$
   (e) $P_4 + O_2 \longrightarrow$
   (f) $Li + N_2 \longrightarrow$
   (g) $P_4O_6 + O_2 \longrightarrow$
   (h) $Al(s) + HCl(g) \longrightarrow$

(i) $K(s) + H_2O(l) \longrightarrow$

(j) $In(s) + HI(aq) \longrightarrow$

(k) $Mg(s) + CH_3CO_2H(aq) \longrightarrow$

(l) $Mg(s) + SnCl_4(g) \overset{\Delta}{\longrightarrow}$

(m) $Al(s) + PbO(s) \overset{\Delta}{\longrightarrow}$

(n) $H_2S + O_2 \overset{\Delta}{\longrightarrow}$

(o) $Li_3P + O_2 \overset{\Delta}{\longrightarrow}$

(p) $CaSO_3 + O_2 \overset{\Delta}{\longrightarrow}$

27. When heated to 700–800°C, diamonds, which are pure carbon, are oxidized by atmospheric oxygen (they burn!). Write the balanced equation for this reaction.

28. Military lasers use the very intense light produced when fluorine combines explosively with hydrogen. What is the balanced equation for this reaction?

29. The product of the reaction of phosphorus with an excess of sulfur is used in matches. Predict the formula of this compound.

30. Small amounts of hydrogen gas and oxygen gas are often produced by electrolysis of water containing a little sulfuric acid as a catalyst. Write the balanced equation for this reaction.

31. Elemental bromine is prepared from sea water, in which is found sodium bromide, NaBr, by treating the sea water with chlorine gas. Write a balanced equation for the process.

32. Metallic copper is produced by roasting ore containing Cu_2S in air, followed by reduction of the product with excess carbon. The gas escaping from the reduction process is burned to provide part of the heat for the roasting process. Write the balanced equation for each of these three steps.

33. Complete and balance the equations for the following acid-base reactions. If the reactions are run in water as a solvent, write the reactants and products as solvated ions. In some cases there may be more than one correct answer depending on the amounts of reactants used.

(a) $HCl(g) + In_2O_3(s) \longrightarrow$

(b) $Mg(OH)_2 + HNO_3 \overset{H_2O}{\longrightarrow}$

(c) $Al(OH)_3(s) + HClO_4(aq) \longrightarrow$

(d) $Li_2O(s) + H_2O(l) \longrightarrow$

(e) $Na_2O(s) + CH_3CO_2H(l) \longrightarrow$

(f) $SrO(s) + H_3PO_4(l) \longrightarrow$

(g) $SO_2(g) + H_2O(l) \longrightarrow$

(h) $NH_3(g) + HI(g) \longrightarrow$

(i) $NH_3(aq) + HBr(aq) \longrightarrow$

(j) $CaO(s) + SO_3(g) \longrightarrow$

(k) $Li_2O(s) + Cl_2O_7(l) \longrightarrow$

(l) $NH_3(g) + NaHSO_4(s) \longrightarrow$

(m) $KCl(s) + H_3PO_4(l) \overset{\Delta}{\longrightarrow}$

34. The following salts contain anions of weak acids. Write a balanced equation for the reaction of a solution of each with an excess of the indicated acid.

(a) $Li[CH_3CO_2] + H_2SO_4 \longrightarrow$

(b) $SrF_2 + HCl \longrightarrow$

(c) $NaCN + HBr \longrightarrow$

(d) $Na_2CO_3 + HClO_4 \longrightarrow$

(e) $CaHPO_4 + HNO_3 \longrightarrow$

(f) $KHCO_3 + HCl \longrightarrow$

(g) $Ca(NO_2)_2 + HI \longrightarrow$

(h) $CaSO_3 + HCl \longrightarrow$

35. Complete and balance the equations of the following reactions, each of which is used to remove hydrogen sulfide from natural gas:

(a) $Ca(OH)_2(s) + H_2S(g) \longrightarrow$

(b) $NaOH(s) + H_2S(g) \longrightarrow$

(c) $Na_2CO_3(aq) + H_2S(g) \longrightarrow$

(d) $K_3PO_4(aq) + H_2S(g) \longrightarrow$

36. The medical laboratory test for cyanide ion, CN^-, involves separation of the cyanide ion from a blood, urine, or tissue sample by the addition of sulfuric acid. The gaseous product is absorbed in a sodium hydroxide solution and then analyzed. Write balanced equations that describe the separation and absorption.

37. Calcium propionate is sometimes added to bread to retard spoilage. This compound can be prepared by the reaction of calcium carbonate, $CaCO_3$, with propionic acid, $C_2H_5CO_2H$, an acid with properties similar to those of acetic acid. Write the balanced equation for the formation of calcium propionate.

38. A teaspoon of baking soda, $NaHCO_3$, in a glass of water can be used to treat acid stomach. Stomach acid is hydrochloric acid. Write a balanced equation for the reaction that occurs. Milk of magnesia, $Mg(OH)_2$, is a somewhat milder antacid. Write a balanced equation describing the action of milk of magnesia. Other antacids contain aluminum hydroxide, $Al(OH)_3$. Write the balanced equation that describes its action.

39. Derivatives of fatty acids occur in plant and animal cells. A fatty acid is a carboxylic acid, like

acetic acid, in which a large hydrocarbon group replaces the CH_3 group of acetic acid. Write a balanced equation for each of the following reactions, which involve fatty acids:

(a) stearic acid, $C_{17}H_{35}CO_2H$, with NaOH
(b) palmitic acid, $C_{15}H_{31}CO_2H$, with Zn
(c) myristic acid, $C_{13}H_{27}CO_2H$, with CaO
(d) sodium palmitate, $C_{15}H_{31}CO_2Na$, with HCl
(e) capric acid, $C_9H_{19}CO_2H$, with LiH
(f) myristic acid, $C_{13}H_{27}CO_2H$, with Na_2CO_3

Additional Exercises

40. Complete and balance the following equations. If the reactions are run in water as a solvent, write the reactants and products as solvated ions. In some cases there may be more than one correct answer depending on the amounts of reactants used.

(a) $Mg(OH)_2(s) + HCl(g) \longrightarrow$

(b) $Sr(OH)_2 + HNO_3 \xrightarrow{H_2O}$

(c) $Ca + P_4 \xrightarrow{\triangle}$
(d) $Cs + S_8 \longrightarrow$
(e) $LiH + H_2O \longrightarrow$

(f) $Mg + SnSO_4 \xrightarrow{H_2O}$
(g) $Si + F_2 \longrightarrow$
(h) $Na(s) + H_2O(l) \longrightarrow$

(i) $Ca(CH_3CO_2)_2 + H_2SO_4 \xrightarrow{H_2O}$
(j) $Li_2O + SO_3 \longrightarrow$
(k) $H_2 + S_8 \longrightarrow$
(l) $Sb + F_2 \longrightarrow$
(m) $NH_3(g) + H_2S(g) \longrightarrow$
(n) $As + S_8 \longrightarrow$
(o) $Al(s) + H_2SO_4(aq) \longrightarrow$
(p) $Cl_2 + I_2 \longrightarrow$

(q) $NaF + HNO_3 \xrightarrow{H_2O}$

41. Hydrogen sulfide, H_2S, is removed from natural gas by passing the raw gas through a solution of ethanolamine, $HOCH_2CH_2NH_2$, whose behavior is similar to that of NH_3 in its acid-base reactions. After the solution is saturated, the H_2S can be recovered by heating the solution to reverse the reaction. Part of the recovered H_2S is burned with air, and the product is reacted with the remaining H_2S to produce sulfur, S_8, which is easier to ship than H_2S. Write the reaction of ethanola-

mine with hydrogen sulfide, showing the Lewis structures of the reactants and products. Write the balanced equations for the reactions that lead to the formation of sulfur from hydrogen sulfide.

42. The amino acid glycine can act either as an acid or as a base. Explain why. The Lewis structure of glycine in neutral solution is

The behavior of the $C-NH_3^+$ group is similar to that of the NH_4^+ ion.

43. Write balanced chemical equations for the preparation of each of the following compounds (a) by a metathetical reaction, (b) by an oxidation-reduction reaction, and (c) by an acid-base reaction: KBr, $AlPO_4$, $Cs(HSO_4)_2$. Write different reactions for (a), (b), and (c) in each case.

44. Calcium cyclamate, $Ca(C_6H_{11}NHSO_3)_2$, an artificial sweetener now banned by the FDA, was purified industrially by converting it to the barium salt through reaction of the acid $C_6H_{11}NHSO_3H$ with barium carbonate, treatment with sulfuric acid (barium sulfate is very insoluble), and then neutralization with calcium hydroxide. Write the balanced equations for these reactions.

45. Calcium hydrogen sulfite, $Ca(HSO_3)_2$, is used in the paper industry in the "cooking liquor" that is important in converting cellulose into paper. $Ca(HSO_3)_2$ is prepared from sulfur, calcium carbonate, air, and water. Write balanced equations for the three steps in its preparation.

46. Write a balanced chemical equation for the reaction used to prepare each of the following compounds from the given starting material(s). Several steps and additional reactants may be required.

(a) H_2SO_4 from ZnS
(b) $(NH_4)NO_3$ from N_2
(c) HCl from Cl_2
(d) KOH from KCl
(e) LiH from H_2SO_4
(f) H_2S from Zn and S

9

OXYGEN, OZONE, AND HYDROGEN

OXYGEN

Oxygen is the most abundant element in the earth's crust and atmosphere (see Table 1-1), and it forms compounds with almost all other elements. It is essential to combustion and to respiration in most plants and animals.

The discovery of oxygen marked the beginning of modern chemistry. Credit for the discovery of oxygen is usually given to Joseph Priestley, an English clergyman and scientist, who prepared oxygen in 1774 by focusing the sun's rays on mercury(II) oxide by means of a lens (see Fig. 9-1). He tested the gaseous product with a burning candle and noted that it caused the candle to burn more brightly than did ordinary air. Shortly thereafter, Antoine-Laurent Lavoisier recognized that the products of the combustion of metals weighed more than the pure metals because oxygen combined with the metals during combustion.

9.1 Chemical Properties of Oxygen

You should expect oxygen to exhibit pronounced nonmetallic behavior since it is located at the top of Group VIA in the Periodic Table, and it does. It forms ionic compounds with metals, and covalent compounds with nonmetals. It is a very electronegative nonmetal (electronegativity of 3.5; see Table 5-2); only fluorine is

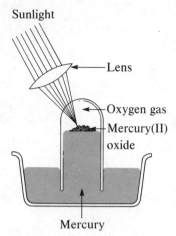

Figure 9-1. Apparatus used by Priestley in his preparation of oxygen. When HgO is heated, it decomposes, yielding gaseous oxygen and mercury.

more electronegative (electronegativity of 4.1). The electron configuration of an oxygen atom is $1s^2 2s^2 2p^4$, with two of the p orbitals singly occupied. Like other nonmetals in the second period, oxygen fills its valence shell and completes an octet of electrons by picking up electrons. Thus it can form the oxide ion, O^{2-}, in many compounds with metals, or it can form two single covalent bonds as in water, H_2O; a double bond as in formaldehyde, $H_2C=O$; or a triple bond as in carbon monoxide, $C\equiv O$.

Oxygen forms compounds with all of the elements except the noble gases helium, neon, argon, and krypton. Except when combined with the more electronegative element fluorine, oxygen has a negative oxidation number, in its compounds, most commonly -2 as expected for an element in Group VIA (Section 8.5). Binary compounds of oxygen in which it has an oxidation number of -2 are called **oxides.** In compounds that contain O—O single bonds, **peroxides,** oxygen exhibits an oxidation number of -1. In the few metal compounds that contain the **superoxide ion,** O_2^-, oxygen has an oxidation number of $-\frac{1}{2}$.

Elemental oxygen is a strong oxidizing agent and will eventually oxidize most organic substances, although fortunately at a slow rate. A fast oxidation rate would make life on earth impossible; all organic matter, including animal life, would burn up.

The ease with which elemental oxygen picks up electrons is mirrored by the difficulty of removing electrons from oxygen in most oxides. Of the elements, only the very reactive fluorine molecule will oxidize oxides to form oxygen.

Soluble ionic metal oxides react with water to form metal hydroxides (Section 8.1) because the oxide ion is a very strong base and readily picks up a hydrogen ion from a water molecule:

$$O^{2-} + H_2O \longrightarrow 2OH^-$$

The oxides of the more nonmetallic elements react with water to form oxyacids. For example, sulfur trioxide and water react to form sulfuric acid:

$$SO_3 + H_2O \longrightarrow H_2SO_4$$

The chemical behavior of oxides, oxyacids, and their salts depends on what elements other than oxygen are present in these compounds and will therefore be described in the chapters dealing with those other elements.

9.2 Occurrence and Preparation of Oxygen

Oxygen is the most abundant, widely distributed element on the earth. As the free element, it forms about 23% of the air; in combination with hydrogen, about 89% of water. About 50% of the mass and 90% of the volume of the earth's crust consist of oxygen combined with other elements, principally silicon. In combination with carbon, hydrogen, and nitrogen, oxygen forms a large part of the weight of the bodies of plants and animals.

Air and water are abundant, cheap, and simple to use, so approximately 97% of commercial oxygen is produced from air and 3% from the electrolysis of water. In 1982, 14.8 million tons of oxygen were produced; it ranked fourth among all chemicals in amount produced.

1. FRACTIONAL EVAPORATION OF LIQUID AIR. Cooling and compressing air until it liquefies and then evaporating off the lower-boiling nitrogen and some other elements yields oxygen in commercial quantities (see Section 22.2). Oxygen

is often stored and shipped in Dewar flasks (see Fig. 22-2) as a liquid (boiling point = −183°C). The evaporation of some of the oxygen makes these flasks self-refrigerating. Oxygen is also stored and shipped as a compressed gas in steel cylinders.

2. ELECTROLYSIS OF WATER.   Pure water is a very poor conductor of electricity; but when a direct current of electricity is passed through water containing a small amount of an electrolyte such as H_2SO_4, NaOH, or Na_2SO_4 (Section 8.1), bubbles of hydrogen are formed at one electrode, the cathode, and oxygen is evolved at the other electrode, the anode (Fig. 9-2). The volume of hydrogen produced is twice that of the oxygen. The net reaction can be summarized by the equation

$$2H_2O(l) + \text{electrical energy} \longrightarrow 2H_2(g) + O_2(g)$$

The use of an electric current to produce a chemical reaction is called **electrolysis.** Because electricity is so expensive, electrolytic oxygen is usually used only when very pure oxygen is required.

Figure 9-2. Laboratory apparatus for the electrolysis of water.

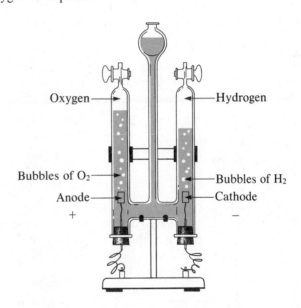

3. DECOMPOSITION OF METAL OXIDES AND PEROXIDES.   A few metal oxides are not thermally stable and thus lose oxygen when heated. For example, when red mercury(II) oxide, HgO, is heated (Fig. 9-1), metallic mercury and oxygen are formed.

$$2HgO \longrightarrow 2Hg + O_2$$

This is the method used by Priestley and Lavoisier, but it is too expensive to be used to prepare oxygen commercially. Similar decomposition reactions that give the metallic element and oxygen occur when silver oxide (Ag_2O), gold(I) oxide (Au_2O), gold(III) oxide (Au_2O_3), and the platinum(II) and platinum(IV) oxides (PtO and PtO_2) are heated. As we shall see in Section 9.17, these are oxides of metals located near the bottom of the activity series.

   In some metal oxides·the metal has a high oxidation number and is relatively unstable. Such metal ions are strong enough oxidizing agents to oxidize oxide ions to oxygen gas, when heated. These metal oxides undergo thermal decomposition reactions in which part of the oxygen is liberated. The remainder is combined in a metal oxide in which the metal has a lower, more stable oxidation number than it did in the original oxide.

$$2PbO_2 \xrightarrow{\Delta} 2PbO + O_2$$
$$\text{Lead(IV) oxide} \qquad \text{Lead(II) oxide}$$

$$4CrO_3 \xrightarrow{\Delta} 2Cr_2O_3 + 3O_2$$
$$\text{Chromium(VI) oxide} \qquad \text{Chromium(III) oxide}$$

Ionic peroxides (compounds containing $O_2{}^{2-}$ ions with an O—O single bond), such as barium peroxide, BaO_2, and cesium peroxide, Cs_2O_2, decompose to the corresponding oxides with evolution of oxygen.

$$2Ba^{2+}(:\overset{..}{O}:\overset{..}{O}:{}^{2-}) \xrightarrow{\Delta} 2Ba^{2+}(:\overset{..}{O}:{}^{2-}) + O_2$$

$$2(Cs^+)_2(:\overset{..}{O}:\overset{..}{O}:{}^{2-}) \xrightarrow{\Delta} 2(Cs^+)_2(:\overset{..}{O}:{}^{2-}) + O_2$$

4. HEATING CERTAIN SALTS THAT CONTAIN OXYGEN. The nitrate salts of certain metals yield oxygen when heated.

$$2NaNO_3 \xrightarrow{\Delta} 2NaNO_2 + O_2$$
$$\text{Sodium nitrate} \qquad \text{Sodium nitrite}$$

$$2Cu(NO_3)_2 \xrightarrow{\Delta} 2CuO + 4NO_2 + O_2$$
$$\text{Copper(II) nitrate} \qquad \text{Copper(II) oxide} \quad \text{Nitrogen dioxide}$$

Oxygen can be prepared on a small scale by heating potassium chlorate above its melting point. If manganese(IV) oxide is mixed with the chlorate, the latter decomposes quite rapidly at about 270°C, nearly 100° below its melting point (Fig. 9-3).

$$2KClO_3 \xrightarrow[MnO_2]{\Delta} 2KCl + 3O_2$$
$$\text{Potassium chlorate} \qquad \text{Potassium chloride}$$

The manganese(IV) oxide may be reclaimed in chemically unchanged form after the reaction is completed; it has served to catalyze the reaction by causing it to take place more rapidly at a lower temperature. A **catalyst** is a substance that changes the rate of a reaction without undergoing a permanent chemical change itself. Many other metal oxides, such as Fe_2O_3 and Cr_2O_3, also serve as catalysts for the decomposition of potassium chlorate. Catalysts are usually specific in their action; a substance that will catalyze one reaction will often have no effect on another. Catalysis is discussed in more detail in Section 15.15.

The preparation of oxygen by heating potassium chlorate can be dangerous if done carelessly. Explosions can occur when combustible materials such as carbon,

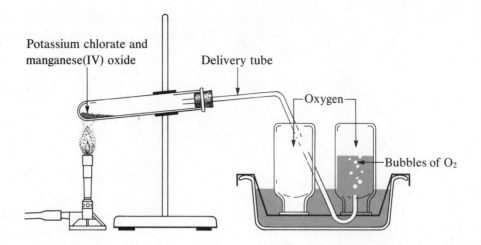

Figure 9-3. In the laboratory, oxygen is usually prepared as shown. The manganese(IV) oxide catalyst and potassium chloride remain after the potassium chlorate has decomposed.

Potassium chlorate and manganese(IV) oxide

Delivery tube

Oxygen

Bubbles of O_2

sulfur, rubber, or a glowing wood splint come in contact with hot potassium chlorate. The potential danger in this experiment cannot be overemphasized. Proceed with caution when preparing oxygen by this method.

9.3 Physical Properties of Oxygen

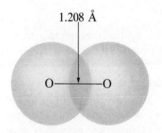

1.208 Å

O——O

Figure 9-4. A representation of the oxygen molecule, O_2. Each sphere represents an oxygen atom.

Oxygen is a colorless, odorless, and tasteless gas at ordinary temperatures. It is slightly more dense than air. At 0°C and a pressure of 1 atmosphere, 1 liter of oxygen weighs 1.429 grams, whereas 1 liter of air under the same conditions weighs 1.292 grams. Although it is only slightly soluble in water (49 milliliters dissolve in 1 liter at 0°C), its solubility is very important to aquatic life and in the decomposition of organic matter present in streams and lakes. Oxygen is pale blue in the liquid state and boils at −183°C at atmospheric pressure. Solid oxygen, also pale blue, melts at −218.4°C. The oxygen molecule, O_2, is diatomic (Fig. 9-4) and **paramagnetic,** that is, attracted by a magnetic field. This paramagnetism is an indication that the diatomic molecule contains unpaired electrons.

Since an oxygen atom has six valence electrons, we might expect two pairs of electrons to be shared between the atoms in the diatomic oxygen molecule, with all electrons paired. However, the size of the paramagnetism of the oxygen molecule indicates two unpaired electrons per molecule. Their presence is consistent with the molecular orbital description of the bonding in O_2 in which the unpaired electrons are distributed singly in the two π_p^* antibonding molecular orbitals (Section 6.10).

9.4 Chemical Reactivity of Oxygen

1. REACTION WITH ELEMENTS. Oxygen reacts directly at room temperature or at elevated temperatures with all other elements except the noble gases, the halogens, and a few second and third row transition metals of low reactivity (those below copper in the activity series whose oxides decompose upon heating; see Section 9.17). The most active metals, the metals located at the bottom of Group IA of the Periodic Table (Cs, Rb, and K), react with oxygen giving superoxides

(RbO$_2$, for example). Their neighbors in the Periodic Table (Na, Ca, Sr, and Ba) give peroxides when they react with oxygen. Less active metals and the nonmetals give oxides.

$$Rb + O_2 \longrightarrow RbO_2$$

Rubidium
superoxide

$$Ba + O_2 \longrightarrow BaO_2$$

Barium
peroxide

$$4Li + O_2 \longrightarrow 2Li_2O$$

Lithium oxide

$$4Ga + 3O_2 \longrightarrow 2Ga_2O_3$$

Gallium oxide

$$C + O_2 \longrightarrow CO_2$$

Carbon dioxide

$$P_4 + 5O_2 \longrightarrow P_4O_{10}$$

Phosphorus(V)
oxide

Oxides of the halogens, of some of the noble gases, and of the metals at the bottom of the activity series can be prepared, but not by direct reaction with oxygen.

As was shown in Chapter 8, the formulas of binary oxides, peroxides, and superoxides generally can be determined from the position of the combining element in the Periodic Table and the oxidation number of oxygen. Thus the reaction of the Group IA element lithium with oxygen at ordinary pressures gives Li$_2$O (oxidation number of Li = +1, oxidation number of O = −2); gallium, a Group IIIA element, gives Ga$_2$O$_3$ (oxidation number of Ga = +3, oxidation number of O = −2); and phosphorus, a Group VA element, gives P$_4$O$_6$ or P$_4$O$_{10}$ (oxidation number of P = +3 or +5, oxidation number of O = −2), depending on the amount of oxygen available. As we also saw in Chapter 8, the oxides of the metals are basic, whereas the oxides of the nonmetals are acidic.

2. REACTION WITH COMPOUNDS.

Elemental oxygen also reacts with some compounds. If a compound is composed of elements that will combine with oxygen when free, the compound may be expected to react with oxygen to form oxides of the constituent elements, *provided* that at least one of the atoms in the compound does not have its maximum oxidation number. For example, hydrogen sulfide, H$_2$S, contains sulfur with an oxidation number of −2. Since the sulfur does not have its maximum oxidation number and since free sulfur will react with oxygen, we expect hydrogen sulfide to react with oxygen. It does, giving water and sulfur dioxide.

$$2H_2S + 3O_2 \longrightarrow 2H_2O + 2SO_2$$

The following equations provide additional examples of compounds that react with oxygen:

$$CS_2 + 3O_2 \longrightarrow CO_2 + 2SO_2$$

Carbon
disulfide

$$C_{12}H_{22}O_{11} + 12O_2 \longrightarrow 12CO_2 + 11H_2O$$
Sugar

$$CH_4 + 2O_2 \longrightarrow CO_2 + 2H_2O$$
Methane

$$2ZnS + 3O_2 \longrightarrow 2ZnO + 2SO_2$$
Zinc sulfide

Oxides containing an element having a lower oxidation number than is usually the case when the element combines with an excess of oxygen will react with additional oxygen. Examples are the following:

$$2CO + O_2 \longrightarrow 2CO_2$$
Carbon
monoxide

$$P_4O_6 + 2O_2 \longrightarrow P_4O_{10}$$
Phosphorus(III)
oxide

Oxides such as carbon dioxide, CO_2, silicon dioxide, SiO_2, sulfur trioxide, SO_3, and magnesium oxide, MgO, do not react with oxygen, because in these compounds the element combined with oxygen already exhibits the maximum oxidation number possible.

An important point that needs to be made here is that an equation often represents only the *main* reaction that takes place under a given set of conditions. There may be by-products, but these are not included in writing the equation for the principal reaction. We write the equation for burning sulfur in oxygen, for example, as follows:

$$S + O_2 \longrightarrow SO_2$$

However, when the reaction takes place, small amounts of sulfur trioxide are produced. When hydrogen sulfide is burned, the chief products are water and sulfur dioxide, but we may also get small amounts of elemental sulfur or of sulfur trioxide. Possible products from the burning of carbon disulfide, in addition to carbon dioxide and sulfur dioxide, are elemental carbon, carbon monoxide, elemental sulfur, and sulfur trioxide.

9.5 Combustion

Combustion is a chemical reaction that is accompanied by the evolution of light and heat. The burning of wood, natural gas, and magnesium in air are examples of combustion. Combustion can also take place in other atmospheres; for example, hydrogen will burn in an atmosphere of chlorine, forming hydrogen chloride.

$$H_2 + Cl_2 \longrightarrow 2HCl$$

Before combustion can take place, the substances involved must be heated to the **kindling temperature,** which is the temperature at which the reaction is rapid enough to proceed without further addition of heat from the outside. The kindling temperature for many substances, especially solids, is not fixed, but depends on how finely divided the substance is. For example, a fine iron wire will burn readily when heated and then placed in pure oxygen, but an iron rod a few millimeters in diameter will not.

When the heat evolved during a slow reaction remains and raises the temperature of the reacting substances to the kindling temperature, **spontaneous combustion** may occur. For example, linseed oil unites with oxygen at ordinary temperatures in an exothermic reaction. Rags soaked with linseed oil and stored without enough circulation of air to carry off the heat produced by slow oxidation may ignite spontaneously; that is, the heat of reaction may accumulate and raise the temperature of the system above the kindling temperature. Many costly and disastrous fires have been started by spontaneous combustion of such materials as coal in large piles, uncured hay stored in unventilated barns, and rags containing paints or drying oils.

Whenever combustion takes place very rapidly, the heat of reaction is liberated almost instantly, and usually a large increase in gaseous volume results (gases expand when heated). An explosion may result if the gases cannot escape. The mixture of gasoline vapors and air in the cylinder of an automobile engine explodes when ignited by a spark. Disastrous explosions sometimes occur in flour mills, grain elevators, and coal mines when finely divided particles are ignited. Any process that leads to a sudden, large increase in gas pressure in a confined space can create an explosion.

9.6 Heat of Reaction

Combustion reactions produce heat and are called exothermic reactions (Section 2.7). Once started, exothermic reactions usually continue without addition of energy from the outside. For example, coal (which is primarily carbon) will burn in air, evolving heat, until the coal is exhausted.

Reactions accompanied by the absorption of heat are called endothermic reactions. They require a continuous supply of energy from the outside to keep them going. For example, the decomposition of mercury(II) oxide into mercury and oxygen will continue only as long as the compound is heated.

The quantity of heat liberated or absorbed during a chemical change occurring at a constant pressure is referred to as the **enthalpy of reaction,** or the **heat of reaction.** As noted in Section 2.7, the heat of reaction is given as a ΔH value following the equation for the reaction. A negative value of ΔH indicates an exothermic reaction; a positive value of ΔH, an endothermic reaction. When hydrogen and oxygen at 25°C and 1 atmosphere of pressure form 1 mole of water at 25°C and 1 atmosphere of pressure, 285.8 kilojoules of heat is produced. The equation is

$$H_2(g) + \tfrac{1}{2}O_2(g) \longrightarrow H_2O(l) \qquad \Delta H = -285.8 \text{ kJ/mol}$$

The negative value of ΔH indicates that the reaction is exothermic. The energy involved in the formation of one mole of a compound from the constituent elements is known as the **enthalpy of formation,** or **heat of formation,** of the compound. The heat of formation of 1 mole of water is −285.8 kilojoules, that of 1 mole of carbon dioxide is −393 kilojoules, and that of 1 mole of magnesium oxide is −601.83 kilojoules. The same amount of heat is involved when a mole of a compound decomposes into its constituent elements as when it is formed. This follows from the Law of Conservation of Energy (Section 1.3). Thus when 1 mole of water is decomposed into hydrogen and oxygen at 25°C and 1 atmosphere of pressure, 285.8 kilojoules of heat is needed for this endothermic process.

$$H_2O(l) \longrightarrow H_2(g) + \tfrac{1}{2}O_2(g) \qquad \Delta H = +285.8 \text{ kJ/mol}$$

Compounds such as water and carbon dioxide that are formed by highly exothermic reactions are stable toward heat and are said to be **thermally stable.** A very high temperature is required to decompose them. On the other hand, compounds such as hydrogen peroxide, H_2O_2, that result from endothermic reactions, are thermally unstable; that is, only moderate heating is necessary to break the bonds holding the atoms together, and the molecule may decompose even at room temperature.

Additional points concerning the energy relationships in chemical reactions will be brought out in the discussion of enthalpy, free energy, and entropy in Chapter 18.

9.7   Rate of Reaction

Oxygen combines slowly with finely divided soft coal at ordinary temperatures. After ignition a piece of coal will burn quietly. Coal dust and oxygen may explode when ignited by a spark. It is evident, then, that the speed of the reaction of coal with oxygen varies greatly.

The rate at which substances are used up or are formed during a chemical change is known as the **rate of reaction.** Among the factors that influence the rate of reaction are the temperature, the concentration of the reactants, whether a catalyst is present, and how finely divided the reactants are (area of contact between the reactants).

1. TEMPERATURE.   Almost all reactions are speeded up when the temperature is increased, although for very fast reactions, such as

$$Ag^+(aq) + Cl^-(aq) \longrightarrow AgCl(s)$$

the effect is negligible. There is a rough rule that, for reactions that do not appear to be instantaneous, a rise in temperature of $10°$ will double the rate of reaction.

2. CONCENTRATION.   The fact that a heated iron wire will burn in pure oxygen but not in air, which is only 21% oxygen by volume, shows the effect of concentration on the rate of oxidation of iron. In general, the rate of reaction increases as the concentration of the reactants is increased. The reaction of zinc with hydrochloric acid to produce zinc chloride and hydrogen proceeds more rapidly if the concentration of hydrochloric acid is increased.

3. CATALYSTS.   The rate of a reaction can be increased if a catalyst is introduced. The catalytic effect of manganese(IV) oxide on the thermal decomposition of potassium chlorate was mentioned in Section 9.2. The commercial feasibility of many industrial processes is due to the use of catalysts for increasing the rates of reactions that would otherwise occur too slowly to be profitably utilized.

4. STATE OF SUBDIVISION.   Reactions between substances take place only when they are in contact, and the more extensive the contact, the more rapid the reaction. As a given amount of a solid substance is more and more finely divided, its surface area increases so that more of the substance can come in contact with a reactant. Wood shavings burn much more rapidly than a massive wood block. Powdered zinc reacts more rapidly with hydrochloric acid than large pieces of zinc.

9.8   The Importance of Oxygen to Life

In human beings and other animals, the energy required for maintenance of normal functions is derived from the slow oxidation of materials in the body. Oxygen passes from the lungs into the blood, where it combines with hemoglobin, producing oxyhemoglobin. In this form oxygen is carried by the blood to the various tissues of the body, where it is released and consumed in reactions with oxidizable materials. The products are mainly carbon dioxide and water. The blood carries the carbon dioxide through the veins to the lungs. There it gives up the carbon dioxide and collects another supply of oxygen. The digestion and assimilation of food regenerate the materials consumed by oxidation in the body, with the same amount of energy being liberated as if the food had been burned outside the body.

The oxygen in the atmosphere is continually replenished through the action of algae and higher plants by a process called **photosynthesis.** The products of photosynthesis vary, but the process can be described in general as the conversion of carbon dioxide and water to glucose (a sugar) and oxygen by chlorophyll using the energy of light.

$$6CO_2 + 6H_2O \xrightarrow[\text{Light}]{\text{Chlorophyll}} C_6H_{12}O_6 + 6O_2$$

Carbon   Water   Glucose   Oxygen
dioxide

Thus the oxygen that is converted to carbon dioxide and water by the metabolic processes of plants and animals is returned to the atmosphere by photosynthetic reactions.

9.9   Uses of Oxygen

Oxygen in the air is essential in combustion processes such as the burning of fuels for the production of heat; oxygen is also required for the decay of organic matter. Plants and animals use oxygen from the air in respiration. Oxygen-enriched air is used in medical practice when a patient is getting an inadequate supply of oxygen due to shock, pneumonia, or some other illness. Health services use about 13% of the oxygen produced commercially.

Approximately 30% of the oxygen produced commercially is used to remove carbon from iron in steel production. Large quantities of pure oxygen are also consumed in metal fabrication and in the cutting and welding of metals with oxyhydrogen and oxyacetylene blowtorches. The chemical industry uses oxygen for oxidizing many substances.

Liquid oxygen is used as an oxidizing agent in the rocket engines on the space shuttle. It is also used to provide gaseous oxygen for life support in space.

OZONE

9.10   Preparation of Ozone

When dry oxygen is passed between the two electrically charged plates of an apparatus called an **ozonizer** (Fig. 9-5), a decrease in the volume of the gas occurs

Figure 9-5. Laboratory apparatus for preparing ozone.

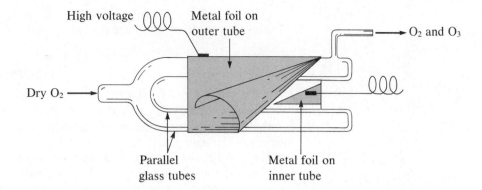

and a substance possessing a distinctive odor is formed. This substance is **ozone.** Its molecules are composed of three oxygen atoms; that is, ozone, O_3, is triatomic.

Ozone and oxygen are **allotropes,** or two different forms of the same element in the same state. Ordinary oxygen, O_2, and ozone, O_3, are both gases, but oxygen molecules are diatomic while ozone molecules are triatomic.

The formation of ozone from oxygen is an endothermic reaction in which the energy is furnished in the form of an electrical discharge, heat, or ultraviolet light.

$$3O_2 \xrightarrow[\text{discharge}]{\text{Electric}} 2O_3 \qquad \Delta H = 286 \text{ kJ}$$

Ozone is prepared commercially by an electrical ozonizer that is more complicated and more efficient than the laboratory apparatus. An alternative method of commercial production is by electrolysis of very cold, concentrated sulfuric acid. Ozone is also formed by processes not suitable for commercial production; for example, by the action of ultraviolet light on oxygen, by lightning, and by most other electrical discharges in air. The sharp odor associated with sparking electrical equipment is due, in part, to ozone. Ozone produced in the lower atmosphere reacts with engine exhaust, and this contributes to smog (see Section 22.4). A great deal of the deterioration of automobile tires is due to contact with ozone in the air.

Significant amounts of ozone are formed in the upper atmosphere by the action of ultraviolet light from the sun on the oxygen there. Considerable discussion has resulted from the concern that residual chlorofluorocarbon propellants (Freons) from aerosol cans and other sources may catalyze the decomposition of ozone in the upper atmosphere, decreasing its concentration. This would lessen the effectiveness of this layer as a protective barrier to the harmful rays of the sun. It is not yet known whether this may happen; but if it does, it could have a drastic effect on the processes involved in life on the earth.

9.11 Properties and Uses of Ozone

Ozone is pale blue as a gas and deep blue as a liquid, with a sharp, irritating odor. It produces headaches and is poisonous. Energy is absorbed when ozone is formed from oxygen, so ozone is more active chemically than oxygen and decomposes readily into oxygen by the exothermic reaction

$$2O_3 \longrightarrow 3O_2 \qquad \Delta H = -286 \text{ kJ}$$

As was mentioned in Section 9.6, in general, substances formed by endothermic reactions tend to be less stable than those formed by exothermic reactions.

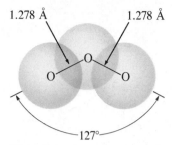

1.278 Å   1.278 Å

Figure 9-6. A representation of the ozone molecule, O_3. The arrangement of spheres indicates the angular nature (127° angle) of the molecule.

The presence of ozone in gas mixtures can readily be detected by passing the gas through a solution of potassium iodide, KI, containing some starch.

$$[2K^+(aq)] + 2I^-(aq) + O_3(g) + H_2O(l) \longrightarrow$$
$$[2K^+(aq)] + 2OH^-(aq) + I_2 + O_2(g)$$

The O_3 oxidizes I^- to I_2, elemental iodine, which imparts a blue color to the starch. Although O_2 reacts with almost all other elements, it usually does so only at elevated temperatures. Ozone is a powerful oxidizing agent and forms oxides with many elements when oxygen will not react.

The molecular structure of ozone is shown in Fig. 9-6. The ozone molecule is angular, and the two bonds are indistinguishable. The electronic structure of ozone can be described by the following resonance forms (Section 5.7):

The uses of ozone depend on its readiness to react with other substances. As a bleaching agent for oils, waxes, fabrics, and starch, it reacts with the colored compounds in these substances, oxidizing them to colorless compounds. It is sometimes used instead of chlorine to purify water.

HYDROGEN

Early in the sixteenth century the Swiss-German physician Paracelsus noted that a flammable gas was formed by the reaction of sulfuric acid with iron. However, it was not until 1766 that Cavendish, an Englishman, recognized this gas as a distinct substance and prepared it by the quaint method of passing steam through a red-hot gun barrel and also by the action of various acids on certain metals. Lavoisier, a French chemist, named the gas hydrogen, meaning "water producer," because water was formed when it was burned in air.

An uncombined hydrogen atom, which consists of one proton and one electron, has the simplest structure of all of the atoms. It has one valence electron in an s orbital and is usually classified with the alkali metals in Group IA of the Periodic Table. However, the chemical properties of hydrogen generally are those of a nonmetal, so it is sometimes placed with the halogens in Group VIIA. For example, a hydrogen atom can fill its valence shell by picking up one electron and forming a hydride ion, H^-, or by sharing one electron and forming a single covalent bond with another atom. Ionic hydrides form with some metals of Groups IA and IIA. Hydrogen forms single covalent bonds with itself and other nonmetals. Many of these covalent compounds with nonmetals exhibit acidic behavior (Section 8.1, Part 2).

9.12   Occurrence and Preparation of Hydrogen

Hydrogen is found in the free state on the earth in negligible quantities. It occurs in elemental form in very small amounts in the atmosphere, in the gases of active volcanoes, in natural gas, in the gases present in some coal mines, and trapped in meteorites. The sun and other stars, as well as the gas found in interstellar space, appear to be composed largely of hydrogen. It is estimated that 90% of the atoms in the universe are hydrogen atoms.

The number of known compounds containing hydrogen is greater than the number for any other element. Hydrogen comprises nearly 11% of the weight of water, its most abundant compound. Hydrogen is found combined with carbon and oxygen in the tissues of all plants and animals. It is an important part of petroleum, many minerals, cellulose and starch, sugar, fats, oils, alcohols, acids, and thousands of other substances.

Hydrogen is more expensive to isolate than is oxygen since useful amounts of free hydrogen are not found in nature. Hydrogen must be obtained from compounds by breaking chemical bonds, and this requires much more energy than does simply condensing air to separate one element from other elements found free in it. When small quantities of hydrogen are required, such as in the laboratory, the gas is usually generated from acids. When commercial quantities of hydrogen are needed, water or petroleum is generally used as the raw material. The most common methods of preparing hydrogen follow.

1. FROM HYDROCARBONS. Hydrogen is produced commercially in large quantities from the hydrocarbons in oil and natural gas. **Hydrocarbons** are compounds such as methane, CH_4, and octane, C_8H_{18}, which contain only carbon and hydrogen. When a mixture of methane (the principal component of natural gas) and steam is heated to a high temperature in the presence of catalysts, a gaseous mixture of carbon monoxide, carbon dioxide, and hydrogen is produced.

$$CH_4 + H_2O \xrightarrow[\text{Catalyst}]{\triangle} CO + 3H_2$$

$$CO + H_2O \xrightarrow[\text{Catalyst}]{\triangle} CO_2 + H_2$$

These are typical reactions, and other hydrocarbons may be substituted for methane.

At high temperatures in the presence of catalysts, hydrocarbons can decompose and rearrange in what are called cracking reactions. These reactions, which are used in the refining of petroleum, may produce hydrogen as a by-product. One example of a simple cracking reaction is

$$2CH_3CH_2CH_3 \xrightarrow{600°} CH_4 + CH_2{=}CH_2 + CH_3CH{=}CH_2 + H_2$$

2. ELECTROLYSIS. Hydrogen is liberated at the cathode when water containing a small amount of an electrolyte is electrolyzed (Section 9.2).

$$2H_2O(l) + \text{electrical energy} \longrightarrow 2H_2(g) + O_2(g) \tag{1}$$

If water is replaced by a concentrated solution of sodium chloride, NaCl, in the electrolysis, hydrogen is still formed, but, instead of oxygen, sodium hydroxide and chlorine are formed.

$$[2Na^+(aq)] + 2Cl^-(aq) + 2H_2O + \text{electrical energy} \longrightarrow$$
$$[2Na^+(aq)] + 2OH^-(aq) + Cl_2(g) + H_2(g) \tag{2}$$

This is a commercial method for producing NaOH (caustic soda) and chlorine; hydrogen is a by-product (Section 8.7, Part 5). If the salt solution used is not fairly concentrated, oxygen is liberated at the anode by the reaction shown in Equation (1). This fact points up an important aspect of electrolysis: the concentration of the electrolyte can be an important determining factor in product formation.

3. REACTION OF METALS WITH ACIDS.

Hydrogen is conveniently produced in the laboratory by the reaction of an active metal with an acid. The metal reduces the hydrogen in the acid to H_2. An apparatus like that shown in Fig. 9-7 is used. The acid solution is added to the metal, and the hydrogen is collected by the downward displacement of water. Although any metal above hydrogen in the activity series (Section 9.17) can be used, zinc and iron are frequently used with a dilute solution of either sulfuric acid or hydrochloric acid.

$$Zn(s) + 2H^+(aq) + [SO_4{}^{2-}(aq)] \longrightarrow Zn^{2+}(aq) + [SO_4{}^{2-}(aq)] + H_2(g) \quad (3)$$
$$\text{Zinc sulfate}$$

$$Fe(s) + 2H^+(aq) + [2Cl^-(aq)] \longrightarrow Fe^{2+}(aq) + [2Cl^-(aq)] + H_2(g) \quad (4)$$
$$\text{Iron(II) chloride}$$

$H^+(aq)$, $SO_4{}^{2-}(aq)$ and $Cl^-(aq)$ are used in these equations to indicate that H_2SO_4 and HCl undergo ionization in water, giving $H_3O^+(aq)$ and the corresponding anions. In a **net ionic equation,** an equation that shows only those species that change during a reaction, the ions enclosed in brackets above would not be shown. For example, the net ionic equation for Equation (3) is

$$Zn(s) + 2H^+(aq) \longrightarrow Zn^{2+}(aq) + H_2(g)$$

It is the hydrogen ion of the dilute aqueous acid that reacts with the active metal. The negative ion does not enter the reaction. However, if the solution is evaporated to dryness after the reaction is complete, a crystalline salt composed of the metal cation and the negative ion of the acid is obtained. In the reactions described by Equations (3) and (4), the salts are zinc sulfate and iron(II) chloride.

4. REACTION OF ACTIVE METALS OR CARBON WITH WATER.

The very active metals of Group IA and IIA, such as sodium and calcium, rapidly displace hydrogen from water at room temperature, producing a solution of the corresponding base.

$$2Na + 2H_2O \longrightarrow 2Na^+(aq) + 2OH^-(aq) + H_2(g)$$
$$Ca + 2H_2O \longrightarrow Ca^{2+}(aq) + 2OH^-(aq) + H_2(g)$$

The reaction of a small piece of sodium with water produces sufficient heat to ignite the metal as well as the hydrogen being produced. It is dangerous to bring large pieces of very active metals into contact with water because explosions may result.

Less active metals will displace hydrogen from water at higher temperatures. Magnesium reacts only slowly with boiling water, but when steam is passed over magnesium or red-hot iron, hydrogen is liberated rapidly (Fig. 9-8).

$$Mg + H_2O \longrightarrow H_2 + MgO$$
$$3Fe + 4H_2O \longrightarrow 4H_2 + Fe_3O_4$$

White-hot carbon (1500–1600°C) will react with steam in an endothermic reaction, producing a mixture of carbon monoxide and hydrogen. This mixture is commonly known as **water gas.**

$$C(s) + H_2O(g) \longrightarrow CO(g) + H_2(g)$$

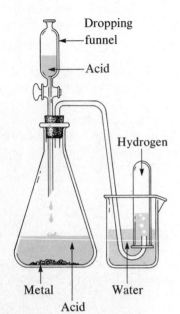

Dropping funnel

Acid

Hydrogen

Metal

Water

Acid

Figure 9-7. Apparatus used in the laboratory preparation of hydrogen.

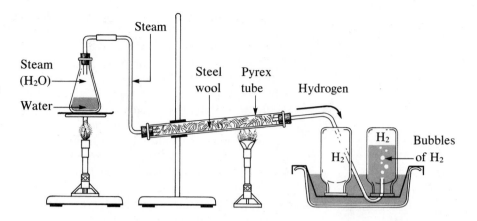

Figure 9-8. Some metals, such as magnesium and iron, reduce the hydrogen in steam, $H_2O(g)$, forming H_2 and metallic oxides. The apparatus for carrying out the reaction of iron with steam to produce hydrogen and iron(II, III) oxide, Fe_3O_4, is illustrated.

Since both carbon monoxide and hydrogen will burn in oxygen or in air and produce heat, water gas is a valuable industrial fuel. When water gas is mixed with steam and passed over a catalyst such as iron oxide or thorium oxide at a fairly high temperature (500°C), carbon dioxide and additional hydrogen are produced. This is called the water gas shift reaction.

$$CO + H_2O + [H_2] \xrightarrow{\text{Catalyst}} CO_2 + H_2 + [H_2]$$

5. REACTION OF CERTAIN ELEMENTS WITH STRONG BASES. Some elements, such as aluminum, zinc, and silicon, react with sodium hydroxide or other strong bases in concentrated aqueous solution, resulting in the liberation of hydrogen.

$$2Al(s) + [2Na^+(aq)] + 2OH^-(aq) + 6H_2O(l) \longrightarrow$$
$$[2Na^+(aq)] + 2Al(OH)_4^-(aq) + 3H_2(g)$$
Sodium aluminate

$$Zn(s) + [2Na^+(aq)] + 2OH^-(aq) + 2H_2O(l) \longrightarrow$$
$$[2Na^+(aq)] + Zn(OH)_4^{2-}(aq) + H_2(g)$$
Sodium zincate

$$Si(s) + 2Na^+(aq) + 2OH^-(aq) + H_2O(l) \longrightarrow Na_2SiO_3(s) + 2H_2(g)$$
Sodium silicate

Aluminum or zinc alone does not displace hydrogen from water, because the oxide that forms produces a film on the metal surface preventing further contact between the metal and the water. Acids and strong bases react with aluminum and zinc to produce hydrogen, because they dissolve the oxide film and allow the metal to come in direct contact with the acid or base.

6. REACTION OF IONIC METAL HYDRIDES WITH WATER. Hydrides of active metals contain the very strongly basic H^- anion and therefore react with water, liberating hydrogen and forming basic solutions.

$$NaH(s) + H_2O(l) \longrightarrow Na^+(aq) + OH^-(aq) + H_2(g)$$
$$CaH_2(s) + 2H_2O(l) \longrightarrow Ca^{2+}(aq) + 2OH^-(aq) + 2H_2(g)$$

Metal hydrides are expensive but convenient sources of hydrogen, especially where space and weight are important factors, as they are in the provision for the inflation of life jackets, life rafts, and military balloons.

9.13   Physical Properties of Hydrogen

Hydrogen is composed of three isotopes. They are ordinary hydrogen, or **protium,** ^1_1H; heavy hydrogen, or **deuterium,** ^2_1H; and **tritium,** ^3_1H. In a sample of hydrogen there is only 1 atom of deuterium for every 7000 atoms of ordinary hydrogen and only 1 atom of tritium for every 10^7 atoms of ordinary hydrogen. The chemical properties of the three isotopes are very similar, because they have identical electronic structures. The physical properties are different, because they have different atomic masses. Deuterium and tritium have lower vapor pressures than does ordinary hydrogen. Consequently, when liquid hydrogen evaporates, the heavier isotopes are somewhat concentrated in the last portions to evaporate.

At ordinary temperatures hydrogen is a colorless, odorless, and tasteless gas consisting of diatomic molecules, H_2 (Fig. 9-9). A pair of electrons in a σ_s bonding molecular orbital (Section 6.3) bonds the two atoms.

With a density of 0.08987 grams/liter at 0°C and 1 atmosphere, hydrogen is the lightest known substance. Because of its low density, it can be collected by the downward displacement of air. The fact that hydrogen is so much lighter than air (which has a density of 1.293 g/L) makes it useful as the lifting agent in balloons.

If sufficiently cooled and compressed, hydrogen changes to a liquid that boils at −252.7°C at atmospheric pressure. The low temperature of liquid hydrogen makes it useful in cooling other materials to low temperatures. It may be used to transform all other gases into solids, with the single exception of helium. Hydrogen freezes at −259.14°C.

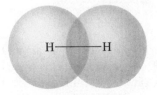

H———H

Figure 9-9.   A representation of the hydrogen molecule, H_2. Each sphere represents a hydrogen atom.

9.14   Chemical Properties of Hydrogen

At ordinary temperatures hydrogen is relatively inactive chemically. When heated, however, it enters into many chemical reactions.

1. REACTION WITH NONMETALS.   When a mixture of hydrogen and oxygen is ignited, water is formed in a vigorous exothermic reaction (negative ΔH).

$$2H_2 + O_2 \longrightarrow 2H_2O \qquad \Delta H = -571.5 \text{ kJ}$$

Because of the violence of the reaction, which often results in an explosion, great caution should be used in handling hydrogen (or any other combustible gas). Hydrogen will burn without explosion under some conditions. The fact that water is formed when hydrogen is burned in either air or oxygen can be demonstrated by allowing a jet of hydrogen to burn inside an inverted, cold, dry beaker (Fig. 9-10). The very high heat of combustion of hydrogen in pure oxygen makes it possible to achieve temperatures up to 2800°C with the oxyhydrogen blowtorch. The hot flame of this torch can be used to cut thick sheets of many metals.

In addition to reacting with oxygen, hydrogen reacts directly with the halogens, sulfur, and nitrogen. The reactions, except for that with fluorine, are very slow at room temperature, but the rates increase with heating. Under some conditions hydrogen reacts explosively with fluorine and chlorine.

$$H_2 + X_2 \longrightarrow 2HX \qquad (X = F, Cl, Br, \text{ or } I)$$
$$8H_2 + S_8 \longrightarrow 8H_2S$$
$$3H_2 + N_2 \longrightarrow 2NH_3$$

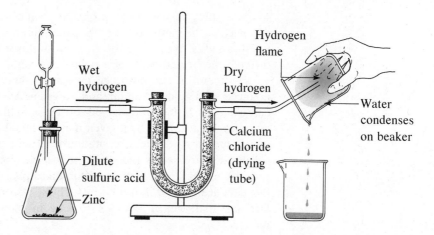

Figure 9-10. When hydrogen burns in air, water vapor is formed. The large beaker here serves as a condenser.

These hydrides are acidic. Ammonia, however, is a very weak acid, and it can also function as a base. These hydrides (as well as hydrides of elements that do not react *directly* with hydrogen) can also be formed by acid-base reactions.

$$KF + H_2SO_4 \longrightarrow KHSO_4 + \underset{\substack{\text{Hydrogen} \\ \text{fluoride}}}{HF}$$

$$Li_3N + 3H_2O \longrightarrow 3LiOH + \underset{\text{Ammonia}}{NH_3}$$

$$Na_3P + 3H_2O \longrightarrow 3NaOH + \underset{\text{Phosphine}}{PH_3}$$

With the exception of the halide ions, all monoatomic negative ions, such as the nitride ion, N^{3-}, and phosphide ion, P_3^{-}, are very strong bases and pick up hydrogen ions from very weak acids, such as water.

2. REACTION WITH METALS. When heated, hydrogen reacts with the elements of Group IA and with calcium, strontium, and barium (the more active elements in Group IIA). The compounds formed are crystalline ionic hydrides that contain hydrogen in the form of the anion H^-.

$$2Li + H_2 \longrightarrow \underset{\substack{\text{Lithium} \\ \text{hydride}}}{2LiH}$$

$$Ca + H_2 \longrightarrow \underset{\substack{\text{Calcium} \\ \text{hydride}}}{CaH_2}$$

The reactions are exothermic, and the resulting hydrides are relatively stable toward heat but react vigorously with water (Section 9.12, Part 6).

Hydrogen is **absorbed** (dissolved) by many transition metals, lanthanide metals, and actinide metals. The materials thus produced are called metallic hydrides because they exhibit many metallic properties. For example, they conduct electricity. In most cases these materials can be described as solutions of hydrogen atoms in the metal. They do not contain hydride ions; instead hydrogen atoms are located in the spaces between the metal atoms. Since these hydrides behave like solutions, they have no fixed molecular composition, and the amount of hydrogen dissolved in the metal can vary. In palladium hydride, for example, the ratio of H to Pd can vary from 0.4 to 0.7. Because of this solubility, hydrogen readily passes through the heated walls of a palladium container.

Hydrogen is also **adsorbed** by certain metals, the process being most pronounced with gold, platinum, and tungsten. **Adsorption** is the adhesion of molecules of a gas, liquid, or dissolved substance to the surface of a solid. The quantity of hydrogen adsorbed by a given mass of metal depends on the condition of the metal, such as the extent of subdivision (the surface area of a given mass of a substance increases as the particles become smaller), and the temperature and pressure under which the adsorption takes place. Adsorbed hydrogen is very active chemically, as is indicated by its rapid union with oxygen when in this state. Ordinary hydrogen and oxygen do not combine significantly unless the mixture is ignited.

3. REACTION WITH COMPOUNDS.   Hydrogen reduces the heated oxides of many metals, with the formation of the metal and water. For example, when hydrogen is passed over heated Fe_3O_4, iron and water are formed.

$$Fe_3O_4 + 4H_2 \longrightarrow 3Fe + 4H_2O$$

In this oxidation-reduction reaction (Section 8.2, Part 4), hydrogen reduces iron from an oxidation number of $+2\frac{2}{3}$ in Fe_3O_4 to an oxidation number of zero in Fe. The hydrogen molecule acts as the reducing agent and is oxidized in the process; each atom loses one electron. The iron in the oxide gains electrons from the hydrogen and hence causes the hydrogen to be oxidized; Fe_3O_4 is the oxidizing agent.

Hydrogen may also reduce some metal oxides to oxides in which the metal has a lower oxidation number.

$$MnO_2 + H_2 \longrightarrow MnO + H_2O$$

9.15   Chemical Equilibrium

The reaction of hydrogen with Fe_3O_4

$$Fe_3O_4(s) + 4H_2(g) \xrightarrow{\triangle} 3Fe(s) + 4H_2O(g) \qquad (1)$$

is the reverse of the reaction used by Cavendish to prepare hydrogen from the reaction of steam with a red-hot (iron) gun barrel

$$3Fe(s) + 4H_2O(g) \xrightarrow{\triangle} Fe_3O_4(s) + 4H_2(g) \qquad (2)$$

When these reactions are run in the laboratory, gas is passed over a heated solid in a tube. In the reaction of Equation (1), the gas is hydrogen, and the steam is swept out of the reaction tube by the current of hydrogen. Thus this reaction can go to completion. In the reaction of Equation (2), the gas is steam, and hydrogen is swept out of the tube by it. Thus this reaction also can go to completion. These reactions are reversible (see Section 8.2, Part 6); that is, each can proceed in either direction.

If a mixture of iron and steam is heated in a closed tube, neither the steam nor the hydrogen that forms can escape. At first only the reaction of Equation (2) will

1. Cobalt(II) chloride, $CoCl_2$
2. Sodium chloride, NaCl
3. Lead sulfide, PbS
4. Sulfur, S
5. Zinc, Zn
6. Marble chips, $CaCO_3$
7. Logwood chips
8. Charcoal, C
9. Mercury(II) iodide, HgI_2
10. Pyrite, FeS_2
11. Chromium(III) oxide, Cr_2O_3
12. Iron(II) sulfate, $FeSO_4$
13. Sodium sulfite, Na_2SO_3
14. **Rosin**
15. Sodium thiosulfate, $Na_2S_2O_3$
16. Iron, Fe
17. Aluminum, Al
18. Potassium hexacyanoferrate, $K_3Fe(CN)_6$
19. Potassium chromium sulfate, $KCr(SO_4)_2$
20. Menthol, $C_{10}H_{19}OH$
21. Potassium permanganate, $KMnO_4$
22. Ammonium nickel sulfate, $(NH_4)_2Ni(SO_4)_2$
23. Copper(II) sulfate 5-hydrate, $CuSO_4 \cdot 5H_2O$
24. Sodium chromate, Na_2CrO_4
25. Trilead tetraoxide, Pb_3O_4
26. Hydroquinone, $C_6H_4(OH)_2$
27. Copper, Cu

Plate 1. All substances are unique in that each has a particular set of properties—such as color, density, odor, composition, melting point, and boiling point—that differentiate it from all other substances. Although color alone would not necessarily identify a substance, we can see how important it is when we consider it along with other distinguishing properties. *Courtesy of Sargent Welch Scientific Company.*

Plate 2. (a) Two solutions, one containing lead nitrate and the other containing sodium iodide. (b) As one solution is added to the other, a yellow solid composed of lead iodide, PbI_2, is formed.
© *Yoav/Phototake*

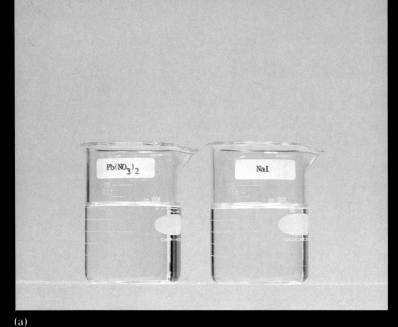

(a)

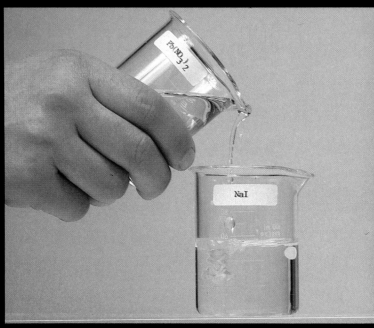

(b)

Plate 3. (*Facing page.*) The light emitted by selenium is separated into distinct colored lines by a prism. Each color (wave length) in the spectrum can be seen in three different positions because the light is very intense.
Color photograph by Fritz Goro. Courtesy of Scientific American, February, 1973.

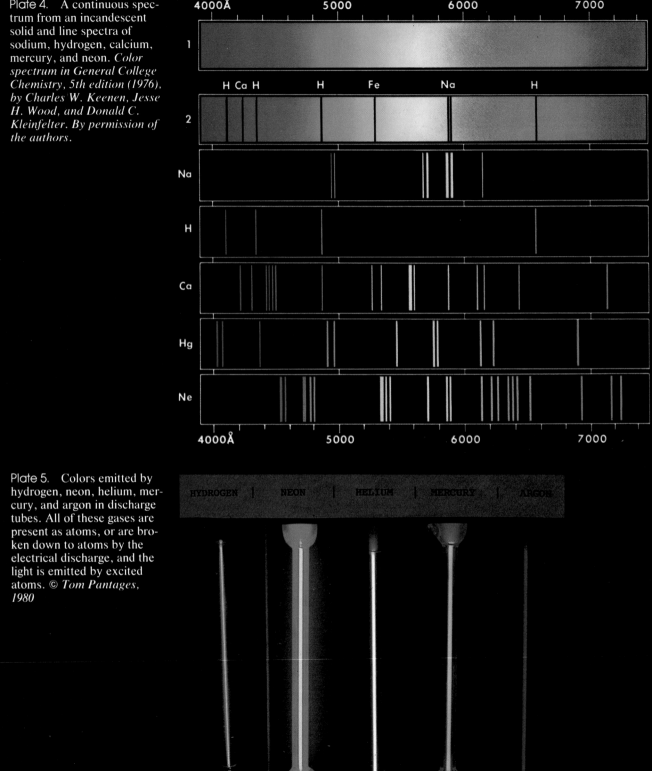

Plate 4. A continuous spectrum from an incandescent solid and line spectra of sodium, hydrogen, calcium, mercury, and neon. *Color spectrum in General College Chemistry, 5th edition (1976), by Charles W. Keenen, Jesse H. Wood, and Donald C. Kleinfelter. By permission of the authors.*

Plate 5. Colors emitted by hydrogen, neon, helium, mercury, and argon in discharge tubes. All of these gases are present as atoms, or are broken down to atoms by the electrical discharge, and the light is emitted by excited atoms. © *Tom Pantages, 1980*

Plate 6. Neon signs are actually special discharge tubes containing various gases that give the different colors. © *Mitchel L. Osborne/The Image Bank*

Plate 7. Excited sodium atoms emit yellow light. Sodium metal can thus be used in lamps to produce high intensity lighting. *Mitchell Funk/The Image Bank*

Plate 8.   A mixture of aluminum powder and ammonium perchlorate is the principal ingredient in the boosters used to launch the space shuttle. The large clouds of smoke contain aluminum oxide and aluminum chloride. The large central fuel tank contains liquid hydrogen and liquid oxygen, which are burned in the engines at the rear of the space craft, giving the pale blue flame and invisible water vapor. *Ron Lindsey/Phototake*

Plate 9. Silver crystals form on a copper wire placed in a solution of silver nitrate. Copper metal is more active than silver and reduces the silver ions of the silver nitrate. *© Tom Pantages, 1980*

Plate 10. This crucible is made of platinum (an inert element located near the bottom of the activity series); thus it can be used at high temperatures without damage. *Courtesy of Ferro Corporation.*

Plate 11. Molten bronze is poured from a crucible made of metal oxides, ionic compounds with high melting points. *Courtesy of Ferro Corporation.*

Plate 12. Colors of indicators at various pH values. Note that each indicator changes color over a pH of 1 to 2 units. (a) Methyl red, (b) Bromothymol blue, and (c) Phenolphthalein. (d) The Universal Indicator is actually a mixture of indicators. Thus it gives a different color at each pH. Such a mixture is contained in pH paper. © *Tom Pantages, 1983. Courtesy of Salem State College.*

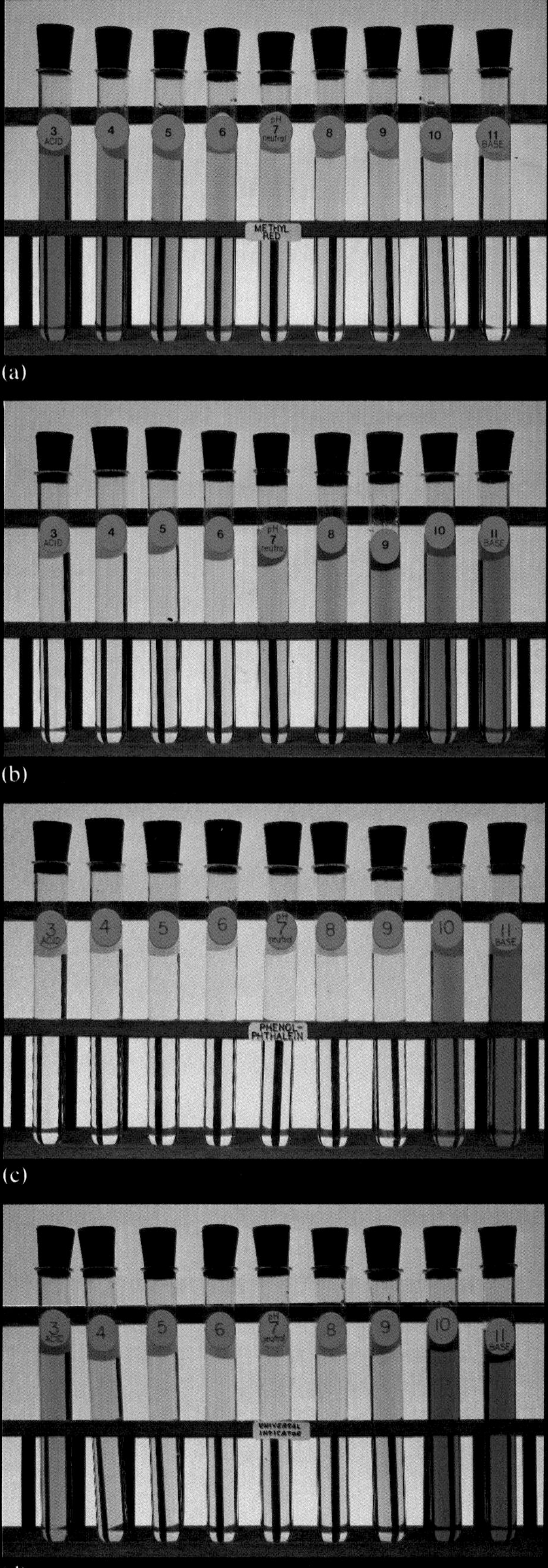

(a)

(b)

(c)

(d)

occur because only iron and steam are present. As this reaction proceeds, however, iron and steam are used up while Fe_3O_4 and hydrogen are formed. It follows that the rate of this reaction decreases while that of the reaction of Equation (1) increases, and eventually the two rates will become equal. At this point the reactions will have reached a state of **chemical equilibrium.** The quantity of each of the four substances in the closed tube remains unchanged at equilibrium because each is constantly being formed at the same rate as it is being consumed. It is customary to represent a state of equilibrium by means of a double arrow:

$$3Fe + 4H_2O \rightleftharpoons Fe_3O_4 + 4H_2$$

A more complete account of the very important phenomenon of chemical equilibrium will be given in Chapter 15.

9.16   Uses of Hydrogen

Hydrogen was once used as the lifting agent in lighter-than-air craft such as blimps and dirigibles, but it has been replaced by helium. Inert helium is much safer than highly combustible hydrogen (Fig. 9-11).

Two-thirds of the world's hydrogen production is used to manufacture ammonia, which is used primarily as a fertilizer and in the manufacture of nitric acid. Ammonia is manufactured by the Haber Process, the reaction of hydrogen with nitrogen (Section 8.7):

$$3H_2(g) + N_2(g) \underset{500°C}{\overset{\substack{\text{High} \\ \text{pressure}}}{\rightleftharpoons}} 2NH_3(g)$$

Figure 9-11. Explosion of hydrogen on the airship *Hindenburg* while landing at Lakehurst, New Jersey, May 6, 1937. Following the ignition of hydrogen in one compartment, ignition of the other two compartments took place within seconds, resulting in the complete destruction of the giant airliner. (*Official United States Navy Photograph*)

Hydrogen is used extensively in the process of **hydrogenation,** in which vegetable oils are changed from liquids to solids. Crisco is an example of a hydrogenated oil. The change results from the addition of H_2 to carbon-carbon double bonds, producing single bonds. Ethane is hydrogenated in a similar reaction:

$$\underset{H}{\overset{H}{>}}C=C\underset{H}{\overset{H}{<}} + H-H \longrightarrow H-\underset{\underset{H}{|}}{\overset{\overset{H}{|}}{C}}-\underset{\underset{H}{|}}{\overset{\overset{H}{|}}{C}}-H$$

Methyl alcohol is produced commercially by the catalyzed reaction of hydrogen with carbon monoxide.

$$2H_2 + CO \xrightarrow{\text{Catalyst}} CH_3OH$$

Water gas, a mixture of hydrogen with carbon monoxide, is an important industrial fuel. Hydrogen is also used in the reduction of certain oxides to obtain the free metal.

9.17   The Activity, or Electromotive, Series

Certain metals, such as sodium and potassium, react readily with cold water, displacing hydrogen and forming metal hydroxides (Section 9.12). Other metals, such as magnesium and iron, react with water only when heated. Sodium and potassium react much more vigorously with acids than do magnesium and iron. Experimental observations like these are used to arrange the metals in order by their chemical activities and thereby to establish an **activity,** or **electromotive, series.** A brief form of the activity series containing only the common elements is given in Table 9-1. Elements exhibiting more metallic behavior are at the top of the series (Groups IA and IIA). In general, the less the metallic character of an element, the lower it appears in the series.

Potassium is the most reactive of the common metals and therefore heads the activity series. Each succeeding metal in the series is less reactive, and gold, the least reactive of all, is found at the bottom of the series. In theory, any metal in the series will displace any other below it from a dilute aqueous solution of a soluble compound, if simple ions are involved. That is, any metal in its elemental form will reduce the ion, in water, of any metal below it in the series. For example, the copper ion in an aqueous solution of a copper(II) salt is reduced to copper metal by iron, which is above copper in the series (Fig. 9-12).

$$Fe(s) + Cu^{2+}(aq) \longrightarrow Cu(s) + Fe^{2+}(aq)$$

In a similar manner aqueous silver ions are reduced by metallic copper, and aqueous mercury ions are reduced by both copper and silver. Any metal above hydrogen in the series will liberate hydrogen from aqueous acids (solutions of acids in water); those from cadmium to the top will liberate hydrogen from hot water or stream; and those from sodium to the top will liberate hydrogen even from cold water. The metals below hydrogen do not displace hydrogen from water or from aqueous acids.

The reactivity of the metals toward oxygen decreases down the series, as do the heat of formation and the stability of the compounds formed. It is evident

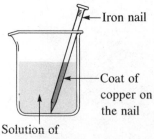

Iron nail

Coat of copper on the nail

Solution of copper sulfate

Figure 9-12. When an iron nail is immersed in a solution of copper sulfate, copper (a less active metal than iron) is deposited as the iron reduces the Cu^{2+} ion to metallic copper and is itself oxidized to Fe^{2+}.

Table 9-1   Activity, or Electromotive, Series of Common Metals

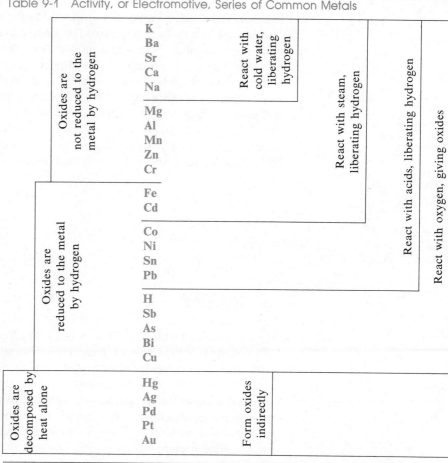

then that the activity series is very useful since it indicates the possibility of a reaction of a given metal with water, acids, salts of other metals, and oxygen. In addition, the series provides some indication of the stability of the compounds formed. It should be noted, however, that the order in which the elements are placed in the series depends somewhat on the conditions under which the activity is observed. A series determined from observations of the activities of the metals with respect to their ions in water will be slightly different from that reflecting the order of activity at high temperatures and in the absence of a solvent. The series shown in Table 9-1 is applicable to reactions in water.

The activity series of metals will be discussed in more detail in Chapter 20.

9.18   Hydrogen as a Secondary Energy Source

Hydrogen, like electricity, is a secondary source of energy, because it must be produced by using energy from a primary source such as coal, nuclear fusion, or the sun. In a few years mass-produced hydrogen might cost no more than gasoline and natural gas. Unlike electricity, hydrogen can be stored easily and then used when needed. It is pollution-free; the only product formed when it is burned is water. Hydrogen could conceivably be a substitute for natural gas, if safety problems can be solved. It could be transported in the nationwide network of underground pipelines that already exists for the transport of gases. This network now links over 60% of all homes and industries in the United States.

As an internal combustion fuel hydrogen is in many ways superior to gasoline and diesel oil. Engines using hydrogen start more easily, particularly in cold weather, and experience less wear. The greatest advantage is that almost no pollution results. A number of hydrogen-powered cars developed for research purposes are already on the roads. The greatest problem is carrying the fuel safely and practically. In recent years hydrogen has been used as a fuel in fuel cells (Section 20.18) that produce electricity much more efficiently than do generators powered by steam.

FOR REVIEW

SUMMARY

Oxygen, a nonmetal found in Group VIA of the Periodic Table, forms compounds with almost all of the elements. Except in a few compounds with the most electronegative element (fluorine), oxygen exhibits negative oxidation numbers in its compounds since it is the second most electronegative element. The most common oxidation number for oxygen is -2, found in **oxides.** The oxidation number of -1 is found in compounds that contain O—O bonds, **peroxides.** The oxidation number of $-\frac{1}{2}$ is found in compounds containing the O_2^- ion, **superoxides.** Oxygen is a strong oxidizing agent that reacts with many metals, nonmetals, and compounds. Many of the reactions between oxygen and a second element are exothermic; that is, heat is produced as the reaction proceeds. The **enthalpy of reaction,** or **heat of reaction,** is negative for an exothermic reaction.

The **rate of reaction** increases with increasing temperature, increasing concentration of reactants, increasing contact between reactants, and the presence of a catalyst.

Ozone, O_3, is an **allotrope** of oxygen, O_2. Ozone forms from oxygen in an endothermic reaction; that is, heat is absorbed from the surroundings as the reaction proceeds. It is a stronger oxidizing agent than oxygen.

Hydrogen has the chemical properties of a nonmetal with a relatively low electronegativity. Thus hydrogen forms ionic hydrides containing the H^- ion with active metals (such as those of Groups IA and IIA), covalent compounds in which it has an oxidation number of -1 with less electronegative nonmetals, and covalent compounds in which it has an oxidation number of $+1$ with more electronegative nonmetals. It will react explosively with oxygen, fluorine, and chlorine, less readily with bromine, and much less readily with iodine, sulfur, and nitrogen. Hydrogen will reduce the oxides of those metals lying below chromium in the **activity series** to form the metal and water. Some of these oxidation-reduction reactions are readily reversible and can proceed under conditions in which **chemical equilibrium** is established. A state of equilibrium is established when the net quantities of reactants and products involved in a reversible reaction occurring in a closed system no longer change; at equilibrium each reactant and product is produced by one reaction at the same rate as it is used up by the opposing reaction.

KEY TERMS AND CONCEPTS

Activity series (9.17)

Absorption (9.14)

Adsorption (9.14)

Allotropes (9.10)

Catalyst (9.2)

Chemical equilibrium (9.15)

Combustion (9.5)

Electrolysis (9.2)

Enthalpy of formation (9.6)

Enthalpy of reaction (9.6)

Equilibrium (9.15)

Heat of formation (9.6)

Heat of reaction (9.6)

Hydrocarbons (9.12)

Kindling temperature (9.5)

Net ionic equation (9.12)

Oxides (9.1)

Paramagnetism (9.3)

Peroxides (9.1)

Photosynthesis (9.8)

Rate of reaction (9.7)

Reversible reaction (9.15)

Spontaneous combustion (9.5)

Superoxide ion (9.1)

EXERCISES

1. Both hydrogen and oxygen are colorless, odorless, and tasteless gases. How may we distinguish between them using physical properties? using chemical properties?

2. Both H_2 and O_2 are diatomic molecules. How do the chemical bonds in the molecules differ, and how are they similar?

3. Why is it easier and cheaper to prepare commercial quantities of liquid oxygen than to prepare commercial quantities of liquid hydrogen?

Oxygen

4. Write the reaction describing the preparation of oxygen by using a lens to focus the sun's rays on a sample of mercury(II) oxide.

5. Which of the various laboratory techniques for the preparation of oxygen should be used if a sample of oxygen free of traces of water (dry oxygen) is required for an experiment?

6. Write balanced equations for the preparation of oxygen from:
   (a) $Cu(NO_3)_2$ (c) H_2O
   (b) $NaNO_3$ (d) BaO_2

7. Write balanced equations showing the release of oxygen gas upon heating the following oxides:
   (a) Au_2O (c) PtO_2 (e) CrO_3
   (b) Ag_2O (d) PbO_2

8. Describe the oxygen-oxygen bonding in the peroxide ion in terms of the Lewis structure of O_2^{2-}.

9. Using Molecular Orbital Theory, describe the bonding in O_2, O_2^-, and O_2^{2-}. Arrange these species in decreasing order of bond energy.

10. Write a balanced equation for the reaction of an excess of oxygen with each of the following (keep in mind that an element tends to reach its highest oxidation number when it combines with an excess of oxygen):
    (a) Ca   (d) Ge   (g) C_2H_6
    (b) Cs   (e) Na_2SO_3 (h) CO
    (c) As   (f) AlP

11. Define *combustion*. Give examples of combustion in an O_2 atmosphere and in a Cl_2 atmosphere.

12. Under what conditions may spontaneous combustion occur?

13. Which of the following materials will burn in O_2: $SiH_4(g)$, SiO_2, CO, CO_2, Mg, CaO? Why won't some of these materials burn in oxygen?

14. Why will magnesium ribbon burn more rapidly in pure oxygen than in air?

15. Why will wood shavings ignite and burn more rapidly than a log of the same kind of wood?

16. The heat of formation of ozone is 143 kJ and that of H(g) (atomic hydrogen) is 218 kJ. Are the reactions in which these compounds are formed endothermic or exothermic?

17. Is the heat of formation of MgO positive or negative?

18. What is the function of a catalyst in a chemical reaction? Under what conditions is it useful to use a catalyst in a chemical reaction?

19. Assume that when O_2 dissolves in water the volume of the solution is the same as the volume of water from which the solution was made. What is the molar concentration of O_2 in a solution at $0°C$ that is saturated with O_2, that is, one that contains the maximum possible amount of dissolved O_2?   *Ans. 2.2×10^{-3} M*

20. What mass and volume of $O_2(g)$ (density = 1.429 g/L) are required to prepare 41.3 g of MgO?   *Ans. 16.4 g; 11.5 L*

21. What mass of oxygen can be obtained by the thermal decomposition of 48.19 g of $Cu(NO_3)_2$? What is the volume of O_2 produced if the density of O_2 is 1.429 g/L?   *Ans. 4.111 g; 2.878 L*

22. Elements generally exhibit their highest oxidation numbers in compounds containing oxygen or fluorine. What is the oxidation number of osmium in 0.3789 g of an osmium oxide prepared by the reaction of 0.2827 g of Os with O_2? Write the equation for the reaction that gives this compound.   *Ans. +8; $Os + 2O_2 \rightarrow OsO_4$*

Hydrogen

23. Write the equation that describes the preparation of hydrogen when steam is passed through a red-hot gun barrel.

24. Write balanced equations for the preparation of hydrogen from $C + H_2O$; H_2SO_4; H_2O + electrical energy; NaH.

25. Which of the various laboratory techniques for the preparation of hydrogen should be used if a sample of hydrogen free of traces of water is required for an experiment?

26. Write balanced equations for and name the compounds formed in the reaction, if any, of each of the following metals with water and with hydrobromic acid (HBr in water): K, Mg, Ga, Bi, Fe, Pt.

27. Write a balanced equation for the reaction of an excess of hydrogen with each of the following: PtO, Na, Fe_2O_3, Cl_2, O_3.

28. Steam reacts with iron to give Fe_3O_4 and hydrogen. The Fe_3O_4 may be reduced to iron by reaction with CO giving CO_2. The overall change in this sequence is the conversion of carbon monoxide and steam to hydrogen and carbon dioxide. Write a single balanced equation for the overall change.

29. What is meant by a reversible reaction? What is meant by chemical equilibrium?

30. Explain: "Metal hydrides are convenient and portable sources of hydrogen."

31. The reaction of NaH with H_2O may be characterized as an acid-base reaction; identify the acid and the base of the reactants. The reaction is also an oxidation-reduction reaction. Identify the oxidizing agent, the reducing agent, and the changes in oxidation number that occur in the reaction.

32. Hydrogen is not very soluble. At $25°C$ a saturated solution of H_2 in water is 7.86×10^{-5} M. How many milliliters of H_2 at $25°$ (density = 0.0823 g/L) are dissolved in a liter of solution?   *Ans. 1.93 mL*

33. What mass of hydrogen would result from the reaction of 27.2 g of BaH_2 with water?   *Ans. 0.787 g*

34. What mass of LiH is required to provide enough hydrogen by reaction with water to fill a balloon at $0°C$ and 1 atm pressure with a volume of 3.50 L?   *Ans. 1.24 g*

35. How many grams of hydrogen and of zinc acetate, $Zn(CH_3CO_2)_2$, can be prepared by the reaction of 22 g of zinc with acetic acid, CH_3CO_2H?   *Ans. 0.68 g H_2; 62 g $Zn(CH_3CO_2)_2$*

36. Deuterium, ^2_1H, usually indicated as D, may be separated from ordinary hydrogen by repeated distillation of water. Ultimately, pure D_2O can be isolated. What is the molecular weight of D_2O to three significant figures. (The atomic weight of D may be found in Table 4-2.)   *Ans. 20.0*

The Activity Series

37. Under what chemical conditions may the activity series be relied on to predict chemical behavior?

38. One of the elements in the activity series (Table 9-1) will not produce hydrogen when it reacts with steam, but it will form hydrogen when it reacts with hydrochloric acid. Reaction of this element with a solution of $Ni(NO_3)_2$ produces $Ni(s)$. What is this element?

39. Zinc and mercury are both members of Group IIB, but ZnS and HgS behave differently when roasted in air. Write the equations for the reactions of ZnS and HgS with O_2 when roasted.

40. With the aid of the activity (electromotive) series (Table 9-1), predict whether or not the following reactions will take place:

(a) $Mg + Co^{2+} \longrightarrow Mg^{2+} + Co$

(b) $Al_2O_3 + 3H_2 \longrightarrow 2Al + 3H_2O$

(c) $2Au + Fe^{2+} \longrightarrow 2Au^+ + Fe$

(d) $Ni + 2H^+ \longrightarrow Ni^{2+} + H_2$

(e) $PtO_2 \xrightarrow{\Delta} Pt + O_2$

(f) $K_2O + H_2 \longrightarrow 2K + H_2O$

(g) $3Cd + 2Bi^{3+} \longrightarrow 2Bi + 3Cd^{2+}$

(h) $3Pd + 2Au^{3+} \longrightarrow 2Au + 3Pd^{2+}$

41. Lithium and beryllium do not appear in the activity series given in Table 9-1. From the position of these elements in the Periodic Table, locate their approximate position in the activity series and write equations for their reactions, if any, with water, steam, oxygen, and acids. Will the oxides of Li and Be be reduced by hydrogen? Are the oxides reduced by heat alone?

42. What mass of copper can be recovered from a solution of copper(II) sulfate by addition of 2.11 kg of aluminum?   *Ans. 7.45 kg*

Additional Exercises

43. Which of the following compounds will react with water to give an acid solution? Which will give a basic solution? Write a balanced equation for the reaction of each with water.

(a) NaH   (d) SO_3 (f) Na_2O_2

(b) H_2S (e) CaO   (g) P_4O_{10}

(c) HCl

44. Write the balanced equation or equations necessary to carry out the following transformations (H_2O, H_2, and/or O_2 may be used as needed):

(a) $KAl(OH)_4$ from Al and K

(b) $Zn_3(PO_4)_2$ from Zn and P

(c) NaCl from Na_2O_2 and Cl_2

(d) $ZnSO_4$ from Zn and H_2S

(e) Fe from Fe_3O_4

45. Increases in the price of petroleum result in increases in the price of fertilizers. Why?

46. The bond length in the O_2 molecule is 1.208 Å, while that in the O_3 molecule is 1.278 Å. Why does ozone have a longer bond?

47. What is the concentration of the barium hydroxide solution formed when 1.38 g of barium reacts with 483 mL of water? Assume that the volume of the solution is the same as that of the water used in its preparation.   *Ans. 2.08×10^{-2} M*

48. What is the concentration of hydrogen ion in a solution that contains 31.9 g of H_2SO_4 per liter of solution? Assume that the H_2SO_4 completely dissociates to hydrogen ion and sulfate ion.
          Ans. 0.650 M

49. The combustion of 9.180 g of sodium to sodium oxide is exothermic; 100.74 kJ of heat is evolved. What is the heat of formation of sodium oxide?
          Ans. −504.6 kJ/mol

50. The heat of formation of $CS_2(l)$ is +21.4 kcal/mol. How much heat is required for the reaction of 10.00 g of S_8 with carbon according to the following equation?

$$4C(s) + S_8(s) \longrightarrow 4CS_2(l)$$

Is the reaction exothermic or endothermic?
          Ans. 3.34 kcal; endothermic

51. What mass of PD_3 may be prepared by the reaction of 0.498 g of Na_3P with D_2O? (See Exercise 36.)   *Ans. 0.184 g*

52. The equation for the combustion of methane in oxygen is

$$CH_4(g) + 2O_2(g) \longrightarrow CO_2(g) + 2H_2O(l)$$
$$\Delta H = -890.4 \text{ kJ}$$

ΔH for this reaction is sometimes called the heat of combustion of methane. The combustion of hydrogen proceeds by the following equation:

$$2H_2(g) + O_2(g) \longrightarrow 2H_2O(l)$$
$$\Delta H = -571.5 \text{ kJ}$$

Which will produce more heat, the combustion of one cubic foot of methane (density = 0.7143 g/L) or the combustion of one cubic foot of hydrogen (density = 0.08987 g/L)? How much more?

          Ans. methane; 761.2 kJ more

10

THE GASEOUS STATE AND THE KINETIC-MOLECULAR THEORY

Those elements and compounds that are gases at room temperature have played an important part in the history of the development of chemistry. The identification of oxygen as a component of air was crucial to the development of the atomic theory. Water was shown to be a compound rather than an element when it was prepared by the reaction of the gases oxygen and hydrogen. Atomic weights were determined in part from a study of the densities of gases. Conclusions regarding chemical stoichiometry and the molecular nature of matter followed from observations of the volumes of gases that combined in chemical reactions. The variations in the pressure and volume of a gas with temperature led to the discovery of the concept of absolute zero and the development of the kinetic-molecular theory of gaseous behavior. Studies of gases also led to the first quantitative models for the description of the behavior of matter.

In this chapter we shall consider the equations that relate the temperature, pressure, volume, and mass of a gas present in a container. These equations are the tools used to describe the quantitative behavior of gases. We will see how to use them to convert physical measurements to the moles of gas present, to the molecular weight of a gas, or to the quantities of gases involved in chemical changes. We will examine the development of the kinetic-molecular theory of gases and compare the experimental behavior of gases with the predictions of the theoretical model. This will help us to determine if the assumptions used in the theory to describe the nature of molecules of gases lead to a useful model for the behavior of gases.

THE PHYSICAL BEHAVIOR OF GASES

10.1 Behavior of Matter in the Gaseous State

We are surrounded by an ocean of gas, the atmosphere, so many of the properties of gases are familiar to us from our daily activities. We know that squeezing a balloon decreases the volume of the gas inside. If we fill a tire pump with air and then cap the tube, the volume of the gas in the pump can be decreased by pushing the pump handle down. In fact, the volume of any gas, unlike that of a solid or a liquid, can be decreased greatly by increasing the pressure on the gas. This property is known as **compressibility.** Increasing the weight on a piston, as is shown in Fig. 10-1, causes the gas confined in the cylinder to decrease in volume until the gas exerts enough pressure to support the greater weight.

Figure 10-1. These figures illustrate what happens when a fixed quantity of gas at constant temperature is confined in a cylinder and subjected to more and more pressure by placing heavier and heavier weights on a movable, gas-tight piston. The number of molecules represented is but a minute fraction of the actual number in such a volume. The molecules are shown lined up in the diagram for clarity, but are actually randomly distributed in the gaseous state.

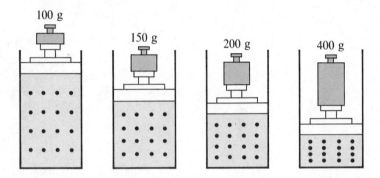

If we heat a gas in a closed container, such as a soda bottle, its pressure increases. Heating a gas in a cylinder such as that shown in Fig. 10-1(a) will increase the pressure inside the cylinder unless the piston moves. Raising the piston increases the volume and keeps the pressure constant, thus allowing the gas to expand. **Expansibility** is characteristic of all gases.

A small quantity of a pungent gas released into the corner of a room in which the air is still can, in time, be smelled all over the room. When a sample of gas is introduced into an evacuated container, it almost instantly fills the container uniformly. These are examples of the **diffusion** of gases. Diffusion results because the molecules of a gas are in constant high-speed motion. Two gases introduced into the same container mix by diffusion. Each gas is said to permeate the other. Thus we say that gases possess the properties of **diffusibility** and **permeability.**

10.2 Measurement of Gas Pressures

The pressure exerted by the atmosphere (a mixture of gases) may be measured by a simple mercury **barometer** (Fig. 10-2). A barometer can be made by filling an 80-centimeter long glass tube, closed at one end, with mercury and inverting it in a container of mercury. The mercury in the tube falls until the pressure exerted by the atmosphere on the surface of the mercury in the container is just sufficient to support the mercury in the tube. Because the pressure of the atmosphere is proportional to the height of the mercury column (the vertical distance between the surface of the mercury in the tube and that in the open vessel), pressure is some-

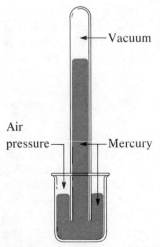

Vacuum

Air pressure — Mercury

Figure 10-2. A mercury barometer.

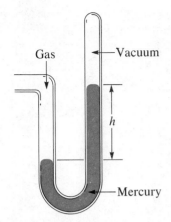

Figure 10-3. A simple two-armed mercury barometer. Such barometers are sometimes called manometers.

times expressed in terms of millimeters of mercury (**mmHg**). A pressure of 1 millimeter of mercury is generally referred to as a **torr** after Evangelista Torricelli, the inventor of the barometer. A mercury column 1 millimeter high exerts a pressure of 1 torr.

A second type of barometer, a **manometer,** has two arms, one closed and one open to the atmosphere or connected to a container filled with gas (Fig. 10-3). The pressure exerted by the atmosphere or the gas in the container is proportional to the height of the column of mercury, that is, the distance between the mercury levels in the two arms of the tube (h in the diagram).

Some portable **aneroid barometers** (Fig. 10-4) are made with scales graduated in feet for measuring elevations. Mountain climbers and pilots use such barometers to determine their altitude above sea level.

The average pressure of the atmosphere at sea level at a latitude of 45° will support a column of mercury 760 millimeters in height. The average sea-level pressure at the latitude of 45° is thus 760 millimeters of mercury, which is defined as **1 atmosphere (atm).** The pressure of the atmosphere varies with the distance above sea level and with climatic changes. At a higher elevation the air is less dense, and thus the pressure is less. The atmospheric pressure at 20,000 feet is only half of that at sea level because about half of the atmosphere is below this elevation.

Figure 10-4. Diagram of an aneroid barometer. Changes in air pressure cause changes in the thickness of the vacuum box to which the coupling rod is attached. The pointer reacts to changes in position of the coupling rod.

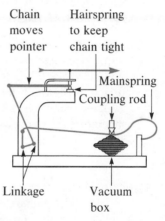

The pressure of a gas is actually the force that it exerts on a unit area of surface. The pressure unit recommended for use with the International System of Units (SI) is the **pascal (Pa),** based on the newton unit for force. A **newton (N)** is the force that, when applied for 1 second, will give a 1-kilogram mass a speed of 1 meter per second. Hence a newton is expressed in units of kg m/s² or kg m s⁻². Since pressure is force per unit area, the pressure in SI units is expressed in terms of newtons per square meter, or pascals.

$$Pa = N/m^2 = kg \ m \ s^{-2}/m^2 = kg \ s^{-2}/m = kg/s^2 \ m$$

In many cases units of **kilopascals, kPa,** are more convenient to use than units of pascals (1 kPa = 1000 Pa).

The conversions between the various pressure units in this section are as follows:

$$1 \ atm = 760 \ mmHg = 760 \ torr = 101.325 \ kPa$$

EXAMPLE 10.1   Weather reports in the United States often give barometric pressures in inches of mercury rather than millimeters of mercury or torr (29.92 in Hg = 760 torr). Convert a pressure of 29.20 in Hg to torr, atmospheres, and kilopascals.

First convert the pressure in inches of mercury to torr using the conversion factor given in the example.

$$29.20 \text{ in Hg} \times \frac{760 \text{ torr}}{29.92 \text{ in Hg}} = 742 \text{ torr}$$

Now convert torr into atmospheres and kilopascals using the conversion factors given in this section.

$$742 \text{ torr} \times \frac{1 \text{ atm}}{760 \text{ torr}} = 0.976 \text{ atm}$$

$$742 \text{ torr} \times \frac{101.325 \text{ kPa}}{760 \text{ torr}} = 98.9 \text{ kPa}$$

10.3 Relation of Volume to Pressure at Constant Temperature; Boyle's Law

Experiment shows that if the temperature of a sample of gas is not changed, its volume is reduced to half when the pressure on the gas is doubled. Conversely, its volume doubles when the pressure is halved. Observations of this sort were summarized by the English physicist and chemist Robert Boyle in 1660 in a statement now known as **Boyle's Law: The volume of a given mass of gas held at constant temperature is inversely proportional to the pressure under which it is measured,** or

$$V \propto \frac{1}{P}$$

where V is the volume of the gas, P, the pressure, and $\propto$ means "is proportional to." The proportionality may be changed to an equality by including a constant, k, referred to as a proportionality constant.

$$V = \text{constant} \times \frac{1}{P}, \quad \text{or } V = k \times \frac{1}{P}$$

Hence

$$PV = \text{constant}, \quad \text{or } PV = k \tag{1}$$

The value of the constant changes if the mass of gas or the temperature changes, but it does not vary if only the pressure or volume is varied for a particular mass of gas at constant temperature. The meaning of Equation (1) is that **the product of the pressure of a given mass of gas times its volume is always constant if the temperature does not change.** Thus, as the pressure of gas increases at constant temperature, the volume must decrease, or as the pressure decreases at constant temperature, the volume must increase, in order for the product to remain constant. A graph showing this relationship is given in Fig. 10-5. The following examples illustrate the application of Boyle's Law.

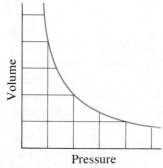

Volume

Pressure

Figure 10-5. A graphical illustration of Boyle's Law. Note how the volume decreases as the pressure increases and, conversely, increases as the pressure decreases in an inversely proportional relationship.

EXAMPLE 10.2   A balloon contains 14.0 L of air at a pressure of 760 torr. What will the volume of the air be when the balloon is taken to a depth of 10 ft in a swimming pool, where the pressure is 981 torr? The temperature of the air does not change.

This problem gives us the volume (14.0 L) of a gas at one pressure (760 torr) and asks for the volume of the gas at a second pressure (981 torr). From Boyle's Law we know that $PV = $ constant. We can find the constant from the volume (14.0 L) at a pressure of 760 torr.

$$PV = k$$

$$760 \text{ torr} \times 14.0 \text{ L} = k = 10,640 \text{ torr L}$$

Using this value of the constant, we can find the volume at 981 torr by rearranging the equation $PV = k$ to solve for V.

$$V = \frac{k}{P}$$

$$V = \frac{10,640 \text{ torr L}}{981 \text{ torr}} = 10.8 \text{ L}$$

There is a less formal way of solving this problem. From Boyle's Law, Equation (1), we know that an increase in pressure from 760 torr to 981 torr at constant temperature will result in a decrease in volume. The new volume can be determined by multiplying the original volume by the ratio of the two pressures. The ratio must have the smaller pressure in the numerator and the larger pressure in the denominator, so that when the original volume is multiplied by this ratio the calculation will show a change to a smaller volume in accord with our prediction.

$$14.0 \text{ L} \times \frac{760 \text{ torr}}{981 \text{ torr}} = 10.8 \text{ L}$$

Thus the volume is less than the original volume. Note that the unit torr appears in both numerator and denominator of the expression; thus it cancels out, leaving L as the unit of measurement in the answer to the problem.

EXAMPLE 10.3   A sample of Freon gas used in an air conditioner has a volume of 325 L and a pressure of 96.3 kPa at 20°C. What will the pressure of the gas be when its volume is 975 L at 20°C?

First find the constant in Boyle's Law, $PV = k$, at the initial pressure and volume.

$$96.3 \text{ kPa} \times 325 \text{ L} = 3.130 \times 10^4 \text{ kPa L}$$

Then find the second pressure from the value of k and the second volume.

$$PV = k$$

$$P = \frac{k}{V}$$

$$P = \frac{3.130 \times 10^4 \text{ kPa L}}{975 \text{ L}} = 32.1 \text{ kPa}$$

There is a less formal way of solving this problem. At constant temperature the pressure of a gas decreases as the volume of the gas increases. The new pressure is determined by the ratio of the two volumes. If the smaller volume is the numerator and the larger volume, the denominator, multiplying the original pressure by this ratio will give the new (lower) pressure.

$$96.3 \ \text{kPa} \times \frac{325 \ \cancel{L}}{975 \ \cancel{L}} = 32.1 \ \text{kPa}$$

The new pressure must be one-third of the initial pressure to permit the volume to increase threefold.

10.4 Relation of Volume to Temperature at Constant Pressure; Charles's Law

If a partially filled balloon is placed in sunlight so that the gas inside becomes warmer, the balloon will expand. This is an example of the effect of temperature on the volume of a confined gas. Studies of the effect of temperature on the volume of confined gases at constant pressure by the French physicist S. A. C. Charles in 1787 led to the generalization known as **Charles's Law: The volume of a given mass of gas is directly proportional to its temperature on the Kelvin scale when the pressure is held constant,** or

$$V \propto T$$

If we use a proportionality constant that depends on the mass of gas and its pressure, we get the equation

$$V = \text{constant} \times T, \quad \text{or} \quad V = k \times T$$

Hence

$$\frac{V}{T} = \text{constant}, \quad \text{or} \quad \frac{V}{T} = k \tag{1}$$

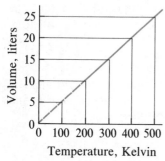

Figure 10-6. A graphical illustration of Charles's Law. Note how the volume increases with increasing Kelvin temperature at constant pressure in a directly proportional relationship.

This constant is different from that in Boyle's Law. It does not vary unless the mass of the gas or the pressure changes. Equation (1) means that, as the Kelvin temperature of a given mass of gas changes with no change in pressure, its volume also changes so that the ratio V/T remains the same. A decrease in T results in a decrease in V, and an increase in T results in an increase in V. The relationship of volume and temperature is shown in Fig. 10-6. (See also Fig. 10-8.)

Charles's Law applies to the volume of a gas and *its Kelvin temperature*. In Section 1.14 we saw that the relationship between the Kelvin and Celsius temperature scales is $K = °C + 273.15$. (Recall that temperatures on the Kelvin scale are by convention reported without a degree sign.)

The following examples illustrate the application of Charles's Law.

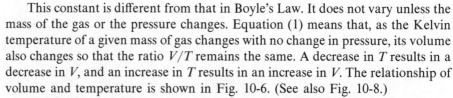

EXAMPLE 10.4   A sample of carbon dioxide, CO_2, occupies 300 mL at 10°C and 750 torr. What volume will the gas have at 30°C and 750 torr? (Assume all figures are significant.)

First we convert the Celsius temperatures to the Kelvin scale.

$$10°C + 273.15 \text{ K} = 283 \text{ K}$$
$$30°C + 273.15 = 303 \text{ K}$$

Using the first Kelvin temperature (283 K) and the corresponding volume (300 mL), we find the constant in the equation for Charles's Law.

$$\frac{V}{T} = k$$

$$\frac{300 \text{ mL}}{283 \text{ K}} = k = 1.06 \text{ mL/K}$$

Now we solve for V at the second temperature.

$$V = T \times k$$
$$V = 303 \text{ K} \times 1.06 \text{ mL/K} = 321 \text{ mL}$$

This problem can also be solved in a less formal way. From Charles's Law we know that when the temperature of a gas is raised at constant pressure, the gas expands. The final volume may be found by multiplying the initial volume by the ratio of Kelvin temperatures that will give the larger volume (high temperature on top in the ratio).

$$300 \text{ mL} \times \frac{303 \text{ K}}{283 \text{ K}} = 321 \text{ mL}$$

EXAMPLE 10.5   A particular hydrogen gas thermometer has a volume of 150.0 cm³ when immersed in a mixture of ice and water (0.00°C). When immersed in boiling liquid ammonia, the volume of the hydrogen, at the same pressure, is 131.7 cm³. Find the temperature of the boiling ammonia on the Kelvin and Celsius scales.

This problem asks us to find the temperature of a 131.7-cm³ sample of hydrogen gas that has a volume of 150.0 cm³ at a temperature of 0.00°C and the same pressure. Since the initial temperature of 0.00°C is 273.15 K, the constant for this sample of hydrogen in the Charles's Law equation is

$$\frac{V}{T} = \frac{150.0 \text{ cm}^3}{273.15 \text{ K}} = 0.54915 \text{ cm}^3/\text{K}$$

The temperature at which the volume is 131.7 cm³ can then be determined.

$$\frac{V}{T} = k$$

$$T = \frac{V}{k} = \frac{131.7 \text{ cm}^3}{0.54915 \text{ cm}^3/\text{K}} = 239.8 \text{ K}$$

Subtracting 273.15 from 239.8 K, we find that the temperature of the boiling ammonia on the Celsius scale is $-33.3°C$.

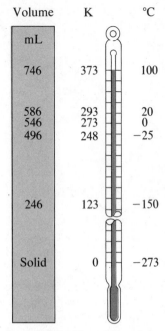

Volume	K	°C
mL		
746	373	100
586	293	20
546	273	0
496	248	−25
246	123	−150
Solid	0	−273

Figure 10-7. The volume of a gas at constant pressure is directly proportional to its Kelvin temperature.

10.5 The Kelvin Temperature Scale

In Section 1.14 we discussed the Kelvin temperature scale and its relationship to the Celsius temperature scale. Now, with some knowledge of the effect of temperature on gases, we are ready to examine the basis for the Kelvin scale.

As we have already seen, experiments show that when a gas is heated, it must expand for its pressure to remain constant. Likewise, cooling any gas at constant pressure results in a decrease in its volume. When the pressure remains constant and 546 milliliters of gas at 0°C is warmed to 1°C, its volume increases by 2 milliliters to 548 milliliters; at 20°C its volume is 586 milliliters; at 273°C its volume is 1092 milliliters, twice its volume at 0°C; and so on (Fig. 10-7). This means that the volume of the gas increases by $\frac{1}{273}$ of its volume at 0° for each rise of 1° on the Celsius scale. The volume of a gas decreases in the same proportion when the temperature falls. If the temperature of 273 milliliters of a gas could be lowered from 0°C to −273°C, then the gas should have no volume at −273°C, because its volume should decrease at the rate of $\frac{1}{273}$ of its volume at 0°C for each fall of 1°C in temperature. Before the temperature of −273°C is reached, however, all gases become liquids or solids.

A plot of data obtained from several such volume and temperature experiments is shown in Fig. 10-8, where the volume of a sample of each of several gases at constant pressure is plotted against the temperature in degrees Celsius. The straight lines reflect the direct proportionality between the volume and temperature of each sample, as described by Charles's Law. The data stop at those temperatures at which the gases condense. If the straight lines are extended back (extrapolated) to the temperatures at which the gases would have a zero volume if they did not condense, the lines all intersect the temperature axis at −273.15°C. Regardless of the gas, these measurements yield −273.15°C. Below this temperature gases would theoretically have a negative volume. Since negative volumes are impossible, −273.15°C must be the lowest temperature possible. This temperature, called **absolute zero,** is taken as the zero point of the Kelvin temperature

Figure 10-8. Charles's Law behavior for several gases, each at constant pressure. All gases extrapolate to a volume of 0 at −273.15°C. The dashed lines represent the extrapolated portion of each line for temperatures below the boiling point.

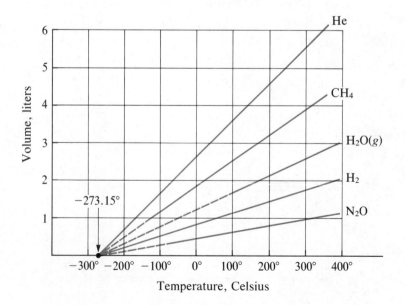

scale. The freezing point of water (0.00°C) is therefore 273.15 kelvin, and the boiling point (100.00°C) is 373.15 kelvin (Fig. 10-7).

To lower the temperature of a substance, we remove heat. Absolute zero (0 K) is the temperature reached when all possible heat has been removed from a substance. Obviously, a substance cannot be cooled any further after all possible heat has been removed from it.

10.6 Reactions Involving Gases; Gay-Lussac's Law

Gases combine, or react, in definite and simple proportions by volume. For example, it has been determined by experiment that one volume of nitrogen will combine with three volumes of hydrogen to give two volumes of ammonia gas, provided that the volumes of the reactants and product are measured under the same conditions of temperature and pressure.

$$N_2(g) + 3H_2(g) \longrightarrow 2NH_3(g)$$

1 volume   3 volumes   2 volumes

The units, of course, must be the same: if the volume of nitrogen is measured in liters, the volumes of hydrogen and ammonia must also be measured in liters. Experimental observations on the volumes of combining gases were summarized by Joseph Louis Gay-Lussac (1788–1850) in **Gay-Lussac's Law: The volumes of gases involved in a reaction, at constant temperature and pressure, can be expressed as a ratio of small whole numbers.** It has been shown that the small whole numbers in such a ratio are equal to the coefficients in the balanced equation that describes the reaction.

EXAMPLE 10.6 What volume of $O_2(g)$ measured at 25°C and 760 torr is required to react with 1.0 L of methane [$CH_4(g)$, the principal component of natural gas] measured under the same conditions of temperature and pressure?

The ratio of the volumes of CH_4 and O_2 will be equal to the ratio of their coefficients in the balanced equation for the reaction.

$$CH_4(g) + 2O_2(g) \longrightarrow CO_2(g) + 2H_2O(g)$$

1 volume   2 volumes   1 volume   2 volumes

From the equation we see that one volume of CH_4, in this case 1 L, will react with two volumes of O_2.

$$1.0 \text{ L } CH_4 \times \frac{2 \text{ L } O_2}{1 \text{ L } CH_4} = 2.0 \text{ L } O_2$$

A volume of 2.0 L of O_2 will be required to react with 1.0 L of CH_4.

It is important to remember that this law applies only to substances in the gaseous state measured at the same temperature and pressure. The volumes of any solids or liquids involved in the reactions are not considered.

EXAMPLE 10.7   What volume of $CO_2(g)$, measured at the same temperature and pressure, is formed by the reaction of 150 mL of $O_2(g)$ with carbon, a solid.

$$C(s) \; + \; O_2(g) \; \longrightarrow \; CO_2(g)$$

A solid   1 volume   1 volume

We can say nothing about the volume of carbon involved in this reaction since it is a solid. However, the coefficients in the equation indicate that one volume of O_2 (in units of mL in this case) reacts to give one volume of CO_2.

$$150 \; \text{mL } O_2 \times \frac{1 \; \text{mL } CO_2}{1 \; \text{mL } O_2} = 150 \; \text{mL } CO_2$$

Thus 150 mL of O_2 produces 150 mL of CO_2 in this reaction.

EXAMPLE 10.8   During 1980 a volume of 876 billion cubic feet of gaseous ammonia, $NH_3(g)$, measured at 25°C and 1 atm, was manufactured in the United States. What volume of $H_2(g)$, measured under the same conditions, was required to prepare this amount of ammonia by reaction with N_2?

The ratio of the volumes of H_2 and N_2 will be equal to the ratio of the coefficients in the equation

$$N_2(g) \; + \; 3H_2(g) \; \longrightarrow \; 2NH_3(g)$$

1 volume   3 volumes   2 volumes

We see that two volumes of NH_3, in this case in units of billion ft³, will be formed from three volumes of H_2. Thus

$$876 \; \text{billion ft}^3 \; NH_3 \times \frac{3 \; \text{billion ft}^3 \; H_2}{2 \; \text{billion ft}^3 \; NH_3} = 1.31 \times 10^3 \; \text{billion ft}^3 \; H_2$$

The manufacture of 876 billion ft³ of NH_3 required 1310 billion ft³ of H_2. (This is the volume of a cube with an edge length of approximately 2 miles.)

10.7   An Explanation of Gay-Lussac's Law; Avogadro's Law

Gay-Lussac's Law concerning combining volumes of gases is explained by the molecular nature of gases if **equal volumes of all gases, measured under the same conditions of temperature and pressure, contain the same number of molecules.** The Italian physicist Amadeo Avogadro advanced this hypothesis in 1811 to account for the behavior of gases. His hypothesis, which has since been experimentally proven, is now accepted as fact and is known as **Avogadro's Law.**

The explanation of Gay-Lussac's Law by Avogadro's Law may be illustrated by the reaction of the gases hydrogen and oxygen at 200°C to give gaseous water (water vapor), with the reactants and product at the same temperature and pressure. Gay-Lussac's Law indicates that two volumes of H_2 and one volume of O_2 will combine to give two volumes of H_2O.

$$2H_2(g) \; + \; O_2(g) \; \longrightarrow \; 2H_2O(g)$$

2 volumes   1 volume   2 volumes

According to Avogadro's Law equal volumes of gaseous H_2, O_2, and H_2O contain the same number of molecules. Since two molecules of water are formed from two molecules of hydrogen and one molecule of oxygen, the volume of hydrogen required and the volume of gaseous water produced are both twice as great as the volume of oxygen required. This relationship is illustrated in Fig. 10-9.

Figure 10-9. Two volumes of hydrogen combine with one volume of oxygen to yield two volumes of water vapor.

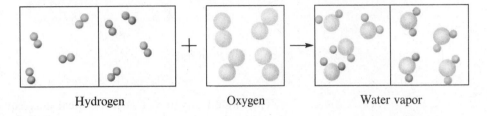

Hydrogen   Oxygen   Water vapor

10.8   The Ideal Gas Equation

Boyle's Law, Charles's Law, and Avogadro's Law are special cases that can be covered by one equation that relates pressure, volume, temperature, and number of moles of an **ideal gas** (a gas that follows these gas laws perfectly). This equation is referred to as the **Ideal Gas Equation.** Although it is derived (Section 10.18) for an ideal gas, under normal conditions of temperature and pressure it also applies very well to real gases because real gases behave very nearly like ideal gases under normal conditions. In this text we will approximate the behavior of all gases as ideal. Although we will not attempt to describe deviations from this behavior quantitatively, we shall consider them qualitatively (Section 10.19).

The Ideal Gas Equation is

$$PV = nRT$$

where P is the pressure of a gas; V, its volume; n, the number of moles of the gas; T, its temperature on the Kelvin scale; and R, a constant called the **gas constant.** The following paragraphs show how the Ideal Gas Equation reduces to either Boyle's Law, Charles's Law, or Avogadro's Law under the appropriate conditions.

Boyle's Law (Section 10.3) states that for a given amount of an ideal gas the product of its volume, V, and its pressure, P, is a constant at constant temperature. Since n, R, and T do not change under the conditions for which Boyle's Law holds, their product is a constant. Therefore the Ideal Gas Equation, $PV = nRT$, becomes

$$PV = \text{constant}$$

which is the mathematical expression for Boyle's Law.

Charles's Law (Section 10.4) states that the volume, V, of n moles of an ideal gas divided by its temperature, T, on the Kelvin scale is equal to a constant at constant pressure. If we rearrange the Ideal Gas Equation so that the terms that do not vary (n, R, and P) are on the right side and V and T are on the left, we obtain the equation expressing Charles's Law.

$$\frac{V}{T} = \frac{nR}{P} = \text{constant}$$

If equal volumes of all gases, under the same conditions of temperature and pressure, contain the same number of molecules (Avogadro's Law, Section 10.7), then they contain the same number of moles, n, of gas. This may be written in the form of an equation as follows:

$$V = \text{constant} \times n$$

The Ideal Gas Equation may be rearranged to yield this expression of Avogadro's Law by placing all the constant quantities (R, T, and P) together:

$$V = \frac{RT}{P} \times n$$

At constant P and T,

$$V = \text{constant} \times n$$

The numerical value of R in the Ideal Gas Equation can be determined by substituting actual experimental values for P, V, n, and T in the equation. It has been shown experimentally that exactly 1 mole of any gas that very closely approximates ideal behavior occupies a volume of 22.414 liters at 273.15 K and a pressure of exactly 1 atmosphere. Substituting these values in $PV = nRT$ gives

$$(1 \text{ atm})(22.414 \text{ L}) = (1 \text{ mol})(R)(273.15 \text{ K})$$

$$R = \frac{22.414 \text{ L} \times 1 \text{ atm}}{1 \text{ mol} \times 273.15 \text{ K}}$$

$$R = 0.08206 \text{ L atm/mol K}$$

The numerical value for R using other units for P, V, n, and T will be different. For example, if we take the pressure in kilopascals instead of atmospheres (1 atm = 101.325 kPa), we calculate another value of R as follows:

$$R = \frac{22.414 \text{ L} \times 101.325 \text{ kPa}}{1 \text{ mol} \times 273.15 \text{ K}}$$

$$R = 8.314 \text{ L kPa/mol K}$$

The units of pressure, volume, temperature, and amount of gas used in the Ideal Gas Equation must always be the same as the units for R.

The following problems illustrate some of the applications of the Ideal Gas Equation.

EXAMPLE 10.9   Calculate the volume occupied by 0.54 mol of N_2 at 15°C and 0.967 atm. Use the value of R with the units of L atm/mol K.

The Ideal Gas Equation contains five terms (P, V, n, R, and T). If we know any four of these, we can rearrange the equation and find the fifth. In this problem we are given 0.54 mol, 15°C, and 0.967 atm and told to use 0.08206 L atm/mol K to find V. Thus we must rearrange $PV = nRT$ to solve for V.

$$V = \frac{nRT}{P}$$

Since the equation requires that the temperature be on the Kelvin scale, we convert the temperature from °C to K.

$$T = 15°C + 273.15 = 288 \text{ K}$$

Substitution gives

$$V = \frac{nRT}{P} = \frac{(0.54 \text{ mol})(0.08206 \text{ L atm/mol K})(288 \text{ K})}{0.967 \text{ atm}}$$

$$V = 13 \text{ L (to two significant figures, as justified by the data)}$$

Note that the units cancel to liters, an appropriate unit for volume. Cancellation of units is a very simple way of checking to see that the units for R are consistent with those for P, V, n, and T.

EXAMPLE 10.10   Methane, CH_4, can be used as fuel for an automobile; however, it is a gas at normal temperatures and pressures, which causes some problems with storage. One gallon of gasoline could be replaced by 655 g of CH_4. What is the volume of this much methane at 25°C and 745 torr? Use the value of R with the units of L atm/mol K.

As shown in Example 10.9 the volume of an ideal gas may be obtained from the equation

$$V = \frac{nRT}{P}$$

The units of R (liters, atmospheres, moles, and Kelvin degrees) require that the pressure be expressed in atmospheres, the amount of gas in moles, and the temperature in Kelvin degrees. Once these conversions are completed, the values can be substituted into the equation and the volume calculated.

$$P = 745 \text{ torr} \times \frac{1 \text{ atm}}{760 \text{ torr}} = 0.9803 \text{ atm}$$

$$n = 655 \text{ g } CH_4 \times \frac{1 \text{ mol}}{16.04 \text{ g } CH_4} = 40.84 \text{ mol}$$

$$T = 25°C + 273.15 = 298.2 \text{ K}$$

Now we substitute into the rearranged equation.

$$V = \frac{nRT}{P}$$

$$V = \frac{(40.84 \text{ mol})(0.08206 \text{ L atm/mol K})(298.2 \text{ K})}{0.9803 \text{ atm}}$$

$$V = 1020 \text{ L}$$

Thus it would require 1020 L (269 gal) of gaseous methane at about 1 atm of pressure to replace 1 gal of gasoline. Obviously, the methane would have to be carried in a tank under pressure to reduce the volume.

EXAMPLE 10.11   While resting, the average human male consumes 200 mL of O_2 per hour at 25°C and 1.0 atm for each kilogram of weight. How many moles of O_2 are consumed by a 70-kg man while resting for 1 h?

For the purpose of this problem, let us assume the oxygen consumption per kilogram per hour is good to two significant figures. The volume of O_2 consumed by a resting 70-kg male is then 14,000 mL/h, or 14 L/h.

To solve for moles, rearrange the Ideal Gas Equation, $n = PV/RT$. Since the pressure and volume are given in atmospheres and liters, use $R = 0.08206$ L atm/mol K. The temperature is $273.15 + 25 = 298$ K. These values can now be used to find n.

$$n = \frac{PV}{RT} = \frac{1.0 \text{ atm} \times 14 \text{ L}}{0.08206 \text{ L atm/mol K} \times 298 \text{ K}} = 0.57 \text{ mol } O_2$$

EXAMPLE 10.12   What is the pressure in kilopascals in a 35.0-L balloon at 25°C filled with dried hydrogen gas produced by the reaction of 38.9 g of NaH with water?

To use the Ideal Gas Equation, $PV = nRT$, for determining P, we must know V, n, R, and T ($P = nRT/V$). Values for V and T are given; R must be 8.314 L kPa/mol K in order for the pressure units to be kilopascals; and n can be calculated from the weight of NaH and the equation for the reaction.

$$NaH + H_2O \longrightarrow NaOH + H_2$$

$$38.9 \text{ g NaH} \times \frac{1 \text{ mol NaH}}{24.00 \text{ g NaH}} = 1.62 \text{ mol NaH}$$

$$1.62 \text{ mol NaH} \times \frac{1 \text{ mol } H_2}{1 \text{ mol NaH}} = 1.62 \text{ mol } H_2$$

so $n = 1.62$ mol

Now we convert T to the Kelvin scale.

$$T = 25°C + 273.15 = 298 \text{ K}$$

Using the Ideal Gas Equation gives

$$P = \frac{nRT}{V} = \frac{1.62 \text{ mol} \times 8.314 \text{ L kPa/mol K} \times 298 \text{ K}}{35.0 \text{ L}} = 115 \text{ kPa}$$

The pressure in the balloon (115 kPa) is about 14% greater than atmospheric pressure (101 kPa).

EXAMPLE 10.13   What volume of hydrogen at 27°C and 723 torr may be prepared by the reaction of 8.88 g of gallium with an excess of hydrochloric acid?

$$2Ga(s) + 6HCl(aq) \longrightarrow 2GaCl_3(aq) + 3H_2(g)$$

This problem requires the following steps:

$$\boxed{\text{Mass of Ga}} \longrightarrow \boxed{\text{Moles of Ga}} \longrightarrow \boxed{\text{Moles of } H_2} \longrightarrow \boxed{\text{Volume of } H_2}$$

$$88.8 \text{ g Ga} \times \frac{1 \text{ mol Ga}}{69.72 \text{ g Ga}} = 0.1274 \text{ mol Ga}$$

$$0.1274 \text{ mol Ga} \times \frac{3 \text{ mol } H_2}{2 \text{ mol Ga}} = 0.1911 \text{ mol } H_2$$

Now we use the Ideal Gas Equation to get the volume of H_2.

$$V = \frac{nRT}{P}$$

$$V = \frac{(0.1911 \text{ mol } H_2)(0.08206 \text{ L atm/mol K})(300 \text{ K})}{0.951 \text{ atm}}$$

$$V = 4.95 \text{ L } H_2$$

10.9 Standard Conditions of Temperature and Pressure

As the foregoing discussion shows, the volume and density (mass per unit volume) of a gas vary with changes in pressure and temperature. To compare densities of gases, we must adopt a set of **standard conditions of temperature and presssure (STP)**. Accordingly, 0°C (273.15 K) and 1 atmosphere (760 torr or 101.325 kilopascals) are universally used as standard conditions for measurements of gases.

EXAMPLE 10.14   Calculate the volume, in liters, occupied by 1.00 mol of an ideal gas at STP.

If we choose the value of R in L atm/mol K, T is exactly 273.15 K and P is exactly 1 atm. Thus

$$V = \frac{nRT}{P}$$

$$V = \frac{(1.00 \text{ mol})(0.08206 \text{ L atm/mol K})(273.15 \text{ K})}{1.00 \text{ atm}}$$

$$V = 22.4 \text{ L}$$

Note that the units cancel to liters, an appropriate unit for volume.

The volume of 1 mole of any gas at 0°C and 1 atmosphere of pressure is 22.4 liters and is referred to as the **molar volume** (see Fig. 10-10).

Figure 10-10. A mole of any gas occupies a volume of approximately 22.4 L at 0°C and 1 atm of pressure. A cube with an edge length of 28.2 cm contains 22.4 L.

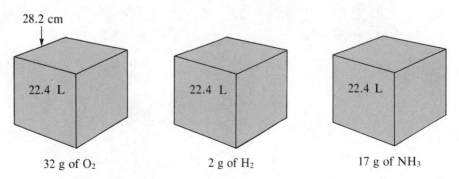

28.2 cm

| 22.4 L | 22.4 L | 22.4 L |

32 g of O_2 2 g of H_2 17 g of NH_3

10.10 Correction of the Volume of a Gas to Standard Conditions

EXAMPLE 10.15   A sample of gaseous ammonia is found to occupy 250 mL under laboratory conditions of 27°C and 740 torr. Correct the volume to standard conditions of 0°C and 760 torr. (Assume all figures in the data are significant.)

The problem can be solved by using either the Ideal Gas Equation or a combination of Boyle's and Charles's Laws.

1. *Using the Ideal Gas Equation.*   Put all constant terms in the equation on the right side and the variable terms on the left. In this example the moles of ammonia and R do not vary, so

$$\frac{PV}{T} = nR = \text{constant}$$

We evaluate the constant term, nR, at the initial conditions (250 mL, $27 + 273.15 = 300$ K, and 740 torr).

$$\frac{740 \text{ torr} \times 250 \text{ mL}}{300 \text{ K}} = nR = 616.7 \text{ mL torr/K}$$

Now we calculate the volume at standard conditions ($0°C = 273.15$ K, 760 torr = 1 atm). Since the amount of ammonia does not change, the equation

$$\frac{PV}{T} = \text{constant}$$

still holds. Rearranging the equation gives

$$V = \text{constant} \times \frac{T}{P}$$

$$V = \frac{616.7 \text{ mL torr} / \text{K} \times 273.15 \text{ K}}{760 \text{ torr}}$$

$$V = 222 \text{ mL}$$

Note that the units cancel to milliliters, a correct unit for volume.

2. *Using a combination of Boyle's and Charles's Laws.*   First we convert the Celsius temperatures to the Kelvin scale: $27°C = 300$ K, and $0°C = 273.15$ K. A decrease in temperature from 300 K to 273.15 K will in itself cause a decrease in the volume of ammonia. Therefore we multiply the original volume by a fraction made up of the two temperatures and having a value less than 1 to correct for the temperature change:

$$250 \text{ mL} \times \frac{273.15 \text{ K}}{300 \text{ K}}$$

The increase in pressure from 740 torr to 760 torr will also in itself decrease the volume, and this factor may be included in the same expression with the temperature factor to obtain the corrected volume:

$$250 \text{ mL} \times \frac{273.15 \cancel{\text{K}}}{300 \cancel{\text{K}}} \times \frac{740 \cancel{\text{torr}}}{760 \cancel{\text{torr}}} = 222 \text{ mL}$$

These procedures can be used to correct to any second set of temperature and pressure values. They are not restricted to correction to standard conditions.

10.11 Densities of Gases

The **density** of a gas is usually expressed as the mass of the gas per liter of gas. Since molecules of different gases have different masses, and since equal volumes of different gases at the same temperature and pressure contain the same number of molecules, the densities of different gases are not the same. For example, the density of oxygen at STP is 1.429 grams per liter, whereas that of hydrogen at STP is 0.08987 grams per liter. Note that you must specify both the temperature and pressure of the gas when reporting its density, since the number of moles of gas (and thus the mass of the gas) in a liter changes with both temperature and pressure. Gas densities are commonly reported at STP.

The density of a gas that exhibits ideal behavior can be calculated in several ways. Two of these are shown in the following examples.

EXAMPLE 10.16   The density of oxygen at STP is 1.429 g/L. Calculate the density of carbon dioxide at STP.

Since equal volumes of two gases at STP contain the same number of molecules, the ratio of the densities of these gases is the same as the ratio of their molecular weights.

$$\frac{\text{Density of } CO_2}{\text{Density of } O_2} = \frac{\text{mol. wt } CO_2}{\text{mol. wt } O_2}$$

$$\text{Density of } CO_2 = \text{density of } O_2 \times \frac{\text{mol. wt } CO_2}{\text{mol. wt } O_2}$$

$$\text{Density of } CO_2 = 1.429 \text{ g/L} \times \frac{44.01}{32.00} = 1.965 \text{ g/L}$$

EXAMPLE 10.17   Calculate the density of ethane, C_2H_6, at a pressure of 183.4 kPa and a temperature of 25°C.

The density of C_2H_6 under the given conditions is the mass of 1 L of C_2H_6. The moles of C_2H_6 in exactly 1 L can be determined using the Ideal Gas Equation and can then be converted to the mass of C_2H_6.

$$n = \frac{PV}{RT}$$

$$n = \frac{183.4 \text{ kPa} \times 1 \text{ L}}{8.314 \text{ L kPa/mol K} \times 298 \text{ K}} = 0.0740 \text{ mol}$$

$$\text{g } C_2H_6 = 0.0740 \text{ mol } C_2H_6 \times \frac{30.1 \text{ g } C_2H_6}{1 \text{ mol } C_2H_6} = 2.23 \text{ g } C_2H_6$$

$$\text{Density} = \frac{2.23 \text{ g}}{1.00 \text{ L}} = 2.23 \text{ g/L}$$

Since the pressure of C_2H_6 was expressed in kilopascals, the value of R with kPa in the units was used.

10.12   Determination of the Molecular Weights of Gases or Volatile Compounds

The molecular weight of a gas can be found experimentally by determining the mass of a given volume of the gas at a known temperature and pressure. The number of moles of gas is determined by using the Ideal Gas Equation, and the molar mass and the molecular weight are determined from the number of moles of gas and its mass (Section 2.4).

EXAMPLE 10.18   Cyclopropane is a commonly used anesthetic. If a 2.00-L flask contains 3.11 g of cyclopropane gas at 684 torr and 23°C, what is the molecular weight of cyclopropane ($R = 0.08206$ L atm/mol K)?

To use the Ideal Gas Equation, we must express temperature on the Kelvin scale, and the units of P and V must correspond to the units of R. Using 0.08206 L atm/mol K as the value of R requires that P be expressed in atmospheres.

$$P = 684 \text{ torr} \times \frac{1 \text{ atm}}{760 \text{ torr}} = 0.900 \text{ atm}$$

$$T = 23°C + 273.15 = 296 \text{ K}$$

Now we calculate the number of moles of cyclopropane using the Ideal Gas Equation.

$$n = \frac{PV}{RT}$$

$$n = \frac{0.900 \text{ atm} \times 2.00 \text{ L}}{0.08206 \text{ L atm/mol K} \times 296 \text{ K}} = 0.0741 \text{ mol}$$

We use the mass of cyclopropane and the number of moles to calculate the molar mass.

$$\text{molar mass of } C_3H_6 = \frac{3.11 \text{ g}}{0.07041 \text{ mol}} = 42.0 \text{ g/mol}$$

A molar mass of 42.0 g/mol corresponds to a molecular weight of 42.0 (Section 2.4).

EXAMPLE 10.19   A sample of phosphorus vapor weighing 3.243×10^{-2} g at 550°C exerted a pressure of 31.89 kPa in a 56.0-mL bulb. What is the molecular weight and molecular formula of phosphorus vapor?

Since the pressure is given in kilopascals, we take 8.314 L kPa/mol K as the value of R. This means that the units of volume must be in liters and that the temperature must be expressed on the Kelvin scale.

$$T = 550 + 273.15 = 823 \text{ K}$$

$$V = 56.0 \text{ mL} \times \frac{1 \text{ L}}{1000 \text{ mL}} = 0.0560 \text{ L}$$

$$n = \frac{PV}{RT}$$

$$n = \frac{31.89 \text{ kPa} \times 0.0560 \text{ L}}{8.314 \text{ L kPa/mol K} \times 823 \text{ K}} = 2.61 \times 10^{-4} \text{ mol}$$

Using the mass of the phosphorus vapor,

$$\text{molar mass} = \frac{3.243 \times 10^{-2} \text{ g}}{2.61 \times 10^{-4} \text{ mol}} = 124 \text{ g/mol}$$

A molar mass of 124 g/mol corresponds to a molecular weight of 124.

Since the molecular weight of phosphorus vapor is 124 and the atomic weight of a single phosphorus atom is 31, there must be $\frac{124}{31}$, or 4, phosphorus atoms in a molecule of phosphorus vapor. The molecular formula is therefore P_4. The P_4 molecule is a tetrahedron (Fig. 10-11).

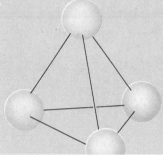

Figure 10-11. The molecular structure of gaseous phosphorus, P_4.

10.13 The Pressure of a Mixture of Gases; Dalton's Law

If gas A is at a pressure of 100 torr in a 1-liter flask, and gas B is at a pressure of 100 torr in a second 1-liter flask, transfer of gas B to the 1-liter flask containing gas A gives a mixture with a total pressure of 200 torr (provided that the temperature does not change). In the absence of chemical interaction between the components of a mixture of gases, the individual gases do not affect one another's pressures, and each gas exerts the same pressure before and after it is mixed. The pressure exerted by each gas is called the **partial pressure** of that gas, and the total pressure of the mixture of gases is the sum of the partial pressures of all the gases present. This law is known as **Dalton's Law of Partial Pressures: The total pressure of a**

mixture of ideal gases is equal to the sum of the partial pressures of the component gases.

$$P_T = P_A + P_B + P_C + \cdots$$

In this equation P_T is the total pressure of a mixture of gases; P_A is the pressure of gas A; P_B, the pressure of gas B; etc.

EXAMPLE 10.20   What is the total presssure in atmospheres in a 10.0-L vessel containing 2.50×10^{-3} mol of H_2, 1.00×10^{-3} mol of He, and 3.00×10^{-4} mol of Ne at 35°C?

The total pressure in the 10.0-L vessel is the sum of the partial pressures of the gases since they do not react with each other.

$$P_T = P_{H_2} + P_{He} + P_{Ne}$$

The partial pressure of each gas can be determined from the Ideal Gas Equation, using $P = nRT/V$.

$$P_{H_2} = \frac{(2.50 \times 10^{-3} \text{ mol})(0.08206 \text{ L atm/mol K})(308 \text{ K})}{10.0 \text{ L}} = 6.32 \times 10^{-3} \text{ atm}$$

$$P_{He} = \frac{(1.00 \times 10^{-3} \text{ mol})(0.08206 \text{ L atm/mol K})(308 \text{ K})}{10.0 \text{ L}} = 2.53 \times 10^{-3} \text{ atm}$$

$$P_{Ne} = \frac{(3.00 \times 10^{-4} \text{ mol})(0.08206 \text{ L atm/mol K})(308 \text{ K})}{10.0 \text{ L}} = 7.58 \times 10^{-4} \text{ atm}$$

$$P_T = (0.00632 + 0.00253 + 0.00076) \text{ atm} = 9.61 \times 10^{-3} \text{ atm}$$

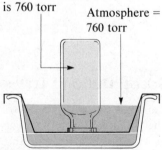

Pressure of gas and water vapor is 760 torr

Atmosphere = 760 torr

Figure 10-12. Method by which the pressure on a confined mixture of gases can be made equal to the atmospheric pressure. The level of water inside and outside the vessel is made the same. In the diagram a typical atmospheric pressure of 760 torr is indicated. The actual pressure depends on temperature and altitude.

The pressure of a sample of a gas that does not react with water can be measured by collecting the gas in a container over water and making its pressure equal to air pressure (which can be measured by a barometer). This is easily accomplished by adjusting the water level (Fig. 10-12) so that it is the same both inside and outside the container. The gas is then at the existing atmospheric pressure.

However, another factor must be considered when determining pressure by this method. There is always a little gaseous water (water vapor) above a sample of water. Thus when a gas is collected over water, it soon becomes saturated with water vapor. The total pressure of the mixture equals the sum of the partial pressures of the gas and the water vapor, according to Dalton's Law. The pressure of the pure gas is therefore equal to the total pressure minus the pressure of the water vapor.

The pressure of the water vapor above a sample of water in a closed container like that in Fig. 10-12 depends on the temperature, as shown in Fig. 10-13 and Table 10-1. At 23°C, the vapor pressure of water is 21.1 torr (Table 10-1). Thus, at 23°C the total pressure of 760 torr in the bottle in Fig. 10-2 consists of 21 torr due to water vapor and 739 torr due to the other gases. At 29°C, water has a vapor pressure of 30 torr. At 20°C the bottle would contain water vapor at 30 torr, and the other gases at 730 torr.

Figure 10-13. A graph illustrating change of water vapor pressure with change in temperature.

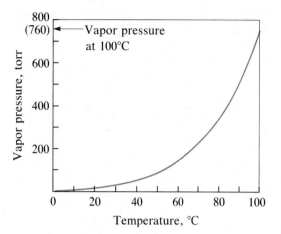

Table 10-1   Vapor Pressure of Ice and Water at Various Temperatures

Temperature, °C	Pressure, torr	Temperature, °C	Pressure, torr	Temperature, °C	Pressure, torr
−10	1.95	12	10.5	27	26.7
−5	3.0	13	11.2	28	28.3
−2	3.9	14	12.0	29	30.0
−1	4.2	15	12.8	30	31.8
0	4.6	16	13.6	35	42.2
1	4.9	17	14.5	40	55.3
2	5.3	18	15.5	50	92.5
3	5.7	19	16.5	60	149.4
4	6.1	20	17.5	70	233.7
5	6.5	21	18.7	80	355.1
6	7.0	22	19.8	90	525.8
7	7.5	23	21.1	95	633.9
8	8.0	24	22.4	99	733.2
9	8.6	25	23.8	100.0	760.0
10	9.2	26	25.2	101.0	787.6
11	9.8				

EXAMPLE 10.21   If 0.200 L of argon is collected over water at a temperature of 26°C and a pressure of 750 torr in a system like that shown in Fig. 10-12, what is the partial pressure of argon?

According to Dalton's Law, the total pressure in the bottle (750 torr) is the sum of the pressure of argon and the pressure of gaseous water.

$$P_T = P_{Ar} + P_{H_2O}$$

Rearranging this equation to solve for the pressure of argon gives

$$P_{Ar} = P_T - P_{H_2O}$$

The pressure of water vapor above a sample of liquid water at 26°C is given as 25.2 torr in Table 10-1. So,

$$P_{Ar} = 750 \text{ torr} - 25.2 \text{ torr} = 725 \text{ torr}$$

EXAMPLE 10.22   A 1.34-g sample of calcium carbide, CaC_2, was allowed to react with water. The acetylene, C_2H_2, produced by the reaction

$$CaC_2(s) + 2H_2O(l) \longrightarrow C_2H_2(g) + Ca(OH)_2(s)$$

was collected over water in a system like that shown in Fig. 10-14. If 471 mL of acetylene gas was collected at a temperature of 23°C and a pressure of 743 torr, what was the percent yield of the reaction?

Figure 10-14. Acetylene from the reaction of CaC_2 with water is collected over water.

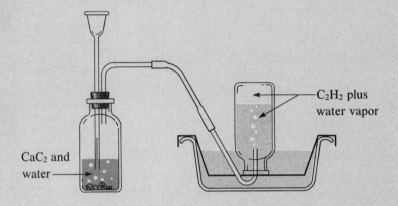

CaC$_2$ and water

C$_2$H$_2$ plus water vapor

To determine the percent yield (Section 2.11), we need the actual yield and the theoretical yield. We determine the theoretical yield as follows:

$$1.34 \text{ g CaC}_2 \times \frac{1 \text{ mol CaC}_2}{64.1 \text{ g CaC}_2} = 2.09 \times 10^{-2} \text{ mol CaC}_2$$

$$2.09 \times 10^{-2} \text{ mol CaC}_2 \times \frac{1 \text{ mol C}_2\text{H}_2}{1 \text{ mol CaC}_2} = 2.09 \times 10^{-2} \text{ mol C}_2\text{H}_2$$

To determine the actual yield we need to calculate the moles of C_2H_2 produced. This is available from the Ideal Gas Equation, $PV = nRT$, and the data in the problem. However, the pressure in the equation must be the pressure of $C_2H_2(g)$, not the pressure of a mixture of $H_2O(g)$ and $C_2H_2(g)$. So we use Dalton's Law to get that partial pressure. The vapor pressure of water at 23°C is 21.1 torr (Table 10-1). Therefore,

$$P_{C_2H_2} = P_T - P_{H_2O}$$

$$P_{C_2H_2} = 743 \text{ torr} - 21.1 \text{ torr} = 722 \text{ torr (or 0.950 atm)}$$

Using the Ideal Gas Equation with P in atmospheres , T on the Kelvin scale, and V in liters, we find n, the moles of C_2H_2 produced.

$$n_{C_2H_2} = \frac{PV}{RT}$$

$$n_{C_2H_2} = \frac{0.950 \text{ atm} \times 0.471 \text{ L}}{0.08206 \text{ L atm/mol K} \times 296 \text{ K}} = 1.34 \times 10^{-2} \text{ mol}$$

Now we calculate the percent yield.

$$\text{Percent yield} = \frac{\text{actual yield}}{\text{theoretical yield}} \times 100$$

$$\text{Percent yield} = \frac{1.84 \times 10^{-2} \text{ mol}}{2.09 \times 10^{-2} \text{ mol}} \times 100 = 88.0\%$$

10.14 Diffusion of Gases; Graham's Law

When a sample of gas is set free from one part of a closed container, it very quickly diffuses throughout the container (see Fig. 10-15).

Figure 10-15. When the stopcock is opened, the molecules intermingle; that is, the gases mix by diffusing together. The rapid motion of the molecules and the relatively large spaces between them explain why diffusion occurs. Note that H_2, the lighter of the two gases, diffuses faster than O_2.

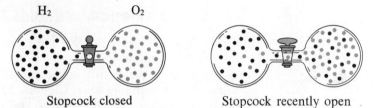

| H₂ | O₂ |

Stopcock closed   Stopcock recently open

If a mixture of gases is placed in a container with porous walls, diffusion of the gases through the walls occurs. The lighter gases diffuse through the small openings of the porous walls more rapidly than the heavier ones. In 1832 Thomas Graham studied the rates of diffusion of different gases and formulated **Graham's Law: The rates of diffusion of gases are inversely proportional to the square roots of their densities (or their molecular weights).**

$$\frac{\text{Rate of diffusion of gas A}}{\text{Rate of diffusion of gas B}} = \frac{\sqrt{\text{density B}}}{\sqrt{\text{density A}}} = \frac{\sqrt{\text{mol. wt B}}}{\sqrt{\text{mol. wt A}}}$$

EXAMPLE 10.23   Calculate the ratio of the rate of diffusion of hydrogen to the rate of diffusion of oxygen.

Using densities. The density of hydrogen is 0.08987 g/L, and that of oxygen is 1.429 g/L.

$$\frac{\text{Rate of diffusion of hydrogen}}{\text{Rate of diffusion of oxygen}} = \frac{\sqrt{1.429 \text{ g/L}}}{\sqrt{0.08987 \text{ g/L}}} = \frac{1.20}{0.30} = \frac{4}{1}$$

Using molecular weights.

$$\frac{\text{Rate of diffusion of hydrogen}}{\text{Rate of diffusion of oxygen}} = \frac{\sqrt{32}}{\sqrt{2}} = \frac{\sqrt{16}}{1} = \frac{4}{1}$$

Hydrogen diffuses four times as rapidly as oxygen.

Rates of diffusion depend on the speeds of the molecules—the faster the molecules move, the faster they diffuse. Molecules of smaller mass move with faster speeds than do molecules of larger mass; thus the former diffuse more rapidly. Differences in diffusion rates of gaseous substances are used in the separation of light isotopes from heavy ones. Diffusion through a porous barrier from a region of higher pressure to one of lower pressure is used for the large-scale separation of gaseous $^{235}_{92}UF_6$ from $^{238}_{92}UF_6$ at the atomic energy installation in Oak Ridge, Tennessee. It is said that for separation to be complete, a given volume of UF_6 must be diffused some two million times.

THE MOLECULAR BEHAVIOR OF GASES

10.15 The Kinetic-Molecular Theory

Bernoulli in 1738, Poule in 1851, and Kronig in 1856 suggested that properties of gases such as compressibility, expansibility, and diffusibility could be explained by assuming that they consist of continuously moving molecules. During the latter half of the nineteenth century, Clausius, Maxwell, Boltzmann, and others developed this hypothesis into the detailed **kinetic-molecular theory** of gases. This theory was developed for an ideal gas using the following assumptions about the nature of the molecules of such a gas:

1. Gases are considered to be composed of molecules that are in continuous, completely random motion in all directions. The molecules move in straight lines and change direction only when they collide with other molecules or with the walls of a container.

2. The pressure of a gas in a container results from the bombardment of the walls of the container by the gas molecules.

3. At a given temperature, the pressure in a container does not change with time. Thus the collisions between molecules and between molecules and walls must be elastic; that is, the collisions involve no loss of energy due to friction.

4. At relatively low pressures the average distance between gas molecules is large compared to the size of the molecules.

5. At relatively low pressures the attractive forces between gas molecules, which depend on the separation of the molecules, are of no significance and can be ignored since the molecules are relatively widely separated.

6. Since the molecules are small compared to the distances between them, they may be considered to have no volume.

7. The average kinetic energy of the molecules (the energy due to their motion) is proportional to the temperature of the gas on the Kelvin scale and is the same for all gases at the same temperature.

As we will show in the next section, the assumptions of the kinetic-molecular theory explain the gas laws in a qualitative way. The true test of the theory, however, is its ability to describe the behavior of a gas quantitatively. We shall show that the Ideal Gas Equation can be derived from the assumptions of the theory; this has led chemists to believe that the assumptions of the theory represent the properties of gas molecules.

10.16 Relation of the Behavior of Gases to the Kinetic-Molecular Theory

The gas laws may be readily explained in light of the kinetic-molecular theory.

1. BOYLE'S LAW. The pressure exerted by a gas in a container is caused by the bombardment of the walls of the container by rapidly moving molecules of the gas. The pressure varies directly with the number of molecules hitting the walls per unit of time. Reducing the volume of a given mass of gas by half will double the number of molecules per unit of volume. The number of impacts per unit of time on the same area of wall surface will also be doubled. Doubling the number of impacts per unit of area doubles the pressure. These results are in accordance with Boyle's Law relating volume and pressure.

2. CHARLES'S LAW. At constant volume an increase in temperature increases the pressure of a gas. This increase in pressure reflects the increase in average speed and kinetic energy of the molecules as the temperature is raised. An increase in the average speed results in more frequent and harder impacts on the walls of the container, that is, in greater pressure.

   In order for the pressure to remain constant with the increasing temperature, the volume must increase so that each molecule travels farther, on the average, before hitting a wall. The smaller number of molecules striking any wall at a given time exactly offsets the greater force with which each molecule hits, making it possible for the pressure to remain constant. This may be visualized by picturing a cylinder fitted with a piston. As the temperature increases, the molecules hit the piston harder. The piston must move up, thereby increasing the volume, if the pressure is to remain the same.

3. DALTON'S LAW. The molecules of one component of a mixture of gases will bombard the walls of the container with the same frequency in the presence of other kinds of molecules as in their absence. Thus the total pressure of a mixture of gases will be the sum of the partial pressures of the individual gases.

4. GRAHAM'S LAW. The facts that the molecules of a gas are in rapid motion and that the space between the molecules is relatively large explains the phenomenon of diffusion. Gas molecules can move past each other easily.

   The rate of diffusion depends on the mass of the molecules of a gas since, at the same temperature, molecules of all gases have the same average kinetic energy. The kinetic energy, KE, of a moving body, in joules, is equal to $\frac{1}{2}mu^2$ where m is its mass in kilograms and u is its speed in meters per second. The average kinetic energy, KE_{avg}, of the molecules of a gas consisting of one type of molecule is equal to $\frac{1}{2}m(u^2)_{avg}$, where $(u^2)_{avg}$ is the average of the squares of the speeds of the molecules. Thus at a given temperature, lighter molecules, like hydrogen, must move at higher speeds than heavier molecules, like oxygen, since both have the same average kinetic energy (Fig. 10-16).

   For two different gases at the same temperature and pressure,

$$\text{Average kinetic energy for first gas} = \tfrac{1}{2}m_1(u_1^2)_{avg}$$

$$\text{Average kinetic energy for second gas} = \tfrac{1}{2}m_2(u_2^2)_{avg}$$

Figure 10-16. The distribution of the molecular speeds of different gases. The lighter gases move with faster average speeds.

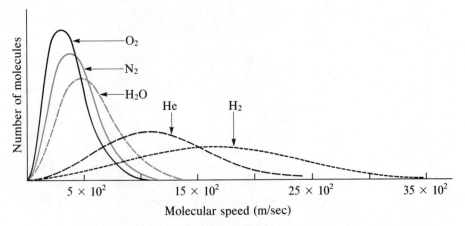

where m_1 and m_2 are the masses of individual molecules of the two gases and $(u_1{}^2)_{avg}$ and $(u_2{}^2)_{avg}$ are the averages of the squares of their speeds.

The kinetic energies of the two gases are equal according to the kinetic-molecular theory. Hence

$$\tfrac{1}{2}m_1(u_1{}^2)_{avg} = \tfrac{1}{2}m_2(u_2{}^2)_{avg}$$

Dividing both sides of the equation by $\tfrac{1}{2}$ gives

$$m_1(u_1{}^2)_{avg} = m_2(u_2{}^2)_{avg}$$

On rearranging, we have

$$\frac{(u_1{}^2)_{avg}}{(u_2{}^2)_{avg}} = \frac{m_2}{m_1}$$

Taking square roots gives

$$\frac{(u_1)_{rms}}{(u_2)_{rms}} = \frac{\sqrt{m_2}}{\sqrt{m_1}}$$

where $(u_1)_{rms}$ and $(u_2)_{rms}$ are the root-mean-square speeds of the respective molecules, that is, the square roots of the averages of the squares of the speeds:

$$(u_1)_{rms} = \sqrt{(u_1{}^2)_{avg}}, \qquad (u_2)_{rms} = \sqrt{(u_2{}^2)_{avg}}$$

The rate at which a gas will diffuse, R, is proportional to u_{rms}, the root-mean-square speed of its molecules. The molecular weights, M_1 and M_2, of two gases are proportional to the masses of the individual molecules. Thus

$$\frac{R_1}{R_2} = \frac{\sqrt{M_2}}{\sqrt{M_1}}$$

This equation states that the ratio of the rates of diffusion of two gases, held at the same temperature and pressure, equals the inverse ratio of the square roots of their molecular weights.

As shown in Section 10.11, the ratio of the densities (d_1 and d_2) of two gases is the same as the ratio of their molecular weights. Since $d_2/d_1 = M_2/M_1$,

$$\frac{R_1}{R_2} = \frac{\sqrt{d_2}}{\sqrt{d_1}}$$

Therefore Graham's Law can equally well be expressed in terms of the square roots of the molecular weights and in terms of the square roots of the densities of the gases.

Hydrogen is the lightest gas and therefore diffuses the most rapidly. The root-mean-square speed of its molecules at room temperature is about 1 mile per second; that of oxygen molecules is about $\frac{1}{4}$ mile per second. Collisions between molecules make diffusion rates much lower than would be expected from such speeds. A molecule in hydrogen gas at standard conditions has about 11 billion collisions per second. Hydrogen molecules travel about 1.7×10^{-5} centimeter between collisions and, on the average, collide about 60,000 times in traveling 1 linear centimeter.

10.17 The Distribution of Molecular Velocities

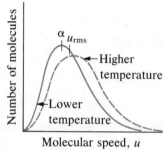

The individual molecules of a gas travel at different speeds. Because of the enormous number of collisions, the speeds of the molecules vary from practically zero to thousands of meters per second. The exchange of energy accompanying each collision can change the speed of a given molecule over a wide range. However, since a large number of molecules is involved, the distribution of the molecular speeds of the total number of molecules is constant. The nature of the distribution of molecular speeds is described by the **Maxwell-Boltzmann Distribution Law.**

The distribution of molecular speeds is shown in Fig. 10-17. The vertical axis represents the number of molecules, and the horizontal axis represents the molecular speed, u. The most probable speed is α, and the root-mean-square speed of the molecules is u_{rms}. The graph indicates that very few molecules move at very low or very high speeds. The number of molecules with intermediate speeds increases rapidly up to a maximum and then drops off rapidly.

At the higher temperature the whole curve is shifted to higher speeds. Thus at a higher temperature more molecules have higher speeds and fewer molecules have lower speeds. This is in accord with expectations based on kinetic-molecular theory.

Figure 10-17. Distributions of molecular speed versus number of molecules of a gas at different temperatures.

10.18 Ideal Gas Equation Derived from the Kinetic-Molecular Theory

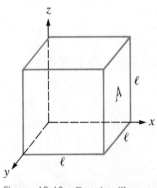

The mathematical expression of the Ideal Gas Equation, $PV = nRT$, may be derived from the kinetic-molecular theory of gases. Consider **N** molecules, each having a mass m, confined within a cubical container, as shown in Fig. 10-18, with an edge length of l centimeters. Although the molecules are moving in all possible directions, we may assume that $\frac{1}{3}$ of the molecules ($\frac{1}{3}$N) are moving in the direction of the x axis, $\frac{1}{3}$ in the direction of the y axis, and $\frac{1}{3}$ in the direction of the z axis. This assumption is valid since the motion of the molecules is entirely random and no particular direction is preferred.

Initially, let us assume that all molecules are moving with the same speed, and consider the collisions of molecules with wall A (Fig. 10-18). A molecule, on the average, travels $2l$ centimeters between two consecutive collisions with wall A as it moves back and forth across the container. If the speed of the molecule is u centimeters per second, it will collide $u/2l$ times per second with wall A. Since the collisions are perfectly elastic, the molecule will rebound with a speed of $-u$,

Figure 10-18. Drawing illustrating the container used in the derivation of the Ideal Gas Equation from the kinetic-molecular theory.

having lost no kinetic energy as a result of the collision. Because momentum is defined as the product of mass and speed, the average momentum before a collision is mu and after a collision is $-mu$. Thus the average change in momentum per molecule per collision is $2mu$. There are $u/2\ell$ collisions per second, making the average change in momentum per molecule per second

$$2mu \times \frac{u}{2\ell} = \frac{mu^2}{\ell}$$

The total change in momentum per second for the $\mathbf{N}/3$ molecules that can collide with wall A is

$$\frac{\mathbf{N}}{3} \times \frac{mu^2}{\ell}$$

This is the average force on wall A, because force is equal to change of momentum per second. Pressure is force per unit area. Since the area of A is ℓ^2, the pressure P on A is

$$P = \frac{\text{force}}{\text{area}} = \frac{1}{\ell^2} \times \frac{\mathbf{N}}{3} \times \frac{mu^2}{\ell} = \frac{\mathbf{N}}{3} \times \frac{mu^2}{\ell^3}$$

However, ℓ^3 is equal to the volume, V, of the cubical container; thus we have

$$P = \frac{\mathbf{N}}{3} \times \frac{mu^2}{V}$$

or
$$PV = \tfrac{1}{3}\mathbf{N}mu^2$$

Actually, molecules move with different speeds, so we must use the average of the squares of their speeds. Thus

$$PV = \tfrac{1}{3}\mathbf{N}m(u^2)_{\text{avg}}$$

This is the fundamental equation of the kinetic-molecular theory of gases. It describes the behavior of an ideal gas exactly, and it is a very good approximation for the behavior of real gases at ordinary pressures.

The equation above may be written as follows:

$$PV = \tfrac{2}{3}\mathbf{N}[\tfrac{1}{2}m(u^2)_{\text{avg}}]$$

The term in the set of brackets, $\tfrac{1}{2}m(u^2)_{\text{avg}}$, is the expression for the average kinetic energy of the gas, KE_{avg} (Section 10.16, Part 4). Thus

$$PV = \tfrac{2}{3}\mathbf{N}(\text{KE}_{\text{avg}})$$

The average kinetic energy of the molecules, KE_{avg}, is directly proportional to the Kelvin temperature, T.

$$\text{KE}_{\text{avg}} = kT$$

The number of molecules, $\mathbf{N}$, is proportional to the number of moles of molecules, n.

$$\mathbf{N} = k'n$$

Substituting the expressions for KE_{avg} and $\mathbf{N}$ in the equation for PV gives

$$PV = \tfrac{2}{3}(k'n)(kT) = n(\tfrac{2}{3}k'k)T$$

The expression $\frac{2}{3}k'k$ is the constant (being made up entirely of constants) that we write as R and call the gas constant. Therefore,

$$PV = nRT$$

Thus from the assumptions of the kinetic-molecular theory (Section 10.15), we can derive the Ideal Gas Equation. Since this equation describes the behavior of gases very well, the assumptions used in its derivation are strongly supported.

10.19 Deviations from Ideal Gas Behavior

When a gas at ordinary temperature and pressure is compressed, the volume is reduced by crowding the molecules closer together. This reduction in volume is really a reduction in the amount of empty space between the molecules. At high pressures the molecules are crowded so close together that the actual volume of the molecules, the molecular volume, is a relatively large fraction of the total volume, including the empty space between molecules. Because the volume of the molecules themselves cannot be compressed, only a fraction of the entire volume is affected by a further increase in pressure. Thus at very high pressures the whole volume is not inversely proportional to the pressure, as predicted by Boyle's Law. It should be noted, nevertheless, that even at moderately high pressures most of the volume is not occupied by molecules.

The molecules in a real gas at relatively low pressures have practically no attraction for one another because they are far apart. Thus they behave almost like molecules of ideal gases. However, as the molecules are crowded closer together at high pressures, the force of attraction between the molecules increases. This attraction has the same effect as does an increase in external pressure. Consequently, when external pressure is applied to a volume of gas, especially at low temperatures, there is a slightly greater decrease in volume than should be achieved by the pressure alone. This slightly greater decrease in volume caused by intermolecular attraction is more pronounced at low temperatures because the molecules move more slowly, their kinetic energy is smaller relative to the attractive forces, and they fly apart less easily after collisions with one another.

In 1879 van der Waals expressed the deviations of real gases from the ideal gas laws quantitatively in what is now known as the **van der Waals Equation:**

$$\left(P + \frac{n^2a}{V^2}\right)(V - nb) = nRT$$

The constant a represents the attraction between molecules, and van der Waals assumed that this force varies inversely as the square of the total volume of the gas. (See Section 11.13 for a discussion of the nature of the van der Waals force.) Since this force augments the pressure and thus tends to make the volume smaller, it is added to the term P.

The term b in the equation represents the total volume of all of the molecules themselves and is subtracted from V, the total volume of the gas. When V is large, both nb and n^2a/V^2 are negligible, and the van der Waals Equation reduces to the simple gas equation, $PV = nRT$.

At low pressures the correction for intermolecular attraction, a, is more important than the correction for molecular volume, b. At high pressures and small volumes the correction for the volume of the molecules becomes important because the molecules themselves are incompressible and constitute an appreciable fraction of the total volume. At some intermediate pressure the two corrections cancel one another, and the gas appears to follow the relationship given by $PV = nRT$ over a small range of pressures.

Strictly speaking, then, the Ideal Gas Equation applies exactly only to gases whose molecules do not attract one another and do not occupy an appreciable part of the whole volume. Because no real gases have these properties, we speak of such hypothetical gases as ideal gases. Under ordinary conditions, however, the deviations from the gas laws are so slight that they may be neglected.

FOR REVIEW

SUMMARY

A gas takes the shape and volume of its container. Gases are readily compressible and capable of infinite expansion. The pressure of a gas may be expressed (in terms of force per unit area) in the SI units of **pascals** or **kilopascals.** Other units used for pressure include **torr, atmospheres,** and **millimeters of mercury** (1 atm = 760 mmHg = 760 torr = 101.325 kPa).

The behavior of gases can be described by a number of laws based on experimental observations of their properties. For example, the volume of a given quantity of a gas is inversely proportional to the pressure of the gas, provided that the temperature does not change **(Boyle's Law).** The volume of a given quantity of gas is directly proportional to its temperature on the Kelvin scale, provided that the pressure does not change **(Charles's Law).** Gases react in simple proportions by volume **Gay-Lussac's Law),** since under the same conditions of temperature and pressure, equal volumes of all gases contain the same number of molecules **(Avogadro's Law).** The rates of diffusion of gases are inversely proportional to the square roots of their densities or to the square roots of their molecular weights **(Graham's Law).** In a mixture of gases the total pressure is equal to the sum of the partial pressures of the gases present. The **partial pressure** of a gas is the pressure that the gas would exert if it were the only gas present. The volumes and densities of gases are often reported under **standard conditions of temperature and pressure (STP):** 0°C and 1 atmosphere.

The equations describing Boyle's Law, Charles's Law, and Avogadro's Law are special cases of the **Ideal Gas Equation,** $PV = nRT$, where P is the pressure of the gas; V, its volume; n, the number of moles of the gas; and T, its Kelvin temperature. R is the **gas constant.** The value used for the gas constant when applying the equation depends on the units used for P and V.

The Ideal Gas Equation has been derived for a gas that obeys the hypotheses of the **kinetic-molecular theory** of gases. This theory assumes that gases consist of molecules of negligible volume, which are widely separated with no attraction between them. The molecules move randomly and change direction only when they collide with one another or with the walls of the container. These collisions are elastic. The pressure of a gas results from the bombardment of the walls of the container by its molecules. The average kinetic energy of the molecules is proportional to the temperature of the gas and is the same for all gases at a given temperature.

All molecules of a gas are not moving at the same speed at the same instant of time. Some move relatively slowly, while others move relatively rapidly. The speeds are distributed over a wide range. The average of the squares of the molecular speeds, $(u^2)_{avg}$, is directly proportional to the temperature of the gas on the Kelvin scale and inversely proportional to the masses of the individual molecules. Heating a gas increases both the value of $(u^2)_{avg}$ and the number of molecules moving at higher speeds. As shown by Graham's Law, heavier molecules diffuse more slowly than do lighter ones. Heavier molecules move more slowly, on the average.

The molecules in a real gas (in contrast to those in an ideal gas) possess a finite volume and attract each other slightly. At relatively low pressures the molecular volume can be neglected. At temperatures well above the temperature at which the gas liquefies, the attractions between molecules can be neglected. Under these conditions the Ideal Gas Equation is a good approximation for the behavior of a real gas. However, at lower temperatures or higher pressures or both, corrections for molecular volume and molecular attractions are required. The **van der Waals Equation** can be used to describe real gases under these conditions.

KEY TERMS AND CONCEPTS

Absolute zero (10.5)
Avogadro's Law (10.7)
Barometer (10.2)
Boyle's Law (10.3, 10.16)
Charles's Law (10.4, 10.16)
Dalton's Law (10.13, 10.16)
Density (10.11)
Diffusion (10.1, 10.14, 10.16)

Gas constant (10.8)
Gay-Lussac's Law (10.6, 10.7)
Graham's Law (10.14, 10.16)
Kinetic energy (10.16)
Kinetic-molecular theory (10.15)
Ideal gas (10.8, 10.15)
Ideal Gas Equation (10.8, 10.18)
Molecular volume (10.19)

Partial pressure (10.13)
Pressure (10.2)
Rate of diffusion (10.14, 10.16)
Real gas (10.19)
Standard conditions, STP (10.9)
Temperature (10.5)
van der Waals Equation (10.19)

EXERCISES

Pressure

⌐S⌐ 1. A typical barometric pressure in Kansas City is 740 torr. What is this pressure in atmospheres, millimeters of mercury, and kilopascals?
            Ans. 0.974 atm; 740 mmHg; 98.7 kPa

⌐S⌐ 2. European tire gauges are marked in units of kilopascals. What reading on such a gauge corresponds to $25 \, lb/in^2$ ($1 \, atm = 14.7 \, lb/in^2$)?
                                    Ans. 170 kPa

3. A medical laboratory catalog describes the pressure in a cylinder of a gas as 14.82 MPa ($1 \, MPa = 10^6 \, Pa$). What is the pressure of this

gas in atmospheres and torr?
            Ans. 146.3 atm; 1.112 × 10⁵ torr

4. Arrange the following gases in increasing order by pressure: H_2 at 375 torr, N_2 at 1.4 atm, O_2 at 98 kPa.   *Ans. $H_2 < O_2 < N_2$*

⌐S⌐ 5. A biochemist adds carbon dioxide, CO_2, to an evacuated bulb like the one shown on the following page and stops when the difference in the heights of the mercury columns, h, is 3.56 cm. What is the pressure of CO_2 in the bulb in atmospheres and in kilopascals?

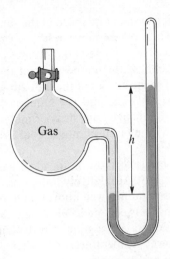

Ans. 0.0468 atm; 4.75 kPa

The Gas Laws

6. Using Boyle's Law, set up a mathematical equation relating the initial volume, V_1, and the final volume, V_2, of a given quantity of an ideal gas at constant temperature as the pressure changes from the initial pressure, P_1, to the final pressure, P_2.

7. Explain in your own words what it means to say that "the volume of a gas is inversely proportional to its pressure at constant temperature."

Ⓢ8. The volume of a sample of carbon monoxide, CO, is 405 mL at 10.0 atm and 467 K. What volume will it occupy at 4.29 atm and 467 K?
Ans. 944 mL

9. The volume of a sample of ethane, C_2H_6, is 3.24 L at 477 torr and 27°C. What volume will it occupy at 27°C and 831 torr?   *Ans. 1.86 L*

10. A sample of oxygen, O_2, occupies 38.9 L at STP. What volume will it occupy at 0°C and 917 torr?
Ans. 32.2 L

Ⓢ11. A typical scuba tank has a volume of 13.2 L. What volume of air in liters at 0.950 atm is required to fill such a scuba tank to a pressure of 153 atm, assuming no change in temperature?
Ans. 2130 L

12. A cylinder of oxygen for medical use contains 35.4 L of oxygen gas at a pressure of 149.6 atm. What is the volume of this oxygen in liters at 1.00 atm of pressure and the same temperature?
Ans. 5.30 × 10³ L

13. Using Charles's Law, set up a mathematical equation relating the initial volume, V_1, and temperature, T_1, of a given quantity of an ideal gas at constant pressure to some other volume, V_2, and temperature, T_2.

14. Explain in your own words what it means to say that "the volume of a gas is directly proportional to its Kelvin temperature at constant pressure."

Ⓢ15. What is the volume of a sample of ethane at 467 K and 1.1 atm if it occupies 405 mL at 298 K and 1.1 atm?   *Ans. 635 mL*

16. What is the volume of a sample of carbon monoxide, CO, at 3°C and 744 torr if it occupies 13.3 L at 55°C and 744 torr?   *Ans. 11.2 L*

17. The gas in a 1.00-L bottle at 25°C can be put into a 1-qt (0.946-L) bottle at the same pressure if the temperature is reduced. What temperature is required?   *Ans. 9°C*

Ⓢ18. A 2.50-L volume of hydrogen measured at the normal boiling point of nitrogen, −210.0°C, is warmed to the normal boiling point of water, 100°C. Calculate the new volume of the gas, assuming ideal behavior and no change in pressure.   *Ans. 14.8 L*

19. A gas occupies 275 mL at 0°C and 305 Pa. What final temperature would be required to increase the pressure to 380 Pa if the volume is held constant?   *Ans. 340 K or 67°C*

Ⓢ20. A high-altitude balloon is filled with 1.41 × 10⁴ L of hydrogen at a temperature of 21°C and a pressure of 745 torr. What is the volume of the balloon at a height of 20 km where the temperature is −48°C and the pressure is 63.1 torr?   *Ans. 1.27 × 10⁵ L*

21. If the CO_2 in the bulb in Exercise 5 is warmed from 21.2°C to 35.2°C, what will the height, h, of the mercury column be? Ignore the change in volume due to the motion of the mercury; that is, assume that the volume remains constant.
Ans. 3.73 cm

Ⓢ22. A spray can is used until it is empty, except for the propellant gas, which has a pressure of 1.1 atm at 23°C. If the can is thrown into a fire ($T = 475°C$), what will be the pressure in the hot can?   *Ans. 2.8 atm*

23. (a) When a gas is allowed to expand, its temperature will decrease if a change of the pressure on it occurs. A 2.00-L sample of a gas at 25°C and a pressure of 875 torr was allowed to expand to a volume of 5.00 L at a pressure of 345 torr. What was the new temperature of the gas?   *Ans. 21°C*

(b) Suppose a gas expands under conditions such that the pressure remains constant. How must the temperature of the gas change? Illustrate by calculating the temperature of the gas in part (a) if its expansion occurs at a constant pressure of 875 torr.   *Ans. 472°C*

S 24. How many moles of hydrogen sulfide, H_2S, are contained in a 327.3-mL bulb at 48.1°C if the pressure is 149.3 kPa?

Ans. 1.830 × 10⁻² mol

25. How many moles of chlorine gas, Cl_2, are contained in a 10.3-L tank at 21.2°C if the pressure is 633 torr?   *Ans. 0.355 mol*

26. How many moles of oxygen gas, O_2, are contained in a cylinder of medical oxygen with a volume of 35.4 L, a pressure of 149 atm, and a temperature of 19.5°C?   *Ans. 220 mol*

27. A small cylinder of helium for use in chemistry lectures has a volume of 334 mL. How many moles of helium are contained in such a cylinder at a pressure of 154 atm and a temperature of 23°C?   *Ans. 2.12 mol*

S 28. What is the temperature of a 0.274-g sample of methane, CH_4, confined in a 300.0-mL bulb at a pressure of 198.7 kPa?

Ans. 420 K or 147°C

29. What is the volume of a bulb that contains 8.17 g of neon at 13°C and a pressure of 8.73 atm?   *Ans. 1.09 L*

S 30. Assume that 453.6 g of dry ice (solid CO_2) is placed in an evacuated 50.0-L closed tank. What is the pressure in the tank in atmospheres at a temperature of 45°C after all the $CO_2(s)$ has been converted to gas?   *Ans. 5.38 atm*

31. How many grams of gas are present in each of the following cases?

S (a) 0.100 L of NO at 703 torr and 62°C

Ans. 0.101 g

(b) 8.75 L of CH_4 at 278.3 kPa and 843 K

Ans. 5.57 g

(c) 73.3 mL of I_2 at 0.462 atm and 125°C

Ans. 0.263 g

(d) 4341 L of BF_3 at 2.22 atm and 788 K

Ans. 10.1 kg

(e) 221 mL of Ne at 0.23 torr and −54°C

Ans. 7.5 × 10⁻⁵ g

32. Calculate the volume in liters of each of the following quantities of gas at STP.

(a) 6.72 g of C_2H_2 *Ans. 5.78 L*

S (b) 0.588 g of NH_3 *Ans. 0.774 L*

(c) 1.47 kg of C_2H_6 *Ans. 1.10 × 10³ L*

(d) 0.720 mol of BF_3 *Ans. 16.1 L*

(e) 13.5 mol of SO_2 *Ans. 303 L*

(f) 0.027 mol of PH_3 *Ans. 0.61 L*

33. (a) What mass of nitrosyl chloride, NOCl, occupies a volume of 0.250 L at a temperature of 325 K and a pressure of 113.0 kPa?

Ans. 0.684 g

(b) What is the density of NOCl under these conditions?   *Ans. 2.74 g/L*

S 34. Calculate the density of fluorine gas, F_2, at STP, and at 30.0°C and 725 torr.

Ans. 1.70 g/L; 1.46 g/L

35. Calculate the density of Freon 12, CF_2Cl_2, at 30.0°C and 0.954 atm.   *Ans. 4.64 g/L*

36. Calculate the density of nitrogen, N_2, at a temperature of 273 K and a pressure of 101.3 kPa.

Ans. 1.25 g/L

37. A liter of methane gas, CH_4, at STP contains more atoms of hydrogen than does a liter of pure hydrogen gas, H_2, at STP. Using Avogadro's Law as a starting point, explain why.

38. For 1 mol of H_2 showing ideal gas behavior, draw labeled graphs of:

(a) the variation of P with V at $T = 273$ K

(b) the variation of V with T at $P = 1.00$ atm

(c) the variation of P with T at $V = 22.4$ L

(d) the variation of the average velocity of the gas molecules with T

39. A cylinder of a standard gas for calibration of blood gas analyzers contains 5.0% CO_2, 12.0% O_2, and the remainder N_2 at a total pressure of 146 atm. What is the partial pressure of each component of this gas? (The percentages given indicate the percent of the total pressure due to each component.)   *Ans. CO_2, 7.3 atm; O_2, 17.5 atm; N_2, 121 atm*

S 40. Most mixtures of hydrogen gas with oxygen gas are explosive. However, a mixture that contains less than 3.0% O_2 is not. If enough O_2 is added to a cylinder of H_2 at 33.2 atm to bring the total pressure to 34.5 atm, is the mixture explosive?   *Ans. Yes (P_{O_2} = 1.3 atm, 3.8% of the total pressure)*

S 41. A 5.73-L flask at 25°C contains 0.0388 mol of N_2, 0.147 mol of CO, and 0.0803 mol of H_2. What is the pressure in the flask in atmospheres, torr, and kilopascals?   *Ans. 1.14 atm; 863 torr; 115 kPa*

42. A mixture of 0.200 g of H_2, 1.00 g of N_2, and 0.820 g of Ar is stored in a closed container at

STP. What is the volume of the container assuming that the gases exhibit ideal behavior?   *Ans. 3.50 L*

S 43. A sample of carbon monoxide was collected over water at a total pressure of 764 torr and a temperature of 23°C. What is the pressure of the carbon monoxide? (See Table 10-1 for the vapor pressure of water.)   *Ans. 743 torr*

S 44. A sample of oxygen collected over water at a temperature of 29.0°C and a pressure of 764 torr has a volume of 0.560 L. What volume would the dry oxygen have under the same conditions of temperature and pressure?
Ans. 0.538 L

45. The volume of a sample of a gas collected over water at 32.0°C and 752 torr is 627 mL. What will the volume of the gas be when dried and measured at STP?   *Ans. 529 mL*

46. A 265-mL sample of gaseous nitric oxide, NO, was collected over mercury at 19°C and 714.0 torr. If the same sample of nitric oxide were collected over water at 19°C and 714.0 torr, what would its volume be?
Ans. 271 mL

47. Which is denser at the same temperature and pressure, dry air or air saturated with water vapor? Explain.   *Ans. Dry air*

48. (a) What is the concentration of the atmosphere in molecules per milliliter at STP?
Ans. 2.69×10^{19} molecules/mL
   (b) At a height of 150 km (about 94 mi) the atmospheric pressure is about 3.0×10^{-6} torr with a temperature of 420 K. What is the concentration of the atmosphere in molecules per milliliter at 150 km?
Ans. 6.9×10^{10} molecules/mL

Combining Volumes, Molecular Weights, Stoichiometry

49. What volume of O_2 at STP is required to oxidize 14.0 L of CO at STP to CO_2? What volume of CO_2 is produced at STP?   *Ans. 7.00 L O_2; 14.0 L CO_2*

50. Methanol (sometimes called wood alcohol), CH_3OH, is produced industrially by the following reaction.

$$CO(g) + 2H_2(g) \xrightarrow[\text{300°C, 300 atm}]{\text{Copper catalyst}} CH_3OH(g)$$

Assuming that the gases behave as ideal gases, what is the ratio of the total volume of the reactants to the final volume?   *Ans. 3 to 1*

S 51. Calculate the volume of oxygen required to burn 7.00 L of propane gas, C_3H_8, to produce carbon dioxide and water, if the volumes of both the propane and oxygen are measured under the same conditions.   *Ans. 35.0 L*

52. A 2.50-L sample of a colorless gas at STP decomposed to give 2.50 L of N_2 and 1.25 L of O_2 at STP. What is the colorless gas?
Ans. N_2O

53. An acetylene tank for an oxyacetylene welding torch will provide 9340 L of acetylene gas, C_2H_2, at STP. How many tanks of oxygen, each providing 7.00×10^3 L of O_2 at STP, will be required to burn the acetylene?

$$2C_2H_2 + 5O_2 \longrightarrow 4CO_2 + 2H_2O$$

Ans. 3.34 tanks

54. How could you show by experiment that the molecular formula of acetylene is C_2H_2 and not CH?

S 55. What is the molecular weight of methylamine if 0.157 g of methylamine gas occupies 125 mL with a pressure of 99.5 kPa at 22°C?
Ans. 31.0

56. In 1894 one of the noble gases of Group VIIIA was first isolated by allowing a large volume of air to react with magnesium so that only the noble gas, contaminated with traces of other noble gases, remained. A 32-mL sample of this gas weighs 0.054 g at 9°C and 748 torr. Determine the apparent molecular weight of this monatomic gas. (The traces of other gases are negligible.) Which noble gas is it?
Ans. 40; Ar

57. A sample of an oxide of nitrogen isolated from the exhaust of an automobile was found to weigh 0.571 g and occupied 1.00 L at 356 torr and 27°C. Calculate the molecular weight of this oxide, and determine if it was NO, NO_2, or N_2O_5.   *Ans. Mol. wt = 30.0; NO*

S 58. The density of a certain gaseous fluoride of phosphorus is 3.93 g/L at STP. Calculate the molecular weight of this fluoride, and determine its molecular formula.   *Ans. 88.1; PF_3*

S 59. Cyclopropane, a gas containing only carbon and hydrogen, is often used as an anaesthetic for major surgery. Taking into account that 0.45 L of cyclopropane at 120°C and 0.72 atm reacts with O_2 to give 1.35 L of CO_2 and 1.35 L of $H_2O(g)$ at the same temperature and pressure, determine the molecular formula of cyclopropane.   *Ans. C_3H_6*

60. Joseph Priestley first prepared pure oxygen by heating mercuric oxide, HgO.

$$2HgO(s) \xrightarrow{\triangle} 2Hg(l) + O_2(g)$$

What volume of O_2 at 15°C and 0.980 atm is produced by the decomposition of 2.36 g of HgO?   *Ans. 131 mL*

61. Cavendish prepared hydrogen in 1766 by the rather quaint method of passing steam through a red-hot gun barrel.

$$4H_2O(g) + 3Fe(s) \xrightarrow{\triangle} Fe_3O_4(s) + 4H_2(g)$$

What volume of H_2 at a pressure of 783 torr and a temperature of 21°C can be prepared from the reaction of 10.0 g of H_2O?   *Ans. 13.0 L*

62. Hydrogen gas will reduce Fe_3O_4 in the reverse of the reaction shown in Exercise 61.
    (a) What volume of H_2 at 195 atm and 35°C is required to reduce 1.00 metric ton (1000 kg = 1 metric ton) of Fe_3O_4 to Fe?   *Ans. 2.24×10^3 L*
    (b) If the reduction is run at 500°C and 1.0 atm, what volume of water vapor is produced?   *Ans. 1.1×10^6 L*

S 63. Gaseous hydrogen chloride, $HCl(g)$, is prepared commercially by the reaction of NaCl with H_2SO_4 (Section 8.7, Part 1). What mass of NaCl is required to prepare enough $HCl(g)$ to fill a 35.4-L cylinder to a pressure of 125 atm at 0°C?   *Ans. 11.5 kg*

64. If the oxygen consumed by a resting human male (Section 10.8, Example 10.11) is used to produce energy by the oxidation of glucose,

$$C_6H_{12}O_6 + 6O_2 \longrightarrow 6CO_2 + 6H_2O$$

what is the mass of glucose required per hour for a resting 70-kg male?   *Ans. 17 g*

65. Sulfur dioxide is an intermediate in the preparation of sulfuric acid (Section 8.7). What volume of SO_2 at 343°C and 1.21 atm is produced by the combustion of 1.00 kg of sulfur?   *Ans. 1.30×10^3 L*

66. What volume of oxygen at 423.0 K and a pressure of 127.4 kPa will be produced by the decomposition of 129.7 g of BaO_2 to BaO and O_2?   *Ans. 10.57 L*

S 67. In a common freshman laboratory experiment, $KClO_3$ is decomposed by heating to give KCl and O_2. What mass of $KClO_3$ must be decomposed to give 638 mL of O_2 at a temperature of 18°C and a pressure of 752 torr?   *Ans. 2.16 g*

68. In a laboratory determination a 0.1009-g sample of a compound containing boron and chlorine gave 0.3544 g of silver chloride upon reaction with silver nitrate. A 0.06237-g sample of the compound exerted a pressure of 6.52 kPa at 27°C in a volume of 147 mL. What is the molecular formula of the compound?   *Ans. B_2Cl_4*

S 69. As 1 g of the radioactive element radium decays over 1 year, it produces 1.16×10^{18} α particles (helium nuclei). Each α particle becomes an atom of helium gas. What volume of helium gas at a pressure of 56.0 kPa and a temperature of 25°C is produced?   *Ans. 8.52×10^{-2} mL*

Kinetic-Molecular Theory, Graham's Law

70. Using the postulates of the kinetic-molecular theory, explain why a gas will uniformly fill a container of any shape.

71. Show how Boyle's Law, Charles's Law, and Dalton's Law follow from the assumptions of the kinetic-molecular theory.

72. Describe what happens to the average kinetic energy of ideal gas molecules when the conditions are changed as follows:
    (a) The pressure of the gas is increased by decreasing the volume at constant temperature.
    (b) The pressure of the gas is increased by increasing the temperature at constant volume.
    (c) The average velocity of the molecules is increased by a factor of 2.

73. Can the speed of a given molecule in a gas double at constant temperature? Explain your answer.

74. What is the ratio of the kinetic energy of a helium atom to that of a hydrogen molecule in a mixture of the two gases? What is the ratio of the root-mean-square speeds, u_{rms}, of the two gases?

75. A 1-L sample of CO initially at STP is heated to 546°C, and its volume is increased to 2 L.
    (a) What effect do these changes have on the

number of collisions of the molecules of the gas per unit area of the container wall?

(b) What is the effect on the average kinetic energy of the molecules?

(c) What is the effect on the root-mean-square speed of the molecules?

S 76. The root-mean-square speed of H_2 molecules at 25°C is about 1.6 km/s. What is the root-mean-square speed of a CH_4 molecule at 25°C?

Ans. 0.57 km/s

77. Show that the ratio of the rate of diffusion of gas 1 to the rate of diffusion of gas 2, R_1/R_2, is the same at 0°C and 100°C.

78. Heavy water, D_2O (mol. wt = 20.03), can be separated from ordinary water, H_2O (mol. wt = 18.01), as a result of the difference in the relative rates of diffusion of the molecules in the gas phase. Calculate the relative rates of diffusion of H_2O and D_2O.

Ans. H_2O diffuses 1.055 times faster than D_2O

79. Show by calculation which of the following gases diffuse more slowly than oxygen does: F_2, Ne, N_2O, C_2H_2, NO, Cl_2, H_2S.

Ans. F_2, Cl_2, N_2O, H_2S

80. Calculate the relative rate of diffusion of HF compared to HCl and of O_2 compared to O_3.

Ans. 1.35; 1.22

S 81. A gas of unknown identity diffuses at the rate of 83.3 mL/s in a diffusion apparatus in which a second gas, whose molecular weight is 44.0, diffuses at the rate of 102 mL/s. Calculate the molecular weight of the first gas. *Ans. 66.0*

S 82. (a) When two cotton plugs, one moistened with ammonia and the other with hydrochloric acid, are simultaneously inserted into opposite ends of a glass tube 87.0 cm long, a white ring of NH_4Cl forms where gaseous NH_3 and gaseous HCl first come into contact.

$$NH_3(g) + HCl(g) \longrightarrow NH_4Cl(s)$$

At what distance from the ammonia-moistened plug does this occur?

Ans. 51.7 cm

(b) A student is trying to identify an amine that is one of the three possible compounds, CH_3NH_2, $(CH_3)_2NH$, or $(CH_3)_3N$. In this case, a white ring due to the amine hydrochloride forms at a distance of 41.2 cm from

the amine-moistened plug. Which is the correct formula for the amine?

Ans. $(CH_3)_2NH$

Nonideal Behavior of Gases

83. Graphs showing the behavior of several different gases are given below. Which of these gases exhibit behavior significantly different from that expected for ideal gases?

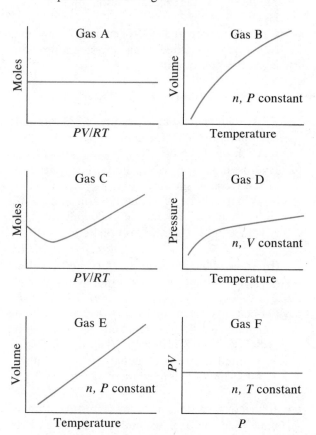

84. Describe the factors responsible for the deviation of the behavior of real gases from that of an ideal gas.

85. For which of the following sets of conditions will a real gas behave most like an ideal gas and for which conditions would a real gas be expected to deviate from ideal behavior? Explain.

(a) high pressure, small volume

(b) high temperature, low pressure

(c) low temperature, high pressure

86. For which of the following gases should the cor-

rection for the molecular volume be largest: CO, CO_2, H_2, He, NH_3, SF_6?

Additional Exercises

87. A commercial mercury vapor analyzer can detect concentrations of gaseous Hg atoms (which are poisonous) in air as low as 2×10^{-3} mg/m^3 of air (1 m^3 = 1000 L). What is the partial pressure of gaseous mercury if the atmospheric pressure is 755 torr at 21°C? *Ans.* 2×10^{-10} *torr*

[S] 88. (a) What is the total volume of the $CO_2(g)$ and $H_2O(g)$ at 600°C and 735 torr produced by the combustion of 1.00 L of $C_3H_8(g)$ measured at STP? *Ans. 23.1 L*

(b) What is the partial pressure of CO_2 in the product gases? *Ans. 315 torr*

89. Butene, an important petrochemical used in the production of synthetic rubber, is composed of carbon and hydrogen. A 0.0124-g sample of this compound produces 0.0390 g of CO_2 and 0.0159 g of H_2O when burned in an oxygen atmosphere. If the volume of a 0.125-g gaseous sample of the compound at 735 torr and 45°C is 60.2 mL, what is the molecular formula of the compound? *Ans.* C_4H_8

[S] 90. Ethanol, C_2H_5OH, is produced industrially from ethylene, C_2H_4, by the following sequence of reactions:

$$3C_2H_4 + 2H_2SO_4 \longrightarrow C_2H_5HSO_4 + (C_2H_5)_2SO_4$$

$$C_2H_5HSO_4 + (C_2H_5)_2SO_4 + 3H_2O \longrightarrow 3C_2H_5OH + 2H_2SO_4$$

What volume of ethylene at STP is required to produce 1.000 metric ton (1000 kg) of ethanol if the overall yield of ethanol is 90.1%? *Ans.* 5.40×10^5 *L*

91. Thin films of amorphous silicon for electronic applications are prepared by decomposing silane gas, SiH_4, on a hot surface at low pressures.

$$SiH_4(g) \xrightarrow{\Delta} Si(s) + 2H_2(g)$$

What volume of SiH_4 at 130 Pa and 800 K is required to produce a 10.0-cm-by-10.0-cm film that is 200 Å thick (1 Å = 10^{-8} cm)? The density of amorphous silicon is 1.9 g/cm^3. *Ans. 0.69 L*

92. One molecule of hemoglobin will combine with four molecules of oxygen. If 1.0 g of hemoglobin combines with 1.53 mL of oxygen at body temperature (37°C) and a pressure of 743 torr, what is the molecular weight of hemoglobin? *Ans. 68,000*

[S] 93. Ethanol, C_2H_5OH, is often produced by the fermentation of sugars. For example, the preparation of ethanol from the sugar glucose is represented by the equation

$$C_6H_{12}O_6(aq) \xrightarrow{\text{Yeast}} 2C_2H_5OH(aq) + 2CO_2(g)$$

What volume of CO_2 at STP is produced by the fermentation of 125 g of glucose if the reaction has a yield of 97.5%? *Ans. 30.3 L*

94. A sample prepared by pumping 10.0 L of air at STP and 1.0 L of CO at STP into a 2.0-L closed flask was heated until all the CO had been converted to CO_2. Assuming that air is 20% oxygen and 80% nitrogen, calculate the partial pressures of N_2, O_2, and CO_2 in the flask at 0°C after the reaction has taken place. *Ans.* N_2, *4.0 atm;* O_2, *0.75 atm;* CO_2, *0.50 atm*

[S] 95. One method of analysis of amino acids is the van Slyke method. The characteristic amino groups ($—NH_2$) in protein material are allowed to react with nitrous acid, HNO_2, to form N_2 gas. From the volume of the gas, the amount of amino acid can be determined. A 0.0604-g sample of a biological material containing glycine, $CH_2(NH_2)COOH$, was analyzed by the van Slyke method giving 3.70 mL of N_2 collected over water at a pressure of 735 torr and 29°C. What was the percentage of glycine in the sample?

$$CH_2(NH_2)COOH + HNO_2 \longrightarrow CH_2(OH)COOH + H_2O + N_2$$

Ans. 17.2%

96. Natural gas often contains hydrogen sulfide, H_2S, a gas that is itself a pollutant and that produces another pollutant, sulfur dioxide, upon combustion. Hydrogen sulfide is removed from raw natural gas by the reaction

$$HOC_2H_4NH_2 + H_2S \longrightarrow (HOC_2H_4NH_3)HS$$

How much ethanolamine, $HOC_2H_4NH_2$, is required to remove the H_2S from 1.00×10^3 ft^3

(1 ft^3 = 28.3 L) of natural gas at STP if the partial pressure of H_2S is 233 Pa?   *Ans. 177 g*

97. One step in the production of sulfuric acid (Section 8.7) involves oxidation of $SO_2(g)$ to $SO_3(g)$ using air at 400°C with vanadium (V) oxide as a catalyst. Assuming that air is 21% oxygen by volume, what volume of air at 19°C and 754 torr is required to produce enough SO_3 to give 1.00 metric ton (1000 kg = 1 metric ton) of sulfuric acid, H_2SO_4?   *Ans. 5.9 × 10^5 L*

S 98. A sample of a compound of xenon and fluorine was confined in a bulb with a pressure of 24 torr. Hydrogen was added to the bulb until the pressure was 96 torr. Passage of an electric spark through the mixture produced Xe and HF. After the HF was removed by reaction with solid KOH, the final pressure of xenon and unreacted hydrogen in the bulb was 48 torr. What is the empirical formula of the xenon fluoride in the original sample? *Note:* Xenon fluorides contain only one xenon atom per molecule.

Ans. XeF$_4$

99. A 0.250-L bottle contains 75.0 mL of a 3.0% by mass solution of hydrogen peroxide. How much will the pressure (in torr) in the bottle increase if the H_2O_2 decomposes to $H_2O(l)$ and $O_2(g)$ at 34°C? Assume that the density of the solution is 1.00 g/cm^3, that the solubility of the oxygen can be neglected, and that the volume of the liquid does not change during the decomposition.

Ans. 3.6 × 10^3 torr, or 4.8 atm

100. The pressure in a sample of hydrogen collected above water in a 425-mL bottle at 35°C is 763 torr. What is the volume of the sample when the temperature falls to 23°C, assuming its pressure does not change?   *Ans. 397 mL*

11

THE LIQUID AND
SOLID STATES

Liquids and solids are familiar forms of matter. We know that liquids flow and assume the shape of any container into which they are poured. We also know that solids are rigid. Most liquids will evaporate; most solids will not. These and other properties of liquids and solids can be explained in terms of the kinetic-molecular theory, which we used to describe the properties of gases. In this chapter we will examine several properties of liquids and solids and see how the kinetic-molecular theory can help us understand them. We will also examine the way metal atoms or ions in compounds can arrange themselves to give crystalline solids.

11.1 The Kinetic-Molecular Description of Liquids and Solids

In Section 10.15 we saw that the molecules of a substance in the gaseous state are in constant and very rapid motion and that the spaces between them are large compared to the sizes of the molecules themselves. As a consequence, the forces of attraction between the molecules are small. However, as the molecules of a gas are brought closer together by increased pressure, the average distance between them is decreased, and the attractions between molecules become stronger. Moreover, as a gas is cooled, the average speed of the molecules decreases, and they tend to pair up for short periods after they collide. If the pressure is sufficiently high and the temperature sufficiently low, the intermolecular attraction overcomes the tendency of the molecules to move apart, and the gas changes to a liquid. Even though molecules in the liquid state are close together or in contact with each other, they still retain a limited amount of motion and move about in the liquid. Since the molecules in a liquid move, the liquid can change its shape, take the shape of a container, diffuse, and evaporate.

The molecules in a liquid are held closely together by their mutual attractive forces. An increase in the pressure on a liquid can reduce the distance between the molecules only slightly, so the volume of a liquid decreases very little with increased pressure. Liquids are relatively incompressible, compared to gases. The molecules in a liquid are able to move past one another in a random fashion (diffusion), but, because of the much shorter distance that any molecule can move before it collides with another molecule, liquids diffuse much more slowly than do gases.

Cooling a liquid causes the average speed of the molecules to decrease. At some point the great majority of molecules no longer have sufficient kinetic energy to move past one another, and they take up comparatively fixed and, in most cases, ordered positions relative to each other. The liquid thus changes to a rigid solid. When the molecules of a liquid lose kinetic energy and assume ordered positions, the liquid is said to freeze—that is, to change from the liquid to the solid state. The only motion remaining is due principally to the vibration of the molecules. The resistance of the molecules in a solid to changes in position causes the rigidity and small compressibility of crystalline solids. However, the facts that diffusion takes place to a slight extent in the solid state and that crystalline compounds show some vapor pressure indicate that the molecules are not motionless. As the temperature of a solid is lowered, the motion of its molecules gradually decreases to a minimum at absolute zero ($-273\,°C$).

A comparison of the behavior of molecules of gases, liquids, and solids based on the kinetic-molecular theory is presented in Fig. 11-1.

Figure 11-1. The arrangement of molecules in a gas, a liquid, and a solid. The density of the gaseous molecules is exaggerated for the purpose of illustration.

Gas | Liquid | Solid

Widely separated, disordered molecules in continuous motion.

Closely spaced, disordered molecules in continuous motion.

Ordered molecules in contact and in relatively fixed positions.

PROPERTIES OF LIQUIDS

11.2 Evaporation of Liquids

Water in an open vessel disappears upon standing as **evaporation** occurs. Some liquids, such as ether, alcohol, and gasoline, evaporate more rapidly than water under the same conditions. Other liquids evaporate more slowly. Motor oil and

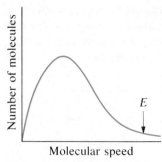

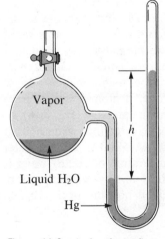

Figure 11-2. Distribution of the molecular speeds of molecules of a liquid. Molecules with speeds greater than E possess sufficient kinetic energy to escape through the surface of the liquid into the gas phase.

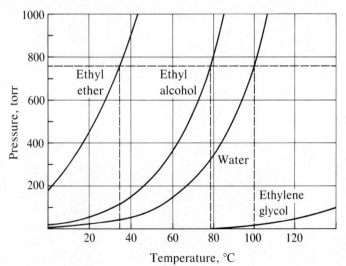

Figure 11-3. A closed vessel at 25°C containing only liquid water and water vapor. The vapor pressure is proportional to the difference in height, h, of the mercury in the two columns of the manometer.

Figure 11-4. The vapor pressures of four common substances at various temperatures. The intersection of a curve with the dashed line at 760 torr, referred to the temperature axis, indicates the normal boiling point of that substance. (The normal boiling point of ethyl ether is 34.6°C; of ethyl alcohol, 78.4°C; of water, 100°C; of ethylene glycol, 198°C, which is off the scale in this figure.)

ethylene glycol (antifreeze) evaporate so slowly that they seem not to evaporate at all.

Evaporation may be explained in terms of the motion of molecules. Just as in a gas, at any given temperature the molecules of a liquid move—some rapidly, many at intermediate rates, and some very slowly. A rapidly moving molecule near the surface of a liquid may possess sufficient kinetic energy to overcome the attraction of its neighbors and escape—that is, evaporate into the space above the liquid (Fig. 11-2), becoming a gas-phase molecule. Such gaseous molecules diffuse away, and more fast-moving molecules leave the liquid phase and appear in the gas phase above the liquid as the process of evaporation continues.

If a liquid is in a closed container, however, molecules that evaporate from it cannot escape from the container. They strike the walls and rebound, and some eventually strike the surface of the liquid and move back into it. The change from vapor to liquid is called **condensation.** As evaporation proceeds, the number of molecules in the vapor state increases, and in turn the rate of condensation increases. When the rate of condensation equals the rate of evaporation, the vapor in the container is in **equilibrium** with the liquid. At equilibrium neither the amount of the liquid nor the amount of the vapor in the container changes, although some molecules from the vapor condense at the same time that an equal number of molecules from the liquid evaporate. When both processes operate and counterbalance each other, the equilibrium is a **dynamic equilibrium.**

$$\text{Liquid} \xrightleftharpoons[\text{Condensation}]{\text{Evaporation}} \text{Vapor}$$

At equilibrium the space above a liquid is saturated with molecules of its vapor. The pressure exerted by a vapor in equilibrium with its liquid at a given temperature is called the **vapor pressure** of the liquid. Neither the area of the surface of the liquid in contact with the vapor nor the size of the vessel has any effect on the vapor pressure. Vapor pressures are commonly measured by introducing a liquid into a closed container and measuring the increase in pressure due to the vapor in equilibrium with the liquid by means of a manometer (Section 10.2 and Fig. 11-3).

Figure 11-4 depicts the vapor pressures of four liquids with quite different vapor pressures. Such differences are related to the intermolecular attractive forces in the liquids, which are least for ether and greatest for ethylene glycol, in this case. Thus the vapor pressure of a liquid is dependent on the particular kind of molecule that composes it. The two liquids that evaporate more rapidly, ethyl ether and ethyl alcohol, have the smaller intermolecular forces, as is evident from their higher vapor pressures. The very low vapor pressure of ethylene glycol (too low even to be shown on the graph at room temperature—about 23°C) is a reflection of the strong forces between its molecules.

11.3 Boiling Points of Liquids

The data in Fig. 11-4 show that the vapor pressure of each liquid increases as the temperature is raised. This is due to the increase in the rate of molecular motion that accompanies an increase in temperature and is observed for all liquids. As shown in Fig. 11-5, at a higher temperature more molecules move rapidly enough to escape from a liquid. The escape of more molecules from the surface of the liquid per unit of time and a greater speed for each molecule that escapes contribute to a higher vapor pressure.

A liquid boils when bubbles of vapor form within it and then rise to the surface where they burst and release the vapor. A liquid will boil when its vapor pressure equals the pressure of the gas above it. The **normal boiling point** of a liquid is the temperature at which its vapor pressure equals 1 atmosphere (760 torr; see Fig. 11-4). A liquid will boil at temperatures higher than its normal boiling point when the external pressure is greater than 1 atmosphere; alternatively, the boiling point may be lowered by decreasing the pressure (Fig. 11-6). At high altitudes, where the atmospheric pressure is less than 760 torr, water boils at temperatures below its normal boiling point of 100°C. Food in boiling water cooks more slowly at high altitudes because the temperature of boiling water is lower there than nearer sea level. The temperature of boiling water in a pressure cooker is higher than normal because of the increased pressure in the cooker. Thus foods cook faster there than in open vessels.

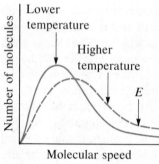

Figure 11-5. Distribution of molecular speeds in a liquid at two different temperatures. At the higher temperature, more molecules have the necessary speed, E, to escape from the liquid into the gas phase.

Figure 11-6. A liquid such as ethyl alcohol that (a) will not boil at room temperature under 1 atmosphere of pressure (b) may boil at that same temperature when the pressure is reduced by placing the liquid in a bell jar and pumping out the air.

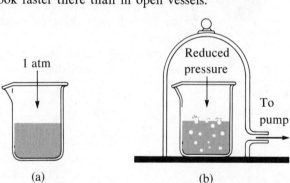

EXAMPLE 11.1   From the graph in Fig. 11-4, determine the boiling point of water contained in a bell jar such as that shown in Fig. 11-6(b) if the pressure is 400 torr.

The graph of the vapor pressure of water versus temperature indicates that the vapor pressure of water is 400 torr at about 83°C. Thus at about 83°C the vapor pressure of the water will equal that of the atmosphere in the bell jar, and the water will boil.

11.4 Heat of Vaporization

Evaporation occurs as molecules with high kinetic energies (faster-moving molecules) escape from the surface of a liquid (Section 11.2). The loss of such molecules leaves behind molecules with lower average kinetic energies and, consequently, a lower temperature ($KE_{avg} \propto T$; Section 10.17). The cooling effect due to the evaporation of water is very evident to a swimmer emerging from the water. In this case some of the water on the swimmer's skin evaporates, leaving behind cooler water, which causes the skin to feel cold.

For a liquid to evaporate at a constant temperature, heat must be supplied in sufficient quantity to offset the cooling effect brought about by the escape of the molecules possessing high kinetic energies. The energy that must be supplied to evaporate a given quantity of liquid at a constant temperature is known as the **heat of vaporization.** The heat of vaporization of water at 25°C, for example, is 2443 joules per gram, or 44.01 kilojoules per mole. That of ammonia is 1368 joules per gram, or 23.31 kilojoules per mole, at its boiling point, −33°C. At higher temperatures less heat is required per mole of evaporated liquid. At its boiling point (100°C) the heat of vaporization of water is 40.67 kilojoules per mole.

$$H_2O(l) \longrightarrow H_2O(g) \qquad \Delta H \text{ (at 25°C)} = 44.01 \text{ kJ/mol}$$

EXAMPLE 11.2   One mechanism for the removal of excess heat generated by the body's metabolic processes is by evaporation of the water in sweat. In a hot, dry climate a person may lose as much as 1.5 L of water (1500 g) per day through evaporation of sweat. Although sweat is not pure water, we can get an approximate value of the amount of heat removed by assuming that it is. How much heat is required to evaporate 1.5 L of water? Assume that the temperature is 46°C (115°F); $\Delta H = 43.02 \text{ kJ/mol}$ at 46°C.

To evaporate 1 mol of water at 46°C requires 43.02 kJ, or 10.28 kcal. So, we must determine the number of moles of water in 1500 g and then determine the amount of heat necessary to vaporize that much water.

$$1500 \text{ g } H_2O \times \frac{1 \text{ mol } H_2O}{18 \text{ g } H_2O} = 83.3 \text{ mol } H_2O$$

$$83.3 \text{ mol } H_2O \times \frac{43.02 \text{ kJ}}{1 \text{ mol } H_2O} = 3600 \text{ kJ (to two significant figures)}$$

Thus 3600 kJ (or 860 kcal) of heat are removed by the evaporation of 1.5 L of water.

Note that 1 kilocalorie of heat in scientific terms is equal to 1 Calorie in nutritional terms (1000 cal of heat = 1 nutritional Cal). The 810 kcal lost by evaporation corresponds to a nutritional intake of 810 Cal, about one-third of the daily dietary requirement.

The quantity of heat evolved as a liquid condenses equals that absorbed as it evaporates

$$H_2O(g) \longrightarrow H_2O(l) \qquad \Delta H \text{ (at 100°C)} = -40.67 \text{ kJ/mol}$$

$$NH_3(g) \longrightarrow NH_3(l) \qquad \Delta H \text{ (at −33°C)} = -23.31 \text{ kJ/mol}$$

A refrigerator cools by evaporating a liquid refrigerant. Heat leaves the refrigerator when the circulating refrigerant (usually CCl_2F_2, one of the Freons) absorbs the energy needed to change it to a gas. In the gaseous state the refrigerant is then circulated through a compressor outside the refrigerated compartment and reliquefied by combined cooling and compression. To be an effective refrigerant a substance must be readily convertible from the gaseous to the liquid state at the refrigerator's temperature and must have a high heat of vaporization.

11.5   Critical Temperature and Pressure

If a sample of water is sealed in an evacuated tube, at 25°C the tube will contain liquid water and water vapor with a pressure of 23.8 torr (Table 10-1). At this temperature there is a clear boundary between the two phases [Fig. 11-7(a)]. As the temperature is increased, the pressure of the vapor increases. At 100°C, for example, the pressure is 760 torr, but liquid water and water vapor are both present and the boundary between them is obvious [Fig. 11-7(b)]. However, at 374°C the boundary disappears [Fig. 11-7(c)], and all of the water in the tube is physically identical. No distinction can be made between liquid and vapor. The temperature above which it is no longer possible to distinguish a liquid from its vapor is called the **critical temperature**. Above this temperature a gas cannot be liquefied, *no matter how much pressure is applied*. The pressure required to liquefy a gas at its critical temperature is called the **critical pressure**. The critical pressure of water is 217.7 atmospheres (165,500 torr). The critical temperatures and critical pressures of some common substances are given in Table 11-1.

Figure 11-7. A sample of liquid water and water vapor (a) at 25°C, (b) at 100°C, and (c) at 374°C. At the critical temperature, 374°C, the boundary between liquid and vapor disappears.

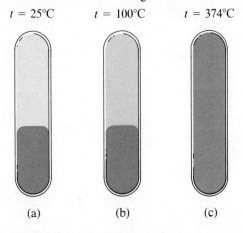

| | $t = 25°C$ | $t = 100°C$ | $t = 374°C$ |
| | (a) | (b) | (c) |

Table 11-1   Critical Temperatures and Pressures of Some Common Substances

	Critical Temperature, K	Critical Pressure, atm
Hydrogen	33.24	12.8
Nitrogen	126.0	33.5
Oxygen	154.3	49.7
Carbon dioxide	304.2	73.0
Ammonia	405.5	111.5
Sulfur dioxide	430.3	77.7
Water	647.1	217.7

EXAMPLE 11.3   If a carbon dioxide fire extinguisher is shaken on a cool day (temperature = 18°C), liquid CO_2 can be heard sloshing around in the cylinder. However, the same cylinder appears to contain no liquid on a hot summer day (temperature = 35°C). Explain these observations.

On the cool day the temperature of the CO_2 is below its critical temperature, which is 304 K, or 31°C (Table 11-1), so liquid CO_2 is present in the cylinder. On the hot day the temperature is greater than the critical temperature, so no liquid CO_2 can form.

Above the critical temperature of a substance, the average kinetic energy of its molecules is sufficient to overcome their mutually attractive forces. Therefore they will not cling together enough to form a liquid, no matter how great the pressure. If the temperature is decreased, the average kinetic energy of the molecules decreases. At the critical temperature the intermolecular forces are just large enough, relative to the average kinetic energy, to liquefy the gas, provided that the pressure is equal to or greater than the critical pressure. The pressure helps the intermolecular forces bring the molecules close enough to one another to liquefy. Below the critical temperature the pressure required for liquefaction decreases with decreasing temperature until it reaches 1 atmosphere at the normal boiling temperature. Substances with strong intermolecular attractions, like water and ammonia, have high critical temperatures; substances with weak intermolecular attractions, like hydrogen and nitrogen, have low critical temperatures.

11.6 Distillation

Dissolved materials in a liquid may make it unsuitable for a particular purpose. For example, water containing dissolved minerals should not be used in batteries because it shortens their life. Water and other liquids may be purified by the process of **distillation.** When impure water is heated in a distilling flask (Fig. 11-8), it vaporizes and passes into the condenser. The vapor is condensed to the

Figure 11-8. Laboratory distillation apparatus. When impure water is distilled with this apparatus, nonvolatile substances remain in the distilling flask. The water is vaporized, condensed, and finally collected as distillate in the receiving flask.

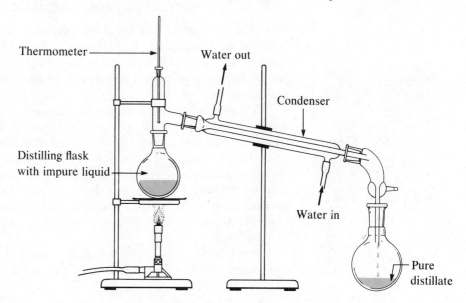

liquid in the water-cooled condenser, and the liquid flows into the receiving vessel. Dissolved mineral matter, such as calcium sulfate or magnesium chloride, is not volatile at the boiling point of water and stays in the flask. Distillation takes advantage of the properties that the addition of heat to a liquid speeds up its rate of evaporation (an endothermic change) and that cooling a vapor favors its condensation (an exothermic change). The distillation of two or more volatile substances from a mixture is described more fully in Section 13.21.

11.7   Surface Tension

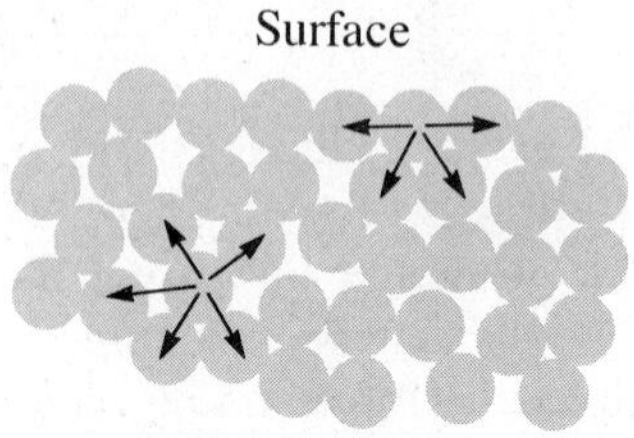

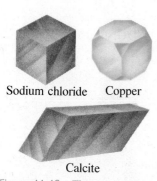

Figure 11-9. Attractive forces experienced by a molecule at the surface and by a molecule in the bulk of a liquid.

The molecules within a liquid are attracted equally in all directions by neighboring molecules; the resultant force on any one molecule within the liquid is therefore zero. However, the molecules on the surface of a liquid are attracted only into the liquid and to either side (Fig. 11-9). This unbalanced molecular attraction tends to pull the surface molecules back into the liquid, and a condition of equilibrium is reached only when as many molecules as possible have been pulled into the liquid. At this point the minimum number of molecules possible are on the surface, and the surface area is reduced to a minimum. A small drop of liquid tends to assume a spherical shape because in a sphere the ratio of surface area to volume is at a minimum. **Surface tension** is defined as the force that causes the surface of a liquid to contract. The surface of a liquid acts as if it were a stretched membrane. A steel needle carefully placed on water will float. Some insects, even though they are heavier than water, move on its surface because they are supported by the surface tension. One of the forces causing water to rise in capillary tubes (tubes with a very small bore) is the surface tension. Water is brought from the soil up through the roots and into the portion of a plant above the soil by this capillary action.

PROPERTIES OF SOLIDS

11.8   Crystalline and Amorphous Solids

When the temperature of a liquid becomes sufficiently low or the pressure on the liquid becomes sufficiently high, the molecules of the liquid can no longer move past one another and a solid is formed. Solids are rigid. They cannot be compressed like gases or poured like liquids because their molecules cannot change position easily.

Most solid substances are crystalline in nature. Although some solids, such as diamonds, sugar, and table salt, are composed of single crystals, most common crystalline solids are aggregates of many small crystals. Some examples are sandstone, chunks of ice, and metal objects. **A crystal is a homogeneous solid in which the atoms, ions, or molecules are arranged in a definite repeating pattern.** Crystals are often bounded by plane surfaces. The angles at which the surfaces of a crystal intersect are always the same for, and are characteristic of, a given substance. Figure 11-10 illustrates single crystals of sodium chloride, copper metal, and calcite ($CaCO_3$).

Many liquids can be cooled well below their freezing points before crystallization begins. This is **supercooling,** and a liquid below its freezing point is said to be

Sodium chloride   Copper

Calcite

Figure 11-10. The appearance of single crystals of sodium chloride, copper metal, and calcite ($CaCO_3$).

supercooled. A supercooled liquid is not stable; it may crystallize spontaneously. Crystallization can often be induced by introduction of a seed crystal of the substance, which provides an ordered structure to which molecules can become attached.

Certain liquid materials such as fused glass, melted butter, or molten asphalt, which contain large cumbersome particles that cannot move readily into the positions of a regular crystal lattice, often show great tendencies to supercool. As the temperature is lowered, the large and irregular structural units composing the material become less and less mobile, until the supercooled liquid finally becomes rigid. Such a material is an **amorphous solid,** or a **glass.** True solids have a crystalline structure, with a definite, ordered arrangement of their internal building units; they have sharp melting points and resist change of shape under pressure. Amorphous solids, on the other hand, lack an ordered internal structure (Fig. 11-11). They do not melt at a definite temperature, but gradually soften and become less viscous when heated. They are really supercooled liquids; the term *amorphous solid* is often used but is not strictly correct.

Figure 11-11. (a) A two-dimensional illustration of the ordered arrangement of atoms in a crystal of boric oxide. (b) An illustration showing the disorder in amorphous boric oxide.

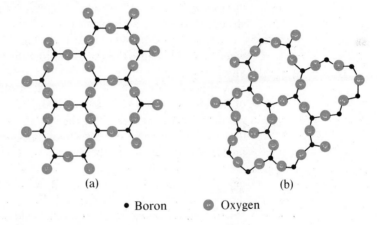

(a)   (b)

• Boron   ⬤ Oxygen

11.9 Crystal Defects

Crystals are never actually perfect; several types of **crystal defects** occur. It is common for some positions that should contain atoms or ions to be vacant [Fig. 11-12(a)]. Less commonly, some atoms or ions in a crystal may be located in positions called **interstitial sites,** which are located between the regular positions [Fig. 11-12(b)]. Certain distortions occur in some impure crystals. An example is when the cations, anions, or molecules of the impurity are too large to fit into the regular positions without distorting the structure. Minute amounts of impurities are sometimes introduced to cause imperfections in a crystalline structure in order to change the electrical conductivity (Section 27.20) or other physical properties. This has practical applications in the manufacture of semiconductors and printed circuits.

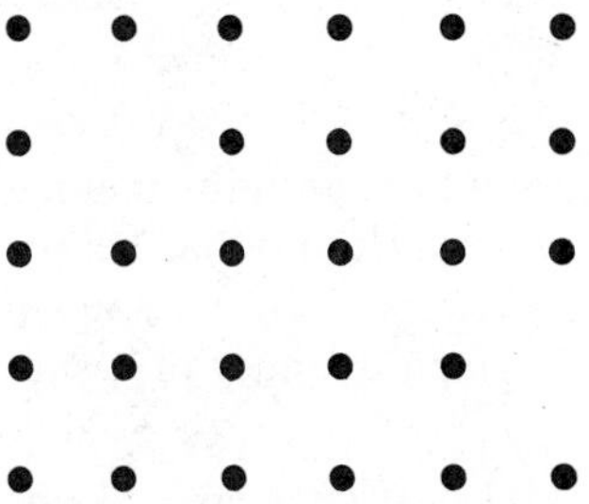

Lattice vacancies
(a)

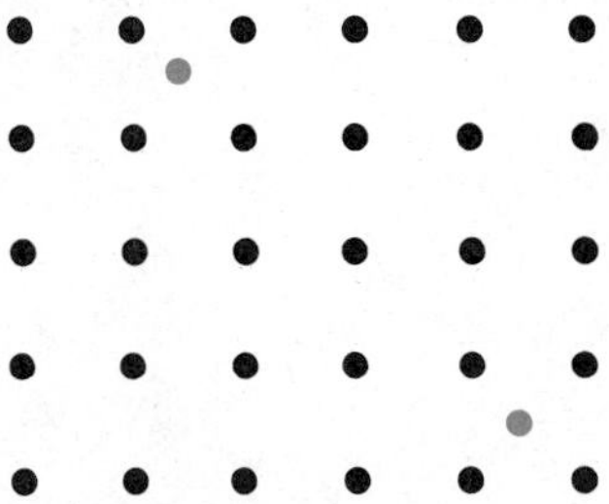

Interstitial atoms
(b)

Figure 11-12. Representations of two types of crystal defects.

11.10 Melting of Solids

When a crystalline solid is heated, at some temperature the kinetic energy of some of the molecules becomes great enough to overcome the intermolecular forces holding them in their fixed positions, and the solid begins to melt. If heating is continued, all of the solid will melt even though the temperature does not rise. If, however, the heating is stopped and no heat is allowed to escape, both solid and liquid phases are present, in equilibrium. The changes continue, but the rate of melting is just equal to the rate of freezing, and the quantities of solid and liquid remain constant. The temperature at which the solid and liquid phases of a given substance are in equilibrium is the **melting point** of the solid, and also the **freezing point** of the liquid.

The melting point of a crystalline solid is determined by the strength of the attractive forces between the units present in the crystal. Small symmetrical molecules, such as H_2, N_2, O_2, and F_2, have weak attractive forces (Section 11.13) and form crystals with low melting points. Crystals consisting of larger, nonpolar molecules have stronger attractive forces and melt at higher temperatures. Crystalline solids built up of unsymmetrical polar molecules melt at still higher temperatures; two examples are ice and sugar. Diamond is an atomic crystal in which carbon atoms are held together by strong covalent bonds (Fig. 25-1); the melting point of diamond is very high. The atoms in the crystals of most metals are joined by strong, modified covalent bonds, and most metals have high melting points. The electrostatic forces of attraction between the ions in ionic solids can be quite strong; thus many ionic crystals also have high melting points.

Crystalline solids such as aluminum, sodium chloride, or iodine melt abruptly when heated because the forces holding the atoms, ions, or molecules together are all of the same strength, and thus the units all break apart at once. The gradual softening of glasses, on the other hand, results from the structural nonequivalence of the atoms. When a glass is heated, the weakest bonds break first. As the temperature is further increased, the stronger bonds are broken. There is a gradual decrease in the size and a corresponding increase in the mobility of the structural units as the temperature is raised.

11.11   Heat of Fusion

When heat is applied to a crystalline solid, its temperature rises until the melting point is reached. The temperature then remains constant as additional heat is applied, until all of the solid has melted. After that the temperature rises again (Fig. 11-13). The quantity of heat needed to change a given quantity of a sub-

Figure 11-13. Heating curve for water. When ice is heated, the temperature increases until the melting point is reached. The temperature then remains constant until all of the ice is melted, then it increases again. When the water begins to boil, the temperature again remains constant until all of the water has vaporized.

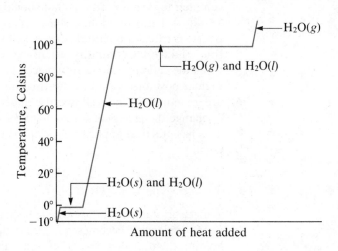

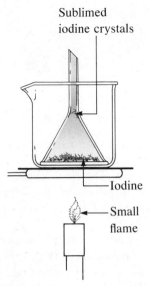

Sublimed iodine crystals

Iodine

Small flame

Figure 11-14. A simple method of demonstrating the sublimation of iodine.

stance from the solid state to the liquid state at constant temperature is known as the **heat of fusion** of the substance. The heat of fusion of ice is 333.6 joules per gram, or 6.01 kilojoules per mole, at 0°C.

$$H_2O(s) \longrightarrow H_2O(l) \qquad \Delta H = 6.01 \text{ kJ/mol}$$

The quantity of heat liberated during crystallization (freezing) equals that absorbed during fusion.

$$H_2O(l) \longrightarrow H_2O(s) \qquad \Delta H = -6.01 \text{ kJ/mol}$$

11.12 Vapor Pressure of Solids

Snow and ice evaporate at temperatures below their melting point. Moth balls have a characteristic odor. When iodine is heated in a container to just below its melting point, the solid evaporates rapidly, and the beautiful purple vapor condenses to crystals in a cooler part of the container (Fig. 11-14). These observations provide evidence that molecules of some solids pass directly from the solid into the vapor state.

The pressure of a vapor in equilibrium with its solid in a closed container is called the **vapor pressure** of the solid. Those solids in which the intermolecular forces are weak exhibit measurable vapor pressures at room temperature. As might be expected, the vapor pressure of a solid increases with an increase in temperature. The vapor pressure of solid iodine is 0.2 torr at 20°C and 90 torr at 114°C, its melting point. The combined process by which a solid passes directly into the vapor state without melting and recondenses into the solid state is called **sublimation.** Many substances, such as iodine, may be purified by sublimation if the impurities have low vapor pressures.

ATTRACTIVE FORCES BETWEEN MOLECULES

11.13 Intermolecular Forces

We saw in Section 11.1 that the liquefaction of a gas or the freezing of a liquid is due to forces of attraction between its constituent molecules. The strengths of these intermolecular attractive forces vary widely, although the *inter*molecular forces between small molecules are weak compared to the *intra*molecular forces that bond the atoms together within molecules. For example, to overcome the intermolecular forces in a mole of liquid HCl and convert it to gaseous HCl requires only about 17 kilojoules. However, to break all the bonds between the hydrogen and chlorine atoms in a mole of hydrogen chloride and to separate the hydrogen atoms from the chlorine atoms requires about 430 kilojoules.

Intermolecular attractive forces, collectively called **van der Waals forces,** are electrical in nature and result from the attraction of charges of opposite sign. One type of van der Waals force results from the electrostatic attraction of the positive end of one polar molecule for the negative end of another. Molecules whose centers of positive and negative electrical charge do not coincide are polar (Section 5.8). Having a positive and a negative end, they possess a dipole moment.

Figure 11-15. Two arrangements of polar molecules that allow an attractive interaction between the dipoles. These arrangements place the negative end of one molecule close to the positive end of another.

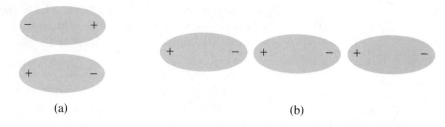

(a)   (b)

The CO molecule, because oxygen is more electronegative than carbon, has a partial negative charge on the oxygen atom and a partial positive charge on the carbon atom. Thus the CO molecule has a dipole moment. The electrostatic attraction of the positive end of one CO molecule for the negative end of another constitutes an attractive force, a **dipole-dipole attraction** (Fig. 11-15), between CO molecules.

Another component of the van der Waals forces, found with both nonpolar and polar molecules, is called the **dispersion force**, or **London force** (after Fritz London, who in 1930 first explained it). Because of the constant motion of its electrons, an atom that on average is nonpolar sometimes becomes polar, and while it is, it has a temporary dipole. A second atom, in turn, is distorted by the appearance of the dipole in the first atom. Its nucleus is attracted toward the negative end of the first atom (Fig. 11-16). Thus a negative end on the second atom is formed away from the negative end on the first atom, and the two rapidly fluctuating, temporary dipoles attract each other. If the two atoms are located in different molecules, the two molecules may attract each other. London forces between two atoms are weak compared to dipole-dipole attractions and are significant only when the atoms are very close together. Atoms must almost touch for London forces to be significant. These forces cause nonpolar substances such as the noble gases and the halogens to condense into liquids and to freeze into solids when the temperature is lowered sufficiently.

For molecules with similar shapes, the magnitude of London forces increases with the size of the molecule. For example, about 8 kilojoules will convert a mole of liquid CH_4 (mol. wt. = 16) to a gas, but about 13 kilojoules are needed to

Figure 11-16. A representation of the formation of the temporary dipoles that give rise to London forces. The temporary dipole on the atom on the right in (a) produces a dipole in the atom on the left. The attraction of the two resulting dipoles (b) results in the attractive force.

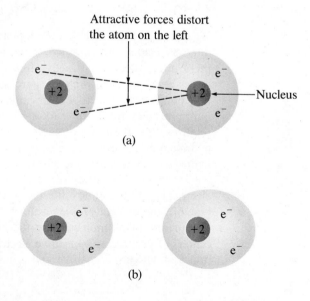

convert a mole of liquid SiH_4 (mol. wt. = 32) to a gas. This suggests that the larger and heavier SiH_4 molecule has stronger London forces. The size dependence is readily explained by London's theory. The valence electrons in a larger molecule are, on the average, farther from the nucleus than those in a smaller molecule. Thus they are less tightly held and can more easily form the temporary dipoles that produce the attraction. The increase in intermolecular attraction with increasing size is reflected in the rise in boiling point in the series of related substances shown in Table 11-2.

Table 11-2   Molecular Weights and Boiling Points

Substance	He	Ne	Ar	Kr	Xe	Rn
Molecular weight	4.0	20.18	39.95	83.8	131.3	222
Boiling point, °C	−268.9	−245.9	−185.7	−152.9	−107.1	−61.8

Substance	H_2	F_2	Cl_2	Br_2	I_2
Molecular weight	2.016	38.0	70.91	159.8	253.8
Boiling point, °C	−252.7	−187	−34.6	58.78	184.4

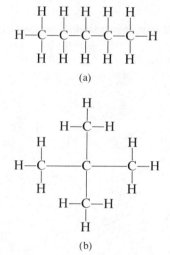

(a)

(b)

Figure 11-17. The Lewis structures of (a) *n*-pentane and (b) neopentane.

The shapes of molecules also affect the magnitudes of the London forces between them. For example, two compounds with the same chemical formula, C_5H_{12}, boil at different temperatures. A sample of *n*-pentane [Fig. 11-17(a)] boils at 36°C, while one of neopentane [Fig. 11-17(b)] boils at 9.5°C. Boiling points increase as the forces of attraction between molecules increase. Thus the London forces in *n*-pentane are larger than those in neopentane. The neopentane molecule is shaped so that the valence electrons in the carbon-carbon bonds are well inside the molecule. *n*-Pentane is shaped so that the valence electrons in the carbon-carbon bonds are closer to the surface of the molecule. Thus any London forces that involve electrons in carbon-carbon bonds must act over a longer distance with neopentane, and this results in weaker total attractive forces.

In general, liquids composed of discrete molecules with no permanent dipole moments have low boiling points relative to their molecular weights, because only the weak London forces must be overcome during vaporization. In such molecules the centers of positive and negative electrical charge coincide. Examples are the molecules of the noble gases and the halogens and other symmetrical molecules such as CH_4, SiH_4, CF_4, SiF_4, SF_6, and UF_6. Substances such as H_2O, HF, and C_2H_5OH, whose molecules have permanent dipole moments, have rather high boiling points relative to their molecular weights.

11.14 Hydrogen Bonding

Nitrosyl fluoride (ONF, mol. wt. = 49) has a boiling point of −56°C. Water (H_2O, mol. wt. = 18) has a much higher boiling point of 100°C, even though it has a lower molecular weight. The difference between the boiling points of these two compounds cannot be attributed to London forces; both molecules have about the same shape, and ONF is the heavier and larger molecule. It cannot be attributed to differences in the dipole moments; both molecules have the same dipole moment. The large difference in the boiling points of these two molecules is due to a special kind of dipole-dipole attraction called **hydrogen bonding.**

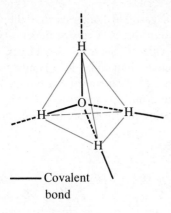

— Covalent bond

----- Hydrogen bond

Figure 11-18. A tetrahedral arrangement of two covalent bonds (solid lines) and two hydrogen bonds (broken lines) about an oxygen atom in water. Two hydrogen atoms participate in covalent bonds and two participate in hydrogen bonds to the oxygen. Each bond from a hydrogen atom and away from the tetrahedron leads to an oxygen of another water molecule.

Figure 11-19. Illustrations of (a) the boiling points and (b) the heats of vaporization of several binary hydrides of Groups IVA-VIIA. Each line connects compounds of the elements in a given group. H_2O, HF, and NH_3 have hydrogen bonding and exhibit abnormally high boiling points and heats of vaporization. CH_4 does not have hydrogen bonding. Plots for four of the noble gases are included to indicate typical behavior due to increasing London forces resulting from increasing mass.

Hydrogen bonds can form when hydrogen is bonded to one of the more electronegative elements, such as fluorine, oxygen, nitrogen, or chlorine. The large difference between the electronegativity of hydrogen (2.1) and that of the second element (4.1 for fluorine, 3.5 for oxygen, 3.0 for nitrogen, and 2.8 for chlorine) leads to a highly polar covalent bond in which the hydrogen atom bears a large fractional positive charge and the second atom bears a large fractional negative charge. The electrostatic attraction between the partially positive hydrogen atom in one molecule and the partially negative atom of the more electronegative element in another molecule gives rise to the strong dipole-dipole attraction called a **hydrogen bond.** Examples include the following: $HF\cdots HF$, $H_2O\cdots HOH$, $H_3N\cdots HNH_2$, $H_3N\cdots HOH$, and $H_2O\cdots HOCH_3$. Figure 11-18 illustrates the hydrogen bonding about a water molecule in liquid water.

Hydrogen bonds are stronger than other dipole-dipole attractions and London forces. The strengths are about 5–10% of those of ordinary covalent bonds. Figure 11-19(a) is a plot of boiling points, and Fig. 11-19(b), a plot of heats of vaporization; both are plotted against the number of the period of the Periodic Table, for the hydrides of elements in Groups IVA, VA, VIA, and VIIA and for several noble gases. In general, boiling points and heats of vaporization decrease with decreasing molecular or atomic weight, with the exceptions of H_2O, HF, and NH_3, compounds that have appreciable hydrogen bonding. Energy is required to break these hydrogen bonds in order to bring the material to the gaseous (vapor) state. Hence those substances forming hydrogen bonds have abnormally high boiling points and heats of vaporization. Methane, CH_4, however, does not have hydrogen bonding and hence has a lower boiling point and heat of vaporization than those of the heavier Group IVA hydrides. Similarly, the lightest noble gas shown, Ne, has lower values than do the heavier ones.

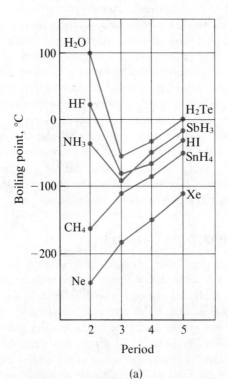

(a)

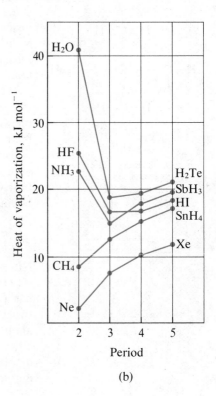

(b)

THE STRUCTURE OF CRYSTALLINE SOLIDS

Layer A

(a)

Layer A

(b)   Layer B

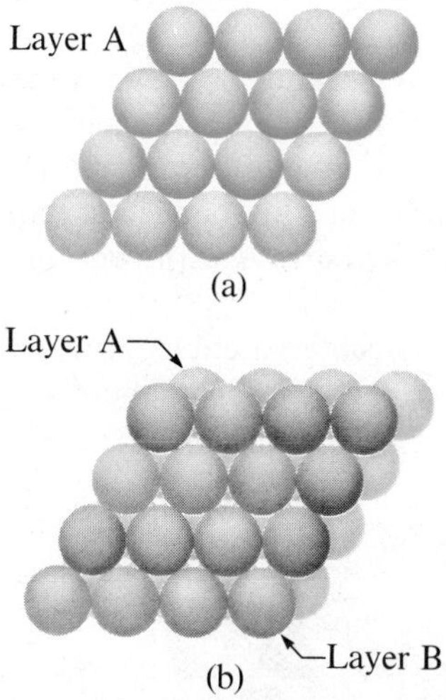

Figure 11-20. (a) A portion of a layer of closest packed spheres in the same plane. Each sphere touches 6 others. (b) Spheres in two closest packed layers. Each sphere in layer B touches 3 spheres in layer A (the colored layer). In an actual crystal many more than two planes would exist. Each sphere contacts 6 spheres in its own layer, 3 spheres in the layer below, and 3 spheres in the layer above, making a total contact for each sphere of 12 other spheres.

11.15 The Structures of Metals

A pure metal is a crystalline solid in which the metal atoms are packed closely together in a repeating array, or pattern. In most metals the atoms pack together as if they were spheres. If spheres of equal size are packed together as closely as possible in a plane, they arrange themselves as shown in Fig. 11-20(a), with each sphere in contact with six others. This arrangement, called **closest packing,** can extend indefinitely in a single layer. Crystals of many metals can be described as stacks of such closest packed layers.

Two types of stacking of closest packed layers are observed in simple metallic crystalline structures. In both types a second layer (B) is placed on the first layer (A) so that each sphere in the second layer is in contact with three spheres in the first layer, as shown in Fig. 11-20(b). A third layer can be positioned in one of two ways.

In one positioning each sphere in the third layer lies directly above a sphere in the first layer [Fig. 11-21(a)]. The third layer is also type A. The stacking continues with alternate type B and type A close packed layers (ABABAB···). This arrangement is called **hexagonal closest packing.** Metals that crystallize this way have a **hexagonal closest packed structure.** Examples include Be, Cd, Co, Li, Mg, Na, and Zn. (Those elements or compounds that crystallize with the same structure have **isomorphous structures.**)

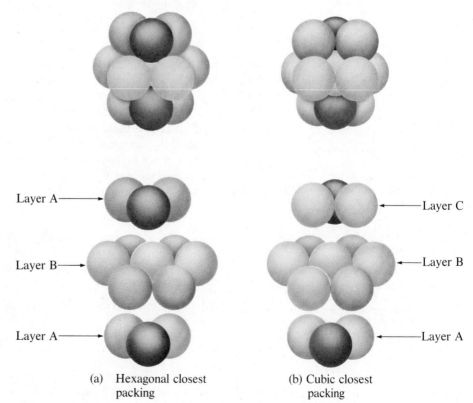

Figure 11-21. A portion of two types of crystal structures in which spheres are packed as compactly as possible. The lower diagrams show the structures expanded for clarification. Note that the first and third layers have identical orientations in (a). The first and third layers have different orientations in (b). In both structures, each sphere is surrounded by 12 others in an infinite extension of the structure and is said to have a coordination number of 12.

Layer A

Layer B

Layer A

(a) Hexagonal closest packing

Layer C

Layer B

Layer A

(b) Cubic closest packing

In the second positioning the third layer is located so that its spheres are not directly above those in either layer A or layer B [Fig. 11-21(b)]. This layer is type C. The stacking continues with alternating layers of type A, type B, and type C (ABCABC···), an arrangement called **cubic closest packing.** Metals crystallizing this way have a **cubic closest packed,** or **face-centered cubic, structure.** Examples include Ag, Al, Ca, Cu, Ni, Pb, and Pt.

In crystals of metals with either hexagonal closest packing or cubic closest packing, each atom touches 12 equidistant neighbors, 6 in its own plane and 3 in each adjacent plane. This gives each atom a coordination number of 12. The **coordination number** of an atom or ion is the number of neighbors nearest to it. About two-thirds of all metals crystallize in closest packed arrays with coordination numbers of 12.

Most of the remaining metals crystallize in a **body-centered cubic structure,** which contains planes of spheres that are not closest packed. Each sphere in a plane is surrounded by four neighbors [Fig. 11-22(a)], rather than six, as in closest

Figure 11-22. (a) A portion of a plane of spheres found in a body-centered cubic structure. Note that the spheres do not touch. (b) Spheres in two layers of a body-centered cubic structure. Each sphere in one layer touches four spheres in the adjacent layer but none in its own layer.

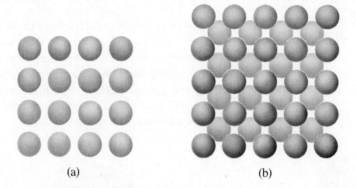

(a) (b)

packed planes. The spheres in such a plane *do not touch.* The structure consists of repeating layers of these planes. The second layer is stacked on top of the first so that a sphere in the second layer *touches* four spheres in the first layer [Fig. 11-22(b)]. The spheres of the third layer are positioned directly above the spheres of the first layer (Fig. 11-23); those of the fourth, above the second; etc. Any atom in this structure touches four atoms in the layer above it and four atoms in the layer below it. An atom in a body-centered cubic structure thus has a coordination number of 8. Isomorphous metals with a body-centered cubic structure include Ba, Cr, Mo, W, and Fe at room temperature.

Figure 11-23. (a) A portion of a body-centered cubic structure, showing parts of three layers. (b) An expanded view of a body-centered cubic structure. In an infinite extension of this structure, each sphere touches four spheres in the layer above it and four spheres in the layer below and is said to have a coordination number of 8.

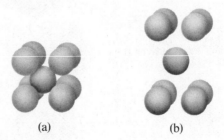

(a) (b)

Polonium (Po) crystallizes in the **simple cubic structure,** which is rare for metals. It contains planes in which each sphere *touches* its four nearest neighbors

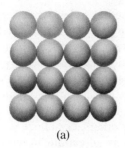

(a)

(b)

Figure 11-24. (a) A portion of a plane of spheres found in a simple cubic structure. Note that the spheres are in contact. (b) A portion of a simple cubic structure, showing two of the planes.

[Fig. 11-24(a)]. Thus the structure is not closest packed. The planes are stacked directly above each other, so that an atom in the second layer touches only one atom in the first layer [Fig. 11-24(b)]. The coordination number of a polonium atom in a simple cubic array is 6; an atom touches four other atoms in its own layer, one atom in the layer above, and one atom in the layer below.

11.16 The Structures of Ionic Crystals

Ionic crystals consist of two or more different kinds of ions, usually having different sizes. The packing of these ions into a crystal structure is more complex than the packing of metal atoms that are the same size and kind.

Most monoatomic ions behave as charged spheres; their attraction for ions of opposite charge is the same in every direction. Consequently, stable structures result (1) when ions of one charge are surrounded by as many ions as possible of the opposite charge and (2) when the cations and anions are in contact with each other. The structures are determined by two factors: the relative sizes of the ions and the relative numbers of positive and negative ions required to maintain the electrical neutrality of the crystal as a whole.

In simple ionic structures the anions, which are normally larger than the cations, are usually arranged in a closest packed array. The spaces remaining between the anions, called **holes,** or **interstices,** are occupied by the smaller cations. Figures 11-25 and 11-26 illustrate the two most common types of holes. The smaller of these is found between three spheres in one plane and one sphere in an adjacent plane [Fig. 11-25(a)]. The four spheres that bound this hole are arranged at the corners of a tetrahedron [Fig. 11-25(b)]; the hole is called a **tetrahedral hole.** The larger type of hole is found at the center of six spheres (three in one layer and three in an adjacent layer) located at the corners of an octahedron [Fig. 11-26]. Such a hole is called an **octahedral hole.**

Depending on the relative sizes of the cations and anions, the cations of an ionic compound can occupy tetrahedral or octahedral holes. As will be discussed

Figure 11-25. (a) Spheres in two adjacent closest packed layers that form a tetrahedral hole. (b) A cation (smaller sphere) located in a tetrahedral hole surrounded by four anions (larger spheres) from a different perspective. The structure has been expanded to show the geometrical relationships.

Figure 11-26. (a) Spheres in two adjacent closest packed layers that form an octahedral hole. (b) A cation (smaller sphere) located in an octahedral hole surrounded by six anions (larger spheres) from a different perspective. The structure has been expanded to show the geometrical relationships.

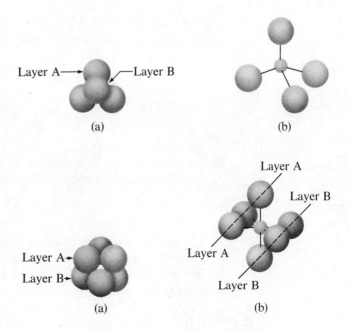

Figure 11-27. A cation in the cubic hole in a simple cubic array of anions.

in Section 11.17, relatively small cations occupy tetrahedral holes, while larger cations occupy octahedral holes. If the cations are too large to fit into the octahedral holes, the packing of the anions may change to give a more open structure, such as a simple cubic array (Fig. 11-24). The larger cations can then occupy the larger cubic holes made possible by the more open spacing (Fig. 11-27).

In either a hexagonal closest packed or a cubic closest packed array of anions, there are two tetrahedral holes for each anion in the array. The isomorphous compounds Li_2O, Na_2O, Li_2S, Na_2S, and Li_2Se, among others, crystallize with a cubic closest packed array of anions with the relatively small cations in tetrahedral holes. The ratio of tetrahedral holes to anions is 2 to 1; thus all of these holes must be filled by cations since the cation-to-anion ratio is 2 to 1 in these compounds. A compound that crystallizes in a closest packed array of anions with cations in the tetrahedral holes can have a maximum cation-to-anion ratio of 2 to 1; all of the tetrahedral holes are filled at this ratio. Compounds with a ratio of less than 2 to 1 may also crystallize in a closest packed array of anions with cations in the tetrahedral holes, if the ionic sizes fit. In these compounds, however, some of the tetrahedral holes remain vacant.

EXAMPLE 11.4   Zinc sulfide crystallizes with zinc ions occupying $\frac{1}{2}$ of the tetrahedral holes in a closest packed array of sulfide ions. What is the formula of zinc sulfide?

Since there are 2 tetrahedral holes per anion (sulfide ion) and $\frac{1}{2}$ of these are occupied by zinc ions, there must be $\frac{1}{2} \times 2$, or 1, zinc ion per sulfide ion. Thus the formula is ZnS.

The ratio of octahedral holes to anions in either a hexagonal or a cubic closest packed structure is 1 to 1. Thus compounds with cations in octahedral holes in a closest packed array of anions can have a maximum cation-to-anion ratio of 1 to 1. In NiO, MnS, NaCl, and KH, for example, all of the octahedral holes are filled. Ratios of less than 1 to 1 are observed when some of the octahedral holes remain empty.

EXAMPLE 11.5   Aluminum oxide crystallizes with aluminum ions in $\frac{2}{3}$ of the octahedral holes in a closest packed array of oxide ions. What is the formula of aluminum oxide?

Since there is 1 octahedral hole per anion (oxide ion) and only $\frac{2}{3}$ of these holes are occupied, the ratio of aluminum to oxygen must be $\frac{2}{3}$ to 1, which would give $Al_{2/3}O$. The simplest whole-number ratio is 2 to 3, so the formula is Al_2O_3.

In a simple cubic array of anions (Fig. 11-27), there is one cubic hole that can be occupied by a cation for each anion in the array. In CsCl, and in other compounds with the same structure, all of the cubic holes are occupied. Half of the cubic holes are occupied in SrH_2, UO_2, $SrCl_2$, and CaF_2.

Different types of ionic compounds crystallize in the same structure because the relative sizes of the ions and the stoichiometry (the features that determine structure) are similar.

11.17 The Radius Ratio Rule for Ionic Compounds

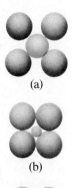

(a)

(b)

(c)

Figure 11-28. Packing of anions (black spheres) around cations of varying size (colored spheres). Decreasing size of cations is illustrated successively in (a), (b), and (c).

The structure assumed by an ionic compound is largely the result of stoichiometry and of simple geometric and electrostatic relationships that depend on the relative sizes of the cation and anion. A relatively large cation can touch a large number of anions and so occupies a cubic or an octahedral hole, whereas a relatively small cation can touch only a few anions and so occupies a tetrahedral hole.

Consider a cation, M^+, with a coordination number of 6. As shown in Fig. 11-28(a), the M^+ ion touches four X^- ions in a plane. In addition, although they are not shown in the figure, there is an X^- ion above the M^+ ion and touching it and another below and touching it. Note that the M^+ ion is large enough to expand the array of X^- ions so the X^- ions are not in contact with one another. As long as the expansion is not great enough to allow still another anion to touch the cation, this is a stable situation; the cation-anion contacts are maintained. Figure 11-28(b) illustrates what happens when the size of the M^+ ion is decreased somewhat. Here the X^- ions touch each other, as do the M^+ and X^- ions. If the size of M^+ is further decreased, it becomes impossible to get a structure with a coordination number of 6. The anions touch [Fig. 11-28(c)], but there is no contact between the M^+ and X^- ions—this is an unstable structure. In this case a more stable structure would be formed with only four anions about the cation. The limiting condition for the formation of a structure with a coordination number of 6 is illustrated in Fig. 11-28(b): the X^- ions touch one another, and the M^+ and X^- ions touch. This occurs when the sizes of the ions are such that the **radius ratio** (the radius of the positive ion, r^+, divided by the radius of the negative ion, r^-) is equal to 0.414.

There is a minimum radius ratio (r^+/r^-) for each coordination number. Below this value an ionic structure having that coordination number is generally not stable. The approximate limiting values for the radius ratios for ionic compounds are given in Table 11-3.

Table 11-3 Limiting Values for the Radius Ratio for Ionic Compounds (r^+ is radius of cation; r^- is radius of anion)

Coordination Number	Type of Hole Occupied	Approximate Limiting Values of r^+/r^-
8	Cubic	Above 0.732
6	Octahedral	0.414 to 0.732
4	Tetrahedral	0.225 to 0.414

EXAMPLE 11.6   Predict the coordination number of Cs^+ ($r^+ = 1.69$ Å) and of Na^+ ($r^+ = 0.95$ Å) in CsCl and NaCl, respectively. The radius of a chloride ion is 1.81 Å.

For CsCl:

$$\frac{r^+}{r^-} = \frac{1.69 \,\text{Å}}{1.81 \,\text{Å}} = 0.934$$

The radius ratio is greater than 0.732 (Table 11-3), which indicates that a coordination number of 8 is likely for Cs^+ in CsCl.

For NaCl:

$$\frac{r^+}{r^-} = \frac{0.95 \,\text{Å}}{1.81 \,\text{Å}} = 0.52$$

This radius ratio is between 0.414 and 0.732 and so indicates that a coordination number of 6 is likely for Na^+ in NaCl (Table 11-3).

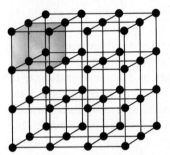

Figure 11-29. A portion of a simple cubic space lattice. One unit cell is shaded in color.

Figure 11-30. A portion of the structure of NaCl illustrating three ways to position a space lattice in the structure. Small spheres represent Na^+; large spheres, Cl^-. (a) Lattice points in the center of the Na^+ ions. (b) Lattice points in the center of the Cl^- ions. (c) Lattice points with a chloride ion to the left and a sodium ion to the right. Black lines connect points that define a unit cell in each lattice. The unit cell of NaCl is usually described as indicated in (a) or (b). Note that this structure has been expanded for clarity; the ions touch in the actual structure.

The radius ratio rule is only a guide to the type of structure that may form. It applies strictly only to ionic crystals and in some cases fails with them. In compounds in which the bonds are covalent, the rule may not hold. In spite of its limitations, however, the radius ratio rule is a useful guide for predicting many structures. It also underlines one of the most significant features responsible for the structures of ionic solids: the relative sizes of the cations and anions.

11.18   Crystal Systems

The atoms in a crystal are arranged in a definite repeating pattern, so any one point in the crystal matches innumerable other points having identical environments. The collection of all of the points within the crystal that have identical environments is called a **space lattice.** A simple three-dimensional cubic space lattice is shown in Fig. 11-29. (The atoms around the points have been omitted so you can see the lattice.) Figure 11-30 illustrates a portion of the structure of sodium chloride. This structure will be described more fully later in this section, but for now, let us consider only the space lattice associated with it. There are an infinite number of ways to construct a space lattice in the sodium chloride structure. The points of one possible space lattice are located at the centers of the sodium ions, as shown in Fig. 11-30(a). Alternatively, points with identical environments could be located at the centers of chloride ions [Fig. 11-30(b)] or between sodium and chloride ions [Fig. 11-30(c)]. In each case the resulting space lattice is the same; only the locations of the points of the lattice differ.

That part of a space lattice that will generate the entire lattice if repeated in three dimensions is called a **unit cell.** The cube shaded in Fig. 11-29 is a unit cell for the simple cubic space lattice. The cubes outlined in Fig. 11-30 illustrate three

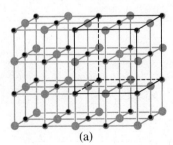

(a)

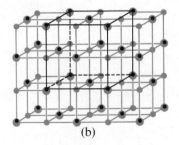

(b)

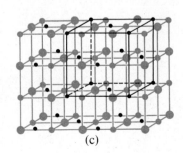

(c)

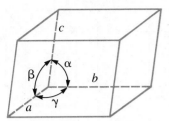

Figure 11-31. A unit cell.

ways in which unit cells may be selected for the sodium chloride structure. The structure of a crystal is specified by describing the size and shape of the unit cell and indicating the arrangement of its contents. The crystal can be built up by repeating the unit cell in three dimensions.

Thus far we have considered only unit cells shaped like a cube, but there are others. In general, a unit cell is a parallelepiped for which the size and shape are defined by the lengths of three axes (a, b, and c) and the angles (α, β, and γ) between them (Fig. 11-31). The axes are defined as being the lengths between points in the space lattice. Consequently, **unit cell axes join points with identical environments.** Unit cells must have one of the seven shapes indicated in Table 11-4.

Table 11-4   Unit Cells of the Seven Crystal Systems

System	Axes	Angles
Cubic	$a = b = c$	$\alpha = \beta = \gamma = 90°$
Tetragonal	$a = b \neq c$	$\alpha = \beta = \gamma = 90°$
Orthorhombic	$a \neq b \neq c$	$\alpha = \beta = \gamma = 90°$
Monoclinic	$a \neq b \neq c$	$\alpha = \gamma = 90°; \beta \neq 90°$
Triclinic	$a \neq b \neq c$	$\alpha \neq \beta \neq \gamma \neq 90°$
Hexagonal	$a = b \neq c$	$\alpha = \beta = 90°; \gamma = 120°$
Rhombohedral	$a = b = c$	$\alpha = \beta = \gamma \neq 90°$

Variations in the number and location of lattice points in these seven unit cells give rise to 14 types of space lattices, shown in Fig. 11-32. In some of these, points with identical surroundings (lattice points) are found only at the corners of the unit cell. In others the points of the lattice occur both at the corners and in the centers of some or all of the faces of the unit cell. In still others the points of the lattice are found both at the corners and in the center of the unit cell.

Students sometimes have trouble reconciling the number of points or atoms given as the number per unit cell with the diagram of the unit cell. Some of the points or atoms shown in a diagram of the unit cell are actually shared by other unit cells and therefore do not lie completely within the unit cell. Keep in mind the following rules:

1. A point or atom lying completely within a unit cell belongs to that unit cell only and is therefore counted as 1 when totaling the number of points or atoms in the unit cell.
2. A point or atom lying on a face of a unit cell is shared equally by two unit cells and is therefore counted as $\frac{1}{2}$ when totaling the number of points or atoms in a unit cell.
3. A point or atom lying on an edge is shared by four unit cells and is therefore counted as $\frac{1}{4}$.
4. A point or atom lying at a corner is shared by eight unit cells and is therefore counted as $\frac{1}{8}$.

Now let us look more closely at the contents of some cubic unit cells. The lattice points associated with the space lattice of each of the three cubic unit cells are indicated in Fig. 11-33. There is 1 lattice point ($8 \times \frac{1}{8}$) associated with the unit cell of the simple cubic lattice [Fig. 11-33(a)]. Since a unit cell containing one lattice point is called a **primitive cell,** the simple cubic lattice is sometimes called a **primitive cubic lattice.** The second cell [Fig. 11-33(b)] has 2 lattice points, 1 at the

Figure 11-32. Unit cells of the 14 types of space lattice.

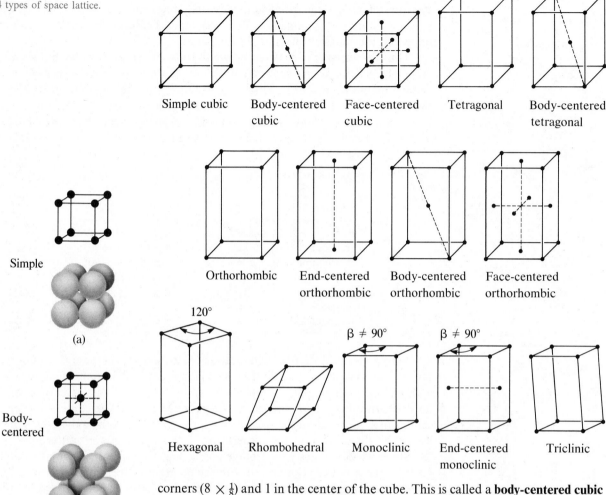

Figure 11-33. Cubic unit cells showing the locations of lattice points in the upper figures and the locations of metal atoms in the lower figures.

corners ($8 \times \frac{1}{8}$) and 1 in the center of the cube. This is called a **body-centered cubic cell.** Such a cell has points with identical surroundings at the corners and at its center. The third cell [Fig. 11-33(c)] has 4 lattice points (points with identical environments), one at the corners ($8 \times \frac{1}{8}$) and 3 from the 6 face centers ($6 \times \frac{1}{2}$). This is called a **face-centered cubic cell.**

The structure of polonium (Section 11.15) may be described as consisting of a simple cubic space lattice with an atom located at each of the lattice points [Fig. 11-33(a)]. The length of the unit cell edge is 3.36 Å. Since the polonium atoms touch along the edges of the cell, the nearest distance between the centers of polonium atoms is 3.36 Å, and the radius of each polonium atom is thus 1.68 Å. (Remember that the distance between two atoms is measured between their nuclei.)

A metal with a body-centered cubic structure consists of a space lattice composed of body-centered cubic unit cells [Fig. 11-33(b)]. One metal atom is located on each lattice point, so there are two identical metal atoms (atoms with identical environments) in the unit cell. The atoms touch along the diagonal of the cubic unit cell.

A metal with a cubic closest packed structure may be described as consisting of a space lattice made up of face-centered cubic unit cells [Fig. 11-33(c)]. One metal atom is located on each lattice point, so there are four equivalent metal atoms in each unit cell. The structure in Fig. 11-33(c) is the same as that in Fig. 11-21(b), but the perspective is different. Note that the atoms touch along the diagonals of the faces of the cell.

Ionic compounds can also crystallize with cubic unit cells; CsCl, NaCl, and one form of ZnS (zinc blende) crystallize with cubic space lattices. Another form of ZnS (wurtzite) crystallizes with a hexagonal space lattice. The assumption of two or more crystal structures by the same substance is called **polymorphism.**

The structure of cesium chloride, CsCl, is simple cubic. The unit cell is shown in Fig. 11-34. Chloride ions are located on the lattice points at the corners of the

Figure 11-34. The unit cells of some ionic compounds of the general formula MX. The black spheres represent positive ions (cations), and the colored spheres represent negative ions (anions). These structures have been expanded to show the geometrical relationships. In the crystal the cations and anions touch.

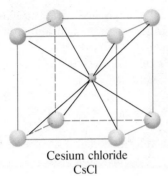

Cesium chloride
CsCl

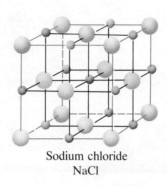

Sodium chloride
NaCl

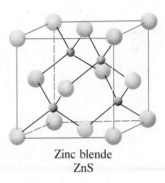

Zinc blende
ZnS

cell, and the cesium ion is located at the center of the cell. The cesium ion and the chloride ions touch along the diagonals of the cubic cell. There is no lattice point in the center of the cell because a cesium ion is not identical to a chloride ion. There are one cesium ion and one chloride ion per unit cell, matching the 1 to 1 stoichiometry reflected in the formula for cesium chloride.

The unit cell of sodium chloride, NaCl, is also illustrated in Fig. 11-34. Chloride ions are located on the lattice points of a face-centered cubic unit cell. Sodium ions are located in the octahedral holes in the middle of the cell edges and in the center of the cell. The sodium and chloride ions touch each other. The unit cell contains four sodium ions and four chloride ions, reflecting the 1 to 1 stoichiometry called for by the formula for sodium chloride.

A unit cell of zinc blende, ZnS, is illustrated in Fig. 11-34. This cubic structure contains sulfide ions on the lattice points of a face-centered cubic cell. Zinc ions are located in alternate tetrahedral holes ($\frac{1}{2}$ of the tetrahedral holes). There are four zinc ions and four sulfide ions in the unit cell, making the unit cell neutral in net charge.

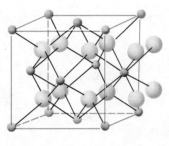

$\bullet$ = Ca^{2+}

$\bigcirc$ = F^-

Figure 11-35. The unit cell of CaF_2. The black spheres represent calcium ions, Ca^{2+}, and the colored spheres, fluoride ions, F^-. Note the face-centered cubic array of Ca^{2+} and the simple cubic array of F^-. This structure has been expanded to show the geometrical relationships. In the crystal the cations and anions touch.

A unit cell of calcium fluoride, CaF_2, is shown in Fig. 11-35. This cubic structure has equivalent calcium ions located on the lattice points of a face-centered cubic cell. All of the tetrahedral holes in the face-centered cubic array of calcium ions are occupied by fluoride ions. There are four calcium ions and eight fluoride ions in a unit cell, giving a calcium-to-fluorine ratio of 1 to 2, as in the formula of calcium fluoride, CaF_2. Close examination of Fig. 11-35 will reveal the simple cubic array of fluoride ions with calcium ions in half of the cubic holes. The structure cannot be described in terms of a space lattice of points on the fluoride ions since the fluoride ions do not all have identical environments. The orientations of the four calcium ions about each of the fluoride ions can differ.

11.19 Calculation of Ionic Radii

If the edge length of a unit cell and the positions of the constituent ions are known, ionic radii for the ions in the crystal lattice can be calculated, making certain assumptions. The following examples illustrate the method and assumptions for cubic structures.

EXAMPLE 11.7   The edge length of the unit cell of LiCl (NaCl-like structure, face-centered cubic) is 5.14 Å. Assuming anion-anion contact, calculate the ionic radius for the chloride ion.

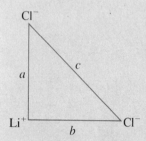

The NaCl structure (Fig. 11-34) contains a right triangle involving two chloride ions and one sodium ion. In the isomorphous LiCl structure, the lithium ion is so small that all ions in the structure touch, as in Fig. 11-28(b).

Because a, the distance between the center of a chloride ion and the center of a lithium ion, is $\frac{1}{2}$ of the edge length of the cubic unit cell,

$$a = \frac{5.14 \text{ Å}}{2}$$

Similarly, b is the distance between the center of a chloride ion and the center of a lithium ion and hence is also $\frac{1}{2}$ of the edge length of the cubic unit cell.

$$b = \frac{5.14 \text{ Å}}{2}$$

By the Pythagorean Theorem, c, the distance between the centers of two chloride ions, can be calculated.

$$c^2 = a^2 + b^2$$

$$c^2 = \left(\frac{5.14}{2}\right)^2 + \left(\frac{5.14}{2}\right)^2 = 13.21$$

$$c = \sqrt{13.21} = 3.63 \text{ Å}$$

Since the anions are assumed to touch each other, c is twice the radius of one chloride ion. Hence the radius of the chloride ion is $\frac{1}{2}c$.

$$r_{Cl^-} = \tfrac{1}{2}c = \tfrac{1}{2} \times 3.63 \text{ Å} = 1.81 \text{ Å}$$

EXAMPLE 11.8   The edge length of the unit cell of KCl (NaCl structure, face-centered cubic) is 6.28 Å. Assuming anion-cation contact, calculate the ionic radius for the potassium ion.

Inspection of the structure in Fig. 11-34 shows that the distance between the center of a potassium ion and the center of a chloride ion is $\frac{1}{2}$ of the edge length of the cubic unit cell for KCl, or

$$\tfrac{1}{2} \times 6.28\text{Å} = 3.14 \text{ Å}$$

Assuming anion-cation contact, 3.14 Å is the sum of the ionic radii for K^+ and Cl^-.

$$r_{K^+} + r_{Cl^-} = 3.14 \text{ Å}$$

In Example 11.7, r_{Cl^-} was calculated as 1.81 Å. Therefore

$$r_{K^+} = 3.14\,\text{Å} - 1.81\,\text{Å} = 1.33\,\text{Å}$$

Note that the chloride ions do not touch in solid potassium chloride.

It is important to realize that values for ionic radii calculated from the edge lengths of unit cells depend on numerous assumptions, such as a perfect spherical shape for ions, which are approximations at best. Hence such calculated values are themselves approximate, and comparisons cannot be pushed too far. Nevertheless, this is one method that has proved useful for calculating ionic radii from experimental measurements such as X-ray crystallographic determinations.

11.20 The Lattice Energies of Ionic Crystals

The **lattice energy** of an ionic compound may be defined as the energy required to separate the ions in a mole of the compound by infinite distances. For the ionic solid MX, the lattice energy is the energy involved in the chemical change

$$\text{Energy} + MX(s) \longrightarrow M^{n+}(g) + X^{n-}(g)$$

The same amount of energy is released when a mole of an ionic compound is formed by bringing positive and negative ions together from infinite distances. Lattice energies can be calculated from basic principles, or they can be measured experimentally.

The calculation of lattice energies is based primarily on the work of Born, Lande, and Mayer. They derived the following equation to express the lattice energy, U, of an ionic crystal:

$$U = \frac{C(Z^+Z^-)}{R_0}$$

where C is a constant that depends on the type of crystal structure and the electronic structures of the ions; Z^+ and Z^- are the charges on the ions; and R_0 is the interionic distance (the sum of the radii of the positive and negative ions). For a given structure the principal factors determining the lattice energy are Z^+, Z^-, and R_0. The lattice energy increases rapidly with the charges on the ions. Keeping all other parameters constant, doubling the charge on both cation and anion quadruples the lattice energy. For example, the lattice energy of LiF (Li$^+$ and F$^-$; $Z^+ = 1$, $Z^- = -1$) is 1023 kilojoules per mole, while that of MgO (Mg^{2+} and O^{2-}; $Z^+ = 2$, $Z^- = -2$) is 3900 kilojoules per mole (R_0 is nearly the same for both compounds).

The lattice energy also increases rapidly with decreasing interionic distance in the crystal lattice, which may result from decreasing either the cation radius, the anion radius, or both. Examples of crystals that exhibit a large difference in interionic distances and lattice energies are lithium fluoride and rubidium chloride.

	Interionic Distance, R_0	Lattice Energy, U
LiF	2.008 Å	1023 kJ/mol
RbCl	3.28 Å	680 kJ/mol

The large lattice energies of many ionic compounds result from strong electrostatic forces between the ions in the crystal. These strong interionic forces are also responsible for the relatively large heats of fusion, sublimation, and vaporization and for the high melting and boiling temperatures of these compounds, as compared to most molecular compounds. These forces must be overcome in order to vaporize a crystal to separated gaseous ions, for example. However, in the case of fusion, the energy requirement is less than the total lattice energy because of the interactions of neighboring ions in the liquid phase. The same strong forces are responsible for the fact that ionic crystals are generally hard, dense, rigid, and nonvolatile. While ionic compounds have such properties, some covalent substances, such as diamond and silicon carbide, also have them (see Sections 25.1 and 27.13). In such cases the very strong forces in the crystal are due to strong covalent bonds, which extend in three dimensions throughout the crystal, rather than to the weaker London forces usually found in molecular compounds.

11.21 The Born-Haber Cycle

The lattice energies of only a few ionic crystals have been measured because it is not possible to measure most lattice energies directly. However, a cyclic process can be used to relate the lattice energy to other quantities and make its calculation possible. This **Born-Haber cycle** involves ΔH_f, the enthalpy of formation of the compound (Section 9.6); I, the ionization potential of the metal (Section 4.15, Part 3); $E.A.$, the electron affinity of the nonmetal (Section 4.15, Part 4); ΔH_s, the enthalpy of sublimation of the metal; D, the bond dissociation energy of the nonmetal; and U, the lattice energy of the compound. The Born-Haber cycle for sodium chloride analyzes the formation of $NaCl(s)$ from $Na(s)$ and $\frac{1}{2}Cl_2(g)$ as a step-by-step process, which may be expressed diagrammatically as follows with the overall change presented in color and the individual steps in black:

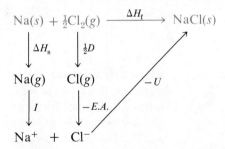

This diagram indicates hypothetical steps in the formation of sodium chloride from 1 mole of sodium metal and $\frac{1}{2}$ mole of chlorine gas. We can assume first that the sodium metal is vaporized and the bonds in the diatomic chlorine molecules are broken. Then the sodium atoms are ionized, and the electrons from them are transferred to the chlorine atoms to form chloride ions. The gaseous sodium ions and chloride ions thus formed come together to give solid sodium chloride. The energy change in this step is the negative of the lattice energy, which is the amount of energy required to produce 1 mole of gaseous sodium ions and 1 mole of gaseous chloride ions from 1 mole of solid sodium chloride. The total energy evolved in the preceding hypothetical preparation of sodium chloride is equal to the experimentally determined enthalpy of formation, ΔH_f, of the compound from its elements.

The calculation of the lattice energy of sodium chloride may be carried out using experimental data as follows:

1. Enthalpy of formation of sodium chloride:

$$Na(s) + \tfrac{1}{2}Cl_2(g) \longrightarrow NaCl(s) \qquad \Delta H = \Delta H_f = -411 \text{ kJ}$$

2. Enthalpy of sublimation of sodium metal:

$$Na(s) \longrightarrow Na(g) \qquad \Delta H = \Delta H_s = 109 \text{ kJ}$$

3. Bond dissociation energy of molecular chlorine:

$$\tfrac{1}{2}Cl_2(g) \longrightarrow Cl(g) \qquad \Delta H = \tfrac{1}{2}D = 121 \text{ kJ}$$

4. Ionization energy of sodium atom:

$$Na(g) \longrightarrow Na^+(g) + e^- \qquad \Delta H = I = 496 \text{ kJ}$$

5. Electron affinity of chlorine atom:

$$Cl(g) + e^- \longrightarrow Cl^-(g) \qquad \Delta H = -E.A. = -368 \text{ kJ}$$

6. Lattice energy of sodium chloride:

$$Na^+(g) + Cl^-(g) \longrightarrow Na^+Cl^-(s) \qquad \Delta H = -U = ? \text{ (to be calculated)}$$

The total enthalpy of formation of the crystal from its elements is given by the equation

$$\Delta H_f = \Delta H_s + \tfrac{1}{2}D + I - E.A. - U$$

The value of ΔH_f is accurately known for many substances. If the other thermochemical values are available, we can solve for the lattice energy, U, by rearranging the preceding equation as follows:

$$U = -\Delta H_f + \Delta H_s + \tfrac{1}{2}D + I - E.A.$$

For sodium chloride, using the above data, the lattice energy is

$$U = (411 + 109 + 121 + 496 - 368) \text{ kJ} = 769 \text{ kJ}$$

The Born-Haber cycle may be used to calculate any one of the quantities in the equation for lattice energy, provided that all of the others are known. Usually, ΔH_f, ΔH_s, I, and D are known. The direct measurement of electron affinity is rather difficult, so really accurate values of electron affinity have been determined only for the halogens. For halogen compounds, then, the Born-Haber cycle can be used to calculate lattice energies, which are found to agree well with those obtained by other means. The electron affinity of oxygen, an important quantity, cannot be measured directly. It has been obtained by using the Born-Haber cycle on ionic oxides using the values of U calculated by the method of Born, Lande, and Mayer (Section 11.20).

11.22 X-Ray Diffraction

The size of the unit cell and the arrangement of atoms in a crystal can be determined from measurements of the **diffraction** of X rays by the crystal. X rays are electromagnetic radiation with wavelengths (Section 4.8) about as long as the distance between neighboring atoms in crystals (about 2 Å).

W. H. Bragg, an English physicist, showed that the diffraction of X rays by a crystal can be interpreted if the crystal is thought of as a reflection grating. The reflection of X rays from planes of lattice points in the crystal may be compared with the reflection of a beam of light from a stack of thin glass plates of equal thickness. It is known that light of a single wavelength is reflected from such a stack only at definite angles, which depend on the wavelength of the light and the thickness of the plates. The planes within a crystal correspond to the plates of glass.

There is a simple mathematical relation between the wavelength, λ, of the X rays, the distance between the planes in the crystal, and the angle of diffraction (reflection). When a beam of monochromatic X rays strikes two planes of a crystal at an angle θ, it is reflected (Fig. 11-36). The rays from the lower plane travel farther than those from the upper plane by a distance equal to the sum of the distances BC and CD. If this distance is equal to 1, 2, 3, or any other integral number of wavelengths (so $n\lambda = BC + CD$), the two beams emerging from the crystal will be in phase and will reinforce each other. From the right triangle ABC of Fig. 11-36,

$$\sin BAC = \frac{BC}{AC}$$

Rearranging gives

$$BC = AC \sin BAC$$

The distance AC is the distance between the crystal planes, which we will call d. Angle BAC is equal to θ. Then

$$BC = d \sin \theta$$

But the difference in the path lengths of the two beams is equal to $BC + CD$. Because $BC = CD$, the total difference in path length is $2BC$. X rays will be reflected, provided that, as indicated previously,

$$n\lambda = BC + CD = 2BC$$

Hence

$$n\lambda = 2d \sin \theta$$

which is the **Bragg equation.**

If the X rays strike the crystal at any angle other than θ, the extra distance $BC + CD$ will not be equal to an integral number of wavelengths, and there will be interference of the reflected rays. This will destroy the intensity of the reflected rays.

Figure 11-36. Diffraction of a monochromatic beam of X rays by two planes of a crystal.

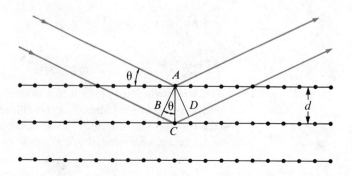

The reflection corresponding to $n = 1$ is called the first-order reflection; that corresponding to $n = 2$ is the second-order reflection; and so forth. Each successive order has a larger angle.

By rotating a crystal and varying the angle θ at which the X-ray beam strikes it, the values of θ at which diffraction occurs can be determined by looking for the maximum diffracted intensities. If the wavelength of the X rays is known, the spacing, d, of the planes within the crystal can be determined from these values. On the other hand, if the X rays strike planes for which d is known, their wavelength may be determined by measuring the angles.

FOR REVIEW

SUMMARY

The physical properties of liquids and solids can be explained in terms of the kinetic-molecular theory, which was also used to explain the properties of gases. In a liquid the intermolecular attractive forces hold the molecules in contact, although they still have sufficient kinetic energy to move past one another (diffuse). Since the molecules are in contact, liquids are not very compressible. The speeds of the molecules in a liquid vary; so some molecules have enough kinetic energy to escape from the liquid, resulting in **evaporation.** The pressure of a vapor in dynamic equilibrium with a liquid is called the **vapor pressure** of the liquid. The vapor pressure of a liquid increases with increasing temperature, and the **boiling point** of the liquid is reached when the vapor pressure equals the external pressure on the liquid. The **normal boiling point** of a liquid is the temperature at which its vapor pressure is equal to 1 atmosphere. The amount of energy necessary to evaporate a unit mass of a liquid at a constant temperature is called its **heat of vaporization.**

When the temperature of a liquid becomes sufficiently low or the pressure on the liquid becomes sufficiently high, the molecules can no longer move past each other, and a rigid solid forms. Many solids are **crystalline;** that is, they are composed of a repeating pattern of atoms, molecules, or ions. Others are supercooled liquids, sometimes called **amorphous solids,** in which the disordered arrangement of molecules found in liquids is retained. Crystalline solids melt when enough energy is added so that the kinetic energy of their molecules just overcomes the intermolecular attractive forces holding the molecules in fixed positions in the crystal. The temperature at which a solid and its liquid are in equilibrium is called the **melting point** of the solid or the **freezing point** of the liquid. Amorphous solids do not have a melting point but soften over a range of temperatures. The **heat of fusion** of a crystalline solid is the amount of energy that must be added to a given mass to convert the solid to a liquid at the same temperature. Some solids exhibit a measurable vapor pressure and will **sublime.**

Intermolecular attractive forces, collectively referred to as **van der Waals forces,** are responsible for the behavior of liquids and solids and are electrostatic in nature. **Dipole-dipole attractions** result from the electrostatic attraction between the negative end of one polar molecule and the positive end of another. The temporary dipole that forms as a result of the motion of the electrons in an atom can induce a dipole in an adjacent atom and give rise to the **dispersion force,** or

London force. London forces increase with increasing molecular size. **Hydrogen bonding** is a special type of dipole-dipole attraction that results when hydrogen is found in compounds with very electronegative elements such as fluorine, oxygen, nitrogen, and chlorine. These various intermolecular forces are responsible for the critical temperature and critical pressure of a substance. The **critical temperature** of a substance is that temperature above which the substance cannot be liquefied, no matter how much pressure is applied. The **critical pressure** is the pressure required to liquefy a gas at its critical temperature.

The structures of crystalline metals and simple ionic compounds can be described in terms of the packing of spheres. Metal atoms can be arrayed in **hexagonal closest packed structures, cubic closest packed structures, body-centered structures,** and **simple cubic structures.** The anions in simple ionic structures commonly adopt one of these structures with the cations occupying the spaces remaining between the anions. Small cations may occupy **tetrahedral holes** in a closest packed array of anions. Larger cations can occupy **octahedral holes.** Still larger cations can occupy cubic holes in a simple cubic array of anions. The **radius ratio rule** serves as a guide to the **coordination number** of the cation. The structure of a crystalline solid can be described by indicating the size, shape, and internal arrangement of its **unit cell.** The dimensions of a unit cell can be determined by a study of the **diffraction** of X rays from a crystal.

The energy required to separate the ions in a mole of an ionic compound by infinite distances is called the **lattice energy** of the compound. The lattice energy is proportional to the product of the charges on the ions and inversely proportional to the distance between their centers. Lattice energies may be calculated or may be determined using the **Born-Haber cycle.**

KEY TERMS AND CONCEPTS

Amorphous solid (11.8)
Body-centered cubic structure (11.15)
Body-centered cubic unit cell (11.18)
Boiling point (11.13)
Born-Haber cycle (11.21)
Bragg equation (11.22)
Condensation (11.2)
Coordination number (11.15)
Critical pressure (11.5)
Critical temperature (11.5)
Crystal (11.8)
Crystal defects (11.9)
Cubic closest packed structure (11.15)
Dipole-dipole attraction (11.14)
Distillation (11.6)

Dynamic equilibrium (11.2)
Evaporation (11.2)
Face-centered cubic structure (11.15)
Face-centered cubic unit cell (11.18)
Freezing point (11.10)
Glass (11.8)
Heat of fusion (11.11)
Heat of vaporization (11.4)
Hexagonal closest packed structure (11.15)
Hydrogen bonding (11.14)
Intermolecular forces (11.13)
Interstitial sites (11.9)
Ionic radius (11.17, 11.19)
Isomorphous structures (11.15)
Lattice energy (11.20)

London force (11.13)
Melting point (11.10)
Normal boiling point (11.3)
Octahedral hole (11.16)
Polymorphism (11.18)
Radius ratio (11.17)
Simple cubic structure (11.15)
Space lattice (11.18)
Sublimation (11.12)
Supercooling (11.8)
Surface tension (11.7)
Tetrahedral hole (11.16)
Unit cell (11.18)
van der Waals forces (11.13)
Vapor pressure (11.2)
X-ray diffraction (11.22)

EXERCISES

Note: The following data for water may be useful in several of the following exercises.

Heat of vaporization of water (100°C)	2258 J/g
Heat of vaporization of water (25°C)	2443 J/g
Heat of fusion of water (0°C)	333.6 J/g
Specific heat of water (solid)	2.04 J/g K
Specific heat of water (liquid)	4.18 J/g K
Specific heat of water (gas)	2.00 J/g K

Kinetic-Molecular Theory of Solids and Liquids

1. Describe the variations in the motion of the molecules of a substance as it changes from a solid to a liquid and from a liquid to a gas.
2. Explain why a liquid will assume the shape of any container into which it is poured, while a solid is rigid and retains its shape.
3. The types of intermolecular forces in a substance are identical whether the substance is a solid, a liquid, or a gas. Why then does a substance change phase from a solid to a liquid or to a gas?
4. What is the relationship between the intermolecular forces in liquids and their boiling points?
5. Ethyl chloride (b.p. = 12°C) is used as a local anaesthetic. When the liquid is sprayed on the skin, it cools the skin enough to freeze and numb it. Explain the cooling effect of liquid ethyl chloride.
6. What are the origins of the van der Waals forces of attraction?
7. Silane, SiH_4, phosphine, PH_3, and hydrogen sulfide, H_2S, melt at $-185°C$, $-133°C$, and $-85°C$, respectively. What does this suggest about the polar character and intermolecular attractions in the three compounds?
8. The heat of vaporization of $CO_2(l)$ is 54 cal/g. Would you expect the heat of vaporization of $CS_2(\ell)$ to be 89 cal/g or 19 cal/g? Explain your answer.
9. The heat of vaporization of a gram of water is larger than its heat of fusion. Explain why.

Properties of Liquids

10. What is the difference between evaporation and boiling?
11. Explain why increasing the temperature of a liquid increases its vapor pressure.
12. How does the boiling point of a liquid differ from its normal boiling point?
13. What is the approximate boiling point of water in Denver when the atmospheric pressure is 625 torr? See Fig. 11-4 for data.
14. A closed piston like that shown in Fig. 10-1 is filled with liquid ether so that there is no space for any vapor. Describe what will happen if the plunger is withdrawn somewhat, forming a volume that can be occupied by vapor. If the system is insulated so that no heat can enter or leave, will the temperature of the liquid increase, decrease, or remain constant? Explain.
15. Why do the boiling points of the halogens increase as their atomic weights increase?
16. Account for the fact that steam produces much more severe burns than does the same mass of boiling water.
17. How much heat in kilojoules is required to evaporate 1 mol of water at 25°C? *Ans. 44.01 kJ*
⟨S⟩18. How much heat is required (a) to change 3.00 mol of water at 100°C to steam at 100°C? (b) to change 0.2500 mol of ice at 0°C to water at 0°C? *Ans. 122 kJ; 1.502 kJ*
⟨S⟩19. In hot, dry climates water is cooled by allowing some of it to evaporate slowly. How much water must evaporate to cool 1.00 kg from 37°C to 21°C? Assume that the heat of vaporization of water is constant between 21° and 37° and is equal to the value at 25°C. *Ans. 27 g*
⟨S⟩20. The specific heat of copper is 0.0931 cal/g K. What weight of steam at 100°C must be converted to water at 100°C to raise the temperature of a 1.00×10^2-g copper block from 20°C to 100°C? *Ans. 1.38 g*
21. How much water at 0°C could be converted to ice at 0°C by the cooling action of vaporizing 26.0 g of liquid ammonia? The heat of vaporization of liquid ammonia is 1368 J/g. *Ans. 107 g*
⟨S⟩22. During the fermentation step in the production of beer, 4.5×10^6 kcal of heat are evolved per

1000 gal of beer produced. How many liters of cooling water are required to maintain the optimum fermentation temperature of 58°F for 1000 gal of beer? how many gallons? The cooling water enters with a temperature of 5°C and is discharged with a temperature of 13°C.

Ans. 6×10^5 L; 2×10^5 gal

23. Is it possible to liquefy sulfur dioxide at room temperature (about 20°C)? Is it possible to liquefy nitrogen at room temperature? Explain your answers.

24. Draw a rough graph showing how the pressure, initially at 145 atm, inside a cylinder of oxygen decreases as $O_2(g)$ is released from the cylinder at 25°C. Draw a similar curve for a cylinder of ammonia at 25°C, which contains $NH_3(l)$ and $NH_3(g)$ at a pressure of 10 atm.

Properties of Solids

25. A sample of fused (amorphous) boric oxide, B_2O_3, (Fig. 11-11) looks like a solid. It is rigid and will shatter if struck with a hammer. Why do chemists regard fused boric oxide as a supercooled liquid?

26. Contrast crystalline and amorphous solids (supercooled liquids) with regard to melting points.

27. What types of liquid materials tend to supercool readily and form glasses?

28. Describe the types of intermolecular forces that can be present in a solid.

29. How do we know that certain solids such as ice and naphthalene have vapor pressures?

30. Why does iodine sublime more rapidly at 110°C than at 25°C?

31. What feature characterizes the equilibrium between a solid and its liquid as dynamic?

32. The melting point of $N_2(s)$ is −210°C. Would you expect the melting point of $O_2(s)$ to be −202°C or −218°C? Explain your answer.

33. The melting point of $H_2O(s)$ is 0°C. Would you expect the melting point of $H_2S(s)$ to be −85°C or 185°C? Explain your answer.

34. Which one of each of the following pairs will have the higher vapor pressure at the same temperature?
(a) $Br_2(s)$ or $I_2(s)$
(b) $HCl(s)$ or $Ar(s)$
(c) $CaCl_2(s)$ or $Cl_2(s)$
(d) $LiF(s)$ or $MgO(s)$

Properties of Crystalline Solids

S 35. Cobalt metal crystallizes in a hexagonal closest packed structure. What is the coordination number of a cobalt atom?

36. Chromium metal crystallizes in a body-centered cubic structure. What is the coordination number of a chromium atom?

37. Copper crystallizes with four equivalent metal atoms in a cubic unit cell. Describe the crystal structure of copper.

38. The free space in a lattice of spheres may be found by subtracting the volume of the spheres in a unit cell from the volume of the cell. Calculate the percentage of free space in each of the three cubic lattices if all atoms in each lattice are of equal radius and touch their nearest neighbors. Which of these lattice structures represents the most efficient packing of spheres into a volume?

S 39. One form of tungsten crystallizes in a body-centered cubic unit cell with a tungsten atom on each lattice point. If the edge length of the unit cell is 3.165 Å, what is the atomic radius of tungsten in this structure? *Ans. 1.370 Å*

40. The atomic radius of lead is 1.75 Å. If lead crystallizes in a face-centered cubic unit cell with nearest neighbors in contact, what is the edge length of the unit cell? *Ans. 4.95 Å*

S 41. Silver crystallizes in a face-centered cubic unit cell with a silver atom on each lattice point. (a) If the edge length of the unit cell is 4.0862 Å, what is the atomic radius of silver? (b) Calculate the density of silver.
Ans. (a) 1.4447 Å; (b) 10.50 g/cm³

S 42. Rutile is a mineral that contains titanium and oxygen. The structure of rutile may be described as a closest packed array of oxygen atoms with titanium in $\frac{1}{2}$ of the octahedral holes. What is the formula of rutile? What is the oxidation number of titanium? *Ans. TiO_2; +4*

43. Thallous iodide crystallizes in a simple cubic array of iodide ions with thallium ions in all of the cubic holes. What is the formula of thallous iodide? *Ans. TlI*

44. Cadmium sulfide, sometimes used as a yellow pigment by artists, crystallizes with cadmium occupying $\frac{1}{2}$ of the tetrahedral holes in a closest packed array of sulfide ions. What is the formula of cadmium sulfide? *Ans. CdS*

45. Magnetite, a magnetic oxide of iron used on recording tapes, crystallizes with iron atoms occu-

pying both $\frac{1}{8}$ of the tetrahedral holes and $\frac{1}{2}$ of the octahedral holes in a closest packed array of oxide ions. What is the formula of magnetite?

Ans. Fe_3O_4

46. Although it is a covalent compound, silicon carbide can be described as a cubic closest packed array of silicon atoms, with carbon atoms in $\frac{1}{2}$ of the tetrahedral holes. What is the empirical formula of silicon carbide? What is the coordination number of carbon in silicon carbide?

Ans. SiC; 4

47. A compound containing zinc, aluminum, and sulfur crystallizes with a closest packed array of sulfide ions. Zinc ions are found in $\frac{1}{8}$ of the tetrahedral holes, and aluminum ions in $\frac{1}{2}$ of the octahedral holes. What is the empirical formula of the compound?

Ans. $ZnAl_2S_4$

48. Why are compounds that are isomorphous with ZnS generally not observed when the radius ratio for the compound is greater than 0.42?

49. The chemically similar alkali metal chlorides NaCl and CsCl have different crystal structures, while chemically different NaCl and MnS have the same structure. Explain.

50. Classify the following compounds according to the type of crystal structure they form, assuming that they can be treated as ionic compounds:

(a) NaCl   (d) RbH   (g) CaTe   (j) AlP
(b) CsCl   (e) MgO   (h) TlBr   (k) BeO
(c) ZnS   (f) ZnTe   (i) CsI

S 51. Each of the following compounds crystallizes in a structure matching that of NaCl, CsCl, ZnS, or CaF_2. From the radius ratio, predict which structure is formed by each. Show your work.

S (a) ZnTe	*Ans. ZnS str.*	
S (b) BaF_2	*Ans. CaF_2 str.*	
S (c) KBr	*Ans. NaCl str.*	
S (d) AlP	*Ans. ZnS str.*	
S (e) CaS	*Ans. NaCl str.*	
(f) SrF_2	*Ans. CaF_2 str.*	
(g) NiO	*Ans. NaCl str.*	
(h) CsBr	*Ans. CsCl str.*	

52. A cubic unit cell contains iodide ions at the corners and a cesium ion in the center. What is the formula of the compound? *Ans. CsI*

53. A cubic unit cell contains rhenium ions at the corners and oxide ions at the center of each edge. What is the empirical formula of the oxide?

Ans. ReO_3

54. A cubic unit cell contains oxide ions in the center of each face, a titanium ion in the center of the cell, and calcium ions at the corners of the cell. What is the empirical formula of this oxide? What is the oxidation number of the titanium ion? *Ans. $CaTiO_3$; +4*

55. TlI crystallizes in the same structure as CsCl. The edge length of the unit cell of TlI is 4.20 Å. Assuming cation-anion contact, calculate the ionic radius of Tl^+. (The ionic radius of I^- may be found on the inside of the back cover.)

Ans. 1.48 Å

S 56. LiH crystallizes with the same crystal structure as NaCl. The edge length of the cubic unit cell of LiH is 4.08 Å. Assuming anion-anion contact, calculate the ionic radius of H^-. *Ans. 1.44 Å*

57. The unit cell edge length of CaF_2 is 5.46295 Å. The density of CaF_2 is 3.1805 g/cm^3. From these data and the atomic weights of calcium and fluorine, calculate Avogadro's number.

Ans. 6.023×10^{23}

S 58. The lattice energy of LiF is 1023 kJ/mol, and the Li–F distance is 2.008 Å. NaF crystallizes in the same structure as LiF but with a Na–F distance of 2.31 Å. Which of the following values most closely approximates the lattice energy of NaF: 510, 890, 1023, 1175, or 4090 kJ/mol? Explain your choice. *Ans. 890 kJ/mol*

59. The lattice energy of KF is 794 kJ/mol, and the interionic distance is 2.69 Å. The Sr–O distance in SrO, which has the same structure as KF, is 2.53 Å. Is the lattice energy of SrO about 745, 844, 2987, 3176, or 3375 kJ/mol? Explain your answer. *Ans. 3375 kJ/mol*

S 60. What X-ray wavelength would give a second order reflection ($n = 2$) with a θ angle of 10.40° from planes with a spacing of 4.00 Å?

Ans. 0.722 Å

61. What is the spacing between crystal planes that diffract X rays with a wavelength of 1.541 Å at an angle, θ, of 15.55° (first-order reflection)?

Ans. 2.87 Å

62. Gold crystallizes in a face-centered cubic unit cell. The second-order reflection ($n = 2$) of X rays for the planes that make up the tops and bottoms of the unit cells is at $\theta = 22.20°$. The wavelength of the X rays is 1.54 Å. What is the density of metallic gold? *Ans. 19.3 g/cm^3*

Additional Exercises

63. Most reactions in a chemical laboratory are run in the gaseous or the liquid state. If molecules

have to come in contact in order to react, explain why a mixture of solid reagents does not, in general, react very rapidly.

64. Explain why some molecules in a solid may have enough energy to sublime away from the solid even though the solid does not contain enough energy to melt.

[S] 65. How much energy is released when 75 g of steam at 135°C is converted to ice at −40°C? *Ans. 240 kJ*

66. How much energy is required to convert 35.0 g of ice at −8.0°C to steam at 105°C? *Ans. 106 kJ*

[S] 67. If 135 g of steam at 100°C and 475 g of ice at 0°C are combined, what is the temperature of the resultant water, assuming that no heat is lost? *Ans. 79.5°C*

68. How much more energy is contained in 5.0×10^2 kg of steam at 100°C than in the same quantity of ice at −10°C? *Ans.* 1.5×10^6 *kJ*

[S] 69. A river is 30 ft wide and has an average depth of 5 ft and a current of 2 mi/h. A power plant dissipates 2.1×10^5 kJ of waste heat into the river every second. What is the temperature difference between the water upstream and downstream from the plant? *Ans. 4°C*

70. If the heat dissipated by the power plant in Exercise 69 is to be removed by evaporating water at 25°C, how many gallons of water per day would be needed? *Ans.* 2.0×10^6 *gal/day*

71. By referring to Fig. 11-4, determine the boiling point of ethyl ether at 500 torr, the boiling point of ethyl alcohol at 0.5 atm, and the boiling point of water at 55 kPa. At what temperature will the vapor pressure of ethyl alcohol be equal to the vapor pressure of ethyl ether at 20°C? *Ans. 24°C; 64°C; 84°C; 66°C*

72. The melting points of NaCl, KCl, and RbCl decrease with increasing atomic number of the alkali metal. Suggest an explanation.

73. Compare the features that determine the geometry of the nearest neighbors around the nonmetal in a nonmetal halide (Chapter 7) with those that determine the geometry of the nearest neighbors about a metal in an ionic metal halide.

74. The carbon atoms in the unit cell of diamond occupy the same positions as both the zinc and sulfur atoms in cubic zinc sulfide. How many carbon atoms are found in the unit cell of diamond? The bonds between the carbon atoms are covalent. What is the hybridization of a carbon atom?

75. In terms of its internal structure, explain the high melting point (and hardness) of diamond. The structure of diamond is described in Exercise 74.

[S] 76. When an electron in an excited molybdenum atom falls from the L to the K shell, an X ray is emitted. These X rays are diffracted at an angle of 7.75° by planes with a separation of 2.64 Å. What is the difference in energy between the K and the L shell in molybdenum? *Ans.* 2.79×10^{-8} *erg or* 1.74×10^4 *eV*

12

WATER AND HYDROGEN PEROXIDE

Almost everyone knows "aitch-two-oh" (H_2O), the formula of water. This is appropriate since none of the hundreds of thousands of other chemical substances is more important than water.

Water covers nearly three-fourths of the earth's surface. It is present in the atmosphere and the earth's crust and composes a large part of all living plant and animal matter. Nutrients are transported into the roots of plants as solutions in water. Biological reactions occur in solution in the water in cells. Many industrial reactions are run in water, and much of the chemistry that is discussed in a general chemistry course occurs in water. Water has its own chemical behavior, and it has been used as a standard for many physical constants and units.

The yearly use of water in the United States averages about 600,000 gallons per person. It is estimated that in the year 2000 the world's consumption of fresh water will be 9700 cubic kilometers (2300 cubic miles) per year.

In this chapter we will consider the properties of water and examine the underlying electronic and molecular structure responsible for its behavior. We will also examine natural waters and the effect of society's activities on their quality. Finally, hydrogen peroxide, H_2O_2, another compound composed only of hydrogen and oxygen, will be discussed and compared to water.

THE PROPERTIES OF WATER

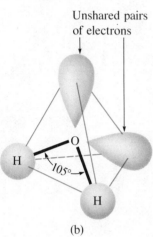

(a)

(b)

Figure 12-1. (a) The Lewis structure and (b) the bent molecular structure of the water molecule. Note that the Lewis structure shows only the distribution of electrons and does not represent the three-dimensional structure of the molecule.

12.1 The Structures of Gaseous, Liquid, and Solid Water

Until the latter part of the eighteenth century, water was thought to be an element, but in 1781 Henry Cavendish showed that water is formed when hydrogen burns in air. A few years later, Antoine-Laurent Lavoisier determined that the relative masses of hydrogen and oxygen in water are 1 and 8, respectively. The formula for water is H_2O.

Water has an abnormally low vapor pressure, a high boiling point, a high heat of vaporization, and remarkable solvent properties when compared to molecules of similar formula such as H_2S, H_2Se, and H_2Te. These unusual properties are all closely related to the structure of the water molecule. An oxygen atom has the electronic structure $1s^2 2s^2 2p^4$ (Section 4.12). In a water molecule, the two unpaired electrons in its valence shell pair with the $1s$ valence electrons of the hydrogen atoms, as indicated by the Lewis structure [Fig. 12-1(a)]. The water molecule contains two bonding electron pairs and two lone pairs. These four regions of high electron density are arranged tetrahedrally about the oxygen atom, giving the water molecule a bent structure [Fig. 12-1(b)]. The reduction of the H—O—H angle from 109.5° (the bond angle in a regular tetrahedral molecule) to 105° is due to repulsions between the lone pairs and bond pairs. The electron pairs in the hydrogen-oxygen bonds are pushed closer together by the stronger lone pair–bond pair repulsions (Section 7.1). Since the distortion from a regular tetrahedral arrangement is small, the hybridization of the oxygen atom can be described as essentially sp^3 (Section 7.3). Two of the four hybrid orbitals form σ bonds with the hydrogen atoms. The two remaining hybrid orbitals contain lone pairs.

The atoms in a molecule of water are held together by covalent bonds, which are distinctly polar in character since oxygen has a higher electronegativity (3.5) than hydrogen (2.1). (See Table 5-2.) Thus the oxygen atom possesses a partial negative charge, and each hydrogen atom, a partial positive charge. The molecule is highly polar since it has a bent structure. If the molecule were linear, the two bond polarities would cancel each other, and the molecule would be nonpolar.

Water molecules in the liquid and solid states are bound together by hydrogen bonds (Section 11.14). Although these hydrogen bonds between molecules are only about 5% as strong as the covalent hydrogen-oxygen bonds within a molecule, they are an important factor in the distinctive properties of both solid and liquid water. Molecules of water in the gaseous state exist largely as individual molecules; the larger distances between molecules and their larger kinetic energy (due to their faster motion) make hydrogen bonding negligible.

X-ray diffraction studies have shown that the water molecules in ice are so arranged that each oxygen atom has four hydrogen atoms as nearest neighbors; two are attached by electron-pair bonds and two by hydrogen bonds in an approximately tetrahedral arrangement (Fig. 12-2). Each hydrogen atom participates in a covalent bond to one oxygen atom and a hydrogen bond to another. This arrangement leads to an open structure, making ice (Fig. 12-3) a substance with a relatively low density. When ice is heated, some of the hydrogen bonds are broken, the open structure is destroyed, and the water molecules pack more

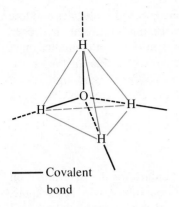

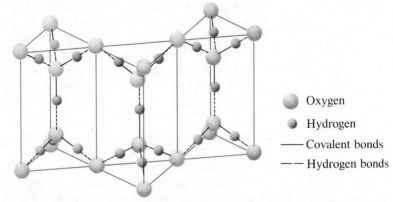

Figure 12-2. The tetrahedral arrangement of four hydrogen atoms about an oxygen atom in ice. Two hydrogen atoms participate in covalent bonds and two in hydrogen bonds to the oxygen. Each bond leading away from a hydrogen atom is to an oxygen of another water molecule.

Figure 12-3. The arrangement of water molecules in ice, showing each oxygen atom with four hydrogen atoms as nearest neighbors, two attached by electron-pair bonds and two by hydrogen bonds. The oxygen atoms at the corners of a prism are also at the corners of adjoining prisms, as this is only a portion of a continuous array.

closely together in the liquid. Thus an unusual property of water is that the solid is less dense than the liquid.

Although the structure of liquid water is still not completely understood, it appears that clusters of hydrogen-bonded water molecules exist throughout the liquid. Some uncombined water molecules are probably present as well (Fig. 12-4). One theory of the structure of liquid water suggests that the clusters contin-

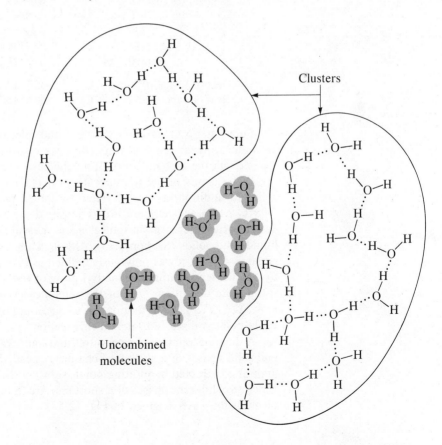

Figure 12-4. The structure of liquid water, showing hydrogen-bonded clusters and uncombined molecules.

ually vary in size and shape because they pick up uncombined molecules and lose molecules that become uncombined. As the temperature is raised, hydrogen bonds are disrupted, and the larger clusters are broken down into smaller ones. The association of water molecules in clusters may be responsible for the abnormally low vapor pressure, high boiling point, high heat of fusion, and high heat of vaporization of water. These properties are due to the energy required to break the hydrogen bonds holding the clusters together.

12.2 Physical Properties of Water

Water is the most widely used substance and is a standard for the determination of a number of physical constants and units. In addition, it has many remarkable, almost unique, properties. These properties include the expansion of the liquid when cooled below 4°C and the abnormally high melting point, boiling point, heat of fusion, and heat of vaporization for a molecule of such small molecular weight (Table 12-1). The vapor pressures of ice and water at various temperatures are listed in Table 10-1.

Table 12-1.   Physical Properties of Water

Freezing point:	0°C, 273.15 K
Boiling point:	100°C, 373.15 K
Density: ice (0°C)	0.9168 g/cm³
liquid (0°C)	0.99984 g/cm³
(3.98°C)	0.99997 g/cm³ (maximum density)
(25°C)	0.99704 g/cm³
(100°C)	0.95837 g/cm³
Heat of fusion (0°C):	6.010 kJ/mol
Heat of vaporization (100°C):	40.67 kJ/mol

Pure water is an odorless, tasteless, and colorless liquid. Large bodies of natural waters appear bluish-green. The taste of drinking water is due to dissolved gases (from the air) and salts (from the earth).

Water is one of the few substances that is more dense as a liquid than as a solid. In addition, as liquid water warms, it reaches its maximum density at 3.98°C. Above this temperature its density decreases with increasing temperature. Water is the reference substance in the measurement of the specific gravity of liquids and solids (Section 1.13). Although the calorie is now defined as being equal to 4.184 J, it was previously defined as being the amount of heat necessary to raise the temperature of exactly 1 gram of water from 14.5°C to 15.5°C. This is still a close definition. One of the reference points on the Kelvin temperature scale, 273.15 K, is by definition the temperature of the triple point of water. The triple point is discussed later in this section.

A diagram relating the pressures and temperatures at which gaseous, liquid, and solid phases of a substance can exist is called a **phase diagram.** Such a diagram is constructed by plotting points showing the pressures and temperatures at which two different phases of a substance are in equilibrium. A part of the phase diagram for water is given in Fig. 12-5.

Figure 12-5. Phase diagram for water.

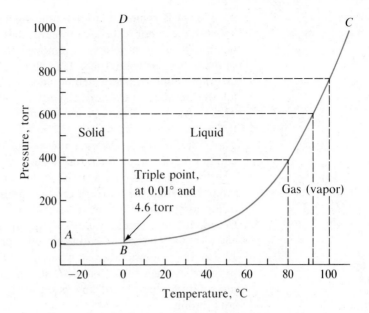

Figure 12-6. Syringes containing water at 760 torr. (a) At 80°C and 760 torr, only liquid water is contained in the syringe. (b) If the water is warmed to 100°C and then a little additional heat is added, the syringe will contain both liquid water and water vapor at 100°C and 760 torr.

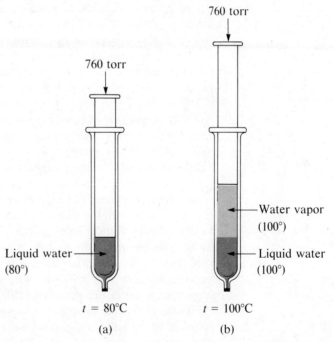

If water at 80°C is confined in a syringe such as that shown in Fig. 12-6(a), with a pressure of 760 torr on the plunger, only liquid water will be present. As heat is added, the temperature of the water will increase. After the water has been warmed to 100°C (at 760 torr), addition of more heat will produce water vapor with a pressure of 760 torr [Fig. 12-6(b)]. However, the temperature will remain constant at 100°C until all of the liquid water has changed to vapor. If the addition of heat is stopped and no heat is allowed to escape when both liquid water and water vapor are present, the mixture will remain at equilibrium.

If heat is removed from water vapor at 110°C and a constant pressure of 760 torr, the vapor will cool until it reaches 100°C, at which point liquid water will begin to form. If more heat is removed, the temperature will not change until all of the vapor has condensed. Thus the point at 100°C and 760 torr on the phase diagram is one point at which liquid water and water vapor are in equilibrium. At a lower pressure of 600 torr, the equilibrium between liquid and vapor occurs at 93.5°C. At 380 torr, the equilibrium occurs at 80.3°C. These points are located on the colored curve *BC* in Fig. 12-5. This boundary gives the vapor pressure of water (Section 11.2), the pressure at which liquid water and water vapor are in equilibrium as the temperature changes.

Water in a syringe at 10°C and a pressure of 380 torr will cool as heat is removed. When the temperature reaches 0.005°C, further loss of heat will result in the formation of ice, with no change in temperature until all of the liquid is frozen. Ice and liquid water are in equilibrium at 0.005°C at a pressure of 380 torr. At 760 torr the equilibrium occurs at 0°C. The colored line *BD* in Fig. 12-5 gives the temperature at which ice and liquid water are in equilibrium as the pressure changes. The colored line *AB* gives the vapor pressure of ice as the temperature changes; at any point along this line, ice and water vapor are in equilibrium.

At any combination of pressure and temperature within the portion of the diagram labeled "Solid," water exists only in the solid form (ice). Similarly, at values of pressure and temperature within the portion labeled "Liquid," water exists only as a liquid. At any combination of pressure and temperature in the region labeled "Gas," water exists only in the gaseous state. For example, the point at −5°C and 760 torr is in the region on the diagram where only solid water is stable. Moving across the diagram horizontally at a constant pressure of 760 torr, we see that as the temperature increases water becomes a liquid at 0°C, which is its *normal melting temperature*. At this temperature and pressure both solid and liquid water can exist in equilibrium. Water is stable only as a liquid at a pressure of 760 torr and temperatures between 0°C and 100°C. It changes to a gas at 100°C, its *normal boiling temperature,* at which liquid water at 760 torr and water vapor at 760 torr can exist in equilibrium. At temperatures above 100°C at 760 torr, water is stable only as a gas.

The almost vertical colored line *BD* in Fig. 12-5 shows that, as pressure increases, the melting temperature remains almost constant (actually decreasing very slightly). On the other hand, the diagram indicates clearly that the boiling temperature increases markedly with increasing pressure. Representative vapor pressures for water are given in Table 10-1.

The lines that separate the various regions in Fig. 12-5 represent collections of points at which an equilibrium exists between two states. Combinations of pressure and temperature that fall on the line between the solid and liquid regions, for example, represent conditions under which the solid and liquid are in equilibrium with each other. One point exists where all three phases are in equilibrium. This point, called the **triple point,** occurs at point *B* on the diagram in Fig. 12-5, where all three lines intersect. At the pressure and temperature of the triple point (4.6 torr and 0.01°C, or 273.15 K), all three states are in equilibrium with each other. At temperatures and pressures below those of the triple point, water changes directly from solid to gas without going through the liquid state (sublimation; see Section 11.12). Frost forms under these conditions as water vapor condenses to ice.

12.3 Chemical Properties of Water

Water is a medium in which many chemical reactions take place. Most biochemical reactions, many geological processes, some industrial reactions, and the majority of the chemical changes studied in a course in general chemistry take place in aqueous solutions.

1. IONIZATION OF WATER. Water can act as a base and accept hydrogen ions from an acid to form the hydronium ion (Section 8.1).

$$HCl(g) + H_2O(l) \longrightarrow H_3O^+(aq) + Cl^-(aq)$$

Water can also act as an acid and provide hydrogen ions to a base such as calcium oxide.

$$CaO(s) + H_2O(l) \longrightarrow Ca^{2+}(aq) + 2OH^-(aq)$$

Thus water can act either as a base or as an acid (depending on the conditions) and can either accept or donate hydrogen ions. In pure water there is a very limited transfer of hydrogen ion from one water molecule to another, giving a very low concentration of hydronium ion and hydroxide ion.

$$2H_2O(l) \rightleftharpoons H_3O^+(aq) + OH^-(aq)$$

Even in completely pure water, there are very small, *but equal,* amounts of H_3O^+ and OH^-. Many of the chemical properties of water depend on its content of hydronium ions and hydroxide ions.

2. REACTION OF WATER WITH CERTAIN ELEMENTS TO GIVE HYDROGEN. Water is a very weak acid, and this is reflected in its reactions with metals. Only those metals that lie above cobalt in the activity series (Section 9.17) react with water and produce hydrogen. The more active metals react with cold water (producing the hydroxides and hydrogen), but the less active ones must be heated to high temperatures in steam (generally producing the oxides and hydrogen).

$$2Na + 2H_2O(cold) \longrightarrow 2Na^+ + 2OH^- + H_2$$
$$3Fe + 4H_2O(steam) \longrightarrow Fe_3O_4 + 4H_2$$

Magnesium reacts with hot water to give the hydroxide and hydrogen.

$$Mg + 2H_2O(hot) \longrightarrow Mg(OH)_2 + H_2$$

When magnesium burns in steam, the oxide rather than the hydroxide results.

$$Mg + H_2O(steam) \longrightarrow MgO + H_2$$

Although it is not a metal, white-hot carbon also reacts with steam to produce a mixture of carbon monoxide and hydrogen called **water gas.**

$$C(s) + H_2O(g) \longrightarrow CO(g) + H_2(g)$$

Because coal can be used as the source of carbon, water gas is used extensively as a fuel gas, as a source of hydrogen for the manufacture of ammonia, and as a source of carbon monoxide for the production of methanol, other alcohols, and plastics.

3. ACID-BASE REACTIONS OF WATER WITH OXIDES. The oxide ion, O^{2-}, is a

very strong base that will remove a hydrogen ion from the weakly acidic water molecule.

$$O^{2-} + H_2O \longrightarrow 2OH^-$$

Soluble ionic metal oxides react with water to give hydroxides. The oxides of the alkali metals (lithium, sodium, potassium, rubidium, and cesium) and several of the alkaline earth metals (calcium, strontium, and barium) react readily with water in exothermic reactions, forming hydroxides.

$$Na_2O(s) + H_2O(l) \longrightarrow 2Na^+(aq) + 2OH^-(aq)$$

$$SrO(s) + H_2O(l) \longrightarrow Sr^{2+}(aq) + 2OH^-(aq)$$

The oxides of the alkaline earth metals beryllium and magnesium are insoluble in water and react only slowly and incompletely. Most other metal oxides are also insoluble and do not react with water. Metal oxides that react with water to form bases are sometimes called **basic anhydrides.**

Oxides of nonmetals, such as CO_2, SO_2, and P_4O_{10}, are covalent molecules that do not contain oxide ions. Most of these molecules react with water to form acids and are sometimes called **acidic anhydrides.** The amount of carbon dioxide, CO_2, in the atmosphere is controlled, in part, by its reaction with the water in the oceans to give carbonic acid.

$$CO_2(g) + H_2O(l) \longrightarrow \underset{\text{Carbonic acid}}{H_2CO_3(aq)}$$

Some acid rain results from the reaction of sulfur dioxide, SO_2, formed during the burning of sulfur-containing coal, with rain water.

$$SO_2(g) + H_2O(l) \longrightarrow \underset{\text{Sulfurous acid}}{H_2SO_3(aq)}$$

The reaction of phosphorus(V) oxide, P_4O_{10}, with water is used to manufacture phosphoric acid for fertilizers or beverages.

$$P_4O_{10}(s) + 6H_2O(l) \longrightarrow \underset{\text{Phosphoric acid}}{4H_3PO_4(l)}$$

4. HYDROLYSIS.   **Hydrolysis** is a general term for any reaction in which the water molecule is split and a hydrogen atom, an oxygen atom, and/or a hydroxide group from the water is incorporated into one or more of the products. Hydrolysis is generally observed for compounds with ionic or polar bonds. The positive end of the bond is attracted to the negative oxygen in a water molecule, and the negative end of the bond, to the positive hydrogen.

Compounds that undergo hydrolysis include almost all polar nonmetal halides (except those of carbon) as illustrated by the following equations:

$$SiBr_4(l) + 4H_2O(l) \longrightarrow Si(OH)_4(s) + 4HBr(aq)$$

$$PCl_5(l) + 4H_2O(l) \longrightarrow (HO)_3PO(aq) + 5HCl(aq)$$

One product of the second reaction, phosphoric acid, is more commonly written as H_3PO_4.

Acid-base reactions between water and the salts of weak acids or weak bases are also hydrolysis reactions. Examples include the following:

$$Li_3P(s) + 3H_2O(l) \longrightarrow 3Li^+(aq) + 3OH^-(aq) + PH_3(g)$$

$$MgS(s) + 2H_2O(l) \longrightarrow Mg(OH)_2(s) + H_2S(g)$$

Reactions of this type will be considered more fully in Chapter 16.

Chlorine and bromine also undergo hydrolysis to a limited extent. The reaction of chlorine with water at room temperature is

$$Cl_2(g) + 2H_2O(l) \rightleftharpoons \underset{\text{Hydrochloric acid}}{H_3O^+(aq) + Cl^-(aq)} + \underset{\substack{\text{Hypochlorous} \\ \text{acid}}}{HOCl(aq)}$$

Hydrochloric acid, a strong acid, is completely ionized, while hypochlorous acid, a weak acid, is only slightly ionized, most of it being present as molecules. Bromine reacts similarly; but fluorine oxidizes water, producing a mixture of ozone, O_3, oxygen, O_2, and oxygen difluoride, OF_2.

5. HYDRATION.   The combination of water with other substances without any splitting of the water molecule is called **hydration.**

When aqueous solutions of salts are evaporated, the salt often separates as crystals that contain the salt and molecules of water combined in definite proportions. Such compounds are known as **hydrates,** and the water is called **water of hydration.** General chemistry laboratories often use solid samples of the hydrates copper sulfate 5-hydrate (sometimes referred to as copper sulfate pentahydrate), $CuSO_4 \cdot 5H_2O$, sodium sulfate 10-hydrate, $Na_2SO_4 \cdot 10H_2O$, and cobalt nitrate 6-hydrate, $Co(NO_3)_2 \cdot 6H_2O$.

Acids, bases, and even some free elements also form hydrates. Oxalic acid, which is often used in oxidation-reduction titrations, crystallizes as a hydrate, $H_2C_2O_4 \cdot 2H_2O$. Evaporation of a cold solution of chlorine and water may yield crystals of chlorine 8-hydrate, $Cl_2 \cdot 8H_2O$.

Many hydrates lose water upon heating. These reactions are usually reversible. For example, when blue crystals of $CuSO_4 \cdot 5H_2O$, are heated, water is given off, the crystalline structure characteristic of the hydrated salt breaks down, and the white anhydrous copper sulfate salt, $CuSO_4$, remains.

$$CuSO_4 \cdot 5H_2O \overset{\triangle}{\rightleftharpoons} CuSO_4 + 5H_2O$$

This reaction is reversible—the addition of water to the anhydrous salt produces the hydrated salt.

Water of hydration may be bound in a crystal in a number of ways. X-ray diffraction has shown that the copper(II) ion in $CuSO_4 \cdot 5H_2O$ is combined with four water molecules by coordinate covalent bonds. The remaining water molecule is hydrogen-bonded to the sulfate anion. Water molecules may also hydrogen-bond to other water molecules in a hydrate. The structure of $Cl_2 \cdot 8H_2O$ consists of a cage of hydrogen-bonded water molecules surrounding a chlorine molecule. Water can also hydrogen-bond to other molecules in a solid, as in $H_2C_2O_4 \cdot 2H_2O$. Finally, water may occupy relatively open spaces in a crystal, such as the holes in the three-dimensional network of the SiO_4 tetrahedra in zeolites (Section 12.11).

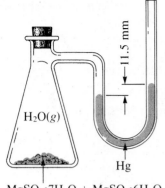

MgSO$_4$·7H$_2$O + MgSO$_4$·6H$_2$O

Figure 12-7. The vapor pressure of MgSO$_4$·7H$_2$O at 25°C. The pressure of water vapor above a sample of MgSO$_4$·7H$_2$O is 11.5 torr.

12.4 Efflorescence and Deliquescence

Many hydrates exhibit measurable vapor pressures. For example, if a sample of Epsom salts, MgSO$_4$·7H$_2$O, is put in a flask like the one shown in Fig. 12-7, at 25°C the pressure in the flask will increase by 11.5 torr due to the reversible reaction

$$MgSO_4 \cdot 7H_2O(s) \rightleftharpoons MgSO_4 \cdot 6H_2O(s) + H_2O(g)$$

The pressure of the water vapor is called the **vapor pressure of the hydrate.** This pressure increases with a rise in temperature and decreases as the temperature is lowered.

When the vapor pressure of a hydrate is higher than the partial pressure of the water vapor in the atmosphere, the hydrate will lose part or all of its water of hydration if exposed to the air. Hydrates that lose water of hydration in this way are said to **effloresce,** and **efflorescence** is said to have taken place. Na$_2$SO$_4$·10H$_2$O, which has a vapor pressure of 14 torr at 25°C, effloresces if the partial pressure of the water vapor in the air is less than 14 torr. It is stable if the partial pressure is more than 14 torr. Some hydrates are stable when the humidity is high but effloresce when it is low. A salt may form more than one hydrate, each of which possesses its own definite vapor pressure at a given temperature. For example, the following are known hydrates of copper(II) sulfate: CuSO$_4$·5H$_2$O, CuSO$_4$·3H$_2$O, and CuSO$_4$·H$_2$O.

Some compounds, such as CaCl$_2$, form hydrates when exposed to moist air. Such a compound is said to be **hygroscopic.** Hygroscopic compounds such as CaCl$_2$, H$_2$SO$_4$, and MgSO$_4$ are used as drying agents, or **desiccants.** Some water-soluble hygroscopic solids remove enough water from the air to dissolve completely in this water and form solutions. Such substances are said to be **deliquescent,** and the process is termed **deliquescence.** Very soluble salts, such as CaCl$_2$·6H$_2$O, are often extremely deliquescent.

12.5 Heavy Water

Water composed of deuterium, $_1^2$H, and oxygen is known as **heavy water,** or deuterium oxide, D$_2$O. Deuterium is present in light (ordinary) water in the ratio of 1 part D for every 6900 parts H. Heavy water was first obtained by the electrolysis of ordinary water. During electrolysis molecules containing deuterium migrate to a cathode more slowly than light water molecules; therefore the heavy water molecules tend to remain behind while the light ones are decomposed. Heavy water is now produced by an exchange equilibrium process, which is less costly and more efficient than the electrolytic method. In the exchange process liquid water and gaseous hydrogen sulfide are mixed. The deuterium atoms exchange between these two compounds, preferentially combining with oxygen as D$_2$O at low temperatures and with sulfur as D$_2$S at elevated temperatures.

$$H_2S + D_2O \underset{\text{Cold}}{\overset{\text{Hot}}{\rightleftharpoons}} D_2S + H_2O$$

Hydrogen sulfide, bubbled through water at an elevated temperature, is enriched with D$_2$S through this exchange. After the enriched hydrogen sulfide is drawn off, the reverse exchange at lower temperatures produces water that contains a larger proportion of D$_2$O than it did. Many repetitions of the process yield water with a D_2O concentration of about 15%, which is then enriched to 99.8% by fractional distillation and electrolysis. To produce 1 ton of D_2O a plant must process 45,000 tons of water and must cycle 150,000 tons of hydrogen sulfide. Several hundred tons of heavy water are produced each year.

Heavy water resembles ordinary water in appearance but differs from it slightly in other physical properties (Table 12-2). The chemical behavior of the two is virtually identical, although salts have slightly lower solubilities in D_2O than in H_2O, and reactions in D_2O are somewhat slower than in H_2O.

Table 12-2   Some Physical Properties of Ordinary Water and Heavy Water

Property	Ordinary Water	Heavy Water
Density at 20°, g/mL	0.997	1.108
Boiling point, °C	100.00	101.41
Melting point, °C	0.00	3.79
Heat of vaporization, kJ/mol	40.7	41.61
Heat of fusion, kJ/mol	6.01	6.3

Heavy water is one of the moderators used in nuclear reactors to reduce the speed of neutrons to that required for the fission process to take place (Chapter 28).

It is possible to detect the presence of as little as 1 part of deuterium in 100,000 parts of hydrogen. For this reason heavy water and deuterium serve as valuable tracers in the study of both chemical and physiological changes. By replacing ordinary hydrogen with deuterium in molecules of food, the processes of digestion and metabolism in the body can be studied.

NATURAL WATERS

12.6   Water, an Important Natural Resource

Although there is enough water in the oceans to cover the entire earth to a depth of about two miles, human needs are generally for fresh water. Unfortunately, however, the supply of fresh water is becoming critically low in many parts of the United States as the population increases.

About a half-gallon of water per day is required to satisfy the biological needs of a human being. Nearly 150 gallons per day per person are used in the United States in maintaining cleanliness, in the cooking of food, and in heating and air-conditioning homes. These quantities are dwarfed by the 750 gallons per person per day used by industries in the United States—5 gallons of water are used to produce 1 gallon of milk, 10 gallons for 1 gallon of gasoline, 80 gallons for 1 kilowatt of electricity, 65,000 gallons for 1 ton of steel, and 3,000,000 gallons for 1 ton of nylon. More water is used for irrigation in the United States than for any other single purpose, the average quantity used being over 750 gallons per person per day.

Fresh water, unlike most mineral resources, is renewable because it is a part of a gigantic cycle. Water evaporates from the earth (the land and the oceans) into the atmosphere; it is transported in the atmosphere by the winds; and it finally falls back to the earth as some form of precipitation. A large part of the water that precipitates is lost to human use, however, due to evaporation and transpiration by plants.

At the present time the oceans are becoming increasingly important as potential sources of fresh water because of recent advances in devising economical and practical desalting processes. It has been estimated that about two billion gallons of pure drinking water are produced per day worldwide by desalting sea water.

12.7   Naturally Occurring Water

All natural waters, even when not polluted by people, are impure since they contain many dissolved substances. Rain water is relatively pure; its chief impurities are dust and dissolved gases. After rain has fallen for some time, the air will have been washed free of polluting gases, dust, and bacteria, and any subsequent rain will be quite free of such impurities. Sea water contains about 3.6% dissolved solids, principally sodium chloride (Table 12-3). Sea water is known to contain at least 72 elements, and it is possible that all naturally occurring elements exist in the sea.

The impurities in the fresh water on the earth's surface vary with the nature of the soil and rocks that it has passed over or through. Increasingly, fresh water also contains impurities added by people. The natural impurities may include dissolved gases such as oxygen, nitrogen, and carbon dioxide (from the air) and ammonia and hydrogen sulfide; dissolved salts; dissolved organic substances from the decay of plant and animal matter; and suspended solids such as sand, clay, silt, organic material; and microorganisms.

During the normal life cycle of a body of fresh water such as a lake, organic sediment slowly accumulates from algae, bacteria, aquatic plants, and animal by-products. Bacteria and other organisms in the lake feed on this sediment and, in the presence of oxygen, decompose the sediment to various molecules and ions, including nitrates and phosphates. This decomposition of organic matter by bacteria in the presence of oxygen is called **aerobic decomposition.** If the supply of oxygen is reduced or if the amount of organic sediment increases to the point where aerobic decomposition cannot keep up with it, the aerobic bacteria die, and other bacteria that live in the absence of oxygen thrive. They too decompose the sediment, but the products are much less pleasant. Decomposition of organic matter in the absence of oxygen is called **anaerobic decomposition.** Anaerobic decomposition is accompanied by the smell of rotten eggs due to hydrogen sulfide (which in fact is the substance responsible for the unpleasant odor of rotten eggs); bubbles of methane are usually visible; and the water is black and often filled with slime.

The process by which a lake grows rich in nutrients (phosphates and nitrates, primarily) and gradually fills with organic sediment and aquatic plants is called **eutrophication.** As a lake ages naturally, it becomes shallower because it fills with sediment, and plant life abounds on the banks and within the lake itself. The lake is slowly transformed into a marsh and ultimately into dry land by the processes accompanying eutrophication.

Table 12-3   The Most Abundant Elements Dissolved in Sea Water (excluding dissolved gases)

Element	Concentration, g/L
Cl	18.98
Na	10.56
Mg	1.272
S	0.884
Ca	0.400
K	0.380
Br	0.065
C[a]	0.028
Sr	0.013
B[b]	0.0045
F	0.0014

[a] In CO_3^{2-} and HCO_3^-
[b] In H_3BO_3

12.8 Water Pollution

In addition to the natural contaminants in water, many other pollutants are added to our water supplies by people. The very existence of this problem went unrecognized for many years, and its magnitude may not be realized yet. Unfortunately, very little is known of the long-term effects on human health of the many, largely unidentified chemical compounds that enter water supplies. A whole new field of chemistry is developing rapidly in which pollutants in our *overall* environment, not just in our water supplies, are studied in connection with their short-range and long-range effects on human health and with their efficient removal.

Several water pollutants are currently causing serious concern, and their effects are worthy of our attention.

1. PHOSPHATES AND OTHER NUTRIENTS. Phosphates have been used widely in laundry detergents and hence have been released in large quantities in municipal wastes into rivers and streams. Furthermore, many fertilizers contain phosphates (Section 8.7). A number of phosphate compounds are also present in certain rocks. Thus additional phosphates reach streams through natural runoff. A high concentration of phosphates in streams and lakes presents a problem because such substances are significant nutrients for algae, bacteria, and aquatic plants. Other nutrients include nitrogen (in the form of ammonia, ammonium salts such as ammonium sulfate, and nitrate salts such as calcium nitrate) and potassium (in potassium salts such as potassium nitrate and potassium sulfate), which are released in municipal wastes, in the runoff from fertilizer application, and in the wastes from animals.

The growth of algae and other aquatic plants is usually controlled by the amount of nutrients available. Excess nutrients lead to rapid plant growth with a consequent rapid increase in the rate of eutrophication of natural waters. In the worst cases the deposition of organic sediment can become so extensive that anaerobic bacteria replace the aerobic bacteria and the lake is no longer able to support life.

Development of effective long-term controls of eutrophication will require a better understanding of the significant nutrients in streams and lakes, the natural plant and animal populations in them, and how these are affected by externally applied factors such as the added nutrients. In the short range, substitutes for phosphates in detergents are now in use, and methods of removing phosphates from and controlling their undesirable effects in municipal wastes are increasingly being used.

2. SEWAGE AND OTHER WASTE MATERIALS. In addition to nutrients, municipal sewage contains microorganisms as well as the by-products from many industrial processes. In some communities raw sewage is dumped directly into natural waters that serve as sources of public water supplies. Contamination may also result from breakdowns in sewage treatment plants or from flooding of such plants. Chlorination of municipal water supplies (Section 12.9) has largely eliminated dangers from bacteria and viruses in drinking water. However, many bodies of natural water are still contaminated with microorganisms and certain chemical by-products, which may pose hazards in domestic water supplies.

The presence of organic matter, nutrients, and microorganisms in water is measured by three tests: **coliform count, algal count,** and **biological oxygen de-**

mand (BOD). A coliform count determines the number of *E. coli* (the characteristic bacteria in animal wastes) present. The algal count is a test for microorganisms other than bacteria and viruses. The BOD measures the volume of oxygen gas taken up by a given amount of water in five days at 20°C. Since the oxygen is consumed by microorganisms in the water, the BOD is a measure of the total quantity of microorganisms and the nutrients available to them. Pure water has a BOD of 1 part per million (equivalent to 1 milligram of O_2 per liter of water) or less. A BOD of 20 parts per million or less is acceptable for municipal or industrial sewage. Raw sewage exhibits a BOD of 100–400 parts per million.

The coliform count, the algal count, and the BOD do not determine the presence of small amounts of organic compounds, often resulting from industrial wastes, but specialized tests for compounds such as benzene, acetone, toluene, chloroform, and carbon tetrachloride have shown small amounts (1–40 parts per billion) of these materials in the drinking water of cities in industrial river basins. The normal water purification procedures do not remove these compounds. In fact, the presence of chloroform, $CHCl_3$, in domestic water supplies may result from the reaction of chlorine (used in water purification) with naturally occurring organic compounds that result from the decay of plant matter in the soil. Some of these organic pollutants can be removed by filtering domestic water through activated charcoal.

Other sources of industrial chemical wastes include improperly constructed waste storage or treatment facilities and the random (and often illegal) discharge or dumping of such wastes.

3. TOXIC POLLUTANTS.

Many waters contain pesticides (organic compounds that are used to kill insects, weeds, and fungi). Pesticides can concentrate in the food chain of fish and wildlife and kill these species or decrease their ability to reproduce. Some of these compounds, such as Aldrin, DDT, Dieldrin, and Kepone, have proved so toxic that their manufacture and use has been prohibited.

The very poisonous compound methylmercury is produced in inland waters by the action of microorganisms in the bottom mud on mercury, which may remain from prior discharge of waste mercury and mercury compounds. The discharge of mercury and mercury compounds is now prohibited. Methylmercury attacks the central nervous system. Furthermore, methylmercury is retained in the human body for long periods so that, even with small rates of intake, toxic amounts can be accumulated.

Microorganisms in the intestines of some animals, including chickens, can methylate mercury. It also is known that fish take in methylmercury both through their food and through their gills to the extent that the concentration in their flesh may be several thousand times that in the surrounding water. Hence, the possibility of mercury poisoning in humans from eating fish (or chicken and other animals) is greater than might be expected from the actual concentration of the element in water, air, or soil.

Considerable attention has recently been devoted to pollution by cadmium, an element in the same periodic family as mercury. The discharge of cadmium as well as other toxic pollutants such as cyanide, benzidine, and polybromobiphenyl (PBB) compounds, has also been prohibited, but it may be some time before these materials are absent from natural waters.

4. ACID RAIN. Rain with an acid concentration of 0.0001–0.001 M is falling over the eastern United States and in Minnesota, Colorado, Oregon and Washington, southeastern Canada, Scandinavia, Japan, and China. Sulfuric acid appears to be the principal component of this acid rain. Such rain increases the acid content of many lakes to the point where plants and animals cannot live in the water. It may also be stripping the soil of elements such as calcium, magnesium, and potassium, which are nutrients necessary for plant life. Some of this loss, however, appears to be made up by the sulfates, nitrates, and ammonium salts present in the rain.

Natural sources of acid in rain include lightning, volcanic eruptions, and anaerobic decomposition of organic matter. However, many scientists believe that most acid rain is a result of the combustion of fossil fuels. These fuels, particularly coal, contain sulfur, which forms gaseous sulfur dioxide when burned. Although the mechanism of the transformation is not fully clear, the sulfur dioxide ends up as the sulfuric acid in acid rain.

It has been suggested that acid rain could be significantly reduced if emission control technology and/or low-sulfur coal were used. However, a switch to low-sulfur coal could depress the midwestern coal mining industry, and consumers of electricity in many states would face substantial rate hikes to pay for the emission control equipment. Some scientists claim that such controls would have very little effect on the acidity of rainfall. The acid rain problem illustrates the conflict between the need to respond promptly to a problem and the need to gather enough information to act effectively.

5. THERMAL POLLUTION. Discharges that raise the temperature of a body of water may harm fish and increase the growth of algae and other microorganisms. A few degrees rise in water temperature due to thermal pollution can decrease the solubility of oxygen in the water to a level too low for many forms of marine life to survive. Cooling towers and other facilities to reduce the heat transferred to natural waters are increasingly employed in industrial situations.

It is conceivable that thermal pollution could be beneficial under certain conditions. For example, the Ralston Purina Company and the Florida Power Company, which has replaced a conventional power plant with a nuclear power plant, cooperated in a study to determine if selected saltwater fish and shellfish might thrive in warmer water, thereby improving the food supply from the sea.

12.9 Purification of Water

Water is purified for city water supplies by first allowing it to stand in large reservoirs where most of the mud, clay, and silt settle out, a process called **sedimentation,** and then filtering it through beds of sand and gravel. Prior to filtration, lime and aluminum sulfate are often added to the water. These chemicals react to form aluminum hydroxide as a gelatinous precipitate that settles slowly, carrying with it much of the suspended matter, including most of the bacteria. The bacteria that remain in the water after filtration are killed by adding chlorine.

Relatively pure water for laboratory use is commonly prepared by the distillation of tap water (Section 11.6). Because the basic constituents of glass slowly dissolve in water, glass equipment is not satisfactory for the preparation and storage of very pure water. Gases from the air, particularly carbon dioxide, often contaminate distilled water and must be removed for some laboratory operations.

12.10   Hard Water

Water containing dissolved calcium, magnesium, and iron salts is known as **hard water.** The negative ions present in hard water are usually chloride, sulfate, and hydrogen carbonate. Hardness in water is objectionable for two reasons:

1. The calcium, magnesium, and iron ions form insoluble soaps such as calcium stearate, $(C_{17}H_{35}CO_2)_2Ca$, by reaction with soluble soaps such as sodium stearate, $C_{17}H_{35}CO_2Na$. Insoluble soaps have no cleansing power, and they adhere to fabrics, giving them a dingy appearance. They also make up the ring in the bathtub. Soap in excess of that needed to precipitate the calcium and magnesium must be added to hard water in order to obtain cleansing action.

2. Hard water is responsible for the formation of boiler scale. At high temperatures much of the mineral matter precipitates as *scale* [substances such as calcium carbonate, magnesium carbonate, iron(II) carbonate, and calcium sulfate]. A major component of some scale is calcium sulfate, which is *less* soluble in hot water than in cold and precipitates partly because of that fact. The scale is a poor conductor of heat and thus causes a waste of fuel. Furthermore, scale may cause boiler explosions. Since it is not a good heat conductor, the metal under it must be very hot (often red hot) in order to boil the water in the boiler. If the scale cracks, the water seeps through and comes in contact with the hot metal.

$$4H_2O + 3Fe(hot) \longrightarrow Fe_3O_4 + 4H_2$$

As it is formed, the hydrogen gas breaks the scale loose and more water gets in, producing still more hydrogen and setting the stage for a violent explosion.

12.11   Water-Softening

It is important that the substances responsible for hardness in water be removed before the water is used for washing or in boilers. The removal of the metallic ions responsible for the hardness is **water-softening.**

Water that contains Ca^{2+}, Mg^{2+}, or Fe^{2+} as hydrogen carbonate salts can be softened by boiling. This drives off carbon dioxide, and the hardness is removed as the metal carbonates precipitate.

$$Ca^{2+} + 2HCO_3^- \xrightarrow{\triangle} CaCO_3(s) + CO_2 + H_2O$$

Carbonates of calcium, magnesium, and iron form some of the deposits found in teakettles and boilers. The hydrogen carbonate ion may also be converted to the carbonate ion by the addition of a base. Commercially, calcium hydroxide is added in the quantity needed to react with the hydrogen carbonate ions to form insoluble calcium carbonate.

$$Ca^{2+} + 2OH^- + Ca^{2+} + 2HCO_3^- \longrightarrow 2CaCO_3(s) + 2H_2O$$

Similar reactions remove Mg^{2+} or Fe^{2+} as insoluble magnesium or iron(II) carbonates. The precipitated carbonates are readily removed from the water by filtration.

Unlike hydrogen carbonate salts, chloride and sulfate salts are not removed by boiling. Crude sodium hydroxide (caustic soda) is often used in large installations to remove both carbonate and noncarbonate salts.

$$Ca^{2+} + 2HCO_3^- + [2Na^+] + 2OH^- \longrightarrow$$
$$CaCO_3(s) + [2Na^+] + CO_3^{2-} + 2H_2O$$

The sodium carbonate produced then reacts to remove any remaining noncarbonate salts. For example,

$$Ca^{2+} + [SO_4^{2-}] + [2Na^+] + CO_3^{2-} \longrightarrow CaCO_3(s) + [2Na^+] + [SO_4^{2-}]$$

The sodium sulfate or sodium chloride produced is relatively soluble, and does not interfere with the cleansing action of soaps or form boiler scale.

Ion exchange provides another useful and highly important method of softening water. When water containing calcium, magnesium, and iron(II) ions filters slowly through layers of sodium aluminosilicates, those ions are replaced by sodium ions and the water is softened. Aluminosilicates are known as **zeolites.**

$$2NaAlSi_2O_6 \quad + \quad Ca^{2+} \longrightarrow \quad Ca(AlSi_2O_6)_2 \quad + \quad 2Na^+$$
Sodium aluminosilicate   Calcium aluminosilicate
(insoluble)   (insoluble)

Calcium aluminosilicate can be reconverted into the sodium compound using a concentrated solution of sodium chloride, thus regenerating the system to remove more calcium ions.

$$Ca(AlSi_2O_6)_2 + 2Na^+ \longrightarrow 2NaAlSi_2O_6 + Ca^{2+}$$

This reaction is the reverse of the preceding one, and the direction the reaction takes is controlled by an excess of either Ca^{2+} or Na^+.

$$2Na(AlSi_2O_6) + Ca^{2+} \rightleftharpoons Ca(AlSi_2O_6)_2 + 2Na^+$$

After the sodium zeolite is regenerated, it is ready for use again. The ion-exchange method of water-softening is effective for the removal of both carbonate and noncarbonate salts (Fig. 12-8).

Synthetic **ion-exchange resins** have also been developed to soften hard water. These insoluble resins remove both cations (positive ions) and anions (negative

Figure 12-8. Industrial water-softening by the ion-exchange method. When regeneration is necessary, the soft water outlet valve is closed, and a concentrated NaCl solution is run through the calcium zeolite and out the waste line until regeneration of the sodium zeolite is complete.

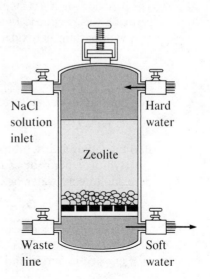

NaCl solution inlet

Hard water

Zeolite

Waste line

Soft water

ions) from hard water; that is, they demineralize the water completely. Examples of such resins are Amberlite and Zeo-Carb. Metal ions in the hard water displace hydrogen ions from one type of ion-exchange resin, RCO_2H (R represents a complex hydrocarbon portion of the resin that does not enter into the exchange reaction).

$$2RCO_2H + Ca^{2+} + [SO_4^{2-}] + 2H_2O \longrightarrow (RCO_2)_2Ca + 2H_3O^+ + [SO_4^{2-}]$$

Another type of ion-exchange resin then removes the ions of the acid from the water.

$$2RNH_2 + 2H_3O^+ + SO_4^{2-} \longrightarrow (RNH_3^+)_2SO_4^{2-}(s) + 2H_2O$$

Hence, if hard water (or a salt solution) is passed first through a resin that replaces all metal ions with H_3O^+ and then through a resin that combines with the resulting acid, it can be completely demineralized. The result is essentially the same as that achieved through distillation of water. In recent years the use of ion-exchange resins in the treatment of water for industrial, laboratory, and domestic purposes has grown tremendously.

HYDROGEN PEROXIDE

Hydrogen forms another compound with oxygen, **hydrogen peroxide,** H_2O_2, in which there is twice as much oxygen for the same weight of hydrogen as there is in water. Hydrogen peroxide was first prepared in 1818 by Thénard, who obtained it by treating barium peroxide with hydrochloric acid. Very small quantities of hydrogen peroxide are present in dew, rain, and snow, probably as a result of the action of ultraviolet light on oxygen mixed with water vapor.

12.12   Preparation of Hydrogen Peroxide

The peroxide ion is a strongly basic ion. In the nineteenth century dilute hydrogen peroxide was produced on a commercial scale exclusively by the reaction of an acid with barium peroxide, which is readily formed by heating barium oxide in air.

$$2BaO + O_2 \longrightarrow 2BaO_2$$

For the reaction of sulfuric acid with barium peroxide, the equation is

$$BaO_2(s) + 2H_3O^+(aq) + SO_4^{2-}(aq) \longrightarrow BaSO_4(s) + H_2O_2(aq) + 2H_2O(l)$$

Barium sulfate is insoluble and can be separated by filtration.

Hydrogen peroxide can also be prepared in the laboratory by adding sodium peroxide to cold water or cold dilute hydrochloric acid.

$$Na_2O_2 + 2H_2O \longrightarrow 2Na^+ + 2OH^- + H_2O_2$$

Hydrogen peroxide can be produced by the electrolysis of a solution of sulfuric acid and ammonium hydrogen sulfate to produce peroxydisulfuric acid, $H_2S_2O_8$, at the anode and hydrogen at the cathode.

$$2H_2SO_4(aq) + \text{electrical energy} \longrightarrow H_2S_2O_8(aq) + H_2(g)$$

The peroxydisulfuric acid is subsequently hydrolyzed to produce hydrogen peroxide.

$$H_2S_2O_8 + 2H_2O \longrightarrow 2H_2SO_4 + H_2O_2$$

The sulfuric acid formed is used again in the production of more hydrogen peroxide.

For many years the electrolytic process accounted for 80–90% of U.S. production of hydrogen peroxide. However, chemical processes have largely replaced it. In the most popular chemical process, an organic compound, 2-ethyl anthraquinone, is reduced by hydrogen in a benzene solution, using palladium as a catalyst, to 2-ethyl dihydroanthraquinone. When the dihydroanthraquinone is allowed to react with oxygen, the original anthraquinone is obtained along with hydrogen peroxide. The anthraquinone is then reused in the production of more hydrogen peroxide.

$$\underset{\substack{\text{2-Ethyl} \\ \text{anthraquinone}}}{C_{16}H_{12}O_2} + H_2 \xrightarrow{\text{Pd catalyst}} \underset{\substack{\text{2-Ethyl} \\ \text{dihydroanthraquinone}}}{C_{16}H_{12}(OH)_2}$$

$$C_{16}H_{12}(OH)_2 + O_2 \longrightarrow C_{16}H_{12}O_2 + H_2O_2$$

When the hydrogen peroxide reaches a concentration of about 0.15 mole per liter, it is extracted with water to produce an 18% aqueous solution, which may be further concentrated by vacuum distillation. Hydrogen peroxide decomposes when heated, necessitating vacuum distillation, which takes place at lower temperatures (lower boiling points at reduced pressures; see Section 11.3).

Very pure hydrogen peroxide is quite stable. Solutions whose concentration is 90% or higher are used routinely in a number of applications. However, when impurities are present, the concentrated solutions are dangerously explosive. For storage purposes a stabilizer is added.

12.13   Properties of Hydrogen Peroxide

Pure hydrogen peroxide is a colorless, syrupy liquid with a sharp odor and an astringent taste. Its relatively low vapor pressure and high boiling point ($150.2\,°C$) are caused by hydrogen bonding among its molecules. It is miscible (mutually soluble) with water, alcohol, and ether in all proportions. Hydrogen peroxide is thermally unstable and decomposes spontaneously in an exothermic reaction.

$$2H_2O_2(l) \longrightarrow 2H_2O(l) + O_2(g) \qquad \Delta H = -196 \text{ kJ}$$

Its instability is due, in part, to the fact that its formation is a highly endothermic process.

Hydrogen peroxide is an oxidizing agent. It oxidizes sulfurous acid, H_2SO_3, to sulfuric acid, H_2SO_4, and sulfides to sulfates.

$$H_2SO_3 + H_2O_2 \longrightarrow H_2SO_4 + H_2O$$
$$PbS + 4H_2O_2 \longrightarrow PbSO_4 + 4H_2O$$

It is useful as a bleach and an antiseptic because, like chlorine, it is a strong oxidizing agent. It has the advantage that the only by-product of its oxidation reactions is water.

Hydrogen peroxide also acts as a reducing agent. For example, it will reduce silver oxide to the metal and chlorine to chloride ion.

$$Ag_2O + H_2O_2 \longrightarrow 2Ag + H_2O + O_2$$

$$Cl_2(aq) + H_2O_2(aq) + 2H_2O(l) \longrightarrow O_2(g) + 2H_3O^+(aq) + 2Cl^-(aq)$$

Hydrogen peroxide reduces many other substances with the evolution of oxygen. These substances include the oxides of mercury, gold, and platinum and the strong oxidizing agents manganese dioxide and potassium permanganate. The titration of hydrogen peroxide with potassium permanganate in acid solution is a common general chemistry laboratory procedure. The reaction proceeds according to the equation

$$6H_3O^+ + 2MnO_4^- + 5H_2O_2 \longrightarrow 5O_2 + 2Mn^{2+} + 14H_2O$$

Hydrogen peroxide is a weak acid; in water it ionizes to a very slight extent in two steps.

$$H_2O_2(aq) \rightleftharpoons H^+(aq) + \quad HO_2^-(aq)$$
$$\text{Hydrogen peroxide ion}$$

$$HO_2^-(aq) \rightleftharpoons H^+(aq) + \quad O_2^{2-}(aq)$$
$$\text{Peroxide ion}$$

Its acidic character shows in its reaction with barium hydroxide.

$$Ba^{2+}(aq) + 2OH^-(aq) + H_2O_2(l) \longrightarrow BaO_2(s) + 2H_2O(l)$$

12.14 The Structure of Peroxides

The Lewis structure and molecular structure of hydrogen peroxide are shown in Fig. 12-9. Its molecular structure can be compared to a sheet of paper folded to an angle of 94°, with the oxygen atoms located in the fold and one hydrogen atom located in each plane.

Peroxides contain an oxygen-oxygen single bond; see, for example, Fig. 12-9 and the following Lewis structures of peroxymonosulfuric acid, H_2SO_5, peroxydisulfuric acid, $H_2S_2O_8$, and the ionic peroxide barium peroxide, BaO_2.

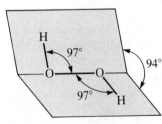

Figure 12-9. The Lewis structure and molecular structure of hydrogen peroxide, H_2O_2.

Peroxymonosulfuric acid   Peroxydisulfuric acid   Barium peroxide

The charge on the *peroxide* ion, which contains two oxygen atoms, is −2, while the usual charge for the single oxygen in an *oxide* ion is −2, as indicated for barium oxide and manganese dioxide:

$$Ba^{2+}[\,\ddot{\underset{..}{O}}\,{}^{2-}] \qquad Mn^{4+}[\,\ddot{\underset{..}{O}}\,{}^{2-}]_2$$

Barium oxide   Manganese dioxide

Manganese dioxide, unlike barium peroxide, does not yield hydrogen peroxide when treated with acids. This fact indicates that MnO_2 is a dioxide rather than a peroxide; i.e., the two oxygen atoms are not joined together but are present as oxide ions. There is no oxygen-oxygen bond in MnO_2.

12.15 Uses of Hydrogen Peroxide

A 3% solution of hydrogen peroxide in water is used as a mild antiseptic, a deodorizer, and a germicide; its effectiveness as a germicide is questionable, however. A somewhat more concentrated solution is a bleach for hair. A 30% solution of hydrogen peroxide is commonly used in the general chemistry laboratory as an oxidizing agent. In commerce the principal consumption of hydrogen peroxide is as an oxidizing agent, particularly in bleaching. It is often used to bleach substances of animal origin, such as wool and hair, which are harmed by most other bleaching agents. Most cotton cloth and paper pulp are bleached with hydrogen peroxide because of the innocuous character of the decomposition products and the permanency of the whiteness produced.

A 90% solution of hydrogen peroxide is used as an oxidizing agent in certain high explosives and in rockets because it releases a lot of energy on reaction, is relatively stable to shock, is noncorrosive and nontoxic, and has a high boiling point and low freezing point.

New uses are rapidly increasing in pollution control applications, such as treatment of sewage for dechlorination and the control of hydrogen sulfide odors.

FOR REVIEW

SUMMARY

The water molecule contains two bonding electron pairs and two lone pairs arranged tetrahedrally about the oxygen atom. The sp^3-hybridized oxygen forms covalent bonds to two hydrogen atoms with a bond angle of 105°. Water molecules are bound together in the liquid and solid states by hydrogen bonds, which give ice its open structure and make solid water less dense than liquid water. The hydrogen bonds give water its relatively high heats of fusion and of vaporization, its high boiling point, and its low vapor pressure. The conditions under which liquid water and solid water, solid water and water vapor, or liquid water and water vapor are in equilibrium are described by a **phase diagram.** At the **triple point** (0.01°C and 4.6 torr), all three phases of water are in equilibrium.

Water behaves both as a weak acid and a weak base. Pure liquid water contains small but equal amounts of hydronium ion and hydroxide ion due to the reversible reaction $2H_2O \rightleftharpoons H_3O^+ + OH^-$. Like other weak acids, water reacts with some metals, giving hydrogen (heating may be required). It will provide hydrogen ions to bases such as the oxide ion and other strongly basic anions. Covalent oxides such as sulfur dioxide react with water to form acids. Water will react with many covalent halides in **hydrolysis** reactions. In a hydrolysis reaction the water molecule is split, and a hydrogen atom, oxygen atom, and/or hydroxide group from the water is incorporated into the products. Some ionic halides, as well as other substances, form **hydrates,** which have water molecules incorporated into the compound. Many hydrates exhibit a measurable vapor pressure and may **effloresce** (lose water) or **deliquesce** (absorb water), depending on the relative humidity.

Heavy water, which is chemically very similar to water, consists of two deuterium atoms ($_1^2H$) bonded to an oxygen atom.

Water is a vital natural resource, satisfying important biological and industrial requirements. Fresh water is available from rivers, lakes, and wells, but it may require purification to remove sediment and microorganisms. It may also require softening in order to remove the Ca^{2+}, Mg^{2+}, and Fe^{2+} ions that produce **hard water.** Pollution of natural waters is a serious problem. This pollution is the result of excess nutrients that find their way into natural waters (leading to increased rates of **eutrophication**), toxic chemicals from sewage or chemical waste dumps, acid rain, or simply the discharge of heat into the waters.

Hydrogen and oxygen form another compound, **hydrogen peroxide,** H_2O_2, which contains an oxygen-oxygen single bond. Hydrogen peroxide may be prepared in the laboratory by acidification of ionic peroxides, compounds containing the peroxide ion O_2^{2-}, a strong base. It may also be prepared by an electrolytic method and by a chemical process beginning with the reduction of 2-ethyl anthraquinone. Hydrogen peroxide is a weak acid, and it will act as either an oxidizing agent or a reducing agent, depending on the chemical conditions. The principal uses of hydrogen peroxide are based on its oxidizing ability.

KEY TERMS AND CONCEPTS

Acidic anhydrides (12.3)
Aerobic decomposition (12.7)
Anaerobic decomposition (12.7)
Basic anhydrides (12.3)
Biological oxygen demand, BOD (12.8)
Chemical properties of hydrogen peroxide (12.13)
Chemical properties of water (12.3)

Deliquescence (12.4)
Efflorescence (12.4)
Eutrophication (12.7)
Hard water (12.10)
Heavy water (12.5)
Hydrate (12.3)
Hydrolysis (12.3)
Ion exchange (12.11)
Peroxides (12.14)
Phase diagram (12.2)

Physical properties of water (12.2)
Preparation of hydrogen peroxide (12.12)
Purification of water (12.9)
Structure of water (12.1)
Triple point (12.2)
Water pollution (12.8)

EXERCISES

Physical Properties of Water

1. Explain why water is a polar molecule.
2. Why does the H—O—H angle in water deviate from the angle expected for a tetrahedral molecule?
3. Under what conditions does water exist as single H_2O molecules?
4. Explain why water, H_2O, has a higher boiling point than hydrogen sulfide, H_2S, a molecule with a similar structure.
5. Using information about the structure of water

molecules and their interaction with one another, discuss the fact that the heat of fusion of water is only 334 J/g but the heat of vaporization of water is 2258 J/g.
6. Assume that the heat of fusion of water is the energy required to break about 10% of the hydrogen bonds in ice and that the heat of vaporization is the energy required to break the remaining hydrogen bonds. With these assumptions, about 2600 J are required to break the hydrogen bonds

in one gram of water. How much energy is required to break one mole of hydrogen bonds between water molecules?   *Ans. 23 kJ/ mol*

7. Explain why ice is less dense than liquid water.

8. From the phase diagram for water (Fig. 12-5) determine its physical state at:
   (a) 600 torr and 40°C
   (b) 400 torr and −25°C
   (c) 400 torr and 50°C
   (d) 200 torr and 90°C

9. (a) Can water be converted from a solid to a liquid at 3 torr? from a gas to a liquid? from a solid to a gas? from a gas to a solid?
   (b) If the pressure is 10 torr, can the changes in state in part (a) be accomplished?
   (c) Explain both (a) and (b) in terms of the phase diagram for water (Fig. 12-5).

10. Explain what is meant by the triple point of a substance.

11. Using the phase diagram for water (Fig. 12-5), determine the following:
    (a) At 400 torr what is the approximate temperature necessary to convert water from a solid to a liquid? from a liquid to a gas?
    (b) At 20°C what is the approximate pressure at which water changes from a gas to a liquid? at 66°C?

12. Using the phase diagram for water (Fig. 12-5), answer the following questions:
    (a) Holding the temperature constant, how may water be changed from gas to liquid and from liquid to solid?
    (b) At a temperature of −10°C, can water exist as a gas? as a liquid? as a solid? Explain.
    (c) Can each of the changes in part (a) be accomplished at −5°C? at +20°C? Explain.

13. What is heavy water, and how is it prepared?

14. Explain the difference in vapor density of H_2O and D_2O at 1 atm and 115°C.

15. The density of water vapor at 100°C is 0.58 g/L. What is the density of D_2O vapor at 100°C?
    Ans. 0.64 g/L

16. What mass of steam at 100°C must be converted to water at 100°C in order to convert 1.00 kg of $D_2O(l)$ at 101.4°C to $D_2O(g)$ at 101.4°C?
    Ans. 0.920 kg

Chemical Properties of Water

17. Write Lewis structures for water, the hydronium ion, and the hydroxide ion.

18. Write balanced chemical equations illustrating how water can function as an acid in a reaction with a metal and with a metal oxide.

19. The phosphoric acid added to some cola soft drinks is prepared by adding water to the product prepared from burning phosphorus in an excess of oxygen. Write balanced chemical equations for these two reactions.

20. During World War II, buckets of sand for fighting fires due to incendiary bombs (composed of burning magnesium) were kept in many buildings. Explain why water could not be used on such bombs.

21. Slaked lime, $Ca(OH)_2$, is prepared by the addition of water to calcium oxide, CaO. Write the balanced chemical equation for the reaction, and calculate what volume of water (density = 1.00 g/mL) is required to produce 1.000 metric ton (1000 kg) of slaked lime.
    Ans. $CaO + H_2O \longrightarrow Ca(OH)_2$; 243 L

22. Complete and balance the following equations:
    (a) $Cs + H_2O \longrightarrow$
    (b) $SO_2 + H_2O \longrightarrow$
    (c) $SiCl_4 + D_2O \longrightarrow$
    (d) $Mg_3P_2 + D_2O \longrightarrow$
    (e) $Al_2S_3 + H_2O \longrightarrow$

23. What mass of hydrogen is produced by the reaction of 29.6 g of hot magnesium with steam to produce magnesium oxide and hydrogen?
    Ans. 2.45 g

24. Plaster of paris, $(CaSO_4)_2 \cdot H_2O$, is prepared from gypsum, $CaSO_4 \cdot 2H_2O$. When mixed with water or exposed to a humid atmosphere, plaster of paris sets (hardens) by reacting with the water to reform gypsum.
    (a) What forces might be responsible for binding the water molecules into the gypsum structure?
    (b) How would you prepare plaster of paris from gypsum?

25. If an excess of $Na_2SO_4 \cdot 10H_2O$ is introduced into an evacuated flask, what will the pressure of water vapor be after equilibrium has been established at 25°C?

26. Relate the vapor pressure of hydrates to temperature.

27. Explain how P_4O_{10} acts as a desiccant and converts $CuSO_4 \cdot 5H_2O$ to $CuSO_4$.

28. Which of the following two reactions will occur when samples of the reactants are mixed together

in a closed container?

$$2(Na_2SO_4 \cdot 10H_2O) + 3(CuSO_4 \cdot 3H_2O) \longrightarrow$$
$$2(Na_2SO_4 \cdot 7H_2O) + 3(CuSO_4 \cdot 5H_2O)$$

$$2(Na_2SO_4 \cdot 7H_2O) + 3(CuSO_4 \cdot 5H_2O) \longrightarrow$$
$$2(Na_2SO_4 \cdot 10H_2O) + 3(CuSO_4 \cdot 3H_2O)$$

29. Distinguish between hygroscopic and deliquescent compounds.

Natural Waters

30. What naturally occurring impurities are typically present in unpolluted rain water? in unpolluted river water?
31. What is the effect of excess nutrients in a lake?
32. Define the following terms:
    (a) BOD
    (b) eutrophication
    (c) aerobic decomposition
    (d) anaerobic decomposition
    (e) hard water
    (f) zeolites
33. What are some of the primary substances responsible for water pollution?
34. What is thermal pollution of water? Is it necessarily undesirable?
35. Can the purification of water lead to its pollution? Can the pollution of water produce any beneficial results?
36. What metal ions are commonly responsible for the hardness of water?
37. Water that is hard due to the presence of hydrogen carbonate salts may be softened by boiling. Explain, using an equation.
38. Distinguish between softening and demineralizing of water. Does demineralization soften water?
39. One of the reasons that so much slaked lime (calcium hydroxide) and aluminum sulfate are produced industrially (Section 8.7) is that these compounds are used in water treatment. Explain how they are used.
40. What are the molarities of Na^+ and Cl^- in sea water (see Table 12-3)? *Ans. Na^+, 0.4593 M; Cl^-, 0.5354 M*
41. Assuming 95% recovery, what volume in liters of sea water must be processed to isolate 1.000 metric ton (1000 kg) of magnesium? *Ans. 8.3×10^5 L*

Hydrogen Peroxide

42. Write the Lewis structure of hydrogen peroxide; of peroxymonosulfuric acid, H_2SO_5; of peroxydisulfuric acid, $H_2S_2O_8$.
43. Distinguish between dioxides and peroxides in terms of electronic structure.
44. What is the oxidation number of oxygen in hydrogen peroxide?
45. How is hydrogen peroxide prepared in the laboratory? How is it prepared commercially? Write balanced equations.
46. Why is hydrogen peroxide purified by distillation under reduced pressure?
47. Complete and balance the following reactions. (Note that in some cases, depending on the stoichiometry, different products may result from the reaction of the same reagents.)
    (a) $SrO + H_2O_2 \longrightarrow$
    (b) $Na_2O_2 + H_3PO_4 \longrightarrow$
    (c) $BaS + H_2O_2 \longrightarrow$
    (d) $H_3PO_3 + H_2O_2 \longrightarrow$
48. Under what conditions does hydrogen peroxide act as a reducing agent? as an oxidizing agent? as an acid? Illustrate with balanced equations.
49. What mass of a solution of hydrogen peroxide (25.0% by mass) is required to convert 1.50 mol of sulfurous acid to sulfuric acid if the yield of the reaction is 100%? *Ans. 204 g*
50. Calculate the number of moles of hydrogen peroxide that can be prepared from 2.5×10^2 g of barium peroxide and an excess of sulfuric acid, assuming the yield of the reaction is 92%. *Ans. 1.4 mol*

Additional Exercises

51. (a) To break each hydrogen bond in ice requires 3.5×10^{-20} J. The measured heat of fusion of ice is 334 J/g. Essentially all of the energy involved in the heat of fusion goes to break hydrogen bonds. What percentage of the hydrogen bonds are broken when ice is converted to liquid water? *Ans. 14%*
    (b) How much additional heat would be required, per gram, to break the remaining hydrogen bonds? *Ans. 2.0 kJ*
52. Explain why a hydrogen bond between two water molecules, $HOH \cdots OH_2$, is stronger than a hydro-

gen bond between two ammonia molecules, $H_2NH\cdots NH_3$.

53. The hydrogen bonds in liquid hydrogen fluoride, HF, are stronger than those in liquid water, yet the heat of vaporization of liquid hydrogen fluoride is less than that of water. Explain.

54. It is believed that Mammoth Cave formed when water containing dissolved carbon dioxide trickled through cracks in a thick bed of limestone ($CaCO_3$) and dissolved away the limestone to form the cave. Explain the chemical reactions between carbon dioxide and water and between that solution and calcium carbonate that lead to the formation of a cave.

55. Will a sample of carbon dioxide in a cylinder at a

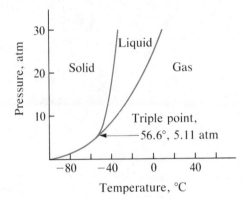

pressure of 20 atm and a temperature of $-30°C$ exist as a solid, a liquid, or a gas? Explain. (A portion of the phase diagram of carbon dioxide is shown.)

56. Dry ice, $CO_2(s)$, does not melt at atmospheric pressure. It sublimes at a temperature of $-78°C$. What is the lowest pressure at which $CO_2(s)$ will melt to give $CO_2(l)$. What is the approximate temperature at which this will occur? (A portion of the phase diagram for carbon dioxide is given in Exercise 55.)

57. How many liters of dry hydrogen at 25.0°C and 760 torr are produced by the reaction of 23.0 g of sodium with water?   *Ans. 12.2 L*

58. What volume of dry oxygen is produced when 68.00 g of hydrogen peroxide decomposes to give oxygen and water at 25°C and 740.0 torr?
          Ans. 25.1 L

59. What mass of sulfuric acid can be prepared by the reaction of 81.27 L of sulfur trioxide at 50.0°C and 750.0 torr with enough water to produce pure sulfuric acid?   *Ans. 296.7 g*

60. What mass of water in kilograms is required to produce enough hydrogen to make 1.000 metric ton (1000 kg) of ammonia?   *Ans. 1587 kg*

13

SOLUTIONS; COLLOIDS

Most chemical reactions take place in solution. In a gaseous or liquid solution, molecules or ions can move freely, come into contact with each other, and react. In solids, molecules or ions cannot move freely, and chemical reactions between solids, if they occur at all, are generally very slow.

Solutions are critical to life. When food is converted into a form that the body can use, the useful nutrients go into solution to pass through the walls of the digestive tract into the blood. There, they are carried throughout the body in solution.

The dissolution of substances from the air and the earth are important in the conversion of rocks to soil, in altering the fertility of the soil, and in changing the form of the earth's surface. Deposits of many minerals are the result of reactions that have taken place in solution followed by the evaporation of the solvent.

In this chapter we will consider what kind of solution may form when a solute and a solvent are mixed, what factors determine whether or not a solution will form, and the resulting properties of the solution. In addition, we will examine colloids, systems that resemble solutions but consist of dispersions of particles somewhat larger than ordinary molecules or ions.

13.1 The Nature of Solutions

When sugar is stirred with enough water, the sugar disappears and a solution of sugar in water is formed. The sugar is said to have dissolved (or gone into solution) in the water. The solution consists of the **solute** (the substance that dissolves, in this case, sugar) and the **solvent** (the substance in which a solute dissolves, in this case, water). The molecules of sugar are uniformly distributed among the molecules of water; that is, the solution is a **homogeneous mixture** of solute and solvent molecules. The molecules of sugar diffuse continuously throughout the water, and even though sugar molecules are heavier than water molecules, the sugar does not settle out on standing.

When sodium chloride dissolves in water, the sodium ions and chloride ions become uniformly distributed throughout the water. This solution is a homogeneous mixture of water molecules, sodium ions, and chloride ions. The solute particles (ions in this case) diffuse throughout the water just as molecular solutes do; they do not settle on standing.

A solution with a small proportion of solute to solvent is **dilute;** one with a large proportion of solute to solvent is **concentrated.** A solution is **saturated** when the dissolved solute exists in equilibrium with excess undissolved solute. In other words, if more solute is added to a saturated solution, it will not dissolve. The **solubility** of a given solute is the quantity that will dissolve in a given amount of solvent to produce a saturated solution. Since solutions can have different concentrations, it is evident that the composition of a solution may be varied between certain limits. Thus solutions are not compounds, because compounds always have the same composition.

All solutions—whether they contain dissolved molecules or dissolved ions—exhibit certain properties: (1) homogeneity, (2) absence of settling, (3) the molecular or ionic state of subdivision of the components, and (4) a composition that can be varied continuously within limits.

In chemistry we usually use liquid solvents. Water is so often used as a solvent that the word *solution* has come to imply a water solution. However, almost any gas, liquid, or solid can act as a solvent for other gases, liquids, or solids. For example, many alloys are solid solutions of one metal dissolved in another: nickel coins contain nickel dissolved in copper. Air is a homogeneous mixture of gases, or a gaseous solution. Oxygen (a gas), alcohol (a liquid), and sugar (a solid) will each dissolve in water (a liquid) to form liquid solutions.

Although it is easy to distinguish solute and solvent in most cases (as when 1 gram of sugar is dissolved in 100 milliliters of water), sometimes this is difficult. For example, in a gaseous solution of equal amounts of oxygen and nitrogen, it is not possible to tell which gas is solute or solvent. In such cases, the choice is arbitrary and, in fact, unimportant.

13.2 Solutions of Gases in Liquids

The amount of gas that will dissolve in a liquid depends on the nature of the gas, the nature of the solvent, the pressure, and the temperature. The solubility of a gas is a specific property of that gas; its solubility differs from that of other gases in a given liquid. For example, at $0\,°C$ and 1 atmosphere, 0.0489 liter of oxygen, 1.713 liters of carbon dioxide, 79.789 liters of sulfur dioxide, or 1176.0 liters of ammonia dissolve in 1 liter of water.

The solubility of a gas in a liquid can be increased by increasing the pressure of the gas. This relationship is quantitatively expressed by **Henry's Law: The mass of a gas that dissolves in a definite volume of liquid is directly proportional to the pressure of the gas.** This means that if 1 gram of a gas dissolves in 1 liter of water at 1 atmosphere of pressure, 5 grams will dissolve at 5 atmospheres of pressure.

The effect of pressure does not follow Henry's Law when a chemical reaction takes place between the gas and the solvent. Thus the solubility of ammonia in water does not increase as rapidly with increasing pressure as predicted by the law because ammonia, being a base, reacts to some extent with water to form ammonium ions and hydroxide ions.

$$NH_3 + H_2O \rightleftharpoons NH_4^+ + OH^-$$

You can see the effect of pressure on solubility in bottled carbonated beverages. Pressure forces carbon dioxide into solution in the beverage, and the bottle is tightly capped to maintain this pressure. When you open the bottle, the pressure decreases, and some of the gas escapes from the solution. The escape of bubbles of a gas from a liquid is known as **effervescence.**

Finally, the solubility of gases in liquids decreases with an increase in temperature. For example, 1.713 liters of carbon dioxide dissolve in 1 liter of water at 760 torr and 0°C, but only 1.194 liters dissolve at 10°C, 0.878 liter at 20°C, and 0.359 liter at 60°C. This relationship is not one of inverse proportions, however, and the solubility of a gas in a liquid at a given temperature must be determined experimentally. The decreasing solubility of gases with increasing temperature is a very important factor in thermal pollution, particularly in connection with the quantity of dissolved oxygen in water (see Section 12.8, Part 5). Most gases can be expelled from solvents by boiling their solutions. The gases oxygen, nitrogen, carbon dioxide, and sulfur dioxide can be removed from water by boiling it for a few minutes. The fact that a gas is expelled from a solution during boiling is due not only to the rise in temperature of the solution. When a solution is boiled in an open container, part of the vapor of the solvent escapes, carrying with it some of the solute gas above the solution. This lowers the partial pressure of that gas above the solution, and in accord with Henry's Law more gas escapes.

13.3 Solutions of Liquids in Liquids (Miscibility)

Some liquids will mix with water in all proportions. Such liquids are usually polar substances (see Section 5.8) or substances that form hydrogen bonds (Section 11.14). For these liquids the dipole-dipole attractions or hydrogen bonds of the solute molecules with the solvent molecules are at least as strong as those between molecules in the pure solute or the pure solvent. Hence the two kinds of molecules mix easily. Two liquids that mix with each other in all proportions are said to be completely **miscible** (Fig. 13-1). Ethyl alcohol, sulfuric acid, and ethylene glycol (antifreeze) are completely miscible with water.

Figure 13-1. Immiscible, partially miscible, and miscible liquids.

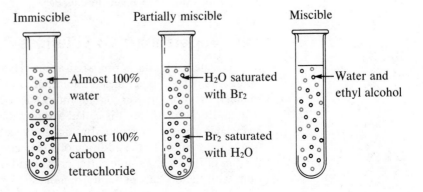

Immiscible | Partially miscible | Miscible

—Almost 100% water
—Almost 100% carbon tetrachloride

—H₂O saturated with Br₂
—Br₂ saturated with H₂O

—Water and ethyl alcohol

Gasoline, carbon disulfide, carbon tetrachloride and many other nonpolar liquids are very nearly insoluble in water. There is no effective attraction between the molecules of such liquids and the polar molecules of water. Thus the only appreciable attractions in such a mixture are between the water molecules; they effectively squeeze out the molecules of the nonpolar liquid. Such liquids are said to be **immiscible** with water (Fig. 13-1). Nonpolar liquids may be miscible with each other, however, because of the absence of an appreciable tendency of molecules to attract either other molecules like themselves or molecules of another nonpolar liquid. The solubility of polar molecules in polar solvents and of nonpolar molecules in nonpolar solvents is an illustration of the old chemical axiom "like dissolves like."

Some liquids, such as ether and bromine, are slightly soluble in water and are said to be partially miscible. Two layers are formed when two immiscible liquids are in the same container. In cases of partial miscibility each layer is a solution of one liquid in the other (Fig. 13-1).

13.4   Saturated and Unsaturated Solutions

If sugar is dissolved in water until no more will dissolve, the solution is saturated. When more sugar is added to a saturated solution, it falls to the bottom, and no more seems to dissolve. Actually, molecules of sugar continue to leave the solid and go into solution. But as the molecules of sugar in solution move about by diffusion, some of them collide with the solid and take up positions in the crystal. Enough molecules return to the solid from the solution that the process of crystallization counterbalances that of dissolution, and a state of dynamic equilibrium exists. The same dynamic equilibrium occurs with saturated solutions of partially miscible liquids, such as bromine and water. At equilibrium bromine molecules leave the bromine layer and enter the water layer, while the same number of bromine molecules leave the water layer and return to the bromine layer. A **saturated solution** is one in which dissolved solute is in equilibrium with undissolved solute. If a crystal of imperfect shape is placed in a saturated solution of the crystalline material, it will slowly mend its shape by losing particles to the solution and gaining particles from the solution. This is experimental evidence that a saturated solution in contact with its solid solute is a system in dynamic equilibrium.

An **unsaturated solution** contains less solute dissolved in a given quantity of solvent than does a saturated solution of the same solute. There can be no undissolved solute in contact with an unsaturated solution; it dissolves. As the solute dissolves, the concentration of such a solution increases until it becomes saturated or until all solute has been dissolved.

13.5   Supersaturated Solutions

If a saturated solution is prepared at an elevated temperature and undissolved solute is removed, sometimes the solution can be cooled without crystallization of solute. If the cool solution contains more solute than it would when in equilibrium with undissolved solute, it is called a **supersaturated solution.** Such solutions are **metastable systems** (unstable, but not so unstable that they cannot exist). Agitation of the solution or the addition of a seed crystal of the solute may start crystallization of the excess solute; after crystallization a saturated solution in equilibrium with the crystals of solute remains.

Gases also form supersaturated solutions. If a solution of a gas in a liquid is prepared either at a low temperature or under pressure (or both), as the solution warms or as the gas pressure is reduced, the solution may become supersaturated. For example, the dissolved carbon dioxide in a bottle of a carbonated beverage might not be liberated when the carbon dioxide pressure above the solution is reduced by opening the bottle, but it will be if the beverage is shaken or stirred.

13.6   The Effect of Temperature on the Solubility of Solids in Liquids

The dependence of solubility on temperature for a number of inorganic substances in water is shown by the solubility curves in Fig. 13-2. Generally, the solubility of these solids increases with increasing temperature, although there are important exceptions (calcium sulfate, Section 12.10, for example). A sharp break in a solubility curve indicates that a new compound with a different solubility has formed. For example, when solid $Na_2SO_4 \cdot 10H_2O$ (Glauber's salt) in equilibrium with a saturated solution is heated to 32.4°C, it loses its water of hydration and forms the anhydrous salt Na_2SO_4. The curve up to 32.4°C shows the effect of increasing temperature on the solubility of $Na_2SO_4 \cdot 10H_2O$ (it increases with increasing temperature), and the curve above that point shows the effect of increasing temperature on the solubility of Na_2SO_4 (it decreases with increasing temperature).

Figure 13-2. Graph showing the effect of temperature on the solubility of several inorganic substances.

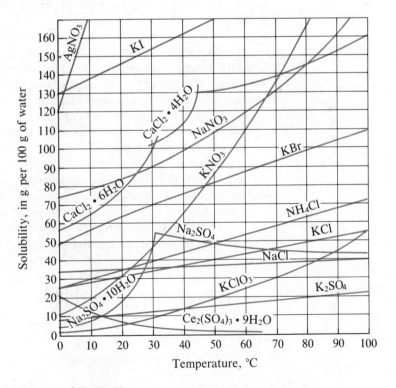

13.7   Rate of Solution

The amount of solute entering a solution per unit of time is called the **rate of solution** of the solute. This rate depends on several factors. First, more soluble substances dissolve more rapidly than do less soluble ones. Second, a solid dissolves only at its surface; so the more finely divided the solute, the greater is the surface area per unit of mass and the more rapidly it dissolves. Third, dissolved molecules diffuse away from the solid solute relatively slowly; so the solution around the solid approaches saturation. Stirring, shaking, or heating the mixture brings unsaturated solution in contact with the solute and thus increases the rate of solution. Fourth, heating often increases the solubility of the solid; if the solute and solvent react during the solution process, heating accelerates the reaction and hence increases the rate of solution.

13.8 Electrolytes and Nonelectrolytes

Substances such as acids, bases, and salts that give solutions that conduct an electric current are called **electrolytes** (Section 8.1). Substances that conduct an electric current when molten are also called electrolytes. Other substances, such as sugar and alcohol, form solutions that do not conduct and are called **nonelectrolytes.**

Electrolytes and nonelectrolytes may be classified experimentally by setting up a simple conductivity apparatus, as shown in Fig. 13-3. The terminals of a storage battery or a 110-volt circuit are connected through an electric lamp to two electrodes in a beaker. When the beaker is filled with pure water, the lamp does not light because very little current flows. When a nonelectrolyte such as sugar, alcohol, or glycerin is added to the water, the lamp still does not light. If, however, an electrolyte such as hydrochloric acid is dissolved in the water, the lamp glows brightly.

With a 0.1 M solution of an acid such as hydrochloric, nitric, or sulfuric acid, of a base such as potassium, sodium, or barium hydroxide, or of most salts, the lamp in the circuit will glow brightly, showing that these solutions are good conductors of electricity. Substances whose aqueous solutions are good conductors of electricity are known as **strong electrolytes.** When a 0.1 M solution of acetic acid or ammonia is placed in the beaker, the lamp dims, showing that these solutions are poor conductors. Substances whose solutions are poor conductors of electricity are called **weak electrolytes.**

Svante Arrhenius, a Swedish chemist, first successfully explained electrolytic conduction. Although his theory has been modified by subsequent knowledge of atomic structure and chemical bonding, the current theory of electrolytes embodies most of the principal postulates of Arrhenius's theory.

According to the current theory, a solution of an electrolyte contains positive and negative ions that move independently. When a solution conducts an electric

Figure 13-3. Apparatus for demonstrating the conductivity of solutions.

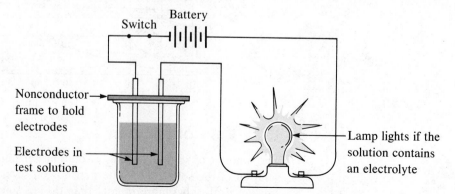

Switch   Battery

Nonconductor frame to hold electrodes

Electrodes in test solution

Lamp lights if the solution contains an electrolyte

current, the positive ions (cations) move toward the negative electrode, while the negative ions (anions) move toward the positive electrode. The movement of ions toward the electrode of opposite charge accounts for electrolytic conduction. A solution of a nonelectrolyte contains *molecules* of the nonelectrolyte rather than ions and thus cannot conduct an electric current.

According to the theory of electrolytes, if a solution is a good conductor of electricity, the solute consists principally of ions, and if a solution is a poor conductor, the solute consists principally of molecules. Thus compounds that dissolve in water and form solutions composed principally of ions are strong electrolytes. Compounds that exist principally as unreacted molecules are weak electrolytes. The partial ionization of acetic acid, a weak acid (and thus a weak electrolyte), is such that the acid in a 0.1 M solution is only about 1% ionized. This means that 99% of the acid is in the molecular form.

Pure water is an extremely poor conductor of electricity, indicating very slight ionization (actually about 0.0000002%). As indicated in Section 12.3, water ionizes when one molecule of water gives up a proton to another molecule of water, yielding hydronium and hydroxide ions.

$$H_2O + H_2O \rightleftharpoons H_3O^+ + OH^-$$

To show how ions migrate during electrolytic conduction, partially fill a U-tube (Fig. 13-4) with a colorless solution of potassium nitrate, acidified with a few drops of sulfuric acid. Then carefully introduce a solution of copper(II) permanganate into the bottom of the U-tube without mixing the two solutions. When a current is passed through the U-tube, blue hydrated copper(II) ions, $Cu^{2+}(aq)$, begin to move through the colorless potassium nitrate solution toward the negative electrode, and the purple permanganate ions, MnO_4^-, toward the positive electrode. The separation of colors is visual evidence that the electric current is carried by ions and that ions of opposite charge move independently in the solution.

Figure 13-4. When an electric current flows in the system shown, blue Cu^{2+} ions and purple MnO_4^- ions travel toward different electrodes, visual evidence that the current is carried by the ions.

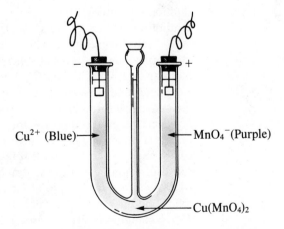

Cu²⁺ (Blue)

MnO₄⁻ (Purple)

Cu(MnO₄)₂

13.9 Generalizations on the Solubilities of Common Metal Compounds

Knowledge of the solubilities of metallic compounds is very useful to the student and chemist. Memorizing solubilities of individual compounds is unnecessary; it is simpler to learn the following general rules. (Remember that these rules are for simple compounds of the more common metals; there are exceptions for less common metals and complex compounds.)

1. Most nitrates and acetates are soluble in water; silver acetate, chromium(II) acetate, and mercury(I) acetate are slightly soluble; bismuth acetate hydrolyzes to bismuth oxyacetate, $BiO(CH_3CO_2)$, which is insoluble in water.

2. All chlorides are soluble except those of mercury(I), silver, lead(II), and copper(I); lead(II) chloride is soluble in hot water.

3. All sulfates, except those of strontium, barium, and lead(II) are soluble; calcium sulfate and silver sulfate are slightly soluble.

4. Carbonates, phosphates, borates, arsenates, and arsenites are insoluble, except those of the ammonium ion and the alkali metals.

5. The hydroxides of the alkali metals and of barium and strontium are soluble, and other hydroxides are insoluble; calcium hydroxide is slightly soluble.

6. Most sulfides are insoluble. However, the sulfides of the ammonium ion and of the alkali metals are soluble, but they hydrolyze (react with water) to give solutions of the hydroxide and hydrogen sulfide ion, HS^-; the sulfides of the alkaline earth metals and of aluminum also hydrolyze to give the hydroxide and HS^- (or H_2S if the hydroxide is insoluble).

13.10   Solid Solutions

If a mixture consisting of equal amounts of lithium chloride and sodium chloride is melted, mixed well, and allowed to cool, the resulting crystalline solid contains an array of chloride ions with a random distribution of lithium ions and sodium ions in holes in the array (Fig. 13-5). The crystal is a **solid solution** of LiCl and NaCl. It is homogeneous, just like a liquid solution. Its composition can be varied (from pure LiCl to pure NaCl), and neither NaCl nor LiCl separate out on standing.

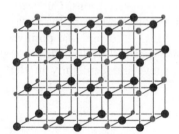

Figure 13-5. A portion of the structure of a solid solution of LiCl and NaCl. The chloride ions (large black spheres) form a face-centered cubic array with the octahedral holes occupied by a random distribution of lithium ions (smaller colored spheres) and sodium ions (larger colored spheres).

Solid solutions of ionic compounds result when ions of one type randomly replace other ions of about the same size in a crystal. Ruby, for example, is a solid solution containing about 1% Cr_2O_3 in Al_2O_3. The mineral powellite is a solid solution of $CaMoO_4$ and $CaWO_4$ with a random distribution of MoO_4^{2-} and WO_4^{2-} ions. Application of fluoride to tooth enamel, which is mostly hydroxyapatite, forms a decay-resistant solid solution of hydroxyapatite, $Ca_5(PO_4)_3OH$, and fluoroapatite, $Ca_5(PO_4)_3F$.

Some ionic substances appear to be **nonstoichiometric compounds;** that is, their chemical formulas deviate from ideal ratios or are variable. In other respects, however, these substances resemble compounds. They are not heterogeneous mixtures but are instead homogeneous throughout. These so-called nonstoichiometric compounds are, in fact, solid solutions of two or more compounds. A sample of ruby, for example, with the formula $Cr_{0.02}Al_{1.98}O_3$ is a solid solution of Cr_2O_3 in Al_2O_3 with a 1 to 99 ratio of Cr to Al. Some nonstoichiometric compounds are solid solutions containing one ion in two different oxidation states. For example, $TiO_{1.8}$ contains both Ti^{3+} and Ti^{4+} ions and may be considered to be a solid solution of Ti_2O_3 and TiO_2.

Some **alloys** are solid solutions composed of two or more metals. In such an alloy atoms of one of the component metals take up positions in the crystal lattice of the other. The solute atoms may randomly replace some of the atoms of the solvent crystal to form what are referred to as **substitutional solid solutions** [Fig. 13-6(a)]. For example, chromium dissolves in nickel to form a solid solution in which the chromium atoms replace nickel atoms in the face-centered cubic struc-

Figure 13-6. Two-dimensional representations of alloys. (a) A substitutional solid solution in which solute atoms (black spheres) replace atoms of the solvent crystal (colored spheres). (b) An interstitial solid solution in which small solute atoms (black spheres) occupy holes in the lattice of the solvent crystal (colored spheres).

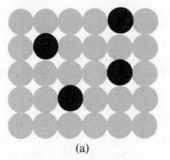
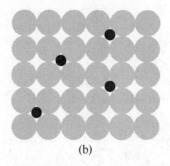

(a)   (b)

ture of nickel. The solubility may be limited (zinc and copper; chromium and nickel) or practically infinite (nickel and copper).

Small atoms (hydrogen, carbon, boron, and nitrogen) may occupy the holes in the lattice of a metal, forming another class of solid solutions called **interstitial solid solutions** [Fig. 13-6(b)]. The solid solution of carbon in gamma-iron (austenite) is an example; the iron atoms are on the lattice points of a face-centered cubic lattice, and the carbon atoms occupy the interstitial positions.

It should be noted that not all alloys are solid solutions. Some are *heterogeneous* mixtures in which the component metals are mutually insoluble and the solid alloy is composed of an intimate mixture of crystals of each metal. For example, tin and lead (in plumber's solder) are insoluble in each other in the solid state. Other alloys, such as Cu_5Zn_8 and Ag_3Al, are actually compounds that form with only one specific stoichiometric composition. In general, the formulas of such intermetallic compounds are not those that might be predicted on the basis of the usual valence rules.

THE PROCESS OF DISSOLUTION

13.11 Dissolution of Nonelectrolytes

When a sample of ethanol is added to water, a solution forms without the input of outside energy. The solution forms by a **spontaneous process.** Chemists recognize two factors involved in such a spontaneous process: (1) changes in energy, and (2) changes in the amount of disorder of the components of the solution. In the process of dissolution the change in disorder results from the mixing of the solute and solvent. The energy changes occur because the intermolecular attractions change from solute-solute and solvent-solvent attractions to solute-solvent attractions. First, let us consider formation of a solution in which energy changes are insignificant, so we can concentrate on how changes in disorder contribute to the formation of solutions.

When the strengths of the intermolecular forces of attraction between solute and solvent molecules (or ions) are the same as those between the molecules in the separate components, a solution is formed with no accompanying energy change. Such a solution is called an **ideal solution.** An ideal solution obeys Raoult's Law exactly (Section 13.19). Solutions of ideal gases (or of gases such as helium and argon, which closely approach ideal behavior) contain molecules with no significant intermolecular attractions, and thus these solutions behave as ideal solutions.

If the stopcock is opened between 500-milliliter bulbs containing helium and

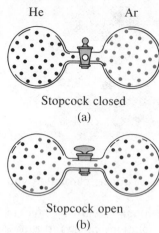

He   Ar

Stopcock closed
(a)

Stopcock open
(b)

Figure 13-7. The spontaneous mixing of helium and argon to give a solution. When the stopcock between samples of the two pure gases (a) is opened, they mix by diffusion to give a solution (b) in which the disorder of the molecules of the two gases is increased.

argon under conditions where they exhibit ideal gas behavior [Fig. 13-7(a)], the gases will spontaneously diffuse together and form a solution of helium and argon [Fig. 13-7(b)]. This solution forms because the disorder of the helium and argon molecules increases when they mix. They occupy a volume twice as large as that which each occupied before mixing, and the molecules are randomly distributed among one another.

Other solutions that closely approximate ideal behavior include solutions of pairs of chemically similar substances, such as the liquids methanol, CH_3OH, and ethanol, C_2H_5OH, or the liquids chlorobenzene, C_6H_5Cl, and bromobenzene, C_6H_5Br. If the samples of helium and argon in Fig. 13-7(a) were replaced with methanol and ethanol (or with chlorobenzene and bromobenzene), when the stopcock was opened, the molecules of the liquids would diffuse together spontaneously (although at a much slower rate than the gases), giving a solution with a disorder greater than that of the pure liquids. These examples show that **processes in which the disorder of the system increases tend to occur spontaneously.** Moving molecules will become randomly distributed among one another unless something holds them back.

Intermolecular forces of attraction can keep molecules from mixing. These forces are small and negligible in gases; so gases are mutually soluble in all proportions. However, the molecules in liquids and solids are relatively close together, and their intermolecular attractions are stronger and much more important in determining their behavior. Sometimes, solute molecules attract one another strongly but attract solvent molecules weakly. The solute molecules thus remain in contact with one another and do not dissolve, even though formation of a solution would increase their disorder. If the solvent molecules attract one another strongly but do not attract the solute molecules, the solvent molecules will not separate to let the solute dissolve. A solution forms only when the attractions between solute and solvent molecules are about equal to the combination of the attractions between solute molecules and the attractions between solvent molecules.

To see why gasoline and water don't mix, let us consider what happens when octane, a typical hydrocarbon in gasoline, is added to water. Water molecules are held in contact by hydrogen bonds (Section 11.14); octane molecules are held in contact by London forces (Section 11.13). When the two liquids are brought together, the forces of attraction between the octane molecules and the water molecules are not strong enough to overcome the hydrogen bonds between the water molecules. A great deal of stirring can put some molecules of water among the octane molecules. This stirring breaks hydrogen bonds between water molecules and destroys some of the London forces between the octane molecules. Relatively weak London forces then form between octane and water molecules. When the stirring stops, however, diffusion of the water molecules brings them back in contact with one another, and their relatively strong hydrogen bonds hold them clustered together. Eventually, water and octane separate into two layers.

On the other hand, if we add methanol to water, a solution forms readily. Methanol molecules hydrogen-bond about equally well to other methanol molecules and to water molecules. Water-water and water-methanol hydrogen bonds are of about equal strengths. Thus there is no stronger hydrogen bonding to cause the water molecules or the methanol molecules to cluster together, and a solution forms because of the increase in disorder.

Now let us consider the second factor involved in a spontaneous solution process, changes in energy. If the solute-solvent attractions are stronger than the solute-solute and solvent-solvent attractions, then heat is released during dissolution (the dissolution process is exothermic) as the stronger attractions form. If the solute-solvent attractions are weaker than the solute-solute and solvent-solvent attractions, heat is absorbed (the dissolution process is endothermic). For example, when 1 mole of sulfuric acid is dissolved in 9 moles of water, the resulting acid-water interactions are stronger than the combination of the acid-acid attractive forces in pure sulfuric acid and the water-water attractive forces in pure water. The solution becomes very hot because 63.2 kilojoules of heat is produced. The loss of energy, as heat, when sulfuric acid dissolves in water indicates that the solution contains less energy than did the separate components before mixing. The **heat of solution** is the amount of heat absorbed or evolved when a solute dissolves in a specified amount of solvent.

Processes in which the energy content of the system decreases tend to occur spontaneously. Thus both a loss of energy and an increase in disorder favor a spontaneous process such as the formation of a solution. For a solution to form, however, it is not necessary for both the energy change and the change in disorder to favor a spontaneous process. When ammonium nitrate is added to water, the ammonium nitrate dissolves even though the solution cools as heat is absorbed from the water. In this case the solute-solute attractions and solvent-solvent attractions combined are larger than the solute-solvent attractions. A solution forms, in spite of the endothermic nature of the process, because the increase in disorder is large enough that it more than compensates for the increase in the energy content. (Incidentally, ammonium nitrate is used to make instant "ice" packs for treatment of athletic injuries. A thin-walled plastic bag of NH_4NO_3 is sealed inside a larger bag filled with water. When the smaller bag is broken, a cold solution of NH_4NO_3 forms.)

13.12 Dissolution of Ionic Compounds

When ionic compounds dissolve in water, the associated ions in the solid separate because water reduces the strong electrostatic forces between them. Let us consider the dissolution of potassium chloride in water. The hydrogen (positive) end of a polar water molecule is attracted to a negative chloride ion at the surface of the solid, and the oxygen (negative) end is attracted to a positive potassium ion. The water molecules surround individual K^+ and Cl^- ions at the surface of the crystal and penetrate between them, reducing the strong interionic forces that bind them together and letting them move off into solution as hydrated ions (Fig. 13-8). Several water molecules associate with each ion in solution as a result of the electrostatic attraction between the charged ion and the dipole of the water. Such an attraction is called an **ion-dipole attraction.** The increase in the distance between oppositely charged ions due to the layer of water molecules around each ion reduces the electrostatic attraction between them. In addition, the layers of water act as insulators, which further reduces the electrostatic attraction. This permits the independent motion of each hydrated ion in a dilute solution, resulting in an increase in the disorder of the system as the ions change from their ordered positions in the crystal to a much more disordered and mobile state in solution. This increased disorder is responsible for the dissolution of many ionic compounds, including potassium chloride, that dissolve with absorption of heat. In other cases the electrostatic attractions between the ions in a crystal are so large

Figure 13-8. The dissolution of potassium chloride in water and the hydration of its ions. Water molecules in front of and behind the ions are not shown. The positive spheres represent potassium ions; the negative spheres, chloride ions.

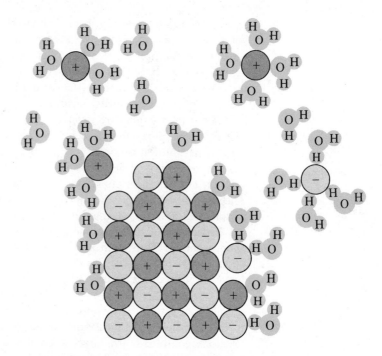

that an increase in disorder could not compensate for the energy required to separate the ions, and the crystal is insoluble.

Ionic compounds dissolve in polar solvents only when the polar solvent molecules can solvate and insulate the ions. In general, the larger the dielectric constant of a solvent, the greater its ability to insulate solvated ions and to reduce the electrostatic attractions between them, and the greater the solubility of an ionic compound in the solvent. This phenomenon is strikingly illustrated by the data of Table 13-1. Ionic substances in general do not dissolve appreciably in nonpolar solvents, such as benzene or carbon tetrachloride, because the nonpolar solvent molecules are not strongly attracted to ions and because nonpolar solvents have low dielectric constants.

Table 13-1   The Solubility of Sodium Chloride and the Dielectric Constant of the Solvent

Solvent	Solubility of NaCl (grams per 100 g of solvent, 25°C)	Dielectric Constant of the Solvent
Water, H_2O	36.12	80.0
Methyl alcohol, CH_3OH	1.3	33.1
Carbon tetrachloride, CCl_4	0.00	2.2

13.13   Dissolution of Molecular Electrolytes

Hydrogen chloride consists of covalent HCl molecules and contains no ions; pure gaseous hydrogen chloride does not conduct an electric current. A solution of hydrogen chloride in a nonpolar solvent such as benzene is also a nonelectrolyte and, hence, does not contain ions. However, a solution of hydrogen chloride in water is a strong electrolyte. The water molecules play an important part in causing ionization; hydrogen chloride reacts with water to form hydronium ions, H_3O^+, and chloride ions, Cl^-. As shown by the Lewis structures, a hydrogen ion (proton) shifts from a hydrogen chloride molecule to a lone pair of electrons on the water molecule.

$$H:\overset{..}{\underset{H}{O}}: \;+\; H:\overset{..}{\underset{..}{Cl}}: \;\longrightarrow\; H:\overset{..}{\underset{H}{O}}:H^+ \;+\; :\overset{..}{\underset{..}{Cl}}:^-$$

The hydronium ions and chloride ions conduct the current in the solution. All common strong acids react with water when they dissolve and are therefore strong electrolytes.

Many other compounds dissolve in water as hydrated molecules. In many cases these molecules are weak electrolytes, and a small fraction of the molecules undergo ionization, causing the solution to conduct electricity weakly. For example, cyanic acid (HOCN, a weak acid) dissolves in water principally as hydrated molecules. A few of these dissolved molecules ionize under ordinary conditions.

$$H:\overset{..}{\underset{H}{O}}: \;+\; H:\overset{..}{\underset{..}{O}}:C:::N: \;\rightleftharpoons\; H:\overset{..}{\underset{H}{O}}:H^+ \;+\; :\overset{..}{\underset{..}{O}}:C:::N:^-$$

Because the cyanate ion, OCN^-, binds a hydrogen ion more strongly than does a water molecule, the reaction gives only a 5.6% yield of ions in a 0.1 M solution of HOCN at 25°C. Other weak acids such as acetic acid, CH_3CO_2H, nitrous acid, HNO_2, and hydrogen cyanide, HCN, are also soluble in water, and only a small fraction of their hydrated molecules ionize at any one time.

Weak bases also dissolve to give solutions of hydrated molecules, some of which undergo ionization. For example, a solution of ammonia in water consists primarily of hydrated ammonia molecules, $NH_3(aq)$, with small amounts of ammonium ions, $NH_4^+(aq)$, and hydroxide ions, $OH^-(aq)$, which result from the reaction of ammonia with water.

$$H:\overset{H}{\underset{H}{N}}: \;+\; H:\overset{..}{\underset{H}{O}}: \;\rightleftharpoons\; H:\overset{H}{\underset{H}{N}}:H^+ \;+\; :\overset{..}{\underset{..}{O}}:H^-$$

The halides and cyanides of mercury, cadmium, and zinc are also weak electrolytes. These molecules have no dipole moments even though the bonds are of the polar covalent type. Their centers of positive and negative charge coincide because of a high degree of molecular symmetry, as in the linear molecule Cl—Hg—Cl. In this molecule, however, the mercury atom has a net positive electric charge, and the chlorine atoms, net negative charges. Water dipoles are attracted to these charge concentrations, and the $HgCl_2$ molecule becomes hydrated. A small fraction of the mercury-chlorine bonds break, and a few hydrated mercury(II) ions and hydrated chloride ions form.

EXPRESSING CONCENTRATION

13.14 Percent Composition

To express the concentration of a solute, we can state the mass of solute in a given mass of solvent, for example, 1 gram of NaCl in 100 grams of water, or we can give its composition as **percent by mass** (Section 3.4). A 10% NaCl solution by mass may contain 10 grams of NaCl in 100 grams of solution (10 grams of NaCl and 90 grams of water), or it may contain any other ratio of grams of NaCl to grams of solution for which the mass of NaCl is 10% of the total mass, for example, 20 grams of NaCl in 200 grams of solution, 1.5 grams of NaCl in 15 grams of solution, or 7.4 grams of NaCl in 74 grams of solution.

$$\% \text{ solute} = \frac{\text{mass of solute}}{\text{mass of solution}} \times 100$$

EXAMPLE 13.1   A bottle of a certain ceramic tile cleanser, which is essentially a solution of hydrogen chloride, contains 130 g of HCl and 750 g of water. What is the percent by mass of HCl in this cleanser?

The percent by mass of the solute is

$$\% \text{ solute} = \frac{\text{mass solute}}{\text{mass solution}} \times 100$$

$$= \frac{130 \text{ g}}{130 \text{ g} + 750 \text{ g}} \times 100 = 15\%$$

When using percent composition by mass in calculations, it is convenient to use grams of solute per 100 grams of solution (g solute/100 g solution), since the number of grams of solute in 100 grams of solution is equal to the percent by mass of solute. To calculate the mass of solute in a given *volume* of solution, given the percent composition by mass, you must know the specific gravity (or density) of the solution.

EXAMPLE 13.2   Concentrated hydrochloric acid, a saturated solution of hydrogen chloride, HCl, in water, is often used in the general chemistry laboratory. It has a specific gravity of 1.19 and contains 37.2% HCl by mass. What mass of HCl is contained in exactly 1 L of this concentrated acid?

This problem requires the following steps.

| Volume of solution | → | Mass of solution | → | Mass of HCl |

From the specific gravity we know that the mass of 1 mL of concentrated hydrochloric acid is 1.19 g. Since 1 L equals 1000 mL,

$$\frac{1.19 \text{ g solution}}{1 \text{ mL}} \times 1000 \text{ mL} = 1190 \text{ g solution}$$

The solution contains 37.2% HCl by mass (37.2 g HCl/100.0 g solution).

$$1190 \text{ g solution} \times \frac{37.2 \text{ g HCl}}{100.0 \text{ g solution}} = 443 \text{ g HCl}$$

EXAMPLE 13.3   A student needs 125 g of HCl to prepare a metal chloride. What volume of concentrated hydrochloric acid with a specific gravity of 1.19 and containing 37.2% HCl by mass contains 125 g of HCl?

This problem requires the following steps.

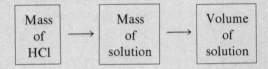

Since the solution contains 37.2% HCl, there are 37.2 g of HCl per 100.0 g of solution.

$$125 \text{ g HCl} \times \frac{100.0 \text{ g solution}}{37.2 \text{ g HCl}} = 336 \text{ g solution}$$

The mass of 1 mL of solution is 1.19 g (1 mL/1.19 g).

$$336 \text{ g solution} \times \frac{1 \text{ mL solution}}{1.19 \text{ g solution}} = 282 \text{ mL solution}$$

Thus 282 mL of the concentrated hydrochloric acid contains 125 g of HCl.

13.15 Molarity

We have already considered the concept of molarity. In Section 3.1 the **molarity, M,** of a solution was defined as the number of moles of solute in exactly 1 liter of solution. Molarity may be calculated by dividing the moles of solute in a solution by the volume of the solution.

$$M = \frac{\text{moles of solute}}{\text{liters of solution}}$$

Because 1 mole of any substance contains the same number of molecules as 1 mole of any other substance, equal volumes of 1 M solutions contain the same number of molecules of solute. The molar method of expressing the concentration of a solution makes it easy to select a desired number of moles, molecules, or ions of the solute by measuring out the appropriate volume of solution. For example, if 1 mole of sodium hydroxide is needed for a given reaction, we can use 40 grams of solid sodium hydroxide (1 mole), or 1 liter of a 1 M solution of the base, or 2 liters of a 0.5 M solution.

To prepare a solution of known molarity, measure out the required amount of the solute, in moles, and add enough solvent to give the desired volume of solution. To prepare 1 liter of a 1.00 M sodium bicarbonate solution, you should dissolve 1.00 mole of pure sodium bicarbonate (84.0 grams of $NaHCO_3$) in enough water to form 1 liter of solution (Fig. 13-9).

Examples of the use of molar concentrations in stoichiometry calculations were presented in Chapter 3. However, here is one more as a reminder.

Figure 13-9. In one method of preparation of a 1.00 M solution of sodium bicarbonate, 84.0 g of NaHCO$_3$ (1 mol) is added to a flask that is calibrated to hold 1.000 L (a) and enough water is added to make 1.000 L of solution (b).

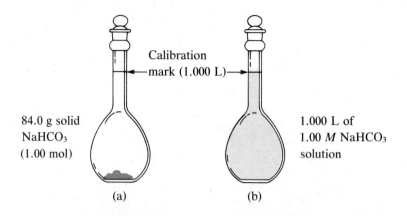

84.0 g solid
NaHCO$_3$
(1.00 mol)

Calibration
mark (1.000 L)

1.000 L of
1.00 M NaHCO$_3$
solution

(a)   (b)

EXAMPLE 13.4   The contact process can be used to produce concentrated sulfuric acid, a solution with a specific gravity of 1.84 and containing 98.3% H$_2$SO$_4$ by mass. What is the molarity of this acid?

Calculating the moles of H$_2$SO$_4$ in 1.00 L of concentrated sulfuric acid requires the following steps.

$$\boxed{\begin{array}{c}\text{Volume of} \\ \text{solution} \\ (1.00\ L)\end{array}} \longrightarrow \boxed{\begin{array}{c}\text{Mass} \\ \text{of} \\ \text{solution}\end{array}} \longrightarrow \boxed{\begin{array}{c}\text{Mass} \\ \text{of} \\ H_2SO_4\end{array}} \longrightarrow \boxed{\begin{array}{c}\text{Moles} \\ \text{of} \\ H_2SO_4\end{array}}$$

The mass of 1 mL of solution is 1.84 g.

$$\frac{1.84\ \text{g solution}}{1.00\ \text{mL solution}} \times 1000\ \text{mL solution} = 1.84 \times 10^3\ \text{g solution}$$

There are 98.3 g of H$_2$SO$_4$ per 100 g of solution since the solution is 98.3% H$_2$SO$_4$ by mass.

$$1.84 \times 10^3\ \text{g solution} \times \frac{98.3\ \text{g H}_2\text{SO}_4}{100\ \text{g solution}} = 1.81 \times 10^3\ \text{g H}_2\text{SO}_4$$

$$1.81 \times 10^3\ \text{g H}_2\text{SO}_4 \times \frac{1\ \text{mol H}_2\text{SO}_4}{98.0\ \text{g H}_2\text{SO}_4} = 18.5\ \text{mol H}_2\text{SO}_4$$

$$\frac{18.5\ \text{mol H}_2\text{SO}_4}{1.00\ L} = 18.5\ M$$

The calculations can be carried out in one step as follows:

$$\left(\frac{1.84\ \text{g solution}}{1\ \text{mL solution}} \times 1000\ \text{mL solution} \times \frac{98.3\ \text{g H}_2\text{SO}_4}{100\ \text{g solution}}\right.$$
$$\left.\times \frac{1\ \text{mol H}_2\text{SO}_4}{98.0\ \text{g H}_2\text{SO}_4}\right)/1.00\ L = 18.5\ M$$

Knowing the molar concentration of a solution enables us to measure out a definite number of moles of solute by volume. However, solutions change in volume as the temperature changes, and thus molarity changes with temperature. For a solution's molarity to be accurate, the solution must be used at or near the temperature at which it was prepared.

A concentration unit called normality is sometimes used to describe the concentrations of acids and bases or of oxidizing and reducing agents. It is related to molarity but differs from it in one important way: the normality of a solute is dependent on the reaction that the solute undergoes. Normality will be described in the chapters dealing with acid-base reactions (Section 14.14) and with electrochemical reactions (Section 20.7).

13.16 Molality

The **molality, m,** of a solution is the number of moles of solute in exactly 1 kilogram of solvent.

$$m = \frac{\text{moles of solute}}{\text{kilograms of solvent}}$$

Note that a *kilogram of solvent* rather than a liter of solution is specified. This is the difference between molality and molarity. Molality is a useful method for expressing concentration because it does not change its value as the temperature changes.

EXAMPLE 13.5   What is the molality of a solution that contains 0.850 g of ammonia, NH_3, in 125 g of water?

After the mass of NH_3 has been converted to moles of NH_3, the molality may be determined by dividing by the kilograms of water, the solvent.

$$0.850 \text{ g NH}_3 \times \frac{1 \text{ mol NH}_3}{17.0 \text{ g NH}_3} = 5.00 \times 10^{-2} \text{ mol NH}_3$$

$$m = \frac{\text{moles of solute}}{\text{kilograms of solvent}} = \frac{\text{mol NH}_3}{\text{kg H}_2\text{O}}$$

$$= \frac{5.00 \times 10^{-2} \text{ mol NH}_3}{0.125 \text{ kg H}_2\text{O}} = 0.400 \, m$$

EXAMPLE 13.6   Calculate the molality of an aqueous solution of sodium chloride if 0.250 kg of the solution contains 40.0 g of NaCl.

The mass of solvent is the difference between 0.250 kg and 0.0400 kg, or 0.210 kg. The molality is determined by converting the mass of NaCl to moles of NaCl and dividing by the mass of the solvent, water, in kilograms.

$$40.0 \text{ g NaCl} \times \frac{1 \text{ mol NaCl}}{58.5 \text{ g NaCl}} = 0.684 \text{ mol NaCl}$$

$$m = \frac{0.684 \text{ mol NaCl}}{0.210 \text{ kg H}_2\text{O}} = 3.26 \, m$$

13.17 Mole Fraction

The **mole fraction, X,** of each component in a solution is the number of moles of the component divided by the total number of moles of all components present. The mole fraction of substance A in a solution (or other mixture) of substances A, B, C, etc., is expressed as follows:

$$\text{Mole fraction of A} = X_A = \frac{\text{moles A}}{\text{moles A} + \text{moles B} + \text{moles C} + \cdots}$$

The sum of the mole fractions of all components of a system always equals 1. Like molality, the mole fraction of a component of a solution does not change with temperature.

EXAMPLE 13.7   Calculate the mole fraction of each component in a solution containing 42.0 g CH_3OH, 35 g C_2H_5OH, and 50.0 g C_3H_7OH.

The number of moles of each component is calculated first.

$$42.0 \text{ g } CH_3OH \times \frac{1 \text{ mol } CH_3OH}{32.0 \text{ g } CH_3OH} = 1.31 \text{ mol } CH_3OH$$

$$35 \text{ g } C_2H_5OH \times \frac{1 \text{ mol } C_2H_5OH}{46.0 \text{ g } C_2H_5OH} = 0.76 \text{ mol } C_2H_5OH$$

$$50.0 \text{ g } C_3H_7OH \times \frac{1 \text{ mol } C_3H_7OH}{60.0 \text{ g } C_3H_7OH} = 0.833 \text{ mol } C_3H_7OH$$

The mole fractions are

$$X_{CH_3OH} = \frac{1.31}{1.31 + 0.76 + 0.833} = \frac{1.31}{2.90} = 0.452$$

$$X_{C_2H_5OH} = \frac{0.76}{1.31 + 0.76 + 0.833} = \frac{0.76}{2.90} = 0.26$$

$$X_{C_3H_7OH} = \frac{0.833}{2.90} = 0.287$$

Note that the sum of the mole fractions, $0.452 + 0.26 + 0.287$, is 1.00.

EXAMPLE 13.8   Calculate the mole fraction of solute and solvent for a 3.0 m solution of sodium chloride.

A 3.0 m solution of sodium chloride contains 3.0 mol of NaCl dissolved in exactly 1 kg, or 1000 g, of water.

$$1000 \text{ g } H_2O \times \frac{1 \text{ mol } H_2O}{18.0 \text{ g } H_2O} = 55.6 \text{ mol } H_2O$$

$$X_{NaCl} = \frac{3.0}{3.0 + 55.6} = 0.051$$

$$X_{H_2O} = \frac{55.6}{3.0 + 55.6} = 0.949$$

Note that the sum of the two mole fractions, $0.051 + 0.949$, is 1.000.

13.18 Applications of Concentration Calculations

When a solution is diluted, the volume is increased by adding more solvent. Although the concentration is decreased, the total amount of solute is constant.

EXAMPLE 13.9   If 0.750 L of a 5.00 M solution of silver nitrate, $AgNO_3$, is diluted to a volume of 1.80 L by adding water, what is the molarity of the resulting diluted solution?

Since the number of moles of silver nitrate does not change on dilution, the problem can be solved by the following steps.

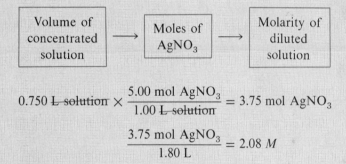

$$0.750 \text{ L solution} \times \frac{5.00 \text{ mol AgNO}_3}{1.00 \text{ L solution}} = 3.75 \text{ mol AgNO}_3$$

$$\frac{3.75 \text{ mol AgNO}_3}{1.80 \text{ L}} = 2.08 \text{ M}$$

The solution was diluted from 5.00 M to 2.08 M.

EXAMPLE 13.10   How many milliliters of water will be required to dilute 11 mL of a 0.45 M acid solution to a concentration of 0.12 M?

Again the number of moles of solute does not change. The following steps are necessary to solve this problem.

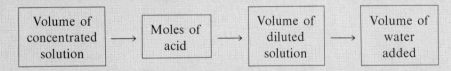

We convert the volume to liters and solve for the moles of acid present.

$$11 \text{ mL} \times \frac{1 \text{ L}}{1000 \text{ mL}} = 1.1 \times 10^{-2} \text{ L}$$

$$1.1 \times 10^{-2} \text{ L solution} \times \frac{0.45 \text{ mol acid}}{1.00 \text{ L solution}} = 4.95 \times 10^{-3} \text{ mol acid}$$

Rearrangement of the expression for molarity (molarity = moles/liters) gives the following expression.

$$\text{Liters} = \frac{\text{moles}}{\text{molarity}}$$

$$\text{Liters of dilute solution} = \frac{4.95 \times 10^{-3} \, \cancel{\text{mol acid}}}{0.12 \, \cancel{\text{mol acid}}/1.00 \text{ L solution}}$$

$$= 4.12 \times 10^{-2} \text{ L solution}$$

We convert to milliliters.

$$4.12 \times 10^{-2} \, \cancel{\text{L}} \times \frac{1000 \text{ mL}}{1.00 \, \cancel{\text{L}}} = 41.2 \text{ mL solution}$$

The volume of water added in the dilution is equal to the final volume of the solution minus the original volume.

$$41.2 \text{ mL} - 11 \text{ mL} = 30 \text{ mL}$$

EXAMPLE 13.11   A sulfuric acid solution containing 571.6 g of H_2SO_4 per liter of solution at 20°C has a density of 1.3294 g/mL. Calculate (a) the molarity, (b) the molality, (c) the percent by mass of H_2SO_4, and (d) the mole fractions for the solution.

(a) $$571.6 \, \cancel{\text{g } H_2SO_4} \times \frac{1 \text{ mol } H_2SO_4}{98.08 \, \cancel{\text{g } H_2SO_4}} = 5.828 \text{ mol } H_2SO_4$$

$$M = \frac{5.828 \text{ mol } H_2SO_4}{1.000 \text{ L}} = 5.828 \, M$$

(b) Since we know the number of moles of H_2SO_4 in 1 L of solution, to calculate the molality we need to find the mass of water in 1 L of solution. The following steps are required.

$$\boxed{\begin{array}{c} \text{Density} \\ \text{of} \\ \text{solution} \end{array}} \longrightarrow \boxed{\begin{array}{c} \text{Mass} \\ \text{of} \\ \text{solution} \end{array}} \longrightarrow \boxed{\begin{array}{c} \text{Mass} \\ \text{of} \\ \text{water} \end{array}} \longrightarrow \boxed{\text{Molality}}$$

The mass of 1 L of solution is given by rearranging the expression for density (density = mass/volume).

$$\text{Density} \times \text{volume} = \text{mass}$$

$$\frac{1.3294 \text{ g}}{1 \, \cancel{\text{mL}}} \times 1000 \, \cancel{\text{mL}} = 1329.4 \text{ g}$$

Thus 1 L of solution weighs 1329.4 g and contains 571.6 g of H_2SO_4. The mass of water is therefore

$$1329.4 \text{ g} - 571.6 \text{ g} = 757.8 \text{ g (or } 0.7578 \text{ kg)}$$

$$\frac{5.828 \text{ mol } H_2SO_4}{0.7578 \text{ kg } H_2O} = 7.691 \, m$$

(c) The solution contains 571.6 g of H_2SO_4 in 1329.4 g of solution.

$$\frac{571.6 \text{ g } H_2SO_4}{1329.4 \text{ g solution}} \times 100 = 43.00\% \ H_2SO_4 \text{ by mass}$$

(d) The number of moles of water present in 1 L of the solution is given by

$$757.8 \text{ g } H_2O \times \frac{1 \text{ mol}}{18.02 \text{ g } H_2O} = 42.05 \text{ mol } H_2O$$

$$X_{H_2SO_4} = \frac{\text{mol } H_2SO_4}{\text{mol } H_2SO_4 + \text{mol } H_2O}$$

$$= \frac{5.828}{5.828 + 42.05} = \frac{5.828}{47.88} = 0.1217$$

$$X_{H_2O} = \frac{42.05}{47.88} = 0.8782$$

Note that the sum of the mole fractions, $0.1217 + 0.8782$, is 0.9999 (or 1.0000 within rounding error of 1 in the last significant figure).

COLLIGATIVE PROPERTIES OF SOLUTIONS

13.19 Lowering the Vapor Pressure of the Solvent

When a nonvolatile substance (or one with such a low vapor pressure that we can disregard it) is dissolved in a liquid, the liquid's vapor pressure is lowered. Sugar (and most other solids) are nonvolatile; thus the vapor pressure of an aqueous sugar solution at 20°C is less than that of pure water at 20°C.

The vapor pressure of a liquid is determined by the frequency of escape of molecules from its surface. In a sugar solution both sugar and water molecules are found at the surface. Consequently, the number of water molecules at the surface is less than in pure water; thus the frequency of escape of water molecules, and hence the vapor pressure, is lower than in pure water.

The amount that the vapor pressure of a solvent is lowered has nothing to do with the kind or size of solute molecules, provided that the solution approximates ideal behavior (Section 13.11). The decrease in vapor pressure depends only on the *concentration* of solute molecules. The lowering of the vapor pressure of a solvent due to the presence of a solute is one of several properties of solutions that depend only on the concentration of the solute species present and that are called **colligative properties.**

For an ideal solution the decrease in vapor pressure is proportional to the ratio of the number of solute molecules to the total number of solute and solvent molecules. The greater the number of solute molecules, the lower the vapor pressure of the solution. These considerations are summed up in **Raoult's Law: The vapor pressure of the solvent in an ideal solution, P_{solv}, is equal to the mole fraction** of the solvent, X_{solv}, times the vapor pressure of the pure solvent, P^0_{solv}.

$$P_{solv} = X_{solv}P^0_{solv} \qquad (1)$$

As was noted in Section 13.11, solutions of chemically similar pairs of substances often closely approximate ideal behavior. Dilute solutions also tend to exhibit ideal behavior. The decrease in vapor pressure, ΔP, of an ideal solution, compared to that of the pure solvent, is equal to the mole fraction of solute, X_{solute}, times the vapor pressure of pure solvent, P^0_{solv}.

$$\Delta P = X_{solute}P^0_{solv} \qquad (2)$$

EXAMPLE 13.12   Calculate the vapor pressure of a solution of 92.1 g of glycerin, $C_3H_8O_3$, in 184 g of ethanol (ethyl alcohol), C_2H_5OH, at 40°C. The vapor pressure of pure ethanol at 40°C is 135.3 torr. Glycerin is essentially nonvolatile at this temperature and forms a solution with ethanol that approximates ideal behavior.

We can find the vapor pressure of ethanol from Equation (1), $P_{solv} = X_{solv}P^0_{solv}$, but first we need to determine its mole fraction.

$$92.1 \text{ g } C_3H_8O_3 \times \frac{1 \text{ mol } C_3H_8O_3}{92.1 \text{ g } C_3H_8O_3} = 1.00 \text{ mol } C_3H_8O_3$$

$$184 \text{ g } C_2H_5OH \times \frac{1 \text{ mol } C_2H_5OH}{46.0 \text{ g } C_2H_5OH} = 4.00 \text{ mol } C_2H_5OH$$

$$X_{C_2H_5OH} = \frac{4.00}{1.00 + 4.00} = 0.800$$

Now we can calculate the vapor pressure of ethanol.

$$P_{C_2H_5OH} = X_{C_2H_5OH}P^0_{C_2H_5OH} = 0.800 \times 135.5 \text{ torr} = 108 \text{ torr}$$

The vapor pressure of the solvent (ethanol) has been lowered from 135.3 torr to 108 torr by adding the glycerin. This change of 27 torr can also be calculated from Equation (2).

$$\Delta P = X_{solute}P^0_{solv} = 0.200 \times 135.3 \text{ torr} = 27 \text{ torr}$$

13.20 Elevation of the Boiling Point of the Solvent

A liquid is at its boiling point when its vapor pressure equals the external pressure on its surface (Section 11.3). Adding a solute lowers the vapor pressure of a liquid; so a higher temperature is needed to increase the vapor pressure of a solution and make it boil. According to Raoult's Law, the lowering of the vapor pressure of the solvent in a dilute solution is directly proportional to the mole fraction of the solute. It follows, then, that the elevation of the boiling point of the solvent is also proportional to the mole fraction of the solute. The elevation of the boiling point does not depend on the kind of solute, provided that it is a nonvolatile substance that does not dissociate into ions. The elevation of the boiling point is the same when solutions of the same mole fraction are considered. For example, 1 mole of sucrose, $C_{12}H_{22}O_{11}$, and 1 mole of glucose, $C_6H_{12}O_6$, each dissolved in 1 kilogram of water, have the same mole fraction and form solutions that have the same boiling point, 100.512°C at 760 torr, an increase of 0.512° above the boiling point of pure water. One mole of any nonelectrolyte dissolved in 1 kilogram of water raises the boiling point by 0.512°C. The difference between the boiling point of a very dilute solution of a nonelectrolyte and the boiling point of the pure solvent is also directly proportional to the *molal* concentration of the solute. Thus 0.200 mole of a nonelectrolyte dissolved in 1 kilogram of water (a 0.200 *m* solution) will increase the boiling point by

$$0.200 \times 0.512°C = 0.102°C$$

The change in boiling point of a dilute solution, ΔT, from that of the pure solvent is given by the expression

$$\Delta T = K_b m \tag{1}$$

where m is the molal concentration of the solute in the solvent and K_b is the increase in boiling point for a 1 *m* solution. K_b is called the **molal boiling-point elevation constant.** The values of K_b for several solvents are listed in Table 13-2; you can see that the value of K_b varies for each solvent.

Table 13-2 Boiling Points, Freezing Points, and Molal Boiling-Point Elevation and Freezing-Point Depression Constants for Several Solvents

Solvent	Boiling Point, °C (760 torr)	$K_b(°C/m)$	Freezing Point, °C	$K_f(°C/m)$
Water	100.0	0.512	0	1.86
Acetic acid	118.1	3.07	16.6	3.9
Benzene	80.1	2.53	5.48	5.12
Chloroform	61.26	3.63	−63.5	4.68
Nitrobenzene	210.9	5.24	5.67	8.1

It should be emphasized that the extent to which the vapor pressure of a solvent is lowered and the boiling point is elevated depends on the number of solute particles present in a given amount of solvent and not on the mass or size of the particles. A mole of sodium chloride forms 2 moles of ions in solution and causes nearly twice as great a rise in boiling point as does 1 mole of nonelectrolyte. One mole of sugar contains 6.022×10^{23} particles (as molecules), whereas 1 mole of sodium chloride contains $2 \times 6.022 \times 10^{23}$ particles (as ions). Calcium chloride, $CaCl_2$, which consists of three ions, causes nearly three times as great a rise in boiling point as does sugar. Section 13.27 explains why the elevation is not exactly twice (for NaCl) or exactly three times (for $CaCl_2$) that of the boiling-point elevation for a nonelectrolyte.

EXAMPLE 13.13   How much does the boiling point of water change when 1.00 g of glycerin, $C_3H_8O_3$, is dissolved in 47.8 g of water?

Since the change in boiling point is proportional to the molal concentration of glycerin, we first calculate its molal concentration.

$$1.00 \text{ g } C_3H_8O_3 \times \frac{1 \text{ mol } C_3H_8O_3}{92.1 \text{ g } C_3H_8O_3} = 1.09 \times 10^{-2} \text{ mol } C_3H_8O_3$$

$$\frac{1.09 \times 10^{-2} \text{ mol } C_3H_8O_3}{0.0478 \text{ kg } H_2O} = 0.228 \text{ } m$$

The change in boiling point is equal to the molal concentration of the glycerin multiplied by K_b (0.512°C/m for water, Table 13-2).

$$\Delta T = K_b m = 0.512°C/m \times 0.228 \text{ } m = 0.117°C$$

EXAMPLE 13.14   What is the boiling point of a solution of 92.1 g of iodine, I_2, in 800.0 g of chloroform, $CHCl_3$, assuming that the iodine is nonvolatile?

First we calculate the molality of iodine in the solution since the change in boiling point of a dilute solution is proportional to m.

$$92.1 \text{ g } I_2 \times \frac{1 \text{ mol } I_2}{253.8 \text{ g } I_2} = 0.3629 \text{ mol } I_2$$

$$\frac{0.3629 \text{ mol } I_2}{0.8000 \text{ kg } CHCl_3} = 0.4536 \text{ } m$$

The value of the molal boiling-point elevation constant for chloroform (Table 13-2) is 3.63°C/m. Thus, for a 0.4536 m solution,

$$\Delta T = K_b m = 3.63°C/m \times 0.4536 \text{ } m = 1.65°C$$

Since the boiling point of chloroform, 61.26°C, is raised by 1.65°C, the boiling point of the solution will be

$$61.26°C + 1.65°C = 62.91°C$$

13.21 Fractional Distillation

In many cases pure solvent may be recovered from a solution by distillation (Section 11.6). If the solute is nonvolatile, an apparatus like the one in Fig. 11-8 may be used. Distillation makes use of the facts that heating a liquid speeds up its rate of evaporation and increases its vapor pressure and that cooling a vapor favors condensation.

The boiling point of a solution is the temperature at which the total vapor pressure of the mixture is equal to the atmospheric pressure. The total vapor pressure of a solution composed of two volatile substances depends on the concentration and the vapor pressure of each substance and will be

1. between the vapor pressures of the pure components,
2. less than the vapor pressure of either pure component, or
3. greater than that of either pure component.

Usually the total vapor pressure, and thus the boiling point, of a mixture of two liquids lies between those of the two components.

When a solution of type 1 is distilled, the vapor produced at the boiling point is richer in the lower-boiling component (the component with the higher vapor pressure) than was the original mixture. When this vapor is condensed, the resulting distillate contains more of the lower-boiling component than did the original mixture. This means that the composition of the boiling mixture constantly changes, the amount of the lower-boiling component of the mixture constantly decreases, and the boiling point rises as distillation continues. The vapor (and the distillate) contains more and more of the higher-boiling component and less and less of the lower-boiling component as distillation proceeds. Changing the receiver at intervals yields successive fractions, each one increasingly richer in the less volatile (higher-boiling) component. If this process of **fractional distillation** is repeated several times, relatively pure samples of the two liquids may be obtained.

Fractionating columns have been devised that achieve separation of liquids that would require a great number of simple fractional distillations of the type just described. Crude oil, a complex mixture of hydrocarbons, is separated into its components by fractional distillation on an enormous scale (Fig. 13-10).

The vapor pressure of a solution of nitric acid and water is less than that of either component (a solution of type 2). When a dilute solution is distilled, the first fraction of distillate consists mostly of water because water has a higher vapor pressure than nitric acid under distillation conditions. As distillation is continued,

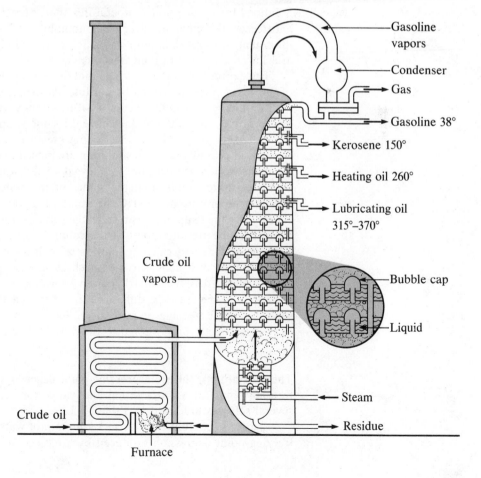

Figure 13-10. Fractional distillation of crude oil. Oil heated to about 425°C in the furnace vaporizes when it enters the tower at the right. The vapors rise through bubble caps in a series of trays in the tower. As the vapors gradually cool, fractions of higher, then of lower, boiling points condense to liquids and are drawn off. The fraction of highest boiling point is drawn off at the bottom as a residue. It is heavy fuel oil. In modern refineries these fractions, which still consist of mixtures of hydrocarbons, are further processed.

the solution remaining in the distilling flask becomes richer in nitric acid. When a concentration of 68% HNO_3 by mass is reached, the solution boils at a constant temperature of 120.5°C (at 760 torr). At this temperature, the solution and the vapor have the same composition, and the solution distills without any further change in composition.

If a nitric acid solution more concentrated than 68% is distilled, the vapor first formed contains a large amount of HNO_3. The solution that remains in the distilling flask contains a greater percentage of water than at first, and the concentration of the nitric acid in the distilling flask decreases as the distillation is continued. Finally, a concentration of 68% HNO_3 is reached, and the solution again boils at the constant temperature of 120.5°C (at 760 torr). The 68% solution of nitric acid is referred to as a **constant boiling solution.**

Solutions of both type 2 and type 3 produce constant boiling solutions when distilled. A constant boiling solution forms a vapor with the same composition as the solution; thus it distills without a change in concentration. Constant boiling solutions are also called **azeotropic mixtures.** Other common and important substances that form azeotropic mixtures with water are HCl (20.24%, 110°C at 760 torr) and H_2SO_4 (98.3%, 338°C at 760 torr).

13.22 Depression of the Freezing Point of the Solvent

Solutions freeze at lower temperatures than pure liquids. We use solutions of antifreezes like ethylene glycol in automobile radiators because they freeze at lower temperatures than pure water. Sea water, with its large salt content, freezes at a lower temperature than fresh water.

The depression of the freezing point of a solvent is a reflection of the lowering of the vapor pressure caused by a solute. At its freezing point a pure liquid is in equilibrium with its solid (Section 11.10), and they have the same vapor pressure. For example, pure water and ice have the same vapor pressure at 0°C. However, the water *in a solution* has a lower vapor pressure than that of ice at 0°C. Consequently, if ice and an aqueous solution are placed in contact and kept at 0°C, the ice melts. If the temperature is allowed to fall, the vapor pressure of the ice decreases more rapidly than does that of the water in the solution, and at a specific temperature below 0°C the ice and the water have the same vapor pressure. This is the temperature at which the solution and the ice are again in equilibrium; it is the freezing point of the solution. This property permits the use of sodium chloride and calcium chloride to melt ice on streets and highways.

One mole of a nonelectrolyte, such as sucrose, glycerin, ethylene glycol, or alcohol, when dissolved in 1 kilogram of water, gives a solution that freezes at −1.86°C. In general, the difference, ΔT, between the freezing point of a pure solvent and the freezing point of a solution of a nonelectrolyte dissolved in that solvent is directly proportional to the *molal* concentration of the solute.

$$\Delta T = K_f m$$

The constant K_f, the **molal freezing-point depression constant,** is the change in freezing point for a 1 m solution and varies for each solvent. Values of K_f for several solvents are listed in Table 13-2.

A mole of sodium chloride in 1 kilogram of water will show nearly twice the freezing-point depression produced by a molecular compound. Each individual ion produces the same effect on the freezing point as a single molecule does. In Section 13.27 we shall consider why the lowering produced by sodium chloride is not exactly twice that produced by a similar amount of a nonelectrolyte.

EXAMPLE 13.15   What is the freezing point of the solution of 92.1 g of I_2 in 800.0 g of $CHCl_3$ described in Example 13.14?

The molal concentration of I_2 was shown to be 0.4536 m. The value of the molal freezing-point depression constant, K_f, in Table 13-2 is 4.68°C/m. Thus for a 0.4536 m solution

$$\Delta T = K_f m = 4.68°C/m \times 0.4536\, m = 2.12°C$$

The freezing point of the solution will be 2.12° lower than that of pure $CHCl_3$, or

Freezing point of solution = −63.5°C − 2.12°C = −65.6°C

13.23 Phase Diagram for an Aqueous Solution of a Nonelectrolyte

Figure 13-11 is a phase diagram for an aqueous solution of a nonelectrolyte, such as sucrose, $C_{12}H_{22}O_{11}$. The phase diagram for pure water (Section 12.2) is included as a dashed line for comparison. You can see that the freezing point for the solution is lower than that for pure water (the solid line separating solid and liquid states is displaced to the left of the dashed one). Correspondingly, the higher boiling point for the solution is shown by the displacement of the solid line separating the liquid and gas states to the right of the dashed one. The decrease in vapor pressure of the solution, at any given temperature, is indicated by the vertical distance between the dashed line and the solid line. On the diagram the freezing-point depression and the boiling-point elevation are the horizontal dis-

Figure 13-11. Phase diagram for a 1 m aqueous solution of a nonelectrolyte (solid lines) compared to that for pure water (dashed lines). The diagram is not to scale in order to show more clearly the differences between the solution and pure water.

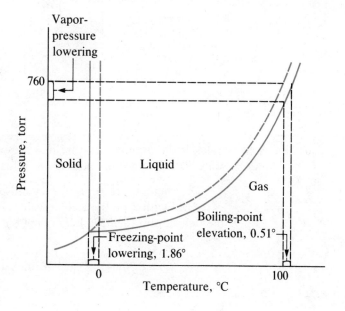

tances between the broken line and the solid line near 0°C and near 100°C, respectively, at a pressure of 760 torr.

13.24   Osmosis and Osmotic Pressure of Solutions

When a solution and its pure solvent are separated by a semipermeable membrane (one through which the solvent but not the solute can pass), the pure solvent will diffuse through the membrane and dilute the solution. This process is known as **osmosis.** Actually, the solvent diffuses through the membrane in both directions simultaneously; however, its rate of diffusion is greater from the pure solvent to the solution than in the opposite direction, so there is an increase in the number of solvent molecules in the solution.

When a solution and a pure solvent are separated by a semipermeable membrane as shown in Fig. 13-12, the volume of the solution increases because the net movement of solvent molecules is from the pure solvent to the solution. The liquid in the tube above the solution slowly rises. If the membrane is strong enough to withstand the pressure and if the tube is long enough, the liquid will rise until its hydrostatic pressure (due to the weight of the column of solution in the tube) is great enough to prevent the further osmosis of solvent molecules into the solution. The pressure required to stop the osmosis from a pure solvent into a solution is called the **osmotic pressure, π,** of the solution. The osmotic pressure of a dilute solution can be calculated from the expression

$$\pi = MRT$$

where M is the molar concentration of the solute, R is the gas constant, and T is the temperature of the solution on the Kelvin scale.

Figure 13-12.   Apparatus for demonstrating osmosis. The levels are equal at the start (a) but at equilibrium (b) the level of the sugar solution is higher due to the net transfer of water molecules into it.

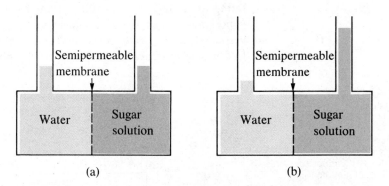

(a)   (b)

EXAMPLE 13.16   What is the osmotic pressure in atmospheres of a 0.30 M solution of glucose in water that is used for intravenous infusion at body temperature, 37°C?

The osmotic pressure, π, in atmospheres can be found using the formula $\pi = MRT$ with T on the Kelvin scale (310 K) and with a value of R that includes the unit of atm (0.08206 L atm/mol K).

$$\pi = MRT$$
$$\pi = 0.30 \text{ mol/L} \times 0.08206 \text{ L atm/mol K} \times 310 \text{ K} = 7.6 \text{ atm}$$

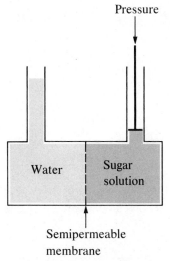

Figure 13-13. Application of a pressure greater than the osmotic pressure will reverse the osmosis.

If a solution is placed in an apparatus like the one in Fig. 13-13, applying pressure greater than the osmotic pressure of the solution reverses the osmosis and increases the volume of the pure solvent. This technique of **reverse osmosis** is used for desalting sea water. A plant using this principle and capable of producing about three million gallons of fresh water per day was opened in Key West, Florida, in 1981.

The effects of osmosis are particularly evident in biological systems since cells are surrounded by semipermeable membranes. Carrots and celery that have become limp due to loss of water to the atmosphere can be made crisp again by placing them in water. Water moves into the carrot or celery cells by osmosis. A cucumber placed in a concentrated salt solution loses water by osmosis and becomes a pickle. Osmosis can also affect animal cells. Solute concentrations are particularly important when solutions are injected into the body. Solutes in body cell fluids and blood serum give these solutions an osmotic pressure of approximately 7.7 atmospheres. Solutions injected into the body must have the same osmotic pressure as blood serum; that is, they should be **isotonic** with blood serum. If a less concentrated solution, a **hypotonic solution,** is injected in sufficient quantity to dilute the blood serum, water from the diluted serum will pass into the blood cells by osmosis, causing the cells to expand and rupture. If a more concentrated solution, a **hypertonic solution,** is injected, the cells will lose water to the more concentrated solution, shrivel, and, possibly, die.

13.25 Determination of Molecular Weights of Substances in Solution

The effect of a solute on the freezing point, boiling point, vapor pressure, or osmotic pressure of a solution can be measured. Changes in these properties are directly proportional to the concentration of solute present, so the molecular weight of the solute can be determined from the change.

EXAMPLE 13.17 A solution of 35.7 g of an organic nonelectrolyte in 220.0 g of chloroform has a boiling point of 64.5°C. What is the molecular weight of this organic compound?

This problem requires the following steps.

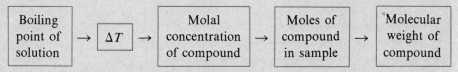

From the boiling point of pure chloroform (61.26°C, Table 13-2), we can calculate ΔT, the increase in boiling temperature.

$$\Delta T = 64.5°C - 61.26°C = 3.2°C$$

From K_b for chloroform (3.63°C/m, Table 13-2), we can calculate the molal concentration of the electrolyte (mol solute/mass solvent) by rearranging the equation $\Delta T = K_b m$.

$$m = \frac{\Delta T}{K_b} = \frac{3.2°C}{3.63°C/m} = 0.88\ m$$

The number of moles of solute in 0.2200 kg (220.0 g) of solvent is then calculated as follows:

$$\text{Moles of solute} = \frac{0.88 \text{ mol solute}}{1.00 \text{ kg solvent}} \times 0.2200 \text{ kg solvent}$$

$$= 0.19 \text{ mol solute}$$

We can then calculate the molar mass.

$$\text{Molar mass} = \frac{35.7 \text{ g}}{0.19 \text{ mol}} = 180 \text{ g/mol, or } 1.8 \times 10^2 \text{ g/mol}$$

A molar mass of 180 g/mol corresponds to a molecular weight of 180. (Note that only two significant figures are justified, despite the fact that all data are expressed to at least three significant figures. Why is this?)

EXAMPLE 13.18 If 4.00 g of a certain nonelectrolyte is dissolved in 55.0 g of benzene, the resulting solution freezes at 2.32°C. Calculate the molecular weight of the nonelectrolyte.

The steps for this problem are

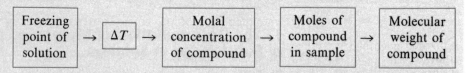

According to Table 13-2, K_f for benzene is 5.12°C/m and its freezing point is 5.48°C. Since the freezing point of the solution is 2.32°C, the 4.00 g of solute has lowered the freezing point from 5.48°C to 2.32°C. Thus $\Delta T = 3.16$°C.

The molality of the solution is

$$m = \frac{\Delta T}{K_f} = \frac{3.16°\cancel{C}}{5.12°\cancel{C}/m} = 0.617 \, m$$

The number of moles of solute in 0.0550 kg (55.0 g) of solvent is

$$\text{Moles of solute} = \frac{0.617 \text{ mol}}{1.000 \text{ kg solvent}} \times 0.0550 \text{ kg solvent}$$

$$= 0.0339 \text{ mol}$$

The molar mass can then be calculated:

$$\text{Molar mass} = \frac{4.00 \text{ g}}{0.0339 \text{ mol}} = 118 \text{ g/mol}$$

$$\text{Mol. wt} = 118$$

EXAMPLE 13.19 One liter of an aqueous solution containing 20.0 g of hemoglobin has an osmotic pressure of 5.9 torr at 22°C. What is the molecular weight of hemoglobin?

The steps for this problem are

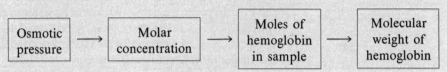

From $\pi = MRT$, we can relate the osmotic pressure to the molar concentration provided the pressure is expressed in atmospheres to match the units in R (0.08206 L atm/mol K) and the temperature is expressed on the Kelvin scale ($T = 273 + 22 = 295$ K).

$$\pi = 5.9 \text{ torr} \times \frac{1 \text{ atm}}{760 \text{ torr}} = 7.76 \times 10^{-3} \text{ atm}$$

Rearrangement of $\pi = MRT$ gives $M = \pi/RT$.

$$M = \frac{7.76 \times 10^{-3} \text{ atm}}{(0.08206 \text{ L atm/mol K})(295 \text{ K})} = 3.2 \times 10^{-4} \text{ mol/L} = 3.2 \times 10^{-4} \text{ } M$$

Since the volume of the solution containing the 20.0 g of hemoglobin is 1 L, 20.0 g of hemoglobin is equal to 3.2×10^{-4} mol.

$$\text{Molar mass} = \frac{20.0 \text{ g}}{3.2 \times 10^{-4} \text{ mol}} = 62,000 \text{ g/mol}$$

$$\text{Mol. wt} = 62,000$$

13.26 The Effect of Electrolytes on the Colligative Properties of Solutions

The effect of nonelectrolytes on the colligative properties (changes in vapor pressure, boiling point, freezing point, and osmotic pressure) of a solution is dependent only on the number, not on the kind, of particles dissolved. For example, 1 mole of any nonelectrolyte in solution in 1 kilogram of water produces the same lowering of the freezing point as does 1 mole of any other nonelectrolyte because 1 mole of any nonelectrolyte contains 6.022×10^{23} molecules. However, the lowering of the freezing point produced by 1 mole of an electrolyte is much greater than that from 1 mole of a nonelectrolyte since an electrolyte ionizes when it dissolves. When 1 mole of sodium chloride is dissolved in 1 kilogram of water, the solution freezes at $-3.37°C$. The freezing point is lowered 1.81 times as much as for a nonelectrolyte. This illustrates the fact that 1 mole of an electrolyte produces more than 6.022×10^{23} solute particles. Almost all acids, bases, and salts behave this way when dissolved.

13.27 Ion Activities

If sodium chloride were completely dissociated in aqueous solution, it would lower the freezing point and raise the boiling point of the solvent twice as much as would an equal molal concentration of a nonelectrolyte, since it gives two moles of ions per mole of compound. However, a mole of sodium chloride lowers the freezing point of water only 1.81 times as much. A similar discrepancy occurs for the boiling-point elevation. Apparently, sodium chloride (and other strong electrolytes) are not completely dissociated in solution.

Peter J. W. Debye and Erich Hückel proposed a theory to explain the apparent incomplete ionization of strong electrolytes. They suggested that although interionic attraction in an aqueous solution is very greatly reduced by hydration of the ions and the insulating action of the polar solvent, it is not completely nullified. The residual attractions prevent the ions from behaving as totally independent particles (Fig. 13-14). In some cases a positive and negative ion may actually touch, giving a solvated unit called an **ion pair.** Thus the **activity** of any particular kind of ion, or the effective concentration, is less than indicated by the actual concentration. This is evident in the colligative properties of the solution and in the amount of electric current conducted by it. Ions become more and more widely separated the more dilute the solution, and the residual interionic attractions become less and less. Thus in extremely dilute solutions the effective concentrations of the ions (their activities) are essentially equal to the actual concentrations.

Figure 13-14. Diagrammatic representation of the various species thought to be present in a solution of potassium chloride.

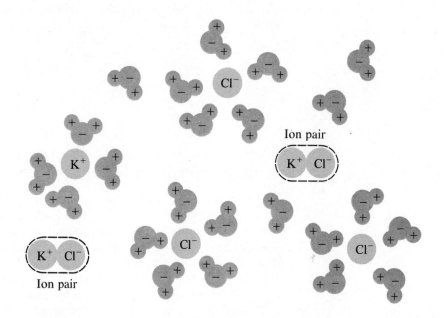

COLLOID CHEMISTRY

Solutions are dispersions of discrete molecules or ions of a solute. We shall now consider dispersions of particles that are significantly larger than simple molecules yet not large enough to be seen using an ordinary microscope.

13.28 Colloidal Matter

Although the mixture obtained when powdered starch is heated with water is not homogeneous, the particles of insoluble starch do not settle out but remain in suspension indefinitely. Such a system is called a **colloidal dispersion;** the finely divided starch is called the **dispersed phase,** and the water is called the **dispersion medium.**

The term **colloid**—from the Greek words *kolla,* meaning glue, and *eidos,* meaning like—was first used in 1861 by Thomas Graham to classify substances such as starch and gelatin, which usually exist in an amorphous condition. We now know that colloidal properties are not limited to a special class of substances; any substance can exist in colloidal form. Many colloidal particles are aggregates of hundreds, or even thousands, of molecules, but others, such as viruses and polymer molecules, consist of a single large molecule. Viruses are giant molecules with molecular weights ranging from several hundred thousands to billions. Tobacco mosaic virus has a molecular weight of about 40,000,000. Proteins and synthetic polymer molecules may have weights ranging from a few thousands to many millions.

Colloids may be dispersed in a gas, a liquid, or a solid, and the dispersed phase may be a gas, a liquid, or a solid. However, a gas dispersed in another gas is not a colloidal system because the particles are of molecular dimensions. A classification of colloidal systems is given in Table 13-3.

Table 13-3   Colloidal Systems

Dispersed Phase	Dispersion Medium	Examples	Common Name
Solid	Gas	Smoke, dust	Solid aerosol
Solid	Liquid	Starch suspension, some inks, paints, milk of magnesia	Sol
Solid	Solid	Colored gems, some alloys	Solid sol
Liquid	Gas	Clouds, fogs, mists, sprays	Liquid aerosol
Liquid	Liquid	Milk, mayonnaise, butter	Emulsion
Liquid	Solid	Jellies, gels, opal (SiO_2 and H_2O), pearl ($CaCO_3$ and H_2O)	Solid emulsion
Gas	Liquid	Foams, whipped cream, beaten egg whites	Foam
Gas	Solid	Pumice, floating soaps	

13.29 Preparation of Colloidal Systems

A colloidal system is prepared by producing particles of colloidal dimensions and distributing these particles through the dispersion medium. Particles of colloidal size are formed by two methods:

1. **dispersion methods,** that is, the subdivision of larger particles or masses. For example, paint pigments are produced by dispersing large particles by grinding in special mills.
2. **condensation methods,** that is, growth from smaller units, such as molecules or ions. For example, clouds form when water molecules condense and form very small droplets.

A few solid substances, when brought into contact with water, disperse spontaneously and form colloidal systems. Gelatin, glue, and starch behave in this manner, and are said to undergo **peptization.** The particles are already of colloidal size; the water simply disperses them. Some atomizers produce colloidal dispersions. Powdered milk with particles of colloidal size is produced by dehydrating milk spray.

An **emulsion** may be prepared by shaking together two immiscible liquids. Agitation breaks one liquid into droplets of colloidal size, which then disperse throughout the other liquid. The droplets of the dispersed phase, however, tend to coalesce, forming large drops, and separation of the liquids into two layers follows. Therefore, emulsions are usually stabilized by substances called **emulsifying agents.** For example, a little soap will stabilize an emulsion of kerosene in water. Milk is an emulsion of butterfat in water with casein as the emulsifying agent. Mayonnaise is an emulsion of oil in vinegar with egg yolk as the emulsifying agent. Oil spills in the ocean are particularly difficult to clean up because the oil and the water form an emulsion.

Condensation methods form colloidal particles by aggregation of smaller particles. If the particles grow beyond the colloidal size range, precipitates form, and no colloidal system results.

Condensation methods often employ chemical reactions. A dark red colloidal suspension of iron(III) hydroxide may be prepared by mixing a concentrated solution of iron(III) chloride with hot water.

$$Fe^{3+} + 3Cl^- + 6H_2O \longrightarrow Fe(OH)_3 + 3H_3O^+ + 3Cl^-$$

A colloidal suspension of arsenic(III) sulfide is produced by the reaction of hydrogen sulfide with arsenic(III) oxide dissolved in water.

$$As_2O_3 + 3H_2S \longrightarrow As_2S_3 + 3H_2O$$

A colloidal gold sol results from the reduction of a very dilute solution of gold chloride by such reducing agents as formaldehyde, tin(II) chloride, or iron(II) sulfate.

$$Au^{3+} + 3e^- \longrightarrow Au$$

Some gold sols prepared by Faraday in 1857 are still perfectly clear.

13.30 The Cleansing Action of Soaps and Detergents

Soaps are made by boiling either fats or oils with a strong base, such as sodium hydroxide. Pioneers made soap by boiling fat with a strongly basic solution made by leaching potassium carbonate, K_2CO_3, from wood ashes with hot water. When animal fat is treated with sodium hydroxide, glycerol and sodium salts of fatty acids such as palmitic, oleic, and stearic acid are formed. The sodium salt of stearic acid, sodium stearate, has the formula $C_{17}H_{35}CO_2Na$.

Detergents (soap substitutes) all contain a hydrocarbon chain, such as $—C_{12}H_{25}$, which is nonpolar, and an ionic group, such as a sulfate, $—OSO_3^-$, or a sulfonate, $—SO_3^-$. Soaps form insoluble calcium and magnesium compounds in hard water; detergents form water-soluble products—a definite advantage for detergents.

The cleansing action of soaps and detergents can be explained in terms of the structures of the molecules involved. The molecules of both consist of a long hydrocarbon chain attached to an ionic group.

Hydrocarbon chain

Ionic end

$$CH_3 \diagdown CH_2 \diagup CH_2 \diagdown CH_2 \diagup CH_2 \diagdown CH_2 \diagup CH_2 \diagdown CH_2 \diagup CH_2 \diagdown CH_2 \diagup CH_2 \diagdown CH_2 \diagup CH_2 \diagdown CH_2 \diagup CH_2 \diagdown CH_2 \diagup CH_2 \diagdown CO_2^- Na^+$$

Sodium stearate (soap)

$$CH_3 \diagdown CH_2 \diagup CH_2 \diagdown CH_2 \diagup CH_2 \diagdown CH_2 \diagup CH_2 \diagdown CH_2 \diagup CH_2 \diagdown CH_2 \diagup CH_2 \diagdown CH_2 \diagup OSO_3^- Na^+$$

Sodium lauryl sulfate (detergent)

The hydrocarbon end of such a molecule is attracted by dirt, oil, or grease particles, and the ionic end is attracted by water (Fig. 13-15). The result is an orientation of the molecules at the interface between the dirt particles and the water in such a way that the dirt particles become suspended as colloidal particles and are readily washed away.

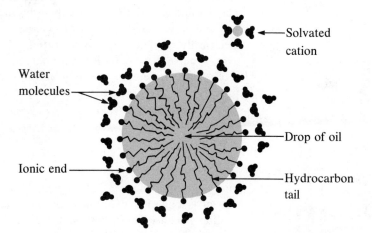

Figure 13-15. Diagrammatic cross section of an emulsified drop of oil in water with soap or detergent as the emulsifier. The negative ions of the emulsifier are oriented at the interface between the oil particle and water. The positive ions are solvated and move independently through the water. The nonpolar hydrocarbon end of the ion is in oil, and the ionic end (—CO_2^- for soap, —SO_3^- for detergent) is in water.

13.31 Brownian Movement

When a colloidal system against a black background in a darkened space is viewed with a powerful microscope at right angles to an intense beam of light, the colloidal particles look like tiny bright flashes of light in irregular rapid, dancing motion (Fig. 13-16). This motion is called **Brownian movement** after the botanist Robert Brown, who first observed it in 1828. He could not explain it, but we now know that the colloidal particles are so small that bombardment by molecules of the dispersion medium makes them move irregularly. Any particle in suspension is bombarded on all sides by the moving molecules of the dispersion medium. Particles larger than colloidal size do not move because bombardment on one side is likely to be counterbalanced by an equal bombardment on the opposite side. For particles of colloidal size, however, the probability of equal and simultaneous bombardments on opposite sides is slight, so they move irregularly. This movement explains why dispersed particles in colloidal systems do not settle, even though they are denser than the dispersion medium. The kinetic-molecular theory received one of its earliest confirmations as a result of studies of Brownian movement.

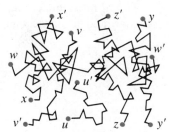

Figure 13-16. A diagram showing how six particles in suspension might move due to unequal bombardment by molecules of the dispersion medium.

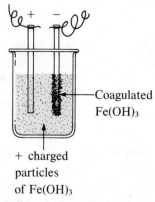

Coagulated
Fe(OH)₃

+ charged
particles
of Fe(OH)₃

Figure 13-17.   Colloidal iron(III)
hydroxide is coagulated in an
electrolytic cell.

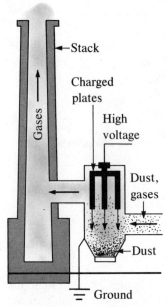

Stack

Charged
plates

High
voltage

Gases

Dust,
gases

Dust

Ground

Figure 13-18.   In a Cottrell pre-
cipitator, positively and negatively
charged particles in smoke are
precipitated as they pass over the
electrically charged plates.

13.32   Electrical Properties of Colloidal Particles

Dispersed colloidal particles are usually electrically charged. A particle of an iron(III) hydroxide sol, for example, does not contain enough hydroxide ions to compensate exactly for the positive charges on the iron(III) ions. Thus each individual colloidal particle bears a positive charge, and the colloidal dispersion consists of charged colloidal particles and hydrated ions of opposite charge, which keep the dispersion electrically neutral. In an electrolytic cell (Fig. 13-17), the dispersed particles move to the negative electrode, the cathode. Since opposite charges attract each other, this is good evidence that the iron(III) hydroxide particles are positively charged. At the cathode the colloidal particles lose their charge and coagulate as a precipitate. All colloidal particles in any one system have charges of the same sign. This helps to keep them dispersed, since like charges repel each other. Most metal hydroxide colloids have positive charges, while most metal sulfides and the metals themselves form negatively charged dispersions.

The carbon and dust particles in smoke are often colloidally dispersed and electrically charged. Frederick Cottrell, an American chemist, developed a process to remove these particles. The charged particles are attracted to highly charged electrodes, where they are neutralized and deposited as dust (Fig. 13-18). The process is also important in the recovery of valuable products from the smoke of smelters, furnaces, and kilns.

13.33   Gels

Under certain conditions, the dispersed phase in a colloidal system coagulates so that the whole mass, including the liquid, sets to an extremely viscous body known as a **gel.** For example, a hot aqueous "solution" of gelatin sets to a gel on cooling. Because the formation of a gel is accompanied by the taking up of water or some other solvent, the gel is said to be hydrated or solvated. Apparently the fibers of the dispersed substance form a complex three-dimensional network, the interstices being filled with the liquid medium or a dilute solution of the dispersed phase.

Pectin, a carbohydrate from fruit juices, is a gel-forming substance important in jelly making. Silica gel, a colloidal dispersion of hydrated silicon dioxide, is formed when dilute hydrochloric acid is added to a dilute solution of sodium silicate. Canned Heat is a gel made by mixing alcohol and a saturated aqueous solution of calcium acetate. The wall of a living cell is colloidal in character and within the cell is a gel. In fact, all living tissue is colloidal, and life processes depend on the chemistry of colloids.

FOR REVIEW

SUMMARY

A **solution** forms when one substance, a **solute,** dissolves in a second substance, a **solvent,** giving a homogeneous mixture of atoms, molecules, or ions. The solute may be a solid, a liquid, or a gas. The solvent is usually a liquid, although solutions of gases in other gases as well as solutions of gases, liquids, and solids in solids are possible. Substances that dissolve and give solutions that contain ions and conduct electricity are called **electrolytes.** Electrolytes may be molecular compounds that react with the solvent to give ions, or they may be ionic compounds. **Nonelectrolytes** are substances that dissolve to give solutions of molecules. These solutions do not conduct electricity.

The extent to which a gas dissolves in a liquid solvent is proportional to the pressure of the gas, provided that the gas does not react with the solvent. Generally, the solubility of a gas decreases with increasing temperature. Polar liquids tend to be soluble **(miscible)** in water while nonpolar liquids tend to be insoluble **(immiscible)** in water. Generally, the solubility of solids in water increases with increasing temperature; however, there are exceptions.

Solutions that obey **Raoult's Law** exactly, **ideal solutions,** form when the average strength of the intermolecular forces of attraction between the solute and solvent is equal to the average strength of the forces of attraction between pure solute molecules and between pure solvent molecules. Ideal solutions form by a **spontaneous process** because the disorder of the molecules of solute and solvent increases when the solution forms. An increase in the average strength of the solute-solvent intermolecular attractive forces, compared to those in the pure solute and solvent, also favors spontaneous formation of a solution. The relative strengths of the solute-solute and solvent-solvent attractions compared to the solute-solvent attractions can be measured by the **heat of solution** of the solute in the solvent. In many cases, the increase in disorder compensates for the decrease in the average strength of the solvent-solute attractions compared to those of the pure components of a solution.

The relative amounts of solute and solvent in a solution can be described quantitatively as the **concentration** of the solution. Units of concentration include **percent composition, molar concentration, molal concentration,** and **mole fraction.**

Properties of a solution that depend only on the concentration of solute particles are called **colligative properties.** They include changes in the vapor pressure, boiling point, and freezing point of the solvent in the solution. The magnitudes of these properties depend only on the total concentration of solute particles in solution and not on the type of particles. The total concentration of solute particles in a solution also affects its **osmotic pressure.** This is the pressure that must be applied to the solution in order to prevent diffusion of molecules of pure solvent through a semipermeable membrane into the solution.

A **colloid** is a suspension of small, insoluble particles, which are usually larger than molecules but small enough not to settle from the dispersion medium. Examples of colloids include smoke, clouds, paints, mayonnaise, and foams.

KEY TERMS AND CONCEPTS

Alloys (13.10)
Azeotropic mixtures (13.21)
Boiling-point elevation (13.20)
Brownian movement (13.31)
Colligative properties (13.19)
Colloid (13.28)
Constant boiling solution (13.21)

Electrolytes (13.8)
Emulsion (13.29)
Fractional distillation (13.21)
Freezing-point depression (13.22)
Gel (13.33)
Heat of solution (13.11)

Henry's Law (13.2)
Ideal solution (13.11)
Ion activity (13.27)
Ion-dipole attraction (13.12)
Miscible liquids (13.3)
Molality (13.16)
Molarity (13.15)

Mole fraction (13.17)
Nonelectrolytes (13.8)
Nonstoichiometric compounds
   (13.10)
Osmosis (13.24)
Osmotic pressure (13.24)

Percent composition (13.14)
Raoult's Law (13.19)
Saturated solution (13.4)
Solid solution (13.10)
Solubility (13.1)
Strong electrolytes (13.8)

Supersaturated solution (13.5)
Unsaturated solution (13.4)
Vapor-pressure lowering (13.19)
Weak electrolytes (13.8)

EXERCISES

The Nature of Solutions

1. How do solutions differ from compounds? from ordinary mixtures?
2. Show how solutions of (a) carbon dioxide in air, (b) ethanol, C_2H_5OH, in water, and (c) potassium chloride in water exhibit the principal characteristics of solutions.
3. Why are the majority of chemical reactions most readily carried out in solutions?
4. Explain the following terms as applied to solutions:
   (a) solute
   (b) solvent
   (c) weak electrolyte
   (d) strong electrolyte
   (e) nonelectrolyte
   (f) saturated
   (g) supersaturated
   (h) unsaturated
5. Describe the effect of (a) the nature of the gas and its solvent, (b) the pressure on the gas, (c) the temperature, and (d) the reaction of the gas with the solvent, on the solubility of a gas.
$\boxed{S}$6. At 0°C, 3.36 g of CO_2 at 1.00 atm will dissolve in exactly 1 L of water. How many grams of CO_2 will dissolve in exactly 1 L of water at 0°C and 4.00 atm?   *Ans. 13.4 g*
7. In order to prepare supersaturated solutions of most solids, saturated solutions are cooled. Supersaturated solutions of gases are prepared by warming saturated solutions. Explain the reasons for the difference in the two procedures.
8. Why is a solution of nitric acid, HNO_3, in water a good conductor of an electric current when both pure nitric acid and water are not?
9. Why are solid ionic compounds nonconductors, whereas ionic compounds that are fused (melted) are good conductors?

10. Using the general rules given in Section 13.9, answer the following questions:
    (a) When HCl is added to an unknown solution, a white precipitate forms. What common cations might be present in the solution?
    (b) The addition of a solid metal sulfide to water produces H_2S and an insoluble solid. What common metal sulfides would exhibit this behavior?
    (c) When H_2SO_4 is added to a solution, no precipitate forms. What common metal cations are not present in the solution?
    (d) A solution of sodium phosphate was added to a second solution, and no precipitate formed. What common metal cations might be present in the second solution?
    (e) A solution of $Pb(NO_3)_2$ was added to a second solution, and no precipitate formed. What common anions might be present in the second solution?
11. Explain why HBr is a nonelectrolyte when dissolved in benzene and an electrolyte when dissolved in water.
12. Suppose you are presented with a clear solution of sodium thiosulfate, $Na_2S_2O_3$. How could you determine whether the solution is unsaturated, saturated, or supersaturated?

The Process of Dissolution

13. Explain why the ions Na^+ and Cl^- are strongly solvated in water but not in benzene.
14. Why are nonpolar compounds such as benzene and carbon tetrachloride generally poor solvents for ionic compounds?
15. Indicate the type of intermolecular attractions

(Sections 11.13, 11.14, 13.11, and 13.12) that should be most important in each of the following solutions:

(a) NaBr in water
(b) methanol, CH_3OH, in ethanol, C_2H_5OH
(c) HNO_3 in water
(d) HCl in benzene, C_6H_6
(e) carbon tetrachloride, CCl_4, in Freon, CF_2Cl_2

16. Which of the following processes, which can proceed spontaneously, occurs with an increase in disorder?
(a) evaporation of water
(b) precipitation of AgCl from solution
(c) condensation of water vapor to ice
(d) mixing of $CH_4(g)$ and $C_2H_6(g)$
(e) dissolution of NH_4NO_3 in water

17. Compare the processes that occur when glycerin, hydrogen chloride, ammonia, and sodium hydroxide dissolve in water. Write equations and prepare sketches showing the form in which each of these compounds is present in its respective solution.

18. Solid water (ice) is soluble in ethanol, C_2H_5OH. Compare the process that occurs when ice dissolves in ethanol with the one that occurs when sodium chloride dissolves in water. (See Section 12.1 for a description of the structure of ice.)

19. Explain, in terms of the intermolecular attractions between solute and solvent, why methanol, CH_3OH, is miscible with water in all proportions but only 0.11 g of butanol, $CH_3CH_2CH_2CH_2OH$, will dissolve in 100 g of water.

20. Suggest an explanation for the observations that ethanol, C_2H_5OH, is completely miscible with water and that ethanethiol, C_2H_5SH, is soluble only to the extent of 1.5 g/100 mL of water.

Units of Concentration

21. Distinguish between 1 M and 1 m solutions.
22. Is the following statement true or false? In all cases, the molality of a solution has a value that is larger than that of the molarity of the same solution. Explain your answer.
[S] 23. There are about 10 g of calcium, as Ca^{2+}, in 1.0 L of milk. What is the molarity of Ca^{2+} in milk?   *Ans. 0.25 M*
24. Calculate the number of moles and the mass of the solute in each of the following solutions:
[S] (a) 2.00 L of 18.5 M H_2SO_4, concentrated sulfuric acid   *Ans. 37.0 mol; 3.63 kg*
(b) 100 mL of 3.8×10^{-5} M NaCN, the minimum lethal concentration of sodium cyanide in blood serum
                     Ans. 3.8×10^{-6} mol; 1.9×10^{-4} g
[S] (c) 500. mL of 0.30 M glucose, $C_6H_{12}O_6$, used for intravenous injection
                     Ans. 0.15 mol; 27 g
(d) 5.50 L of 13.3 M formaldehyde, H_2CO, used to "fix" tissue samples   *Ans. 73.2 mol; 2.20 kg*
(e) 325 mL of 1.8×10^{-6} M $FeSO_4$, the minimum concentration of iron sulfate detectable by taste in drinking water
                     Ans. 5.8×10^{-7} mol; 8.8×10^{-5} g

25. Calculate the molarity of each of the following solutions:
(a) 0.195 g of cholesterol, $C_{27}H_{46}O$, in 0.100 L of serum, the average concentration of cholesterol in human serum
                     Ans. 5.04×10^{-3} M
[S] (b) 4.25 g of NH_3 in 0.500 L of solution, the concentration of NH_3 in household ammonia   *Ans. 0.499 M*
(c) 222.0 g of ethylene glycol, $C_2H_4(OH)_2$, in 393.6 mL of an antifreeze solution
                     Ans. 9.087 M
[S] (d) 595 g of isopropanol, C_3H_8OH, in 1.00 L of solution, the concentration of isopropanol in rubbing alcohol   *Ans. 9.74 M*
(e) 0.029 g of I_2 in 0.100 L of solution, the solubility of I_2 in water at 20°C
                     Ans. 1.1×10^{-3} M

26. Calculate the molality of each of the following solutions:
[S] (a) 71.0 g of sodium carbonate (washing soda), Na_2CO_3, in 1.00 kg of water, a saturated solution at 0°C   *Ans. 0.670 m*
[S] (b) 583 g of H_2SO_4 in 1.50 kg of water, the acid solution used in an automobile battery
                     Ans. 3.96 m
(c) 0.86 g of NaCl in 100 g of water, a solution of sodium chloride for intravenous injection
                     Ans. 0.15 m
(d) 46.85 g of codeine, $C_{18}H_{21}NO_3$, in 125.5 g of ethanol, C_2H_5OH *Ans. 1.247 m*
(e) 0.372 g of histamine, C_5H_9N, in 125 g of chloroform, $CHCl_3$ *Ans. 3.58×10^{-2} m*

27. Calculate the mole fractions of solute and solvent in each of the solutions in Exercise 26.
Ans. [S] *(a)* Na_2CO_3, *0.0119;* H_2O, *0.9881*
[S] *(b)* H_2SO_4, *0.0666;* H_2O, *0.933*
(c) $NaCl$, *0.0027;* H_2O, *0.997*
(d) $C_{18}H_{21}NO_3$, *0.05433;* C_2H_5OH, *0.9457*
(e) C_5H_9N, *0.00426;* $CHCl_3$, *0.996*

[S] 28. What mass of nitric acid (68.0% HNO_3 by mass) is needed to prepare 400.0 g of a 10.0% solution of HNO_3 by mass? *Ans. 58.8 g*

29. What mass of sulfuric acid (95% by mass) is needed to prepare 200.0 g of a 20.0% solution of the acid by mass? *Ans. 42 g*

[S] 30. What mass of a 4.00% NaOH solution by mass contains 15.0 g of NaOH? *Ans. 375 g*

[S] 31. What mass of HCl is contained in 45.0 mL of an HCl solution with a specific gravity of 1.19 and containing 37.21% HCl by mass? *Ans. 19.9 g*

32. What mass of solid NaOH (97.0% NaOH by mass) is required to prepare 1.00 L of a 10.0% solution of NaOH by mass? The density of the 10.0% solution is 1.109 g/cm^3. *Ans. 114 g*

[S] 33. The hardness of water (hardness count) is usually expressed as parts per million (by mass) of $CaCO_3$, which is equivalent to milligrams of $CaCO_3$ per liter of water. What is the molar concentration of Ca^{2+} ions in a water sample with a hardness count of 175? *Ans.* 1.75×10^{-3} *M*

34. The Safe Drinking Water Act sets the maximum permissible amount of cadmium in drinking water at 0.01 mg/L. What is the maximum permissible molar concentration of cadmium in drinking water? *Ans.* 9×10^{-8} *M*

35. The concentration of glucose, $C_6H_{12}O_6$, in normal spinal fluid is 75 mg/100 g. What is the molal concentration? *Ans.* 4.2×10^{-3} *m*

[S] 36. What volume of a 0.20 *M* K_2SO_4 solution contains 57 g of K_2SO_4? *Ans. 1.6 L*

37. A 0.553 *M* solution of $NaHCO_3$ has a density of 1.032 g/cm^3. What is the molality of the solution? *Ans. 0.561 m*

[S] 38. Calculate what volume of a sulfuric acid solution (specific gravity = 1.070, and containing 10.00% H_2SO_4 by mass) contains 18.50 g of pure H_2SO_4 at a temperature of 25°C. *Ans. 172.9 mL*

39. A 13.0% solution of K_2CO_3 by mass has a density of 1.09 g/cm^3. Calculate the molarity of the solution. *Ans. 1.03 M*

40. Equal volumes of 0.050 *M* $Ca(OH)_2$ and 0.400 *M* HCl are mixed. Calculate the molarity of each ion present in the final solution.
Ans. 0.025 M Ca^{2+}; *0.15 M* H_3O^+; *0.200 M* Cl^-

[S] 41. What volume of 0.600 *M* HCl is required to react completely with 2.50 g of sodium hydrogen carbonate?

$$NaHCO_3 + HCl \longrightarrow NaCl + CO_2 + H_2O$$

Ans. 49.6 mL

[S] 42. Calculate the volume of 0.050 *M* HCl necessary to precipitate the silver contained in 12.0 mL of 0.050 *M* $AgNO_3$.

$$Ag^+ + Cl^- \longrightarrow AgCl(s)$$

Ans. 12.0 mL

43. What volume of a 0.33 *M* solution of hydrobromic acid would be required to neutralize completely 1.00 L of 0.15 *M* barium hydroxide? *Ans. 0.91 L*

44. To 10.0 mL of a 0.100 *M* $K_2Cr_2O_7$ solution was added 10.0 mL of a 0.100 *M* $Pb(NO_3)_2$ solution. What mass of $PbCr_2O_7$ forms? *Ans. 0.423 g*

[S] 45. A gaseous solution was found to contain 15% H_2, 10.0% CO, and 75% CO_2 by mass. What is the mole fraction of each component?
Ans. H_2, *0.78; CO, 0.038;* CO_2, *0.18*

46. A sample of lead glass is prepared by melting together 20.0 g of silica, SiO_2, and 80.0 g of lead(II) oxide, PbO. What is the mole fraction of SiO_2 and of PbO in the glass?
Ans. SiO_2, *0.482; PbO, 0.518*

47. Calculate the mole fractions of methanol, CH_3OH, ethanol, C_2H_5OH, and water in a solution that is 40% methanol, 40% ethanol, and 20% water by mass. (Assume the data are good to two significant figures.) *Ans. 0.39; 0.27; 0.34*

[S] 48. Concentrated hydrochloric acid is 37.0% HCl by mass and has a specific gravity of 1.19. Calculate (a) the molarity, (b) the molality, and (c) the mole fraction of HCl and of H_2O.
Ans. (a) 12.1 M; (b) 16.1 m;
(c) HCl, 0.225; H_2O, *0.775*

49. Calculate (a) the percent composition and (b) the molality of an aqueous solution of

NaNO$_3$, if the mole fraction of NaNO$_3$ is 0.20.
Ans. (a) 54% NaNO$_3$, 46% H$_2$O; (b) 14 m

Colligative Properties

50. A solution of ethanol (a nonelectrolyte) and a solution of sodium chloride (an electrolyte) both freeze at −0.5°C. What other physical properties of the two solutions are identical?

51. Which will evaporate faster under the same conditions, 50 mL of distilled water or 50 mL of sea water?

52. The triple point of air-free water (Section 12.2) is defined as 273.15 K. Why is it important that the water be air-free?

53. Why will a mole of sodium chloride depress the freezing point of 1 kg of water by almost twice as much as a mole of glycerin will?

54. (a) Such substances as sodium chloride, NaCl, calcium chloride, CaCl$_2$, and the nonelectrolyte urea, NH$_2$CONH$_2$, are frequently used to melt ice on streets and highways. Rank the three from high to low in terms of effectiveness per pound in lowering the freezing point. Explain your ranking.
    (b) Would these substances be as effective in melting ice on a day when the maximum temperature is −20°F? Explain why or why not.

55. Explain what is meant by osmotic pressure. How is the osmotic pressure of a solution related to its concentration?

56. The cell walls of red and white blood cells are semipermeable membranes. The concentration of solute particles in the blood is about 0.6 M. What will happen to blood cells that are placed in pure water? in a 1 M sodium chloride solution?

57. A 1 m solution of HCl in benzene has a freezing point of 0.4°C. Is HCl an electrolyte in benzene? Explain.

58. Arrange the following solutions in decreasing order by their freezing points: 0.1 M Na$_3$PO$_4$, 0.1 M C$_2$H$_5$OH, 0.01 M CO$_2$, 0.15 M NaCl, and 0.2 M CaCl$_2$.

59. Some mammals, including humans, excrete hypertonic urine in order to conserve water. In humans, some parts of the ducts of the kidney (which contain the urine) pass through a fluid that contains a much more concentrated salt so-lution than that normally found in the body. Explain how this could help conserve water in the body.

S 60. A solution of 5.00 g of an organic compound in 25.00 g of carbon tetrachloride (bp 76.8°C; $K_b = 5.02° C/m$) boils at 81.5°C at 1 atm. What is the molecular weight of the compound?
Ans. 2.1 × 10^2

S 61. A solution contains 5.00 g of urea, CO(NH$_2$)$_2$, per 0.100 kg of water. If the vapor pressure of pure water at 25°C is 23.7 torr, what is the vapor pressure of the solution?   *Ans. 23.4 torr*

62. A 12.0-g sample of a nonelectrolyte is dissolved in 80.0 g of water. The solution freezes at −1.94°C. Calculate the molecular weight of the substance.   *Ans. 144*

63. A sample of an organic compound (nonelectrolyte) weighing 1.350 g lowered the freezing point of 10.0 g of benzene by 3.66°C. Calculate the molecular weight of the organic compound.   *Ans. 189*

S 64. Calculate the boiling-point elevation of 0.100 kg of water containing 0.010 mol of NaCl, 0.020 mol of Na$_2$SO$_4$, and 0.030 mol of MgCl$_2$, assuming complete dissociation of these electrolytes.   *Ans. 0.87°C*

65. What is the approximate freezing point of a 0.27 m aqueous solution of sodium bromide? Assume complete dissociation of this electrolyte.   *Ans. −1.0°C*

S 66. How could you prepare a 3.08 m aqueous solution of glycerin, C$_3$H$_8$O$_3$? What is the freezing point of this solution?   *Ans. Dissolve 284 g of glycerin in 1.00 kg of water; −5.73°C*

67. If 26.4 g of the nonelectrolyte dibromobenzene, C$_6$H$_4$Br$_2$, is dissolved in 0.250 kg of benzene, what is (a) the freezing point of the solution and (b) the boiling point of the solution at 1 atm?   *Ans. (a) 3.19°C; (b) 81.2°C*

68. A sample of sulfur weighing 0.210 g was dissolved in 17.8 g of carbon disulfide, CS$_2$ ($K_b = 2.34°C/m$). If the boiling-point elevation was 0.107°C, what is the formula of a sulfur molecule in carbon disulfide?   *Ans. S$_8$*

69. What is the boiling point, at 1 atm, of a solution containing 140.0 g of sucrose, C$_{12}$H$_{22}$O$_{11}$, in 400.0 g of water?   *Ans. 100.524°C*

S 70. Lysozyme is an enzyme that cleaves cell walls. A 0.100-L sample of a solution of lysozyme that contains 0.0750 g of the enzyme exhibits an osmotic pressure of 1.32×10^{-3} atm at 25°C. What is the molecular weight of lysozyme?

Ans. 13,900

71. The osmotic pressure of a solution containing 7.0 g of insulin per liter is 23 torr at 25°C. What is the molecular weight of insulin?

Ans. 5700

72. The osmotic pressure of human blood is 7.6 atm at 37°C. What mass of glucose, $C_6H_{12}O_6$, is required to make 1.00 L of aqueous solution for intravenous feeding if the solution must have the same osmotic pressure as blood at body temperature, 37°C?

Ans. 54 g

Colloids

73. Distinguish between dispersion methods and condensation methods for preparing colloidal systems.

74. How do colloidal "solutions" differ from true solutions with regard to dispersed particle size and homogeneity?

75. Identify the dispersed phase and the dispersion medium in each of the following colloidal systems: starch dispersion, smoke, fog, pearl, whipped cream, floating soap, jelly, milk, and ruby.

76. Explain the cleansing action of soap.

77. How can it be demonstrated that colloidal particles are electrically charged?

78. What is the structure of a gel?

79. Explain the phenomenon of Brownian movement. What is its relationship to the stability of the dispersed particles of a colloidal system?

Additional Exercises

80. For which of the following gases is solubility in water not directly proportional to the pressure of the gas: HCl, H_2, SO_2, NH_3, CH_4, Ne? Explain.

81. Why will ice melt when placed in an aqueous solution that is kept at 0°C?

82. What is a constant boiling solution? Describe two ways to prepare a constant boiling solution of hydrochloric acid.

83. The salt lithium nitrate, $LiNO_3$, does not depress the freezing point of water twice as much as nonelectrolytes do. Explain.

84. The approximate radius of a water molecule, assuming a spherical shape, is 1.40 Å (1 Å = 10^{-8} cm). Assume that water molecules cluster around each metal ion in a solution so that the water molecules essentially touch both the metal ion and each other. On this basis, and assuming that 4, 6, 8, and 12 are the only possible coordination numbers, what is the maximum number of water molecules that can hydrate each of the following ions?
(a) Mg^{2+} (radius 0.65 Å)   (c) Rb^+ (1.48 Å)
(b) Al^{3+} (0.50 Å)   (d) Sr^{2+} (1.13 Å)

85. Hydrogen gas dissolves in the metal palladium with hydrogen atoms going into the holes between metal atoms. Determine the molarity, molality, and percent by mass of hydrogen atoms in a solution (specific gravity = 10.8) of 0.94 g of hydrogen gas in 215 g of palladium metal.   *Ans. 47 M; 4.3 m; 0.44%*

86. How many liters of $NH_3(g)$ at 25°C and 1.46 atm are required to prepare 3.00 L of a 2.50 M solution of NH_3?   *Ans. 126 L*

Ⓢ 87. A 1.80-g sample of an acid, H_2X, required 14.00 mL of KOH solution for neutralization of all the hydrogen ion. Exactly 14.2 mL of this same KOH solution was found to neutralize 10.0 mL of 0.750 M H_2SO_4. Calculate the molecular weight of H_2X.   *Ans. 243*

88. A 0.200-L volume of gaseous ammonia measured at 30.0°C and 764 torr was absorbed in 0.100 L of water. How many milliliters of 0.0100 M hydrochloric acid are required in the neutralization of this aqueous ammonia? What is the molarity of the aqueous ammonia solution? (Assume no change in volume when the gaseous ammonia is added to the water.)   *Ans. 808 mL; 0.0808 M*

89. Calculate the percent by mass and the molality in terms of $CuSO_4$ for a solution prepared by dissolving 11.5 g of $CuSO_4 \cdot 5H_2O$ in 0.100 kg of water. *Remember to consider the water released from the hydrate.*   *Ans. 6.59%; 0.442 m*

90. It is desired to produce 1.000 L of 0.050 M nitric acid by diluting 10.00 M nitric acid. Calculate the volume of the concentrated acid and the volume of water required in the dilution.

Ans. 5.0 mL of HNO_3; 995 mL of H_2O

Ⓢ 91. What is the molarity of H_3PO_4 in a solution that is prepared by dissolving 10.0 g of P_4O_{10} in sufficient water to make 0.500 L of solution?

Ans. 0.282 M

92. A solution of sodium carbonate having a volume of 0.400 L was prepared from 4.032 g of $Na_2CO_3 \cdot 10H_2O$. Calculate the molarity of this solution.   *Ans. 0.0352 M*

93. A solution of $Ba(OH)_2$ has a molarity of 0.1055. What volume of 0.211 *M* nitric acid is required in the neutralization of 15.5 mL of the $Ba(OH)_2$ solution?   *Ans. 15.5 mL*

Ⓢ94. The sulfate in 50.0 mL of dilute sulfuric acid was precipitated using an excess of barium chloride. The mass of $BaSO_4$ formed was 0.482 g. Calculate the molarity of the sulfuric acid solution.   *Ans. 0.0413 M*

Ⓢ95. A sample of $HgCl_2$ weighing 9.41 g is dissolved in 32.75 g of ethanol, C_2H_5OH ($K_b = 1.20°C/m$). The boiling-point elevation of the solution is 1.27°C. Is $HgCl_2$ an electrolyte in ethanol? Show your calculations.   *Ans. No*

96. A salt is known to be an alkali metal fluoride. A quick approximate freezing point determination indicates that 4 g of the salt dissolved in 100 g of water produces a solution that freezes at about −1.4°C. What is the formula of the salt? Show your calculations.   *Ans. RbF*

Ⓢ97. A solution of 0.045 g of an unknown organic compound in 0.550 g of camphor melts at 158.4°C. The melting point of pure camphor is 178.4°C. K_f for camphor is 37.7°C/*m*. The solute contains 93.46% C and 6.54% H by mass. What is the molecular formula of the solute? Show your calculations.   *Ans. $C_{12}H_{10}$*

98. How many liters of HCl gas, measured at 30.0°C and 745 torr are required to prepare 2.50 L of a 1.60 *M* solution of hydrochloric acid?

Ans. 101 L

99. The sugar fructose contains 40.0% C, 6.7% H, and 53.3% O by mass. A solution of 11.7 g of fructose in 325 g of ethanol has a boiling point of 78.59°C. The boiling point of ethanol is 78.35°C, and K_b for ethanol is 1.20°C/*m*. What is the molecular formula of fructose?

Ans. $C_6H_{12}O_6$

100. The vapor pressure of methanol, CH_3OH, is 94 torr at 20°C. The vapor pressure of ethanol, C_2H_5OH, is 44 torr at the same temperature. Calculate the mole fraction of methanol and of ethanol in a solution of 50.0 g of methanol and 50.0 g of ethanol. Ethanol and methanol form a solution that behaves like an ideal solution. Calculate the vapor pressure of methanol and of ethanol above the solution at 20°C. Calculate the mole fraction of methanol and of ethanol in the vapor above the solution.

Ans. In solution: CH_3OH, 0.590; C_2H_5OH, 0.410; P_{CH_3OH}, 55 torr; $P_{C_2H_5OH}$, 18 torr In vapor: CH_3OH, 0.75; C_2H_5OH, 0.25

14

ACIDS AND BASES

Acids and bases have been defined in a number of ways. When Boyle first characterized them in 1680, he noted that acids dissolve many substances, change the color of certain natural dyes (for example, litmus from blue to red), and lose these characteristic properties after coming in contact with alkalies (bases). In the eighteenth century it was recognized that acids have a sour taste, react with limestone with the liberation of a gaseous substance (CO_2), and interact with alkalies to form neutral substances. In 1787 Lavoisier proposed that acids are binary compounds of oxygen and considered oxygen to be responsible for their acidic properties. The hypothesis that oxygen is an essential component of acids was disproved by Davy in 1811 when he showed that hydrochloric acid contains no oxygen. Davy contributed greatly to the development of the acid-base concept by concluding that hydrogen, rather than oxygen, is the essential constituent of acids. In 1814 Gay-Lussac concluded that acids are substances that can neutralize alkalies and that these two classes of substances can be defined only in terms of each other. Davy and Gay-Lussac provided the foundation for our concepts of acids and bases in aqueous solutions.

The significance of hydrogen was reemphasized in 1884 when Arrhenius defined an acid as a compound that dissolves in water to yield hydrogen ions, and a base as a compound that dissolves in water to yield hydroxide ions. In 1923 the close relationship between acids and bases was pointed out by both Johannes

Brönsted, a Danish chemist, and Thomas Lowry, an English chemist. Their models for acid-base behavior define acids as proton donors and bases as proton acceptors. An even broader view of the relationship between acids and bases was developed by the American chemist G. N. Lewis. The Lewis theory defines acids as electron-pair acceptors and bases as electron-pair donors.

In this chapter we shall first consider the description of acid-base behavior presented by Brönsted and Lowry and then the more generalized concept suggested by Lewis. Finally, we shall examine the properties of acids and bases in aqueous solutions, because a great many of their reactions occur there.

THE BRÖNSTED-LOWRY CONCEPT OF ACIDS AND BASES

14.1 The Protonic Concept of Acids and Bases

A compound that donates a proton (a hydrogen ion, H^+) to another compound is called a Brönsted acid, and a compound that accepts a proton is called a Brönsted base (Section 8.1). Stated simply, an acid is a proton donor, and a base is a proton acceptor.

An **acid-base reaction** is the transfer of a proton from a proton donor (acid) to a proton acceptor (base). A proton may be donated by a molecular acid, such as HCl, H_2SO_4, H_3PO_4, CH_3CO_2H, H_2S, or H_2O, or it may be donated by either an anion or a cation. Anions that can exhibit acidic behavior include HSO_4^-, $H_2PO_4^-$, HPO_4^{2-}, and HS^-. Cations that can exhibit acidic behavior include H_3O^+, NH_4^+, $[Cu(H_2O)_4]^{2+}$, and $[Fe(H_2O)_6]^{3+}$. A proton may be accepted by a molecular base, such as H_2O, NH_3, or CH_3NH_2, or by an anion, such as OH^-, HS^-, S^{2-}, HCO_3^-, CO_3^{2-}, HSO_4^-, SO_4^{2-}, HPO_4^{2-}, Cl^-, F^-, NO_3^-, or PO_4^{3-}, or by a cation, such as $[Fe(H_2O)_5OH]^{2+}$ or $[Cu(H_2O)_3OH]^+$.

The most familiar bases are ionic compounds such as $NaOH$ and $Ca(OH)_2$, which contain the hydroxide ion, OH^-. The hydroxide ion accepts protons from acids, forming water.

$$H_3O^+ + OH^- \longrightarrow 2H_2O$$

In our discussions of acid-base properties, we will sometimes identify specific ions as being acids or bases. However, it must be reemphasized that no solid or solution can contain only cations or only anions. For every ion in a solid or solution there must be an ion of the opposite charge, a **counter ion,** so that the solid or solution is electrically neutral. If the counter ions do not affect the acid-base properties of the system, we may leave them out of equations, but they will be present in the actual solution.

When an acid loses a proton, it forms what is called the **conjugate base** of the acid. The conjugate base is a base because it can pick up a proton (and reform the acid).

Acid		Proton	Conjugate Base
HCl	$\longrightarrow$	H^+	$+ \ Cl^-$
H_2SO_4	$\longrightarrow$	H^+	$+ \ HSO_4^-$
H_2O	$\longrightarrow$	H^+	$+ \ OH^-$
HSO_4^-	$\longrightarrow$	H^+	$+ \ SO_4^{2-}$
NH_4^+	$\longrightarrow$	H^+	$+ \ NH_3$
$[Fe(H_2O)_6]^{3+}$	$\longrightarrow$	H^+	$+ \ [Fe(H_2O)_5(OH)]^{2+}$

When a base accepts a proton, it forms what is called the **conjugate acid** of the base. The conjugate acid is an acid because it can give up a proton (and reform the base).

Base		Proton		Conjugate Acid
H_2O	$+$	H^+	$\longrightarrow$	H_3O^+
NH_3	$+$	H^+	$\longrightarrow$	NH_4^+
OH^-	$+$	H^+	$\longrightarrow$	H_2O
S^{2-}	$+$	H^+	$\longrightarrow$	HS^-
CO_3^{2-}	$+$	H^+	$\longrightarrow$	HCO_3^-
F^-	$+$	H^+	$\longrightarrow$	HF
$[Fe(H_2O)_5OH]^{2+}$	$+$	H^+	$\longrightarrow$	$[Fe(H_2O)_6]^{3+}$

In order for an acid to act as a proton donor, a base (proton acceptor) must be present to receive the proton. An acid does not form its conjugate base unless a second base is present to accept the proton. When the second base accepts the proton, it forms its conjugate acid, the second acid. When hydrogen chloride, HCl, reacts with anhydrous ammonia, NH_3, forming ammonium ions, NH_4^+, and chloride ions, Cl^-, hydrogen chloride (acid$_1$) gives up a proton forming chloride ion, its conjugate base (base$_1$). Ammonia acts as the proton acceptor and therefore is a base (base$_2$). The proton combines with ammonia to give its conjugate acid, the ammonium ion (acid$_2$). The equations for this and several other acid-base reactions are as follows:

Acid$_1$		Base$_2$		Acid$_2$		Base$_1$
HCl	$+$	NH_3	$\rightleftharpoons$	NH_4^+	$+$	Cl^-
HNO_3	$+$	F^-	$\rightleftharpoons$	HF	$+$	NO_3^-
HSO_4^-	$+$	CO_3^{2-}	$\rightleftharpoons$	HCO_3^-	$+$	SO_4^{2-}
NH_4^+	$+$	S^{2-}	$\rightleftharpoons$	HS^-	$+$	NH_3

In all of these acid-base reactions, the forward reaction is the transfer of a proton from acid$_1$ to base$_2$. The reverse reaction is the transfer of a proton from acid$_2$ to base$_1$. In effect, base$_1$ and base$_2$ are in competition for the proton, and the species that are present in the greatest concentrations will depend on which base competes most effectively for the proton. In all of the examples given in the table above, base$_2$ is the stronger base. As a result, at equilibrium the system will consist primarily of a mixture of acid$_2$ and base$_1$.

When an acid such as hydrogen chloride dissolves in water, the proton acceptor is a water molecule. Thus water functions as a base in this reaction.

$$HCl + H_2O \rightleftharpoons H_3O^+ + Cl^-$$

$$\underset{acid_1}{} \quad \underset{base_2}{} \qquad \underset{acid_2}{} \quad \underset{base_1}{}$$

The hydronium ion is the conjugate acid of water.

When a base such as ammonia dissolves in water, water functions as a proton donor and hence as an acid.

$$H_2O + NH_3 \rightleftharpoons NH_4^+ + OH^-$$

$$\underset{acid_1}{} \quad \underset{base_2}{} \qquad \underset{acid_2}{} \quad \underset{base_1}{}$$

The hydroxide ion is the conjugate base of water. Under the appropriate conditions, then, water can function either as an acid or as a base. Note that the hydroxide ion is a stronger base than ammonia, so this reaction gives only a small amount of NH_4^+ and OH^-; the majority of ammonia molecules do not react.

14.2 Amphiprotic Species

Certain molecules and ions may exhibit either acidic or basic behavior under the appropriate conditions. A species that, like water, may either gain or lose a proton is said to be **amphiprotic.** Water may lose a proton to a base, such as NH_3, or gain a proton from an acid, such as HCl, and so is classified as amphiprotic.

The proton-containing anions given in Section 14.1 are also amphiprotic, which can be readily seen from the following equations:

Acid$_1$	Base$_2$		Acid$_2$	Base$_1$
HS^-	$+ OH^-$	$\rightleftharpoons$	H_2O	$+ S^{2-}$
HBr	$+ HS^-$	$\rightleftharpoons$	H_2S	$+ Br^-$
HCO_3^-	$+ CN^-$	$\rightleftharpoons$	HCN	$+ CO_3^{2-}$
H_3O^+	$+ HCO_3^-$	$\rightleftharpoons$	H_2CO_3	$+ H_2O$

The hydroxides of certain metals, especially those near the boundary between metals and nonmetals in the Periodic Table, are amphiprotic and so react either as acids or bases.

Acid$_1$	Base$_2$		Acid$_2$	Base$_1$
$[Al(H_2O)_3(OH)_3] +$	OH^-	$\rightleftharpoons$	H_2O	$+ [Al(H_2O)_2(OH)_4]^-$
H_3O^+	$+ [Al(H_2O)_3(OH)_3]$	$\rightleftharpoons$	$[Al(H_2O)_4(OH)_2]^+ +$	H_2O

In the first reaction one of the water molecules of the hydrated aluminum hydroxide loses a proton to the hydroxide ion. In the second reaction the aluminum hydroxide receives a proton from the hydronium ion. In both cases the insoluble hydrated aluminum hydroxide dissolves.

14.3 The Strengths of Acids and Bases

The fundamental acid-base relationship for the reaction of any Brönsted acid with water is

$$HA + H_2O \rightleftharpoons H_3O^+ + A^-$$

where A^- is the conjugate base of the acid HA and the hydronium ion is the conjugate acid of water. Water is the base that reacts with the acid HA. **Strong acids** react with water and ionize in dilute solution, giving a 100% yield (or very nearly a 100% yield) of hydronium ion and the conjugate base of the acid. **Weak acids** ionize in aqueous solution, giving small yields of hydronium ion (ordinarily 10% or less). Thus the strength of an acid can be measured by the extent of its tendency to form hydronium ions in aqueous solution by donation of protons to water molecules.

The extent to which protons will be donated to water molecules depends on the strength of the conjugate base, A^-, of the acid. If A^- is a strong base and accepts protons readily, there will be relatively little A^- and H_3O^+ in solution, and the acid, HA, is a weak one. If A^- is a weak base and does not readily accept protons, the solution will contain primarily A^- and H_3O^+, and the acid is a strong one. Strong acids form weak conjugate bases, and weak acids form relatively strong conjugate bases. The first six acids in Table 14-1 are the common strong acids, which are completely dissociated in aqueous solution. The conjugate bases of these acids are weaker bases than water. When one of these acids dissolves in water, the proton is transferred to water, the stronger base.

Those acids that lie between the hydronium ion and water in Table 14-1 form conjugate bases that can compete with water for possession of a proton. Both

Table 14-1   The Relative Strengths of Conjugate Acid-Base Pairs

	Acid			(conj) Base
Perchloric acid	$HClO_4$		ClO_4^-	Perchlorate ion
Sulfuric acid	H_2SO_4		HSO_4^-	Hydrogen sulfate ion
Hydrogen iodide	HI	Stronger acids than	I^-	Iodide ion
Hydrogen bromide	HBr	H_3O^+; form H_3O^+ in	Br^-	Bromide ion
Hydrogen chloride	HCl	100% yield in H_2O.	Cl^-	Chloride ion
Nitric acid	HNO_3		NO_3^-	Nitrate ion
Hydronium ion	H_3O^+		H_2O	Water
Hydrogen sulfate ion	HSO_4^-		SO_4^{2-}	Sulfate ion
Phosphoric acid	H_3PO_4		$H_2PO_4^-$	Dihydrogen phosphate ion
Hydrogen fluoride	HF		F^-	Fluoride ion
Nitrous acid	HNO_2		NO_2^-	Nitrite ion
Acetic acid	CH_3CO_2H		$CH_3CO_2^-$	Acetate ion
Carbonic acid	H_2CO_3		HCO_3^-	Hydrogen carbonate ion
Hydrogen sulfide	H_2S		HS^-	Hydrogen sulfide ion
Ammonium ion	NH_4^+		NH_3	Ammonia
Hydrogen cyanide	HCN		CN^-	Cyanide ion
Hydrogen carbonate ion	HCO_3^-		CO_3^{2-}	Carbonate ion
Hydrogen sulfide ion	HS^-		S^{2-}	Sulfide ion
Water	H_2O weak		OH^-	Hydroxide ion strongest·est
Ethanol	C_2H_5OH		$C_2H_5O^-$	Ethoxide ion
Ammonia	NH_3	Stronger bases than	NH_2^-	Amide ion
Hydrogen	H_2	OH^-; form OH^- in	H^-	Hydride ion
Methane	CH_4	100% yield in H_2O.	CH_3^-	Methide ion

Increasing Acid Strength (left margin, pointing up)

Increasing Base Strength (right margin, pointing down)

weakest (handwritten note above Perchlorate ion)

hydronium ions and nonionized acid molecules are present in equilibrium in a solution of one of these acids. Their relative strengths can be measured by their percent ionizations in aqueous solutions of equal concentration, with the stronger acid forming the higher yield of hydronium ion. From the data given below, the order of acid strength for three of these acids is $HSO_4^- > HNO_2 > CH_3CO_2H$.

$$HSO_4^- + H_2O \rightleftharpoons H_3O^+ + SO_4^{2-} \qquad \text{(29\% in 0.10 } M \text{ solution)}$$
$$HNO_2 + H_2O \rightleftharpoons H_3O^+ + NO_2^- \qquad \text{(6.5\% in 0.10 } M \text{ solution)}$$
$$CH_3CO_2H + H_2O \rightleftharpoons H_3O^+ + CH_3CO_2^- \qquad \text{(1.3\% in 0.10 } M \text{ solution)}$$

Compounds that lie below water in the column of acids in Table 14-1 exhibit no observable acidic behavior when dissolved in water. Their conjugate bases are stronger than the hydroxide ion, and if any conjugate base were formed, it would react with water to reform the acid.

The fundamental reaction of a Brönsted base with water is given by the equation

$$H_2O + B \rightleftharpoons HB^+ + OH^-$$

where HB^+ is the conjugate acid of the base B, and the hydroxide ion is the conjugate base of water. Water is the acid that reacts with the base. **Strong bases** react with water giving a 100% yield (or very nearly a 100% yield) of hydroxide ion and the conjugate acid of the base. **Weak bases** react with water giving small yields of hydroxide ion (ordinarily 10% or less). Soluble ionic hydroxides are strong bases since they dissolve in water and give a 100% yield of hydroxide ion.

The extent to which a base will form hydroxide ion in aqueous solution depends on the strength of the base relative to that of the hydroxide ion, as shown in the column of bases in Table 14-1. A strong base, such as one of those lying below hydroxide ion in the table, will accept a proton from water, forming the conjugate acid and hydroxide ion in 100% yield. Those bases lying between water and hydroxide ion in Table 14-1 will accept protons from water, but the result will be a mixture of the hydroxide ion and the base. Since stronger bases give higher concentrations of hydroxide ion, the relative strengths of these bases can be measured by the yield of hydroxide ion that each produces in aqueous solution. From the data given below, the order of base strength for three of these bases is $NH_3 > CH_3CO_2^- > NO_2^-$.

$$NH_3 + H_2O \rightleftharpoons NH_4^+ + OH^- \qquad \text{(1.3\% in 0.10 } M \text{ solution)}$$
$$CH_3CO_2^- + H_2O \rightleftharpoons CH_3CO_2H + OH^- \qquad \text{(0.0075\% in 0.10 } M \text{ solution)}$$
$$NO_2^- + H_2O \rightleftharpoons HNO_2 + OH^- \qquad \text{(0.0015\% in 0.10 } M \text{ solution)}$$

Bases that are weaker than water show no observable basic behavior in dilute aqueous solution.

According to the Brönsted-Lowry model, then, the strongest acids have the weakest conjugate bases, and the strongest bases have the weakest conjugate acids. Nitrous acid, HNO_2, a stronger acid than acetic acid, CH_3CO_2H, forms the nitrite ion, NO_2^-, as its conjugate base. The nitrite ion is a weaker base than the acetate ion, $CH_3CO_2^-$, the conjugate base of acetic acid. Table 14-1 illustrates the relationships between conjugate acid-base pairs. An acid will donate a proton to a conjugate base of any acid lying below it in the table. Thus, nitrous acid will react with acetate ion to form nitrite ion and acetic acid.

$$HNO_2 + CH_3CO_2^- \rightleftharpoons NO_2^- + CH_3CO_2H$$

An acid-base neutralization reaction occurs when equivalent quantities of an acid and a base are mixed. This statement does not mean, however, that the resulting solution is always chemically neutral; it may contain either excess hydronium ions or excess hydroxide ions. The nature of the particular acid and base involved determines whether the resulting aqueous solution is acidic or basic. The following equations illustrate some reactions that take place when equivalent quantities of acids and bases are brought together in aqueous solution:

1. *A strong acid plus a strong base gives a neutral solution.*

$$H_3O^+(aq) + [Cl^-(aq)] + [Na^+(aq)] + OH^-(aq) \longrightarrow$$
$$2H_2O(l) + [Na^+(aq)] + [Cl^-(aq)]$$

Both acid and base are completely ionized. The reaction goes to completion, and water molecules, sodium ions, and chloride ions are the only products. The solution is neutral.

2. *A strong acid plus a weak base gives a weakly acidic solution.*

$$H_3O^+(aq) + [Cl^-(aq)] + NH_3(aq) \rightleftharpoons NH_4^+(aq) + [Cl^-(aq)] + H_2O(l)$$

Because ammonia is a moderately weak base, its conjugate acid (the ammonium ion) is a weak acid. A small fraction of the ammonium ions give up protons to water molecules, giving an acidic solution.

3. *A weak acid plus a strong base gives a weakly basic solution.*

$$CH_3CO_2H(aq) + [Na^+(aq)] + OH^-(aq) \rightleftharpoons$$
$$H_2O(l) + [Na^+(aq)] + CH_3CO_2^-(aq)$$

Because acetic acid is a moderately weak acid, its conjugate base, the acetate ion, is a moderately weak base. A small fraction of the acetate ions pick up protons from water, producing hydroxide ions and giving a basic solution.

4. *A weak acid plus a weak base.*

$$CH_3CO_2H(aq) + NH_3(aq) \rightleftharpoons NH_4^+(aq) + CH_3CO_2^-(aq)$$

This is the most complex of the four types of reactions and will be discussed in detail in Chapter 16. However, it is appropriate to note here that if the weak acid and the weak base have the same strengths, the solution will be neutral. Ammonia, for example, takes up about the same amount of protons to form ammonium ion as acetic acid gives up, and, although the reaction does not go to completion, the solution is neutral. If the weak acid and the weak base are of *unequal* strengths, the solution will be either acidic or basic, depending on the relative strengths of the two substances.

14.4 The Relative Strengths of Strong Acids and Bases

The strongest acids, such as HCl, HBr, and HI, appear to have about the same strength in water. The water molecule is such a strong base compared to the conjugate bases Cl^-, Br^-, and I^- that ionization of the strong acids HCl, HBr, and HI is nearly complete in aqueous solutions. Hence in water these acids are all strong and appear to have equal strengths. In solvents less strongly basic than water, HCl, HBr, and HI differ markedly in their tendency to give up a proton to the solvent. In ethanol, a weaker base than water, the extent of ionization increases in the order HCl < HBr < HI, and it is evident that HI is the strongest of these acids. Water tends to equalize any differences in strength among strong acids, and this is known as the **leveling effect** of water.

Water also exerts a leveling effect on the strengths of very strong bases. For example, the oxide ion, O^{2-}, and the amide ion, NH_2^-, are such strong bases that they react completely with water.

$$O^{2-} + H_2O \longrightarrow OH^- + OH^-$$
$$NH_2^- + H_2O \longrightarrow NH_3 + OH^-$$

Thus O^{2-} and NH_2^- appear to have the same base strength.

In binary compounds of hydrogen with nonmetals, the acid strength (the tendency to lose a proton) increases as the H—A bond strength decreases down a group in the Periodic Table. For Group VIIA the order of increasing acidity is $HF < HCl < HBr < HI$, in the absence of any leveling effect due to the solvent. Likewise, for Group VIA the order of increasing acid strength is $H_2O < H_2S < H_2Se < H_2Te$.

Across a row in the Periodic Table, the acid strength of binary hydrogen compounds increases with increasing electronegativity of the nonmetal atom as the polarity of the H—A bond increases. Thus the order of increasing acidity (for removal of one proton) across the second row is $CH_4 < NH_3 < H_2O < HF$, and across the third row is $SiH_4 < PH_3 < H_2S < HCl$.

Compounds containing oxygen and one or more hydroxide groups can be acidic, basic, or amphoteric depending on the position in the Periodic Table (and thus the electronegativity) of E, the element bonded to the oxygen and hydroxide group. Such compounds have the general formula $O_nE(OH)_m$ and include sulfuric acid, $O_2S(OH)_2$, sulfurous acid, $OS(OH)_2$, nitric acid, O_2NOH, perchloric acid, O_3ClOH, and aluminum hydroxide, $Al(OH)_3$. If the central atom, E, has a low electronegativity, its attraction for electrons is low, little tendency exists for it to form a strong covalent bond with the oxygen atom, and the bond between the element and oxygen is weaker than that between oxygen and hydrogen. Hence, bond a is ionic, hydroxide ions are released to the solution, and the material behaves as a base. Large atomic size, small nuclear charge, and low oxidation number operate to lower the electronegativity of E and are characteristic of the more metallic elements.

$$-\overset{|}{\underset{|}{E}}-O-H$$

Bond b

Bond a

If, on the other hand, the element E has a relatively high electronegativity, it strongly attracts the electrons it shares with the oxygen atom, giving rise to a relatively strong bond between E and oxygen. The oxygen-hydrogen bond is thereby weakened, since electrons are displaced toward E. Bond b is polar and readily releases hydrogen ions to the solution, so the material behaves as an acid. Small atomic size, large nuclear charge, and high oxidation number operate to increase the electronegativity of E and are characteristic of the more nonmetallic elements. As the electronegativity of E increases, the O—H bond becomes weaker, and the acid strength increases.

Increasing the oxidation number of E also increases the acidity of an oxyacid, since this increases the attraction of E for the electrons it shares with oxygen and thereby weakens the O—H bond. Sulfuric acid, $\overset{+6}{O_2S(OH)_2}$, is more acidic than sulfurous acid, $\overset{+4}{OS(OH)_2}$; likewise nitric acid, $\overset{+5}{O_2NOH}$, is more acidic than nitrous acid, $\overset{+3}{ONOH}$. In each of these pairs the oxidation number of the central atom (indicated by the small superior number) is larger for the stronger acid.

The hydroxides of elements with intermediate electronegativities and relatively high oxidation numbers (for example, elements near the diagonal line separating the metals from the nonmetals in the Periodic Table) are usually **amphoteric.** This means that the hydroxides act as acids toward strong bases and as bases toward strong acids. The amphoterism of aluminum hydroxide is reflected in its solubility in both strong acids and strong bases. In strong bases the relatively insoluble hydrated aluminum hydroxide, $[Al(H_2O)_3(OH)_3]$, is converted to the soluble ion, $[Al(H_2O)_2(OH)_4]^-$, by reaction with hydroxide ion (Section 14.2). In strong acids it is converted in the first step to the soluble ion $[Al(H_2O_4)(OH)_2]^+$ by reaction with hydronium ion and in two additional steps to the soluble ion $[Al(H_2O)_6]^{3+}$.

$$[Al(H_2O)_3(OH)_3] + 3H_3O^+ \rightleftharpoons [Al(H_2O)_6]^{3+} + 3H_2O$$

The different direction in the trend of acid strengths for the binary hydrogen halides arises from the fact that the halogen is attached directly to a hydrogen in the binary acid (instead of to an oxygen, which is in turn attached to a hydrogen, as in the oxyacids). In either kind of compound, as we go from a larger halogen to a smaller (i.e., from I to Br to Cl), a progressive shrinking of the electron shells takes place, including a pulling closer of the adjacent atom, whether the adjacent atom be hydrogen (as in the hydrogen halides) or oxygen (as in the oxyacids). As the adjacent atom is pulled closer, it is brought into a region of higher electron density. If you picture the valence electrons as being inside a smaller volume for the smaller halogen, you can visualize the higher electron density. If the adjacent atom being pulled closer is a hydrogen atom (as is the case with the hydrogen halides), it is held more tightly by the higher electron density and is pulled loose with greater difficulty. Conversely, the hydrogen atom is in a region of lower electron density when attached to a larger halogen, so it is attracted less tightly and is pulled away more easily.

THE LEWIS CONCEPT OF ACIDS AND BASES

14.5 Definitions and Examples

In 1923 G. N. Lewis proposed a generalized model of acid-base behavior in which acids and bases are not restricted to proton donors and proton acceptors. **According to the Lewis concept, an acid is any species (molecule or ion) that can accept a pair of electrons, and a base is any species (molecule or ion) that can donate a pair of electrons.** An acid-base reaction occurs when a base donates a pair of electrons to an acid. An acid-base adduct, a compound that contains a coordinate covalent bond between the Lewis acid and the Lewis base, forms.

The following equations show the general application of the Lewis model. In the first reaction two ammonia molecules, **Lewis bases,** each donate a pair of electrons to a silver ion, the **Lewis acid.**

$$2H\colon\!\!\overset{..}{\underset{H}{N}}\colon\; +\; Ag^+ \longrightarrow \left[\; H\colon\!\!\overset{H}{\underset{H}{N}}\colon Ag\colon\!\!\overset{H}{\underset{H}{N}}\colon H\;\right]^+$$

$$\;\;\;\;\;\;\text{Base}\;\;\;\;\;\;\;\text{Acid}\;\;\;\;\;\;\;\;\;\;\;\;\;\;\;\;\;\;\text{Acid-base}$$
$$\;\text{adduct}$$

The boron atom in boron trifluoride, BF_3, has only six electrons in its valence shell. Consequently, BF_3 is a very good Lewis acid and reacts with many Lewis bases; fluoride ion is the Lewis base in this reaction:

$$\left[\colon\!\!\overset{..}{\underset{..}{F}}\colon\right]^- +\; \overset{\colon\!\overset{..}{F}\colon}{\underset{\colon\!\overset{..}{F}\colon}{B}}\colon\!\!\overset{..}{\underset{..}{F}}\colon \longrightarrow \left[\;\overset{\colon\!\overset{..}{F}\colon}{\underset{\colon\!\overset{..}{F}\colon}{\colon\!\!\overset{..}{F}\colon B}}\colon\!\!\overset{..}{\underset{..}{F}}\colon\;\right]^-$$

$$\;\;\;\;\text{Base}\;\;\;\;\;\;\;\;\text{Acid}\;\;\;\;\;\;\;\;\;\;\;\;\;\text{Acid-base}$$
$$\;\text{adduct}$$

Nonmetal oxides act as Lewis acids and react with oxide ions, the Lewis base, to form oxyanions.

$$Ca^{2+}\left[\colon\!\!\overset{..}{\underset{..}{O}}\colon\right]^{2-} +\; \overset{\colon\!\overset{..}{O}\colon}{\underset{\colon\!\overset{..}{O}\colon}{S}}\colon\colon\!\!\overset{..}{O}\colon \longrightarrow Ca^{2+}\left[\;\overset{\colon\!\overset{..}{O}\colon}{\underset{\colon\!\overset{..}{O}\colon}{\colon\!\!\overset{..}{O}\colon S}}\colon\!\!\overset{..}{O}\colon\;\right]^{2-}$$

$$\;\;\;\;\;\;\;\text{Base}\;\;\;\;\;\;\;\;\;\;\text{Acid}\;\;\;\;\;\;\;\;\;\;\;\;\;\;\;\;\text{Acid-base}$$
$$\;\text{adduct}$$

Many Lewis acid-base reactions are displacement reactions in which one Lewis base displaces another Lewis base from an acid-base adduct or in which one Lewis acid displaces another Lewis acid.

$$\left[\; H\colon\!\!\overset{H}{\underset{H}{N}}\colon Ag\colon\!\!\overset{H}{\underset{H}{N}}\colon H\;\right]^+ +\; 2\colon\!N\!\equiv\!C\colon^- \longrightarrow [\colon\!N\!\equiv\!C\colon Ag\colon C\!\equiv\!N\colon]^- +\; 2\colon\!\!\overset{..}{\underset{H}{N}}\colon H$$

$$\text{Acid-base adduct}\;\;\;\;\;\;\;\;\;\;\;\;\;\;\;\text{Base}\;\;\;\;\;\;\;\;\;\;\;\;\;\text{New adduct}\;\;\;\;\;\;\;\;\;\;\;\text{New base}$$

$$Ca^{2+}\left[\;\overset{\colon\!\overset{..}{O}\colon}{\underset{\colon\!\overset{..}{O}\colon}{C}}\colon\colon\!\!\overset{..}{O}\colon\;\right]^{2-} +\; SO_3 \longrightarrow Ca^{2+}\left[\;\overset{\colon\!\overset{..}{O}\colon}{\underset{\colon\!\overset{..}{O}\colon}{\colon\!\!\overset{..}{O}\colon S}}\colon\!\!\overset{..}{O}\colon\;\right]^{2-} +\; CO_2$$

$$\;\;\text{Acid-base adduct}\;\;\;\;\;\;\;\;\;\text{Acid}\;\;\;\;\;\;\;\;\;\;\;\;\;\;\;\text{New adduct}\;\;\;\;\;\;\;\;\;\;\;\text{New acid}$$

$$H\colon\!\!\overset{..}{\underset{..}{Cl}}\colon +\; \colon\!\!\overset{..}{\underset{H}{O}}\colon H \longrightarrow H\colon\!\!\overset{..}{\underset{H}{O}}\colon H^+ +\; \colon\!\!\overset{..}{\underset{..}{Cl}}\colon^-$$

$$\;\text{Acid-base}\;\;\;\;\;\;\;\text{Base}\;\;\;\;\;\;\;\text{New adduct}\;\;\;\;\text{New base}$$
$$\;\;\text{adduct}$$

The last displacement reaction shows how the reaction of a Brönsted acid with a base fits into the Lewis model. A Brönsted acid such as HCl is an acid-base adduct according to the Lewis concept, and proton transfer occurs because a more stable acid-base adduct is formed.

ACIDS AND BASES IN AQUEOUS SOLUTION

14.6 Brönsted Acids in Aqueous Solution

When a Brönsted acid stronger than water dissolves in water, protons are transferred from the acid molecules to water molecules, giving hydronium ions, H_3O^+. Thus the properties that are common to Brönsted acids in aqueous solution are those of the hydronium ion. The hydronium ion is the species actually formed, but the symbol H^+ is often used instead of H_3O^+ for the sake of convenience. It is important to remember, however, that H^+ is an abbreviation and that hydrogen ion is always hydrated in water solution. In a strict sense even the formula H_3O^+ is an abbreviation. For example, good evidence has been accumulated for the existence of the species $H_9O_4^+$, corresponding to $[H_3O(H_2O)_3]^+$.

Since aqueous solutions of acids contain higher concentrations of hydronium ions than pure water, these solutions all exhibit the following properties, which are due to the presence of the hydronium ion:

1. They have a sour taste.
2. They change the color of certain indicators. For example, they change litmus from blue to red and phenolphthalein from red to colorless.
3. They react with metals above hydrogen in the activity series (Section 9.17) and liberate hydrogen; for example,

$$2H_3O^+(aq) + Zn(s) \longrightarrow Zn^{2+} + H_2(g) + 2H_2O(l)$$

4. They react with many basic metal oxides and hydroxides forming salts and water.

$$2H_3O^+ + FeO \longrightarrow Fe^{2+} + 3H_2O$$
$$2H_3O^+ + Fe(OH)_2 \longrightarrow Fe^{2+} + 4H_2O$$

5. They react with the salts of either weaker or more volatile acids, such as carbonates or sulfides, to give a new salt and a new acid.

$$2H_3O^+ + CaCO_3 \longrightarrow H_2CO_3 + Ca^{2+} + 2H_2O$$
$$\longrightarrow H_2O + CO_2(g)$$

$$2H_3O^+ + FeS \longrightarrow H_2S(g) + Fe^{2+} + 2H_2O$$

14.7 Formation of Brönsted Acids

Brönsted acids may be prepared by any of the following methods:

1. *By the direct union of the constituent elements.* Since binary compounds of hydrogen with the more electronegative nonmetals are acids (Section 14.4), the direct reaction of hydrogen with such nonmetals as F_2, Cl_2, Br_2, and S_8 yields an acid. For example,

$$H_2 + Br_2 \longrightarrow 2HBr$$
$$8H_2 + S_8 \overset{\triangle}{\rightleftharpoons} 8H_2S$$

2. *By the reaction of water with an oxide of a nonmetal.* Most oxides of nonmetals

are acidic (Section 8.4). The action of water on such a nonmetal oxide forms an acid.

$$SO_3 + H_2O \longrightarrow H_2SO_4$$
$$P_4O_{10} + 6H_2O \longrightarrow 4H_3PO_4$$
$$CO_2 + H_2O \rightleftharpoons H_2CO_3$$

3. *By the metathetical reaction (Section 8.2):*
   a. *of a salt of a volatile acid with a nonvolatile or slightly volatile acid.*

   $$NaF(s) + H_2SO_4(l) \longrightarrow NaHSO_4(s) + HF(g)$$

   b. *of a salt with an acid to produce a second acid and an insoluble precipitate.*

   $$3H_2SO_4(l) + Ca_3(PO_4)_2(s) \longrightarrow 2H_3PO_4(l) + 3CaSO_4(s)$$

   c. *of a salt of a weak acid with a strong acid.*

   $$[Na^+(aq)] + CH_3CO_2^-(aq) + H_3O^+(aq) + [Cl^-(aq)] \longrightarrow$$
   $$CH_3CO_2H(aq) + [Na^+(aq)] + [Cl^-(aq)] + H_2O$$

4. *By hydrolysis (Section 12.3) of certain nonmetal halides containing polar bonds.*

   $$PBr_3 + 3H_2O \longrightarrow H_3PO_3 + 3HBr$$
   $$PCl_5 + 4H_2O \longrightarrow H_3PO_4 + 5HCl$$
   $$SiI_4 + 4H_2O \longrightarrow Si(OH)_4 + 4HI$$

14.8   Polyprotic Acids

Acids can be classified by the number of protons per molecule that can be given up in a reaction. Acids such as HCl, HNO_3, and HCN that contain one ionizable hydrogen atom in each molecule are called **monoprotic acids.** Their reactions with water are as follows:

$$HCl + H_2O \longrightarrow H_3O^+ + Cl^-$$
$$HNO_3 + H_2O \longrightarrow H_3O^+ + NO_3^-$$
$$HCN + H_2O \rightleftharpoons H_3O^+ + CN^-$$

Acetic acid, CH_3CO_2H, is monoprotic because only one of the four hydrogen atoms in each molecule is given up as a proton in reactions with bases.

$$H-\underset{\underset{H}{|}}{\overset{\overset{H}{|}}{C}}-CO_2H + H_2O \rightleftharpoons H_3O^+ + H-\underset{\underset{H}{|}}{\overset{\overset{H}{|}}{C}}-CO_2^-$$

Diprotic acids contain two ionizable hydrogen atoms per molecule; ionization of such acids occurs in two stages. The primary ionization always takes place to a greater extent than the secondary ionization. For example, carbonic acid ionizes as follows:

$$H_2CO_3 + H_2O \rightleftharpoons H_3O^+ + HCO_3^- \qquad \text{(primary ionization)}$$
$$HCO_3^- + H_2O \rightleftharpoons H_3O^+ + CO_3^{2-} \qquad \text{(secondary ionization)}$$

Triprotic acids, such as phosphoric acid, ionize in three steps:

$$H_3PO_4 + H_2O \rightleftharpoons H_3O^+ + H_2PO_4^- \quad \text{(primary ionization)}$$

$$H_2PO_4^- + H_2O \rightleftharpoons H_3O^+ + HPO_4^{2-} \quad \text{(secondary ionization)}$$

$$HPO_4^{2-} + H_2O \rightleftharpoons H_3O^+ + PO_4^{3-} \quad \text{(tertiary ionization)}$$

The terms monobasic, dibasic, and tribasic are often used instead of monoprotic, diprotic, and triprotic.

14.9 Brönsted Bases in Aqueous Solution

When a Brönsted base stronger than water dissolves in water, a solution containing a higher concentration of hydroxide ions than found in pure water is formed. Strong bases give a 100% yield of hydroxide ion, whereas weak bases give only a small yield of hydroxide ion.

The following properties are common to strong Brönsted bases in aqueous solution and are due to the presence of the hydroxide ion:

1. They have a bitter taste.
2. They change the colors of certain indicators. For example, they change litmus from red to blue, and phenolphthalein from colorless to red.
3. They neutralize aqueous acids, forming water and a solution of a salt.

$$[M^+] + OH^- + H_3O^+ + [A^-] \longrightarrow [M^+] + [A^-] + 2H_2O$$

Some weak Brönsted bases also form enough hydroxide ion to exhibit these properties.

14.10 Formation of Hydroxide Bases

Bases containing hydroxide ion (hydroxide bases) may be prepared by the following methods:

1. *By the reaction of active metals with water.* The active metals that lie above magnesium in the activity series (Section 9.17) react directly with a stoichiometric amount of water to give strong bases.

$$2K + 2H_2O \longrightarrow 2KOH + H_2$$

$$Ca + 2H_2O \longrightarrow Ca(OH)_2 + H_2$$

If an excess of water is present, aqueous solutions of the bases are formed.

2. *By the reaction of oxides of active metals with water.* The oxides of active metals react with water to give bases (Section 8.3).

$$Li_2O + H_2O \longrightarrow 2LiOH$$

$$SrO + H_2O \longrightarrow Sr(OH)_2$$

If an excess of water is present, aqueous solutions of the bases may form.

3. *By the metathetical reaction of a salt with a base to give a solution of a second base and an insoluble precipitate.*

$$[Ca^{2+}] + SO_4^{2-} + Ba^{2+} + [2OH^-] \longrightarrow BaSO_4(s) + [Ca^{2+}] + [2OH^-]$$

4. *By electrolysis of certain salt solutions.*

$$[2Na^+] + 2Cl^- + 2H_2O \xrightarrow{\text{Electrolysis}} [2Na^+] + 2OH^- + H_2(g) + Cl_2(g)$$

14.11   Acid-Base Neutralization

Neutralization is the reaction of an acid with a stoichiometric amount of a base. When aqueous solutions of hydrochloric acid and sodium hydroxide are mixed in the proper proportions, the acidic and basic properties disappear. The hydronium ion, which is responsible for the acidic properties, reacts with the hydroxide ion, which is responsible for the basic properties, producing water.

$$H_3O^+ + [Cl^-] + [Na^+] + OH^- \longrightarrow 2H_2O + [Na^+] + [Cl^-]$$

The sodium and chloride ions undergo no chemical change and appear as the crystalline salt sodium chloride if the solution is evaporated. Since the only change that occurs in the solution is the reaction of hydronium and hydroxide ions, the neutralization may be represented simply by the net equation

$$H_3O^+ + OH^- \longrightarrow 2H_2O$$

This equation may be used to represent the reaction of an aqueous solution of any strong acid with an aqueous solution of any strong base.

14.12   Salts

The products of a neutralization reaction in aqueous solution are water and a salt. As described in Section 8.1, a salt is an ionic compound. Among the cations most frequently found in salts are simple metal ions such as Na^+, K^+, Ca^{2+}, and Ba^{2+}. Hydrated metal ions such as $[Al(H_2O)_6]^{3+}$ and more complex ions such as NH_4^+ and $(CH_3)_3NH^+$ are also found. The simple anions of salts include F^-, Cl^-, Br^-, and I^-. Many salts contain polyatomic negative ions such as NO_3^-, SO_4^{2-}, PO_4^{3-}, and CN^-.

Salts can be formed by other reactions as well. For example, Brönsted acid-base reactions, metathetical reactions, and the direct reactions between metals and nonmetals can all produce salts.

A salt that contains neither a replaceable hydrogen nor a hydroxide group is sometimes called a **normal salt;** examples are NaCl, K_2SO_4, and $Ca_3(PO_4)_2$. When only part of the acidic hydrogens of a polyprotic acid are replaced by a cation, the compound is known as a **hydrogen salt;** examples are sodium hydrogen carbonate, $NaHCO_3$ (often called sodium bicarbonate, or baking soda), and sodium dihydrogen phosphate, NaH_2PO_4.

When only part of the hydroxide groups of an ionic metal hydroxide are replaced by another anion, the compound is known as a **hydroxysalt;** examples are barium hydroxychloride, Ba(OH)Cl, and bismuth dihydroxychloride, $Bi(OH)_2Cl$. Some hydroxysalts readily lose the elements of one or more molecules of water and form **oxysalts.** For example, bismuth dihydroxychloride breaks down to give bismuth oxychloride, BiOCl.

$$Bi(OH)_2Cl \longrightarrow BiOCl + H_2O$$

14.13 Quantitative Reactions of Acids and Bases

Using chemical reactions to determine the concentration of a solution (to **standardize** a solution) or to determine the amount of a substance present in a sample using a **standard solution** (a solution of known concentration) makes up the branch of quantitative analysis called **volumetric analysis.** Since the end point of a titration (Section 3.2) involving a neutralization reaction can be detected readily by use of an indicator, solutions of acids and bases are often used in volumetric analyses.

EXAMPLE 14.1 Titration of a 40.00-mL sample of a solution of H_3PO_4 requires 35.00 mL of 0.1500 M KOH to reach the end point. Determine the molar concentration of H_3PO_4 if the reaction is

$$2KOH + H_3PO_4 \longrightarrow K_2HPO_4 + 2H_2O$$

The following steps are required for this problem.

Volume of KOH	$\longrightarrow$	Moles of KOH	$\longrightarrow$	Moles of H_3PO_4	$\longrightarrow$	Molarity of H_3PO_4

$$0.03500 \text{ L KOH} \times \frac{0.1500 \text{ mol KOH}}{1 \text{ L KOH}} = 5.250 \times 10^{-3} \text{ mol KOH}$$

$$5.250 \times 10^{-3} \text{ mol KOH} \times \frac{1 \text{ mol } H_3PO_4}{2 \text{ mol KOH}} = 2.625 \times 10^{-3} \text{ mol } H_3PO_4$$

$$\frac{2.625 \times 10^{-3} \text{ mol } H_3PO_4}{0.04000 \text{ L}} = 0.06562 \text{ } M$$

EXAMPLE 14.2 You may have used crystalline potassium hydrogen phthalate, $KHC_8H_4O_4$, as a standard acid in the laboratory because it is easy to purify and weigh. If you titrate 1.5024 g of this acid with 37.28 mL of NaOH solution, what is the concentration of the NaOH solution?

$$KHC_8H_4O_4 + NaOH \longrightarrow KNaC_8H_4O_4 + H_2O$$

You can determine concentration of NaOH from the number of moles of NaOH in 37.28 mL (0.03728 L) of the solution. To determine the concentration requires the following steps.

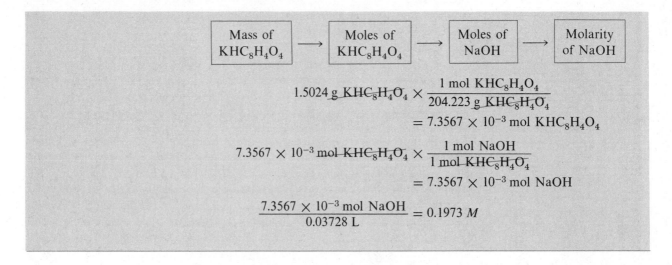

$$1.5024 \text{ g } \cancel{\text{KHC}_8\text{H}_4\text{O}_4} \times \frac{1 \text{ mol KHC}_8\text{H}_4\text{O}_4}{204.223 \text{ g } \cancel{\text{KHC}_8\text{H}_4\text{O}_4}}$$

$$= 7.3567 \times 10^{-3} \text{ mol KHC}_8\text{H}_4\text{O}_4$$

$$7.3567 \times 10^{-3} \text{ mol } \cancel{\text{KHC}_8\text{H}_4\text{O}_4} \times \frac{1 \text{ mol NaOH}}{1 \text{ mol } \cancel{\text{KHC}_8\text{H}_4\text{O}_4}}$$

$$= 7.3567 \times 10^{-3} \text{ mol NaOH}$$

$$\frac{7.3567 \times 10^{-3} \text{ mol NaOH}}{0.03728 \text{ L}} = 0.1973 \text{ } M$$

We can solve titration problems using mole relationships as described here and in Section 3.2. However, there is an alternate method, which uses equivalents, that is described in the next section.

14.14 Equivalents of Acids and Bases

An **equivalent of a Brönsted acid** is the mass of the acid in grams that will provide 1 mole of protons (H^+) in a reaction. An **equivalent of a base** is the mass of the base in grams that will provide 1 mole of hydroxide ions in a reaction or will react with 1 mole of protons. An equivalent is sometimes called a **gram-equivalent weight**. In an acid-base reaction 1 equivalent of an acid will react with 1 equivalent of a base.

In an acid-base reaction protons (H^+) are transferred from the acid to the base, and the number of protons transferred is used in the calculation of the mass of an equivalent of the acid and the base. When hydrochloric acid reacts with sodium hydroxide, a proton is transferred to the hydroxide ion.

$$\text{HCl} + \text{NaOH} \longrightarrow \text{NaCl} + \text{H}_2\text{O}$$

When 1 mole of hydrochloric acid reacts with 1 mole of sodium hydroxide, 1 mole of protons (6.022×10^{23} protons) is transferred to 1 mole of hydroxide ions (6.022×10^{23} hydroxide ions). It follows, then, that 1 equivalent (equiv) of hydrochloric acid is the same as 1 mole (36.5 g) of HCl (1 equiv HCl/1 mol HCl), and 1 equivalent of sodium hydroxide is the same as 1 mole (40.0 g) of NaOH (1 equiv NaOH/1 mol NaOH). However, 1 mole of sulfuric acid reacts with 2 moles of sodium hydroxide, and 2 moles of protons are transferred from the acid to the base.

$$\text{H}_2\text{SO}_4 + 2\text{NaOH} \longrightarrow \text{Na}_2\text{SO}_4 + 2\text{H}_2\text{O}$$

Hence 1 equivalent of sulfuric acid is equal to $\frac{1}{2}$ mole of the acid (2 equiv H_2SO_4/1 mol H_2SO_4).

Do not forget that the number of protons *actually transferred* in an acid-base reaction determines the mass of an equivalent of the acid or the base. For example, phosphoric acid may react with potassium hydroxide in any one of three ways:

$$H_3PO_4 + KOH \longrightarrow KH_2PO_4 + H_2O \tag{1}$$

$$H_3PO_4 + 2KOH \longrightarrow K_2HPO_4 + 2H_2O \tag{2}$$

$$H_3PO_4 + 3KOH \longrightarrow K_3PO_4 + 3H_2O \tag{3}$$

The numbers of moles of protons transferred, per mole of H_3PO_4, in the three reactions are 1, 2, and 3 moles, respectively. Thus an equivalent of H_3PO_4 in the three cases is equal to 1 mole, $\frac{1}{2}$ mole, and $\frac{1}{3}$ mole, respectively. *It must be emphasized that the mass of an equivalent of an acid or a base must be deduced from the reaction, not merely from the formula of the substance.*

EXAMPLE 14.3   Calculate the mass of an equivalent of H_3PO_4 in Equations (1), (2), and (3).

The molecular weight of H_3PO_4 is 97.99. In Equation (1), 1 mol of H_3PO_4 provides 1 mol of H^+ (1 equiv H_3PO_4/1 mol H_3PO_4).

$$\frac{97.99 \text{ g } H_3PO_4}{1 \text{ mol } H_3PO_4} \times \frac{1 \text{ mol } H_3PO_4}{1 \text{ equiv } H_3PO_4} = \frac{97.99 \text{ g } H_3PO_4}{1 \text{ equiv } H_3PO_4}$$

Thus the mass of an equivalent of H_3PO_4 in Equation (1) is 97.99 g.

In Equation (2), 1 mol of H_3PO_4 provides two mol of H^+ (2 equiv H_3PO_4/1 mol H_3PO_4).

$$\frac{97.99 \text{ g } H_3PO_4}{1 \text{ mol } H_3PO_4} \times \frac{1 \text{ mol } H_3PO_4}{2 \text{ equiv } H_3PO_4} = \frac{48.99 \text{ g } H_3PO_4}{1 \text{ equiv } H_3PO_4}$$

The mass of an equivalent of H_3PO_4 in Equation (2), therefore, is 48.99 g.

In Equation (3), 1 mol of H_3PO_4 provides 3 mol of H^+ (3 equiv H_3PO_4/1 mol H_3PO_4).

$$\frac{97.99 \text{ g } H_3PO_4}{1 \text{ mol } H_3PO_4} \times \frac{1 \text{ mol } H_3PO_4}{3 \text{ equiv } H_3PO_4} = \frac{32.66 \text{ g } H_3PO_4}{1 \text{ equiv } H_3PO_4}$$

Hence the mass of an equivalent of H_3PO_4 in Equation (3) is 32.66 g.

The **normality, N,** of a solution is the number of equivalents of solute per liter of solution. (Compare this with the definition of molarity in Section 3.1.)

$$N = \frac{\text{equivalents of solute}}{\text{liters of solution}} = \frac{\text{equiv}}{L}$$

Normality may also be determined from the molarity of the solution, provided that the reaction is known.

$$N = \frac{\text{moles of solute}}{\text{liters of solution}} \times \frac{\text{equivalents of solute}}{\text{moles of solute}} = \frac{\text{equiv}}{L}$$

A solution that contains 1 equivalent of solute in 1 liter of solution is a 1 N (one normal) solution. Provided that both hydrogens react, a 1 N solution of sulfuric acid contains 1 equivalent (98 g/2 = 49 g) of H_2SO_4 per liter; a 2 N solution of H_2SO_4 contains 2 equivalents, or 98 g; and a 0.01 N solution contains 0.49 g. A

1 N solution of sulfuric acid is identical to a 0.5 M solution of sulfuric acid; each contains 49 g of solute per liter of solution.

It should be evident that a 1 M solution of hydrochloric acid is also 1 N, since 1 mole of HCl is equal to 1 equivalent of HCl. However, a 1 M solution of sulfuric acid is 2 N, since 1 mole of H_2SO_4 is equal to 2 equivalents of H_2SO_4, if both hydrogens react.

EXAMPLE 14.4   Calculate the normality of a solution containing 3.65 g of HCl per 0.500 L of solution.

The mass of an equivalent of HCl is 36.5 g.

$$3.65 \text{ g HCl} \times \frac{1 \text{ equiv HCl}}{36.5 \text{ g HCl}} = 0.100 \text{ equiv HCl}$$

$$\frac{0.100 \text{ equiv}}{0.500 \text{ L}} = 0.200 \ N \text{ HCl}$$

EXAMPLE 14.5   How many grams of H_2SO_4 are contained in 1.2 L of 0.50 N H_2SO_4 solution that reacts according to the equation

$$H_2SO_4 + Ba(OH)_2 \longrightarrow BaSO_4 + 2H_2O$$

This problem may be solved by the following steps.

Volume of solution	$\longrightarrow$	Equivalents of H_2SO_4	$\longrightarrow$	Moles of H_2SO_4	$\longrightarrow$	Mass of H_2SO_4

A 0.50 N H_2SO_4 solution contains 0.50 equiv/L.

$$1.2 \text{ L} \times \frac{0.50 \text{ equiv } H_2SO_4}{1.0 \text{ L}} = 0.60 \text{ equiv } H_2SO_4$$

Since each mole of H_2SO_4 provides two protons in the acid-base reaction, 1 mol of H_2SO_4 is equal to 2 equiv of H_2SO_4.

$$0.60 \text{ equiv } H_2SO_4 \times \frac{1 \text{ mol } H_2SO_4}{2 \text{ equiv } H_2SO_4} = 0.30 \text{ mol } H_2SO_4$$

$$0.30 \text{ mol } H_2SO_4 \times \frac{98.0 \text{ g } H_2SO_4}{1 \text{ mol } H_2SO_4} = 29 \text{ g } H_2SO_4$$

EXAMPLE 14.6   What is the normality of a solution of $Ba(OH)_2$ that contains 17.14 mg of $Ba(OH)_2$ per 0.1000 L of solution?

$$Ba(OH)_2 + 2HCl \longrightarrow BaCl_2 + 2H_2O$$

Since both of the hydroxide ions react, 1 mol of $Ba(OH)_2$ provides 2 mol of OH^-. Hence there is 1 mol of $Ba(OH)_2$ per 2 equiv of $Ba(OH)_2$. This problem requires the following steps.

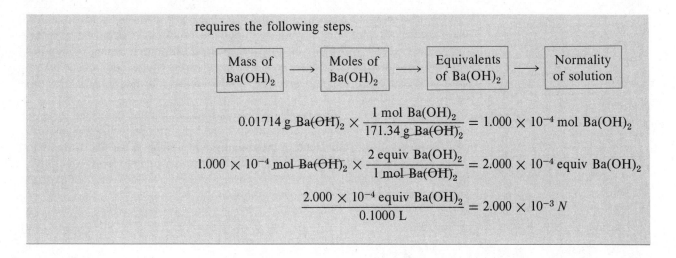

$$0.01714 \text{ g Ba(OH)}_2 \times \frac{1 \text{ mol Ba(OH)}_2}{171.34 \text{ g Ba(OH)}_2} = 1.000 \times 10^{-4} \text{ mol Ba(OH)}_2$$

$$1.000 \times 10^{-4} \text{ mol Ba(OH)}_2 \times \frac{2 \text{ equiv Ba(OH)}_2}{1 \text{ mol Ba(OH)}_2} = 2.000 \times 10^{-4} \text{ equiv Ba(OH)}_2$$

$$\frac{2.000 \times 10^{-4} \text{ equiv Ba(OH)}_2}{0.1000 \text{ L}} = 2.000 \times 10^{-3} \ N$$

FOR REVIEW

SUMMARY

A compound that can donate a proton to another compound is called a **Brönsted acid.** The compound that accepts the proton is called a **Brönsted base.** The species remaining after a Brönsted acid has lost a proton is the **conjugate base** of the acid. When a Brönsted base gains a proton, its **conjugate acid** forms. Thus an **acid-base reaction** occurs when a proton is transferred from a reactant acid to a reactant base, with formation of the conjugate base of the reactant acid and the conjugate acid of the reactant base. **Amphiprotic species** can act as both a proton donor and a proton acceptor. Water is the most important amphiprotic species. It can form both the hydronium ion, H_3O^+, and the hydroxide ion, OH^-.

Strong acids form weak conjugate bases, and weak acids form strong conjugate bases. Thus **strong acids** are completely ionized in aqueous solution because their conjugate bases are weaker bases than water. **Weak acids** are only partially ionized because their conjugate bases are strong enough to compete successfully with water for possession of protons. **Strong bases** react with water to give a 100% yield of hydroxide ion. **Weak bases** give only small yields of hydroxide ion. The strengths of the binary acids increase from left to right across a period of the Periodic Table ($CH_4 < NH_3 < H_2O < HF$), and they increase down a group of the Table ($HF < HCl < HBr < HI$). The strengths of oxyacids containing the same central element increase as the oxidation number of the element increases ($H_2SO_3 < H_2SO_4$). The strengths of oxyacids also increase as the electronegativity of the central element increases [$H_2SeO_4 < H_2SO_4$; $Al(OH)_3 < Si(OH)_4 < H_3PO_4 < H_2SO_4 < HClO_4$].

A **Lewis acid** is a molecule or ion that can accept a pair of electrons. A **Lewis base** is the molecule or ion that donates the pair of electrons. Thus a Lewis acid-base reaction results in the formation of a coordinate covalent bond.

The characteristic properties of aqueous solutions of Brönsted acids are due to the presence of hydronium ions; those of aqueous solutions of Brönsted bases are due to the presence of hydroxide ions. The **neutralization** that occurs when aqueous solutions of acids and bases are mixed results from the reaction of the hydronium and hydroxide ions to form water. Salts are also formed in neutralization reactions.

The concentration of an acid or a base in solution may be described in terms of the number of moles per liter (molarity) or of the number of equivalents per liter (normality). An **equivalent,** or **gram-equivalent weight,** of an acid is the mass of the acid in grams that will provide 1 mole of protons in a reaction. An equivalent of a base is the mass of the base in grams that will provide 1 mole of hydroxide ions or will accept 1 mole of protons.

KEY TERMS AND CONCEPTS

Amphiprotic (14.2)	Hydrogen salt (14.12)	Oxysalt (14.12)
Amphoteric (14.4)	Hydroxysalt (14.12)	Polyprotic acid (14.8)
Brönsted acid (14.1)	Lewis acid (14.5)	Salt (14.12)
Brönsted base (14.1)	Lewis base (14.5)	Strong acids (14.3)
Conjugate acid (14.1)	Monoprotic acid (14.8)	Strong bases (14.3)
Conjugate base (14.1)	Neutralization (14.11)	Triprotic acid (14.8)
Diprotic acid (14.8)	Normal salt (14.12)	Weak acids (14.3)
Equivalent (14.14)	Normality (14.14)	Weak bases (14.3)

EXERCISES

Brönsted Acids and Bases

1. According to the Brönsted-Lowry concept, what is an acid? a base?
2. Show by suitable equations that each of the following species can act as a Brönsted acid:
   (a) H_3O^+ (c) H_2O (e) NH_4^+
   (b) HCl (d) CH_3CO_2H (f) HSO_4^-
3. Show by suitable equations that each of the following species can act as a Brönsted base:
   (a) H_2O (c) NH_3 (e) S^{2-}
   (b) OH^- (d) CN^- (f) $H_2PO_4^-$
4. What is the conjugate acid formed when each of the following reacts as a base?
   (a) OH^- (d) HSO_4^- (g) NH_2^-
   (b) F^- (e) HCO_3^- (h) H^-
   (c) H_2O (f) NH_3 (i) N^{3-}
   Is each a strong or weak acid?
5. What is the conjugate base formed when each of the following reacts as an acid?

   (a) OH^- (d) HBr (g) HS^-
   (b) H_2O (e) HSO_4^- (h) PH_3
   (c) HCO_3^- (f) NH_3 (i) H_2O_2
6. What is the conjugate base of each of the following?
   (a) NH_4^+ (c) HNO_3 (e) CH_3CO_2H
   (b) HCl (d) $HClO_4$ (f) HCN
   Is each a strong or weak base?
7. Write equations for the reaction of each of the following with water:
   (a) HCl (d) NH_2^- (f) F^-
   (b) HNO_3 (e) $HClO_4$ (g) NH_4^+
   (c) NH_3
   What is the role played by water in each of these acid-base reactions?
8. (a) Identify the strong Brönsted acids and strong Brönsted bases in the list of important industrial inorganic compounds in Table 8-6.

(b) List those compounds in Table 8-6 that can behave as Brönsted acids with strengths lying between those of H_3O^+ and H_2O.

(c) List those compounds in Table 8-6 that can behave as Brönsted bases with strengths lying between those of H_2O and OH^-.

9. Hydrogen chloride can be prepared by the acid-base reaction of pure sulfuric acid with solid sodium chloride (Section 8.7). Write the balanced chemical equation for the reaction, and identify the conjugate acid-base pairs in this reaction.

10. Identify and label the Brönsted acid, its conjugate base, the Brönsted base, and its conjugate acid in each of the following equations:
    (a) $HNO_3 + H_2O \longrightarrow H_3O^+ + NO_3^-$
    (b) $CN^- + H_2O \rightleftharpoons HCN + OH^-$
    (c) $H_2SO_4 + Cl^- \longrightarrow HCl + HSO_4^-$
    (d) $HSO_4^- + OH^- \longrightarrow SO_4^{2-} + H_2O$
    (e) $O^{2-} + H_2O \longrightarrow 2OH^-$
    (f) $[Cu(H_2O)_3(OH)]^+ + [Al(H_2O)_6]^{3+} \rightleftharpoons$
        $[Cu(H_2O)_4]^{2+} + [Al(H_2O)_5(OH)]^{2+}$
    (g) $H_2S + NH_2^- \longrightarrow HS^- + NH_3$

11. Gastric juice, the digestive fluid produced in the stomach, contains hydrochloric acid, HCl. Milk of magnesia, a suspension of solid $Mg(OH)_2$ in an aqueous medium, is sometimes used to neutralize excess stomach acid. Write a complete balanced equation for the neutralization reaction, and identify the conjugate acid-base pairs.

12. Nitric acid reacts with insoluble copper(II) oxide to form soluble copper(II) nitrate, $Cu(NO_3)_2$. Write the balanced chemical equation for the reaction of an aqueous solution of HNO_3 with CuO, and indicate the conjugate acid-base pairs.

13. What are amphiprotic species? Illustrate with suitable equations.

14. State which of the following species are amphiprotic and write chemical equations illustrating the amphiprotic character of the species:
    (a) H_2O (c) S^{2-} (e) HSO_4^-
    (b) $H_2PO_4^-$ (d) CH_4 (f) H_2CO_3

Strengths of Acids and Bases

15. Predict which acid in each of the following pairs is the stronger:
    (a) H_2O or H_2Te (d) HSO_4^- or $HSeO_4^-$
    (b) NH_3 or PH_3 (e) HSO_3^- or HSO_4^-
    (c) $B(OH)_3$ or $Al(OH)_3$

Explain your reasoning for each.

16. Predict which acid in each of the following pairs is the stronger:
    (a) NH_3 or H_2O (c) NH_3 or H_2S
    (b) H_2O or HF (d) PH_3 or HI
    Explain your reasoning for each.

17. Rank the compounds in each of the following groups in order of increasing acidity, and explain the order you assign:
    (a) HOCl, HOBr, HOI
    (b) HCl, HBr, HI
    (c) $HOCl, HOClO, HOClO_2, HOClO_3$
    (d) HF, H_2O, NH_3, CH_4
    (e) $Mg(OH)_2, Si(OH)_4, ClO_3(OH)$
    (f) $NaHSO_3, NaHSeO_3, NaHSO_4$

18. Rank the bases in each of the following groups in order of increasing base strength. Explain.
    (a) H_2O, OH^-, H^-, Cl^-
    (b) $ClO_4^-, ClO^-, ClO_2^-, ClO_3^-$
    (c) $NH_2^-, HS^-, HTe^-, PH_2^-$
    (d) BrO_2^-, ClO_2^-, IO_2^-

19. What is the relationship between the extent of ionization of an acid in aqueous solution and the strength of the acid?

20. Both HF and HCN ionize in water to a limited extent. Which of the conjugate bases, F^- or CN^-, is the stronger base? See Table 14-1.

21. Soaps are sodium and potassium salts of a family of acids called fatty acids, which are isolated from animal fats. These acids, which are related to acetic acid, CH_3CO_2H, all contain the carboxyl group, $-CO_2H$, and have about the same strength as acetic acid. Examples include palmitic acid, $C_{15}H_{31}CO_2H$, and stearic acid, $C_{17}H_{35}CO_2H$.
    (a) Write a balanced chemical equation indicating the formation of sodium palmitate, $C_{15}H_{31}CO_2Na$, from palmitic acid and sodium carbonate, and a corresponding equation for the formation of sodium stearate.
    (b) Is a soap solution acidic, basic, or neutral?

22. Some porcelain cleansers contain sodium hydrogen sulfate, $NaHSO_4$. Is a solution of $NaHSO_4$ acidic, neutral, or basic? These cleansers remove the deposits due to hard water (primarily calcium carbonate, $CaCO_3$) very effectively. Write a balanced chemical equation for one reaction between $NaHSO_4$ and $CaCO_3$.

23. What is meant by the leveling effect of water on strong acids and strong bases?

24. How can the relative strengths of strong acids be measured?

Lewis Acids and Bases

25. According to the Lewis concept, what is an acid? a base?

26. Write the Lewis structures of the reactants and product of each of the following equations, and identify the Lewis acid and the Lewis base in each:
    (a) $O^{2-} + SO_3 \longrightarrow SO_4{}^{2-}$
    (b) $AlCl_3 + Cl^- \longrightarrow AlCl_4{}^-$
    (c) $B(OH)_3 + OH^- \longrightarrow B(OH)_4{}^-$
    (d) $CO_2 + OH^- \longrightarrow HCO_3{}^-$
    (e) $I^- + I_2 \longrightarrow I_3{}^-$

27. Calcium oxide is prepared by the decomposition of calcium carbonate at elevated temperatures (Section 8.7). This is the reverse of the Lewis acid-base reaction between CaO and CO_2. Write the chemical equation for the reaction of CaO with CO_2 showing the Lewis structures of the reactants and product, and identify the Lewis acid and Lewis base.

28. The reaction of SO_3 with H_2SO_4 to give pyrosulfuric acid, $H_2S_2O_7$ (Section 8.7), is a Lewis acid-base reaction that results in the formation of S—O—S bonds. Write the chemical equation for this reaction showing the Lewis structures of the reactants and product, and identify the Lewis acid and Lewis base.

29. The dissolution of solid $Al(NO_3)_3$ in water is accompanied by both a Lewis and a Brönsted acid-base reaction. Write balanced chemical equations for the reactions.

30. Each of the species given below may be considered to be a Lewis acid-base adduct of a proton with a Lewis base. For each of the following pairs, indicate which adduct is the stronger acid:
    (a) HCl or HCN   (d) H_2O or NH_3
    (b) HBr or H_3O^+ (e) $HSO_4{}^-$ or $HCO_3{}^-$
    (c) H_3O^+ or $NH_4{}^+$

31. Write a balanced chemical equation for the Lewis acid-base reaction between $Al(OH)_4{}^-$ and CO_2, which causes the precipitation of $Al(OH)_3$ when CO_2 is added to an aqueous solution of $Al(OH)_4{}^-$.

Solutions of Brönsted Acids and Bases

32. Write the equation for the essential reaction between aqueous solutions of strong acids and of strong bases.

33. Write equations to show the stepwise ionization of the following polyprotic acids: H_2S, H_2CO_3, and H_3PO_4.

34. Containers of $NaHCO_3$, sodium hydrogen carbonate (sometimes called sodium bicarbonate), are often kept in chemical laboratories for use in neutralizing any acids or bases spilled. Write balanced chemical equations for the neutralization by $NaHCO_3$ of (a) a solution of HCl and (b) a solution of KOH.

35. In aqueous solution arsenic acid, H_3AsO_4, ionizes in three steps. Write an equation for each step, and label the conjugate acid-base pairs. Which arsenic-containing species are amphiprotic?

36. Define the term *salt*, and give examples of normal salts, hydrogen salts, hydroxysalts, and oxysalts.

37. Write equations to illustrate three typical and characteristic reactions of aqueous acids.

38. Contrast the action of water on oxides of metals with that of water on oxides of nonmetals.

39. Give four methods of preparing hydroxide bases and of preparing Brönsted acids.

40. Describe the chemical reaction, using equations where appropriate, that you would expect from the combination of a solution of $HClO_4$ with each of the following:
    (a) aluminum
    (b) calcium carbonate
    (c) a potassium hydroxide solution
    (d) a blue-colored litmus solution
    (e) solid sodium oxide
    (f) solid aluminum sulfide, Al_2S_3
    (g) ammonia gas

41. Define acid-base *neutralization*. Does your definition imply that the resulting solution is always neutral? Explain.

Titration; Normality

42. Calculate the mass of an equivalent of each of the reactants in the following equations to one decimal place:

[S] (a) $Sr(OH)_2 + 2HNO_3 \longrightarrow Sr(NO_3)_2 + 2H_2O$
Ans. $Sr(OH)_2$, 60.8 g; HNO_3, 63.0 g

(b) $LiOH + LiH_2PO_4 \longrightarrow Li_2HPO_4 + H_2O$
Ans. LiOH, 23.9 g; LiH_2PO_4, 103.9 g

[S] (c) $2Al(OH)_3 + 3H_2SO_4 \longrightarrow Al_2(SO_4)_3 + 6H_2O$
Ans. $Al(OH)_3$, 26.0 g; H_2SO_4, 49.0 g

(d) $NiCO_3 + 2HF \longrightarrow NiF_2 + H_2O + CO_2$
Ans. $NiCO_3$, 59.4 g; HF, 20.0 g

[S] (e) $KOH + H_2SO_4 \longrightarrow KHSO_4 + H_2O$
Ans. KOH, 56.1; H_2SO_4, 98.1 g

(f) $H_2S + Li_3P \longrightarrow Li_2S + H_3P$ (not balanced)
Ans. H_2S, 17.0 g; Li_3P, 17.3 g

43. Calculate the normality of each of the following solutions:

[S] (a) 5.0 equiv of HCl in 2.0 L of solution
Ans. 2.5 N

(b) 0.050 equiv of $Ba(OH)_2$ in 0.67 L of solution
Ans. 0.075 N

[S] (c) 0.0015 equiv of HCl in 100.0 mL of solution
Ans. 0.015 N

(d) 0.04733 equiv of $HClO_4$ in 98.02 mL of solution
Ans. 0.4829 N

[S] 44. A 0.244 *N* solution of KOH is titrated with H_2SO_4, producing K_2SO_4. If a 48.0-mL sample of the KOH solution is used, what volume of 0.244 *M* H_2SO_4 is required to reach the end point? What volume of 0.244 *N* H_2SO_4? (Assume both protons react.)
Ans. 24.0 mL of 0.244 M H_2SO_4; 48.0 mL of 0.244 N H_2SO_4

[S] 45. What volume of 0.421 *N* HCl is required to titrate 47.00 mL of 0.204 *N* KOH? *Ans. 22.8 mL*

[S] 46. Titration of 0.1500 g of an acid requires 47.00 mL of 0.0120 *N* NaOH to reach the end point. What is the mass of an equivalent of the acid?
Ans. 266 g

47. Titration of 10.0 mL of a 0.444 *N* HI solution with NH_3 requires 23.2 mL of the NH_3 solution to reach the end point. What is the normality of the NH_3 solution? *Ans. 0.191 N*

48. What is the normality of a H_2SO_4 solution if 24.8 mL of the solution is required to titrate 2.50 g of $NaHCO_3$?

$$H_2SO_4 + 2NaHCO_3 \longrightarrow Na_2SO_4 + 2H_2O + 2CO_2$$

Ans. 1.20 N

49. A sample of solid calcium hydroxide, $Ca(OH)_2$, is allowed to stand in water until a saturated solution is formed. A titration of 50.00 mL of this solution with 5.00×10^{-2} *N* HCl requires 24.4 mL of the acid to reach the end point.

$$Ca(OH)_2 + 2HCl \longrightarrow CaCl_2 + 2H_2O$$

What is the normality of a saturated $Ca(OH)_2$ solution? the molarity? What is the solubility of $Ca(OH)_2$ in grams per liter of solution?
Ans. 2.44×10^{-2} N; 1.22×10^{-2} M; 0.904 g/L

50. A 0.1824-g sample of an acid is titrated with 0.1236 *N* NaOH, and 23.14 mL is required to reach the end point. What is the mass of an equivalent of the acid? *Ans. 63.77 g*

Additional Exercises

51. In papermaking the use of wood pulp as a source of cellulose and of alum-rosin sizing (which keeps ink from soaking into the pages and blurring) produces a paper that deteriorates in 25–50 years. This is quite satisfactory for most purposes, but it is not suitable for the needs of libraries and archives. The deterioration is due to formation of acids. Alum-rosin slowly produces sulfuric acid under humid conditions. Also, wood pulp contains lignin, which forms carboxylic acids as it ages. These acids are similar to acetic acid in that they contain the acidic $—CO_2H$ group. In order to stop this acidification, books are sometimes soaked in magnesium hydrogen carbonate solution; treated with cyclohexylamine, $C_6H_{11}NH_2$, a base like ammonia but with one hydrogen atom replaced by a $—C_6H_{11}$ group; and treated with gaseous diethyl zinc, $(C_2H_5)_2Zn$, a covalent molecule containing Zn—C bonds. Diethyl zinc is a source of $C_2H_5^-$, which is very much like CH_3^- in its properties. Write balanced equations for these reactions. Use the formula RCO_2H for the carboxylic acids formed from lignin; the exact nature of R is not important in these reactions.

52. In dilute aqueous solution HF acts as a weak acid. However, pure liquid HF (b.p. = 19.5°C) is a strong acid. In liquid HF, HNO_3 acts like a base and accepts protons. The acidity of liquid HF can be increased by adding one of several inorganic fluorides that are Lewis acids and accept F^- ion

(for example, BF_3 or SbF_5). Write balanced chemical equations for the reaction of pure HNO_3 with pure HF and of pure HF with BF_3. Write the Lewis structures of the reactants and products, and identify the conjugate acid-base pairs.

53. Amino acids are biologically important molecules that are the building blocks of proteins. The simplest amino acid is glycine, $H_2N\,CH_2CO_2H$. The common feature of amino acids is that they contain the groups indicated in color: amino, $-NH_2$, and carboxyl, $-CO_2H$. An amino acid can function as either an acid or a base. For glycine the acid strength of the carboxyl group is about the same as that of acetic acid, CH_3CO_2H, and the base strength of the amino group is slightly greater than that of ammonia, NH_3.

    (a) Write the Lewis structures of the ions that form when glycine is dissolved in 1 M HCl and in 1 M KOH.

    (b) Write the Lewis structure of glycine when this amino acid is dissolved in water. (*Hint:* Consider the relative base strengths of the $-NH_2$ and $-CO_2^-$ groups.)

54. Using Lewis structures, write balanced equations for the following reactions:

    (a) $HF + NH_3 \longrightarrow$

    (b) $H_3O^+ + CN^- \longrightarrow$

    (c) $Na_2O + SO_3 \longrightarrow$

    (d) $Li_3N + H_2O \longrightarrow$

    (e) $NH_3 + CH_3^- \longrightarrow$

55. Predict the products of the following reactions, and write balanced equations for each. There may be more than one reasonable equation depending on the stoichiometry assumed for the reactants.

    (a) $Fe_2O_3(s) + H_2SO_4(l) \longrightarrow$

    (b) $Li_2O(s) + SiO_2(s) \xrightarrow{\triangle}$

    (c) $HCN(g) + Na \longrightarrow$

    (d) $NaHCO_3(s) + NaNH_2(s) \longrightarrow$

    (e) $Li_3N(s) + NH_3(l) \longrightarrow$

    (f) $BaO(s) + Cl_2O_7(l) \longrightarrow$

    (g) $NaF(s) + H_3PO_4(l) \longrightarrow$

    (h) $MgH_2(s) + H_2S(g) \longrightarrow$

    (i) $NaCH_3(s) + NH_3(g) \longrightarrow$

    (j) $KHCO_3(s) + KHS(s) \longrightarrow$

    (k) $H_2SO_4(l) + NaCH_3CO_2(s) \longrightarrow$

56. Trichloroacetic acid, CCl_3CO_2H, is amphiprotic.

    $$CCl_3CO_2H + B \longrightarrow CCl_3CO_2^- + BH$$

$$HA + CCl_3CO_2H \longrightarrow CCl_3CO_2H_2^+ + A^-$$

Write equations for the reaction of pure $CCl_3CO_2H(l)$ with $H_2O(l)$, $HClO_4(l)$, $HBr(g)$, $NH_3(g)$, and $CH_3CO_2H(l)$. Trichloroacetic acid has an acid strength between those of H_3O^+ and HSO_4^-.

57. Write equations for three methods of preparing the insoluble salt $BaSO_4$.

58. What mass of magnesium when treated with an excess of dilute sulfuric acid will produce the same volume of hydrogen gas as produced by 2.00 g of aluminum under the same conditions?

    Ans. 2.70 g

[S] 59. The reaction of 0.871 g of sodium with an excess of liquid ammonia containing a trace of $FeCl_3$ as a catalyst produced 0.473 L of pure H_2 measured at 25°C and 745 torr. What is the equation for the reaction of sodium with liquid ammonia? Show your calculations.

    Ans. $2Na + 2NH_3 \longrightarrow 2NaNH_2 + H_2$

60. The reaction of WCl_6 with Al at about 400°C gives black crystals of a compound containing only tungsten and chlorine. A sample of this compound, when reduced with hydrogen, gives 0.2232 g of tungsten metal and hydrogen chloride, which is absorbed in water. Titration of the hydrochloric acid thus produced requires 46.2 mL of 0.1051 M NaOH to reach the end point. What is the empirical formula of the black tungsten chloride?   *Ans. WCl_4*

61. A 0.3420 g sample of potassium acid phthalate, $KHC_8H_4O_4$, reacts with 35.73 mL of a NaOH solution in a titration. What is the concentration of the NaOH?

    $$KHC_8H_4O_4 + NaOH \longrightarrow KNaC_8H_4O_4 + H_2O$$

    Ans. 0.04691 M

62. A 21.34 mL sample of a solution of sulfuric acid, H_2SO_4, is required to titrate 25.00 mL of a 0.1079 M LiOH solution. What is the concentration of the sulfuric acid solution?

    $$H_2SO_4 + 2LiOH \longrightarrow Li_2SO_4 + 2H_2O$$

    Ans. 0.06321 M

63. How many milliliters of a 0.1500 M solution of KOH will be required to titrate 20.00 mL of a 0.1312 M solution of H_3PO_4?

    $$H_3PO_4 + 2KOH \longrightarrow K_2HPO_4 + 2H_2O$$

    Ans. 34.99 mL

15

CHEMICAL KINETICS AND CHEMICAL EQUILIBRIUM

When we are planning to run a chemical reaction, we should ask whether or not it will produce the desired products in useful quantities. We could answer this question through the use of equilibrium and thermodynamic calculations, as discussed in the latter part of this chapter and in Chapters 16–18, or we could use qualitative considerations, such as the fact that the reaction of a strong acid with a strong base gives a salt or that an active metal (a reducing agent) generally reacts with a nonmetal (an oxidizing agent). We should also ask whether or not the reaction will proceed rapidly enough to be useful. A reaction that takes 50 years to produce a product is almost as useless as one that will never produce a product.

This chapter considers two ways to answer these questions. The first part deals with the rates at which chemical reactions yield products **(chemical kinetics).** In it we examine the factors influencing the rates of chemical reactions, the mechanisms by which reactions proceed, and the quantitative techniques used to determine and to describe the rates at which reactions occur. The second part of the chapter deals with chemical equilibrium and describes how it is related to kinetics.

CHEMICAL KINETICS

15.1 The Rate of Reaction

All rates involve a division by some unit of time. For example, the rate of production of a well is measured in gallons *per minute*. The rate of use of coal by an electric generating plant is measured in tons *per hour.* **The rate of a chemical reaction can be defined as the decrease in concentration of a reactant or the increase in concentration of a product during a given unit of time.** When pure hydrogen peroxide is dissolved in water, it slowly decomposes according to the following equation:

$$2H_2O_2(aq) \longrightarrow 2H_2O(l) + O_2(g)$$

The change in concentration with time at 40°C of a solution that is initially 1 M in hydrogen peroxide is shown in Fig. 15-1 and in Table 15-1. From these data, which must be determined experimentally, we can determine the rate, or speed, at which the hydrogen peroxide decomposes.

$$\text{Rate} = \frac{\text{change in concentration of reactant}}{\text{time interval}}$$

$$= -\frac{[H_2O_2]_{t_2} - [H_2O_2]_{t_1}}{t_2 - t_1} = -\frac{\Delta[H_2O_2]}{\Delta t}$$

In these equations brackets are used to signify molar concentrations, and the symbol delta (Δ), to signify "the change in." Thus $[H_2O_2]_{t_1}$ represents the molar concentration of hydrogen peroxide at some time t_1; $[H_2O_2]_{t_2}$, the molar concentration of hydrogen peroxide at a later time t_2; and $\Delta[H_2O_2]$, the change in molar concentration of hydrogen peroxide during the time interval Δt (i.e., $t_2 - t_1$). The minus sign preceding the whole expression indicates that the concentration of the reactant is decreasing during the course of the reaction.

Figure 15-1. The decomposition of H_2O_2 ($2H_2O_2 \rightarrow 2H_2O + O_2$) at 40°C. The intensity of the color represents the concentration of H_2O_2 at the indicated time after the reaction begins. The color fades to indicate that the concentration of H_2O_2 decreases with time.

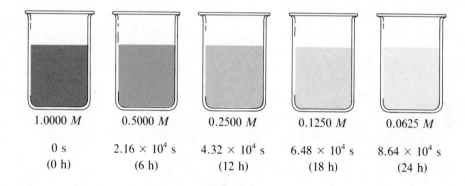

1.0000 M	0.5000 M	0.2500 M	0.1250 M	0.0625 M
0 s	2.16×10^4 s	4.32×10^4 s	6.48×10^4 s	8.64×10^4 s
(0 h)	(6 h)	(12 h)	(18 h)	(24 h)

Table 15-1   The Variation in the Rate of Decomposition of H_2O_2 at 40°C

Time, s	$[H_2O_2]$, mol L^{-1}	$\Delta[H_2O_2]$, mol L^{-1}	Δt, s	Rate, mol L$^-$ s^{-1}
0	1.0000			
2.16×10^4	0.5000	-0.5000	2.16×10^4	2.31×10^{-5}
4.32×10^4	0.2500	-0.2500	2.16×10^4	1.16×10^{-5}
6.48×10^4	0.1250	-0.1250	2.16×10^4	0.579×10^{-5}
8.64×10^4	0.0625	-0.0625	2.16×10^4	0.289×10^{-5}

EXAMPLE 15.1   Calculate the rate of decomposition of a 1.0000 M solution of hydrogen peroxide at 40°C for the period of the reaction from 0 to 21,600 s, using the data in Table 15-1.

In Table 15-1 we find that for the period of the reaction from 0 to 21,600 s,

$$[H_2O_2]_{t_1} = 1.0000 \text{ mol L}^{-1} \qquad (t_1 = 0 \text{ s})$$

and

$$[H_2O_2]_{t_2} = 0.5000 \text{ mol L}^{-1} \qquad (t_2 = 21{,}600 \text{ s})$$

Thus

$$\Delta t = t_2 - t_1 = 21{,}600 \text{ s} - 0 \text{ s} = 21{,}600 \text{ s}$$
$$\Delta[H_2O_2] = [H_2O_2]_{t_2} - [H_2O_2]_{t_1}$$
$$= 0.5000 \text{ mol L}^{-1} - 1.0000 \text{ mol L}^{-1}$$
$$= -0.5000 \text{ mol L}^{-1}$$

$$\text{Rate} = -\left(\frac{\Delta[H_2O_2]}{\Delta t}\right) = -\left(\frac{-0.5000 \text{ mol L}^{-1}}{21{,}600 \text{ s}}\right)$$

$$= 2.31 \times 10^{-5} \text{ mol L}^{-1} \text{ s}^{-1}$$

The unit mol/L (moles per liter, or molarity) can be written as mol L^{-1} since $1/X = X^{-1}$, according to the mathematical rules for exponents. The unit s^{-1} (per second) is similarly equivalent to 1/sec.

Since the rate of a chemical reaction changes with time, the reaction rates presented in Table 15-1 are average rates for the time intervals indicated. The rate of decomposition of hydrogen peroxide is faster at the beginning of the decomposition ($t = 0$ seconds) than it is at 21,600 seconds. The rate of 2.31×10^{-3} mol L^{-1} is the average rate of decomposition between 0 seconds and 21,600 seconds. If we calculate the rate for the time interval from 0 seconds to 86,400 seconds, we find

$$\text{Rate} = -\left(\frac{0.0625 \text{ mol L}^{-1} - 1.0000 \text{ mol L}^{-1}}{86{,}400 \text{ s} - 0 \text{ s}}\right)$$

$$= -\left(\frac{-0.9375 \text{ mol L}^{-1}}{86{,}400 \text{ s}}\right) = 1.08 \times 10^{-5} \text{ mol L}^{-1} \text{ s}^{-1}$$

This value is the average of the rates for the four time intervals in Table 15-1.

A graph of the concentration of hydrogen peroxide versus time shows the rate of the reaction at any instant of time. This rate is given by the slope of a straight line tangent to the curve at that time. Such a graph is shown in Fig. 15-2, with a tangent drawn at $t = 40{,}000$ seconds. The slope of this line (-8.90×10^{-6}) is $\Delta[H_2O_2]/\Delta t$. Thus the rate of decomposition at 40,000 seconds is

$$\text{Rate} = 8.90 \times 10^{-6} \text{ mol L}^{-1} \text{ s}^{-1}$$

The initial rate of the reaction is equal numerically (but opposite in sign) to the slope of the tangent at time zero.

Figure 15-2. A graph of concentration versus time for the 1.0000 M solution of H_2O_2 described in Fig. 15-1. The rate at any instant of time is equal to the slope of a line tangent to this curve at that time. A tangent is shown for $t = 40,000$ s.

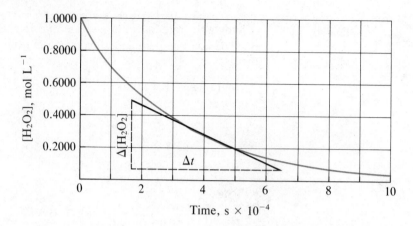

The rates of chemical reactions vary greatly. The initial rate for the decomposition of 1 M hydrogen peroxide at 40°C is 3.2×10^{-5} mol L^{-1} s^{-1}, and the reaction takes many hours to go to completion. The initial rate of formation of water when 1 M hydrochloric acid and 1 M sodium hydroxide react at 25°C is 1.4×10^{11} mol L^{-1} s^{-1}. This acid-base reaction occurs as rapidly as the two solutions can be mixed—almost instantaneously. The initial rate for the reaction of hydrochloric acid with sodium hydroxide is 10^{16} times faster than that for the decomposition of hydrogen peroxide.

The rates at which reactants disappear during reactions are rarely constant. As the reactants are consumed, rates usually decrease until they reach zero. In an irreversible reaction, such as the decomposition of hydrogen peroxide, the rate of the reaction reaches zero when all of the reactant has been consumed. The decomposition of hydrogen peroxide is effectively irreversible because the oxygen produced by the reaction escapes from the open beaker. In a reversible reaction, a reaction in which the products react to reform the reactants, the rate becomes zero when equilibrium is established. The reaction of nitrogen with hydrogen to form ammonia in a closed system is an example of a reversible reaction (Section 8.7). In order to establish the rate of a reversible chemical reaction, the reaction is studied at the beginning, before equilibrium is reached.

15.2   Rate of Reaction and the Nature of the Reactants

The rate of a reaction is strongly influenced by the nature of the participating substances. The reaction of an acid with a base, a process involving the simple combination of two ions of opposite charge,

$$H_3O^+ + OH^- \longrightarrow 2H_2O$$

is much faster than the decomposition of hydrogen peroxide, a process involving the rearrangement of molecules,

$$2H_2O_2 \longrightarrow 2H_2O + O_2$$

Even similar reactions have different rates under the same conditions if different reactants are involved. The active metals calcium and sodium both react with water to form hydrogen gas and the metal hydroxide. But calcium reacts at a moderate rate, while sodium reacts so rapidly that its reaction is almost explosive.

15.3 Rate of Reaction and the State of Subdivision of the Reactants

Except for those reactions involving substances in the gaseous state or in solution, reactions occur at the boundary, or interface, between two phases. The rate of a reaction between two phases depends to a great extent on the area of surface contact between them. Reactions involving solids occur on their surfaces. Finely divided solids, because of the greater surface area available, react more rapidly than do large pieces of the same substances. For example, large pieces of wood smolder slowly, smaller pieces burn rapidly, and grain dust (which, like wood, is composed of cellulose) may burn at an explosive rate (Fig. 15-3).

Figure 15-3. Grain dust (very finely divided grain) can burn at an explosive rate. If the dust is suspended in a confined space, such as a grain elevator, an explosion can result. A grain dust explosion in this storage elevator destroyed several silos, killed one worker, and injured several dozen others. (*Courtesy of United Press International, Inc.*)

15.4 Rate of Reaction and Temperature

Chemical reactions are accelerated by increases in temperature. The initial rate of decomposition of a 1 M hydrogen peroxide solution is 3.2×10^{-5} mol L^{-1} s^{-1} at 40°C and is slightly more than twice as fast, 7.2×10^{-5} mol L^{-1} s^{-1}, at 50°C. The oxidation of either iron or coal is very slow at ordinary temperatures but proceeds rapidly at high temperatures. Foods cook faster at higher temperatures than at lower ones. We use a burner, a hot plate, or a device called a heating mantle in the laboratory to increase the rates of reactions that proceed slowly at ordinary temperatures. In many cases the rate of a reaction in a homogeneous system is *approximately* doubled by an increase in temperature of only 10°C (Fig. 15-4). This rule, however, is a rough approximation and applies only to reactions that last longer than a second or two.

Figure 15-4. The effect of temperature on reaction rate, shown graphically. The rate of reaction is often approximately doubled for each rise in temperature of 10°C.

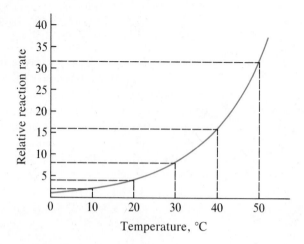

15.5 Rate of Reaction and Concentration; Rate Equations

At a fixed temperature the rate of a given reaction depends on the concentrations of the reactants. The rates at which limestone buildings deteriorate due to reaction with the air pollutant sulfur dioxide depend on the amount of sulfur dioxide in the air, for example. In a highly polluted atmosphere where the concentration is high, they crumble more rapidly than they would in less polluted air. The rate of decomposition of hydrogen peroxide (Table 15-1) decreases with decreasing concentration. The rate at any instant of time has been found to be directly proportional to the concentration of hydrogen peroxide at that time.

$$\text{Rate} = k[\text{H}_2\text{O}_2] \tag{1}$$

The proportionality constant, k, is called the **rate constant.** The brackets signify the molar concentration of the reactant. The rate constant is independent of the concentration of the reactant but does vary with temperature. At 40°C, k is $3.2 \times 10^{-5}\ \text{s}^{-1}$; at 50°C, $7.2 \times 10^{-5}\ \text{s}^{-1}$. The rate for the reaction of hydrochloric acid with sodium hydroxide is described by a similar expression:

$$\text{Rate} = k[\text{H}_3\text{O}^+][\text{OH}^-] \tag{2}$$

The value of k in this equation differs from that in Equation (1) since the rate constant is dependent on the nature of the reacting substances. The value of k in Equation (2) is $1.4 \times 10^{11}\ \text{L mol}^{-1}\ \text{s}^{-1}$.

Equations (1) and (2) are quantitative descriptions based on experimental measurements and are examples of **rate equations,** or **rate laws.** These equations give the relationship between reaction rate and concentrations of reactants.

In general, a rate equation has the form

$$\text{Rate} = k[\text{A}]^m[\text{B}]^n[\text{C}]^p \cdots$$

in which [A], [B], and [C] represent molar concentrations of reactants (or sometimes products or other substances), k is the rate constant for the particular reaction, and the exponents m, n, and p are usually positive integers (although fractions and negative numbers sometimes appear). *Both k and the exponents m, n, and p must be determined experimentally from the variation of the rate of a reaction as the concentrations of the reactants are varied.*

EXAMPLE 15.2   One of the reactions occurring in the ozone layer in the upper atmosphere is the combination of nitric oxide, NO, with ozone, O_3.

$$NO + O_3 \longrightarrow NO_2 + O_2$$

This reaction has been studied in the laboratory, and the following rate data obtained at 25°C:

[NO], mol L^{-1}	[O_3], mol L^{-1}	$\dfrac{\Delta[NO_2]}{\Delta t}$, mol L^{-1} s^{-1}
1.00×10^{-6}	3.00×10^{-6}	0.660×10^{-4}
1.00×10^{-6}	6.00×10^{-6}	1.32×10^{-4}
1.00×10^{-6}	9.00×10^{-6}	1.98×10^{-4}
2.00×10^{-6}	9.00×10^{-6}	3.96×10^{-4}
3.00×10^{-6}	9.00×10^{-6}	5.94×10^{-4}

Determine the rate equation for the reaction at 25°C.

The rate equation will have the form

$$Rate = k[NO]^m[O_3]^n$$

The values of m and n cannot be determined from the chemical equation, but must be found from the experimental data. In the first three lines in the data table, [NO] is held constant and [O_3] is allowed to vary. The reaction rate changes in direct proportion to the change in [O_3]. When [O_3] doubles, the rate doubles; when [O_3] increases by a factor of 3, the rate increases by a factor of 3. Thus the rate is directly proportional to [O_3], or

$$Rate = k'[O_3]$$

Hence n in the rate equation must be equal to 1. Writing [O_3] is equivalent to writing [O_3]^1.

In the last three lines in the data table, [NO] varies as [O_3] is held constant. When [NO] doubles the rate doubles, and when [NO] triples, the rate also triples. Thus the rate is also directly proportional to [NO] and m in the rate equation is also equal to 1.

The rate equation is thus

$$Rate = k[NO][O_3]$$

EXAMPLE 15.3   Acetaldehyde decomposes when heated to yield methane and carbon monoxide according to the equation

$$CH_3CHO(g) \longrightarrow CH_4(g) + CO(g)$$

Determine the rate equation for the reaction from the following experimental data:

[CH_3CHO], mol L^{-1}	$\dfrac{\Delta[CH_3CHO]}{\Delta t}$, mol L^{-1} s^{-1}
1.75×10^{-3}	2.06×10^{-11}
3.5×10^{-3}	8.25×10^{-11}
7.0×10^{-3}	3.30×10^{-10}

The rate equation will have the form

$$Rate = k[CH_3CHO]^n$$

From the experimental data we can see that when $[CH_3CHO]$ doubles, the rate increases by a factor of 4; when $[CH_3CHO]$ increases by a factor of 4, the rate increases by a factor of 16. Thus the rate is proportional to the square of the concentration, or

$$Rate = k[CH_3CHO]^2$$

It is not unusual to find that a rate equation does not reflect the overall stoichiometry of a chemical reaction. In some cases one or more of the reactants may not even appear in the rate equation.

15.6   Order of a Reaction

The order of a reaction is determined from the exponents found in its rate equation. The order can be given as the order with respect to a specific reactant or as the overall order of the reaction. **The order of a reaction with respect to one of the reactants is equal to the power to which the concentration of that reactant is raised.**

Consider a reaction for which the rate equation is

$$Rate = k[A]^m[B]^n$$

If the exponent m is 1, the reaction is said to be **first order** with respect to A. If m is 2, the reaction is said to be **second order** with respect to A. If n is 2, the reaction is said to be second order with respect to B. If a reaction is **zero order** with respect to a reactant, the rate of the reaction does not change as the concentration of that reactant changes. If this reaction were zero order with respect to B, n would be 0. Note that $[B]^0$ is equal to 1, no matter what the value of $[B]$.

The overall order of a reaction is the sum of the orders with respect to each reactant. If m is 1 and n is 1, the reaction is first order in A and first order in B. The overall order of the reaction is given by the sum of m and n; therefore the overall order of this particular reaction is second order.

EXAMPLE 15.4   Experiment shows that the reaction of nitrogen dioxide with carbon monoxide

$$NO_2 + CO \longrightarrow NO + CO_2$$

is second order in NO_2 and zero order in CO at 200°C. What is the rate equation for the reaction?

The rate equation will have the form

$$Rate = k[NO_2]^m[CO]^n$$

The reaction is second order in NO_2; thus $m = 2$. The reaction is zero order in CO; thus $n = 0$. The rate equation is

$$\text{Rate} = k[NO_2]^2[CO]^0 = k[NO_2]^2$$

Remember that a number raised to the zero power is equal to 1, so $[CO]^0 = 1$.

EXAMPLE 15.5   What are the orders with respect to each reactant and the overall order of the reaction

$$NO + O_3 \longrightarrow NO_2 + O_2$$

in Example 15.2, Section 15.5?

The rate equation for the reaction was found to be

$$\text{Rate} = k[NO][O_3]$$

Thus the reaction is first order with respect to both NO and O_3. The overall reaction is second order since the sum of the exponents is 2.

15.7 Half-Life of a Reaction

1. FIRST-ORDER REACTION.   An equation relating the rate constant k to the initial concentration $[A_0]$ and the concentration $[A]$ present after any given time t can be derived for a first-order reaction. The derivation requires more advanced mathematics than we have been using so we will not present it here, but the resulting equation is

$$\log \frac{[A_0]}{[A]} = \frac{kt}{2.303} \tag{1}$$

The following example demonstrates the use of this equation.

EXAMPLE 15.6   The rate constant for the first-order decomposition of cyclobutane, C_4H_8 (Section 15.11), at 500°C is $9.2 \times 10^{-3}\ s^{-1}$. How long will it take for 90.0% of a sample of 0.100 M C_4H_8 to decompose, that is, for the concentration of C_4H_8 to decrease to 0.0100 M?

The initial concentration of C_4H_8, $[A_0]$, is 0.100 mol L^{-1}; $[A]$, the concentration at time t, is 0.0100 mol L^{-1}; and k is $9.2 \times 10^{-3}\ s^{-1}$.

$$\log \frac{[A_0]}{[A]} = \frac{kt}{2.303}$$

$$t = \log \frac{[A_0]}{[A]} \times \frac{2.303}{k}$$

$$t = \log \left(\frac{0.100\ \text{mol L}^{-1}}{0.0100\ \text{mol L}^{-1}} \right) \times \frac{2.303}{9.2 \times 10^{-3}\ \text{sec}^{-1}} = 2.5 \times 10^2\ \text{sec}$$

The half-life of a reaction, $t_{1/2}$, is the time required for half of the original concentration of the limiting reactant to be consumed. In each succeeding half-life, half of the remaining concentration of the reactant will be used up. The decomposition of hydrogen peroxide illustrated in Fig. 15-1 displays the concentration after each of several successive half-lives. During the first half-life (from 0 seconds to 2.16×10^4 seconds), the concentration decreases from 1.0000 M to 0.5000 M. During the second half-life (from 2.16×10^4 seconds to 4.32×10^4 seconds), it decreases from 0.5000 M to 0.2500 M; during the third half-life, from 0.2500 M to 0.1250 M. The concentration decreases by half during each successive period of 2.16×10^4 seconds. The decomposition of hydrogen peroxide is a first-order reaction, and, as shown below, the half-life of a first-order reaction is independent of the concentrations of the reactants. However, half-lives of higher-order reactions are not independent of the concentrations of the reactants.

The equation for determining the half-life for a first-order reaction can be derived from Equation (1) as follows.

$$\log \frac{[A_0]}{[A]} = \frac{kt}{2.303}$$

$$t = \log \frac{[A_0]}{[A]} \times \frac{2.303}{k}$$

If the time t is the half-life, $t_{1/2}$, the concentration of the limiting reactant at the end of this time, $[A]$, is equal to $\frac{1}{2}$ of the initial concentration. Hence, at time $t_{1/2}$, $[A] = \frac{1}{2}[A_0]$.

Therefore

$$t_{1/2} = \log \frac{[A_0]}{\frac{1}{2}[A_0]} \times \frac{2.303}{k}$$

$$= \log 2 \times \frac{2.303}{k}$$

$$= 0.301 \times \frac{2.303}{k} = \frac{0.693}{k}$$

Thus

$$t_{1/2} = \frac{0.693}{k} \qquad (2)$$

You can see in Equation (2) that the half-life of a first-order reaction is inversely proportional to the rate constant k. Hence, a fast reaction with a short half-life has a large k; a slow reaction with a longer half-life has a smaller k.

EXAMPLE 15.7   Calculate the rate constant for the first-order decomposition of hydrogen peroxide in water at 40°C using the data from Fig. 15-1.

The half-life for the decomposition of H_2O_2 is 2.16×10^4 s.

$$t_{1/2} = \frac{0.693}{k}$$

$$k = \frac{0.693}{t_{1/2}} = \frac{0.693}{2.16 \times 10^4 \text{ s}} = 3.21 \times 10^{-5} \text{ s}^{-1}$$

2. SECOND-ORDER REACTION. Let us consider two types of second-order reactions: (1) those that are second order with respect to one reactant and (2) those that are first order with respect to each of two reactants *and* in which the initial concentration of each reactant is the same. In such cases the equation relating the rate constant k to the initial concentration $[A_0]$ of any of the reactants and to the concentration $[A]$ present after any given time t is

$$\frac{1}{[A]} - \frac{1}{[A_0]} = kt \tag{3}$$

EXAMPLE 15.8   The reaction of an organic ester (a compound with the general formula $RCOOR'$, where R and R' are alkyl groups; see Chapter 25) with a strong base is second order with a rate constant equal to $4.50 \text{ L mol}^{-1} \text{ min}^{-1}$. If the initial concentrations of ester and base are both $0.0200 \ M$ what is the concentration remaining after 10.0 min?

$$\frac{1}{[A]} - \frac{1}{[A_0]} = kt$$

$$[A_0] = 0.0200 \text{ mol L}^{-1}$$

$$k = 4.5 \text{ L mol}^{-1} \text{ min}^{-1}$$

$$t = 10.0 \text{ min}$$

Therefore, $\dfrac{1}{[A]} - \dfrac{1}{0.0200 \text{ mol L}^{-1}} = 4.50 \text{ L mol}^{-1} \text{ min}^{-1} \times 10.0 \text{ min}$

or $\dfrac{1}{[A]} = 4.50 \text{ L mol}^{-1} \text{ min}^{-1} \times 10.0 \text{ min} + \dfrac{1}{0.0200 \text{ mol L}^{-1}}$

$$= 45.0 \text{ L mol}^{-1} + \frac{1}{0.0200} \text{ mol}^{-1} \text{ L}$$

$$= 45.0 \text{ L mol}^{-1} + 50.0 \text{ L mol}^{-1}$$

$$= 95.0 \text{ L mol}^{-1}$$

Then $[A] = \dfrac{1}{95.0 \text{ L mol}^{-1}} = 0.0105 \text{ mol L}^{-1}$

Hence 0.0105 mol of both the ester and the strong base remains per liter at the end of 10.0 min, compared to the 0.0200 mol of each originally present.

The equation used for calculating the half-life of a second-order reaction in which the initial concentration of each reactant is the same can be derived from Equation (3) as follows.

$$\frac{1}{[A]} - \frac{1}{[A_0]} = kt$$

If $t = t_{1/2}$, then $[A] = \frac{1}{2}[A_0]$

Therefore,

$$\frac{1}{\frac{1}{2}[A_0]} - \frac{1}{[A_0]} = kt_{1/2}$$

$$\frac{1 - \frac{1}{2}}{\frac{1}{2}[A_0]} = kt_{1/2}$$

$$\frac{\frac{1}{2}}{\frac{1}{2}[A_0]} = kt_{1/2}$$

$$\frac{1}{[A_0]} = kt_{1/2}$$

$$t_{1/2} = \frac{1}{k[A_0]} \tag{4}$$

For a second-order reaction $t_{1/2}$ depends on the initial concentration, whereas it is independent of the initial concentration for a first-order reaction. Consequently, the use of the half-life concept is more complex and less useful for second-order reactions. For example, the rate constant of a second-order reaction cannot be calculated directly from the half-life, or vice versa, unless the initial concentration is known; this is not true for first-order reactions.

EXAMPLE 15.9   Calculate the half-life for the second-order reaction described in Example 15.8 ($k = 4.50 \text{ L mol}^{-1} \text{ min}^{-1}$) if the initial concentrations of both reactants are 0.0200 M.

The calculations in Example 15.8 showed that slightly more than half of each of the reactants (0.0105 mol L^{-1}) remains after 10.0 min. The half-life, therefore, must be a little bit longer than 10.0 min. This estimate provides us with a useful order of magnitude to check the calculated answer.

$$t_{1/2} = \frac{1}{k[A_0]} = \frac{1}{(4.50 \text{ L mol}^{-1} \text{ min}^{-1})(0.0200 \text{ mol L}^{-1})} = 11.1 \text{ min}$$

In accord with our estimate the half-life is a little longer than 10.0 min when the initial concentrations are 0.0200 M.

EXAMPLE 15.10   What is the half-life for the reaction in Examples 15.8 and 15.9 if the initial concentrations of both reactants are 0.0300 M?

$$t_{1/2} = \frac{1}{k[A_0]} = \frac{1}{(4.50 \text{ L mol}^{-1} \text{ min}^{-1})(0.0300 \text{ mol L}^{-1})} = 7.41 \text{ min}$$

Example 15.9 and this one clearly show the dependence of half-life on initial concentrations for a second-order reaction. The increase in the initial concentrations from 0.0200 M to 0.0300 M shortens the half-life for this particular reaction from 11.1 min to 7.41 min.

15.8   Collision Theory of the Reaction Rate

Chemists believe that before two atoms, molecules, or ions can react, they must collide with one another. In a few reactions every collision between reactants leads to products. The rate of such reactions is determined solely by how rapidly the reactants can diffuse together, and they are thus called **diffusion-controlled reactions.** Diffusion-controlled reactions are very fast; hence they have large rate constants. For a typical diffusion-controlled second-order gas-phase reaction at 25°C, such as the reaction of an oxygen atom with a nitrogen molecule

$$O + N_2 \longrightarrow NO + N$$

the rate constant falls in the range 10^{10}–10^{12} L mol^{-1} s^{-1}. The diffusion-controlled reaction between hydronium ions and hydroxide ions in water at 25°C

$$H_3O^+ + OH^- \longrightarrow 2H_2O$$

has a rate constant of 1.4×10^{11} L mol^{-1} s^{-1}. At these rates over 95% of the reactants would be consumed in 10^{-11} second.

Almost all other reactions occur at a much slower rate because only a very small fraction of collisions result in the formation of products. In the majority of collisions the reactants simply bounce away unchanged. For a collision to lead to a reaction, the following must occur:

1. The reacting species usually must have a particular orientation during the collision so that the atoms that are bonded together in the product come in contact.
2. The collision must occur with enough energy for the valence shells of the reacting species to penetrate into each other so the electrons can rearrange and form new bonds (and new chemical species).

The gas-phase reaction of nitrogen dioxide with carbon monoxide

$$NO_2 + CO \longrightarrow NO + CO_2$$

above 225°C illustrates the factors necessary for effective collision. During the course of the reaction, an oxygen atom is transferred from an NO_2 molecule to a CO molecule. There are many orientations of the NO_2 and CO molecules during collision that will not place an oxygen atom of the NO_2 molecule close to the carbon atom of the CO molecule. Three of these are indicated in Fig. 15-5. These collisions will not be effective in producing a chemical reaction. Only a collision in which an oxygen atom strikes the carbon atom [Fig. 15-5(d)] can produce a reaction.

Even if the orientation is correct, a collision may still not lead to reaction. As the oxygen atom of an NO_2 molecule approaches the carbon atom of a CO molecule, the electrons in the two molecules begin to repel each other. Unless the molecules possess a kinetic energy greater than a certain minimum value, the two

Figure 15-5. Some possible collisions between NO_2 and CO molecules. Only in (d) are the molecules correctly oriented for transfer of an oxygen atom from NO_2 to CO to give NO and CO_2.

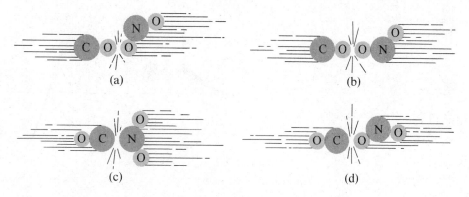

(a)

(b)

(c)

(d)

molecules will bounce away from each other before they can get close enough to react. If the molecules are moving fast enough that their kinetic energy exceeds this minimum, then the repulsion between their electrons is not strong enough to keep them apart, and the molecules can get close enough for a C—O bond to begin to form as the N—O bond begins to break. Using dots to represent partially formed or broken bonds, we can write the resulting species as follows:

$$NO_2 + CO \longrightarrow \left[O{=}N\cdots O \cdots C{\equiv}O \right]$$

The species $O{=}N\cdots O\cdots C{\equiv}O$, which contains the partially formed $C\cdots O$ and partially broken $N\cdots O$ bonds, is called the **activated complex,** or **transition state.** An activated complex is a combination of reacting molecules that is intermediate between reactants and products and in which some bonds have weakened and new bonds have begun to form. Ordinarily, an activated complex cannot be isolated. It breaks down to give either reactants or products, depending on the conditions under which the reaction takes place.

$$NO_2 + CO \longrightarrow \left[O{=}N\cdots O\cdots C{\equiv}O \right] \xrightarrow[\text{or}]{} \begin{array}{l} NO_2 + CO \\ NO + CO_2 \end{array}$$

The collision theory shows why reaction rates decrease as concentrations decrease. With decreased concentration of any of the reacting substances, the chances for collisions between molecules are decreased because fewer molecules are present per unit of volume. Fewer collisions mean a slower reaction rate. The concentration of a substance, when expressed in moles per liter, provides a direct measure of the number of molecules per unit of volume and is therefore appropriately used in rate equations (Sections 15.5 and 15.6).

15.9 Activation Energy and the Arrhenius Equation

The minimum energy necessary to form an activated complex during a collision between reactants is called the **activation energy,** E_a (Fig. 15-6). In a slow reaction the activation energy is much larger than the average energy content of the molecules. The energies of a large fraction of the molecules in a system are close to the average value. However, a few of the molecules, the fast-moving ones, have relatively high energies. The collisions between the fast-moving molecules are most

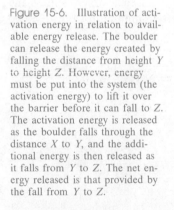

Figure 15-6. Illustration of activation energy in relation to available energy release. The boulder can release the energy created by falling the distance from height Y to height Z. However, energy must be put into the system (the activation energy) to lift it over the barrier before it can fall to Z. The activation energy is released as the boulder falls through the distance X to Y, and the additional energy is then released as it falls from Y to Z. The net energy released is that provided by the fall from Y to Z.

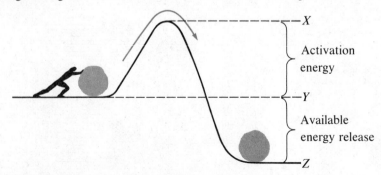

apt to result in reactions. In very fast reactions the fraction of molecules possessing the necessary activation energy is large, and most collisions between molecules result in reaction.

The energy relationships for the general reaction of a molecule of A with a molecule of B to form molecules of C and D are shown in Fig. 15-7.

$$A + B \rightleftharpoons C + D$$

The figure shows that after the activation energy, E_a, is exceeded, and as C and D begin to form, the system loses energy until its total energy is lower than that of the initial mixture. The forward reaction (that between molecules A and B) therefore tends to take place readily if sufficient energy is available in any one collision to exceed the activation energy, E_a. In Fig. 15-7, ΔE represents the difference in energy between the two molecules of reactants (A and B) and the two molecules of products (C and D). The sum of E_a and ΔE represents the activation energy for the reverse reaction

$$C + D \longrightarrow A + B$$

For a given reaction the rate constant is related to the activation energy by a relationship known as the **Arrhenius equation.**

$$k = A \times 10^{-E_a/2.303\,RT}$$

R is a constant with the value 8.314 J mol^{-1} K^{-1}, T is temperature on the Kelvin scale, E_a is the activation energy in joules per mole, and A is a constant called the **frequency factor,** which is related to the frequency of collisions and the orientation of the reacting molecules. A indicates how many collisions have the correct orientation to lead to products. The remainder of the equation, $10^{-E_a/2.303\,RT}$, gives the fraction of the collisions in which the energy of the reacting species is greater than E_a, the activation energy for the reaction.

The Arrhenius equation describes quantitatively much of what we have discussed about reaction rates. Reactions with similar frequency factors but different activation energies, E_a, will have different rate constants because the rate constants are proportional to $10^{-E_a/2.303\,RT}$. For two reactions at the same temperature, the reaction with the higher activation energy will have the lower rate constant and the slower rate. The larger value of E_a decreases the size of $10^{-E_a/2.303\,RT}$,

Figure 15-7. Potential energy relationships for the reaction $A + B \rightleftharpoons C + D$. The energy represented by the broken curve is that for the system with a molecule of A and a molecule of B present; the energy represented by the solid curve is that for the system with a molecule of C and a molecule of D present. The activation energy for the forward reaction is represented by E_a, the activation energy for the reverse reaction by $(E_a + \Delta E)$. The species present at the peak maximum corresponds to the transition state.

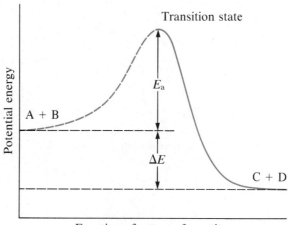

reflecting the smaller fraction of molecules with sufficient energy to react. Alternatively, the reaction with the smaller E_a will have a larger fraction of molecules with the necessary energy to react (Fig. 15-8). This will be reflected as a larger value of $10^{-E_a/2.303\,RT}$, a larger rate constant, and a faster rate for the reaction. An increase in temperature has the same effect as a decrease in activation energy. A larger fraction of molecules has the necessary energy to react (Fig. 15-9), as indicated by an increase in the value of $10^{-E_a/2.303\,RT}$. The rate constant is also directly proportional to the frequency factor, A. Hence a change in conditions or reactants that increases the fraction of collisions in which the orientation of the molecules is right for reaction results in an increase in A and, consequently, an increase in k.

Figure 15-8. As the activation energy of a reaction decreases, the number of molecules with at least this much energy increases, as shown by the colored areas.

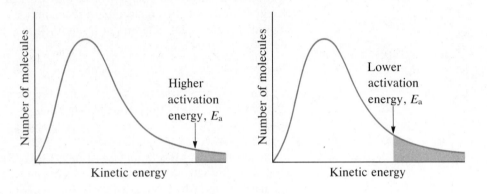

Figure 15-9. At a higher temperature, T_2, more molecules have an energy greater than E_a, as shown by the colored area.

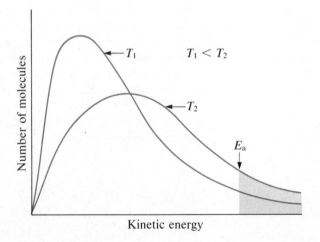

In order to determine E_a for a reaction, it is necessary to measure k at different temperatures and evaluate E_a from the Arrhenius equation. The Arrhenius equation may be rewritten as follows:

$$\log k = \log A - \frac{E_a}{2.303\,RT}$$

A plot of $\log k$ against $1/T$ gives a straight line whose slope is $-E_a/2.303\,R$, from which E_a may be determined.

EXAMPLE 15.11   The variation of the rate constant with temperature for the decomposition of $HI(g)$ to $H_2(g)$ and $I_2(g)$ is given in the table. What is the activation energy for the reaction?

T, K	$1/T$, K^{-1}	k, L mol^{-1} s^{-1}	$\log k$
555	1.80×10^{-3}	3.52×10^{-7}	-6.453
575	1.74×10^{-3}	1.22×10^{-6}	-5.913
645	1.55×10^{-3}	8.59×10^{-5}	-4.066
700	1.43×10^{-3}	1.16×10^{-3}	-2.936
781	1.28×10^{-3}	3.95×10^{-2}	-1.403

A graph of $\log k$ versus $1/T$ is given in Fig. 15-10. The slope of this line is given by the following expression:

$$\text{Slope} = \frac{\Delta(\log k)}{\Delta(1/T)}$$

$$= \frac{(-6.003) - (-1.633)}{(1.75 \times 10^{-3}\,K^{-1}) - (1.30 \times 10^{-3}\,K^{-1})}$$

$$= \frac{-4.370}{0.45 \times 10^{-3}\,K^{-1}} = -9.71 \times 10^3\,K$$

$$\text{Slope} = -\frac{E_a}{2.303\,R}$$

Thus

$$-E_a = \text{slope} \times 2.303\,R = -9.71 \times 10^3\,K \times 2.303 \times 8.314\,J\,mol^{-1}\,K^{-1}$$

$$E_a = 186\,kJ\,mol^{-1}$$

Figure 15-10. A graph of the linear relationship between $\log k$ and $1/T$ for the reaction $2HI \rightarrow H_2 + I_2$ according to the Arrhenius equation.

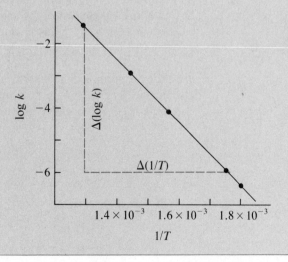

15.10 Elementary Reactions

A balanced equation for a chemical reaction indicates what is reacting and what is produced, but it says nothing about how the reaction actually takes place. The process, or pathway, by which a reaction occurs is called the **reaction mechanism,** or the **reaction path.**

Reactions often occur in steps. The decomposition of ozone, for example, is believed to follow a mechanism with two steps.

$$O_3 \longrightarrow O_2 + O \tag{1}$$

$$O + O_3 \longrightarrow 2O_2 \tag{2}$$

The two steps add up to the overall reaction for the decomposition,

$$2O_3 \longrightarrow 3O_2$$

The oxygen atom produced in the first step is used in the second and thus does not appear as a final product. Species that are produced in one step and consumed in another are called **intermediates.**

Each of the steps in a reaction mechanism is called an **elementary reaction.** Elementary reactions occur exactly as they are written and cannot be broken down into simpler steps. An overall reaction can often be broken down into a number of steps. Thus, although the overall reaction indicates that two molecules of ozone react to give three molecules of oxygen, the reaction path does not involve the collision of two ozone molecules. Instead, a molecule of ozone decomposes to an oxygen molecule and an intermediate oxygen atom, then the oxygen atom reacts with a second ozone molecule to give two oxygen molecules. These two elementary reactions occur exactly as they are written in Equations (1) and (2).

15.11   Unimolecular Reactions

An elementary reaction is **unimolecular** if the rearrangement of a single molecule (or ion) produces one or more molecules of product. A unimolecular reaction may be one of several elementary reactions in a complex mechanism, or it may be the only reaction in a mechanism. Equation (1) in Section 15.10,

$$O_3 \longrightarrow O_2 + O$$

illustrates a unimolecular elementary reaction occurring in a two-step reaction mechanism. The gas-phase decompositions of dinitrogen pentaoxide, N_2O_5, and cyclobutane, C_4H_8, occur with one-step unimolecular mechanisms.

All that is required for each of these three reactions to occur is the separation of the single reactant molecule into two parts. The latter two reactions also show that an overall reaction may be an elementary reaction.

Chemical bonds do not simply fall apart during chemical reactions. The decomposition of C_4H_8, for example, requires the input of 261 kilojoules of energy per mole to distort the molecules into activated complexes that can decompose into products. (Thus the activation energy for this reaction is 261 kilojoules per mole of C_4H_8 reacting.)

In a sample of C_4H_8, very few molecules contain enough energy to form activated complexes. Those that do, react by the process indicated above. Others collide with rapidly moving molecules and pick up additional energy. (The kinetic energy of the rapidly moving molecules is converted into activation energy in the other molecules.) If this energy is sufficient to form the activated complex, then the activated molecules too can undergo reaction. In effect, a particularly energetic collision will knock the reacting molecules into the geometry of the activated complex. However, only a small fraction of gas molecules travel at sufficiently high speeds with large kinetic energies (Fig. 15-8). Hence at any one time only a few molecules pick up enough energy from collisions to climb the activation energy barrier and react.

Doubling the concentration of C_4H_8 molecules in a sample gives twice as many molecules per liter. Although the fraction of molecules with enough energy to react will be the same, the total number of such molecules will be twice as great. Consequently, the change in the amount of C_4H_8 per liter and thus the reaction rate ($-\Delta[C_4H_8]/\Delta t$) will be twice as great. The reaction rate is directly proportional to the concentration of C_4H_8.

$$Rate = k[C_4H_8]$$

This relationship applies to any *unimolecular elementary reaction;* the reaction rate is directly proportional to the concentration of the reactant, and the reaction exhibits first-order behavior. The proportionality constant is the rate constant for the particular unimolecular reaction.

15.12 Bimolecular Reactions

The collision *and combination* of two reactants to give an activated complex in an elementary reaction is called a **bimolecular reaction.** Equation (2) in Section 15.10,

$$O + O_3 \longrightarrow 2O_2$$

is an example of a bimolecular elementary reaction that occurs in a two-step reaction mechanism. The reaction of nitrogen dioxide with carbon monoxide (Section 15.8) and the decomposition of two hydrogen iodide molecules to give hydrogen, H_2, and iodine, I_2 (Fig. 15-11), are examples of reactions whose mechanisms consist of a single bimolecular elementary reaction.

Figure 15-11. Probable mechanism for the dissociation of two HI molecules to produce one molecule of H_2 and one molecule of I_2.

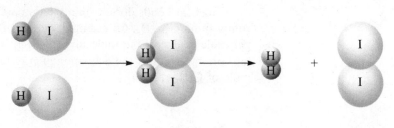

Two HI molecules   Transition state   Hydrogen molecule   Iodine molecule

For the bimolecular elementary reaction

$$A + B \longrightarrow \text{products}$$

the rate equation is first order in A and first order in B.

$$\text{Rate} = k[A][B]$$

Doubling the concentration of either A or B doubles the total number of molecular collisions and the number of effective collisions between molecules A and B. Thus if the concentration of A is doubled (and that of B is left the same), the rate of reaction is doubled. Doubling the concentrations of both A and B quadruples the number of collisions and the rate of reaction (Fig. 15-12). If the initial concentrations of both A and B are tripled, then the reaction proceeds nine times as fast. For the bimolecular elementary reaction

$$A + A \longrightarrow \text{products}$$

the rate equation is second order in A.

$$\text{Rate} = k[A][A] = k[A]^2$$

Figure 15-12. A schematic representation of the effect of concentration on the number of possible collisions and hence the rate of reaction.

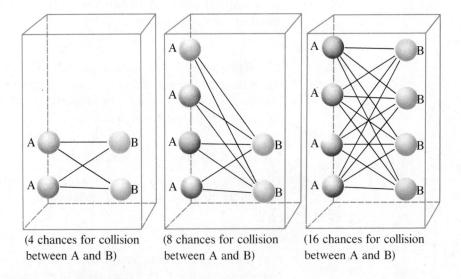

(4 chances for collision between A and B)   (8 chances for collision between A and B)   (16 chances for collision between A and B)

15.13   Termolecular Reactions

An elementary **termolecular reaction** involves the simultaneous collision of any combination of three atoms, molecules, or ions. Termolecular elementary reactions are uncommon because the probability of three particles colliding simultaneously is less than a thousandth of that of two particles colliding. There are, however, a few established termolecular elementary reactions. The reaction of nitric oxide with oxygen,

$$2NO + O_2 \longrightarrow 2NO_2$$

the reaction of nitric oxide with chlorine,

$$2NO + Cl_2 \longrightarrow 2NOCl$$

and the reaction of hydrogen with iodine

$$H_2 + I_2 \longrightarrow 2HI$$

all involve termolecular steps.

The reaction of hydrogen with iodine might appear to involve a bimolecular reaction of H_2 with I_2. In fact, two iodine atoms, produced by the dissociation of an iodine molecule, $(I_2 \rightarrow 2I)$, collide with a hydrogen molecule to give a transition state, which then splits to give two hydrogen iodide molecules (Fig. 15-13).

For the termolecular elementary reaction

$$A + B + C \longrightarrow products$$

the rate equation is first order in A, B, and C.

$$Rate = k[A][B][C]$$

For the termolecular elementary reaction

$$2A + B \longrightarrow products$$

the rate equation is second order in A and first order in B.

$$Rate = k[A]^2[B]$$

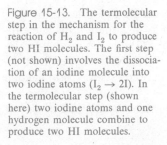

Figure 15-13. The termolecular step in the mechanism for the reaction of H_2 and I_2 to produce two HI molecules. The first step (not shown) involves the dissociation of an iodine molecule into two iodine atoms $(I_2 \rightarrow 2I)$. In the termolecular step (shown here) two iodine atoms and one hydrogen molecule combine to produce two HI molecules.

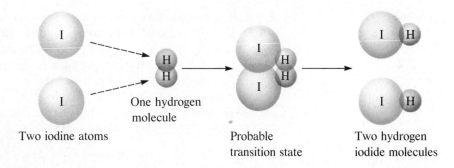

Two iodine atoms   One hydrogen molecule   Probable transition state   Two hydrogen iodide molecules

15.14 Reaction Mechanisms

The stepwise sequence of elementary reactions that converts reactants into products is called the reaction mechanism, or reaction path (Section 15.10). The decomposition of C_4H_8 $(C_4H_8 \rightarrow 2C_2H_4)$, for example, has a one-step mechanism (Section 15.11); the decomposition of ozone $(2O_3 \rightarrow 3O_2)$, a two-step mechanism. Since elementary reactions involving three or more reactants are rare, it is reasonable to expect that complex reactions take place in several steps. For a reaction such as the following one to take place in one step,

$$2MnO_4^- + 10Cl^- + 16H_3O^+ \longrightarrow 2Mn^{2+} + 5Cl_2 + 24H_2O$$

2 permanganate ions (MnO_4^-), 10 chloride ions, and 16 hydronium ions would have to collide simultaneously in an effective collision. This, of course, is very unlikely.

Even an apparently simple reaction such as the formation of ethylene, C_2H_4, from ethane, C_2H_6, may proceed by a complex mechanism. Ethylene is one of the most important reagents used by the chemical industry for manufacture of polyethylene, polyester fiber, synthetic rubber, and some detergents. Ethylene may be prepared by the pyrolysis (decomposition by heating) of ethane, a component of natural gas, at temperatures of 500°–800°C.

$$C_2H_6 \longrightarrow C_2H_4 + H_2$$

The actual mechanism is much more complex than suggested by this equation and involves the following elementary reactions:

1. *Initiation.* The reaction begins with a unimolecular reaction in which a C_2H_6 molecule splits at its weakest point, the C—C bond.

$$C_2H_6 \overset{\triangle}{\longrightarrow} 2CH_3 \qquad (1)$$

The CH_3 group abstracts a hydrogen atom from another C_2H_6 molecule in a bimolecular reaction.

$$CH_3 + C_2H_6 \longrightarrow CH_4 + C_2H_5 \qquad (2)$$

2. *Propagation.* The C_2H_5 group undergoes a unimolecular dissociation.

$$C_2H_5 \longrightarrow C_2H_4 + H \qquad (3)$$

The hydrogen atom produced in Equation (3) reacts with another C_2H_6 molecule in a bimolecular reaction.

$$H + C_2H_6 \longrightarrow C_2H_5 + H_2 \qquad (4)$$

The C_2H_5 group produced in Equation (4) reacts according to Equation (3), producing another hydrogen atom to undergo reaction as in Equation (4) and produce another C_2H_5. The reactions given by Equations (3) and (4) produce C_2H_4 and H_2, which are the principal products of the pyrolysis. Normally, each initiation [Equations (1) and (2)] is followed by about a hundred cycles of propagation [Equations (3) and (4)] before termination.

3. *Termination.* The chain of propagation reactions may be stopped by one of the following bimolecular reactions:

$$2C_2H_5 \longrightarrow C_4H_{10}$$
$$2C_2H_5 \longrightarrow C_2H_4 + C_2H_6$$

The C_2H_5 produced in Equation (4) can react according to Equation (3) to produce another hydrogen atom for Equation (4). This reinitiates the series of elementary reactions represented by Equations (3) and (4), and they repeat themselves over and over. Such a mechanism, involving repeating reactions, is known as a **chain mechanism.**

Some of the elementary reactions in a reaction path are relatively slow. The slowest reaction step determines the maximum rate, since a reaction can proceed no faster than its slowest step. The slowest step, therefore, is the **rate-determining step** of the reaction.

We can write the rate equation for each elementary reaction in a reaction mechanism (Sections 15.11–15.13), *but we cannot ordinarily write a correct rate equation or establish the reaction order for an overall reaction that involves several steps simply by inspection of the overall balanced equation.* We must determine the overall rate equation from experimental data. The reaction of nitrogen dioxide and carbon monoxide (Example 15.4 and Section 15.8) is an excellent example.

$$NO_2 + CO \longrightarrow CO_2 + NO$$

For temperatures above 225°C the reaction proceeds by a single one-step bimolecular elementary reaction (Section 15.12). For temperatures above 225°C, therefore, the reaction is first order with respect to NO_2 and first order with respect to CO. The rate equation is

$$\text{Rate} = k[NO_2][CO]$$

The overall order of the reaction is second order.

At temperatures below 225°C the reaction proceeds by two elementary reactions, the first of which is slow and is therefore the rate-determining step.

$$NO_2 + NO_2 \longrightarrow NO_3 + NO \qquad \text{(slow)}$$
$$NO_3 + CO \longrightarrow NO_2 + CO_2 \qquad \text{(fast)}$$

(A summation of the two equations represents the net overall reaction.) At lower temperatures the rate equation, based on the first step as the rate-determining step, is

$$\text{Rate} = k[NO_2]^2$$

Note that the rate equation for the reaction at lower temperatures does not even include the concentration of CO. At temperatures below 225°C, therefore, the reaction is second order with respect to NO_2 and zero order with respect to CO; the overall order of the reaction is still second order.

In general, when the rate-determining step is the first step, it provides the rate equation for the overall mechanism. However, when the rate-determining step occurs later in the mechanism, the rate equation may be complex. The oxidation of iodide ion by hydrogen peroxide illustrates this point.

$$H_2O_2 + 3I^- + 2H^+ \longrightarrow 2H_2O + I_3^-$$

In a solution with a high concentration of acid, one reaction pathway has the following rate equation:

$$\text{Rate} = k[H_2O_2][I^-][H_3O^+]$$

This rate equation is consistent with several mechanisms, three of which follow:

Mechanism A

$$H_2O_2 + H_3O^+ + I^- \longrightarrow 2H_2O + HOI \qquad \text{(slow)}$$
$$HOI + H_3O^+ + I^- \longrightarrow 2H_2O + I_2 \qquad \text{(fast)}$$
$$I_2 + I^- \longrightarrow I_3^- \qquad \text{(fast)}$$

Mechanism B

$$H_3O^+ + I^- \longrightarrow HI + H_2O \qquad \text{(fast)}$$
$$H_2O_2 + HI \longrightarrow H_2O + HOI \qquad \text{(slow)}$$
$$HOI + H_3O^+ + I^- \longrightarrow 2H_2O + I_2 \qquad \text{(fast)}$$
$$I_2 + I^- \longrightarrow I_3^- \qquad \text{(fast)}$$

Mechanism C

$$H_3O^+ + H_2O_2 \longrightarrow H_3O_2^+ + H_2O \qquad \text{(fast)}$$
$$H_3O_2^+ + I^- \longrightarrow H_2O + HOI \qquad \text{(slow)}$$
$$HOI + H_3O^+ + I^- \longrightarrow 2H_2O + I_2 \qquad \text{(fast)}$$
$$I_2 + I^- \longrightarrow I_3^- \qquad \text{(fast)}$$

In Mechanism A the slow step is the first elementary reaction, so the rate equation for the overall reaction is equal to that for this step. In Mechanisms B and C the rate-determining step is the second elementary reaction, and the overall rate equation is not simply the rate equation for this step. To derive the rate equations for Mechanisms B and C requires some familiarity with equilibrium constants, which will be introduced later in this chapter. However, since Mechanisms A, B, and C all have the same overall rate equation, it is not possible to distinguish among them only on this basis. Additional experimental information is required to distinguish which reaction pathway actually leads to product. Moreover, the rate equation provides no information about what happens after the rate-determining step. The subsequent steps must be worked out from other chemical knowledge or from other measurements.

The determination of the mechanism of a reaction is important in selecting conditions that provide a good yield of the desired product. Knowing a reaction's mechanism sometimes helps a chemist to prepare a previously unknown compound. Compared to the number of known chemical reactions, rather few reaction mechanisms have been completely characterized. The study of reaction mechanisms, or the **kinetics of reaction,** is a very active research area.

15.15 Catalysts

The rate of many reactions can be accelerated by **catalysts** (Section 9.2), substances that are not themselves used up by the reaction. Such substances may be divided into two general classes: **homogeneous catalysts** and **heterogeneous catalysts.**

1. HOMOGENEOUS CATALYSTS. A homogeneous catalyst is present in the same phase as the reactants. It interacts with a reactant forming an intermediate substance, which then decomposes or reacts with another reactant to regenerate the original catalyst and give product.

The ozone in the stratosphere (the upper atmosphere) that protects the earth from ultraviolet radiation is formed by the following mechanism when ultraviolet light (hv) interacts with oxygen molecules.

$$O_2(g) \xrightarrow{hv} 2O(g) \qquad (1)$$
$$O(g) + O_2(g) \longrightarrow O_3(g) \qquad (2)$$

As shown in Section 15.10, ozone decomposes by the following mechanism.

$$O_3(g) \longrightarrow O_2(g) + O(g) \qquad (3)$$
$$O(g) + O_3(g) \longrightarrow 2O_2(g) \qquad (4)$$

The rate of the decomposition of ozone is influenced by the presence of nitric oxide, NO, which acts as a catalyst in the decomposition.

$$NO(g) + O_3(g) \longrightarrow NO_2(g) + O_2(g) \qquad (5)$$

$$O_3(g) \longrightarrow O_2(g) + O(g) \qquad (6)$$

$$NO_2(g) + O(g) \longrightarrow NO(g) + O_2(g) \qquad (7)$$

The overall chemical change for Equations (5) through (7) is the same as for Equations (3) and (4):

$$2O_3 \longrightarrow 3O_2$$

Since the nitric oxide is not permanently used up in these reactions, it is a catalyst. The rate of the decomposition of ozone is greater in the presence of nitric oxide because of its catalytic activity. Under normal conditions the rate at which ozone forms in the upper atmosphere according to Equations (1) and (2) is equal to the rate at which it decomposes according to Equations (3) and (4) and (5) through (7). A state of equilibrium exists, and thus the total concentration of ozone does not change significantly over long periods. However, if additional nitric oxide were introduced into the stratosphere by atmospheric nuclear explosions or from the exhaust of high-flying supersonic aircraft, the rate of decomposition would increase and the total amount of ozone present would decrease. (Certain compounds that contain chlorine also catalyze the decomposition of ozone.)

Many biological reactions are catalyzed by enzymes, complex substances produced by living organisms. A pure solution of sugar in water will not react with oxygen, even if it stands for years. However, the enzyme zymase, produced by yeast cells, catalyzes the reaction between sugar and oxygen to form alcohol in a matter of hours. A part of the body's energy is produced by the oxidation of sugar in a multistep process that is catalyzed by enzymes. Many other reactions in living organisms are also catalyzed by enzymes. There may be as many as 30,000 different enzymes in the human body, each of which is a protein constructed to be effective as a catalyst for a specific chemical reaction useful to the body. The subject of enzymes and their kinetic activity will be examined in greater detail in Chapter 26.

2. HETEROGENEOUS CATALYSTS. Heterogeneous catalysts act by furnishing a surface at which a reaction can occur. Gas-phase and liquid-phase reactions catalyzed by heterogeneous catalysts typically occur on the surface of the catalyst rather than within the gas or liquid phase. For this reason heterogeneous catalysts are sometimes called *contact catalysts*.

Heterogeneous catalysis has at least four steps: (1) adsorption of the reactant onto the surface of the catalyst, (2) activation of the adsorbed reactant, (3) reaction of the adsorbed reactant, and (4) diffusion of the product from the surface into the gas or liquid phase (desorption). Any one of these steps may be slow and thus be the rate-determining step. In general, however, the overall rate of the reaction is faster than the rate would be if the reactants were in the gas or liquid phase. The steps that are believed to occur in the reaction of compounds containing a carbon-carbon double bond with hydrogen on a nickel catalyst are illustrated in Fig. 15-14. This is the catalyst used in the hydrogenation of edible fats and oils containing several carbon-carbon double bonds (polyunsaturated fats

Figure 15-14. Steps in the catalysis by nickel of the reaction $C_2H_4 + H_2 \rightarrow C_2H_6$. Both molecules are adsorbed by weak attractive forces. Activation occurs when the bonding electrons in the molecules rearrange to form bonds to metal atoms. Following the reaction of the activated atoms, the weakly adsorbed C_2H_6 molecule escapes from the surface.

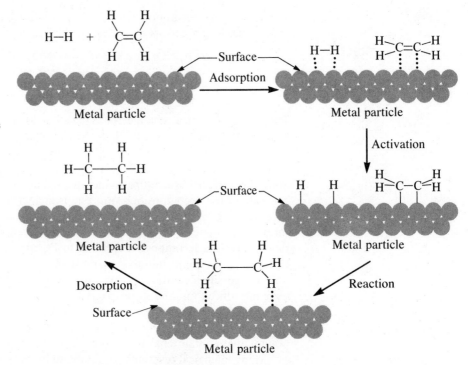

and oils) to produce edible fats and oils containing only single carbon-carbon bonds (saturated fats and oils).

Other significant industrial processes involving the use of contact catalysts include the preparation of sulfuric acid (Section 8.7, Part 1), the preparation of ammonia (Section 8.7, Part 2), the oxidation of ammonia to nitric acid (Section 23.13), and the synthesis of methanol, CH_3OH (Section 25.19).

Heterogeneous catalysts are also used in the catalytic converters found on most gasoline-powered automobiles. The exhaust gases are mixed with air and passed over a mixture of metal oxides. The catalyst promotes oxidation of carbon monoxide and unburned hydrocarbons to relatively harmless carbon dioxide and water. The newer catalysts also promote the decomposition of nitrogen oxides to nitrogen and oxygen. However, these catalysts also catalyze the oxidation of SO_2 to SO_3, which reacts with water vapor and forms a sulfuric acid mist. Since sulfur-containing fuels produce SO_2, these fuels must be avoided.

Both homogeneous and heterogeneous catalysts function by providing a reaction path with a lower activation energy than would be found in the absence of the catalyst (Fig. 15-15). This lower activation energy results in an increase in rate (Section 15.9). Note that a catalyst decreases the activation energy for both the forward and the reverse reactions and hence *accelerates both the forward and the reverse reactions.*

Some substances, called **inhibitors,** decrease the rate of a chemical reaction. In many cases these are substances that react with and "poison" some catalyst in the system and thereby prevent its action. Many biological poisons are inhibitors that reduce the catalytic activity of an organism's enzymes. Catalytic converters are poisoned by lead, so lead-free fuels must be used in automobiles equipped with such converters.

Figure 15-15. Potential energy diagram showing the effect of a catalyst on the activation energy of a reaction.

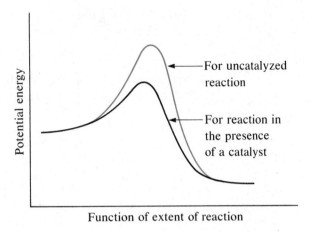

CHEMICAL EQUILIBRIUM

15.16 The State of Equilibrium

Equilibrium is reached in a system when the conversion of reactants into products and the conversion of products back into reactants occur simultaneously at the same rate. Consequently, at equilibrium there is no change in the amounts of reactants and products present in a system. For example, we have defined the vapor pressure of a liquid (Section 11.2) as the pressure exerted by the vapor in equilibrium with its liquid. The conversions in this case are evaporation and condensation, and at equilibrium there is no change in the quantity of liquid or of vapor.

The point at which equilibrium occurs in a reversible reaction varies with the conditions under which the reaction takes place. Knowing the effect of a change in conditions on an equilibrium enables chemists to select conditions and to control the relative amounts of substances present at equilibrium. For example, it was known for many years that nitrogen and hydrogen react to form ammonia.

$$N_2 + 3H_2 \rightleftharpoons 2NH_3$$

However, ammonia was manufactured in useful quantities by this reaction only after the factors that influence its equilibrium were understood (Section 23.4).

15.17 Law of Mass Action

Whenever the products of a chemical reaction can react to reform the reactants, the two reactions occur simultaneously and make up a reversible reaction (Section 8.2, Part 6). A reversible reaction does not go to completion as long as the products do not escape, the reactants are not completely converted into products, and a state of equilibrium is attained. Most chemical reactions that occur in a closed system are reversible and do not go to completion.

Let us consider the oxidation of carbon monoxide by nitrogen dioxide in a closed flask at 300°C.

$$NO_2 + CO \rightleftharpoons NO + CO_2$$

At the beginning of the reaction the flask contains only NO_2 and CO. Above 225°C the rate of the forward reaction (Section 15.14) is given by the expression

$$Rate_1 = k_1[NO_2][CO]$$

At first no molecules of NO and CO_2 are present, so there can be no reverse reaction to reform NO_2 and CO. The rate of the reverse reaction, $Rate_2$, is zero (see Fig. 15-16). However, as soon as some of the products (NO and CO_2) are formed, they begin to react. Initially the rate is relatively slow because their concentrations are low. However, as the forward reaction between NO_2 and CO proceeds, the concentrations of NO and CO_2 increase, and the rate of the reverse reaction, $Rate_2$, increases. The rate of the reverse reaction above 225°C is given by

$$Rate_2 = k_2[NO][CO_2]$$

Meanwhile, the concentrations of NO_2 and CO decrease, so the rate of the forward reaction, $Rate_1$, falls off. Consequently, the two reaction rates approach each other and finally become equal. When $Rate_2$ equals $Rate_1$, a condition of **dynamic equilibrium** is established. The opposing reactions are in full operation but at the same rate. Thus at equilibrium we may write

$$Rate_2 = Rate_1$$

$$k_2[NO][CO_2] = k_1[NO_2][CO]$$

or, by rearranging,

$$\frac{k_1}{k_2} = \frac{[NO][CO_2]}{[NO_2][CO]}$$

Because k_1 and k_2 are constants, the ratio k_1/k_2 is also a constant, and the expression may be written

$$K = \frac{[NO][CO_2]}{[NO_2][CO]}$$

K is called the **equilibrium constant** for the reaction. Just as the rate constants k_1 and k_2 are specific for each reaction at a definite temperature, K is a constant

Figure 15-16. Rates of reaction of forward and reverse reactions for $NO_2 + CO \rightleftharpoons NO + CO_2$, assuming only NO_2 and CO are present initially.

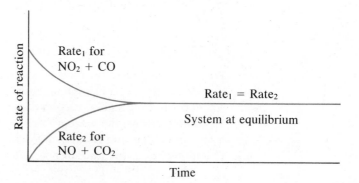

specific to the system in equilibrium at a given temperature. The values for the molar concentrations used to evaluate K from the above expression must always be the concentrations present after the reaction has reached equilibrium.

It is important to remember that while the rates of reaction, Rate_1 and Rate_2, are equal at equilibrium, the molar concentrations of the reactants and products in the equilibrium mixture are usually not equal. However, the individual concentration of each reactant and product remains constant *at equilibrium* because the rate at which any one reactant is being used up in one reaction is equal to the rate at which it is being formed by the opposite reaction.

A general equation for a chemical reaction may be written

$$m\text{A} + n\text{B} + \cdots \rightleftharpoons x\text{C} + y\text{D} + \cdots$$

When this reaction has reached equilibrium, the ratio

$$\frac{[\text{C}]^x[\text{D}]^y \cdots}{[\text{A}]^m[\text{B}]^n \cdots}$$

called the **reaction quotient,** is constant; that is,

$$\frac{[\text{C}]^x[\text{D}]^y \cdots}{[\text{A}]^m[\text{B}]^n \cdots} = K$$

This is a mathematical expression of the law of chemical equilibrium, or the **law of mass action: When a reversible reaction has attained equilibrium at a given temperature, the reaction quotient (the product of the molar concentrations of the substances to the right of the arrow divided by the product of the molar concentrations of the substances to the left, with each concentration raised to a power equal to the number of moles of that substance appearing in the equation) is a constant.**

The law of mass action is more precisely expressed in terms of the activities of the reactants and products (Section 13.27) rather than their concentrations. However, as we have seen, the activity of a dilute solute is closely approximated by its molar concentration, so concentrations are commonly used in equilibrium constants for dissolved species. The activity of a gas is approximated by its pressure (in atmospheres), so pressures are commonly used in equilibrium constants. Also, since the molar concentration of a gas is directly proportional to its pressure, molar concentrations are sometimes used in equilibrium constants. The activity of a pure solid, a pure liquid, or a solvent is 1.

The mathematical expression of the law of mass action means that, regardless of how we might change individual concentrations and temporarily upset an equilibrium, the composition of the system will always adjust itself to a new condition of equilibrium for which the reaction quotient, $[\text{C}]^x[\text{D}]^y \cdots / [\text{A}]^m[\text{B}]^n \cdots$, will again have the value K, provided that the temperature does not change. When a mixture of A, B, C, and D is prepared in such proportions that the reaction quotient is not equal to K, then the system is not in equilibrium, and its composition will change until equilibrium is established. If the reaction quotient is less than K, the rate of reaction of A with B will be greater than that of C with D, so that A and B will be used up faster than they are formed until the reaction quotient becomes equal to K. Conversely, if the reaction quotient is greater than K, the rate of reaction of C with D will be greater than that of A with B until the reaction quotient becomes equal to K.

The value of an equilibrium constant is a measure of the completeness of a reversible reaction. A large value for K indicates that equilibrium is attained only after the reactants A and B have been largely converted into the products C and D. When K is very small—much less than 1—equilibrium is attained when only small amounts of A and B have been converted to C and D.

EXAMPLE 15.12   Write the equilibrium constant expressions for the reactions $3O_2 \rightleftharpoons 2O_3$ and $3H_2 + N_2 \rightleftharpoons 2NH_3$.

$$3O_2 \rightleftharpoons 2O_3 \qquad K = \frac{[O_3]^2}{[O_2]^3}$$

$$3H_2 + N_2 \rightleftharpoons 2NH_3 \qquad K = \frac{[NH_3]^2}{[H_2]^3[N_2]}$$

The use of square brackets indicates that molar concentrations (units of mol L^{-1}) are used in these equations.

Using the law of mass action we can determine the concentrations of reactants and products at equilibrium.

EXAMPLE 15.13   At 2000°C the equilibrium constant for the reaction

$$N_2(g) + O_2(g) \rightleftharpoons 2NO(g)$$

is 4.1×10^{-4}. What is the concentration of $NO(g)$ in a mixture of $NO(g)$, $N_2(g)$, and $O_2(g)$ at equilibrium with $[N_2] = 0.036$ mol L^{-1} and $[O_2] = 0.0089$ mol L^{-1}?

At equilibrium the reaction quotient $[NO]^2/[N_2][O_2]$ is equal to the equilibrium constant. Thus we have

$$K = \frac{[NO]^2}{[N_2][O_2]}$$

Since we know K, $[N_2]$, and $[O_2]$, we can solve for $[NO]^2$ by rearranging the equation.

$$[NO]^2 = K[N_2][O_2]$$
$$[NO] = \sqrt{K[N_2][O_2]}$$
$$= \sqrt{4.1 \times 10^{-4} \times 0.036 \text{ mol L}^{-1} \times 0.0089 \text{ mol L}^{-1}}$$
$$= \sqrt{1.31 \times 10^{-7} (\text{mol L}^{-1})^2}$$
$$= 3.6 \times 10^{-4} \text{ mol L}^{-1}$$

Thus the concentration of $NO(g)$ is 3.6×10^{-4} mol L^{-1} at equilibrium under these conditions.

EXAMPLE 15.14   Acetic acid, CH_3CO_2H, reacts with ethanol, C_2H_5OH, to form water and ethyl acetate, $CH_3CO_2C_2H_5$, the solvent responsible for the odor in some nail polish removers.

$$CH_3CO_2H + C_2H_5OH \rightleftharpoons CH_3CO_2C_2H_5 + H_2O$$

The equilibrium constant for this reaction (when run in dioxane as a solvent) is 4.0. What mass of ethyl acetate is formed by the reaction of 0.10 mol of CH_3CO_2H and 0.15 mol of C_2H_5OH in enough dioxane to make 1 L of solution?

From the law of mass action, we know that at equilibrium

$$\frac{[CH_3CO_2C_2H_5][H_2O]}{[CH_3CO_2H][C_2H_5OH]} = 4.0$$

We need to solve this equation for the concentration of $CH_3CO_2C_2H_5$ at equilibrium.

When we are not given equilibrium concentrations, it is sometimes helpful to set up a table of initial concentrations (the concentrations of reactants before any reaction occurs) and equilibrium concentrations (the concentrations of reactants and products after equilibrium has been established). For this problem we have the following table. Entries in color are given in the problem statement. Entries in black are calculated as described below.

	$[CH_3CO_2H]$	$[C_2H_5OH]$	$[CH_3CO_2C_2H_5]$	$[H_2O]$
Initial concentrations, M	0.10	0.15	0	0
Equilibrium concentrations, M	$0.10 - x$	$0.15 - x$	x	x

The problem tells us the initial concentrations of CH_3CO_2H and C_2H_5OH, the concentrations before any reaction occurs. The concentration of the products, $CH_3CO_2C_2H_5$ and H_2O, is zero before the reaction starts.

Now let us consider the situation after equilibrium has been reached. We do not know the concentration of $CH_3CO_2C_2H_5$ at equilibrium, so we will set this concentration equal to x. From the chemical equation we can see that one mole of H_2O is formed for each mole of $CH_3CO_2C_2H_5$ produced, so the concentration of H_2O at equilibrium can also be set equal to x. From the equation we can see that one mole of CH_3CO_2H (or of C_2H_5OH) reacts for each mole of $CH_3CO_2C_2H_5$ that forms. Thus if x mol of $CH_3CO_2C_2H_5$ appears as product, x mol of CH_3CO_2H is consumed, and $(0.10 - x)$ mol remains. Also, $(0.15 - x)$ mol of C_2H_5OH remains at equilibrium.

If we substitute the equilibrium concentrations into the expression of the law of mass action, we have

$$K = \frac{[CH_3CO_2C_2H_5][H_2O]}{[CH_3CO_2H][C_2H_5OH]}$$

$$4.0 = \frac{(x)(x)}{(0.10 - x)(0.15 - x)}$$

This equation contains only one unknown and can be solved for x, the concentration of $CH_3CO_2C_2H_5$. Let us expand the equation, and solve the resulting quadratic using the technique shown in Appendix A.4.

$$4.0 = \frac{x^2}{x^2 - 0.25x + 0.015}$$

$$4.0(x^2 - 0.25x + 0.015) = x^2$$

$$4.0x^2 - x + 0.060 = x^2$$

$$3x^2 - x + 0.060 = 0$$

Now we use the quadratic formula.

$$x = \frac{-(-1) \pm \sqrt{(-1)^2 - 4(3)(0.060)}}{2(3)}$$

$$= \frac{1 \pm \sqrt{1 - 0.72}}{6}$$

$$= 0.25 \text{ and } 0.078$$

Only one of these solutions, 0.078, is physically reasonable. If $x = 0.25$, then $[CH_3CO_2C_2H_5] = 0.25$ M, but the maximum concentration can be only 0.10 M since 0.10 mol of CH_3CO_2H, the amount present before reaction, can only form 0.10 mol of product. Thus

$$[CH_3CO_2C_2H_5] = 0.078 \ M \text{ (at equilibrium)}$$

$$\text{Mass } CH_3CO_2C_2H_5 = \frac{0.078 \text{ mol}}{1 L} \times 1.0 L \times \frac{88.1 \text{ g } CH_3CO_2C_2H_5}{1 \text{ mol}}$$

$$= 6.9 \text{ g}$$

15.18 Determination of Equilibrium Constants

Although the values of equilibrium constants can be determined from the forward and reverse rate constants for a reaction, they are more commonly determined by measuring, *at equilibrium*, the concentrations of reactants and products of a reaction.

EXAMPLE 15.15 An equimolar mixture of hydrogen and iodine was heated at 400°C until no further change in the concentration of H_2, I_2, or HI was observed. At this point it was assumed that equilibrium had been reached, and it was found by analysis that $[H_2] = 0.221$ M, $[I_2] = 0.221$ M, and $[HI] = 1.563$ M. Calculate the equilibrium constant for the reaction at 400°.

$$H_2(g) + I_2(g) \rightleftharpoons 2HI(g)$$

The reaction quotient for this reaction is

$$\frac{[HI]^2}{[H_2][I_2]}$$

At equilibrium this ratio is a constant equal to the equilibrium constant.

$$K = \frac{[HI]^2}{[H_2][I_2]}$$

Substitution of the concentrations determined at equilibrium gives

$$K = \frac{[HI]^2}{[H_2][I_2]} = \frac{(1.563 \; mol \; L^{-1})^2}{(0.221 \; mol \; L^{-1})(0.221 \; mol \; L^{-1})} = 50.0 \quad \text{(units cancel out)}$$

If we start with pure HI at any molar concentration, or with any mixture of H_2 and I_2, or with any mixture of H_2, I_2, and HI, and hold the temperature of the system at 400°C until equilibrium is established, the molar concentrations of the three substances will have changed so that the reaction quotient $[HI]^2/[H_2][I_2]$ is equal to 50.0.

The following example illustrates that it is not always necessary to measure the concentration of each species present at equilibrium. The chemical equation can be used to relate concentrations.

EXAMPLE 15.16   Iodine molecules react reversibly with iodide ions to produce triiodide ions.

$$I_2(aq) + I^-(aq) \rightleftharpoons I_3^-(aq)$$

If a liter of solution prepared from 0.1000 mol of I_2 and 0.1000 mol of the strong electrolyte KI contains 9.33×10^{-2} mol of iodine at equilibrium, what is the equilibrium constant for the reaction?

From the law of mass action, we know that the equilibrium constant is given by the equation

$$K = \frac{[I_3^-]}{[I_2][I^-]}$$

In order to determine the value of K, we need to know the values of $[I_3^-]$, $[I_2]$, and $[I^-]$ at equilibrium.

We can set up the following table of initial concentrations and equilibrium concentrations. Entries in color are those given in the problem.

	$[I_2]$	$[I^-]$	$[I_3^-]$
Initial concentrations, M	0.1000	0.1000	0
Equilibrium concentrations, M	9.33×10^{-2}	9.33×10^{-2}	6.7×10^{-3}

First consider the table entries for initial concentrations. The data tell us that the solution was prepared from 0.1000 mol of I_2 per liter and 0.1000 mol of KI per liter. Thus, before the reaction begins, the concentration of I_2 is 0.1000 mol/1 L or 0.1000 M. Since KI is a strong electrolyte, the concentration of I^- is equal to the concentration of KI (1 mol I^-/1 mol KI), or 0.1000 M. The initial concentration of I_3^- is zero since no I_3^- has yet been formed.

At equilibrium the concentration of I_2 is given as 9.33×10^{-2} M. Before reaction (when the solution was first prepared) the concentration of I_2, $[I_2]_i$, was 0.1000 M; however, it decreased to 9.33×10^{-2} M at equilibrium. The change in concentration of I_2 was

$$[I_2]_i - [I_2] = 0.1000\ M - (9.33 \times 10^{-2}\ M) = 6.7 \times 10^{-3}\ M$$

Thus 6.7×10^{-3} mol of I_2 per liter of solution was consumed as the reaction proceeded to equilibrium. From the chemical equation we know that one mole of I_3^- is formed for each mole of I_2 that reacts. Thus the concentration of I_3^- at equilibrium, $[I_3^-]$, must be

$$\frac{6.7 \times 10^{-3}\ \text{mol I}_2}{1\ \text{L}} \times \frac{1\ \text{mol I}_3^-}{1\ \text{mol I}_2} = \frac{6.7 \times 10^{-3}\ \text{mol I}_3^-}{1\ \text{L}} = 6.7 \times 10^{-3}\ M$$

Before reaction the concentration of I^-, $[I^-]_i$, was also 0.1000 M. The concentration of I^- at equilibrium, $[I^-]$, is that amount that did not react to form I_3^-. Since the chemical equation tells us that one mole of I^- reacts for each mole of I_3^- formed, the amount of I^- that reacts is equal to the amount of I_3^- formed, or 6.7×10^{-3} M. Therefore

$$[I^-] = [I^-]_i - [I_3^-] = 0.1000\ M - (6.7 \times 10^{-3}\ M) = 9.33 \times 10^{-2}\ M$$

Now we simply substitute these equilibrium concentrations into the expression for the equilibrium constant and evaluate it.

$$K = \frac{[I_3^-]}{[I_2][I^-]} = \frac{(6.7 \times 10^{-3}\ \text{mol L}^{-1})}{(9.33 \times 10^{-2}\ \text{mol L}^{-1})(9.33 \times 10^{-2}\ \text{mol L}^{-1})} = 0.770\ \text{mol}^{-1}\ \text{L}$$

Units for equilibrium constants (mol^{-1} L in Example 15.16) differ depending on the form of the particular mathematical expression used to evaluate the equilibrium constant. In some cases the units may even cancel out, as in Example 15.15. It is common practice to omit the units of K, so we will do so in most examples.

15.19 Effect of Change of Concentration on Equilibrium

A chemical system may be shifted out of equilibrium by increasing the rate of the forward or the reverse reaction. When the system

$$A + B \rightleftharpoons C + D$$

is in equilibrium and an additional quantity of A is added, the rate of the forward reaction increases because the concentration of the reacting molecules increases (Fig. 15-17). The rate of the forward reaction becomes faster than that of the reverse reaction, so the system is out of equilibrium. However, as the concentrations of C and D increase, the rate of the reverse reaction increases, while the decrease in the concentrations of A and B causes the rate of the forward reaction to decrease. The rates of the two reactions become equal again, and a second state of equilibrium is reached. Although the concentrations of A, B, C, and D have

Figure 15-17. The effect of adding reactant A on the rates of the forward and reverse reactions for $A + B \rightleftharpoons C + D$ when the system is initially at equilibrium.

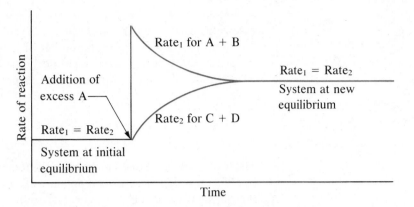

changed, the reaction quotient [C][D]/[A][B] is again equal to its original value of K.

The equilibrium is said to have been shifted to the right, since in the new state of equilibrium the products C and D are present in greater concentrations than they were before the addition of A. Also, B is present in smaller concentration and A is present in greater concentration than they were before. Increasing the concentration of B would also shift the equilibrium to the right. Increasing the concentration of either C or D, or both, would shift the equilibrium to the left and increase the concentrations of A and B. The equilibrium may also be shifted to the right by removing either C or D, or both. This causes a reduction in the rate of the reaction to the left, which proceeds more slowly than the reaction to the right until the reaction rates again become equal as equilibrium is attained.

The effect of a change in concentration on a system in equilibrium is illustrated by the equilibrium $H_2 + I_2 \rightleftharpoons 2HI$, for which K was found to be 50.0 at $400°C$ with $[H_2] = [I_2] = 0.221\ M$ and $[HI] = 1.563\ M$ (Example 15.15). If H_2 is introduced into the system quickly so that its concentration doubles before it begins to react, the rate of the reaction of H_2 with I_2 to form HI increases. When equilibrium is again reached, $[H_2] = 0.374\ M$, $[I_2] = 0.153\ M$, and $[HI] = 1.692\ M$. If these new values are substituted in the expression for the equilibrium constant for this system, we have

$$\frac{[HI]^2}{[H_2][I_2]} = \frac{(1.692)^2}{(0.374)(0.153)} = 50.0 = K$$

Hence, by doubling the concentration of H_2, we have caused the formation of more HI, used up about one-third of the I_2 present at the first equilibrium, and used up some, but not all, of the excess H_2 added.

The effect of a change in concentration on a system in equilibrium is an important application of **Le Châtelier's principle: If a stress (such as a change in concentration, pressure, or temperature) is applied to a system in equilibrium, the equilibrium shifts in a way that tends to undo the effect of the stress.**

15.20   Effect of Change in Pressure on Equilibrium

Changes in pressure measurably affect systems in equilibrium only when gases are involved, and then only when the chemical reaction produces a change in the total number of gaseous molecules in the system. As the pressure on a gaseous system increases, the gases are compressed, the total number of molecules per unit of volume increases, and their molar concentrations increase. The stress in this case is an increase in the number of molecules per unit of volume. A chemical reaction that reduces the total number of molecules per unit of volume relieves the stress and will, in accord with the Le Châtelier's principle, be favored.

Consider the effect of an increase in pressure on the system in which one molecule of nitrogen and three molecules of hydrogen interact to form two molecules of ammonia.

$$N_2(g) + 3H_2(g) \rightleftharpoons 2NH_3(g)$$

The formation of ammonia decreases the total number of molecules in the system by 50%, thus reducing the total pressure exerted by the system. Experiment shows that an increase in pressure does drive the reaction to the right. On the other hand, a decrease in the pressure on the system favors decomposition of ammonia into hydrogen and nitrogen. These observations are fully in accord with Le Châtelier's principle. (Table 23-2 in the chapter on nitrogen and its compounds shows equilibrium concentrations of ammonia at several temperatures and pressures.)

Let us now consider the reaction in which a molecule of nitrogen interacts with a molecule of oxygen with the formation of two molecules of nitric oxide.

$$N_2(g) + O_2(g) \rightleftharpoons 2NO(g)$$

Because there is no change in the total number of molecules in the system during reaction, a change in pressure does not favor either formation or decomposition of gaseous nitric oxide.

Whenever gases are involved in a system in equilibrium, the pressure of each gas can be substituted for its concentration in the expression for the equilibrium constant because the concentration of a gas at constant temperature varies directly with the pressure. Thus for the system

$$N_2(g) + 3H_2(g) \rightleftharpoons 2NH_3(g)$$

we may write

$$\frac{P_{NH_3}^2}{P_{N_2} P_{H_2}^3} = K_p$$

Remember that if partial pressures of gases are substituted for their concentrations, the equilibrium constant, K_p, is still a constant, but its numerical value and units may change. An equilibrium constant using gas pressures will be indicated with a subscript p in this text.

The units for K_p for the reaction to produce ammonia, if the three partial pressures are expressed in atmospheres, would be

$$K_p = \frac{atm^2}{atm \times atm^3} = \frac{1}{atm^2} = atm^{-2}$$

Although equilibrium constants have units (unless they happen to cancel out as they do for the reaction $N_2 + O_2 \rightleftharpoons 2NO$), their values are often given without units.

EXAMPLE 15.17   A vessel contains gaseous carbon monoxide, carbon dioxide, hydrogen, and water in equilibrium at 980°C. The pressures of these gases are CO, 0.150 atm; CO$_2$, 0.200; H$_2$, 0.090 atm; H$_2$O, 0.200 atm. Hydrogen is pumped into the vessel, and the equilibrium pressure of CO changes to 0.230 atm. Calculate the partial pressures of the other substances at the new equilibrium, assuming no change in temperature.

$$CO_2(g) + H_2(g) \rightleftharpoons CO(g) + H_2O(g)$$

Before we can calculate the pressures after the addition of hydrogen, we need the value of the equilibrium constant. This can be determined from the initial equilibrium pressures of the reactants and products.

$$K_p = \frac{P_{CO}P_{H_2O}}{P_{CO_2}P_{H_2}} = \frac{(0.150 \text{ atm})(0.200 \text{ atm})}{(0.200 \text{ atm})(0.090 \text{ atm})} = 1.67$$

(Units cancel out, and the equilibrium constant in this case has no units.)

Now we set up a table showing what we know about the initial pressures in the system (the pressures *after* addition of H$_2$ *but before* the additional H$_2$ began to react) and about the pressures at the new equilibrium.

	P_{CO_2}	P_{H_2}	P_{CO}	P_{H_2O}
Initial pressures, atm	0.200	?	0.150	0.200
Equilibrium pressures, atm	?	?	0.230	?

At the new equilibrium P_{CO} is 0.230 atm. Therefore P_{CO} has increased by 0.080 atm (from 0.150 atm to 0.230 atm). The balanced equation tells us that for each increase of 1 mol of CO, the concentration of H$_2$O must also increase by 1 mol; hence the pressure increase for H$_2$O from its initial pressure of 0.200 atm will also be 0.080 atm. Thus P_{H_2O} at the new equilibrium is

$$0.200 \text{ atm} + 0.080 \text{ atm} = 0.280 \text{ atm}$$

The balanced equation also tells us that for every mole of CO produced, one mole of CO$_2$ must be consumed, and hence the pressure of CO$_2$ must decrease from its initial pressure of 0.200 atm by the same amount as the increase of pressure for CO (0.080 atm). Thus P_{CO_2} at the new equilibrium is

$$0.200 \text{ atm} - 0.080 \text{ atm} = 0.120 \text{ atm}$$

Now we have all of the pressures at the new equilibrium except that of H$_2$, which is our unknown and can be designated as x. (Note that it is not necessary to calculate the initial pressure of H$_2$ because this value is not required in the final calculation.)

	P_{CO_2}	P_{H_2}	P_{CO}	P_{H_2O}
Initial pressures, atm	0.200	—	0.150	0.200
Equilibrium pressures, atm	0.120	x	0.230	0.280

Now we can calculate the partial pressure of hydrogen at the new equilibrium by substituting P_{CO}, P_{H_2O}, and P_{CO_2} at the new equilibrium into the expression for

K_p and solving for P_{H_2}:

$$K_p = \frac{P_{CO}P_{H_2O}}{P_{CO_2}P_{H_2}} = \frac{(0.230 \text{ atm})(0.280 \text{ atm})}{(0.120 \text{ atm})(x)} = 1.67$$

$$x = \frac{(0.230 \text{ atm})(0.280 \text{ atm})}{(0.120 \text{ atm})(1.67)} = 0.321 \text{ atm}$$

Therefore the partial pressures for CO, H_2O, CO_2, and H_2 at the new equilibrium are 0.230 atm, 0.280 atm, 0.120 atm, and 0.321 atm, respectively.

EXAMPLE 15.18   In a 3.0-L vessel the following partial pressures are measured at equilibrium: N_2, 0.380 atm; H_2, 0.400 atm; NH_3, 2.000 atm. Hydrogen is removed from the vessel until the partial pressure of nitrogen, at a new equilibrium, is equal to 0.450 atm. Calculate the partial pressures of the other substances under the new conditions.

$$N_2(g) + 3H_2(g) \rightleftharpoons 2NH_3(g)$$

$$K_p = \frac{P_{NH_3}^2}{P_{N_2}P_{H_2}^3}$$

$$K_p = \frac{(2.000 \text{ atm})^2}{(0.380 \text{ atm})(0.400 \text{ atm})^3} = 1.645 \times 10^2 \text{ atm}^{-2}$$

Since the amount of H_2 removed from the system is unknown, we let P_{H_2} at the new equilibrium be x.

The removal of H_2 shifts the equilibrium to the left, producing additional N_2 and therefore increasing the pressure of N_2 in proportion to its increased molar concentration. At the new equilibrium $P_{N_2} = 0.450$ atm. Hence the pressure of N_2 has increased by 0.070 atm. The balanced equation tells us that 2 mol of NH_3 must be used up to produce 1 mol of N_2. The pressure of NH_3 must decrease in proportion to the decrease in its molar concentration. The pressure of NH_3 must therefore decrease by twice the amount that the pressure of N_2 increases; thus the pressure of NH_3 must decrease by 0.140 atm. At the new equilibrium

$$P_{NH_3} = 2.000 - 0.140 = 1.860 \text{ atm}$$

Now we can set up a table containing the initial pressures (*after* H_2 was removed, *but before* the reaction begins to shift to restore the equilibrium) and the new equilibrium pressures. Values given in the problem statement are shown in color.

	P_{N_2}	P_{H_2}	P_{NH_3}
Initial pressures, atm	0.380	—	2.000
Equilibrium pressures, atm	0.450	x	1.860

To calculate P_{H_2} at the new equilibrium, we substitute P_{N_2} and P_{NH_3} in the expression for K_p and solve for x.

$$K_p = \frac{P_{NH_3}^2}{P_{N_2}P_{H_2}^3} = \frac{(1.860 \text{ atm})^2}{(0.450 \text{ atm})(x^3)} = 1.645 \times 10^2 \text{ atm}^{-2}$$

$$x^3 = \frac{(1.860 \text{ atm})^2}{(0.450 \text{ atm})(1.645 \times 10^2 \text{ atm}^{-2})}$$

$$= 4.674 \times 10^{-2} \text{ atm}^3 = 46.74 \times 10^{-3} \text{ atm}^3$$

$$x = \sqrt[3]{46.74 \times 10^{-3} \text{ atm}^3} = 0.360 \text{ atm}$$

At the new equilibrium the partial pressures of NH_3 and H_2 are 1.860 atm and 0.360 atm, respectively.

15.21 Effect of Change in Temperature on Equilibrium

All chemical changes involve either the evolution or the absorption of energy. In every system in equilibrium an endothermic and an exothermic reaction are taking place simultaneously. The endothermic reaction ($\Delta H > 0$) is favored by an increase in temperature, that is, an increase in energy, and the exothermic reaction ($\Delta H > 0$) is favored by a decrease in temperature. Changing the temperature changes the value of the equilibrium constant.

The effect of temperature changes on systems in equilibrium is summarized by **van't Hoff's law: When the temperature of a system in equilibrium is raised, the equilibrium is displaced in such a way that heat is absorbed.** This generalization is a special case of Le Châtelier's principle.

In the reaction between gaseous hydrogen and gaseous iodine, heat is evolved.

$$H_2(g) + I_2(g) \rightleftharpoons 2HI(g) \qquad \Delta H = -9.4 \text{ kJ}$$

Thus lowering the temperature of this system favors formation of hydrogen iodide; raising it favors decomposition. Raising the temperature decreases the value of the equilibrium constant since the concentration of HI at the new equilibrium decreases while the molar concentrations of H_2 and I_2 increase. The value of the equilibrium constant

$$K = \frac{[HI]^2}{[H_2][I_2]}$$

decreases from 67.5 at 357°C to 50.0 at 400°C.

The equation for the formation of ammonia from hydrogen and nitrogen is

$$N_2 + 3H_2 \rightleftharpoons 2NH_3 \qquad \Delta H = -92.2 \text{ kJ}$$

The reaction is exothermic, and the equilibrium can be shifted to the right to favor the formation of more ammonia by lowering the temperature (see Table 23-2). However, equilibrium is reached more slowly because of the large decrease of reaction rate with decreasing temperature. In the commercial production of ammonia, it is not feasible to use temperatures much lower than 500°C. At lower temperatures, even in the presence of a catalyst, the reaction proceeds too slowly to be practical.

15.22   Effect of a Catalyst on Equilibrium

Iron powder is used as a catalyst in the production of ammonia from nitrogen and hydrogen to increase the rate of reaction of these two elements.

$$N_2 + 3H_2 \xrightarrow[\text{Fe}]{\triangle} 2NH_3$$

However, this same catalyst serves equally well to increase the rate of the reverse reaction, that is, the decomposition of ammonia into its constituent elements.

$$2NH_3 \xrightarrow[\text{Fe}]{\triangle} N_2 + 3H_2$$

Thus the net effect of iron in the reversible reaction

$$N_2 + 3H_2 \rightleftharpoons 2NH_3$$

is to cause equilibrium to be reached more rapidly. *A catalyst has no effect on the value of the equilibrium constant.* It merely increases the rate of both the forward and the reverse reactions to the same extent, so that equilibrium is reached more rapidly.

15.23   Homogeneous and Heterogeneous Equilibria

A **homogeneous equilibrium** is an equilibrium within a single phase, such as a mixture of gases or a solution. Most of the equilibria that have been considered in this chapter are homogeneous equilibria involving reversible changes in only one phase, the gas phase.

A **heterogeneous equilibrium** is an equilibrium between two or more different phases. Liquid water in equilibrium with ice, liquid water in equilibrium with water vapor, and a solid in contact with its saturated solution are examples of heterogeneous equilibria. Each of these equilibria involves some kind of boundary surface between two phases.

The equilibrium between calcium carbonate, calcium oxide, and carbon dioxide is a heterogeneous equilibrium.

$$CaCO_3(s) \rightleftharpoons CaO(s) + CO_2(g)$$

The expression for the equilibrium constant for this reversible reaction is

$$\frac{[CaO][CO_2]}{[CaCO_3]} = K$$

The concentration of a solid substance is proportional to its density and remains constant as long as the temperature does not change. Since the concentrations of solids involved in an equilibrium are constant, they can be included in the equilibrium constant and need not appear in the reaction quotient obtained by the law of mass action. We may write

$$[CO_2] = \frac{[CaCO_3]}{[CaO]} \times K$$

The right-hand side of this equation contains only constant terms. These can be combined into a single constant, K.

$$[CO_2] = K$$

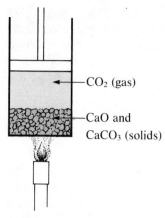

CO₂ (gas)

CaO and
CaCO₃ (solids)

Figure 15-18. The thermal decomposition of calcium carbonate in a closed system is an example of a heterogeneous equilibrium.

You can assume that all equilibrium constants for heterogeneous equilibria contain such constant terms.

The expression for the equilibrium constant for the preceding reaction in terms of pressure is

$$P_{CO_2} = K_p$$

This equation means that at a given temperature there is only one pressure at which gaseous carbon dioxide can be in equilibrium with solid calcium carbonate and calcium oxide. A sample of $CaCO_3$ placed in a cylinder with a movable piston (Fig. 15-18) and heated at 900°C will decompose into CaO and CO_2. At equilibrium the pressure of carbon dioxide in the cylinder will equal K_p (actually 1.04 atm), if the piston is held stationary.

$$P_{CO_2} = K_p = 1.04 \text{ atm} \qquad \text{(at 900°C)}$$

At this pressure, since equilibrium has been reached, $CaCO_3$ decomposes into CaO and CO_2 at the same rate that CaO and CO_2 react to produce $CaCO_3$. If we increase the pressure by pushing the piston down and compressing the gas, more CO_2 will combine with CaO to form $CaCO_3$, and the pressure will again drop to 1.04 atm. If, on the other hand, we decrease the pressure by raising the piston, just enough $CaCO_3$ will decompose to bring the pressure exerted by the CO_2 back to its equilibrium value of 1.04 atm.

In the commercial production of quicklime, CaO, from limestone, the carbon dioxide is continuously removed as fast as it is formed by means of a stream of air; this causes the reaction to proceed essentially only to the right. Since the pressure of carbon dioxide never reaches the equilibrium pressure, equilibrium is never established.

FOR REVIEW

SUMMARY

The **rate** at which a chemical reaction proceeds can be defined as either the decrease in concentration of a reactant or the increase in concentration of a product per unit of time. In general, the rate of a given reaction increases as the temperature or the concentration of a reactant increases. The rate of a reaction can also be increased by a **catalyst,** a substance that is not permanently changed by the reaction. Reactions involving two phases proceed more rapidly, the more finely divided the condensed phase. The rate of a given reaction can be described by an experimentally determined **rate equation** of the form

$$\text{Rate} = k[A]^m[B]^n[C]^p \cdots$$

in which [A], [B], and [C] represent molar concentrations of reactants (or sometimes products or other substances); m, n, and p are usually, but not always, positive integers; and k is the specific rate constant. The exponents n, m, p, etc., describe the **order of the reaction** with respect to each specific reactant. The **overall order of the reaction** is the sum of the exponents. The **half-life** of a reaction is the time required for half of the original reactant to be consumed.

Before atoms, molecules, or ions can react, they must collide. When every collision leads to reaction, the rate is controlled by how rapidly the reactants diffuse together. These **diffusion-controlled reactions** are very fast. Most reactions occur more slowly because the reacting species must be oriented correctly when they collide and because they must possess a certain minimum energy, the **activation energy,** in order to form an activated complex, or transition state. The rate constant for a reaction is related to these effects by the **Arrhenius equation:**

$$k = A \times 10^{-E_a/2.303\,RT}$$

The collection of individual steps, or **elementary reactions,** by which reactants are converted into products during the course of an overall reaction is called the **reaction mechanism,** or **reaction path.** The overall rate of a reaction is determined by the rate of the slowest step, the **rate-determining step.** Although it is not possible to write a rate equation for an overall reaction without data from experimental observations, once the elementary reactions have been determined, rate equations can be written by inspection for each elementary reaction. **Unimolecular elementary reactions** have first-order rate equations; **bimolecular elementary reactions,** second-order rate equations; and **termolecular elementary reactions** (which are uncommon), third-order rate equations.

If, at some point during the course of a reaction, the rate of formation of products becomes equal to the rate at which the products reform reactants, the reaction is said to have reached **dynamic equilibrium.** At equilibrium there is no net change in the concentrations of reactants or products. For any reaction

$$m\mathrm{A} + n\mathrm{B} + \cdots \rightleftharpoons x\mathrm{C} + y\mathrm{D} + \cdots$$

at equilibrium the **reaction quotient** is equal to a constant K, the equilibrium constant for the reaction.

$$K = \frac{[\mathrm{C}]^x[\mathrm{D}]^y \cdots}{[\mathrm{A}]^m[\mathrm{B}]^n \cdots}$$

At equilibrium the values of [A], [B], [C], [D], etc., may vary, but the reaction quotient will always be equal to K. Addition or removal of a reactant or product may shift a reaction out of equilibrium, but the reaction will proceed so that the concentrations change and equilibrium is reestablished. The effect of a change in conditions is described by **Le Châtelier's principle:** If a stress such as a change in temperature, pressure, or concentration is applied to a system at equilibrium, the equilibrium shifts in a way that relieves the effects of the stress.

KEY TERMS AND CONCEPTS

Activated complex (15.8)
Activation energy (15.9)
Arrhenius equation (15.9)
Bimolecular reaction (15.12)
Catalysis (15.15)
Chain mechanism (15.14)
Collision theory of reaction (15.8)

Diffusion-controlled reactions (15.8)
Elementary reaction (15.10)
Equilibrium (15.16)
Equilibrium constant (15.17)
Frequency factor (15.9)
Half-life of a reaction (15.7)
Heterogeneous catalysts (15.15)

Heterogeneous equilibrium (15.23)
Homogeneous catalysts (15.15)
Homogeneous equilibrium (15.23)
Law of mass action (15.17)
Le Châtelier's principle (15.19)
Order of a reaction (15.6)

Pressure effects on equilibrium (15.20)

Rate constant (15.5)

Rate-determining step (15.14)

Rate equations (15.5)

Rate of reaction (15.1)

Reaction mechanism (15.10)

Reaction quotient (15.17)

Temperature effects on equilibrium (15.21)

Termolecular reaction (15.13)

Unimolecular reaction (15.11)

van't Hoff's law (15.21)

EXERCISES

Reaction Rates and Rate Equations

1. Explain how each of the factors that determine the rate of a reaction is responsible for changing the rate.

2. How do the rate of a reaction and its rate constant differ?

3. Doubling the concentration of a reactant increases the rate of a reaction four times. What is the order of the reaction with respect to that reactant? *Ans. Second order*

4. Nitric oxide gas reacts with chlorine gas according to the equation

$$2NO(g) + Cl_2(g) \longrightarrow 2NOCl(g)$$

The following initial rates of reaction have been observed for certain reactant concentrations:

NO, mol L^{-1}	Cl$_2$, mol L^{-1}	Rate, mol L^{-1} h^{-1}
0.50	0.50	1.14
1.00	0.50	4.56
1.00	1.00	9.12

What is the rate equation describing the rate dependence on the concentrations of NO and Cl$_2$?
Ans. Rate = k[NO]²[Cl];
k = 9.12 L² mol⁻² h⁻¹

☒ 5. The rate constant for the decomposition at 45°C of dinitrogen pentaoxide, N$_2$O$_5$, dissolved in chloroform, CHCl$_3$, is 6.2×10^{-4} min^{-1}.

$$2N_2O_5 \longrightarrow 4NO_2 + O_2$$

The decomposition is first order in N$_2$O$_5$.
(a) What is the rate of the reaction when [N$_2$O$_5$] = 0.40 M?
Ans. 2.5 × 10⁻⁴ mol L⁻¹ min⁻¹
(b) What is the concentration of N$_2$O$_5$ remaining at the end of 1 h if the initial concentration of N$_2$O$_5$ was 0.40 M? *Ans. 0.38 M*

6. One of the reactions involved in the formation of photochemical smogs is

$$O_3(g) + NO(g) \longrightarrow O_2(g) + NO_2(g)$$

The rate constant for this reaction is 1.2×10^7 L mol^{-1} s^{-1}. The reaction is first order in O$_3$ and first order in NO. Calculate the rate of formation of NO$_2$ in air in which the O$_3$ concentration is 2×10^{-8} M and the NO concentration 7×10^{-8} M. *Ans. 2 × 10⁻⁸ mol L⁻¹ s⁻¹*

☒ 7. Most of the 17.1 billion pounds of HNO$_3$ produced in the United States during 1980 was prepared by the following sequence of reactions, each run in a separate reaction vessel.

$$4NH_3(g) + 5O_2(g) \longrightarrow$$
$$4NO(g) + 6H_2O(g) \quad (1)$$
$$2NO(g) + O_2(g) \longrightarrow 2NO_2(g) \quad (2)$$
$$3NO_2(g) + H_2O(l) \longrightarrow$$
$$2HNO_3(aq) + NO(g) \quad (3)$$

The first reaction is run by burning ammonia in air over a platinum catalyst. This reaction is fast. The reaction in Equation (3) is also fast. The second reaction limits the rate at which nitric acid can be prepared from ammonia. If Equation (2) is second order in NO and first order in O$_2$, what is the rate of formation of NO$_2$ when the oxygen concentration is 0.50 M and the nitric oxide concentration is 0.75 M? The rate constant for the reaction is 5.8×10^{-6} L^2 mol^{-2} s^{-1}.
Ans. 1.6 × 10⁻⁶ mol L⁻¹ s⁻¹

8. Nitrosyl chloride, NOCl, decomposes to NO and Cl$_2$.

$$2NOCl(g) \longrightarrow 2NO(g) + Cl_2(g)$$

Determine the rate equation and the rate constant for this reaction from the following data:

[NOCl], M	0.20	0.40	0.60
Rate, mol L^{-1} s^{-1}	1.60×10^{-9}	6.40×10^{-9}	1.44×10^{-8}

Ans. Rate $= k[NOCl]^2$;
$k = 4.0 \times 10^{-8}$ L mol^{-1} s^{-1}

S 9. Hydrogen reacts with nitric oxide to form nitrous oxide, laughing gas, according to the equation

$$H_2(g) + NO(g) \longrightarrow N_2O(g) + H_2O(g)$$

Determine the rate equation and the rate constant for the reaction from the following data:

[NO], M	0.30	0.60	0.60
[H$_2$], M	0.35	0.35	0.70
Rate, mol L^{-1} s^{-1}	2.835×10^{-3}	1.134×10^{-2}	2.268×10^{-2}

Ans. Rate $= k[NO]^2[H_2]$;
$k = 9.0 \times 10^{-2}$ L^2 mol^{-2} s^{-1}

10. The following data have been determined for the reaction

$$I^- + OCl^- \longrightarrow IO^- + Cl^-$$

[I$^-$], M	0.10	0.20	0.30
[OCl$^-$], M	0.050	0.050	0.010
Rate, mol L^{-1} s^{-1}	3.05×10^{-4}	6.10×10^{-4}	1.83×10^{-4}

Determine the rate equation and rate constant for this reaction. *Ans.* Rate $= k[I^-][OCl^-]$;
$k = 6.1 \times 10^{-2}$ L mol^{-1} s^{-1}

11. A liter of a 1 M solution of H$_2$O$_2$ slowly decomposes into H$_2$O and O$_2$. If 0.50 mol of H$_2$O$_2$ decomposes during the first 6 h of the reaction, explain why only 0.25 mol decomposes during the next 6-h period.

12. Radioactive materials decay by a first-order process. The very dangerous isotope strontium-90 decays with a half-life of 28 years.

$$^{90}_{38}Sr \longrightarrow {}^{90}_{39}Y + e^-$$

What is the rate constant for this decay?
Ans. 2.5×10^{-2} L mol^{-1} yr^{-1}

13. Determine the rate constant for the decomposition of H$_2$O$_2$ shown in Fig. 15-1 from the data given in the figure. *Ans.* 3.21×10^{-5} s^{-1}

S 14. The decomposition of SO$_2$Cl$_2$ to SO$_2$ and Cl$_2$ is a first-order reaction with $k = 2.2 \times 10^{-5}$ s^{-1} at 320°C. Determine the half-life of this reaction.

At 320°C, how much SO$_2$Cl$_2$(g) would remain in a 1.00-L flask 90.0 min after the introduction of 0.0238 mol of SO$_2$Cl$_2$? Assume the rate of the reverse reaction is so slow that it can be ignored.
Ans. $t_{1/2} = 3.2 \times 10^4$ s; 2.11×10^{-2} mol

Collision Theory of Reaction Rates

15. Chemical reactions occur when reactants collide. For what reasons may a collision fail to produce a chemical reaction?

16. If every collision between reactants leads to a reaction, what determines the rate at which the reaction will occur?

17. What is the activation energy of a reaction, and how is this energy related to the activated complex of the reaction?

18. Account for the relationship between the rate of a reaction and its activation energy.

19. What is the effect on the rate of many reactions of an increase in temperature of 10°C? Explain this effect in terms of the collision theory of reaction rate.

S 20. If the rate of a reaction doubles for every 10°C rise in temperature, how much faster would the reaction proceed at 45°C than at 25°C? at 95°C than at 25°C? *Ans.* 4 times faster; 128 times faster

21. In an experiment, a sample of NaClO$_3$ was 90% decomposed in 48 min. Approximately how long would this decomposition have taken if the sample had been heated 20° higher? *Ans.* 12 min

S 22. The rate constant at 325°C for the reaction C$_4$H$_8$ → 2C$_2$H$_4$ (Section 15.11) is 6.1×10^{-8} s^{-1}, and the activation energy is 261 kJ per mole of C$_4$H$_8$. Determine the frequency factor for the reaction. *Ans.* 3.8×10^{15} s^{-1}

23. The rate constant for the decomposition of acetaldehyde, CH$_3$CHO, to methane, CH$_4$, and carbon monoxide, CO, in the gas phase is 1.1×10^{-2} L mol^{-1} s^{-1} at 703 K and 4.95 L mol^{-1} s^{-1} at 865 K. Determine the activation energy for this decomposition. *Ans.* 190 kJ

24. An elevated level of the enzyme alkaline phosphatase (ALP) in the serum is an indication of possible liver or bone disorder. The level of serum ALP is so low that it is very difficult to measure directly. However, ALP catalyzes a number of reactions, and its relative concentra-

tion can be determined by measuring the rate of one of these reactions under controlled conditions. One such reaction is the conversion of *p*-nitrophenyl phosphate (PNPP) to *p*-nitrophenoxide ion (PNP) and phosphate ion. Control of temperature during the test is very important; the rate of the reaction increases 1.47 times if the temperature changes from 30 °C to 37 °C. What is the activation energy for the ALP-catalyzed conversion of PNPP to PNP and phosphate?

Ans. $43\ kJ\ mol^{-1}$

Elementary Reactions, Reaction Mechanisms, Catalysts

25. Define the following:
(a) unimolecular reaction
(b) bimolecular reaction
(c) elementary reaction
(d) overall reaction
26. What is the rate equation for the elementary termolecular reaction $A + 2B \rightarrow$ products? for $3A \rightarrow$ products?
27. Why are elementary reactions involving three or more reactants very uncommon?
28. In general, can we predict the effect of doubling the concentration of A on the rate of the overall reaction $A + B \rightarrow C$? Can we predict the effect if the reaction is known to be an elementary reaction?
29. Which of the following equations, as written, could describe elementary reactions?
(a) $Cl_2 + CO \longrightarrow Cl_2CO$; rate $= k[Cl_2]^{3/2}[CO]$
(b) $PCl_3 + Cl_2 \longrightarrow PCl_5$; rate $= k[PCl_3][Cl_2]$
(c) $2NO + 2H_2 \longrightarrow N_2 + 2H_2O$; rate $= k[NO][H_2]$
(d) $2NO + O_2 \longrightarrow 2NO_2$; rate $= k[NO]^2[O_2]$
(e) $NO + O_3 \longrightarrow NO_2 + O_2$; rate $= k[NO][O_3]$

Ans. (b), (d), and (e)
30. Account for the increase in reaction rate brought about by a catalyst.
31. Describe how a homogeneous catalyst and a heterogeneous catalyst function.
32. (a) It has been suggested that chlorine atoms resulting from decomposition of chlorofluoromethanes, such as CCl_2F_2, catalyze the decomposition of ozone in the ozone layer of the earth's atmosphere. One simplified mechanism for the decomposition is

$$O_3 \xrightarrow{\text{Sunlight}} O_2 + O$$
$$O_3 + Cl \longrightarrow O_2 + ClO$$
$$ClO + O \longrightarrow Cl + O_2$$

Explain why chlorine atoms are catalysts in the gas-phase transformation

$$2O_3 \rightleftharpoons 3O_2$$

(b) Nitric oxide is also involved in the decomposition of ozone by the mechanism

$$O_3 \xrightarrow{\text{Sunlight}} O_2 + O$$
$$O_3 + NO \longrightarrow NO_2 + O_2$$
$$NO_2 + O \longrightarrow NO + O_2$$

Is NO a catalyst for the decomposition? Explain your answer.
33. Write the rate equation for each of the elementary reactions given in both parts of Exercise 32.

Equilibrium Constants

34. Using the law of mass action, derive the mathematical expression of the equilibrium constant for the following reversible reactions:
(a) $N_2(g) + 3H_2(g) \rightleftharpoons 2NH_3(g)$
(b) $CH_4(g) + Cl_2(g) \rightleftharpoons CH_3Cl(g) + HCl(g)$
(c) $N_2(g) + O_2(g) \rightleftharpoons 2NO(g)$
(d) $2SO_2(g) + O_2(g) \rightleftharpoons 2SO_3(g)$
(e) $4NH_3(g) + 5O_2(g) \rightleftharpoons 4NO(g) + 6H_2O(g)$
(f) $N_2O_4(g) \rightleftharpoons 2NO_2(g)$
(g) $CO_2(g) + H_2(g) \rightleftharpoons CO(g) + H_2O(g)$
(h) $NH_4Cl(s) \rightleftharpoons NH_3(g) + HCl(g)$
(i) $BaSO_3(s) \rightleftharpoons BaO(s) + SO_2(g)$
(j) $2Pb(NO_3)_2(s) \rightleftharpoons 2PbO(s) + 4NO_2(g) + O_2(g)$
(k) $2H_2(g) + O_2(g) \rightleftharpoons 2H_2O(l)$
(l) $S_8(g) \rightleftharpoons 8S(g)$

[S] 35. The rate of the reaction $H_2(g) + I_2(g) \rightarrow 2HI(g)$ at 25 °C is given by

$$\text{Rate} = 1.7 \times 10^{-18}\,[H_2][I_2]$$

The rate of decomposition of gaseous HI to $H_2(g)$

and $I_2(g)$ at 25°C is given by

$$Rate = 2.4 \times 10^{-21} [HI]^2$$

What is the equilibrium constant for the formation of gaseous HI from the gaseous elements at 25°C?   *Ans. 7.1 × 10²*

S 36. A sample of $NH_3(g)$ was formed from $H_2(g)$ and $N_2(g)$ at 500°C. If the equilibrium mixture was found to contain 1.35 mol H_2 per liter, 1.15 mol N_2 per liter and 4.12×10^{-1} mol NH_3 per liter, what is the value of the equilibrium constant for the formation of NH_3?
Ans. 6.00 × 10⁻² (M⁻², or mol⁻² L²)

37. Most of the 81 billion pounds of sulfuric acid produced in the United States during 1980 resulted from the reaction sequence

$$S_8(g) + 8O_2(g) \longrightarrow 8SO_2(g) \qquad (1)$$

$$2SO_2(g) + O_2(g) \xrightarrow{V_2O_5} 2SO_3(g) \qquad (2)$$

$$SO_3(g) + H_2O(l) \longrightarrow H_2SO_4(l) \qquad (3)$$

V_2O_5 is required in Equation (2) because the oxidation of SO_2 to SO_3 is slow in the absence of a catalyst. Each of these reactions has been studied extensively. Under a specific set of conditions at 500°C, equilibrium pressures of SO_2, O_2, and SO_3 in the reaction of Equation (2) have been determined to be 0.342 atm, 0.173 atm, and 0.988 atm, respectively. What is K_p for this reaction?   *Ans. 48.2 atm⁻¹*

S 38. A 0.72-mol sample of PCl_5 is put into a 1.00-L vessel and heated. At equilibrium the vessel contains 0.40 mol of $PCl_3(g)$ as well as Cl_2 and undissociated $PCl_5(g)$. What is the equilibrium constant for the decomposition of PCl_5 to PCl_3 and Cl_2 at this temperature?   *Ans. 0.50 M*

S 39. The vapor pressure of water is 0.196 atm at 60°C. What is the equilibrium constant for the transformation $H_2O(l) \rightleftharpoons H_2O(g)$?
Ans. K_p = 0.196 atm

40. A sample of ammonium chloride was heated in a closed container.

$$NH_4Cl(s) \rightleftharpoons NH_3(g) + HCl(g)$$

At equilibrium the pressure of $NH_3(g)$ was found to be 1.75 atm. What is the equilibrium constant for the decomposition at this temperature?
Ans. K_p = 3.06 atm²

41. In general, the equilibrium constant for a reaction in the gas phase has different units and a

different numerical value when pressures rather than concentrations are used to evaluate it. Show that such is not the case for the decomposition of HI into H_2 and I_2; show that the numerical value of the equilibrium constant for this particular reaction is independent of the units in which concentration is expressed.

S 42. Sodium sulfate 10-hydrate, $Na_2SO_4 \cdot 10H_2O$, dehydrates according to the equation

$$Na_2SO_4 \cdot 10H_2O(s) \rightleftharpoons$$
$$Na_2SO_4(s) + 10H_2O(g)$$

with $K = 4.08 \times 10^{-25}$ at 25°C. What is the pressure of water vapor in equilibrium with a sample of $Na_2SO_4 \cdot 10H_2O$?   *Ans. 3.64 × 10⁻³ atm*

S 43. Calculate the equilibrium concentration of NO_2 in 1 L of a solution prepared from 0.129 mol of N_2O_4 with chloroform as the solvent. For the reaction $N_2O_4 \rightleftharpoons 2NO_2$, in chloroform, $K = 1.07 \times 10^{-5}$.   *Ans. 1.17 × 10⁻³ M*

S 44. The equilibrium constant K for the reaction

$$PCl_5(g) \rightleftharpoons PCl_3(g) + Cl_2(g)$$

is 0.0211 at a certain temperature. What are the equilibrium concentrations of PCl_5, PCl_3, and Cl_2 starting with a concentration of PCl_5 of 1.00 M?
Ans. PCl₅, 0.865 M; PCl₃, 0.135 M; Cl₂, 0.135 M

45. The equilibrium constant for the reaction

$$H_2(g) + CO_2(g) \rightleftharpoons H_2O(g) + CO(g)$$

is 1.6 at 990°C. Calculate the number of moles of each component in the final equilibrium mixture obtained from adding 1.00 mol of H_2, 2.0 mol of CO_2, 0.75 mol of H_2O, and 1.0 mol of CO to a 5.00-L reactor at 990°C.   *Ans. 0.60 mol H₂; 1.6 mol CO₂; 1.2 mol H₂O; 1.4 mol CO*

S 46. The equilibrium constant K_p for the decomposition of nitrosyl bromide

$$2NOBr(g) \rightleftharpoons 2NO(g) + Br_2(g)$$

is 1.0×10^{-2} atm at 25°C. What percentage of NOBr is decomposed at 25°C and a total pressure of 0.25 atm?   *Ans. 34%*

Effect of Changes in Pressure, Temperature, or Concentration on Equilibrium

47. State Le Châtelier's principle as it applies to chemical equilibria.

48. Under what conditions do changes in pressure affect systems in equilibrium?

49. How will an increase in temperature affect each of the following equilibria? an increase in pressure?

   (a) $N_2(g) + 3H_2(g) \rightleftharpoons 2NH_3(g)$
   $$\Delta H = -92.2 \text{ kJ}$$
   (b) $H_2O(l) \rightleftharpoons H_2O(g)$ $\quad \Delta H = 41 \text{ kJ}$
   (c) $N_2(g) + O_2(g) \rightleftharpoons 2NO(g)$ $\quad \Delta H = 181 \text{ kJ}$
   (d) $3O_2(g) \rightleftharpoons 2O_3(g)$ $\quad \Delta H = 285 \text{ kJ}$
   (e) $CaCO_3(s) \rightleftharpoons CaO(s) + CO_2(g)$
   $$\Delta H = 176 \text{ kJ}$$

50. For each of the following reactions between gases at equilibrium, determine the effect on the equilibrium concentrations of the products when the temperature is decreased and when the external pressure on the system is decreased.

   (a) $2H_2O(g) \rightleftharpoons 2H_2(g) + O_2(g)$
   $$\Delta H = 484 \text{ kJ}$$
   (b) $N_2(g) + O_2(g) \rightleftharpoons 2NO(g)$ $\quad \Delta H = 181 \text{ kJ}$
   (c) $N_2(g) + 3H_2(g) \rightleftharpoons 2NH_3(g)$
   $$\Delta H = -92.2 \text{ kJ}$$
   (d) $2O_3(g) \rightleftharpoons 3O_2(g)$ $\quad \Delta H = -285 \text{ kJ}$
   (e) $H_2(g) + F_2(g) \rightleftharpoons 2HF(g)$ $\quad \Delta H = 541 \text{ kJ}$

51. Suggest four ways in which the equilibrium concentration of ammonia can be increased for the reaction in Exercise 50(c).

52. For the reaction in Exercise 50(b), the equilibrium constant at 2000°C is 6.2×10^{-4}. Consider each of the following situations, and decide whether or not a reaction will occur and, if so, in which direction it will predominantly proceed.

   (a) A 5.0-L box contains 0.26 mol of N_2, 0.0062 mol of O_2, and 0.0010 mol of NO at 2000°C.
   (b) A 2.5-L box contains 0.26 mol of N_2, 0.0062 mol of O_2, and 0.0010 mol of NO at 2500°C.

53. Under what conditions will the reversible decomposition

   $$CaCO_3(s) \rightleftharpoons CaO(s) + CO_2(g)$$

   proceed to completion in a closed container so that no $CaCO_3$ remains?

54. The binding of oxygen by hemoglobin (Hb), giving oxyhemoglobin (HbO_2), is partially regulated by the concentration of H^+ and CO_2 in the blood. Although the equilibrium is complicated, it can be summarized as

   $$HbO_2 + H^+ + CO_2 \rightleftharpoons CO_2-Hb-H^+ + O_2$$

(a) Write the equilibrium constant expression for this reaction.

(b) Explain why the production of lactic acid and CO_2 in a muscle during exertion stimulates release of O_2 from the oxyhemoglobin in the blood passing through the muscle.

S 55. The equilibrium constant for the reaction

   $$CO + H_2O \rightleftharpoons CO_2 + H_2$$

   is 5.0 at a given temperature.

   (a) Upon analysis, an equilibrium mixture of the substances present at the given temperature was found to contain 0.20 mol of CO, 0.30 mol of water vapor, and 0.90 mol of H_2 in a liter. How many moles of CO_2 were there in the equilibrium mixture? *Ans. 0.33 mol*

   (b) Maintaining the same temperature, additional H_2 was added to the system, and some water vapor was removed by drying. A new equilibrium mixture was thereby established containing 0.40 mol of CO, 0.30 mol of water vapor, and 1.2 mol of H_2 in a liter. How many moles of CO_2 were in the new equilibrium mixture? Compare this with the quantity in part (a) and discuss whether the second value is reasonable. Explain how it is possible for the water vapor concentration to be the same in the two equilibrium solutions even though some vapor was removed before the second equilibrium was established.
   Ans. 0.50 mol

S 56. A 1.00-L vessel at 400°C contains the following equilibrium concentrations: N_2, 1.00 M; H_2, 0.50 M; and NH_3, 0.50 M. How many moles of hydrogen must be removed from the vessel in order to increase the concentration of nitrogen to 1.2 M?

   Ans. 0.94 (Note that more hydrogen is removed than was originally present as elemental H_2. Is this possible?)

57. (a) Using the law of mass action, write the expression for the equilibrium constant for the reversible reaction

   $$N_2 + O_2 \rightleftharpoons 2NO \qquad \Delta H = 181 \text{ kJ}$$

   (b) What will happen to the concentration of NO at equilibrium if (1) more O_2 is added? (2) N_2 is removed? (3) the pressure on the system is increased? (4) the temperature of the system is increased?

(c) This reaction produces nitrogen oxide pollutants during the operation of an internal combustion engine. NO forms in the cylinders during combustion and is swept out in the exhaust. The reaction for the decomposition of NO ($2NO \rightarrow N_2 + O_2$) has an equilibrium constant of $K_p = 3.0 \times 10^{31}$ at 25°C (the temperature of a warm summer day). At equilibrium what would be the atmospheric NO pressure ($P_{O_2} = 0.2$ atm, $P_{N_2} = 0.8$ atm)?
Ans. 7×10^{-17} *atm*

(d) The actual concentration of NO along an expressway during rush-hour traffic is higher than this value. Suggest an explanation.

Additional Exercises

58. Some bacteria are resistant to the antibiotic penicillin because they produce penicillinase, an enzyme with a molecular weight of 30,000, which converts penicillin into inactive molecules. Although the kinetics of enzyme-catalyzed reactions can be complex, at low concentrations this reaction can be described by a rate equation that is first order in the catalyst (penicillinase) and that also involves the concentration of penicillin. From the following data for a solution containing 0.15 μg (0.15×10^{-6} g) of penicillinase, determine the order of the reaction with respect to penicillin and the value of the rate constant.

[Penicillin], M	Rate, mol L^{-1} min^{-1}
2.0×10^{-6}	1.0×10^{-10}
3.0×10^{-6}	1.5×10^{-10}
4.0×10^{-6}	2.0×10^{-10}

Ans. Rate $= k$[*penicillinase*][*penicillin*]; $k = 1.0 \times 10^7\ L\ mol^{-1}\ min^{-1}$

[S] 59. The hydrolysis of the sugar sucrose to the sugars glucose and fructose

$$C_{12}H_{22}O_{11} + H_2O \longrightarrow C_6H_{12}O_6 + C_6H_{12}O_6$$

follows a first-order rate equation for the disappearance of sucrose.

$$\text{Rate} = k[C_{12}H_{22}O_{11}]$$

(The products of the reaction, glucose and fructose, have the same molecular formulas but differ in the arrangement of the atoms in their molecules.)

(a) In neutral solution, $k = 2.1 \times 10^{-11}\ s^{-1}$ at 27°C and $8.5 \times 10^{-11}\ s^{-1}$ at 37°C. Determine the activation energy, the frequency factor, and the rate constant for this equation at 47°C.
Ans. $E_a = 108$ *kJ;* $A = 1.3 \times 10^8\ s^{-1}$; $k = 3.1 \times 10^{-10}\ s^{-1}$
(*An alternative and equally acceptable method of working the problem produces the answers* $E_a = 109$ *kJ;* $A = 2.0 \times 10^8\ s^{-1}$; $k = 3.2 \times 10^{-10}\ s^{-1}$.)

(b) The equilibrium constant for the reaction is 1.36×10^5 at 27°C. What are the concentrations of glucose, fructose, and sucrose after a 0.150 M aqueous solution of sucrose has reached equilibrium? Remember that the activity of a solvent (the effective concentration) is 1.
Ans. [*Sucrose*] $= 1.65 \times 10^{-7}$ *M;* [*glucose*] $=$ [*fructose*] $= 0.150$ *M*

(c) How long will the reaction of a 0.150 M solution of sucrose require to reach equilibrium at 27°C in the absence of a catalyst? Since the concentration of sucrose at equilibrium is so low, assume that the reaction is irreversible.
Ans. 4.29×10^{10} *s (1360 years!)*

60. Ethanol and acetic acid interact to form ethyl acetate, an important industrial solvent, and water, according to the equation

$$C_2H_5OH + CH_3CO_2H \rightleftharpoons CH_3COOC_2H_5 + H_2O$$

When 1 mol each of C_2H_5OH and CH_3CO_2H are allowed to react at 100°C in a sealed tube, equilibrium is established when $\frac{1}{3}$ mol of each of the reactants remains. Calculate the equilibrium constant for the reaction. (Note: Water is not a solvent in this reaction.)
Ans. 4

61. Assume that the rate equations given below apply to each reaction at equilibrium. Determine the rate equation for each reverse reaction.

[S] (a) $NO(g) + O_3(g) \longrightarrow NO_2(g) + O_2(g)$
$\text{Rate}_1 = k_1[NO][O_3]$
Ans. $Rate_2 = k_2[NO_2][O_2]$

(b) $2NO(g) + O_2(g) \longrightarrow 2NO_2(g)$
$\text{Rate}_1 = k_1[NO]^2[O_2]$
Ans. $Rate_2 = k_2[NO_2]^2$

(c) $2NO_2(g) + F_2(g) \longrightarrow 2NO_2F(g)$
$\text{Rate}_1 = k_1[NO_2][F_2]$
Ans. $Rate_2 = k_2[NO_2F]^2[NO_2]^{-1}$

[S] (d) $2NO(g) + 2H_2(g) \longrightarrow N_2(g) + 2H_2O(g)$
Rate$_1 = k_1[NO][H_2]$
Ans. $Rate_2 = k_2[N_2][H_2O]^2[NO]^{-1}[H_2]^{-1}$

62. The reaction between hydrogen, $H_2(g)$, and sulfur, $S_8(g)$, to produce H_2S is exothermic at 25 °C. How should the pressure and temperature be adjusted in order to improve the equilibrium yield of H_2S, assuming that all reactants and products are in the gaseous state? How will these conditions affect the rate of attainment of equilibrium?

63. Write the equilibrium constant expression for the acid-base reaction of each of the following substances with water:
(a) CH_3CO_2H (c) CN^- (e) F^-
(b) NH_3 (d) HSO_4^- (f) HCO_3^-

64. One possible mechanism for the reaction of H_2O_2 with I^- in acid solution (Section 15.14) is

$H_3O^+ + I^- \longrightarrow HI + H_2O$ (fast)
$H_2O_2 + HI \longrightarrow H_2O + HOI$ (slow)
$HOI + H_3O^+ + I^- \longrightarrow 2H_2O + I_2$ (fast)
$I_2 + I^- \longrightarrow I_3^-$ (fast)

(a) Write the rate equation for the slow elementary reaction. *Ans.* $Rate = k[H_2O_2][HI]$
(b) Write the equilibrium constant expression for the first elementary reaction.
Ans. $K = \frac{[HI]}{[H_3O^+][I^-]}$
(c) Solve the equilibrium constant expression from part (b) for [HI], and substitute this concentration into the rate equation from part (a) to obtain the overall rate equation for this mechanism.
Ans. $Rate = kK[H_3O^+][I^-][H_2O_2] = k'[H_3O^+][I^-][H_2O_2]$

65. Using the expression for the equilibrium constant for the fast reaction of H_3O^+ with H_2O_2 and the rate equation for the slow elementary reaction of $H_3O_2^+$ with I^-, derive the overall rate equation based on Mechanism C in Section 15.14 for the reaction of H_2O_2 with I^- in acid solution.

16

IONIC EQUILIBRIA
OF WEAK ELECTROLYTES

Water is probably the most important solvent on our planet. Our very lives depend on reactions that occur in solution in our cells. Many aqueous solutions contain ions that result from complete ionization of strong electrolytes or from partial ionization of weak electrolytes. The concentrations of these ions are very important in determining the rates, mechanisms, and yields of reactions in such solutions.

The concentrations of ions in a solution of a strong electrolyte can be calculated from its concentration since it is 100% ionized in solution. However, the concentrations of ions in a solution of a weak electrolyte cannot be determined so easily. Both ions and nonionized molecules of the weak electrolyte are present, and their concentrations must be calculated from the concentration of the weak electrolyte and the extent to which it ionizes. These calculations involve the concepts of equilibria introduced in Chapter 15.

Weak acids and weak bases are probably the most significant weak electrolytes. They are important in many purely chemical processes and in those of biological interest. For example, amino acids are both weak acids and weak bases. In this chapter we will consider some ways of expressing concentrations of hydronium ion and hydroxide ion in solutions of weak acids and weak bases and will then examine equilibria involving these weak electrolytes.

16.1   The Ionization of Water

Water is an extremely weak electrolyte that undergoes self-ionization (Section 12.3) with the formation of very small, *but equal,* amounts of hydronium and hydroxide ions.

$$2H_2O \rightleftharpoons H_3O^+ + OH^-$$

This slight ionization plays an important role in acid-base reactions in water. Since this reaction is reversible, we can apply the law of mass action and write

$$\frac{[H_3O^+][OH^-]}{[H_2O]^2} = K$$

The brackets stand for "molar concentration of"; $[H_3O^+]$ means molar concentration of hydronium ion, for example. For dilute solutions the molar concentration of water may be considered to be constant ($[H_2O] = 55.6\ M$), and we can write

$$[H_3O^+][OH^-] = K[H_2O]^2 = K_w$$

K_w is the **ion product for water;** at $25°C$ it has the value 1.0×10^{-14}. As you will see, this is an important equilibrium equation; you should memorize it. The degree of ionization of water and the resulting concentrations of hydronium ion and hydroxide ion increase as the temperature increases. At $100°C$, K_w is about 1×10^{-12}, 100 times larger than at $25°C$.

The ionization of water yields equal numbers of hydronium and hydroxide ions. Therefore, in pure water at $25°C$, $[H_3O^+] = [OH^-]$, and

$$[H_3O^+]^2 = [OH^-]^2 = 1.0 \times 10^{-14}$$
$$[H_3O^+] = [OH^-] = \sqrt{1.0 \times 10^{-14}} = 1.0 \times 10^{-7}$$

All aqueous solutions contain both hydronium ions and hydroxide ions. If an acid is added to pure water, $[H_3O^+]$ becomes larger than 1.0×10^{-7}, and $[OH^-]$ becomes less than 1.0×10^{-7}, but not 0. If a base is added to pure water, $[OH^-]$ becomes greater than 1.0×10^{-7}, and $[H_3O^+]$ becomes less than 1.0×10^{-7}, but not 0. The product of $[H_3O^+]$ and $[OH^-]$ at a given temperature is always a constant, so the value of either concentration can never equal 0.

One of the problems students have in chemistry is that chemists sometimes use different names or symbols to represent the same thing. For example, most American chemists call element 104 rutherfordium; others call it kurchatovium; radiochemists generally call it element-104 or simply 104. Very few chemists use the recommended name, unnilquadium (Section 4.13).

There is a similar lack of consistency in acid-base chemistry. Although almost all chemists recognize that a hydrogen ion is present in water as the hydronium ion, H_3O^+, many of them write it as H^+ or $H^+(aq)$ and talk about solutions of hydrogen ions or protons (which are, after all, hydrogen ions). In most of the literature dealing with equilibria, H^+ or $H^+(aq)$ is used instead of H_3O^+. The self-ionization of water is often written

$$H_2O \rightleftharpoons H^+ + OH^-$$

for which the equilibrium equation is

$$[H^+][OH^-] = K_w = 1.0 \times 10^{-14} \text{ (at } 25°C)$$

You can see this is the equation for the ion product of water, except that $[H^+]$ replaces $[H_3O^+]$. In fact, all equilibrium expressions for reactions in water can be written using $[H^+]$ to replace $[H_3O^+]$.

There is nothing wrong with using hydrogen ion rather than hydronium ion, provided that you remember that a hydrogen ion is bound to at least one molecule of water in aqueous solutions. (However, some instructors will want you to use hydronium ion to show that you remember.) From this point, we will usually talk about hydrogen ions and use the symbol H^+ in dealing with equilibrium problems. If you want to use hydronium ions in place of hydrogen ions, you can replace H^+ with H_3O^+ in equilibria and chemical equations and add one water molecule for each hydronium ion on the other side of the chemical equations.

16.2 pH and pOH

The concentration of hydrogen ion in a solution is a measure of its acidity or basicity. This concentration is often expressed in terms of the **pH of the solution, the negative logarithm of the hydrogen ion concentration.**

$$\text{pH} = -\log[H^+] \quad \text{or} \quad \text{pH} = \log\frac{1}{[H^+]}$$

The pH value is the negative power to which 10 must be raised to give the hydrogen ion concentration.

$$[H^+] = 10^{-\text{pH}}$$

A neutral solution, one in which $[H^+] = [OH^-]$, has a pH of 7 at 25°C. The pH of an acidic solution ($[H^+] > [OH^-]$) is less than 7; the pH of a basic solution ($[OH^-] > [H^+]$) is greater than 7. The pH of a solution can be calculated from its hydrogen ion concentration or measured with a **pH meter** (Fig. 16-1. See also Fig. 16-2.) Table 16-1 shows some common solutions and their pH values.

Figure 16-1. A pH meter. The pH of a solution is determined electronically from the difference in electrical potentials between the two electrodes when they are dipped into the solution and is displayed on the scale for convenient reading. (*Photograph courtesy of Corning Scientific*)

Figure 16-2. Some pairs of pH electrodes are combined into a single unit for convenience.

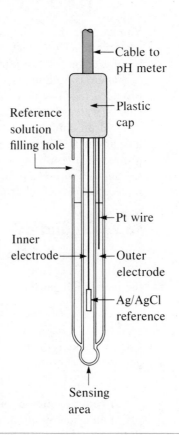

Table 16-1   Relationships of [H⁺], [OH⁻], pH, and pOH

$[H^+]$	$[OH^-]$	pH	pOH	Sample Solution	
10^1	10^{-15}	-1	15		Strongly acidic
10^0 or 1	10^{-14}	0	14	1 M HCl	
10^{-1}	10^{-13}	1	13		
10^{-2}	10^{-12}	2	12	Lime juice	
10^{-3}	10^{-11}	3	11	Stomach acid	
				Wine	
10^{-4}	10^{-10}	4	10		
10^{-5}	10^{-9}	5	9	Coffee	
10^{-6}	10^{-8}	6	8		
10^{-7}	10^{-7}	7	7	Pure water	Neutral
				Blood	
10^{-8}	10^{-6}	8	6		
10^{-9}	10^{-5}	9	5		
10^{-10}	10^{-4}	10	4		
10^{-11}	10^{-3}	11	3	Milk of magnesia	
10^{-12}	10^{-2}	12	2	Household ammonia	
10^{-13}	10^{-1}	13	1		
10^{-14}	10^0 or 1	14	0	1 M NaOH	Strongly basic

EXAMPLE 16.1   Calculate the pH of a 0.01 M solution of nitric acid, a strong acid.

Nitric acid is completely ionized in dilute solutions, so the concentration of the hydrogen ion is 0.01 M, or 1×10^{-2} M. Substituting, we have

$$pH = -\log[H^+] = -\log(1 \times 10^{-2}) = -(-2.0) = 2.0$$

(The use of logarithms is explained in Appendix A.3. On a calculator simply take the logarithm of 1×10^{-2}; the pH is equal to the negative of this logarithm.)

EXAMPLE 16.2   The concentration of the hydrochloric acid secreted by the stomach after a meal is about 1.2×10^{-3} M. What is the pH of stomach acid?

Because hydrochloric acid is a strong acid, the hydrogen ion concentration of stomach acid is the same as its molar concentration. Thus we have

$$pH = -\log(1.2 \times 10^{-3}) = -(\log 1.2 + \log 10^{-3})$$

$$= -(0.079 - 3.00) = -(-2.92) = 2.92$$

(On a calculator take the logarithm of 1.2×10^{-3}; the pH is equal to its negative.)

The answer 2.92 represents *two* significant figures rather than three. The number to the left of the decimal in a logarithm is the *characteristic*, which merely establishes the decimal place in the number that the logarithm represents. The only significant figures in a logarithm are those to the right of its decimal. The number of significant figures in a logarithm should be equal to the number of significant figures in the number from which the logarithm is obtained. There are two significant figures in the hydrogen ion concentration (1.2×10^{-3}) and two in the pH.

EXAMPLE 16.3   Water in equilibrium with air contains $4.4 \times 10^{-5}\%$ carbon dioxide. The resulting carbonic acid, H_2CO_3, gives the solution a hydrogen ion concentration of 2.0×10^{-6} M, about 20 times larger than that of pure water. Calculate the pH of the solution.

$$pH = -\log[H^+] = -\log(2.0 \times 10^{-6}) = -(\log 2.0 + \log 10^{-6})$$

$$= -(0.30 - 6.00) = -(-5.70) = 5.70$$

(On a calculator take the logarithm of 2.0×10^{-6}; the pH is equal to its negative.)

Thus we see that water in contact with air is acidic, rather than neutral, due to dissolved carbon dioxide.

The following examples illustrate the conversion of pH values to hydrogen ion concentrations.

EXAMPLE 16.4   Calculate the hydrogen ion concentration of a solution whose pH is 9.0.

$$pH = -\log [H^+] = 9.0$$
$$\log [H^+] = -9.0$$
$$[H^+] = 10^{-9.0} \quad \text{or} \quad [H^+] = \text{antilog of } -9.0$$
$$[H^+] = 1 \times 10^{-9} \, M$$

(On a calculator take the antilog, or the "inverse" log, of −9, or calculate 10^{-9}.)

EXAMPLE 16.5   Calculate the hydrogen ion concentration of blood, the pH of which is 7.3 (slightly alkaline).

$$pH = -\log [H^+] = 7.3$$
$$\log [H^+] = -7.3$$
$$[H^+] = 10^{-7.3} \quad \text{or} \quad [H^+] = \text{antilog of } -7.3$$
$$[H^+] = 5 \times 10^{-8} \, M$$

(On a calculator simply take the antilog, or the "inverse" log, of −7.3, or calculate $10^{-7.3}$.)

We may define **pOH of a solution as the negative logarithm of the hydroxide ion concentration.**

$$pOH = -\log [OH^-]$$

Similarly, we may define pK_w as the negative logarithm of the ion product for water (K_w). Now we can write the ion product equation for water in terms of pH, pOH, and pK_w.

$$[H^+][OH^-] = K_w$$
$$(-\log [H^+]) + (-\log [OH^-]) = -\log K_w$$
$$pH + pOH = pK_w$$

Since K_w has the value 1.0×10^{-14},

$$pK_w = -\log (1.0 \times 10^{-14}) = -(-14.00) = 14.00$$

it follows that

$$pH + pOH = 14.00$$
$$pH = 14.00 - pOH$$
$$pOH = 14.00 - pH$$

Several examples of hydrogen ion and hydroxide ion concentrations and the corresponding pH and pOH values are given in Table 16.1.

EXAMPLE 16.6   What are the pOH and pH of a 0.0125 M solution of potassium hydroxide, KOH?

Potassium hydroxide is completely ionized in solution, so $[OH^-] = 0.0125$ M.

$$pOH = -\log [OH^-] = -\log 0.0125$$
$$= -(-1.903) = 1.903$$

The pH can be found from the pOH.

$$pH + pOH = 14.00$$
$$pH = 14.00 - pOH = 14.00 - 1.903 = 12.10$$

Alternatively, we can calculate the pH from the ion-product equation.

$$[H^+][OH^-] = 1.0 \times 10^{-14}$$

$$[H^+] = \frac{1.0 \times 10^{-14}}{[OH^-]} = \frac{1.4 \times 10^{-14}}{0.0125}$$

$$= 8.0 \times 10^{-13}$$

$$pH = -\log (8.0 \times 10^{-13})$$
$$= -(-12.10) = 12.10$$

EXAMPLE 16.7   Calculate the pOH and pH of a 0.00030 M solution of strontium hydroxide, $Sr(OH)_2$.

$Sr(OH)_2$ is a strong base so it is completely ionized, and 1 mol of $Sr(OH)_2$ gives 2 mol of OH^-. Thus

$$[OH^-] = \frac{0.00030 \text{ mol } Sr(OH)_2}{1 \text{ L}} \times \frac{2 \text{ mol } OH^-}{1 \text{ mol } Sr(OH)_2} = 0.00060 \text{ } M$$

$$pOH = -\log [OH^-] = -\log 0.00060$$
$$= -(-3.22) = 3.22$$

$$pH + pOH = 14.00$$
$$pH = 14.00 - pOH = 14.00 - 3.22 = 10.78$$

IONIC EQUILIBRIA OF WEAK ACIDS AND WEAK BASES

16.3   Ion Concentrations in Solutions of Strong Electrolytes

Strong electrolytes are completely ionized in aqueous solution, so the ion concentrations may be found directly from the molar concentration. For example, in a 0.01 M solution of hydrochloric acid (HCl → H^+ + Cl^-), both the hydrogen ion concentration, $[H^+]$, and the chloride ion concentration, $[Cl^-]$, are equal to the concentration of HCl, [HCl]. From the chemical equation we can find the conversions 1 mol H^+/1 mol HCl and 1 mol Cl^-/1 mol HCl. Thus a HCl concentration of 0.01 M (0.01 mol HCl/1 L) can be converted to $[H^+]$ and $[Cl^-]$ as follows:

$$[H^+] = \frac{0.01 \text{ mol HCl}}{1 \text{ L}} \times \frac{1 \text{ mol } H^+}{1 \text{ mol HCl}} = \frac{0.01 \text{ mol } H^+}{1 \text{ L}} = 0.01 \ M$$

$$[Cl^-] = \frac{0.01 \text{ mol HCl}}{1 \text{ L}} \times \frac{1 \text{ mol } Cl^-}{1 \text{ mol HCl}} = \frac{0.01 \text{ mol } Cl^-}{1 \text{ L}} = 0.01 \ M$$

Because 1 mole of the strong electrolyte potassium sulfate, K_2SO_4, contains 2 moles of potassium ions and 1 mole of sulfate ions, $[K^+]$ and $[SO_4^{2-}]$ in a 0.03 M solution of K_2SO_4 are given by

$$[K^+] = \frac{0.03 \text{ mol } K_2SO_4}{1 \text{ L}} \times \frac{2 \text{ mol } K^+}{1 \text{ mol } K_2SO_4} = 0.06 \ M$$

$$[SO_4^{2-}] = \frac{0.03 \text{ mol } K_2SO_4}{1 \text{ L}} \times \frac{1 \text{ mol } SO_4^{2-}}{1 \text{ mol } K_2SO_4} = 0.03 \ M$$

16.4 The Ionization of Weak Acids

We can tell by measuring the pH of an aqueous solution of a weak acid such as acetic acid, CH_3CO_2H, that only a fraction of the molecules are ionized into hydrogen cations and acetate anions. The remaining acid is present as the nonionized, or molecular, form. Alternatively, we can calculate the concentrations of nonionized acetic acid, of hydrogen cations, and of acetate anions from the equilibrium constant for the ionization of acetic acid. For acetic acid the ionization reaction is

$$CH_3CO_2H(aq) \rightleftharpoons H^+(aq) + CH_3CO_2^-(aq)$$

Since equilibrium is attained almost immediately when an acid dissolves in water, we can use the law of mass action and write the equation for K_a, **the acid dissociation constant, or acid ionization constant.**

$$K_a = \frac{[H^+][CH_3CO_2^-]}{[CH_3CO_2H]} \tag{1}$$

Before we use this equilibrium expression, let's show that it is the same as the one using hydronium ion. The chemical equation using hydronium ion is written

$$CH_3CO_2H(aq) + H_2O(l) \rightleftharpoons H_3O^+(aq) + CH_3CO_2^-(aq)$$

For this equation the law of mass action gives

$$\frac{[H_3O^+][CH_3CO_2^-]}{[CH_3CO_2H][H_2O]} = K$$

For dilute solutions the number of moles of water consumed in the formation of hydronium ions is negligible compared to the number of moles of water present, and the molar concentration of water remains practically constant. Therefore we can write

$$\frac{[H_3O^+][CH_3CO_2^-]}{[CH_3CO_2H]} = [H_2O]K$$

$$\frac{[H_3O^+][CH_3CO_2^-]}{[CH_3CO_2H]} = K_a$$

Except for the replacement of $[H^+]$ with $[H_3O^+]$, this last equation is identical to Equation (1).

Concentrations of ions and molecules in the equation for K_a are generally expressed in moles per liter. The data in Table 16-2 show that K_a for acetic acid remains practically constant over a considerable range of concentrations. Note that the extent to which acetic acid ionizes increases with decreasing concentration. For example, 0.100 M acetic acid is 1.34% ionized (98.66% is in the molecular form), while 0.0100 M acetic acid is 4.15% ionized (95.85% is in the molecular form). The reason for this is that in a more dilute solution at equilibrium the ions are farther apart and have less tendency to combine to form undissociated acetic acid molecules.

Table 16-2   Ionization Constants for Acetic Acid at Different Concentrations

Molarity	% Ionized	$[H^+]$ and $[CH_3CO_2^-]$	$[CH_3CO_2H]$	K_a
0.100	1.34	0.00134	0.09866	1.82×10^{-5}
0.0800	1.50	0.00120	0.07880	1.83×10^{-5}
0.0300	2.45	0.000735	0.02927	1.85×10^{-5}
0.0100	4.15	0.000415	0.009585	1.80×10^{-5}

The ionization constants of a number of weak acids are given in Table 16-3; a more complete list is given in Appendix G. The acids in the table are listed in order of decreasing strength, as indicated by the decreasing size of the ionization constant (compare with Table 14-1 in Section 14.3). As the fraction of acid existing in the ionized form decreases, the ionization constant decreases. Thus HF with its K_a of 7.2×10^{-4} is a much stronger acid than HCN, which has a K_a of 4×10^{-10}.

Table 16-3   Ionization Constants of Some Weak Acids

Ionization Reaction	K_a at 25°C
$HF \rightleftharpoons H^+ + F^-$	7.2×10^{-4}
$HNO_2 \rightleftharpoons H^+ + NO_2^-$	4.5×10^{-4}
$HNCO \rightleftharpoons H^+ + NCO^-$	3.46×10^{-4}
$HCO_2H \rightleftharpoons H^+ + HCO_2^-$	1.8×10^{-4}
$CH_3CO_2H \rightleftharpoons H^+ + CH_3CO_2^-$	1.8×10^{-5}
$HClO \rightleftharpoons H^+ + ClO^-$	3.5×10^{-8}
$HBrO \rightleftharpoons H^+ + BrO^-$	2×10^{-9}
$HCN \rightleftharpoons H^+ + CN^-$	4×10^{-10}

The following exemplify the types of problems that are based on the partial ionization of weak acids.

EXAMPLE 16.8   In a 0.0800 M solution acetic acid is 1.50% ionized. Calculate $[H^+]$, $[CH_3CO_2^-]$, and $[CH_3CO_2H]$ in the solution.

First, we write the equation for the ionization of acetic acid.

$$CH_3CO_2H \rightleftharpoons H^+ + CH_3CO_2^-$$

The equation shows that, for each mole of CH_3CO_2H that ionizes, one mole of H^+ and one mole of $CH_3CO_2^-$ are formed. Since 1.50% of 0.0800 mol/L of acid ionizes,

$$[H^+] = [CH_3CO_2^-] = 0.0800 \ M \times 0.0150 = 0.00120 \ M$$

and

$$[CH_3CO_2H] = 0.0800 \ M - 0.00120 \ M = 0.0788 \ M$$

EXAMPLE 16.9   Using the concentrations found in Example 16.8, calculate the ionization constant of acetic acid.

First, we write the expression for the ionization constant of acetic acid and then substitute in it the values found above to solve for K_a.

$$K_a = \frac{[H^+][CH_3CO_2^-]}{[CH_3CO_2H]} = \frac{0.00120 \times 0.00120}{0.0788} = 1.83 \times 10^{-5}$$

EXAMPLE 16.10   The pH of a 0.0495 M solution of nitrous acid, HNO_2, is 2.34. What is K_a for nitrous acid?

The equation for the ionization of nitrous acid is

$$HNO_2 \rightleftharpoons H^+ + NO_2^-$$

for which

$$K_a = \frac{[H^+][NO_2^-]}{[HNO_2]}$$

Thus we need to know the concentrations of H^+, NO_2^-, and HNO_2 at equilibrium in order to calculate K_a.

In many equilibrium problems we are given concentrations of reactants (and sometimes of products) before equilibrium is established and are asked to determine concentrations at equilibrium. It is often helpful to set up a table such as the one below showing the initial concentrations (the concentrations before equilibrium is established) and the concentrations at equilibrium. The concentration given in the problem is shown in color; the entries given in black are calculated as described following the table.

	$[HNO_2]$	$[H^+]$	$[NO_2^-]$
Initial concentrations, M	0.0495	~0	0
Final concentrations, M	0.0450	0.0045	0.0045

We are given the pH (from which we can get [H⁺] at equilibrium) and the total concentration of HNO_2 (0.0495 M), which may be considered to be that before any ionizes. Thus before equilibrium is established, $[HNO_2]_i = 0.0495\ M$. The subscript letter i is used to indicate an initial concentration, the concentration before any reaction occurs. As shown in Table 16-1, [H⁺] in pure water (the concentration of H⁺ before the reaction begins) is not zero. However, in most cases involving the dissolution of a weak acid in water, the initial concentration of H⁺ is sufficiently small that it may be neglected, so we take $[H^+]_i = \sim0$. (The symbol $\sim$ means "approximately," so ~0 means "approximately zero.") The concentration of NO_2^- is zero before any HNO_2 reacts with water.

Since pH $= -\log[H^+]$, we can calculate [H⁺] at equilibrium.

$$pH = -\log[H^+] = 2.34$$
$$\log[H^+] = -2.34$$
$$[H^+] = 10^{-2.34} = 0.0045\ M$$

From the chemical equation we see that one mole of NO_2^- forms for each mole of H⁺ formed. Thus $[NO_2^-]$ also equals 0.0045 M.

For each mole of H⁺ that forms, one mole of HNO_2 ionizes. Thus the equilibrium concentration of HNO_2 is equal to the initial concentration (0.0495 M) minus the amount that ionizes (0.0045 M).

$$[HNO_2] = 0.0495\ M - 0.0045\ M = 0.0450\ M$$

At equilibrium

$$K_a = \frac{[H^+][NO_2^-]}{[HNO_2]}$$

$$= \frac{0.0045 \times 0.0045}{0.0495}$$

$$= 4.5 \times 10^{-4}$$

EXAMPLE 16.11   Taking K_a for acetic acid to be 1.8×10^{-5}, calculate the percent ionization in a 0.100 M solution of the acid.

The equation for the ionization of acetic acid is

$$CH_3CO_2H \rightleftharpoons H^+ + CH_3CO_2^-$$

Here is a table of initial concentrations (concentrations before the reaction occurs) and equilibrium concentrations, with the one given in the problem in color.

	$[CH_3CO_2H]$	$[H^+]$	$[CH_3CO_2^-]$
Initial concentrations, M	0.100	~0	0
Equilibrium concentrations, M	$0.100 - x$	x	x

If we let x be the number of moles of H⁺ formed by the ionization of the acid in 1 L of solution, the concentration of H⁺ at equilibrium, [H⁺], is equal to x mol/1 L or x M. Since the ionization of acetic acid yields one mole of $CH_3CO_2^-$ for each mole of H⁺ produced, the concentration of acetate ion at equilibrium,

$[CH_3CO_2^-]$, is also equal to x M. The equilibrium concentration of nonionized acetic acid is equal to the initial concentration of CH_3CO_2H minus the amount that ionizes to give H^+ and $CH_3CO_2^-$.

$$[CH_3CO_2H] = [CH_3CO_2H]_i - x = 0.100 - x$$

At equilibrium

$$\frac{[H^+][CH_3CO_2^-]}{[CH_3CO_2H]} = K_a = 1.8 \times 10^{-5}$$

Substitution of equilibrium concentrations yields

$$\frac{(x)(x)}{0.100 - x} = \frac{x^2}{0.100 - x} = 1.8 \times 10^{-5}$$

The small value of K_a indicates that the ratio $x^2/(0.100 - x)$ is small, so x will be small relative to 0.100, from which it is to be subtracted. Thus $(0.100 - x)$ is virtually equal to 0.100, and the preceding equation may be simplified for an approximate solution as follows:

$$\frac{x^2}{0.100} = 1.8 \times 10^{-5}$$

$$x^2 = 1.8 \times 10^{-6}$$

$$x = \sqrt{1.8 \times 10^{-6}} = 1.3 \times 10^{-3} \text{ mol L}^{-1}$$

Thus the concentration of H^+ and of $CH_3CO_2^-$ at equilibrium is 1.3×10^{-3} M.

We were justified in neglecting x in the expression $(0.100 - x)$, since $0.100 - 0.0013 = 0.0987$ which is almost 0.100. When x is less than about 5% of the total concentration of the acid, use this approximation method; if x is greater than 5%, then the complete quadratic equation should be solved (see Appendix A.4 for the solution of quadratic equations).

To calculate the percent ionization, divide the concentration of the acid in the ionized form, which is equal to $[H^+]$, by the total concentration of the acid, and then multiply by 100. Thus we have

$$\frac{\text{Ionized } CH_3CO_2H}{\text{Total } CH_3CO_2H} \times 100 = \frac{[H^+]}{[CH_3CO_2H]_i} \times 100$$

$$= \frac{1.3 \times 10^{-3}}{0.100} \times 100 = 1.3\% \text{ ionized}$$

The values calculated in this example do not exactly match the values given in Table 16-2 because only two significant figures were used here.

16.5   The Ionization of Weak Bases

The most common weak base is aqueous ammonia. When gaseous ammonia is dissolved in water, the solution becomes basic because of the reaction

$$NH_3(aq) + H_2O(l) \rightleftharpoons NH_4^+(aq) + OH^-(aq)$$

Application of the law of mass action to this system gives the expression

$$\frac{[NH_4^+][OH^-]}{[NH_3][H_2O]} = K$$

Only a small fraction of the water is consumed in the reaction because we are working with a dilute solution; thus the concentration of water is practically constant and we may write

$$\frac{[NH_4^+][OH^-]}{[NH_3]} = K[H_2O] = K_b = 1.8 \times 10^{-5}$$

Aqueous ammonia is about as strong a base as acetic acid is an acid; the ionization constants for the two substances are the same (to two significant figures).

Equilibria involving the ionization of weak bases may be treated mathematically in the same manner as those of weak acids. The ionization constants of several weak bases are given in Table 16-4 and in Appendix H.

Table 16-4 Ionization Constants of Some Weak Bases

Base	Ionization	K_b at 25°C
$(CH_3)_2NH$ Dimethylamine	$+ H_2O \rightleftharpoons (CH_3)_2NH_2^+ + OH^-$	7.4×10^{-4}
CH_3NH_2 Methylamine	$+ H_2O \rightleftharpoons CH_3NH_3^+ + OH^-$	4.4×10^{-4}
$(CH_3)_3N$ Trimethylamine	$+ H_2O \rightleftharpoons (CH_3)_3NH^+ + OH^-$	7.4×10^{-5}
NH_3 Ammonia	$+ H_2O \rightleftharpoons NH_4^+ + OH^-$	1.8×10^{-5}
$C_6H_5NH_2$ Aniline	$+ H_2O \rightleftharpoons C_6H_5NH_3^+ + OH^-$	4.6×10^{-10}

EXAMPLE 16.12   The sedative Veronal, $C_8H_{12}N_2O_3$ (Fig. 16-3), is a weak base with an ionization constant, K_b, of 1.1×10^{-8}. What is $[OH^-]$ in a 0.010 M aqueous solution of Veronal?

Using Vr as the abbreviation for Veronal, the equation for the reaction with water is

$$Vr + H_2O \rightleftharpoons VrH^+ + OH^-$$

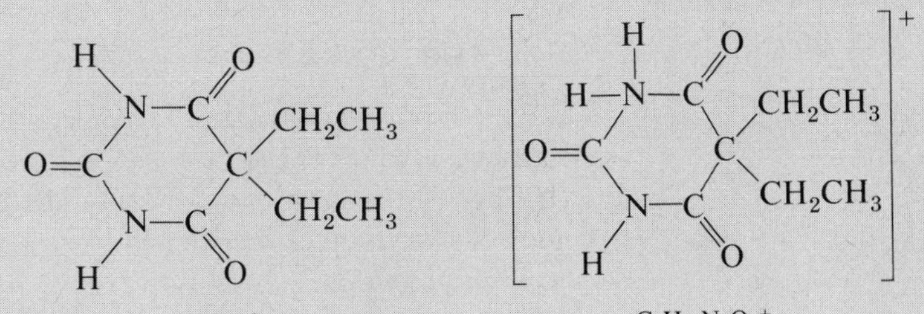

Figure 16-3. The structure of the base Veronal, $C_8H_{12}N_2O_3$, and of the cation formed when a proton is added to it.

Before any Veronal reacts with water, [Vr] is 0.010 M. As shown in Table 16-1, the concentration of OH^- in pure water ([OH^-] before the reaction) is not zero. However, in most cases $[OH^-]_i$ can be taken to be approximately zero (~ 0). The concentration of VrH^+ is zero before any Veronal reacts with water.

The concentrations of VrH^+ and OH^- at equilibrium are equal since the reaction of 1 mol of Vr produces 1 mol of VrH^+ and 1 mol of OH^-. Thus $[VrH^+] = [OH^-] = x$, and the concentration of Vr at equilibrium is $0.010 - x$, where x is the number of moles of Veronal that accept a proton per liter of solution.

The following table shows initial concentrations (concentrations before the reaction occurs) and equilibrium concentrations, with the one given in the problem shown in color.

	[Vr]	[VrH$^+$]	[OH$^-$]
Initial concentrations, M	0.010	0	~ 0
Equilibrium concentrations, M	$0.010 - x$	x	x

At equilibrium

$$\frac{[VrH^+][OH^-]}{[Vr]} = K_b = 1.1 \times 10^{-8}$$

or

$$\frac{(x)(x)}{0.010 - x} = \frac{x^2}{0.010 - x} = 1.1 \times 10^{-8}$$

If we assume that x is small relative to 0.010 (a reasonable assumption in view of the small K_b), then

$$x^2 = 0.010 \times 1.1 \times 10^{-8}$$
$$x = \sqrt{1.1 \times 10^{-10}} = 1.0 \times 10^{-5}$$

Thus, the concentration of OH^- is $1.0 \times 10^{-5} M$.

To check our assumption about the magnitude of x relative to 0.010, we need to check the percent ionization of Veronal, or the value of $[VrH^+]/[Vr]_i$ multiplied by 100. If the percent ionization is less than 5%, the approximation is justified.

$$\frac{[VrH^+]}{[Vr]_i} \times 100 = \frac{1.0 \times 10^{-5}}{0.010} \times 100 = 0.10\%$$

The approximation is justified.

16.6 The Salt Effect

The expression for the ionization constant holds for pure, dilute solutions of weak electrolytes. However the extent of ionization of a weak electrolyte may be influenced by the presence of other ions. The addition of a salt slightly increases the degree of ionization of a weak acid. For example, the ionization constant of acetic acid in dilute aqueous solution is 1.8×10^{-5}. This increases to 2.2×10^{-5} when the solution is made 0.1 M in sodium chloride. This effect, the **salt effect,** occurs because the activities of the ions (Section 13.27) of a weak electrolyte decrease due to interionic attractions for the ions of the strong electrolyte. An ionization constant is, therefore, not strictly constant, but varies slightly with the ionic strength of the solution. For most purposes, and for ours, the expression for the ionization constant may be used without regard to the slight errors that may arise due to the salt effect.

16.7 The Common Ion Effect

The acidity of an aqueous solution of acetic acid decreases when the strong electrolyte sodium acetate, $NaCH_3CO_2$, is added. This can be explained by Le Châtelier's principle (Section 15.19); the addition of acetate ions causes the equilibrium to shift to the left, increasing the concentration of CH_3CO_2H and reducing the concentration of H^+.

$$CH_3CO_2H \rightleftharpoons H^+ + CH_3CO_2^-$$

Since sodium acetate and acetic acid have the acetate ion in common, this influence on the equilibrium is known as the **common ion effect.** Since the equilibrium is disturbed by the addition of $CH_3CO_2^-$, a shift occurs that reduces the concentration of $CH_3CO_2^-$. This is accomplished by the combination of H^+ with $CH_3CO_2^-$, forming molecular acetic acid and reducing $[H^+]$ of the solution.

The common ion effect is important in the adjustment of the hydrogen ion concentration for many of the precipitations and separations in a qualitative analysis scheme, and it also plays an essential role in blood and other biological fluids.

The extent to which the concentration of the hydrogen ion is decreased may be calculated from the expression for the ionization constant.

EXAMPLE 16.13   Calculate $[H^+]$ in a 0.10 M solution of CH_3CO_2H that is 0.50 M with respect to $NaCH_3CO_2$.

	$[CH_3CO_2H]$	$[H^+]$	$[CH_3CO_2^-]$
Initial concentrations, M	0.10	~0	0.50
Equilibrium concentrations, M	$0.10 - x$	x	$0.50 + x$

Before reaction $[CH_3CO_2H]_i$ is 0.10 M, $[H^+]_i$ is approximately zero, and $[CH_3CO_2^-]_i$ is 0.50 M. $NaCH_3CO_2$ is an ionic compound, and thus is completely ionized in solution. A 0.50 M solution of $NaCO_3CO_2$ is 0.50 M in Na^+ and 0.50 M in $CH_3CO_2^-$.

At equilibrium the unknown concentration of H^+ may be taken as x. Since one mole of $CH_3CO_2^-$ is formed for each mole of H^+ formed, the concentration of $CH_3CO_2^-$ at equilibrium is equal to $0.50 + x$, the sum of the amount of $CH_3CO_2^-$ present before the reaction and the amount of $CH_3CO_2^-$ produced as equilibrium is reached. The concentration of nonionized acetic acid at equilibrium is $0.10 - x$. Substituting in the expression for the ionization constant for acetic acid, we have

$$K_a = \frac{[H^+][CH_3CO_2^-]}{[CH_3CO_2H]} = \frac{x(0.50 + x)}{0.10 - x} = 1.8 \times 10^{-5}$$

Even in the absence of $NaCH_3CO_2$, the concentration of the acetate ion derived from the ionization of 0.10 M acetic acid would be small (0.0013 mol L^{-1}) relative to that from $NaCH_3CO_2$ (0.50 mol L^{-1}). Since the degree of ionization is even smaller in the presence of sodium acetate, the concentration of acetate ion may be taken as equal to the concentration of the sodium acetate; that is, x may be neglected in the term $0.50 + x$. Likewise, the concentration of the nonionized acetic acid is very nearly equal to 0.10 M, and x in the term $0.10 - x$ may be dropped. Thus an approximate solution to the problem is

$$\frac{[H^+][CH_3CO_2H]}{[CH_3CO_2H]} = \frac{x(0.50)}{0.10} = 1.8 \times 10^{-5}$$

$$x = \frac{0.10}{0.50} \times 1.8 \times 10^{-5} = 3.6 \times 10^{-6} \, M = [H^+]$$

Table 16-2 shows that the concentration of the hydrogen ion in 0.100 M CH_3CO_2H is 0.00134 M. This is reduced to 0.0000036 M when the solution is also 0.50 M in sodium acetate. The simplifying assumption that x is negligible compared to either 0.50 or 0.10 is justified.

EXAMPLE 16.14   A 0.10 M solution of aqueous ammonia, also containing ammonium chloride, has a hydroxide ion concentration of 2.8×10^{-6} M. What is the concentration of ammonium ion in the solution?

	$[NH_3]$	$[NH_4^+]$	$[OH^-]$
Initial concentrations, M	0.10	—	—
Equilibrium concentrations, M	$0.10 - (2.8 \times 10^{-6})$	x	2.8×10^{-6}

Since we are calculating the equilibrium concentration of NH_4^+, we set $[NH_4^+] = x$. In order to determine the other concentrations in the table, we first write the equation for the ionization of ammonia.

$$NH_3 + H_2O \rightleftharpoons NH_4^+ + OH^-$$

This equation indicates that for each mole of OH^- formed, one mole of NH_3 is converted to NH_4^+, and the concentration of NH_3 decreases by one mole. Since we are given the equilibrium concentration of OH^-,

$$[OH^-] = 2.8 \times 10^{-6} \, M$$

the initial concentration of NH_3 must have decreased by that amount to produce 2.8×10^{-6} mol of OH^- per liter. Thus

$$[NH_3] = [NH_3]_i - (2.8 \times 10^{-6}) = 0.10 - (2.8 \times 10^{-6})$$

We now have all the concentrations necessary to use the equilibrium constant expression and need not find $[NH_4^+]_i$ or $[OH^-]_i$.

Since 2.8×10^{-6} is small relative to 0.10, we can approximate $[NH_3]$ as 0.10, and

$$K_b = \frac{[NH_4^+][OH^-]}{[NH_3]} = \frac{x(2.8 \times 10^{-6})}{0.10} = 1.8 \times 10^{-5}$$

Solving for x:

$$x = \frac{(0.10)(1.8 \times 10^{-5})}{2.8 \times 10^{-6}} = 0.64$$

Thus we find the concentration of the ammonium ion is 0.64 M at equilibrium. Of this amount, 2.8×10^{-6} mol is formed by the ionization of ammonia, and the remainder comes from the added ammonium chloride.

EXAMPLE 16.15   Exactly 10 mL of 0.100 M sodium hydroxide solution is added to 25 mL of 0.100 M acetic acid solution. Calculate the hydrogen ion concentration of the resulting 35 mL of solution.

Sodium hydroxide, a strong base, reacts with acetic acid to form sodium acetate. A volume of 10 mL (0.010 L) of 0.100 M sodium hydroxide contains

$$0.010 \, \cancel{L} \times 0.100 \, \text{mol}/\cancel{L} = 1.0 \times 10^{-3} \, \text{mol NaOH}$$

A volume of 25 mL (0.025 L) of 0.100 M acetic acid contains

$$0.025 \, \cancel{L} \times 0.100 \, \text{mol}/\cancel{L} = 2.5 \times 10^{-3} \, \text{mol CH}_3\text{CO}_2\text{H}$$

The 1.0×10^{-3} mol of NaOH neutralizes 1.0×10^{-3} mol of CH_3CO_2H according to the equation

$$NaOH(aq) + CH_3CO_2H(aq) \longrightarrow Na^+(aq) + CH_3CO_2^-(aq) + H_2O(l)$$

producing 1.0×10^{-3} mol of $NaCH_3CO_2$ and leaving 1.5×10^{-3} mol of unreacted CH_3CO_2H. Thus a solution containing sodium acetate and acetic acid results.

The problem is a common ion problem like that in Example 16.13. In order to find the equilibrium concentration of H^+, we need to calculate the initial concentrations of 1.5×10^{-3} mol of CH_3CO_2H and 1.0×10^{-3} mol of $NaCH_3CO_2$ dissolved in the total 35 mL of solution.

$$[CH_3CO_2H]_i = \frac{1.5 \times 10^{-3} \, \text{mol}}{0.035 \, \text{L}} = 4.29 \times 10^{-2} \, M$$

$$[CH_3CO_2^-]_i = [NaCH_3CO_2] = \frac{1.0 \times 10^{-3} \, \text{mol}}{0.035 \, \text{L}} = 2.86 \times 10^{-2} \, M$$

	[CH₃CO₂H]	[H⁺]	[CH₃CO₂⁻]
Initial concentrations, M	4.29×10^{-2}	~ 0	2.86×10^{-2}
Equilibrium concentrations, M	$(4.29 \times 10^{-2}) - x$	x	$(2.86 \times 10^{-2}) + x$

$$\frac{[H^+][CH_3CO_2^-]}{[CH_3CO_2H]} = \frac{x[(2.86 \times 10^{-2}) + x]}{(4.29 \times 10^{-2}) - x} = 1.8 \times 10^{-5}$$

Assuming that x is small relative to either 2.86×10^{-2} or 4.29×10^{-2},

$$\frac{x(2.86 \times 10^{-2})}{4.29 \times 10^{-2}} = 1.8 \times 10^{-5}$$

$$x = \frac{4.29 \times 10^{-2}}{2.86 \times 10^{-2}} \times 1.8 \times 10^{-5}$$

$$= 2.7 \times 10^{-5}$$

Thus $[H^+] = 2.7 \times 10^{-5}\,M$, a value low enough to justify our approximation based on neglecting x.

16.8   Buffer Solutions

Mixtures of weak acids and their salts or mixtures of weak bases and their salts are called **buffer solutions,** or **buffers.** They resist a change in hydrogen ion concentration when small amounts of acid or base are added. An example of a buffer solution is a mixture of acetic acid and sodium acetate.

EXAMPLE 16.16   (a) Calculate the pH of a buffer that is a mixture containing 0.10 M acetic acid and 0.10 M sodium acetate. (b) Calculate the change in pH when 1.0 mL of 0.10 M NaOH is added to 100 mL of this buffer.

(a) Determination of the pH of the buffer solution requires the calculation of $[H^+]$ in a typical common ion equilibrium, as illustrated in Example 16.13.

	$[CH_3CO_2H]$	$[H^+]$	$[CH_3CO_2^-]$
Initial concentrations, M	0.10	~0	0.10
Equilibrium concentrations, M	$0.10 - x$	x	$0.10 + x$

Assuming that x is negligible compared to 0.10,

$$K_a = \frac{[H^+][CH_3CO_2^-]}{[CH_3CO_2H]} = \frac{x(0.10)}{0.10} = 1.8 \times 10^{-5}$$

$$[H^+] = x = \frac{0.10}{0.10} \times 1.8 \times 10^{-5} = 1.8 \times 10^{-5}$$

$$pH = -\log[H^+] = -\log(1.8 \times 10^{-5}) = 4.74$$

(b) On addition of 1.0 mL (0.0010 L) of 0.10 M sodium hydroxide to 100 mL (0.100 L) of this buffer solution, an equivalent amount of acetic acid is neutralized by the sodium hydroxide, and sodium acetate is formed.

$$CH_3CO_2H + OH^- \longrightarrow H_2O + CH_3CO_2^-$$

Now we will calculate the new concentrations of acetic acid and sodium acetate and then use them to calculate the concentration of hydrogen ion.

Before reaction, 0.100 L of the buffer solution contains

$$0.100\,\cancel{L} \times \frac{0.10\,\text{mol CH}_3\text{CO}_2\text{H}}{1\,\cancel{L}} = 1.0 \times 10^{-2}\,\text{mol CH}_3\text{CO}_2\text{H}$$

and 1.0×10^{-2} mol of NaCH$_3$CO$_2$. Also, 1 mL (0.0010 L) of 0.10 M sodium hydroxide contains

$$0.0010\,\cancel{L} \times \frac{0.10\,\text{mol NaOH}}{1\,\cancel{L}} = 1.0 \times 10^{-4}\,\text{mol NaOH}$$

The 1.0×10^{-4} mol of NaOH neutralizes 1.0×10^{-4} mol of CH$_3$CO$_2$H leaving

$$(1.0 \times 10^{-2}) - (0.01 \times 10^{-2}) = 0.99 \times 10^{-2} \text{ mol of } CH_3CO_2H$$

and producing 1.0×10^{-4} mol of $NaCH_3CO_2$. This makes a total of

$$(1.0 \times 10^{-2}) + (0.01 \times 10^{-2}) = 1.01 \times 10^{-2} \text{ mol of } NaCH_3CO_2$$

After reaction, 0.99×10^{-2} mol of CH_3CO_2H and 1.01×10^{-2} mol of $NaCH_3CO_2$ are contained in 101 mL of solution, so the concentrations are 9.9×10^{-3} mol/0.101 L = 0.098 M CH_3CO_2H and 1.01×10^{-2} mol/0.101 L = 0.100 M $NaCH_3CO_2$. Now we calculate the pH of the solution, which is 0.098 M in CH_3CO_2H and 0.100 M in $NaCH_3CO_2$.

	$[CH_3CO_2H]$	$[H^+]$	$[CH_3CO_2^-]$
Initial concentrations, M	0.098	~0	0.100
Equilibrium concentrations, M	$0.098 - x$	x	$0.100 + x$

Assuming that x is small relative to 0.100 or 0.098,

$$K_a = \frac{[H^+][CH_3CO_2^-]}{[CH_3CO_2H]} = \frac{x(0.100)}{0.098} = 1.8 \times 10^{-5}$$

$$[H^+] = x = \frac{0.098}{0.100} \times 1.8 \times 10^{-5} = 1.76 \times 10^{-5}$$

$$pH = -\log[H^+] = -\log(1.76 \times 10^{-5}) = 4.75$$

Thus the addition of the base barely changes the pH of the solution.

An acetic acid–sodium acetate mixture acts as a buffer to keep the hydrogen ion concentration almost constant because, as sodium hydroxide is added, the hydroxide ions react with the few hydrogen ions present, and then more of the acetic acid ionizes, restoring the hydrogen ion concentration to near its original value. If a small amount of hydrochloric acid were added to the buffer solution, most of the hydrogen ions from the hydrochloric acid would combine with acetate ions, forming acetic acid molecules.

$$H^+ + CH_3CO_2^- \longrightarrow CH_3CO_2H$$

Thus there would be very little increase in the concentration of the hydrogen ion, and the pH would remain practically unchanged.

Buffers play a very important role in chemical processes; they keep hydrogen ion and hydroxide ion concentrations approximately constant over a range of changes in condition, such as the addition of acid, base, or water. Common buffer pairs besides CH_3CO_2H and $CH_3CO_2^-$ include NH_3 and NH_4^+, H_2CO_3 and HCO_3^-, and $H_2PO_4^-$ and HPO_4^{2-}.

Blood is an important example of a buffered solution, with the principal acid and ion responsible for the buffering action being carbonic acid, H_2CO_3, and the hydrogen carbonate ion, HCO_3^-. When an excess of hydrogen ion enters the blood stream, it is removed primarily by the reaction

$$H^+ + HCO_3^- \longrightarrow H_2CO_3$$

When an excess of the hydroxide ion is present, it is removed by the reaction

$$OH^- + H_2CO_3 \longrightarrow H_2O + HCO_3^-$$

The pH of human blood thus remains very near 7.35, slightly alkaline.

16.9   The Ionization of Weak Diprotic Acids

Polyprotic acids, those from which more than one proton may be removed, ionize in water in successive steps. An example of a **diprotic acid** is hydrogen sulfide. In aqueous solution the first step in the ionization of hydrogen sulfide yields hydrogen ions and hydrogen sulfide ions.

$$H_2S(aq) \rightleftharpoons H^+(aq) + HS^-(aq)$$

The expression for the ionization constant for the first ionization of H_2S is

$$\frac{[H^+][HS^-]}{[H_2S]} = K_{H_2S} = 1.0 \times 10^{-7}$$

The hydrogen sulfide ion in turn ionizes and forms hydrogen ions and sulfide ions.

$$HS^-(aq) \rightleftharpoons H^+(aq) + S^{2-}(aq)$$

The expression for the ionization constant for the hydrogen sulfide ion is

$$\frac{[H^+][S^{2-}]}{[HS^-]} = K_{HS^-} = 1.3 \times 10^{-13}$$

Note that K_{HS^-} is smaller than K_{H_2S} by a factor of almost 10^6. This means that very little of the HS^- formed by the ionization of H_2S ionizes to give H^+ and S^{2-}. Thus the concentrations of H^+ and HS^- are practically equal in a pure aqueous solution of H_2S.

When the first ionization constant of a polyprotic acid is much larger than the second, it is convenient and appropriate to treat the first ionization step separately and to calculate concentrations of species resulting from it before calculating concentrations of species resulting from subsequent ionization steps.

EXAMPLE 16.17   The concentration of H_2S in a saturated aqueous solution of the gas at room temperature is approximately 0.1 M. Calculate $[H^+]$, $[HS^-]$, and $[S^{2-}]$ of the solution.

We have the following table for the first ionization step

$$H_2S \rightleftharpoons H^+ + HS^-$$

	$[H_2S]$	$[H^+]$	$[HS^-]$
Initial concentrations, M	0.1	~0	0
Equilibrium concentrations, M	0.1 − x	x	x

If we let x equal $[H^+]$, then $[HS^-]$ will also be equal to x, and $[H_2S]$ will be equal to $0.1 - x$. Since x will be quite small compared to 0.1, it may be dropped from the term $0.1 - x$. Solving for x, we have

$$K_{H_2S} = \frac{[H^+][HS^-]}{[H_2S]} = \frac{x^2}{0.1} = 1.0 \times 10^{-7}$$

$$x^2 = 1.0 \times 10^{-8}$$

$$x = 1 \times 10^{-4}\, M = [H^+] = [HS^-]$$

Now we treat the second step in the ionization of H_2S

$$HS^- \rightleftharpoons H^+ + S^{2-}$$

and let $[S^{2-}] = y$. Since, in any solution, all possible equilibria must be satisfied simultaneously, $[H^+]$ and $[HS^-]$ must be the same in the equilibria involved in the ionizations of both HS^- and H_2S.

	$[HS^-]$	$[H^+]$	$[S^{2-}]$
Initial concentrations, M	1×10^{-4}	1×10^{-4}	0
Equilibrium concentrations, M	$(1 \times 10^{-4}) - y$	$(1 \times 10^{-4}) + y$	y

Substituting in the expression for the ionization constant of HS^-, we have

$$K_{HS^-} = \frac{[H^+][S^{2-}]}{[HS^-]} = \frac{[(1 \times 10^{-4}) + y]y}{(1 \times 10^{-4}) - y} = 1.3 \times 10^{-13}$$

Assuming that y is small relative to 1×10^{-4} and thus may be neglected,

$$[S^{2-}] = y = \frac{1 \times 10^{-4}}{1 \times 10^{-4}} \times 1.3 \times 10^{-13}$$

$$= 1.3 \times 10^{-13}$$

Since y is in fact much smaller than 1×10^{-4}, the approximation that y can be neglected is justifiable.

(Note that in a pure aqueous solution of H_2S $[S^{2-}]$ is equal to K_{HS^-}. In fact, for any weak diprotic acid the concentration of the divalent anion is equal to the secondary ionization constant.)

By multiplying the expressions for the ionization constants K_{H_2S} and K_{HS^-}, we obtain

$$\frac{[H^+][HS^-]}{[H_2S]} \times \frac{[H^+][S^{2-}]}{[HS^-]} = K_{H_2S} \times K_{HS^-}$$

$$\frac{[H^+]^2[S^{2-}]}{[H_2S]} = K_{H_2S} K_{HS^-} = 1.3 \times 10^{-20}$$

This is the equilibrium constant for the reaction

$$H_2S(aq) \rightleftharpoons 2H^+(aq) + S^{2-}(aq)$$

EXAMPLE 16.18   Calculate the concentration of sulfide ion in a saturated solution of hydrogen sulfide to which hydrochloric acid has been added to make the hydrogen ion concentration of the solution 0.10 M at equilibrium. (A saturated solution of H_2S is 0.10 M.)

To calculate the sulfide ion concentration, we can use the equilibrium

$$H_2S \rightleftharpoons 2H^+ + S^{2-}$$

	$[H_2S]$	$[H^+]$	$[S^{2-}]$
Initial concentrations, M	0.10	—	0
Equilibrium concentrations, M	$0.10 - x$	0.10	x

$[H^+]$ is 0.10 M at equilibrium since enough hydrochloric acid has been added to make it 0.10 M. Since K_a for the reaction

$$H_2S \rightleftharpoons 2H^+ + S^{2-}$$

is very small, the amount of H^+ added to the solution by the ionization of H_2S and the amount of H_2S that ionizes will be very small. Thus we can neglect x. Substituting in the ionization constant equation gives

$$\frac{[H^+]^2[S^{2-}]}{[H_2S]} = \frac{(0.10)^2 x}{0.10} = 1.3 \times 10^{-20}$$

$$[S^{2-}] = x = \frac{(1.3 \times 10^{-20})(0.10)}{(0.10)^2} = 1.3 \times 10^{-19}\ M$$

(We could use the same equation to calculate the hydrogen ion concentration necessary to produce a given sulfide ion concentration.)

16.10   The Ionization of Weak Triprotic Acids

A typical example of a **triprotic acid** is phosphoric acid, which ionizes in water in three steps.

STEP 1	$H_3PO_4 \rightleftharpoons H^+ + H_2PO_4^-$
STEP 2	$H_2PO_4^- \rightleftharpoons H^+ + HPO_4^{2-}$
STEP 3	$HPO_4^{2-} \rightleftharpoons H^+ + PO_4^{3-}$

Each step in the ionization has a corresponding ionization constant. The expressions for the ionization constants of the three steps are

$$\frac{[H^+][H_2PO_4^-]}{[H_3PO_4]} = K_{H_3PO_4} = 7.5 \times 10^{-3}$$

$$\frac{[H^+][HPO_4^{2-}]}{[H_2PO_4^-]} = K_{H_2PO_4^-} = 6.3 \times 10^{-8}$$

$$\frac{[H^+][PO_4^{3-}]}{[HPO_4^{2-}]} = K_{HPO_4^{2-}} = 3.6 \times 10^{-13}$$

In each successive step the ionization is less. This is a general characteristic of polyprotic acids; successive ionization constants differ by a factor of about 10^5 to 10^6. (See Section 16.9 for the values for H_2S.)

This set of three dissociation reactions appears to make calculations of equilibrium concentrations very complicated. However, because the successive ionization constants differ by a factor of 10^4 to 10^6, the calculations can be broken down into a series of steps similar to those for monoprotic acids.

EXAMPLE 16.19   Calculate the concentrations of all species—H_3PO_4, $H_2PO_4^-$, HPO_4^{2-}, PO_4^{3-}, H^+, and OH^-—present at equilibrium in a solution containing a total phosphoric acid concentration of 0.100 M.

Let x be the concentrations of H^+ and $H_2PO_4^-$ produced by the first step in this equilibrium.

$$H_3PO_4 \rightleftharpoons H^+ + H_2PO_4^-$$

The concentration of H_3PO_4 will be $0.100 - x$. A small amount of the $H_2PO_4^-$ produced in the first step will disappear by dissociation into H^+ and HPO_4^{2-} in the second step, and a small amount of additional H^+ will be produced in both the second and third steps. However, because K_2 and K_3 are very small compared to K_1, it is logical to assume that the decrease in $[H_2PO_4^-]$ in the second step and the increase in $[H^+]$ in the second and third steps are negligible compared to the quantities of each ion (x) produced in the first step.

Hence at equilibrium:

	$[H_3PO_4]$	$[H^+]$	$[H_2PO_4^-]$
Initial concentrations, M	0.100	~0	0
Equilibrium concentrations, M	$0.100 - x$	x	x

$$K_1 = \frac{[H^+][H_2PO_4^-]}{[H_3PO_4]} = 7.5 \times 10^{-3}$$

$$\frac{(x)(x)}{0.100 - x} = \frac{x^2}{0.100 - x} = 7.5 \times 10^{-3}$$

If we neglect x in the term $0.100 - x$, we have

$$\frac{x^2}{0.100} = 7.5 \times 10^{-3}$$

$$x^2 = 7.5 \times 10^{-4}$$

$$x = 2.7 \times 10^{-2}$$

This value of x corresponds to $(2.7 \times 10^{-2}/0.100) \times 100$, or 27% of the value of the initial concentration of H_3PO_4. Thus it is much too large to neglect. A relatively large value of K_a, such as 7.5×10^{-3}, should also serve to alert us that x probably should not be neglected. Hence an accurate value of x can be calculated by means of the quadratic formula (see Appendix A.4).

$$x^2 = (7.5 \times 10^{-4}) - (7.5 \times 10^{-3})x$$

$$x^2 + (7.5 \times 10^{-3})x - (7.5 \times 10^{-4}) = 0$$

$$x = \frac{-(7.5 \times 10^{-3}) \pm \sqrt{(7.5 \times 10^{-3})^2 - 4(-7.5 \times 10^{-4})}}{2(1)}$$

$$= 2.4 \times 10^{-2} \, M$$

An alternative, and very useful, method for solving the equation is by **successive approximations.**

First approximation. On a trial basis, we will neglect the x term in the denominator:

$$K_1 = \frac{x^2}{0.100} = 7.5 \times 10^{-3}$$

$$x^2 = 7.5 \times 10^{-4}$$

$$x = 2.7 \times 10^{-2} \, M$$

Second approximation. The value of x obtained from the first approximation is not negligible compared to $0.100 \, M$. A better estimate of the equilibrium concentration of phosphoric acid, using the approximate value of x just obtained, is

$$0.100 - x = 0.100 - 0.027 = 0.073 \, M$$

$$K_1 = \frac{x^2}{0.073} = 7.5 \times 10^{-3}$$

$$x^2 = (7.5 \times 10^{-3})(7.3 \times 10^{-2}) = 5.5 \times 10^{-4}$$

$$x = 2.3 \times 10^{-2} \, M$$

Since the values of x obtained from the first and second approximations differ by 15%, a third approximation is necessary.

Third approximation. A still better estimate of the equilibrium concentration of phosphoric acid, using the value of x obtained from the second approximation, is

$$0.100 - x = 0.100 - 0.023 = 0.077 \, M$$

$$K_1 = \frac{x^2}{0.077} = 7.5 \times 10^{-3}$$

$$x^2 = (7.5 \times 10^{-3})(0.077) = 5.8 \times 10^{-4}$$

$$x = 2.4 \times 10^{-2} \, M$$

(Note that this value of x is the same as that calculated using the quadratic formula.)

The values of x from the last two approximations agree to within approximately 4%. Hence we can safely conclude that

$$[H^+] = x = 0.024 \, M$$

$$[H_2PO_4^-] = x = 0.024 \, M$$

$$[H_3PO_4] = 0.100 - x = 0.100 - 0.024 = 0.076 \, M$$

$$[OH^-] = \frac{K_w}{[H^+]} = \frac{1.0 \times 10^{-14}}{2.4 \times 10^{-2}} = 4.2 \times 10^{-13} \, M$$

To calculate the concentration of HPO_4^{2-}, we must use the second step of the acid dissociation.

$$H_2PO_4^- \rightleftharpoons H^+ + HPO_4^{2-}$$

Let y be the concentration of HPO_4^- produced in the second step.

	$[H_2PO_4{}^-]$	$[H^+]$	$[HPO_4{}^-]$
Initial concentrations, M	0.024	0.024	0
Equilibrium concentrations, M	$0.024 - y$	$0.024 + y$	y

$$K_2 = \frac{(0.024 + y)y}{0.024 - y} = 6.3 \times 10^{-8}$$

Since K_2 is so small, we can neglect the additive and subtractive y terms. Then the preceding equation becomes

$$\frac{0.024\, y}{0.024} = y = 6.3 \times 10^{-8}\, M = [HPO_4{}^-]$$

Note that this provides a check for our earlier assumption that the increase in $[H^+]$ and the decrease in $[H_2PO_4{}^-]$ (now known to be $6.3 \times 10^{-8}\, M$) are negligible compared to the quantity of each ion produced in the first step (0.024 mol).

Now we consider the third dissociation step

$$HPO_4{}^{2-} \rightleftharpoons H^+ + PO_4{}^{3-}$$

and determine the concentration of $PO_4{}^{3-}$. Let z be the number of moles $PO_4{}^{3-}$ formed by this step.

	$[HPO_4{}^{2-}]$	$[H^+]$	$[PO_4{}^{3-}]$
Initial concentrations, M	6.3×10^{-8}	0.024	0
Equilibrium concentrations, M	$(6.3 \times 10^{-8}) - z$	$0.024 + z$	z

$$K_3 = \frac{(0.024 + z)(z)}{(6.3 \times 10^{-8}) - z} = 3.6 \times 10^{-13}$$

Since K_3 is very small compared to K_2, the extent of the third dissociation is insignificant, and we can neglect the additive and subtractive z terms in the preceding equation. Thus we can easily solve for z:

$$\frac{0.024\, z}{6.3 \times 10^{-8}} = 3.6 \times 10^{-13}$$

$$z = [PO_4{}^{3-}]$$

$$= \frac{(3.6 \times 10^{-13})(6.3 \times 10^{-8})}{2.4 \times 10^{-2}} = 9.5 \times 10^{-19}\, M$$

The quantity of H^+ added to the solution and the decrease in $[HPO_4{}^{2-}]$, z in the third step, are indeed negligible compared to the amounts previously existing in the solution.

16.11   Hydrolysis

The conjugate base of a weak acid reacts with water to increase the concentration of hydroxide ion (Section 14.3). A salt of a weak acid and a strong base, such as $NaCH_3CO_2$, contains the conjugate base ($CH_3CO_2{}^-$) of a weak acid and so forms aqueous solutions that are basic. The conjugate acid of a weak base reacts with water to increase the concentration of hydrogen ion. A salt of a weak base and a strong acid, such as NH_4Cl, contains the conjugate acid (NH_4^+) of a weak base and so forms aqueous solutions that are acidic. Since both the conjugate base of a strong acid and the conjugate acid of a strong base are very weak, a salt such as NaCl or KNO_3 that is the product of a strong acid and a strong base forms aqueous solutions that are neutral. A salt of a weak acid and a weak base can react with water to form neutral, basic, or acidic solutions, depending on the relative strengths of the conjugate acid and conjugate base present in the salt.

A reaction of this type in which water is one of the reactants is called **hydrolysis** (Section 12.3, Part 4). The concentrations of ions and molecules that result from hydrolysis of salts of weak acids and weak bases will be discussed in the sections that follow.

Ions known to be hydrated in solution are often indicated in abbreviated form without showing the hydration. For example, Al^{3+} is often written instead of $[Al(H_2O)_6]^{3+}$. However, when water participates in a reaction, as in hydrolysis, the hydration becomes important. In dealing with hydrolysis, therefore, we use formulas that show the extent of hydration.

16.12 Hydrolysis of a Salt of a Strong Base and a Weak Acid

A salt formed from a strong base and a weak acid contains the conjugate base of the weak acid. For instance, sodium acetate, $NaCH_3CO_2$, is the salt of the strong base sodium hydroxide and the weak acid acetic acid. Such salts dissolve in water to give basic solutions.

$$CH_3CO_2^- + H_2O \rightleftharpoons CH_3CO_2H + OH^-$$

Applying the law of mass action to the reaction, we have

$$\frac{[CH_3CO_2H][OH^-]}{[CH_3CO_2^-][H_2O]} = K$$

Because the number of moles of water involved in the reaction is negligible compared with the total number of moles of water present in dilute solution, we may assume that the concentration of water remains unchanged during the hydrolysis. Hence the preceding equation can be written

$$\frac{[CH_3CO_2H][OH^-]}{[CH_3CO_2^-]} = K[H_2O] = K_b \text{ (or } K_h)$$

This equilibrium constant is simply the ionization constant for the base $CH_3CO_2^-$ (Section 16.5). (Since the acid-base reaction occurs during hydrolysis, the equilibrium constant is sometimes called a **hydrolysis constant** and designated by K_h.) When the value for K_h is small, the degree of hydrolysis is slight; when it is large, hydrolysis is extensive. It should be emphasized that a hydrolysis reaction is simply a Brönsted-Lowry acid-base reaction, which may be treated by the same equilibrium concepts developed earlier in this chapter.

If the value of the hydrolysis constant for the base $CH_3CO_2^-$ cannot be found in a table, its value may be calculated from the values for the ionization constants of water and of acetic acid. For every aqueous solution

$$[H^+][OH^-] = K_w \quad \text{or} \quad [OH^-] = \frac{K_w}{[H^+]} \quad \text{(See Section 16.1)}$$

Substituting $K_w/[H^+]$ for $[OH^-]$ in the expression for the hydrolysis constant, we obtain

$$\frac{[CH_3CO_2H][OH^-]}{[CH_3CO_2^-]} = \frac{[CH_3CO_2H]K_w}{[CH_3CO_2^-][H^+]} = K_h$$

But the expression

$$\frac{[CH_3CO_2H]}{[CH_3CO_2^-][H^+]}$$

is the reciprocal of K_a, or $1/K_a$, for acetic acid. Therefore

$$K_h = \frac{K_w}{K_a} = \frac{1.0 \times 10^{-14}}{1.8 \times 10^{-5}} = 5.6 \times 10^{-10}$$

For any salt of a strong base and a weak acid, the hydrolysis constant (or equivalently, the ionization constant of the conjugate base of the weak acid) is given by the expression

$$K_h = \frac{K_w}{K_a}$$

where K_a is the ionization constant of the weak acid.

Calculations involving hydrolysis are identical to those in other acid-base systems.

EXAMPLE 16.20 Calculate the hydroxide ion concentration, the percent hydrolysis, and the pH of a 0.050 M solution of sodium acetate.

The equation for the hydrolysis of sodium acetate is

$$CH_3CO_2^- + H_2O \rightleftharpoons CH_3CO_2H + OH^-$$

and the expression for the hydrolysis constant is

$$\frac{[CH_3CO_2H][OH^-]}{[CH_3CO_2^-]} = K_h = \frac{K_w}{K_a} = 5.6 \times 10^{-10}$$

	$[CH_3CO_2^-]$	$[CH_3CO_2H]$	$[OH^-]$
Initial concentrations, M	0.050	0	~0
Equilibrium concentrations, M	0.050 − x	x	x

In this table, the information given in the problem is indicated in color.

If we let x equal the concentration of acetic acid formed by the hydrolysis of the acetate ion, then the concentration of hydroxide ion will also be equal to x, and the concentration of acetate ion at equilibrium will be 0.050 − x. Substituting in the hydrolysis constant expression, we have

$$K_h = \frac{x^2}{0.050 - x} = 5.6 \times 10^{-10}$$

Since x is very small relative to 0.050, we may drop it from the denominator. Then

$$x^2 = 0.050 \times 5.6 \times 10^{-10} = 28 \times 10^{-12}$$

$$x = 5.3 \times 10^{-6} = [CH_3CO_2H] = [OH^-]$$

Thus the concentration of hydroxide ion is 5.3×10^{-6} M, and the percent hydrolysis is equal to

$$\frac{[CH_3CO_2H]}{[CH_3CO^-]} \times 100 = \frac{5.3 \times 10^{-6}}{0.050} \times 100 = 0.011\%$$

We can calculate the pH of the solution by first finding the hydronium ion concentration:

$$[H_3O^+] = \frac{K_w}{[OH^-]} = \frac{1.0 \times 10^{-14}}{5.3 \times 10^{-6}} = 1.9 \times 10^{-9}$$

and then using that value.

$$pH = -\log[H_3O^+] = -\log(1.9 \times 10^{-9}) = 8.72$$

This pH corresponds to a basic solution, as expected.

EXAMPLE 16.21   Calculate the hydrolysis constant for sulfide ion, the sulfide ion concentration, and the extent of hydrolysis in a 0.0010 M solution of sodium sulfide.

The sulfide ion hydrolyzes according to the equation

$$S^{2-} + H_2O \rightleftharpoons HS^- + OH^-$$

The hydrogen sulfide ion thus formed also undergoes hydrolysis, but the extent of this is so slight compared with that of the sulfide ion that it can be neglected. The expression for the hydrolysis constant of the sulfide ion is

$$\frac{[HS^-][OH^-]}{[S^{2-}]} = K_h = \frac{K_w}{K_a \text{ (for } HS^-)} = \frac{1.0 \times 10^{-14}}{1.3 \times 10^{-13}} = 7.7 \times 10^{-2}$$

The large value for the hydrolysis constant indicates that most of the sulfide ion undergoes hydrolysis. For this reason we let x equal the S^{2-} concentration at equilibrium.

	$[S^{2-}]$	$[HS^-]$	$[OH^-]$
Initial concentrations, M	0.0010	0	~0
Equilibrium concentrations, M	x	$0.0010 - x$	$0.0010 - x$

Substituting these values in the expression for K_h, we have

$$K_h = \frac{[HS^-][OH^-]}{[S^{2-}]} = \frac{(0.0010 - x)(0.0010 - x)}{x} = 7.7 \times 10^{-2}$$

Since hydrolysis is nearly complete, x is small compared to 0.0010 and may be neglected in the numerator. The expression then becomes

$$\frac{(0.0010)^2}{x} = 7.7 \times 10^{-2}$$

$$x = \frac{(0.0010)^2}{7.7 \times 10^{-2}} = 1.3 \times 10^{-5} = [S^{2-}]$$

The degree of hydrolysis (the fraction of the sulfide ion that undergoes hydrolysis) may be found by dividing the amount of S^{2-} hydrolyzed,

$$0.0010 \; M - 0.000013 \; M = 0.00099 \; M$$

by the total amount of S^{2-} originally present.

$$\frac{0.00099}{0.0010} = 0.99$$

The percent hydrolysis can be obtained by multiplying the degree of hydrolysis by 100.

$$\frac{0.00099}{0.0010} \times 100 = 0.99 \times 100$$

$$= 99\%$$

16.13 Hydrolysis of a Salt of a Weak Base and a Strong Acid

Ammonium chloride, NH_4Cl, is a salt of a weak base and a strong acid. The ammonium ion, the conjugate acid of the weak base ammonia, is acidic and reacts with water according to the equation

$$NH_4^+ + H_2O \rightleftharpoons NH_3 + H_3O^+$$

From the law of mass action, we obtain

$$\frac{[NH_3][H_3O^+]}{[NH_4^+]} = [H_2O]K = K_a \; (\text{or } K_h)$$

The equilibrium constant is simply the ionization constant for the acid NH_4^+ (Section 16.4). Since the acid-base reaction occurs during the hydrolysis of a salt, the constant is also called the hydrolysis constant, K_h.

The value of K_h for NH_4^+ may be determined from K_w and the ionization constant, K_b, for NH_3. The expression for the hydrolysis constant for the ammonium ion, based on the equation

$$NH_4^+ + H_2O \rightleftharpoons NH_3 + H_3O^+$$

is written

$$\frac{[NH_3][H_3O^+]}{[NH_4^+]} = K_h$$

Substituting $K_w/[\text{OH}^-]$ for $[\text{H}_3\text{O}^+]$ in the above equation, we obtain

$$\frac{[\text{NH}_3]K_w}{[\text{NH}_4{}^+][\text{OH}^-]} = K_h$$

Since the expression $\dfrac{[\text{NH}_3]}{[\text{NH}_4{}^+][\text{OH}^-]}$ is equal to the reciprocal for K_b for NH_3 and hence equal to $\dfrac{1}{K_b \text{ (for NH}_3)}$, we may write

$$K_h = \frac{K_w}{K_b} = \frac{1.0 \times 10^{-14}}{1.8 \times 10^{-5}} = 5.6 \times 10^{-10}$$

The hydrolysis constant for any salt formed from a weak base and a strong acid (or the ionization constant of the conjugate acid of the weak base) may be calculated from the expression

$$K_h = \frac{K_w}{K_b}$$

where K_b is the ionization constant of the weak base.

16.14 Hydrolysis of a Salt of a Weak Base and a Weak Acid

When we dissolve ammonium acetate, $\text{NH}_4\text{CH}_3\text{CO}_2$, in water, we have a solution of ammonium ion (an acid) and acetate ion (a base), both of which undergo hydrolysis.

$$\text{NH}_4{}^+ + \text{H}_2\text{O} \rightleftharpoons \text{NH}_3 + \text{H}_3\text{O}^+$$
$$\text{CH}_3\text{CO}_2{}^- + \text{H}_2\text{O} \rightleftharpoons \text{CH}_3\text{CO}_2\text{H} + \text{OH}^-$$

The extent to which these reactions take place is about the same because ammonium ion and acetate ion have nearly equal hydrolysis constants. Therefore the products are present in nearly equal concentrations at equilibrium. As H^+ and OH^- are produced by these reactions, they react to form water because the product of their concentrations cannot exceed 1.0×10^{-14}. Thus the solution remains neutral as hydrolysis proceeds.

The net reaction involved in the hydrolysis may be seen as being the sum of three equations:

$$\begin{aligned}
\text{NH}_4{}^+ + \text{H}_2\text{O} &\rightleftharpoons \text{NH}_3 + \text{H}_3\text{O}^+ \\
\text{CH}_3\text{CO}_2{}^- + \text{H}_2\text{O} &\rightleftharpoons \text{CH}_3\text{CO}_2\text{H} + \text{OH}^- \\
\underline{\text{H}_3\text{O}^+ + \text{OH}^- \rightleftharpoons 2\text{H}_2\text{O}} & \\
\text{NH}_4{}^+ + \text{CH}_3\text{CO}_2{}^- &\rightleftharpoons \text{CH}_3\text{CO}_2\text{H} + \text{NH}_3
\end{aligned}$$

Applying the law of mass action to the net reaction, we obtain

$$\frac{[\text{CH}_3\text{CO}_2\text{H}][\text{NH}_3]}{[\text{NH}_4{}^+][\text{CH}_3\text{CO}_2{}^-]} = K_h$$

To express the hydrolysis constant in terms of the ion product for water and the ionization constants for aqueous ammonia and acetic acid, we multiply the numerator and denominator of the above equation by $[H_3O^+][OH^-]$.

$$\frac{[NH_3]}{[NH_4^+][OH^-]} \times \frac{[CH_3CO_2H]}{[H_3O^+][CH_3CO_2^-]} \times \frac{[H_3O^+][OH^-]}{1} = K_h$$

Thus it may readily be seen that

$$K_h = \frac{K_w}{K_b \text{ (for NH}_3) \times K_a \text{ (for CH}_3CO_2H)}$$
$$= \frac{1.0 \times 10^{-14}}{(1.8 \times 10^{-5})(1.8 \times 10^{-5})} = 3.1 \times 10^{-5}$$

When the ions of a salt of a weak base and a weak acid do not have the same hydrolysis constants, the aqueous solution of the salt is either basic or acidic, depending on which ion has the larger hydrolysis constant. For example, ammonium cyanide, NH_4CN, hydrolyzes to give a basic solution, since K_b for CN^- is larger than K_a for NH_4^+. The hydrolysis of the cyanide ion,

$$CN^- + H_2O \rightleftharpoons HCN + OH^-$$

is more extensive than that of the ammonium ion,

$$NH_4^+ + H_2O \rightleftharpoons NH_3 + H_3O^+$$

and an excess of hydroxide ions accumulates in the solution.

EXAMPLE 16.22   Calculate the extent of hydrolysis and the pH of a solution that is $1.0\,M$ in NH_4CN.

When ammonium cyanide is dissolved in water, both the ammonium ion and the cyanide ion undergo hydrolysis. The equations are

$$NH_4^+ + H_2O \rightleftharpoons NH_3 + H_3O^+$$
$$CN^- + H_2O \rightleftharpoons HCN + OH^-$$
$$H_3O^+ + OH^- \rightleftharpoons 2H_2O$$

The net reaction is given by the equation

$$NH_4^+ + CN^- \rightleftharpoons NH_3 + HCN$$

The expression for the hydrolysis constant may be written

$$\frac{[NH_3][HCN]}{[NH_4^+][CN^-]} = K_h = \frac{K_w}{K_b \text{ (for NH}_3) \times K_a \text{ (for HCN)}}$$

$$= \frac{1.0 \times 10^{-14}}{(1.8 \times 10^{-5})(4.0 \times 10^{-10})} = 1.4$$

It can be shown that, although the hydrolysis of cyanide ion is somewhat more extensive than that of ammonium ion, the difference in the extent of hydrolysis of the two ions is relatively small compared to the total quantity of each that undergoes hydrolysis. It is justifiable to assume that for every NH_4^+ that hydrolyzes, a CN^- also hydrolyzes, provided that the NH_4CN concentration is not extremely low. Let x equal both the number of moles of NH_4^+ and the number of moles of CN^- undergoing hydrolysis.

	$[NH_4^+]$	$[CN^-]$	$[NH_3]$	$[HCN]$
Initial concentrations, M	1.0	1.0	0	0
Equilibrium concentrations, M	$1.0 - x$	$1.0 - x$	x	x

$$K_h = \frac{[NH_3][HCN]}{[NH_4^+][CN^-]} = \frac{x^2}{(1.0 - x)^2} = 1.4$$

Extracting the square root of both sides of the equation, we find

$$\frac{x}{1.0 - x} = 1.18$$

$$x = 1.18 - 1.18x$$

$$2.18x = 1.18$$

$$x = 0.54 = [NH_3] = [HCN]$$

The extent of hydrolysis is the amount of NH_4CN hydrolyzed divided by the total amount of NH_4CN originally present. The percent hydrolysis is

$$\frac{0.54}{1.0} \times 100 = 54\%$$

At equilibrium,

$$[NH_4^+] = [CN^-] = 1.0 - x = 1.0 - 0.54 = 0.46$$

Substituting in the expression for the ionization constant for HCN, we find

$$\frac{[H^+][CN^-]}{[HCN]} = \frac{[H^+](0.46)}{0.54} = 4 \times 10^{-10}$$

$$[H^+] = \frac{0.54}{0.46} \times 4 \times 10^{-10} = 4.7 \times 10^{-10}$$

$$pH = -\log(4.7 \times 10^{-10}) = 9.3 \quad \text{(a basic solution)}$$

16.15   The Ionization of Hydrated Metal Ions

A large number of metal ions behave as acids in solution. For example, the aluminum(III) ion of aluminum nitrate reacts with water to give a hydrated aluminum ion, $[Al(H_2O)_6]^{3+}$, which is an acid.

$$Al(NO_3)_3(s) + 6H_2O(l) \longrightarrow [Al(H_2O)_6]^{3+}(aq) + 3NO_3^-(aq)$$

$$[Al(H_2O)_6]^{3+} \rightleftharpoons [Al(OH)(H_2O)_5]^{2+} + H^+$$

An excess of hydrogen ions is produced in the solution. Because such reactions involve the reaction of a salt [in this case, $Al(NO_3)_3$] with water, they too are sometimes called hydrolysis reactions.

Just like a polyprotic acid, the hydrated aluminum ion ionizes in stages, as shown by

Plate 15. A mixture of nitrogen oxides, ozone, and unburned hydro-carbons form smog when exposed to sunlight. *Tom McHugh/Photo Researchers, Inc.*

Plate 16. (*Facing page.*) An oil drilling platform in the North Sea. Oil is an important source of fuel and raw materials for the organic chemicals industry. *Courtesy*

Plate 17. Ions having no *d* electrons or a full *d* subshell are usually colorless. Ions with partially filled *d* levels, transition metal ions, are usually colored. © *Yoav/Phototake*

Plate 18. Colors of various compounds of chromium, with different oxidation states and ligands. © *Yoav/Phototake*

(a)

Plate 19. (a) Colors of complex ions of cobalt(III). (b) Colors of complex ions of cobalt(II). © *Tom Pantages, 1983. Chemicals courtesy of William R. Robinson and Alfa Products.*

(b)

Plate 20. The colors of many gems and minerals are due to the presence of small amounts of transition metal ions. *Nino Mascardi/The Image Bank*

Plate 21. Stalactites and stalagmites are formed by the crystallization of calcium carbonate. © *Harald Sund/ The Image Bank*

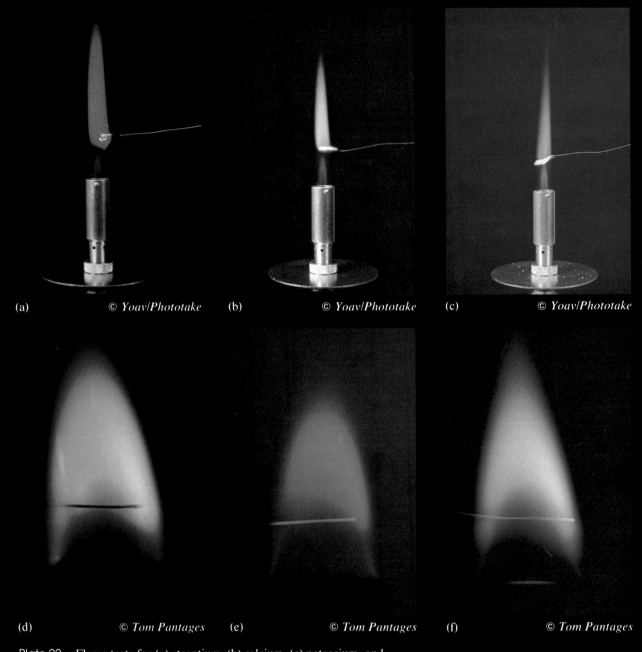

(a)　　　© *Yoav/Phototake*　(b)　　　© *Yoav/Phototake*　(c)　　　© *Yoav/Phototake*

(d)　　　© *Tom Pantages*　(e)　　　© *Tom Pantages*　(f)　　　© *Tom Pantages*

Plate 22. Flame tests for (a) strontium, (b) calcium, (c) potassium, and in closeup for (d) copper, (e) lithium, and (f) sodium.

Plate 23. The same effect that produces a flame test colors many fireworks; certain colors are produced by heating metal ions. *Norman Owen Tomalin/Bruce Coleman, Inc.*

$$[Al(H_2O)_6]^{3+} \rightleftharpoons [Al(OH)(H_2O)_5]^{2+} + H^+$$

$$[Al(OH)(H_2O)_5]^{2+} \rightleftharpoons [Al(OH)_2(H_2O)_4]^+ + H^+$$

$$[Al(OH)_2(H_2O)_4]^+ \rightleftharpoons [Al(OH)_3(H_2O)_5] + H^+$$

The ionization of a cation carrying more than one charge is not extensive beyond the first stage. Additional examples of the first stage in the ionization of hydrated metal ions are

$$[Fe(H_2O)_6]^{3+} \rightleftharpoons [Fe(OH)(H_2O)_5]^{2+} + H^+$$

$$[Cu(H_2O)_4]^{2+} \rightleftharpoons [Cu(OH)(H_2O)_3]^+ + H^+$$

$$[Zn(H_2O)_6]^{2+} \rightleftharpoons [Zn(OH)(H_2O)_5]^+ + H^+$$

EXAMPLE 16.23   Let us calculate the pH of a 0.10 M solution of aluminum chloride that dissolves to give the hydrated aluminum ion, $[Al(H_2O)_6]^{3+}$, in solution. The ionization constant, K_a, is 1.4×10^{-5} for the reaction

$$[Al(H_2O)_6]^{3+} \rightleftharpoons [Al(OH)(H_2O)_5]^{2+} + H^+$$

The expression for the ionization constant is

$$\frac{[Al(OH)(H_2O)_5{}^{2+}][H^+]}{[Al(H_2O)_6{}^{3+}]} = K_a = 1.4 \times 10^{-5}$$

Let x equal the equilibrium concentration of hydrogen ion, which is equal to the concentration of $[Al(OH)(H_2O)_5]^{2+}$. Then $0.10 - x$ will equal the equilibrium concentration of $[Al(H_2O)_6]^{3+}$.

	$[Al(H_2O)_6{}^{3+}]$	$[H^+]$	$[Al(OH)(H_2O)_5{}^{2+}]$
Initial concentrations, M	0.10	~0	0
Equilibrium concentrations, M	$0.10 - x$	x	x

Substituting in the expression for the ionization constant, we have

$$K_a = \frac{[Al(OH)(H_2O)_5{}^+][H^+]}{[Al(H_2O)_6{}^{3+}]} = \frac{x^2}{0.10 - x} = 1.4 \times 10^{-5}$$

Since K_a is small, x will also be small, and we may drop x from the term $0.10 - x$ in an approximate solution of the problem. Thus we obtain

$$\frac{x^2}{0.10} = 1.4 \times 10^{-5}$$

$$x = 1.2 \times 10^{-3} = [H^+]$$

$$pH = -\log(1.2 \times 10^{-3}) = 2.9 \quad \text{(an acid solution)}$$

The constants for the different stages of ionization are not known for many metal ions, so we cannot calculate the extent of their ionization. However, if the hydroxide of a metal ion is insoluble in water, the hydrated metal ion will ionize extensively to give a very acidic solution. In fact, practically all metal ions other than those of the alkali metals ionize to give acidic solutions. The ionization constant increases as the charge of the metal ion increases or as the size of the metal ion decreases.

16.16   Acid-Base Indicators

Certain organic substances change color in dilute solution when the hydrogen ion concentration reaches a particular value. For example, phenolphthalein is a colorless substance in any aqueous solution with a hydrogen ion concentration greater than 5.0×10^{-9} M (pH less than 8.3). In solutions with a hydrogen ion concentration less than 5.0×10^{-9} M (pH greater than 8.3), phenolphthalein is red or pink. Substances like phenolphthalein, which can be used to determine the pH of a solution, are called **acid-base indicators.** Acid-base indicators are either weak organic acids, HIn, or weak organic bases, InOH, where the letters In stand for a complex organic group.

The equilibrium in a solution of the acid-base indicator **methyl orange,** a weak acid, can be represented by the equation

$$HIn \rightleftharpoons H^+ + In^-$$

Red   Yellow

The anion of methyl orange is yellow, and the nonionized form is red. If acid is added to the solution, the increase in the hydrogen ion concentration shifts the equilibrium toward the red form in accordance with the law of mass action.

$$K_a = \frac{[H^+][In^-]}{[HIn]}$$

The indicator's color is the visible result of the ratio of the concentrations of the two species HIn and In$^-$. For methyl orange,

$$\frac{[In^-]}{[HIn]} = \frac{[\text{substance with yellow color}]}{[\text{substance with red color}]} = \frac{K_a}{[H^+]}$$

When [H$^+$] has the same numerical value as K_a, the ratio of [In$^-$] to [HIn] is equal to 1, meaning that 50% of the indicator is present in the red acid form and 50% in the yellow ionic form, and the solution appears orange in color. When the hydrogen ion concentration increases to a pH of 3.1, about 90% of the indicator is present in the red form and 10% in the yellow form, and the solution turns red. No change in color is visible for any further increase in the hydrogen ion concentration.

Addition of a base to the system reduces the hydrogen ion concentration and shifts the equilibrium toward the yellow form. At a pH of 4.4 about 90% of the indicator is in the yellow ionic form, and a further decrease in the hydrogen ion concentration does not produce a visible color change. The pH range between 3.1 (red) and 4.4 (yellow) is the **color-change interval** of methyl orange; the pronounced color change takes place between these pH values.

The large number of known acid-base indicators cover a wide range of pH values and can be used to determine the approximate pH of an unknown solution by a process of elimination. Table 16-5 lists some of these indicators, their colors, and their color-change intervals. The use of these indicators for specific purposes will be discussed in Section 16.17.

Table 16-5 Some Acid-Base Indicators

Indicator	Color in the More Acid Range	pH Range	Color in the More Basic Range
Methyl violet	Yellow	0–2	Violet
Thymol blue	Pink	1.2–2.8	Yellow
Bromophenol blue	Yellow	3.0–4.7	Violet
Methyl orange	Pink	3.1–4.4	Yellow
Bromocresol green	Yellow	4.0–5.6	Blue
Bromocresol purple	Yellow	5.2–6.8	Purple
Litmus	Red	4.7–8.2	Blue
Phenolphthalein	Colorless	8.3–10.0	Pink
Thymolphthalein	Colorless	9.3–10.5	Blue
Alizarin yellow G	Colorless	10.1–12.1	Yellow
Trinitrobenzene	Colorless	12.0–14.3	Orange

16.17 Titration Curves

Titration curves, plots of the pH versus the volume of acid or base added in a titration (Section 3.2), show the point where equivalent quantities of acid and base are present (the equivalence point), which aids in the choice of a proper indicator.

The simplest acid-base reactions are those of a strong acid with a strong base. Let us consider the titration of a 25.0-milliliter sample of 0.100 M hydrochloric acid with 0.100 M sodium hydroxide. The values of the pH measured after successive additions of small amounts of NaOH are shown in Fig. 16-4 and listed in Table 16-6. The pH increases very slowly at first, increases very rapidly in the middle portion of the curve, and then increases very slowly again. The point of inflection (the midpoint of the vertical part of the curve) is the **equivalence point** for the titration. It indicates when equivalent quantities of acid and base are present. For the titration of a strong acid and a strong base, the equivalence point occurs at a pH of 7 [Table 16-6, column (a), and Fig. 16-4]. A solution with a pH of 7 is neutral; the hydrogen ion concentration is equal to hydroxide ion concentration.

The titration of a weak acid with a strong base (or of a weak base with a strong acid) is somewhat more complicated than that just discussed, but follows the same basic principles. Let us consider the titration of 25.0 milliliters of 0.100 M acetic acid with 0.100 M sodium hydroxide and compare the titration curve with that of the strong acid (Fig. 16-4). Table 16-6, column (b), gives the pH values during the titration; Fig. 16-5 shows the titration curve.

The similarities in the two titration curves are readily apparent, but there are also important differences. The titration curve for the weak acid begins at a higher pH value (less acidic) and maintains higher pH values up to the equivalence point. This is because acetic acid is a weak acid and releases only a small quantity of hydrogen ions. The pH at the equivalence point is also higher (8.72, rather than 7.00) because the solution contains sodium acetate. The acetate ion is a weak base and raises the pH as described in Section 16.12.

$$CH_3CO_2^- + H_2O \rightleftharpoons CH_3CO_2H + OH^-$$

Table 16-6   pH Values (a) in the Titration of a Strong Acid with a Strong Base
and (b) in the Titration of a Weak Acid with a Strong Base

| | | pH Values | |
Volume of 0.100 M NaOH Added, mL	Moles of NaOH Added	(a) Titration of 25.00 mL of 0.100 M HCl	(b) Titration of 25.00 mL of 0.100 M CH$_3$CO$_2$H
0.0	0.0	1.00	2.87
5.0	0.00050	1.18	4.14
10.0	0.00100	1.37	4.57
15.0	0.00150	1.60	4.92
20.0	0.00200	1.95	5.35
22.0	0.00220	2.20	5.61
24.0	0.00240	2.69	6.13
24.5	0.00245	3.00	6.44
24.9	0.00249	3.70	7.14
25.0	0.00250	7.00	8.72
25.1	0.00251	10.30	10.30
25.5	0.00255	11.00	11.00
26.0	0.00260	11.29	11.29
28.0	0.00280	11.75	11.75
30.0	0.00300	11.96	11.96
35.0	0.00350	12.22	12.22
40.0	0.00400	12.36	12.36
45.0	0.00450	12.46	12.46
50.0	0.00500	12.52	12.52

(a) Titration of 25.00 mL of 0.100 M HCl (0.00250 mol of HCl) with 0.100 M NaOH
(b) Titration of 25.00 mL of 0.100 M CH$_3$CO$_2$H (0.00250 mol of CH$_3$CO$_2$H) with 0.100 M NaOH

Figure 16-4.   Titration curve for the titration of 25.00 mL of 0.100 M HCl (strong acid) with 0.100 M NaOH (strong base). The pH ranges for the color change of phenolphthalein, litmus, and methyl orange are indicated by the shaded areas.

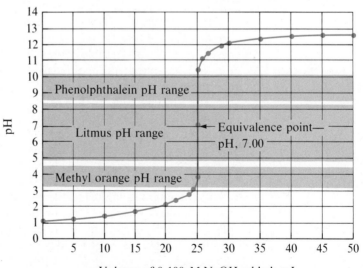

After the equivalence point the two curves are identical (Table 16-6 and Fig. 16-5) because the pH is dependent on the excess of hydroxide ion in both cases. After an acid is neutralized, whether it is a strong acid (like HCl) or a weak acid (like

Figure 16-5. Titration curve for the titration of 25.00 mL of 0.100 M CH$_3$CO$_2$H (weak acid) with 0.100 M NaOH (strong base). The pH ranges for the color change of phenolphthalein, litmus, and methyl orange are indicated by the shaded areas.

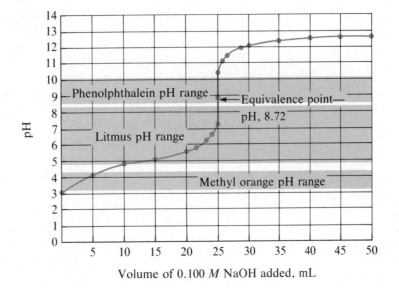

Volume of 0.100 M NaOH added, mL

CH$_3$CO$_2$H), it has no further significant effect on the concentration of the hydroxide ion.

Titration curves help us pick an indicator that will provide a sharp color change at the equivalence point. The pH ranges for the color changes of three indicators from Table 16-5 are indicated in Figs. 16-4 and 16-5.

Phenolphthalein has a sharp color change over a small added volume of NaOH and hence is suitable for titrations that are either strong acid–strong base or weak acid–strong base. The equivalence point is located in the steeply rising portion of the titration curve, where the curve passes through the color-change interval of the indicator.

Litmus is a suitable indicator for the HCl titration because the titration curve shows that it changes color within 0.10 milliliter of the equivalence point. However, litmus would be a poor choice for the CH$_3$CO$_2$H titration because the pH is within the color change interval of litmus when only about 12 milliliters of NaOH have been added and does not leave the range until 25 milliliters have been added. The end of the color change would be at about the equivalence point (a good feature), but the color change would be very gradual, taking place during the addition of 13 milliliters of NaOH, making it almost useless as an indicator of the equivalence point.

Methyl orange could be used for the HCl titration (Fig. 16-4), although its use would have two disadvantages: (1) it completes its color change slightly before the equivalence point is reached (but very close to it, so this is not too serious), and (2) it changes color, as the plot shows, during the addition of nearly 0.5 milliliter of NaOH, which is not as sharp a color change as that of litmus or phenolphthalein. The titration curve shows that methyl orange would be completely useless as an indicator for the CH$_3$CO$_2$H titration. Its color change is not sharp, beginning after only about 1 milliliter of NaOH has been added and ending when about 8 milliliters have been added. More serious is the fact that the color change is completed long before the equivalence point (which occurs when 25.0 milliliters of NaOH have been added) is reached and hence provides no indication of the equivalence point.

The pH range in which an equivalence point will be observed can also be determined by calculating a pH curve. For example, the hydrogen ion concentration of a 0.100 M solution of acetic acid can be calculated as described in Example 16.11 and then converted to pH as described in Section 16.2. The determination of the hydrogen ion concentration in a solution resulting from the addition of a known amount of sodium hydroxide to 0.100 M acetic acid has been described in Example 16.15. At the equivalence point equimolar amounts of sodium hydroxide and acetic acid have been mixed. The problem therefore becomes the calculation of the pH of a solution of sodium acetate, a calculation described in Section 16.12. Beyond the equivalence point the pH is controlled by the excess sodium hydroxide present and may be determined from the concentration of hydroxide ion, as described in Section 16.2.

FOR REVIEW

SUMMARY

Water is an extremely weak electrolyte that undergoes **self-ionization,**

$$H_2O \rightleftharpoons H^+ + OH^-$$

Thus pure water contains equal concentrations of hydrogen ion and hydroxide ion. The **ion product for water, K_w,** is the equilibrium constant for the self-ionization reaction; at 25°C

$$K_w = [H^+][OH^-] = 1.0 \times 10^{-14}$$

All aqueous solutions contain both H^+ and OH^-, with the product $[H^+][OH^-]$ equal to 1.0×10^{-14} at 25°C.

The concentration of H^+ in a solution may be expressed as the pH of the solution; **pH $= -$log $[H^+]$.** The concentration of OH^- may be expressed as the pOH of the solution: **pOH $= -$log $[OH^-]$.** In pure water pH $= 7$ and pOH $= 7$. In any aqueous solution at 25°C, pH + pOH = 14.00.

Although strong electrolytes ionize completely in solution, a solution of a weak electrolyte contains a mixture of molecular and ionic species. The concentration of the nonionized molecules of a weak acid (HA), of hydrogen ion, and of the conjugate base (A^-) of the weak acid, can be determined from K_a, the **ionization constant of the weak acid.** K_a is the equilibrium constant for the reaction

$$HA \rightleftharpoons H^+ + A^-; \qquad K_a = \frac{[H^+][A^-]}{[HA]}$$

The **ionization constant for a weak base** (B) is the equilibrium constant for the reaction

$$B + H_2O \rightleftharpoons BH^+ + OH^-; \qquad K_b = \frac{[BH^+][OH^-]}{[B]}$$

The extent of the ionization of a weak acid in solution can be reduced by adding a compound containing the conjugate base of the weak acid **(the common** ion effect) or by adding another acid. Likewise, the extent of the ionization of a weak base can be reduced by adding the conjugate acid of the weak base or by adding hydroxide. A solution containing a mixture of an acid and its conjugate base, or of a base and its conjugate acid, is called a **buffer solution.** Unlike solutions of acids, bases, or most salts, the hydrogen ion concentration of a buffer solution does not change greatly when a small amount of acid or base is added to it.

The conjugate base of a weak acid reacts with water to give a basic solution. The conjugate acid of a weak base gives an acid solution. A reaction of this type in which water is one of the reactants is called **hydrolysis.** The equilibrium constant for the hydrolysis of a salt of a weak acid, or the **hydrolysis constant K_h,** can be calculated from K_w and K_a for the weak acid

$$K_h = \frac{K_w}{K_a}$$

For a salt of a weak base,

$$K_h = \frac{K_w}{K_b}$$

where K_b is the ionization constant of the weak base.

The pH of a solution may be determined by using a **pH meter** or with **acid-base indicators.** Acid-base indicators are also used to determine the equivalence point in acid-base titrations. The correct indicator for the titration of an acid and a base is selected by considering the **titration curve** for the acid-base pair.

KEY TERMS AND CONCEPTS

Acid-base indicators (16.16)
Buffer solutions (16.8)
Common ion effect (16.7)
Equivalence point (16.17)
Hydrolysis (16.11)
Hydrolysis constant (16.12)
Ionization constant (16.4)

Ionization of hydrated metal
   ions (16.15)
Ionization of water (16.1)
Ionization of weak acids (16.4)
Ionization of weak bases (16.5)
Ion product for water (16.1)
pH (16.2)

pH meter (16.2)
pOH (16.2)
Polyprotic acid (16.9)
Salt effect (16.6)
Titration curves (16.17)

EXERCISES

Values of K not given may be found in Appendixes G and H.

Ionic Equilibria of Acids and Bases

1. Compare the extent of ionization of strong electrolytes and weak electrolytes in water.
2. Equilibrium calculations are not necessary when

dealing with ionic concentrations in solutions of strong electrolytes such as NaOH and HCl. Why not?

§ 3. What ionic and molecular species are present in an aqueous solution of hydrogen fluoride, HF? a solution of sulfuric acid? a solution of SO_2 in water (sulfurous acid)?

4. Calculate the concentration of each of the ions in the following solutions of strong electrolytes:

Ⓢ(a) 0.085 M HNO_3

Ans. $[H^+] = 0.085\ M;$ $[NO_3^-] = 0.085\ M$

(b) 1.248 M $CoCl_2$

Ans. $[Co^{2+}] = 1.248\ M;$ $[Cl^-] = 2.496\ M$

Ⓢ(c) 0.107 M $[Fe(H_2O)_6]_2(SO_4)_3$

Ans. $[Fe(H_2O)_6^{3+}] = 0.214\ M;$
$[SO_4^{2-}] = 0.321\ M$

5. From the equilibrium concentrations given, calculate K for each of the following weak acids or weak bases:

Ⓢ(a) CH_3CO_2H: $[H^+] = 1.34 \times 10^{-3}\ M;$
$[CH_3CO_2^-] = 1.34 \times 10^{-3}\ M;$
$[CH_3CO_2H] = 9.866 \times 10^{-2}\ M$

Ans. 1.82×10^{-5}

(b) HNO_2: $[H^+] = 0.011\ M;$
$[NO_2^-] = 0.0438\ M;$ $[HNO_2] = 1.07\ M$

Ans. 4.5×10^{-4}

(c) NH_3: $[OH^-] = 3.1 \times 10^{-3}\ M;$
$[NH_4^+] = 3.1 \times 10^{-3}\ M;$ $[NH_3] = 0.533\ M$

Ans. 1.8×10^{-5}

(d) ClO^-: $[OH^-] = 4.0 \times 10^{-4}\ M;$
$[HClO] = 2.38 \times 10^{-5}\ M;$
$[ClO^-] = 0.273\ M$ Ans. 3.5×10^{-8}

6. Using the ionization constants in Appendix G, calculate the hydrogen ion concentration and the percent ionization in each of the following solutions:

Ⓢ(a) 0.0092 M $HClO$

Ans. $1.8 \times 10^{-5}\ M;$ 0.20%

(b) 0.0810 M HCN

Ans. $6 \times 10^{-6}\ M;$ 7×10^{-3}%

(c) 0.138 M CH_3CO_2H

Ans. $1.6 \times 10^{-3}\ M;$ 1.2%

(d) 0.0992 M $HC_2O_4^-$

Ans. $2.5 \times 10^{-3}\ M;$ 2.5%

7. Formic acid, HCO_2H, is the irritant that causes the body's reaction to an ant's sting. What is the concentration of hydrogen ion and the percent ionization in a 0.417 M solution of formic acid?

Ans. $8.7 \times 10^{-3}\ M;$ 2.1%

8. Calculate the hydrogen ion concentration and percent ionization of the weak acid in each of the following solutions. Note that the ionization constants may be such that the change in electrolyte concentration cannot be neglected and the quadratic formula or successive approximations may be required.

Ⓢ(a) 0.0184 M $HCNO$ ($K_a = 3.46 \times 10^{-4}$)

Ans. $2.36 \times 10^{-3}\ M;$ 12.8%

(b) 0.100 M HF ($K_a = 7.2 \times 10^{-4}$)

Ans. $8.1 \times 10^{-3}\ M;$ 8.1%

(c) 0.0655 M $[Fe(H_2O)_6]^{3+}$ ($K_a = 4.0 \times 10^{-3}$
for $[Fe(H_2O)_6]^{3+} + H_2O \rightleftharpoons$
$[Fe(OH)(H_2O)_5]^{2+} + H_3O^+)$

Ans. $1.4 \times 10^{-2}\ M;$ 22%

Ⓢ(d) 0.02173 M CH_2ClCO_2H ($K_a = 1.4 \times 10^{-3}$)

Ans. $4.9 \times 10^{-3}\ M;$ 22%

(e) 0.10 M HSO_3NH_2 ($K_a = 1.0 \times 10^{-1}$)

Ans. $0.062\ M;$ 62%

9. Sodium hydrogen sulfate, $NaHSO_4$, is used as a solid acid in some porcelain cleansers because it reacts with calcium carbonate precipitates. What are the hydrogen ion concentration and the percent ionization of the HSO_4^- ion ($K_a = 1.2 \times 10^{-2}$) in a 0.50 M solution of $NaHSO_4$?

Ans. $7.2 \times 10^{-2}\ M;$ 14%

10. Propionic acid, $C_2H_5CO_2H$ ($K_a = 1.34 \times 10^{-5}$), is used in the manufacture of calcium propionate, a food preservative. What is the hydrogen ion concentration in a 0.717 M solution of $C_2H_5CO_2H$?   Ans. $3.10 \times 10^{-3}\ M$

Ⓢ11. The ionization constant of lactic acid, $CH_3CHOHCO_2H$, is 1.36×10^{-4}. If 20.0 g of lactic acid is used to make a solution with a volume of 1.00 L, what is the concentration of hydrogen ion in the solution?   Ans. $5.49 \times 10^{-3}\ M;$ note that with three significant figures the answer is different ($5.43 \times 10^{-3}\ M$) if the simplifying assumption is not made.

12. Calculate the percent ionization of a 0.50 M solution of H_3PO_4. (Consider the second and third ionization steps to be negligible compared to the first.)   Ans. 12%

13. What is the fluoride ion concentration in 0.750 L of a solution that contains 0.1500 g of HF?

Ans. $2.3 \times 10^{-3}\ M$

14. Calculate the hydroxide ion concentration and the percent ionization of the weak base in each of the following solutions:

Ⓢ(a) 0.0784 M $C_6H_5NH_2$ ($K_b = 4.6 \times 10^{-10}$)

Ans. $6.0 \times 10^{-6}\ M;$ 7.6×10^{-3}%

(b) 0.1098 M $(CH_3)_3N$ ($K_b = 7.4 \times 10^{-5}$)

Ans. $2.8 \times 10^{-3}\ M;$ 2.6%

Ⓢ(c) 0.222 M CN^- ($K_b = 2.5 \times 10^{-5}$)

Ans. $2.4 \times 10^{-3}\ M;$ 1.1%

(d) 0.300 M N_2H_4 ($K_b = 3 \times 10^{-6}$ for
$N_2H_4 + H_2O \rightleftharpoons N_2H_5^+ + OH^-$)

Ans. $9 \times 10^{-4}\ M;$ 0.3%

(e) 1.13 M BrO^- ($K_b = 5 \times 10^{-6}$)

Ans. $2 \times 10^{-3}\ M;$ 0.2%

15. What is the hydroxide ion concentration in a dilute solution of household ammonia that contains

0.3124 mol of NH_3 per liter?

Ans. 2.4 × 10⁻³ M

16. The artificial sweetener sodium saccharide (often referred to as sodium saccharin on soft drink bottles) contains the basic ion $C_7H_4NSO_3^-$ ($K_b = 4.8 \times 10^{-3}$). What is the hydroxide ion concentration in a liter of water sweetened with 0.100 g of $Na(C_7H_4NSO_3)$?   Ans. 1.5 × 10⁻³ M

17. Calculate the hydroxide ion concentration and the percent ionization of the weak base in each of the following solutions. Note that the ionization constants are sufficiently large that it may be necessary to use the quadratic formula or successive approximations.

S (a) 4.113 × 10⁻³ M CH_3NH_2
($K_b = 4.4 \times 10^{-4}$)
Ans. 1.1 × 10⁻³ M; 28%

(b) 0.11 M $(CH_3)_2NH$ ($K_b = 7.4 \times 10^{-4}$)
Ans. 8.7 × 10⁻³ M; 7.9%

(c) 0.00253 M $C_6H_5O^-$ ($K_b = 7.81 \times 10^{-5}$)
Ans. 4.07 × 10⁻⁴ M; 16.1%

(d) 0.437 M S^{2-} ($K_b = 7.7 \times 10^{-2}$)
Ans. 0.15 M; 34%

18. Calculate the ionization constants for each of the following solutes from the percent ionization and the concentration of the solute:

S (a) 0.050 M HClO, 8.4 × 10⁻²% ionized
Ans. 3.5 × 10⁻⁸

S (b) 0.010 M HNO_2, 19% ionized
Ans. 4.4 × 10⁻⁴

(c) 1.0 M NH_3, 0.42% ionized
Ans. 1.8 × 10⁻⁵

(d) 0.10 M HF, 8.1% ionized
Ans. 7.1 × 10⁻⁴

(e) 0.300 M CH_2ClCO_2H, 6.60% ionized
Ans. 1.40 × 10⁻³

(f) 1.0 × 10⁻⁴ M H_2O_2, 1.5 × 10⁻²% ionized
Ans. 2.2 × 10⁻¹²

19. What is the concentration of acetic acid in a solution that is 0.30% ionized?   Ans. 2.0 M

20. A 0.010 M solution of HF is 23% ionized. Calculate $[H^+]$, $[F^-]$, $[HF]$, and the ionization constant for HF.   Ans. $[H^+] = [F^-] = 2.3 \times 10^{-3}$ M;
$[HF] = 7.7 \times 10^{-3}$ M; $K_a = 6.9 \times 10^{-4}$

Ionization of Water; pH and pOH

21. Describe the self-ionization of water.

22. Define pH both in words and mathematically.

S 23. Calculate the pH and pOH of pure water at 25°C.

24. The ionization constant for water (K_w) is 9.614 × 10⁻¹⁴ at 60°C. Calculate $[H^+]$, $[OH^-]$, pH, and pOH for pure water at 60°C.
Ans. $[H^+] = [OH^-] = 3.101 \times 10^{-7}$ M;
pH = pOH = 6.5085

25. Calculate the pH and pOH of each of the following solutions:

S (a) 0.200 M HCl
Ans. pH = 0.699; pOH = 13.30

(b) 0.0143 M NaOH
Ans. pH = 12.16; pOH = 1.84

S (c) 0.0071 M $Ca(OH)_2$
Ans. pH = 12.15; pOH = 1.85

(d) 3 M HNO_3
Ans. pH = −0.5; pOH = 14.5

26. Calculate the pH and pOH of each of the acidic aqueous solutions in Exercise 6.
Ans. (a) pH = 4.74, pOH = 9.26;
(b) 5.2, 8.8; (c) 2.80, 11.20;
(d) 2.60, 11.40

27. Calculate the pH and pOH of each of the basic aqueous solutions in Exercise 14.
Ans. (a) pH = 8.78, pOH = 5.22;
(b) 11.45, 2.55; (c) 11.38, 2.62;
(d) 11.0, 3.0; (e) 11.3, 2.7

28. Calculate the pH and pOH of a 0.50 M solution of $NaHSO_4$ (see Exercise 9).
Ans. pH = 1.14; pOH = 12.86

29. Calculate the pH and pOH of a 0.3124 M solution of NH_3 (see Exercise 15).
Ans. pH = 11.38; pOH = 2.62

30. Calculate the pH and pOH of each of the following solutions:

(a) 0.0407 M HNO_3
Ans. pH = 1.390; pOH = 12.610

(b) 0.0407 M $HC_2O_4^-$
Ans. pH = 2.79; pOH = 11.21

S (c) 0.407 M $HC_2O_4^-$
Ans. pH = 2.29; pOH = 11.71

(d) 0.19 M HNO_2
Ans. pH = 2.04; pOH = 11.96

(e) 1.4 × 10⁻² M HCNO
Ans. pH = 2.69; pOH = 11.31

31. Calculate the pH and pOH of each of the following solutions:

(a) 0.127 M NH_3
Ans. pH = 11.18; pOH = 2.82

(b) 0.244 M $(CH_3)_3N$
Ans. pH = 11.63; pOH = 2.37

(c) 0.0244 M $(CH_3)_3N$
Ans. pH = 11.12; pOH = 2.88

(d) $0.104\ M\ H_2BO_3^-$ $(K_b = 1.7 \times 10^{-6})$

Ans. pH = 10.62; pOH = 3.38

(e) $0.479\ M\ SO_3^{2-}$ $(K_b = 1.6 \times 10^{-7})$

Ans. pH = 10.44; pOH = 3.56

32. Calculate the pH, pOH, and total methylamine concentration in a solution of methylamine that is 6.90% ionized.

Ans. pH = 11.77; pOH = 2.23;
total CH_3NH_2 = 0.086 M

33. White vinegar is a 5.0% by mass solution of acetic acid in water. If the density of white vinegar is $1.007\ g/cm^3$, what is the pH?   *Ans. 2.41*

34. Calculate the pH of each of the following solutions, containing two solutes in the concentrations indicated:

Ⓢ (a) $0.50\ M\ CH_3CO_2H$, $0.50\ M\ NaCH_3CO_2$

Ans. 4.74

(b) $0.100\ M\ HN_3$, $0.200\ M\ NaN_3$ *Ans. 4.3*

(c) $0.225\ M\ HCNO$, $0.315\ M\ NaCNO$

Ans. 3.607

(d) $0.400\ M\ HNO_2$, $1.875\ M\ Ca(NO_2)_2$

Ans. 4.32

Ⓢ(e) $0.125\ M\ NH_3$, $1.00\ M\ NaOH$ *Ans. 14.00*

Ⓢ(f) $0.400\ M\ NaHSO_4$, $0.400\ M\ Na_2SO_4$

Ans. 1.95

Ionic Equilibria of Polyprotic Acids and Bases

35. Explain why successive ionization constants for polyprotic acids are progressively smaller.

36. Evaluate the ionization constants $K_{H_2CO_3}$ and $K_{HCO_3^-}$ as well as the constant for the equilibrium

$$H_2CO_3 \rightleftharpoons 2H^+ + CO_3^{2-}$$

from the following concentrations found at equilibrium in a solution of H_2CO_3 and $NaHCO_3$:

$$[H^+] = 8.0 \times 10^{-6}\ M$$
$$[HCO_3^-] = 1.6 \times 10^{-3}\ M$$
$$[CO_3^{2-}] = 1.4 \times 10^{-8}\ M$$
$$[H_2CO_3] = 3.0 \times 10^{-2}\ M$$

Ans. $K_{H_2CO_3} = 4.3 \times 10^{-7}$;
$K_{HCO_3^-} = 7.0 \times 10^{-11}$;
$K_1 \cdot K_2 = 3.0 \times 10^{-17}$

Ⓢ 37. Calculate the concentration of each species present in a 0.050 M solution of H_2S.

Ans. $[H^+] = 7.1 \times 10^{-5}\ M$;
$[OH^-] = 1.4 \times 10^{-10}\ M$; $[H_2S] = 0.050\ M$;
$[HS^-] = 7.1 \times 10^{-5}\ M$; $[S^{2-}] = 1.3 \times 10^{-13}\ M$

38. Calculate the concentration of each species present in a 0.010 M solution of phthalic acid, $C_6H_4(CO_2H)_2$.

$$C_6H_4(CO_2H)_2 \rightleftharpoons C_6H_4(CO_2H)(CO_2^-) + H^+$$
$$(K_a = 1.1 \times 10^{-3})$$

$$C_6H_4(CO_2H)(CO_2^-) \rightleftharpoons C_6H_4(CO_2^-)_2 + H^+$$
$$(K_a = 3.9 \times 10^{-6})$$

Ans. $[H^+] = 2.8 \times 10^{-3}\ M$;
$[OH^-] = 3.6 \times 10^{-12}\ M$;
$[C_6H_4(CO_2H)_2] = 7.2 \times 10^{-3}\ M$;
$[C_6H_4(CO_2H)(CO_2^-)] = 2.8 \times 10^{-3}\ M$;
$[C_6H_4(CO_2^-)_2] = 3.9 \times 10^{-6}\ M$

39. Nicotine, $C_{10}H_{14}N_2$, is a base that will accept two protons $(K_1 = 7 \times 10^{-7},\ K_2 = 1.4 \times 10^{-11})$. What is the concentration of each species present in a 0.050 M solution of nicotine?

Ans. $[H^+] = 5 \times 10^{-11}\ M$;
$[OH^-] = 2 \times 10^{-4}\ M$;
$[C_{10}H_{14}N_2] = 0.050\ M$;
$[C_{10}H_{14}N_2H^+] = 2 \times 10^{-4}\ M$;
$[C_{10}H_{14}N_2H_2^{2+}] = 1.4 \times 10^{-11}\ M$

40. Calculate the concentration of each species present in a 0.0250 M solution of Na_3PO_4.

$$PO_4^{3-} + H_2O \rightleftharpoons HPO_4^{2-} + OH^-$$
$$(K_b = 2.8 \times 10^{-2})$$

$$HPO_4^{2-} + H_2O \rightleftharpoons H_2PO_4^- + OH^-$$
$$(K_b = 1.6 \times 10^{-7})$$

$$H_2PO_4^- + H_2O \rightleftharpoons H_3PO_4 + OH^-$$
$$(K_b = 1.3 \times 10^{-12})$$

Ans. $[H^+] = 6.3 \times 10^{-13}\ M$;
$[OH^-] = 1.6 \times 10^{-2}\ M$;
$[PO_4^{3-}] = 9.1 \times 10^{-3}\ M$;
$[HPO_4^{2-}] = 1.6 \times 10^{-2}\ M$;
$[H_2PO_4^-] = 1.6 \times 10^{-7}\ M$;
$[H_3PO_4] = 1.3 \times 10^{-17}\ M$; $[Na^+] = 0.0750\ M$

Buffers

41. Explain why the pH does not change significantly when a small amount of an acid or a base is added to a solution containing equal amounts of H_3PO_4 and NaH_2PO_4.

42. Calculate the pH of each of the following buffer solutions, containing equal volumes of two solutes in the concentrations indicated:

(a) $0.50\ M\ CH_3CO_2H$, $0.50\ M\ NaCH_3CO_2$

Ans. 4.74

(b) 0.50 M CH_3CO_2H, 0.20 M $NaCH_3CO_2$ *Ans. 4.35*

(c) 0.40 M NH_3; 0.40 M NH_4Cl *Ans. 9.26*

[S] (d) 0.25 M H_3PO_4; 0.15 M NaH_2PO_4 *Ans. 1.90*

(e) 0.10 M H_2S; 0.20 M NaHS *Ans. 7.30*

43. In venous blood the following equilibrium is set up by dissolved carbon dioxide:

$$H_2CO_3 \rightleftharpoons H^+ + HCO_3^-$$

If the pH of the blood is 7.4, what is the ratio of $[HCO_3^-]$ to $[H_2CO_3]$? *Ans. 11 to 1*

44. Calculate the pH of 0.500 L of a 0.0880 M solution of $NaHCO_3$ to which has been added 1.10×10^{-2} mol of HCl. *Ans. 6.84*

45. Calculate the pH of a solution resulting from mixing 0.10 L of 0.10 M NaOH with 0.40 L of 0.025 M HF. *Ans. 7.72*

46. Calculate the pH of a solution made by mixing equal volumes of 0.30 M NH_3 and 0.030 M HNO_3. *Ans. 10.21*

47. Calculate the pH of a solution prepared by mixing 10.0 mL of 0.10 M LiOH and 20.0 mL of a 0.500 M benzoic acid solution. K_a for benzoic acid is 6.7×10^{-5}. *Ans. 3.22*

48. Calculate the pH of 0.500 L of a 0.120 M solution of HClO to which 0.010 mol of $Ca(OH)_2$ has been added. *Ans. 7.15*

49. Given a 0.1 M solution of ammonia that is also 0.1 M in ammonium chloride (a buffer solution) and a 1.8×10^{-5} M solution of NaOH, calculate (a) the initial pH in each solution, (b) the pH when a liter of each of the original solutions is treated with 0.03 mol of solid NaOH, and (c) the pH when a liter of each of the original solutions is treated with 0.03 mol of HCl. (Assume no volume change with the addition of the NaOH or the HCl.) Is the buffer solution effective in serving its intended function of holding $[H^+]$ relatively constant compared to the unbuffered solution?
Ans. (a) 9.3 for each solution; (b) 9.5; 12.5; (c) 9.0; 1.5

50. A buffer solution is prepared from 5.0 g of NH_4NO_3 and 0.100 L of 0.10 M NH_3. What is the pH of the buffer? *Ans. 8.46*

51. Calculate the pH of 0.750 L of a solution that contains 87.5 g of $Ba(N_3)_2$ and 0.250 mol of HN_3. *Ans. 4.5*

52. How many moles of sodium acetate must be added to 1.0 L of a 1.0 M acetic acid solution to prepare a buffer solution with a pH of 5.08? of 4.20? *Ans. 2.2 mol; 0.29 mol*

[S] 53. How many moles of NH_4Cl must be added to 1.0 L of a 1.0 M NH_3 solution to prepare a buffer solution with a pH of 9.00? of 9.50? *Ans. 1.8 mol; 0.57 mol*

54. Addition of 1.0 mL of 0.10 M NaOH to 0.100 L of the buffer solution discussed in Section 16.8 changed the pH from 4.74 to 4.75. What is the new pH when 1.0 mL of 0.10 M NaOH is added to 0.100 L of an unbuffered solution of hydrochloric acid with a pH of 4.74? *Ans. 10.99*

[S] 55. A buffer solution is made up of equal volumes of 0.100 M acetic acid and 0.500 M sodium acetate. (a) What is the pH of this solution? (b) What is the pH that results from adding 1.00 mL of 0.100 M HCl to 0.200 L of the buffer solution? (Use 1.80×10^{-5} for the ionization constant of acetic acid.) *Ans. 5.444; 5.438*

Titration Curves and Indicators

56. Calculate the pH after the addition of 20.0, 24.9, 25.0, 25.1, and 30.0 mL of NaOH in the titration of HCl described in Table 16-6 in Section 16.17.

57. Calculate the pH after the addition of 20.0, 24.9, 25.0, 25.1, 25.5, and 30.0 mL of NaOH in the titration of CH_3CO_2H described in Table 16-6 in Section 16.17.

58. Determine a theoretical titration curve for the titration of 20.0 mL of 0.100 M $HClO_4$, a strong acid, with 0.200 M KOH.

59. Determine a theoretical titration curve for the titration of 25.0 mL of 0.150 M NH_3 with 0.150 M HCl.

60. Which of the indicators in Table 16-5 in Section 16.16 would be appropriate for the titration described in Exercise 58? in Exercise 59?

61. Why does an acid-base indicator change color over a range of pH values rather than at a specific pH?

62. Using the data in Table 16-5, arrange the following indicators (which are weak acids) in increasing order of acid strength: methyl orange, litmus, thymol blue.
Ans. litmus < methyl orange < thymol blue

63. The indicator dinitrophenol is an acid with a K_a

of 1.1×10^{-4}. In a 1.0×10^{-4} M solution, it is colorless in acid and yellow in base. Calculate the pH range over which it goes from 10% ionized (colorless) to 90% ionized (yellow).

Ans. 3.0–4.9

64. A 0.5000-g sample of an impure monobasic amine is titrated to the equivalence point with 75.00 mL of 0.100 M HCl. The pH after 40.00 mL of acid is added is 10.65. The molecular weight of the pure base is known to be 59.1. (a) Calculate the ionization constant of this base and (b) its percent purity. (c) Which indicator from the following list would be most suitable for this titration? State your reasoning.

Orange IV (red–yellow; pH 1.2–2.6)
Methyl orange (red–yellow; pH 3.1–4.4)
Methyl red (red–yellow; pH 4.8–6.0)
Bromthymol blue (yellow–blue; pH 6.0–7.6)
Phenolphthalein (colorless–pink; pH 8.3–10.0)
Thymolphthalein (colorless–blue; pH 9.3–10.5)

Ans. (a) 5.1×10^{-4}; (b) 88.7%; (c) methyl red

65. The hydrogen ion concentration at the half-equivalence point (the point at which half of the amount of base necessary to reach the equivalence point has been added) in the titration of a weak acid with a strong base is equal to K_a for the weak acid. Explain.

Hydrolysis

66. Why does the salt of a strong acid and a weak base give an acidic solution?
67. Why does the salt of a weak acid and a strong base give a basic solution?
68. Why does the salt of a strong acid and a strong base give an approximately neutral solution?
69. Does the salt of a weak acid and a weak base give an acidic solution, a basic solution, or a neutral solution? Explain your answer.
70. Neutralization is defined in Section 14.11 as the reaction of an acid with a stoichiometric amount of a base. Why do some neutralization reactions give nonneutral solutions?
71. Why is a hydrolysis reaction simply an acid-base reaction?
72. Arrange the conjugate bases of the acids in Table 16-3 in order of increasing base strength.

73. Arrange the conjugate acids of the bases in Table 16-4 in order of decreasing acid strength.
74. Calculate the ionization constant (hydrolysis constant) for each of the following acids or bases:
$\boxed{s}$(a) F^- *Ans. 1.4×10^{-11}*
   (b) AsO_4^{3-} *Ans. 1×10^{-1}*
   (c) NO_2^- *Ans. 2.2×10^{-11}*
$\boxed{s}$(d) NH_4^+ *Ans. 5.6×10^{-10}*
   (e) $(CH_3)_2NH_2^+$ *Ans. 1.4×10^{-11}*
   (f) $HC_2O_4^-$ (as a base)   *Ans. 1.7×10^{-13}*
   (g) S^{2-} *Ans. 7.7×10^{-2}*
75. Calculate the pH of each of the following solutions:
$\boxed{s}$(a) 0.4735 M NaCN   *Ans. 11.5*
   (b) 0.1050 M NH_4NO_3 *Ans. 5.12*
   (c) 0.0270 M $Ca(N_3)_2$ *Ans. 8.4*
$\boxed{s}$(d) 0.333 M $[(CH_3)_2NH_2]_2SO_4$ [Note that $(CH_3)_2NH_2^+$ is the conjugate acid of the weak base $(CH_3)_2NH$, just as NH_4^+ is the conjugate acid of the weak base NH_3.]

Ans. 5.52

   (e) 0.100 M NH_4F *Ans. 6.20*
76. Calculate the pH of a solution made by mixing equal volumes of 0.040 M aqueous aniline, $C_6H_5NH_2$, and 0.040 M nitric acid. (K_b for $C_6H_5NH_2 = 4.6 \times 10^{-10}$.)   *Ans. 3.18*
77. (a) Write the equation for the hydrolysis reaction that makes potassium cyanide, KCN, behave as a base.
   (b) What is K_h for this reaction?
   (c) What is the pH of a 0.255 M solution of KCN?

Ans. (a) $CN^- + H_2O \rightleftharpoons HCN + OH^-$;
(b) $K_h = 2 \times 10^{-5}$; (c) 11.4

78. The ion HTe^- is an amphiphrotic species; it can act as either an acid or a base. What is K_a for the acid reaction of HTe^- with H_2O? What is K_b for the reaction in which HTe^- functions as a base in water?   *Ans. 1×10^{-5}; 4.3×10^{-12}*
79. Sodium nitrite, $NaNO_2$, is used in meat processing. What are $[H^+]$ and the pH of a 0.225 M $NaNO_2$ solution? Is the solution acidic or basic?

Ans. 4.5×10^{-9} M; 8.35; basic

80. Novocaine, $C_{13}H_{21}O_2N_2Cl$, is the salt of the base procaine and hydrochloric acid. The ionization constant for procaine is 7×10^{-6}. What is the pH of a 2.0% solution by mass of novocaine assuming that the density of the solution is 1.0 g/cm³?

Ans. 5.0

81. A 0.0010 M solution of KCN is hydrolyzed to the extent of 14.0%. Calculate the value of the ionization constant of HCN.   *Ans. 4.4 × 10⁻¹⁰*

82. Calculate the pH of a 0.470 M solution of lithium carbonate, Li_2CO_3.   *Ans. 11.9*

83. What concentration of calcium acetate, $Ca(CH_3CO_2)_2$, will give a solution in which the percent hydrolysis is 0.10%?
Ans. 2.8 × 10⁻⁴ M

Additional Exercises

84. By calculation, check the values given in Section 14.3 for the percent ionization of 0.10 M solutions of the weak acids HSO_4^-, HNO_2, and CH_3CO_2H.

85. By calculation, check the values given in Section 14.3 for the percent ionization of 0.10 M solutions of the weak bases NH_3, $CH_3CO_2^-$, and NO_2^-.

86. The pH of a sample of acid rain (Section 12.8, Part 4) is 4.07. Assuming that the acidity is due to the presence of HNO_2 in the rain, what is the total concentration of HNO_2?
Ans. 1.6 × 10⁻⁵ M

87. Using the K_a values in Appendix G and in Section 16.15, place $[Al(H_2O)_6]^{3+}$ in the correct location in Table 14-1 in Section 14.3, the table of relative strengths of conjugate acid-base pairs.

88. A typical urine sample contains 2.3% by mass of the base urea, $CO(NH_2)_2$

$$CO(NH_2)_2 + H_2O \rightleftharpoons$$
$$CO(NH_2)(NH_3)^+ + OH^-$$
$$K_b = 1.5 \times 10^{-14}$$

If the density of the urine sample is 1.06 g/cm³ and the pH of the sample is 6.35, calculate the concentration of $CO(NH_2)_2$ and of $CO(NH_2)(NH_3)^+$ in the sample.
Ans. [CO(NH₂)₂] = 0.41 M;
[CO(NH₂)(NH₃)⁺] = 2.7 × 10⁻⁷ M

89. The solutions of ammonia used for cleaning windows usually contain about 10% NH_3 by mass. If a solution that is exactly 10% by mass has a density of 0.99 g/cm³, what are the pH and the hydroxide ion concentration?
Ans. 12.01; 1.0 × 10⁻² M

90. In many detergents phosphates have been replaced with silicates as water conditioners. If 125 g of a detergent that contains 8.0% Na_2SiO_3 by weight is used in 4.0 L of water, what are the pH and the hydroxide ion concentration in the wash water?

$$SiO_3^{2-} + H_2O \rightleftharpoons SiO_3H^- + OH^-$$
$$K_b = 1.6 \times 10^{-3}$$
$$SiO_3H^- + H_2O \rightleftharpoons SiO_3H_2 + OH^-$$
$$K_b = 3.1 \times 10^{-5}$$
Ans. pH = 11.70; [OH⁻] = 5.0 × 10⁻³ M

91. Saccharin, $C_7H_4NSO_3H$, is a weak acid (K_a = 2.1 × 10⁻¹²). If 0.250 L of diet cola with a pH of 5.48 was prepared from 2.00 × 10⁻³ g of sodium saccharide (often referred to as sodium saccharin on soft drink bottles), $Na(C_7H_4NSO_3)$, what are the final concentrations of saccharin and sodium saccharide in the solution?
Ans. [C₇H₄NSO₃H] = 3.9 × 10⁻⁵ M;
[Na(C₇H₄NSO₃)] = 2.5 × 10⁻¹¹ M

92. The ionization constant for water, K_w, is 5.474 × 10⁻¹⁴ at 50°C. At 50°C, K_b for NH_3 is 1.892 × 10⁻⁵. What is the ionization constant for NH_4^+ at 50°C?   *Ans. 2.893 × 10⁻⁹*

93. If abominable snowmen were to exist in the world in the same proportion to humans as do hydrogen ions to water molecules in pure water, how many abominable snowmen would there be in the world? Assume a world population of 4 billion people.   *Ans. 7*

94. How many moles of hydrogen chloride must be added to 1.50 L of a 0.450 M solution of CH_3NH_2 to give a pH of 10.95?   *Ans. 0.22 mol*

95. What is the pH of 1.000 L of a solution of 100.0 g of glutamic acid ($C_5H_9NO_4$, a diprotic acid; K_1 = 8.5 × 10⁻⁵, K_2 = 3.39 × 10⁻¹⁰) to which has been added 20.0 g of NaOH during the preparation of monosodium glutamate, the flavoring agent? What is the pH when exactly 1 mol of NaOH per mole of acid has been added?
Ans. 4.51; 4.82

96. Lime juice is among the most acidic of fruit juices, with a pH of 1.92. If the acidity is due to citric acid, which we can abbreviate as H_3Cit, what is the ratio of each of the following to $[Cit^{3-}]$: $[H_3Cit]$; $[H_2Cit^-]$; $[HCit^{2-}]$?

$$H_3Cit \rightleftharpoons H^+ + H_2Cit^-   K_a = 8.4 \times 10^{-4}$$
$$H_2Cit^- \rightleftharpoons H^+ + HCit^{2-}   K_a = 1.8 \times 10^{-5}$$
$$HCit^{2-} \rightleftharpoons H^+ + Cit^{3-}   K_a = 4.0 \times 10^{-6}$$

Ans. 2.9 × 10⁷; 2.0 × 10⁶; 3.0 × 10³ to 1

97. Sulfuric acid is a strong acid. The extent of the

ionization, $H_2SO_4 \rightleftharpoons H^+ + HSO_4^-$, is effectively 100%. The hydrogen sulfate ion, HSO_4^-, is a weaker acid with $K_a = 1.2 \times 10^{-2}$.

(a) What is the concentration of sulfate ion, SO_4^{2-}, in a 0.102 M solution of $NaHSO_4$?

Ans. 0.029 M

(b) What is the concentration of sulfate ion in a 0.102 M solution of H_2SO_4?

Ans. 9.9 $\times$ 10^{-3} M

98. Sulfuric acid ionizes in two steps:

$$H_2SO_4 \rightleftharpoons H^+ + HSO_4^-$$
$$HSO_4^- \rightleftharpoons H^+ + SO_4^{2-}$$

Since sulfuric acid is a strong acid, the first step is effectively 100% complete ($K_{H_2SO_4}$ is large). The ionization constant for the second step, $K_{HSO_4^-}$, is 1.2×10^{-2}. Check the assumption that the first step goes completely to the right, based on the approximation that for a polyprotic acid, successive ionization constants often differ by a factor of 1×10^6 and calculate (a) $K_{H_2SO_4}$, (b) the concentration of H_2SO_4, (c) the percent ionization of H_2SO_4 into H^+ and HSO_4^- in a 0.100 M solution of H_2SO_4.

Ans. (a) 1 $\times$ 10^4; (b) 1 $\times$ 10^{-6} M; (c) 99.999%

99. Show that the rate of reaction of acetic acid with sodium hydroxide is given by the following expression:

$$\text{Rate} = \frac{kK_a[CH_3CO_2H][OH^-]}{[CH_3CO_2^-]}$$

17

THE SOLUBILITY
PRODUCT PRINCIPLE

In chemical reactions that involve the formation of solids (precipitation reactions), we often work with slightly soluble substances that are completely ionized in solution. Equilibrium systems consisting of slightly soluble ionic substances in contact with their saturated aqueous solutions are particularly important in analytical chemistry. In this chapter we shall be concerned with the theory and application of the precipitation and dissolution of slightly soluble ionic substances.

THE FORMATION OF PRECIPITATES

17.1 The Solubility Product

A slightly soluble electrolyte dissolves until a saturated solution, a solution of its ions in equilibrium with undissolved solute, is formed. The use of radioactive tracers shows that at equilibrium the solid (the undissolved solute) continues to dissolve, the process of crystallization occurs, and the opposing processes have equal rates. The equilibrium of a solution of silver chloride with solid silver chloride can be expressed by

$$AgCl(s) \underset{\text{Precipitation}}{\overset{\text{Dissolution}}{\rightleftharpoons}} Ag^+(aq) + Cl^-(aq) \quad \text{(saturated solution)}$$

Applying the law of mass action to this system, we obtain

$$\frac{[Ag^+][Cl^-]}{[AgCl]} = K$$

Because the concentration of the solid silver chloride remains constant, we can write

$$[Ag^+][Cl^-] = K[AgCl] = K_{sp}$$

K_{sp} **is called the solubility product,** or sometimes the **solubility product constant.** The product of the concentrations must be equal to the solubility product when a saturated solution is in equilibrium with undissolved solute.

Calcium phosphate dissolves according to the equation

$$Ca_3(PO_4)_2(s) \rightleftharpoons 3Ca^{2+}(aq) + 2PO_4^{3-}(aq)$$

The expression for its solubility product is

$$[Ca^{2+}]^3[PO_4^{3-}]^2 = K_{sp}$$

For a salt like magnesium ammonium phosphate, which dissolves according to the equation

$$MgNH_4PO_4(s) \rightleftharpoons Mg^{2+}(aq) + NH_4^+(aq) + PO_4^{3-}(aq)$$

we have

$$[Mg^{2+}][NH_4^+][PO_4^{3-}] = K_{sp}$$

Solubility products have units, as do all equilibrium constants (Section 15.18); however, we will follow the common practice and omit them.

The solubility product is constant only for solutions of *slightly* soluble electrolytes. It is not constant for moderately or highly soluble salts, such as NaCl and KNO_3. Furthermore, high concentrations of ions not involved in the equilibrium increase the solubility of a slightly soluble substance—the salt effect (Section 16.6). The salt effect varies with the concentrations and with the charges of the additional ions. Hence the solubility product is constant *only* for saturated solutions with extremely small ionic concentrations. However, the solubility product concept is very useful for calculating approximate concentrations of slightly soluble electrolytes.

17.2 Calculation of Solubility Products from Solubilities

The solubility product of a slightly soluble electrolyte can be calculated from its solubility in moles per liter. When concentrations are given in other units, such as grams per liter, they must be converted to moles per liter.

EXAMPLE 17.1   Calculate the solubility product, K_{sp}, of silver chromate, Ag_2CrO_4, whose solubility is 1.3×10^{-4} mol L^{-1}.

From the equation for the dissolution of silver chromate

$$Ag_2CrO_4(s) \rightleftharpoons 2Ag^+(aq) + CrO_4^{2-}(aq)$$

we note that K_{sp} is given by

$$[Ag^+]^2[CrO_4^{2-}] = K_{sp}$$

From the equation for the dissolution of Ag_2CrO_4, we can see that 1 mol of $CrO_4^{2-}(aq)$ is formed for each mole of Ag_2CrO_4 that dissolves (1 mol CrO_4^{2-}/1 mol Ag_2CrO_4), and 2 mol of $Ag^+(aq)$ are formed (2 mol Ag^+/1 mol Ag_2CrO_4).

$$\frac{1.3 \times 10^{-4} \text{ mol } Ag_2CrO_4}{1 \text{ L}} \times \frac{2 \text{ mol } Ag^+}{1 \text{ mol } Ag_2CrO_4} = \frac{2.6 \times 10^{-4} \text{ mol } Ag^+}{1 \text{ L}}$$

$$\frac{1.3 \times 10^{-4} \text{ mol } Ag_2CrO_4}{1 \text{ L}} \times \frac{1 \text{ mol } CrO_4^{2-}}{1 \text{ mol } Ag_2CrO_4} = \frac{1.3 \times 10^{-4} \text{ mol } CrO_4^{2-}}{1 \text{ L}}$$

Thus $[Ag^+] = 2.6 \times 10^{-4}$ M and $[CrO_4^{2-}] = 1.3 \times 10^{-4}$ M.

A table of concentrations before the reaction (before any Ag_2CrO_4 dissolves) and at equilibrium contains the following entries:

	$[Ag^+]$	$[CrO_4^{2-}]$
Initial concentrations, M	0	0
Equilibrium concentrations, M	2.6×10^{-4}	1.3×10^{-4}

Substituting the concentrations of Ag^+ and CrO_4^{2-} in the expression for the solubility product gives

$$[Ag^+]^2[CrO_4^{2-}] = K_{sp}$$
$$(2.6 \times 10^{-4})^2(1.3 \times 10^{-4}) = 8.8 \times 10^{-12} = K_{sp}$$

Note that $[Ag^+]$ is squared in this expression.

In this example it was not really necessary to include the initial concentrations of the ions in the table since they were both zero. However, it is good practice to do so since initial concentrations will be helpful later in this chapter.

EXAMPLE 17.2   The solubility of lead chromate, $PbCrO_4$, the artist's pigment chrome yellow, is 4.3×10^{-5} g L^{-1}. Calculate the solubility product for $PbCrO_4$.

We need the solubility of $PbCrO_4$ in moles per liter. From this solubility we can find the equilibrium concentrations, and from them, K_{sp}. We can use the formula weight of $PbCrO_4$, 323.2, to find the molar solubility.

$$\frac{4.3 \times 10^{-5} \text{ g } PbCrO_4}{1 \text{ L}} \times \frac{1 \text{ mol } PbCrO_4}{323.2 \text{ g } PbCrO_4}$$

$$= \frac{1.33 \times 10^{-7} \text{ mol } PbCrO_4}{1 \text{ L}} = 1.33 \times 10^{-7} \text{ } M$$

$$PbCrO_4(s) \rightleftharpoons Pb^{2+}(aq) + CrO_4^{2-}(aq)$$

Since 1 mol of $PbCrO_4$ gives 1 mol of $Pb^{2+}(aq)$ and 1 mol of $CrO_4^{2-}(aq)$, both $[Pb^{2+}]$ and $[CrO_4^{2-}]$ are equal to the molar solubility of $PbCrO_4$.

Now we set up a table of concentrations.

	$[Pb^{2+}]$	$[CrO_4^{2-}]$
Equilibrium concentrations, M	1.33×10^{-7}	1.33×10^{-7}

We substitute the equilibrium concentrations into the expression for K_{sp}.

$$[Pb^{2+}][CrO_4{}^{2-}] = K_{sp}$$
$$(1.33 \times 10^{-7})(1.33 \times 10^{-7}) = 1.8 \times 10^{-14} = K_{sp}$$

A table of solubility products of some slightly soluble electrolytes at 25°C is given in Appendix E.

17.3 Calculation of Molar Solubilities from Solubility Products

The solubility of a slightly soluble electrolyte can be readily calculated from its solubility product.

EXAMPLE 17.3 The solubility product for silver chloride is 1.8×10^{-10} (Appendix E). Calculate the molar solubility of silver chloride.

Silver chloride forms a saturated solution according to the equation

$$AgCl(s) \rightleftharpoons Ag^+(aq) + Cl^-(aq)$$

Let x be the solubility of AgCl in moles per liter. The molar concentrations of both Ag^+ and Cl^- will also be x since the salt is completely dissociated. Thus we have this table:

	$[Ag^+]$	$[Cl^-]$
Equilibrium concentrations, M	x	x

Substituting in the solubility product expression for AgCl gives

$$[Ag^+][Cl^-] = K_{sp} = 1.8 \times 10^{-10}$$
$$x^2 = 1.8 \times 10^{-10}$$
$$x = \sqrt{1.8 \times 10^{-10}} = 1.3 \times 10^{-5}$$

The molar solubility of silver chloride is $1.3 \times 10^{-5}\ M$.

EXAMPLE 17.4 Although most mercury compounds are very poisonous, sixteenth century physicians used calomel, Hg_2Cl_2, as a medication. Their patients (usually) did not die of mercury poisoning because calomel is quite insoluble. Calculate the molar solubility of Hg_2Cl_2 ($K_{sp} = 1.1 \times 10^{-18}$; see Appendix E).

The equation for the dissolution of Hg_2Cl_2 is

$$Hg_2Cl_2(s) \rightleftharpoons Hg_2{}^{2+}(aq) + 2Cl^-(aq)$$

If the solubility of Hg_2Cl_2 is x mol L^{-1}, then

$$[Hg_2^{2+}] = \frac{x \text{ mol } Hg_2Cl_2}{1 \text{ L}} \times \frac{1 \text{ mol } Hg_2^{2+}}{1 \text{ mol } Hg_2Cl_2} = \frac{x \text{ mol}}{1 \text{ L}}$$

$$[Cl^-] = \frac{x \text{ mol } Hg_2Cl_2}{1 \text{ L}} \times \frac{2 \text{ mol } Cl^-}{1 \text{ mol } Hg_2Cl_2} = \frac{2x \text{ mol}}{1 \text{ L}}$$

Thus x mol of Hg_2Cl_2 gives $2x$ mol of Cl^-.

	$[Hg_2^{2+}]$	$[Cl^-]$
Equilibrium concentrations, M	x	$2x$

We substitute these values into the expression for K_{sp} and calculate the value of x.

$$[Hg_2^{2+}][Cl^-]^2 = K_{sp} = 1.1 \times 10^{-18}$$
$$(x)(2x)^2 = 4x^3 = 1.1 \times 10^{-18}$$
$$x = 6.5 \times 10^{-7}$$

The molar solubility of Hg_2Cl_2 is 6.5×10^{-7} M; $[Hg_2^{2+}] = x = 6.5 \times 10^{-7}$ M, and $[Cl^-] = 2x = 1.3 \times 10^{-6}$ M.

17.4 The Precipitation of Slightly Soluble Electrolytes

A slightly soluble electrolyte will precipitate if its ions are present in solution in such quantities that the product of their concentrations, raised to the appropriate powers, *exceeds the solubility product of the electrolyte*. We can show that silver chloride will precipitate if we mix equal volumes of a 2×10^{-4} M solution of silver nitrate and a 2×10^{-4} M solution of sodium chloride. The volume doubles when we mix equal volumes; therefore each concentration is reduced to half its initial value. Consequently, immediately upon mixing, $[Ag^+]$ and $[Cl^-]$ will both be equal to $\frac{1}{2}(2 \times 10^{-4}$ $M) = 1 \times 10^{-4}$ M. The product $[Ag^+][Cl^-]$ will, *momentarily*, be greater than K_{sp} for AgCl.

$$K_{sp} = 1.8 \times 10^{-10}$$
$$[Ag^+][Cl^-] = (1 \times 10^{-4})(1 \times 10^{-4}) = 1 \times 10^{-8} > K_{sp}$$

The system will not be at equilibrium, and silver chloride will precipitate until the product $[Ag^+][Cl^-]$ equals the solubility product; that is, until

$$[Ag^+][Cl^-] = K_{sp}$$

The ions Ag^+ and Cl^- are present in equal concentrations in a saturated solution of AgCl in pure water in contact with solid AgCl, and the product $[Ag^+][Cl^-]$ just equals K_{sp}. If we add NaCl to such a solution, the product of $[Ag^+]$ and $[Cl^-]$ will be larger than the solubility product, and AgCl will form until the ion product again equals K_{sp}. At the new equilibrium $[Ag^+]$ will be less and $[Cl^-]$ greater than they were in the solution of AgCl in pure water. The greater we make $[Cl^-]$, the smaller $[Ag^+]$ will become. However, we can never reduce $[Ag^+]$ to zero because the product $[Ag^+][Cl^-]$ will always be equal to the solubility product.

17.5 Calculation of the Concentration of an Ion Necessary to Form a Precipitate

When the concentration of one of the ions of a slightly soluble electrolyte and the value for the solubility product of the electrolyte are known, we can calculate the concentration that the other ion must exceed before precipitation will occur.

EXAMPLE 17.5   If a solution contains 0.001 mol of CrO_4^{2-} per liter, what concentration of Ag^+ ion must be exceeded (by adding solid $AgNO_3$ to the solution) before Ag_2CrO_4 will begin to precipitate, neglecting any increase in volume due to the addition of silver nitrate? ($K_{sp} = 9 \times 10^{-12}$; see Appendix E.)

The equilibrium involved in the problem is

$$Ag_2CrO_4(s) \rightleftharpoons 2Ag^+(aq) + CrO_4^{2-}(aq)$$

There is no equilibrium between Ag_2CrO_4, Ag^+, and CrO_4^{2-} until the first trace of solid Ag_2CrO_4 appears. Equilibrium is established only when enough $AgNO_3$ has been added to form the first trace of $Ag_2CrO_4(s)$, and only then can the relationship between $[Ag^+]$ and $[CrO_4^{2-}]$ be determined from the solubility product expression. Let x equal $[Ag^+]$ at this point.

	$[Ag^+]$	$[CrO_4^{2-}]$
Initial concentrations, M	0	0.001
Equilibrium concentrations, M	x	0.001

Since enough $AgNO_3$ to form only a trace of solid Ag_2CrO_4 has been added, $[CrO_4^{2-}]$ has not changed significantly. Substitution into the solubility product expression gives

$$[Ag^+]^2[CrO_4^-] = K_{sp} = 9 \times 10^{-12}$$

$$(x^2)(0.001) = 9 \times 10^{-12}$$

$$x^2 = \frac{9 \times 10^{-12}}{0.001} = 9 \times 10^{-9}$$

$$x = 9.5 \times 10^{-5} \quad \text{or} \quad 1 \times 10^{-4}\, M \text{ (rounded)}$$

A concentration of Ag^+ greater than $1 \times 10^{-4}\, M$ is necessary to cause precipitation of Ag_2CrO_4 under these conditions.

17.6 Calculation of the Concentration of an Ion in Solution after Precipitation

It is often useful to know the concentration of an ion remaining in solution after precipitation. This, too, can be determined using the solubility product.

EXAMPLE 17.6   Clothing washed in water with a manganese concentration exceeding 0.1 mg/L $(1.8 \times 10^{-6}\ M)$ may be stained by the manganese. A laundry wishes to add a base to their water to precipitate the manganese as the hydroxide $Mn(OH)_2$ $(K_{sp} = 4.5 \times 10^{-14})$. At what pH will $[Mn^{2+}]$ be equal to $1.8 \times 10^{-6}\ M$? The dissolution of $Mn(OH)_2$ is described by the equation

$$Mn(OH)_2(s) \rightleftharpoons Mn^{2+}(aq) + 2OH^-(aq)$$

At equilibrium $[OH^-]$ has been adjusted so that $[Mn^{2+}]$ is $1.8 \times 10^{-6}\ M$.

	$[Mn^{2+}]$	$[OH^-]$
Equilibrium concentrations, M	1.8×10^{-6}	x

Substitution into the expression for K_{sp} gives

$$[Mn^{2+}][OH^-]^2 = K_{sp} = 4.5 \times 10^{-14}$$

$$(1.8 \times 10^{-6})(x)^2 = 4.5 \times 10^{-14}$$

$$x^2 = \frac{4.5 \times 10^{-14}}{1.8 \times 10^{-6}} = 2.5 \times 10^{-8}$$

$$x = 1.6 \times 10^{-4}\ M$$

Thus

$$[OH^-] = 1.6 \times 10^{-4}\ M$$

$$pOH = -\log[OH^-] = -\log(1.6 \times 10^{-4}) = 3.8$$

$$pH = 14 - pOH = 10.20$$

If the laundry adds a base to their water until the pH rises to 10.20, the manganese ion will be reduced to a concentration of $1.8 \times 10^{-6}\ M$; at or below that concentration, it will not stain.

17.7 Supersaturation of Slightly Soluble Electrolytes

When the product of the concentrations of the ions raised to their proper powers exceeds the solubility product, precipitation usually occurs. However, a supersaturated solution may be formed, and then precipitation does not occur immediately. For example, if magnesium is being precipitated as magnesium ammonium phosphate from a solution with a low concentration of magnesium ion, the solution may stand several hours before a visible precipitate appears. Furthermore, even though precipitation may begin immediately, it may not be complete. The first crystals to precipitate are often very small and more soluble than are the larger ones that form as the solution stands in contact with the precipitate (see Section 17.8). To ensure maximum precipitation you should let a solution stand before filtering or centrifuging it.

17.8   Solubility and Crystal Size

The solubility of fine crystals of barium sulfate has been found to be about twice as great as that of coarse crystals. In fact, small crystals of any substance are more soluble than large ones because there are proportionately more ions inside the larger crystal. Ions in the interior of a crystal are bound more tightly than are those at the faces and edges. The fraction of ions occupying surface positions is greater in a small crystal than in a large crystal, making the tendency for ions to enter a solution greater for the small crystal. Because small crystals are more soluble than large ones, the smallest crystals gradually dissolve while larger ones grow still larger. True equilibrium is not reached until large, perfect crystals are formed. Solubility products are determined for solutions in contact with relatively large crystals.

Very small crystals pass through filters and resist centrifugation; therefore precipitates are often digested (warmed in the solution) to increase the size of the crystals. Heat increases the rate at which large crystals grow from the smaller ones. Simply allowing a precipitate to stand in solution may also yield larger crystals.

17.9   Calculations Involving Both Ionization Constants and Solubility Products

In dealing with precipitation we often work with systems that involve more than one equilibrium. The following examples illustrate how to treat multiple equilibria.

EXAMPLE 17.7   Calculate the hydrogen ion concentration required to prevent the precipitation of ZnS in a solution that is 0.16 M in $ZnCl_2$ and saturated with H_2S (0.10 M in H_2S).

We need to consider two reactions in this system.

$$ZnS(s) \rightleftharpoons Zn^{2+}(aq) + S^{2-}(aq) \qquad (1)$$
$$H_2S(aq) \rightleftharpoons 2H^+(aq) + S^{2-}(aq) \qquad (2)$$

ZnS will not precipitate if the product $[Zn^{2+}][S^{2-}]$ is less than K_{sp} for ZnS (1.1×10^{-21}). Let us find the conditions under which the first trace of ZnS will appear.

	$[Zn^{2+}]$	$[S^{2-}]$
Equilibrium concentrations, M	0.16	x

$$[Zn^{2+}][S^{2-}] = K_{sp} = 1.1 \times 10^{-21}$$

$$(0.16)(x) = 1.1 \times 10^{-21}$$

$$x = \frac{1.1 \times 10^{-21}}{0.16} = 6.9 \times 10^{-21} \, M$$

Hence, when $[Zn^{2+}]$ is 0.16 M, ZnS will begin to appear unless $[S^{2-}]$ is less than $6.9 \times 10^{-21} M$.

The concentration of S^{2-} can be adjusted by changing the concentration of H^+ (see Equation 2 and Section 16.9).

$$\frac{[H^+]^2[S^{2-}]}{[H_2S]} = K_a = 1.3 \times 10^{-20}$$

Now we let $[H^+]$ be equal to y and calculate $[H^+]$ for a solution in which $[S^{2-}]$ is $6.9 \times 10^{-21}\ M$.

	$[H^+]$	$[S^{2-}]$	$[H_2S]$
Equilibrium concentrations, M	y	6.9×10^{-21}	0.10

$$\frac{[H^+]^2[S^{2-}]}{[H_2S]} = K_a = 1.3 \times 10^{-20}$$

$$\frac{(y)^2(6.9 \times 10^{-21})}{0.10} = 1.3 \times 10^{-20}$$

$$y^2 = \frac{0.10}{6.9 \times 10^{-21}} \times 1.3 \times 10^{-20} = 1.9 \times 10^{-1}$$

$$y = 4.3 \times 10^{-1} = 0.43\ M$$

Thus $[S^{2-}]$ is $6.9 \times 10^{-21}\ M$ when $[H^+]$ is 0.43 M. If $[H^+]$ is greater than 0.43 M, $[S^{2-}]$ will be less than $6.9 \times 10^{-21}\ M$, and precipitation of ZnS will not occur.

EXAMPLE 17.8   Calculate the concentration of ammonium ion, supplied by NH_4Cl, required to prevent the precipitation of $Mg(OH)_2$ in a liter of solution containing 0.10 mol of NH_3 and 0.10 mol of Mg^{2+}.

Two equilibria are involved in this system.

$$Mg(OH)_2(s) \rightleftharpoons Mg^{2+}(aq) + 2OH^-(aq)$$

$$NH_3(aq) + H_2O(l) \rightleftharpoons NH_4^+(aq) + OH^-(aq)$$

If $Mg(OH)_2$ is not to form, then the product $[Mg^{2+}][OH^-]^2$ must be less than K_{sp} for $Mg(OH)_2$. $[OH^-]$ can be reduced by the addition of NH_4^+, which shifts the second reaction to the left and thus reduces $[OH^-]$.

First we determine the hydroxide ion concentration at which the first trace of $Mg(OH)_2$ will form when $[Mg^{2+}]$ is 0.10 M. This is the hydroxide concentration for which $[Mg^{2+}][OH^-]^2$ equals K_{sp}.

	$[Mg^{2+}]$	$[OH^-]$
Equilibrium concentrations, M	0.10	x

$$[Mg^{2+}][OH^-]^2 = K_{sp} = 1.5 \times 10^{-11}$$

$$(0.10)(x)^2 = 1.5 \times 10^{-11}$$

$$x^2 = \frac{1.5 \times 10^{-11}}{0.10} = 1.5 \times 10^{-10}$$

$$x = 1.2 \times 10^{-5}\ M$$

$Mg(OH)_2$ will not begin to precipitate if $[OH^-]$ is less than $1.2 \times 10^{-5}\ M$.

Now we must find what $[NH_4^+]$ is necessary to reduce $[OH^-]$ to 1.2×10^{-5} M.

	$[NH_3]$	$[NH_4^+]$	$[OH^-]$
Equilibrium concentrations, M	0.10	y	1.2×10^{-5}

$$\frac{[NH_4^+][OH^-]}{[NH_3]} = K_b = 1.8 \times 10^{-5}$$

$$\frac{(y)(1.2 \times 10^{-5})}{0.10} = 1.8 \times 10^{-5}$$

$$y = \frac{0.10}{1.2 \times 10^{-5}} \times 1.8 \times 10^{-5} = 0.15 \ M$$

If $[NH_4^+]$ equals 0.15 M, $[OH^-]$ will be 1.2×10^{-5} M. Any $[NH_4^+]$ greater than 0.15 M will reduce $[OH^-]$ below 1.2×10^{-5} M and prevent formation of $Mg(OH)_2$.

17.10   Fractional Precipitation

When two anions form slightly soluble compounds with the same cation or when two cations form slightly soluble compounds with the same anion, the less soluble compound will precipitate first as the precipitating agent (the precipitant) is added to a solution containing both ions. Almost all of the less soluble compound will precipitate before any of the more soluble one does; however, any remaining less soluble compound will precipitate along with the more soluble one (**coprecipitate**) as more precipitant is added.

For example, a solution containing both sodium iodide and sodium chloride is treated with silver nitrate. Silver iodide, being less soluble than silver chloride, is precipitated first. Only after most of the iodide is precipitated will the chloride begin to precipitate.

EXAMPLE 17.9   A solution contains 0.010 mol of KI and 0.10 mol of KCl per liter. $AgNO_3$ is gradually added to this solution. Which will be precipitated first, AgI or AgCl?

We must find $[Ag^+]$ at which AgCl will begin to precipitate and $[Ag^+]$ at which AgI will begin to precipitate. The salt that forms at the lower $[Ag^+]$ will precipitate first.

For AgI. Let x be equal to $[Ag^+]$ at which precipitation begins.

$$[Ag^+][I^-] = K_{sp} = 1.5 \times 10^{-16}$$

$$(x)(0.010) = 1.5 \times 10^{-16}$$

$$x = \frac{1.5 \times 10^{-16}}{0.010} = 1.5 \times 10^{-14} \ M$$

Thus AgI will begin to precipitate when $[Ag^+]$ is 1.5×10^{-14} M.

For AgCl. Let y be equal to [Ag$^+$] at which precipitation begins.

$$[Ag^+][Cl^-] = K_{sp} = 1.8 \times 10^{-10}$$

$$(y)(0.10) = 1.8 \times 10^{-10}$$

$$y = \frac{1.8 \times 10^{-10}}{0.10} = 1.8 \times 10^{-9} \, M$$

Thus [Ag$^+$] must be 1.8×10^{-9} M for AgCl to precipitate.

AgI begins to precipitate at a lower silver ion concentration than that at which AgCl begins to precipitate, so AgI will begin to precipitate first.

EXAMPLE 17.10   (a) What is the concentration of I$^-$ in the solution described in Example 17.9 when AgCl begins to precipitate, and what fraction of the original I$^-$ remains in solution at this point?

When [Ag$^+$] is 1.8×10^{-9} M, AgCl will begin to precipitate. Let x be equal to [I$^-$] when [Ag$^+$] is 1.8×10^{-9} M.

$$[Ag^+][I^-] = K_{sp} = 1.5 \times 10^{-16}$$

$$(1.8 \times 10^{-9})(x) = 1.5 \times 10^{-16}$$

$$x = \frac{1.5 \times 10^{-16}}{1.8 \times 10^{-9}} = 8.3 \times 10^{-8} \, M$$

[I$^-$] will be 8.3×10^{-8} M when AgCl begins to precipitate.

The fraction of I$^-$ remaining at this point may be determined as follows:

$$\frac{[I^-] \text{ when precipitation of AgCl begins}}{[I^-] \text{ originally present}} = \frac{8.3 \times 10^{-8}}{0.010} = 8.3 \times 10^{-6}$$

Only about 8 parts in a million of the original I$^-$ remains in solution when AgCl begins to precipitate.

(b) What will the concentration of I$^-$ in the solution be after half of the Cl$^-$ initially present is precipitated as AgCl?

When half of the Cl$^-$ initially present has been precipitated, [Cl$^-$] will have been reduced to 0.050 M.

$$[Ag^+] = \frac{K_{AgCl}}{[Cl^-]} = \frac{1.8 \times 10^{-10}}{0.050} = 3.6 \times 10^{-9} \, M$$

$$[I^-] = \frac{K_{AgI}}{[Ag^+]} = \frac{1.5 \times 10^{-16}}{3.6 \times 10^{-9}} = 0.417 \times 10^{-7} = 4.2 \times 10^{-8} \, M$$

Half of the very small amount of I$^-$ remaining in solution when AgCl begins to precipitate (8.3×10^{-8} M) is coprecipitated as half of the Cl$^-$ is precipitated.

THE DISSOLUTION OF PRECIPITATES

When the product of the initial molar concentrations of the ions raised to their appropriate powers is less than ($<$) the solubility product, a solid either dissolves completely or dissolves until the ion concentration product is equal to the solubility product. For example, calcium carbonate dissolves in water as long as

$$[Ca^{2+}][CO_3^{2-}] < K_{sp}$$

For a precipitate to dissolve completely, it is necessary to make the concentration of at least one of its ions low enough that the ion concentration product is less than the solubility product. Ions can be removed (1) by the formation of a weak electrolyte, (2) by the conversion of an ion to another species, or (3) by the formation of a complex ion.

17.11   Dissolution of a Precipitate by the Formation of a Weak Electrolyte

Slightly soluble solids derived from weak acids are often soluble in strong acids. For example, $CaCO_3$, FeS, and $Ca_3(PO_4)_2$ dissolve in HCl because their anions react to form weak acids (H_2CO_3, H_2S, and $H_2PO_4^-$).

When hydrochloric acid is added to solid calcium carbonate, the hydrogen ion from the acid combines with the carbonate ion and forms the hydrogen carbonate ion, a weak acid.

$$H^+(aq) + CO_3^{2-}(aq) \rightleftharpoons HCO_3^-(aq)$$

Additional hydrogen ion reacts with the hydrogen carbonate ion according to the equation

$$H^+(aq) + HCO_3^-(aq) \rightleftharpoons H_2CO_3(aq)$$

Finally, the solution becomes saturated with the slightly ionized and unstable carbonic acid, and carbon dioxide gas is evolved.

$$H_2CO_3(aq) \rightleftharpoons H_2O(l) + CO_2(g)$$

These reactions reduce the carbonate ion concentration to such a low level that the product of the calcium ion and carbonate ion concentrations is less than the solubility product of calcium carbonate.

$$[Ca^{2+}][CO_3^{2-}] < K_{sp}$$

Consequently, the calcium carbonate dissolves. Even acetic acid provides enough hydrogen ion to dissolve calcium carbonate because, although weak, acetic acid is still a stronger acid than either the hydrogen carbonate ion or carbonic acid.

Lead sulfate dissolves in solutions of ammonium acetate because the formation of slightly ionized (but soluble) lead acetate reduces the product of the lead ion and sulfate ion concentrations below the value of the solubility product for lead sulfate.

$$PbSO_4(s) \rightleftharpoons Pb^{2+}(aq) + SO_4^{2-}(aq)$$
$$Pb^{2+}(aq) + 2CH_3CO_2^-(aq) \rightleftharpoons Pb(CH_3CO_2)_2(aq)$$
$$[Pb^{2+}][SO_4^{2-}] < K_{sp}$$

The formation of water causes most metal hydroxides, such as $Al(OH)_3$, $Mg(OH)_2$, and $Fe(OH)_3$, to dissolve in acid solutions.

$$Al(OH)_3(s) \rightleftharpoons Al^{3+}(aq) + 3OH^-(aq)$$

$$3H^+(aq) + 3OH^-(aq) \rightleftharpoons 3H_2O\,(l)$$

Net: $Al(OH)_3(s) + 3H^+(aq) \longrightarrow Al^{3+}(aq) + 3H_2O(l)$

$$[Al^{3+}][OH^-]^3 < K_{sp}$$

$$Mg(OH)_2(s) \rightleftharpoons Mg^{2+}(aq) + 2OH^-(aq)$$

$$2NH_4^+(aq) + 2OH^-(aq) \rightleftharpoons 2NH_3(aq) + 2H_2O\,(l)$$

Net: $Mg(OH)_2(s) + 2NH_4^+(aq) \longrightarrow Mg^{2+}(aq) + 2H_2O(l) + 2NH_3(aq)$

$$[Mg^{2+}][OH^-]^2 < K_{sp}$$

17.12 Dissolution of a Precipitate by the Conversion of an Ion to Another Species

The solubility products for some metal sulfides are sufficiently large so that the hydrogen ion from a strong acid lowers the sulfide ion concentration (by forming the weak electrolyte hydrogen sulfide) enough to dissolve the sulfide. For example, iron(II) sulfide is readily dissolved by hydrochloric acid according to the equation

$$FeS(s) \rightleftharpoons Fe^{2+}(aq) + S^{2-}(aq)$$

$$2H^+(aq) + S^{2-}(aq) \rightleftharpoons H_2S\,(g)$$

Net: $FeS(s) + 2H^+(aq) \rightleftharpoons Fe^{2+}(aq) + H_2S(g)$

$$[Fe^{2+}][S^{2-}] < K_{sp}$$

However, the solubility products for other metal sulfides, such as lead sulfide, are so small that even very high concentrations of hydrogen ion from strong acids are not sufficient to lower the sulfide ion concentration enough to dissolve them. To dissolve these sulfides, the sulfide ion concentration must be decreased by oxidizing the sulfide to elemental sulfur.

$$3S^{2-}(aq) + 2NO_3^-(aq) + 8H^+(aq) \longrightarrow 3S(s) + 2NO(g) + 4H_2O(l)$$

Lead sulfide dissolves in nitric acid, then, because the sulfide ion is oxidized to sulfur, making the product of the lead ion and sulfide ion concentrations less than the solubility product:

$$[Pb^{2+}][S^{2-}] < K_{sp}$$

17.13 Dissolution of a Precipitate by the Formation of a Complex Ion

Many slightly soluble electrolytes dissolve as a result of the formation of complex ions. Several examples that are important to qualitative analysis follow.

$$AgCl(s) + 2NH_3(aq) \rightleftharpoons Ag(NH_3)_2^+(aq) + Cl^-(aq)$$

$$CuCN(s) + CN^-(aq) \rightleftharpoons Cu(CN)_2^-(aq)$$

$$Zn(OH)_2(s) + 2OH^-(aq) \rightleftharpoons Zn(OH)_4^{2-}(aq)$$

$$Sn(OH)_2(s) + OH^-(aq) \rightleftharpoons Sn(OH)_3^-(aq)$$

$$Al(OH)_3(s) + OH^-(aq) \rightleftharpoons Al(OH)_4^-(aq)$$
$$As_2S_3(s) + 3S^{2-}(aq) \rightleftharpoons 2AsS_3^{3-}(aq)$$
$$HgS(s) + S^{2-}(aq) \rightleftharpoons HgS_2^{2-}(aq)$$
$$Sb_2S_3(s) + 3S^{2-}(aq) \rightleftharpoons 2SbS_3^{3-}(aq)$$

The stability of a complex ion is described by an equilibrium constant for the formation of the complex ion from its components in solution. The larger the equilibrium constant, the more stable the complex. For example, the complex ion $Cu(CN)_2^-$ forms by the reaction

$$Cu^{2+} + 2CN^- \rightleftharpoons Cu(CN)_2^-$$

The equilibrium constant for this reaction is the **formation constant, K_f.**

$$K_f = \frac{[Cu(CN)_2^-]}{[Cu^{2+}][CN^-]^2} = 1 \times 10^{16}$$

This constant is sometimes referred to as the **stability constant** or **association constant.** Appendix F contains a table of formation constants. The larger the formation constant, the more stable the complex.

Alternatively, the stability of a complex ion can be described by its **dissociation constant, K_d,** the equilibrium constant for the decomposition of the complex ion into its components in solution. For $Cu(CN)_2^-$ the dissociation is

$$Cu(CN)_2^- \rightleftharpoons Cu^{2+} + 2CN^-$$

and

$$K_d = \frac{[Cu^{2+}][CN^-]^2}{[Cu(CN)_2^-]} = 1 \times 10^{-16}$$

It should be apparent that

$$K_d = \frac{1}{K_f}$$

The smaller the dissociation constant, the more stable the complex.

As an example of dissolution by complex ion formation, let us consider the dissolution of silver chloride in aqueous ammonia. Silver chloride dissolves in water giving a small concentration of Ag^+ ($[Ag^+] = 1.3 \times 10^{-5}$ M; see Example 17.3). However, if NH_3 is present in the water, the complex diamminesilver ion, $Ag(NH_3)_2^+$, can form according to the equation

$$Ag^+(aq) + 2NH_3(aq) \rightleftharpoons Ag(NH_3)_2^+(aq)$$

The formation constant, K_f, for $Ag(NH_3)_2^+$ is 1.6×10^7.

$$\frac{[Ag(NH_3)_2^+]}{[Ag^+][NH_3]^2} = K_f = 1.6 \times 10^7$$

The large value for this formation constant indicates that most of the free silver ions produced by the dissolution of AgCl combine with NH_3 to form $Ag(NH_3)_2^+$. As a consequence, the concentration of silver ions, $[Ag^+]$, is reduced, and the product $[Ag^+][Cl^-]$ becomes less than the solubility product of silver chloride.

$$[Ag^+][Cl^-] < K_{sp}$$

More silver chloride then dissolves. If the concentration of ammonia is great enough, all of the silver chloride will dissolve.

17.14 Calculations Involving Complex Ions

The following examples illustrate calculations of equilibrium concentrations in reactions involving complex ions.

EXAMPLE 17.11   Calculate the concentration of silver ions in a solution that is 0.10 M with respect to $Ag(NH_3)_2^+$.

The formation of the complex ion may be represented by the equation

$$Ag^+(aq) + 2NH_3(aq) \rightleftharpoons Ag(NH_3)_2^+(aq)$$

The reverse reaction shows the dissociation of $Ag(NH_3)_2^+$. Each mole of $Ag(NH_3)_2^+$ that dissociates gives 1 mol of Ag^+ and 2 mol of NH_3. If x is equal to $[Ag^+]$ at equilibrium,

$$[NH_3] = 2x \quad \text{and} \quad [Ag(NH_3)_2^+] = 0.10 - x$$

	$[Ag^+]$	$[NH_3]$	$[Ag(NH_3)_2^+]$
Initial concentrations, M	0	0	0.10
Equilibrium concentrations, M	x	$2x$	$0.10 - x$

$$\frac{[Ag(NH_3)_2^+]}{[Ag^+][NH_3]^2} = K_f = 1.6 \times 10^7$$

$$\frac{0.10 - x}{(x)(2x)^2} = 1.6 \times 10^7$$

The formation constant of $Ag(NH_3)_2^+$ is large, so we can assume that $[Ag(NH_3)_2^+]$ is close to 0.10 M and that x is negligible in the term $0.10 - x$. Therefore

$$\frac{0.10}{(x)(2x)^2} = K_f = 1.6 \times 10^7$$

$$0.10 = (4x^3)(1.6 \times 10^7)$$

$$x^3 = \frac{0.10}{(4)(1.6 \times 10^7)} = 1.6 \times 10^{-9}$$

$$x = 1.2 \times 10^{-3} \, M = [Ag^+]$$

$$2x = 2.4 \times 10^{-3} \, M = [NH_3]$$

(Since only 1.2% of the $Ag(NH_3)_2^+$ dissociates to Ag^+ and NH_3, neglecting x is justified.)

EXAMPLE 17.12   Unexposed silver halides are removed from photographic film when they react with a solution of $Na_2S_2O_3$ (hypo) and form the complex ion $Ag(S_2O_3)_2^{3-}$ ($K_f = 4.7 \times 10^{-13}$). What mass of $Na_2S_2O_3$ is required to prepare 1 L of a solution that will dissolve 1.00 g of silver bromide, $AgBr$, by the formation of $Ag(S_2O_3)_2^{3-}$?

Two equilibria are involved when $AgBr$ dissolves in a solution containing the thiosulfate ion, $S_2O_3^{2-}$.

$$AgBr(s) \rightleftharpoons Ag^+(aq) + Br^-(aq)$$

$$Ag^+(aq) + 2S_2O_3^{2-}(aq) \rightleftharpoons Ag(S_2O_3)_2^{3-}(aq)$$

In order for the $AgBr(s)$ to dissolve completely, the maximum $[Ag^+]$ must be low enough that the product $[Ag^+][Br^-]$ is less than K_{sp} for $AgBr(s)$.

Let us first calculate the concentration of Br^- produced by the dissolution of 1.00 g of AgBr and then determine the maximum possible concentration of Ag^+ that will not result in the reprecipitation of AgBr.

$$1.00 \text{ g AgBr} \times \frac{1 \text{ mol AgBr}}{187.77 \text{ g AgBr}} = 5.33 \times 10^{-3} \text{ mol AgBr}$$

$$5.33 \times 10^{-3} \text{ mol AgBr} \times \frac{1 \text{ mol Br}^-}{1 \text{ mol AgBr}} = 5.33 \times 10^{-3} \text{ mol Br}^-$$

$$\frac{5.33 \times 10^{-3} \text{ mol Br}^-}{1.000 \text{ L}} = 5.33 \times 10^{-3} \, M = [Br^-]$$

Now find the maximum $[Ag^+]$ without reprecipitation of AgBr taking place. At this point $[Br^-]$ is $5.33 \times 10^{-3} \, M$, and $[Ag^+]$ is set equal to x.

$$[Ag^+][Br^-] = K_{sp} = 3.3 \times 10^{-13}$$

$$(x)(5.33 \times 10^{-3}) = 3.3 \times 10^{-13}$$

$$x = \frac{3.3 \times 10^{-13}}{5.33 \times 10^{-3}} = 6.19 \times 10^{-11} \, M$$

The maximum $[Ag^+]$ is $6.19 \times 10^{-11} \, M$. Below this concentration, when $[Br^-]$ is $5.33 \times 10^{-3} \, M$, AgBr will not form because

$$[Ag^+][Br^-] < K_{sp}$$

Now we determine $[S_2O_3^{2-}]$ when $[Ag^+]$ is $6.19 \times 10^{-11} \, M$. Since 5.33×10^{-3} mol of AgBr dissolves,

$$(5.33 \times 10^{-3}) - (6.19 \times 10^{-11}) = 5.33 \times 10^{-3} \text{ mol Ag}(S_2O_3)_2^{3-}$$

must form. Thus $[Ag(S_2O_3)_2^{3-}]$ is equal to $5.33 \times 10^{-3} \, M$. To form 5.33×10^{-3} mol of $Ag(S_2O_3)_2^{3-}$ would require $2 \times (5.33 \times 10^{-3})$ mol of $S_2O_3^{2-}$. However, the total number of moles of $S_2O_3^{2-}$ that must be added to the solution is equal to the number of moles that react in forming the complex *plus* the number of moles of free $S_2O_3^{2-}$ in solution at equilibrium. At equilibrium $[S_2O_3^{2-}]$ is set equal to y, $[Ag^+]$ is $6.19 \times 10^{-11} \, M$, and $[Ag(S_2O_3)_2^{3-}]$ is $5.33 \times 10^{-3} \, M$.

$$\frac{[Ag(S_2O_3)_2^{3-}]}{[Ag^+][S_2O_3^{2-}]^2} = K_f = 4.7 \times 10^{13}$$

$$\frac{5.33 \times 10^{-3}}{(6.19 \times 10^{-11})(y)^2} = 4.7 \times 10^{13}$$

$$y^2 = \frac{5.33 \times 10^{-3}}{6.19 \times 10^{-11}} \times \frac{1}{4.7 \times 10^{13}} = 1.83 \times 10^{-6}$$

$$y = 1.35 \times 10^{-3} \, M = [S_2O_3^{2-}]$$

If $[S_2O_3^{2-}]$ is increased above $1.35 \times 10^{-3} \, M$, $[Ag^+]$ will decrease below $6.19 \times 10^{-11} \, M$, and no solid AgBr will remain. The total amount of $S_2O_3^{2-}$ that must be added to the solution by adding $Na_2S_2O_3$ is

$$2 \times (5.33 \times 10^{-3} \, mol) + 1.35 \times 10^{-3} \, mol$$
$$= 1.20 \times 10^{-2} \, mol \, S_2O_3{}^{2-} = 1.20 \times 10^{-2} \, mol \, Na_2S_2O_3$$

$$1.20 \times 10^{-2} \, \text{mol Na}_2\text{S}_2\text{O}_3 \times \frac{158.1 \, g \, Na_2S_2O_3}{1 \, \text{mol Na}_2\text{S}_2\text{O}_3} = 1.9 \, g \, Na_2S_2O_3$$

Thus 1 L of a solution prepared from 1.9 g of $Na_2S_2O_3$ will dissolve 1.0 g of AgBr.

17.15 Effect of Hydrolysis on the Dissolution of Precipitates

Many salts hydrolyze when they dissolve in water, some with the formation of precipitates, the evolution of gases, or both. For example, hydrated aluminum hydroxide precipitates and hydrogen sulfide is produced when aluminum sulfide, Al_2S_3, dissolves in water. The aluminum ions and the sulfide ions combine with water.

$$Al_2S_3(aq) \rightleftharpoons 2Al^{3+}(aq) + 3S^{2-}(aq)$$
$$2Al^{3+}(aq) + 12H_2O(l) \longrightarrow 2[Al(H_2O)_6]^{3+}(aq)$$
$$3S^{2-}(aq) + 6H_2O(l) \rightleftharpoons 3H_2S(aq) + 6OH^-(aq)$$
$$2[Al(H_2O)_6]^{3+}(aq) + 6OH^-(aq) \rightleftharpoons 2[Al(H_2O)_3(OH)_3](s) + 6H_2O(l)$$

The net reaction is given by

$$Al_2S_3(s) + 12H_2O(l) \rightleftharpoons 2[Al(H_2O)_3(OH)_3](s) + 3H_2S(aq)$$

If the initial amount of aluminum sulfide is high enough, the solubility product of hydrogen sulfide may be exceeded and it will evolve as a gas. When a relatively insoluble sulfide such as lead sulfide, PbS, dissolves in water, an appreciable amount of the sulfide ion hydrolyzes to form the hydrogen sulfide ion and, in some cases, hydrogen sulfide.

$$PbS(s) \rightleftharpoons Pb^{2+}(aq) + S^{2-}(aq)$$
$$S^{2-}(aq) + H_2O(l) \rightleftharpoons HS^-(aq) + OH^-(aq)$$
$$HS^-(aq) + H_2O(l) \rightleftharpoons H_2S(aq) + OH^-(aq)$$

Under these conditions, the lead ion concentration is not the same as that of the sulfide ion but equals the sum of the concentrations of sulfide ion, hydrogen sulfide ion, and hydrogen sulfide.

$$[Pb^{2+}] = [S^{2-}] + [HS^-] + [H_2S]$$

This means that in calculating the solubility of a relatively insoluble sulfide from the solubility product, we must consider the hydrolysis of the sulfide ion.

EXAMPLE 17.13   Calculate the solubility of lead sulfide ($K_{sp} = 8.4 \times 10^{-28}$) in water, neglecting the hydrolysis of the sulfide ion.

Let x equal the molar solubility of PbS in water. Then x will be equal to $[Pb^{2+}]$ and, if we assume the hydrolysis of the sulfide ion is negligible, to $[S^{2-}]$.

Substituting in the expression for the solubility product, we obtain

$$[Pb^{2+}][S^{2-}] = K_{sp}$$
$$(x)(x) = x^2 = 8.4 \times 10^{-28}$$
$$x = 2.9 \times 10^{-14} \text{ mol L}^{-1}$$

If the hydrolysis of the sulfide ion is neglected, we find the solubility of lead sulfide in water to be 2.9×10^{-14} mol L^{-1}.

EXAMPLE 17.14   Now solve the problem of Example 17.13 correctly, taking into consideration the hydrolysis of the sulfide ion.

We shall consider three steps in the dissolution of PbS in water.

$$PbS(s) \rightleftharpoons Pb^{2+} + S^{2-} \qquad\qquad K_{sp} = 8.4 \times 10^{-28}$$

$$S^{2-} + H_2O \rightleftharpoons HS^- + OH^- \qquad\qquad K_{h_1} = \frac{[HS^-][OH^-]}{[S^{2-}]} = \frac{K_w}{K_a \text{ (for HS}^-)}$$

$$= 7.7 \times 10^{-2}$$

$$HS^- + H_2O \rightleftharpoons H_2S + OH^- \qquad\qquad K_{h_2} = \frac{[H_2S][OH^-]}{[HS^-]} = \frac{K_w}{K_a \text{ (for H}_2S)}$$

$$= 1.0 \times 10^{-7}$$

Since K_{h_2} is quite small compared to K_{h_1}, we might be tempted to consider only the first stage of the hydrolysis and neglect the second. This is often a valid way to proceed, as shown in Example 16.21 of Section 16.12. We found in that example that the sulfide ion in a 0.0010 M solution of sodium sulfide is 99% hydrolyzed. Thus the first stage of that hydrolysis produced 0.99×0.0010 mol, or 9.9×10^{-4} mol of OH$^-$. Since this quantity of OH$^-$ is large compared to the 1.0×10^{-7} mol furnished by the water, the total concentration of OH$^-$ in solution was found to be 9.9×10^{-4} mol L^{-1} (1.0×10^{-7} is negligible compared to 9.9×10^{-4}, to several significant figures).

We then calculated the ratio of [H$_2$S] to [HS$^-$], from K_{h_2}, to establish the relative amounts of HS$^-$ and H$_2$S produced in the first and second stages of hydrolysis, respectively.

$$\frac{[H_2S][OH^-]}{[HS^-]} = \frac{[H_2S](9.9 \times 10^{-4})}{[HS^-]} = K_{h_2} = 1.0 \times 10^{-7}$$

$$\frac{[H_2S]}{[HS^-]} = 1.0 \times 10^{-4}$$

We found that [H$_2$S] was only 0.0001 times [HS$^-$], indicating that the second stage of the hydrolysis was negligible compared to the first stage and could be neglected *in that calculation* (see Example 16.21).

However, in this example, we have a much smaller amount of sulfide ion. Our rough calculation for the solubility of lead sulfide, neglecting hydrolysis, in Example 17.13 indicated only 2.9×10^{-14} mol of lead sulfide per liter. The hydroxide ion concentration due to the hydrolysis of the sulfide ion

$$S^{2-} + H_2O \rightleftharpoons HS^- + OH^-$$

is of the order of 3×10^{-14} M. However, this hydroxide ion concentration is so small relative to that provided by the water (1.0×10^{-7} M) that $[OH^-]$ of the solution actually has the value of 1.0×10^{-7} M. Even some increase in the solubility of PbS and some additional OH^- produced in the second stage of hydrolysis would not likely be enough to add sufficient OH^- to be significant compared with the amount furnished by the water. We can now calculate the ratio of $[H_2S]$ to $[HS^-]$ from K_{h_2}, for a solution in which $[OH^-]$ is 1×10^{-7} M (a neutral solution).

$$\frac{[H_2S][OH^-]}{[HS^-]} = \frac{[H_2S](1.0 \times 10^{-7})}{[HS^-]} = K_{h_2} = 1.0 \times 10^{-7}$$

Therefore, $$\frac{[H_2S]}{[HS^-]} = 1$$

Hence, for our neutral solution, $[H_2S]$ is equal to $[HS^-]$, indicating that the second stage of hydrolysis *is* important in this calculation and cannot be neglected. For $[HS^-]$ to equal $[H_2S]$, half of the HS^- ions produced in the first stage of hydrolysis must hydrolyze to produce H_2S in the second stage of hydrolysis.

Now, consider the three equilibria listed earlier for this system. As PbS dissolves, equal quantities of Pb^{2+} and S^{2-} are produced. Essentially all of the S^{2-} produced is hydrolyzed and hence converted to HS^- in the first stage of hydrolysis. At this point, not yet taking the second stage of hydrolysis into account, $[Pb^{2+}]$ is equal to $[HS^-]$. However, we have established that half of the HS^- produced is hydrolyzed in the second stage of hydrolysis, thereby reducing the concentration of HS^- to half of its former value and thus to half of the concentration of Pb^{2+}. Hence, at final equilibrium,

$$[HS^-] = \tfrac{1}{2}[Pb^{2+}]$$

Using 1.0×10^{-7} M for $[OH^-]$, we can calculate the ratio of $[HS^-]$ to $[S^{2-}]$ by substituting in the expression for K_{h_1} and then solving for $[S^{2-}]$.

$$\frac{[HS^-][OH^-]}{[S^{2-}]} = \frac{[HS^-](1.0 \times 10^{-7})}{[S^{2-}]} = K_{h_1} = 7.7 \times 10^{-2}$$

$$\frac{[HS^-]}{[S^{2-}]} = 7.7 \times 10^5, \quad \text{or} \quad [S^{2-}] = \frac{[HS^-]}{7.7 \times 10^5}$$

Substituting the above value for $[S^{2-}]$ in the expression for the solubility product for lead sulfide, we have

$$[Pb^{2+}][S^{2-}] = [Pb^{2+}]\frac{[HS^-]}{7.7 \times 10^5} = K_{sp} = 8.4 \times 10^{-28}$$

Because $[HS^-]$ is equal to $\tfrac{1}{2}[Pb^{2+}]$, we can write

$$[Pb^{2+}]\left(\frac{\tfrac{1}{2}[Pb^{2+}]}{7.7 \times 10^5}\right) = 8.4 \times 10^{-28}$$

$$\frac{\tfrac{1}{2}[Pb^{2+}]^2}{7.7 \times 10^5} = 8.4 \times 10^{-28}$$

$$[Pb^{2+}]^2 = 12.94 \times 10^{-22}$$

$$[Pb^{2+}] = 3.6 \times 10^{-11} = \text{molar solubility of PbS}$$

We have found, then, that the solubility of lead sulfide in water is 3.6×10^{-11} mol L^{-1} when the hydrolysis of the sulfide ion is considered. This value is more than a thousand times greater than that calculated $(2.9 \times 10^{-14}$ mol L$^{-1})$ by neglecting hydrolysis. If we were to take into account the fact that the lead ion hydrolyzes slightly to give Pb(OH)$^{+}$ and Pb(OH)$_2$, the solubility of lead sulfide in water would be found to be even greater.

Finally, note that the hydrolysis of the sulfide ion would produce 3.6×10^{-11} mol of OH^{-} per liter in the first stage of hydrolysis and half of that amount, or 1.8×10^{-11} mol of OH^{-} per liter, in the second stage of hydrolysis, for a total of 5.4×10^{-11} mol of OH^{-} per liter. This amount is indeed negligible compared to the 1.0×10^{-7} mol of OH^{-} per liter furnished by the water. Hence the assumption that [OH^{-}] is equal to 1.0×10^{-7} M for this particular case is entirely justified.

The insoluble carbonates behave similarly to the sulfides. When a relatively insoluble carbonate such as barium carbonate, BaCO$_3$, is dissolved in water, its solubility is increased due to hydrolysis of the carbonate ion.

$$CO_3^{2-} + H_2O \rightleftharpoons HCO_3^{-} + OH^{-}$$
$$HCO_3^{-} + H_2O \rightleftharpoons H_2CO_3 + OH^{-}$$

The concentration of the barium ion is equal to the sum of the concentrations of the carbonate ion, CO$_3^{2-}$, the hydrogen carbonate ion, HCO$_3^{-}$, and carbonic acid, H$_2$CO$_3$.

FOR REVIEW

SUMMARY

An equilibrium involving the precipitation or dissolution of a slightly soluble electrolyte is described by the **solubility product, K_{sp},** the equilibrium constant for the equilibrium between the solid and its ions in solution. The solubility product of a compound can be calculated from its solubility, and the solubility of the compound can be calculated from its solubility product.

A slightly soluble electrolyte will begin to precipitate when the product of the concentrations of its ions raised to their appropriate powers exceeds the solubility product of the electrolyte. Precipitation will continue until this ion concentration product equals the solubility product. Consequently, for a solution containing one of the ions of a slightly soluble electrolyte, we can calculate how much of the other ion of the electrolyte must be added either to cause precipitation of the electrolyte to begin or to reduce the concentration of the first ion to any desired value.

Many slightly soluble electrolytes contain anions of weak acids. These electrolytes often dissolve in acidic solutions because **hydrolysis** of the anion to give the weak acid reduces the concentration of the anion to the point where the ion concentration product is less than the solubility product. The ion concentration product can also be reduced by the conversion of an ion to another species or by the formation of a complex ion. The stability of a complex ion is described by its **formation constant, K_f.**

Hydrolysis of the ions of a precipitate can have a significant effect on its solubility in water.

KEY TERMS AND CONCEPTS

Complex ion (17.13)
Coprecipitation (17.10)
Dissociation constant (17.13)
Dissolution of precipitates
  (17.11, 17.12, 17.13)

Formation constant (17.13)
Fractional precipitation (17.10)
Hydrolysis (17.15)
Molar solubility (17.3)
Precipitation (17.4)

Solubility product (17.1)
Supersaturated solution (17.7)

EXERCISES

Solubility Products

1. A saturated solution of a slightly soluble electrolyte in contact with some of the solid electrolyte is said to be a system in equilibrium. Explain. Why is such a system called a heterogeneous equilibrium?

2. How do the concentrations of Ag^+ and CrO_4^{2-} in a liter of water above 1.0 g of solid Ag_2CrO_4 change when 100 g of solid Ag_2CrO_4 is added to the system? Explain.

3. How do the concentrations of Pb^{2+} and S^{2-} change when K_2S is added to a saturated solution of PbS?

4. Under what circumstances, if any, will a sample of solid AgCl completely dissolve in pure water?

5. Refer to Appendix E for solubility products for calcium salts. Determine which of the calcium salts listed is most soluble in moles per liter. Determine which is most soluble in grams per liter.

6. Solid silver bromide is in equilibrium with a saturated solution of its ions, Ag^+ and Br^-. How, if at all, will this equilibrium be affected if (a) more solid silver bromide is added? (b) silver nitrate is added? (c) sodium bromide is added? (d) the temperature is raised? (The solubility increases with temperature.)

7. Write the expression for the solubility product of each of the following slightly soluble electrolytes:

(a) $PbCl_2$
(b) AgBr
(c) $Ba_3(PO_4)_2$
(d) Ag_2S
(e) $MgNH_4AsO_4$

8. Which of the following compounds will precipitate from a solution initially containing the indicated concentrations of ions (see Appendix E for K_{sp} values)?

(a) $CaCO_3$: $[Ca^{2+}] = 0.003\ M$,
            $[CO_3^{2-}] = 0.003\ M$
            Ans. Precipitates

(b) $Co(OH)_2$: $[Co^{2+}] = 0.01\ M$,
            $[OH^-] = 1 \times 10^{-7}\ M$
            Ans. Does not precipitate

(c) $CaHPO_4$: $[Ca^{2+}] = 0.01\ M$,
            $[HPO_4^{2-}] = 2 \times 10^{-6}\ M$
            Ans. Does not precipitate

(d) $Pb_3(PO_4)_2$: $[Pb^{2+}] = 0.01\ M$,
            $[PO_4^{3-}] = 1 \times 10^{-13}\ M$
            Ans. Precipitates

(e) Ag_2S: $[Ag^+] = 1 \times 10^{-10}\ M$,
            $[S^{2-}] = 1 \times 10^{-13}\ M$
            Ans. Precipitates

9. Calculate the solubility product of each of the following from the solubility given:

(a) AgBr, 5.7×10^{-7} mol L^{-1} *Ans. 3.2×10^{-13}*
(b) $CaCO_3$, 6.9×10^{-3} g L^{-1} *Ans. 4.8×10^{-9}*

$\boxed{S}$ (c) PbF_2, 2.1×10^{-3} mol L^{-1} *Ans. 3.7 × 10⁻⁸*

$\boxed{S}$ (d) Ag_2CrO_4, 4.3×10^{-2} g L^{-1}

Ans. 8.7 × 10⁻¹²

(e) Ag_2SO_4, 4.47 g L^{-1} *Ans. 1.18 × 10⁻⁵*

10. The *Handbook of Chemistry and Physics* gives solubilities for the compounds listed below in grams per 100 mL of water. Since these compounds are only slightly soluble, assume that the volume does not change on dissolution, and calculate the solubility product for each.
    (a) TlCl, 0.29 g/100 mL   *Ans. 1.5 × 10⁻⁴*
    (b) $Ce(IO_3)_4$, 1.5×10^{-2} g/100 ml

Ans. 4.6 × 10⁻¹⁷

(c) $Gd_2(SO_4)_3$, 3.98 g/100 mL

Ans. 1.36 × 10⁻⁴

(d) InF_3, 4.0×10^{-2} g/100 mL

Ans. 7.9 × 10⁻¹⁰

11. Calculate the concentrations of ions in a saturated solution of each of the following (see Appendix E for solubility products):
    $\boxed{S}$ (a) AgI   *Ans. [Ag⁺] = [I⁻] = 1.2 × 10⁻⁸ M*
    $\boxed{S}$ (b) Ag_2SO_4 *Ans. [Ag⁺] = 2.8 × 10⁻² M;*
    [SO₄²⁻] = 1.4 × 10⁻² M
    (c) $Mn(OH)_2$ *Ans. [Mn²⁺] = 2.2 × 10⁻⁵ M;*
    [OH⁻] = 4.5 × 10⁻⁵ M
    $\boxed{S}$ (d) $Sr(OH)_2 \cdot 8H_2O$

Ans. [Sr²⁺] = 4.3 × 10⁻² M;
    [OH⁻] = 8.6 × 10⁻² M

12. Calculate the molar solubility of each of the following minerals from its K_{sp}:
    (a) alabandite, MnS: $K_{sp} = 5.6 \times 10^{-16}$
    Ans. 2.4 × 10⁻⁸ M
    (b) anglesite, $PbSO_4$: $K_{sp} = 1.8 \times 10^{-8}$
    Ans. 1.3 × 10⁻⁴ M
    (c) brucite, $Mg(OH)_2$: $K_{sp} = 1.5 \times 10^{-11}$
    Ans. 1.6 × 10⁻⁴ M
    (d) fluorite, CaF_2: $K_{sp} = 3.9 \times 10^{-11}$
    Ans. 2.1 × 10⁻⁴ M

13. Most barium compounds are very poisonous; however, barium sulfate is often administered internally as an aid in the X-ray examination of the lower intestinal track. This use of $BaSO_4$ is possible because of its insolubility. Calculate the molar solubility of $BaSO_4$ ($K_{sp} = 1.08 \times 10^{-10}$) and the mass of barium present in 1 L of water saturated with $BaSO_4$.
    Ans. 1.04 × 10⁻⁵ M; 1.43 × 10⁻³ g

14. Public Health Service standards for drinking water set a maximum of 250 mg/L of SO_4^{2-}, or $[SO_4^{2-}] = 2.60 \times 10^{-3}$ M, because of its cathartic

action (it is a laxative). Does natural water that is saturated with $CaSO_4$ ("gyp" water) as a result of passing through soil containing gypsum, $CaSO_4 \cdot 2H_2O$, meet these standards? What is $[SO_4^{2-}]$ in such water?
    Ans. No; [SO₄²⁻] = 4.9 × 10⁻³ M

15. The first step in the preparation of magnesium metal is the precipitation of $Mg(OH)_2$ from sea water by the addition of $Ca(OH)_2$. The concentration of $Mg^{2+}(aq)$ in sea water is 5.37×10^{-2} M. Using the solubility product for $Mg(OH)_2$, calculate the pH at which $[Mg^{2+}]$ is reduced to 5.37×10^{-5} M by the addition of $Ca(OH)_2$.   *Ans. 10.72*

16. Calculate the concentration of Tl⁺ when TlCl ($K_{sp} = 1.9 \times 10^{-4}$) just begins to precipitate from a solution that is 0.0250 M in Cl⁻.
    Ans. 7.6 × 10⁻³ M

$\boxed{S}$ 17. Calculate the concentration of Sr^{2+} when SrF_2 ($K_{sp} = 3.7 \times 10^{-12}$) starts to precipitate from a solution that is 0.0025 M in F⁻.
    Ans. 5.9 × 10⁻⁷ M

18. Calculate the concentration of F⁻ required to begin precipitation of CaF_2 in a solution that is 0.010 M in Ca²⁺.   *Ans. 6.2 × 10⁻⁵ M*

19. Iron concentrations greater than 5.4×10^{-6} M in water used for laundry purposes can cause staining. What $[OH^-]$ is required to reduce $[Fe^{2+}]$ to this level by precipitation of $Fe(OH)_2$?
    Ans. 3.8 × 10⁻⁵ M

20. (a) Calculate $[Ag^+]$ in a saturated aqueous solution of AgBr ($K_{sp} = 3.3 \times 10^{-13}$).
    Ans. 5.7 × 10⁻⁷ M
    (b) What will $[Ag^+]$ be when enough KBr has been added to make $[Br^-] = 0.050$ M?
    Ans. 6.6 × 10⁻¹² M
    (c) What will $[Br^-]$ be when enough $AgNO_3$ has been added to make $[Ag^+] = 0.020$ M?
    Ans. 1.6 × 10⁻¹¹ M

$\boxed{S}$ 21. The solubility product of $CaSO_4 \cdot 2H_2O$ is 2.4×10^{-5}. What mass of this salt will dissolve in 1.0 L of 0.010 M K_2SO_4?   *Ans. 0.34 g*

22. Calculate the maximum concentration of lead(II) ion in a solution of lead(II) sulfate in which the concentration of sulfate ion is 0.0045 M.
    Ans. 4.0 × 10⁻⁶ M

23. In one experiment a precipitate of $BaSO_4$ ($K_{sp} = 1.08 \times 10^{-10}$) was washed with 0.100 L of distilled water; in another experiment a precipitate of $BaSO_4$ was washed with 0.100 L of

$0.010\ M\ H_2SO_4$. Calculate the quantity of $BaSO_4$ that dissolved in each experiment, assuming that the wash liquid became saturated with $BaSO_4$. *Ans. 1.04×10^{-6} mol; 1.08×10^{-9} mol*

24. A volume of 0.800 L of a $2 \times 10^{-4}\ M\ Ba(NO_3)_2$ solution is added to 0.200 L of $5 \times 10^{-4}\ M$ Li_2SO_4. Will $BaSO_4$ precipitate? Explain your answer. *Ans. Yes*

Solubility Products and Ionization Constants

25. Which of the following compounds, when dissolved in a 0.01 M solution of $HClO_4$, will have a solubility greater than in pure water: $CuCl$, $CaCO_3$, MnS, $PbBr_2$, CaF_2? Explain your answers. *Ans. $CaCO_3$, MnS, CaF_2*

26. A solution of 0.075 $M\ CoBr_2$ is saturated with H_2S ($[H_2S] = 0.10\ M$). What is the minimum pH at which CoS ($K_{sp} = 5.9 \times 10^{-21}$) will precipitate? *Ans. 0.89*

27. (a) What are the concentrations of Ca^{2+} and CO_3^{2-} in a saturated solution of $CaCO_3$ ($K_{sp} = 4.8 \times 10^{-9}$)? *Ans. $[Ca^{2+}] = [CO_3^{2-}] = 6.9 \times 10^{-5}\ M$*
    (b) What are the concentrations of Ca^{2+} and CO_3^{2-} in a buffer solution with a pH of 4.55 in contact with an excess of $CaCO_3$? *Ans. $[Ca^{2+}] = 0.35\ M$; $[CO_3^{2-}] = 1.4 \times 10^{-8}\ M$*

28. To a 0.10 M solution of $Pb(NO_3)_2$ is added enough $HF(g)$ to make $[HF] = 0.10\ M$. (a) Will PbF_2 precipitate from this solution? (b) What is the minimum pH at which PbF_2 will precipitate? *Ans. (a) Yes; (b) 0.93*

29. Using the solubility product, calculate the molar solubility of $AgCN$ (a) in pure water and (b) in a buffer solution with a pH of 3.00. *Ans. (a) $1.1 \times 10^{-8}\ M$; (b) $2 \times 10^{-5}\ M$*

30. Calculate the molar solubility of $Sn(OH)_2$ (a) in pure water (pH = 7) and (b) in a buffer solution containing equal concentrations of NH_3 and NH_4^+. *Ans. (a) $5 \times 10^{-12}\ M$; (b) $2 \times 10^{-16}\ M$*

31. A 0.125 M solution of $Mn(NO_3)_2$ is saturated with H_2S ($[H_2S] = 0.10\ M$). At what pH will MnS ($K_{sp} = 5.6 \times 10^{-16}$) begin to precipitate? *Ans. 3.27*

32. Calculate the concentration of Cd^{2+} resulting from the dissolution of $CdCO_3$ in a solution that

is 0.250 M in CH_3CO_2H, 0.375 M in $NaCH_3CO_2$, and 0.010 M in H_2CO_3. *Ans. $1 \times 10^{-5}\ M$*

33. A volume of 50 mL of 1.8 $M\ NH_3$ is mixed with an equal volume of a solution containing 0.95 g of $MgCl_2$. What mass of NH_4Cl must be added to the resulting solution to prevent the precipitation of $Mg(OH)_2$? *Ans. 7.1 g*

34. Calculate the molar solubility of CaF_2 in a 0.100 M solution of HNO_3. *Ans. $1.2 \times 10^{-2}\ M$*

Separation of Ions

35. (a) With what volume of water must a precipitate containing $NiCO_3$ be washed to dissolve 0.100 g of this compound? Assume that the wash water becomes saturated with $NiCO_3$ ($K_{sp} = 1.36 \times 10^{-7}$). *Ans. 2.28 L*
    (b) If the $NiCO_3$ were a contaminant in a sample of $CoCO_3$ ($K_{sp} = 1.0 \times 10^{-12}$), what mass of $CoCO_3$ would have been lost? Keep in mind that both $NiCO_3$ and $CoCO_3$ dissolve in the same solution. *Ans. 7.3×10^{-7} g*

36. A solution is 0.010 M in both Cu^{2+} and Cd^{2+}. What percentage of Cd^{2+} remains in the solution when 99.9% of the Cu^{2+} has been precipitated as CuS by addition of sulfide? *Ans. 100%*

37. A solution is 0.15 M in both Pb^{2+} and Ag^+. If Cl^- is added to this solution, what is $[Ag^+]$ when $PbCl_2$ begins to precipitate? *Ans. $1.7 \times 10^{-8}\ M$*

38. The maximum allowable chloride ion concentration in drinking water is 0.25 g/L ($7.1 \times 10^{-3}\ M$). A commercial kit for the analysis of chloride ion in water contains K_2CrO_4, which is dissolved in a water sample as an indicator, and a standard solution of $AgNO_3$, which is used as a titrant. As the $AgNO_3$ solution is added to the water sample drop by drop, insoluble white $AgCl$ is formed. After "all" of the chloride ion has been precipitated, the next drop of $AgNO_3$ solution reacts with the K_2CrO_4 to form an orange-colored precipitate of Ag_2CrO_4, which indicates the end of the titration. What percentage of the initial chloride ion content remains unprecipitated when the first trace of Ag_2CrO_4 forms in a solution for which initial $[Cl^-] = 7.1 \times 10^{-3}\ M$ and $[CrO_4^{2-}] = 1.0 \times 10^{-4}\ M$? Assume that the titration does not change the volume of the solution. *Ans. 0.008%*

39. A solution that is 0.10 M in both Pb^{2+} and Fe^{2+} and 0.30 M in HCl is saturated with H_2S ($[H_2S] = 0.10$ M). What concentrations of Pb^{2+} and Fe^{2+} remain in the solution?

   Ans. $[Pb^{2+}] = 5.8 \times 10^{-8}$ M; $[Fe^{2+}] = 0.1$ M

40. What reagent might be used to separate the ions in each of the following mixtures, which are 0.1 M with respect to each ion? In some cases it may be necessary to control the pH. (*Hint:* Consider the K_{sp} values in Appendix E.)
   (a) Hg_2^{2+} and Cu^{2+}
   (b) SO_4^{2-} and Cl^-
   (c) Hg^{2+} and Co^{2+}
   (d) Zn^{2+} and Sr^{2+}
   (e) Ba^{2+} and Mg^{2+}
   (f) CO_3^{2-} and OH^-

Formation Constants of Complex Ions

S 41. Calculate the cadmium ion concentration, $[Cd^{2+}]$, in a solution prepared by mixing 0.100 L of 0.0100 M $Cd(NO_3)_2$ with 0.150 L of 0.100 M $NH_3(aq)$. *Ans.* 2.7×10^{-4} M

42. Calculate the silver ion concentrations, $[Ag^+]$, of a solution prepared by dissolving 1.00 g of $AgNO_3$ and 10.0 g of KCN in sufficient water to make 1.00 L of solution. *Ans.* 3×10^{-21} M

43. Sometimes equilibria for complex ions are described in terms of dissociation constants, K_d. For the complex ion AlF_6^{3-} the dissociation reaction is

$$AlF_6^{3-} \rightleftharpoons Al^{3+} + 6F^-$$

   and

$$K_d = \frac{[Al^{3+}][F^-]^6}{[AlF_6^{3-}]} = 2 \times 10^{-24}$$

   (a) Calculate the value of the formation constant, K_f, for AlF_6^{3-}. *Ans.* 5×10^{23}
   (b) Using the value of the formation constant for the complex ion $Co(NH_3)_6^{2+}$, calculate the dissociation constant. *Ans.* 1.2×10^{-5}

44. Calculate the concentration of Ni^{2+} in a 1.0 M solution of $[Ni(NH_3)_6](NO_3)_2$. *Ans.* 0.014 M

45. Calculate the concentration of Zn^{2+} in a 0.30 M solution of $[Zn(CN)_4]^{2-}$. *Ans.* 4×10^{-5} M

46. What are the concentrations of Ag^+, CN^-, and $Ag(CN)_2^-$ in a saturated solution of AgCN?

   Ans. $[Ag^+] = 5.5 \times 10^{-9}$ M;
   $[CN^-] = 1 \times 10^{-10}$ M;
   $[Ag(CN)_2^-] = 5.5 \times 10^{-9}$ M

47. The equilibrium constant for the reaction

$$Hg^{2+}(aq) + 2Cl^-(aq) \rightleftharpoons HgCl_2(aq)$$

   is 1.6×10^{13}. Is $HgCl_2$ a strong electrolyte or a weak electrolyte? What are the concentrations of Hg^{2+} and Cl^- in a 0.015 M solution of $HgCl_2$?

   Ans. *Weak electrolyte;* $[Hg^{2+}] = 6.2 \times 10^{-6}$ M;
   $[Cl^-] = 1.2 \times 10^{-5}$ M

Dissolution of Precipitates

48. (a) Which of the following slightly soluble compounds will have a solubility greater than that calculated from its solubility product due to hydrolysis of the anion present: $CoSO_3$, CuI, $PbCO_3$, $PbCl_2$, Tl_2S, $KClO_4$?

   Ans. $CoSO_3$, $PbCO_3$, Tl_2S
   (b) For which compound in part (a) will hydrolysis be most extensive? *Ans.* Tl_2S

49. Both AgCl and AgI dissolve in NH_3.
   S (a) What mass of AgCl will dissolve in 1.0 L of 1.0 M NH_3? *Ans.* 6.9 g
   (b) What mass of AgI will dissolve in 1.0 L of 1.0 M NH_3? *Ans.* 1.1×10^{-2} g

50. Calculate the minimum number of moles of cyanide ion that must be added to 100 mL of solution to dissolve 2×10^{-2} mol of silver cyanide, AgCN. *Ans.* 2×10^{-2} mol

51. Calculate the minimum concentration of ammonia needed in 1.0 L of solution to dissolve 3.0×10^{-3} mol of silver bromide. *Ans.* 1.3 M

52. A roll of 35-mm black and white film contains about 0.27 g of unexposed AgBr before developing. What mass of $Na_2S_2O_3 \cdot 5H_2O$ (hypo) in 1.0 L of developer is required to dissolve the AgBr as $Ag(S_2O_3)_2^{3-}$ ($K_f = 4.7 \times 10^{13}$)?

   Ans. 0.81 g

53. Calculate the solubility of NiS ($K_{sp} = 3 \times 10^{-21}$) in a 0.25 M solution of $NH_3(aq)$. For the reaction

$$Ni^{2+}(aq) + 6NH_3(aq) \rightleftharpoons Ni(NH_3)_6^{2+}$$

   K_f is 1.8×10^8. *Ans.* 1×10^{-8} M

54. Calculate the volume of 1.50 M CH_3CO_2H required to dissolve a precipitate composed of 350 mg each of $CaCO_3$, $SrCO_3$, and $BaCO_3$.

   Ans. 10.2 mL

55. (a) The K_{sp} for CdS is 3.6×10^{-29}. Calculate the maximum concentration of sulfide ion that can exist in a solution that is 0.0010 M in $CdCl_2$. *Ans.* 3.6×10^{-26} M
   (b) Calculate the solubility of CdS in water. Do

not neglect the hydrolysis of sulfide ion.

Ans. 7.4 × 10⁻¹² M

56. Calculate the solubility of $Al(OH)_3$ in water ($K_{sp} = 1.9 \times 10^{-33}$). Do not neglect the hydrolysis of the aluminum ion (Section 16.15).

Ans. 2.6 × 10⁻¹⁰ M

Additional Exercises

57. Even though $Ca(OH)_2$ is an inexpensive base, its limited solubility restricts its use. What is the pH of a saturated solution of $Ca(OH)_2$?

Ans. 12.40

58. What mass of NaCN must be added to 1 L of 0.010 M $Mg(NO_3)_2$ in order to produce the first trace of $Mg(OH)_2$? *Ans. 4.8 × 10⁻³ g*

Ⓢ 59. The calcium ions in human blood serum are necessary for coagulation. In order to prevent coagulation when a blood sample is drawn for laboratory tests, an anticoagulant is added to the sample. Potassium oxalate, $K_2C_2O_4$, can be used as an anticoagulant because it removes the calcium as a precipitate of $CaC_2O_4 \cdot H_2O$. In order to prevent coagulation, it is necessary to remove all but 1.0% of the Ca^{2+} in serum. If normal blood serum with a buffered pH of 7.40 contains 9.5 mg of Ca^{2+} per 100 mL of serum, what mass of $K_2C_2O_4$ is required to prevent the coagulation of a 10-mL blood sample that is 55% serum by volume? [All volumes are accurate to two significant figures. Note that the volume of fluid (serum) in a 10-mL blood sample is 5.5 mL. Assume that the K_{sp} value for CaC_2O_4 in serum is the same as in water.] *Ans. 2.2 × 10⁻³ g*

Ⓢ 60. About 50% of urinary calculi (kidney stones) consist of calcium phosphate, $Ca_3(PO_4)_2$. The normal midrange calcium content excreted in the urine is 0.10 g of Ca^{2+} per day. The normal midrange amount of urine passed may be taken as 1.4 L per day. What is the maximum concentration of phosphate ion possible in urine before a calculus begins to form? *Ans. 4 × 10⁻⁹ M*

Ⓢ 61. The pH of a normal urine sample is 6.30, and the total phosphate concentration ($[PO_4^{3-}]$ + $[HPO_4^{2-}]$ + $[H_2PO_4^-]$ + $[H_3PO_4]$) is 0.020 M.

What is the minimum concentration of Ca^{2+} necessary to induce calculus formation? (See Exercise 60 for additional information.)

Ans. 3 × 10⁻³ M

62. Magnesium metal (a component of alloys used in aircraft and a reducing agent used in the production of uranium, titanium, and other active metals) is isolated from sea water by the following sequence of reactions:

$$Mg^{2+}(aq) + Ca(OH)_2(aq) \longrightarrow$$
$$Mg(OH)_2(s) + Ca^{2+}(aq)$$

$$Mg(OH)_2(s) + 2HCl \longrightarrow MgCl_2(s) + 2H_2O$$

$$MgCl_2(l) \xrightarrow{\text{Electrolysis}} Mg(s) + Cl_2(g)$$

Sea water has a density of 1.026 g/cm³ and contains 1272 parts per million of magnesium as $Mg^{2+}(aq)$ by mass. What mass, in kilograms, of $Ca(OH)_2$ is required to precipitate 99.9% of the magnesium in 1.00×10^3 L of sea water?

Ans. 3.97 kg

63. Calculate $[HgCl_4^{2-}]$ in a solution prepared by adding 8.0×10^{-3} mol of NaCl to 0.100 L of a 0.040 M $HgCl_2$ solution. *Ans. 0.040 M*

64. In a titration of cyanide ion, 28.72 mL of 0.0100 M $AgNO_3$ is added before precipitation begins. (The reaction of Ag^+ with CN^- goes to completion, producing the $[Ag(CN)_2]^-$ complex. Precipitation of solid AgCN takes place when excess Ag^+ is added to the solution, above the amount needed to complete the formation of $[Ag(CN)_2]^-$.) How many grams of NaCN were in the original sample? *Ans. 0.0281 g*

65. A 0.010-mol sample of solid AgCN is rendered soluble in 1 L of solution by adding just enough excess cyanide ion to form $[Ag(CN)_2]^-$. When all of the solid silver cyanide has just dissolved, the concentration of free cyanide ion is 1.125×10^{-7} M. Neglecting hydrolysis of the cyanide ion, determine the concentration of free, uncomplexed silver ion in the solution. If more cyanide ion is added (without changing the volume) until the equilibrium concentration of free cyanide ion is 1.0×10^{-6} M, what will be the equilibrium concentration of free silver ion?

Ans. 8 × 10⁻⁹ M; 1 × 10⁻¹⁰ M

18

CHEMICAL THERMODYNAMICS

We have pointed out that the total quantity of energy in the universe is constant and that energy can be transformed from one form to another (Section 1.2). There are many examples, both large and small scale, of such energy transformations, for example, an electric light bulb (electrical energy to light and heat, which are forms of radiant energy), a gas stove (potential energy of the fuel to heat), and an automobile (potential energy of gasoline to kinetic energy of motion and heat).

In this chapter we will consider some energy changes that accompany chemical changes. We will see that these energy changes can be predicted quantitatively and that their values can be used to predict whether or not a chemical reaction will occur. Finally, we will see that the equilibrium constant for a reaction can be calculated from these values.

When we are trying to predict the occurrence of a chemical reaction, such as the oxidation of glucose in the body to carbon dioxide and water, we often have several questions. Will the substances react when put together? If they do react, what will the final concentrations of reactants and products be when equilibrium is established? If a reaction occurs, how much heat, if any, will be produced? How fast will the reaction go? What is its mechanism? The last two questions can be answered from a consideration of the kinetics of the reaction (Chapter 15). The first three are the basis of this chapter. Chemical thermodynamics is helpful in predicting whether or not a reaction will proceed, the amounts of reactants and products present at equilibrium, and the energy changes associated with the reaction.

18.1 Introduction

Chemical thermodynamics is the study of the energy transformations and transfers that accompany chemical and physical changes. The part of the universe that undergoes a change and that is being studied is called the **system,** and the rest of the universe, the **surroundings. Chemical thermodynamics is the study of the energy transformations that occur in a system and of any transfers of energy that may occur between the system and the surroundings.**

Two fundamental laws of nature are particularly important in thermodynamics: (1) systems tend toward a state of minimum potential energy, and (2) systems tend toward a state of maximum disorder. For example, if you drop a box, it falls to the floor. During this change in the box's position, part of its potential energy (due to its height above the floor) is converted to kinetic energy. When the box hits the floor, the kinetic energy is converted to heat. The loss of heat indicates that the box has gone to a lower energy state. It certainly has a lower potential energy since it can no longer fall as far as it could fall before the change in its position. If the box contained a jig-saw puzzle that had been assembled before being dropped, almost certainly the puzzle would be partly disassembled (more disordered) after the fall. The latter fact may not seem very important or fundamental, but most people would agree that it would be foolish to try to assemble a jig-saw puzzle by dropping the separated pieces on the floor, whereas the reverse process is readily accomplished by dropping the assembled puzzle on the floor. A system tends to become less orderly because there are many more ways to be disorderly than ways to be orderly. The probability for a system's becoming more disordered or more random is greater than that for its becoming more ordered.

THERMOCHEMISTRY

18.2 Internal Energy, Heat, and Work

The study of thermochemistry focuses on *changes* in the internal energy of a system due to heat effects associated with chemical changes in it. The **internal energy, E,** is simply the total of all possible kinds of energy present in that portion of the universe that is being studied; that is, the total of heat, kinetic energy, bond energy, chemical energy, etc., present in the system.

Even though we cannot measure or calculate the value of E for a system, we can determine the change in internal energy, ΔE, that accompanies a change in the system. (The symbol Δ stands for the difference between two quantities—in this case, two energies.) As an example, let us consider a system that consists of a mole of hydrogen molecules. Under certain conditions, 1 mole of hydrogen molecules will dissociate into 2 moles of hydrogen atoms upon addition of 436 kilojoules of energy. Although we cannot determine the internal energy of the system,

E_1, when it exists in its initial state as a mole of hydrogen molecules or its energy, E_2, when it exists in its final state as 2 moles of hydrogen atoms, we can find the difference in internal energy between the two states. The internal energy of 2 moles of hydrogen atoms is 436 kilojoules greater than the internal energy of 1 mole of hydrogen molecules because 436 kilojoules of energy has been added to the system. The change in internal energy is equal to the difference between E_2 and E_1.

$$\Delta E = E_2 - E_1$$

The value of ΔE is positive when energy is transferred from the surroundings to the system. When energy is transferred from the system to the surroundings, the value of ΔE is negative.

Energy is transferred into or out of a system in one, or both, of two ways: by heat transfer and/or by work. The symbol for the amount of **heat** transferred during a change in the system is q. If heat flows from the surroundings into the system, q is positive ($q > 0$). If heat flows from the system to the surroundings, the value of q is negative ($q < 0$). For example, when a system consisting of 18 grams of steam at 100°C condenses to liquid water at 100°C, doing no work in the process, the internal energy of the system decreases because 40.7 kilojoules of heat flows from the system ($q = -40.7$ kJ).

The other means of changing the internal energy of a system is (1) to let the system do work on the surroundings, in which case the internal energy of the system decreases by the amount of work it does, or (2) to let work be done on the system by the surroundings, in which case the internal energy of the system increases by the amount of work done on it. The symbol for the amount of **work** done by or on a system is w. When energy is transferred from the system to the surroundings as work, work is done on the surroundings and the value of w is positive ($w > 0$). When energy is transferred from the surroundings to the system as work, the surroundings do work on the system and the value of w is negative ($w < 0$). In this chapter we will limit our consideration of work to that of expansion against a constant pressure (called expansion work).

Gases expanding against a restraining pressure are capable of doing work, as in a steam engine or an automobile engine. A system initially in the energy state E_1 with volume V_1 that goes at a constant pressure to a new energy state E_2 with a larger volume V_2 does expansion work on the surroundings. The amount of work done is equal to $P(V_2 - V_1)$, since the pressure, P, remains constant. The work term $P(V_2 - V_1)$ is an energy term with energy units. Pressure is defined as force divided by area and therefore has units of force/length^2; volume has units of length^3. Thus the units for the work term are

$$P(V_2 - V_1) = \frac{\text{force}}{\cancel{\text{length}^2}} \times \text{length}^{\cancel{3}} = \text{force} \times \text{length}$$

Units of force $\times$ length are units of work, or energy. If the pressure is expressed in pascals (newtons/meter^2) and the volume change in meters^3, the resulting product will have units of joules (newton meters).

A system can contract as well as expand. When a system contracts, the volume decreases ($V_2 - V_1 < 0$), and work is done on the system by the surroundings. A change in volume is usually referred to in thermodynamics as an expansion; an increase in volume is a positive expansion, and a decrease in volume, a negative expansion.

18.3   The First Law of Thermodynamics

The **First Law of Thermodynamics** is actually the Law of Conservation of Energy (Section 1.3): **The total amount of energy in the universe is constant.** The First Law is often considered in a rather special form. If an amount of heat, q, is added to a system with an internal energy E_1, and the system then does some work, w, on the surroundings, the system ends up with a new internal energy, E_2. The Law of Conservation of Energy requires that the final internal energy of the system be equal to its initial internal energy plus the energy added as heat from the surroundings minus the energy lost as work done on the surroundings.

$$E_2 = E_1 + q - w$$

or

$$E_2 - E_1 = q - w$$

$E_2 - E_1$ is the change in the internal energy of the system, or ΔE.

$$\Delta E = E_2 - E_1 = q - w$$

In other words, the change in the internal energy of a given system equals the heat transferred to the system from the surroundings minus the work transferred from the system to the surroundings. The value of ΔE may be either positive or negative depending on the relative values of q and w.

If ΔE is the energy change for the system and ΔE_{sur} is the energy change for the surroundings

$$\Delta E + \Delta E_{sur} = 0$$

This emphasizes the fact expressed by the First Law that the total amount of energy in the universe is constant.

EXAMPLE 18.1   (a) If 600 J of heat is added to a system in energy state E_1 and the system does 450 J of work on the surroundings, what is the energy change of the system?

To calculate ΔE we use the expression relating ΔE, q, and w.

$$\Delta E = q - w$$

Since heat is added to the system, q is positive ($q > 0$), so q is +600 J. The value of w is also positive ($w > 0$) since work is done on the surroundings; w is +450 J.

$$\Delta E = q - w = (+600 \text{ J}) - (+450 \text{ J}) = +150 \text{ J}$$

The internal energy of the system increases by 150 J.

(b) What is the energy change of the surroundings?

$$\Delta E + \Delta E_{sur} = 0$$
$$150 \text{ J} + \Delta E_{sur} = 0$$
$$\Delta E_{sur} = -150 \text{ J}$$

The energy of the surroundings decreases by 150 J.

(c) What is the internal energy of the system in the new energy state, E_2?

Since we cannot determine the internal energy of the system, all we can say is that its final internal energy, E_2, is 150 J greater than its initial internal energy, E_1.

$$\Delta E = E_2 - E_1 = 150 \text{ J}$$
$$E_2 = E_1 + 150 \text{ J}$$

EXAMPLE 18.2   A system consisting of 18.015 g (one mole) of $H_2O(g)$ at 100°C and 1 atm with a volume of 30.12 L (0.03012 m^3) condenses at a constant pressure of 1 atm to $H_2O(l)$ at 100°C and 1 atm with a density of 0.9584 g/cm^3. There is 40,668 J of heat released to the surroundings. Calculate ΔE for this process.

Since

$$\Delta E = q - w$$

we need two values to determine ΔE: the amount of heat transferred to or from the system and the amount of work done on or by the system. The problem tells us that 40,668 J of heat is transferred from the water to the surroundings, so q is $-40,668$ J. The sign is negative since heat leaves the system.

The amount of work (in joules) involved in this change can be calculated from the expression

$$w = P(V_2 - V_1)$$

if P is expressed in pascals (1 atm = 101,325 Pa; Section 10.2) and V_2 and V_1, in cubic meters.

The value of V_1 is given as 0.03012 m^3. The value of V_2 may be calculated as follows:

$$V_2 = 18.015 \text{ g} \times \frac{1 \text{ cm}^3}{0.9584 \text{ g}} = 18.80 \text{ cm}^3$$

$$= 18.80 \text{ cm}^3 \times \left(\frac{1 \text{ m}}{100 \text{ cm}}\right)^3 = 1.880 \times 10^{-5} \text{ m}^3$$

Now the value of w can be determined.

$$w = P(V_2 - V_1)$$
$$= 101,325 \text{ Pa} \times (1.880 \times 10^{-5} \text{ m}^3 - 3012 \times 10^{-5} \text{ m}^3)$$
$$= -3050 \text{ Pa m}^3 = -3050 \text{ J}$$

As indicated in Section 18.2, units of Pa m^3 are equivalent to joules. Note that the value of w is negative, indicating that work is done on the system as its volume decreases from 0.03012 m^3 to 0.0000188 m^3 at a constant pressure of 101,325 Pa (1 atm). Now that we know both q ($-40,668$ J) and w (-3050 J), we can calculate ΔE.

$$\Delta E = q - w$$
$$= (-40,668 \text{ J}) - (-3050 \text{ J})$$
$$= -37,618 \text{ J} = -37.618 \text{ kJ}$$

The internal energy of the system decreases by 37,618 J ($\Delta E < 0$) as the system gives up 40,668 J to the surroundings ($q < 0$) and the surroundings do 3050 J of work on the system ($w < 0$).

18.4   State Functions

A mole of water that condenses from a gas to a liquid at 100°C (as in Example 18.2) undergoes a change in internal energy, ΔE, that is simply the difference in the internal energies of 1 mole of $H_2O(l)$ at 100°C and 1 mole of $H_2O(g)$ at 100°C. This difference does not depend on how the steam is converted to liquid water. We could convert the gas directly to liquid at 100°C and 1 atmosphere, or we could convert the mole of $H_2O(g)$ to 1 mole of $H_2(g)$ and $\frac{1}{2}$ mole of $O_2(g)$ and then reconvert the $H_2(g)$ and $O_2(g)$ to 1 mole of $H_2O(l)$ at 100°C and 1 atmosphere. Either way, ΔE for the process would be -37.618 kilojoules.

When a property of a system (its internal energy, for example) is not dependent on how the system gets from one state to another but is only dependent on the state itself, the property is said to be a **state function.** How the system goes from the initial state to the final state is irrelevant to the final value of a state function. To put it another way, state functions have no memory—they forget the path used for the transformation. The change in the value of a state function equals its value at the final state minus its value at the initial state.

An illustration of a state function is the change in the potential energy of a tennis ball as it is moved from a second-floor window ledge to a third-floor window ledge. (The potential energy is a measure of how hard the ball would hit the ground if it fell.) The change in potential energy is the same whether the ball is taken directly from the second to the third floor or it is carried to the roof and then back to the third floor. The potential energy of the tennis ball is a state function, since a difference in the path by which it is moved from the second to the third floor will make no difference in how hard the ball will hit the ground if it falls from the third-floor window.

In contrast to state functions, there are other functions whose values depend on the paths followed. The distinction between the two types of functions can be clarified by the expression

$$\Delta E = E_2 - E_1 = q - w$$

The internal energy, E, is a state function; ΔE of a reaction is the same regardless of how that reaction is carried out. On the other hand, the values of q and w are not constant; they vary with the process used. As an example of how an energy change can be independent of the path used to cause it, consider the tennis ball again. More work is required to carry the tennis ball from the second floor to the roof and back to the third floor of a ten-story building than to carry it directly from the second floor to the third floor, but the change in its potential energy is the same for either path. Similarly, the change in energy, ΔE, associated with the transformation of $H_2O(g)$ to $H_2O(l)$ at constant temperature and pressure is always the same and is a state function. However, q and w depend on the way in which the transformation is carried out, and thus they are not state functions. Quantities that are state functions are designated by capital letters; those that are not, by lower-case letters.

18.5   Enthalpy Changes; Heat of Reaction

The amount of heat, q, absorbed or released by a system undergoing a chemical change is usually not a state function. However, if the change occurs in such a way that the only work done is due to a change in volume of the system, q is a state function and is called the **enthalpy change, ΔH,** of the system.

Most chemical reactions occur under the essentially constant pressure of the atmosphere; consequently, the enthalpy change of these, or of any other reactions that occur at constant pressure, can be determined by adding the work term to the value of ΔE. For such a reaction, the change in internal energy can be calculated from the equation

$$\Delta E = q - w$$

Since the reaction occurs at constant pressure and the only work done results from the change in volume of the system, the work term in this case is equal to $P(V_2 - V_1)$, or $P\Delta V$.

$$\Delta E = q - P\Delta V$$

Rearrangement gives the expression for q.

$$q = \Delta E + P\Delta V$$

When the process occurs at constant pressure with all work given by $P\Delta V$, q is called ΔH.

$$\Delta H = q = \Delta E + P\Delta V$$

The enthalpy change, ΔH, is the value of the total amount of heat a chemical reaction can provide to or absorb from the surroundings at constant pressure. At constant pressure *and* constant volume,

$$P\Delta V = 0 \quad \text{and} \quad \Delta H = \Delta E$$

If ΔH for a chemical reaction conducted at constant pressure (state 1 = reactants, state 2 = products) is negative, the system *evolves* heat to the surroundings, and the reaction is **exothermic.** If ΔH for a chemical reaction conducted at constant pressure is positive, heat is *absorbed* by the system from the surroundings, and the reaction is **endothermic.**

We have seen enthalpy changes before. In Section 9.6 we described the enthalpy, or heat, of reaction as the quantity of heat liberated or absorbed during a chemical change. Heat of vaporization was identified in Section 11.4 as the heat necessary to convert 1 mole of a liquid to 1 mole of a gas. The heat of vaporization of ammonia at $-33°C$, for example, is the enthalpy change for the process

$$NH_3(l) \longrightarrow NH_3(g) \qquad \Delta H = 23.3 \text{ kJ}$$

The heat of fusion of a substance was described in Section 11.11 as the enthalpy change for conversion of 1 mole of the solid substance to the liquid state.

Enthalpy changes for a large number of chemical reactions have been measured and tabulated. The National Bureau of Standards has been particularly active in the compilation of such data. Since the enthalpy change of a reaction can vary with the temperature or pressure, data are measured and tabulated under a specific set of conditions called a **standard state.** One common standard state is $25°C$ (298.15 K) and 1 atmosphere of pressure. The state of the substance (gas, liquid, or solid) must be specified. The symbol ΔH_{298}° indicates an enthalpy change for a chemical reaction that occurs under these standard state conditions.

Probably the most useful tabulation of enthalpy changes is for a particular type of chemical reaction in which *1 mole* of a pure substance is formed from the free elements in their most stable states under standard state conditions. This enthalpy change is referred to as the **standard molar enthalpy of formation** of the substance formed and is designated by ΔH_{f298}°. (Note the subscript letter f, which identifies an enthalpy of *formation*.) The standard molar enthalpy of formation of $CO_2(g)$ is -393.5 kilojoules per mole, which is the enthalpy change for the reaction

$$C(s) + O_2(g) \longrightarrow CO_2(g)$$

with the reactants and products at 1 atmosphere of pressure and 25°C and with the carbon present as graphite, the most stable form of carbon under these conditions. For silver oxide, Ag_2O, ΔH°_{f298} is -31.0 kilojoules per mole; that is, it is equal to ΔH°_{298} for the reaction

$$2Ag(s) + \tfrac{1}{2} O_2(g) \longrightarrow Ag_2O(s)$$

The reaction of $\tfrac{1}{2}$ mole of O_2 and 2 moles of Ag to produce 1 mole of Ag_2O is the correct one to use for this purpose since the enthalpy of formation refers to 1 mole of product. By convention, the standard molar enthalpy of formation of an element in its most stable form is equal to zero (see Table 18-1).

Standard molar enthalpies of formation of some common substances are given in Table 18-1 and in Appendix J. The units in these tables are joules and

Table 18-1 Standard Molar Enthalpies of Formation, Standard Molar Free Energies of Formation, and Absolute Standard Molar Entropies (298.15 K, 1 atm). (See Appendix J for additional values.)

Substance	ΔH°_{f298}, kJ mol^{-1}	ΔG°_{f298}, kJ mol^{-1}	S°_{298}, J mol^{-1} K^{-1}
Carbon			
C(s) (graphite)	0	0	5.740
C(g)	716.68	671.29	157.99
CO(g)	-110.5	-137.2	197.56
$CO_2(g)$	-393.5	-394.4	213.6
$CH_4(g)$	-74.81	-50.75	186.15
Chlorine			
$Cl_2(g)$	0	0	222.96
Cl(g)	121.7	105.7	165.09
Copper			
Cu(s)	0	0	33.15
CuS(s)	-53.1	-53.6	66.5
Hydrogen			
$H_2(g)$	0	0	130.57
H(g)	218.0	203.3	114.60
$H_2O(g)$	-241.8	-228.6	188.71
$H_2O(l)$	-285.8	-237.2	69.91
HCl(g)	-92.31	-95.30	186.80
$H_2S(g)$	-20.6	-33.6	205.7
Oxygen			
$O_2(g)$	0	0	205.03
O(g)	249.2	231.8	160.95
Silver			
$Ag_2O(s)$	-31.0	-11.2	121
$Ag_2S(s)$	-32.6	-40.7	144.0

kilojoules, but many older tables have calories (1 cal = 4.184 J) and kilocalories. If you have occasion to use other tables, be sure to check the units. Making use of these tabulated values and the fact that the enthalpy is a state function, we can obtain the enthalpy changes for a large number of chemical reactions. Several other properties of interest can be calculated from the enthalpy changes.

EXAMPLE 18.3   In Section 9.10 the heat of the reaction

$$3O_2(g) \longrightarrow 2O_3(g)$$

was given as 286 kJ. Assuming that this is the standard enthalpy change, ΔH°_{298} (with both reactant and product in their standard states), what is the standard molar enthalpy of formation, ΔH°_{f298}, of $O_3(g)$?

ΔH°_{f298} is the enthalpy change for the formation of one mole of a substance in its standard state from the elements in their standard states. Thus ΔH°_{f298} for $O_3(g)$ is the enthalpy change for the reaction

$$\tfrac{3}{2}O_2(g) \longrightarrow O_3(g)$$

Since 286 kJ of heat are absorbed when 2 mol of $O_3(g)$ are formed, to form 1 mol of $O_3(g)$ will require half as much heat, or 143 kJ. Thus the enthalpy change is 143 kJ, and

$$\Delta H^\circ_{f298} = 143 \text{ kJ mol}^{-1}$$

EXAMPLE 18.4   What is the enthalpy change for the reaction of 1 mol of $H_2(g)$ with 1 mol of $Cl_2(g)$ to produce 2 mol of $HCl(g)$ at standard state conditions?

$$H_2(g) + Cl_2(g) \longrightarrow 2HCl(g)$$

Since $H_2O(g)$ and $Cl_2(g)$ are the most stable states of the elements hydrogen and chlorine under standard state conditions and a pure substance is being formed at standard state conditions, ΔH° must be proportional to the standard molar enthalpy of formation of $HCl(g)$. In this example 2 mol of $HCl(g)$ are being formed; therefore,

$$\Delta H^\circ = 2 \, \Delta H^\circ_{fHCl(g)}$$

From Table 18-1

$$\Delta H^\circ_{fHCl(g)} = -92.31 \text{ kJ mol}^{-1}$$

Therefore,

$$\Delta H^\circ = 2 \text{ mol } HCl \times (-92.31 \text{ kJ mol}^{-1}) = -184.6 \text{ kJ}$$

Thus the reaction evolves 184.6 kJ of heat.

$$H_2(g) + Cl_2(g) \longrightarrow 2HCl(g) \qquad \Delta H^\circ = -184.6 \text{ kJ}$$

18.6 Hess's Law

Standard molar enthalpies of formation can be used to calculate other enthalpy changes because of a principle summed up in **Hess's law: If a process can be** considered to be the sum of several stepwise processes, the enthalpy change for the total process equals the sum of the enthalpy changes for the various steps. Hess's law tells us that the enthalpy change for a reaction that can be considered to be a sequence of reaction steps can be calculated by adding the enthalpy changes of all of the steps. Hess's law can be used even if the steps are hypothetical.

Before we apply Hess's law, let us look briefly at two important features of enthalpy changes:

1. ΔH for a reaction in one direction is equal in magnitude and opposite in sign to ΔH for the reaction in the reverse direction.
2. ΔH is directly proportional to the quantities of reactants or products.

The first statement tells us that if we know ΔH for the reaction

$$H_2O(s) \longrightarrow H_2O(l)$$

is 6.0 kilojoules, then ΔH for the reverse reaction,

$$H_2O(l) \longrightarrow H_2O(s)$$

is simply the negative of that for the forward reaction, or -6.0 kilojoules. The second statement indicates that if the heat of fusion of 1 mole of water is 6.0 kilojoules,

$$H_2O(s) \longrightarrow H_2O(l) \qquad \Delta H = 6.0 \text{ kJ}$$

then the heat of fusion of 2 moles of water is twice as great,

$$2H_2O(s) \longrightarrow 2H_2O(l) \qquad \Delta H = 12 \text{ kJ}$$

EXAMPLE 18.5 Calculate the standard molar enthalpy of formation of $CO_2(g)$ from the following standard enthalpy changes:

$$2C(s) + O_2(g) \longrightarrow 2CO(g) \qquad \Delta H° = -221.0 \text{ kJ}$$
$$2CO(g) + O_2(g) \longrightarrow 2CO_2(g) \qquad \Delta H° = -566.0 \text{ kJ}$$

The standard molar enthalpy of formation of $CO_2(g)$ is the enthalpy change for the reaction

$$C(s) + O_2(g) \longrightarrow CO_2(g) \qquad \Delta H° = H°_{fCO_2}$$

This equation and the enthalpy change can be obtained by multiplying the first equation above and its enthalpy change by $\frac{1}{2}$ and then adding $\frac{1}{2}$ of the second equation and its enthalpy change.

$$\frac{1}{2}[2C(s) + O_2(g) \longrightarrow 2CO(g)] \qquad \Delta H° = \frac{1}{2}(-221.0) = -110.5 \text{ kJ}$$
$$\frac{1}{2}[2CO(g) + O_2(g) \longrightarrow 2CO_2(g)] \qquad \Delta H° = \frac{1}{2}(-566.0) = -283.0 \text{ kJ}$$
$$\overline{C(s) + O_2(g) \longrightarrow CO_2(g) \qquad \Delta H° = (-110.5) + (-283.0)}$$
$$= -393.5 \text{ kJ}$$

The enthalpy of formation of $CO_2(g)$ at 25°C and 1 atm is -393.5 kJ mol^{-1}. Hence $\Delta H°_{fCO_2(g)} = -393.5$ kJ.

EXAMPLE 18.6   What is $\Delta H°$ for the following reaction?

$$CH_4(g) + 2O_2(g) \longrightarrow CO_2(g) + 2H_2O(l)$$

This reaction can be viewed as occurring in several steps:

STEP 1 $CH_4(g) \longrightarrow C(s) + 2H_2(g)$ $\Delta H_1° = -\Delta H_{fCH_4(g)}°$

STEP 2 $2O_2(g) \longrightarrow 2O_2(g)$ $\Delta H_2° = 2\,\Delta H_{fO_2(g)}° = 0$

STEP 3 $2H_2(g) + O_2(g) \longrightarrow 2H_2O(l)$ $\Delta H_3° = 2\,\Delta H_{fH_2O(l)}°$

STEP 4 $C(s) + O_2(g) \longrightarrow CO_2(g)$ $\Delta H_4° = \Delta H_{fCO_2(g)}°$

Adding the reactions of Steps 1, 2, 3, and 4 gives

$$CH_4(g) + \overset{2}{\cancel{4}}O_2(g) + \cancel{2H_2(g)} + \cancel{C(s)} \longrightarrow \cancel{C(s)} + \cancel{2O_2(g)}$$
$$+ \cancel{2H_2(g)} + 2H_2O(l) + CO_2(g)$$

The net chemical change is

$$CH_4(g) + 2O_2(g) \longrightarrow 2H_2O(l) + CO_2(g)   \Delta H° = ?$$

In either case (by direct reaction or by the combination of reactions in Steps 1 through 4) we end up with the same result: state 1 [1 mol of $CH_4(g)$ at 1 atm and 25°C and 2 mol of $O_2(g)$ at 1 atm and 25°C] is converted to state 2 [2 mol of $H_2O(l)$ at 1 atm and 25°C and 1 mol of $CO_2(g)$ at 1 atm and 25°C]. The enthalpy change is independent of how the conversion from state 1 to state 2 is accomplished (H is a state function). Thus,

$$\Delta H° = \Delta H_1° + \Delta H_2° + \Delta H_3° + \Delta H_4°$$

or $\Delta H° = -\Delta H_{fCH_4(g)}° + 2\,\Delta H_{fO_2(g)}° + 2\,\Delta H_{fH_2O(l)}° + \Delta H_{fCO_2(g)}°$

$= -1\,\text{mol } \cancel{CH_4(g)} \times (-74.81\,\text{kJ } \cancel{\text{mol}^{-1}}) + 2\,\text{mol } \cancel{O_2(g)}$

$\times (0\,\text{kJ } \cancel{\text{mol}^{-1}}) + 2\,\text{mol } \cancel{H_2O(l)} \times (-285.8\,\text{kJ } \cancel{\text{mol}^{-1}})$

$+ 1\,\text{mol } \cancel{CO_2(g)} \times (-393.5\,\text{kJ } \cancel{\text{mol}^{-1}})$

$= -890.3\,\text{kJ}$

Hence 890.3 kJ of heat are evolved during the combustion of 1 mol of methane.

$$CH_4(g) + 2O_2(g) \longrightarrow 2H_2O(l) + CO_2(g)   \Delta H° = -890.3\,\text{kJ}$$

There is a more useful statement of Hess's law for the preceding type of calculation: *The standard enthalpy change for a chemical reaction is equal to the sum of the standard molar enthalpies of formation of all the products, each one multiplied by the number of moles of the product in the balanced chemical equation, minus the corresponding sum for the reactants.*

$$\Delta H° = \Sigma\,\Delta H_{f\text{products}}° - \Sigma\,\Delta H_{f\text{reactants}}°$$

For the reaction $mA + nB \rightarrow xC + yD$,

$$\Delta H = [(x \times \Delta H_{fC}°) + (y \times \Delta H_{fD}°)] - [(m \times \Delta H_{fA}°) + (n \times \Delta H_{fB}°)]$$

EXAMPLE 18.7 Calculate $\Delta H°$ for the reaction

$$2Ag_2S(s) + 2H_2O(l) \longrightarrow 4Ag(s) + 2H_2S(g) + O_2(g)$$

The standard molar enthalpies of formation, $\Delta H°_{f_{298}}$, of the compounds involved may be found in Table 18-1.

$$\Delta H° = 4\,\Delta H°_{f_{Ag(s)}} + 2\,\Delta H°_{f_{H_2S(g)}} + \Delta H°_{f_{O_2(g)}} - 2\,\Delta H°_{f_{Ag_2S(s)}} - 2\,\Delta H°_{f_{H_2O(l)}}$$

The standard molar enthalpy of formation of an element in its most stable state is zero. Thus

$$\Delta H° = 4\,\text{mol Ag(s)} \times (0 \text{ kJ mol}^{-1}) + 2\,\text{mol H}_2\text{S(g)} \times (-20.6 \text{ kJ mol}^{-1})$$
$$+ 1\,\text{mol O}_2\text{(g)} \times (0 \text{ kJ mol}^{-1}) - 2\,\text{mol Ag}_2\text{S(s)} \times (-32.6 \text{ kJ mol}^{-1})$$
$$- 2\,\text{mol H}_2\text{O(l)} \times (-285.8 \text{ kJ mol}^{-1})$$
$$= [-41.2 - (-65.2) - (-571.6)] \text{ kJ} = 595.6 \text{ kJ}$$

Hence the reaction is endothermic.

$$2Ag_2S(s) + 2H_2O(l) \longrightarrow 4Ag(s) + 2H_2S(g) + O_2(g) \qquad \Delta H° = 595.6 \text{ kJ}$$

EXAMPLE 18.8 Calculate $\Delta H°$ for the reaction

$$2Na(s) + 2H_2O(l) \longrightarrow 2NaOH(s) + H_2(g)$$

The standard molar enthalpies of formation for the reactants and products involved are as follows: $H_2O(l)$, -285.8 kJ mol^{-1}; NaOH(s), -426.8 kJ mol^{-1}; Na and H_2, 0 kJ mol^{-1}.

$$\Delta H° = 2\,\Delta H°_{f_{NaOH(s)}} + \Delta H°_{f_{H_2(g)}} - 2\,\Delta H°_{f_{Na(s)}} - 2\,\Delta H°_{f_{H_2O(l)}}$$
$$\Delta H° = 2\,\text{mol NaOH(s)} \times (-426.8 \text{ kJ mol}^{-1}) + 1\,\text{mol H}_2\text{(g)} \times (0 \text{ kJ mol}^{-1})$$
$$- 2\,\text{mol Na(s)} \times (0 \text{ kJ mol}^{-1}) - 2\,\text{mol H}_2\text{O(l)} \times (-285.8 \text{ kJ mol}^{-1})$$
$$= [-853.5 + 571.6] \text{ kJ} = -281.9 \text{ kJ}$$

Hence the reaction evolves heat.

In terms of bond strengths within molecules, a positive value of ΔH indicates that the bonds are stronger in the reactants than in the products, and hence heat must be absorbed for the reaction to take place. A negative value of ΔH, on the other hand, indicates that the bonds are stronger in the products than in the reactants, and hence less energy is required to break the bonds in the reactants than is evolved in the formation of the products. The net effect is the evolution of heat.

18.7 Bond Energies

The strength of a chemical bond is measured by the energy required to break it, that is, to separate the atoms in a molecule and leave them as distinct isolated gaseous atoms. For a diatomic molecule this energy, called the **bond energy, D,** is equal to the standard enthalpy change for the reaction

$$XY(g) \longrightarrow X(g) + Y(g)$$

Bond energies for commonly occurring diatomic molecules range from 946 kilojoules per mole for N_2 (triple bond) to 150 kilojoules per mole for I_2 (single bond) and from 569 kilojoules per mole for HF to 295 kilojoules per mole for HI.

Molecules of three or more atoms necessarily have two or more bonds. The sum of all of the bond energies in such a molecule is equal to the heat of formation of the molecule from the isolated gaseous atoms. For example, the heat of formation of the S_8 ring from eight sulfur atoms is eight times the energy of a sulfur-sulfur single bond.

Average bond energies of some common bonds appear in Table 18-2. The energies of double or triple bonds between two atoms are generally higher than those of single bonds between the same two atoms. However, a double bond is usually not quite twice as strong as a single bond between the same two atoms, and a triple bond is not quite three times as strong as a single bond.

Table 18-2   Some Average Bond Energies (kJ mol^{-1})

Single Bonds

H	C	N	O	F	Si	P	S	Cl	Br	I	
436	415	390	464	569	295	320	340	432	370	295	H
	345	290	350	439	290	265	260	330	275	240	C
		160	200	270	—	210	—	200	245(?)	—	N
			140	185	370	350	—	205	—	200	O
				160	540	489	285	255	235	—	F
					175	215	225	359	290	215	Si
						215	230	330	270	215	P
							215	250	215	—	S
								243	220	210	Cl
									190	180	Br
										150	I

Multiple Bonds

C=C,	611	C=N,	615	C=O,	741	N=N, 418
C≡C,	837	C≡N,	891	C≡O,	1080	N≡N, 946

Tabulated values for enthalpy changes can be used to obtain bond energies, or bond energies can be used to obtain approximate values for enthalpy changes.

EXAMPLE 18.9   Evaluate the bond energy of a H—H single bond from the data in Table 18-1.

The energy of a mole of single bonds between hydrogen atoms, D_{H-H}, is equal to the standard enthalpy of the reaction

$$H_2(g) \longrightarrow 2H(g) \qquad D_{H-H} = \Delta H°$$

Referring to Table 18-1, we find the enthalpy of formation of 1 mole of hydrogen atoms, $\Delta H^\circ_{f_{H(g)}}$, is 218.0 kJ mol^{-1}. Therefore,

$$D_{H-H} = \Delta H^\circ = 2\,\Delta H^\circ_{f_{H(g)}} - \Delta H^\circ_{f_{H_2(g)}}$$
$$= 2 \text{ mol H}(g) \times (218.0 \text{ kJ mol}^{-1}) - 1 \text{ mol H}_2(g) \times (0 \text{ kJ mol}^{-1})$$
$$= 436.0 \text{ kJ}$$

Thus 436.0 kJ is required to break the H—H bonds in 1 mol of H_2. The rounded value is given in Table 18-2.

EXAMPLE 18.10   Evaluate the bond energy of the C≡O bond in CO(g) from the data in Table 18-1.

The energy of a mole of C≡O bonds, $D_{C≡O}$, is equal to the standard enthalpy of the reaction

$$CO(g) \longrightarrow C(g) + O(g) \qquad D_{C≡O} = \Delta H^\circ$$

Using Hess's law, we have

$$D_{C≡O} = \Delta H^\circ = \Delta H^\circ_{f_{C(g)}} + \Delta H^\circ_{f_{O(g)}} - \Delta H^\circ_{f_{CO(g)}}$$
$$= 1 \text{ mol C}(g) \times (716.68 \text{ kJ mol}^{-1}) + 1 \text{ mol O}(g)$$
$$\times (249.2 \text{ kJ mol}^{-1}) - 1 \text{ mol CO}(g) \times (-110.5 \text{ kJ mol}^{-1})$$
$$= 1076.4 \text{ kJ}$$

Thus 1076.4 kJ is required to break the C≡O bond in CO(g). The rounded value is reported in Table 18-2.

EXAMPLE 18.11   Methanol, CH_3OH, is manufactured by the reaction

$$CO(g) + 2H_2(g) \longrightarrow CH_3OH(g)$$

Calculate the approximate enthalpy change of the reaction from the bond energies in Table 18-2.

The enthalpy change for this reaction can be calculated as the sum of the enthalpy changes of three reactions, each of which involves bond breaking or bond formation.

C≡O(g) $\longrightarrow$ C(g) + O(g)	$\Delta H = D_{C≡O}$
2H—H(g) $\longrightarrow$ 4H(g)	$\Delta H = 2D_{H-H}$

$$C(g) + O(g) + 4H(g) \longrightarrow \underset{\underset{\displaystyle H}{|}}{\overset{\overset{\displaystyle H}{|}}{H-C-O-H}} \qquad \Delta H = -3D_{C-H} - D_{C-O} - D_{O-H}$$

$$CO(g) + 2H_2(g) \longrightarrow CH_3OH(g) \qquad \Delta H = D_{C≡O} + 2D_{H-H}$$
$$- 3D_{C-H} - D_{C-O} - D_{O-H}$$

In Table 18-2 we find

$$D_{C≡O} = 1080 \text{ kJ mol}^{-1} \qquad D_{H-H} = 436 \text{ kJ mol}^{-1} \qquad D_{C-H} = 415 \text{ kJ mol}^{-1}$$
$$D_{C-O} = 350 \text{ kJ mol}^{-1} \qquad D_{O-H} = 464 \text{ kJ mol}^{-1}$$

Thus
$$\Delta H = [1080 + (2 \times 436) - (3 \times 415) - 350 - 464]\,kJ$$
$$= -107\,kJ$$

ENTROPY AND FREE ENERGY

18.8 The Spontaneity of Chemical and Physical Changes

A **spontaneous change,** in a thermodynamic sense, is any change in a system that proceeds in the absence of any outside influence on the system. For example, when two containers of pure gases are connected, the gases form a solution (Fig. 13-7) without any outside influence—they mix spontaneously. Liquid water freezes to ice spontaneously at $-1°C$; excess hydrogen ions and hydroxide ions combine spontaneously in aqueous solution; carbon combines spontaneously with oxygen to give carbon monoxide or carbon dioxide. Since the definition of a spontaneous change does not mention time, a change may be spontaneous even though it proceeds very slowly. For those who own diamonds (which are essentially pure carbon) it is fortunate that the spontaneous reaction of carbon with oxygen is very slow at room temperature.

Exothermic changes for which ΔH is large (a large amount of heat is given off) are frequently spontaneous. The following reaction is exothermic (Example 18.8):

$$2Na(s) + 2H_2O(l) \longrightarrow 2NaOH(s) + H_2(g) \qquad \Delta H° = -281.9\,kJ$$

Because of the large negative value of ΔH, we should expect this reaction to occur spontaneously at 25°C and 1 atmosphere, and it does. The reaction in Example 18.7

$$2Ag_2S(s) + 2H_2O(l) \longrightarrow 4Ag(s) + 2H_2S(g) + O_2(g) \qquad \Delta H° = 595.6\,kJ$$

is endothermic and we should not expect it to occur spontaneously. However, the reverse reaction ($\Delta H = -595.6\,kJ$) should be expected to occur spontaneously, and it does. Silver exposed to hydrogen sulfide and air tarnishes rapidly.

Predictions based solely on the value of $\Delta H°$ are not always valid, particularly if $\Delta H°$ has a small value. For example, a reaction that proceeds spontaneously at one temperature and pressure may not proceed spontaneously at another temperature and pressure. At 25°C and 1 atmosphere of pressure, the reaction

$$2Ag(s) + \tfrac{1}{2}O_2(g) \longrightarrow Ag_2O(s)$$

proceeds spontaneously; $\Delta H°$ is -31 kilojoules. However, above about 200°C this reaction is not spontaneous, even though ΔH changes very little with the change in temperature. Above its melting temperature a crystalline solid melts spontaneously, even though melting is an endothermic change ($\Delta H > 0$). Below its melting temperature it does not, yet the ΔH value changes very little. Whether chemical reactions and physical changes proceed spontaneously or not can depend on temperature and pressure. Yet temperature and pressure changes often have virtually no effect on the value of ΔH for a reaction.

It is evident, then, that another factor in addition to ΔH must be considered when determining whether or not a given reaction will proceed spontaneously. This factor is the entropy change that occurs as the change in the system takes place.

18.9   Entropy and Entropy Changes

Consider two changes that are spontaneous but not exothermic: the melting of ice at room temperature,

$$H_2O(s) \longrightarrow H_2O(l) \qquad \Delta H = 6.0 \text{ kJ}$$

and the decomposition of calcium carbonate at high temperature,

$$CaCO_3(s) \longrightarrow CaO(s) + CO_2(g) \qquad \Delta H = 178 \text{ kJ}$$

The disorder of the system increases when either of these reactions occurs. Water molecules are held in fixed positions in a regular, repeating array in an ice crystal (Section 12.1). When the ice melts, the water molecules are free to move through the liquid as well as to change their orientations. The molecules in the liquid are much more randomly distributed than those in the solid. Thus the amount of order is higher in the solid than in the liquid. As calcium carbonate decomposes, the system changes from an ordered array of calcium ions and carbonate ions in solid calcium carbonate to an ordered array of calcium and oxide ions in solid calcium oxide plus a disordered collection of carbon dioxide molecules in the gas phase. The arrangement of the carbon dioxide molecules in the gas phase is even more random (the disorder is greater) than that of an equal number of water molecules in the liquid phase.

The randomness, or the amount of disorder, of a system can be determined quantitatively and is referred to as the entropy, S, of the system. The greater the randomness, or disorder, in a system, the higher is its entropy. Entropy is one of the most important of scientific concepts; as we shall see, an entropy increase, corresponding to an increase in disorder, is the major driving force in many chemical and physical processes.

Every substance has an entropy as one of its characteristic properties, just as it has color, hardness, volume, melting point, density, and enthalpy. Like E and H, S is a state function. The entropy of a system is equal to the sum of the entropies of its components. The **entropy change, ΔS,** for a chemical change is equal to the sum of the entropies of the products of that change minus the sum of the entropies of the reactants. For the reaction $mA + nB \rightarrow xC + yD$ the entropy change is

$$\Delta S = \Sigma S_{products} - \Sigma S_{reactants}$$
$$= [(x \times S_C) + (y \times S_D)] - [(m \times S_A) + (n \times S_B)]$$

where S_A is the entropy of A, S_B is the entropy of B, etc.

A positive value for ΔS ($\Delta S > 0$) indicates an increase in randomness, or disorder, during the change; a negative value for ΔS ($\Delta S < 0$), a decrease in randomness, or an increase in order. The change to a more random and less ordered system when ice melts or when calcium carbonate decomposes corresponds to an increase in entropy ($\Delta S > 0$). Similarly, as liquid molecules go to the still more disordered and random state characteristic of gases, entropy increases. The amount of change in the entropy value is a measure of the increase in the randomness, or disorder, of a system.

It is possible to determine the absolute entropy content of a substance using the technique described in Section 18.15. Absolute standard molar entropy values for several substances are listed in Table 18-1 and in Appendix J. Each of these S_{298}° values represents the actual entropy content of 1 mole of a substance. The superscript symbol in S_{298}° indicates that the value is for a substance in a standard state (298.15 K and 1 atm). Note that solids (most ordered) tend to have lower entropies than liquids (less ordered) and liquids tend to have lower entropies than gases (least ordered). In the same physical state, substances with simple molecules tend to have lower molar entropies than those of substances with more complicated molecules. Since the latter have larger numbers of atoms that can move about, they can exhibit greater randomness or disorder; thus they have higher entropies. Hard substances, such as diamond, tend to be more ordered and have lower entropies than do softer materials, such as graphite or sodium.

The following examples show how absolute standard molar entropies can be used to calculate entropy changes.

EXAMPLE 18.12   Determine the entropy change for the process $H_2O(l) \rightarrow H_2O(g)$ when both $H_2O(l)$ and $H_2O(g)$ are in their standard states at 25°C.

The entropy change for this reaction is the sum of the entropies of the products, only $H_2O(g)$ in this case, minus the sum of the entropies of the reactants, $H_2O(l)$ in this case. Entropy values are listed in Table 18-1.

$$\Delta S^\circ = S_{H_2O(g)}^\circ - S_{H_2O(l)}^\circ$$
$$= 1 \text{ mol } H_2O(g) \times (188.71 \text{ J mol}^{-1} \text{ K}^{-1}) - 1 \text{ mol } H_2O(l)$$
$$\times (69.91 \text{ J mol}^{-1} \text{ K}^{-1})$$
$$= 118.80 \text{ J K}^{-1}$$

The value for ΔS° is positive, indicating greater disorder in gaseous H_2O than in liquid H_2O. This is in accord with the fact that the molecules are in much more rapid and random motion in the gaseous state than they are in the liquid state.

EXAMPLE 18.13   Calculate the entropy change for the following reaction when the reactants and products are in their standard states at 25°C:

$$2H_2(g) + O_2(g) \longrightarrow 2H_2O(l)$$

Note that absolute standard molar entropies of the elements are not zero at 25°C. Thus

$$\Delta S^\circ = 2S_{H_2O(l)}^\circ - 2S_{H_2(g)}^\circ - S_{O_2(g)}^\circ$$
$$= 2 \text{ mol } H_2O(l) \times (69.91 \text{ J mol}^{-1} \text{ K}^{-1}) - 2 \text{ mol } H_2(g)$$
$$\times (130.57 \text{ J mol}^{-1} \text{ K}^{-1}) - 1 \text{ mol } O_2(g) \times (205.03 \text{ J mol}^{-1} \text{ K}^{-1})$$
$$= -326.35 \text{ J K}^{-1}$$

As this reaction proceeds, the entropy *of the system* decreases. We will see in a subsequent section that the entropy of the surroundings increases.

18.10 Free Energy Changes

To connect the concepts of randomness and enthalpy, we can say that reactions tend to proceed toward a state of minimum energy (negative ΔH) and maximum disorder (positive ΔS). The difference between ΔH, the enthalpy change of a reaction, and $T \Delta S$, the product of the temperature and the entropy change, equals the change in another state function, the **free energy, G.** The relationship of the **free energy change, ΔG,** the enthalpy change, and the entropy change for a reaction is given by the equation

$$\Delta G = \Delta H - T \Delta S$$

where T is the temperature of the reaction on the Kelvin scale. The free energy change is the thermodynamic quantity that is used to predict the spontaneity of a chemical reaction run at constant temperature and pressure. **Reactions for which the value of ΔG is negative ($\Delta G < 0$) are spontaneous.**

In the free energy equation, $\Delta G = \Delta H - T \Delta S$, if $T \Delta S$ is small compared to ΔH, then ΔG and ΔH have nearly the same value and each predicts the spontaneity of a given reaction. However, when the value of $T \Delta S$ is significant compared to that of ΔH, only the ΔG value can be used to predict spontaneity in a reliable way. The ΔH term of the free energy equation represents the energy, in the form of heat, involved in a reaction at constant pressure; hence it corresponds to a difference in energy between the initial and final states in a process. The ΔS term of the free energy equation (and hence the $T \Delta S$ term), on the other hand, corresponds to the difference in the amount of order of the atoms in the products and reactants.

When the initial and final states have the same enthalpy (ΔH is zero) and ΔS is not zero, the entropy change, indicated by the $T \Delta S$ term, controls whether or not the reaction will occur. When the amounts of order of the initial and final states are the same (ΔS, and hence $T \Delta S$, is zero) and ΔH is not zero, the enthalpy change, indicated by the ΔH term, determines whether or not the reaction will occur. If ΔH and ΔS both differ from zero, a combination of the two determines whether or not the reaction will occur.

The free energy equation $\Delta G = \Delta H - T \Delta S$ tells us that a reaction that is exothermic (negative ΔH) and produces more numerous molecules (more disorder, positive ΔS) will proceed spontaneously (both ΔH and ΔS contribute toward a more negative value for ΔG). A reaction that is endothermic (positive ΔH) and produces a more ordered system (negative ΔS) will not proceed spontaneously (both ΔH and ΔS contribute toward a more positive value for ΔG). If a reaction is exothermic (negative ΔH) but produces a more ordered system (negative ΔS) or if a reaction is endothermic (positive ΔH) but produces a less ordered system (positive ΔS), then ΔH and ΔS are working in opposite directions, and the relative sizes of ΔH and $T \Delta S$ will determine the prediction.

Note that T must always be positive; therefore ΔS (which can be either positive or negative) determines the sign of the $T \Delta S$ term. Also, as T increases, the value of $T \Delta S$ gets larger, so ΔS plays an increasingly important role in the value of ΔG.

Again, it should be emphasized that predicting whether or not a given reaction will proceed spontaneously says *nothing whatever about the rate.* A reaction with a negative ΔG value will proceed spontaneously, but it may do so at a very rapid rate or at an incredibly slow rate.

18.11 Determination of Free Energy Changes

Free energy is a state function. Hence values of $\Delta G°$, the free energy change for a reaction run under standard state conditions, can be calculated from standard molar free energies of formation, $\Delta G°_{f298}$, in the same way that $\Delta H°$ values are calculated from standard molar enthalpies of formation. The free energy change of a chemical process equals the sum of the free energies of the products minus the sum of the free energies of the reactants. For the reaction

$$mA + nB \longrightarrow xC + yD$$

the free energy change is

$$\Delta G° = \Sigma \Delta G°_{products} - \Sigma \Delta G°_{reactants}$$
$$= [(x \times \Delta G°_{fC}) + (y \times \Delta G°_{fD})] - [(m \times \Delta G°_{fA}) + (n \times \Delta G°_{fB})]$$

Standard molar free energy of formation values for several substances are given in Table 18-1 and in Appendix J for the standard state condition of 25°C and 1 atmosphere. Note that the standard molar free energy of formation of any free element in its most stable state is zero.

EXAMPLE 18.14   Calculate the standard free energy change, $\Delta G°$, for the reaction

$$2Ag_2S(s) + 2H_2O(l) \longrightarrow 4Ag(s) + 2H_2S(g) + O_2(g)$$

Standard molar free energies of formation of the compounds involved may be found in Table 18-1 or Appendix J; for $Ag_2S(s)$ $\Delta G°_{f298}$ is -40.7 kJ mol^{-1}; for $H_2O(l)$, -237.2 kJ mol^{-1}; for $H_2S(g)$, -33.6 kJ mol^{-1}; and for $Ag(s)$ and $O_2(g)$, 0 kJ mol^{-1}. Thus

$$\Delta G° = 4\,\Delta G°_{fAg(s)} + 2\,\Delta G°_{fH_2S(g)} + \Delta G°_{fO_2(g)} - 2\,\Delta G°_{fAg_2S(s)} - 2\,\Delta G°_{fH_2O(l)}$$

$$= 4 \text{ mol Ag(s)} \times (0 \text{ kJ mol}^{-1}) + 2 \text{ mol H}_2\text{S(g)} \times (-33.6 \text{ kJ mol}^{-1})$$
$$+ 1 \text{ mol O}_2\text{(g)} \times (0 \text{ kJ mol}^{-1}) - 2 \text{ mol Ag}_2\text{S(s)} \times (-40.7 \text{ kJ mol}^{-1})$$
$$- 2 \text{ mol H}_2\text{O(l)} \times (-237.2 \text{ kJ mol}^{-1})$$

$$= 488.6 \text{ kJ}$$

The positive value of $\Delta G°$ indicates that the forward reaction should not occur spontaneously but that the reverse reaction ($\Delta G° = -488.6$ kJ) should at 25°C and 1 atm.

EXAMPLE 18.15   The reaction of calcium oxide with the pollutant, sulfur trioxide,

$$CaO(s) + SO_3(g) \longrightarrow CaSO_4(s)$$

has been proposed as one way of removing SO_3 from the smoke resulting from burning high-sulfur coal. Using the following $\Delta H°_f$ and $S°_f$ values, calculate the standard free energy change for the reaction at 298 K.

$CaO(s)$: $\Delta H°_f = -635.5$ kJ mol^{-1}; $S°_f = 40$ J mol^{-1} K^{-1}

$SO_3(g)$: $\Delta H°_f = -395.7$ kJ mol^{-1}; $S°_f = 256.6$ J mol^{-1} K^{-1}

$CaSO_4(s)$: $\Delta H°_f = -1432.7$ kJ mol^{-1}; $S°_f = 107$ J mol^{-1} K^{-1}

This calculation can be approached in two ways.

1. ΔH_f° and ΔS_f° for the reaction can be calculated from the data, as shown in Sections 18.6 and 18.9, and then ΔG° can be calculated from the free energy equation,

$$\Delta G^\circ = \Delta H_f^\circ - T \Delta S_f^\circ$$

2. Using S_f° and ΔH_f° for each compound, ΔG_f° for each compound can be determined, and ΔG° for the reaction can be calculated from these data using the technique illustrated in Example 18.14.

We will use the first approach.

$$
\begin{aligned}
\Delta H^\circ &= \Delta H_{f_{CaSO_4(s)}}^\circ - \Delta H_{f_{CaO(s)}}^\circ - \Delta H_{f_{SO_3(g)}}^\circ \\
&= 1 \text{ mol CaSO}_4(s) \times (-1432.7 \text{ kJ mol}^{-1}) - 1 \text{ mol CaO}(s) \\
&\quad \times (-635.5 \text{ kJ mol}^{-1}) - 1 \text{ mol SO}_3(g) \times (-395.7 \text{ kJ mol}^{-1}) \\
&= -401.5 \text{ kJ} = -401,500 \text{ J} \\
\Delta S^\circ &= S_{CaSO_4(s)}^\circ - S_{CaO(s)}^\circ - S_{SO_3(g)}^\circ \\
&= 1 \text{ mol CaSO}_4(s) \times (107 \text{ J mol}^{-1} \text{ K}^{-1}) - 1 \text{ mol CaO}(s) \\
&\quad \times (40 \text{ J mol}^{-1} \text{ K}^{-1}) - 1 \text{ mol SO}_3(g) \times (256.6 \text{ J mol}^{-1} \text{ K}^{-1}) \\
&= -189.6 \text{ J K}^{-1} \\
\Delta G^\circ &= \Delta H^\circ - T \Delta S^\circ \\
&= -401,500 \text{ J} - [298 \text{ K} \times (-189.6 \text{ J K}^{-1})] \\
&= -401,500 \text{ J} + 56,500 \text{ J} \\
&= -345,000 \text{ J} = -345 \text{ kJ}
\end{aligned}
$$

Note that the units of ΔH° were initially in *kilojoules* and the units of ΔS°, in *joules*. It is necessary to use the same units when ΔH° and ΔS° are used to calculate ΔG°.

18.12   The Second Law of Thermodynamics

If a process occurs at a constant pressure in a closed system, a system in which no heat can get in or out, ΔH must be zero. If the process is spontaneous, the free energy change, ΔG, is negative, and since ΔH is zero, the entropy change, ΔS, must be positive. Thus the entropy increases in a closed system when a spontaneous process occurs. The universe is a closed system; the universe includes everything; nothing can get in or out. **The Second Law of Thermodynamics states that any spontaneous change that occurs in the universe must be accompanied by an increase in the entropy of the universe.**

It is difficult to apply the Second Law in its pure form. However, a useful modification can be developed from the free energy equation presented in Section 18.10.

$$\Delta G = \Delta H - T \Delta S \quad \text{(all values pertaining to the system)}$$

Dividing by T gives

$$\frac{\Delta G}{T} = \frac{\Delta H}{T} - \Delta S \quad \text{(for the system)}$$

It can be shown that

$$\Delta S_{sur} = -\frac{\Delta H_{sys}}{T} \quad \text{(at constant temperature and pressure)}$$

where ΔS_{sur} is the change in entropy for the surroundings, and ΔH_{sys} is the change in enthalpy for the system.

Thus,

$$\frac{\Delta G_{sys}}{T} = -\Delta S_{sur} - \Delta S_{sys}$$

or

$$-\frac{\Delta G_{sys}}{T} = \Delta S_{sur} + \Delta S_{sys}$$

$$\Delta S_{univ} = \Delta S_{sur} + \Delta S_{sys}$$

Hence,

$$-\frac{\Delta G_{sys}}{T} = \Delta S_{univ}$$

This equation is valid only at constant temperature and pressure, because the relationships from which it is derived are valid only at constant temperature and pressure.

This equation indicates that if ΔG_{sys} is negative, corresponding to a spontaneous change, the entropy of the universe must increase, that is, ΔS_{univ} must be positive. Thus it can be seen that **for a spontaneous change to take place at constant temperature and pressure, the entropy of the universe must increase.**

18.13 Free Energy Changes and Nonstandard States

Many chemical reactions occur when the reactants and products are not in their standard states. The free energy change, ΔG, of such a reaction can be determined from the free energy change, $\Delta G°$, of the same reaction when both the reactants and the products are in standard states by using the equation

$$\Delta G = \Delta G° + 2.303RT \log Q$$

In this equation T is the temperature on the Kelvin scale (298.15 K), R is a constant (8.314 J K^{-1}), and Q is the reaction quotient of the chemical reaction. For the chemical reaction

$$m\text{A} + n\text{B} + \cdots \longrightarrow x\text{C} + y\text{D} + \cdots$$

the reaction quotient is

$$Q = \frac{[\text{C}]^x[\text{D}]^y \cdots}{[\text{A}]^m[\text{B}]^n \cdots}$$

Thus the reaction quotient has the same form as the equilibrium constant, K, for the reaction. However, Q has no fixed value; its magnitude is determined by the concentrations of reactants and products at whatever stage in the reaction we choose to evaluate Q. When A and B are first mixed, no products are present, and

Q is equal to zero. The value of Q increases as the reaction proceeds. Only when the reaction has reached equilibrium is Q equal to K. As with the equilibrium constant (Section 15.17), strictly speaking, the activities of the reactants and products should be used to evaluate Q. However, we will continue to use pressure (in atmospheres) for gases, concentrations (in moles per liter) for dissolved species, and unity for pure solids and liquids as good approximations of the activities of these species.

EXAMPLE 18.16 The reaction of calcium oxide with sulfur trioxide (Example 18.15) rarely occurs under standard state conditions. Calculate ΔG for this reaction at 25°C when the pressure of SO_3 is 0.15 atm.

To determine ΔG under these nonstandard state conditions, we use the equation

$$\Delta G = \Delta G° + 2.303RT \log Q$$

For the reaction $CaO(s) + SO_3(g) \rightarrow CaSO_4(s)$

$$Q = \frac{[CaSO_4]}{[CaO][SO_3]}$$

CaO and $CaSO_4$ are solids, so their concentrations (activities) are 1. The concentration (activity) of the gas SO_3 is taken as its pressure in atmospheres, 0.15.

$$Q = \frac{1}{(1) \times (0.15)} = 6.67$$

In Example 18.15 $\Delta G°$ was shown to be -345 kJ. Thus

$$\Delta G = \Delta G° + 2.303RT \log Q$$

$$\Delta G = -345 \text{ kJ} + (2.303)(8.314 \text{ J K}^{-1})\left(\frac{1 \text{ kJ}}{10^3 \text{ J}}\right)(298.15 \text{ K})(\log 6.67)$$

$$= -345 \text{ kJ} + 4.7 \text{ kJ}$$

$$= -340 \text{ kJ}$$

The change of conditions caused a change in ΔG from -345 kJ to -340 kJ, but the reaction is still spontaneous under these conditions. Note that R had units of J K^{-1}, which had to be converted to kJ K^{-1}.

18.14 The Relationship Between Free Energy Changes and Equilibrium Constants

If ΔG for any reaction (A + B → C + D, for example) is negative, the reaction will occur spontaneously as written, and the quantities of products will increase. If ΔG is positive, the reverse reaction will occur spontaneously; that is, the species on the right will react to give increased quantities of the species on the left. What is the situation at equilibrium, which, by definition, is the state when the quantities of reactants and products do not change? At equilibrium, ΔG can be neither positive nor negative; thus it can only be zero. At equilibrium, we therefore have

$$\Delta G = 0 = \Delta G^\circ + 2.303 RT \log Q$$

Since the reaction is at equilibrium, the value of the reaction quotient must be equal to that of the equilibrium constant for the reaction (Q is equal to K), so we have

$$\Delta G = 0 = \Delta G^\circ + 2.303 RT \log K$$
$$\Delta G^\circ = -2.303 RT \log K \tag{1}$$

This derivation shows us that the value of the standard state free energy change, ΔG°, for a reaction can be used to determine the equilibrium constant for that reaction and that values of ΔG° can be determined experimentally from equilibrium constants.

EXAMPLE 18.17   Calculate the standard state free energy change for the ionization of acetic acid at 25°C.

$$CH_3CO_2H(aq) \longrightarrow H^+(aq) + CH_3CO_2^-(aq)$$

The equilibrium constant for the reaction has been found to be 1.8×10^{-5}.

At equilibrium, $\Delta G^\circ = -2.303 RT \log K$, where

$$K = \frac{[H^+][CH_3CO_2^-]}{[CH_3CO_2H]} = 1.8 \times 10^{-5}$$

Thus $\Delta G^\circ = -(2.303)(8.314 \text{ J K}^{-1})(298.15 \text{ K})[\log (1.8 \times 10^{-5})]$
$$= -(2.303)(8.314 \text{ J K}^{-1})(298.15 \text{ K})(-4.74) = 2.7 \times 10^4 \text{ J}$$
$$= 27 \text{ kJ}$$

Therefore the free energy change for the *complete* transformation of 1 mol of acetic acid in its standard state in water into 1 mol of hydrogen ion and 1 mol of acetate ion, both in their standard states in water, is 27 kJ. Since ΔG° is positive, the reaction is not spontaneous. However, the reverse reaction,

$$H^+(aq) + CH_3CO_2^-(aq) \longrightarrow CH_3CO_2H(aq)$$

is spontaneous ($\Delta G^\circ = -27 \text{ kJ mol}^{-1}$).

EXAMPLE 18.18   Using data in Appendix J, calculate the value of K at 298.15 K for the reaction

$$H_2(g) + \tfrac{1}{2}O_2(g) \longrightarrow H_2O(g)$$

The equilibrium constant for this reaction can be evaluated using the equation

$$\Delta G^\circ = -2.303 RT \log K$$

Since this reaction involves formation of 1 mol of $H_2O(g)$ from the elements, ΔG° for the reaction is equal to the standard molar free energy of formation of water vapor, $-228.59 \text{ kJ mol}^{-1}$ (Appendix J) or $-228,590 \text{ J mol}^{-1}$. Therefore

$$-228,590 \text{ J} = (-2.303)(8.314 \text{ J K}^{-1})(298.15 \text{ K})(\log K)$$

$$\log K = \frac{-228,590 \text{ J}}{(-2.303)(8.314 \text{ J K}^{-1})(298.15 \text{ K})} = 40.042$$

$$K = 1.10 \times 10^{40}$$

Note that when reactants are gases, the equilibrium constant calculated from Equation (1) involves the pressures of the gases in atmospheres. Thus

$$K = \frac{P_{H_2O(g)}}{P_{H_2(g)} \times P_{O_2(g)}^{1/2}} = 1.10 \times 10^{40}$$

Now we see one reason why calculation of the value of $\Delta G°$ is useful. It allows us to predict the value of the equilibrium constant for a reaction and thus to determine whether or not a reaction will give significant amounts of products.

If we combine Equation (1) with the equation relating $\Delta G°$, $\Delta H°$, and $\Delta S°$, we have

$$\Delta G° = \Delta H° - T \Delta S° = -2.303 RT \log K \qquad (2)$$

As the temperature varies, $\Delta H°$ and $\Delta S°$ change only slightly, so if we assume that they are indeed independent of temperature, we can obtain a simple equation relating the equilibrium constant and temperature.

Consider an equilibrium reaction at two different temperatures, T_1 and T_2, with equilibrium constants K_{T_1} and K_{T_2}.

$$\Delta H° - T_1 \Delta S° = -2.303 RT_1 \log K_{T_1} \qquad (3)$$

and $\qquad \Delta H° - T_2 \Delta S° = -2.303 RT_2 \log K_{T_2} \qquad (4)$

Equations (3) and (4) may be rearranged to

$$\frac{\Delta H°}{T_1} - \Delta S° = -2.303 R \log K_{T_1} \qquad (5)$$

and $\qquad \dfrac{\Delta H°}{T_2} - \Delta S° = -2.303 R \log K_{T_2} \qquad (6)$

by dividing Equation (3) by T_1 and Equation (4) by T_2. Subtracting Equation (6) from Equation (5) gives

$$\frac{\Delta H°}{T_1} - \frac{\Delta H°}{T_2} - \Delta S° - (-\Delta S°) = -2.303 R \log K_{T_1} - (-2.303 R \log K_{T_2})$$

or $\qquad \Delta H° \left(\dfrac{1}{T_1} - \dfrac{1}{T_2} \right) = 2.303 R \log \dfrac{K_{T_2}}{K_{T_1}}$

or, by multiplying the left side by $T_1 T_2 / T_1 T_2$ and dividing both sides by $2.303 R$,

$$\frac{\Delta H°(T_2 - T_1)}{2.303 RT_1 T_2} = \log \frac{K_{T_2}}{K_{T_1}} \qquad (7)$$

Equations (1) and (7) prove to be quite useful. If we know $\Delta G°$ at 298 K, we can determine the equilibrium constant, K, at 298 K using Equation (1). Then, if we know $\Delta H°$ (which is essentially independent of temperature) at 298 K, we can obtain K for any other temperature by means of Equation (7) (within the limitation that both $\Delta H°$ and $\Delta S°$ are independent of T). The value of $\Delta G°$ at a temperature other than 298 K can be calculated if $\Delta G°$ and $\Delta H°$ are known at 298 K because the value of $\Delta S°$ at 298 K can be determined using the equation

$$\Delta G° = \Delta H° - T\Delta S° \qquad (T = 298 \text{ K})$$

Then, since both $\Delta H°$ and $\Delta S°$ are essentially independent of temperature, $\Delta G°$ at other temperatures can be calculated.

EXAMPLE 18.19   (a) For the reaction

$$\text{CuS}(s) + \text{H}_2(g) \longrightarrow \text{Cu}(s) + \text{H}_2\text{S}(g)$$

calculate $\Delta G°$ and $\Delta H°$ at 298.15 K and 1 atm.

$\Delta G° = \Delta G°_{f\text{Cu}(s)} + \Delta G°_{f\text{H}_2\text{S}(g)} - \Delta G°_{f\text{CuS}(s)} - \Delta G°_{f\text{H}_2(g)}$

$\Delta G° = 1 \text{ mol Cu}(s) \times (0 \text{ kJ mol}^{-1}) + 1 \text{ mol H}_2\text{S}(g) \times (-33.6 \text{ kJ mol}^{-1})$
$\qquad\qquad - 1 \text{ mol CuS}(s) \times (-53.6 \text{ kJ mol}^{-1}) - 1 \text{ mol H}_2(g) \times (0 \text{ kJ mol}^{-1})$

$\Delta G° = 20.0 \text{ kJ} \qquad$ (reaction is not spontaneous)

$\Delta H° = \Delta H°_{f\text{Cu}(s)} + \Delta H°_{f\text{H}_2\text{S}(g)} - \Delta H°_{f\text{CuS}(s)} - \Delta H°_{f\text{H}_2(g)}$

$\Delta H° = 1 \text{ mol Cu}(s) \times (0 \text{ kJ mol}^{-1}) + 1 \text{ mol H}_2\text{S}(g) \times (-20.6 \text{ kJ mol}^{-1})$
$\qquad\qquad - 1 \text{ mol CuS}(s) \times (-53.1 \text{ kJ mol}^{-1}) - 1 \text{ mol H}_2(g) \times (0 \text{ kJ mol}^{-1})$

$\Delta H° = 32.5 \text{ kJ} \qquad$ (reaction is endothermic)

(b) Calculate the value for the equilibrium constant, K, at 298.15 K and 1 atm.

$$K = \frac{P_{\text{H}_2\text{S}}}{P_{\text{H}_2}} \qquad \text{(where P is the partial pressure of a gas in atm)}$$

$$\Delta G° = -2.303RT \log K \qquad (R = 8.314 \text{ J K}^{-1})$$

Then
$$\log K = \frac{\Delta G°}{-2.303RT \log K}$$

$$= \frac{20.0 \times 10^3 \text{ J}}{-(2.303)(8.314 \text{ J K}^{-1})(298.15 \text{ K})}$$

$$= -3.5038$$

$$K = 3.13 \times 10^{-4} \qquad \text{(at 298.15 K)}$$

Note that at 298.15 K the equilibrium constant has a value less than 1, indicating that the elements in the reaction are present in larger quantities as reactants than as products and that the equilibrium therefore lies far to the left.

(c) Estimate the value for K at 798 K and 1 atm.

$$\frac{\Delta H°(T_2 - T_1)}{2.303RT_1T_2} = \log \frac{K_{T_2}}{K_{T_1}}$$

Then
$$\frac{32{,}500 \text{ J} (798 \text{ K} - 298 \text{ K})}{(2.303)(8.314 \text{ J K}^{-1})(798 \text{ K})(298 \text{ K})} = \log \frac{K_{798}}{K_{298}}$$

$$3.5689 = \log \frac{K_{798}}{K_{298}}$$

$$\log K_{798} - \log K_{298} = 3.5689$$
$$\log K_{298} = -3.5038 \qquad \text{[calculated in part (b)]}$$

Therefore
$$\log K_{798} = 3.5689 + (-3.5038) = 0.0651$$
$$K = 1.16 \qquad \text{(at 798 K)}$$

Note that the equilibrium constant at 798 K is greater than 1, indicating that the elements are present in greater quantities as products than as reactants and that the equilibrium has been displaced to the right.

(d) Calculate ΔS° at 298.15 K and 1 atm.

Although we could calculate ΔS° from absolute standard molar entropy values (Section 18.9), let us use a different method since we already have values for ΔG and ΔH.

$$\Delta G^\circ = \Delta H^\circ - T \Delta S^\circ$$

$$\Delta S^\circ = \frac{\Delta H^\circ - \Delta G^\circ}{T} = \frac{32{,}500 \text{ J} - 20{,}000 \text{ J}}{298.15 \text{ K}}$$

$$= 41.9 \text{ J K}^{-1}$$

Note that this value for ΔS° is positive and favors a spontaneous reaction, whereas the positive value of ΔH° is unfavorable for reaction. This is, therefore, a case where ΔS° is sufficiently large that ΔH° is not a valid indicator of spontaneity and ΔG° must be considered. In particular, a determination as to whether or not the reaction is spontaneous is expected to depend on the temperature.

We found in part (a) that $\Delta G^\circ = +20.0$ kJ at 298.15 K, indicating that at this temperature the reaction is not spontaneous. Let us now calculate ΔG° at a higher temperature assuming that ΔH° and ΔS° do not change significantly as the temperature increases.

(e) Estimate ΔG° at 798 K and 1 atm.

$$\Delta G^\circ = \Delta H^\circ - T \Delta S^\circ$$
$$= 32{,}500 - (798 \text{ K})(41.9 \text{ J K}^{-1})$$
$$= -936 \text{ J} = -0.936 \text{ kJ}$$

At the higher temperature ΔG° is negative, showing that at 798 K the reaction *is* spontaneous. ΔH°, being relatively independent of temperature, still has a value of about 32.5 kJ at 798 K and hence does not indicate the change to spontaneity.

(f) Estimate the temperature at which ΔG° is equal to zero at 1 atm, assuming that ΔH° and ΔS° do not change significantly as the temperature increases.

$$\Delta G^\circ = \Delta H^\circ - T \Delta S^\circ \qquad [\Delta S^\circ = 41.9 \text{ J K}^{-1}, \text{ as calculated in part (d)}]$$
$$0 = 32{,}500 \text{ J} - T(41.9 \text{ J K}^{-1})$$

$$T = \frac{32,500 \text{ J}}{41.9 \text{ J K}^{-1}} = 776 \text{ K}$$

Hence $\Delta G° = 0$ at 776 K. At temperatures below 776 K the values of $\Delta G°$ are positive; at temperatures above 776 K they are negative. Therefore, above 776 K the reaction is spontaneous; below 776 K the reaction as written is not spontaneous—indeed, it is spontaneous in the opposite direction.

EXAMPLE 18.20   (a) Determine the temperature at which liquid water and gaseous water are in equilibrium with each other at 1 atm.

Since the two states are in equilibrium, the free energy change, $\Delta G°$, in going from the liquid to the gas is zero. Assuming again that $\Delta H°$ and $\Delta S°$ are independent of temperature, for the reaction $H_2O(l) \rightarrow H_2O(g)$,

$$\begin{aligned}
\Delta H° &= \Delta H°_{fH_2O(g)} - \Delta H°_{fH_2O(l)} = 1 \text{ mol } H_2O(g) \times (-241.8 \text{ kJ mol}^{-1}) \\
&\quad - 1 \text{ mol } H_2O(l) \times (-285.8 \text{ kJ mol}^{-1}) \\
&= 44.0 \text{ kJ} = 44,000 \text{ J}
\end{aligned}$$

$$\begin{aligned}
\Delta S° &= S°_{H_2O(g)} - \Delta S°_{H_2O(l)} \\
&= 1 \text{ mol } H_2O(g) \times (188.71 \text{ J mol}^{-1} \text{ K}^{-1}) \\
&\quad - 1 \text{ mol } H_2O(l) \times (69.91 \text{ J mol}^{-1} \text{ K}^{-1}) \\
&= 118.80 \text{ J K}^{-1}
\end{aligned}$$

$$\Delta G° = \Delta H° - T\Delta S° = 0 \qquad (\Delta G° = 0 \text{ at equilibrium})$$

$$T = \frac{\Delta H°}{\Delta S°} = \frac{44,000 \text{ J}}{118.80 \text{ J K}^{-1}}$$

$$= 370 \text{ K} = 97°\text{C}$$

The correct answer, of course, is 373 K (100°C), the boiling point of water at 1 atm, but the calculation of the value 370 K was based on the assumptions that $\Delta H°$ and $\Delta S°$ are independent of temperature. This is a good place to reemphasize that these assumptions are only approximately correct. They are sufficiently close to be highly useful, but for really exact calculations values of $\Delta H°$ and $\Delta S°$ at the temperature in question must be used.

(b) Recalculate the temperature at which liquid water and gaseous water are in equilibrium with each other at 1 atm, this time using the values of $\Delta H°$ and $\Delta S°$ at the temperature 97°C.

$$\Delta H°_{H_2O} \text{ at } 97°\text{C} = 40,720 \text{ J mol}^{-1}; \ \Delta S°_{H_2O} \text{ at } 97°\text{C} = 109.1 \text{ J mol}^{-1} \text{ K}^{-1}$$

Note that the values of $\Delta H°$ (44,000 J) and of $\Delta S°$ (118.80 J K^{-1}) at 25°C are close enough, as stated earlier, to the values at 97°C for an approximate calculation but are not sufficiently close for an exact calculation.

Again using

$$\Delta G° = \Delta H° - T\Delta S° = 0 \qquad (\Delta G° = 0 \text{ at equilibrium})$$

$$T = \frac{\Delta H°_{370}}{\Delta S°_{370}} = \frac{40,720 \text{ J}}{109.1 \text{ J K}^{-1}} = 373.2 \text{ K}$$

or $373.2 - 273.2 = 100.0\,°C$

In this part of the example we obtained the experimental value of 100°C. If the calculation is repeated with the values of $\Delta H°$ and $\Delta S°$ at 100°C (40,656 J and 108.95 J K^{-1}), a value of $T = 100.0\,°C$ is again obtained, indicating that the values for $\Delta H°$ and $\Delta S°$ at 97°C are close enough to give the correct temperature of 100.0°C (to four significant figures).

At temperatures below 373 K the value of $T\Delta S°$ becomes smaller, and hence a smaller term is subtracted from $\Delta H°$ to give $\Delta G°$ ($\Delta G° = \Delta H° - T\Delta S°$). Thus $\Delta G°$ is positive (it equals zero at the equilibrium state at 373 K), indicating that *at temperatures below 373 K (100°C) at 1 atm of pressure the spontaneous change is from gaseous H$_2$O to liquid H$_2$O*. At temperatures higher than 373 K, $T\Delta S°$ becomes larger, and hence a larger term is subtracted from $\Delta H°$. Thus $\Delta G°$ is negative, indicating that *at temperatures above 373 K (100°C) at 1 atm of pressure the spontaneous change is from liquid H$_2$O to gaseous H$_2$O*.

Here the heat effect ($\Delta H°$) and the entropy effect ($\Delta S°$) work at cross purposes, since $\Delta S°$ is positive (maximum disorder, indicating favorable conditions for the change), and $\Delta H°$ is positive (endothermic, indicating unfavorable conditions for the change). Neither $\Delta H°$ nor $\Delta S°$ alone is sufficient to predict spontaneity. $\Delta G°$ (which combines $\Delta H°$, $\Delta S°$, and T) is the quantity on which to base such predictions. As shown above, $\Delta G°$ is positive at temperatures below 373 K (100°C) and negative at temperatures above 373 K.

18.15   The Third Law of Thermodynamics

In the preceding sections we have considered the changes in state functions that accompany chemical or physical changes; we have not tried to obtain values for the functions in any particular state. In fact, it is not possible to obtain absolute values for E, H, and G. However, it is possible to determine the **absolute entropy** of a pure substance at any given temperature. The reason for this is found in the **Third Law of Thermodynamics: The entropy of any pure, perfect crystalline substance at absolute zero (0 K) is equal to zero.** At absolute zero all molecular motion is at a minimum, and at that temperature a pure crystalline substance has no disorder (its entropy is zero). All molecular motion is also at a minimum in an impure substance at absolute zero, but the impurity can be distributed in different ways, giving rise to disorder (and a nonzero value for entropy).

If we take a pure crystalline substance at absolute zero (with an entropy of zero) and measure the entropy change as its temperature is increased to any temperature T, then we have measured the absolute entropy of the substance at the temperature T. Hence we can find *absolute entropies* of pure substances; in contrast, for free energy and enthalpy, we can find only differences between two values, not the absolute values. Absolute entropies allow us to compare the relative amounts of disorder present in different pure substances, and they can be used to determine entropy changes (Section 18.9). Caution must be exercised when using absolute entropies, however, since the absolute entropy of a pure elemental substance at standard state conditions will *not* be equal to zero. Table 18-1 and Appendix J contain absolute standard molar entropies, $S°_{298}$, of some common substances at standard state conditions.

FOR REVIEW

SUMMARY

Chemical thermodynamics is the study of the energy transformations and transfers that accompany chemical and physical changes. From such a study we can determine if a change will occur spontaneously and how far it will proceed—the equilibrium position of the change. The spontaneity of a chemical change is determined by the changes in potential energy and in disorder that accompany the change.

The **First Law of Thermodynamics** states that in any change that occurs in nature the total energy of the universe remains constant. Thus by measuring the energy lost or gained by the system we can determine the energy change within it. Energy can be lost from a system as the system does **work, w,** on the surroundings ($w > 0$) or as the surroundings do work on the system ($w < 0$). The amount of work due to expansion is given by the expression $P(V_2 - V_1)$, when the expansion is carried out at constant pressure. Energy can also be transferred as **heat, q.** In an **endothermic process** heat is transferred into the system from the surroundings ($q > 0$). In an **exothermic process** heat is transferred from the system to the surroundings ($q < 0$).

Although the amounts of heat and work accompanying a change may vary depending on how the change is carried out, the change in internal energy of the system is independent of how the change is accomplished. A property such as internal energy that does not depend on how the system gets from one state to another is called a **state function.** If a chemical change is carried out at constant pressure such that the only work done is expansion work, q is a state function and is called the **enthalpy change, ΔH,** of the reaction. The enthalpy change accompanying the formation of 1 mole of a substance from the elements in their most stable states, all at 298.15 K and 1 atm (a **standard state**), is called the **standard molar enthalpy of formation of the substance, ΔH°_{f298}.** Using **Hess's law** the enthalpy change of any reaction can be described as the sum of the standard molar enthalpies of all of the products minus those of all of the reactants. When using Hess's law, remember that ΔH for a reaction in one direction is equal to magnitude and opposite in sign to ΔH for the reaction in the opposite direction, that ΔH is directly proportional to the quantities of reactants, and that the standard molar enthalpy of formation of any element in its most stable state is zero. **Bond energies** can be obtained from enthalpies of formation.

A **spontaneous change,** a change in a system that will proceed without any outside influence, is favored when the change is exothermic or when the change leads to an increase in the randomness, or disorder, of the system. The randomness of a system can be determined quantitatively and is called the **entropy, S,** of the system. S is a state function, and values of S°_{298} for many substances in a standard state have been measured and tabulated. The entropy change ΔS°, for many reactions can be calculated as the sum of the **absolute standard molar entropies** of all products minus the sum of those of all reactants. A positive value of ΔS° ($\Delta S^{\circ} > 0$) indicates that the disorder of the system has increased, while a negative value of ΔS° ($\Delta S^{\circ} < 0$) indicates that the disorder of the system has decreased. The **Second Law of Thermodynamics** states that any spontaneous change that occurs in the universe must be accompanied by an increase in the entropy of the universe.

The **free energy change, ΔG,** of a reaction is the difference between the enthalpy change of the reaction and the product of the temperature and entropy change of the reaction.

$$\Delta G = \Delta H - T \Delta S$$

For a reaction at constant temperature and pressure, a negative value of ΔG ($\Delta G < 0$) indicates that the reaction is spontaneous. Values of $\Delta G°$ can be calculated from **standard molar free energies of formation, $\Delta G°_{f_{298}}$,** or from the values of $\Delta H°$ and $\Delta S°$ for the reaction.

The free energy change of a reaction that occurs in a nonstandard state can be determined using the equation

$$\Delta G = \Delta G° + 2.303 RT \log Q$$

where T is equal to 298.15 K, R is equal to $8.314\ J\ K^{-1}$, and Q is the reaction quotient. When a reaction has reached equilibrium, Q is equal to K, the equilibrium constant for the reaction, and ΔG is equal to zero. Thus at equilibrium

$$\Delta G° = -2.303 RT \log K$$

The **Third Law of Thermodynamics** states that the entropy of a pure crystalline substance at 0 K is equal to zero. Thus a measurement of the entropy change of such a substance as it is heated from 0 K to a higher temperature gives the absolute entropy of the substance at the higher temperature. The entropy change for a chemical reaction can be determined by subtracting the sum of the absolute molar entropies of all of the reactants from the sum of those of the products.

$$\Delta S° = \Sigma S°_{\text{products}} - \Sigma S°_{\text{reactants}}$$

KEY TERMS AND CONCEPTS

Absolute entropy (18.15)
Bond energy (18.7)
Chemical thermodynamics (18.1)
Endothermic process (18.5)
Enthalpy change (18.5)
Entropy (18.9)
Entropy change (18.9)
Equilibrium constant (18.14)
Exothermic process (18.5)
First Law of Thermodynamics (18.3)

Free energy change (18.10)
Heat (18.2)
Hess's law (18.6)
Internal energy (18.2)
Reaction quotient (18.13)
Second Law of Thermodynamics (18.12)
Spontaneous change (18.8)
Standard molar enthalpy of formation (18.5)
Standard molar entropy (18.9)
Standard state (18.5)

State function (18.4)
Surroundings (18.1)
System (18.1)
Thermochemistry (18.1)
Third Law of Thermodynamics (18.15)
Work (18.2)

EXERCISES

Heat, Work, and Internal Energy

1. State the First Law of Thermodynamics in two forms.
2. Calculate the missing value of ΔE, q, or w for a system given the following data:
   (a) $q = 570$ J; $w = 300$ J   *Ans.* $\Delta E = 270$ J
   Ⓢ(b) $\Delta E = -7500$ J; $w = -4500$ J
                              Ans. $q = -12{,}000$ J
   (c) $\Delta E = 250$ J; $q = 300$ J   *Ans.* $w = 50$ J
   Ⓢ(d) The system absorbs 1.000 kJ of heat energy and does 650 J of work on the surroundings.
                              Ans. 350 J
3. In which of the following changes at constant pressure is work done by the surroundings on the system? by the system on the surroundings? Is no work done in any of these changes? What is the value of w in each case: $w > 0$, $w < 0$, or $w = 0$?

Initial State	*Final State*
(a) $H_2O(g)$	$H_2O(l)$
(b) $H_2O(s)$	$H_2O(g)$
(c) $2Na(s) + Cl_2(g)$	$2NaCl(s)$
(d) $H_2(g) + Cl_2(g)$	$2HCl(g)$
(e) $3H_2(g) + N_2(g)$	$2NH_3(g)$
(f) $Na_2SO_4 \cdot 10H_2O(s)$	$Na_2SO_4(s) + 10H_2O(g)$
(g) $NO_2(g) + CO(g)$	$NO(g) + CO_2(g)$

   Ⓢ4. Calculate the work involved in compressing a system consisting of exactly 1 mol of H_2O as it changes from a gas at 373 K (volume $= 30.6$ L) to a liquid at 373 K (volume $= 18.9$ mL) under a constant pressure of 1 atm. Does this work increase or decrease the internal energy of the system?   *Ans. 3.10 J; increase*
5. A gas in expanding against a constant pressure of 0.50 atm from 10.0 L to 16.0 L absorbs 125 J of heat. What is the change in internal energy of the gas?   *Ans. -180 J*

Enthalpy Changes; Hess's Law

6. What is the distinction between ΔE and ΔH for a system undergoing a change at constant pressure?
7. The enthalpy, H, of a system has been referred to as the heat content of the system. Why?
8. What is the difference between ΔH and ΔH_{298}° for a reaction?
Ⓢ9. Does $\Delta H_{f_{H_2O(g)}}^\circ$ differ from ΔH_{298}° for the reaction $2H_2(g) + O_2(g) \rightarrow 2H_2O(g)$? If so, how?
10. For the conversion of graphite to diamond, $C(s,\ graphite) \rightarrow C(s,\ diamond)$, $\Delta H = 1.90$ kJ mol^{-1}. Do the heats of combustion of graphite and carbon differ; that is, are the enthalpy changes for the following reactions the same or different?

    $$C(s,\ graphite) + O_2(g) \longrightarrow CO_2(g)$$
    $$C(s,\ diamond) + O_2(g) \longrightarrow CO_2(g)$$

11. (a) Using the data in Appendix J, calculate the standard enthalpy change, ΔH°, for each of the following reactions:

    Ⓢ(1) $Fe_2O_3(s) + 13CO(g) \longrightarrow$
    $$2Fe(CO)_5(g) + 3CO_2(g)$$
    Ans. -387.6 kJ

    (2) $2LiOH(s) + CO_2(g) \longrightarrow$
    $$Li_2CO_3(s) + H_2O(g)$$
    Ans. -89.4 kJ

    Ⓢ(3) $CH_4(g) + N_2(g) \longrightarrow$
    $$HCN(g) + NH_3(g)$$
    Ans. 164 kJ

    (4) $CS_2(g) + 3Cl_2(g) \longrightarrow$
    $$CCl_4(g) + S_2Cl_2(g)$$
    Ans. -238 kJ

    (5) $N_2(g) + O_2(g) \longrightarrow 2NO(g)$
    Ans. 180.5 kJ

    (b) Which of these reactions are exothermic?
    Ans. (1), (2), and (4)

12. The decomposition of hydrogen peroxide, H_2O_2, has been used to provide thrust in the control jets of various space vehicles. How much heat is produced by the decomposition of exactly 1 mol of H_2O_2 under standard conditions?

    $$2H_2O_2(l) \longrightarrow 2H_2O(g) + O_2(g)$$

    Ans. -54.0 kJ

13. In 1774 oxygen was prepared by Joseph Priestley by heating red mercury(II) oxide with the light from the sun focused through a lens (Chapter 9). How much heat is required to decompose 1 mol of $HgO(s)$ to $Hg(l)$ and $O_2(g)$ under standard conditions?   *Ans. 90.83 kJ*
Ⓢ14. How many kilojoules of heat energy will be liberated when 49.70 g of manganese are burned to

form $Mn_3O_4(s)$ at standard state conditions? ΔH°_{f298} of Mn_3O_4 is equal to -1388 kJ mol^{-1}. *Ans. 418.5 kJ*

[S] 15. The heat of formation of $OsO_4(s)$, $\Delta H^\circ_{f_{OsO_4(s)}}$, is -391 kJ mol^{-1} at 298 K and the heat of sublimation is 56.4 kJ mol^{-1}. What is ΔH°_{298} for the process $Os(s) + 2O_2(g) \rightarrow OsO_4(g)$ under standard state conditions? *Ans. −335 kJ*

16. The oxidation of the sugar glucose, $C_6H_{12}O_6$, is described by the following equation:

$$C_6H_{12}O_6(s) + 6O_2(g) \longrightarrow 6CO_2(g) + 6H_2O(l) \qquad \Delta H^\circ = -2816 \text{ kJ}$$

Metabolism of glucose gives the same products, although the glucose reacts with oxygen in a series of steps in the body. (a) How much heat in kilojoules is produced by the metabolism of 1.0 g of glucose? (b) How many nutritional Calories (1 cal = 4.184 J; 1 nutritional Cal = 1000 cal) are produced by the metabolism of 1.0 g of glucose? *Ans. 16 kJ; 3.7 Cal*

17. The white pigment TiO_2 is prepared by the hydrolysis of titanium tetrachloride, $TiCl_4$, in the gas phase.

$$TiCl_4(g) + 2H_2O(g) \longrightarrow TiO_2(s) + 4HCl(g)$$

How much heat is evolved in the production of exactly 1 mol of $TiO_2(s)$ under standard state conditions of 25 °C and 1 atm? *Ans. 67.1 kJ*

[S] 18. Calculate the standard molar enthalpy of formation of NO(*g*) from the following data:

$$N_2(g) + 2O_2(g) \longrightarrow 2NO_2(g) \quad \Delta H^\circ = 66.4 \text{ kJ}$$
$$2NO(g) + O_2(g) \longrightarrow 2NO_2(g) \quad \Delta H^\circ = -114.1 \text{ kJ}$$

Ans. 90.3 kJ mol^{-1}

19. A sample of $WO_2(s)$ with a mass of 0.9745 g was "burned" in oxygen at constant pressure giving $WO_3(s)$ and 1.143 kJ of heat. The enthalpy of formation of $WO_3(s)$ under these conditions is -842.91 kJ mol^{-1}. Calculate the enthalpy of formation of $WO_2(s)$. *Ans. −589.7 kJ mol^{-1}*

[S] 20. The heat of combustion of a hydrocarbon (a compound of carbon and hydrogen) is the standard state enthalpy change for the reaction of the compound with oxygen to give $CO_2(g)$ and $H_2O(l)$. Determine the heats of combustion of (a) octane (C_8H_{18}, $\Delta H^\circ_f = -208.4$ kJ mol^{-1}), a component of gasoline and (b) methane ($CH_4(g)$, $\Delta H^\circ_f = -74.81$ kJ mol^{-1}), the major component of natural gas. Which has the higher heat content per gram? *Ans. −5512 kJ; −890.3 kJ; CH_4*

[S] 21. Using the data in Appendix J, calculate the bond energies of F_2, Cl_2, and FCl. All are gases in their most stable form at standard state conditions.
Ans. F—F = 158.0 kJ per mole of bonds; Cl—Cl = 243.36 kJ per mole of bonds; F—Cl = 255.15 kJ per mole of bonds

22. Using the data in Appendix J, calculate the bond energies of N_2, O_2, and NO. All are gases in their most stable form at standard state conditions.
Ans. N≡N = 945.408 kJ per mole of bonds; O═O = 498.34 kJ per mole of bonds; N═O = 631.6 kJ per mole of bonds

23. Using the data in Appendix J, calculate the Ti—Cl single bond energy in $TiCl_4$. *Ans. 430.0 kJ*

[S] 24. (a) Using the bond energies given in Table 18-2 in Section 18.7, determine the approximate enthalpy change for the formation of ethylene from ethane, which is described in Section 15.14.

$$C_2H_6(g) \longrightarrow C_2H_4(g) + H_2(g)$$

(b) Compare this with the standard state enthalpy change. *Ans. 128 kJ; 137.0 kJ*

25. The enthalpy of formation of $AsF_5(g)$ has been determined to be -16.46 kJ g^{-1} of arsenic using the reaction $2As(s) + 5F_2(g) \rightarrow 2AsF_5(g)$. Using this information and the data in Appendix J, calculate the As—F single bond energy in AsF_5. *Ans. 386 kJ*

26. (a) Water gas, a mixture of H_2 and CO, is an important industrial fuel produced by the reaction of steam with red-hot coke (essentially pure carbon).

$$C(s) + H_2O(g) \longrightarrow CO(g) + H_2(g)$$

Assuming that coke has the same enthalpy of formation as graphite, calculate ΔH°_{298} for this reaction. *Ans. 131.30 kJ*

(b) Methanol, a liquid fuel that could possibly replace gasoline, can be prepared from water gas and additional hydrogen at high temperatures and pressures in the presence of a suitable catalyst.

$$2H_2(g) + CO(g) \longrightarrow CH_3OH(g)$$

Under the conditions of the reaction, methanol forms as a gas. Calculate ΔH°_{298} for this reaction and for the condensation of gaseous methanol to liquid methanol.

Ans. $-90.2\ kJ\ mol^{-1}$; $-38.0\ kJ\ mol^{-1}$

(c) Calculate the heat of combustion of 1 mol of liquid methanol to $H_2O(g)$ and $CO_2(g)$.

Ans. $-638.4\ kJ$

Entropy Changes

27. What is the connection between entropy and disorder?
28. What is the absolute standard molar entropy of a pure substance?
29. State the Second Law of Thermodynamics in terms of entropy changes for the universe. State the Second Law of Thermodynamics in terms of free energy changes.
30. State the Third Law of Thermodynamics.
31. Give the sign of the entropy change for each of the following systems going from the initial state to the final state. Explain your answers.

Initial State	Final State
(a) egg in shell	scrambled egg
(b) chicken feed and baby chick	full grown chicken
(c) $NaCl(s)$ at 298 K	$NaCl(s)$ at 10 K
(d) $H_2O(g)$ at 273 K and 1 atm	$H_2O(s)$ at 273 K and 1 atm
(e) 1 mol Sn and 1 mol O_2	1 mol SnO_2
(f) 1 mol $CaCO_3$	1 mol CaO and 1 mol CO_2

S 32. (a) Using the data in Appendix J, calculate the standard entropy change for each reaction in part (a) of Exercise 11.

Ans. (1) -1124.5; (2) -34.9; (3) 16.4; (4) -265.5; (5) $24.8\ J\ K^{-1}$

(b) For which of the reactions in Exercise 11 are the entropy changes favorable for the reaction to proceed spontaneously?

Ans. (3) and (5)

S 33. What is the entropy change accompanying the evaporation of 1 mol of chloroform, $CHCl_3(l) \rightarrow CHCl_3(g)$, under standard state conditions?

Ans. $94\ J\ K^{-1}$

34. What is the entropy change accompanying the reaction $N_2(g) + 3H_2(g) \rightarrow 2NH_3(g)$ under standard state conditions?

Ans. $-198.6\ J\ K^{-1}$

Free Energy Changes

35. In dealing with free energy, why is the emphasis on the change in free energy as a system changes rather than on the values of the free energies of the initial and final states?
36. What is the distinction between ΔH and ΔG for a system undergoing a change at constant temperature and pressure?
37. What is meant by the term *spontaneous reaction?*
38. Explain why the free energy change of a reaction varies with temperature.
39. Under what conditions is ΔG equal to $\Delta G°$ for the reaction $H_2(g) + O_2(g) \rightarrow H_2O(l)$?
40. For the following reactions

$$H_2(g) + Cl_2(g) \longrightarrow 2HCl(g) \qquad (1)$$
$$2HCl(g) \longrightarrow H_2(g) + Cl_2(g) \qquad (2)$$

show that $\Delta H°$, $\Delta S°$, and $\Delta G°$ for the reaction of Equation (1) are equal to the negatives of $\Delta H°$, $\Delta S°$, and $\Delta G°$, respectively, for the reaction of Equation (2) using the data in Table 18-1.

S 41. (a) Using the data in Appendix J, calculate the standard free energy changes for the reactions given in Exercise 11.

Ans. (1) -52.4; (2) -78.8; (3) 159; (4) -159; (5) $173.1\ kJ$

(b) Which of those reactions are spontaneous? Why?

Ans. (1), (2), and (4)

42. The standard enthalpies of formation of $NO(g)$, $NO_2(g)$, and $N_2O_3(g)$ are 90.25 kJ mol^{-1}, 33.2 kJ mol^{-1}, and 83.72 kJ mol^{-1}, respectively. Their standard entropies are 210.65 J mol^{-1} K^{-1}, 239.9 J mol^{-1} K^{-1}, and 312.2 J mol^{-1} K^{-1}, respectively.

S (a) Use the data above to calculate the free energy change for the following reaction at 25.0°C.

$$N_2O_3(g) \longrightarrow NO(g) + NO_2(g)$$

Ans. $-1.6\ kJ$

(b) Repeat the above calculation for 0.00°C and 100.0°C assuming that the enthalpy and entropy changes do not vary with a change in temperature. Is the reaction spontaneous at 0.00°C? at 100.0°C?

Ans. 1.9 kJ; -11.9 kJ.
The reaction is spontaneous at 100.0°C, but not at 0.00°C.

43. For a certain process at 300 K, $\Delta G = -77.0$ kJ

and $\Delta H = -56.9$ kJ. Find the entropy change for this process at this temperature.
Ans. $\Delta S = 67.0\ J\ K^{-1}$

44. Hydrogen chloride, $HCl(g)$, and ammonia, $NH_3(g)$, escape from bottles of their solutions and react to form the white glaze often seen on glass in chemistry laboratories.

$$HCl(g) + NH_3(g) \longrightarrow NH_4Cl(s)$$

(a) Calculate the free energy change, $\Delta G°$, for this reaction.   *Ans. $-89.7\ kJ$*

(b) At what temperature will $\Delta G°$ for the reaction be equal to zero?
Ans. $618.6\ K\ (345.4°C)$

Free Energy Changes and Equilibrium Constants

45. Explain why equilibrium constants change with temperature.

46. No matter what their bond energy, all compounds will decompose if heated to a sufficiently high temperature. Why will the reaction $AB \rightarrow A + B$, where A and B represent atoms, eventually become spontaneous with $K > 1$ as the temperature of the system is increased?

47. Calculate the equilibrium constant for the decomposition of solid NH_4Cl to $HCl(g)$ and $NH_3(g)$. $\Delta G°$ for this reaction is 89.7 kJ (Exercise 44).   *Ans. 1.94×10^{-16}*

S 48. Consider the reaction

$$I_2(g) + Cl_2(g) \longrightarrow 2ICl(g)$$

(a) For this reaction $\Delta H° = -26.9$ kJ and $\Delta S° = 11.3$ J K^{-1}. Calculate $\Delta G°$ for the reaction.   *Ans. $-30.3\ kJ$*

(b) Calculate the equilibrium constant for this reaction at 25.0°C.   *Ans. 2.03×10^5*

49. If the entropy of vaporization of H_2O is equal to 109 J mol^{-1} K^{-1} and the enthalpy of vaporization is 40.62 kJ mol^{-1}, calculate the normal boiling temperature of water in °C.   *Ans. $100°C$*

50. Will the conversion of graphite to diamond become spontaneous at any temperature?
Ans. No. $\Delta H > 0$, $\Delta S < 0$.

51. For the decomposition of $CaCO_3(s)$ into $CaO(s)$ and $CO_2(g)$ at 1 atm,

S (a) estimate the minimum temperature at which you would conduct the reaction.
Ans. $1109\ K\ (836°C)$

(b) calculate the equilibrium vapor pressure of

$CO_2(g)$ above $CaCO_3(s)$ in a closed container at 298 K.   *Ans. $P_{CO_2} = 1.50 \times 10^{-23}$ atm*

S 52. If the enthalpy of vaporization of CH_2Cl_2 is 29.0 kJ mol^{-1} at 25.0°C and the entropy of vaporization is 92.5 J mol^{-1} K^{-1}, calculate a value for the normal boiling temperature of CH_2Cl_2.
Ans. $313.5\ K\ (40.4°C)$

53. The equilibrium constant, K_p, for the reaction $N_2O_4(g) \rightleftharpoons 2NO_2(g)$ is 0.142 at 298 K. What is $\Delta G°$ for the reaction?   *Ans. $4.84\ kJ$*

54. Calculate $\Delta G°$ at 298 K for the reaction of 1 mol of $H^+(aq)$ with 1 mol of $OH^-(aq)$ using the equilibrium constant for the self-ionization of water at 298 K.

$$H_2O \rightleftharpoons H^+(aq) + OH^-(aq)$$
$$K_w = 1.00 \times 10^{-14}$$

Ans. $-79.9\ kJ$

55. Consider the decomposition of dinitrogen trioxide described in Exercise 42.

$$N_2O_3(g) \longrightarrow NO(g) + NO_2(g)$$

At what temperature does this reaction become spontaneous?   *Ans. $287.2\ K\ (14.0°C)$*

56. The pollutant gas hydrogen sulfide is removed from natural gas by the reaction

$$2H_2S(g) + SO_2(g) \longrightarrow 3S(s) + 2H_2O(g)$$

What is the equilibrium constant for this reaction? Is the reaction endothermic or exothermic?
Ans. 5.35×10^{15}; exothermic

Additional Exercises

57. The structure of solid NaCl and a solution of NaCl are illustrated in Fig. 8-1. (a) What is the sign of ΔS for the reaction $NaCl(s) \rightarrow NaCl(aq)$? (b) Sodium chloride spontaneously dissolves in water. What is the sign of ΔG for this reaction? (c) The heat of solution of $NaCl(s)$ is 3.88 kJ mol^{-1}. Is this consistent with a spontaneous reaction? (d) What is the driving force for the reaction?

58. For the vaporization of bromine liquid to bromine gas,

S (a) calculate the change in enthalpy and the change in entropy at standard state conditions.   *Ans. $\Delta H° = 30.91\ kJ\ mol^{-1}$; $\Delta S° = 93.12\ J\ mol^{-1}\ K^{-1}$*

(b) discuss relative disorder in bromine liquid

compared to bromine gas. State what you can about the spontaneity of the vaporization.

[S] (c) estimate the value of $\Delta G°$ for the vaporization of bromine from the data in Appendix J.

Ans. $3.14\ kJ\ mol^{-1}$

[S] (d) state what you can about the spontaneity of the process from the value you obtained for $\Delta G°$ in part (c).

[S] (e) estimate the temperature at which liquid and gaseous Br_2 are in equilibrium with each other at 1 atm (assume $\Delta H°$ and $\Delta S°$ are independent of temperature).

Ans. 331.9 K, or 58.7°C

(f) state in which direction the process would be spontaneous at 298 K and at 398 K, using the temperature value obtained in part (e).

(g) compare $\Delta H°$, $\Delta S°$, and $\Delta G°$ in terms of their usefulness in predicting the spontaneity of the vaporization of bromine.

59. Ethanol, C_2H_5OH, is used as a fuel for motor vehicles, particularly in Brazil. (a) Write the balanced equation for the combustion of ethanol to $CO_2(g)$ and $H_2O(g)$, and, using the data in Appendix J, calculate the heat of combustion of 1 mol of ethanol. (b) The density of ethanol is 0.7893 g mL^{-1}. Calculate the heat of combustion of exactly 1 mL of ethanol. (c) Assuming that the mileage an automobile gets is directly proportional to the heat of combustion of the fuel, calculate how many times farther an automobile could be expected to go on 1 gal of gasoline than on 1 gal of ethanol. Assume that gasoline has the heat of combustion and the density of *n*-octane, C_8H_{18}, ($\Delta H_f° = -208.4$ kJ mol^{-1}; density = 0.7025 g mL^{-1}).

Ans. (a) $C_2H_5OH(l) + 3O_2(g) \rightarrow 2CO_2(g) + 3H_2O(g)$; -1234.8 kJ; (b) -21.16 kJ; (c) 1.487 times farther

60. Which of the following represents the molecular structure of hydroxylamine?

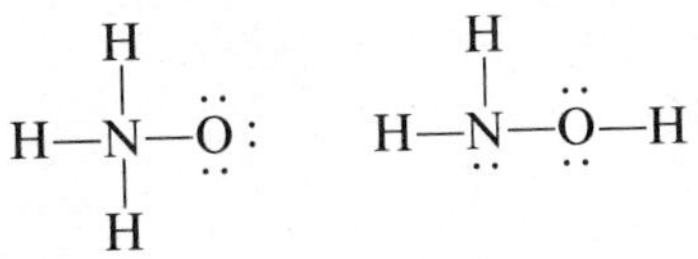

(a) (b)

[*Hint:* Using bond energies, determine if the conversion of structure (a) to structure (b) is endothermic or exothermic.] *Ans. (b)*

61. Carbon tetrachloride, an important industrial solvent, is prepared by the chlorination of methane, $CH_4(g) + 4Cl_2(g) \rightarrow CCl_4(g) + 4HCl(g)$, at 850 K. (a) What is the equilibrium constant for this reaction at 850 K? (b) Will the reaction vessel need to be heated or cooled to keep the temperature of the reaction constant?

Ans. $K_p = 2.05 \times 10^{23}$; cooled, $\Delta H = -397.3$ kJ

[S] 62. Acetic acid, CH_3CO_2H, can form a dimer, $(CH_3CO_2H)_2$, in the gas phase.

$$2CH_3CO_2H(g) \rightleftharpoons (CH_3CO_2H)_2(g)$$

The dimer is held together by two hydrogen bonds

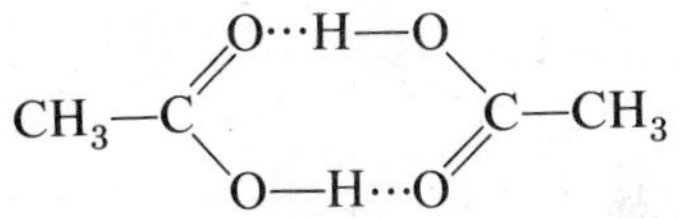

with a total strength of 66.5 kJ per mole of dimer. At 25°C the equilibrium constant for the dimerization is 1.3×10^3 (pressure in atmospheres). What is $\Delta S°$ for the reaction at 25°C?

Ans. $-163\ J\ K^{-1}$

[S] 63. At 1000 K the equilibrium constant for the reaction $Br_2(g) \rightleftharpoons 2Br$ is 2.8×10^4 (pressure in atmospheres). What is $\Delta G°$ for the reaction? Assume that the bond energy of Br_2 does not change between 298 K and 1000 K and calculate $\Delta S°$ for the reaction at 1000 K.

Ans. $\Delta G° = -85$ kJ; $\Delta S° = 278\ J\ K^{-1}$

64. Nitric acid, HNO_3, can be prepared by the following sequence of reactions (Section 8.7):

$$4NH_3(g) + 5O_2(g) \longrightarrow 4NO(g) + 6H_2O(g)$$
$$2NO(g) + O_2(g) \longrightarrow 2NO_2(g)$$
$$3NO_2(g) + H_2O(l) \longrightarrow 2HNO_3(l) + NO(g)$$

How much heat is evolved when 1 mol of $NH_3(g)$ is converted to $HNO_3(l)$? Assume all reactants and products are in their standard states at 25°C and 1 atm. *Ans.* 307.3 kJ

19

THE HALOGENS AND
THEIR COMPOUNDS

The elements of Group VIIA of the Periodic Table are known as the **halogens,** which means *salt formers.* Their binary compounds are called **halides.** The first four halogens—fluorine, chlorine, bromine, and iodine—were all isolated between 1774 and 1886. Salts of these elements are common in nature. Astatine, the fifth halogen, was first prepared artificially in 1940. It is a radioactive element with a limited stability and will not be considered extensively in this chapter.

19.1 Chemical Properties of the Halogens

As can be inferred from their location at the far right-hand side of the Periodic Table, the properties of the halogens reflect their pronounced nonmetallic character, which decreases as the atomic number increases. The valence shell of a halogen atom has the electron configuration ns^2np^5. Thus a halogen atom forms only one single covalent bond with less electronegative nonmetals, although it can form several bonds to more electronegative nonmetals. In a compound with a metal, a halogen atom generally accepts one electron to form a stable univalent negative ion. Like other nonmetals, the free halogens are oxidizing agents (Section 8.2). Fluorine is the strongest oxidizing agent of all of the known elements, and the halogens become progressively weaker oxidizing agents as their nonmetallic character decreases down the group.

Halogens form diatomic molecules in which the atoms are bonded together by a single covalent bond, as indicated in the Lewis structures

$$:\ddot{X}:\ddot{X}: \qquad :\ddot{X}-\ddot{X}:$$

where X represents any one of the halogen atoms. The molecular orbital electron configuration for F_2 (Section 6.11) is

$$KK(\sigma_{2s})^2(\sigma_{2s}{}^*)^2(\sigma_{2p_x})^2(\pi_{2p_y}, \pi_{2p_z})^4(\pi_{2p_y}{}^*, \pi_{2p_z}{}^*)^4$$

A fluorine molecule has eight of its valence electrons in bonding molecular orbitals and six in antibonding molecular orbitals. The valence electrons occupy shells for which n is equal to 3, 4, 5, and 6 for chlorine, bromine, iodine, and astatine, respectively, and these molecules have molecular orbital electron configurations analogous to that for fluorine. Consequently, both the molecular orbital model and the Lewis structure predict an X—X bond order of 1 in each halogen molecule.

All of the halogens except fluorine form some compounds in which they exhibit a positive oxidation number. With the exception that fluorine does not exhibit a positive oxidation number, the chemical properties of the halogens differ in degree rather than in kind. For example, each reacts with hydrogen according to the equation

$$H_2 + X_2 \longrightarrow 2HX$$

The enthalpy of formation, $\Delta H_f°$, of HX (Table 19-5) changes in the order HF < HCl < HBr < HI, indicating that HF is formed by the most exothermic reaction and that the reactions become progressively less exothermic—in fact, the last reaction (with iodine) is endothermic. The high ionization energies make the formation of positive ions highly unlikely, except possibly for iodine. On the other hand, the halogens, except for fluorine, do exhibit positive oxidation numbers in compounds with more electronegative elements. Since fluorine is the most electronegative element, it does not form compounds in which it has a positive oxidation number. Oxidation numbers of $+1$, $+3$, $+5$, and $+7$ (and sometimes other values) are shown when the halogens share electrons with oxygen in their oxides, oxyacids, and oxyacid salts, or with more electronegative halogens in interhalogen compounds (Section 19.6).

The halogens oxidize a variety of metals, nonmetals, and ions, giving ionic or covalent products in which the halogen has an oxidation number of -1. An elemental halogen will oxidize those halide ions that are formed from less electronegative halogens. Thus fluorine oxidizes chloride, bromide, and iodide ions, while chlorine oxidizes only bromide and iodide ions.

THE ELEMENTAL HALOGENS

19.2 Occurrence and Preparation of the Elemental Halogens

The halogens are too reactive to occur in a free state in nature, but their compounds are widely distributed. Chlorides are the most abundant, and although fluorides, bromides, and iodides are less common, they are reasonably available. The principal natural occurrences of the halogens are given in Table 19-1.

Table 19-1   Natural Occurrences of the Halogens

Fluorine	Bromine
Fluorite, or fluorspar, CaF_2	Sea water ($8 \times 10^{-4}\ M$)
Fluoroapatite, $Ca_5(PO_4)_3F$	Underground brines
Cryolite, Na_3AlF_6	Salt deposits
Sea water ($7 \times 10^{-5}\ M$)	Dead Sea
Teeth, bone, blood (small amounts)	

Chlorine	Iodine
Sea water ($0.54\ M$)	Sea water ($4 \times 10^{-7}\ M$)
Salt beds ($NaCl$, $MgCl_2$, $CaCl_2$)	Sea plants (kelp)
Great Salt Lake ($4.5\ M$)	Underground brines
Gastric juice (0.05–$0.1\ M$ HCl)	Human thyroid gland

The best sources of the halogens (except iodine) are salts in which they exist as halide anions with an oxidation number of -1. Halide ions may be oxidized to free diatomic halogen molecules (oxidation number of zero) by various methods that depend on the ease with which they are oxidized. The ease of oxidation increases with increasing atomic size in the order $F^- < Cl^- < Br^- < I^-$. The valence electron removed during oxidation is closest to the nucleus in F^- and farthest from it in I^-. The elemental halogens are prepared by the following techniques:

1. ELECTROLYTIC OXIDATION.   Nonspontaneous reactions can be forced to occur by the input of electrical energy in a process called **electrolysis.** Electrolysis is often used to carry out oxidation-reduction reactions with species that are oxidized or reduced with difficulty. Fluorine is the most powerful oxidizing agent of the known elements. It spontaneously oxidizes most other elements, forming fluorides. The reverse reaction, the oxidation of fluorides to elemental fluorine, cannot be accomplished chemically—electrolytic oxidation is necessary. Since fluorine oxidizes water, the oxidation is commonly carried out in a mixture of three parts potassium hydrogen fluoride, KHF_2, with two parts anhydrous hydrogen fluoride, HF. When electrolysis of the fused mixture (m.p. $= 72°C$) begins, HF is decomposed to form fluorine gas at the anode and hydrogen at the cathode.

$$2HF + \text{electrical energy} \longrightarrow H_2(g) + F_2(g)$$

The two gases are kept apart to prevent their recombination to form hydrogen fluoride. They are drawn from the cell continuously, and more anhydrous hydrogen fluoride is added to regenerate the fused mixture. After purification the fluorine gas is compressed in special steel cylinders at 27 atmospheres of pressure.

Chlorides can be chemically oxidized to chlorine, but the necessary reagents are expensive. Consequently, chlorine is commercially produced by electrolysis of the chloride ion in aqueous solutions of sodium chloride (Fig. 19-1). The overall equation for this reaction is

$$[2Na^+(aq)] + 2Cl^-(aq) + 2H_2O(l) \longrightarrow$$
$$H_2(g) + Cl_2(g) + [2Na^+(aq)] + 2OH^-(aq)$$

Figure 19-1. An electrolytic cell for the preparation of chlorine gas from a solution of sodium chloride.

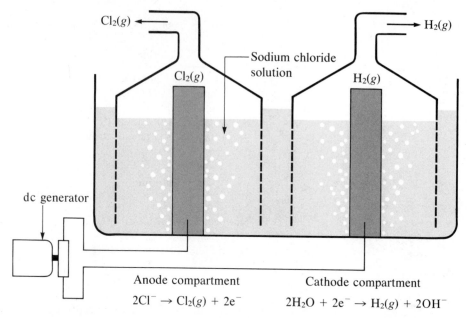

Anode compartment

$$2Cl^- \rightarrow Cl_2(g) + 2e^-$$

Cathode compartment

$$2H_2O + 2e^- \rightarrow H_2(g) + 2OH^-$$

The reactions at the electrodes in the electrolytic cell are given by the equations

At the anode (oxidation): $2Cl^- \longrightarrow Cl_2(g) + 2e^-$

At the cathode (reduction): $2H_2O + 2e^- \longrightarrow 2OH^- + H_2(g)$

Net reaction: $2Cl^- + 2H_2O \longrightarrow Cl_2(g) + 2OH^- + H_2(g)$

Sodium hydroxide and hydrogen, which are valuable by-products of the process, must be kept separate from the chlorine to prevent unwanted secondary reactions.

Chlorine is a by-product when metals such as sodium, calcium, and magnesium are produced by the electrolytic decomposition of their molten chlorides. For example, when molten magnesium chloride is electrolyzed, magnesium and chlorine are formed.

$$MgCl_2(l) + \text{electrical energy} \longrightarrow Mg(s) + Cl_2(g)$$

2. CHEMICAL OXIDATION.   Chloride, bromide, and iodide ions are easier to oxidize than fluoride ions. Thus, chlorine, bromine, and iodine can be prepared by the chemical oxidation of the respective halides.

Fluorine is a strong enough oxidizing agent to oxidize the chloride ion to chlorine, the bromide ion to bromide, and the iodide ion to iodine. However, it is rarely used for these oxidations because it is so reactive that it is difficult to handle. Small quantities of chlorine can be prepared in the laboratory by oxidation of the chloride ion in acid solution with the strong oxidizing agent manganese dioxide, MnO_2, potassium permanganate $KMnO_4$, or sodium dichromate, $Na_2Cr_2O_7$.

$$MnO_2 + 2Cl^- + 4H^+ \longrightarrow Mn^{2+} + Cl_2(g) + 2H_2O$$
$$2MnO_4^- + 10Cl^- + 16H^+ \longrightarrow 2Mn^{2+} + 5Cl_2(g) + 8H_2O$$
$$Cr_2O_7^{2-} + 6Cl^- + 14H^+ \longrightarrow 2Cr^{3+} + 3Cl_2(g) + 7H_2O$$

Sodium chloride and sulfuric acid are usually used as the sources of chloride ion and hydrogen ion.

The methods for the small, laboratory-scale oxidation of bromides to bromine are like those used for the oxidation of chlorides. Bromine is prepared commercially by the oxidation of bromide ion by chlorine.

$$2Br^-(aq) + Cl_2(g) \longrightarrow Br_2(l) + 2Cl^-(aq)$$

Chlorine is a stronger oxidizing agent than bromine, and the equilibrium for this reaction lies well to the right. Essentially all bromine in the United States is produced by chlorine oxidation of bromide ions obtained from Arkansas brines, drawn from as far underground as two miles. These brines contain 3800–5000 parts per million of bromide. Bromine is obtained as a mixture with chlorine and water after treating heated brine with chlorine. The mixture is decanted to separate the crude bromine from the aqueous phase. The bromine is purified by fractional distillation (Section 13.21). About 95% of the bromine in the brine is normally recovered.

Elemental iodine is produced by the oxidation of iodide with chlorine. The reaction is similar to the one used to produce bromine.

$$2I^-(aq) + Cl_2(g) \longrightarrow I_2(s) + 2Cl^-(aq)$$

Most of the iodine appears as a solid, which is separated from the solution by filtration. An excess of chlorine must be avoided; it causes unwanted secondary reactions that form iodine monochloride, ICl, and iodic acid, HIO_3. Considerable iodine is obtained from the iodides concentrated in kelp and other sea plants and from oil field brines.

3. CHEMICAL REDUCTION. Iodine exhibits its +5 oxidation number in sodium iodate, $NaIO_3$, an impurity in deposits of Chile saltpeter, $NaNO_3$. The iodine is liberated by reducing the iodate with sodium hydrogen sulfite.

$$2IO_3^- + 5HSO_3^- \longrightarrow 3HSO_4^- + 2SO_4^{2-} + H_2O + I_2(s)$$

19.3 Physical Properties of the Halogens

The physical properties of the halogens (Table 19-2) reflect the nonpolar nature of their molecules and the relatively weak forces between them (Section 11.13).

Table 19-2   Physical Properties of the Halogens[a]

	Fluorine	Chlorine	Bromine	Iodine	Astatine
Atomic number	9	17	35	53	85
Atomic weight	18.998	35.453	79.904	126.904	(210)[b]
Valence shell	$2s^22p^5$	$3s^23p^5$	$4s^24p^5$	$5s^25p^5$	$6s^26p^5$
Radius of X^-, Å	1.36	1.81	1.95	2.16	. . .
Covalent bond radius, Å	0.64	0.99	1.14	1.33	1.40
Physical state at 25°C	Gas	Gas	Liquid	Solid	Solid
Color	Pale yellow	Greenish-yellow	Reddish-brown	Black (s); violet (g)	. . .
Melting point, °C	−218	−101	−7.3	114	. . .
Boiling point, °C	−188	−34.1	58.78	184	. . .
Electronegativity	4.1	2.8	2.7	2.2	. . .

[a] The halogens are represented by the letter X.
[b] Most stable isotope.

These intermolecular forces increase as the atomic size increases, so there is a progressive change in each physical property. For example, fluorine is a pale yellow gas of low density; chlorine is a greenish-yellow gas; bromine is a deep reddish-brown liquid three times as dense as water; iodine is a grayish-black crystalline solid with a low melting point. Liquid bromine has a high vapor pressure, and the reddish vapor is visible in a bottle containing the liquid. Iodine crystals have a high vapor pressure; when gently heated, these crystals change into a beautiful deep violet vapor without melting. The vapor condenses readily on cooling. This sublimation process (Section 11.12) is used in the purification of iodine.

Bromine is only slightly soluble in water, a polar solvent, but is miscible in all proportions in less polar or nonpolar solvents such as alcohol, ether, chloroform, carbon tetrachloride, and carbon disulfide, forming solutions that vary in color from yellow to reddish-brown, depending on the concentration.

Iodine is soluble in chloroform, carbon tetrachloride, carbon disulfide, and many hydrocarbons, forming solutions of I_2 molecules. The solutions have the same violet color as iodine in the gas phase, which is also molecular. Iodine is only slightly soluble in water but is quite soluble in alcohol, ether, and aqueous solutions of iodides; in all of these it forms brown solutions. These brown solutions result because iodine molecules are weak Lewis acids and combine with solvent molecules that can function as Lewis bases or with the iodide ion, which can also act as a Lewis base. The equation for the reversible reaction of iodine with the iodide ion to give the triiodide ion, I_3^-, is

$$:\!\overset{..}{\underset{..}{I}}\!:^- \; + \; :\!\overset{..}{\underset{..}{I}}\!:\!\overset{..}{\underset{..}{I}}\!: \; \rightleftharpoons \; :\!\overset{..}{\underset{..}{I}}\!:\!\overset{..}{\underset{..}{I}}\!:\!\overset{..}{\underset{..}{I}}\!:^-$$

A negative iodide ion, or other Lewis base, close to an iodine molecule, whose outer electrons are far from the nucleus, disturbs these electronic arrangements enough to form a coordinate-covalent bond. Compounds containing the complexes I_5^-, Br_3^-, Cl_3^-, and ICl_2^-, which also contain coordinate-covalent bonds, are known.

19.4 Chemical Properties of the Elemental Halogens

The elemental (free) halogens are oxidizing agents with strengths decreasing in the order $F_2 > Cl_2 > Br_2 > I_2$. In general, the heavier a halogen, the less is its strength as an oxidizing agent. For example, when fluorine reacts with a second element, the resulting fluoride contains the second element in its most oxidized state. When any other halogen reacts with a second element, the resulting halide may not contain the second element in its most oxidized state.

$$\begin{array}{ll} S + xsF_2 \longrightarrow SF_6 & 2Co + 3F_2 \longrightarrow 2CoF_3 \\ S + xsCl_2 \longrightarrow SCl_2 & Co + xsCl_2 \longrightarrow CoCl_2 \\ S + xsBr_2 \longrightarrow S_2Br_2 & Co + xsBr_2 \longrightarrow CoBr_2 \\ S + xsI_2 \longrightarrow \text{no reaction} & Co + xsI_2 \longrightarrow CoI_2 \end{array}$$

In these reactions xs indicates a stoichiometric excess of a reactant.

The reactions of elemental halogens with a variety of substances are generalized in Table 19-3.

Table 19-3 Chemical Properties of the Elemental Halogens

General Equation	Comments
Reactions with Elements	
$2M + nX_2 \longrightarrow 2MX_n$	With almost all metals
$H_2 + X_2 \longrightarrow 2HX$	With decreasing reactivity in the order $F_2 > Cl_2 > Br_2 > I_2$.
$Xe + \dfrac{n}{2}F_2 \longrightarrow XeF_n$	$n = 2, 4,$ or 6
$nX_2 + X_2' \longrightarrow 2X'X_n$	X' heavier than X; n an odd integer
$S + 3X_2 \longrightarrow SX_6$	With F_2; Se and Te can replace S
$S + X_2 \longrightarrow SX_2$	With Cl_2
$2S + X_2 \longrightarrow S_2X_2$	With Cl_2 or Br_2
$2P + 3X_2 \longrightarrow 2PX_3$	With excess P; also with As, Sb, or Bi
$2P + 5X_2 \longrightarrow 2PX_5$	With excess halogen, except I_2
Reactions with Compounds	
$X_2 + 2X'^- \longrightarrow 2X^- + X_2'$	X' heavier than X
$F_2 + H_2O \longrightarrow O_2 + OF_2 + H_2O_2 + HF$	Unbalanced; actual distribution of products depends on reaction conditions
$X_2 + H_2O \longrightarrow H^+ + X^- + HOX$	Not with F_2
$X_2 + H_2S \longrightarrow 2HX + S$	With Cl_2 or Br_2
$X_2 + CO \longrightarrow COX_2$	With Cl_2 or Br_2
$X_2 + SO_2 \longrightarrow SO_2X_2$	With F_2 or Cl_2
$X_2 + PX_3 \longrightarrow PX_5$	Not with I_2
$-\overset{\mid}{\underset{\mid}{C}}-H + X_2 \longrightarrow -\overset{\mid}{\underset{\mid}{C}}-X + HX$	With Cl_2 or Br_2 and many hydrocarbons
$\overset{}{\underset{}{>}}C=C\overset{}{\underset{}{<} + X_2 \longrightarrow -\overset{X}{\underset{\mid}{C}}-\overset{X}{\underset{\mid}{C}}-$	With Cl_2, Br_2, I_2, and many hydrocarbons containing C=C double bonds

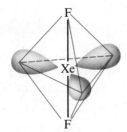

XeF$_2$

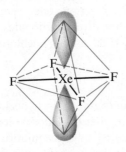

XeF$_4$

Figure 19-2. Structures of the molecules XeF_2 (linear) and XeF_4 (square planar). Note the close similarity to the isoelectronic species ICl_2^- and ICl_4^- in Fig. 19-5.

Fluorine reacts directly with and forms binary fluorides with all of the elements except the lighter noble gases (He, Ne, and Ar). Fluorine is such a strong oxidizing agent that many substances ignite on contact with it. A stream of fluorine directed onto water actually causes the water to "burn," with the formation of O_2, OF_2, H_2O_2, O_3, and HF. Wood and asbestos rapidly ignite and burn in fluorine gas. Heated glass burns in fluorine, producing smoke that looks like wood smoke. Most hot metals burn vigorously in fluorine. However, fluorine can be handled in a copper, iron, magnesium, or nickel container because an adherent film of the metal fluoride protects the surface from further attack.

Fluorine readily displaces chlorine and the other halogens from solid metal halides; with an excess of fluorine, halogen fluorides are formed (Section 19.6). Fluorine and hydrogen react explosively, even at temperatures as low as that of liquid nitrogen (77 K). Fluorine is the only element that will react directly with the noble gas xenon. A variety of compounds, such as XeF_2 and XeF_4 (Fig. 19-2), are formed (Section 22.8).

Although chlorine is a strong oxidizing agent, it is less active than fluorine. For example, fluorine and hydrogen react explosively, but when chlorine and hydrogen are mixed in the dark, the reaction between them is so slow that it is almost imperceptible. However, when the mixture is exposed to light, the reaction is explosive.

$$H_2(g) + Cl_2(g) \longrightarrow 2HCl(g) \qquad \Delta H° = -184.61 \text{ kJ}$$

Chemical reactions whose rates are enhanced by the effects of light are called **photochemical reactions.**

We should note at this point that before Chapter 18, we used the general enthalpy symbol, ΔH, although the heats of reaction given were, in fact, standard molar enthalpies. Since $\Delta H°$ was defined in Chapter 18, we will now use this symbol.

Chlorine is less active than fluorine toward metals; its oxidation reactions usually require higher temperatures. However, some finely divided metals, such as antimony, ignite in an atmosphere of chlorine at room temperature. Chlorine reacts with most nonmetals (C, N_2, and O_2 are notable exceptions), forming covalent compounds. For example, if the stoichiometry is controlled to keep the chlorine at a low level, it reacts with sulfur to give sulfur monochloride, S_2Cl_2 (Fig. 19-3), a liquid used in the vulcanization of rubber. If the chlorine is in excess, sulfur is oxidized to sulfur dichloride, SCl_2. A limited supply of chlorine oxidizes phosphorus to the trichloride, PCl_3. With an excess the pentachloride, PCl_5, is formed.

Chlorine generally reacts with compounds that contain only carbon and hydrogen (hydrocarbons) by adding to multiple bonds

$$CH_2{=}CH_2 + Cl_2 \longrightarrow CH_2Cl{-}CH_2Cl$$

or by substitution. For example, an exothermic stepwise substitution of chlorine for hydrogen occurs when methane, CH_4, reacts with chlorine.

$$CH_4(g) + Cl_2(g) \longrightarrow CH_3Cl(g) + HCl(g) \qquad \Delta H° = -94.14 \text{ kJ}$$
$$CH_3Cl(g) + Cl_2(g) \longrightarrow CH_2Cl_2(g) + HCl(g) \qquad \Delta H° = -132.9 \text{ kJ}$$
$$CH_2Cl_2(g) + Cl_2(g) \longrightarrow CHCl_3(g) + HCl(g) \qquad \Delta H° = -105.3 \text{ kJ}$$
$$CHCl_3(g) + Cl_2(g) \longrightarrow CCl_4(g) + HCl(g) \qquad \Delta H° = -93.26 \text{ kJ}$$

When chlorine is dissolved in water, a **disproportionation,** or **auto-oxidation-reduction, reaction** takes place. In a disproportionation reaction, one type of chemical species is both oxidized and reduced. Chlorine disproportionates in water, giving a solution containing hydrochloric acid, HCl, and hypochlorous acid, HOCl.

$$Cl_2 + H_2O \rightleftharpoons H^+ + Cl^- + HOCl$$

Half the chlorine atoms are oxidized to the $+1$ oxidation number (in hypochlorous acid), and half are reduced to the -1 oxidation number (in chloride ion). This disproportionation reaction is incomplete (as indicated by the equilibrium equation), so chlorine water is a solution of chlorine molecules, hypochlorous acid molecules, hydrogen ions, and chloride ions. When exposed to light, this solution undergoes a photochemical decomposition.

$$2HOCl \xrightarrow{\text{Sunlight}} 2H^+ + 2Cl^- + O_2(g)$$

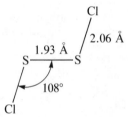

Figure 19-3. The molecular structure of S_2Cl_2. Note the S—S bond, which is responsible for the uncommon oxidation number of $+1$ for sulfur in this compound.

Oxygen is given off as the light-sensitive hypochlorous acid decomposes, HOCl is removed from the equilibrium mixture, and eventually the chlorine completely decomposes. Chlorine water is usually stored in dark bottles to prevent this decomposition.

The chemical properties of bromine are similar to those of chlorine, although bromine is the weaker oxidizing agent and its reactivity is less than that of chlorine. This lower reactivity is reflected in lower enthalpies of reaction with bromine than with chlorine. Bromine dissolved in water disproportionates less than chlorine, but the resulting hypobromous acid, HOBr, is also sensitive to light, decomposing with evolution of oxygen and formation of hydrobromic acid, HBr.

Iodine is the least reactive of the four naturally occurring halogens. It is the weakest oxidizing agent, and its anion is the most easily oxidized. Iodine reacts with metals, but heating is often required. It will not oxidize other halide ions but will oxidize certain other nonmetal anions, such as the sulfide ion, S^{2-}.

$$S^{2-} + I_2 \longrightarrow S + 2I^-$$

Compared with the other halogens, iodine reacts only slightly with water. Traces of iodine in water react with a mixture of starch and iodide ion, resulting in a deep blue color. This reaction is used as a very sensitive test for the presence of iodine in water.

19.5   Uses of the Elemental Halogens

Fluorine gas has been used to fluorinate organic compounds (to replace hydrogen with fluorine) since its initial discovery by Moissan in 1886. The direct reaction of fluorine with an organic compound is difficult to control, so the organic compound is dissolved in liquid hydrogen fluoride and the solution is electrolyzed. The fluorine reacts as rapidly as it is formed, so it never builds up a large concentration, and the reaction is easily controlled. The resulting **fluorocarbons** are quite stable and nonflammable. They are used as lubricants, refrigerants, plastics, and in artificial blood. Freon-12, CCl_2F_2, is widely used as a refrigerant. Teflon is a polymer composed of $-CF_2CF_2-$ units. Perfluorodecalin, $C_{10}F_{16}$, is useful as a blood substitute, in part because oxygen is very soluble in this chemically inert substance. Fluorine gas is used to make uranium hexafluoride, UF_6, which is necessary in the gaseous diffusion process for the separation of isotopes for the production of atomic energy. Fluorine is also used to make sulfur(VI) fluoride, SF_6, a stable gas used in high-voltage equipment. Fluoride ion, in SnF_2 or NaF, is added to water supplies and to some toothpastes to decrease cavities.

Large quantities of chlorine are used to bleach wood pulp and cotton cloth. The chlorine reacts with water to form hypochlorous acid, which oxidizes colored substances to colorless ones. Small amounts of chlorine kill the bacteria in drinking water supplies. Large quantities of chlorine are used in chlorinating hydrocarbons (replacing hydrogen with chlorine) and in producing compounds such as carbon tetrachloride, CCl_4, chloroform, $CHCl_3$, ethyl chloride, C_2H_5Cl, and polyvinyl chloride (PVC) and other polymers.

Bromine is used to produce certain dyes, light-sensitive silver bromide for photographic film, and sodium and potassium bromides for sedatives and soporifics.

Iodine in alcohol solution with potassium iodide is used as an antiseptic (tincture of iodine). Iodine salts are essential for the proper functioning of the thyroid gland; an iodine deficiency may lead to the development of a goiter. Iodized table salt contains 0.023% potassium iodide. Silver iodide is used in photographic film and in the seeding of clouds to induce rain. Iodoform, CHI_3, is an antiseptic.

COMPOUNDS OF THE HALOGENS

19.6   Interhalogens

Compounds formed from two different halogens are called **interhalogens.** Interhalogen molecules consist of one atom of a heavier halogen bonded by single bonds to an odd number of atoms of a lighter halogen. The structures of IF_3, IF_5, and IF_7 are shown in Fig. 19-4. Formulas for other interhalogens, each of which can be prepared by the direct reaction of the respective halogens, are given in Table 19-4. Because smaller halogen atoms are grouped about a larger one, the maximum number of smaller atoms possible increases as the radius of the larger ion increases (see Section 11.17). Many of these compounds are unstable, and most are extremely reactive. The interhalogens react like their related halides; halogen fluorides, for example, are stronger oxidizing agents than halogen chlorides.

Figure 19-4.   Structures of IF_3 (T-shaped), IF_5 (square pyramidal), and IF_7 (pentagonal bipyramidal).

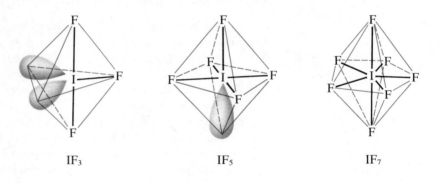

IF₃   IF₅   IF₇

Table 19-4   Interhalogens

XX′	XX′₃	XX′₅	XX′₇
ClF	ClF₃	ClF₅	
BrF	BrF₃	BrF₅	
BrCl			
IF	IF₃	IF₅	IF₇
ICl	ICl₃		
IBr			

The ionic **polyhalides** of the alkali metals, compounds such as KI_3, $KICl_2$, $KICl_4$, $CsIBr_2$, and $CsBrCl_2$ that contain an anion composed of at least three halogen atoms, are closely related to the interhalogens. The formation of the polyhalide anion I_3^- is responsible for the solubility of iodine in aqueous solutions containing iodide ion. Typical structures of **polyhalide ions** are shown in Fig. 19-5.

Figure 19-5. Structures of the ions ICl_2^- (linear) and ICl_4^- (square planar). (See also Fig. 19-2.)

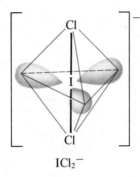 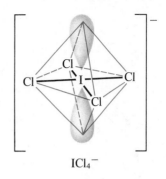

ICl_2^- ICl_4^-

19.7 Hydrogen Halides

Binary compounds containing only hydrogen and a halogen are called **hydrogen halides.** These compounds have the general formula HX, in which X represents the halogen. At room temperature the pure compounds HF, HCl, HBr, and HI are gases composed of covalent molecules in which a hydrogen atom and a halogen atom are joined by a single polar covalent bond. The polarity of these molecules is a function of the electronegativity of the halogen atom; it is greatest for HF and least for HI. Figure 19-6 shows the relative sizes of the hydrogen halide molecules.

Pure hydrogen fluoride differs from the other hydrogen halides in that its molecules are associated through hydrogen bonding (Section 11.14). Hydrogen fluoride crystallizes with chains of hydrogen-bonded molecules extending the entire length of the crystal (Fig. 19-7). Liquid hydrogen fluoride exhibits an anomalously high boiling point for a hydrogen halide (Table 19-5 and Fig. 11-19) due to the extensive association of the molecules. Hydrogen-bonded dimers, $(HF)_2$, are also observed in the vapor.

The tendency for HF to form hydrogen bonds enables many metal fluorides to form stable hydrogen difluoride salts, which contain hydrogen bonds in the anions. Examples include $NaHF_2$ and KHF_2. The hydrogen difluoride ion can be described as a resonance hybrid (Section 5.7).

$$F—H \cdots F^- \longleftrightarrow F^- \cdots H—F$$

Figure 19-6. Approximate relative sizes of hydrogen halide molecules. The bond distance (the distance between the nuclei of two atoms) is indicated for each molecule.

0.92 Å	1.28 Å	1.41 Å	1.62 Å	1.70 Å
H F	H Cl	H Br	H I	H At
HF	HCl	HBr	HI	HAt

Figure 19-7. A chain of hydrogen-bonded HF molecules as found in the solid compound.

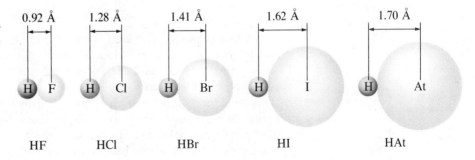

Table 19-5 Some Properties of the Hydrogen Halides

	HF	HCl	HBr	HI
Molecular weight	20.006	36.461	80.917	127.91
Melting point, °C	−83.1	−114.2	−86.8	−50.8
Boiling point, °C	19.9	−85.1	−66.7	−35.36
Solubility in water, g/100 g of water	∞ (0°C)	82.3 (0°C)	221 (0°C)	234 (10°C)
Enthalpy of formation, $\Delta H^\circ_{f_{298}}$, kJ mol^{-1}	−271.1	−92.307	−36	26.5

Only very weak hydrogen bonds can occur in hydrogen chloride, hydrogen bromide, and hydrogen iodide, because the halogen atoms in these molecules have relatively low electronegativities, and, consequently, the polarities of the bonds are small.

The anhydrous hydrogen halides are rather inactive chemically and do not attack dry metals at room temperature. However, they will react with many metals at elevated temperatures, forming metal halides and hydrogen. These reactions are sometimes used to prepare anhydrous metal halides.

$$\text{Fe} + 2\text{HCl}(g) \xrightarrow{300°C} \text{FeCl}_2 + \text{H}_2(g)$$

The hydrogen halides fume in moist air; that is, they combine with water vapor to form small drops of the corresponding acid. All of the hydrogen halides are very soluble in water (Table 19-5). With the exception of hydrogen fluoride, they are strong acids; they ionize completely in dilute aqueous solution.

Since the hydrogen halides are acids, they can be prepared by the general techniques used to prepare acids (Section 14.7); however, each technique is not suitable for every hydrogen halide. Hydrogen halides are prepared as follows:

1. BY DIRECT REACTION OF THE ELEMENTS. As indicated in Table 19-3, fluorine, chlorine, and bromine react directly with hydrogen to form the various hydrogen halides.

$$\text{H}_2 + \text{X}_2 \longrightarrow 2\text{HX}$$

This reaction is used to prepare hydrogen chloride and hydrogen bromide commercially. Both hydrogen and chlorine are readily available from the electrolysis of brine (Section 19.2), and these gases are burned together, forming HCl, in specially designed burners.

Bromine reacts much less vigorously with hydrogen than chlorine does. For bromine to react appreciably, the mixture must be heated to about 200°C in contact with a catalyst such as platinum or carbon. The reaction is reversible and quickly reaches equilibrium, but almost complete conversion of bromine to hydrogen bromide is accomplished by using an excess of hydrogen. The excess hydrogen is recycled by liquefying the hydrogen bromide and pumping away the gaseous hydrogen.

Hydrogen fluoride can also be prepared by the direct reaction of the elements, but elemental fluorine is too expensive to use commercially and too difficult to handle to use in most laboratories.

The direct union of hydrogen and iodine is unsatisfactory for the preparation of hydrogen iodide because the reaction is slow and the equilibrium yield is low. Heat decomposes hydrogen iodide, and thus the rate of reaction cannot be increased by heating.

2. BY ACID-BASE REACTIONS. Although the halide ions, except for the fluoride ion, are only weakly basic, hydrogen halides can be prepared by acid-base reactions in which a nonvolatile strong acid, such as sulfuric acid, is added to a metal halide. The escape of the gaseous hydrogen halide drives the reaction to completion. For example, hydrogen fluoride is usually prepared by heating a mixture of calcium fluoride, CaF_2, and concentrated sulfuric acid in a lead or platinum container.

$$CaF_2 + H_2SO_4 \longrightarrow CaSO_4 + 2HF(g)$$

Hydrogen fluoride is also a by-product in the preparation of phosphate fertilizers by the reaction of apatite, $Ca_5(PO_4)_3F$, with sulfuric acid (Section 8.7). The hydrogen fluoride is evolved as a gas. If absorbed in water, it forms hydrofluoric acid, which attacks glass and is therefore stored in plastic bottles, generally polyethylene ones.

Hydrogen chloride is prepared, both in the laboratory and commercially, by the reaction of concentrated sulfuric acid with sodium chloride, the least expensive of the chlorides.

$$NaCl + H_2SO_4 \longrightarrow NaHSO_4 + HCl(g)$$

This reaction goes to completion because hydrogen chloride is insoluble in the reaction mixture. Since sodium hydrogen sulfate, $NaHSO_4$, is also an acid, adding more sodium chloride and heating produces more hydrogen chloride.

$$NaCl + NaHSO_4 \longrightarrow Na_2SO_4 + HCl(g)$$

The hydrogen chloride thus produced may be sold as a gas in pressurized cylinders or absorbed in water and marketed as hydrochloric acid.

Hydrogen bromide and hydrogen iodide cannot be prepared in similar reactions because sulfuric acid is a strong enough oxidizing agent that it oxidizes both bromide and iodide ions. Sulfuric acid is reduced to sulfur dioxide by bromide ion and to sulfur dioxide, sulfur, or hydrogen sulfide by iodide ion, depending on the relative amounts of acid and iodide ion used in the reaction.

3. BY HYDROLYSIS REACTIONS. Many covalent nonmetal halides, such as PCl_3, PBr_3, PI_3, PCl_5, $SiBr_4$, and SCl_4, react with water (hydrolyze) with the formation of a hydrogen halide and the oxyacid of the nonmetal. Quite often the reaction is a vigorous one.

$$PBr_3 + 3H_2O \longrightarrow 3HBr(g) + H_3PO_3$$
$$SCl_4 + 3H_2O \longrightarrow 4HCl(g) + H_2SO_3$$

This method is rarely used to prepare hydrogen chloride but is commonly used for the preparation of hydrogen bromide or hydrogen iodide.

4. BY HALOGENATION OF HYDROCARBONS. About two-thirds of the hydrogen chloride produced in the United States each year is a by-product of the reactions used to prepare chlorinated hydrocarbons, which are in great demand in

a variety of chemical industries. For example, hydrogen chloride is a by-product of a reaction used to manufacture ethyl chloride from ethane and chlorine:

$$C_2H_6(g) + Cl_2(g) \longrightarrow C_2H_5Cl(g) + HCl(g)$$

Hydrogen bromide is a by-product of the manufacture of brominated hydrocarbons by a similar reaction involving bromine and hydrocarbons.

19.8 Hydrohalic Acids

Aqueous solutions of the hydrogen halides are called **hydrofluoric acid, hydrochloric acid, hydrobromic acid,** and **hydriodic acid.** Hydrofluoric acid is a weak acid, but the other **hydrohalic acids** are very strong. This is due to differences in the H—X bond strengths in the hydrogen halide molecules. In order for ionization to occur, the H—X bond must be broken, and the resulting ions must be hydrated. The H—F bond is the strongest H—X bond (about twice as strong as the H—I bond), and its greater strength is responsible for the weaker acid strength of hydrofluoric acid.

The acidic behavior of hydrohalic acids reflects the presence of hydronium ion. The hydronium ion is reduced by metals above hydrogen in the activity series (Section 9.17); halide salts and hydrogen are formed. The general reaction is illustrated with the metal zinc.

$$Zn(s) + 2H_3O^+(aq) + [2X^-(aq)] \longrightarrow Zn^{2+}(aq) + [2X^-(aq)] + H_2(g) + 2H_2O(l)$$

Hydronium ion acts as an oxidizing agent in the reaction. Hydronium ion also reacts with bases. The reaction with cobalt(II) hydroxide is described by the equation

$$Co(OH)_2(s) + 2H_3O^+(aq) + [2X^-(aq)] \longrightarrow Co^{2+}(aq) + [2X^-(aq)] + 4H_2O(l)$$

Reactions of the hydrohalic acids with metals or with metal hydroxides, oxides, or carbonates are often used to prepare soluble metal salts. In particular, most chloride salts are soluble ($AgCl$, $PbCl_2$, and Hg_2Cl_2 are the common exceptions). The hydronium ion from strong acids also reacts with other bases:

$$NH_3(aq) + H_3O^+(aq) + [X^-(aq)] \longrightarrow NH_4^+(aq) + [X^-(aq)] + H_2O(l)$$
$$CH_3CO_2^-(aq) + H_3O^+(aq) + [X^-(aq)] \longrightarrow CH_3CO_2H(aq) + [X^-(aq)] + H_2O(l)$$

The hydrated halide ions in hydrohalic acids give these substances the properties associated with $X^-(aq)$. The heavier halide ions (Cl^-, Br^-, and I^-) can act as reducing agents and are oxidized by lighter halogens or other oxidizing agents.

$$2H_3O^+ + 2Br^- + Cl_2 \longrightarrow 2H_3O^+ + 2Cl^- + Br_2$$
$$14H_3O^+ + 6Cl^- + Cr_2O_7^{2-} \longrightarrow 2Cr^{3+} + 3Cl_2 + 21H_2O$$

They also serve as precipitating agents for insoluble metal halides, for example:

$$Ag^+ + [NO_3^-] + [H_3O^+] + Cl^- \longrightarrow AgCl(s) + [H_3O^+] + [NO_3^-]$$

One of the most unusual and unique properties of hydrofluoric acid is its reactivity with sand (silicon dioxide) and with glass, which is a mixture of silicates (mainly calcium silicate). The reactions are

$$SiO_2 + 4HF \longrightarrow SiF_4(g) + 2H_2O$$
$$CaSiO_3 + 6HF \longrightarrow CaF_2 + SiF_4(g) + 3H_2O$$

The silicon escapes from these reactions as silicon tetrafluoride, a volatile compound. Because hydrogen fluoride attacks glass, it is used to frost or to etch glass objects. Light bulbs are frosted with hydrogen fluoride, and measurement markings are made on thermometers, burets, and other glassware with it.

The largest use for hydrogen fluoride is in the production of fluorocarbons for refrigerants such as the Freons, plastics such as Teflon, and propellants. The second largest use is in the manufacture of cryolite, K_3AlF_6, which is important in the production of aluminum. The acid is also used in the production of other inorganic fluorides, such as boron trifluoride, BF_3, which are used as catalysts in the industrial synthesis of certain organic compounds.

Extreme care should be used when handling hydrofluoric acid because it causes painful and slow-healing burns. Since it is a local anesthetic, its presence may not be noticed until the damage has been done.

Hydrochloric acid is inexpensive. Thus, it is second only to sulfuric acid in industrial use and is used in the manufacture of metal chlorides, dyes, glue, glucose, and various other chemicals. A considerable amount is also used in the activation of oil wells and as a pickle liquor (an acid used to remove the oxide coating from iron or steel that is to be galvanized, tinned, or enameled).

Compared to hydrochloric acid, the amounts of hydrobromic acid and hydroiodic acid used commercially are insignificant.

19.9   Metal Halides

Metal halides are among the most commonly used of chemical compounds. They are relatively easily prepared, generally soluble salts used as reagents for the preparation of a wide variety of other metal-containing compounds.

Anhydrous metal halides may be prepared by the reaction of an elemental halogen or of a gaseous hydrogen halide with a metal, generally at an elevated temperature. Hydrated metal halides are obtained by the action of hydrohalic acids on metals lying above hydrogen in the activity series and by the reaction of hydrohalic acids with most metal oxides, hydroxides, or carbonates. Several insoluble metal halides may be prepared by metathetical reactions. For example, magnesium fluoride can be prepared by mixing solutions of magnesium nitrate and potassium fluoride.

$$Mg^{2+}(aq) + [2NO_3^-(aq)] + [2K^+(aq)] + 2F^-(aq) \longrightarrow$$
$$MgF_2(s) + [2K^+(aq)] + [2NO_3^-(aq)]$$

This reaction can be expressed more simply as follows:

$$Mg^{2+}(aq) + 2F^-(aq) \longrightarrow MgF_2(s)$$

Other insoluble metal halides include the chlorides, bromides, and iodides of silver, lead, and mercury(I) as well as the fluorides of the Group IIA metals and of lead. The solubilities of the fluorides in water often differ from those of the other halides. For example, AgF is quite water-soluble, while AgCl, AgBr, and AgI are insoluble. The fluorides of the Group IIA metals are insoluble, while the other halide salts of these metals are soluble.

The variation in character among the metal halides is remarkable; their properties range from those of ionic species to those of covalent molecules (Fig. 19-8). The halides of the alkali metals, of the alkaline earth metals, of most fluorides, and of transition metals having lower oxidation numbers ($+2$, $+3$) are ionic in

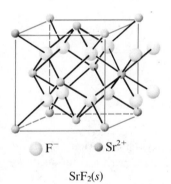

○ F^- ● Sr^{2+}

$SrF_2(s)$

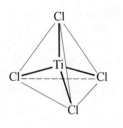

$TiCl_4(l)$

Figure 19-8. The structures of SrF_2, an ionic solid, and $TiCl_4$, a covalent molecule.

their behavior. These ionic compounds tend to have high melting points (above about 400°C) and high electrical conductivities in the molten state. Those that dissolve in water undergo little or no hydrolysis.

The halides of metals with high charge-to-size ratios and/or high oxidation numbers behave more like nonmetals and are predominantly covalent in character because of the attraction of the metal for the valence electrons of the halide ion. These halides are characterized by low melting points, by ready volatility, by solubility in nonpolar solvents, by lack of electrical conductivity in the molten state, and by extensive hydrolysis in water.

As might be expected, some metal halides are intermediate in character between the ionic and covalent ones. For example, iron(III) chloride, $FeCl_3$, is volatile, low-melting, and readily hydrolyzed, but it is also a good conductor of electricity in the molten state.

19.10   Binary Halogen-Oxygen Compounds

Although the halogens do not react directly with oxygen, binary halogen-oxygen compounds (shown in Table 19-6) can be prepared by indirect methods. Oxygen is more electronegative than chlorine, bromine, and iodine; thus their compounds with oxygen are called oxides. On the other hand, fluorine compounds with oxygen are called fluorides because fluorine is the more electronegative element. Most binary halogen-oxygen compounds are extremely reactive and unstable. Iodine(V) oxide, I_2O_5, is the only one of these compounds which does not decompose on heating.

Table 19-6   Binary Halogen-Oxygen Compounds

Oxygen Fluorides	Chlorine Oxides	Bromine Oxides	Iodine Oxides
OF_2	Cl_2O	Br_2O	I_2O_4
O_2F_2	ClO_2	Br_3O_8	I_4O_9
	Cl_2O_6	BrO_2	I_2O_5
	Cl_2O_7		I_2O_7

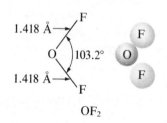

Of the two known oxygen fluorides (Table 19-6), **oxygen difluoride, OF_2,** is the most important. It is prepared by the reaction of fluorine with a solution of sodium hydroxide.

$$2F_2 + [2Na^+] + 2OH^- \longrightarrow OF_2(g) + [2Na^+] + 2F^- + H_2O$$

Oxygen difluoride is a covalent molecule (Fig. 19-9), having oxidation numbers of +2 for oxygen and −1 for fluorine. Although OF_2 is appreciably soluble in water, it is not the anhydride of an acid; it does not react with water.

Chlorine(I) oxide, or simply **chlorine monoxide, Cl_2O,** a yellowish-red gas that is apt to explode violently, has the same basic structure as oxygen difluoride (Fig. 19-9). It is prepared by the reaction of chlorine with freshly prepared mercury(II) oxide.

$$2Cl_2 + 2HgO \longrightarrow Cl_2O(g) + HgCl_2 \cdot HgO$$

Chlorine monoxide is an active oxidizing agent and reacts with water to form hypochlorous acid, HOCl.

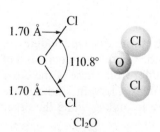

Figure 19-9.   Angular structures of the OF_2 and Cl_2O molecules (drawn approximately to scale). Bond distances are measured between the nuclei of the atoms.

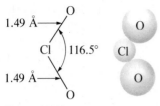

Figure 19-10. Structure of the ClO_2 molecule.

Chlorine dioxide, ClO_2 (Fig. 19-10), is available from the reaction of chlorine with silver chlorate.

$$Cl_2 + 2AgClO_3 \longrightarrow 2ClO_2(g) + 2AgCl + O_2$$

On an industrial scale ClO_2 is prepared by oxidation of sodium chlorite with chlorine.

$$2NaClO_2 + Cl_2 \longrightarrow 2NaCl + 2ClO_2(g)$$

Chlorine dioxide is used industrially for the bleaching of flour and in water treatment. The pure yellow gas is apt to explode violently, but it can be handled safely if diluted with carbon dioxide or air. Chlorine dioxide reacts slowly with water (forming chloric and hydrochloric acids) and with ozone (forming Cl_2O_6, a dark red liquid).

19.11 Oxyacids of the Halogens and Their Salts

The only known oxyacid of fluorine is the very unstable **hypofluorous acid, HOF,** which is prepared by the reaction of gaseous fluorine with ice.

$$F_2(g) + H_2O(s) \longrightarrow HOF(g) + HF(g)$$

Hypofluorous acid, which is so unstable that it decomposes in less than an hour at room temperature, does not ionize in water, and no salts are known. The principal oxyacids of the other halogens are shown in Table 19-7.

Table 19-7   Oxyacids of the Heavier Halogens

Oxidation Number of Halogen	Oxyacids of Chlorine	Oxyacids of Bromine	Oxyacids of Iodine
+1	HClO	HBrO	HIO
+3	$HClO_2$		
+5	$HClO_3$	$HBrO_3$	HIO_3
+7	$HClO_4$	$HBrO_4$	HIO_4
			H_5IO_6

The compounds HXO, HXO_2, HXO_3, and HXO_4, where X represents one of the heavier halogens (Cl, Br, or I), are called **hypohalous, halous, halic,** and **perhalic acids,** respectively (Section 5.12). The acid strengths of these acids are directly proportional to the oxidation numbers of the halogens (Section 14.3), increasing from the hypohalous acids, which are very weak acids, to the perhalic acids, which are very strong acids.

Chlorine and bromine both react with water to a limited extent, forming a mixture of the respective hypohalous and hydrohalic acids (Section 19.4). But to prepare the pure hypohalous acid, a reaction with mercury(II) oxide, similar to that used for chlorine monoxide (Section 19.10), is used.

$$2X_2 + 3HgO + H_2O \longrightarrow HgX_2 \cdot 2HgO(s) + 2HOX(aq)$$

None of the hypohalous acids of the heavier halogens has been isolated in the free state. Their thermal instability means that they are stable only in solution. The

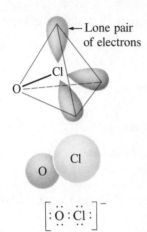

Figure 19-11. Structure of the hypochlorite ion, ClO⁻.

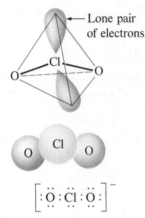

Figure 19-12. Structure of the chlorite ion, ClO₂⁻.

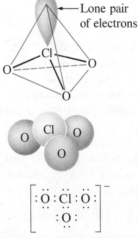

Figure 19-13. Structure of the chlorate ion, ClO₃⁻.

hypohalous acids are all very weak acids, whose acid strengths decrease as the halogen's size increases and as its electronegativity decreases (Section 14.4). Thus HOCl is a stronger acid than HOBr, which in turn is stronger than HOI.

Solutions of salts containing **hypohalite ions, OX⁻,** may be prepared by adding a base to a solution of a hypohalous acid, but only the salts of hypochlorous acid have been isolated as solids. The structure of the hypochlorite ion, ClO⁻, is shown in Fig. 19-11. The hypohalite ions hydrolyze by accepting protons from water, and solutions of metal hypohalites are basic due to the presence of excess hydroxide ions.

$$OX^- + H_2O \rightleftharpoons HOX + OH^-$$

Sodium hypochlorite is used as an inexpensive bleach (Chlorox) and germicide. It is produced commercially by the electrolysis of cold, dilute aqueous sodium chloride solution under conditions where the resulting chlorine and hydroxide ion can react.

$$2Cl^- + 2H_2O \xrightarrow{\text{Electrical energy}} 2OH^- + Cl_2 + H_2$$

$$Cl_2 + 2OH^- \longrightarrow Cl^- + ClO^- + H_2O$$

Net reaction: $Cl^- + H_2O \xrightarrow{\text{Electrical energy}} ClO^- + H_2$

All of the hypohalites are unstable with respect to disproportionation, but the reaction is slow for hypochlorite. Hypobromite and hypoiodite disproportionate rapidly, even at low temperatures.

$$3OX^- \longrightarrow 2X^- + XO_3^-$$

The only known halous acid is **chlorous acid, HClO₂,** obtained by the reaction of barium chlorate with dilute sulfuric acid.

$$Ba(ClO_2)_2(aq) + H_2SO_4(aq) \longrightarrow BaSO_4(s) + 2HClO_2(aq)$$

A solution of HClO₂ is obtained by filtering off the barium sulfate. Chlorous acid is not stable; it slowly decomposes in solution to give chlorine dioxide, hydrochloric acid, and water. Chlorous acid reacts with bases to give salts containing the **chlorite ion, ClO₂⁻** (Fig. 19-12). Metal chlorite salts can also be prepared by the action of chlorine dioxide on a metal peroxide. For example, sodium peroxide reacts as follows:

$$2ClO_2 + Na_2O_2 \longrightarrow 2NaClO_2 + O_2$$

Sodium chlorite is used extensively in the bleaching of paper because it is a strong oxidizing agent that does not damage the paper.

Chloric acid, HClO₃, and **bromic acid, HBrO₃,** are stable only in solution, but **iodic acid, HIO₃,** can be isolated as a stable white solid from the reaction of iodine with concentrated nitric acid.

$$I_2 + 10HNO_3 \longrightarrow 2HIO_3 + 10NO_2(g) + 4H_2O$$

The lighter halic acids can be obtained from their barium salts by reaction with dilute sulfuric acid. The reaction is analogous to that used to prepare chlorous acid. All of the halic acids are strong acids and very active oxidizing agents. Salts containing **halate ions** such as ClO₃⁻ (Fig. 19-13) can be prepared by reaction of

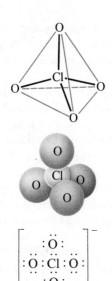

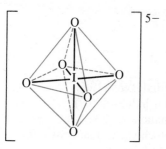

Figure 19-14. Structure of the perchlorate ion, ClO_4^-.

Figure 19-15. The octahedral structure of the paraperiodate ion, IO_6^{5-}.

the acids with bases. Metal chlorates are also prepared by electrochemical oxidation of a hot solution of a metal halide. Bromates can be produced by the oxidation of bromides with hypochlorite ion, whereas oxidation of iodides with chlorates gives iodates. Sodium chlorate is used as a weed killer; potassium chlorate is used in some matches and to prepare oxygen in the laboratory (Section 9.2).

Perchloric acid, HClO₄, may be obtained by heating a **perchlorate,** such as potassium perchlorate, with sulfuric acid under reduced pressure.

$$KClO_4 + H_2SO_4 \longrightarrow HClO_4 + KHSO_4$$

Perchloric acid explodes above 92°C, but it will distill at temperatures below 92°C at reduced pressures and can be removed from the reaction mixture with little loss by decomposition. Dilute aqueous solutions of perchloric acid are quite stable thermally, but at concentrations above 60% they are unstable and dangerous. Perchloric acid is a powerful oxidizing agent, and serious explosions have occurred when concentrated solutions were heated with easily oxidized substances. However, its reactions as an oxidizing agent are slow when it is cold and dilute. The acid is a valuable reagent because it is among the strongest of all common acids; it is a good oxidizing agent; and most perchlorates are soluble. Salts of the **perchlorate ion, ClO_4^-** (Fig. 19-14), are prepared by reactions of bases with perchloric acid and, commercially, by the electrolysis of hot solutions of chlorides.

Perbromic acid, HBrO₄, and its salts were unknown until 1968. This was unexpected and puzzling since analogous compounds of chlorine and iodine had long been known. Salts containing the **perbromate ion, BrO_4^-,** were finally prepared by the oxidation of bromates with xenon difluoride, XeF_2, in aqueous solution. They are now routinely prepared by the reaction of bromates with fluorine in basic solution. Perbromic acid is prepared by the acidification of perbromate salts.

Several different acids containing iodine with its +7 oxidation number are known; these include **metaperiodic acid, HIO₄,** and **paraperiodic acid, H₅IO₆.** The iodine atom is large enough for six oxygen atoms to surround it (Fig. 19-15), whereas the smaller chlorine and bromine atoms only accommodate four oxygen atoms. Salts of the periodic acids can be readily prepared by reactions with bases. In addition, sodium metaperiodate, $NaIO_4$, can be prepared by the oxidation of sodium iodate, $NaIO_3$, by chlorine in a hot alkaline solution. Paraperiodates are also prepared by the electrolytic oxidation of iodates, a method analogous to the preparation of perchlorates. Free paraperiodate can be obtained by evaporating a solution of metaperiodic acid, formed by the reaction of barium metaperiodate with sulfuric acid. Salts of paraperiodic acid, such as $Na_2H_3IO_6$, $Na_3H_2IO_6$, and Ag_5IO_6, have been prepared.

FOR REVIEW

SUMMARY

The elements of Group VIIA of the Periodic Table, the **halogens,** are reactive nonmetals that exist as diatomic molecules, X_2, under ordinary conditions. The halogens (except fluorine) usually form compounds in which they have an oxidation number of −1, +1, +3, +5, or +7, although in some compounds they exhibit other oxidation numbers. Fluorine is the most electronegative of the known elements; thus its oxidation number in compounds is always −1. A halogen exhibits the highest electronegativity of any of the elements in each row of the Periodic Table.

The chemical properties of the halogens change as expected down the group. The oxidizing ability of the elemental halogens and the resistance of the halides to oxidation decrease as the size of the halogen increases. This is reflected in the ease of preparation of the elements from their halide salts. A halogen will oxidize only heavier halide ions. The gaseous **hydrogen halides** form from the direct reaction of halogens and hydrogen, but with decreasing vigor as the halogen becomes heavier. Hydrogen halides can also be prepared by the general techniques used to make acids. They dissolve in water to give solutions of **hydrohalic acids.** Reactions of the halogens with other elements also become less vigorous as the size of the halogen increases.

The halogens form **halides** with less electronegative elements. Halides of the active metals and of transition metals having lower oxidation numbers are essentially ionic. Halides of the metals in Groups IIIA, IVA, and VA and of transition metals having higher oxidation numbers have a significant covalent character while halides of nonmetals are covalent. **Interhalogens** are formed by the combination of two different halogens. **Binary halogen-oxygen compounds,** which are generally of low stability, as well as **hypohalous, halous, halic,** and **perhalic acids** and their salts are known.

KEY TERMS AND CONCEPTS

Binary halogen-oxygen
  compounds (19.10)
Disproportionation reaction
  (19.4)
Electrolysis (19.2)

Halic acids (19.11)
Halous acids (19.11)
Hydrogen halides (19.7)
Hydrohalic acids (19.8)
Hypohalous acids (19.11)

Interhalogens (19.6)
Metal halides (19.9)
Perhalic acids (19.11)
Photochemical reactions (19.4)
Polyhalide ions (19.6)

EXERCISES

Elemental Halogens

1. Arrange the halogens in order of increasing (a) atomic radius, (b) electronegativity, (c) boiling point, and (d) oxidizing activity.
2. Write the Lewis structure of chlorine.
3. Describe the electrochemical preparation of fluorine.
4. Explain why fluorine must be prepared by an electrochemical technique. Why is a molten mixture of KF and HF used for the reaction medium instead of an aqueous solution of a fluoride?

5. Why can fluorine, which reacts with all of the elements except a few of the lighter noble gases, be stored in metal cylinders?
6. Write the equation that describes the electrochemical preparation of chlorine (a) from an aqueous solution of sodium chloride and (b) from molten iron(III) chloride.
7. Describe the chemistry of the extraction of bromine from brine.
8. Suggest two reasons why fluorine is not used in

the extraction of bromine from brine.

9. Describe the production of iodine from sodium iodate.

10. Common bleaches such as chlorine, hydrogen peroxide, and sodium hypochlorite act to whiten materials by oxidizing traces of colored substances to colorless ones. Since fluorine is an oxidizing agent, why is it not used as a bleach?

11. Write balanced chemical equations that describe the reaction of chlorine with each of the following:
    (a) lithium   (d) an excess of phosphorus
    (b) magnesium   (e) iodine
    (c) hydrogen   (f) sulfur

12. Using the reactions of the halogens with sulfur and with phosphorus as examples, show that the oxidizing ability of the halogens decreases as their size increases.

13. Show that the reaction of bromine with water is a disproportionation reaction by considering the changes in oxidation number that occur.

14. Why is iodine more soluble in a solution of calcium iodide than in pure water?

15. Why is iodine monofluoride more polar than iodine monochloride?

16. Why will chlorine added to an indoor swimming pool last longer than the same amount of chlorine added to an outdoor pool? Write the chemical equations for the reactions that are responsible for this behavior.

17. Write chemical equations describing the changes that occur in each of the following cases:
    (a) sulfur is burned in a fluorine atmosphere
    (b) molten sodium fluoride is electrolyzed
    (c) potassium bromide is treated with chlorine
    (d) phosphorus is heated with an excess of iodine
    (e) hydrogen sulfide is bubbled into a solution of iodine
    (f) the barrier in a hydrogen fluoride production cell breaks and hydrogen and fluorine mix
    (g) aluminum is heated with iodine
    (h) 1 mol of methane reacts with 2 mol of chlorine

Halogen Compounds

18. Arrange the halide ions in order of increasing (a) natural abundance, (b) ionic radius, (c) ease of oxidation, and (d) strength as a reducing agent.

19. Identify the oxidation numbers that fluorine and bromine commonly exhibit in their compounds.

20. With what elements can chlorine form binary compounds in which it exhibits a positive oxidation number?

21. Write the Lewis structure for each of the iodine fluorides.

22. Describe the molecular geometry of each of the iodine fluorides.

23. Write an equation describing a convenient laboratory preparation for each of the hydrogen halides.

24. When a hydrogen halide functions as a reducing agent, which element in the molecule is reduced? Compare HF, HCl, HBr, and HI with respect to their strengths as reducing agents.

25. Hydrohalic acids act as oxidizing agents with some metals. Write an equation that is an example of such behavior.

26. How do hydrogen fluoride and hydrofluoric acid differ in their chemical behavior from the other hydrogen halides and hydrohalic acids?

27. Write a balanced chemical equation that describes the reaction that occurs in each of the following cases:
    (a) calcium is added to hydrobromic acid
    (b) potassium hydroxide is added to hydrofluoric acid
    (c) ammonia is bubbled through hydrofluoric acid
    (d) sodium acetate is added to hydroiodic acid
    (e) silver oxide is added to hydrobromic acid
    (f) chlorine is bubbled through hydrobromic acid

28. Give two examples of both ionic and covalent binary metal halides. Contrast the chemical and physical behavior of these two classes of halides.

29. Why is OF_2 called oxygen difluoride rather than fluorine oxide?

30. In what kind of compounds can chlorine exhibit an oxidation number of $+1$?

31. Write chemical equations that show that Cl_2O and I_2O_5 are acid anhydrides.

32. What is the oxidation number of bromine in each of its oxides listed in Table 19-6?

33. Write the chemical equation for the net reaction that occurs during the preparation of each of the oxyhalide anions that can be prepared electrochemically.

34. Write the Lewis structure of each of the oxyacids

of chlorine. Calculate the oxidation number of chlorine in each of these acids.

35. Write a chemical equation for the reaction that occurs in each of the following cases:
    (a) sodium hydroxide is added to a solution of chloric acid
    (b) metallic zinc is added to a solution of perbromic acid
    (c) ammonia is bubbled through a solution of chlorous acid
    (d) an excess of calcium hydroxide reacts with paraperiodic acid
    (e) bromine is added to a solution of lithium hydroxide
    (f) solid sodium hypochlorite is added to water

36. Which is the stronger acid, $HClO_2$ or $HClO_3$? Why?

37. Which is the stronger acid, $HClO_3$ or $HBrO_3$? Why?

38. Predict the products of each of the following:
    (a) aluminum reacts with oxygen difluoride
    (b) a solution of hypochlorous acid is heated
    (c) hydrochloric acid is added to a solution of hypochlorous acid

Additional Exercises

39. Write the balanced equation for the chemical reaction that occurs when ICl dissolves in water. (*Hint:* Consider which atom is more readily oxidized.)

40. Write the chemical reaction for the preparation of ethylene dibromide, $C_2H_4Br_2$, a constituent of leaded gasoline, from ethylene and bromine. Use Lewis structures instead of chemical formulas in your equation.

41. Explain the changes in the first ionization energies of the halogens (Table 4-9) as a function of their atomic sizes.

42. A sample of cadmium chloride with a mass of 1.766 g gave 1.089 g of cadmium on electrolysis. Determine the formula of cadmium chloride.
    Ans. $CdCl_2$

43. Iodine reacts with liquid chlorine at $-40°C$ to give an orange compound containing 54.3% iodine and 45.7% chlorine. Write the balanced equation for the reaction.
    Ans. $I_2 + 3Cl_2 \rightarrow 3ICl_3$

44. The reaction of V_2O_3 with Cl_2 gives a yellow liq-

uid that contains 29.42% vanadium, 61.3% chlorine, and the remainder oxygen. At $19°C$ a sample of the liquid with a mass of 0.433 g vaporized in a 115-mL flask, giving a gas with a pressure of 390 torr. What is the molecular formula of this vanadium oxychloride? *Ans. $VOCl_3$*

45. The reaction of titanium metal with $HF(g)$ at $250°C$ for 2 days yielded a titanium fluoride that contained 45.67% titanium. Write the chemical equation that describes the reaction.
    Ans. $2Ti + 6HF \rightarrow 2TiF_3 + 3H_2$

46. What mass of PCl_3 can be prepared by the reaction of chlorine with 13.55 g of phosphorus?
    Ans. 60.08 g

47. How many grams of sodium iodate are required to prepare 0.100 kg of iodine by reduction with sodium hydrogen sulfite? *Ans. 156 g*

48. Predict and draw diagrams for the molecular structures of the following molecules and ions: IO_6^{5-}, ClO_3^-, $SnBr_4$, IF_4^+, ClF_2^+.
    Ans. Octahedral; pyramidal; tetrahedral; seesaw; bent (109.5°)

49. Which of the following compounds or ions have the same structure: $ClICl^-$, Cl_2O, OF_2, BrF_3, ClO_3^-, ClO_4^-, ICl_4^-, IF_4^+, XeF_4?
    Ans. Cl_2O and OF_2; ICl_4^- and XeF_4

50. What volume of $HCl(g)$ measured at $100°C$ and 1 atm can be prepared from 50.0 g of NaCl?
    Ans. 26.2 L

51. What volume of $HBr(g)$ at $25°C$ and 0.977 atm is required to prepare 0.3037 L of 0.1191 M HBr solution? *Ans. 0.906 L*

52. The density of $HF(g)$, 0.991 g/L at $19°C$ and 1.00 atm, is unexpectedly high. Calculate the apparent molecular weight of HF and explain this result. *Ans. 23.7; due to HF dimers*

53. What mass of HI can be prepared by the complete hydrolysis of 10.0 g of PI_3? *Ans. 9.32 g*

54. What volume (in liters) of 0.1191 M HBr solution is required to react with 0.736 g of ZnO to give $ZnBr_2$? *Ans. 0.152 L*

55. From the data in Appendix J, calculate the free energy change for the reaction of hydrogen with each of the halogens. Determine which of these reactions is not spontaneous at $25°C$.
    Ans. $H_2 + I_2 \rightarrow 2HI$

56. Calculate the bond energies of the F—F, Cl—Cl, Br—Br, and I—I bonds from the data in Appendix J, and arrange these molecules in increasing

order of their X—X bond energy.

Ans. $I_2 < F_2 < Br_2 < Cl_2$

57. A gas mixture at equilibrium at 1375 K exhibits a partial pressure of HBr of 0.998 atm, of H_2 of 3.82×10^{-3} atm, and of Br_2 of 3.82×10^{-3} atm. What is K_p for the decomposition of HBr to H_2 and Br_2?   *Ans. 1.46×10^{-5}*

58. Thallium(I) chloride is one of the few insoluble chlorides. What is the molar solubility of TlCl, whose solubility product is 1.8×10^{-4}?

Ans. 1.3×10^{-2} M

20

ELECTROCHEMISTRY AND OXIDATION-REDUCTION

Electrochemistry deals with the chemical changes produced by an electric current and with the production of electricity by chemical reactions. In **electrolytic cells,** electrical energy brings about desired chemical changes. In **voltaic cells,** chemical reactions produce electrical energy. In both types of cells, oxidation occurs at one electrode (the anode) and reduction at the other electrode (the cathode).

Energy changes during chemical reactions (chemical thermodynamics; see Chapter 18) are often studied electrochemically because the quantity of electrical energy produced or consumed during electrochemical changes can be measured very accurately. Furthermore, an understanding of the reactions that take place at the electrodes of electrochemical cells throws much light on the processes of oxidation and reduction and chemical activity.

Many electrochemical processes are important in science and industry. Electrical energy is used in the manufacture of hydrogen, oxygen, ozone, hydrogen peroxide, chlorine, sodium hydroxide, and halogen-oxygen compounds (Chapter 19). Electrical energy is also used in the production of many other chemicals and in the electrorefining of metals, the electroplating of metals and alloys, and the production of metal articles by electrodeposition. The generation of electricity by batteries is an electrochemical process.

20.1 The Conduction of Electricity

The movement of charged particles, either electrons or positive or negative ions, through a conductor produces a current of electricity. Metallic conduction occurs when electrons move through a metal with no changes in the metal and no movement of the metal atoms. **Electrolytic, or ionic, conduction** occurs when ions move through a molten substance, a solution, or, occasionally, a solid.

Compounds that conduct an electric current by the movement of ions are called **electrolytes** (Section 13.8). Two electrodes, which are good conductors, dip into the electrolyte. Electrons are forced onto one of the electrodes and withdrawn from the other by a battery or a generator (Fig. 20-1). The electrode onto which the electrons are forced becomes negatively charged, and the electrode from which electrons are withdrawn becomes positively charged. The positive ions (cations) in the electrolyte are attracted by the negatively charged cathode and move toward it, and the negative ions (anions) are attracted by the positively charged anode and move toward it. The motion of the ions results in the passage of a current.

Figure 20-1. An electrolytic cell used to determine the conductivity of a solution.

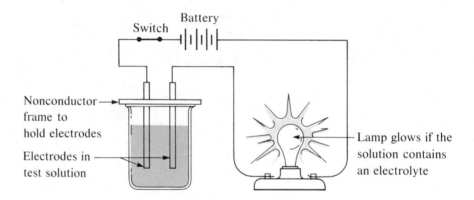

ELECTROLYTIC CELLS

20.2 The Electrolysis of Molten Sodium Chloride

When an electric current passes between two electrodes placed in molten sodium chloride, chemical reactions are observed at each electrode. Molten sodium chloride contains equal numbers of sodium ions and chloride ions that move about with considerable freedom. When an electrochemical cell containing molten sodium chloride is connected to a battery, the positive sodium ions are attracted to the cathode, where they combine with electrons to form sodium atoms, that is, metallic sodium. The reaction at the cathode, called a **half-reaction,** is

$$Na^+ + e^- \longrightarrow Na$$

The oxidation number of sodium decreases from $+1$ to 0 in this reduction reaction, sometimes called cathodic reduction. Thus the **cathode** is the electrode at which reduction occurs. The negative chloride ions are attracted to the anode, where they give up one electron each and become chlorine atoms, which then combine to form molecules of chlorine gas. The half-reaction at the anode is

$$2Cl^- \longrightarrow Cl_2 + 2e^-$$

The oxidation number of chlorine increases from -1 to 0, in a reaction known as anodic oxidation. The **anode** is the electrode at which oxidation occurs.

The net reaction for the electrolytic decomposition of molten sodium chloride is the sum of the two electrode half-reactions. The number of electrons produced by the oxidation must equal the number of electrons consumed in the reduction, so the cathode reaction must be multiplied by 2 to balance the complete equation.

$$
\begin{array}{ll}
\textit{Cathode:} & 2Na^+ + 2e^- \longrightarrow 2Na \\
\textit{Anode:} & \underline{2Cl^- \longrightarrow Cl_2 + 2e^-} \\
& 2Na^+ + 2Cl^- \longrightarrow 2Na + Cl_2 \\
\text{or} & 2NaCl \longrightarrow 2Na + Cl_2
\end{array}
$$

The production of an oxidation-reduction reaction by a direct current (dc) of electricity is known as **electrolysis.**

20.3 The Electrolysis of Aqueous Hydrochloric Acid

When an aqueous solution of fairly concentrated hydrochloric acid is electrolyzed, the positive hydrogen ions are attracted to the cathode, accept electrons from it, and are reduced to hydrogen gas.

$$\textit{Cathodic reduction:} \quad 2H^+ + 2e^- \longrightarrow H_2$$

The negative chloride ions are attracted to the anode, lose electrons, and are oxidized to chlorine gas.

$$\textit{Anodic oxidation:} \quad 2Cl^- \longrightarrow Cl_2 + 2e^-$$

The overall cell reaction may be obtained by adding the electrode reactions.

$$
\begin{array}{ll}
\textit{Cathode:} & 2H^+ + 2e^- \longrightarrow H_2 \\
\textit{Anode:} & \underline{2Cl^- \longrightarrow Cl_2 + 2e^-} \\
& 2H^+ + 2Cl^- \longrightarrow H_2 + Cl_2
\end{array}
$$

20.4 The Electrolysis of an Aqueous Sodium Chloride Solution

When a concentrated aqueous solution of sodium chloride is electrolyzed (Fig. 20-2), chlorine gas is formed at the anode, and hydrogen gas and hydroxide ion are formed at the cathode. In order to account for these products, we need to consider all of the possible electrode reactions.

The reductions possible at the cathode are the reduction of sodium ion and the reduction of water:

$$Na^+ + e^- \longrightarrow Na \qquad (1)$$
$$2H_2O + 2e^- \longrightarrow H_2 + 2OH^- \qquad (2)$$

Water is more easily reduced than is sodium ion, so hydroxide ions and hydrogen gas form [Equation (2)]. The hydroxide ions begin to migrate toward the anode.

Figure 20-2. The electrolysis of an aqueous solution of sodium chloride. The reduction of water molecules at the cathode produces hydrogen gas and hydroxide ions, and the oxidation of chloride ions at the anode produces chlorine gas. A solution of sodium hydroxide remains.

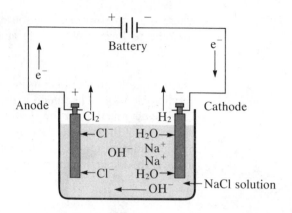

The oxidations possible at the anode are the oxidation of chloride ion and the oxidation of water:

$$2Cl^- \longrightarrow Cl_2 + 2e^- \tag{3}$$

$$2H_2O \longrightarrow O_2 + 4H^+ + 4e^- \tag{4}$$

Chloride ion and water are oxidized with almost equal ease. Hence the concentration of the salt solution plays the dominant role in determining the product. With a high chloride ion concentration, the reaction of Equation (3) occurs, and chlorine gas is formed at the anode. With a low chloride ion concentration, the reaction of Equation (4) occurs as well, and oxygen is formed in addition to the chlorine. Little chlorine is produced from a very dilute solution.

The net reaction for the electrolysis of a concentrated aqueous solution of sodium chloride is obtained by adding the electrode reactions.

Cathode: $2H_2O + 2e^- \longrightarrow H_2 + 2OH^-$
Anode: $\underline{2Cl^- \longrightarrow Cl_2 + 2e^-}$
$$2H_2O(l) + 2Cl^- \longrightarrow H_2(g) + Cl_2(g) + 2OH^-$$

Hydroxide ions are formed during the electrolysis, while chloride ions are removed from solution. Because the sodium ions remain unchanged, sodium hydroxide accumulates as electrolysis proceeds.

The electrolysis of aqueous sodium chloride is an important commercial process for the production of hydrogen, chlorine, and sodium hydroxide (Sections 8.7 and 19.2).

20.5 The Electrolysis of an Aqueous Sulfuric Acid Solution

Sulfuric acid ionizes in water in two steps.

$$H_2SO_4 \longrightarrow H^+ + HSO_4^- \quad \text{(essentially complete)}$$
$$HSO_4^- \rightleftharpoons H^+ + SO_4^{2-}$$

In dilute solution the sulfate ion is the principal anion present. During electrolysis two oxidations are possible at the anode: oxidation of the sulfate ion and oxidation of water.

$$2SO_4^{2-} \longrightarrow S_2O_8^{2-} + 2e^-$$
$$2H_2O \longrightarrow O_2(g) + 4H^+ + 4e^-$$

Oxygen gas and hydrogen ions form because water is more easily oxidized than is sulfate ion.

The only reduction possible at the cathode is

$$2H^+ + 2e^- \longrightarrow H_2$$

The net reaction is the sum of the anode and cathode reactions. (The cathode reaction is multiplied by 2 to balance the number of electrons in the complete oxidation and reduction.)

Anode: $2H_2O \longrightarrow O_2 + 4H^+ + 4e^-$
Cathode: $4H^+ + 4e^- \longrightarrow 2H_2$
$$\overline{2H_2O(l) \longrightarrow 2H_2(g) + O_2(g)}$$

The electrolysis of an aqueous solution of sulfuric acid produces hydrogen and oxygen. Although the hydrogen ion from the sulfuric acid is consumed by the cathodic reduction, it is regenerated at the same rate by the anodic oxidation of water. Water is used up, and the sulfuric acid becomes more concentrated as electrolysis proceeds.

20.6 Electrolytic Refining of Metals

The electrodes used in the electrolytic cells considered thus far are conductive but unreactive materials such as graphite or platinum. However, when a more reactive metal is used as the anode of an electrolytic cell, the anode reaction may involve oxidation of the metal. For example, when a strip of metallic copper is used as the anode in the electrolysis of a solution of copper(II) sulfate (Fig. 20-3), the following anodic oxidation occurs:

$$Cu \longrightarrow Cu^{2+} + 2e^-$$

The oxidation of copper is easier than the oxidation of water, so the electrode goes into solution as copper(II) ions. The copper(II) ions formed at the anode migrate to the cathode, where they are more readily reduced than is the water.

$$Cu^{2+} + 2e^- \longrightarrow Cu$$

Thus pure copper plates out on the cathode.

Figure 20-3. The electrolytic refining of copper.

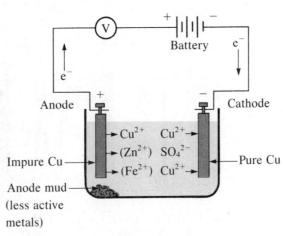

$$Cu \rightarrow Cu^{2+} + 2e^- \qquad Cu^{2+} + 2e^- \rightarrow Cu$$

Traces of impurities can reduce copper's electrical conductivity by more than 10%, so impure copper wires may get dangerously hot when they conduct electricity. Crude copper is refined electrochemically to improve its electrical conductivity. When impure copper is used as an anode, less active impurities in the copper, such as gold and silver, do not dissolve (are not oxidized) and fall to the bottom of the cell, forming a mud from which they are readily recovered. More active impurities, such as zinc and iron, are oxidized and go into the solution as ions. If the electrical potential between the electrodes is carefully regulated, these ions are not reduced at the cathode; only copper is reduced and deposited. The refined copper is deposited either on a thin sheet of pure copper, which serves as a cathode, or on some other metal from which the deposit may be stripped.

20.7   Faraday's Law of Electrolysis

In 1832–1833 Michael Faraday performed experiments that resulted in the statement of **Faraday's law: The amount of a substance undergoing a chemical change at each electrode during electrolysis is directly proportional to the quantity of electricity that passes through the electrolytic cell** (Fig. 20-4).

The quantity of electricity can be expressed as the number of electrons. The quantity of a substance undergoing chemical change can be expressed in terms of moles of the substance or in terms of equivalents of the substance. An **equivalent** of a substance is the mass of the substance, in grams, that combines with or releases 1 mole of electrons. The **normality, N,** of a solution is the number of equivalents of solute per liter of solution (equiv solute/L solution). A liter of solution that contains 1 equivalent of a substance is a $1N$ solution (Section 14.14).

Experiments show that one electron reduces one silver ion, while two electrons are required to reduce one copper(II) ion.

$$Ag^+ + e^- \longrightarrow Ag$$
$$Cu^{2+} + 2e^- \longrightarrow Cu$$

Figure 20-4. Amounts of various elements discharged at the cathode by 1 faraday of electricity (96,487 C, or 1 mol of electrons).

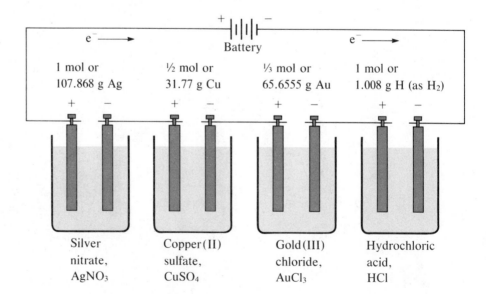

Therefore 6.022×10^{23} electrons reduce 6.022×10^{23} silver ions (1 mole) to silver atoms. The mass of silver produced is 107.868 g; thus the mass of an equivalent of silver is the same as the mass of one mole. Since two electrons are required to reduce one copper(II) ion to a copper atom, 6.022×10^{23} electrons reduce only $\frac{1}{2}$ mole of copper. Thus the mass of an equivalent of copper is $\frac{1}{2}$ of the mass of one mole.

A **faraday** is 1 mole of electrons; 1 faraday will reduce 1 equivalent of a substance or is produced by the oxidation of 1 equivalent of a substance; 1 mole of electrons, or 6.022×10^{23} electrons is lost or gained.

$$1 \text{ faraday} = 6.022 \times 10^{23} \text{ electrons} = 1 \text{ mole of electrons}$$
$$= 96{,}487 \text{ coulombs (C)}$$

1 coulomb (C) = quantity of electricity involved when a current of 1 ampere (A) flows past a given point for 1 second = 1 A s

Calculations involving the passage of an electric current (which is measured in amperes) through a cell may be handled like any other stoichiometry problem. Convert the current in amperes to coulombs, then to faradays, and, finally, to moles of electrons. Use the moles of electrons as you would moles of any other reactant.

EXAMPLE 20.1   Calculate the mass of copper produced by the cathodic reduction of Cu^{2+} ion during the passage of 1.600 A of current through a solution of copper(II) sulfate for 1.000 h.

Copper(II) ion is reduced at the cathode.

$$Cu^{2+} + 2e^- \longrightarrow Cu$$

Here 1 mol of copper is produced for each 2 mol of electrons, e^-, that pass through the solution (1 mol Cu/2 mol e^-). The number of moles of electrons passing through the cell is obtained in the following manner:

$$1.600 \text{ A} \times 1.000 \text{ h} \times \frac{60 \text{ min}}{1 \text{ h}} \times \frac{60 \text{ s}}{1 \text{ min}} = 5760 \text{ A s} = 5760 \text{ C}$$

$$5760 \text{ C} \times \frac{1 \text{ faraday}}{96{,}487 \text{ C}} = 5.970 \times 10^{-2} \text{ faraday}$$

$$5.970 \times 10^{-2} \text{ faraday} \times \frac{1 \text{ mol } e^-}{1 \text{ faraday}} = 5.970 \times 10^{-2} \text{ mol } e^-$$

The above half-reaction at the cathode is used to relate moles of electrons to the mass of copper produced.

$$5.970 \times 10^{-2} \text{ mol } e^- \times \frac{1 \text{ mol Cu}}{2 \text{ mol } e^-} = 2.985 \times 10^{-2} \text{ mol Cu}$$

$$2.985 \times 10^{-2} \text{ mol Cu} \times \frac{63.55 \text{ g Cu}}{1 \text{ mol Cu}} = 1.897 \text{ g Cu}$$

ELECTRODE POTENTIALS

20.8 Single Electrodes and Electrode Potentials

The **electrical potential** is the energy change that accompanies transfer of 1 coulomb from one location to another. The unit of electrical potential is the **volt (V)**, which is 1 joule per coulomb. If an energy change of 1 joule accompanies the transfer of 1 coulomb, the potential difference is 1 volt (1 V = 1 J/C). The energy changes associated with the operation of electrolytic and voltaic cells are expressed in volts.

When a strip of a metal such as zinc is placed in water, a few of the atoms leave the metal and pass into solution as ions. The valence electrons lost in forming the ions remain on the metal. The change may be expressed for zinc by

$$Zn \longrightarrow Zn^{2+}(aq) + 2e^- \text{ (on the strip)}$$

The zinc strip becomes negatively charged, and the solution becomes positively charged. Thus a difference in potential is established between the metal and the solution—the metal is negative with respect to the solution. The negative charge generated on the zinc strip attracts the positive zinc ions and the electrons on the strip can reduce them back to the metal

$$Zn^{2+} + 2e^- \longrightarrow Zn$$

Thus the change is reversible, and an equilibrium is established.

$$Zn \rightleftharpoons Zn^{2+} + 2e^-$$

An increase in the concentration of zinc ions in the solution, brought about by the addition of a zinc salt, tends to reverse the process.

At equilibrium the difference in potential between the metal and the solution depends on (1) the tendency of the metal to form ions, (2) the concentration of the metal ions, and (3) the tendency of the metal ions to be reduced to the metal. More active (more easily oxidized) metals have greater tendencies to form ions and thus are more negative electrodes. A zinc electrode is more negative than a copper electrode, because zinc is more easily oxidized than copper. The electrode potential also varies with the concentration of the metal ion. The potential of the zinc electrode can be varied by changing the concentrations of the zinc ions in solution. With an increase in the zinc ion concentration, the equilibrium

$$Zn \rightleftharpoons Zn^{2+} + 2e^-$$

shifts to the left, and the difference in potential between the metal and the solution decreases. A decrease in the zinc ion concentration shifts the equilibrium to the right, and the difference in potential increases. Since the zinc electrode is a solid, its concentration is constant, and the size of the electrode does not influence the electrode potential.

To be precise, we should use the *activity* rather than the *concentration* of the zinc ion in this discussion (Section 13.27). However, the approximation that the activity of an ionic species is equal to its concentration is acceptable for our purposes.

20.9   Standard Electrode Potentials

There is no satisfactory method for measuring the difference in electrical potential between a single metal electrode and a solution of its ions. However, we can measure the difference between the potentials of two electrodes. Thus if we want to compare two different electrodes, we must compare the potentials of two cells, each containing one of the electrodes coupled with a third. The third electrode is used as a standard with which the potentials of the other two are compared. The **standard hydrogen electrode** (Fig. 20-5) is the electrode customarily chosen for such comparison. The voltage between this electrode and its solution is arbitrarily taken to be zero—its potential is defined as exactly zero volts. (This is similar to arbitrarily establishing sea level as zero elevation.)

Electrodes involving gases can be set up by bubbling the gas around an inert metallic conductor that conducts electrons but does not enter the electrode reaction. The hydrogen electrode is constructed so that hydrogen gas is bubbled around a platinum foil or wire covered with very finely divided platinum (platinized) and immersed in a solution containing hydrogen ions. There is a tendency for hydrogen ions to be reduced to hydrogen by acquiring electrons from the electrode. Thus a difference in potential between the electrode and the solution is established, as it was for zinc.

$$2H^+(aq) + 2e^- \text{ (on the electrode)} \rightleftharpoons H_2$$

The platinum acts as an electrode and as a catalyst that quickly brings the reaction to equilibrium.

The potential of a gas electrode increases with increasing pressure (activity) of the gas. Therefore a standard hydrogen electrode is prepared by bubbling hydrogen gas at a temperature of 25°C and a pressure of 1 atmosphere around platinized platinum immersed in a solution containing a 1 M concentration of hydrogen ions. *We arbitrarily assign the value of exactly zero volt to the potential for the standard hydrogen electrode.*

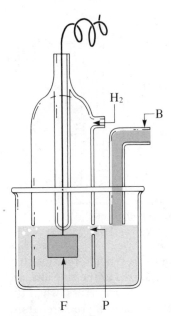

Figure 20-5.   Diagram of a simple standard hydrogen electrode. F indicates the platinized platinum foil; P, the port for escape of hydrogen bubbles; B, part of the salt bridge.

20.10   Measurement of Electrode Potentials

To measure the electrode potential of zinc, for example, an electrochemical cell is set up consisting of a zinc strip in contact with a 1 M solution of zinc ions as half of the cell and a standard hydrogen electrode as the other half (Fig. 20-6). The two halves of the cell are connected by a **salt bridge** consisting of a concentrated solution of an electrolyte, such as potassium chloride. This bridge allows positive and negative ions, such as K^+ and Cl^-, to migrate from cell to cell and thus allows a current to flow while preventing the two solutions from mixing. The zinc strip in the solution of zinc ions and the platinum strip of the hydrogen electrode are connected to a voltmeter, which shows the potential difference, in volts, between the two electrodes.

Before the circuit is closed, that is, before the zinc and platinum strips are connected by a wire, we have the following electrode half-reactions, each at equilibrium.

$$2H^+ + 2e^- \rightleftharpoons H_2$$
$$Zn^{2+} + 2e^- \rightleftharpoons Zn$$

Figure 20-6.   Measurement of the electrode potential of zinc, using the standard hydrogen electrode as the reference.

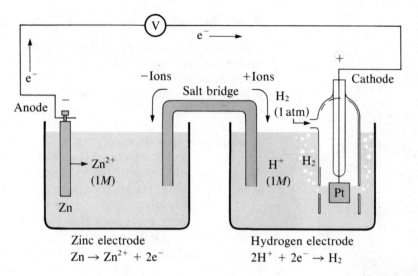

Because zinc has a greater tendency to form ions than hydrogen has, the zinc strip acquires a higher electron density (from the electrons released as the zinc ions are formed) than does the platinum of the hydrogen electrode. Therefore, when the electrical circuit is closed, the greater electron density, or electron pressure, at the zinc electrode causes electrons to flow from the zinc electrode to the hydrogen electrode. The electron density increases at the hydrogen electrode, and its equilibrium is shifted to the left, causing hydrogen ions to be reduced to hydrogen gas. The reaction that occurs at the hydrogen electrode as current flows is

$$2H^+ + 2e^- \longrightarrow H_2$$

At the same time the electron density at the zinc electrode decreases because electrons are flowing to the hydrogen electrode. The equilibrium at the zinc electrode is shifted to the right, and more zinc is oxidized to zinc ions. The reaction that occurs at the zinc electrode as current flows is

$$Zn \longrightarrow Zn^{2+} + 2e^-$$

The overall reaction that takes place when the zinc-hydrogen cell is operating is an oxidation-reduction reaction.

Anodic oxidation:	$Zn \longrightarrow Zn^{2+} + 2e^-$
Cathodic reduction:	$2H^+ + 2e^- \longrightarrow H_2$
Cell reaction:	$Zn + 2H^+ \longrightarrow Zn^{2+} + H_2(g)$

This same reaction takes place when zinc metal is immersed in a solution of hydrochloric acid. However, in the zinc-hydrogen cell, the reaction takes place without any contact between the reactants. The electrons involved in the oxidation-reduction reaction are transferred from the zinc to the hydrogen ions through a wire. This is important from a practical point of view because the current can be used to do work in an electric motor, to light an electric lamp, or to produce some other form of energy.

The voltmeter shows the difference in potential between the two half-cells, the **electromotive force (emf)**, which in this case is 0.76 volt. This 0.76-volt difference causes electrons to move from the zinc electrode to the hydrogen electrode. Since the potential of the hydrogen electrode is assumed to be zero volt, we assign the entire electromotive force of the cell to the zinc electrode. We say that the potential of the zinc electrode for the oxidation of zinc metal to zinc ion,

$$Zn \longrightarrow Zn^{2+} + 2e^-$$

is 0.76 volt above that of the hydrogen electrode and is positive ($+0.76$ V). For the reverse reaction, the reduction of zinc ion to metallic zinc,

$$Zn^{2+} + 2e^- \longrightarrow Zn$$

the electrode potential has the same absolute value but is of opposite sign (-0.76 V).

When copper metal in contact with a 1 M solution of copper(II) ions is the second electrode in a cell containing a standard hydrogen electrode, we find that hydrogen has a greater tendency to form positive ions than copper has. The formation of hydrogen ions from hydrogen leaves more electrons on the platinum than are left on the copper when the latter forms positive ions. Consequently, electrons flow from the hydrogen electrode to the copper electrode where they reduce copper(II) ions.

Anodic oxidation:	$H_2 \rightleftharpoons 2H^+ + 2e^-$
Cathodic reduction:	$Cu^{2+} + 2e^- \rightleftharpoons Cu$
Cell reaction:	$H_2 + Cu^{2+} \rightleftharpoons Cu + 2H^+$

The emf of the cell as measured by a voltmeter is 0.337 volt. Since copper ion is reduced to the free element more easily than is hydrogen ion, the potential of the copper electrode for the reduction

$$Cu^{2+} + 2e^- \longrightarrow Cu$$

has a positive value ($+0.337$ V) with respect to the hydrogen electrode.

The values of electrode potentials are customarily given for the reduction process. Standard electrode potentials, or standard reduction potentials, $E°$, are potentials at electrodes in which all components are in a standard state at 25°C with ion concentrations (activities) of 1 M and gas pressures (activities) of 1 atmosphere. The standard reduction potential, $E°$, for zinc is -0.76 volt. The standard reduction potential for any metal electrode having an electron density greater than the hydrogen electrode has a negative sign. An electrode that has less electron density than the hydrogen electrode (copper, for example) is positive with reference to the hydrogen electrode and hence has a positive potential. The standard reduction potential for copper is $+0.337$ volt. The assignment of positive or negative values to electrode potentials is arbitrary, and some scientists follow the opposite convention, defining electrode potentials in terms of the oxidation process and hence assigning the opposite signs to the potentials.

The standard potential of an electrode is its potential with respect to the standard hydrogen electrode, measured at 25°C when the concentration of each ion in the solution is 1 M and the pressure of any gas involved is 1 atmosphere. Compared with the standard hydrogen electrode potential of 0.000 volt, the standard reduction potential of sodium is -2.71 volt, that of zinc is -0.76 volt, that of copper is $+0.337$ volt, and that of silver is $+0.7991$ volt. As the standard reduction potential for a series of half-cells becomes more negative, it becomes progressively more difficult to carry out the reduction. Thus Na^+ is more difficult to reduce than Zn^{2+}, which is more difficult to reduce than Cu^{2+}, which is more difficult to reduce than Ag^+. Also, the more negative the standard reduction potential, the more positive is the potential for the reverse reaction, the oxidation

$$M \longrightarrow M^{n+} + ne^-$$

and the more easily oxidized is the metal.

Any cell that generates an electric current by an oxidation-reduction reaction is called a **voltaic cell.** The cell composed of zinc and hydrogen electrodes is a voltaic cell that can be diagrammed as follows:

$$\overset{e^-}{\longrightarrow}$$
$$\text{Zn} \,|\, \text{Zn}^{2+}, M = 1 \quad \| \quad \text{H}^+, M = 1 \,|\, \text{H}_2, 1 \text{ atm (Pt)}$$

This diagram indicates zinc metal in contact with a 1 M solution of zinc ions. The solution of zinc ions is connected by a salt bridge, represented by ‖, to a 1 M solution of hydrogen ions in a hydrogen electrode with a gaseous hydrogen pressure of 1 atm. A single vertical line, |, is used to separate two different phases; two different species in the same phase would be separated by a semicolon. The anode (the electrode at which oxidation occurs) is customarily written on the left in such a diagram.

The purpose of the salt bridge in a cell is to establish a path for electrolytic conduction between the solutions of the two electrodes. An excess of zinc ions accumulates in the half-cell where zinc goes into solution. Similarly, an excess of anions accumulates in the half-cell where hydrogen ions are reduced. Migration of ions through the salt bridge (negative ions into the zinc half-cell solution and positive ions into the hydrogen half-cell solution) compensates for this imbalance and maintains neutral solutions. Without the salt bridge no current would flow in the external circuit, and the cell reactions would not take place. The ions that migrate through the salt bridge to maintain neutrality do not need to be the ions participating in the electrode reactions. In the zinc-hydrogen cell, they are in fact largely chloride ions and potassium ions supplied by the salt bridge.

20.11 The Activity, or Electromotive, Series

The positions of the metals in the activity series (Section 9.17) can be determined from their standard reduction potentials. When the standard reduction potentials of a number of metals are arranged in order from the most negative to the most positive, we have the activity series, shown in Table 20-1. This table also gives the reduction potentials of a few nonmetals. (Additional reduction potentials are given in Table 20-2 and in Appendix I.)

The activity series correlates many chemical properties of the elements; some of the more important ones are given in the following list.

1. The metals with large negative reduction potentials at the top of the series are good reducing agents in the free state. They are the metals most easily oxidized to their ions by the removal of electrons (active metals).

2. The elements with large positive reduction potentials at the bottom of the series are good oxidizing agents when in the oxidized form, that is, when the metals are in the form of ions and the nonmetals are in the elemental state.

3. The reduced form of any element will reduce the oxidized form of any element below it. For example, metallic zinc will reduce copper(II) ions according to the equation

$$\text{Zn} + \text{Cu}^{2+} \rightleftharpoons \text{Cu} + \text{Zn}^{2+}$$

Table 20-1 Standard Reduction Potentials

	Electrode Reaction		Standard Reduction Potential, $E°$, V
	Oxidized Form	Reduced Form	
Potassium	$K^+ + e^-$	$\rightleftharpoons K$	−2.925
Calcium	$Ca^{2+} + 2e^-$	$\rightleftharpoons Ca$	−2.87
Sodium	$Na^+ + e^-$	$\rightleftharpoons Na$	−2.714
Magnesium	$Mg^{2+} + 2e^-$	$\rightleftharpoons Mg$	−2.37
Aluminum	$Al^{3+} + 3e^-$	$\rightleftharpoons Al$	−1.66
Manganese	$Mn^{2+} + 2e^-$	$\rightleftharpoons Mn$	−1.18
Zinc	$Zn^{2+} + 2e^-$	$\rightleftharpoons Zn$	−0.763
Chromium	$Cr^{3+} + 3e^-$	$\rightleftharpoons Cr$	−0.74
Iron	$Fe^{2+} + 2e^-$	$\rightleftharpoons Fe$	−0.440
Cadmium	$Cd^{2+} + 2e^-$	$\rightleftharpoons Cd$	−0.40
Cobalt	$Co^{2+} + 2e^-$	$\rightleftharpoons Co$	−0.277
Nickel	$Ni^{2+} + 2e^-$	$\rightleftharpoons Ni$	−0.250
Tin	$Sn^{2+} + 2e^-$	$\rightleftharpoons Sn$	−0.136
Lead	$Pb^{2+} + 2e^-$	$\rightleftharpoons Pb$	−0.126
Hydrogen	$2H^+ + 2e^-$	$\rightleftharpoons H_2$	0.00
Copper	$Cu^{2+} + 2e^-$	$\rightleftharpoons Cu$	+0.337
Iodine	$I_2 + 2e^-$	$\rightleftharpoons 2I^-$	+0.5355
Mercury	$Hg_2^{2+} + 2e^-$	$\rightleftharpoons 2Hg$	+0.789
Silver	$Ag^+ + e^-$	$\rightleftharpoons Ag$	+0.7991
Bromine	$Br_2(l) + 2e^-$	$\rightleftharpoons 2Br^-$	+1.0652
Platinum	$Pt^{2+} + 2e^-$	$\rightleftharpoons Pt$	+1.2
Oxygen	$O_2 + 4H^+ + 4e^-$	$\rightleftharpoons 2H_2O$	+1.23
Chlorine	$Cl_2 + 2e^-$	$\rightleftharpoons 2Cl^-$	+1.3595
Gold	$Au^+ + e^-$	$\rightleftharpoons Au$	+1.68
Fluorine	$F_2 + 2e^-$	$\rightleftharpoons 2F^-$	+2.87

Increasing ease of reduction of the reactant

Increasing ease of oxidation of the product

20.12 Calculation of Cell Potentials

The net cell reaction of a voltaic cell made up of a standard zinc electrode and a standard copper electrode (Fig. 20-7), is obtained by adding the two half-reactions together.

Anode half-reaction:	$Zn \rightleftharpoons Zn^{2+} + 2e^-$
Cathode half-reaction:	$Cu^{2+} + 2e^- \rightleftharpoons Cu$
Net cell reaction:	$Zn + Cu^{2+} \rightleftharpoons Zn^{2+} + Cu$

The emf of the cell can be calculated by adding the standard potential of the copper electrode to that of the zinc electrode, taking into account the appropriate sign on each potential, based on whether the half-reaction taking place is oxidation or reduction. The standard reduction potential for zinc is −0.76 volt. This is the potential for the half-reaction

$$Zn^{2+} + 2e^- \longrightarrow Zn$$

The oxidation of Zn to Zn^{2+}, the reverse of that half-reaction, has a potential of +0.76 volt. The standard reduction potential for copper is +0.337 volt. Since

Figure 20-7. A zinc-copper voltaic cell.

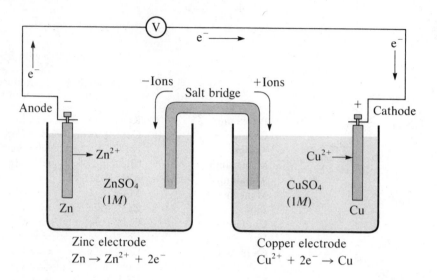

Zinc electrode
$$Zn \rightarrow Zn^{2+} + 2e^-$$

Copper electrode
$$Cu^{2+} + 2e^- \rightarrow Cu$$

Cu^{2+} is being reduced to Cu, the potential for the copper half-reaction is $+0.337$ volt. The emf for the cell is

$$(+0.76 \text{ V}) + (+0.337 \text{ V}) = 1.10 \text{ V}$$

The positive emf value calculated for the zinc-copper cell indicates that the net cell reaction will proceed spontaneously to the right and that the cell will deliver an electric current. A negative emf value for a cell indicates that the net cell reaction will not proceed spontaneously to the right but will proceed spontaneously to the left.

EXAMPLE 20.2   Calculate the emf of a cell composed of copper and silver electrodes in 1 M solutions of their respective ions.

Because copper has a lower $E°$ than silver (Cu is higher in the activity series than Ag), copper metal will reduce silver ion.

$$Cu + 2Ag^+ \rightleftharpoons Cu^{2+} + 2Ag$$

To obtain this net cell reaction, we add the half-reactions. To obtain the emf we add the half-cell potentials.

$$
\begin{array}{ll}
Cu \rightleftharpoons Cu^{2+} + 2e^- & E° = -0.337 \text{ V} \\
\underline{2Ag^+ + 2e^- \rightleftharpoons 2Ag} & \underline{E° = +0.799 \text{ V}} \\
Cu + 2Ag^+ \rightleftharpoons Cu^{2+} + 2Ag & E° = (-0.337 \text{ V}) + (+0.779 \text{ V}) \\
& = +0.442 \text{ V}
\end{array}
$$

The emf of the cell is 0.442 V. Since it is positive, the reaction proceeds spontaneously to the right as written, and the cell delivers current. Note that the potential of the copper half-reaction used in this solution, an oxidation half-reaction, is equal in magnitude but of opposite sign to that listed in Table 20-1 for the reduction half-reaction

$$Cu^{2+} + 2e^- \longrightarrow Cu$$

EXAMPLE 20.3 Calculate the emf of a cell made up of standard bromine and chlorine electrodes.

The half-reactions and standard potentials (Table 20-1) are

$$
\begin{array}{ll}
Cl_2 + 2e^- \rightleftharpoons 2Cl^- & E^\circ = +1.3595 \text{ V} \\
\underline{2Br^- \rightleftharpoons Br_2 + 2e^-} & E^\circ = -1.0652 \text{ V} \\
2Br^- + Cl_2 \rightleftharpoons Br_2 + 2Cl^- & E^\circ = +0.2943 \text{ V}
\end{array}
$$

The emf of the cell is thus found to be 0.2943 V, meaning that the net cell reaction proceeds spontaneously to the right as written (Cl_2 oxidizing Br^-). If the half-reactions are added as follows:

$$
\begin{array}{ll}
2Cl^- \rightleftharpoons Cl_2 + 2e^- & E^\circ = -1.3595 \text{ V} \\
\underline{Br_2 + 2e^- \rightleftharpoons 2Br^-} & E^\circ = +1.0652 \text{ V} \\
2Cl^- + Br_2 \rightleftharpoons Cl_2 + 2Br^- & E^\circ = -0.2943 \text{ V}
\end{array}
$$

the negative emf value indicates that the reaction proceeds spontaneously to the left, rather than to the right. Chlorine oxidizes bromide ion to free bromine, but bromine will not oxidize chloride ion to free chlorine (Section 19.1).

Standard reduction potentials (such as those in Table 20-1) refer to standard conditions with concentrations of 1 M in aqueous solution, pressures of 1 atm, and a temperature of 25°C. If the conditions change, the potentials change, and some members may even change places in the table. The effect of the concentration on potential is illustrated by the reduction potentials for the hydrogen half-reaction at two different concentrations.

$$
\begin{array}{ll}
2H^+(1\ M) + 2e^- \rightleftharpoons H_2(1\ atm) & E^\circ = 0.00 \text{ V} \\
2H^+(10^{-7}\ M) + 2e^- \rightleftharpoons H_2(1\ atm) & E^\circ = -0.41 \text{ V}
\end{array}
$$

The values given in Table 20-1 do not apply to nonaqueous solutions or to molten salts, although similar electromotive series can be constructed for them. The electrode potentials are different from those for water, and the members of the series fall in a different order because solvation energies vary from solvent to solvent.

20.13 Electrode Potentials for Other Half-Reactions

In addition to reactions between elements, there are many other oxidation-reduction reactions that can take place at electrodes. Table 20-2 and Appendix I contain standard reduction potentials for a number of half-reactions in which the electrode is an inactive substance such as carbon, which undergoes no change itself and merely serves as a carrier of electrons.

The standard reduction potential of the Pt, Sn^{4+}, Sn^{2+} electrode (Table 20-2) is that of an inert platinum strip immersed in a solution that is 1 M in both Sn^{4+} and Sn^{2+} ions. Similarly, the standard reduction potential of a Pt, Fe^{3+}, Fe^{2+} electrode is that when Fe^{3+} and Fe^{2+} are both 1 M. The emf of a cell composed of these two electrodes

$$
\text{(Pt) } Sn^{2+},\ M = 1;\ Sn^{4+},\ M = 1 \quad \overset{e^-}{\longrightarrow} \quad \| \quad Fe^{3+},\ M = 1;\ Fe^{2+},\ M = 1 \text{ (Pt)}
$$

Table 20-2 Standard Reduction Potentials for Some Selected Half-Reactions

| Electrode | Electrode Reaction | | Standard Reduction Potential, V |
	Oxidized Form	Reduced Form	
Fe, Fe(OH)$_2$, OH$^-$	Fe(OH)$_2$ + 2e$^-$ $\rightleftharpoons$ Fe + 2OH$^-$		-0.877
Pb, PbSO$_4$, SO$_4{}^{2-}$	PbSO$_4$ + 2e$^-$ $\rightleftharpoons$ Pb + SO$_4{}^{2-}$		-0.356
Pt, Sn^{4+}, Sn^{2+}	Sn^{4+} + 2e$^-$ $\rightleftharpoons$ Sn^{2+}		$+0.15$
Ag, AgCl, Cl$^-$	AgCl + e$^-$ $\rightleftharpoons$ Ag + Cl$^-$		$+0.222$
Hg, Hg$_2$Cl$_2$, Cl$^-$	Hg$_2$Cl$_2$ + 2e$^-$ $\rightleftharpoons$ 2Hg + 2Cl$^-$		$+0.27$
NiO$_2$, Ni(OH)$_2$, OH$^-$	NiO$_2$ + 2H$_2$O + 2e$^-$ $\rightleftharpoons$ Ni(OH)$_2$ + 2OH$^-$		$+0.49$
Pt, Fe^{3+}, Fe^{2+}	Fe^{3+} + e$^-$ $\rightleftharpoons$ Fe^{2+}		$+0.771$
Pt, Cr$_2$O$_7{}^{2-}$, H$^+$, Cr^{3+}	Cr$_2$O$_7{}^{2-}$ + 14H$^+$ + 6e$^-$ $\rightleftharpoons$ 2Cr^{3+} + 7H$_2$O		$+1.33$
Pt, MnO$_4{}^-$, H$^+$, Mn^{2+}	MnO$_4{}^-$ + 8H$^+$ + 5e$^-$ $\rightleftharpoons$ Mn^{2+} + 4H$_2$O		$+1.51$
PbO$_2$, PbSO$_4$, H$_2$SO$_4$	PbO$_2$ + SO$_4{}^{2-}$ + 4H$^+$ + 2e$^-$ $\rightleftharpoons$ PbSO$_4$ + 2H$_2$O		$+1.685$

can be calculated as follows:

$$\begin{array}{ll} \text{Sn}^{2+} \rightleftharpoons \text{Sn}^{4+} + 2e^- & E° = -0.15 \text{ V} \\ \underline{2\text{Fe}^{3+} + 2e^- \rightleftharpoons 2\text{Fe}^{2+}} & \underline{E° = +0.771 \text{ V}} \\ \text{Sn}^{2+} + 2\text{Fe}^{3+} \rightleftharpoons \text{Sn}^{4+} + 2\text{Fe}^{2+} & E° = +0.62 \text{ V} \end{array}$$

The emf of the cell is 0.62 V when standard electrodes are used. This means that the reaction will proceed spontaneously to the right as written and will deliver a current. It also means that when solutions containing tin(II) and iron(III) are mixed, the Sn^{2+} will reduce the Fe^{3+}.

20.14 The Nernst Equation

Standard reduction potentials (Section 20.11), $E°$, are tabulated for the standard state condition with 1 M solutions of ions, 1 atmosphere of pressure for gases, and a temperature of 25°C. However, as pointed out in Section 20.12, the reduction potential of a half-reaction changes when the concentrations change. The **Nernst equation** can be used to calculate emf values at other concentrations. The Nernst equation for the emf of a half-cell at 25°C is

$$E = E° - \frac{0.05915}{n} \log Q$$

where E = the emf for the reaction at the new concentration
   $E°$ = the *standard* electrode potential
   n = the number of moles of electrons stated in the half-reaction
   Q = the reaction quotient (Section 18.13) for the half-reaction, an expression that takes the same form as the equilibrium constant for the half-reaction, neglecting the electrons.

For example, a typical half-reaction and the corresponding Nernst equation is

$$\text{Sn}^{4+} + 2e^- \rightleftharpoons \text{Sn}^{2+}$$

$$E = E° - \frac{0.05915}{2} \log \frac{[\text{Sn}^{2+}]}{[\text{Sn}^{4+}]}$$

or, with the value of the standard reduction potential, $E°$, from Table 20-2 inserted,

$$E = 0.15 - \frac{0.05915}{2} \log \frac{[Sn^{2+}]}{[Sn^{4+}]}$$

where the brackets, as usual, mean *molar concentration of*, and log signifies the common logarithm. If the concentrations of both Sn^{2+} and Sn^{4+} are 1 M (standard state concentrations),

$$E = E° - \frac{0.05915}{2} \log \frac{1}{1}$$

Since $\log 1 = 0$,

$$E = E° - \frac{0.05915}{2}(0) \quad \text{or} \quad E = E°$$

This of course shows that the reduction potential, E, is equal to the *standard reduction potential*, $E°$, when the concentrations are the standard state concentration, 1 M. (Note also that E is equal to $E°$ if $[Sn^{2+}]$ equals $[Sn^{4+}]$, even if the concentrations are not 1 M.)

EXAMPLE 20.4   Calculate E for the half-reaction

$$Sn^{4+} + 2e^- \rightleftharpoons Sn^{2+}$$

at 25°C if the concentration of Sn^{2+} is 0.40 M and the concentration of Sn^{4+} is 0.10 M.

$$E = 0.15 - \frac{0.05915}{2} \log \frac{[Sn^{2+}]}{[Sn^{4+}]}$$

$$= 0.15 - \frac{0.05915}{2} \log \frac{0.40}{0.10}$$

$$= 0.15 - \frac{0.05915}{2}(0.6021)$$

$$= 0.13 \text{ V}$$

In Section 20.12 we noted that the standard reduction potential, $E°$, for the reduction of $2H^+$ to H_2 at standard state conditions is 0.00 V, but the emf value, E, for the same reduction at 10^{-7} M is -0.41 V. The following example shows the calculation of this value.

EXAMPLE 20.5   Calculate the emf value for the reduction of H^+ at 1.0×10^{-7} M to H_2 at 1.0 atm.

$$2H^+(1.0 \times 10^{-7} \text{ } M) + 2e^- \rightleftharpoons H_2(1.0 \text{ atm}) \quad E° = 0.00 \text{ V}$$

$$E = E° - \frac{0.05915}{n} \log \frac{P_{H_2}}{[H^+]^2}$$

$$= 0.00 - \frac{0.05915}{2} \log \frac{1.0}{(1.0 \times 10^{-7})^2}$$

$$= 0.00 - \frac{0.05915}{2} \log \frac{1.0}{1.0 \times 10^{-14}}$$

$$= 0.00 - \frac{0.05915}{2} \log (1.0 \times 10^{14})$$

$$= 0.00 - \frac{0.05915}{2} (14)$$

$$= -0.41 \text{ V}$$

Hence for the half-reaction

$$2H^+(1.0 \times 10^{-7}\, M) + 2e^- \rightleftharpoons H_2(1.0 \text{ atm}) \qquad E = -0.41 \text{ V}$$

20.15 Relationship of the Free Energy Change, the Cell Potential, and the Equilibrium Constant

The standard free energy change of an electrochemical reaction, $\Delta G°$ (Section 18.10), is related to the standard cell potential, $E°$, and the equilibrium constant, K (Section 18.14), by the following two equations:

$$\Delta G° = -nFE° \qquad \text{and} \qquad \Delta G° = -RT \ln K$$

where n = the number of moles of electrons exchanged in the reaction,

F = a constant known as the Faraday constant, whose value is 96.487 kJ V^{-1}, or 96,487 coulombs,

$E°$ = the standard cell potential,

R = the gas constant (8.314 J K^{-1}),

T = the Kelvin temperature,

K = the equilibrium constant,

and ln indicates a natural logarithm (base e).

Combining the two expressions for $\Delta G°$, we have

$$-nFE° = -RT \ln K$$

Then

$$E° = \frac{RT}{nF} \ln K$$

At 25°C (298.15 K), after conversion to the common logarithm (base 10) and with substitution for R, the equation becomes

$$E° = \frac{0.05915}{n} \log K$$

Thus the standard free energy change and the equilibrium constant of a reaction can be determined from the standard reduction potentials of two half-cell reactions, which add up to give the reaction.

EXAMPLE 20.6   Calculate the standard free energy change at 25°C for the reaction

$$Cd + Pb^2 \rightleftharpoons Cd^{2+} + Pb$$

The half-reactions and calculation of the emf of the cell are

$$
\begin{array}{lll}
Cd \rightleftharpoons Cd^{2+} + 2e^- & E° = +0.40 \text{ V} \\
Pb^{2+} + 2e^- \rightleftharpoons Pb & E° = -0.13 \text{ V} \\
\hline
Cd + Pb^{2+} \rightleftharpoons Cd^{2+} + Pb & E° = +0.27 \text{ V}
\end{array}
$$

Hence the standard cell potential is 0.27 V.

$$\Delta G° = -nFE° = -2(96.487 \text{ kJ V}^{-1})(0.27 \text{ V}) = -52 \text{ kJ}$$

The negative value for $\Delta G°$ indicates that the net cell reaction proceeds spontaneously to the right (Section 18.10). A positive $E°$ means a negative $\Delta G°$, so a positive $E°$ also indicates a spontaneous forward reaction.

EXAMPLE 20.7   Calculate the equilibrium constant at 25°C for the reaction in Example 20.6.

$$E° = \frac{0.05915}{n} \log K$$

$$0.27 = \frac{0.05915}{2} \log K$$

$$\log K = 0.27 \times \frac{2}{0.05915} = 9.13$$

$$K = \text{antilog } 9.13 = 1.3 \times 10^9$$

At equilibrium the molar concentration of cadmium ion in the solution is more than a billion times that of lead ion!

EXAMPLE 20.8   Calculate the free energy change and the equilibrium constant for the following reaction at 25°C:

$$Zn + Cu^{2+}(0.20 \, M) \rightleftharpoons Zn^{2+}(0.50 \, M) + Cu$$

Under standard state conditions

$$
\begin{array}{lll}
Zn \rightleftharpoons Zn^{2+} + 2e^- & E° = +0.76 \text{ V} \\
Cu^{2+} + 2e^- \rightleftharpoons Cu & E° = +0.34 \text{ V} \\
\hline
Zn + Cu^{2+} \rightleftharpoons Zn^{2+} + Cu & E° = +1.10 \text{ V}
\end{array}
$$

Hence for the cell reaction *under standard state conditions* (Cu^{2+} and Zn^{2+} at concentrations of 1 M or, more accurately, at unit activity), $E°$ is +1.10 V.

The Nernst equation lets us calculate the potential, E, when [Zn^{2+}] and [Cu^{2+}] are 0.50 M and 0.20 M, respectively.

Zinc half-reaction:

$$E = E° - \frac{0.05915}{n} \log \frac{[Zn^{2+}]}{[Zn°]} = 0.76 - \frac{0.05915}{2} \log \frac{0.50}{1}$$

$$= 0.76 - \frac{0.05915}{2}(0.699 - 1) = 0.76 + 0.01 = +0.77 \text{ V}$$

Copper half-reaction:

$$E = E° - \frac{0.05915}{n} \log \frac{[Cu°]}{[Cu^{2+}]} = +0.34 - \frac{0.05915}{2} \log \frac{1}{0.20}$$

$$= +0.34 - \frac{0.05915}{2}(0.699) = +0.34 - 0.02 = +0.32 \text{ V}$$

Thus, for the cell reaction under these conditions,

$$E = (+0.77) + (+0.32) = +1.09 \text{ V}$$

(compared to the $E°$ value of $+1.10$ V for standard state conditions).

$$\Delta G = -nFE = -2(96.487 \text{ kJ V}^{-1})(1.09 \text{ V}) = -210 \text{ kJ}$$

The negative value of ΔG indicates that the reaction proceeds spontaneously to the right.

The value of K at a given temperature (25°C in this case) is calculated from $E°$ and is constant for the various concentrations.

$$E° = \frac{0.05915}{n} \log K$$

$$1.10 = \frac{0.05915}{2} \log K$$

$$\log K = 1.10 \times \frac{2}{0.05915} = 37.19$$

$$K = 1.6 \times 10^{37}$$

VOLTAIC CELLS

20.16 Primary Voltaic Cells

A voltaic cell is an electrochemical cell in which a chemical reaction is used to produce electrical energy—that is, a battery. **Primary cells,** or **irreversible cells,** cannot be recharged because the electrodes and electrolytes are such that they cannot be restored to their original states by an external electrical potential. The Daniell cell and the dry cell are examples of primary cells.

1. THE DANIELL CELL. A Daniell cell (Fig. 20-8) consists of a copper electrode (A) with crystals of copper(II) sulfate in a saturated solution of copper(II) sulfate

below a zinc electrode (B) suspended near the top of the cell in a dilute solution of zinc sulfate. The zinc sulfate solution floats on the denser solution of copper(II) sulfate. When the two metals are connected by a wire, electrons flow from the more easily oxidized zinc to the less easily oxidized copper. The zinc is oxidized to zinc ions, while the copper(II) ions are reduced to metallic copper.

Anodic oxidation: $Zn \longrightarrow Zn^{2+} + 2e^-$
Cathodic reduction: $Cu^{2+} + 2e^- \longrightarrow Cu$

Cell reaction: $Zn + Cu^{2+} \longrightarrow Zn^{2+} + Cu$

As the zinc ion concentration increases, the anode reaction shifts to the left, and a (slightly) lower voltage is produced.

A cell similar to the Daniell cell, but using cadmium and nickel instead of zinc and copper, has a longer cell life.

2. THE DRY CELL.   A flashlight battery, or dry cell, (Fig. 20-9) consists of a zinc container, which serves as the anode, a carbon (graphite) cathode, and a moist mixture of ammonium chloride, manganese dioxide, zinc chloride, and some inert filler such as sawdust. This mixture is separated from the zinc anode by a porous paper liner. When the cell delivers a current the zinc goes into solution as zinc ions and leaves electrons on the zinc container so that it becomes the negative electrode. This electrode, then, is the anode, since this is where oxidation occurs. The oxidation half-reaction is

$$Zn \rightleftharpoons Zn^{2+} + 2e^-$$

The reaction at the cathode involves reduction of manganese dioxide, in a process that is not completely understood. One possible reaction involves the reduction of the manganese from an oxidation number of $+4$ in MnO_2 to $+3$ in $MnO(OH)$.

$$e^- + NH_4^+ + MnO_2 \rightleftharpoons NH_3 + MnO(OH)$$

The ammonia formed at the cathode unites with some of the zinc ions, forming the complex ion $[Zn(NH_3)_4]^{2+}$. This reaction helps to hold down the concentration of zinc ions, thereby keeping the potential of the zinc electrode more nearly constant. It also prevents the accumulation of an insulating layer of ammonia molecules on the cathode, a condition called **polarization,** which stops the action of the cell.

3. CELLS FOR SPECIAL PURPOSES.   Cells are often used in batteries that must meet special requirements for a specific purpose. For example, a small battery for torpedoes produces a lot of power, but only once over a period of just 30 seconds. A battery cell suitable for a camera light meter must produce a small amount of current but must last for a long period. A wristwatch battery must produce a very small current for at least 1 year. A special battery used in weather balloons meets the requirements of small size, minimal weight, and an operating lifetime of a few hours. It is made by laying a mixture of dry magnesium powder and dry silver chloride on a strip of paper, which is then rolled up. When this roll is dipped into water, the reaction begins.

The small Rubin-Mallory cell is used in hearing aids. It utilizes a zinc container as the anode and a mercury–mercury oxide electrode with a carbon rod as the cathode. The electrolyte is a small amount of sodium, or potassium, hydroxide. The cell produces a potential of 1.35 volts. The half-reactions and net cell reaction are

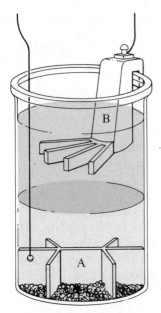

Figure 20-8.   The Daniell, or gravity, cell. (A) sheet copper surrounded by saturated copper(II) sulfate solution and crystals of copper sulfate; (B) zinc plate surrounded by zinc sulfate solution.

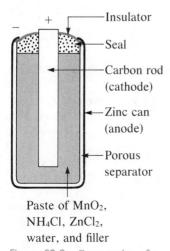

Insulator

Seal

Carbon rod (cathode)

Zinc can (anode)

Porous separator

Paste of MnO_2, NH_4Cl, $ZnCl_2$, water, and filler

Figure 20-9.   Cross section of a flashlight battery, or dry cell.

Anode:	$Zn + 2OH^- \longrightarrow Zn(OH)_2 + 2e^-$
Cathode:	$HgO + H_2O + 2e^- \longrightarrow Hg + 2OH^-$
Net cell reaction:	$Zn + HgO + H_2O \longrightarrow Zn(OH)_2 + Hg$

20.17 Secondary Voltaic Cells

In **secondary,** or **reversible, cells,** the substances consumed in producing electricity may be regenerated in their original forms by passing a current of electricity through the cell in the reverse direction of that of the discharge. This recharges the cell. The lead storage battery and the Edison storage battery both consist of several secondary cells connected together.

1. THE LEAD STORAGE BATTERY.   The electrodes of the cells in a lead storage battery consist of lead alloy plates in the form of grids. The openings of one set of grids are filled with lead dioxide, and the openings of the other, with spongy lead metal. Dilute sulfuric acid serves as the electrolyte. When the battery discharges (delivers a current), the spongy lead is oxidized to lead ions, and the lead plates become negatively charged.

$$Pb \longrightarrow Pb^{2+} + 2e^-   \text{(oxidation)}$$

The lead ions combine with the sulfate ions of the electrolyte and coat the lead electrode with lead sulfate.

$$Pb^{2+} + SO_4^{2-} \longrightarrow PbSO_4(s)$$

The lead sulfate is insoluble and remains on the electrode for regeneration to lead during recharging. The net reaction at the lead electrode when the battery discharges is

$$Pb + SO_4^{2-} \longrightarrow PbSO_4(s) + 2e^-$$

The negatively charged lead electrode is the anode since it is the electrode at which oxidation occurs.

Electrons can flow from the lead electrode through an external circuit and enter the lead dioxide electrode. The lead dioxide is reduced to lead(II) ions, and water is formed.

$$PbO_2 + 4H^+ + 2e^- \longrightarrow Pb^{2+} + 2H_2O   \text{(reduction)}$$

Again lead ions combine with sulfate ions of the electrolyte, and the lead dioxide plate becomes coated with lead sulfate.

$$Pb^{2+} + SO_4^{2-} \longrightarrow PbSO_4(s)$$

The net reaction at the positive lead dioxide electrode when the battery is discharging is

$$PbO_2 + 4H^+ + SO_4^{2-} + 2e^- \longrightarrow PbSO_4(s) + 2H_2O$$

The positively charged lead dioxide electrode is the cathode since it is the electrode at which reduction occurs.

The lead storage battery can be recharged by passing electrons in the reverse direction through each cell by applying an external potential. The cell becomes an electrolytic cell during recharging. The half-reactions are just the reverse of those that occur when the cell is operating as a voltaic cell producing a current. Electrons forced into the lead electrode reduce lead ions from the lead sulfate so this electrode becomes the cathode during recharging. Electrons are withdrawn from the lead dioxide electrode, making it the anode, and lead ions from the lead sulfate are oxidized.

The charge and discharge at the two plates may be summarized as follows:

$$Pb + SO_4{}^{2-} \underset{\text{Charge}}{\overset{\text{Discharge}}{\rightleftharpoons}} PbSO_4(s) + 2e^- \quad \text{(at the lead plate)}$$

$$PbO_2 + 4H^+ + SO_4{}^{2-} + 2e^- \underset{\text{Charge}}{\overset{\text{Discharge}}{\rightleftharpoons}} PbSO_4(s) + 2H_2O$$

(at the lead dioxide plate)

During recharging the sulfate ion of the electrolyte is regenerated, and lead sulfate is converted back to spongy lead at the lead electrode. Hydrogen ions and sulfate ions of the electrolyte are regenerated, and lead sulfate is converted back to lead dioxide at the lead dioxide electrode.

The net cell reaction of the lead storage battery is obtained by adding the two electrode reactions together and is:

$$Pb + PbO_2 + 4H^+ + 2SO_4{}^{2-} \underset{\text{Charge}}{\overset{\text{Discharge}}{\rightleftharpoons}} 2PbSO_4(s) + 2H_2O$$

This equation indicates that the amount of sulfuric acid decreases as the cell discharges. Conversely, charging the cell regenerates the acid. Sulfuric acid is much heavier than water, so a lead storage battery may be tested by determining the specific gravity of the electrolyte. When the specific gravity is 1.25–1.30, the battery is charged; when the specific gravity falls below 1.20, the battery needs recharging. As can be calculated from the potentials in Table 20-2, a single standard lead cell has a potential of

$$(+0.36 \text{ V}) + (+1.69 \text{ V}) = 2.05 \text{ V}$$

The potential falls off slowly as the cell is used. A 12-volt automobile battery contains six lead storage cells.

2. THE EDISON STORAGE BATTERY.   The negative electrode of the Edison storage battery consists of steel plates packed with finely divided iron; the positive electrode, of steel plates packed with hydrated nickel dioxide. The electrolyte is a 21% potassium hydroxide solution containing some lithium hydroxide. When the cell delivers a current, iron is oxidized at the anode,

Anode: $Fe + 2OH^- \longrightarrow Fe(OH)_2(s) + 2e^-$

and NiO_2 is reduced at the cathode,

Cathode: $NiO_2 + 2H_2O + 2e^- \longrightarrow Ni(OH)_2(s) + 2OH^-$

These reactions are reversed during charging. The net reaction is

$$Fe + NiO_2 + 2H_2O \underset{\text{Charge}}{\overset{\text{Discharge}}{\rightleftharpoons}} Fe(OH)_2(s) + Ni(OH)_2(s)$$

The emf of each cell in an Edison storage battery is about 1.4 V. The cell must be sealed, otherwise potassium hydroxide and lithium hydroxide would form carbonates by reaction with carbon dioxide from the air. The cell can be sealed because no gases are produced during reaction. This battery is lighter and more durable than the lead storage battery but more expensive.

Note that the anode is positive and the cathode negative in an electrolytic cell (see Fig. 20-2, for example), whereas the anode is negative and the cathode positive in a voltaic cell (Fig. 20-6). *For all types of cells, however, the electrode where oxidation occurs is the anode and the electrode where reduction occurs is the cathode.* Figure 20-10 shows diagrammatically the relationship between a voltaic cell and an electrolytic cell. The sign convention is such that when a voltaic cell is supplying the current for an electrolytic cell, the negative electrode of one cell is connected to the negative electrode of the other cell, and likewise the positive electrodes of the two cells are connected.

Figure 20-10. Illustration of the relationship between a voltaic cell and an electrolytic cell. The voltaic cell (*left*) provides the electric current to run the electrolytic cell (*right*). Note that the signs for the electrodes are opposite for the two types of cell, resulting in electrodes of like sign being connected. Note also that regardless of whether the cell is voltaic or electrolytic, oxidation occurs at the anode and reduction at the cathode.

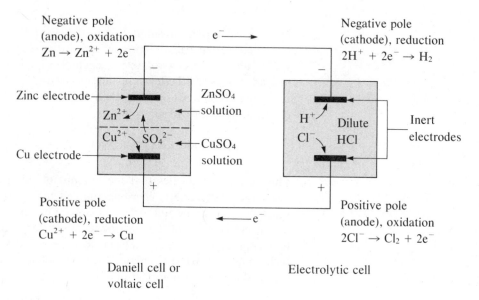

Negative pole (anode), oxidation
$Zn \rightarrow Zn^{2+} + 2e^-$

Negative pole (cathode), reduction
$2H^+ + 2e^- \rightarrow H_2$

Zinc electrode

$ZnSO_4$ solution

Inert electrodes

Cu electrode

$CuSO_4$ solution

Zn^{2+} Cu^{2+} SO_4^{2-}

H^+ Cl^- Dilute HCl

Positive pole (cathode), reduction
$Cu^{2+} + 2e^- \rightarrow Cu$

Positive pole (anode), oxidation
$2Cl^- \rightarrow Cl_2 + 2e^-$

Daniell cell or voltaic cell

Electrolytic cell

20.18 Fuel Cells

Fuel cells are voltaic cells in which electrode materials, usually in the form of gases, are supplied continuously to an electrochemical cell and consumed to produce electricity.

A typical fuel cell, of the type currently used in the space shuttle, is based on the reaction of hydrogen (the fuel) and oxygen (the oxidizer) to form water. Hydrogen gas is diffused through the anode, a porous electrode with a catalyst such as finely divided platinum or palladium on its surface. Oxygen is diffused through the cathode, a porous electrode impregnated with cobalt oxide, platinum, or silver as catalyst. The two electrodes are separated by a concentrated solution of sodium hydroxide or potassium hydroxide (Fig. 20-11) as an electrolyte.

Hydrogen diffuses through the anode and is absorbed on the surface in the form of hydrogen atoms, which react with hydroxide ions of the electrolyte to form water.

$$H_2 \xrightarrow{\text{Catalyst}} 2H$$
$$2H + 2OH^- \longrightarrow 2H_2O + 2e^-$$

Net anode reaction: $H_2 + 2OH^- \longrightarrow 2H_2O + 2e^-$

The electrons produced at the anode flow through the external circuit to the

Figure 20-11. Diagram of a hydrogen-oxygen fuel cell. In actual operation it would take several such cells to light an ordinary 110-V bulb.

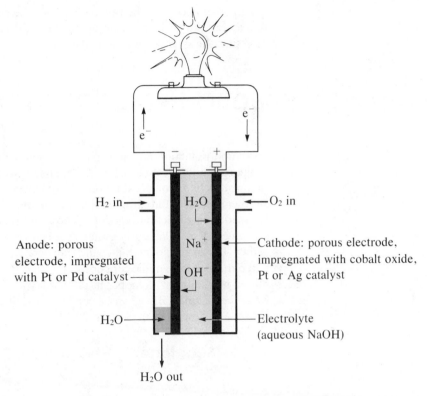

H₂ in →

H₂O

← O₂ in

Anode: porous electrode, impregnated with Pt or Pd catalyst

Na⁺

OH⁻

Cathode: porous electrode, impregnated with cobalt oxide, Pt or Ag catalyst

H₂O

Electrolyte (aqueous NaOH)

H₂O out

cathode. The oxygen absorbed on the cathode's surface is reduced to hydroxide ions.

Cathode reaction: $O_2 + 2H_2O + 4e^- \longrightarrow 4OH^-$

The hydroxide ions replace those that react at the anode. As in all voltaic cells, the electrical output of the cell results from the flow of electrons through the external circuit from anode to cathode.

The overall cell reaction is the combination of hydrogen and oxygen to produce water.

Anodic oxidation: $2H_2 + 4OH^- \longrightarrow 4H_2O + 4e^-$
Cathodic reduction: $O_2 + 2H_2O + 4e^- \longrightarrow 4OH^-$
Cell reaction: $2H_2 + O_2 \longrightarrow 2H_2O$

A great deal of effort is being expended in investigating other fuels, such as methane and other hydrocarbons, and other electrode systems.

The efficiency of the fuel cell is better than that of conventional generating equipment, because electric current is produced directly from the reaction of fuel and oxidizer without the inherently wasteful intermediate conversion of chemical energy to heat. In conventional power plants this conversion can waste 60–80% of the available chemical energy. Fuel cells can, at the present stage of development, increase conversion efficiencies from about 30% in conventional power plants to over 40%. It is estimated that sufficient power for 20,000 people can be produced in a unit that is 18 feet high and covers less than ½ acre of ground. An additional advantage of the fuel cell is that it produces no pollution. The chemical products of a cell such as described here would be carbon dioxide and water vapor. However, power needs for the process, such as the energy required to decompose the fuel and to pump the cell components into the cell, must be taken into account in assessing the overall efficiency of fuel cells for producing power commercially.

OXIDATION-REDUCTION REACTIONS

Oxidation-reduction reactions (redox reactions) involve oxidation and reduction of the reactants (Section 8.2, Part 4). The chemical changes occurring in electrochemical cells involve oxidation of one species at the anode and reduction of another species at the cathode; thus the net reactions of most electrochemical cells are oxidation-reduction reactions. Those net reactions with a positive cell potential proceed spontaneously either in a cell or when the reactants are mixed together. For example, metallic zinc transfers electrons directly to a copper(II) ion and reduces it to metallic copper leaving zinc(II) ions when a piece of zinc metal is placed in a solution of copper nitrate, or it transfers electrons through an external wire when a zinc half-cell and a copper half-cell are combined in an electrochemical cell, as described in Section 20.12.

20.19 Spontaneity, Free Energy Change, and Equilibrium Constants of Redox Reactions

Since a spontaneous oxidation-reduction reaction will occur either in an electrochemical cell or on direct mixing of the reactants, standard reduction potentials and the Nernst equation can be used to predict the spontaneity of an oxidation-reduction reaction and to calculate its free energy change and its equilibrium constant (Section 20.15). The oxidation-reduction reaction, however, must be written as the sum of two half-reactions with known standard reduction potentials.

EXAMPLE 20.9   Chloride ion can be oxidized to chlorine in an acid solution by permanganate ion.

$$2MnO_4^- + 10Cl^- + 16H^+ \longrightarrow 2Mn^{2+} + 5Cl_2 + 8H_2O$$

Show that this reaction is spontaneous at standard state conditions and calculate $\Delta G°$ and K for it.

In order to solve this problem, we need to write the reaction as the sum of two half-reactions, which are found in Appendix I:

$$Cl_2 + 2e^- \longrightarrow 2Cl^- \qquad E° = 1.3595 \text{ V}$$
$$MnO_4^- + 8H^+ + 5e^- \longrightarrow Mn^{2+} + 4H_2O \qquad E° = 1.51 \text{ V}$$

The oxidation-reduction reaction and its emf can be written as the following sum:

$$
\begin{array}{ll}
5(2Cl^- \longrightarrow Cl_2 + 2e^-) & E° = -1.3595 \text{ V} \\
2(MnO_4^- + 8H^+ + 5e^- \longrightarrow Mn^{2+} + 4H_2O) & E° = +1.51 \text{ V} \\
\hline
2MnO_4^- + 16H^+ + 10Cl^- \longrightarrow 2Mn^{2+} + 5Cl_2 + 8H_2O & E° = +0.15 \text{ V}
\end{array}
$$

Since the potential for a cell involving this net reaction is positive, the reaction is spontaneous.

$\Delta G°$ and K can be determined as described in Section 20.15.

$$\Delta G° = -nFE°$$
$$= -10 \times (96.487 \text{ kJ V}^{-1}) \times (0.15 \text{ V})$$
$$= -1.4 \times 10^2 \text{ kJ}$$

$$E° = \frac{RT}{nF} \ln K = \frac{0.05915}{n} \log K$$

$$\log K = E° \times \frac{n}{0.05915} = 0.15 \times \frac{10}{0.05915}$$

$$= 25.359$$

$$K = 2.3 \times 10^{25}$$

20.20 Balancing Redox Equations by the Half-Reaction Method

Equations for oxidation-reduction reactions can become quite complicated and difficult to balance by trial and error. The half-reaction method for balancing redox equations provides a systematic approach. In this method the overall reaction is broken down into two half-reactions, one representing the oxidation step and the other the reduction step. Each half-reaction is balanced separately, then the two half-reactions are added to give the balanced equation.

EXAMPLE 20.10   Iron(II) is oxidized to iron(III) by chlorine. Use the half-reaction method to balance the equation for the reaction

$$Fe^{2+} + Cl_2 \longrightarrow Fe^{3+} + Cl^-$$

The oxidation half-reaction is oxidation of Fe^{2+} to Fe^{3+}.

$$Fe^{2+} \longrightarrow Fe^{3+}$$

To balance this half-reaction in terms of ion charges and electrons, we add 1 electron to the right side of the equation.

$$Fe^{2+} \longrightarrow Fe^{3+} + e^-$$
$$+2 \quad = \quad +3 \quad -1$$

The half-reaction for the reduction of Cl_2 to Cl^- has 2 electrons to the left side of the equation.

$$Cl_2 + 2e^- \longrightarrow 2Cl^-$$
$$-2 \quad = \quad -2$$

In the oxidation half-reaction 1 electron is lost, while in the reduction half-reaction 2 electrons are gained. The oxidation half-reaction is multiplied by 2 to balance the electrons.

$$2Fe^{2+} \longrightarrow 2Fe^{3+} + 2e^-$$

When the balanced half-reactions are added together, the two electrons on either side of the equation cancel one another, and the balanced equation is obtained.

$$2Fe^{2+} \longrightarrow 2Fe^{3+} + 2e^-$$
$$\underline{Cl_2 + 2e^- \longrightarrow 2Cl^-}$$
$$2Fe^{2+} + Cl_2 \longrightarrow 2Fe^{3+} + 2Cl^-$$

EXAMPLE 20.11 The dichromate ion will oxidize iron(II) to iron(III) in acid solution. Balance the equation

$$Cr_2O_7^{2-} + H^+ + Fe^{2+} \longrightarrow Cr^{3+} + Fe^{3+} + H_2O$$

by the half-reaction method.

The oxidation half-reaction is balanced as shown in Example 20.10.

$$Fe^{2+} \longrightarrow Fe^{3+} + e^-$$

The reduction half-reaction involves the change

$$Cr_2O_7^{2-} \longrightarrow 2Cr^{3+}$$

In acid solution, excess oxygen combines with hydrogen ion to give water; thus this half-reaction also involves H^+ and H_2O. By inspection we can see that the 7 atoms of oxygen from the $Cr_2O_7^{2-}$ ion require 14 H^+ ions to produce 7 molecules of water.

$$Cr_2O_7^{2-} + 14H^+ \longrightarrow 2Cr^{3+} + 7H_2O$$

To balance this reduction half-reaction in terms of ion charges and electrons, we add 6 electrons on the left side of the equation.

$$Cr_2O_7^{2-} + 14H^+ + 6e^- \longrightarrow 2Cr^{3+} + 7H_2O$$
$$-2 \qquad +14 \qquad -6 \quad = \quad +6$$

Now we multiply the oxidation half-reaction by 6 so both half-reactions involve the same number of electrons.

$$6Fe^{2+} \longrightarrow 6Fe^{3+} + 6e^-$$

On addition of the balanced half-reactions, the 6 electrons on either side of the equation cancel one another, and we have

$$6Fe^{2+} \longrightarrow 6Fe^{3+} + 6e^-$$
$$\underline{Cr_2O_7^{2-} + 14H^+ + 6e^- \longrightarrow 2Cr^{3+} + 7H_2O}$$
$$Cr_2O_7^{2-} + 14H^+ + 6Fe^{2+} \longrightarrow 2Cr^{3+} + 6Fe^{3+} + 7H_2O$$

EXAMPLE 20.12 In acidic solution hydrogen peroxide oxidizes Fe^{2+} to Fe^{3+}. Use the half-reaction method to balance the equation.

$$H_2O_2 + H^+ + Fe^{2+} \longrightarrow H_2O + Fe^{3+}$$

The oxidation half-reaction is

$$Fe^{2+} \longrightarrow Fe^{3+} + e^-$$

The reduction half-reaction involves the reduction of H_2O_2 to H_2O in acid solution.

$$H_2O_2 \longrightarrow ?H_2O$$

By inspection we can see there is sufficient oxygen in 1 molecule of H_2O_2 to form 2 molecules of water. This requires 2 hydrogen ions.

$$H_2O_2 + 2H^+ \longrightarrow 2H_2O$$

Then 2 electrons are needed to balance the positive charges.

$$H_2O_2 + 2H^+ + 2e^- \longrightarrow 2H_2O$$

So that both half-reactions involve the same number of electrons, we multiply the oxidation half-reaction by 2.

$$2Fe^{2+} \longrightarrow 2Fe^{3+} + 2e^-$$

Now we add the two half-reactions.

$$
\begin{array}{r}
2Fe^{2+} \longrightarrow 2Fe^{3+} + 2e^- \\
H_2O_2 + 2H^+ + 2e^- \longrightarrow 2H_2O \\
\hline
H_2O_2 + 2H^+ + 2Fe^{2+} \longrightarrow 2H_2O + 2Fe^{3+}
\end{array}
$$

When using the half-reaction method in basic solution, you must remember that in *a basic solution* excess oxygen combines with water to give hydroxide ion,

$$O^{2-} + H_2O \longrightarrow 2OH^-$$

and excess hydrogen combines with hydroxide ion to give water,

$$H^+ + OH^- \longrightarrow H_2O$$

EXAMPLE 20.13   Hydrogen peroxide reduces MnO_4^- to MnO_2 in basic solution. Write a balanced equation for the reaction.

Initially we have

$$H_2O_2 + MnO_4^- \longrightarrow MnO_2$$

and we know that the reaction is in basic solution. The reduction half-reaction involves the change

$$MnO_4^- \longrightarrow MnO_2$$

Since it is in basic solution, the excess oxygen combines with water to give hydroxide ion.

$$MnO_4^- + 2H_2O \longrightarrow MnO_2 + 4OH^-$$

Adding electrons to balance the charge gives us the following half-reaction:

$$MnO_4^- + 2H_2O + 3e^- \longrightarrow MnO_2 + 4OH^-$$

The oxidation half-reaction involves oxidation of hydrogen peroxide. Since oxygen already has an oxidation number of -1 in H_2O_2, it can only be oxidized to O_2, with an oxidation number of zero. The oxidation half-reaction must involve the change

$$H_2O_2 \longrightarrow O_2$$

In basic solution excess hydrogen combines with hydroxide ion to give water.

$$H_2O_2 + 2OH^- \longrightarrow O_2 + 2H_2O$$

Addition of electrons to balance the charge gives us

$$H_2O_2 + 2OH^- \longrightarrow O_2 + 2H_2O + 2e^-$$

After multiplying to get the same number of electrons in both half-reactions, we add them to give the overall equation.

$$2MnO_4^- + 4H_2O + 6e^- \longrightarrow 2MnO_2 + 8OH^-$$
$$3H_2O_2 + 6OH^- \longrightarrow 3O_2 + 6H_2O + 6e^-$$
$$\overline{2MnO_4^- + 4H_2O + 3H_2O_2 + 6OH^- \longrightarrow 2MnO_2 + 8OH^- + 3O_2 + 6H_2O}$$

Cancellation of identical species on the left and right sides of this equation gives

$$2MnO_4^- + 3H_2O_2 \longrightarrow 2MnO_2 + 3O_2 + 2OH^- + 2H_2O$$

20.21   Some Half-Reactions

When redox equations are balanced by the half-reaction method, the correct half-reactions must be used. The half-reactions in Table 20-3 should be studied carefully, for they are frequently encountered in the balancing of redox equations. The products listed are the *major products* under the conditions given but often are not the sole products.

Table 20-3   Important Half-Reactions

$H_2O_2 + 2H^+ + 2e^- \longrightarrow 2H_2O$	Hydrogen peroxide as an oxidizing agent in acid solution
$H_2O_2 \longrightarrow O_2 + 2H^+ + 2e^-$	Hydrogen peroxide as a reducing agent in acid solution
$MnO_4^- + 8H^+ + 5e^- \longrightarrow Mn^{2+} + 4H_2O$	Permanganate ion as an oxidizing agent in acid solution
$MnO_4^- + 2H_2O + 3e^- \longrightarrow MnO_2 + 4OH^-$	Permanganate ion as an oxidizing agent in neutral or basic solution
$Cr_2O_7^{2-} + 14H^+ + 6e^- \longrightarrow 2Cr^{3+} + 7H_2O$	Dichromate ion as an oxidizing agent in acid solution
$NO_3^- + 2H^+ + e^- \longrightarrow NO_2 + H_2O$	Concentrated nitric acid as an oxidizing agent toward less active metals, such as Cu, Ag, and Pb, and certain anions, such as Br^-
$NO_3^- + 4H^+ + 3e^- \longrightarrow NO + 2H_2O$	Dilute nitric acid as an oxidizing agent toward less active metals and certain nonmetals, such as Br^-

Table 20-3   (continued)

$2NO_3^- + 10H^+ + 8e^- \longrightarrow N_2O + 5H_2O$	Dilute nitric acid as an oxidizing agent toward moderately active metals, such as Zn and Fe
$H_2SO_4 + 2H^+ + 2e^- \longrightarrow SO_2 + 2H_2O$	Concentrated sulfuric acid as an oxidizing agent toward HBr, C, and Cu
$H_2SO_4 + 6H^+ + 6e^- \longrightarrow S + 4H_2O$	Concentrated sulfuric acid as an oxidizing agent toward H_2S
$H_2SO_4 + 8H^+ + 8e^- \longrightarrow H_2S + 4H_2O$	Concentrated sulfuric acid as an oxidizing agent toward HI
$NO_2^- + 2H^+ + e^- \longrightarrow NO + H_2O$	Nitrous acid as an oxidizing agent
$NO_2^- + H_2O \longrightarrow NO_3^- + 2H^+ + 2e^-$	Nitrous acid as a reducing agent
$HClO + H^+ + 2e^- \longrightarrow Cl^- + H_2O$	Hypochlorous acid as an oxidizing agent
$X_2 + 2e^- \longrightarrow 2X^-$	The halogens F_2, Cl_2, Br_2, and I_2 as oxidizing agents
$O_2 + 4H^+ + 4e^- \longrightarrow 2H_2O$	Oxygen as an oxidizing agent in acid solution
$2H^+ + 2e^- \longrightarrow H_2$	The hydrogen ion as an oxidizing agent
$H_2 \longrightarrow 2H^+ + 2e^-$	Molecular hydrogen as a reducing agent in aqueous solution
$M \longrightarrow M^{n+} + ne^-$	Metals as reducing agents
$M^{n+} + ne^- \longrightarrow M$	Metal ions as oxidizing agents
$M^{n+} \longrightarrow M^{n+x} + xe^-$	Oxidation of a metal ion to a higher oxidation number; for example, $Cr^{2+} \longrightarrow Cr^{3+} + e^-$ $Fe^{2+} \longrightarrow Fe^{3+} + e^-$ $Sn^{2+} \longrightarrow Sn^{4+} + 2e^-$
$M^{n+} + xe^- \longrightarrow M^{n-x}$	Reduction of a metal ion to a lower oxidation number; for example, $Fe^{3+} + e^- \longrightarrow Fe^{2+}$

20.22   Balancing Redox Equations by the Change in Oxidation Number Method

Oxidation and reduction always occur together and to the same extent; the total increase in oxidation number due to oxidation must equal the total decrease in oxidation number due to reduction. This makes it possible to use changes in oxidation number in balancing redox equations.

EXAMPLE 20.14   Use the changes in oxidation number to balance the equation for the reaction of antimony and chlorine to form antimony(III) chloride.

First we write the reactants and products of the reaction and then indicate the oxidation number of each atom in the reactants and products (Sections 5.9 and 8.5). (Remember that elements in the free state are always assigned an oxidation number of zero.)

$$\overset{0}{Sb} + \overset{0}{Cl_2} \longrightarrow \overset{+3\,-1}{SbCl_3}$$

We indicate the changes in oxidation number that occur during the reaction. An ascending arrow ($\uparrow$) denotes an increase in oxidation number, and a descending arrow ($\downarrow$) denotes a decrease in oxidation number.

$$\overset{0}{Sb} \longrightarrow \overset{+3}{Sb} \uparrow 3 \qquad \text{(total increase of 3 in oxidation number per formula unit, Sb)}$$

$$\overset{0}{Cl_2} \longrightarrow \overset{-1}{2Cl} \downarrow 2 \qquad \text{(total decrease of 2 in oxidation number per formula unit, } Cl_2)$$

To balance the equation we must select the proper numbers of antimony and chlorine atoms so that the total increase in oxidation number will equal the total decrease. If we use 2 atoms of antimony and 3 molecules (6 atoms) of chlorine, then there will be both an increase and a decrease of 6 units of oxidation number.

$$\overset{0}{Sb} \longrightarrow \overset{+3}{Sb} \uparrow \qquad 3 \times 2 = 6$$

$$\overset{0}{Cl_2} \longrightarrow \overset{-1}{2Cl} \downarrow \qquad 2 \times 3 = 6$$

The balanced equation, then, is

$$2Sb + 3Cl_2 \longrightarrow 2SbCl_3$$

EXAMPLE 20.15 The chloride ion is oxidized by the permanganate ion in acid solution to give the manganese(II) ion, molecular chlorine, and water. Balance the equation

$$MnO_4^- + Cl^- + H^+ \longrightarrow Mn^{2+} + Cl_2 + H_2O$$

using the changes in oxidation number.

After we assign oxidation numbers, it becomes evident that manganese is reduced from +7 to +2 and chlorine is oxidized from −1 to 0.

$$\overset{+7}{MnO_4^-} + Cl^- + H^+ \longrightarrow Mn^{2+} + \overset{0}{Cl_2} + H_2O$$

$$2Cl^- \longrightarrow \overset{0}{Cl_2} \uparrow \qquad 2 \times 5 = 10$$

$$\overset{+7}{Mn} \longrightarrow Mn^{2+} \downarrow \qquad 5 \times 2 = 10$$

We can see that 10 chloride ions increase by 10 units of oxidation number in going to 5 chlorine molecules (10 chlorine atoms), and 2 manganese atoms decrease by 10 units of oxidation number.

$$2MnO_4^- + 10Cl^- + \text{?}H^+ \longrightarrow 2Mn^{2+} + 5Cl_2 + \text{?}H_2O$$

Now we balance the ion charges on both sides of the equation. On the right side of the equation, the only charged particles are the 2 manganese ions with a total ionic charge of +4. The sum of the charges on the left must also be equal to +4. The sum of the charges on the 2 permanganate ions and the 10 chloride ions is

−12, so 16 hydrogen ions are necessary to give a sum of +4.

$$-2 - 10 + 16 = +4$$

The 16 hydrogen ions produce 8 molecules of water.

$$2MnO_4^- + 10Cl^- + 16H^+ \longrightarrow 2Mn^{2+} + 5Cl_2 + 8H_2O$$

To check the balancing, note there are 8 oxygen atoms on each side of the arrow.

EXAMPLE 20.16   Very active reducing agents such as zinc can reduce dilute nitric acid to ammonium ion. Using the changes in oxidation number, balance the equation for this reduction with zinc.

First we write the reactants and products of the reaction and assign oxidation numbers to the atoms that undergo a change in oxidation number.

$$\overset{0}{Zn} + \overset{+5}{NO_3^-} + H^+ \longrightarrow Zn^{2+} + \overset{-3}{NH_4^+} + H_2O$$

Then we balance the equation with regard to the atoms that change in oxidation number.

$$\overset{0}{Zn} \longrightarrow Zn^{2+} \uparrow \qquad 2 \times 4 = 8$$

$$\overset{+5}{N} \longrightarrow \overset{-3}{N} \downarrow \qquad 8 \times 1 = 8$$

$$4Zn + NO_3^- + ?H^+ \longrightarrow 4Zn^{2+} + NH_4^+ + ?H_2O$$

Next, we balance the ion charges to find the number of hydrogen ions required.

$$NO_3^- + ?H^+ \longrightarrow 4Zn^{2+} + NH_4^+$$

$$4Zn + \underset{-1}{NO_3^-} + \underset{+10}{10H^+} \underset{=}{\longrightarrow} \underset{+8}{4Zn^{2+}} + \underset{+1}{NH_4^+} + ?H_2O$$

Of the 10 hydrogen atoms on the left, 4 are found in the ammonium ion on the right. Thus the other 6 hydrogen atoms form 3 molecules of water on the right.

$$4Zn + NO_3^- + 10H^+ \longrightarrow 4Zn^{2+} + NH_4^+ + 3H_2O$$

To check the balancing of the equation, we note there are 3 oxygen atoms on each side of the arrow. We also note that the net charge on each side of the arrow is 9+. Charges and atoms thus balance respectively on each side.

FOR REVIEW

SUMMARY

In an **electrolytic cell,** electrical energy is used to produce a chemical change; in a **voltaic cell,** chemical changes are used to produce electrical energy. In either type of cell the electrode at which oxidation occurs is the **anode,** and the electrode at which reduction occurs, the **cathode.**

Typical changes carried out in electrochemical cells include electrolysis of molten sodium chloride producing sodium and chlorine; of an aqueous solution of hydrogen chloride producing hydrogen and chlorine; of an aqueous solution of sodium chloride producing hydrogen, chlorine, and a solution of sodium hydroxide; and of a solution of sulfuric acid producing hydrogen and oxygen. The extent of the chemical changes in these processes can be related to the amount of electricity that passes through the cell. A **faraday,** 1 mole of electrons, reduces 1 **equivalent** of a substance at the cathode or is produced by the oxidation of 1 equivalent of a substance at the anode. The amount of electrical charge possessed by a mole of electrons is 96,487 coulombs; 1 **coulomb** is the quantity of electricity involved when a current of 1 ampere flows for 1 second.

The potential that causes electrons from a voltaic cell to move from one electrode to the other through an external circuit is called the **electromotive force,** or **emf,** of the cell. The emf of a cell can be regarded as the sum of the electrode potential at the anode plus the electrode potential at the cathode. A positive emf value indicates that the cell reaction will occur spontaneously as written. **Standard electrode potentials,** or **standard reduction potentials,** are tabulated for a variety of reduction (cathode) **half-reactions** with ion concentrations of 1 M, gas pressures of 1 atm, and a temperature of 25 °C. The potential of an oxidation half-cell (the anode half-reaction) is the same magnitude as and opposite in sign to the reduction potential. The more positive the potential associated with a half-reaction, the greater is the tendency of that reaction to occur as written. Electrode potentials vary with the concentrations of the reactants and products involved in the half-reaction and can be calculated from standard electrode potentials using the **Nernst equation:**

$$E = E° - \frac{0.05915}{n} \log Q$$

where n is the number of moles of electrons involved in the cell reaction.

The standard free energy change, $\Delta G°$, for a cell reaction can be calculated from the emf of a cell based on the reaction. At standard conditions

$$\Delta G° = -nFE°$$

($F = 96.487$ kJ V^{-1}, or 96,487 coulombs.) The equilibrium constant for the reaction is related to the standard free energy change and thus to $E°$.

$$E° = \frac{0.05915}{n} \log K$$

Examples of commercial voltaic cells include the Daniell cell and the familiar flashlight battery, or dry cell. These cells are **primary cells,** which cannot be recharged. The lead storage battery is an example of a **secondary cell,** which is rechargeable.

Oxidation-reduction (redox) reactions can be considered to consist of two half-reactions. The standard free energy change and equilibrium constant of an oxidation-reduction reaction can be calculated from the standard electrode potentials of the two half-reactions. After the half-reactions are balanced, they can be added to obtain the overall balanced equation for the reaction. Redox equations can also be balanced by considering the changes in oxidation number that accompany the reaction.

KEY TERMS AND CONCEPTS

Activity series (20.11)
Anode (20.2)
Cathode (20.2)
Change in oxidation number method for balancing redox equations (20.22)
Coulomb (20.7)
Electrical potential (20.8)
Electrolysis (20.2)
Electrolytes (20.1)
Electrolytic cell (20.2)
Electrolytic conduction (20.1)
Electrolytic refining (20.6)
Electromotive force, emf (20.10)
Equilibrium constants (20.19)
Equivalent (20.7)
Faraday (20.7)
Faraday's law (20.7)
Free energy change (20.15)
Fuel cell (20.18)
Half-reaction (20.2)
Half-reaction method for balancing redox equations (20.20)
Nernst equation (20.14)
Normality (20.7)
Primary cells (20.16)
Salt bridge (20.10)
Secondary cells (20.17)
Spontaneity of redox reactions (20.19)
Standard electrode potential (20.10)
Standard hydrogen electrode (20.9)
Volt (20.8)
Voltaic cell (20.10)

EXERCISES

1. How does conduction of electricity in a metal differ from conduction of electricity in a solution of an electrolyte?
2. How does a voltaic cell differ from an electrolytic cell?
3. Define an anode and a cathode.
4. Complete and balance the following half-reactions. (In each case indicate whether oxidation or reduction occurs.)
   (a) $Sn^{4+} \longrightarrow Sn^{2+}$
   (b) $[Ag(NH_3)_2]^+ \longrightarrow Ag$
   (c) $Hg_2Cl_2 \longrightarrow Hg$
   (d) $O_2 \longrightarrow H_2O$ (in acid)
   [S] (e) $O_2 \longrightarrow OH^-$ (in base)
   (f) $SO_3^{2-} \longrightarrow SO_4^{2-}$ (in acid)
   [S] (g) $MnO_4^- \longrightarrow Mn^{2+}$ (in acid)
   (h) $Cl^- \longrightarrow ClO_3^-$ (in base)

Electrolytic Cells and Faraday's Law of Electrolysis

5. By means of description and chemical equations, explain the electrolytic purification of copper from impurities such as silver, zinc, and gold.
6. In this and previous chapters we have noted commercial electrolytic processes for the preparation of several substances, some of which are hydrogen, oxygen, chlorine, and sodium hydroxide. Write equations for the net cell reaction and the individual electrode half-reactions describing these commercial processes.
7. Using a diagram of the type

$$\text{Anode}\,|\,\text{anode soln} \overset{e^-}{\underset{\rightarrow}{}} \| \text{ cathode soln}\,|\,\text{cathode}$$

   diagram the electrolytic cell for each of the following cell reactions:
   (a) $MgCl_2 \longrightarrow Mg + Cl_2$ (using Mg and inert carbon as electrodes)
   [S] (b) $2NaCl(aq) + 2H_2O \longrightarrow 2NaOH(aq) + H_2 + Cl_2$ (using inert electrodes of specially treated titanium, Ti)
   (c) $Fe + Cu^{2+} \longrightarrow Fe^{2+} + Cu$
   (d) $Cl_2 + 2Br^- \longrightarrow Br_2 + 2Cl^-$ (using platinum electrodes)
8. Write the anode half-reaction, the cathode half-reaction, and the cell reaction for the following electrolytic cells:
   [S] (a) (C) $NaCl(l)\,|\,Cl_2 \overset{e^-}{\rightarrow} \| NaCl(l)\,|\,Na(l)$ (Fe)
   (b) (Pt) $Cl_2\,|\,Cl^- \overset{e^-}{\rightarrow} \| H^+\,|\,H_2$ (Pt)
   (c) (C) $CO_2\,|\,O^{2-} \overset{e^-}{\rightarrow} \| Al^{3+}\,|\,Al$
   [S] (d) (Pb) $PbSO_4$, $PbO_2\,|\,H_2SO_4(aq) \overset{e^-}{\rightarrow} \| H_2SO_4(aq)\,|\,PbSO_4$, Pb
9. Tarnished silverware is coated with Ag_2S. The

tarnish can be removed by placing the silverware in an aluminum pan and covering it with a solution of an inert electrolyte such as NaCl. Explain the electrochemical basis for this procedure.

10. Soon after a copper metal rod is placed in a silver nitrate solution, copper ions are observed in the solution and silver metal has been deposited on the rod. Can the copper rod be considered to be an electrode? If so, is it an anode or a cathode?

11. State Faraday's law. What is a faraday of electricity?

S 12. Calculate the value of the Faraday constant, F, from the charge on a single electron, 1.6021×10^{-19} C. *Ans. 9.648×10^4 C*

13. How many moles of electrons are involved in the following electrochemical changes?

 S (a) 0.800 mol of I_2 is converted to I^-
 Ans. 1.60 mol

 (b) 118.7 g of Sn^{2+} is converted to Sn^{4+}
 Ans. 2.000 mol

 S (c) 27.6 g of SO_3 is converted to SO_3^{2-}
 Ans. 0.690 mol

 (d) 0.174 g of MnO_4^- is converted to Mn^{2+}
 Ans. 7.31×10^{-3} mol

 (e) 1.0 L of O_2 at STP is converted to H_2O in acid solution *Ans. 0.18 mol*

 (f) 100 mL of 0.50 M Cu^{2+} is converted to Cu
 Ans. 0.10 mol

 S (g) 15.80 mL of 0.1145 M MnO_4^- is converted to Mn^{2+} *Ans. 9.046×10^{-3} mol*

S 14. How many faradays of electricity are involved in the electrochemical changes described in Exercise 13? *Ans. (a) 1.60; (b) 2.000; (c) 0.690; (d) 7.31×10^{-3}; (e) 0.18; (f) 0.10; (g) 9.046×10^{-3}*

S 15. How many coulombs of electricity are involved in the electrochemical changes described in Exercise 13? *Ans. (a) 1.54×10^5; (b) 1.930×10^5; (c) 6.66×10^4; (d) 7.05×10^2; (e) 1.7×10^4; (f) 9.6×10^3; (g) 8.728×10^2*

16. Aluminum is manufactured by the electrolysis of a molten mixture of Al_2O_3 and Na_3AlF_6. How many moles of electrons are required to convert 1.0 mol of Al^{3+} to Al? How many faradays? How many coulombs? *Ans. 3.0 mol; 3.0 faraday; 2.9×10^5 C*

17. How many moles of electrons are required to prepare 1.000 metric ton (1000 kg) of chlorine gas by electrolysis of an aqueous solution of sodium

chloride as described in Section 20.4? How many faradays? How many coulombs? Assume that the efficiency of the electrochemical cell is 100%; that is, every electron involved results in the production of a chlorine atom. (In an actual commercial cell the efficiency is about 65%.)
Ans. 2.821×10^4 mol; 2.821×10^4 faraday; 2.722×10^9 C

S 18. Ammonium perchlorate, NH_4ClO_4, used in the solid fuel in the booster rockets on the space shuttle, is prepared from sodium perchlorate, $NaClO_4$, which is produced commercially by the electrolysis of a hot, stirred solution of sodium chloride.

$$NaCl + 4H_2O \longrightarrow NaClO_4 + 4H_2$$

How many moles of electrons are required to produce 1.00 kg of sodium perchlorate? How many faradays? How many coulombs?
Ans. 65.3 mol; 65.3 faraday; 6.30×10^6 C

19. Electrolysis of a sulfuric acid solution with a certain amount of current produces 0.3718 g of hydrogen. What mass of silver would be produced by the same amount of current? What mass of copper? *Ans. 39.79 g Ag; 11.72 g Cu*

S 20. How many grams of zinc will be deposited from a solution of zinc(II) sulfate by 3.40 faraday of electricity? *Ans. 111 g*

21. An experiment is conducted using the apparatus depicted in Fig. 20-4. How many grams of gold would be plated out of solution by the current required to plate out 4.97 g of copper? How many moles of hydrogen would simultaneously be released from the hydrochloric acid solution? How many moles of oxygen would be freed at each anode in the copper sulfate and silver nitrate solutions?
Ans. 10.3 g Au; 0.0782 mol H_2; 0.0391 mol O_2

22. How many moles of electrons flow through a lamp that draws a current of 2.0 A for 1.0 min?
Ans. 1.2×10^{-3} mol

S 23. How many grams of cobalt will be deposited from a solution of cobalt(II) chloride electrolyzed with a current of 20.0 A for 54.5 min?
Ans. 20.0 g

24. How many faradays of electricity would be required to reduce 21.0 g of $Na_2[CdCl_4]$ to metallic cadmium? How long would this take (in minutes) with a current of 7.5 A? *Ans. 0.140 faraday; 30 min*

25. Chromium metal can be plated electrochemically

from an acidic aqueous solution of CrO_3. (a) What is the half-reaction for the process? (b) What mass of chromium, in grams, will be deposited by a current of 2.50 A passing for 20.0 min? (c) How long will it take to deposit 1.0 g of chromium using a current of 10.0 A?

Ans. (a) $CrO_3 + 6H^+ \rightarrow Cr + 3H_2O$; *(b) 0.269 g; (c) 1100 s*

26. A single commercial electrolytic cell for the production of chlorine draws a current of 150,000 A. Assume that the efficiency of the cell is 100%, and calculate the mass of chlorine in kilograms produced by such a cell in 1.0 h. *Ans. 198 kg*

27. Which metals in Table 20-1 could be purified by an electrolysis similar to that used to purify copper?

28. Why are different products obtained when molten $ZnCl_2$ is electrolyzed than when a solution of $ZnCl_2$ is electrolyzed? Why do a solution of $CuCl_2$ and molten $CuCl_2$ give the same products on electrolysis?

Standard Reduction Potentials and Electromotive Force

29. Diagram voltaic cells having the following net reactions:
   §(a) $Mn + 2Ag^+ \longrightarrow Mn^{2+} + Ag$
   (b) $Sn^{4+} + H_2 \longrightarrow Sn^{2+} + 2H^+$
   (c) $Cr_2O_7^{2-} + 14H^+ + 6I^- \longrightarrow$
   $$3I_2 + 2Cr^{3+} + 7H_2O$$

30. Define *standard reduction potential*.

31. Why is the potential for the standard hydrogen electrode listed as 0.00?

32. Which is the better oxidizing agent in each of the following pairs at standard conditions?
   (a) Cr^{3+} or Co^{2+}
   (b) I_2 or Mn^{2+}
   (c) MnO_4^- or $Cr_2O_7^{2-}$ (in acid solution)
   (d) Pb^{2+} or Sn^{4+}

33. Which is the better reducing agent in each of the following pairs at standard conditions?
   (a) F^- or Cu   (c) Sn^{2+} or Fe^{2+}
   (b) H_2 or I^- (d) Fe^{2+} or Cr

34. Calculate the emf of each cell at standard conditions based on the following reactions:
   §(a) $Zn + I_2 \longrightarrow Zn^{2+} + 2I^-$ *Ans. +1.298 V*
   (b) $Cd + Pb^{2+} \longrightarrow Cd^{2+} + Pb$

   Ans. +0.27 V
   (c) $Pt^{2+} + 2Cl^- \longrightarrow Pt + Cl_2$

   Ans. −0.16 V

(d) $Mn + 2AgCl \longrightarrow Mn^{2+} + 2Cl^- + 2Ag$
Ans. +1.40 V

§(e) $MnO_4^- + 8H^+ + 5Au \longrightarrow$
$$Mn^{2+} + 4H_2O + 5Au^+$$
Ans. −0.17 V

(f) $Fe + NiO_2 + 2H_2O \longrightarrow$
$$Fe(OH)_2 + Ni(OH)_2$$
Ans. +1.37 V

35. Determine the standard emf for each of the following cells:

§(a) $Co|Co^{2+}, M = 1 \overset{e^-}{\underset{\rightarrow}{\|}} Cr^{3+}, M = 1|Cr$
Ans. −0.46 V

(b) $Ni|Ni^{2+}, M = 1 \overset{e^-}{\underset{\rightarrow}{\|}} Br^-, M = 1|Br_2(l)$ (Pt)
Ans. +1.315 V

(c) $Pb, PbSO_4(s)|SO_4^{2-}, M = 1 \overset{e^-}{\underset{\rightarrow}{\|}} H^+,$
$$M = 1|H_2, 1 \text{ atm (Pt)}$$
Ans. +0.356 V

(d) (Pt) $Mn^{2+}, M = 1; MnO_4^-, M = 1; H^+,$
$$M = 1 \overset{e^-}{\underset{\rightarrow}{\|}} Fe^{3+}, M = 1; Fe^{2+}, M = 1 \text{ (Pt)}$$
Ans. −0.74 V

36. Write the cell reaction for a voltaic cell based on each of the following pairs of half-reactions, and calculate the emf of the cell under standard conditions:
   §(a) $Sc^{3+} + 3e^- \longrightarrow Sc$
   $Ag^+ + e^- \longrightarrow Ag$
   Ans. +2.88 V

   (b) $S + 2e^- \longrightarrow S^{2-}$
   $Cl_2 + 2e^- \longrightarrow 2Cl^-$
   Ans. +1.84 V

   (c) $ZnS + 2e^- \longrightarrow Zn + S^{2-}$
   $CdS + 2e^- \longrightarrow Cd + S^{2-}$
   Ans. +0.23 V

   (d) $Co(OH)_3 + e^- \longrightarrow Co(OH)_2 + OH^-$
   $Cr(OH)_4^- + 3e^- \longrightarrow Cr + 4OH^-$
   Ans. +1.4 V

   §(e) $HClO_2 + 2H^+ + 2e^- \longrightarrow HClO + H_2O$
   $ClO_3^- + 3H^+ + 2e^- \longrightarrow HClO_2 + H_2O$
   Ans. +0.43 V

§ 37. Rechargeable nickel-cadmium cells are used in calculators and other battery-powered devices. The Telstar communication satellite also uses these cells. The cell reaction is

$$NiO_2 + Cd + 2H_2O \longrightarrow Ni(OH)_2 + Cd(OH)_2$$

Calculate the emf of such a cell using the following half-cell potentials:

$$NiO_2 + 2H_2O + 2e^- \longrightarrow Ni(OH)_2 + 2OH^-$$
$$E° = +0.49 \text{ V}$$

$$Cd(OH)_2 + 2e^- \longrightarrow Cd + 2OH^-$$
$$E° = -0.81 \text{ V}$$

Ans. 1.30 V

38. (a) Calculate the emf at standard conditions of an Edison storage cell; of a single lead storage cell. *Ans. 1.37 V; 2.041 V*

  (b) A lead storage battery, such as used in an automobile, contains six lead storage cells. What maximum voltage would be expected of such a battery? *Ans. 12.25 V*

39. Under standard conditions the potential of the cell diagrammed below is +1.05 V. What is the metal M? Show your calculations.

$$M \,|\, M^{n+} \,\|\, Cu^+ \,|\, Cu$$

Ans. Ga

40. What is the cell with the highest emf that could be constructed from the metals, iron, nickel, copper, and silver? *Ans. $Fe \,|\, Fe^{2+} \,\|\, Ag^+ \,|\, Ag$*

The Nernst Equation

41. Why is the Nernst equation important in electrochemistry?

42. Why does the potential of a single electrode change as the concentrations of the species involved in the half-reaction change?

43. Show by suitable equations that the electrode reaction for the zinc electrode is a reversible one.

44. Calculate the emf for each of the following half-reactions:

  S (a) $Sn^{2+}(0.0100 \text{ M}) + 2e^- \longrightarrow Sn$
  Ans. −0.195 V

  (b) $Hg \longrightarrow Hg^{2+}(0.2500 \text{ M}) + 2e^-$
  Ans. −0.836 V

  S (c) $O_2(0.0010 \text{ atm}) + 4H^+(0.100 \text{ M}) + 4e^- \longrightarrow 2H_2O(l)$
  Ans. 1.13 V

  S (d) $Cr_2O_7{}^{2-}(0.150 \text{ M}) + 14H^+(0.100 \text{ M}) + 6e^- \longrightarrow 2Cr^{3+}(0.000100 \text{ M}) + 7H_2O(l)$
  Ans. 1.26 V

  (e) $Mn^{2+}(0.0125 \text{ M}) + 4H_2O(l) \longrightarrow MnO_4{}^-(0.0125 \text{ M}) + 8H^+(0.100 \text{ M}) + 5e^-$
  Ans. −1.42 V

  (f) $Sn^{4+}(0.00010 \text{ M}) + 2e^- \longrightarrow Sn^{2+}(4.0 \text{ M})$
  Ans. 0.01 V

S 45. Hypochlorous acid, HOCl, is a stronger oxidizing agent in acidic solution than in neutral solution. Calculate the potential for the reduction of HOCl to Cl⁻ in a solution with a pH of 7.00 in which [HOCl] and [Cl⁻] both equal 1.00 M. *Ans. +1.28 V*

46. The standard reduction potential of oxygen in acidic solution is 1.23 V ($O_2 + 4H^+ + 4e^- \longrightarrow 2H_2O$). Calculate the standard reduction potential of oxygen in basic solution and compare your result with the value in Appendix I. (*Hint:* What is [H⁺] when [OH⁻] is 1 M?) *Ans. 0.40 V*

47. The Nernst equation for the net reaction of an electrochemical cell is

$$E = E° - \frac{0.05915}{n} \log Q$$

where Q is the reaction quotient for the reaction and n is the total number of moles of electrons transferred in the net reaction. Show that the Nernst equation for the net reaction

$$2Au + 3Cl_2 \longrightarrow 2Au^{3+} + 6Cl^-$$

is identical to that derived by addition of the Nernst equations for the two half-reactions

$$Au \longrightarrow Au^{3+} + 3e^- \qquad E° = -1.50 \text{ V}$$
$$Cl_2 + 2e^- \longrightarrow 2Cl^- \qquad E° = +1.3595 \text{ V}$$

48. Calculate the voltage produced by each of the following cells:

  S (a) $Zn \,|\, Zn^{2+}, M = 0.0100 \overset{e^-}{\rightarrow} \| \; Cu^{2+}, M = 1.00 \,|\, Cu$
  Ans. +1.16 V

  (b) $Al \,|\, Al^{3+}, M = 0.250 \overset{e^-}{\rightarrow} \| \; Co^{2+}, M = 0.0500 \,|\, Co$
  Ans. +1.36 V

  S (c) $(Pt) \, Br_2(l) \,|\, Br^-, M = 0.450 \overset{e^-}{\rightarrow} \| \; Cl^-, M = 0.0500 \,|\, Cl_2(g), 0.900 \text{ atm} \, (Pt)$
  Ans. +0.3494 V

  S (d) $(Pt) \, H_2, 0.790 \text{ atm} \,|\, H^+, M = 0.500 \overset{e^-}{\rightarrow} \| \; Cl^-, M = 0.0500 \,|\, Cl_2(g), 0.100 \text{ atm} \, (Pt)$
  Ans. +1.4217 V

49. What is the theoretical potential required to elec-

trolyze a 0.0300 M solution of copper(II) chloride, producing metallic copper and chlorine at 0.300 atm of pressure? *Ans. −1.124 V*

50. A standard zinc electrode is combined with a hydrogen electrode with H_2 at 1 atm. If the emf of the cell is 0.46 V, what is the pH of the electrolyte in the hydrogen electrode? *Ans. 5.07*

Free Energy Changes and Equilibrium Constants

51. For a cell based on each of the following reactions run at standard conditions, calculate the emf of the cell, the standard free energy change of the reaction, and the equilibrium constant of the reaction:
   S (a) $Mn(s) + Cd^{2+}(aq) \longrightarrow Mn^{2+}(aq) + Cd(s)$
   Ans. +0.78 V; −150 kJ; 2.4 × 10²⁶
   (b) $2Al(s) + 3Co^{2+}(aq) \longrightarrow$
   $2Al^{3+}(aq) + 3Co(s)$
   Ans. +1.38 V; −799 kJ; 9.62 × 10¹³⁹
   S (c) $2Br^-(aq) + I_2(s) \longrightarrow Br_2(l) + 2I^-(aq)$
   Ans. −0.5297 V; 102.2 kJ; 1.229 × 10⁻¹⁸
   (d) $Cr_2O_7^{2-} + 3Fe(s) + 14H^+(aq) \longrightarrow$
   $2Cr^{3+}(aq) + 3Fe^{2+}(aq) + 7H_2O(l)$
   Ans. 1.77 V; −1020 kJ; 3.49 × 10¹⁷⁹
   (e) $2Mn^{2+}(aq) + 8H_2O(l) + 5Fe^{2+}(aq) \longrightarrow$
   $2MnO_4^-(aq) + 16H^+(aq) + 5Fe(s)$
   Ans. −1.95 V; 1880 kJ; 2.14 × 10⁻³³⁰

S 52. Calculate the free energy change and equilibrium constant for the reaction
   $$2Br^- + F_2 \longrightarrow 2F^- + Br_2$$
   Ans. −348 kJ; 1.06 × 10⁶¹

53. Using the standard reduction potentials for the half-reactions in the hydrogen-oxygen fuel cell, calculate the free energy change and equilibrium constant for the combustion of hydrogen:
   $$2H_2 + O_2 \longrightarrow 2H_2O$$
   Ans. −475 kJ; 1.51 × 10⁸³

S 54. Copper(I) salts disproportionate in water to form copper(II) salts and copper metal:
   $$2Cu^+ \longrightarrow Cu^{2+} + Cu$$
   What concentration of Cu^+ remains at equilibrium in 1.00 L of a solution prepared from 1.00 mol of Cu_2SO_4? *Ans. 5.48 × 10⁻⁴ M*

55. Use the emf values of the following half-cells and show that hydrogen peroxide, H_2O_2, is unstable with respect to decomposition into oxygen and water:
   $$H_2O_2 + 2H^+ + 2e^- \longrightarrow 2H_2O$$
   $$E° = +1.77 \text{ V}$$
   $$O_2 + 2H^+ + 2e^- \longrightarrow H_2O_2$$
   $$E° = +0.68 \text{ V}$$

56. Should each of the following compounds be stable in a 1 M aqueous solution? (*Hint:* Check the possibility of oxidation or reduction of the anion by the cation.)
   (a) $Ba(MnO_4)_2$ *Ans. Yes*
   (b) FeI_3 *Ans. No*
   (c) $Pd[HgBr_4]$ *Ans. Yes*
   (d) $[Co(NH_3)_6](ClO)_2$ *Ans. No*
   (e) $Na_2[Cd(CN)_4]$ *Ans. Yes*

Oxidation-Reduction Reactions

57. Balance the following redox equations:
   (a) $P + Cl_2 \longrightarrow PCl_5$
   S (b) $IF_5 + Fe \longrightarrow FeF_3 + IF_3$
   (c) $Sn^{2+} + Cu^{2+} \longrightarrow Sn^{4+} + Cu^+$
   S (d) $H_2S + Hg_2^{2+} \longrightarrow Hg + S + H^+$
   (e) $CN^- + ClO_2 \longrightarrow CNO^- + Cl^-$
   (f) $Fe^{2+} + Ce^{4+} \longrightarrow Fe^{3+} + Ce^{3+}$
   (g) $HBrO \longrightarrow H^+ + Br^- + O_2$
   (h) $ICl + H_2O \longrightarrow H^+ + Cl^- + HOI$

58. Balance the following redox equations:
   S (a) $Zn + BrO_4^- + OH^- + H_2O \longrightarrow$
   $[Zn(OH)_4]^{2-} + Br^-$
   (b) $H_2SO_4 + HBr \longrightarrow SO_2 + Br_2 + H_2O$
   S (c) $MnO_4^- + S^{2-} + H_2O \longrightarrow$
   $MnO_2 + S + OH^-$
   (d) $NO_3^- + I_2 + H^+ \longrightarrow$
   $IO_3^- + NO_2 + H_2O$
   S (e) $Cu + H^+ + NO_3^- \longrightarrow$
   $Cu^{2+} + NO_2 + H_2O$
   (f) $Zn + H^+ + NO_3^- \longrightarrow$
   $Zn^{2+} + N_2O + H_2O$
   S (g) $Cu + H^+ + NO_3^- \longrightarrow$
   $Cu^{2+} + NO + H_2O$
   (h) $MnO_4^- + H_2S + H^+ \longrightarrow$
   $Mn^{2+} + S + H_2O$
   S (i) $MnO_4^- + NO_2^- + H_2O \longrightarrow$
   $MnO_2 + NO_3^- + OH^-$
   (j) $MnO_4^{2-} + H_2O \longrightarrow$
   $MnO_4^- + OH^- + MnO_2$
   S (k) $Br_2 + SO_2 + H_2O \longrightarrow$
   $H^+ + Br^- + SO_4^{2-}$

59. Balance the following redox equations:
    S (a) $Al + [Sn(OH)_4]^{2-} \longrightarrow$
    $$[Al(OH)_4]^- + Sn + OH^-$$
    (b) $Cl^- + H^+ + NO_3^- \longrightarrow$
    $$Cl_2 + NO_2 + H_2O$$
    S (c) $H_2S + H_2O_2 \longrightarrow S + H_2O$
    (d) $MnO_4^- + Se^{2-} + H_2O \longrightarrow$
    $$MnO_2 + Se + OH^-$$
    (e) $MnO_4^{2-} + Cl_2 \longrightarrow MnO_4^- + Cl^-$
    S (f) $OH^- + NO_2 \longrightarrow NO_3^- + NO_2^- + H_2O$
    (g) $Br_2 + CO_3^{2-} \longrightarrow Br^- + BrO_3^- + CO_2$
    (h) $NH_3 + O_2 \longrightarrow NO + H_2O$
    S (i) $C + HNO_3 \longrightarrow NO_2 + H_2O + CO_2$
    (j) $HClO_3 \longrightarrow HClO_4 + ClO_2 + H_2O$
    (k) $ClO_3^- + H_2O + I_2 \longrightarrow IO_3^- + Cl^- + H^+$
    (l) $Cr_2O_7^{2-} + HNO_2 + H^+ \longrightarrow$
    $$Cr^{3+} + NO_3^- + H_2O$$

60. Complete and balance the following equations. (Note that when a reaction occurs in acidic solution, H^+ and/or H_2O may be added on either side of the equation, as necessary, to balance the equation properly; when a reaction occurs in basic solution, OH^- and/or H_2O may be added, as necessary, on either side of the equation. No indication of the acidity of the solution is given if neither H^+ nor OH^- is involved as a reactant or product.)
    S (a) $Zn + NO_3^- \longrightarrow Zn^{2+} + N_2$
    $\qquad$ (acidic solution)
    (b) $Zn + NO_3^- \longrightarrow Zn^{2+} + NH_3$
    $\qquad$ (basic solution)
    S (c) $CuS + NO_3^- \longrightarrow Cu^{2+} + S + NO$
    $\qquad$ (acidic solution)
    (d) $NH_3 + O_2 \longrightarrow NO_2$ (gas phase)
    (e) $H_2SO_4 + HI \longrightarrow I_2 + SO_2$
    $\qquad$ (acidic solution)
    S (f) $Cl_2 + OH^- \longrightarrow Cl^- + ClO_3^-$
    $\qquad$ (basic solution)
    (g) $H_2O_2 + MnO_4^- \longrightarrow Mn^{2+} + O_2$
    $\qquad$ (acidic solution)
    S (h) $NO_2 \longrightarrow NO_3^- + NO_2^-$ (basic solution)
    (i) $KClO_3 \longrightarrow KCl$ (no solvent)
    S (j) $Fe^{3+} + I^- \longrightarrow Fe^{2+} + I_2$
    (k) $P_4 \longrightarrow PH_3 + HPO_3^{2-}$ (basic solution)
    (l) $P_4 \longrightarrow PH_3 + HPO_3^{2-}$ (acidic solution)

61. Complete and balance the following reactions (see the instructions for Exercise 60):
    (a) $Zn + H^+ \longrightarrow$ (acidic solution)
    (b) $MnO_2 + Cl^- \longrightarrow$ (acidic solution)
    (c) $Pb^{4+} + Sn^{2+} \longrightarrow$

(d) $Fe^{2+} + H_2O_2 \longrightarrow$ (acidic solution)
(e) $Cl_2 + SO_2 \longrightarrow$ (acidic solution)

(f) $ZnS + O_2 \xrightarrow{\Delta}$ (no solvent)
(g) $ClO^- + Sn^{2+} \longrightarrow$ (acidic solution)
(h) $PbO_2 + SeO_3^{2-} \longrightarrow$ (basic solution)
(i) $Cr_2O_7^{2-} + Br^- \longrightarrow$ (acidic solution)

Additional Exercises

62. When gold is plated electrochemically from a basic solution of $[Au(CN)_4]^-$, O_2 forms at one electrode and Au is deposited at the other. Write the half-reactions occurring at each electrode and the net reaction for the electrochemical cell. (The cyanide ion, CN^-, is not oxidized or reduced under these conditions.)

63. A lead storage battery is used to electrolyze a solution of hydrogen chloride. Sketch the two cells, label the cathode and anode in each cell, give the sign of each electrode, indicate the direction of flow of electrons through the system, show the movement of ions in the cells, and write the half-reactions occurring at each electrode.

64. A current of 9.0 A flowed for 45 min through water containing a small quantity of sodium hydroxide. How many liters of gas were formed at the anode at 27.0°C and 750 torr of pressure?
    Ans. 1.6 L

65. A lead storage battery has initially 200 g of lead and 200 g of PbO_2, plus excess H_2SO_4. Theoretically, how long could this cell deliver a current of 10.0 A, without recharging, if it were possible to operate it so that the reaction goes to completion?
    Ans. 4.48 h

66. A total of 69,500 coulombs of electricity was required to reduce 37.7 g of M^{3+} to the metal. What is M?
    Ans. Gd

S 67. A current of 10.0 A is applied for 1.0 h to 1.0 L of a solution containing 1.0 mol of HCl. Calculate the pH of the solution at the end of this time.
    Ans. 0.20

68. A current of 10.0 A is applied for 1.0 h to 1.0 L of a solution containing 5.0 mol of NaCl. Calculate the pH of the solution at the end of this time.
    Ans. 13.57

69. Describe briefly how you could determine the solubility product of CuI ($K_{sp} \approx 10^{-7}$) using an electrochemical measurement.

70. In the chapter on halogens (Chapter 19), you learned that when chlorine dissolves in water, it disproportionates, producing chloride ion and hypochlorous acid. At what hydrogen ion concentration does the potential (emf) for the disproportionation of chlorine change from a negative value to a positive value, assuming 1.00 atm of pressure and concentrations of 1.00 M for all species except hydrogen ion? (The standard electrode potential for the reduction of chlorine to chloride ion is 1.36 V, and for hypochlorous acid to chlorine, 1.63 V.) Could chlorine be produced from hypochlorite and chloride ions in solution, through the reverse of the disproportionation reaction, by acidifying the solution with strong acid? Explain.

Ans. 2.7×10^{-5} M; yes

[S] 71. The standard reduction potentials for the reactions

$$Ag^+ + e^- \longrightarrow Ag$$
$$\text{and} \quad AgCl + e^- \longrightarrow Ag + Cl^-$$

are $+0.7991$ V and $+0.222$ V, respectively. From these data and the Nernst equation, calculate a value for the solubility product (K_{sp}) for AgCl. Compare your answer with the value given in Appendix E. *Ans. 1.76×10^{-10}*

72. Calculate the standard reduction (electrode) potential for the reaction $H_2O + e^- \rightarrow \frac{1}{2}H_2 + OH^-$ using the Nernst equation and the fact that the standard reduction potential for the reaction $H^+ + e^- \rightarrow \frac{1}{2}H_2$ is by definition equal to 0.00 V.
Ans. -0.83 V

73. The standard reduction potentials for the reactions

$$Ag^+ + e^- \longrightarrow Ag$$
$$\text{and} \quad [Ag(NH_3)_2]^+ + e^- \longrightarrow Ag + 2NH_3$$

are $+0.7991$ V and $+0.373$ V, respectively. From these values and the Nernst equation, determine K_f for the $[Ag(NH_3)_2]^+$ ion. Compare your answer with the value given in Appendix F.
Ans. 1.60×10^7

21

SULFUR AND ITS COMPOUNDS

The elements sulfur, selenium, tellurium, and polonium follow oxygen in Group VIA of the Periodic Table. These elements, oxygen through polonium, constitute the **chalcogen family.** Oxygen (discussed in Chapter 9) is the first member of this group and exhibits properties that set it apart from the other elements of the chalcogens, just as fluorine, the first member of the halogen family, is different in many respects from the other elements of its group. Polonium is formed only as a product of radioactive change and is highly radioactive.

21.1 Chemical Properties of the Elements of Group VIA

Uncombined atoms of the elements in Group VIA have six valence electrons with the valence shell electron configuration ns^2np^4 ($n = 2$ through 6); two of the p orbitals have only one electron. Thus, these atoms can fill their valence shells by picking up two electrons and forming an anion, such as S^{2-}, with two negative charges or by sharing electrons in covalent bonds.

Oxygen is second only to fluorine (among the known elements) in electronegativity, and so it usually has an oxidation number of -2 in its compounds. The less electronegative elements of the group have both negative and positive oxidation numbers. Going down the group, the electronegativity decreases, the atomic size increases, and the strength of the elements as oxidizing agents decreases. The latter property is reflected very strikingly in the heats of formation of their respective hydrides (Table 21-1). As you should expect by now, metallic character increases down the group. Oxygen is a typical nonmetal, selenium has some metallic character, and polonium exhibits definite metallic characteristics.

TABLE 21-1 Some Properties of the Chalcogens

Property	Oxygen	Sulfur	Selenium	Tellurium
Atomic number	8	16	34	52
Atomic weight	15.9994	32.064	78.96	127.60
Valence electronic structure	$2s^2 2p^4$	$3s^2 3p^4$	$4s^2 4p^4$	$5s^2 5p^4$
Radius of divalent anion, Å	1.40	1.84	1.98	2.21
Radius of covalent atom, Å	0.66	1.04	1.17	1.37
Physical state	Gas	Solid	Solid	Solid
Color	Colorless	Yellow	Red or gray	Silvery
Electronegativity	3.5	2.4	2.5	2.0
Enthalpy of formation of hydride, kJ mol^{-1}	−285.83	−20.6	29.7	99.6
Common oxidation numbers	−2, −1	−2, +2, +4, +6	−2, +4, +6	−2, +4, +6

Oxygen and sulfur are the most commercially useful elements in Group VIA. The largest single use for sulfur is in the production of sulfuric acid, the most important industrial chemical. Sulfur is also heavily used in vulcanizing rubber and in producing gunpowder, sulfites, thiosulfates, fertilizers, and medicines.

Selenium is used in photoelectric cells, since its ionization energy is low enough that light can remove an electron. Ordinary glass contains traces of iron (II) silicate, which makes it green; adding a little sodium selenide makes it colorless. If more sodium selenide is added, the glass turns red. Selenium is added to glass in the form of sodium selenide rather than as elemental selenium because the free element is volatile at the high temperatures used in glassmaking and is very toxic. In the electronics industry selenium of high purity is used as an efficient rectifier, a device for changing alternating current to direct current. Selenium is also used in the production of certain stainless steels and special copper alloys.

Tellurium is used to color glass blue, brown, or red. Alloyed in small percentages with lead, it increases the hardness of things like battery plates and printing type, which are made from lead. Traces of tellurium increase the machinability of stainless steel and give a hard, wear-resistant surface to cast iron.

SULFUR

21.2 Occurrence and Preparation of Elemental Sulfur

Sulfur has been known from very early times because it occurs as a free solid in nature. The principal deposits in the United States are in Texas and Louisiana.

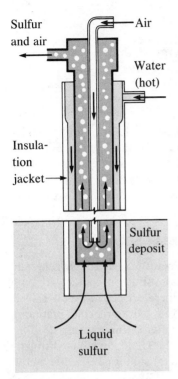

Figure 21-1. Diagram illustrating the Frasch process for mining sulfur.

Figure 21-2. Crystals of orthorhombic sulfur.

Figure 21-3. Crystals of monoclinic sulfur.

There are also extensive deposits in Mexico and in the volcanic regions of Italy and Japan. Minerals containing sulfur in the combined form are numerous and widely distributed. Sulfides of iron, zinc, lead, and copper and sulfates of sodium, calcium, barium, and magnesium are widespread and abundant. Hydrogen sulfide is a common component of natural gas and also occurs in many volcanic gases. Sulfur compounds are also found in coal. Sulfur is a constituent of some proteins and therefore exists in the combined state in animal and vegetable matter.

Sulfur is extracted by the **Frasch process** (Fig. 21-1) from enormous underground deposits in Texas and Louisiana. Superheated water (at 170°C and a pressure of 100 pounds per square inch) is forced down the outermost of three concentric pipes to the sulfur deposit located several hundred feet underground. When the hot water melts the sulfur, compressed air is forced down the innermost pipe. The liquid sulfur mixed with air forms a foam that is less dense than water, and this mixture readily flows up through the middle pipe. The emulsified sulfur is conveyed to large settling vats, where it solidifies on cooling. Sulfur produced by this method is 99.5–99.9% pure and requires no purification for most uses. Sulfur is also obtained, in quantities exceeding those from the Frasch process, from hydrogen sulfide recovered during the purification of natural gas (Section 21.5).

21.3 Physical Properties of Elemental Sulfur

Sulfur exists in several allotropic forms (Section 9.10). Native sulfur is a yellow solid that forms orthorhombic crystals; it is called **rhombic sulfur** (Fig. 21-2). This is soluble in carbon disulfide, CS_2, producing a solution from which well-formed crystals separate when it is allowed to evaporate slowly. Crystals of rhombic sulfur melt at 112.8°C and form a straw-colored liquid known as **λ-sulfur.** When this liquid cools and crystallizes, long transparent needles of **monoclinic sulfur** are formed (Fig. 21-3). This modification of sulfur, the stable form above 96°C, melts at 119.25°C and is soluble in carbon disulfide. At room temperature it gradually changes to the rhombic form.

Rhombic sulfur, monoclinic sulfur, λ-sulfur, and solutions of rhombic sulfur in carbon disulfide all contain S_8 molecules in which the atoms form eight-membered, puckered rings [Fig. 21-4(a)]. Each atom of sulfur is linked to each of its two neighbors in the ring by single electron-pair bonds; each atom thus has a completed octet of electrons.

The straw-colored liquid λ-sulfur is quite mobile. Its viscosity is low because S_8 molecules are essentially spherical and offer relatively little resistance as they move past each other. As the temperature rises, the S_8 rings of the λ-sulfur rupture and form long chains of sulfur atoms [Fig. 21-4(b)]. These chains become entangled with one another and combine end to end to form still longer chains with larger numbers of sulfur atoms, so the viscosity of the liquid increases. The liquid gradually darkens in color and becomes so viscous that finally (at about 230°C) it will not pour. The unpaired electrons at the ends of the chains of sulfur atoms are responsible for the dark red color. This liquid form of the element is known as **μ-sulfur.** When it is cooled rapidly, a rubberlike amorphous mass that is insoluble in carbon disulfide results. This supercooled liquid is known as **plastic sulfur.** After standing for some time at room temperature, plastic sulfur, like all of the other allotropes, changes to the rhombic form.

Figure 21-4. (a) An S_8 molecule and (b) a chain of sulfur atoms.

(a)   (b)

Sulfur boils at 444.6°C and forms a vapor consisting of S_2, S_6, and S_8 molecules; at about 1000°C the vapor density corresponds to the formula S_2.

Although sulfur is stable as an eight-membered ring at room temperature and should be written as S_8 to be perfectly correct, chemists commonly use the symbol S to simplify the coefficients in chemical equations. From this point on we will write S in chemical equations rather than S_8.

21.4 Chemical Properties of Elemental Sulfur

Elemental sulfur is reactive, even at ordinary temperatures, although generally less so than oxygen. Many metals react with solid sulfur on contact at room temperature. Mercury combines with sulfur at temperatures as low as −180°C. Only the noble gases and the elements iodine, nitrogen, tellurium, gold, platinum, and palladium do not combine directly with elemental sulfur.

Sulfur, with the electron configuration $1s^2 2s^2 2p^6 3s^2 3p^4$, exhibits distinctly nonmetallic behavior. It oxidizes metals, giving metal sulfides.

$$2Na + S \longrightarrow Na_2S$$
$$2Na + 4S \longrightarrow Na_2S_4$$
$$Ca + S \longrightarrow CaS$$
$$Fe + S \longrightarrow FeS$$
$$Fe + 2S \longrightarrow FeS_2$$

Many metals produce a variety of binary sulfides. The products of these reactions depend on conditions such as the ratio of reactants and the temperature of the reaction. The simplest metal sulfides contain the sulfide ion, S^{2-}. Adding excess sulfur yields compounds containing polysulfide ions, for example, S_2^{2-}, S_4^{2-}, and S_5^{2-} (Section 21.7).

Sulfur forms compounds with nonmetals in which it participates in two covalent bonds. The oxidation number of sulfur in compounds such as SCl_2, CS_2, P_4S_{10}, CH_3—SH, and H_2S is either +2 or −2, depending on whether the nonmetal is more electronegative (chlorine) or less electronegative (carbon, hydrogen, and phosphorus) than sulfur (Section 5.5). Sulfur also forms compounds containing four, five, or six bonding electron pairs with more electronegative nonmetals. In such compounds, for example, SO_3, H_2SO_4, $SOCl_2$, SF_4, and SF_6, sulfur exhibits an oxidation state of +4 or +6.

Elemental sulfur acts as an oxidizing agent toward hydrogen, forming hydrogen sulfide, H_2S, in low yield; toward carbon at elevated temperatures, forming carbon disulfide, CS_2; and toward phosphorus when heated, forming a variety of phosphorus sulfides depending on the stoichiometry.

$$H_2 + S \longrightarrow H_2S$$
$$C + 2S \longrightarrow CS_2$$

$$P_4 + 10S \longrightarrow P_4S_{10}$$
$$P_4 + 5S \longrightarrow P_4S_5$$

Sulfur acts as a reducing agent toward nonmetals, such as oxygen and the halogens, that are more electronegative than it is. When sulfur is ignited in the air, it burns with a blue flame forming sulfur dioxide and a little sulfur trioxide.

$$S(s) + O_2(g) \longrightarrow SO_2(g) \qquad \Delta H° = -296.83 \text{ kJ}$$
$$2SO_2(g) + O_2(g) \longrightarrow 2SO_3(g) \qquad \Delta H° = -197.8 \text{ kJ}$$

As we shall see later in this chapter, the oxidation of sulfur is important in the production of sulfurous acid, H_2SO_3, and sulfuric acid, H_2SO_4.

Sulfur reduces chlorine in a series of reactions giving products that depend on the temperature and relative quantities of the reactants.

$$2S + Cl_2 \longrightarrow S_2Cl_2 \qquad \text{("sulfur monochloride," a misnomer)}$$
$$S_2Cl_2 + Cl_2 \rightleftharpoons 2SCl_2 \qquad \text{(sulfur dichloride)}$$
$$SCl_2 + Cl_2 \rightleftharpoons SCl_4 \qquad \text{(sulfur tetrachloride, decomposes above } -31°C)$$

The valence electronic structure of S_2Cl_2 is

$$:\!\overset{..}{\underset{..}{Cl}}\!-\!\overset{..}{S}\!-\!\overset{..}{S}\!-\!\overset{..}{\underset{..}{Cl}}\!:$$

Only one bromide of sulfur is known, the monobromide, S_2Br_2, a garnet-red liquid.

Sulfur forms a series of fluorides that includes the hexafluoride, SF_6, a very stable gaseous compound. In SF_6 the fluorine atoms are arranged around the sulfur atom at the corners of a regular octahedron (Fig. 21-5), and the bonding is highly covalent. Twelve electrons, not eight, occupy the valence shell of sulfur in SF_6. This is possible because of the unfilled $3d$ orbitals in the valence shell; sulfur exhibits d^2sp^3 hybridization in SF_6. In binary compounds with the halogens, sulfur has an oxidation number of $+6$ only with fluorine because the fluorine atom is small and highly electronegative compared with the other halogens, and because sulfur with this oxidation number is a strong enough oxidizing agent to oxidize chloride, bromide, and iodide.

Sulfur reduces strong oxidizing agents such as concentrated nitric acid and hot concentrated sulfuric acid. The equations are

$$S + 6H^+ + 6NO_3^- \longrightarrow 2H^+ + SO_4^{2-} + 2H_2O + 6NO_2(g)$$
$$S + 2H_2SO_4 \longrightarrow 3SO_2(g) + 2H_2O(g)$$

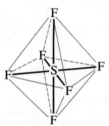

Figure 21-5. The octahedral molecular structure of the SF_6 molecule.

COMPOUNDS OF SULFUR

21.5 Hydrogen Sulfide

Hydrogen sulfide, H_2S, is a colorless gas with an offensive odor. It is responsible for the odor of rotten eggs. Hydrogen sulfide is toxic and great care must be exercised in handling it. It is, in fact, as poisonous as hydrogen cyanide (prussic acid), which has been used in death chambers in some states. Small amounts of hydrogen sulfide in the air cause headaches, while larger amounts cause paralysis in the nerve centers of the heart and lungs, leading to unconsciousness and death. Hydrogen sulfide is particularly deceptive because it paralyzes the olfactory nerves; after a short exposure one does not smell it. Many persons have died as a result of this.

Hydrogen sulfide can be prepared by treating certain metal sulfides with a dilute strong acid; for example,

$$FeS + 2H^+ \longrightarrow Fe^{2+} + H_2S$$

The production of hydrogen sulfide by the direct union of its elements is unsatisfactory because the reaction is reversible and, as it is usually carried out, not more than 2% of the reacting elements are combined at equilibrium.

$$H_2(g) + S(s) \rightleftharpoons H_2S(g) \qquad \Delta H° = -20.63 \text{ kJ}$$

An aqueous solution of hydrogen sulfide may be easily prepared by the hydrolysis of the organic sulfur-containing compound **thioacetamide, CH_3CSNH_2.**

$$CH_3CSNH_2 + 2H_2O \longrightarrow CH_3CO_2^- + NH_4^+ + H_2S(aq)$$

Heating accelerates this hydrolysis considerably.

The sulfur in hydrogen sulfide readily gives up electrons, making it a good reducing agent. In acidic solutions, hydrogen sulfide reduces Fe^{3+} to Fe^{2+}, Br_2 to Br^-, MnO_4^- to Mn^{2+}, $Cr_2O_7^{2-}$ to Cr^{3+}, and HNO_3 to NO_2. The sulfur in hydrogen sulfide is usually oxidized to elemental sulfur, unless a large excess of the oxidizing agent is present. In this case the sulfur may be oxidized to SO_3^{2-} or SO_4^{2-} (or to SO_2 and SO_3 in the absence of water).

Hydrogen sulfide decomposes into hydrogen and sulfur when heated. It burns in air, forming water and sulfur dioxide.

$$2H_2S(g) + 3O_2(g) \longrightarrow 2H_2O(g) + 2SO_2(g) \qquad \Delta H° = -103.6 \text{ kJ}$$

When hydrogen sulfide burns in a limited supply of air or when a mixture of hydrogen sulfide and sulfur dioxide (which can be produced by the combustion of H_2S) is heated, elemental sulfur is formed. In this way tons of sulfur are recovered from the hydrogen sulfide found in the natural gas from many sources and as a significant by-product of oil refineries.

$$2H_2S + SO_2 \xrightarrow{\Delta} 2H_2O + 3S$$

The deposits of sulfur in volcanic regions may be the result of this reaction because both H_2S and SO_2 are constituents of volcanic gases.

Hydrogen sulfide is a weak acid. Thus most of the more reactive metals will displace hydrogen from hydrogen sulfide. Lead sulfide is formed by the action of hydrogen sulfide on metallic lead, according to the equation

$$Pb(s) + H_2S(g) \longrightarrow PbS(s) + H_2(g) \qquad \Delta H° = -79.91 \text{ kJ}$$

Hydrogen sulfide forms black silver sulfide on silver metal. Hence silverware tarnishes when used with eggs and other foods containing certain sulfur compounds.

$$4Ag(s) + 2H_2S(g) + O_2(g) \longrightarrow 2Ag_2S(s) + 2H_2O(l) \qquad \Delta H° = -595.8 \text{ kJ}$$

21.6 Hydrosulfuric Acid and Its Salts

An aqueous solution of hydrogen sulfide is called **hydrosulfuric acid.** The acid is weak and diprotic; that is, it ionizes in two stages. Hydrogen sulfide yields **hydrogen sulfide ions, HS⁻,** in the first stage and **sulfide ions, S²⁻,** in the second.

$$H_2S \rightleftharpoons H^+ + HS^- \qquad K_1 = 1.0 \times 10^{-7}$$
$$HS^- \rightleftharpoons H^+ + S^{2-} \qquad K_2 = 1.3 \times 10^{-13}$$

Because hydrogen sulfide is diprotic, two series of salts are possible. For example, both sodium sulfide, Na_2S, and sodium hydrogen sulfide, $NaHS$, exist. Passing an excess of hydrogen sulfide into a solution of sodium hydroxide yields **sodium hydrogen sulfide, NaHS.**

$$H_2S + [Na^+] + OH^- \rightleftharpoons [Na^+] + HS^- + H_2O$$

Adding a stoichiometric amount of sodium hydroxide to a solution of sodium hydrogen sulfide gives **sodium sulfide, Na₂S.**

$$[Na^+] + OH^- + [Na^+] + HS^- \rightleftharpoons [2Na^+] + S^{2-} + H_2O$$

The sulfide ion and the hydrogen sulfide ion are strong bases, as indicated by the weakness of hydrogen sulfide as an acid. Aqueous solutions of soluble sulfides and hydrogen sulfides are basic.

$$S^{2-} + H_2O \rightleftharpoons HS^- + OH^-$$
$$HS^- + H_2O \rightleftharpoons H_2S + OH^-$$

The sulfide ion is slowly oxidized by oxygen, so sulfur precipitates when a solution of hydrogen sulfide or a sulfide salt is exposed to the air for a time.

$$4H^+ + 2S^{2-} + O_2 \longrightarrow 2H_2O + 2S(s)$$

21.7 Polysulfide Ions

When elemental sulfur is added to a solution of a soluble metal sulfide, the sulfur dissolves by combining with the sulfide ion to form complex **polysulfide ions,** formulated as S_n^{2-} ($n = 2$ through 5). The sulfur atoms of these complex ions are linked together through shared electron pairs. The electronic and geometric structures of S_2^{2-}, S_5^{2-}, and H_2S_5 are shown in Fig. 21-6. **Disulfide ions, S_2^{2-},** are analogous to peroxide ions (Section 12.14).

Polysulfide ions, like peroxide ion and free sulfur, are oxidizing agents. For example, the disulfide ion oxidizes tin(II) sulfide to the thiostannate(IV) ion.

$$SnS + S_2^{2-} \longrightarrow SnS_3^{2-}$$

Figure 21-6. Lewis structures and geometric structures of S_2^{2-}, S_5^{2-}, and H_2S_5.

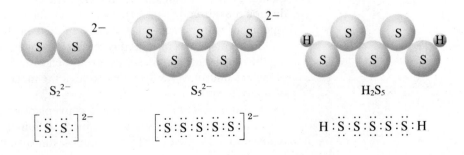

When solutions containing polysulfide ions are acidified, a white, very finely divided form of free sulfur (milk of sulfur) and hydrogen sulfide are produced.

$$S_2^{2-} + 2H^+ \longrightarrow H_2S(g) + S(s)$$

21.8   Sulfur Dioxide

The odor of burning sulfur is due to **sulfur dioxide, SO$_2$**, a colorless gas that is very soluble in water (80 volumes of gas to 1 volume of water at STP.) It readily condenses to a liquid that boils at $-10°C$ and freezes to a white solid that melts at $-75.5°C$. Liquid sulfur dioxide can be stored in steel cylinders or shipped in steel tank cars because its vapor pressure is only 5 atmospheres at 32°C. However, it must be kept perfectly dry because even traces of moisture cause enough sulfurous acid to form to corrode the steel container.

1. THE STRUCTURE OF SULFUR DIOXIDE.

Molecules of sulfur dioxide are bent, with two equal sulfur-oxygen bond distances of 1.43 Å and a bond angle of 119.5° (Fig. 21-7). To indicate that the two sulfur-oxygen bonds are equivalent, the electronic structure of the sulfur dioxide molecule must be described as a resonance hybrid (Section 5.7).

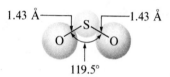

1.43 Å ——— ——— 1.43 Å

119.5°

Figure 21-7. The molecular structure of sulfur dioxide.

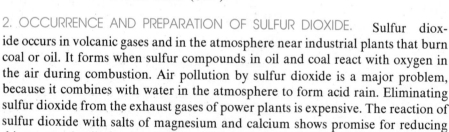

(a)   (b)

The electronic structure is not structure (a) part of the time and structure (b) part of the time, as would be the case in an equilibrium mixture. Instead, the electronic structure of the molecule is an average of the two Lewis structures. A double-headed single arrow ($\longleftrightarrow$) distinguishes the resonance notation from the equilibrium notation of a double arrow ($\rightleftharpoons$).

2. OCCURRENCE AND PREPARATION OF SULFUR DIOXIDE.

Sulfur dioxide occurs in volcanic gases and in the atmosphere near industrial plants that burn coal or oil. It forms when sulfur compounds in oil and coal react with oxygen in the air during combustion. Air pollution by sulfur dioxide is a major problem, because it combines with water in the atmosphere to form acid rain. Eliminating sulfur dioxide from the exhaust gases of power plants is expensive. The reaction of sulfur dioxide with salts of magnesium and calcium shows promise for reducing this source of pollution (see Sections 22.3 through 22.5). It is possible that the sale of sulfur or sulfuric acid extracted from the gases could help offset the costs involved.

Combustion of gasoline containing sulfur compounds also produces small quantities of sulfur dioxide. In the catalytic converters in automobiles, the sulfur dioxide is oxidized to sulfur trioxide, which then reacts with water to form a small amount of sulfuric acid mist that is emitted in the exhaust (see Section 22.5).

Sulfur dioxide is produced commercially by burning free sulfur and by roasting (heating in air) certain sulfide ores, such as ZnS, FeS$_2$, and Cu$_2$S. The roasting of metal sulfide ores (to form the oxide of the metal) is the first step in the refining of many metals.

Sulfur dioxide may be prepared conveniently in the laboratory by the action of sulfuric acid on either sodium sulfite or sodium hydrogen sulfite. Sulfurous acid is first formed, but it quickly decomposes into sulfur dioxide and water.

$$2H^+ + SO_3^{2-} \longrightarrow H_2SO_3 \longrightarrow H_2O + SO_2(g)$$
$$H^+ + HSO_3^- \longrightarrow H_2SO_3 \longrightarrow H_2O + SO_2(g)$$

Sulfur dioxide is also formed when any of a number of reducing agents react with hot concentrated sulfuric acid. Three examples are

$$Cu(s) + 2H_2SO_4(l) \longrightarrow CuSO_4(s) + SO_2(g) + 2H_2O(l) \quad \Delta H° = -11.88 \text{ kJ}$$
$$S(s) + 2H_2SO_4(l) \longrightarrow 3SO_2(g) + 2H_2O(l) \quad \Delta H° = +165.83 \text{ kJ}$$
$$2HBr(aq) + H_2SO_4(aq) \longrightarrow Br_2(l) + SO_2(g) + 2H_2O(l) \quad \Delta H° = +49.20 \text{ kJ}$$

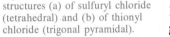

3. REACTIONS OF SULFUR DIOXIDE. Sulfur dioxide is a weak reducing agent. It is slowly oxidized by the oxygen in air to sulfur trioxide, according to the equation

$$2SO_2(g) + O_2(g) \rightleftharpoons 2SO_3(g) \quad \Delta H° = -197.8 \text{ kJ}$$

This oxidation is much faster in the presence of a suitable catalyst, and the reaction is one of the steps in the production of sulfuric acid (Section 21.11).

Chlorine oxidizes sulfur dioxide to **sulfuryl chloride, SO_2Cl_2** [Fig. 21-8(a)].

$$SO_2(g) + Cl_2(g) \longrightarrow SO_2Cl_2(l) \quad \Delta H° = -97.5 \text{ kJ}$$

A related compound, **thionyl chloride, $SOCl_2$** [Fig. 21-8(b)], is formed by the metathetical reaction of sulfur dioxide with phosphorus(V) chloride.

$$SO_2(g) + PCl_5(s) \longrightarrow SOCl_2(l) + POCl_3(l) \quad \Delta H° = -103 \text{ kJ}$$

Like most nonmetal halides, sulfuryl chloride and thionyl chloride hydrolyze, giving hydrogen chloride and sulfur dioxide or sulfur trioxide, respectively.

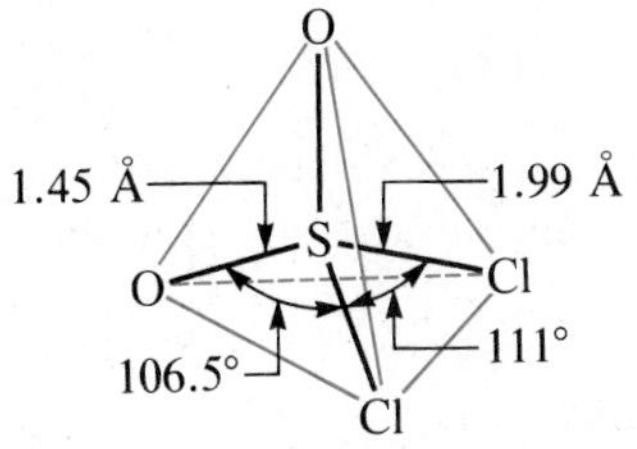

Figure 21-8. The molecular structures (a) of sulfuryl chloride (tetrahedral) and (b) of thionyl chloride (trigonal pyramidal).

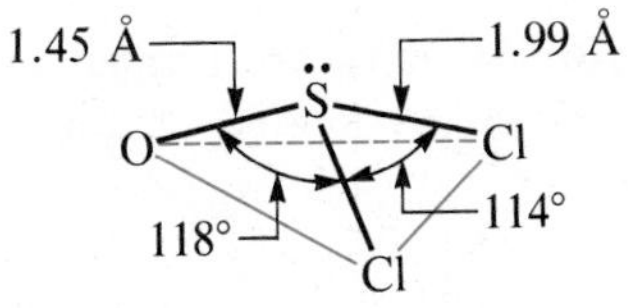

21.9 Sulfurous Acid and Sulfites

Sulfur dioxide dissolves in water to form a weakly acidic solution of **sulfurous acid, H_2SO_3,** as might be expected for the oxide of a nonmetal.

$$H_2O + SO_2 \rightleftharpoons H_2SO_3$$

Sulfurous acid is unstable, and anhydrous H_2SO_3 cannot be isolated. Boiling a solution of sulfurous acid expels the sulfur dioxide. Like other diprotic acids, sulfurous acid ionizes in two steps.

$$H_2SO_3 \rightleftharpoons H^+ + HSO_3^- \quad K_1 = 1.2 \times 10^{-2}$$
$$HSO_3^- \rightleftharpoons H^+ + SO_3^{2-} \quad K_2 = 6.2 \times 10^{-8}$$

Sulfurous acid is a moderately weak acid. Ionization is small in the first stage and much smaller in the second.

Both **sulfites** and **hydrogen sulfites** are formed by sulfurous acid. Sulfites also form when sulfur dioxide reacts with metal oxides, for example,

$$Na_2O + SO_2 \longrightarrow Na_2SO_3$$

Sulfur dioxide functions as a Lewis acid (Section 14.5) in these reactions.

Sulfurous acid reduces strong oxidizing agents. It is oxidized slowly by oxygen in the air to the more stable sulfuric acid.

$$2H_2SO_3 + O_2 \longrightarrow 4H^+ + 2SO_4^{2-}$$

Solutions containing the purple permanganate ion rapidly turn colorless when sulfurous acid is added; the MnO_4^- ion is reduced to the colorless Mn^{2+} ion.

$$2MnO_4^- + 5H_2SO_3 \longrightarrow 2Mn^{2+} + 4H^+ + 5SO_4^{2-} + 3H_2O$$

Sulfite and hydrogen sulfite salts can be prepared by adding a stoichiometric amount of a base to a sulfurous acid solution and then isolated as solids by evaporating the water. Solid sodium hydrogen sulfite forms sodium sulfite, sulfur dioxide, and water when heated.

$$2NaHSO_3 \xrightarrow{\Delta} Na_2SO_3 + SO_2 + H_2O$$

Like sulfurous acid, solutions of sulfites are susceptible to air oxidation, and sulfates are formed. Thus solutions of sulfites always contain sulfates after standing in contact with air.

21.10 Sulfur Trioxide

When sulfur burns in air, small amounts of **sulfur trioxide, SO₃,** are formed in addition to sulfur dioxide. When sulfur dioxide and oxygen are heated together, the trioxide is formed slowly according to the equation

$$2SO_2(g) + O_2(g) \rightleftharpoons 2SO_3(g) \qquad \Delta H° = -197.8 \text{ kJ}$$

If the temperature of the system is raised to about 400°C, equilibrium is reached more rapidly, but the yield is reduced due to a shift in equilibrium to the left with increasing temperature (see van't Hoff's law, Section 15.21). However, even at 400°C the rate is too slow for the reaction to be commercially useful. The presence of a catalyst, such as finely divided platinum or vanadium(V) oxide, increases the rate enough to make commercial production feasible at temperatures that give a reasonable yield.

1. THE STRUCTURES OF SULFUR TRIOXIDE. Sulfur trioxide in the vapor phase is monomeric; that is, its molecules are single SO₃ units, with the sulfur atom at the center and the oxygen atoms at the corners of an equilateral triangle as shown in Fig. 21-9(a). The measured sulfur-oxygen bond distance in sulfur trioxide is less than that of a single sulfur-oxygen bond. Thus the electronic structure of sulfur trioxide is a resonance hybrid of the following three Lewis structures.

There are three distinct solid forms of sulfur trioxide. One is icelike in appearance and is built of trimeric S_3O_9 molecules [Fig. 21-9(b)]. A second is an asbestoslike solid, in which tetrahedral SO₄ units are joined to each other through common oxygen atoms at two corners of each tetrahedron to give long chains that extend the length of the crystal [Fig. 21-9(c)]. In a third (not shown), SO₃ chains are joined together to give a layered arrangement. When it goes from the gaseous state to the solid state, a sulfur trioxide molecule changes its hybridization from sp^2 (trigonal planar) to sp^3 (tetrahedral) while functioning as a Lewis acid to form a bond to a fourth oxygen atom.

Figure 21-9. Sulfur trioxide structures.

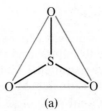

(a)

Structure of the SO₃ vapor molecule (symmetrical planar molecule).

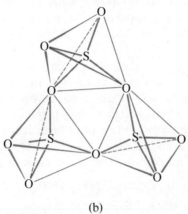

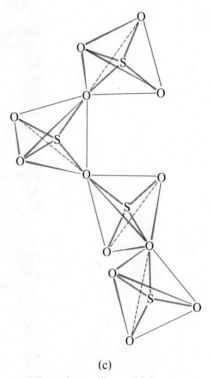

(b)

The orthorhombic icelike solid form with formula S_3O_9 (cyclic with three tetrahedral SO_4 groupings joined through three of the nine oxygen atoms).

(c)

The asbestoslike solid form with tetrahedral SO_4 groupings, each joined to another through a common oxygen atom to form infinite chain molecules.

2. PROPERTIES OF SULFUR TRIOXIDE. Liquid sulfur trioxide boils at 43°C and freezes at 17°C to the icelike solid. When a trace of moisture is added to liquid sulfur trioxide, it changes to the asbestoslike solid. Liquid sulfur trioxide fumes in moist air and dissolves in water in a highly exothermic reaction to form sulfuric acid, H_2SO_4. With a limited amount of water, the molecular acid is formed according to the equation

$$SO_3(l) + H_2O(l) \longrightarrow H_2SO_4(l) \qquad \Delta H° = -87.11 \text{ kJ}$$

Sulfur trioxide dissolves readily in concentrated sulfuric acid and forms **pyrosulfuric acid, $H_2S_2O_7$,** also known as fuming sulfuric acid or oleum.

$$H_2SO_4 + SO_3 \longrightarrow H_2S_2O_7$$

Above about 900°C sulfur trioxide decomposes into the dioxide and oxygen. The trioxide acts as a Lewis acid in Lewis acid-base reactions with many oxides and hydroxides, with the formation of sulfates and hydrogen sulfates, respectively.

$$BaO + SO_3 \longrightarrow BaSO_4$$
$$KOH + SO_3 \longrightarrow KHSO_4$$

21.11 Sulfuric Acid

Sulfuric acid, H_2SO_4, is prepared by oxidizing sulfur to sulfur trioxide and then converting the trioxide to sulfuric acid (Section 8.7). Pure sulfuric acid is a colorless, oily liquid that freezes at 10.5°C. It fumes when heated because it decomposes to water and sulfur trioxide. More sulfur trioxide than water is lost during this heating, until a concentration of 98.33% acid is reached. Acid of this concentration boils at 338°C without further change in concentration and is sold as concentrated H_2SO_4.

Concentrated sulfuric acid dissolves in water with the evolution of a large amount of heat. **Caution! Adding water to concentrated sulfuric acid may cause dangerous spattering of the acid.** Do not add water to concentrated sulfuric acid; add the acid slowly to water while stirring the solution to distribute the heat of dilution.

The sulfur atom in the sulfuric acid molecule is at the center of a tetrahedron surrounded by two oxygen atoms and two hydroxyl groups (Fig. 21-10). The hydrogen atoms form hydrogen bonds between tetrahedra, binding them together and giving the liquid a high boiling point.

The large heat of dilution of sulfuric acid is caused by the formation of hydrates and hydronium ions. The hydrates $H_2SO_4 \cdot H_2O$, $H_2SO_4 \cdot 2H_2O$, and $H_2SO_4 \cdot 4H_2O$ are known. The acid ionizes in two stages.

$$H_2SO_4 \longrightarrow H^+ + HSO_4^- \quad \text{(100\% ionized)}$$
$$HSO_4^- \rightleftharpoons H^+ + SO_4^{2-} \quad K_a = 1.2 \times 10^{-2}$$

In dilute solution, sulfuric acid undergoes almost complete primary ionization. The secondary ionization is less complete, but even so, HSO_4^- is a moderately strong acid.

The strong affinity of concentrated sulfuric acid for water makes it a good dehydrating agent. Gases that do not react with the acid may be dried by being passed through it. So great is the affinity of concentrated sulfuric acid for water that it will remove hydrogen and oxygen, in the form of water, from many compounds containing these elements. Organic substances containing hydrogen and oxygen in the ratio of 2 to 1, such as cane sugar, $C_{12}H_{22}O_{11}$, and cellulose, $(C_6H_{10}O_5)_n$, are charred by concentrated sulfuric acid.

$$C_{12}H_{22}O_{11} \longrightarrow 12C + 11H_2O$$

Concentrated sulfuric acid can cause serious burns because it reacts with organic compounds in the skin.

Sulfuric acid acts as an oxidizing agent, particularly when it is hot and concentrated. Depending on factors such as its concentration, the temperature, and the strength of the reducing agent, sulfuric acid oxidizes many compounds and simultaneously undergoes reduction to SO_2, HSO_3^-, SO_3^{2-}, S, H_2S, or S^{2-}. Its oxidizing action toward hydrogen bromide and hydrogen iodide was noted in Section 19.7, and toward metals and sulfur, in Section 21.8. The displacement of volatile acids from their salts by concentrated sulfuric acid was described in Section 14.7. Aqueous solutions of sulfuric acid exhibit the characteristic properties of strong acids.

The amount of sulfuric acid used industrially exceeds that of any other manufactured compound. Its importance is so great that the amount of it produced from year to year is a fairly accurate index of industrial prosperity. The 32.3

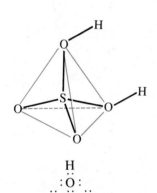

Figure 21-10. The tetrahedral molecular structure and Lewis structure of the sulfuric acid molecule.

million tons produced in the United States in 1982 represented a significant decrease from the 40.7 million tons in 1981 and reflected the general economic problems of the time.

The major uses of sulfuric acid are based largely on its strongly acidic character (Section 8.7). It is used to produce ammonium sulfate (1.8 million tons in the United States in 1982) and soluble phosphate fertilizers; to refine petroleum and to remove impurities from such products as gasoline and kerosene; to pickle steel (clean its surface of rust); to produce dyes, drugs, and disinfectants from coal tar; as an electrolyte itself or to produce metal sulfates for electrolytes; to manufacture other chemicals, such as hydrochloric and nitric acids; and to produce textiles, paints, plastics, explosives, and lead storage batteries.

21.12 Sulfates

Being a diprotic acid, sulfuric acid forms both **sulfates,** such as Na_2SO_4, and **hydrogen sulfates,** such as $NaHSO_4$. The sulfates of barium, strontium, calcium, and lead are only slightly soluble in water. These salts occur in nature as the minerals barite, $BaSO_4$, celestite, $SrSO_4$, gypsum, $CaSO_4 \cdot 2H_2O$, and anglesite, $PbSO_4$. They can be prepared in the laboratory by metathetical reactions. For example, adding barium nitrate to a solution of sodium sulfate causes the precipitation of white barium sulfate.

$$Ba^{2+} + 2NO_3^- + 2Na^+ + SO_4^{2-} \longrightarrow BaSO_4(s) + 2Na^+ + 2NO_3^-$$

This reaction is the basis of a qualitative and quantitative test for the sulfate ion and the barium ion. Although barium and its soluble salts are very toxic, barium sulfate is insoluble enough ($K_{sp} = 1.08 \times 10^{-10}$) that it can be safely put into the alimentary tract to enhance the contrast for taking X rays.

Among the important soluble sulfates are Glauber's salt, $Na_2SO_4 \cdot 10H_2O$, Epsom salts, $MgSO_4 \cdot 7H_2O$, blue vitriol, $CuSO_4 \cdot 5H_2O$, green vitriol, $FeSO_4 \cdot 7H_2O$, and white vitriol, $ZnSO_4 \cdot 7H_2O$. The hydrogen sulfates, such as $NaHSO_4$, are acids as well as salts. Sodium hydrogen sulfate is the primary ingredient in some household cleansers.

21.13 Derivatives of Sulfuric Acid and of Sulfates

Several compounds may be viewed as derivatives of sulfuric acid or of sulfates; that is, compounds in which the oxygen or hydroxide group of a sulfuric acid molecule or a sulfate ion has been replaced by some other atom or combination of atoms. For example, a **thiosulfate ion, $S_2O_3^{2-}$,** may be considered to be a sulfate ion in which one of the oxygen atoms has been replaced by a sulfur atom (Fig. 21-11).

Sulfites are slowly oxidized to sulfates by oxygen (Section 21.9). Sulfur plays a similar role in oxidizing sulfites to thiosulfates. For example, when a mixture of sulfur and a solution of sodium sulfite is boiled, sodium thiosulfate, $Na_2S_2O_3$, is formed.

$$[2Na^+] + SO_3^{2-} + S \longrightarrow [2Na^+] + S_2O_3^{2-}$$

Crystals of the pentahydrate, $Na_2S_2O_3 \cdot 5H_2O$, separate when the solvent is evaporated. Sodium thiosulfate, also called hypo, is used in the photographic process

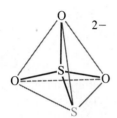

Thiosulfate ion, $S_2O_3^{2-}$

(a)

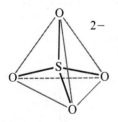

Sulfate ion, SO_4^{2-}

(b)

Figure 21-11. Tetrahedral molecular structures and Lewis structures of (a) the thiosulfate ion and (b) the sulfate ion.

in the fixing solution to dissolve from the plate or film any silver halides that have not been reduced to metallic silver by the developer.

$$AgX + 2S_2O_3^{2-} \rightleftharpoons [Ag(S_2O_3)_2]^{3-} + X^- \quad (X = \text{a halogen})$$

The thiosulfate ion is a reducing agent. It is oxidized by iodine to the **tetrathionate ion, $S_4O_6^{2-}$.**

$$2S_2O_3^{2-} + I_2 \longrightarrow S_4O_6^{2-} + 2I^-$$

This reaction is used extensively in analytical chemistry. For example, in one quantitative analysis for copper ion, iodine is produced by the following reaction and then titrated with a standard thiosulfate solution.

$$Cu^{2+} + 3I^- \longrightarrow I_2 + CuI(s)$$

When a solution of sodium thiosulfate is acidified, unstable **thiosulfuric acid, $H_2S_2O_3$,** is formed.

$$S_2O_3^{2-} + 2H^+ \longrightarrow H_2S_2O_3$$

The acid decomposes immediately into sulfurous acid and sulfur.

$$H_2S_2O_3 \longrightarrow H_2SO_3 + S(s)$$

Both **peroxymonosulfuric acid, H_2SO_5,** and **peroxydisulfuric acid, $H_2S_2O_8$** (Fig. 21-12), may be viewed as peroxide derivatives of sulfuric acid. A hydroxyl group of the sulfuric acid molecule has been replaced by a —OOH group in peroxymonosulfuric acid and by a —OOSO$_3$H group in peroxydisulfuric acid. Peroxydisulfuric acid is produced commercially by the anodic oxidation of hydrogen sulfate ions in 45–55% sulfuric acid at a low temperature.

$$2HSO_4^- \longrightarrow H_2S_2O_8 + 2e^-$$

Electrolysis of potassium hydrogen sulfate in aqueous solution gives potassium peroxydisulfate, $K_2S_2O_8$. Treating this salt at low temperatures with concentrated sulfuric acid produces peroxymonosulfuric acid, commonly called **Caro's acid.** This acid is also produced by the reaction of sulfur trioxide with hydrogen peroxide, H_2O_2.

$$H_2O_2 + SO_3 \longrightarrow SO_2(OH)(OOH)$$

Figure 21-12. Structures of (a) peroxymonosulfuric acid and (b) peroxydisulfuric acid. Each acid possesses an oxygen-oxygen (peroxide) linkage; hence the use of the prefix *peroxy-* in the names.

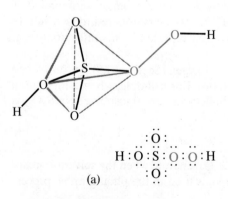

(a)

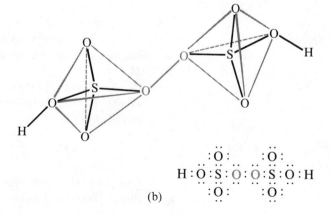

(b)

This reaction is analogous to that of sulfur trioxide with water.

$$H_2O + SO_3 \longrightarrow SO_2(OH)_2$$

The peroxysulfuric acids and their salts are useful as strong oxidizing agents and are used to prepare hydrogen peroxide (Section 12.12).

Chlorosulfonic acid, HSO_3Cl, and sulfamic acid, SO_3NH_3, are related structurally to sulfuric acid, as shown in Fig. 21-13. The formula for sulfamic acid, sometimes written as HSO_3NH_2, is more properly written as SO_3NH_3. Chlorosulfonic acid is a colorless liquid that fumes in moist air, reacts vigorously with water giving sulfuric acid and hydrogen chloride (hydrolysis of a nonmetal halide), and is used to introduce the sulfonate group, $-SO_3H$, into many organic compounds. Sulfamic acid is a white, crystalline, nonhygroscopic solid. It is one of the few strong monoprotic acids that can be weighed without taking special drying precautions. It is of some importance, therefore, in analytical chemistry. Sulfamates are important ingredients in some weed killers.

Figure 21-13. Structure of a sulfuric acid molecule compared to the structures of two related acids. The normal bond angle in a tetrahedral structure is 109°. The O—S—N angle in sulfamic acid is 108°, indicating a slight distortion of the tetrahedron. Such a distortion arises when the structure is not symmetrical—that is, when the atoms at the corners of the tetrahedron are not all the same.

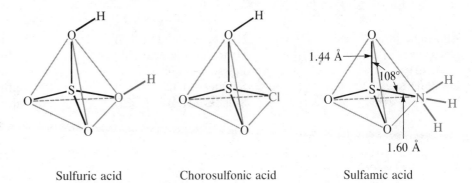

Sulfuric acid   Chorosulfonic acid   Sulfamic acid

FOR REVIEW

SUMMARY

The chemical behavior of **sulfur** is that expected for a nonmetallic element in Group VIA. Sulfur reacts with almost all metals. It readily picks up two electrons to form the **sulfide ion, S^{2-},** in which it has an oxidation number of -2. Sulfur also reacts with most nonmetals. It exhibits an oxidation number of -2 in compounds such as hydrogen sulfide. With more electronegative elements sulfur commonly forms compounds such as sulfur dioxide and sulfur trioxide, in which it exhibits oxidation numbers of $+4$ and $+6$, respectively. In some compounds an oxidation number of $+2$ is observed; sulfur dichloride is an example.

Sulfur is obtained from underground deposits by the **Frasch process** or from hydrogen sulfide obtained from natural gas. The most stable form of elemental sulfur at 25°C is rhombic sulfur, although other **allotropes** may be prepared easily by heating it.

Sulfur burns in air and forms **sulfur dioxide,** which reacts with water to form the weak, unstable acid **sulfurous acid,** H_2SO_3. **Sulfites** may be prepared by the neutralization of sulfurous acid or by Lewis acid-base reactions between sulfur dioxide and metal oxides. The slow reaction of sulfur dioxide with oxygen produces **sulfur trioxide.** This reaction is speeded up by the presence of a catalyst, such as vanadium(V) oxide or platinum. Sulfur trioxide is used to prepare the important industrial chemical **sulfuric acid, H_2SO_4.** Sulfuric acid is a strong acid and an oxidizing agent. Neutralization of sulfuric acid and the reaction of sulfur trioxide with metal oxides produce **sulfates.** Replacement of an OH group or an oxygen atom in a sulfuric acid molecule or a sulfate ion produces derivatives of sulfuric acid and of the sulfate ion. Examples include thiosulfate ion, $S_2O_3{}^{2-}$, peroxymonosulfuric acid, $HOSO_2(O_2H)$, peroxydisulfuric acid, $HOSO_2(O_2)SO_2OH$, chlorosulfonic acid, HSO_3Cl, and sulfamic acid, SO_3NH_3.

KEY TERMS AND CONCEPTS

Allotropic forms of sulfur (21.3)
Derivatives of sulfuric acid (21.13)
Disulfide ions (21.7)
Frasch process (21.2)

Hydrogen sulfide (21.5)
Hydrosulfuric acid (21.6)
Polysulfide ions (21.7)
Sulfates (21.12)
Sulfides (21.4)

Sulfites (21.9)
Sulfur dioxide (21.8)
Sulfur trioxide (21.10)
Sulfuric acid (21.11)
Sulfurous acid (21.9)

EXERCISES

Sulfur

1. Describe five chemical properties of sulfur or of compounds of sulfur that characterize it as a non-metal.
2. What kinds of sulfur compounds are most nearly ionic? Which kinds are covalent? Give two examples of each type.
3. Explain, with the aid of a diagram, the Frasch process for extracting sulfur from underground deposits.
4. What are the allotropic forms of solid sulfur, and how may they be produced?
5. What is the molecular structure of sulfur in the liquid, solid, and gaseous states?
6. Based on the location of sulfur in the Periodic Table, predict the products of and write a balanced equation for each of the following. (There may be more than one correct answer depending on your choice of stoichiometries.)
    (a) $Li + S \longrightarrow$
    (b) $Ga + S \longrightarrow$
    (c) $P + S \longrightarrow$
    (d) $H_2 + S \longrightarrow$
    (e) $F_2 + S \longrightarrow$
7. Write an equation for the reaction of sulfur with each of the following:
    (a) H_2 (b) C   (c) Fe   (d) O_2 (e) HNO_3
8. Write chemical equations describing three chemical reactions in which elemental sulfur acts as an oxidizing agent; three in which it acts as a reducing agent.
9. The reaction of an excess of sulfur with metallic copper at 400°C gives a copper sulfide, which contains 50.2% sulfur. Write the equation for the reaction.   *Ans. $Cu + 2S \longrightarrow CuS_2$*
10. Air contains 20.99% oxygen by volume. What volume of air in cubic meters at 27.0°C and 0.9868 atm is required to oxidize 1.00 metric ton (1000 kg) of sulfur to sulfur dioxide, assuming that all of the oxygen reacts with sulfur?
        Ans. $3.71 \times 10^3 m^3$

Compounds of Sulfur

11. Determine the oxidation number of sulfur in each of the following species:
    (a) HS^- (c) SF_6 (e) $S_2O_3^{2-}$
    (b) $SOCl_2$ (d) Na_2S (f) $S_4O_6^{2-}$

12. Account for the fact that hydrogen sulfide is a gas at room temperature while water, which has a smaller molecular weight, is a liquid.

13. Why are solutions of sulfides and hydrogen sulfides alkaline? Write equations.

14. Write an equation for the reaction of hydrogen sulfide in acidic solution with each of the following:
    (a) Fe^{3+} (c) MnO_4^-
    (b) Br_2 (d) $Cr_2O_7^{2-}$

15. How can sodium sulfite be made in the laboratory?

16. Why do the two sulfur-oxygen bonds in sulfur dioxide have the same length?

17. Describe the hybridization of the sulfur atom in gaseous molecules of SO_2 and SO_3.

18. How does the hybridization of the sulfur atom change when gaseous SO_3 condenses to give the asbestoslike form of solid SO_3?

19. Why is sulfuric acid a stronger acid than sulfurous acid?

20. Which is the stronger acid, $NaHSO_3$ or $NaHSO_4$? Which is the stronger base, Na_2SO_4 or Na_2SO_3?

21. Why do solutions of sulfites usually contain sulfate ions?

22. Write chemical equations describing, respectively, the reaction of water and of sulfuric acid with sulfur trioxide. What are the products of these reactions?

23. What are the possible products of the reduction of sulfuric acid? What is the oxidation number of sulfur in each of these products?

24. Write the Lewis structure of each of the following:
    (a) S^{2-} (d) H_2SO_3 (g) H_2SO_5
    (b) H_2S_2 (e) SO_3 (h) $H_2S_2O_7$
    (c) SO_2 (f) Na_2SO_4 (i) $H_2S_2O_8$

25. In 1981, 2.111×10^6 tons of ammonium sulfate was produced in the United States. If all of this ammonium sulfate resulted from the reaction of ammonia with sulfuric acid, what percentage of the 4.0674×10^7 tons of sulfuric acid produced in 1981 was used in ammonium sulfate manufacture? *Ans. 3.852%*

26. Sulfur(IV) fluoride, SF_4, is an active fluorinating agent; that is, it reacts with many halides and oxides, converting them to fluorides. The reaction of $NaAsO_3$ with SF_4 gives a compound that contains 10.8% sodium, 35.35% arsenic, and 53.80% fluorine. What is the empirical formula of this compound? *Ans. $NaAsF_6$*

27. What volume of hydrogen sulfide at STP can be produced by the reaction of 308 g of aluminum sulfide, Al_2S_3, with an excess of phosphoric acid? *Ans. 138 L*

28. A volume of 22.85 mL of a 0.1023 M standard sodium thiosulfate solution is required to titrate a 25.00-mL sample of a solution containing iodine. What is the iodine concentration in the 25.00-mL sample? *Ans. 0.04675 M*

29. Based on the location of sulfur in the Periodic Table, predict the products of and write a balanced equation for each of the following reactions. (There may be more than one correct answer depending on your choice of stoichiometries.)
    (a) $F_2 + SO_2 \longrightarrow$
    (b) $Na_2O + SO_2 \longrightarrow$
    (c) $KClO + Na_2S \longrightarrow$
    (d) $O_2 + H_2S \longrightarrow$
    (e) $Cl_2 + H_2S \longrightarrow$
    (f) $I^- + SO_2 \longrightarrow$

Additional Exercises

30. Why does SF_6 exist but SCl_6 does not? Why does SCl_4 exist but SBr_4 does not?

31. (a) Which is a stronger acid, H_2O or H_2S?
    (b) Which is a stronger base, OH^- or SH^-?
    (c) Which is a stronger base, SH^- or S^{2-}? Explain each answer.

32. The following compounds can be considered to be derived from sulfurous acid, sulfuric acid, or their salts, by replacement of one or more atoms in these molecules. Suggest structures for these molecules and write their Lewis structures.
    (a) SOF_2 (d) $Na_2S_2O_8$
    (b) $Ca(FSO_3)_2$ (e) $H_2S_2O_7$
    (c) $(CH_3)_2SO_4$ (f) $S_2O_6F_2$

33. The average oxidation number of sulfur is not one of its common ones in Na_2S_2, H_2S_2, K_2S_5, and $Na_2S_2O_3$. Calculate the average oxidation number of sulfur in these compounds. What is the common structural feature in these compounds? In view of your answer to the preceding

question, write a Lewis structure for S_2F_{10}. What is the oxidation number of sulfur in S_2F_{10}?

34. What single chemical test could be used to distinguish sulfides, polysulfides, sulfites, and sulfates from one another?

35. Interpret the following reaction in terms of an acid-base relationship according to the Lewis theory:

$$CaO + SO_3 \longrightarrow CaSO_4$$

36. Show by equations the dehydrating action of sulfuric acid on cane sugar and on cellulose.

37. Write formulas to show that the peroxysulfuric acids may be considered as derivatives of hydrogen peroxide; that they may be considered as derivatives of sulfuric acid.

38. Account for the formation of sulfur in a solution of sodium thiosulfate in contact with air (air contains carbon dioxide, the anhydride of carbonic acid).

39. Devise the best valence electron structure you can for the SCl molecule. How satisfactory is your structure compared to that of S_2Cl_2? Comment on the relative probabilities for the existence of SCl and S_2Cl_2.

40. The reactions involved in the preparation of sulfuric acid are highly exothermic. Using the data in Appendix J, calculate the enthalpy changes for the following reactions:
    (a) $S(s) + O_2(g) \longrightarrow SO_2(g)$
    (b) $2SO_2(g) + O_2(g) \longrightarrow 2SO_3(g)$
    (c) $SO_3(g) + H_2O(l) \longrightarrow H_2SO_4(l)$
    Ans. (a) $-296.83\ kJ$; (b) $-197.8\ kJ$;
    (c) $-132.5\ kJ$

41. Determine if the following reactions are spontaneous at standard conditions using either $\Delta G°$ values or cell potentials. The equations are not necessarily complete or balanced.
    (a) $NiS + Sn \longrightarrow SnS + Ni$
    Ans. Spontaneous
    (b) $S + NO_2^- \longrightarrow S^{2-} + NO_3^-$ (in basic solution)   *Ans. Not spontaneous*
    (c) $H_2S + CrO_4^{2-} \longrightarrow S + Cr(OH)_3$ (in basic solution)   *Ans. Not spontaneous*
    (d) $Al_2O_3(s) + H_2S(g) \longrightarrow Al_2S_3(s) + H_2O(l)$
    Ans. Not spontaneous

42. A compound of sulfur, oxygen, and fluorine was found to contain approximately 27.0% sulfur and 32.3% fluorine. A 0.106-g sample of this compound occupied a volume of 20.2 mL at STP. What is the molecular formula of the compound? Suggest a Lewis structure for the compound. Explain your answer.   *Ans.* SO_3F_2

43. A molten mixture of rubidium fluoride and uranium(IV) fluoride is oxidized with fluorine to produce a uranium compound in which most but not all of the uranium has an oxidation number of +5. The product is found to contain 54.43% uranium. A 1.0357-g sample of the product immersed in 100.0 mL of 0.1007 M acidified potassium iodide solution reacts according to the following equation:

$$2I^- + 2UF_6^- \longrightarrow 2UF_4 + I_2 + 4F^-$$

The iodine produced is titrated with 14.80 mL of 0.1494 M sodium thiosulfate solution. What percentage of the original uranium was oxidized to the +5 oxidation number?   *Ans. 93.36%*

22

THE ATMOSPHERE, AIR POLLUTION, AND THE NOBLE GASES

THE ATMOSPHERE

22.1 The Composition of the Atmosphere

The **atmosphere,** or air, is the mixture of gaseous substances that surrounds the earth. For the most part it consists of almost constant proportions of uncombined nitrogen, oxygen, and the noble gases. The percentage of carbon dioxide in the air varies somewhat and that of dust and water vapor varies widely. There are also traces of hydrogen, ammonia, hydrogen sulfide, nitrogen oxides, sulfur dioxide, and other gases. The average composition of dry air at sea level is given in Table 22-1.

Table 22-1   Composition of Dry Air

Component	Percent by Volume	Component	Percent by Volume
Nitrogen	78.03	Helium	0.0005
Oxygen	20.99	Krypton	0.0001
Argon	0.94	Ozone	0.00006
Carbon dioxide	0.035–0.04	Hydrogen	0.00005
Neon	0.0012	Xenon	0.000009

The composition of dry air does not vary much with location or altitude. However, the density of air, as reflected by its pressure, decreases with increasing altitude. The average pressure at sea level and 45° latitude is 1.0 atmosphere; at 15,000 feet, 0.53 atmosphere; at 10 miles, about 0.05 atmosphere; at 30 miles, only about 1×10^{-4} atmosphere.

22.2 Liquid Air

Before air is liquefied, carbon dioxide and water vapor must be removed because these substances freeze when cooled and clog the pipes of the liquid-air machine (Fig. 22-1). Once free of water and carbon dioxide, the air is compressed to a pressure of about 200 atmospheres and cooled to remove the heat produced by compression. The cold compressed air is next passed into the liquefier where it escapes through a valve and expands to a pressure of about 20 atmospheres. The expanding air absorbs heat from other compressed air in the small inner coil of the liquefier. As the process continues, each quantity of air that escapes from the valve is colder than that preceding it, and finally liquid air is formed.

Figure 22-1. Diagram of the liquefier used in the commercial preparation of liquid air.

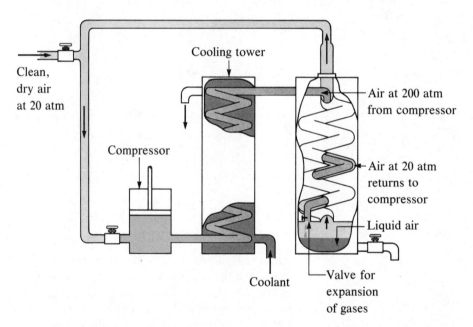

Figure 22-2. Cross section of a Dewar flask.

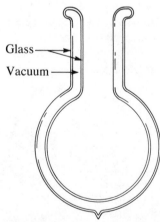

Liquid air is a mobile liquid with a faint blue color. It evaporates rapidly in an open container with the absorption of a large quantity of heat. To reduce the rate of evaporation, liquid air is stored in **Dewar flasks** (Fig. 22-2). These flasks have an evacuated space between their inner and outer walls, because a vacuum is an excellent insulator. Most Dewar flasks are silvered on the inside surface of the evacuated space to reflect heat. Thermos bottles are Dewar flasks.

Liquid air is a mixture so it has no definite boiling point. Quantities of nitrogen, oxygen, and argon for commercial use are obtained by the fractional distillation of liquid air. Oxygen boils at −183.0°C, and nitrogen, at −195.8°C (Fig. 22-3). The low temperature of liquid air makes it very useful in research requiring low temperatures. The study of the effects of very low temperatures on substances is called **cryogenics.**

Figure 22-3. Comparative Celsius temperatures for changes of state of several substances.

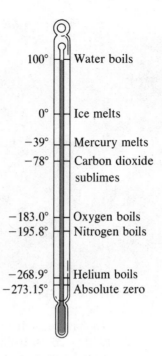

100°   Water boils

0°   Ice melts

−39°   Mercury melts

−78°   Carbon dioxide
sublimes

−183.0°   Oxygen boils
−195.8°   Nitrogen boils

−268.9°   Helium boils
−273.15°   Absolute zero

AIR POLLUTION

22.3 The Air Pollution Problem

Air pollution is not a new problem. William Shakespeare described the problem in *Hamlet* in about 1601:

> This most excellent canopy, the air, look you, this brave o'erhanging firmament, this majestical roof fretted with golden fire, why, it appears no other thing to me than a foul and pestilent congregation of vapors.*

In 1272 Edward I banned the use of smoke-producing coal in an effort to clear the air of London, and the British Parliament ordered a man tortured and hanged for illegally burning the banned coal. Later, Richard III put a high tax on the use of coal. However, these efforts accomplished little, and London remained smoky. In 1952 a combination of smoke and fog killed almost 4000 people in London during a four-day period. In Donora, Pennsylvania, in 1948, twenty people died and 5900 became ill during a period of serious air pollution. These are dramatic examples. Less dramatic, but no less tragic for those involved, were the years 1953, 1963, and 1966, when air pollution in New York City was a serious problem. It is estimated that 700 more people than usual died as a result of smog during each of those years. Polluted air may ultimately result in shortened lives for all of us, but it is a special problem for those suffering from respiratory diseases such as emphysema and bronchitis. Lung cancer is twice as prevalent among people living in air-polluted cities as among those living in rural areas with cleaner air.

Hamlet, Prince of Denmark, William Shakespeare (1601); Act II, Scene 2.

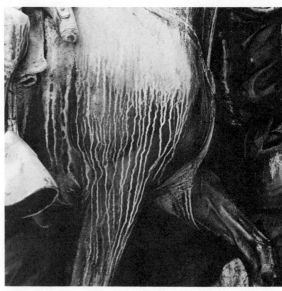

Figure 22-4.   The Robert Gould Shaw Memorial in Boston has been damaged by acid rain. The close-up shows in detail the damage to the forequarters of the horse. (*Courtesy of Rick Stafford*)

The problems due to air pollution reach well beyond those related to human health (Fig. 22-4). To cite only a few, air pollution can damage or ruin flowers and crops, make paint peel and discolor, damage textiles, discolor dyes, add to the expense of cleaning clothes, damage limestone and marble buildings and statues, corrode metals, damage rubber tires, and increase lighting costs by partially blocking out the sun. In the United States alone, an estimated 200 million tons or more of pollutants are emitted into the air annually. Strictly on an economic basis, it has been estimated that directly or indirectly air pollution costs the people in the United States over $12 billion per year. Air pollution is a major problem. It will be advantageous for us to learn more about it and to take steps to solve it.

22.4   The Causes of Air Pollution

The six most common air pollutants are those considered in the federal standards for air quality. The six, with a partial list of the effects attributed to each, are as follows:*

1. *Sulfur oxides* cause acute leaf injury and attack trees, irritate the human respiratory tract, corrode metals and other building materials, ruin hosiery and other textiles, disintegrate book pages and leather, destroy paint pigments, and erode statues.

2. *Particulates* (*solids*) obscure vision, aggravate lung illnesses, deposit grime on buildings and personal belongings, and corrode metals.

3. *Carbon monoxide* causes headaches, dizziness, and nausea, impairs mental processes by reducing oxygen content in the blood, and in sufficient quantity causes death.

---

*Sources: *Air Conservation,* American Association for the Advancement of Science, No. 80; *The Effects of Air Pollution,* Health, Education, and Welfare (HEW)—National Air Pollution Control Administration (NAPCA), No. 1556; *Air Pollution Primer,* National Tuberculosis and Respiratory Disease Association; *Facts and Issues,* League of Women Voters, No. 393; *Air Pollution Injury to Vegetation,* NAPCA, No. AP-71.

4. *Nitrogen oxides* cause leaf damage, stunt plants, irritate eyes and nasal membranes, cause brown pungent haze, corrode metals and other building materials, and damage rubber.

5. *Hydrocarbons* may be carcinogenic (cancer-producing) and can retard plant growth and cause abnormal leaf and bud development.

6. *Photochemical oxidants*

    a. Ozone and resultant chemical products discolor leaves of many crops, trees, and shrubs, damage and fade textiles, reduce physical ability (including athletic performance), speed deterioration of rubber, disturb lung function, irritate eyes, nose, and throat, and induce coughing.

    b. Peroxyacetyl nitrate (PAN) discolors leaves, irritates eyes, and disturbs lung function.

Over 100 specific polluting compounds fall into these six categories. In addition to these compounds many other materials are contaminants in specific locations. These include, to list only a few, fluorides, lead compounds, natural radioactive materials, chlorine, beryllium, mercury, and pesticides.

Most air pollution arises from combustion. Coal is largely carbon; fuel oils, gasoline, and natural gas are largely hydrocarbons (compounds containing only carbon and hydrogen). If combustion is complete, carbon burns and produces only carbon dioxide; hydrocarbons (for example, methane, CH_4) produce only carbon dioxide and water.

$$C + O_2 \longrightarrow CO_2$$
$$CH_4 + 2O_2 \longrightarrow CO_2 + 2H_2O$$

However, combustion is sometimes incomplete because of a deficiency of oxygen, and carbon monoxide forms.

$$2C + O_2 \longrightarrow 2CO$$
$$2CH_4 + 3O_2 \longrightarrow 2CO + 4H_2O$$

Sulfur dioxide is produced when generating plants burn coal containing impurities such as elemental sulfur and pyrite, FeS_2, and this contributes greatly to air pollution.

$$S + O_2 \longrightarrow SO_2$$
$$4FeS_2 + 11O_2 \longrightarrow 2Fe_2O_3 + 8SO_2$$

The sulfur dioxide either reacts with water to form sulfurous acid (which is readily oxidized by oxygen to sulfuric acid) or is oxidized by nitrogen oxides or other oxidizing agents to sulfur trioxide (which unites with water to form sulfuric acid; see Section 8.7). These acids are among the substances chiefly responsible for ruining nylon hosiery and for corroding metals. The catalytic converters of automobile exhaust systems produce small quantities of sulfur dioxide, sulfur trioxide, and sulfuric acid from reactions of the sulfur compounds present in gasoline.

Nitrogen oxides are produced in combustion processes because nitrogen and oxygen in the air react at high temperatures (Sections 23.9 and 23.17).

$$N_2(g) + O_2(g) \longrightarrow 2NO(g) \qquad \Delta H^\circ = 180.5 \text{ kJ}$$
$$2NO(g) + O_2(g) \longrightarrow 2NO_2(g) \qquad \Delta H^\circ = -114.1 \text{ kJ}$$
$$3NO_2 + H_2O \longrightarrow 2HNO_3 + NO$$

Many of the solids in polluted air (carbon particles, ash, etc.) are noncombustible or unburned material from fuels.

Other materials that contribute to air pollution do not come from combustion but result from vaporization of liquids (or sometimes from sublimation of solids). Solid particles are introduced into the air by abrasion or grinding of materials. Examples include metal filings from polishing or grinding metals, dust from dust storms or raised by automobile wheels on dusty roads, and (what is said to be the most common particulate matter in human lungs) finely divided rubber from wear of automobile tires.

Secondary reactions produce additional pollutants. The oxidation of sulfur dioxide to sulfur trioxide and the subsequent reaction with moisture in the atmosphere to produce sulfuric acid have been mentioned. Another example is the photochemical reaction of unburned hydrocarbons and nitrogen oxides from automobile exhaust to form peroxyacetyl nitrate (PAN) and ozone. These compounds are primary constituents of smog, a serious problem in many cities.

The problem of smog is intensified by the air inversion common in Los Angeles and some other cities. Normally, the layer of air nearest the earth's surface becomes warmer than the layers above and rises, allowing fresh air to take its place. Sometimes, however, a warm layer of air develops above the ground layer and traps it at the earth's surface. Pollutants from auto exhaust, industrial smokestacks, and other sources are also trapped, sometimes for days at a time.

Tetraethyl lead is used in some gasolines to provide greater antiknock qualities. Its use has been reduced, however, with the introduction of catalytic converters, because lead compounds poison the catalysts used in the converters and render them ineffective. When tetraethyl lead, $Pb(C_2H_5)_4$, is burned in an engine, lead oxide, which forms harmful deposits, is formed. Therefore, ethylene dibromide, $C_2H_4Br_2$, is added; it reacts in the engine with the lead oxide and forms lead bromide, $PbBr_2$, a volatile substance that is emitted through the exhaust. Lead compounds are known to have serious physiological effects.

22.5 Some Solutions to the Air Pollution Problem

Removal of particulate (solid) pollutants from smokestack gases is easily accomplished using electrostatic Cottrell precipitators (see Section 13.32). Processes for removing sulfur oxides, however, are not well developed, though several methods are being studied. The wet limestone process is one that shows promise.

$$SO_2 + CaCO_3 \xrightarrow{H_2O} CaSO_3 + CO_2$$

$$SO_3 + CaCO_3 \xrightarrow{H_2O} CaSO_4 + CO_2$$

The use of low-sulfur coal, natural gas, or fuel oil, instead of high-sulfur coal, would reduce the problem, but natural gas and fuel oil are becoming more expensive and reserves of low-sulfur coal are located far from where they are needed.

Removing the sulfur is one possible solution. Fuel oil is fairly easy to desulfurize, but only 12% of the sulfur oxides presently emitted by combustion sources originate in fuel oil, whereas 65% originate in coal. Coal contains 2.5–3% sulfur on the average. Of the coal mined in the United States, 25% is such that its sulfur content can be reduced to 1% by existing processes. For the other 75% of the coal mined, it is presently possible to remove only about 40% of its sulfur.

Over 95% of nitrogen oxide emissions result from combustion; 25% of the emissions come from combustion in steam-electric power plants. Some reduction in nitrogen oxide emissions can be made by modifying the combustion process, although this leads to some loss of efficiency and the technology is not yet perfected. The technology for removal of nitrogen oxides from stack gases is also still in the laboratory stage.

Carbon monoxide emissions can be lessened by finer control of combustion, including better tuning of auto engines. The technology to eliminate the carbon monoxide problem completely, either by more efficient combustion or by removal from stacks and exhausts, is not presently available.

Catalytic converters (Fig. 22-5) decrease carbon monoxide, nitrogen oxides, and unburned hydrocarbon pollutants in automobile exhaust to low levels. One system consists of two converters, each containing a different catalytic material. In the first the pollutants react with oxygen from additional added air on the surface of the solid catalyst.

$$2CO + O_2 \longrightarrow 2CO_2$$
$$\underset{\substack{\text{(Typical}\\\text{hydrocarbon}\\\text{in gasoline)}}}{C_7H_{16}} + 11O_2 \longrightarrow 7CO_2 + 8H_2O$$

The catalyst in the second converter promotes the spontaneous decomposition of nitrogen oxides to the elements.

$$2NO \longrightarrow N_2 + O_2 \qquad \Delta G° = -173.1 \text{ kJ}$$
$$2NO_2 \longrightarrow N_2 + 2O_2 \qquad \Delta G° = -102.6 \text{ kJ}$$

Unfortunately, catalytic converters also catalyze reactions involving the sulfur compounds in gasoline. Reaction of hydrocarbons with these sulfur compounds produces the very toxic hydrogen sulfide, but this only occurs with an abnormally rich mixture (high ratio of fuel to air). Careful tuning of the engine can eliminate that problem. More serious is the catalysis of the reaction with oxygen to produce sulfur dioxide and sulfur trioxide. These react with water and ultimately form small quantities of sulfuric acid mist. Fortunately, technology is available for removing sulfur compounds from gasoline during refining. Although this removal

Figure 22-5. Catalytic converters used to decrease pollutants in the exhaust of automobiles.

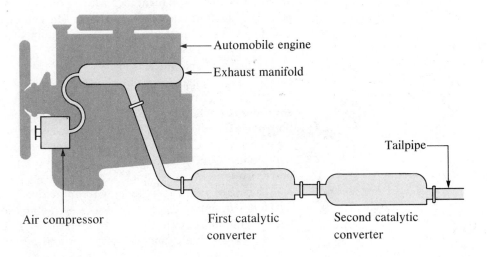

Automobile engine

Exhaust manifold

Tailpipe

Air compressor

First catalytic converter

Second catalytic converter

will increase the cost to the consumer, the extra expense may be necessary if catalytic converters continue to be used extensively.

As was pointed out in Section 22.4, lead compounds poison the catalysts in converters. Hence low-lead or unleaded gasolines must be used in automobiles equipped with converters. The undesirable emission of toxic lead compounds in the exhaust is also eliminated by the use of unleaded gasolines.

Some small automobiles have engines that use a stratified fuel concept based on the idea that good performance requires a fuel-rich mixture only near the spark plug. Each cylinder is divided into a small, high-temperature auxiliary combustion chamber near the spark plug and a larger, lower-temperature main combustion chamber. There is a fuel intake valve in each chamber. The auxiliary chamber is connected to the main chamber by a small opening called a torch nozzle, through which the initial flame can spread. A fuel mixture with a relatively high ratio of fuel to air is introduced into the auxiliary chamber, and an extremely lean fuel mixture is introduced into the main chamber. When the mixture in the auxiliary chamber is ignited, very little nitric oxide is produced even at the high temperature because of the relatively low amount of air, the source of nitrogen and oxygen. The lean mixture is ignited by the flame spreading through the torch nozzle, and the resulting explosion drives the piston. The lower temperature in the main chamber offsets the effect of the larger volume of air and keeps the production of nitric oxide to a low level. In addition, the more complete combustion of the leaner mixture in the main chamber reduces the amounts of carbon monoxide and unburned hydrocarbons produced, and it should improve the gasoline mileage.

Evaporation of gasoline from tanks during filling and during use contributes surprising amounts of hydrocarbon vapors to the air. It has been estimated that in a single large U.S. city, before the introduction of vapor-tight fuel-tank caps, 16 tons of hydrocarbons evaporated into the air each day. Automobile engines in the same city put over 1700 tons of hydrocarbons and approximately 650 tons of nitrogen oxides into the air daily.

Alternatives to the internal combustion engine, as well as modifications of the existing models, are under study. The alternatives include electric automobiles with rechargeable batteries and steam automobiles (or ones using gases other than steam). It should be noted, however, that recharging of batteries requires electrical power, which may require additional generating facilities and bring about a possible subsequent increase in pollution. A change to avoid one type of pollution often introduces another type. The complete resolution of the pollution dilemma will involve the handling of exceedingly complex and intertwined problems.

To many people nuclear energy is the answer to our energy crisis. Nuclear energy is used in many areas. It raises problems of thermal discharge, which raises the temperature of lakes, streams, and the surrounding air, and of radioactive emissions (Chapter 28).Thermal pollution (sometimes called thermal loading) by both nuclear and conventional power plants is a serious problem. Safety measures have been worked out to prevent the escape of radioactive materials, but radioactive wastes must be handled very carefully to prevent any leakage. (Nuclear power is discussed in Chapter 28.)

The use of solar energy (energy from the sun) is in a very rudimentary state of development but holds promise as a clean method for solving some of our power needs. Windmills may once again be used to generate small quantities of electrical power in areas that experience sufficient and reasonably consistent wind velocities. Meanwhile we must solve, at least on a short-term basis, the problems with combustion of fossil fuels.

In 1971 the United States government announced strict air quality standards that were to go into effect by July 1, 1975, but were in part postponed several times. These standards establish limits for the amounts of sulfur oxides, particulate matter, carbon monoxide, photochemical oxidants, nitrogen oxides, and hydrocarbons that can safely exist in the air. Full enforcement of the limits would undoubtedly cause a number of major changes, such as closing some sections of large cities to automobile traffic at certain hours, a greater use of public transportation and car pools, and very significant changes in the fuels used by electric generating plants and other industries. Hence considerable controversy continues.

In summary, air pollution involves the loss of natural resources, degrades the environment, and destroys health. Methods presently known to control pollution cost money, time, and effort. Development of the new technology necessary for really adequate control of pollution carries a high price. It appears, nevertheless, that we really have no choice but to pay the necessary price in dollars and inconvenience.

THE NOBLE GASES

22.6 Discovery and Production of the Noble Gases

Although all of the noble gases are present in the earth's atmosphere, helium was first discovered in the atmosphere around the sun. In 1868 the French astronomer Pierre-Jules-César Janssen went to India to study a total eclipse of the sun. Using a spectroscope, he found one new yellow line, which caused the English chemist Edward Frankland and the English astronomer Sir Norman Lockyer to conclude that the sun contained a new element. They called this new element **helium,** from the Greek word *helios,* which means the sun. For 27 years attempts to find helium on earth were unsuccessful, but in 1895 the Scottish chemist and physicist Sir William Ramsay showed that a gas present in trace amounts in the uranium mineral cleveite has a spectrum identical with that of helium.

Professor Hamilton P. Cady and D. F. McFarland of the University of Kansas discovered helium in natural gas in 1905 after being asked to analyze a sample of natural gas that would not ignite. They found that the gas contained 1.84% helium by mass, and so what seemed at first to be a financial loss turned out to be the beginning of a profitable business in the sale of helium. To isolate the helium the condensable components of the gas are liquefied in a liquid-air machine, leaving only helium as a gas. The United States has most of the world's commercial supply of this element in its helium-bearing gas fields.

In 1894 the British chemist Lord Rayleigh observed that a liter of nitrogen that he had prepared by removing the oxygen, carbon dioxide, and water vapor from air weighed 1.2572 grams, while a liter of nitrogen prepared from ammonia weighed only 1.2506 grams under the same conditions. This difference caused Rayleigh to suspect a previously undiscovered element in the atmosphere. By passing nitrogen obtained from air over red-hot magnesium,

$$3Mg + N_2 \longrightarrow Mg_3N_2$$

he found a small amount of gas that would not combine with any other element. Lord Rayleigh and Sir William Ramsay found that the residual gas showed spectral lines never before observed. In 1894 they announced the isolation of the first noble gas, which they called **argon** (meaning the lazy one). It was later found that this residual gas also contained traces of the other noble gases: helium, neon, krypton, and xenon.

In 1898 Sir William Ramsay and his assistant Morris W. Travers isolated **neon** (meaning the new element) by the fractional distillation of impure liquid oxygen. Shortly thereafter they showed the less volatile fractions from liquid air to contain two other new elements, **krypton** (the hidden element) and **xenon** (the stranger).

In 1900 the German physicist Friedrich E. Dorn discovered that one of the disintegration products of radium is a gas similar in chemical properties to the noble gases. At first this gas was called radium emanation, but later its name was changed to **radon.**

Argon, neon, krypton, and xenon are produced by the fractional distillation of liquid air. Radon is collected from radium salts.

22.7   Physical Properties of the Noble Gases

The principal properties of the noble gases are given in Table 22-2. Their boiling points and melting points are extremely low compared to those of other substances of comparable atomic or molecular weights. The reason for this is that no strong chemical bonds hold the atoms together in the liquid or solid states. Only weak London forces (Section 11.13) are present, and these can hold the atoms together only when molecular motion is very slight, as it is at very low temperatures. Helium is the only substance known that does not solidify on cooling. It remains liquid down to absolute zero ($-273.15\,°C$) at ordinary pressures; however, it will solidify under elevated pressure.

Table 22-2   Properties of the Noble Gases

Property	Helium	Neon	Argon	Krypton	Xenon	Radon
Atomic number	2	10	18	36	54	86
Electrons in outer shell	2	8	8	8	8	8
Atomic weight	4.0	20.18	39.95	83.8	131.3	222
Melting point, °C	-269.7^a	-248.6	-189.3	-157.2	-111.9	-71
Boiling point, °C	-268.9	-246.1	-186	-153.4	-108.1	-62

[a] At a pressure of 26 atm.

22.8   Chemical Properties of the Noble Gases

The elements of Group VIIIA were formerly called the inert gases, since until 1962 it was thought that they were chemically inert. When it was discovered that they are not really inert, the name **noble gases** was adopted. They are sometimes referred to as rare gases. The stability of the noble gas electronic structure has been mentioned frequently in preceding chapters. The two electrons of helium make a complete shell, while in each of the other noble gases, eight electrons in the outer shell make it complete. Many elements form stable ions that have noble gas electronic structures.

Chemists have tried almost since the noble gases were discovered to combine them with other elements. Early efforts were successful to a limited extent. A strong dipole, such as that of the water molecule, may polarize a noble gas atom so that it acts as a dipole itself and thereby attracts the first strong dipole. The larger the noble gas atom, the greater is its susceptibility toward dipole induction. Thus hydrates of argon, krypton, xenon, and radon have been prepared. These compounds have formulas such as $Ar \cdot nH_2O$, $Kr \cdot nH_2O$, and $Xe \cdot nH_2O$, where n is an integer from 1 to 6 but equals 6 only with noble gases of largest radii. These compounds are crystalline solids, which form when water and the noble gases are brought together at low temperatures and high pressures.

The noble gases have higher ionization energies than any of the other elements in their respective horizontal rows of the Periodic Table (see Section 4.15), and this encouraged a widely held belief in their inertness. However, the ionization energies of the larger noble gases are of about the same magnitude as those of some reactive elements and, in some cases, are even smaller. In 1962 Neil Bartlett, then in Canada, reported the preparation of a yellow compound of xenon, which he formulated as $Xe[PtF_6]$. This discovery led a group of chemists at the Argonne National Laboratory to think that perhaps xenon might be oxidized by a highly electronegative element such as fluorine, and they prepared **xenon tetrafluoride, XeF_4.** This was the first stable compound of a noble gas with another single element. The compound was made by the surprisingly simple procedure of mixing xenon gas and fluorine gas at 400°C for 1 hour and cooling to −78°C. The material forms colorless crystals, which melt at about 90°C and are stable at room temperature, showing no reaction or decomposition after prolonged storage in glass vessels. After the preparation of xenon tetrafluoride, the same group prepared XeF_2, $XeOF_4$, and XeO_3. Since that time quite a number of additional compounds have been prepared. Table 22-3 lists the formulas for some of them.

Table 22-3   Formulas for Some Compounds of Three Noble Gases

Xenon			Krypton	Radon
$XePtCl_6$	XeO_3	Na_2XeF_8	KrF_2	RnF_n (where n
$XeCl_2$	XeO_4	K_2XeF_8	$KrF_2 \cdot 2SbF_5$	is not known)
XeF_2	$XeF_2 \cdot 2SbF_5$	Rb_2XeF_8	KrF_4?	
XeF_4	$XeF_2 \cdot 2TaF_5$	Cs_2XeF_8		
XeF_6	$XeF_6 \cdot 2SbF_5$	$(NO^+)_2[XeF_8^{2-}]$		
$XeOF_2$	$XeF_6 \cdot AsF_5$	$Na_4XeO_6 \cdot nH_2O$		
$XeOF_4$	$XeF_6 \cdot BF_3$	(where $n = 6$ and 8)		
XeO_2F_2	$RbXeF_7$	$Ba_2XeO_6 \cdot 1.5H_2O$		
	$CsXeF_7$	$K[XeO_3F]$		

The preparation of compounds of the noble gases not only opened the door to an exciting new area of research but also served to point out that the ideas of science are subject to change. We should always keep our minds open to new possibilities and not be blinded by commonly accepted or entrenched ideas.

22.9 Stability of Noble Gas Compounds

The xenon fluorides are formed in exothermic reactions; thus their heats of formation are negative (see Table 22-4). Xenon-fluorine bonds are somewhat weak, as shown by the average bond energies in Table 22-4. By way of comparison, the energies of other single bonds range from 170–500 kJ/mol. The compounds are stable at room temperature in spite of the moderately weak Xe-F bonding.

Table 22-4   Heats of Formation and Average Bond Energies for the Xenon Fluorides

Compound	Heat of Formation for Gaseous State, kJ/mol	Average Bond Energy, kJ/mol
XeF_2	−108	131
XeF_4	−215	130
XeF_6	−298	126

The oxyfluorides of xenon, in general, have positive heats of formation and are quite unstable. $XeOF_4$ is an exception, with a heat of formation of −38 kJ/mol; it is the only oxyfluoride of a noble gas that is relatively easy to prepare and characterize.

Fluorides of krypton and radon have positive heats of formation. As expected, therefore, they are more difficult to make and less stable than the corresponding xenon fluorides. The heat of formation of KrF_2 is +60.2 kJ/mol.

22.10 Chemical Properties of Noble Gas Compounds

Stable compounds of xenon form when xenon bonds to either of the two highly electronegative elements fluorine and oxygen. The compounds with oxygen are prepared by replacing fluorine atoms in the xenon fluorides with oxygen.

Xenon compounds are very strong oxidizing agents that may disproportionate in water, and they react with strong Lewis acids by donating a fluoride ion. Xenon difluoride is quite soluble in water. It dissolves as undissociated molecules that soon oxidize the water to produce xenon, hydrogen fluoride, and oxygen.

$$2XeF_2 + 2H_2O \longrightarrow 2Xe + 4HF + O_2$$

The hydrolysis of xenon tetrafluoride is more complicated because it is accompanied by disproportionation of the xenon(IV) fluoride and oxidation of water. With a stoichiometric amount of water, $XeOF_4$ is formed, but the oxide, XeO_3, forms in an excess of water.

$$6XeF_4 + 8H_2O \longrightarrow 2XeOF_4 + 4Xe + 16HF + 3O_2$$
$$6XeF_4 + 12H_2O \longrightarrow 2XeO_3 + 4Xe + 24HF + 3O_2$$

Dry solid XeO_3 is an extremely sensitive explosive that must be handled with care.

When XeF_6 reacts with water, the oxidation number of +6 for xenon is retained.

$$XeF_6 + 3H_2O \longrightarrow XeO_3 + 6HF$$

Both XeF_6 and XeO_3 disproportionate in basic solution, producing salts of the perxenate ion, XeO_6^{4-}, for example,

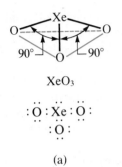

XeO₃

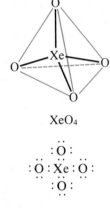

XeO₄

(b)

Figure 22-6. The Lewis structures and the molecular structures of XeO_3 (trigonal pyramidal) and XeO_4 (tetrahedral). See Fig. 19-2 for structures of XeF_2 and XeF_4.

$$2XeF_6 + 4Na^+ + 16OH^- \longrightarrow Na_4XeO_6 + Xe + O_2 + 12F^- + 8H_2O$$

$$2XeO_3 + 4Na^+ + 4OH^- \longrightarrow Na_4XeO_6 + Xe + O_2 + 2H_2O$$

Xenon difluoride reacts with acids such as HSO_3F, F_5TeOH, and $HClO_4$, which are resistant to oxidation, with evolution of hydrogen fluoride and formation of Xe—O bonds. The product in each case depends on the stoichiometry of the reaction.

$$XeF_2 + HSO_3F \longrightarrow F-Xe-O-SO_2F + HF$$

$$XeF_2 + F_5TeOH \longrightarrow F-Xe-O-TeF_5 + HF$$

$$XeF_2 + 2F_5TeOH \longrightarrow F_5Te-O-Xe-O-TeF_5 + 2HF$$

Standard reduction potentials indicate that Xe is reduced more easily than is hydrogen ion. Hence Xe is below hydrogen in the activity series. Reduction takes place more readily in acidic solution than in basic solution. The following equations show the oxidation half-reactions and the reduction potentials.

In acidic solution:

$$XeO_3 + 6H^+ + 6e^- \longrightarrow Xe + 3H_2O \qquad E° = 1.8 \text{ V}$$

$$H_4XeO_6 + 2H^+ + 2e^- \longrightarrow XeO_3 + 3H_2O \qquad E° = 3.0 \text{ V}$$

In basic solution:

$$HXeO_4^- + 3H_2O + 6e^- \longrightarrow Xe + 7OH^- \qquad E° = 0.9 \text{ V}$$

$$HXeO_6^{3-} + 2H_2O + 2e^- \longrightarrow HXeO_4^- + 4OH^- \qquad E° = 0.9 \text{ V}$$

The structures of XeF_2 (linear) and XeF_4 (square planar) are shown in Chapter 19 (Fig. 19-2) and are like those of some isoelectronic polyhalide anions (Fig. 19-5). Oxygen derivatives of XeF_2 also have a linear structure. The structure of XeO_3 [Fig. 22-6(a)] is a trigonal pyramid with the four atoms at the corners. The XeO_4 molecule has a tetrahedral structure [Fig. 22-6(b)]. Evidence indicates that the XeF_6 molecule has a distorted octahedral structure.

22.11 Uses of the Noble Gases

Helium is used for filling balloons and lighter-than-air craft. Since helium does not burn, it is safer to use than hydrogen. Although it is twice as dense as hydrogen, helium has a lifting power that is 92.6% of that of hydrogen. (See Exercise 6 at the end of this chapter.)

Nitrogen at high pressures is a narcotic, but helium is not. Therefore, mixtures of oxygen and helium are used by divers working under high pressures in order to avoid the disoriented mental state known as nitrogen narcosis, the so-called rapture of the deep. This disorientation results from the narcotic effect of nitrogen when air is breathed at high pressures. Mixtures of oxygen and helium are also used in the treatment of certain respiratory conditions such as asthma. The lightness and rapid diffusion of helium decrease the muscular effort involved in breathing.

Helium provides an inert atmosphere for the melting and welding of easily oxidizable metals and for many chemical processes that are sensitive to air. Liquid helium is used to attain low temperatures for cryogenic research.

Neon is used in neon lamps and signs. When an electric spark is passed through a tube containing neon at low pressure, a brilliant orange-red glow is emitted. The color of the light given off by a neon tube may be changed by mixing argon or mercury vapor with the neon and by utilizing tubes made of colored glasses. Neon lamps cost less to operate than ordinary electric lamps, and their light penetrates fog better.

Argon is used in gas-filled electric light bulbs, where its lower heat conductivity and chemical inertness make it preferable to nitrogen for inhibiting vaporization of the tungsten filament and prolonging the life of the bulb. Fluorescent tubes commonly contain a mixture of argon and mercury vapor. Many Geiger-counter tubes are filled with argon.

A krypton-xenon photographic flash tube has been developed for high-speed photographic exposures. An electric discharge through the tube gives a very intense light, which lasts only 0.00002 second.

Radon is collected from disintegrating radium. At one time it was sealed in small tubes that were implanted in tumors for cancer radiotherapy, but it has been replaced by other radioactive materials.

FOR REVIEW

SUMMARY

The **atmosphere** surrounding the earth is a mixture of gases consisting principally of nitrogen (78%) and oxygen (21%) with smaller amounts of water vapor, carbon dioxide, ozone, the noble gases (helium, neon, argon, krypton, xenon, and radon), and certain gases that are usually regarded as air pollutants. The principal classes of pollutants include sulfur oxides, carbon monoxide, nitrogen oxides, hydrocarbons, photochemical oxidants, and particulates, which are not gases but suspended solids.

Carbon monoxide and hydrocarbons in the atmosphere generally result from incomplete combustion. Sulfur oxides result from the combustion of sulfur-containing fuels or from the use of sulfide-containing ores in metallurgical processing. Nitrogen oxides are formed as nitrogen and oxygen in the air combine at the high temperatures accompanying combustion. Nitrogen oxides and hydrocarbons combine in sunlight to produce photochemical oxidants such as peroxyacetyl nitrate (PAN) and ozone, which are components of photochemical smog. Each of these air pollutants can have harmful effects on the health of plants and animals and can damage metals, fibers, and building materials.

A number of techniques have been used to reduce the amounts of pollutants released into the atmosphere. These range from relatively simple devices such as vapor-tight fuel-tank caps, which prevent the evaporation of hydrocarbons, to catalytic converters, desulfurization of coal, and washing of smokestack gases. Present pollution control methods cost money, time, and effort, and, in some cases, cause inconvenience. However, air pollution can result in loss of natural resources, degradation of the environment, and destruction of health, so it may be best to pay the price necessary for its control.

The **noble gases** exhibit a very limited chemistry due to the exceptional stability of their filled valence electron shells. Only xenon will react directly with fluorine, forming the xenon fluorides XeF_2, XeF_4, and XeF_6. Xenon oxides may be prepared by the reaction of water with the fluorides, while the reaction of certain acids with XeF_2 will form derivatives containing Xe—O bonds.

KEY TERMS AND CONCEPTS

Air pollution (22.3–22.5)　　Cryogenics (22.2)　　Liquid air (22.2)
Atmosphere (22.1)　　Dewar flask (22.2)　　Noble gases (22.6–22.12)

EXERCISES

1. What evidence can you cite to show that air is a mixture rather than a compound?

2. Name the six principal air pollutants and identify a source of each. Briefly describe one technique for controlling each of these pollutants.

3. Write balanced equations for each of the following processes:
   (a) the formation of sulfur dioxide from zinc sulfide present in burning coal
   (b) the formation of nitrogen dioxide in an internal combustion engine
   (c) the formation of sulfuric acid mist from sulfur dioxide
   (d) the removal of carbon monoxide from an automobile's exhaust using a catalytic converter

4. A concentration of 750 parts per million by volume of carbon monoxide is lethal. What mass of carbon monoxide is required to provide a lethal uniform concentration in a room 6.8 m long, 5.3 m wide, and 2.2 m high (about 22 × 17 × 7 ft) if the temperature is 22°C and the atmospheric pressure is 740 torr?　*Ans. 67 g*

5. Based on the periodic properties of the elements, write balanced chemical equations for the reactions, if any, between the components of air given in Table 22-1 and hot magnesium.

6. (a) Calculate the density of air, of hydrogen, and of helium at STP in g L^{-1}. (Assume the average molecular weight of air is 29.10.)
   Ans. Air, 1.30 g L^{-1}; hydrogen, 0.0900 g L^{-1}; helium, 0.179 g L^{-1}
   (b) The lifting power of a gas with a density less than that of air is determined by the difference between its mass and that of an equal volume of air. Calculate the lifting power of hydrogen, the lifting power of helium [which by calculation in part (a) is shown to be approximately twice as heavy as hydrogen for a given volume], and the lifting power of helium as a percentage of that of hydrogen.　*Ans. Hydrogen, 1.21; helium, 1.12; 92.6%*

7. Write the Lewis structures, for XeF_2, XeF_4, XeF_6, F—Xe—O—SO_2F, and XeO_3.

8. Write complete and balanced equations for the following reactions:
   (a) XeO_3 + $Ba(OH)_2$ (in water) $\longrightarrow$
   (b) XeF_2 + $HClO_4$ $\longrightarrow$
   (c) XeF_2 + $2HClO_4$ $\longrightarrow$
   (d) XeF_6 + Na $\longrightarrow$

9. A mixture of xenon and fluorine was heated at 400°C for 1 h. A sample of the white solid that formed reacted with hydrogen to give 54.0 mL of xenon at STP and hydrogen fluoride, which was collected in water, giving a solution of hydrofluoric acid. The hydrofluoric acid solution was titrated, and 45.62 mL of 0.2115 M sodium hydroxide was required to reach the equivalence point. Determine the empirical formula for the white solid and write balanced chemical equations for the reactions involving xenon.
   Ans. $Xe + 2F_2 \rightarrow XeF_4$
   $XeF_4 + 2H_2 \rightarrow Xe + 4HF$

10. A 0.492-g sample of the XeF_4 described in Exercise 9 exerted a pressure of 47.8 torr at 127°C in a bulb with a volume of 1238 mL. Determine the molecular formula of xenon tetrafluoride from these data.　*Ans. XeF_4*

11. Basic solutions of Na_4XeO_6 are powerful oxidants. What mass of $Mn(NO_3)_2 \cdot 6H_2O$ will react with 125.0 mL of a 0.1717 M basic solution of Na_4XeO_6 if the products include Xe and a solution of sodium permanganate?　*Ans. 9.857 g*

23

NITROGEN AND
ITS COMPOUNDS

The elements of Group VA of the Periodic Table make up the **nitrogen family.**
Nitrogen is the first member of the family, which also includes phosphorus, arsenic, antimony, and bismuth. Nitrogen and phosphorus are among our most important and useful nonmetals. There is a regular gradation with increasing atomic weight in the nitrogen family, from the characteristics of a true nonmetal (nitrogen) to those of an almost true metal (bismuth).

As observed for fluorine and oxygen, the lightest members of their families, some of the properties of nitrogen are anomalous, while the expected trends are found for the other members of the family (discussed in later chapters). The chemistry of nitrogen is sufficiently different, important, and interesting to study it separately from that of the other members of the family.

23.1 Chemical Properties of Nitrogen

The electronic structure of a nitrogen atom is $1s^2 2s^2 2p^3$, with the three $2p$ electrons distributed in the p_x, p_y, and p_z orbitals. Since nitrogen has no d orbitals in its valence shell, that shell contains a maximum of eight electrons in covalent compounds. Nitrogen forms sp, sp^2, and sp^3 hybrid orbitals. It is one of the most electronegative elements; only fluorine and oxygen have higher electronegativities.

Nitrogen fills its valence shell in several ways. It gains three electrons to form the very strongly basic **nitride ion, N^{3-},** with active metals such as lithium and magnesium. Transition metals also form nitrides, such as vanadium nitride, VN, but they are not ionic. The octet can be completed by the formation of three single bonds as in NH_3 and H_2NOH, or by multiple-bond formation, as in molecular nitrogen, N_2, cyanide ion, CN^-, and nitrous acid, HNO_2.

Since nitrogen is a nonmetal, several of its oxides are acidic and react with water to form oxyacids such as nitric acid, HNO_3, and nitrous acid, HNO_2. Nitrogen-hydrogen bonds are weakly acidic; for example, anhydrous ammonia, NH_3, reacts with very strong bases, such as N^{3-} and H^-, with loss of a proton and formation of the amide ion, NH_2^-. The lone pair of electrons on nitrogen in compounds with three single bonds confers Lewis base character on these molecules.

Compounds of nitrogen in which it exhibits all of the possible oxidation numbers from -3 to $+5$ are known (Table 23-1). Much of the chemistry of nitrogen involves oxidation-reduction reactions, which convert nitrogen from one oxidation number to another.

Table 23-1   Oxidation Numbers of Nitrogen in Some Compounds and Ions

Oxidation Number	Example
-3	N^{3-}, NCl_3, NH_4^+, HN_2^-, $(CH_3)_3N$
-2	N_2H_4 (H_2N-NH_2), $N_2H_5^+$ ($H_2N-NH_3^+$)
-1	H_2NOH, $CH_3N=NCH_3$
0	N_2
$+1$	N_2O
$+2$	NO
$+3$	HNO_2, $ClNO$, NF_3
$+4$	NO_2, N_2O_4
$+5$	N_2O_5, HNO_3, NF_4^+

The various oxidation numbers of nitrogen emphasize the formal nature of this concept (Section 5.9). The distribution of electrons in a molecule is a much more fundamental property than the oxidation numbers of the atoms involved. For example, NF_3 and NCl_3 have identical Lewis structures and similar chemistry, yet nitrogen has an oxidation number of $+3$ in NF_3 and -3 in NCl_3. (This difference results from the convention that the more electronegative atom is assigned a negative oxidation number.) Oxidation numbers are useful, but arbitrary. They are helpful in balancing equations, predicting empirical formulas, and organizing types of chemical behavior, but they do not indicate the charge on an atom.

ELEMENTAL NITROGEN

23.2 Occurrence and Preparation of Elemental Nitrogen

Nitrogen was first recognized as an element by the Scottish botanist Daniel Rutherford in 1772. He demonstrated that this gas supports neither life nor combustion. The atmosphere contains 78% nitrogen by volume and 75% by weight. There are more than 20 million tons of nitrogen over every square mile of the earth's surface. Natural gas also contains some free nitrogen. The most important mineral sources of nitrogen in combined form are the **saltpeter** (KNO_3) deposits in India and other countries of the Far East and the **Chile saltpeter** ($NaNO_3$) deposits in South America. Nitrogen is an essential element of the proteins of all plants and animals.

Nitrogen is obtained industrially by the fractional distillation of liquid air (Section 22.2). Nitrogen prepared in this way usually contains small amounts of oxygen and the noble gases, particularly argon.

Pure nitrogen is prepared in the laboratory by heating a solution of ammonium nitrite.

$$NH_4NO_2(aq) \longrightarrow 2H_2O(l) + N_2(g) \qquad \Delta H° = -305.1 \text{ kJ}$$

In actual practice a mixture of sodium nitrite and ammonium chloride is used to furnish the ions of ammonium nitrite because NH_4NO_2 is so unstable that it cannot be stored. Atmospheric nitrogen (nitrogen plus the noble gases) can be isolated in the laboratory by burning phosphorus in air that is confined over water. The P_4O_{10} formed dissolves in the water, and the residual gas is primarily nitrogen.

23.3   Properties of Elemental Nitrogen

Under ordinary conditions nitrogen is a colorless, odorless, and tasteless gas. It boils at $-195.8°C$ under 1 atmosphere pressure and freezes at $-210.0°C$. It is slightly less dense than air, as air contains heavier molecules of oxygen as well as those of nitrogen. Under standard conditions 100 milliliters of water dissolves 2.4 milliliters of nitrogen.

Nitrogen molecules contain a triple bond, $:N\equiv N:$. The molecular orbital description of the bonding in N_2 was discussed in Section 6.9. The triple bond in a nitrogen molecule is very strong; N_2 is the most stable homonuclear diatomic molecule known. The bond energy is 955 kilojoules per mole of triple bonds. The energy of the oxygen-oxygen double bond in O_2 is 498 kilojoules per mole; the energy of the fluorine-fluorine single bond is 159 kilojoules per mole (Section 6.14).

Nitrogen molecules are very unreactive. The only reactions known at room temperature are with lithium to give lithium nitride, Li_3N, with certain transition-metal complexes, and with nitrogen-fixing bacteria. Nitrogen forms nitrides when heated with active metals and low yields of ammonia when heated with hydrogen. Heating with oxygen followed by rapid cooling (quenching) produces nitric oxide, NO. A great deal of effort has gone into improving the yields and rates of conversion of molecular nitrogen into compounds, primarily ammonia for use in fertilizers and nitric acid for industrial use. Thus the mechanism by which bacteria form nitrogen compounds from nitrogen molecules at room temperature is one of the most challenging and exciting of the unsolved problems in chemistry.

Most uses of elemental nitrogen are based on its inactivity. It is used when a chemical process requires an inert atmosphere. Canned foods such as coffee and hydrogenated vegetable oils cannot oxidize in a pure nitrogen atmosphere, and so they retain a better flavor and color if sealed with nitrogen instead of air. Luncheon meats are often packed in nitrogen to retard spoilage. Large quantities of elemental nitrogen are used in the various processes for fixing atmospheric nitrogen (bringing about its combination with other elements). These processes are discussed later.

AMMONIA AND RELATED COMPOUNDS

As was noted in Section 8.7, ammonia is one of the more important industrial inorganic compounds. It is used extensively in fertilizers and to prepare nitric acid. Ammonia, NH_3, is produced in nature when any nitrogen-containing organic material decomposes in the absence of air. The decomposition may be due to heat or to bacteria that produce decay. The odor of ammonia is common in stables and in sewage where decay is taking place.

23.4   Preparation of Ammonia

Ammonia is usually prepared in the laboratory by heating an ammonium salt with a strong base such as sodium hydroxide. The acid-base reaction with the weakly acidic ammonium ion gives ammonia.

$$NH_4^+(aq) + OH^-(aq) \longrightarrow NH_3(g) + H_2O(l)$$

Sometimes ammonia is prepared in small quantities by the hydrolysis of ionic nitrides. The nitride ion is a much stronger base than the hydroxide ion.

$$Mg_3N_2 + 6H_2O \longrightarrow 3Mg(OH)_2 + 2NH_3(g)$$

Ammonia is produced in commercial quantities by the direct combination of its elements in the **Haber process:**

$$N_2(g) + 3H_2(g) \xrightleftharpoons{\text{Catalyst}} 2NH_3(g) \qquad \Delta H° = -92.2 \text{ kJ}$$

Since this reaction is very slow at room temperature, it is carried out at an elevated temperature and pressure with a catalyst so that the rate is fast enough to be practical. The reaction is exothermic; thus the yield of ammonia becomes less and less as the temperature of the system is raised. However, since four volumes of the reactants ($1N_2$ and $3H_2$) give two volumes of the product ($2NH_3$), high pressure increases the yield at any temperature. Thus the process is carried out at the lowest temperature and highest pressure practicable. In actual practice a pressure of about 200–600 atmospheres and a temperature of 400–600°C are ordinarily used. Of the many catalysts that have been tried, the most efficient is a mixture of iron oxide and potassium aluminate. Nitrogen for the process is obtained from liquid air, and much of the hydrogen is obtained from water gas (Section 9.12). The mixture of hydrogen and nitrogen is compressed, heated, and passed over the catalyst. The ammonia is removed by liquefaction, and the residual hydrogen and nitrogen are recycled through the process.

The equilibrium concentrations of ammonia at several temperatures and pressures are shown in Table 23-2.

Table 23-2   The Equilibrium Concentrations of Ammonia (Percent by Volume) at Several Temperatures and Pressures for the Reaction $N_2 + 3H_2 \rightleftharpoons 2NH_3$

Pressure, atm	Temperature, °C					
	200°	300°	400°	500°	600°	800°
1	15.3	2.18	0.44	0.129	0.05	0.012
100	80.6	52.1	25.1	10.4	4.47	1.15
200	85.8	62.8	36.3	17.6	8.25	2.24
1000	98.3	92.5	80.0	57.5	31.5	—

Fritz Haber, a German chemist, received the 1918 Nobel Prize in chemistry for his success in developing this direct synthesis of ammonia on a commercial scale.

23.5 Properties of Ammonia

Ammonia is a colorless gas with an irritating odor. It is such a powerful heart stimulant that people have died from inhaling it. Gaseous ammonia is easily liquefied by cooling or compressing, giving a colorless liquid that boils at −33°C. The liquid has a vapor pressure of about 10 atmospheres at 25°C; hence ammonia is readily handled in the liquid form. Solid ammonia is white and crystalline; it melts at −78°C. The heat of vaporization of liquid ammonia (1374 J/g) is higher than that of any other liquid except water, so ammonia is used as a refrigerant. Ammonia is quite soluble in water, alcohol, and ether. The fact that 1 liter of water at 0°C dissolves 1185 liters of the gas makes the ammonia fountain (Fig. 23-1) possible; squirting a few drops of water into a flask filled with gaseous ammonia triggers the fountain. Ammonia can be completely expelled from an aqueous solution by boiling.

In the ammonia molecule the three hydrogen atoms are linked to the nitrogen atom by covalent bonds.

$$
\begin{array}{c}
\text{H} \\
| \\
\text{H——N :} \\
| \\
\text{H}
\end{array}
$$

The molecule is a trigonal pyramid in shape (Fig. 23-2), with the unshared pair of electrons occupying the fourth corner of a tetrahedron (by occupying one of the sp^3 hybrid orbitals; Chapter 7). This lone pair can readily be shared by an atom or ion with an incomplete outer shell, forming a coordinate covalent bond. The ammonia molecule is highly polar, a property partly responsible for its physical and chemical behavior. Of the hydrides of the nitrogen family, ammonia is the most highly hydrogen-bonded in the liquid state, the most stable toward decomposition by heat, and the most strongly basic.

The chemical properties of ammonia may be outlined as follows:

1. Ammonia is a Brönsted base since it readily accepts protons and a Lewis base since it acts as an electron-pair donor (Chapter 14). When ammonia dissolves in

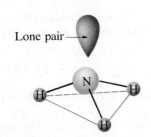

Flask filled with NH₃ (g)

—Water drops

—Dropper

—Air pressure

—Water

Figure 23-1. The ammonia fountain results when air pressure forces the water from the beaker into the space left when the gaseous ammonia dissolves in the water introduced from the dropper.

Lone pair—

Figure 23-2. Geometric structure of the ammonia molecule.

water only about 1% reacts to form ammonium ions and hydroxide ions. Ammonia is a weak base.

$$H\!-\!\overset{H}{\underset{H}{\underset{|}{\overset{|}{N}}}}\!:\; +\; H\!-\!\overset{..}{\underset{..}{O}}\!:\; \rightleftharpoons \left[H\!-\!\overset{H}{\underset{H}{\underset{|}{\overset{|}{N}}}}\!-\!H\right]^{+} +\; [\,:\!\overset{..}{\underset{..}{O}}\!-\!H]^{-}$$

Ammonia accepts protons from hydronium ions and other acids.

$$NH_3 + H_3O^+ \rightleftharpoons NH_4^+ + H_2O$$
$$NH_3 + HCl \rightleftharpoons NH_4^+ + Cl^-$$

The ammonium ion, NH_4^+, is similar in size to the potassium ion, and ionic compounds of these two ions exhibit many similarities in their structures and solubilities. Salts of the ammonium ion are called **ammonium salts.** Gaseous ammonia and gaseous hydrogen chloride react to form a cloud of very small crystals of ammonium chloride, the white deposit so apparent on windowpanes and other glass in chemical laboratories.

Ammonia forms **ammines** by sharing electrons with certain metal ions, just as water forms hydrates (Chapter 29).

$$Cu^{2+} + 4NH_3 \longrightarrow \qquad [Cu(NH_3)_4]^{2+}$$
Tetraamminecopper(II) ion

$$Ag^+ + 2NH_3 \longrightarrow \qquad [Ag(NH_3)_2]^+$$
Diamminesilver ion

In these species, ammonia functions as a Lewis base.

2. Ammonia displays acidic behavior, although it is a much weaker acid than water. It will react with very strong bases, such as CH_3^-, the methanide ion.

$$LiCH_3 + NH_3 \longrightarrow LiNH_2 + CH_4$$

Like other acids, ammonia reacts with metals, although it is so weak that high temperatures are often required. Depending on the stoichiometry, **amides** (salts of NH_2^-), **imides** (salts of NH^{2-}), or **nitrides** (salts of N^{3-}) are formed.

$$2Na + 2NH_3(l) \xrightarrow[\text{catalyst}]{\text{Fe}} 2NaNH_2 + H_2$$

$$2Li + NH_3(g) \xrightarrow{\;\Delta\;} Li_2NH + H_2$$

$$3Mg + 2NH_3(g) \xrightarrow{\;\Delta\;} Mg_3N_2 + 3H_2$$

Compounds that hydrolyze in water often also undergo **ammonolysis,** the equivalent acid-base reaction with ammonia. The equation for the ammonolysis of mercury(II) chloride is

$$HgCl_2 + 2NH_3 \longrightarrow Hg(NH_2)Cl + NH_4Cl$$

Many ammonolysis reactions are complicated by subsequent reactions of the initial product.

3. The nitrogen atom in ammonia has its lowest possible oxidation number and thus is not susceptible to reduction. However, it can be oxidized; hot ammonia and the ammonium ion are active reducing agents.

$$2NH_3(g) + 3CuO(s) \xrightarrow{\Delta} N_2(g) + 3Cu(s) + 3H_2O(g) \qquad \Delta H° = -161.1 \text{ kJ}$$

$$(NH_4)_2Cr_2O_7(s) \xrightarrow{\Delta} N_2(g) + 4H_2O(g) + Cr_2O_3(s) \qquad \Delta H° = -300.4 \text{ kJ}$$

$$NH_4NO_3(s) \xrightarrow{\Delta} N_2O(g) + 2H_2O(g) \qquad \Delta H° = -36.03 \text{ kJ}$$

4. The reaction of hydrogen with nitrogen to form ammonia is reversible. Thus heat (or the action of an electric spark) decomposes ammonia into hydrogen and nitrogen.

About three-fourths of the 15.5 million tons of ammonia produced in the United States during 1982 was used in fertilizers, either as the compound or as an ammonium salt such as the sulfate or nitrate. Application of anhydrous ammonia or of solutions of ammonium compounds is a standard farm procedure. Large quantities of ammonia are used in the production of nitric acid, urea, and other nitrogen compounds. Ammonia is a common refrigerant used in the production of ice and for the maintenance of low temperatures in refrigerating plants. Household ammonia is an aqueous solution of ammonia. It is used to clean windows and to remove the carbonate hardness from hard water.

23.6 Hydrazine

Sodium hypochlorite oxidizes ammonia in a mixture of sodium hydroxide and either gelatin or glue to **hydrazine, N_2H_4.** The reactions are

$$NH_3 + OCl^- \longrightarrow NH_2Cl + OH^-$$
$$NH_2Cl + NH_3 + OH^- \longrightarrow N_2H_4 + Cl^- + H_2O$$

Chloramine, NH_2Cl, is an intermediate in the process. A large excess of ammonia is used, and the reactants must be thoroughly mixed at low temperatures.

A comparison of their Lewis structures shows that hydrazine may be regarded as the nitrogen analogue of hydrogen peroxide.

Hydrazine may also be regarded as an ammonia derivative in which one hydrogen of the ammonia molecule is replaced by an —NH_2 group. Anhydrous hydrazine is thermally stable but very reactive. It burns in air and reacts vigorously with halogens. Like ammonia, hydrazine is a base, although it is weaker than ammonia and forms two series of salts.

$$N_2H_4 + HCl \rightleftharpoons N_2H_5^+ + Cl^-$$
$$N_2H_4 + 2HCl \rightleftharpoons N_2H_6^{2+} + 2Cl^-$$

Hydrazine is readily oxidized to nitrogen gas and water by hydrogen peroxide.

$$N_2H_4(l) + 2H_2O_2(l) \longrightarrow N_2(g) + 4H_2O(g) \qquad \Delta H° = -642.2 \text{ kJ}$$

The reaction products are both gases at the temperature of the reaction, and the reaction is highly exothermic. Hydrazine and hydrogen peroxide have therefore been used as rocket propellants.

23.7 Hydroxylamine

Hydroxylamine, NH₂OH, may be considered to be an ammonia derivative in which an —OH group has replaced one hydrogen atom.

$$\text{H--}\overset{..}{\underset{\underset{\text{H}}{|}}{\text{N}}}\text{--}\overset{..}{\underset{..}{\text{O}}}\text{--H} \qquad \text{H--}\overset{..}{\underset{\underset{\text{H}}{|}}{\text{N}}}\text{--H}$$

Hydroxylamine   Ammonia

The pure substance is an unstable white solid that decomposes to ammonia, water, nitrogen, and nitrous oxide at about 15°C. The decomposition can be explosive at elevated temperatures. Aqueous solutions are more stable. Hydroxylamine is a base with a strength somewhat less than that of ammonia. It is an active reducing agent. It is usually prepared and handled as a **hydroxylammonium salt,** such as the chloride, [NH₃OH⁺]Cl⁻.

OXIDES OF NITROGEN

23.8 Nitrous Oxide

When ammonium nitrate is heated, **nitrous oxide, N₂O,** is formed.

$$NH_4NO_3(s) \xrightarrow{\Delta} N_2O(g) + 2H_2O(g) \qquad \Delta H° = -36.03 \text{ kJ}$$

In this oxidation-reduction reaction the nitrogen in the ammonium ion is oxidized by the nitrogen in the nitrate ion. **Caution! If ammonium nitrate is heated too strongly, an explosion will occur.**

Nitrous oxide is a colorless gas possessing a mild, pleasing odor and a sweet taste. It is used under the name *laughing gas* as an anesthetic for minor operations, especially in dentistry. Its structure and resonance forms are shown in Fig. 23-3.

Nitrous oxide resembles oxygen in its behavior as an oxidizing agent with combustible substances. It decomposes when heated to form nitrogen and oxygen.

$$2N_2O(g) \xrightarrow{\Delta} 2N_2(g) + O_2(g) \qquad \Delta H° = -164.1 \text{ kJ}$$

Because a third of the gas liberated is oxygen, nitrous oxide supports combustion better than air. A glowing splint will burst into flame when thrust into a bottle of this gas.

1.126 Å

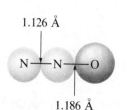

1.186 Å

Figure 23-3. Resonance forms and the geometric structure of the nitrous oxide molecule, N₂O.

23.9 Nitric Oxide

Copper reacts with dilute nitric acid, and **nitric oxide, NO,** is formed as the chief reduction product.

$$3Cu + 8H^+ + 2NO_3^- \longrightarrow 3Cu^{2+} + 2NO(g) + 4H_2O$$

The nitric oxide usually contains other nitrogen oxides. Pure nitric oxide may be obtained by reducing nitric acid with an iron(II) salt in dilute acid solution.

$$3Fe^{2+} + NO_3^- + 4H^+ \longrightarrow 3Fe^{3+} + NO(g) + 2H_2O$$

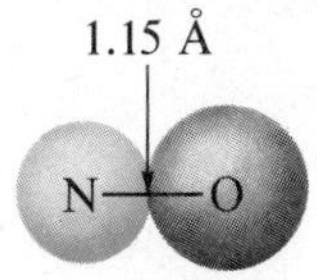

1.15 Å

Figure 23-4. Structure of the nitric oxide molecule, NO.

Nitric oxide is produced commercially by oxidation of ammonia (Section 8.7, Part 2 and Section 23.13, which describe the preparation of nitric acid from ammonia). Nitric oxide is formed by direct union of nitrogen and oxygen in the air by lightning during thunderstorms.

$$N_2(g) + O_2(g) \rightleftharpoons 2NO(g) \qquad \Delta H^\circ = +180.5 \text{ kJ}$$

It has been estimated that 40 million tons of nitrogen are fixed annually this way.

Nitric oxide is a slightly soluble, colorless gas. It is the most thermally stable of the nitrogen oxides. It is one of the air pollutants produced by internal combustion engines by the reaction of nitrogen and oxygen in the air during the combustion process (Section 22.4). Its structure is illustrated in Fig. 23-4.

Nitric oxide can act as an oxidizing agent or as a reducing agent; for example,

Oxidizing agent: $\qquad P_4 + 6NO \longrightarrow P_4O_6 + 3N_2$

Reducing agent: $\qquad Cl_2 + 2NO \longrightarrow 2ClNO$

Two resonance forms can be written for nitric oxide, or the electronic structure can be described by the molecular orbital model (Section 6.13).

$$:\overset{..}{N}=\overset{.}{O}: \quad \longleftrightarrow \quad :\overset{.}{N}=\overset{..}{O}:$$

The molecule contains an unpaired electron, which makes the substance paramagnetic. As is often the case with such molecules, two molecules of nitric oxide can combine to form a dimer by pairing their unpaired electrons. The dimer, N_2O_2, is diamagnetic; it contains no unpaired electrons.

$$2NO \longrightarrow \quad \overset{..}{\underset{O}{N}} - N \overset{O}{\underset{..}{}}$$

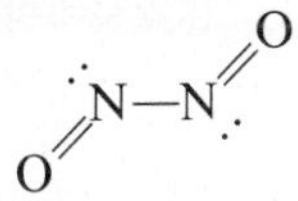

23.10 Dinitrogen Trioxide

When the temperature of a mixture of equal parts of nitric oxide and nitrogen dioxide is lowered to $-21\,^\circ\text{C}$, the gases form a blue liquid consisting of **dinitrogen trioxide, N_2O_3,** molecules. These form when the unpaired electron on a nitric oxide molecule pairs up with the unpaired electron on a nitrogen dioxide molecule. Dinitrogen trioxide exists only in liquid and solid forms. When heated, it forms a gaseous brown equilibrium mixture of NO and NO_2.

$$N_2O_3 \rightleftharpoons NO + NO_2$$

Dinitrogen trioxide is the anhydride of nitrous acid. The resonance and geometric structures of N_2O_3 are shown in Fig. 23-5.

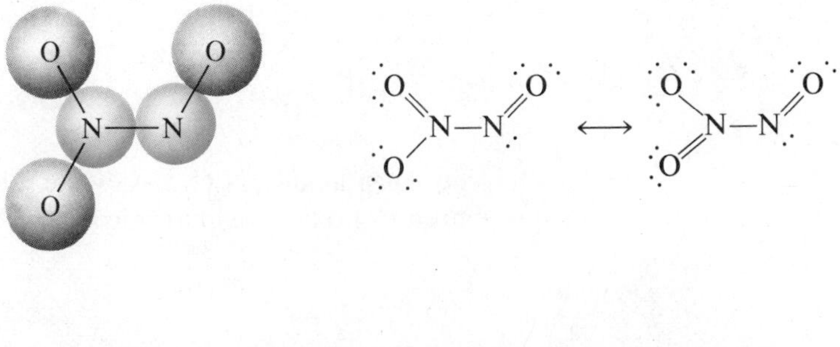

Figure 23-5. Resonance forms and the geometric structure of the dinitrogen trioxide molecule, N_2O_3.

23.11   Nitrogen Dioxide and Dinitrogen Tetraoxide

Nitrogen dioxide, NO₂, is prepared in the laboratory by heating the nitrate of a heavy metal:

$$2Pb(NO_3)_2 \xrightarrow{\triangle} 2PbO + 4NO_2 + O_2$$

or by the reaction of concentrated nitric acid with metallic copper:

$$Cu + 4H^+ + 2NO_3^- \longrightarrow Cu^{2+} + 2NO_2 + 2H_2O$$

Nitrogen dioxide is prepared commercially by oxidizing nitric oxide with air.

$$2NO + O_2 \longrightarrow 2NO_2$$

Depending on the conditions nitrogen dioxide disproportionates in one of two ways when it reacts with water. In cold water a mixture of nitric and nitrous acids is formed.

$$2NO_2 + H_2O \longrightarrow H^+ + NO_3^- + HNO_2$$

At higher temperatures nitric acid and nitric oxide form (Section 23.13).

At 140°C nitrogen dioxide has a deep brown color and a density corresponding to the formula NO₂. At low temperatures the color almost entirely disappears and the density corresponds to the formula N₂O₄, (dinitrogen tetraoxide). At room temperature an equilibrium exists between the two compounds.

$$2NO_2(g) \rightleftharpoons N_2O_4(g) \qquad \Delta H° = -57.20 \text{ kJ}; \Delta G° = -4.77 \text{ kJ}$$

Nitrogen dioxide contains an unpaired electron, which is responsible for its intense color and paramagnetism. When two molecules of nitrogen dioxide share their unpaired electrons, diamagnetic dinitrogen tetraoxide forms, and the color and paramagnetism disappear. The resonance forms and geometric structures of NO₂ and the structure of N₂O₄ are shown in Fig. 23-6.

Figure 23-6. Formulas and geometric structures of nitrogen dioxide and dinitrogen tetraoxide molecules.

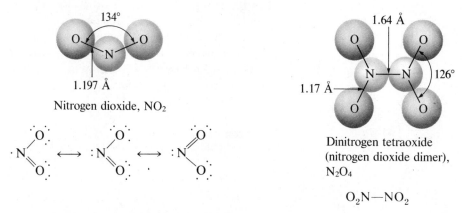

23.12   Dinitrogen Pentaoxide

Dinitrogen pentaoxide, N₂O₅, is a white solid formed by the dehydrating action of phosphorus(V) oxide on pure nitric acid.

$$P_4O_{10} + 4HNO_3 \longrightarrow 4HPO_3 + 2N_2O_5$$

Dinitrogen pentaoxide decomposes at temperatures above 30°C.

$$2N_2O_5 \longrightarrow 4NO_2 + O_2$$

It is the anhydride of nitric acid.

$$N_2O_5 + H_2O \longrightarrow 2HNO_3$$

Figure 23-7 shows the formula and the arrangement of atoms in N_2O_5 in the gaseous state. Crystals are made up of equal numbers of bent NO_2^+ ions (N—O bond length is 1.15 Å) and NO_3^- ions (N—O bond length is 1.24 Å).

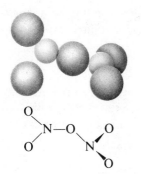

Figure 23-7. Formula and geometric structure of the dinitrogen pentaoxide molecule, N_2O_5.

OXYACIDS OF NITROGEN

23.13 Nitric Acid

Nitric acid, HNO₃, was known to the alchemists of the eighth century as *aqua fortis* (meaning strong water). It was prepared from potassium nitrate and was used in the separation of gold from silver; it dissolves silver but not gold. Traces of nitric acid occur in the atmosphere after thunderstorms, and its salts are widely distributed in nature. Chile saltpeter ($NaNO_3$) is found in tremendous deposits (2 by 200 miles and up to 5 feet thick) in the desert region near the boundary of Chile and Peru. Bengal saltpeter (KNO_3) is found in India and other countries of the Far East.

Nitric acid can be prepared in the laboratory by heating a nitrate salt (such as sodium or potassium nitrate) with concentrated sulfuric acid.

$$NaNO_3 + H_2SO_4 \xrightarrow{\Delta} NaHSO_4 + HNO_3(g)$$

At room temperature the reaction is reversible, and equilibrium is soon established. But nitric acid boils at 86°C, while sulfuric acid boils at 338°C ($NaNO_3$ and $NaHSO_4$ are nonvolatile), and nitric acid is readily removed by distillation. Consequently, the equilibrium shifts to the right, and the reaction goes to completion.

Nitric acid is produced commercially by the **Ostwald process:** oxidation of ammonia to nitric oxide, NO; oxidation of nitric oxide to nitrogen dioxide, NO_2; and conversion of nitrogen dioxide to nitric acid. A mixture of 1 volume of ammonia with ten volumes of air is heated to 600–700°C and then brought into contact with a gauze platinum-rhodium alloy catalyst (90% Pt and 10% Rh). The ammonia literally burns with a flame, and the temperature of the system increases to about 1000°C.

$$4NH_3(g) + 5O_2(g) \longrightarrow 4NO(g) + 6H_2O(g) \qquad \Delta H° = -905.5 \text{ kJ}$$

Admission of additional air then cools the mixture and oxidizes the nitric oxide to nitrogen dioxide.

$$2NO(g) + O_2(g) \longrightarrow 2NO_2(g) \qquad \Delta H° = -114.1 \text{ kJ}$$

Nitrogen dioxide, excess oxygen, and unreactive nitrogen from the air are passed through a water spray in an absorption tower. Nitric acid and nitric oxide form as the nitrogen dioxide disproportionates.

$$3NO_2 + H_2O \longrightarrow 2H^+ + 2NO_3^- + NO$$

The nitric oxide is combined with more oxygen and returned to the absorption tower. The nitric acid is drawn off and concentrated. Most of the 7.6 million tons of domestic nitric acid produced in 1982 came from the Ostwald process.

23.14   Properties of Nitric Acid

The resonance forms and the geometric structure of the nitric acid molecule are shown in Fig. 23-8. Pure nitric acid is a colorless liquid (density = 1.50 g/mL) that boils at 86°C and freezes to a white solid at −42°C. It fumes in moist air, forming a cloud of very small droplets of aqueous nitric acid. A 68% aqueous solution of the acid has a constant boiling point of 120.5°C at 1 atmosphere. This is the commercial concentrated acid; it has a specific gravity of 1.4048 at 20°C.

Figure 23-8.   Resonance forms and the geometric structure showing the arrangement of atoms in the HNO_3 molecule, which in the vapor phase is planar. N—O distances are (a) 1.41 Å and (b) 1.22 Å.

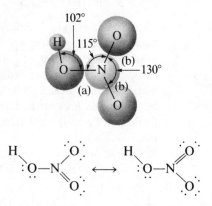

Pure nitric acid is unstable; it decomposes in light or when heated to produce a mixture of nitrogen oxides, of which nitrogen dioxide is predominant.

$$4HNO_3 \xrightarrow{\Delta} 4NO_2 + 2H_2O + O_2$$

It is often yellow or brown in color due to nitrogen dioxide formed by its decomposition. In aqueous solution nitric acid is stable and exhibits both the properties of a strong acid and the properties of a strong oxidizing agent.

The action of nitric acid on a metal rarely produces hydrogen gas (by reduction of hydrogen ion) in more than small amounts. Instead, the nitrogen in the acid is reduced. The products formed depend on the concentration of the acid, the activity of the metal, and the temperature. A mixture of oxides and other reduction products is usually produced, but less active metals, such as copper, silver, and lead, reduce dilute nitric acid primarily to nitric oxide. The following three equations show the oxide that is produced in largest quantity in each particular reaction.

The reaction of dilute nitric acid with copper gives nitric oxide.

$$3Cu + 8H^+ + 2NO_3^- \longrightarrow 3Cu^{2+} + 2NO(g) + 4H_2O$$

With concentrated nitric acid the principal product is nitrogen dioxide.

$$Cu + 4H^+ + 2NO_3^- \longrightarrow Cu^{2+} + 2NO_2(g) + 2H_2O$$

More active metals, such as zinc and iron, give nitrous oxide. With zinc we have

$$4Zn + 10H^+ + 2NO_3^- \longrightarrow 4Zn^{2+} + N_2O(g) + 5H_2O$$

When the acid is very dilute, either nitrogen or ammonium ions may be formed, depending on the conditions

$$5Zn + 12H^+ + 2NO_3^- \longrightarrow 5Zn^{2+} + N_2(g) + 6H_2O$$

$$4Zn + 10H^+ + NO_3^- \longrightarrow 4Zn^{2+} + NH_4^+ + 3H_2O$$

In concentrated nitric acid zinc, as well as copper, produces nitric oxide and nitrogen dioxide. The nitrate salts of these metals separate as crystals when the resulting solutions are evaporated.

Nonmetallic elements, such as sulfur, carbon, iodine, and phosphorus, are oxidized by concentrated nitric acid to their oxyacids (or acid anhydrides), with the formation of nitrogen dioxide.

$$S + 6HNO_3 \longrightarrow H_2SO_4 + 6NO_2(g) + 2H_2O$$

$$C + 4HNO_3 \longrightarrow CO_2 + 4NO_2(g) + 2H_2O$$

Many compounds are oxidized by nitric acid. Hydrochloric acid is readily oxidized by concentrated nitric acid to chlorine and chlorine dioxide. A mixture of nitric acid and hydrochloric acid reacts vigorously with metals. This mixture is particularly useful in dissolving gold, platinum, and other metals that lie below hydrogen in the activity series. A mixture of three parts of concentrated hydrochloric acid and one part concentrated nitric acid by volume is called **aqua regia** (meaning royal water). The action of aqua regia on gold may be represented, in somewhat simplified form, by the equation

$$Au + 4HCl + 3HNO_3 \longrightarrow HAuCl_4 + 3NO_2(g) + 3H_2O$$

Nitric acid reacts with proteins, such as those in the skin, to give a yellow material called xanthoprotein. You may have noticed that if you get nitric acid on your fingers they turn yellow. Nitric acid is used extensively in the laboratory and in industry as a strong acid and an active oxidizing agent. It is used in the manufacture of explosives, dyes, plastics, and drugs. Salts of nitric acid (nitrates) are valuable as fertilizers. Gunpowder is a mixture of potassium nitrate, sulfur, and charcoal. Ammonal, an explosive, is a mixture of ammonium nitrate and aluminum powder.

23.15 Nitrates

Nitrates are salts of nitric acid. The nitrate ion has a trigonal planar structure (Section 7.1), with each bond being a hybrid of two single bonds and a double bond. The three resonance forms are

The salts of nitric acid form when metals or their oxides, hydroxides, or carbonates react with nitric acid. Most nitrates are soluble in water; indeed, one of the significant uses of nitric acid is to prepare soluble metal nitrates.

All nitrates decompose when heated. When sodium nitrate (or potassium nitrate) is heated, the nitrite is formed and oxygen is evolved.

$$2NaNO_3 \xrightarrow{\Delta} 2NaNO_2 + O_2$$

Heating a nitrate of a heavy metal produces the oxide of the metal. With copper(II) nitrate the equation is

$$2Cu(NO_3)_2 \xrightarrow{\Delta} 2CuO + 4NO_2 + O_2$$

Heating ammonium nitrate produces nitrous oxide, N_2O (Section 23.8).

Nitrates, especially ammonium nitrate, can explode.

23.16 Nitrous Acid and Nitrites

Pale blue **nitrous acid, HNO_2,** forms in solution when an acid is added to a cold solution of a nitrite.

$$2Na^+ + 2NO_2^- + 2H^+ + SO_4^{2-} \longrightarrow 2HNO_2 + 2Na^+ + SO_4^{2-}$$

Nitrous acid is very unstable and exists only in solution. It disproportionates slowly at room temperature (rapidly when heated) giving nitric acid and nitric oxide.

$$3HNO_2 \longrightarrow H^+ + NO_3^- + 2NO(g) + H_2O$$

Nitrous acid is a weak acid. It is an active oxidizing agent with strong reducing agents and is oxidized to nitric acid by strong oxidizing agents. Figure 23-9 shows the structure of nitrous acid.

Sodium nitrite is the most important salt of nitrous acid. It is usually made by reducing molten sodium nitrate with lead.

$$NaNO_3 + Pb \longrightarrow NaNO_2 + PbO$$

The nitrites are much more stable than nitrous acid (the salts of all oxyacids are more stable than the acids themselves). Like nitrates, nitrites are soluble in water ($AgNO_2$ is sparingly soluble). The electronic structure of the nitrite ion is described by two resonance forms.

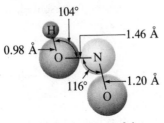

Figure 23-9. Structure of the nitrous acid molecule, HNO_2.

104°
0.98 Å
1.46 Å
116°
1.20 Å

Nitrites, like nitrates, can explode.

When nitrous acid is reduced by hydrazine, **hydrazoic acid, HN_3,** is formed.

$$N_2H_4(aq) + HNO_2(aq) \longrightarrow HN_3(aq) + 2H_2O(l)$$

Hydrazoic acid detonates violently when subjected to shock. It is a weak acid and reacts with both oxidizing and reducing agents. Its salts are called **azides,** which, like the acid, are unstable. Azides of electropositive metals decompose smoothly to nitrogen and the metal when heated; other metal azides explode. Lead azide, $Pb(N_3)_2$, is used as a detonator for explosives. Resonance forms and the molecular

Figure 23-10. Resonance forms and the geometric structure of the hydrazoic acid molecule, HN_3.

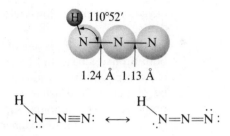

structure of hydrazoic acid are shown in Fig. 23-10. The three nitrogens of the azide ion, N_3^-, lie in a straight line.

23.17 The Nitrogen Cycle

Nitrogen is an essential constituent of all plants and animals. It is present principally in proteins, complex organic materials that also contain carbon, hydrogen, and oxygen (Chapter 26). Most plants obtain the nitrogen necessary for growth through their roots as nitrogen compounds, primarily ammonium and nitrate salts. However, some legumes, such as clover, alfalfa, peas, and beans, are able to obtain their nitrogen from the air by means of nitrogen-fixing bacteria, which live in nodules on their roots. These bacteria convert atmospheric nitrogen into nitrites and nitrates, which are assimilated by the host plant. Generally, the bacteria fix more nitrogen than is used by the plant, so that when the legume dies, the unused nitrogen remains in the soil and is available for other plants. Animals obtain nitrogen compounds by eating plants and other animals.

Tremendous quantities of nitrogen are fixed as nitric oxide by lightning (Section 23.9). The nitric oxide is then oxidized to nitrogen dioxide by atmospheric oxygen, and reactions with water form nitrous and nitric acids in the same way as in the Ostwald process (Section 23.13). These acids are carried by rain to the soil, where they react with oxides and carbonates of metals to form nitrites and nitrates, respectively. Some soil bacteria oxidize nitrites to nitrates, a process called **nitrification.** Other bacteria change ammonia into nitrites. Denitrifying bacteria decompose nitrates and other nitrogen compounds, forming free nitrogen. The decay of both plant and animal matter returns nitrogen to the soil in the form of nitrates, and either ammonia or free nitrogen is produced. Thus the nitrogen cycle is of fundamental importance to all plants and animals.

FOR REVIEW

SUMMARY

Nitrogen, a member of Group VA, forms a diatomic molecule with a particularly strong triple bond. As a consequence, the N_2 molecule is not very reactive. At room temperature it reacts only with lithium (forming an ionic **nitride**), with certain transition-metal complexes, and with nitrogen-fixing bacteria. At elevated temperatures it reacts with active metals forming nitrides and with hydrogen or oxygen in reversible low-yield reactions forming ammonia or nitric oxide, respectively.

Ammonia, NH₃, is one of the more significant industrial chemicals. It exhibits both Brönsted base and Lewis base character and will undergo oxidation. Thus it is a compound from which a variety of other nitrogen-containing compounds can be prepared. **Hydrazine, hydroxylamine, nitric oxide,** and **ammonium salts** are prepared from ammonia. Ammonia is prepared by the direct reaction of hydrogen and nitrogen in the **Haber process.** Because the rate of reaction is slow at low temperatures and the yield of product is low at elevated temperatures, careful study of the effects of temperature, pressure, and catalysts was necessary to make this process practical.

Nitrogen exhibits oxidation numbers ranging from $+1$ to $+5$ in its oxides. The oxides include **nitrous oxide, N₂O, nitric oxide, NO, dinitrogen trioxide, N₂O₃, nitrogen dioxide, NO₂,** and **dinitrogen pentaoxide, N₂O₅.** The most important of these oxides are nitric oxide and nitrogen dioxide since they are intermediates in the preparation of nitric acid by oxidation of ammonia. These two oxides can also be prepared by reduction of nitric acid. Dinitrogen trioxide is the anhydride of nitrous acid; dinitrogen pentaoxide, the anhydride of nitric acid.

The oxidation of ammonia to nitric oxide, followed by the oxidation of nitric oxide to nitrogen dioxide, then the reaction of nitrogen dioxide with water, produces **nitric acid.** Nitric acid can also be distilled from a mixture of concentrated sulfuric acid and a metal nitrate. This acid is one of the commercially important strong acids. It is also a strong oxidizing agent. Metals dissolve in nitric acid by reduction of the nitrate ion, generally with formation of nitrogen dioxide or nitric oxide, rather than by reduction of hydrogen ion with the production of hydrogen gas. **Nitrous acid,** a weak acid, can be prepared by the acidification of a solution of a nitrite salt.

KEY TERMS AND CONCEPTS

Amides (23.5)

Azides (23.16)

Haber process (23.4)

Imides (23.5)

Nitrates (23.15)

Nitride ion (23.1)

Nitrides (23.5)

Nitrites (23.16)

Nitrogen cycle (23.17)

Ostwald process (23.13)

EXERCISES

Nitrogen

1. Why does nitrogen form a maximum of four single covalent bonds?

2. Write balanced equations illustrating three ways in which elemental nitrogen can be converted into ammonia. (Two of these paths involve more than one reaction.)

3. What is the oxidation number of nitrogen in each of the following?

   (a) N_2
   (b) NH_4^+
   (c) $NaNO_2$
   (d) N_2H_4
   (e) NH_2OH
   (f) NO_2
   (g) N_2O_4
   (h) NH_4NO_3
   (i) N_2O
   (j) NCl_3
   (k) NF_3
   (l) Li_3N
   (m) HN_3

4. What volume of nitrogen measured at 75°C and 740 torr is formed by the decomposition of 20.0 g of ammonium nitrite?   *Ans. 9.16 L*

Hydrogen Compounds of Nitrogen

5. Write balanced equations illustrating the acidic and basic character of the two binary compounds of hydrogen and nitrogen.

6. Write electronic structures for the following:
   (a) N_2 (c) NH_4^+ (e) HN_3
   (b) NH_3 (d) N_2H_4 (f) NH_2OH

7. What are the products of the chemical reaction occurring between an ionic nitride and water?

8. Explain the effects of temperature, pressure, and a catalyst on the direct synthesis of ammonia.

9. The salt hydroxylammonium chloride, $[NH_3OH]Cl$, can be formed by the reaction of hydrogen chloride and hydroxylamine. Would you expect the proton from the hydrogen chloride to be found on the oxygen or on the nitrogen of the hydroxylamine? Explain. (*Hint:* Ammonia is more basic than water.)

10. What mass of ammonia is produced by adding water to the product of the reaction of 93.0 g of lithium with elemental nitrogen?   *Ans. 76.1 g*

11. What volume of ammonia measured at 1.20 atm and 25.0°C is required to react with 0.800 L of 0.510 *M* HCl solution?   *Ans. 8.31 L*

12. What is the maximum volume of ammonia that can be collected at 27°C and 750 torr by the treatment of 2.00 g of ammonium chloride with 0.500 L of 0.500 *M* sodium hydroxide?
     Ans. 0.933 L

Nitrogen Oxides and Oxyacids

13. What is the anhydride of nitrous acid? of nitric acid?

14. Write equations for the preparation of each of the oxides of nitrogen.

15. Using the data in Appendix J, determine $\Delta H°$ and $\Delta G°$ values for the conversion

$$2NO_2(g) \rightleftharpoons N_2O_4(g)$$

The values are widely different. Discuss possible reasons for this difference, and discuss the significance of each of the two values as completely as you can. (You may wish to refer to Chapter 18 in answering this question.) Check your answers against the values for $\Delta H°$ and $\Delta G°$ given in Section 23.11.

16. Outline the chemistry of the production of nitric acid from ammonia.

17. List the possible reduction products of nitric acid. What factors determine which product will be formed when nitric acid is reduced?

18. Write equations for the reaction of concentrated HNO_3 with each of the following:
    (a) Cu   (c) C
    (b) S   (d) SO_2

19. Write electronic structures for the following:
    (a) HNO_3 (f) NO_2
    (b) NO_3^- (g) N_2O_3
    (c) HNO_2 (h) N_2O_5
    (d) N_2O (i) NCl_3
    (e) NO   (j) NOCl

20. Write equations for the preparation of nitrous acid starting with sodium nitrate.

21. What would be the result of bubbling SO_2 through a strongly acidic barium nitrate solution? Write equations for the reactions that occur.

22. What quantity of nitric acid could be prepared by the Ostwald process from 300 ft^3 of ammonia, measured at 4.00 atm and 250°C, if a 93.0% yield is obtained based on the original quantity of ammonia.   *Ans. 46.4 kg*

23. The oxidation of ammonia and of nitric oxide are exothermic processes. Using the data in Appendix J, calculate $\Delta H°$ for these reactions.
     Ans. −905.50 kJ; −114.1 kJ

Additional Exercises

24. What evidence is there that the coordinate covalent bond between the proton and an ammonia molecule is stronger than that between the proton and water?

25. Draw a valence electron formula for the azide ion, N_3^-.

26. Describe the nitrogen cycle in nature.

27. Which is the stronger oxidizing agent, hydrazine or hydrogen peroxide? Give evidence for your answer.

28. At 25°C and 1.0 atm, a mixture of $N_2O_4(g)$ and $NO_2(g)$ contains 30.0% $NO_2(g)$ by volume. Calculate the partial pressures of the two gases when they are at equilibrium at 25°C with a total pressure of 9.0 atm.   *Ans. $P_{N_2O_4} = 8.0$ atm;*
     $P_{NO_2} = 1.0$ atm

29. An equilibrium mixture at 400°C contained 0.63 mol of hydrogen, 0.45 mol of nitrogen, and 0.24 mol of ammonia per liter. Calculate the equilibrium constant for the system.   *Ans. 0.51*

24

PHOSPHORUS AND ITS COMPOUNDS

Phosphorus is an active nonmetal. It forms compounds that are used in foods, explosives, matches, agricultural chemicals (including insecticides, herbicides, and fertilizers), soaps and detergents, and special metal alloys.

Phosphorus is essential to both plants and animals. Bones, teeth, and nerve and muscle tissue contain combined phosphorus. The nucleic acids, molecules that provide an organism's genetic information, contain phosphorus. Phosphorus compounds are also important in the metabolism of sugar. Foods like eggs, beans, peas, and milk furnish phosphorus for the body's requirements. Plants obtain phosphorus from soluble phosphates in the soil. Soils that have become deficient in phosphorus may be enriched by the addition of soluble phosphates.

24.1 Comparison of Nitrogen and Phosphorus

Phosphorus and nitrogen are both members of Group VA and have similar valence shell electron configurations, ns^2np^3. However, aside from the stoichiometries of some of their simpler compounds (NH_3 and PH_3, NF_3 and PF_3, for example), marked differences exist between the two elements and their compounds. These differences result from the larger size of the phosphorus atom, its lower electronegativity (2.1) relative to that of nitrogen (3.1), and the presence of empty d orbitals in the valence shell ($n = 3$) of phosphorus. Like nitrogen, phosphorus is a nonmetal, and its chemistry is essentially that of covalent compounds. However, while nitrogen is quite inert except at elevated temperatures, phosphorus is reactive even at low temperatures. The union of nitrogen with oxygen is an endothermic reaction, but phosphorus burns readily in air with the evolution of considerable heat.

Nitric acid and nitrous acid are relatively strong oxidizing agents, while phosphoric acid and phosphorous acid are very weak oxidizing agents. Ammonia, NH_3, is a much weaker reducing agent than phosphine, PH_3. Furthermore, nitrogen forms bonds to only two or three oxygen atoms in its oxyacids and their salts and thus exhibits a coordination number (Chapter 23) of only 2 or 3. Phosphorus exhibits a coordination number of 4 in its oxyacids and their salts. Nitrogen is limited to a maximum coordination number of 4 (in NH_4^+, for example), while phosphorus exhibits higher coordination numbers (in PCl_5 and PF_6^-, for example) because it can use the d orbitals in its valence shell in bonding.

PHOSPHORUS

24.2 Preparation of Phosphorus

Phosphorus is produced commercially by heating calcium phosphate, obtained from phosphate rock (Section 8.7), with sand and coke in an electric furnace.

$$2Ca_3(PO_4)_2 + 6SiO_2 + 10C \longrightarrow 6CaSiO_3 + 10CO(g) + P_4(g)$$

The phosphorus distills out of the furnace and is condensed to a solid, or burned to phosphorus(V) oxide, P_4O_{10}, from which other phosphorus compounds are manufactured. The calcium silicate, along with any impurities, melts and is drawn off as a slag. The elemental phosphorus is shipped to plants at other locations, where it is converted into phosphoric acid and phosphates. In this way freight costs for the oxygen and water used in the manufacture of phosphorus compounds are minimized. The profits of chemical industries often depend on such considerations.

24.3 Physical Properties of Phosphorus

Phosphorus exists in many allotropic forms, but only two are of general interest and importance: **white phosphorus** and **red phosphorus.** Phosphorus prepared as described in Section 24.2 is a white, translucent, waxlike solid. When exposed to light, it slowly turns yellow as a superficial coating of red phosphorus forms. For this reason white phosphorus is sometimes called yellow phosphorus. White phosphorus is very soluble in carbon disulfide, less so in ether, chloroform, and other organic solvents, and very nearly insoluble in water. It melts at 44.2°C and boils at 280°C. As a solid or in solution, white phosphorus exists as P_4 molecules. The four phosphorus atoms are arranged at the corners of a regular tetrahedron, and each atom is covalently bonded to the three other atoms of the molecule (Fig. 24-1). Phosphorus vapor just above the boiling point contains P_4 molecules. At very high temperatures it consists of P_2 molecules; these are assumed to have the electronic structure $:P{\equiv}P:$, which is similar to that of the N_2 molecule.

White phosphorus is very poisonous; about 0.15 gram produces acute pains and convulsions and may result in death. Necrosis, or death, of the bones of the

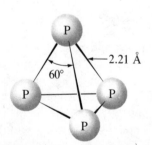

Figure 24-1. The white phosphorus molecule, P_4.

jaw and nose results from continued inhalation of small amounts of phosphorus fumes. In very small doses phosphorus stimulates the nervous system.

Red phosphorus is made by heating white phosphorus to between 230°C and 300°C with air excluded.

$$P(white) \rightleftharpoons P(red) \qquad \Delta H° = -18 \text{ kJ}$$

This change is faster when catalyzed by a trace of iodine. Because the change from white to red phosphorus is exothermic, the heat of combustion of red phosphorus is less than that of white. Red phosphorus sublimes when heated and, like white phosphorus, forms P_4 and P_2 molecules. The red allotrope is insoluble in the solvents that dissolve the white one.

The atoms in a sample of red phosphorus are bonded together as a polymer, with the length of the chain limited only by the size of the given piece. The exact structure of red phosphorus is not definitely known, but the chain is thought to be formed from P_4 molecules by the rupture of one P—P bond and the subsequent joining of the ruptured tetrahedra (Fig. 24-2).

Figure 24-2. One possible structure of red phosphorus. Note that no bond exists between P and P″ in any given tetrahedron. These bonds have been ruptured in the formation of red phosphorus from white phosphorus. The tetrahedra with ruptured bonds join to form a chain structure.

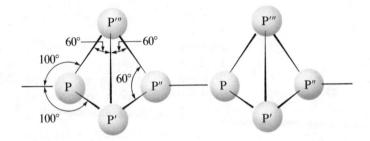

24.4 Chemical Properties of Phosphorus

Phosphorus is an active nonmetal. The most commonly observed oxidation numbers of phosphorus are −3 with metals or less electronegative nonmetals and +3 and +5 with more electronegative nonmetals. In compounds that contain phosphorus-phosphorus bonds, phosphorus also exhibits oxidation numbers that are unusual for a Group VA element; for example, diphosphorus tetrahydride, P_2H_4, and tetraphosphorus trisulfide, P_4S_3 (Fig. 24-3).

The most important chemical property of phosphorus is its reactivity with oxygen. Slow oxidation of white phosphorus at room temperature causes it to get warm, and it spontaneously inflames when its temperature reaches 35–45°C. Because of this, white phosphorus must be stored under water. Burns caused by phosphorus are very painful and slow to heal. *It should be handled only with forceps and put in water immediately after use.* Red phosphorus is less active than the white form; it does not ignite in air unless it is heated to about 250°C. However, the products of the reactions of red phosphorus are the same as those with the white form.

When phosphorus is burned either in an excess of air or in oxygen, covalent phosphorus(V) oxide, P_4O_{10}, is formed. In moist air the cloud of solid phosphorus(V) oxide forms a fog of minute droplets of phosphoric acid. Phosphorus of both forms combines exothermically with the halogens. When heated, it combines with sulfur and many of the metals. Concentrated nitric acid oxidizes phosphorus to orthophosphoric acid.

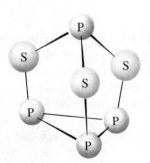

Figure 24-3. The molecular structure of P_4S_3. Note the P—P bonds.

24.5 Uses of Phosphorus

White phosphorus once was used in the manufacture of matches. But the workers who were exposed to the fumes suffered from necrosis of the bones, so it was replaced in the heads of strike-anywhere matches by tetraphosphorus trisulfide, P_4S_3 (Fig. 24-3).

Large quantities of phosphorus are converted into acids and salts to be used in fertilizers, in baking powder, and in the chemical industries. Other uses are in the manufacture of special alloys such as ferrophosphorus and phosphorbronze. Substantial quantities of phosphorus are used in making fireworks, bombs, and rat poisons. Phosphorus has been burned to produce smoke screens during warfare.

BINARY COMPOUNDS OF PHOSPHORUS

24.6 Phosphine

Phosphorus forms a series of hydrides, the most important of which is **phosphine, PH_3,** a gaseous compound analogous to ammonia in formula and structure. Unlike ammonia, phosphine cannot be made by the direct union of the elements. It is prepared by the disproportionation of white phosphorus in a hot concentrated solution of sodium hydroxide.

$$3Na^+ + 3OH^- + P_4 + 3H_2O \longrightarrow 3Na^+ + 3H_2PO_2^- + PH_3(g)$$
$$\text{Sodium hypophosphite} \qquad \text{Phosphine}$$

Pure phosphine is not spontaneously flammable. However, diphosphorus tetrahydride, P_2H_4, a by-product of the reaction, is spontaneously flammable and will ignite the phosphine on contact with air. Consequently, phosphine should be prepared and handled under an atmosphere of nitrogen or some other inert gas.

Phosphine can also be prepared by the hydrolysis of active metal phosphides. The **phosphide ion, P^{3-},** is a very strong base.

$$Ca_3P_2 + 6H_2O \longrightarrow 3Ca(OH)_2 + 2PH_3(g)$$

Phosphine is a colorless, very poisonous gas, which has an odor like that of decaying fish. It is easily decomposed by heat.

$$4PH_3 \xrightarrow{\triangle} P_4 + 6H_2$$

Like ammonia, gaseous phosphine unites with gaseous hydrogen halides to form **phosphonium compounds,** PH_4Cl, PH_4Br, and PH_4I. Unlike ammonium halides, however, phosphonium compounds do not give solutions containing the phosphonium ion, PH_4^+. Instead, phosphine escapes and the hydrohalic acid remains. Phosphine is only very slightly soluble in water and is a much weaker base than ammonia.

Phosphides of several metals may be prepared by reducing the corresponding phosphate with carbon:

$$Ca_3(PO_4)_2 + 8C \longrightarrow Ca_3P_2 + 8CO$$

or by the direct reaction of the metal with phosphorus.

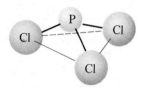

Figure 24-4. The molecular structure of PCl_3.

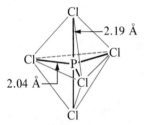

Figure 24-5. The molecular structure of PCl_5 in the gaseous and liquid states.

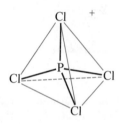

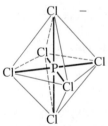

Figure 24-6. The structures of the ions PCl_4^+ and PCl_6^- found in solid PCl_5.

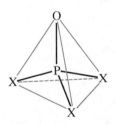

Figure 24-7. The molecular structure of phosphorus(V) oxyhalides, POX_3.

24.7 Phosphorus Halides

With one exception, phosphorus reacts directly with the halogens, forming trihalides, PX_3, and pentahalides, PX_5; the pentaiodide is not known. The trihalides are much more stable than the corresponding nitrogen trihalides; nitrogen pentahalides are not stable.

The Group VA atom in a trihalide uses three orbitals for σ bonds and one for the lone pair, as shown by the Lewis structure of PF_3.

$$:P\!-\!F \qquad F\!-\!P$$

The pentahalides, such as PF_5, require five orbitals on the Group VA atom. Since there are only four valence orbitals on nitrogen, it cannot form pentahalides. The heavier members of Group VA have empty d orbitals in their valence shells. One of these can hybridize with the s and p orbitals to form dsp^3 orbitals to bond five halogen atoms.

The trichloride and the pentachloride are the most important halides of phosphorus. **Phosphorus trichloride, PCl_3,** is prepared by passing chlorine over molten phosphorus.

$$P_4 + 6Cl_2 \longrightarrow 4PCl_3$$

The trichloride is a colorless liquid composed of pyramidal molecules with the phosphorus atom at one corner (Fig. 24-4). Like most other nonmetal halides, phosphorus trichloride irreversibly hydrolyzes, forming phosphorous acid and hydrogen chloride.

$$PCl_3 + 3H_2O \longrightarrow H_3PO_3 + 3HCl(g)$$

Because of this hydrolysis, all of the halides of phosphorus fume in moist air. **Phosphorus pentachloride, PCl_5,** is prepared by oxidizing the trichloride with excess chlorine.

$$PCl_3 + Cl_2 \longrightarrow PCl_5$$

The pentachloride is a straw-colored solid, which sublimes when warmed and decomposes reversibly into the trichloride and chlorine when heated.

$$PCl_5 \rightleftharpoons PCl_3 + Cl_2$$

In both the gaseous and liquid states, the PCl_5 molecule is a trigonal bipyramid (Fig. 24-5).

The pentahalides of phosphorus are Lewis bases. They pick up a halide ion to give the PX_6^- anion. X-ray studies show that solid phosphorus pentachloride is an ionic compound, $[PCl_4^+][PCl_6^-]$, with a tetrahedral cation and an octahedral anion (Fig. 24-6). PCl_5 also forms ions in solution in polar solvents. Phosphorus pentahalides hydrolyze, forming phosphoric acid with excess water.

$$PX_5 + 4H_2O \longrightarrow H_3PO_4 + 5HX(g)$$

Partial hydrolysis produces phosphorus(V) oxyhalides, POX_3, which have tetrahedral structures (Fig. 24-7).

$$PX_5 + H_2O \longrightarrow POX_3 + 2HX(g)$$

24.8   Phosphorus Oxides

Phosphorus forms two common oxides, **phosphorus(III) oxide, P_4O_6**, and **phosphorus(V) oxide, P_4O_{10}**. The oxidation numbers of phosphorus in these compounds are $+3$ and $+5$, respectively. The molecular structures of P_4O_6 and P_4O_{10} are shown in Fig. 24-8. The phosphorus atoms in these compounds are hybridized to form sp^3 orbitals. In the nitrogen(III) and nitrogen(V) oxides N_2O_3 and N_2O_5, nitrogen has sp^2 hybrid orbitals. The combustion of phosphorus in air generally produces a mixture of P_4O_6 and P_4O_{10}. However, it is virtually impossible to get much P_4O_6 by burning phosphorus, even in a limited supply of air.

Figure 24-8. The molecular structures of P_4O_6 and P_4O_{10}. Phosphorus atoms are shown in color.

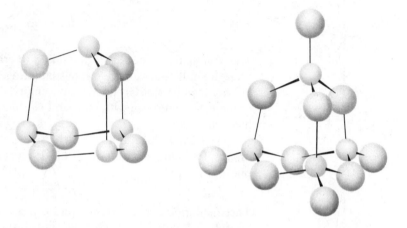

Phosphorus(III) oxide is a white crystalline solid with a garliclike odor. Its vapor is very poisonous. It oxidizes slowly in air and inflames when heated to 70°C, forming phosphorus(V) oxide. Phosphorus(III) oxide dissolves slowly in cold water to form phosphorus acid, H_3PO_3.

Phosphorus(V) oxide, P_4O_{10}, is a white flocculent powder that melts at 420°C. Its enthalpy of formation is very high (-2984 kJ), and for this reason it is quite stable and a very poor oxidizing agent. With a limited amount of water, it forms metaphosphoric acid, $(HPO_3)_n$.

$$nP_4O_{10} + 2nH_2O \longrightarrow 4(HPO_3)_n$$

When more water is added, the metaphosphoric acid slowly changes to orthophosphoric acid, H_3PO_4. The net reaction is given by the equation

$$P_4O_{10}(s) + 6H_2O(l) \longrightarrow 4H_3PO_4(l) \qquad \Delta H° = -368.6 \text{ kJ}$$

When a piece of P_4O_{10} is dropped into water, it reacts with a hissing sound, and much heat is liberated. Because of this great reactivity with water, phosphorus(V) oxide is used extensively for drying gases and removing water from many compounds.

ACIDS OF PHOSPHORUS AND THEIR SALTS

The important oxyacids of phosphorus are orthophosphoric acid, H_3PO_4, diphosphoric (pyrophosphoric) acid, $H_4P_2O_7$, triphosphoric acid, $H_5P_3O_{10}$, metaphosphoric acid, $(HPO_3)_n$, phosphorous acid, H_3PO_3, and hypophosphorous acid, H_3PO_2. The various phosphoric acids all have the same anhydride, P_4O_{10}. Each acid represents a different degree of hydration of this oxide but not a different oxidation number for phosphorus, which is +5 in every case.

24.9 Orthophosphoric Acid

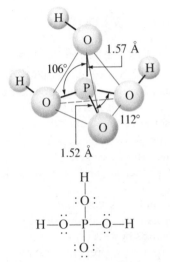

Figure 24-9. The electronic and molecular structures of the orthophosphoric acid molecule, H_3PO_4.

Pure **orthophosphoric acid, H_3PO_4** (commonly called **phosphoric acid**), forms colorless, deliquescent crystals that melt at 42.4°C. However, it is more commonly available as an 82% aqueous solution known as **syrupy phosphoric acid.** The orthophosphoric acid molecule is tetrahedral (Fig. 24-9), and the phosphorus atom has a coordination number of 4.

One commercial method of preparing orthophosphoric acid is to treat calcium phosphate with concentrated sulfuric acid.

$$Ca_3(PO_4)_2 + 3H_2SO_4 \longrightarrow 2H_3PO_4 + 3CaSO_4(s)$$

The products are diluted with water, and the calcium sulfate is removed by filtration. This method gives a dilute acid that is lightly contaminated with calcium dihydrogen phosphate, $Ca(H_2PO_4)_2$, and fluorine compounds formed from the fluoroapatite, $Ca_5(PO_4)_3F$, usually associated with native calcium phosphate. Pure orthophosphoric acid is manufactured by oxidizing phosphorus to P_4O_{10} and dissolving the product in water.

Large quantities of impure orthophosphoric acid are used in the manufacture of fertilizers. The acid is used in medicine as an astringent, an antipyretic, and a stimulant. It is also used in some cola drinks.

24.10 Orthophosphates

Because orthophosphoric acid is a triprotic acid, it forms three series of salts, corresponding to the three stages of ionization.

$$H_3PO_4 \rightleftharpoons H^+ + H_2PO_4^- \qquad \text{(primary ionization, } K_1 = 7.5 \times 10^{-3}\text{)}$$
$$H_2PO_4^- \rightleftharpoons H^+ + HPO_4^{2-} \qquad \text{(secondary ionization, } K_2 = 6.3 \times 10^{-8}\text{)}$$
$$HPO_4^{2-} \rightleftharpoons H^+ + PO_4^{3-} \qquad \text{(tertiary ionization, } K_3 = 3.6 \times 10^{-13}\text{)}$$

The sodium salts are **sodium dihydrogen phosphate, NaH_2PO_4, disodium hydrogen phosphate, Na_2HPO_4,** and **trisodium phosphate, Na_3PO_4.**

Sodium dihydrogen phosphate forms aqueous solutions that are weakly acidic because the dihydrogen phosphate ion is a weak acid.

$$H_2PO_4^- + H_2O \rightleftharpoons HPO_4^{2-} + H_3O^+$$

Disodium hydrogen phosphate solutions are basic because the monohydrogen phosphate ion is a stronger base than it is an acid.

$$HPO_4^{2-} + H_2O \rightleftharpoons H_2PO_4^- + OH^-$$

Trisodium phosphate solutions are strongly basic as a result of hydrolysis.

$$PO_4^{3-} + H_2O \rightleftharpoons HPO_4^{2-} + OH^-$$

Hydrogen and ammonium phosphate salts decompose when heated and volatile products such as water and ammonia are given off. The resulting phosphorus-containing products have P—O—P bonds (Sections 24.11 through 24.13).

Several examples are

$$n NaH_2PO_4 \xrightarrow{\Delta} (NaPO_3)_n + n H_2O(g)$$

$$2 Na_2HPO_4 \xrightarrow{\Delta} Na_4P_2O_7 + H_2O(g)$$

$$n NaNH_4HPO_4 \xrightarrow{\Delta} (NaPO_3)_n + n NH_3(g) + n H_2O(g)$$

$$2 MgNH_4PO_4 \xrightarrow{\Delta} Mg_2P_2O_7 + 2 NH_3(g) + H_2O(g)$$

Sodium dihydrogen phosphate is used as a boiler cleansing compound (to prevent the formation of boiler scale) and as an acid in some baking powders; disodium hydrogen phosphate, as a boiler cleansing compound and in the weighting of silk; and trisodium phosphate, as a water softener and boiler cleansing compound. Calcium dihydrogen phosphate, $Ca(H_2PO_4)_2$, is used as a fertilizer and as a constituent of baking powders. Calcium monohydrogen phosphate, $CaHPO_4$, is added to animal feed for its mineral properties and is used as a polishing agent in toothpastes.

Phosphorus, in the form of soluble orthophosphates, is essential to plant growth. Calcium phosphate is too insoluble to furnish an adequate supply of phosphorus to plants. Calcium dihydrogen phosphate, however, is soluble in water and therefore suitable as a fertilizer. This compound is prepared commercially by treating fluoroapatite with sulfuric acid (Section 8.7).

$$2 Ca_5(PO_4)_3F + 7 H_2SO_4 + 3 H_2O \longrightarrow$$
$$3 Ca(H_2PO_4)_2 \cdot H_2O + 2 HF(g) + 7 CaSO_4$$

The mixture of calcium dihydrogen phosphate and calcium sulfate is sold as superphosphate of lime. A mixture containing a higher percentage of phosphorus, triple superphosphate of lime, is manufactured by treating pulverized calcium phosphate with impure orthophosphoric acid.

$$Ca_3(PO_4)_2 + 4 H_3PO_4 \longrightarrow 3 Ca(H_2PO_4)_2$$

24.11   Diphosphoric Acid

Diphosphoric acid, $H_4P_2O_7$, may be prepared by heating orthophosphoric acid to 250°C.

$$2 H_3PO_4 \xrightarrow{\Delta} H_4P_2O_7 + H_2O$$

This reaction involves the elimination of a molecule of water from two molecules of orthophosphoric acid, as shown by the equation

The structure of the diphosphoric acid molecule is shown in Fig. 24-10. It is a white crystalline solid that melts at 61°C. When dissolved in water, it gradually hydrolyzes to orthophosphoric acid. The hydrogen atoms in diphosphoric acid undergo stepwise neutralization. Thus we can form tetrasodium diphosphate, $Na_4P_2O_7$, trisodium hydrogen diphosphate, $Na_3HP_2O_7$, disodium dihydrogen di-

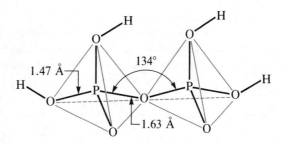

Figure 24-10. The molecular structure of the diphosphoric acid molecule, $H_4P_2O_7$, consists of two tetrahedra joined at a corner.

phosphate, $Na_2H_2P_2O_7$, and sodium trihydrogen diphosphate, $NaH_3P_2O_7$. As you might expect, the acid strengths of the hydrogen diphosphate anions ($H_3P_2O_7^-$, $H_2P_2O_7^{2-}$, and $HP_2O_7^{3-}$) decrease and the base strengths increase with increasing negative charge.

24.12 Triphosphoric Acid

Triphosphoric acid, $H_5P_3O_{10}$ (Fig. 24-11), is formed by the elimination of two molecules of water from three molecules of orthophosphoric acid.

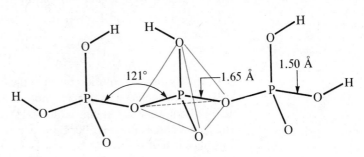

Triphosphoric acid

The reaction is reversible and triphosphoric acid will form orthophosphoric acid upon standing in water.

The sodium salt of triphosphoric acid, $Na_5P_3O_{10}$, is used as a water softener. It may be prepared by fusing equimolar quantities of sodium diphosphate and sodium metaphosphate.

$$nNa_4P_2O_7 + (NaPO_3)_n \longrightarrow nNa_5P_3O_{10}$$

Polyphosphate linkages like those in pyrophosphates and triphosphates have great biochemical importance. The energy required for the contraction of muscles

Figure 24-11. The molecular structure of triphosphoric acid, $H_5P_3O_{10}$.

results from hydrolysis of P—O—P bonds in a complex organic triphosphate known as **adenosine triphosphate.** The enzyme-catalyzed hydrolysis releases 29 kilojoules of energy per mole of adenosine triphosphate, and this energy is used in muscle contraction. The equation for the reaction is given below. (R represents the complex organic portion of the molecule.)

$$\underset{\text{Adenosine triphosphate}}{RO-\overset{\overset{\displaystyle O}{\|}}{\underset{\underset{\displaystyle OH}{|}}{P}}-O-\overset{\overset{\displaystyle O}{\|}}{\underset{\underset{\displaystyle OH}{|}}{P}}-O-\overset{\overset{\displaystyle O}{\|}}{\underset{\underset{\displaystyle OH}{|}}{P}}-OH} + H_2O \longrightarrow \underset{\text{Adenosine diphosphate}}{RO-\overset{\overset{\displaystyle O}{\|}}{\underset{\underset{\displaystyle OH}{|}}{P}}-O-\overset{\overset{\displaystyle O}{\|}}{\underset{\underset{\displaystyle OH}{|}}{P}}-OH} + \underset{\text{Phosphoric acid}}{HO-\overset{\overset{\displaystyle O}{\|}}{\underset{\underset{\displaystyle OH}{|}}{P}}-OH}$$

24.13 Metaphosphoric Acid and Its Salts

Heating orthophosphoric acid above 400°C eliminates 1 mole of water from each mole of orthophosphoric acid, and **metaphosphoric acid, $(HPO_3)_n$,** solidifies from the liquid as a glassy product sometimes called glacial phosphoric acid.

$$nH_3PO_4 \xrightarrow{\Delta} (HPO_3)_n + nH_2O(g)$$

Metaphosphoric acid contains rings and chains built up by oxygen bridges between adjacent phosphorus atoms. There are four oxygen atoms linked to every phosphorus atom. The polymeric HPO_3 units are represented by the formula $(HPO_3)_n$. Metaphosphoric acid dissolves readily in water, in which it slowly the liquid as a glassy product sometimes called glacial phosphoric acid. changes into orthophosphoric acid.

Sodium metaphosphate, $(NaPO_3)_n$, is made by heating sodium dihydrogen orthophosphate.

$$nNaH_2PO_4 \xrightarrow{\Delta} (NaPO_3)_n + nH_2O(g)$$

If the product is heated to about 700°C and then cooled rapidly, a water-soluble glassy polymetaphosphate is formed. This is a long-chain polymer that is best formulated as $(NaPO_3)_n$, although it has been called sodium hexametaphosphate because it once was thought to have the formula $(NaPO_3)_6$. Metaphosphates form soluble complexes with calcium ion and reduce its concentration in solutions so it cannot be precipitated by soaps. For this reason sodium polymetaphosphate and other phosphates have been used extensively as water softeners in detergents. However, their use has been restricted because they enhance the growth of algae in water (Section 12.8).

24.14 Phosphorous Acid and Its Salts

The action of water on P_4O_6, PCl_3, PBr_3, or PI_3 forms **phosphorous acid, H_3PO_3.** Pure phosphorous acid is most readily obtained by hydrolyzing phosphorus trichloride,

$$PCl_3 + 3H_2O \longrightarrow H_3PO_3 + 3HCl(g)$$

heating the resulting solution to expel the hydrogen chloride, and evaporating the water until white crystals of phosphorous acid appear on cooling. The crystals are deliquescent, very soluble in water, and smell like garlic. The solid melts at 70.1°C and decomposes at about 200°C by disproportionation into phosphine and orthophosphoric acid.

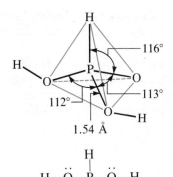

Figure 24-12. Phosphorous acid, H_3PO_3. Note that one hydrogen atom is bonded directly to the phosphorus atom. The other two hydrogen atoms are bonded to oxygen atoms. Only those hydrogen atoms bonded through an oxygen atom are acidic.

$$4H_3PO_3 \longrightarrow PH_3 + 3H_3PO_4$$

The electronic and molecular structures of phosphorous acid are shown in Fig. 24-12.

Phosphorous acid and its salts are active reducing agents that are readily oxidized to phosphoric acid and phosphates, respectively. Phosphorous acid reduces the silver ion to free silver, mercury(II) salts to mercury(I) salts, and sulfurous acid to sulfur.

Phosphorous acid forms only two series of salts, such as sodium dihydrogen phosphite, NaH_2PO_3, and disodium hydrogen phosphite, Na_2HPO_3. The third atom of hydrogen cannot be replaced by a cation; it is bonded to the phosphorus atom rather than to an oxygen atom.

Notice the difference in the spelling of the name of the element (phosphor*us*) and the name of the acid H_3PO_3 (phosphor*ous*). The name of the acid is made of the stem of the element's name plus the *-ous* ending [which indicates a lower oxidation number for phosphorus ($+3$) than that in phosphoric acid, H_3PO_4 ($+5$)]. The name phosphor*ous* acid is analogous to the names chlor*ous* acid, ni-tr*ous* acid, and sulfur*ous* acid. You may find it helpful to review the nomenclature discussed in Section 5.12.

24.15 Hypophosphorous Acid and Its Salts

The solution remaining from the preparation of phosphine from white phosphorus and sodium hydroxide contains sodium hypophosphite, NaH_2PO_2 (Section 24.6). The corresponding barium salt may be obtained by using barium hydroxide in the preparation. When barium hypophosphite is treated with sulfuric acid, barium sulfate precipitates and **hypophosphorous acid, H_3PO_2,** forms in solution.

$$Ba^{2+} + 2H_2PO_2^- + 2H^+ + SO_4^{2-} \longrightarrow BaSO_4(s) + 2H_3PO_2$$

The acid is weak and monoprotic, forming only one series of salts. Two nonacidic hydrogen atoms are bonded directly to the phosphorus atom (Fig. 24-13). Hypophosphorous acid and its salts are strong reducing agents because phosphorus has the unusually low oxidation number of $+1$.

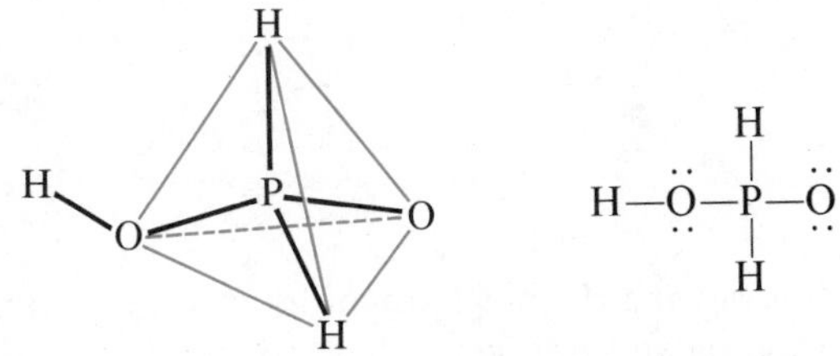

Figure 24-13. Hypophosphorous acid, H_3PO_2. Note that two of the three hydrogen atoms are attached directly to the phosphorus atom. Hence only one hydrogen atom is acidic, the one bonded through an oxygen atom.

FOR REVIEW

SUMMARY

Phosphorus is a nonmetallic element found in Group VA of the Periodic Table. Its common allotropes are the very reactive **white phosphorus,** composed of tetrahedral P_4 molecules, and the less reactive polymeric **red phosphorus.** Although differing in reactivity, both allotropes give the same products in reactions with other elements.

Phosphorus is an active nonmetal. It commonly exhibits an oxidation number of -3 with metals or with less electronegative nonmetals and $+3$ or $+5$ with more electronegative nonmetals. Thus it forms **phosphides,** compounds that contain the P^{3-} ion, with active metals. Hydrolysis of phosphides or disproportionation of phosphorus in basic solution produces **phosphine, PH_3.** Phosphorus is oxidized by halogens and forms phosphorus trihalides and pentahalides, PX_3 and PX_5. The reaction of phosphorus with oxygen is very vigorous; phosphorus(V) oxide, P_4O_{10}, forms. Phosphorus(III) oxide, P_4O_6, can be prepared indirectly.

Orthophosphoric acid, H_3PO_4, is prepared by the acid-base reaction of phosphates with sulfuric acid or by the reaction of water with phosphorus(V) oxide. Orthophosphoric acid is a triprotic acid that forms dihydrogen phosphate, $H_2PO_4^-$, hydrogen phosphate, HPO_4^{2-}, and phosphate, PO_4^{3-} salts. When heated at various temperatures, orthophosphoric acid loses water and forms diphosphoric acid, triphosphoric acid, or metaphosphoric acid.

The hydrolysis of phosphorus trichloride produces **phosphorous acid, H_3PO_3,** a compound containing phosphorus with an oxidation number of $+3$. One hydrogen atom is bonded directly to the phosphorus atom, so this acid contains only two acidic hydrogen atoms. **Hypophosphorous acid, H_3PO_2,** contains two hydrogen atoms bonded directly to the phosphorus atom and one acidic hydrogen.

KEY TERMS AND CONCEPTS

Diphosphoric acid (24.11)
Hypophosphorous acid (24.15)
Metaphosphoric acid (24.13)
Orthophosphoric acid (24.9)

Phosphides (24.6)
Phosphine (24.6)
Phosphorous acid (24.14)
Phosphorus halides (24.7)

Red phosphorus (24.3)
Triphosphoric acid (24.12)
White phosphorus (24.3)

EXERCISES

Phosphorus

1. Compare and contrast the chemical properties of elemental phosphorus and nitrogen; phosphoric acid and nitric acid; phosphine and ammonia.
2. Contrast the properties of white and red phosphorus.
3. Complete and balance the following equations (*xs* indicates that the reactant is present in excess):
   (a) $P_4 + xsS \longrightarrow$
   (b) $P_4 + Li \longrightarrow$
   (c) $P_4 + xsO_2 \longrightarrow$
   (d) $P_4 + Mg \longrightarrow$
   (e) $P_4 + xsF_2 \longrightarrow$
4. Is a phosphorus-phosphorus triple bond stronger or weaker than three P—P single bonds? Explain your answer.
5. How much $Ca_3(PO_4)_2$ would be needed to prepare 1.0 ton of phosphorus in the electric furnace process, if a yield of 94% is obtained?

Ans. 5.3 tons

Compounds of Phosphorus

6. Determine the oxidation number of phosphorus in each of the following:
   (a) PH_3 (e) H_3PO_4
   (b) P_2H_4 (f) $H_4P_2O_7$
   (c) H_3PO_2 (g) PH_4I
   (d) H_3PO_3 (h) PCl_3
7. Write a Lewis structure for each of the following:
   (a) PH_3
   (b) P_2H_4
   (c) H_3PO_2
   (d) H_3PO_3
   (e) H_3PO_4
   (f) $H_4P_2O_7$
   (g) PH_4I (an ionic compound)
   (h) PCl_3
8. Write equations showing the stepwise ionization of phosphoric acid.
9. Write equations for the preparation and hydrolysis of sodium phosphide. Compare the hydrolysis of sodium phosphide to that of sodium nitride.
10. How can the phosphorus in insoluble tricalcium phosphate be made available for plant nutrition?
11. Write equations showing the hydrolysis of PO_4^{3-}, HPO_4^{2-}, and $H_2PO_4^-$.
12. Write equations to show the effect of heat on NaH_2PO_4, $NaNH_4HPO_4$, and $MgNH_4PO_4$.
13. Explain the action of sodium polymetaphosphates as water softeners. Why is the use of such materials in detergents being questioned (see Chapter 12)?
14. Why does phosphorous acid form only two series of salts although its molecule contains three hydrogen atoms?
15. Write equations for the preparation of hypophosphorous acid, starting with white phosphorus.
16. Show that the decomposition of phosphorous acid by heat involves a disproportionation reaction.
17. Explain the difference in the spelling of the name of the element, phosphorus, and the name of H_3PO_3, phosphorous acid.
18. Write equations for each of the following preparations:
    (a) P_4 from $Ca_3(PO_4)_2$
    (b) P_4O_{10} from P_4
    (c) H_3PO_4 from P_4O_{10}
    (d) Na_2HPO_4 from H_3PO_4

(e) $Na_4P_2O_7$ from Na_2HPO_4
19. Draw valence electron structures for diphosphine and P_4.
20. Compare the structures of PCl_4^+, PCl_5, PCl_6^-, and $POCl_3$.
21. Complete and balance the following equations (xs indicates that the reactant is present in excess):
    (a) $NaH_2PO_4 + NH_3 \longrightarrow$
    (b) $PF_5 + KF \longrightarrow$
    (c) $P_4O_6 + CaO \longrightarrow$
    (d) $P_4O_{10} + K_2O \longrightarrow$
    (e) $xsK_3PO_4 + HCl \longrightarrow$
    (f) $Na_3P + xsH_2O \longrightarrow$
    (g) $Na_4P_2O_7 + xsH_2O \longrightarrow$
    (h) $PCl_3 + CH_3OH \longrightarrow$
    (Treat CH_3OH as a derivative of water in which one H is replaced by the $-CH_3$ group.)
22. How much $POCl_3$ can be produced from 50.0 g of PCl_5 and the appropriate amount of P_4O_{10}?
    Ans. 61.4 g
23. What volume of 0.100 M NaOH will be required to neutralize the solution produced by dissolving 1.00 g of PCl_3 in 200 mL of water? Note that when H_3PO_3 is titrated under these conditions, only one proton of the phosphorous acid molecule reacts.
    Ans. 291 mL

Additional Exercises

24. Explain why H_3PO_4 is a stronger acid than H_3PO_3.
25. Name each of the following compounds:
    (a) K_2HPO_4 (e) $Na_4P_2O_7$
    (b) $FePO_4$ (f) $Ca(H_2PO_4)_2$
    (c) NaH_2PO_2 (g) Na_2HPO_3
    (d) $(NaPO_3)_n$
26. A detergent is advertised as containing "only 10% phosphate, expressed as phosphorus pentaoxide." Is the name misleading? If so, explain why. Calculate what this percentage would be if expressed as sodium tripolyphosphate, $Na_5P_3O_{10}$, a common form for detergent phosphate.
    Ans. 26%
27. How many pounds of phosphate ion would be added to the water system of the United States in 1 year if one load of clothes is washed every week for every person in the country? Assume a popu-

lation of 200 million and that 4 oz of detergent that is 20% sodium orthophosphate, Na_3PO_4, is used per washing.   *Ans. 3×10^8 lb*

28. (a) When phosphine burns to give liquid ortho-phosphoric acid, an unusually large amount of energy is involved. Calculate $\Delta H°$ for the reaction using the data in Appendix J.

         Ans. -1272 kJ

   (b) Calculate the value of $\Delta H°$ for production of *solid* orthophosphoric acid by this reaction.

         Ans. -1284 kJ

   (c) Calculate $\Delta G°$ for the reaction giving solid orthophosphoric acid.   *Ans. -1132 kJ*

   (d) Compare the values of $\Delta H°$ and $\Delta G°$, and on the basis of the comparison draw specific conclusions about the reaction.

29. At 422 K the partial pressures of gaseous PCl_5, PCl_3, and Cl_2 in a closed system were 0.453 atm, 6.08×10^{-2} atm, and 6.08×10^{-2} atm, respectively. Determine K for the dissociation reaction

$$PCl_5(g) \rightleftharpoons PCl_3(g) + Cl_2(g)$$

         Ans. 8.16×10^{-3}

30. The solubility product of $Ca_3(PO_4)_2$ is 1×10^{-25}. What is the concentration of $Ca_3(PO_4)_2$ in a saturated solution of $Ca_3(PO_4)_2$?   *Ans. 4×10^{-6} M*

25

CARBON AND ITS COMPOUNDS

ELEMENTAL CARBON

Carbon is the first member of Group IVA; the other members are silicon, germanium, tin, and lead. Carbon is predominantly nonmetallic in character. Silicon and germanium are metalloids exhibiting both nonmetallic and metallic characteristics, and tin and lead are metallic. Each of these elements has four valence electrons with a valence shell electron configuration of ns^2np^2. Each exhibits a maximum oxidation number of $+4$ and each (especially the heavier ones) can also have an oxidation number of $+2$ when combined with other elements.

With few exceptions, carbon forms covalent bonds in compounds. All four of its valence electrons are usually used in bonding, which involves sp, sp^2, or sp^3 hybridization of the carbon atomic orbitals. Thus carbon can have a coordination number of 2, 3, or 4 with a linear, trigonal planar, or tetrahedral geometry, respectively.

Carbon is nineteenth among the elements in abundance; it constitutes only about 0.027% of the earth's crust. It is found in the free state as diamond and graphite and in compounds in natural gas, petroleum, coal, plants and animals, limestone, dolomite, coral, and chalk. Carbon is found in the air as carbon dioxide and in natural waters as carbon dioxide, carbonic acid, and carbonates.

Elemental carbon exists in two allotropic forms: diamond and graphite. Charcoal, coke, and carbon black are microcrystalline, or amorphous, forms of carbon with the graphite structure.

25.1   Diamond

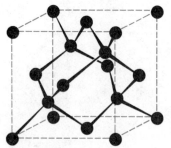

Figure 25-1.   The crystal structure of diamond. Each sphere represents a carbon atom, with the darker spheres indicating the atoms at the corners and the centers of the faces of a cubic unit cell, and the colored spheres, the atoms in tetrahedral sites. The covalent bonds are wedge-shaped to help show the perspective.

Diamonds are found in South Africa, Zaire, the Gold Coast, Brazil, India, Australia, and Arkansas, but most come from South Africa. Diamonds are formed when very pure carbon is subjected to high temperatures and very high pressures.

Many attempts have been made to synthesize diamond, but apparently none were successful until 1954. In 1954 H. Tracy Hall and his co-workers found that small crystals of diamond formed when graphite, in molten iron(II) sulfide, FeS, as a solvent, was subjected to high temperature and pressure under carefully controlled conditions. This method of making diamonds has become increasingly important because synthetic diamonds are superior to natural stones for certain industrial uses.

Diamond is very brittle, and it is the hardest substance known with the possible exception of boron carbide, B_4C. Diamond is inert to all chemicals at ordinary temperatures but burns when heated in air or oxygen. At 1000°C in the absence of air, diamond changes to graphite.

A diamond crystal belongs to the cubic system. Each atom is sp^3 hybridized and covalently bonded to four others at the corners of a tetrahedron (Fig. 25-1). All of the atoms in a diamond crystal are bonded into a single giant molecule. Because the bonds are very strong and extend throughout the crystal in its three dimensions, diamond is very hard and has a high melting point (probably the highest of all elements). Diamond does not conduct electricity, because it has no mobile electrons—they are all used in bond formation.

25.2   Graphite

Natural graphite is mined in Mexico, Madagascar, Ceylon, and Canada. The United States leads in the production of synthetic graphite. To synthesize graphite, a mixture of amorphous carbon with a little sand and iron oxide (as catalysts) is heated in an electric furnace for 24–30 hours at about 3500°C. The carbon vapor that is formed condenses as graphite. Graphite, also called plumbago, or black lead, is distinctly different in crystalline form and physical properties from diamond. It is a soft, grayish black solid that crystallizes in hexagonal plates with a metallic luster and that conducts electricity. The name *graphite* comes from the Greek, meaning *to write*. Graphite is the "lead" in lead pencils. It melts at 3527°C and is inert to most chemical reagents. The internal energy of graphite is less than that of diamond. Thus graphite is the more stable of the two crystalline forms of carbon, in accord with the principle that substances tend to assume the form with the lowest possible energy (Section 18.10).

The comparative softness and electrical conductivity of graphite are related to its structure (Fig. 25-2). A crystal of graphite contains layers of carbon atoms, each of which has three others as nearest neighbors. Each atom forms three σ bonds, one to each of its nearest neighbors in the layer, by formation of sp^2 hybrid orbitals. The unhybridized p orbital of each carbon atom projects above and below the layer (Fig. 25-3) and can overlap with the unhybridized p orbital on each of the adjacent carbon atoms. There is one electron for each unhybridized p orbital, and these electrons pair to form π bonds between adjacent carbon atoms.

Figure 25-2. The crystal structure of graphite.

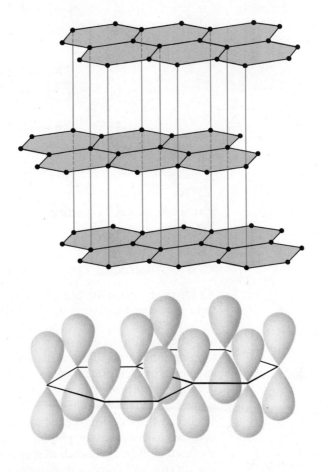

Figure 25-3. The orientation of unhybridized *p* orbitals of the carbon atoms in graphite. Each *p* orbital is perpendicular to the plane of carbon atoms.

Two of the many resonance forms necessary to describe the electronic structure of a graphite layer are shown in Fig. 25-4. The σ and π bonds bind the atoms tightly together within layers. However, the layers can be separated easily because only weak van der Waals forces hold them together. These weak bonds give graphite its flaky, soft character. The loosely held electrons of the resonating double bonds are responsible for the electrical conductivity and the color of graphite.

The π bonding in graphite can also be described by the Molecular Orbital Theory (Chapter 6). The unhybridized *p* orbitals of the carbon atoms overlap to give molecular orbitals that extend over an entire layer so that an electron from any of the *p* orbitals can move over the entire layer. Since the electrons can move,

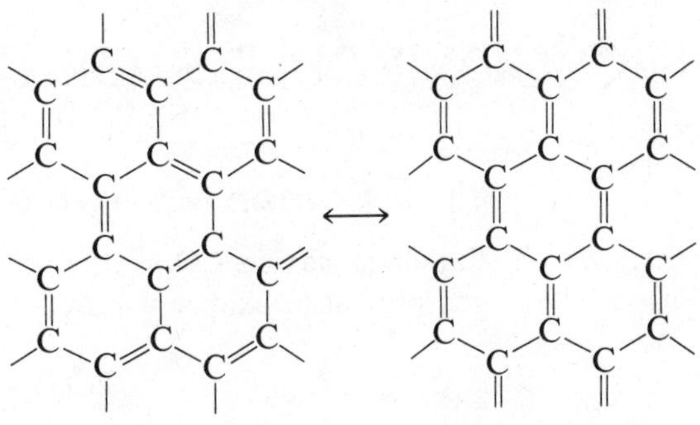

Figure 25-4. Two of the resonance forms of graphite necessary to describe its electronic structure, which is a resonance hybrid.

Figure 25-5. A portion of one molecular orbital in graphite, which contains delocalized π electrons.

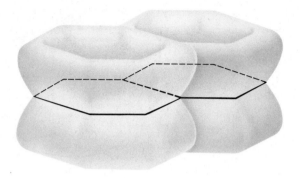

graphite conducts. A portion of one of these molecular orbitals is shown in Fig. 25-5.

25.3   Chemical Properties and Uses of Carbon

All forms of carbon are almost inert toward most reagents at ordinary temperatures. However, graphite is slowly oxidized by a mixture of nitric acid and sodium chlorate. The activity of carbon increases rapidly as the temperature increases, and at elevated temperatures it is very reactive. Hot carbon unites with oxygen, forming either carbon monoxide, CO, or carbon dioxide, CO_2, depending on the amount of oxygen present. With sulfur, carbon forms carbon disulfide, CS_2; with certain metals it forms carbides, such as iron carbide, Fe_3C; and with fluorine it forms carbon tetrafluoride, CF_4.

Because of its resistance to heat and chemical action, its high heat of vaporization, and its conductivity, graphite is used to make electrodes and crucibles. It is also used in the manufacture of paints, commutator brushes, and pencils. Suspensions of colloidal graphite in water or in oil provide excellent lubricants. Diamond is used in cutting, grinding, and polishing operations because of its hardness.

Wood charcoal, a microcrystalline form of carbon, is produced by the destructive distillation of wood, that is, by heating wood in the absence of air. It is used as a fuel. Charcoal is also used as a decolorizing agent for liquids such as sugar solutions, alcohol, and petroleum products; it can adsorb substances that discolor the liquids. **Adsorption** is a surface phenomenon in which the surface forces of the adsorbing agent attract and hold molecules of the substance being adsorbed. Gases as well as solids and liquids are adsorbed on the surface of the charcoal. Charcoal is used to remove objectionable gases from petroleum products and to purify water by adsorbing small amounts of chloroform and other organic molecules that may be present.

INORGANIC COMPOUNDS OF CARBON

25.4   Carbon Monoxide

Carbon monoxide, CO, is produced when carbon is burned without enough oxygen present to oxidize it fully.

$$2C(s) + O_2(g) \xrightarrow{\Delta} 2CO(g)$$

When steam is passed through a bed of red-hot coke, a mixture of hydrogen and carbon monoxide, called **water gas,** is formed.

$$C(s) + H_2O(g) \longrightarrow H_2(g) + CO(g) \qquad \Delta H° = +131.3 \text{ kJ}$$

Because the reaction is endothermic, the coke soon becomes too cool to react. Hence oxygen and steam are passed together through the coke to keep it sufficiently hot. Water gas is used extensively as an industrial fuel and as a source of hydrogen.

Carbon monoxide is prepared in the laboratory by heating crystals of oxalic acid, $H_2C_2O_4$, with concentrated sulfuric acid, which dehydrates the oxalic acid and absorbs the water produced.

$$H_2C_2O_4 \xrightarrow{\text{Conc. } H_2SO_4} H_2O + CO_2 + CO$$

The carbon dioxide in the resulting mixture may be removed by passing the mixture through solid sodium hydroxide, which absorbs the carbon dioxide leaving pure carbon monoxide.

Carbon monoxide is a colorless, odorless, and tasteless gas that is only slightly soluble in water. Liquid carbon monoxide boils at $-192°C$ and freezes at $-205°C$. Its Lewis structure is $:C{\equiv}O:$, and the arrangement of electrons gives a completed shell to each atom. The molecular orbital model for carbon monoxide is mentioned in Section 6.13.

Carbon monoxide burns readily in oxygen, forming carbon dioxide.

$$2CO(g) + O_2(g) \longrightarrow 2CO_2(g) \qquad \Delta H° = -565.97 \text{ kJ}$$

This reaction makes carbon monoxide useful as a gaseous fuel. At high temperatures carbon monoxide will reduce many metal oxides, so it is an important reducing agent in metallurgical processes. Two examples are

$$CuO + CO \longrightarrow Cu + CO_2$$
$$FeO + CO \longrightarrow Fe + CO_2$$

Carbon monoxide is a very dangerous poison; since it is odorless and tasteless, it gives no warning of its presence. It combines with the hemoglobin in blood to form a compound too stable to be broken down by body processes. After the hemoglobin has combined with carbon monoxide, it cannot combine with oxygen, and this destroys the blood's ability to carry oxygen.

25.5 Carbon Dioxide

Carbon dioxide, CO_2 (Fig. 25-6), is produced when any form of carbon or almost any carbon compound is burned in an excess of oxygen.

$$C + O_2 \longrightarrow CO_2$$
$$CH_4 + 2O_2 \longrightarrow CO_2 + 2H_2O$$
$$C_2H_5OH + 3O_2 \longrightarrow 2CO_2 + 3H_2O$$

Many carbonates liberate carbon dioxide when they are heated. Quicklime, CaO, is produced by heating limestone (Section 8.7).

$$CaCO_3 \xrightarrow{\Delta} CaO + CO_2(g)$$

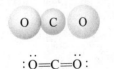

:Ö=C=Ö:

Figure 25-6. The electronic and molecular structures of CO_2.

Large quantities of carbon dioxide are obtained as a by-product of the fermentation of sugar (glucose) during the preparation of alcohol and alcoholic beverages. The net reaction is

$$\underset{\text{Glucose}}{C_6H_{12}O_6} \xrightarrow{\text{Yeast}} \underset{\text{Ethanol}}{2C_2H_5OH} + 2CO_2(g)$$

Carbon dioxide can be produced in the laboratory by the reaction of acid with a carbonate.

$$CuCO_3 + 2H^+ \longrightarrow Cu^{2+} + H_2O + CO_2(g)$$

Carbon dioxide is a colorless, odorless gas that is 1.5 times as heavy as air. It is a component of all carbonated beverages. With water it forms carbonic acid, which has a mildly acid taste. A liter of water at 20°C dissolves 0.9 liter of carbon dioxide. The gas is easily liquefied by compression because its critical temperature is relatively high (31.1°C). When liquid carbon dioxide is allowed to evaporate, the vapor freezes to a snowlike solid at −56.2°C. The solid vaporizes without melting (sublimes) because its vapor pressure is 1 atmosphere at −78.5°C. This property makes solid carbon dioxide, or **dry ice,** valuable as a refrigerant that is always free from liquid.

The Lewis structure of carbon dioxide is $:\overset{..}{O}=C=\overset{..}{O}:$ and shows two double bonds. Since these bonds are strong, the molecule is quite stable. At 2000°C the amount that is dissociated into carbon monoxide and oxygen is only 1.8%.

$$2CO_2 \rightleftharpoons 2CO + O_2$$

When carbon dioxide is heated with carbon, carbon monoxide is formed.

$$CO_2 + C \longrightarrow 2CO$$

Magnesium burns in carbon dioxide and reduces it to carbon.

$$CO_2 + 2Mg \longrightarrow 2MgO + C$$

Carbon dioxide is an efficient fire extinguisher because most substances will not burn in it, it is easily generated, and it is cheap. Air containing as little as 2.5% of carbon dioxide will extinguish a flame.

Carbon dioxide is not toxic, although a large concentration can cause suffocation, that is, lack of oxygen. Breathing air that contains an elevated level of carbon dioxide stimulates the respiratory centers and causes rapid breathing.

The atmosphere contains about 0.04% by volume of carbon dioxide and serves as a huge reservoir of this compound. Green plants absorb carbon dioxide and with the help of sunlight and chlorophyll (as a catalyst) convert the carbon dioxide and water into sugar and oxygen. The reactions that take place during this process, called **photosynthesis,** are complex, but the overall transformation may be summarized by

$$6CO_2 + 6H_2O \longrightarrow C_6H_{12}O_6 + 6O_2$$

The sugar produced during photosynthesis is glucose, $C_6H_{12}O_6$, used by plants to synthesize other sugars, starches, cellulose, and fats.

Carbon dioxide is a product of respiration and is returned to the air by green plants and by animals. The organisms (mostly nongreen plants) that cause plant and animal matter to decay and sugars to ferment also produce carbon dioxide, as does the combustion of carbon-containing fuels. The gases from volcanoes and other geological phenomena are other sources of carbon dioxide. The solubility of carbon dioxide in water makes oceans and lakes great reservoirs of this compound; the oceans keep the carbon dioxide content of the atmosphere almost constant. However, the carbon dioxide content of the atmosphere has increased detectably in the last few years as a result of the burning of fuels.

25.6   Carbonic Acid and Carbonates

Carbon dioxide is the anhydride of **carbonic acid, H_2CO_3,** which forms slowly when carbon dioxide dissolves in water. Carbonic acid is a diprotic acid; the following ionization constants are reported for aqueous carbonic acid:

$$H_2CO_3(aq) \rightleftharpoons H^+(aq) + HCO_3^-(aq) \qquad K_{a_1} = \frac{[H^+][HCO_3^-]}{[H_2CO_3]} = 4.3 \times 10^{-7}$$

$$HCO_3^-(aq) \rightleftharpoons H^+(aq) + CO_3^{2-}(aq) \qquad K_{a_2} = \frac{[H^+][CO_3^{2-}]}{[HCO_3^-]} = 7 \times 10^{-11}$$

The first value is not strictly correct. It is calculated using the assumption that all of the dissolved carbon dioxide has reacted with water and is present as carbonic acid. In fact, most of the carbon dioxide is present as CO_2 molecules rather than H_2CO_3 molecules. The actual ionization constant for carbonic acid, taking into account the true concentration of H_2CO_3 molecules in a carbon dioxide–water solution, is about 2×10^{-4}. In calculations of equilibria involving carbonic acid, it is convenient to assume that all of the carbon dioxide in solution is present as carbonic acid, and thus the value of 4.3×10^{-7} is commonly used for K_a in such calculations. This leads to correct values for the concentration of hydronium ion and hydrogen carbonate ion in solution.

Carbon dioxide reacts with water slowly enough to be visually observed. When an excess of a dilute solution of acetic acid is added to a dilute solution of sodium hydroxide containing a small amount of phenolphthalein indicator, the sodium is neutralized by the acetic acid as rapidly as the two solutions can be mixed (as indicated by the fading or disappearance of the indicator color). Neutralization by a saturated aqueous solution of carbon dioxide under the same conditions takes several seconds.

When a solution of sodium hydroxide is saturated with carbon dioxide, **sodium hydrogen carbonate, $NaHCO_3$,** is formed.

$$[Na^+] + OH^- + CO_2 \longrightarrow [Na^+] + HCO_3^-$$

The compound $NaHCO_3$ is also called **sodium bicarbonate,** or baking soda. The hydrogen carbonate anion is a stronger base than it is an acid, so solutions of salts of hydrogen carbonate are weakly alkaline.

$$HCO_3^- + H_2O \rightleftharpoons H_2CO_3 + OH^-$$

When equivalent amounts of sodium hydroxide and a solution of sodium hydrogen carbonate are mixed, crystals of the hydrate $Na_2CO_3 \cdot 10H_2O$ form on evaporation. The hydrate, called washing soda, is commonly used as a water softener in the home. If heated gently, it forms anhydrous **sodium carbonate, Na_2CO_3,** called soda ash industrially. Solutions of sodium carbonate are strongly basic due to extensive hydrolysis of the carbonate ion.

$$CO_3^{2-} + H_2O \rightleftharpoons HCO_3^- + OH^-$$

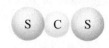

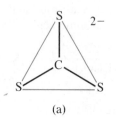

Figure 25-7.   The electronic and molecular structures of CS_2.

25.7   Carbon Disulfide

At high temperatures carbon is oxidized by sulfur to **carbon disulfide, CS_2** (Fig. 25-7).

$$C + 2S \longrightarrow CS_2$$

Air must be excluded because the volatile carbon disulfide is highly flammable, burning according to the equation

$$CS_2 + 3O_2 \longrightarrow CO_2 + 2SO_2$$

Carbon disulfide is produced commercially by heating methane in sulfur vapor at about 700°C in the presence of either silica gel or activated alumina.

Pure carbon disulfide is a colorless, very volatile liquid with a disagreeable odor. The liquid is heavy (specific gravity is in the range 1.27–1.29) and is immiscible with water. The vapor is heavier than air, very poisonous, and highly flammable.

Carbon disulfide, like carbon dioxide, is a Lewis acid. Carbon disulfide reacts with the sulfide ion, giving the thiocarbonate ion, CS_3^{2-} [Fig. 25-8(a)], and with the amide ion, giving the dithiocarbamate ion, $H_2NCS_2^-$ [Fig. 25-8(b)].

$$S^{2-} + CS_2 \longrightarrow CS_3^{2-}$$
$$NH_2^- + CS_2 \longrightarrow H_2NCS_2^-$$

When sulfur replaces part or all of the oxygen in a molecule or ion, the prefix *thio-* is used in the name. Thus CO_3^{2-} is the carbonate ion, and CS_3^{2-} is the thiocarbonate ion.

Large quantities of carbon disulfide are used in making rayon by the viscose process and in the manufacture of cellophane.

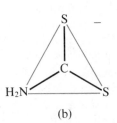

Figure 25-8.   The structures of (a) the thiocarbonate ion and (b) the dithiocarbamate ion.

25.8   Carbon Tetrachloride

Carbon tetrachloride, CCl_4, is manufactured by the reaction of chlorine with methane, CH_4.

$$CH_4 + 4Cl_2 \xrightarrow{\Delta} CCl_4 + 4HCl$$

It is an excellent solvent for fats, oils, and greases, and it was once used for dry-cleaning fabrics. Because its vapor is about five times as heavy as air and it does not burn, it was also used in fire extinguishers. However, carbon tetrachloride has been recognized as a carcinogen, so its use has been sharply curtailed.

25.9   Calcium Carbide

Calcium carbide, CaC_2, is an important commercial substance made by heating calcium oxide with coke in an electric furnace.

$$CaO + 3C \longrightarrow CaC_2 + CO$$

The product is drawn off and solidified by cooling. Calcium carbide is an ionic compound with the Lewis structure $Ca^{2+}[:C{\equiv}C:^{2-}]$. It hydrolyzes in water forming acetylene (ethyne), $HC{\equiv}CH$ (Section 25.14).

$$CaC_2 + 2H_2O \longrightarrow Ca(OH)_2 + C_2H_2(g)$$

25.10 Cyanamides and Cyanides

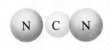

$$: \ddot{N} = C = \ddot{N} :^{2-}$$

Figure 25-9. The structure of the cyanamide ion. Note the similarity to the isoelectronic molecules CO_2 (Fig. 25-6) and CS_2 (Fig. 25-7).

The reaction of calcium carbide with nitrogen at about 1100°C gives **calcium cyanamide, CaCN$_2$.**

$$CaC_2 + N_2 \xrightarrow{1100°C} CaCN_2 + C$$

The linear cyanamide ion, NCN^{2-}, is isoelectronic and isostructural with carbon dioxide and carbon disulfide (compare Fig. 25-9 with Figs. 25-6 and 25-7).

The fusion of calcium cyanamide with carbon and sodium carbonate produces **sodium cyanide, NaCN.**

$$CaCN_2 + C + Na_2CO_3 \longrightarrow CaCO_3 + 2NaCN$$

The soluble sodium cyanide is separated from the insoluble calcium carbonate by washing the products with water. Sodium cyanide can also be prepared by the reaction of sodium amide with carbon at 500–600°C.

$$NaNH_2 + C \xrightarrow{500-600°C} NaCN + H_2$$

The cyanide ion is basic. It combines with acid to form **hydrogen cyanide, HCN.**

$$H^+(aq) + CN^-(aq) \longrightarrow HCN(g)$$

Hydrogen cyanide is a gas that smells like bitter almonds. When it dissolves in water, hydrocyanic acid is formed. This acid is so weak ($K_a = 4 \times 10^{-10}$) that it will not turn litmus red. Its salts, such as sodium cyanide, form solutions that are alkaline by hydrolysis.

Hydrogen cyanide is very poisonous; a dose of about 0.05 gram is fatal to humans.

Both cyanamides and cyanides find extensive use in the manufacturing of plastics and polymers. Calcium cyanamide is used to prepare melamine, a raw material for the production of plastics, such as Melmac. Hydrogen cyanide is used in the preparation of acrylonitrile polymers used in synthetic fibers, such as Dynel and Orlon.

ORGANIC COMPOUNDS OF CARBON

The term *organic* was first used about 1777 to mean occurring in or derived from living organisms. Such substances as starch, alcohol, and urea were classified as organic, for starch is produced by living plants, alcohol is a product of fermentation caused by microorganisms, and urea is contained in urine. In 1828, however, the German chemist Friedrich Wohler synthesized urea from materials obtained from inanimate sources, and the original meaning of organic was no longer applicable. **Organic compounds,** in the modern sense, are compounds of carbon that contain either carbon-carbon bonds or carbon-hydrogen bonds or both. Thousands of carbon compounds not found in or derived from living organisms have been produced by chemists, and about five million organic compounds have been characterized.

That so many organic compounds exist is due primarily to the ability of carbon atoms to combine with other carbon atoms, forming chains of different lengths and rings of different sizes. Other than carbon, the elements most frequently found in organic compounds are hydrogen, oxygen, nitrogen, sulfur, the halogens, phosphorus, and some of the metals.

The simplest organic compounds contain only carbon and hydrogen and are called **hydrocarbons.** Many hydrocarbons are found in plants, animals, and their fossils; other hydrocarbons have been prepared in the laboratory. Several types of hydrocarbons have been characterized. They are distinguished by the type of bonding and the hybridization of the orbitals (Sections 7.3 through 7.5 and 7.8) of their carbon atoms.

25.11 Saturated Hydrocarbons

Saturated hydrocarbons contain only single covalent bonds. All of the carbon atoms in a saturated hydrocarbon have sp^3 hybridization and are bonded to four other carbon or hydrogen atoms. Saturated hydrocarbons of one series have the general molecular formula C_nH_{2n+2}, where n is an integer, and are called **alkanes,** or **paraffins** (from the Latin for having little affinity, or being not very reactive).

A few alkanes and some of their properties are listed in Table 25-1. With the exception of methane, each of the alkanes listed in the table contains a chain of carbon atoms. All of the carbon atoms are sp^3 hybridized and bonded either to other carbon atoms or to hydrogen atoms by single bonds. All of the C—C bond lengths are about 1.54 Å, all of the C—H lengths are about 1.09 Å, and the bond angles are close to 109.5° (the tetrahedral angle). Thus chains of carbon atoms have a staggered, or zigzag, configuration. The Lewis structures and molecular structures of methane, ethane, and pentane are illustrated in Fig. 25-10. In the

Table 25-1 Some Alkanes[a]

	Molecular Formula	Melting Point, °C	Boiling Point, °C	Usual Form	Number of Structural Isomers
Methane	CH_4	−182.5	−161.5	Gas	1
Ethane	C_2H_6	−183.2	−88.6	Gas	1
Propane	C_3H_8	−187.7	−42.1	Gas	1
Butane	C_4H_{10}	−138.3	−0.5	Gas	2
Pentane	C_5H_{12}	−129.7	36.1	Liquid	3
Hexane	C_6H_{14}	−95.3	68.7	Liquid	5
Heptane	C_7H_{16}	−90.6	98.4	Liquid	9
Octane	C_8H_{18}	−56.8	125.7	Liquid	18
Nonane	C_9H_{20}	−53.6	150.8	Liquid	35
Decane	$C_{10}H_{22}$	−29.7	174.0	Liquid	75
Undecane	$C_{11}H_{24}$	−25.6	195.8	Liquid	159
Dodecane	$C_{12}H_{26}$	−9.6	216.3	Liquid	355
Tridecane	$C_{13}H_{28}$	−5.4	235.4	Liquid	802
Tetradecane	$C_{14}H_{30}$	5.9	253.5	Liquid	1858
Octadecane	$C_{18}H_{38}$	28.2	316.1	Solid	60,523

[a]Physical properties for C_4H_{10} and the heavier molecules are those of the *normal* isomer, *n*-butane, *n*-pentane, etc.

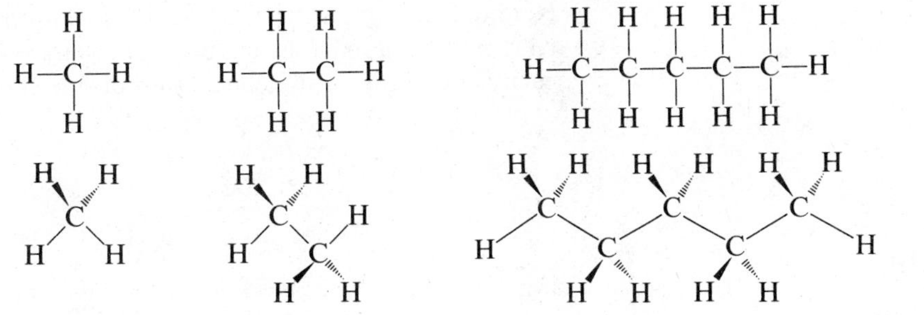

Figure 25-10. The structures of methane, ethane, and pentane. Striped wedges indicate bonds that extend behind the plane of the paper, and solid wedges indicate bonds that extend in front of the plane of the paper.

Methane   Ethane   Pentane

molecular structures the bonds shown as striped wedges extend behind the page, and the bonds shown as solid wedges extend in front of the page. Each carbon atom can rotate about its carbon-carbon single bonds; hence the chain of carbon atoms need not lie in a plane. A long chain can twist itself into many different shapes. For clarity, carbon atoms are usually written in straight lines in Lewis structures, but remember that Lewis structures usually show only electron distribution and bonds, not the geometry about each atom.

Two hydrocarbons have the formula C_4H_{10}; their common names are normal butane (or n-butane) and isobutane. These two butanes are **structural isomers.** They have the same molecular formula and hence the same composition, but different physical and chemical properties because the arrangement of the atoms in their molecules is different. Normal butane, or n-butane, is a straight-chain molecule, and isobutane is a branched-chain molecule.

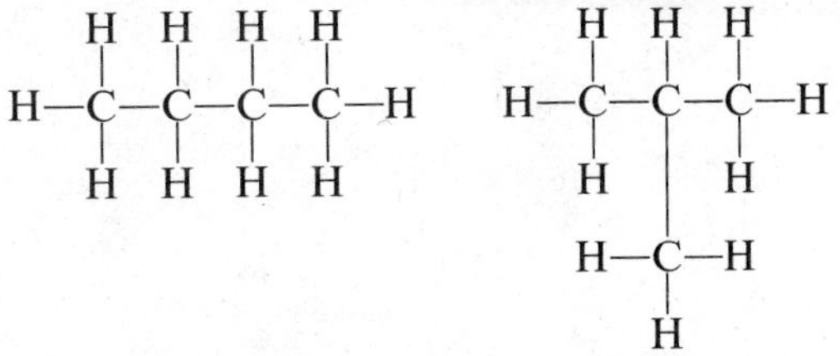

n-Butane   Isobutane

The number of possible isomers increases with the number of carbon atoms in the hydrocarbon. Hydrocarbons with a large number of carbon atoms have a great many isomers (see Table 25-1). The three structural isomers of the hydrocarbon C_5H_{12} have the common names n-pentane, isopentane, and neopentane.

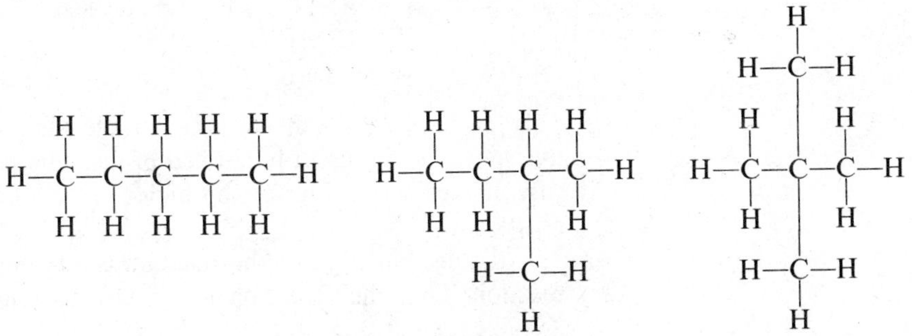

n-Pentane, or
pentane
$CH_3CH_2CH_2CH_2CH_3$

Isopentane, or
2-methylbutane
$CH_3CH_2CH(CH_3)CH_3$

Neopentane, or
2,2-dimethylpropane
$CH_3C(CH_3)_2CH_3$

Note that the following three structures, however, all represent the same molecule, *n*-pentane, and hence are not separate isomers. They are identical—they each contain an unbranched chain of five carbon atoms.

All of the hydrogen atoms in the —CH$_3$ groups at the ends of a hydrocarbon chain (like that in Fig. 25-10, for example) are equivalent, because a carbon atom can rotate freely around a C—C bond.

The members of a second series of saturated hydrocarbons have the general molecular formula C$_n$H$_{2n}$, where *n* is an integer, and are called **cycloalkanes.** As the name implies, the molecules of these substances are cyclic (possessing rings). The smallest member of the series is cyclopropane; next are cyclobutane, cyclopentane, etc.

Cyclopropane   Cyclobutane   Cyclopentane

The reactions of alkanes all involve the breaking of C—H or C—C single bonds. In a **substitution reaction,** a typical reaction of alkanes, a second type of atom is substituted for a hydrogen atom.

Ethane   Ethyl chloride

This reaction transforms an alkane molecule into one that contains a more reactive species or group, a **functional group.** The functional group, a halogen in this case, makes it possible for the molecule to take part in many different kinds of reactions.

Alkanes burn in air; the reaction is a highly exothermic oxidation-reduction reaction. Thus they are excellent fuels. A typical combustion reaction is

$$2C_2H_6(g) + 7O_2(g) \longrightarrow 4CO_2(g) + 6H_2O(g) \qquad \Delta H° = -2856 \text{ kJ}$$

Methane is the principal component of natural gas. The butane used in camping stoves and lighters is an alkane. Gasoline is a mixture of straight and branched chain alkanes containing from five to nine carbon atoms plus various additives. Kerosene, diesel oil, and fuel oil are primarily mixtures of alkanes with higher molecular weights.

25.12 The Nomenclature of Saturated Hydrocarbons

The International Union of Pure and Applied Chemistry (IUPAC) has devised nomenclature for the hydrocarbons, their isomers, and their derivatives (compounds in which one or more hydrogen atoms have been replaced by other atoms or groups of atoms). The nomenclature for saturated hydrocarbons is based on two rules.

1. The longest *continuous* chain of more than four carbon atoms in a saturated hydrocarbon is indicated by one of the following prefixes: five carbons, *penta-;* six carbons, *hexa-;* seven carbons, *hepta-;* eight carbons, *octa-;* nine carbons, *nona-;* and ten carbons, *deca-.* The suffix *-ane* is added at the end (one *a* is dropped when two occur in succession). A two-carbon chain is called ethane; a three-carbon chain, propane; and a four-carbon chain, butane. Thus the names of the hydrocarbons listed in Table 25-1 indicate that these compounds, which are the normal isomers when more than one form exists, contain a single chain of carbon atoms consisting of the same number of carbon atoms as the number of carbon atoms in the molecule.

2. The positions and names of branches from the longest chain or of atoms that replace hydrogen atoms on the chain (substituents) are added as prefixes to the name of the chain. The position of attachment is identified by the number of the carbon atom to which a branch or substituent is attached; the number of the carbon atom is found by counting from the end of the chain nearer to the substituent.

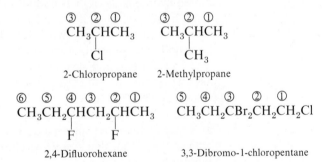

Notice that *-o* replaces *-ide* at the end of the name of an electronegative substituent and that the number of substituents of the same type is indicated by the prefixes *di-* (two), *tri-* (three), *tetra-* (four), etc. (for example, *difluoro-* for two fluoride substituents).

The longest continuous chain of carbon atoms in *n*-butane is four, but in isobutane the longest continuous chain of carbon atoms is only three even though there are a total of four carbon atoms in the molecule. Hence, in IUPAC nomenclature, *n*-butane is called butane, but isobutane is 2-methylpropane.

A substituent that contains one less hydrogen than the corresponding alkane is called an **alkyl group.** The name of an alkyl group is obtained by dropping the suffix *-ane* of the alkane name and adding *-yl.* For example, methane becomes *methyl,* and ethane becomes *ethyl.*

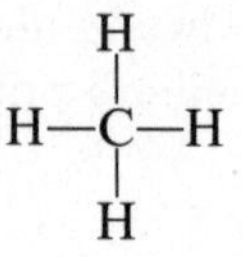
Methane

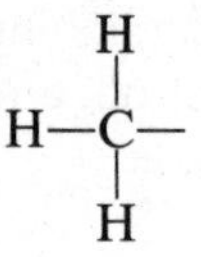
A methyl group

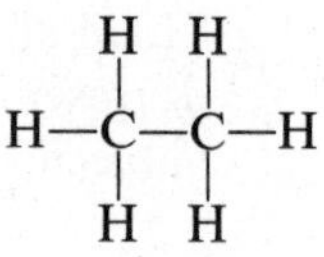
Ethane

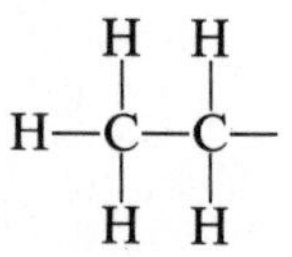
An ethyl group

The open bonds in the methyl and ethyl groups indicate that these groups are bonded to another atom.

Removal of any of the four hydrogen atoms from methane forms a methyl group. These four hydrogen atoms are equivalent. Likewise, removing any one of the six equivalent hydrogen atoms in ethane gives an ethyl group. However, in both propane and 2-methylpropane there are two different types of hydrogen atoms, distinguished by the adjacent atoms or groups of atoms.

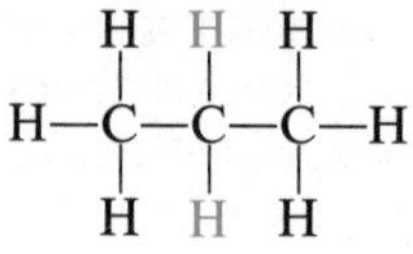
Propane

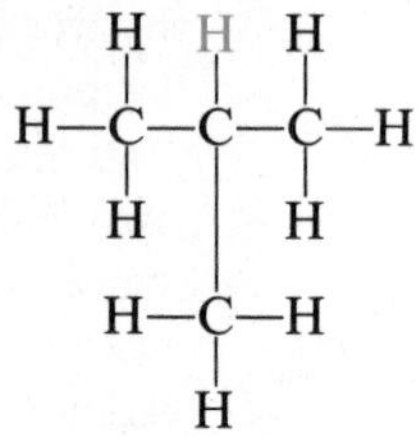
2-Methylpropane

The six equivalent hydrogen atoms of the first type in propane and the nine equivalent hydrogen atoms of that type in 2-methylpropane (all shown in black) are each bonded to a **primary carbon,** a carbon atom bonded to only one other carbon atom. The two colored hydrogen atoms in propane are of a second type. They differ from the six hydrogen atoms of the first type in that they are bonded to a **secondary carbon,** a carbon atom bonded to two other carbon atoms. The colored hydrogen atom in 2-methylpropane differs from the other nine hydrogen atoms in that molecule; it is bonded to a **tertiary carbon,** a carbon atom bonded to three other carbon atoms. Two different alkyl groups can be formed from each of these molecules, depending on which hydrogen atom is removed. The names and structures of these and several other alkyl groups are listed in Table 25-2. Note that alkyl groups do not exist as stable independent entities. They are always part of some larger molecule.

The location of an alkyl group on a hydrocarbon chain is indicated in the same way as any other substituent. If more than one substituent is present on the same carbon atom, they are listed either alphabetically or in order of increasing complexity.

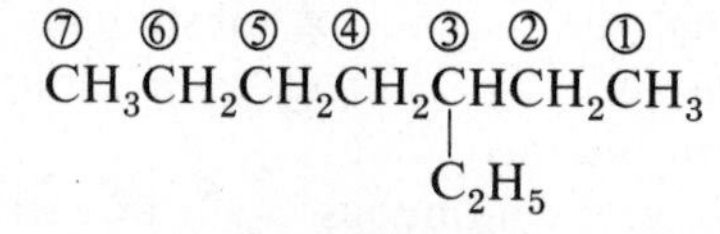
3-Ethylheptane

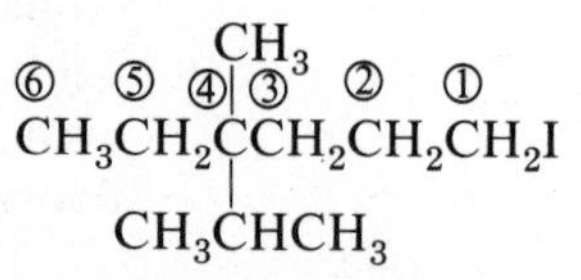
1-Iodo-4-methyl-4-isopropylhexane

According to IUPAC nomenclature, the longest continuous chain of carbon atoms is numbered to produce the lowest number or sum of numbers for the substituted atom(s) and/or group(s). For example, the six carbon atoms of the chain in 1-iodo-4-methyl-4-isopropylhexane (above) could be numbered from left to right, resulting in the name 3-methyl-3-isopropyl-6-iodohexane; however, the

Table 25-2 Some Alkyl Groups

Alkyl Group	Structure
Methyl	CH_3-
Ethyl	CH_3CH_2-
n-Propyl	$CH_3CH_2CH_2-$
Isopropyl	$CH_3\overset{\mid}{C}HCH_3$
n-Butyl	$CH_3CH_2CH_2CH_2-$
sec-Butyl (where sec stands for secondary)	$CH_3CH_2\overset{\mid}{C}HCH_3$
Isobutyl	CH_3CHCH_2- $\quad\underset{\mid}{CH_3}$
t-Butyl (where t stands for tertiary)	$CH_3\overset{\mid}{C}CH_3$ $\quad\underset{\mid}{CH_3}$

latter gives a sum of 12 for the numbers of the substituents, whereas the preferred numbering from right to left gives a sum of 9.

Cycloalkanes are named in the same way as alkanes. The carbons in the ring are numbered, with a substituted carbon being the first.

1-Chloro-2-isobutylcyclohexane

25.13 Alkenes

Hydrocarbon molecules that contain a double bond are members of another series called **alkenes.** The two carbon atoms linked by a double bond are bound together by two bonds, one σ bond and one π bond (Section 7.8). The alkenes have the general molecular formula C_nH_{2n}.

Ethene, C_2H_4, commonly called **ethylene,** is the simplest alkene. Each carbon atom in ethene has a trigonal planar structure, as illustrated in Fig. 7-19 in Section 7.8. The second member of the series is **propene (propylene);** the next members are the butene isomers. The series is analogous to the alkane series.

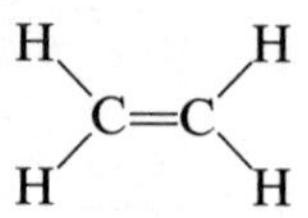

Ethene
(ethylene)

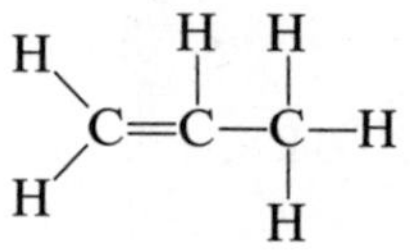

Propene
(propylene)

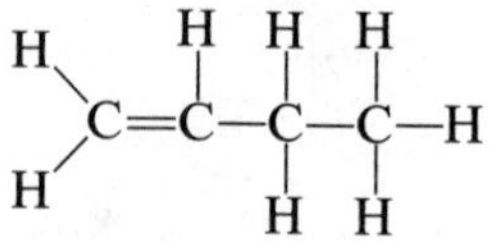

1-Butene

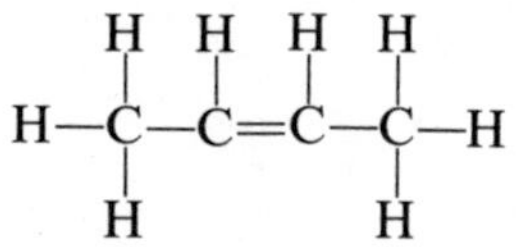

2-Butene

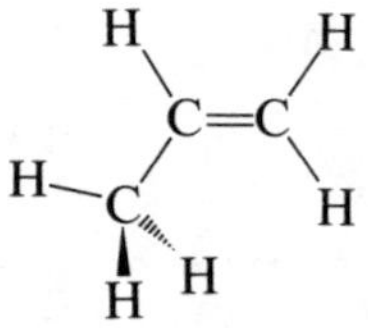

Figure 25-11. The molecular structure of propene.

The IUPAC name of an alkene is derived from that of the alkane with the same number of carbon atoms. The presence of the double bond in the chain is indicated by replacing the suffix -*ane* with the suffix -*ene*. The location of the double bond in the chain is indicated by a number that specifies the position of the first carbon atom in the double bond.

The carbon atoms involved in single bonds in alkenes have sp^3 hybridization, and those in double bonds have sp^2 hybridization. Thus the geometry around the carbon atoms at the double bonds is trigonal planar. The molecular structure of propene is illustrated in Fig. 25-11.

Carbon atoms are free to rotate around a single bond but not around a double bond; a double bond is quite rigid. This makes it possible to have two separate isomers of 2-butene, one with both methyl groups on the same side of the double bond and one with the methyl groups on opposite sides. When formulas of butene are written with 120° bond angles around the doubly bonded carbon atoms, the isomers are apparent.

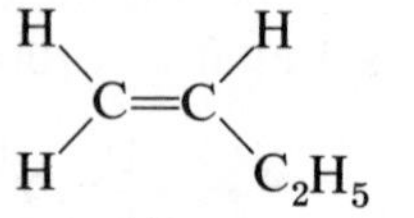

1-Butene

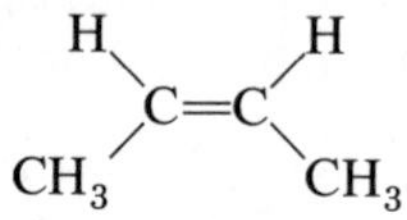

cis isomer 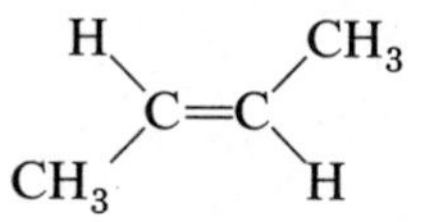   /   *trans* isomer

2-Butene

The 2-butene isomer in which the two methyl groups are on the same side is called the *cis* isomer; the one in which the two methyl groups are opposite is called the *trans* isomer. In these **geometrical isomers,** the same types of atoms are attached to each other, but the geometries of the two molecules differ. The different geometries produce different properties that make separation of the isomers possible.

Alkenes are much more reactive than alkanes. A π bond, being a relatively weaker bond, is disrupted much more easily than a σ bond. Thus the reaction characteristic of alkenes is a type in which the π bond is broken and replaced by two σ bonds. Such a reaction is called an **addition reaction.**

In the presence of a suitable catalyst (Pt, Pd, or Ni, for example), alkenes undergo an addition reaction with hydrogen to form the corresponding alkanes.

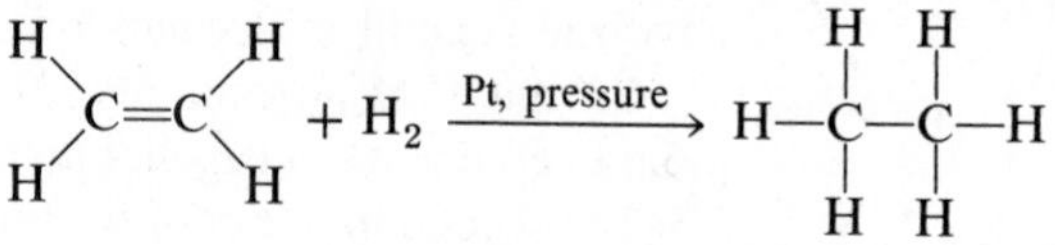

Ethene

Ethane

Chlorine adds to the double bond in an alkene, instead of replacing a hydrogen as it does in an alkane.

$$\underset{\text{Ethene}}{\overset{\text{H}}{\underset{\text{H}}{>}}C=C\overset{\text{H}}{\underset{\text{H}}{<}}} + Cl_2 \longrightarrow \underset{\text{1,2-Dichloroethane}}{H-\overset{\overset{\text{H}}{|}}{\underset{\underset{\text{Cl}}{|}}{C}}-\overset{\overset{\text{H}}{|}}{\underset{\underset{\text{Cl}}{|}}{C}}-H}$$

Many other reagents react with the double bond in alkenes. An example is the acid-catalyzed addition of water to an alkene.

$$\underset{\text{Ethene}}{\overset{\text{H}}{\underset{\text{H}}{>}}C=C\overset{\text{H}}{\underset{\text{H}}{<}}} + H_2O \xrightarrow[\text{catalyst}]{\text{Acid}} \underset{\text{Ethanol}}{H-\overset{\overset{\text{H}}{|}}{\underset{\underset{\text{H}}{|}}{C}}-\overset{\overset{\text{H}}{|}}{\underset{\underset{\text{OH}}{|}}{C}}-H}$$

Alkenes can add to themselves. This very important addition reaction gives polymers.

$$n\left(\overset{\text{H}}{\underset{\text{H}}{>}}C=C\overset{\text{H}}{\underset{\text{H}}{<}}\right) \xrightarrow{\text{Catalyst}} \left(\overset{\overset{\text{H}\ \text{H}}{|\ \ |}}{\underset{\underset{\text{H}\ \text{H}}{|\ \ |}}{C-C}}\right)_n \quad \text{(where } n \text{ is large)}$$

Polyethylene

Most of the ethylene (ethene) used to make polyethylene and in other industrial reactions is produced from alkanes by cracking operations. The preparation of ethylene from ethane is described in Section 15.14. Ethylene is a basic raw material used in the polymer, petrochemical, and plastics industries. Over 12 million tons of ethylene were produced in the United States in 1982. Only five chemicals were produced in larger amounts.

25.14 Alkynes

Hydrocarbon molecules with a triple bond are called **alkynes;** they make up another series of unsaturated hydrocarbons. Two carbon atoms joined by a triple bond are bound together by one σ bond and two π bonds (Section 7.8). The alkynes have the general molecular formula C_nH_{2n-2}.

The simplest and most important member of the alkyne series is **ethyne, C_2H_2,** commonly called **acetylene.** The Lewis structure for ethyne is

$$H-C\equiv C-H$$

Ethyne
(acetylene)

The bonding in the ethyne molecule, which has a linear structure, is illustrated in Fig. 7-22 in Section 7.8.

The IUPAC nomenclature for alkynes is similar to that for alkenes except that the suffix *-yne* is used to indicate a triple bond in the chain.

$$\overset{④\quad③\quad②\ ①}{CH_3CH_2C\equiv CH} \qquad \overset{⑥\quad⑤\quad④\ ③②\ ①}{CH_3CH_2C\equiv CCHCH_3}$$
$$\underset{\qquad\qquad\qquad\qquad\qquad\qquad CH_3}{|}$$

1-Butyne   2-Methyl-3-hexyne

Chemically, the alkynes are similar to the alkenes, except, having two π bonds, they react even more readily, adding twice as much reagent in addition reactions. The reaction of acetylene with bromine is a typical example.

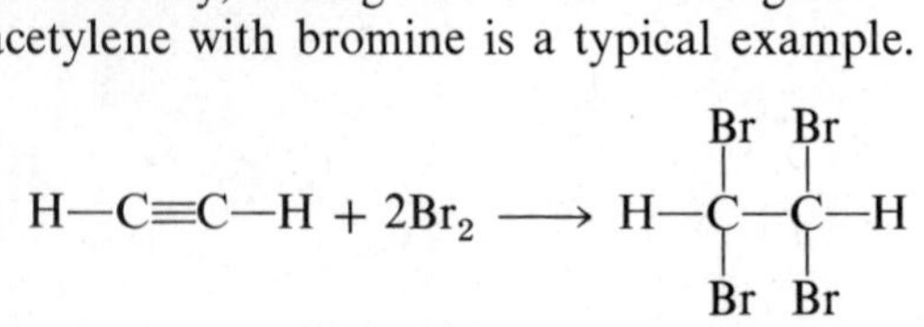

$$H-C\equiv C-H + 2Br_2 \longrightarrow H-\overset{\displaystyle Br}{\underset{\displaystyle Br}{\overset{|}{\underset{|}{C}}}}-\overset{\displaystyle Br}{\underset{\displaystyle Br}{\overset{|}{\underset{|}{C}}}}-H$$

Tetrabromoethane

Acetylene and all the other alkynes burn very easily.

$$2H-C\equiv C-H(g) + 5O_2(g) \longrightarrow 4CO_2(g) + 2H_2O(g) \qquad \Delta H° = -2511.2 \text{ kJ}$$

The flame from burning acetylene is very hot; it is used in welding and cutting metal. When acetylene burns, some of it breaks down a second way according to the equation

$$2C_2H_2 + O_2 \longrightarrow 4C + 2H_2O$$

At the temperature of the flame, the particles of carbon give off a brilliant white light. At one time acetylene was used to light homes and in bicycle and automobile lamps; acetylene lamps are still used by some cavers.

25.15 Aromatic Hydrocarbons

Benzene, C_6H_6, is the simplest member of a large family of hydrocarbons called **aromatic hydrocarbons.** The benzene molecule contains a hexagonal ring of sp^2 hybridized carbon atoms with the unhybridized p orbital of each one perpendicular to the ring. The electronic structure of benzene can be described as a resonance hybrid of two Lewis structures.

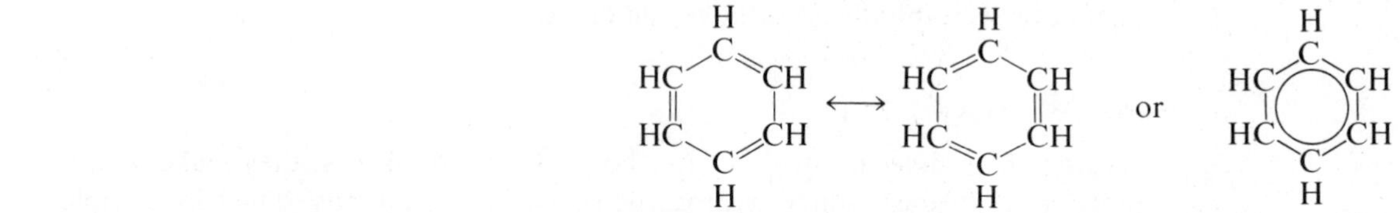

Three valence electrons in the sp^2 hybrid orbitals of each carbon atom and the valence electron of each hydrogen atom form the framework of sigma bonds in the benzene molecule (Fig. 25-12). The unhybridized carbon p orbitals, each with one electron, combine to form π bonds. As shown in Figs. 25-13 and 25-14, the electrons in the π bonds are delocalized around the ring, because the electronic structure is the average of the resonance forms. Since the π electrons are delocalized, benzene does not have alkene character. Each bond between two carbon atoms is neither a single nor a double bond but is intermediate in character (a hybrid). Each bond is equivalent to the others and has a bond order of $1\frac{1}{2}$. Although benzene is often written as one of its resonance forms in chemical formulas, it is important for you to remember that all of the carbon-carbon bonds in benzene are equivalent. The structure of benzene is sometimes written with a circle within the ring (right-hand structure above) to emphasize the fact that the bonds are equivalent and the electrons are delocalized.

The π bonding in benzene can be described by the molecular orbital model. The six unhybridized p orbitals can be combined to give three π bonding molecu-

Figure 25-12. The overlap of sp^2 hybrid orbitals of the carbon atoms and s orbitals of the hydrogen atoms that form σ bonds in a benzene ring.

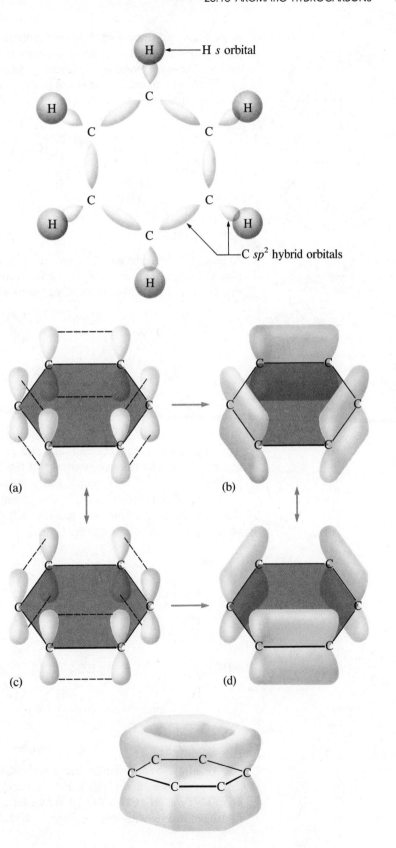

Figure 25-13. The π overlap in a benzene molecule. Hydrogen atoms have been omitted for clarity. The six sp^2 hybridized carbon atoms lie in a plane. The six unhybridized p orbitals, shown in color in (a) and (c), extend above and below the plane and are perpendicular to it. The dashed lines between lobes of p orbitals indicate side-by-side overlap. The overlap can occur in either direction between adjacent lobes as shown in (a) and (c). The resulting π orbitals are shown in (b) and (d), with the orbital portion of each (in color) lying above and below the hexagonal plane. The resonance forms in (a) and (c) and in (b) and (d) are exactly equivalent. The true structure is an average (hybrid) of the two (see Fig. 25-14).

Figure 25-14. The distribution of π electrons in a benzene molecule.

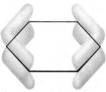

Figure 25-15. The three filled π bonding molecular orbitals in benzene.

lar orbitals and three π antibonding molecular orbitals. The six electrons (as three pairs) from the unhybridized p orbitals occupy the three π bonding molecular orbitals. These π bonding molecular orbitals are shown in Fig. 25-15.

There are many derivatives of benzene. The hydrogen atoms can be replaced by many different species or groups. The following are typical examples.

Toluene   Styrene   Bromobenzene

Xylene is a derivative of benzene in which two hydrogen atoms have been replaced by methyl groups. Xylene has three isomers since the two methyl groups can occupy three different relative positions on the ring.

o-Xylene   *m*-Xylene   *p*-Xylene

Any disubstituted benzene can take one of these three structurally different (isomeric) forms. The relative positions of the two substituents are indicated by the prefixes *o*- (*ortho*-), *m*- (*meta*-), and *p*- (*para*-).

Some important aromatic hydrocarbons and their derivatives contain more than one ring. Examples are naphthalene, $C_{10}H_8$, found in moth balls, anthracene, $C_{14}H_{10}$, and benzpyrene, $C_{20}H_{12}$.

Naphthalene   Anthracene   Benzpyrene

Aromatic hydrocarbons are widely used in manufacturing dyes, synthetic drugs, explosives, and plastics. For example, the important explosive trinitrotoluene (TNT), $C_6H_2(CH_3)(NO_2)_3$, is synthesized from toluene by replacing three hydrogen atoms with nitro groups, $-NO_2$.

$$CH_3$$

Trinitrotoluene

25.16 Isomerism

Isomers are groups of atoms with identical chemical formulas but different arrangements of the atoms. Thus far we have seen two kinds of isomerism: structural and geometrical. Structural isomers (Section 25.11) are molecules with different arrangements of bonds, such as *n*-butane and 2-methylpropane or the three isomers of xylene (Section 25.15). Geometrical isomers have the same atoms bonded in the same order, but some parts of the two molecules have different relative spatial arrangements. In *cis-trans* isomers of organic compounds, for example, groups are arranged either on the same side (*cis-*) or on different sides (*trans-*) of a bond that cannot rotate (Section 25.13).

A third type of isomers, called **optical isomers,** are detected by using polarized light. Optical isomers are detected by their effect on polarized light. Ordinary light is composed of rays that vibrate in all planes parallel to the direction in which the light travels. When ordinary light is passed through a polarizer, such as is used in Polaroid sunglasses or a Nicol prism, only light that vibrates in one plane is transmitted; such light is called **plane-polarized light.** The direction of the plane in which the light vibrates can be detected using another polarizer. When a beam of plane-polarized light passes through a solution containing a single form of an optical isomer, the plane of vibration of the light is rotated. Molecules that rotate the plane of vibration of plane-polarized light are **optically active molecules.** The amount of rotation and thus the optical activity vary from compound to compound.

A compound that forms optical isomers exists as at least two different forms. One of these isomers rotates plane-polarized light clockwise; such an isomer is called a **dextrorotatory isomer,** or a $+$ isomer. The other isomer rotates plane-polarized light counterclockwise by an amount exactly equal to the clockwise rotation of the $+$ isomer. The second isomer is called a **levorotatory isomer,** or a $-$ isomer. A mixture of equal amounts of a dextrorotatory isomer and a levorotatory isomer of the same substance shows no rotation of plane-polarized light because the effects of the two isomers cancel. In most properties, such as melting point, boiling point, and solubility, optical isomers exhibit identical characteristics. However, optical isomers may exhibit different degrees of reactivity in reactions with other optically active compounds.

Optical isomers of a compound have the same relationship to each other as that of a right hand to a left hand. Both have the same shape and the same order of attachment of the various parts to one another; they are almost alike. However, they differ in that they are mirror images of each other. Just as your right and left hands cannot be superimposed, two optical isomers of a molecule cannot be superimposed, no matter how they are rotated. Figure 25-16 illustrates the mirror relationship for a hand, for the molecule CHClBrI, and for the amino acid

Figure 25-16. Mirror images of a hand, the molecule CHClBrI, and the molecule alanine. Although these objects have the same composition and are the same size and shape, their mirror images cannot be superimposed.

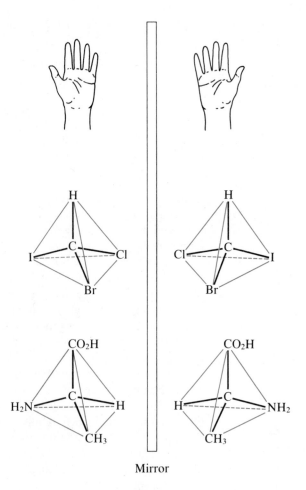

Mirror

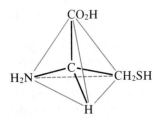

Figure 25-17. The structure of L-cysteine.

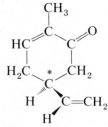

Figure 25-18. The molecule of carvone. The asymmetric carbon is indicated by the asterisk.

alanine. Note that simple molecules that contain sp^3 hybridized carbon atoms form optical isomers only when a carbon atom is bonded to four different groups; such a carbon atom within a molecule is referred to as an **asymmetric carbon.**

Practically all optically active molecules found in nature are found as only one of the two possible isomers. For example, the amino acids found in proteins (Chapter 26) are all of the type designated as L. The structure of L-cysteine, one of these amino acids, is shown in Fig. 25-17.

A striking example of the effect of optical activity is found in the oils responsible for the scents of spearmint and caraway seeds. Both oils are composed of carvone (Fig. 25-18). However, caraway seed oil contains D-carvone, and spearmint oil contains the other optical isomer, L-carvone.

25.17 Petroleum

Petroleum is composed chiefly of saturated hydrocarbons, but may also contain unsaturated hydrocarbons, aromatic hydrocarbons, and their derivatives. The main step in refining petroleum is distilling it into a number of fractions (fractional distillation; Section 13.21), each of which is a complex mixture of hydrocarbons with properties that make it valuable. Additional purification is usually necessary before these fractions can be used. Some of the more important hydrocarbon products from the refining of petroleum are listed in Table 25-3.

Table 25-3 Some Hydrocarbon Products from Petroleum

	Approximate Composition	Boiling Point Range, °C	Uses
Petroleum ether	C_4 to C_{10}	35–80	Solvent
Gasoline	C_4 to C_{13}	40–225	Motor fuel
Kerosene	C_{10} to C_{16}	175–300	Fuel, lighting
Lubricating oils	C_{20} up	350 up	Lubrication
Paraffin	C_{23} to C_{29}	50–60 (m.p.)	Candles, waxed paper
Asphalt		Viscous liquids	Paving, roofing
Coke		Solid	Fuel

Gasoline, a mixture of n-hexane, n-heptane, n-octane, and their isomers, is rated on an arbitrary scale in which isooctane (2,2,4-trimethylpentane), once thought to be the ideal fuel for gasoline engines, is given a rating, or octane number, of 100. Gasolines with octane numbers less than 100 are less efficient than isooctane, and those with higher numbers, more efficient. Until recently, gasoline was a blend of aliphatic (open-chain) hydrocarbons and aromatic hydrocarbons, and tetraethyl lead. However, manufacturers are now making more gasoline without the tetraethyl lead because of the toxic effects of lead compounds and their poisoning effect on catalytic converters (Sections 22.4 and 22.5). Gasoline can be produced without tetraethyl lead by blending in more highly branched hydrocarbon isomers, methanol, and ethanol. At the same time, engines are being designed to run on lower-octane gasolines.

DERIVATIVES OF HYDROCARBONS

25.18 Introduction

Hydrocarbon derivatives such as chloroethane, ethanol, and acetic acid are formed by replacing one or more hydrogen atoms of a hydrocarbon by other atoms or groups of atoms, called functional groups. An alcohol is a derivative of a hydrocarbon that contains an —OH group in place of a hydrogen atom. Thus methanol, CH_3OH [Fig. 25-19(a)], is a derivative of methane; ethanol, C_2H_5OH [Fig. 25-19(b)], is a derivative of ethane.

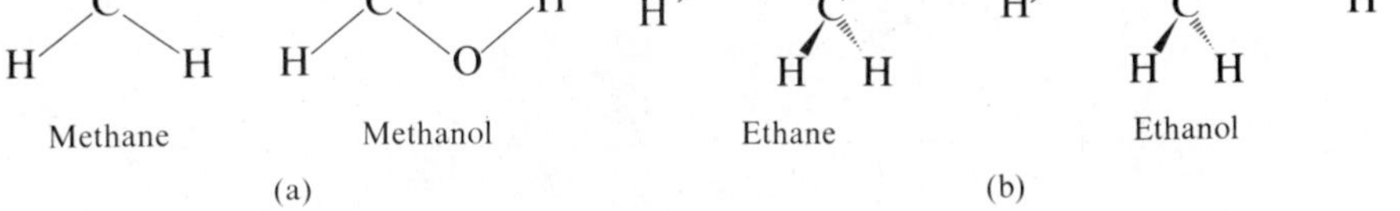

Figure 25-19. (a) The molecular structures of methane and methanol. (b) The molecular structures of ethane and ethanol.

Methane   Methanol   Ethane   Ethanol

(a)   (b)

Table 25-4 lists several important types of compounds derived from hydrocarbons, the corresponding functional groups, and the general formulas. In the general formulas, R stands for the portion of the molecule other than the functional group; R might represent an alkyl group (Section 25.12) or a phenyl group, $C_6H_5—$, for example.

The following sections describe the properties of some specific hydrocarbon derivatives. Their functional groups should be noted carefully.

Table 25-4 Several Types of Derivatives of Hydrocarbons

Derivative Type	Functional Group	General Formula	Typical Examples
Alcohols	—OH	R—OH	CH_3CH_2OH Ethanol (a primary alcohol) OH | CH_3CHCH_3 2-Propanol (a secondary alcohol) OH | CH_3CCH_3 | CH_3 2-Methyl-2-propanol (a tertiary alcohol)
Ethers	—O—	R—O—R	$CH_3—O—C_2H_5$ Methyl ethyl ether $C_2H_5—O—C_2H_5$ Diethyl ether
Amines	| —N—	H / R—N \\ H R | R—N—H R | R—N—R	CH_3NH_2 Methylamine (a primary amine) $(C_2H_5)_2NH$ Diethylamine (a secondary amine) $(C_3H_7)_3N$ Tri-*n*-propylamine (a tertiary amine)
Thiols	—SH	R—SH	C_2H_5SH Ethanethiol
Aldehydes	O ‖ —CH	O ‖ R—CH	O ‖ HCH Formaldehyde O ‖ CH_3CH Acetaldehyde

Table 25-4 (continued)

Derivative Type	Functional Group	General Formula	Typical Examples	
Ketones	$-\overset{\overset{O}{\|\|}}{C}-$	$R-\overset{\overset{O}{\|\|}}{C}-R$	$CH_3-\overset{\overset{O}{\|\|}}{C}-CH_3$ Dimethyl ketone	$CH_3-\overset{\overset{O}{\|\|}}{C}-C_6H_5$ Methyl phenyl ketone
Acids	$-\overset{\overset{O}{\|\|}}{C}OH$	$R\overset{\overset{O}{\|\|}}{C}OH$	$CH_3\overset{\overset{O}{\|\|}}{C}OH$ Acetic acid	$C_6H_5\overset{\overset{O}{\|\|}}{C}OH$ Benzoic acid
Esters	$-\overset{\overset{O}{\|\|}}{C}OR$	$R\overset{\overset{O}{\|\|}}{C}-OR$	$CH_3\overset{\overset{O}{\|\|}}{C}OC_2H_5$ Ethyl acetate	$C_3H_7\overset{\overset{O}{\|\|}}{C}OC_2H_5$ Ethyl butyrate
Salts	$-\overset{\overset{O}{\|\|}}{C}O^-M^+$	$R\overset{\overset{O}{\|\|}}{C}O^-M^+$	$CH_3\overset{\overset{O}{\|\|}}{C}O^-Na^+$ Sodium acetate	$C_2H_5\overset{\overset{O}{\|\|}}{C}O^-K^+$ Potassium pro-pionate
Amides	$-\overset{\overset{O}{\|\|}}{C}-NH_2$	$R\overset{\overset{O}{\|\|}}{C}NH_2$	$CH_3\overset{\overset{O}{\|\|}}{C}NH_2$ Acetamide	$C_6H_5\overset{\overset{O}{\|\|}}{C}NH_2$ Benzamide

25.19 Alcohols, R—OH

All **alcohols** have one or more —OH functional groups, yet they do not behave like bases such as sodium hydroxide and potassium hydroxide. They do not form hydroxide ions in water, nor do they have the other usual properties of bases.

Methanol, or **methyl alcohol, CH_3OH,** is produced commercially from either carbon monoxide or carbon dioxide.

$$CO + 2H_2 \xrightarrow{\text{Ag or Cu}} CH_3OH$$
$$CO_2 + 3H_2 \longrightarrow CH_3OH + H_2O$$

It is a colorless liquid, which boils at 65°C. It smells and tastes like ethyl alcohol. However, methanol is poisonous—breathing the vapor or drinking the liquid may cause blindness or death. Methanol is used in the manufacture of formaldehyde and other organic products; as a solvent for resins, gums, and shellac; and to denature ethyl alcohol (make it unsafe to drink).

Ethanol, C_2H_5OH, called **ethyl alcohol, grain alcohol,** or simply **alcohol,** is the most important of the alcohols. It has long been prepared by fermentation of starch, cellulose, and various sugars.

$$C_6H_{12}O_6 \xrightarrow[\text{yeast}]{\text{Enzymes in}} 2C_2H_5OH + 2CO_2(g)$$

Glucose   Ethanol

Large quantities of ethanol are synthesized from ethylene or acetylene. The synthesis from ethylene is summarized by the equation

$$
\underset{\text{H}}{\overset{\text{H}}{\text{C}}}=\underset{\text{H}}{\overset{\text{H}}{\text{C}}}-\text{H} + \text{HOH} \xrightleftharpoons{\text{H}^+} \underset{\text{H}}{\overset{\text{H}}{\text{C}}}-\underset{\text{OH}}{\overset{\text{H}}{\text{C}}}-\text{H}
$$

Ethanol is a colorless liquid with a characteristic and somewhat pleasant odor. It is miscible with water in all proportions. The boiling point of the pure alcohol is 78.37°C. It forms a constant-boiling mixture with water that contains 95.57% alcohol by weight and boils at 78.15°C. Ethanol is the least toxic of all the alcohols and is found in all alcoholic beverages. It is used as a solvent in tinctures, essences, and extracts; in the preparation of iodoform, ether, dyes, perfumes, and collodion; as a solvent in the lacquer industry; and to some extent as a motor fuel. It is currently used as an additive to gasoline, producing **gasohol** (now more commonly called **superunleaded gasoline**).

Alcohols containing two or more hydroxyl groups can be made. Important examples are 1,2-ethanediol (ethylene glycol), $C_2H_4(OH)_2$, and 1,2,3-propanetriol (glycerol), $C_3H_5(OH)_3$.

Ethylene glycol   Glycerol

Ethylene glycol is used as a solvent and as an antifreeze in automobile radiators. Glycerol (also called glycerin) is produced, along with soap, when either fats or oils are heated with an alkali metal hydroxide.

An alcohol's name comes from the hydrocarbon from which it is derived. The hydrocarbon's final -*e* is replaced by -*ol*, and the carbon atom to which the —OH group is bonded is indicated by a number placed before the name.

25.20   Ethers, R—O—R

Ethers are compounds obtained from alcohols by the elimination of a molecule of water from two molecules of the alcohol. For example, when ethanol is treated with a limited amount of sulfuric acid and heated to 140°C, **diethyl ether** (ordinary ether) and water are formed.

Diethyl ether

In the general formula for ethers, R—O—R, the hydrocarbon groups (R) may be the same or different. Diethyl ether is the most important compound of this class.

It is a colorless volatile liquid (boiling point = 35°C), which is highly flammable. It has been used since 1846 as an anesthetic. Diethyl ether and other ethers are valuable solvents for gums, fats, waxes, and resins.

25.21 Aldehydes, R—C—H

Alcohols represent the first stage of oxidation of hydrocarbons. Further oxidation produces **aldehydes,** compounds containing the group —CHO. Aldehydes are generally made by the oxidation of primary alcohols. When a mixture of methanol and air is passed through a heated tube containing either silver or a mixture of iron powder and molybdenum oxide, the simplest aldehyde, called **formaldehyde, HCHO,** is formed.

$$2H\overset{\underset{\displaystyle H}{|}}{\underset{\underset{\displaystyle H}{|}}{C}}{-}OH + O_2 \xrightarrow{\text{Catalyst}} 2H-C\overset{\displaystyle H}{\underset{\displaystyle O}{\diagdown}} + 2H_2O$$

Methanol   Formaldehyde

Formaldehyde is a colorless gas with a pungent and irritating odor; it is soluble in water in all proportions. Formaldehyde is sold in an aqueous solution called **formalin,** which contains about 37% formaldehyde by weight. The solution also contains 7% methanol, added to inhibit the reaction of formaldehyde molecules with one another to form an insoluble polymer (a compound of high molecular weight). The formaldehyde polymer Bakelite makes formaldehyde industrially important. Formaldehyde causes coagulation of proteins, so it is used to preserve tissue specimens and to embalm. It is also useful as a disinfectant and as a reducing agent in the production of silvered mirrors.

Acetaldehyde, CH₃CHO, boils at 20.2°C. It is colorless, water-soluble, and smells like freshly cut green apples. It is used in the manufacture of aniline dyes, synthetic rubber, and other organic materials.

The systematic names of aldehydes are formed by replacing the *-e* ending of the parent alkane by *-al.* Acetaldehyde is the common name for ethanal. The chain is numbered with the —CHO group as the first carbon. Thus 3-methylbutanal is

$$\overset{④}{CH_3}{-}\overset{\overset{\displaystyle CH_3}{|}}{\underset{③}{CH}}{-}\overset{②}{CH_2}{-}\overset{①}{C}\overset{\displaystyle O}{\underset{\displaystyle H}{\diagup}}$$

25.22 Ketones, R—C—R

Ketones, like aldehydes, contain the **carbonyl group,** $>C{=}O$. In fact, a ketone may be regarded as an aldehyde in which the hydrogen in the aldehyde group is replaced by a hydrocarbon group. In the general structural formula for ketones (see section heading) the R groups may be the same or different. Ketones are often prepared by the oxidation of secondary alcohols.

Dimethyl ketone, CH₃COCH₃, commonly called **acetone,** is the simplest and most important ketone. It is made commercially by fermenting corn or molasses or by dehydrogenation of 2-propanol (isopropyl alcohol).

$$CH_3\overset{\displaystyle \overset{OH}{|}}{C}HCH_3 \xrightarrow[\text{Cu catalyst}]{350°C} CH_3\overset{\displaystyle \overset{O}{\|}}{C}CH_3 + H_2$$

2-Propanol   Dimethyl ketone

It is also a product of the destructive distillation of wood. Acetone is a colorless liquid, boiling at 56.5°C and possessing a characteristic pungent odor and a sweet taste. Among the many uses of acetone are as a solvent for cellulose acetate, cellulose nitrate, acetylene, plastics, and varnishes; as a paint, varnish, and finger-nail polish remover; and as a solvent in the manufacture of drugs, chemicals, smokeless powder, and the explosive cordite.

25.23 Carboxylic Acids, $R-\overset{\displaystyle \overset{O}{\|}}{C}-OH$

Carboxylic acids contain the **carboxyl group, —CO₂H.** They are usually weak acids, yet readily form metal salts. Carboxylic acids may be prepared by the oxidation of primary alcohols or of aldehydes.

$$RCH_2OH \xrightarrow{\text{Oxidation}} R-C\overset{\displaystyle H}{\underset{\displaystyle O}{\diagdown}} \xrightarrow{\text{Oxidation}} R-C\overset{\displaystyle OH}{\underset{\displaystyle O}{\diagdown}}$$

Primary alcohol   Aldehyde   Carboxylic acid

The simplest carboxylic acid is **formic acid, HCO₂H,** known since 1670. Its name comes from the Latin word *formicus,* for ant; it was first isolated by the distillation of red ants. It is partially responsible for the irritation of ant bites and bee stings.

Acetic acid, CH₃CO₂H, constitutes 3–6% of vinegar. Cider vinegar is produced by allowing apple juice to ferment, which changes the sugar present to ethanol. Then bacteria produce an enzyme (Chapter 26) that catalyzes the air oxidation of the alcohol to acetic acid. Pure, anhydrous acetic acid, sometimes called **glacial acetic acid,** boils at 118.1°C and freezes at 16.6°C. Pure acetic acid has a penetrating odor and produces painful burns. It is an excellent solvent for many organic and some inorganic compounds. Acetic acid is essential in the production of cellulose acetate, a component of many synthetic fibers such as rayon.

Oxalic acid is a dicarboxylic acid with the formula (CO₂H)₂, or H₂C₂O₄. Its molecule consists of two carboxyl groups (dicarboxylic) bonded together.

$$\underset{O}{\overset{HO}{\diagdown}}C-C\underset{O}{\overset{OH}{\diagup}}$$

Oxalic acid

Oxalic acid is a colorless crystalline solid, found as a metal hydrogen salt in sorrel, rhubarb, and other plants, to which it imparts a sour taste. In concentrated form it is poisonous.

Benzoic acid, $C_6H_5CO_2H$, is a colorless crystalline solid with a cyclic structure. It is a monocarboxylic acid that occurs in cranberries and coal tar. The sodium salt of benzoic acid, **sodium benzoate,** is used to preserve foods such as tomato ketchup and fruit juices.

Benzoic acid

25.24 Esters, $R-\overset{\overset{\displaystyle O}{\|}}{C}-OR$

Esters are produced by the reaction of acids with alcohols. For example, the ester **ethyl acetate, $CH_3CO_2C_2H_5$,** is formed when acetic acid reacts with ethanol.

The distinctive and attractive odors and flavors of many flowers and ripe fruits are due to the presence of one or more esters. Among the most important of the natural esters are fats (such as lard, tallow, and butter) and oils (such as linseed, cottonseed, and olive oils), which are esters of the trihydroxyl alcohol (triol) glycerol, $C_3H_5(OH)_3$, with large acids such as palmitic acid, stearic acid, and oleic acid.

POLYMERS

25.25 Introduction

Polymers and **plastics,** compounds of very high molecular weights, are built up of a large number of simple molecules, or **monomers,** which have reacted with one another. Cellulose, starch, proteins, and rubber are natural polymers (although most rubber in use is a synthetic polymer). Nylon, rayon, polyethylene, and Dacron are synthetic polymers. Polymers are used in clothing, fibers, insulation, and construction materials; and recently, in cars to save weight, and in turn, gasoline. However, polymers are themselves made from petroleum, so part of the advantage is offset.

25.26 Rubber

Natural rubber comes mainly from latex, the sap of the rubber tree. Rubber consists of very long molecules, which are polymers formed by the union of isoprene units, C_5H_8.

$$\underset{\text{Isoprene}}{CH_2{=}\overset{\overset{\displaystyle CH_3}{|}}{C}{-}CH{=}CH_2} + \underset{\text{Isoprene}}{CH_2{=}\overset{\overset{\displaystyle CH_3}{|}}{C}{-}CH{=}CH_2} \Bigg\downarrow$$

$$\cdots{-}CH_2{-}\overset{\overset{\displaystyle CH_3}{|}}{C}{=}CH{-}CH_2{\vdots}CH_2{-}\overset{\overset{\displaystyle CH_3}{|}}{C}{=}CH{-}CH_2\cdots$$

The number of isoprene units in a rubber molecule is about 2000, giving it a molecular weight of approximately 136,000.

Rubber has the undesirable property of becoming sticky when warmed, but this can be eliminated by **vulcanization.** Rubber is vulcanized by heating it with sulfur to about 140°C. During the process sulfur atoms add at the double bonds in the linear polymer and form bridges that bind one rubber molecule to another. In this way, a linear polymer is converted into a three-dimensional polymer. During vulcanization, fillers are added to increase the wearing qualities of the rubber and to color it. Among the substances used as fillers are carbon black (black), zinc oxide (white), antimony(V) sulfide (orange), and titanium dioxide (white).

Synthetic rubbers resemble natural rubber and are often superior to it in certain respects. For example, neoprene is a synthetic elastomer with rubberlike properties.

$$n CH_2{=}CH{-}\overset{\overset{\displaystyle Cl}{|}}{C}{=}CH_2 \xrightarrow{\text{Polymerization}} \left[{-}CH_2{-}CH{=}\overset{\overset{\displaystyle Cl}{|}}{C}{-}CH_2{-}\right]_n$$

Chloroprene   Neoprene

The monomer chloroprene is similar to isoprene, except that a chlorine atom replaces the methyl group. Neoprene is more elastic than natural rubber, resists abrasion well, and is less affected by oil and gasoline. It is used for making gasoline and oil hoses, automobile and refrigerator parts, and electrical insulation. There are many other synthetic elastomers of this type.

Both natural rubber and neoprene are examples of **addition polymers,** polymers that form by addition reactions (Section 25.13).

25.27 Synthetic Fibers

The synthetic fiber nylon is a **condensation polymer.** Nylon is made from hexamethylenediamine, which contains an amine group (—NH$_2$) at both ends, and adipic acid, which contains a carboxylic acid group (—CO$_2$H) at both ends. During condensation linkages of the type R—NH—CO—R are formed, and water is eliminated. The part shown in color is an **amide linkage.**

$$\underset{\text{Hexamethylenediamine}}{NH_2{-}(CH_2)_6{-}\overset{\overset{\displaystyle H}{|}}{N}\boxed{H} + \boxed{HO}} \underset{\text{Adipic acid}}{OC{-}(CH_2)_4{-}CO_2H} \longrightarrow$$

$$NH_2{-}(CH_2)_6{-}NH{-}CO{-}(CH_2)_4{-}CO_2H + H_2O$$

Because the molecule resulting from the condensation has an $-NH_2$ group at one end and a $-CO_2H$ group at the other, the condensation process can be repeated many times to form a linear polymer of great length.

$$(n + 1)H_2N(CH_2)_6NH_2 + (n + 1)HO_2C(CH_2)_4CO_2H \longrightarrow$$

Hexamethylenediamine $\qquad\qquad\qquad$ Adipic acid

$$H_2N(CH_2)_6NH \left[\overset{\overset{\displaystyle O}{\|}}{C}(CH_2)_4\overset{\overset{\displaystyle O}{\|}}{C}NH(CH_2)_6NH \right]_n \overset{\overset{\displaystyle O}{\|}}{C}(CH_2)_4CO_2H + 2nH_2O$$

Nylon

Fine threads are produced by extruding melted nylon through a spinneret. Nylon is used in hosiery and other clothing, bristles for toothbrushes, surgical sutures, strings for tennis rackets, fishing line leaders, and many other products.

Dacron is made from ethylene glycol and terephthalic acid by a condensation process that forms an ester linkage. Dacron is one of the family of polyesters.

$$(n + 1)HO\overset{\overset{\displaystyle H}{|}}{\underset{\underset{\displaystyle H}{|}}{C}}\overset{\overset{\displaystyle H}{|}}{\underset{\underset{\displaystyle H}{|}}{C}}OH + (n + 1)HO_2C-C \underset{\underset{\displaystyle C=C}{\underset{\displaystyle H \quad H}{|}}}{\overset{\overset{\displaystyle H \quad H}{C-C}}{||}} C-CO_2H \longrightarrow$$

Ethylene $\qquad\qquad\qquad\qquad$ Terephthalic acid
glycol

$$HO\overset{\overset{\displaystyle H}{|}}{\underset{\underset{\displaystyle H}{|}}{C}}\overset{\overset{\displaystyle H}{|}}{\underset{\underset{\displaystyle H}{|}}{C}}O \left[\overset{\overset{\displaystyle O}{\|}}{C}-C \overset{\overset{\displaystyle H \quad H}{C-C}}{\underset{\underset{\displaystyle H \quad H}{C=C}}{||}} C-\overset{\overset{\displaystyle O}{\|}}{C}-O-\overset{\overset{\displaystyle H}{|}}{\underset{\underset{\displaystyle H}{|}}{C}}\overset{\overset{\displaystyle H}{|}}{\underset{\underset{\displaystyle H}{|}}{C}}-O \right]_n \overset{\overset{\displaystyle O}{\|}}{C}-C \overset{\overset{\displaystyle H \quad H}{C-C}}{\underset{\underset{\displaystyle H \quad H}{C=C}}{||}} C-CO_2H + 2nH_2O$$

Dacron

25.28 Polyethylene

Polyethylene is widely used. Polyethylene results from the polymerization of ethylene (Section 25.13). It is a flexible, tough polymer that is very water-resistant and has excellent insulating properties. It is used in plastic bottles, bags for fruits and vegetables, and many other items.

If the ethylene molecules contain substituents, the polymer will also contain the substituents.

$$nCH_2{=}CHR \xrightarrow{\text{Catalysts}} \left[CH_2-\underset{\underset{\displaystyle R}{|}}{CH} \right]_n$$

Polypropylene (R $=$ CH_3) is the polymer of propylene, $CH_3CH{=}CH_2$. **Polyvinylchloride,** or **PVC** (R $=$ Cl), is the polymer of vinyl chloride, $CH_2{=}CHCl$. **Teflon** results from the polymerization of tetrafluoroethylene, $CF_2{=}CF_2$; and **polystyrene** (R $=$ C_6H_5) is the polymer of styrene, $C_6H_5CH{=}CH_2$.

FOR REVIEW

SUMMARY

With the valence shell electron configuration $2s^2 2p^2$, **carbon** is the first member of Group IVA of the Periodic Table. Carbon generally completes its valence shell by sharing electrons in covalent bonds. Since the valence shell of a carbon atom can contain a maximum of eight electrons, hybridization is limited to sp, sp^2, and sp^3. Carbon forms typical inorganic compounds such as carbon monoxide, carbon dioxide (a weakly acidic nonmetal oxide), carbonates (salts containing the $CO_3{}^{2-}$ ion), carbon disulfide, cyanamides, and cyanides. Strong, stable bonds between carbon atoms produce complex molecules containing chains and rings. The chemistry of these compounds is called **organic chemistry.**

Hydrocarbons are organic compounds composed only of carbon and hydrogen. The **alkanes** are **saturated hydrocarbons;** that is, hydrocarbons that contain only single bonds. **Alkenes** are hydrocarbons that contain one or more carbon-carbon double bonds. **Alkynes** contain one or more carbon-carbon triple bonds. **Aromatic hydrocarbons** contain ring structures with delocalized π-electron systems.

Most hydrocarbons have **structural isomers,** compounds with the same chemical formula but different arrangements of atoms. For example, *n*-butane and isobutane are structural isomers. Many aromatic hydrocarbons also have structural isomers; *o*-xylene, *m*-xylene, and *p*-xylene are examples. In addition, molecules of the alkenes and alkynes exhibit isomerism based on the position of the multiple bond in the molecule. Most alkenes also exhibit *cis-trans* isomerism, which is the result of the lack of rotation about a carbon-carbon double bond. Certain organic compounds form **optical isomers,** which contain one or more **asymmetric carbons,** carbon atoms bonded to four different groups.

Alkanes undergo **substitution reactions** in which one of the hydrogen atoms of the alkane is replaced by another type of atom. Alkenes and alkynes undergo **addition reactions** in which a reactant adds to the molecule and converts the carbon-carbon π bond into σ bonds. All hydrocarbons react with oxygen and burn.

Organic compounds that are not hydrocarbons can be considered to be derivatives of hydrocarbons. A hydrocarbon derivative can be formed by replacing one or more hydrogen atoms of a hydrocarbon by a **functional group,** which contains at least one atom of an element other than carbon or hydrogen. The properties of hydrocarbon derivatives are determined largely by the functional group. The —OH group is the functional group of an **alcohol.** The —O— group is the functional group of an **ether.** Other functional groups include the —CHO group of an **aldehyde,** the —CO— group of a **ketone,** and the —CO$_2$H group of a **carboxylic acid.**

Polymers, compounds of very high molecular weights, form when a large number of smaller molecules (**monomers**) react with one another. **Addition polymers** such as rubber and polyethylene form by addition reactions, while **condensation polymers** such as nylon and the polyesters form by condensation reactions.

KEY TERMS AND CONCEPTS

Addition polymer (25.26)
Addition reaction (25.13)
Alcohols (25.19)
Aldehydes (25.21)
Alkanes (25.11)
Alkenes (25.13)
Alkyl group (25.12)
Alkynes (25.14)
Aromatic hydrocarbons (25.15)
Asymmetric carbon (25.16)
Carbonyl group (25.22)

Carboxyl group (25.23)
Carboxylic acid (25.23)
Condensation polymer (25.27)
Cycloalkanes (25.11)
Esters (25.24)
Ethers (25.20)
Functional group (25.11)
Geometrical isomers (25.13)
Ketones (25.22)
Monomers (25.25)
Optical isomers (25.16)

Optically active molecules
  (25.16)
Photosynthesis (25.5)
Polymers (25.25)
Primary carbon (25.12)
Secondary carbon (25.12)
Structural isomers (25.11)
Substitution reaction (25.11)
Tertiary carbon (25.12)

EXERCISES

Carbon

1. Describe the crystal structures of graphite and diamond, and relate the physical properties of each to their structures.
2. Compare and contrast the bonding in graphite, benzene, ethylene, and acetylene.
3. How is coal, a solid fuel, converted into a gaseous fuel?
4. What volume of water gas at 600 K and 750 torr would result from the action of 100 g of water on carbon?   *Ans. 554 L*

Inorganic Compounds of Carbon

5. Write the Lewis structure, including resonance forms where necessary, for each of the following molecules or ions: CS_2; CO_3^{2-}; CCl_4. Identify the hybridization of the carbon atom in each.
6. Write the equation describing the chemical reaction that occurs when carbon dioxide is bubbled through sodium hydroxide solution.
7. Explain the toxicity of carbon monoxide.
8. Compare the heat of combustion of carbon monoxide with that of carbon. Explain.
9. Write the electronic structure for carbon monoxide.
10. Write a balanced equation for the complete combustion of each of the following:
    (a) CH_4
    (b) CH_3OH
    (c) HCO_2H
    (d) C_2H_2

11. Write equations for the production of the following:
    (a) carbon disulfide
    (b) carbon tetrachloride
    (c) calcium carbide
    (d) acetylene
    (e) sodium cyanide
    (f) hydrogen cyanide
12. How much calcium carbonate would be formed from the addition of 10.0 L of carbon dioxide, measured at 27°C and 770 torr, to an excess of a solution of calcium hydroxide?   *Ans. 41.2 g*

Hydrocarbons

13. Write the chemical formula and Lewis structure of an alkane, an alkene, an alkyne, and an aromatic hydrocarbon, each of which contains six carbon atoms.
14. What is the difference between the electronic structures of saturated and unsaturated hydrocarbons?
15. Draw a three-dimensional structure for each of the following, using solid and striped wedge-shaped bonds, where appropriate:
    (a) CH_4
    (b) $n\text{-}C_4H_{10}$
    (c) C_3H_4
    (d) C_3H_6
16. Write Lewis structures for all of the isomers of

C_4H_9Br. Do any of these isomers exist as optical isomers?

17. Write Lewis structures for all of the isomers of the alkene C_5H_{10}.

18. Write Lewis structures for all of the isomers of the alkyne C_5H_8.

19. Write the Lewis structures and the names of all isomers of the alkyl groups $—C_3H_7$ and $—C_4H_9$.

20. Indicate which of the following molecules can form optical isomers:
    (a) 2-butanethiol
    (b) dichlorodifluoromethane
    (c) 1-cyanoethanol [$CH_3CH(CN)OH$]
    (d) *trans*-2-pentene
    (e) 3-ethyloctane
    (f) 4-ethyloctane

21. Draw three-dimensional structures, using solid and striped wedge-shaped bonds, for the optical isomers of phenylalanine, $H_2NCH(R)CO_2H$, where R is $C_6H_5CH_2$.

22. Write a Lewis structure for each of the following molecules:
    (a) propane
    (b) 3-ethyl-1-heptyne
    (c) *trans*-4-chloro-2-octene
    (d) 2,3,4-trimethylpentene
    (e) methylcyclohexane
    (f) *o*-dichlorobenzene
    (g) methyl ethyl ether
    (h) propanethiol
    (i) isopropyl alcohol

23. Give a complete name for each of the following compounds:
    (a) $CH_3CH_2CH_2CH{=}CHCH_3$
    (b) $CH_3CH_2CHBrCHICH_3$
    (c) $CH_3CH_2\underset{\underset{\displaystyle CH_3CHCH_3}{|}}{C}HCH_3$

    (d) $CH_3CH_2CHCl\underset{\underset{\displaystyle CH{=}CH_2}{|}}{C}HCH_2CH_2CH_3$

    (e) $(CH_3)_3CH$
    (f)

    $H_2C\begin{array}{c} CH_2{-}CH_2 \\ \\ CH_2{-}CH_2 \end{array}CHCH_2CH(CH_3)CH_3$

    (g) $\underset{H}{\overset{CH_3}{\diagdown}}C{=}C\underset{C_2H_5}{\overset{H}{\diagup}}$

    (h) $(CH_3)_2CHCH_2C{\equiv}CH$

24. Does the following molecule exist as *cis-trans* isomers? optical isomers?

$\begin{array}{c} CH_2 \\ H_2C \quad CHCl \\ H_2C \quad CHBr \\ CH_2 \end{array}$

25. Write structures for all possible isomers of the alkyl group $—C_5H_{11}$.

26. How much heat will be released by the combustion of exactly 1 g of C_2H_6? of C_2H_2?
    Ans. 47.49 kJ g^{-1}; 48.22 kJ g^{-1}

Hydrocarbon Derivatives

27. Identify the functional groups present in the molecule carvone (Fig. 25-18).

28. Identify each of the following hydrocarbon derivatives in which R is an alkyl group:
    (a) RCO_2H (d) ROR
    (b) RCHO (e) RC(O)R
    (c) ROH (f) RNH_2

29. Write the Lewis structure, indicating resonance structures where appropriate, for each of the following molecules or ions:
    (a) methanol
    (b) formaldehyde
    (c) acetone
    (d) acetate ion
    (e) bromobenzene

30. Draw a three-dimensional structure, using solid and striped wedge-shaped bonds where appropriate, for each of the following molecules:
    (a) CH_3OH (d) CH_3OCH_3
    (b) C_2H_5CHO (e) $C_2H_5NH_2$
    (c) CH_3CO_2H

31. Which alcohol can be oxidized to give the acid $(CH_3)_2CHCH_2CO_2H$? which aldehyde?

32. What are the products formed by the stepwise oxidation of methanol? of ethanol?

33. Write a complete and balanced equation for each of the following reactions:
    (a) ethylene is burned in air
    (b) 2-butene + $H_2O \longrightarrow$
    (c) acetylene (1 mol) + Br_2 (1 mol) $\longrightarrow$
    (d) benzoic acid reacts with propanol
    (e) ethanol is oxidized to an aldehyde by oxygen

Additional Exercises

34. How could the heats of combustion of graphite and diamond be used to show that the conversion

    $$C(s, \textit{graphite}) \longrightarrow C(s, \textit{diamond})$$

    is endothermic?

35. On the basis of hybridization of orbitals of the carbon atom, explain why cyclohexane exists in the form of a "puckered" (bent) ring and benzene in the form of a planar ring.

36. Why is CH_3CH_2OH soluble in water while its isomer CH_3OCH_3 is not?

37. Indicate the types of hybridized orbitals used and the geometry about each carbon atom in each of the following molecules:
    (a) $H_2C=O$ (d) $CH_3C\equiv CH$
    (b) C_2H_4 (e) CH_3CONH_2
    (c) C_2H_6

38. Write a complete and balanced equation for each of the following reactions:
    (a) $(CH_3)_2CHCH_2OH + Na \longrightarrow$
    (b) $CH_3CH_2CO_2H + (CH_3)_3N \longrightarrow$
    (c) $(CH_3)_3CCH_2SH + NaOH \longrightarrow$
    (d) $(CH_3)_2NH + Li \longrightarrow$
    (e) $CH_3Li + H_2O \longrightarrow$
    (f) $CH_3OLi + CS_2 \longrightarrow$

39. (a) What volume of $C_2H_2(g)$ at STP would result from the hydrolysis of 10.0 g of CaC_2?
    Ans. 3.49 L

    (b) How many grams of magnesium would have to react with HCl to provide enough hydrogen to reduce the C_2H_2 of part (a) to ethane?
    Ans. 7.58 g

40. (a) If all of the 24.68 billion pounds of ethylene produced in the United States during 1982 were produced by the pyrolysis of ethane

    $$C_2H_6 \longrightarrow C_2H_4 + H_2$$

    what mass of hydrogen, in pounds, would have been produced?
    Ans. 1.773 billion pounds

    (b) Is this enough hydrogen to prepare the 30.98 billion pounds of ammonia manufactured during 1982?

    $$3H_2 + N_2 \longrightarrow 2NH_3$$

    Ans. No;
    5.500 billion pounds of H_2 was required

41. (a) From the reactions

    $$C_2H_4(g) + 2H_2(g) \longrightarrow 2CH_4(g)$$
    $$C(g) + 2H_2(g) \longrightarrow CH_4(g)$$

    and by considering the bonds that must be broken and formed to get from reactants to products, show that the $C=C$ bond energy, $D_{C=C}$, is given by

    $$D_{C=C} = \Delta H^\circ_{fC(g)} + \Delta H^\circ_{fCH_4(g)} - \Delta H^\circ_{fC_2H_4(g)}$$

    Calculate, using data in Appendix J, a value for $D_{C=C}$.
    Ans. 590 kJ

    (b) By considerations similar to those in part (a) show that the reaction

    $$C_2H_4(g) + H_2(g) \longrightarrow C_2H_6(g)$$

    leads to the conclusion that

    $$D_{C=C} = D_{C-C} + 2D_{C-H}$$
    $$- D_{H-H} + \Delta H^\circ_{fC_2H_6(g)} - \Delta H^\circ_{fC_2H_4(g)}$$

    and calculate, using data in Table 18-2 and Appendix J, a value for $D_{C=C}$.
    Ans. 602 kJ

42. Assuming a value of (n) equal to six in the formula for nylon, calculate how many pounds of hexamethylenediamine would be needed to produce 1.00 ton of nylon. (See Appendix C for conversion factors.)   *Ans. 1.02×10^3 lb*

43. How much acetic acid, in grams, is in exactly 1 qt of vinegar if the vinegar contains 3.00% acetic acid by volume? The density of acetic acid is 1.049 g/mL. (See Appendix C for conversion factors.)   *Ans. 29.8 g*

26

BIOCHEMISTRY

Biochemistry is the study of the chemical composition and structure of living organisms and the reactions that take place within them. Considering the diverse and complex functions of a cell, it is not surprising to find that its molecules are among the most complicated substances known.

Biomolecules are often divided into four major classes—proteins, carbohydrates, lipids, and nucleic acids—according to their structures or functions. Proteins, carbohydrates, and nucleic acids are found in the form of macromolecules with molecular weights in the range 10^4–10^6, or larger, which form by the polymerization of monomers of relatively low molecular weights. Molecules of the fourth major class, the lipids, are considerably smaller than those of the other three classes. However, they can associate to form clusters of molecules of enormous size.

26.1 The Cell

All cells are basically similar. They contain many of the same structures, the same or similar systems of enzymes, and the same kind of genetic material. The principal constituent of living matter is water. Most plants and animals are 60–90% water, most of which is the solvent in which both organic and inorganic substances are dissolved and transported in the cell or through the organism.

An understanding of the role played by water in biological systems requires that a distinction be made between substances that are **hydrophilic** (literally water-loving) and those that are **hydrophobic** (water-hating). Water is a polar solvent that readily dissolves either polar or ionic substances. In general, the less polar a substance is, the more hydrophobic it is. The nonpolar hydrocarbons are excellent examples of hydrophobic compounds; they are almost totally immiscible with water. Some biomolecules contain polar or ionic substituents that impart hydrophilic character, and they are therefore water-soluble. Those biomolecules that do not contain polar or ionic substituents are hydrophobic and are effectively insoluble in water.

Cells are commonly divided into two classes: **eukaryotic cells** such as that represented in Fig. 26-1 contain a nucleus; **prokaryotic cells** do not. The nucleus contains the chromosomes, which are a mixture of two substances: deoxyribonucleic acid (DNA) and protein. The nucleus stores the genetic information that enables the cell to replicate itself and to synthesize the various proteins that it needs for biological activity.

Figure 26-1. Simplified drawing of a eukaryotic cell.

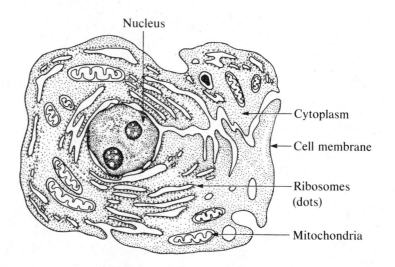

Nucleus

Cytoplasm

Cell membrane

Ribosomes (dots)

Mitochondria

The **cytoplasm** is the entire contents of the cell, except the nucleus. It can be divided into the **cytosol,** the water-soluble contents of the cell, and a variety of cellular organelles such as the ribosomes, mitochondria, and chloroplasts. **Ribosomes,** the site of the directed polymerization (genetic coding) of amino acids to form enzymes and other proteins, are present in all cells. **Mitochondria,** a cell's power plants, are present in eukaryotic cells. A variety of biomolecules (including carbohydrates, amino acids, and fatty acids) are oxidized to carbon dioxide and water in the mitochondria. Some of the energy released during this oxidation is stored as chemical energy through the synthesis of compounds called **nucleoside triphosphates,** for example, adenosine triphosphate (ATP) and guanosine triphosphate (GTP). **Chloroplasts** are present in all plants and algae that are capable of photosynthesis. Light-induced synthesis of carbohydrates from carbon dioxide and water takes place in the chloroplasts. All cells are surrounded by a cell membrane that insulates the contents of the cell from surrounding fluids.

PROTEINS

26.2 Amino Acids

When proteins are heated with aqueous acid or base or are treated with certain hydrolytic enzymes, they undergo hydrolysis (Section 12.3, Part 4), releasing compounds called **amino acids.** All amino acids contain both an amine group, $-NH_2$, and a carboxylic acid group, $-CO_2H$ (Section 25.23). At least 150 naturally occurring amino acids are known; the most important, however, are the 20 genetically coded amino acids used in the synthesis of proteins. The genetically coded amino acids are all **α-amino acids,** which means that the amino group is attached to the carbon atom adjacent, or α, to the carbon of the carboxylic acid group. The primary α-amino acids can be represented by the following structure:

$$H_3N^+-\overset{\overset{\displaystyle R}{|}}{C}H-CO_2^-$$

The structures of these amino acids differ only in the nature of the substituent (R). Thus they are usually grouped on the basis of similarities in the structure of this side chain. The names and structures of the genetically coded amino acids are given in Fig. 26-2 on the following two pages.

The formulas of the amino acids are often written in the form

$$H_2N-\overset{\overset{\displaystyle R}{|}}{C}H-CO_2H$$

However, it is important to recognize that the basic H_2N- end of this molecule acts as a proton acceptor, while the $-CO_2H$ end functions as a proton donor. The proton transfer forms a species called a **zwitterion** (from German, meaning hybrid ion). Zwitterions are molecules that contain ionic charges but that are electrically neutral overall. For example, the amino acid alanine forms a zwitterion.

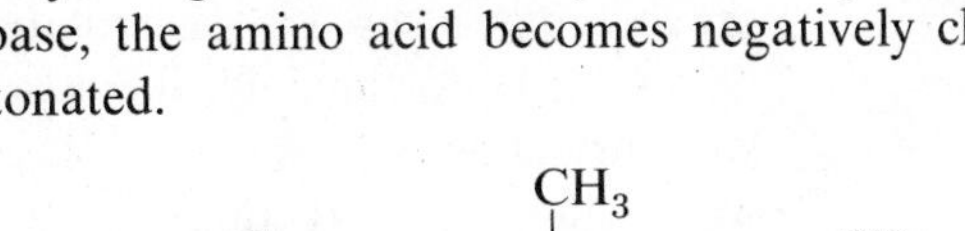

$$H_2\overset{..}{N}-\overset{\overset{\displaystyle CH_3}{|}}{C}H-CO_2\boxed{H} \longrightarrow H_3N^+-\overset{\overset{\displaystyle CH_3}{|}}{C}H-CO_2^-$$

Zwitterion

The zwitterion is the dominant form of an amino acid in aqueous solution at or near neutral pH (pH $= 7$). In the presence of an acid, however, the amino acid becomes positively charged as the carboxylate ion is protonated. Conversely, on addition of a base, the amino acid becomes negatively charged as the amino group is deprotonated.

$$H_3N^+-\overset{\overset{\displaystyle CH_3}{|}}{C}H-CO_2H \underset{OH^-}{\overset{H^+}{\rightleftharpoons}} H_3N^+-\overset{\overset{\displaystyle CH_3}{|}}{C}H-CO_2^- \underset{H^+}{\overset{OH^-}{\rightleftharpoons}} H_2N-\overset{\overset{\displaystyle CH_3}{|}}{C}H-CO_2^-$$

Acidic solution   Neutral solution   Basic solution

With the exception of glycine the α-amino acids are optically active (Section 25.16). Two optical isomers, which differ in their interaction with plane-polarized

Figure 26-2. The 20 genetically coded amino acids at pH = 7.

Nonpolar R Group

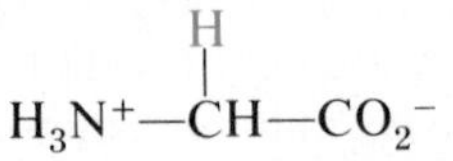

Glycine (GLY)

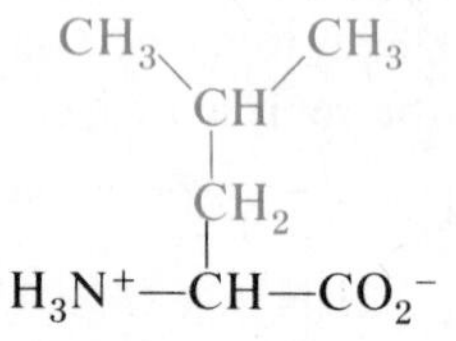

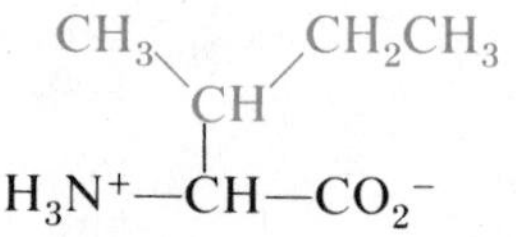

Isoleucine (ILE)

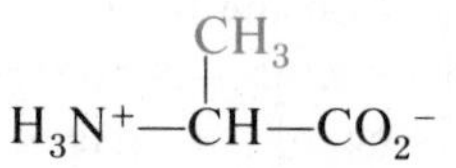

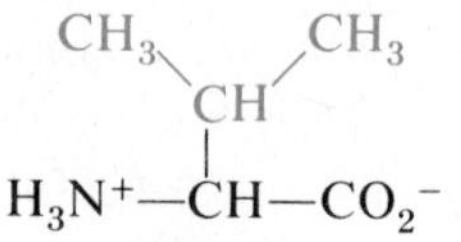

Leucine (LEU)

Valine (VAL)

Alanine (ALA)

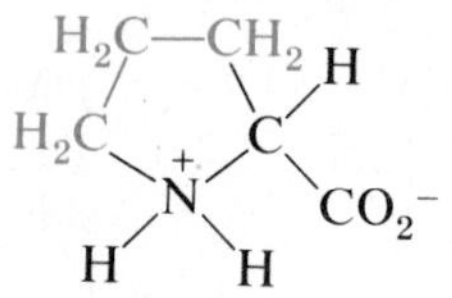

Proline (PRO)

Aromatic R Group

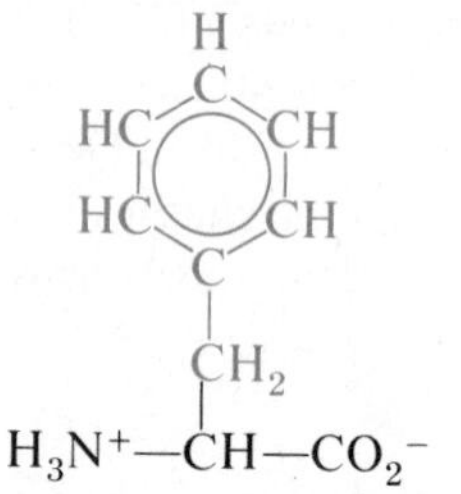

Phenylalanine (PHE)

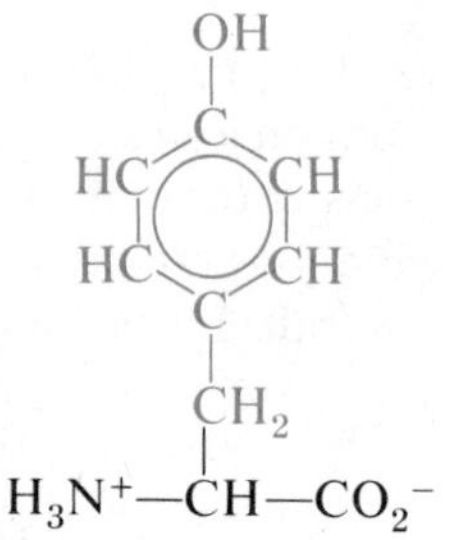

Tyrosine (TYR)

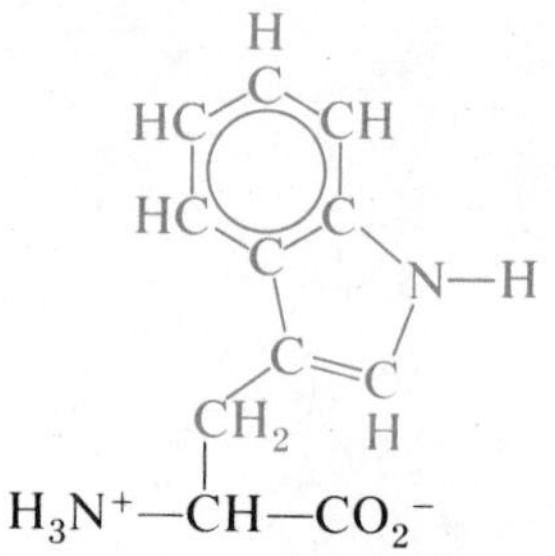

Tryptophan (TRP)

light, are possible for every optically active substance. The dextrorotatory (+) isomer rotates the plane of plane-polarized light in a clockwise direction, whereas the levorotatory (−) isomer rotates the plane of such light in a counterclockwise direction. Unfortunately, there is no direct relationship between the molecular structure of an isomer and its classification as either (+) or (−). Optical isomers whose molecular structures are related to the structure of the (+) isomer of glyceraldehyde, $HOCH_2CH(OH)CHO$, are labeled D isomers, and their mirror images, which are related structurally to the (−) isomer of glyceraldehyde, are labeled L isomers. Every amino acid isolated from a naturally occurring protein is an L amino acid. Looking down the H—C bond of an L amino acid, we see that the $-CO_2^-$, $-R$, and $-NH_3^+$ substituents are arranged in a clockwise fashion. In a D amino acid these substituents are arranged in a counterclockwise manner (Fig. 26-3). Though occasionally found in nature, D amino acids are not genetically coded into proteins.

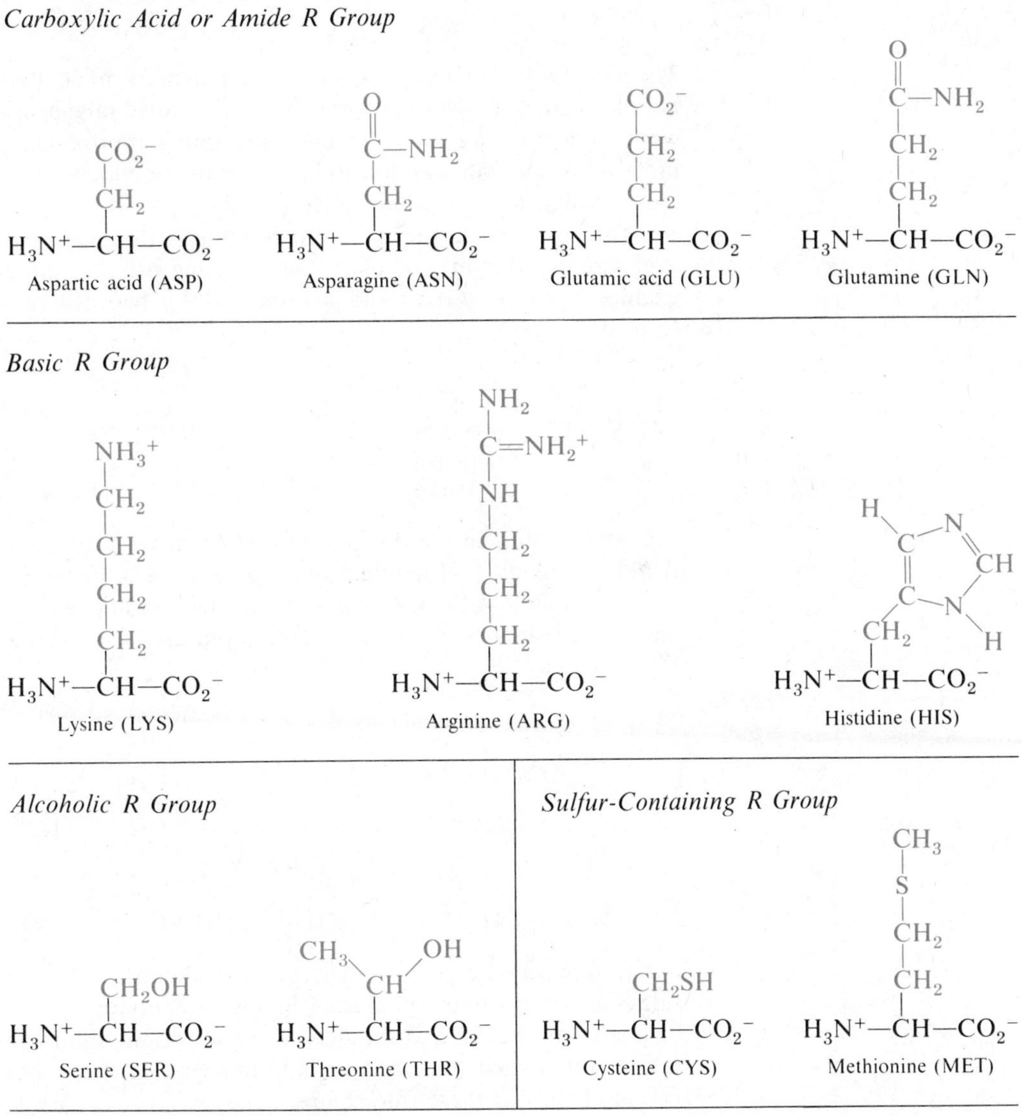

Carboxylic Acid or Amide R Group

$$CO_2^-$$
$$CH_2$$
$$H_3N^+-CH-CO_2^-$$
Aspartic acid (ASP)

$$O$$
$$C-NH_2$$
$$CH_2$$
$$H_3N^+-CH-CO_2^-$$
Asparagine (ASN)

$$CO_2^-$$
$$CH_2$$
$$CH_2$$
$$H_3N^+-CH-CO_2^-$$
Glutamic acid (GLU)

$$O$$
$$C-NH_2$$
$$CH_2$$
$$CH_2$$
$$H_3N^+-CH-CO_2^-$$
Glutamine (GLN)

Basic R Group

$$NH_3^+$$
$$CH_2$$
$$CH_2$$
$$CH_2$$
$$CH_2$$
$$H_3N^+-CH-CO_2^-$$
Lysine (LYS)

$$NH_2$$
$$C=NH_2^+$$
$$NH$$
$$CH_2$$
$$CH_2$$
$$CH_2$$
$$H_3N^+-CH-CO_2^-$$
Arginine (ARG)

$$H$$
$$C-N$$
$$\quad\quad CH$$
$$C-N$$
$$CH_2 \quad H$$
$$H_3N^+-CH-CO_2^-$$
Histidine (HIS)

Alcoholic R Group

$$CH_2OH$$
$$H_3N^+-CH-CO_2^-$$
Serine (SER)

$$CH_3 \quad OH$$
$$CH$$
$$H_3N^+-CH-CO_2^-$$
Threonine (THR)

Sulfur-Containing R Group

$$CH_2SH$$
$$H_3N^+-CH-CO_2^-$$
Cysteine (CYS)

$$CH_3$$
$$S$$
$$CH_2$$
$$CH_2$$
$$H_3N^+-CH-CO_2^-$$
Methionine (MET)

Figure 26-3. The molecular structures of the D and L isomers of an amino acid.

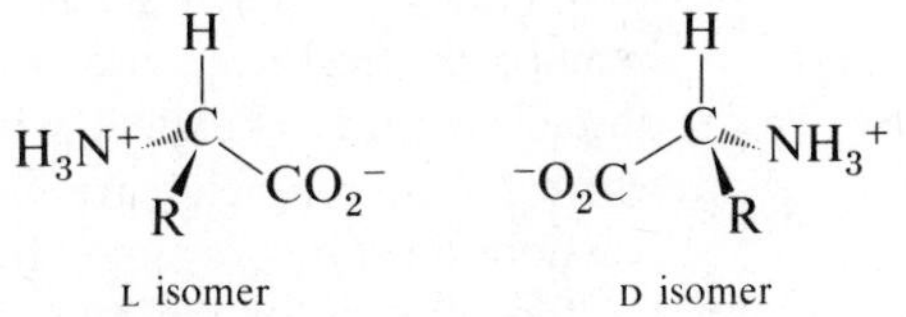

$$H$$
$$H_3N^+\cdots C$$
$$R \quad CO_2^-$$
L isomer

$$H$$
$$C\cdots NH_3^+$$
$$^-O_2C \quad R$$
D isomer

26.3 Peptides and Proteins

The structures of peptides and proteins were first determined by the German chemist Emil Fischer between 1900 and 1910. **Peptides** are polymers formed from two or more amino acids linked by covalent amide linkages given the special name **peptide bonds.** The nature of a peptide bond (in color) can be seen in the following structure for a dipeptide.

$$\text{H}_3\text{N}^+ - \overset{\overset{\textstyle R}{|}}{\text{CH}} - \overset{\overset{\textstyle O}{\|}}{\text{C}} - \text{NH} - \overset{\overset{\textstyle R}{|}}{\text{CH}} - \text{CO}_2^-$$

The name *peptide* is usually restricted to polymers made up of only a small number of amino acids. Such polymers are often called **oligopeptides** (from the Greek word *oligo,* meaning few). Polymers containing approximately 10–40 amino acid units are called **polypeptides.** Relatively large polymers containing 40–10,000 (or more) amino acid units are called **proteins.**

The peptide bond results from combining the α-carboxyl end of one amino acid with the α-amine end of another. Starting with two different amino acids, for example, glycine (GLY) and alanine (ALA), two different dipeptides can be formed.

$$\text{H}_3\text{N}^+ - \text{CH}_2 - \overset{\overset{\textstyle O}{\|}}{\text{C}} - \text{NH} - \overset{\overset{\textstyle CH_3}{|}}{\text{CH}} - \text{CO}_2^- \qquad \text{H}_3\text{N}^+ - \overset{\overset{\textstyle CH_3}{|}}{\text{CH}} - \overset{\overset{\textstyle O}{\|}}{\text{C}} - \text{NH} - \text{CH}_2 - \text{CO}_2^-$$

Glycylalanine (GLY-ALA)   Alanylglycine (ALA-GLY)

The glycine unit in the dipeptide GLY-ALA has a free amine, and this unit is therefore called the **N-terminal amino acid.** In the dipeptide ALA-GLY, however, the glycine unit is the **C-terminal amino acid.** Peptides and proteins are written and named starting with the N-terminal amino acid. The tetrapeptide

is therefore called cysteinylvalyllysylphenylalanine, or CYS-VAL-LYS-PHE. The side chains of the four amino acids in this tetrapeptide are shown in color. Peptide linkages can also form at the side chains of certain amino acids. Some oligopeptides contain such branched peptide linkages. Proteins, however, seldom form peptide bonds at these side chains.

It would be very useful to be able to synthesize any given protein, since the body's inability to synthesize one or more proteins is the cause of such diseases as hemophilia, sickle-cell anemia, and diabetes. Unfortunately, there are three major obstacles to the chemical synthesis of peptides or proteins. First, the order in which the amino acids are incorporated must be rigorously controlled. Second, reactions must be prevented from occurring at the amino acid side chains. Finally, the free energy for the hydrolysis of the peptide bond is negative:

$$\text{H}_3\text{N}^+ - \overset{\overset{\textstyle R}{|}}{\text{CH}} - \overset{\overset{\textstyle O}{\|}}{\text{C}} - \text{NH} - \overset{\overset{\textstyle R}{|}}{\text{CH}} - \text{CO}_2^- + \text{H}_2\text{O} \longrightarrow$$

$$\text{H}_3\text{N}^+ - \overset{\overset{\textstyle R}{|}}{\text{CH}} - \text{CO}_2^- + \text{H}_3\text{N}^+ - \overset{\overset{\textstyle R}{|}}{\text{CH}} - \text{CO}_2^-$$
$$\Delta G^\circ = -17 \text{ kJ/mol}$$

Therefore, the formation of a peptide bond is an uphill battle. Peptides and proteins can be produced by chemical syntheses, but not in large amounts. However, genetic engineering techniques, which use bacterial cells to biosynthesize mammalian peptides or proteins, are improving. Several proteins, including insulin for treating diabetes, can be produced this way.

26.4 The Structures of Proteins

The extraordinary variations possible in the structures of proteins allow for remarkable differences in their types and functions. Among the types of proteins and their functions are the following:

1. *Enzymes.* Enzymes, which catalyze biochemical reactions, invariably contain a protein component.
2. *Hormones.* A number of oligopeptides and proteins, such as insulin and human growth hormone, bind to specific sites in or on the cell to stimulate biological activity.
3. *Structural proteins.* A wide variety of proteins are involved in the formation of various kinds of connective tissue, skin, hair, horn, and hoof.
4. *Energy-storing proteins.* The caseins found in mammalian milk, the albumins in eggs, and a variety of proteins in seeds store energy and serve as food.
5. *Transport proteins.* Proteins are involved in the transport of both organic and inorganic nutrients across cell membranes. The metal-containing proteins, or metalloproteins, such as the cytochromes, are involved in electron transport.
6. *Antibodies.* The antibodies produced by the immune response system are proteins.
7. *Muscle tissue.* Muscles are composed of fibrous bundles of the proteins actin and myosin, which are involved in the contraction and relaxation of muscle tissue.

Proteins can function in many different ways because their three-dimensional structures vary widely. Their structures can vary so much (even though they are formed from only 20 different amino acids) because the amino acids can be used in many different orders in a peptide chain of given length. For example, there are 20^2 or 400 possible dipeptides, 20^3 or 8000 tripeptides, and 20^4 or 160,000 tetrapeptides. Even for a protein as small as insulin, with only 51 amino acid units, there are 2.3×10^{66} possible amino acid sequences. Since the three-dimensional structure of a protein is controlled by the sequence of amino acid side chains and there is an almost infinite variability in this sequence, an essentially infinite number of three-dimensional protein structures is available.

The most fundamental structure, the **primary structure,** of a protein is simply the sequence of amino acids, listed as though printed on ticker tape, starting with the N-terminal amino acid and proceeding through the C-terminal amino acid.

A polypeptide chain can fold upon itself to produce a **secondary structure,** in which the number of hydrogen bonds between peptide linkages is maximized.

$$\diagdown N{-}H\cdots O{=}C\diagup$$

There are only a limited number of secondary structures that a polypeptide chain can adopt that maximize the number of hydrogen bonds between peptide link-

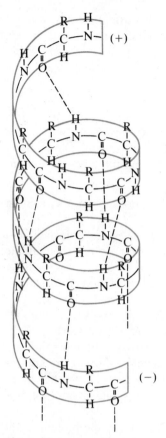

Figure 26-4. The α-helix structure of a peptide chain, one form of secondary structure of protein molecules. The hydrogen bonds that stabilize this structure are indicated by dashed lines.

Figure 26-5. The β-sheet, or pleated sheet, secondary structure of protein molecules. [*From H. D. Springall, The Structural Chemistry of Proteins (Butterworth Scientific Publishers, London, 1954.*)]

ages. Two of the most common forms of secondary structure are the **α-helix** (Fig. 26-4) and the **β-sheet,** or **pleated sheet** (Fig. 26-5). In the α-helix the polypeptide backbone forms a helical coil with approximately 3.6 amino acids per 360° turn. The α-helix is the dominant secondary structure in fibrous proteins such as collagen or the α-keratins of hair and wool. The β-sheet structure results from hydrogen bonds between adjacent strands of a polypeptide backbone. Extended arrays of the β-sheet secondary structure are found in the β-keratins, such as silk.

Proteins often contain regions where a helical or sheet structure is folded into a complex conformation known as the **tertiary structure** (Fig. 26-6). Interactions between amino acid side chains produce this structure.

Many larger proteins contain more than one polypeptide chain. The **quaternary structure** of a protein describes how these independent chains are related to each other. The oxygen-carrying protein hemoglobin (Section 29.5) consists of four separate polypeptide chains, two α subunits and two β subunits, held together in a quaternary structure.

Factors that **denature** a protein disturb its natural conformation, without necessarily cleaving the primary structure, and may lead to the loss of biological activity. A denatured protein has lost some, or all, of its secondary, tertiary, or quaternary structure.

26.5 Enzymes

Enzymes are proteins that catalyze chemical reactions in living systems with remarkable efficiency. The enzyme carbonic anhydrase, for example, which catalyzes the reaction

$$CO_2 + H_2O \rightleftharpoons H^+ + HCO_3^-$$

is such a good catalyst that a single molecule can transform over 36 million substrate molecules per minute into product. This corresponds to an increase of 10^7–10^8 in the rate of this reaction. One explanation for this efficiency is that the binding of the substrate (the molecule undergoing the reaction) to the enzyme may stretch or strain the substrate until its conformation more closely resembles

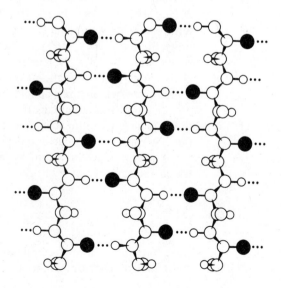

Figure 26-6. The tertiary structure of myoglobin, an oxygen-storing protein found in muscle tissue. The folded tertiary structure is shown as well as the helical secondary structure. The primary structure is not shown in detail.

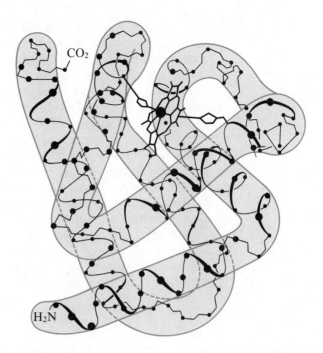

the transition state for the reaction (Section 15.8), thereby reducing the activation energy. Enzymes are not only efficient, but also often remarkably selective. The ability of some enzymes to be specific for a very few substrates in a variety of substances with related structures suggests that the enzyme and substrate fit together somewhat like a lock and key do. The enzyme activates only those substrates that fit, just as a lock is opened only by keys that fit.

In 1913 L. Michaelis and M. L. Menten suggested that enzyme-catalyzed reactions proceed through at least two steps. First, the substrate binds to the enzyme to form an enzyme-substrate complex. Second, the enzyme acts on the substrate to convert it into product. Many substances may bind to an enzyme, but only a limited number of them can react. The enzyme succinate dehydrogenase, for example, not only binds its normal substrate, succinate, but also binds other dicarboxylates, such as malonate, which do not react. By occupying a portion of the available binding sites on the enzyme, malonate slows down the rate of reaction (fewer molecules of enzyme are available to catalyze it). In effect, the activity of the enzyme is **inhibited.** Since the substrate (in this case, succinate) and the inhibitor (malonate) compete for the same binding sites on the enzyme, this form of enzyme inhibition is called **competitive inhibition.** Enzymes are also inhibited by substances that have no resemblance to the substrate, for example, heavy metal ions such as Ag^+, Pb^{2+}, and Hg^{2+}, which bind either in the presence or in the absence of substrate binding. This form of enzyme inhibition is called **noncompetitive inhibition.**

For a cell to function at its maximum efficiency, the rates of its chemical reactions muct be adjusted to its needs. Enzymes alone are not enough; the cell must also have control mechanisms that can turn on enzyme activity when the product of an enzyme-catalyzed reaction is needed and switch off enzyme activity when the product is present in excess. The cell can control the rate of enzyme-catalyzed reactions in four ways. First, it can influence the rate of a reaction by changing the concentration of the substrate. In the absence of substrate, the enzyme is effectively switched off. Second, the cell can produce natural inhibitors of enzyme activity, which operate by either competitive or noncompetitive inhibition mechanisms. Third, the cell can use **allosteric enzymes,** which contain two or more binding sites. Allosteric enzymes bind both a substrate molecule and a small molecule known as the **effector.** Binding of the effector is thought to alter the structure of the enzyme, thereby either switching on or switching off its activity. Binding of the effector usually does not influence the binding of the substrate; it either facilitates or prevents conversion of the substrate into product. There are several forms of control of allosteric enzymes. In so-called **feedback inhibition** an enzyme involved in the synthesis of an end-product is inhibited by binding the end-product as an effector. Alternatively, when the synthesis of a particular product should occur only when the substrate is present in excess, an enzyme may exhibit **substrate activation,** where more than one substrate must bind to the enzyme before it is activated. Fourth, the cell can control the rate of an enzyme-catalyzed reaction by controlling the population of enzyme molecules. This can be done by controlling either the rate at which the enzyme is synthesized or the rate at which the enzyme is degraded or destroyed.

CARBOHYDRATES

26.6 Introduction

The name *carbohydrate* originally meant a compound with the empirical formula CH_2O, literally, a hydrate of carbon. This definition has since been broadened to include all aldehydes and ketones having two or more hydroxyl groups. Among the compounds that fall into the class of **carbohydrates** are the starches, cellulose, glycogen, and the sugars. The carbohydrates are an important source of energy for all organisms and form the supporting tissue of plants and some animals. There are three important classes of carbohydrates: the monosaccharides, the disaccharides, and the polysaccharides.

26.7 Monosaccharides

The simplest carbohydrates contain only a single aldehyde or ketone functional group (Sections 25.21 and 25.22) and are known as **monosaccharides.** They are colorless, crystalline, and frequently possess a sweet taste. A monosaccharide is classified as either an **aldose** or a **ketose,** depending on whether it contains an aldehyde or a ketone functional group, respectively. With the exception of the simplest ketose, the three-carbon monosaccharide dihydroxyacetone, all of the monosaccharides are optically active. Although both D and L isomers are possible, the dominant form of carbohydrates isolated from natural sources is the D isomer. Structures of the D and L forms of the simplest aldose, glyceraldehyde, are

D-Glyceraldehyde   L-Glyceraldehyde

The formulas of monosaccharides are often written using a convention first proposed by Emil Fischer, in which the carbon skeleton is written vertically, with the aldehyde or ketone function toward the top, and with the second to the last —OH group pointed toward the right for the D isomer and toward the left for the L isomer. Fischer projections for the two isomers of glyceraldehyde are

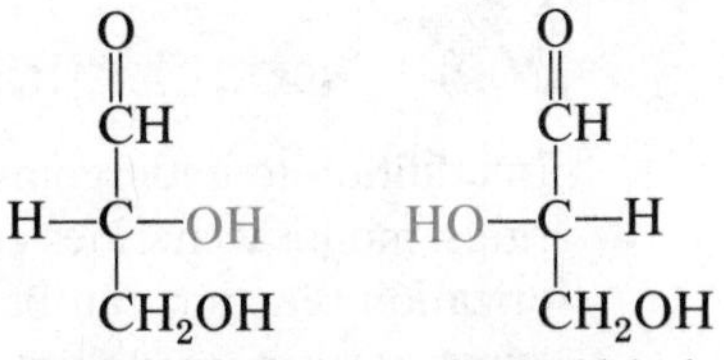

$$
\begin{array}{cc}
\underset{\text{D-Glyceraldehyde}}{\overset{\displaystyle O}{\underset{\displaystyle \text{CH}}{\parallel}} \; \text{H–C–OH} \; \text{CH}_2\text{OH}} &
\underset{\text{L-Glyceraldehyde}}{\overset{\displaystyle O}{\underset{\displaystyle \text{CH}}{\parallel}} \; \text{HO–C–H} \; \text{CH}_2\text{OH}}
\end{array}
$$

Monosaccharides can undergo a reversible intramolecular reaction to form cyclic compounds. An aldose, such as glucose, forms a six-membered **pyranose ring** (Fig. 26-7), while a ketose, such as fructose, forms a five-membered **furanose ring** (Fig. 26-8). In each case two isomers known as the α and β forms are produced; they differ only in orientation at the first carbon atom, or C(1).

Glucose, $C_6H_{12}O_6$ (Fig. 26-7), also called **dextrose,** is the most abundant organic compound in nature. It is the sole component of the polysaccharides cellu-

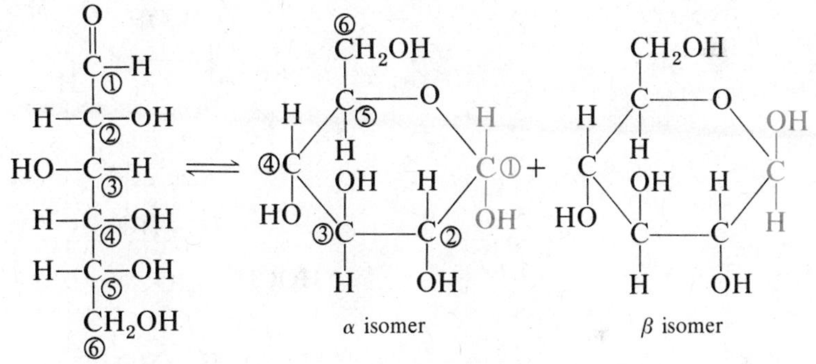

Figure 26-7. Glucose undergoes a reversible reaction leading to the formation of a six-membered glucopyranose ring. Two isomers, differing only in orientation at the C(1) atom, are formed.

D-Glucose   D-Glucopyranose   α isomer   β isomer

lose, starch, and glycogen. In medicine, glucose is often called blood sugar, since it is by far the most abundant carbohydrate in the bloodstream. Human blood normally contains about 1 gram of glucose per liter. Individuals suffering from diabetes are unable to assimilate glucose and eliminate it through the kidneys. In 100 milliliters of a diabetic's urine, there may be as much as 8–10 grams of glucose, and its presence is one symptom of the disease. **Fructose** (Fig. 26-8), also

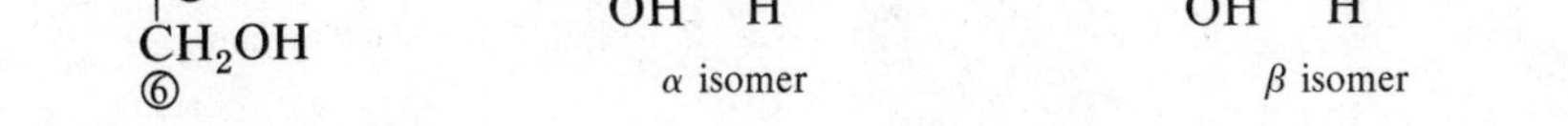

Figure 26-8. Fructose undergoes a reversible reaction leading to the formation of α and β isomers of a five-membered fructofuranose ring.

D-Fructose   D-Fructofuranose   α isomer   β isomer

called **levulose** or **fruit sugar,** has the same molecular formula as glucose, but contains a ketone rather than an aldehyde functional group. Fructose occurs naturally in both fruits and honey and is found combined with glucose in the disaccharide sucrose, called cane sugar. Fructose is the sweetest of all sugars; a given mass of fructose is nearly twice as sweet as the same amount of sucrose.

26.8 Disaccharides and Polysaccharides

In addition to undergoing reversible reactions that lead to pyranose and furanose rings, monosaccharides can also polymerize to form chains. However, the polymerization reaction can be reversed only in the presence of acid or a suitable enzyme catalyst, so polysaccharides are relatively stable toward hydrolysis.

The most important **disaccharide** is **sucrose,** or **cane sugar** (Fig. 26-9). Hydrolysis of sucrose, $C_{12}H_{22}O_{11}$, yields a 50:50 mixture of the monosaccharides fructose and glucose (both $C_6H_{12}O_6$), a mixture often called **invert sugar,** which is a major component of honey. **Lactose,** or **milk sugar,** is a disaccharide that contains an α-D-glucopyranose ring and a β-D-galactopyranose ring (Fig. 26-10). Many individuals of African or Asiatic origin cannot digest milk because they lack enzymes that hydrolyze the β-linkage in lactose.

Figure 26-9. The structure of the disaccharide sucrose, formed from reaction of α-D-glucopyranose with β-D-fructofuranose. (Carbon atoms in the rings are not shown.)

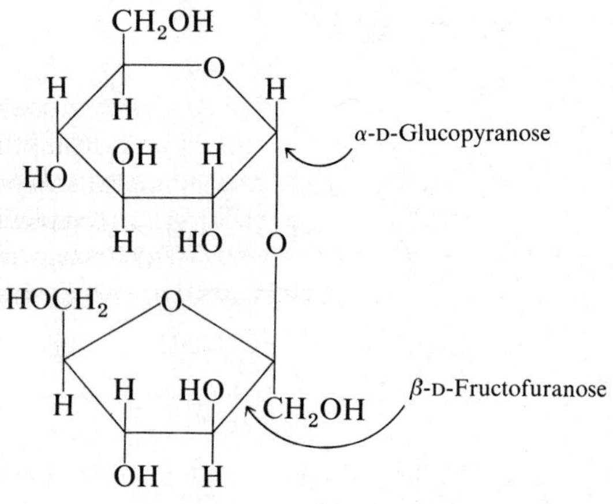

Figure 26-10. The structure of the disaccharide lactose, formed by reaction of β-D-galactopyranose with α-D-glucopyranose. (Carbon atoms in the rings are not shown.)

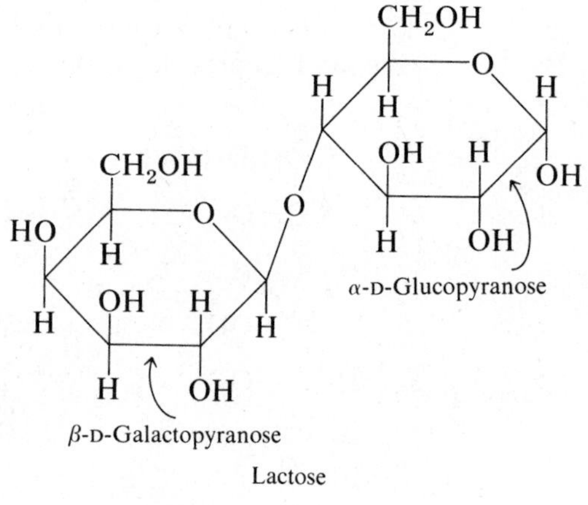

Three major classes of **polysaccharides** are formed by the polymerization of D-glucopyranose rings: starch, glycogen, and cellulose. **Starch** serves primarily as a long-term storage medium for food energy in plants. It accumulates in seeds, tubers, and fruits. Starch is often the main food supply for the young plant until a leaf system has developed and the plant can manufacture its own food through photosynthesis.

Glycogen is a polysaccharide that serves the same function in animals as starch serves in plants. Glycogen is found primarily in the liver and muscle tissue.

Cellulose is a linear polysaccharide of β-D-glucopyranose units (Fig. 26-11) found in plants. It is the principal component of the rigid cell wall of plant cells, and therefore provides structure rather than energy. The lack of an enzyme capable of cleaving the β-linkage in cellulose is the only thing that prevents humans from being able to digest the paper of this text or cotton, the latter of which is at least 90% cellulose by weight. Most animals cannot digest cellulose because they lack suitable enzymes to catalyze its hydrolysis. Termites and the ruminant animals, such as cattle, sheep, and goats, are able to use cellulose as food because bacteria in their digestive tracts contain enzymes that can cleave the β-linkages. Termites are commonly destroyed by killing these bacteria—they then starve to death.

Figure 26-11. The structure of cellulose, a linear polysaccharide formed by the polymerization of β-D-glucopyranose. (Carbon atoms in the rings are not shown.)

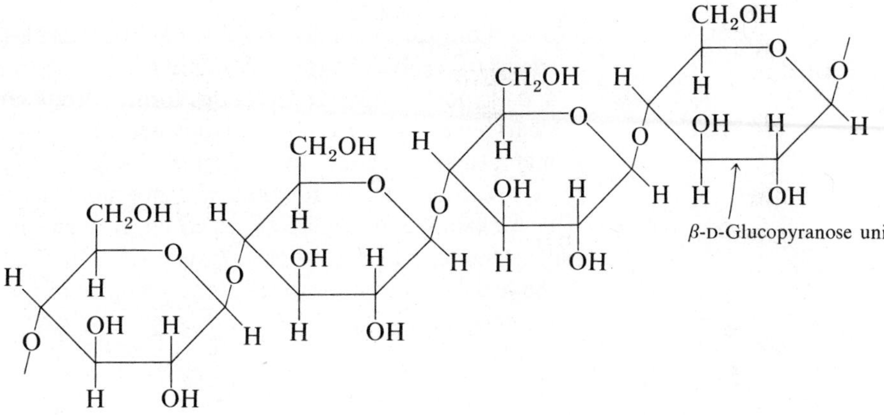

β-D-Glucopyranose unit

Cellulose

LIPIDS

26.9 Complex Lipids

Lipids can be extracted from living systems with nonpolar solvents such as chloroform, ether, or benzene. Lipids are more diverse than either proteins or carbohydrates; this class of biomolecules consists of compounds of widely differing structures. Their only structural similarity is the presence of a hydrophobic hydrocarbon backbone with a hydrophilic head, which may or may not be highly charged. The word *lipid* is derived from the Greek word *lipos* (meaning fat), although not all lipids are fats.

Lipids are often divided into two classes, complex and simple, on the basis of the presence or absence of a fatty acid structure. The **complex lipids** are derivatives of long-chain carboxylic acids, or **fatty acids,** such as stearic acid.

$$CH_3CH_2CH_2CH_2CH_2CH_2CH_2CH_2CH_2CH_2CH_2CH_2CH_2CH_2CH_2CH_2CO_2H$$

Stearic acid

Although fatty acids are available to a cell from a wide variety of sources, the concentration of free fatty acid in the cell is negligible. Most fatty acids occur in plant and animal cells in the form of triesters (Section 25.24) of the alcohol 1,2,3-propanetriol (glycerol), $HOCH_2CH(OH)CH_2OH$.

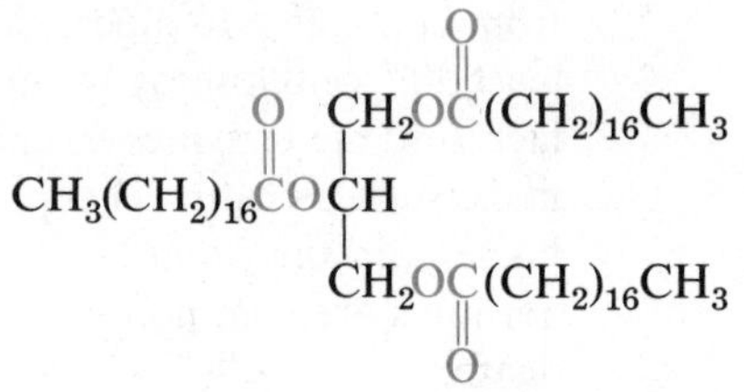

$$
\begin{array}{c}
\qquad\qquad\qquad \overset{\displaystyle O}{\underset{\displaystyle \|}{}} \\
\qquad\qquad CH_2O\overset{\displaystyle O}{\overset{\|}{C}}(CH_2)_{16}CH_3 \\
CH_3(CH_2)_{16}\overset{\displaystyle O}{\overset{\|}{C}}O\,CH \\
\qquad\qquad CH_2O\underset{\displaystyle O}{\underset{\|}{C}}(CH_2)_{16}CH_3
\end{array}
$$

Tristearylglycerol

Such lipids are commonly known as **triglycerides,** although the proper name is **triacylglycerols.** The primary function of these triglycerides is the long-term storage of food energy. Triglycerides formed from saturated fatty acids, such as tristearylglycerol, are typically solids and are known as **fats,** or sometimes as animal fats. Triglycerides formed from unsaturated fatty acids, which contain one or more carbon-carbon double bonds, are more frequently liquids and are known as **oils;** examples include cottonseed oil, linseed oil, and palm oil.

One of the characteristic reactions of fats and oils is **saponification,** which is the hydrolysis of the triester in the presence of base to form glycerol and a salt of the fatty acid.

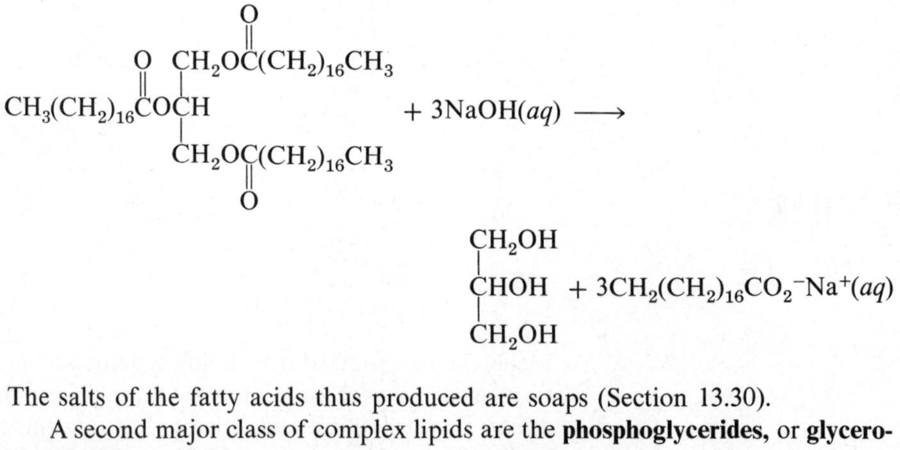

$$
\begin{array}{c}
\qquad\qquad\qquad \overset{\displaystyle O}{\underset{\displaystyle \|}{}} \\
\qquad\qquad CH_2O\overset{\displaystyle O}{\overset{\|}{C}}(CH_2)_{16}CH_3 \\
CH_3(CH_2)_{16}\overset{\displaystyle O}{\overset{\|}{C}}O\,CH \qquad\qquad + 3NaOH(aq) \longrightarrow \\
\qquad\qquad CH_2O\underset{\displaystyle O}{\underset{\|}{C}}(CH_2)_{16}CH_3
\end{array}
$$

$$
\begin{array}{c}
CH_2OH \\
CHOH \quad + 3CH_2(CH_2)_{16}CO_2{}^-Na^+(aq) \\
CH_2OH
\end{array}
$$

The salts of the fatty acids thus produced are soaps (Section 13.30).

A second major class of complex lipids are the **phosphoglycerides,** or **glycerophosphatides,** which are major components of cell membranes. These lipids can be thought of as derivatives of phosphoric acid and glycerol, with the following structure:

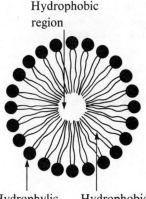

Hydrophobic
region

Hydrophylic   Hydrophobic
head of   tail of
phospholipid   phospholipid
molecule   molecule

Figure 26-12. The phosphatidyl cholines, which possess a charged hydrophilic head and a nonpolar hydrophobic tail, can associate to form micelles in which the hydrophilic heads are oriented toward the surrounding water molecules and the hydrophobic tails share a region of the solution where water molecules are partially excluded.

Figure 26-13. The phosphatidyl cholines can associate to form bilayers that serve as the basis for cell membranes. The nonpolar tails associate to form hydrophobic pockets, and the hydrophilic heads point toward the polar solvent, water. Membranes may also contain other lipids, such as cholesterol, and proteins that are involved in the transport of molecules across the cell membrane. (*From "The Assembly of Cell Membranes," by Harvey F. Lodish and James E. Rothman. Copyright © 1979 by Scientific American, Inc. All rights reserved.*)

$$R-\overset{\overset{\displaystyle O}{\|}}{C}-O-CH_2$$
$$R-\overset{\overset{\displaystyle O}{\|}}{C}-O-CH$$
$$\overset{\|}{O} \quad CH_2-O-\overset{\overset{\displaystyle O}{\|}}{P}-OH$$
$$\overset{|}{OH}$$

Among the most important of the phosphoglycerides are the **phosphatidyl cholines,** or **lecithins,** of which the following is a representative example:

$$CH_3(CH_2)_{16}\overset{\overset{\displaystyle O}{\|}}{C}OCH_2$$
$$CH_3(CH_2)_{16}\overset{\|}{C}OCH \quad O$$
$$\overset{\|}{O} \quad CH_2OPOCH_2CH_2N(CH_3)_3{}^+$$
$$\overset{|}{O^-}$$

Like soaps, phosphatidyl cholines contain a highly charged hydrophilic head and a nonpolar hydrophobic tail. These lipids can spontaneously group to form **micelles** in which the ionic heads face the surrounding water and the nonpolar tails form a hydrophobic pocket (Fig. 26-12). Alternatively, they can form bilayers, which serve as the basis for membranes that separate the cell from its surroundings or segment the interior of the cell (Fig. 26-13).

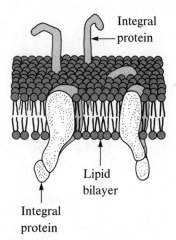

Integral
protein

Lipid
bilayer

Integral
protein

26.10 Simple Lipids

Lipids that do not contain a fatty acid are given the rather misleading name **simple lipids.** The simple lipids include substances such as the **fatty alcohols** (cetyl alcohol and myricyl alcohol, for example), which are long-chain alcohols.

$$CH_3(CH_2)_{14}CH_2OH \qquad CH_3(CH_2)_{29}CH_2OH$$
Cetyl alcohol   Myricyl alcohol

With fatty acids these fatty alcohols form esters (Section 25.24) known as **waxes.** The family of simple lipids also includes the **terpenes,** which include pigments such as β-carotene, fat-soluble vitamins such as vitamin A, and forms of natural rubber (Section 25.26).

A remarkably diverse group of lipids, **steroids,** can be synthesized from terpenes by a complex set of reactions. Among the steroids are such compounds as cholesterol, an important component of cell membranes, and the sex hormones testosterone and estradiol-17 (Fig. 26-14).

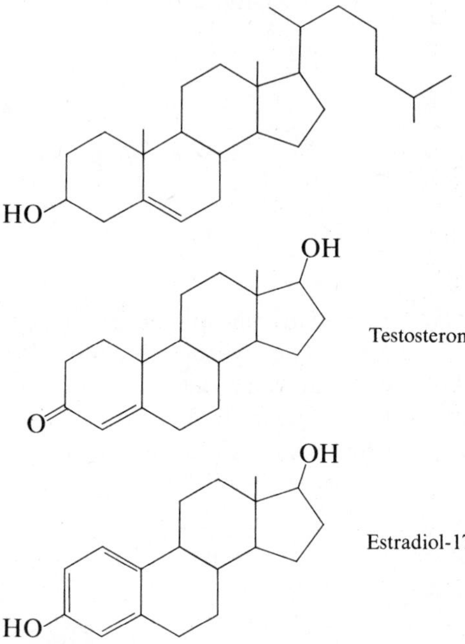

Figure 26-14. The steroids represent a class of compounds that have the same essential fused-ring structure. The steroids function as important components of cell membranes, such as cholesterol; as precursors in the synthesis of vitamins, such as vitamin D_2; as hormones, such as the male and female sex hormones testosterone and estradiol-17; and as ingredients of oral contraceptives.

Cholesterol

Testosterone

Estradiol-17

NUCLEIC ACIDS AND THE GENETIC CODE

26.11 Nucleic Acids

Understanding the chemical nature of nucleic acids is the basis for understanding biochemical genetics. The **nucleic acids** are macromolecules of very high molecular weights whose sole function is the storage and transfer of genetic information. There are two types of nucleic acids: **ribonucleic acids (RNA),** located primarily in the cytoplasm of the cell, and **deoxyribonucleic acids (DNA),** concentrated in the nucleus of the cell.

The ribonucleic acids (RNA) are built on the framework of a β-D-ribofuranose ring, and the deoxyribonucleic acids (DNA) contain a modified ribofuranose ring in which the —OH group on the second, or 2', carbon is replaced by a hydrogen atom.

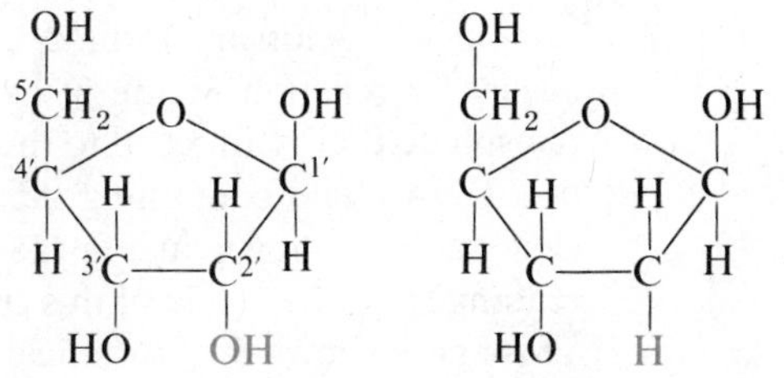

β-D-Ribofuranose   2-Deoxy-β-D-ribofuranose

Nucleosides are formed by combining one of these monosaccharides with one of five nitrogen-containing bases (Fig. 26-15). **Nucleotides** are phosphate esters of nucleosides. For example, the base uracil, which is found only in RNA, not in

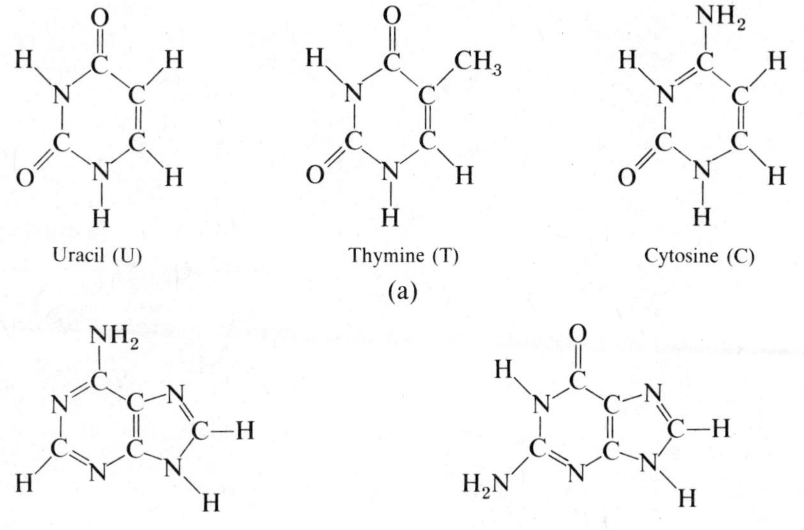

Figure 26-15. The five most common nitrogen-containing bases found in nucleic acids are divided into two families: (a) the pyrimidines uracil (U), thymine (T), and cytosine (C); and (b) the purines adenine (A) and guanine (G).

Uracil (U)   Thymine (T)   Cytosine (C)

(a)

Adenine (A)   Guanine (G)

(b)

DNA, can be joined to a ribofuranose ring at the 1′ carbon to form the nucleoside **uridine.** This nucleoside can then combine with the phosphate ion, PO_4^{3-}, to form the nucleotide **uridine monophosphate.**

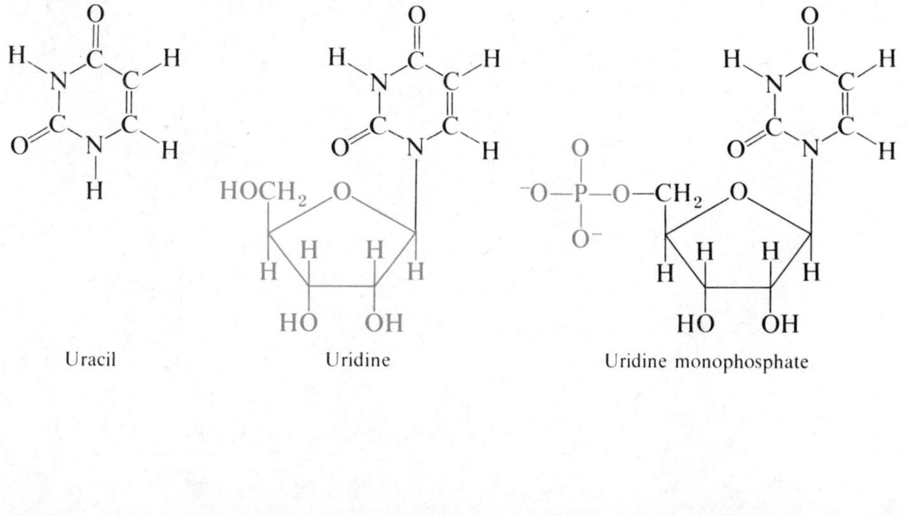

Uracil   Uridine   Uridine monophosphate

Nucleic acids are formed by polymerization of nucleotides. Polymerization occurs by reaction of the 3′ hydroxyl group of one nucleotide with the acidic phosphate that is attached to the 5′ carbon of another to form an ester. A segment of a DNA chain containing the four nucleotides that are found in deoxyribonucleic acids is shown in Fig. 26-16. The sequence of nucleotides in Fig. 26-16, reading from the 5′ end of this chain to the 3′ end, can be symbolized as T-A-C-G. This figure shows the tetranucleotide as it would exist at or near neutral pH. The phosphodiester linkages between adjacent nucleosides in nucleic acids are strong acids. The phosphate groups should therefore be expected to carry a significant

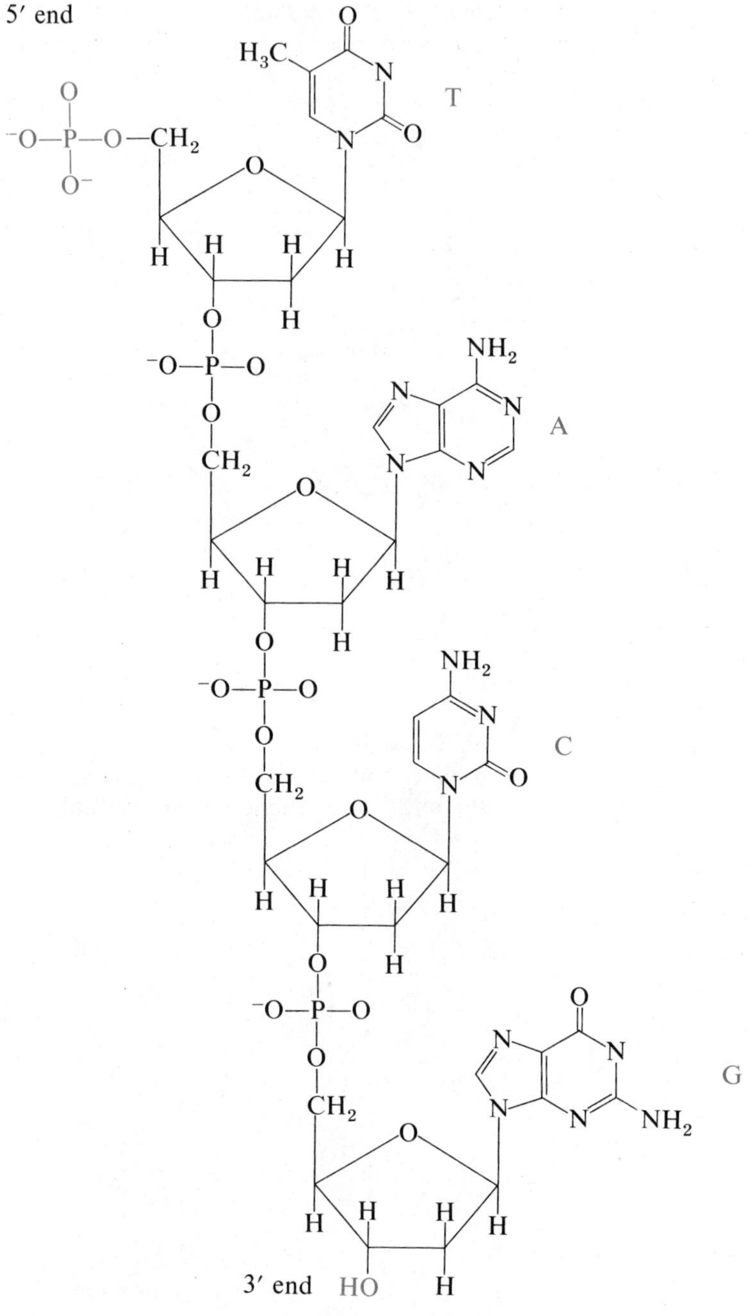

Figure 26-16. The structure of a tetranucleotide segment of a single strand of deoxyribonucleic acid (DNA). Reading from the 5′ to the 3′ end, this tetranucleotide can be symbolized as T-A-C-G. (Carbon atoms in the rings are not shown.)

negative charge in aqueous solution at or near neutral pH. The nucleic acids are therefore normally found associated either with highly charged positive proteins known as **histones** or with cations such as the Mg^{2+} ion.

There are several important structural differences between the nucleic acids DNA and RNA. Each RNA nucleotide contains an —OH group on the 2′ carbon of the furanose ring, but DNA nucleotides do not. Furthermore, the nitrogen-containing bases used to form nucleotides are different. DNA contains exclusively the two purines adenine (A) and guanine (G) and the two pyrimidines cytosine (C) and thymine (T). RNA contains the pyrimidine uracil (U) instead of thymine and often a number of other nucleotides. Evidence suggests, however, that these additional nucleotides are formed by modification of adenine, guanine, cytosine, or uracil after the RNA polymer is produced.

In 1953 J. D. Watson and F. H. C. Crick proposed a mechanism for the storage of genetic information based on the structure of DNA. For this work they subsequently received the Nobel Prize. According to Watson and Crick, DNA is composed of two strands, running in opposite directions, which are bridged by hydrogen bonds between specific pyrimidine groups on one strand and purine groups on the other. These two strands are twisted into a double α-helix structure with approximately 10 nucleotide pairs per 360° turn (Fig. 26-17). Information is

Figure 26-17. The double helix structure of deoxyribonucleic acid (DNA). Schematic drawing (*right*) and model of the double-stranded α-helix (*left*).

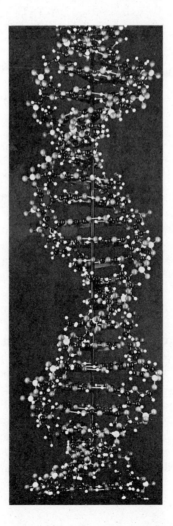

stored in this structure because the strongest hydrogen bonds form between adenine and thymine and between guanine and cytosine (Fig. 26-18). Wherever adenine is found on one strand, thymine will be found on the other, and wherever guanine is found on one strand, cytosine will be found on the other. The two strands of the DNA **double helix** are therefore perfect complements of each other. Knowing the sequence of nucleotides on one strand of the double helix allows the construction of the sequence on the other.

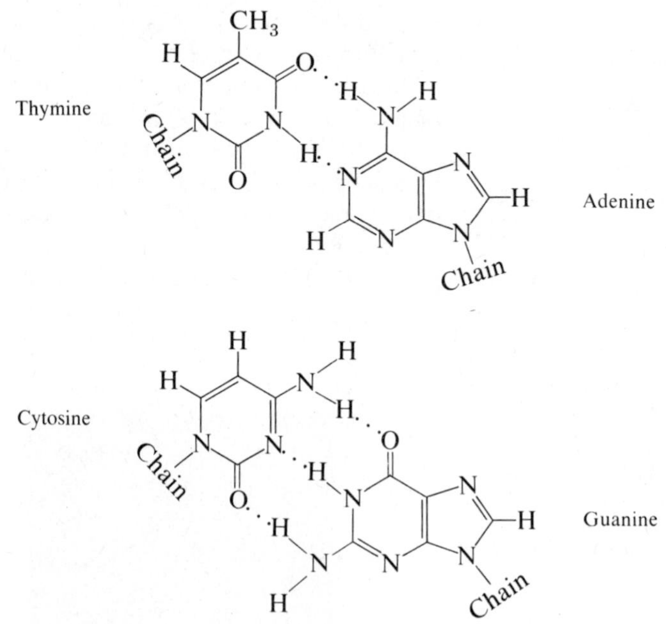

Figure 26-18. The hydrogen bonds in A-T and G-C base pairs. (Carbon atoms in the ring are not shown.)

26.12 The Genetic Code

Modern genetics, and much of modern biochemistry, rests on the theory that the double helix of DNA serves as the medium for storing the information necessary to control the sequence of incorporation of amino acids during protein biosynthesis. To transmit this information during cell division, the DNA molecule must be capable of faultless **replication,** or reproduction, so that each of the new cells gets a perfect copy of the information. To be useful to the cell, the information coded within the structure of the DNA double helix must be capable of being **transcribed,** or read. Finally, a mechanism must exist to **translate** this code into a sequence of amino acids during the assembly of a protein.

The successful replication of the DNA double helix is based on the preference of adenine (A) on one strand of a double helix for thymine (T) on the complementary strand and of guanine (G) for cytosine (C) on the complementary strand. The first step in the replication of DNA during cell division involves the binding of certain proteins that separate a portion of the double helix into two single strands of DNA. Each strand is then copied by the formation of short segments with A in the strand bonded to T in the new segment, G to C, T to A, and C to G. These segments finally join to form the complement of the strand being copied. Replication proceeds until two identical copies of the original DNA molecule are eventu-

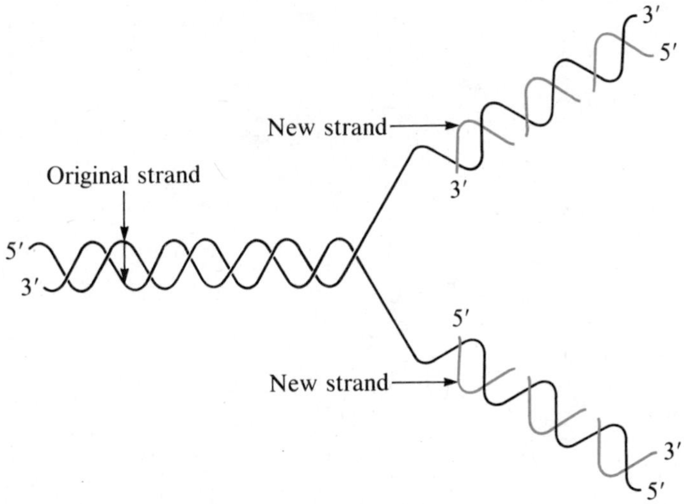

Figure 26-19. The replication of the DNA double helix. Both strands are copied simultaneously, and each strand is copied in small segments that are then joined together. The overall process is semiconservative since each copy retains one strand from the parent molecule.

New strand

Original strand

New strand

DNA double helix

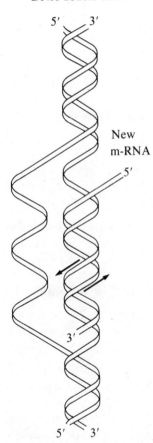

New m-RNA

Figure 26-20. The transcription of a sequence of messenger RNA on the template of one strand of a DNA double helix. Note that the messenger RNA chain is the complement of the DNA strand being copied.

ally formed. Replication is a semiconservative process since each copy retains one strand from the parent DNA molecule (Fig. 26-19).

DNA stores the information that enables a cell to synthesize any of several thousand proteins. RNA processes this information in a two-step mechanism. The first step is called **transcription.** An RNA complement of one strand of the DNA double helix is synthesized under the control of certain enzymes. Since this segment of RNA carries the message contained in the DNA molecule, it is called **messenger RNA,** or **m-RNA.** Just as the two strands of the DNA double helix run in opposite directions, the m-RNA segment synthesized during transcription is a reversed copy of the original DNA strand (Fig. 26-20), with one major difference—RNA contains the nucleotide uracil wherever DNA would contain thymine. Thus the base pairs used during transcription are A-U and G-C.

The second step in the processing of the information contained within the DNA double helix is **translation.** During translation the m-RNA segment serves as the template on which the protein is synthesized. The synthesis of a protein takes place in a ribosome (Section 26.1), which contains both proteins and a second form of RNA known as **ribosomal RNA.** The first step in translation, therefore, is binding of the m-RNA segment to a ribosome. The amino acids are transported to the ribosome by a third form of RNA known as **transfer RNA,** or **t-RNA.** Individual t-RNA molecules, each carrying an amino acid, enter the ribosome and bind to the m-RNA chain at specific positions through hydrogen bonds analogous to those linking the base pairs in the DNA double helix. The amino acids are then joined to form a polypeptide chain.

The sequence of nucleotides in a DNA strand is the basis for coding the sequence of amino acids in proteins. Each sequence of three nucleotides (a **codon**) is specific for a single amino acid. Since there are 64 possible codons and only 20 amino acids to be coded, this genetic code is somewhat redundant (Table 26-1). Among the codons are three that signal chain termination. When a termination codon is encountered during protein biosynthesis, the ribosome releases both the finished polypeptide chain and the m-RNA segment.

The identity of the amino acid coded by any three-nucleotide sequence on a m-RNA segment can be determined from the genetic code in Table 26-1.

Table 26-1   The Amino Acids Coded by Specific Triplets of Nucleotide
Residues on Messenger RNA

First Position (5′ End)	Second Position				Third Position (3′ End)
	U	C	A	G	
U	PHE	SER	TYR	CYS	U
	PHE	SER	TYR	CYS	C
	LEU	SER	Term[a]	Term[a]	A
	LEU	SER	Term[a]	TRP	G
C	LEU	PRO	HIS	ARG	U
	LEU	PRO	HIS	ARG	C
	LEU	PRO	GLN	ARG	A
	LEU	PRO	GLN	ARG	G
A	ILE	THR	ASN	SER	U
	ILE	THR	ASN	SER	C
	ILE	THR	LYS	ARG	A
	MET	THR	LYS	ARG	G
G	VAL	ALA	ASP	GLY	U
	VAL	ALA	ASP	GLY	C
	VAL	ALA	GLU	GLY	A
	VAL	ALA	GLU	GLY	G

[a]The three codons UAA, UGA, and UAG code for the termination of the polypeptide chain.

EXAMPLE 26.1   What is the sequence of amino acids coded by the following sequence of nucleotides on a strand of DNA, reading from the 5′ to the 3′ end: A-T-C-G-C-T-A-C-G-A-A-T?

The m-RNA segment that forms on this DNA template is the complement of this sequence and runs in the opposite direction. Thus the m-RNA chain, reading from the 3′ to the 5′ end, would be U-A-G-C-G-A-U-G-C-U-U-A. Reading this message in the other direction, from the 5′ to the 3′ end, yields the triplets A-U-U, C-G-U, A-G-C, and G-A-U, which code for the sequence ILE-ARG-SER-ASP, reading from the N- to the C-terminal amino acid.

All of the information necessary for the synthesis of the 2000–3000 different proteins found in the simplest cell is contained in the DNA molecules of the cell's **chromosomes.** The portion of the DNA molecule that codes for a single polypeptide chain is known as a **gene.** Each gene is specific for the synthesis of a single kind of protein, although multiple copies of a gene may be found on a chromosome.

26.13   Metabolism

The term **metabolism** is used to describe the chemical reactions that take place in living organisms. Metabolism includes the reactions the cell uses to produce energy, to capture nutrients for the synthesis of biomolecules, to degrade macromolecules to produce the basic building blocks for future construction, and to shuffle intermediates between more abundant and less abundant species. Metabolism is divided into pathways that result in either the synthesis or the degradation of a compound. **Catabolic pathways** lead to the degradation of a metabolic intermediate, and **anabolic pathways** result in the synthesis of a compound.

The energy required for the growth and maintenance of cells is provided by metabolic reactions of two general types: photosynthesis and respiration, or fermentation. **Photosynthesis** is an endothermic process by which the chloroplasts of green plants and algae convert light into chemical energy. The most important set of photosynthetic reactions eventually lead to the reaction of carbon dioxide and water to produce oxygen and a carbohydrate, such as glucose.

$$6CO_2 + 6H_2O \xrightarrow{\text{Light}} C_6H_{12}O_6 + 6O_2 \qquad \Delta G° = 2870 \text{ kJ mol}^{-1}$$

Respiration, or **fermentation,** is a set of reactions that provide energy through the metabolic degradation, or catabolism, of organic compounds. The reactions involved in respiration can be divided into two categories: (1) those that can occur in the absence of molecular oxygen **(anaerobic reactions),** and (2) those that require oxygen **(aerobic reactions).** Anaerobic pathways typically do not lead to a net oxidation or reduction of a metabolic intermediate and thus liberate significantly smaller amounts of energy than do the oxidation-reduction reactions that accompany aerobic pathways.

The sequence of reactions involved in the anaerobic degradation of glucose is known collectively as **glycolysis.** Glycolysis ultimately degrades a single molecule of glucose into two pyruvate ions, with the release of more than 200 kilojoules of energy per mole of glucose consumed.

Glucose   Pyruvate ion

Anaerobic organisms, such as the yeast cells used in the production of alcoholic beverages by the fermentation of sugars, convert the pyruvate produced in glycolysis to either acetaldehyde or ethanol.

Aerobic organisms ultimately convert pyruvate into carbon dioxide and water through a sequence of reactions known as the **citric acid cycle,** or the **Krebs cycle.** The net reaction in aerobic organisms is thus the opposite of the photosynthetic reaction and liberates approximately 2870 kilojoules per mole of glucose consumed.

$$C_6H_{12}O_6 + 6O_2 \longrightarrow 6CO_2 + 6H_2O \qquad \Delta G° = -2870 \text{ kJ mol}^{-1}$$

About 50% of the energy released during the degradation of glucose is lost in the form of heat. The remainder is stored as chemical energy through the synthesis of nucleoside triphosphates such as **adenosine triphosphate, ATP.**

ATP, and the related triphosphates GTP, CTP, and UTP, can react with water to produce adenosine diphosphate, ADP (or GDP, CDP, or UDP), and the phosphate ion, with the release of approximately 31 kilojoules per mole of ATP hydrolyzed. The nucleoside diphosphates, such as ADP, can undergo further hydrolysis to form nucleoside monophosphates, such as AMP, with the release of an additional 30 kilojoules per mole. The nucleoside triphosphates, and to some extent the diphosphates, serve as the cell's short-term energy-storage mechanism. ATP, for example, is synthesized from AMP and ADP during the degradation of carbohydrates, proteins, and lipids. When the cell needs energy either to drive a chemical reaction, such as the synthesis of proteins from amino acids, or for a mechanical process, such as the contraction of a muscle, ATP or its equivalent is consumed.

Glycolysis and the citric acid cycle are important means of producing biological energy in the form of ATP, but these pathways have other functions as well. There are several points where intermediates are produced that can be used in the synthesis of amino acids, or where intermediates from degradation of amino acids and proteins can be inserted for eventual conversion to carbohydrates. The metabolism of carbohydrates and the metabolism of proteins are therefore intimately related as are those of carbohydrates and the complex lipids.

The complete degradation of either carbohydrates or lipids leads eventually to the same products: carbon dioxide and water. However, $2\frac{1}{2}$ times more energy per gram is released during the oxidation of lipids. In times of plenty, that is, when biological energy is abundant, the cell first stores this energy by synthesizing ATP from ADP and/or AMP. Once this short-term storage capacity is exceeded, the cell may store additional energy by synthesizing carbohydrates. These carbohydrates may then be linked to form polysaccharides such as glycogen and starch, which serve as a more efficient means of storing energy for times of shortage. In times of extreme plenty, the cell stores excess energy in the form of complex lipids, or fats. Each of these forms of storage—ATP, carbohydrate, polysaccharide, and lipid—represents a progressively more efficient means of storing excess energy during times of plenty to provide for times of shortage that could come. It is an unfortunate fact, in modern times, that for some humans these times of shortage never come, while for others, the times of plenty are all too infrequent.

FOR REVIEW

SUMMARY

Molecules present in living organisms are called biomolecules and are often divided into four classes—proteins, carbohydrates, lipids, and nucleic acids—each having its own structure and functions.

Proteins are polymers composed of **amino acids,** molecules with the general formula H_3N^+—CHR—CO_2^-. The 20 amino acids that are genetically coded into proteins are distinguished by the nature of the R side chain on the α-carbon. Most amino acids are optically active. The combination of two or more amino acids by formation of **peptide bonds,**

$$\underset{\underset{}{\overset{\overset{O}{\parallel}}{-C-NH-}}}{}$$

produces a **polypeptide.** Proteins are large polypeptides that can fulfill a variety of functions because their structures vary widely. The **primary structure** of a protein is the order in which its amino acids are connected. The **secondary structure** is produced by the folding of the chain of amino acids to maximize hydrogen bonding between peptide bonds; an **α-helix** or a **β-sheet (pleated sheet)** is a common secondary structure. One of the most remarkable classes of proteins is the **enzymes,** which are catalysts of biochemical reactions.

Carbohydrates are aldehydes and ketones that have two or more —OH groups and that can be energy sources or energy stores in plants and animals. The simplest carbohydrates are the **monosaccharides,** such as the sugars glucose and fructose. **Disaccharides** are dimers composed of two monosaccharides; sucrose, for example, consists of glucose and fructose. **Starch,** a food storage carbohydrate found in plants; **glycogen,** a food storage carbohydrate found in animals; and **cellulose,** the carbohydrate that is the structural component of plants, are all **polysaccharides,** large polymers formed from glucose.

Fats and **oils** are members of the class of molecules known as **complex lipids,** esters formed from long-chain carboxylic acids **(fatty acids)** and glycerol. The closely related **phosphoglycerides** serve as the basis for cell membranes. **Simple lipids,** which include vitamin A and the steroids, do not contain fatty acids.

Nucleic acids are polymers of very high molecular weights that store and transfer genetic information. They also determine which proteins are synthesized by a cell. A nucleic acid consists of a polymeric chain of alternating phosphate groups and ribofuranose rings with nitrogen-containing bases bonded to the rings. The **DNA** polymer is a **double helix** that is held together by hydrogen bonds between pairs of specific bases, one on each strand of the helix. This base pairing makes each strand a complement of the other. The sequence of groups of three nucleotides **(codons)** in a DNA strand is the basis for coding the sequence of amino acids in proteins.

KEY TERMS AND CONCEPTS

Adenosine triphosphate, ATP (26.13)
Aerobic reactions (26.13)
Aldose (26.7)
Anabolic pathways (26.13)
Anaerobic reactions (26.13)
Amino acids (26.2)
Carbohydrates (26.6)
Cellulose (26.8)
Chromosomes (26.12)
Codon (26.12)
Complex lipids (26.9)
Deoxyribonucleic acid, DNA (26.11)
Disaccharides (26.8)

Double helix (26.11)
Enzymes (26.5)
Gene (26.12)
Genetic code (26.12)
α-Helix (26.4)
Hydrophilic (26.1)
Hydrophobic (26.1)
Ketose (26.7)
Inhibition (26.5)
Lipid (26.9, 26.10)
Metabolism (26.13)
Monosaccharides (26.7)
Nucleic acids (26.11)
Nucleosides (26.11)
Nucleotides (26.11)

Oligopeptides (26.3)
Peptides (26.3)
Photosynthesis (26.13)
Polypeptides (26.3)
Polysaccharides (26.8)
Primary structure (26.4)
Proteins (26.3)
Quaternary structure (26.4)
Replication (26.12)
Ribonucleic acid, RNA (26.11)
Secondary structure (26.4)
β-Sheet (26.4)
Simple lipids (26.10)
Tertiary structure (26.4)
Zwitterion (26.2)

EXERCISES

1. List the four principal chemical constituents of living cells. Identify the specific functions of each of these constituents.
2. Define *hydrophilic* and *hydrophobic*. What factor or factors determine whether a compound is hydrophilic or hydrophobic? Give an example of each class of compound.

Proteins

3. Into what simpler components may a protein be hydrolyzed?
4. Define the following: amino acid, peptide, and protein.
5. Write symbolic formulas for all of the possible tripeptides that could be formed from the amino acids GLY, ALA, and VAL.
6. Write the structure of the tripeptide CYS-LYS-GLU at neutral pH.
7. Distinguish between the primary, secondary, tertiary, and quaternary structures of a protein.
8. Define what is meant by *denaturation*.
9. Define the following: enzyme, substrate, and product.
10. Describe what happens to the rate of an enzyme-catalyzed reaction from the instant that the enzyme and substrate are mixed until the substrate is consumed.

11. Fungal lactase, a blue protein found in wood-rotting fungi, contains approximately 0.397% copper by mass. If the molecular weight of fungal lactase is approximately 64,000, how many copper atoms are there in each protein molecule?

    Ans. 4 Cu atoms/protein molecule

Carbohydrates

12. What is the literal meaning of *carbohydrate,* and how did this name arise?
13. Write the name and formula for one example each of a monosaccharide, a disaccharide, and a polysaccharide.
14. How does a pyranose differ from a furanose?
15. What is the difference between the α and β isomers of glucopyranose?
16. Which carbon atoms in glucose are asymmetric (Section 25.16)?
17. How many grams of invert sugar can be produced by the hydrolysis of 1.00 g of sucrose, $C_{12}H_{22}O_{11}$?   *Ans. 1.05 g of invert sugar*

Lipids

18. Write the general formula for a neutral fat. How do fats and oils differ in composition?
19. Define saponification and write an equation for

the saponification of tripalmitylglycerol; the triester of palmitic acid, $CH_3(CH_2)_{14}CO_2H$.

20. Describe the common feature or features of complex and simple lipids, and provide an example of each class of compound. In what way do complex and simple lipids differ?

21. The molecular weight of an unknown triglyceride can be estimated by determining the mass of KOH required to saponify a known mass of lipid. What is the molecular weight of a triglyceride if 3.12 mg of the triglyceride requires 0.590 mg of KOH for saponification?

Ans. 890 g/mol (tripalmitylglycerol)

22. Explain why the triacylglycerols do not form micelles.

Nucleic Acids

23. What is the difference between the nucleotides adenosine monophosphate and deoxyadenosine monophosphate?

24. What is the connection between nucleic acids, amino acids, and proteins?

25. Complete the following table.

	DNA	RNA
Sugar unit		
Purine bases		
Pyrimidine bases		

26. Write the complete structure of the following segment of DNA, reading from the 5′ end to the 3′ end: G-T-C-A.

27. If a single strand of DNA contains the nucleotide sequence A-G-G-C-T-C-A-G-C-T-A-G, reading from the 5′ to the 3′ carbon, what would be the sequence of nucleotides on the complementary strand of the DNA double helix, reading in the same direction?

28. What would be (a) the product of replication and (b) the product of transcription of the following fragment of a DNA molecule?

5′ G-T-C-A-A-T-G-G-A 3′

Additional Exercises

29. Distinguish between aerobic and anaerobic processes.

30. The synthesis of organic molecules in nature from carbon dioxide and water is an endothermic process. What is the source of the required energy?

31. Discuss the manner in which the energy requirements of a cell are satisfied.

32. The degradation of both fatty acids, such as palmitic acid, and carbohydrates, such as glucose, leads to the formation of carbon dioxide and water. However, the degradation of palmitic acid produces $2\frac{1}{2}$ times as much energy as that of glucose, per gram of substance consumed. Explain this observation in terms of the average oxidation numbers of the carbon atoms in fatty acids, carbohydrates, and carbon dioxide.

33. Would the alanine in a solution with a pH of 5.0 migrate toward a positive electrode, a negative electrode, or show no preference for either electrode on electrolysis of the solution?

34. In what sense can sucrose be considered an anhydride?

35. What sequence of amino acids would be specified by a single strand of DNA containing the nucleotide sequence A-G-G-C-T-C-A-G-C-T-A-G, reading from the 5′ to the 3′ carbon?

36. What evidence do we have that a twenty-first genetically coded amino acid will not be found?

37. How would you explain the observation that proteins can contain additional amino acids that are structurally related to the 20 genetically coded amino acids?

38. If the degradation of 1 mol of glucose and 1 mol of palmitic acid yields 38 and 129 mol of ATP, respectively, calculate the moles of ATP produced per gram of lipid or carbohydrate consumed.

Ans. 0.21 mol ATP/g glucose; 0.504 mol ATP/g palmitic acid

27

BORON AND SILICON

Boron and silicon are members of different groups of the Periodic Table (Groups IIIA and IVA, respectively), but in many respects their chemical behavior is similar. This is one of several pairs of similar elements, one of which is in both the next family and the next group from the other—a diagonal relationship in the Periodic Table. For example, lithium is similar in chemical properties to magnesium; beryllium is similar to aluminum; and, as noted, boron is similar to silicon. This phenomenon arises because the ratio of the atomic radius to the nuclear charge for the two elements is about the same, so the outer-shell electrons are attracted to the nucleus with about the same force.

A comparison of some of the properties of boron (the first member of Group IIIA) with those of aluminum (the second member of Group IIIA) and those of silicon (the second member of Group IVA) illustrates this behavior. Boron resembles aluminum in forming compounds in which it is trivalent—for example, B_2O_3 and Al_2O_3, $B(OH)_3$ and $Al(OH)_3$, and BCl_3 and $AlCl_3$. However, BCl_3 is more like $SiCl_4$ than like $AlCl_3$ in physical and chemical properties. Boron and silicon chlorides are liquids at $0°C$, aluminum chloride is a solid. Aluminum chloride hydrolyzes reversibly, forming a weak base.

$$AlCl_3 + 3H_2O \rightleftharpoons Al(OH)_3(s) + 3H^+ + 3Cl^-$$

Boron trichloride and silicon tetrachloride hydrolyze almost completely and irreversibly, forming very weak acids.

$$BCl_3 + 3H_2O \longrightarrow B(OH)_3(s) + 3H^+ + 3Cl^-$$
$$SiCl_4 + 4H_2O \longrightarrow Si(OH)_4(s) + 4H^+ + 4Cl^-$$

Boron and silicon are primarily nonmetallic in their chemical properties, but both show some metallic character in the elemental state. For this reason boron and silicon are included in the elements called semimetallic elements, or metalloids (Section 8.3), which are intermediate in nature between metals and nonmetals.

BORON

27.1   Occurrence and Preparation of Boron

Boron and its compounds have been known for a long time. The boron-containing compound borax is mentioned in early Latin works on chemistry. Impure elemental boron was first prepared in 1808 by Gay-Lussac and Thénard in France and by Davy in England, by reducing boric acid with potassium.

Boron, in compounds with oxygen, constitutes less than 0.001% of the earth's crust. It does not occur in the free state in nature. Boron is widely distributed as **boric acid, B(OH)$_3$,** or **H$_3$BO$_3$,** in volcanic regions and as **borates,** such as **borax, Na$_2$B$_4$O$_7 \cdot$ 10H$_2$O, kernite, Na$_2$B$_4$O$_7 \cdot$ 4H$_2$O,** and **colemanite, Ca$_2$B$_6$O$_{11} \cdot$ 5H$_2$O,** in dry lake regions including the desert areas of southern California.

Reduction of boric oxide with powdered magnesium produces impure boron as a brown amorphous powder.

$$B_2O_3 + 3Mg \longrightarrow 2B + 3MgO$$

The magnesium oxide is removed by dissolving it in hydrochloric acid. Pure boron may be obtained by passing a mixture of boron trichloride and hydrogen either through an electric arc or over a hot tungsten filament.

$$2BCl_3(g) + 3H_2(g) \xrightarrow{\text{1500°C}} 2B(s) + 6HCl(g) \qquad \Delta H° = 253.7 \text{ kJ}$$

27.2   The Structure of Elemental Boron

Crystalline boron is transparent and nearly as hard as diamond. Boron crystals have a rather unusual structure; boron atoms occupy the 12 corners of an icosahedron (Fig. 27-1). Icosahedra are packed together, in a manner similar to cubic closest packing of spheres, in the most common form of boron (α-rhombohedral). The length of the bonds between adjacent boron atoms in each icosahedron is 1.76 Å. However, bonds of two different lengths (1.71 Å and 2.03 Å) are present between the icosahedra (Fig. 27-2).

Each of six of the twelve boron atoms of an icosahedron is joined by a regular two-electron B—B bond (1.71 Å in length) to a boron atom of another icosahedron (these bonds are not shown in Fig. 27-2). In Chapters 5 and 6, we discussed such bonds in which two atoms are bonded together by one pair of electrons in a bond that results from the overlap of two atomic orbitals (Section 6.1). These bonds are called **two-electron two-center bonds.** Each of the other six atoms of

Figure 27-1.   An icosahedron, a symmetrical solid shape with 20 faces, each of which is an equilateral triangle. The faces meet at 12 corners.

Figure 27-2. The structure of α-rhombohedral boron. Each icosahedron contains twelve boron atoms. Each of six of these is bonded to a boron atom in another icosahedron by a two-center bond (1.71 Å) and each of the other six of these is bonded to two boron atoms, each in a separate icosahedron, by a three-center bond (2.03 Å). Only the three-center bonds between icosahedra are shown here.

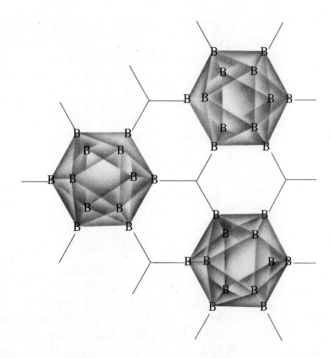

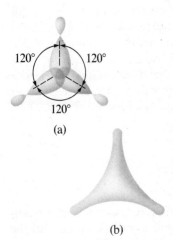

Figure 27-3. The three-center B—B—B bond involving the overlap of three sp^3 atomic orbitals, which share one pair of electrons. (a) The overlap diagram for the atomic orbitals. (b) The overall outline of the bonding molecular orbital.

each icosahedron is bonded to *two* other atoms, one in each of two other icosahedra, by a two-electron three-center bond (2.03 Å in length), as shown in Fig. 27-2.

In a **two-electron three-center bond,** *three* atoms are bonded together by one pair of electrons in a molecular orbital formed from the overlap of three atomic orbitals. The three-center bond is especially likely for atoms that do not have sufficient electrons to satisfy ordinary two-center bond requirements. Boron makes the greatest use of the three-center bond; it exhibits this type of bond not only in elemental boron but also in several boron hydrides (Section 27.4). As was pointed out in Chapter 6, in the regular two-center bond, one bonding molecular orbital and one antibonding molecular orbital result from the overlap of two atomic orbitals (Section 6.1). In the three-center bond, one bonding molecular orbital and two antibonding molecular orbitals (or one bonding orbital, one non-bonding orbital, and one antibonding orbital) result from the overlap of *three* atomic orbitals. The two electrons occupy the bonding molecular orbital. The formation of the bonding orbital is shown in Fig. 27-3.

27.3 The Chemical Behavior of Boron

The valence shell electron configuration of a boron atom is $2s^2 2p^1$. Boron is covalent in its compounds, such as BF_3 and B_2O_3 (Fig. 11-11), with three single covalent bonds to boron. The geometry about boron in such compounds is trigonal planar, and the boron atom is sp^2 hybridized. The p orbital that is not hybridized is empty; consequently, boron compounds with three covalent bonds are strong Lewis acids. The addition of a fourth group in a Lewis acid-base reaction gives a four-coordinated boron atom with a tetrahedral geometry and sp^3 hybridization.

Boron has a large affinity for fluorine and oxygen. It combines with fluorine at room temperature and with chlorine, bromine, oxygen, and sulfur at elevated temperatures, but it does not combine directly with iodine. Boron reacts with carbon in an electric arc to form **boron carbide, B_4C,** which is second only to diamond in hardness and is used in cutting tools. With the exceptions of BF_3 and B_4C, boron compounds hydrolyze or oxidize readily to give boric acid, $B(OH)_3$, or boric oxide, B_2O_3.

Boron acts as a reducing agent when heated with water, sulfur dioxide, nitric oxide, carbon dioxide, or any of many other oxides. It is used to deoxidize fused copper before it is cast. It also reduces both concentrated nitric and sulfuric acids, forming boric acid. Most of boron's chemical properties are characteristic of a nonmetal; its halides are volatile and irreversibly hydrolyzed. However, boron is intermediate in character between nonmetals and metals; this is evidenced in part by the fact that $B(OH)_3$ is a very weak acid.

27.4   Boron Hydrides

Boron forms a series of volatile hydrides that are quite different from the hydrides of carbon and silicon. The hydride having the molecular formula BH_3 is not stable at room temperature, although Lewis acid-base adducts, such as H_3B-CO, are known. The simplest and most important of the boron hydrides is **diborane, B_2H_6,** the dimer of BH_3. However, an electronic structure cannot be written for B_2H_6 that is consistent with the theory of regular covalent bonds as outlined in Chapter 5 and with the properties of the compound. It would require 14 valence electrons (7 electron pairs) to form the covalent structure

$$\begin{array}{ccc} H & & H \\ | & & | \\ H-B & - & B-H \\ | & & | \\ H & & H \end{array}$$

but only 12 electrons are available for bond formation. (Compounds of this sort are sometimes called electron-deficient.) The structure most in keeping with the properties of diborane is one with two hydrogen bridges (Fig. 27-4). Each boron atom is bonded to two hydrogen atoms by regular two-center single covalent bonds. These four bonds are all in the same plane and use 8 of the 12 valence electrons. A hydrogen above and below the plane is connected to the two boron atoms by a two-electron three-center B—H—B bond. The three atoms in each bond are bonded by a molecular orbital that contains one pair of electrons, analogous to the B—B—B three-center bond in elemental boron. The two three-center bonds hold the remaining 4 electrons.

Figure 27-4. The structure of the diborane molecule, B_2H_6. (a) The spatial arrangement of atoms showing the bridging hydrogens (in color). (b) The spatial arrangement of the bonding orbitals.

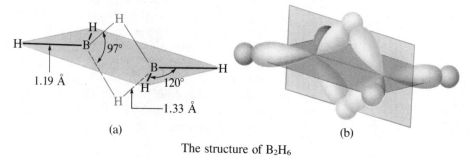

(a)

(b)

The structure of B_2H_6

The reaction of lithium aluminum hydride with boron trifluoride in ether solution yields diborane.

$$4BF_3 + 3LiAlH_4 \longrightarrow 2B_2H_6(g) + 3LiF + 3AlF_3$$

Above 300°C diborane rapidly decomposes into boron and hydrogen. It ignites spontaneously in moist air and forms boric oxide and water. Diborane reacts violently with chlorine to form boron trichloride and hydrogen chloride. It is hydrolyzed by water to boric acid and hydrogen.

$$B_2H_6(g) + 6H_2O(l) \longrightarrow 2H_3BO_3(s) + 6H_2(g) \qquad \Delta H° = -509.2 \text{ kJ}$$

Hydrides of higher molecular weight than B_2H_6 are also possible, and many have been made. Examples include B_4H_{10}, B_5H_9, B_6H_{10}, and $B_{10}H_{14}$.

When an excess of sodium hydride reacts with boron trifluoride, **sodium borohydride, NaBH$_4$,** is formed. The four hydrogen atoms in the tetrahedral anion $[BH_4]^-$ are covalently bonded to the boron atom.

$$Na^+ \begin{bmatrix} & H & \\ & | & \\ H & -B- & H \\ & | & \\ & H & \end{bmatrix}$$

Sodium borohydride

Sodium borohydride and other similar complex hydrides, such as lithium aluminum hydride, $LiAlH_4$, are very useful reducing agents, particularly for organic compounds.

27.5 Boron Halides

Boron trifluoride, boron trichloride, and **boron tribromide** can be prepared by direct union of the elements. These halides are trigonal planar molecules in which boron is sp^2 hybridized. The trifluoride is a colorless gas that hydrolyzes in water to form boric acid and hydrofluoric acid.

$$BF_3(g) + 3H_2O(l) \longrightarrow H_3BO_3(s) + 3HF(g) \qquad \Delta H° = 86.6 \text{ kJ}$$

The hydrofluoric acid combines with excess boron trifluoride and gives **fluoroboric acid, HBF$_4$.**

$$BF_3 + HF \longrightarrow H^+ + BF_4^-$$

In this latter reaction the BF_3 molecule acts as a Lewis acid (electron-pair acceptor) and accepts a pair of electrons from a fluoride ion, as shown by the equation

Fluoroboric acid is a strong acid that is stable only in solution.

The trichloride of boron is a colorless mobile liquid, the tribromide is a viscous liquid, and the triiodide is a white crystalline solid.

27.6 Boric Oxide and Oxyacids of Boron

Boron burns at 700°C in oxygen, forming **boric oxide, B$_2$O$_3$.** Boric oxide is used in the production of chemically resistant glass and certain optical glasses. The oxide dissolves in hot water to form **boric acid, H$_3$BO$_3$.**

$$B_2O_3 + 3H_2O \longrightarrow 2H_3BO_3$$

The boron atom in H_3BO_3 is at the center of an equilateral triangle with oxygen atoms at the corners.

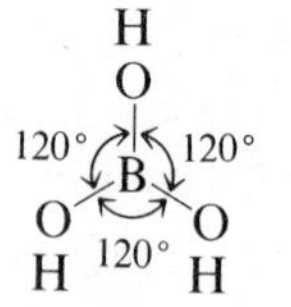

Boric acid

In the solid acid these triangular units are held together by hydrogen bonding.

When boric acid is heated to 100°C, a molecule of water is formed from a pair of adjacent —OH groups, leaving **metaboric acid, HBO₂.** With further heating, at about 150°C, additional B—O—B linkages form, connecting BO_3 groups together through shared oxygen atoms, to form **tetraboric acid, H₂B₄O₇.** At still higher temperatures, boric oxide is formed.

$$H_3BO_3 \longrightarrow HBO_2 + H_2O$$
$$4HBO_2 \longrightarrow H_2B_4O_7 + H_2O$$
$$H_2B_4O_7 \longrightarrow 2B_2O_3 + H_2O$$

27.7 Borates

Borates are salts of the oxyacids of boron. Borates result from the reaction of a base with an oxyacid or from the fusion of boric acid or boric oxide with a metal oxide or hydroxide. Borate anions range from the simple trigonal planar BO_3^{3-} ion to complex species containing chains and rings of three- and four-coordinated boron atoms. The structures of the anions found in CaB_2O_4, $K[B_5O_6(OH)_4] \cdot 2H_2O$ (commonly written $KB_5O_8 \cdot 4H_2O$), and $Na_2[B_4O_5(OH)_4] \cdot 8H_2O$ (commonly written $Na_2B_4O_7 \cdot 10H_2O$) are shown in Fig. 27-5.

Figure 27-5. The borate anions found in (a) CaB_2O_4, (b) $KB_5O_8 \cdot 4H_2O$, and (c) $Na_2B_4O_7 \cdot 10H_2O$. The anion in CaB_2O_4 is an infinite chain.

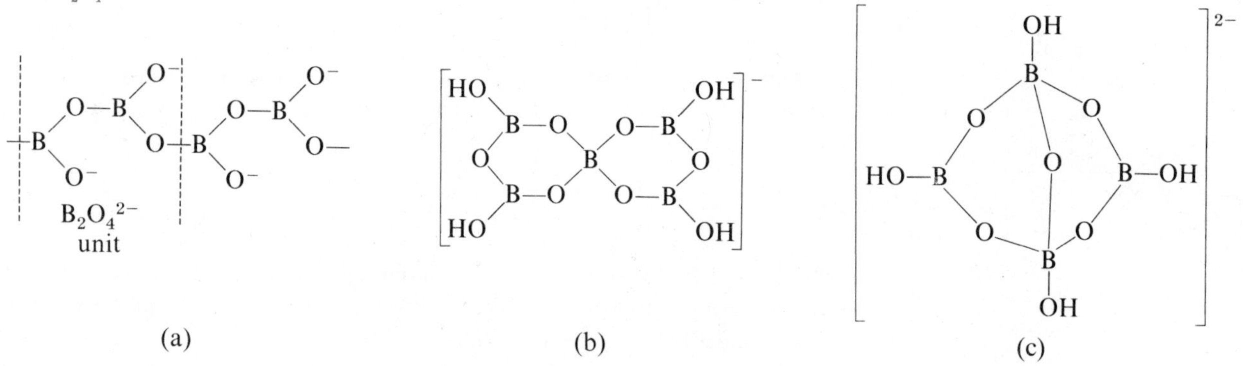

(a)   (b)   (c)

Commercially, the most important borate is **borax, or sodium tetraborate 10-hydrate, Na₂B₄O₇ · 10H₂O.** Most of the supply of borax comes directly from dry lakes, such as Searles Lake in California, or is prepared from kernite, $Na_2B_4O_7 \cdot$

$4H_2O$. Borax is a salt of a strong base and a weak acid, so its aqueous solutions are basic due to the hydrolysis of the tetraborate ion.

$$B_4O_7^{2-} + 7H_2O \rightleftharpoons 4H_3BO_3 + 2OH^-$$

Borax is used to soften water and to make washing compounds. These uses depend on the alkaline character of its solutions and the insolubility of the borates of calcium and magnesium.

When heated, borax fuses to form glass that dissolves metal oxides. An example is given by the equation

$$Na_2B_4O_7 + CuO \longrightarrow 2NaBO_2 + Cu(BO_2)_2$$

Because of this property, borax is used as a flux to remove oxides from metal surfaces in soldering and welding. Different metals form borax glasses of different colors, a fact applied to detecting certain metals. Cobalt borax glass is blue, for example.

27.8 Boron-Nitrogen Compounds

A number of compounds containing boron-nitrogen bonds have been prepared and studied. Two of particular interest are **boron nitride, BN,** and **borazine, $B_3N_3H_6$.**

Boron nitride is the final product of the thermal decomposition of many boron-nitrogen compounds, such as $B(NH_2)_3$ and $BF_3 \cdot NH_3$. It can also be prepared by heating boron with nitrogen or ammonia or by heating borax with ammonium chloride. Crystalline boron nitride is a white solid that sublimes somewhat below 3000 °C; it melts at this temperature under pressure. It is inert to most reagents but can be decomposed by reaction with molten bases. The crystalline structure of boron nitride is analogous to that of graphite (Section 25.2 and Fig. 25-2)—layers of atoms in both are made up of hexagonal rings.

Boron nitride   Graphite
layer   layer

This form of boron nitride is sometimes called **inorganic graphite.**

A second form of boron nitride, **Borazon,** has the same structure as diamond (Section 25.1 and Fig. 25-1) and is almost as hard as diamond. It is now produced commercially.

Borazine, $B_3N_3H_6$, and hydrogen are formed when ammoniates of the boron hydrides, such as $BH_3 \cdot NH_3$, are heated at 180–200 °C. Borazine has been called **inorganic benzene,** because its structure and physical properties are similar to those of benzene; however, its chemistry is different because the B—N bonds are polar and the C—C bonds are not.

$$
\begin{array}{cc}
\text{Borazine} & \text{Benzene}
\end{array}
$$

Since a boron atom and a nitrogen atom bonded together have the same number of electrons as two carbon atoms and the sizes of these two groups are about the same, it is not surprising that the structures of boron nitride are similar to those of graphite and diamond and that borazine resembles benzene.

SILICON

In Chapters 25 and 26 we examined the chemical behavior and compounds of carbon, the first member of Group IVA. The rest of this chapter deals with the second member of the group, silicon. Carbon, with its ability to form compounds containing C—C bonds, plays the dominant structural role in the animal and vegetable worlds, and silicon, which readily forms compounds containing Si—O—Si bonds, is of prime importance in the mineral world. Indeed, the name *silicon* is derived from the Latin word for flint, *silex*.

27.9 Occurrence and Preparation of Silicon

The earth's crust is composed almost entirely of minerals in which silicon atoms are connected by oxygen atoms in complex structures involving chains, layers, and three-dimensional frameworks. These minerals constitute the bulk of most common rocks (except limestone and dolomite), soils, clays, and sands. Materials such as granite, bricks, cement, mortar, ceramics, and glasses are composed of silicon compounds. Silicon is nearly one-fourth of the mass of the earth's crust; only oxygen is more abundant. Its binary compound with oxygen is called **silica, SiO_2.** Sand and sandstone, as well as quartz, amethyst, agate, and flint, are forms of impure silica. Most rocks are built up of common metal cations and silicate anions. A great variety of silicates exist in nature, and many of these have important specific uses. However, they are so stable that it is not economical to use them as sources of metals (except beryl, $Be_3Al_2Si_6O_{18}$, from which beryllium is obtained).

Elemental silicon was first prepared in an impure, amorphous form by Berzelius in 1823 by heating silicon tetrafluoride with potassium. It can also be obtained by the reduction of silicon dioxide with strong reducing agents at high temperatures. Using carbon and magnesium as the reducing agents, the equations are

$$SiO_2(s) + 2C(s) \longrightarrow Si(s) + 2CO(g)$$
$$SiO_2(s) + 2Mg(s) \longrightarrow Si(s) + 2MgO(s)$$

Extremely pure silicon, such as that required for the fabrication of electronic semiconductor devices, is prepared by the decomposition of silicon tetrahalides or silane, SiH_4, at high temperatures.

27.10 Properties and Uses of Silicon

Silicon crystallizes with a diamondlike structure; each silicon atom is covalently bonded to four neighboring silicon atoms at the corners of a regular tetrahedron. A single crystal of silicon is a giant three-dimensional molecule. Thus silicon is hard enough to scratch glass. It is a brittle, gray-black, metallic-appearing solid that melts at 1410°C.

Silicon is unreactive at low temperatures and resists attack by air, water, and acids. However, it is active enough that it does not occur in the free state on the earth's surface. It reacts with strong oxidizing agents and strong bases. It dissolves in hot solutions of sodium hydroxide and potassium hydroxide, forming silicates and hydrogen.

$$Si + 2OH^- + H_2O \longrightarrow SiO_3{}^{2-} + 2H_2(g)$$

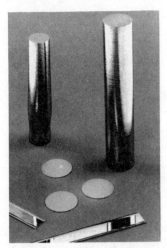

Figure 27-6. Single crystals of silicon. Such crystals are produced for use in solar cells, semiconductors, and other devices. (*Courtesy of Dow Corning Corporation*)

Silicon reacts with the halogens at high temperatures and forms volatile tetrahalides, such as SiF_4. It oxidizes in air at elevated temperatures to give the dioxide, SiO_2.

Elemental silicon is used as a deoxidizer in the production of steel, copper, and bronze, and in the manufacture of certain acid-resistant alloys such as Duriron and Tantiron. Highly purified silicon, containing no more than 1 part of impurity per million parts of silicon, is used in electronic semiconductor devices such as transistors, in microcomputers, and in solar cells (Fig. 27-6).

27.11 The Chemical Behavior of Silicon

Silicon is a metalloid; it exhibits both nonmetallic and metallic character. Although both carbon and silicon have a valence electron shell containing two *s* and two *p* electrons, little of the chemistry of silicon can be inferred from that of carbon except for the stoichiometry of some simple halides and hydrides. Silicon forms compounds in which it is sp^3 or sp^3d^2 hybridized and participates in four or six single covalent bonds, respectively. It does not form double or triple bonds. Silicon forms a series of hydrides and a series of halides, both having the formula Si_nX_{2n+2} (where X is hydrogen or a halide) and containing Si—Si single bonds. However, these compounds are much more reactive, much less stable, and much more limited in number than the corresponding carbon compounds because the Si—Si bond ($D_{Si-Si} = 175\ kJ/mol$) is much weaker than the C—C bond ($D_{C-C} = 350\ kJ/mol$). On the other hand, Si—O, Si—C, and Si—F bonds are particularly stable. Exposure of most silicon compounds to water or oxygen results in their decomposing to compounds containing Si—O bonds, unless the silicon compounds are stabilized by the presence of Si—O, Si—C, or Si—F bonds.

The valence shell of silicon contains *d* orbitals, and it can expand to hold more than the octet. Thus silicon compounds of the general formula SiX_4, where X is a highly electronegative species or group, can act as Lewis acids giving six-coordinated silicon (six bonds) with sp^3d^2 hybridization. For example, silicon tetrafluoride, SiF_4, reacts with sodium fluoride to give $Na_2[SiF_6]$, which contains the octahedral $[SiF_6]^{2-}$ ion.

The presence of *d* orbitals in the valence shell of silicon contributes to the reactivity of silicon compounds. For example, when $SiCl_4$ is exposed to water, it reacts rapidly in a stepwise fashion:

This process is repeated until all four chlorine atoms are replaced by hydroxide groups; $Si(OH)_4$ is the final product of the hydrolysis. The overall reaction is

$$SiCl_4 + 4H_2O \longrightarrow Si(OH)_4 + 4HCl$$

Silicic acid, $Si(OH)_4$ or H_4SiO_4, is unstable and gradually dehydrates to SiO_2.

$$Si(OH)_4 \longrightarrow SiO_2 + 2H_2O$$

Carbon cannot expand its valence shell to accommodate more than an octet. Thus carbon tetrachloride cannot form five-coordinated intermediates of the type involved in the hydrolysis of silicon tetrachloride. Hence tetrahedral carbon compounds are much less reactive than are silicon compounds.

27.12   Silicon Hydrides

Silicon forms a series of hydrides that include SiH_4, Si_2H_6, Si_3H_8, Si_4H_{10}, and Si_6H_{14}. The chemical behavior of the silicon hydrides is decidedly different from that of the hydrocarbons of similar formulas. For example, the silicon hydrides flame spontaneously in air, but the hydrocarbons do not. Only silicon hydrides corresponding to the alkanes are known; there are none that are analogous to the alkenes or alkynes, with double or triple bonds. (Only carbon, nitrogen, and oxygen readily form multiple bonds.)

Acids react with magnesium silicide to form **silane, SiH_4,** which has a structure and formula analogous to methane, CH_4.

$$Mg_2Si + 4H^+ \longrightarrow 2Mg^{2+} + SiH_4(g)$$

Silane is a colorless gas, which is thermally stable at ordinary temperatures but spontaneously flammable in air. The products of its oxidation are silicon dioxide and water.

$$SiH_4(g) + 2O_2(g) \longrightarrow SiO_2(s) + 2H_2O(g) \qquad \Delta H° = -1429 \text{ kJ}$$

Stepwise halogenation of silane is possible, but the reactions are difficult to control. A better method of obtaining partially halogenated silanes involves the use of a hydrogen halide and the corresponding aluminum halide as a catalyst. With HBr, using $AlBr_3$ as the catalyst, the reactions are given by the equations

$$SiH_4 + HBr \longrightarrow SiH_3Br + H_2$$
$$SiH_3Br + HBr \longrightarrow SiH_2Br_2 + H_2$$

$$SiH_2Br_2 + HBr \longrightarrow SiHBr_3 + H_2$$
$$SiHBr_3 + HBr \longrightarrow SiBr_4 + H_2$$

Silane is extremely sensitive to hydroxides, reacting easily to give silicates and hydrogen.

$$SiH_4 + 2OH^- + H_2O \longrightarrow SiO_3^{2-} + 4H_2(g)$$

In contrast, methane is unreactive to hydroxides.

Perhaps the most important reaction of compounds containing a Si—H bond, at least from a commercial standpoint, is the reaction with alkenes (Section 25.13).

$$CH_3CH{=}CH_2 + H_2SiCl_2 \longrightarrow CH_3CH_2CH_2SiHCl_2$$
$$CH_3CH{=}CH_2 + CH_3CH_2CH_2SiHCl_2 \longrightarrow (CH_3CH_2CH_2)_2SiCl_2$$

These reactions are used in the preparation of silicones (Section 27.17).

27.13 Silicon Carbide

When a mixture of sand and a large excess of coke is heated in an electric furnace, **silicon carbide (carborundum), SiC,** is produced according to the equation

$$SiO_2(s) + 3C(s) \longrightarrow SiC(s) + 2CO(g) \qquad \Delta H° = 624.7 \text{ kJ}$$

The product is blue-black, iridescent crystals, nearly as hard as diamonds and very stable at high temperatures. The crystals are crushed, and the particles are screened to uniform size, mixed with a binder of clay or sodium silicate, molded into various shapes such as grinding wheels, and fired. Silicon carbide is used as an abrasive for cutting, grinding, and polishing.

Silicon carbide exists in many different crystalline forms; yet in all of these forms each atom is surrounded tetrahedrally by four of the other kind. One crystalline form has a structure like that of diamond (Section 25.1). To rupture a crystal of silicon carbide, a number of very strong covalent bonds must be broken. The high decomposition temperature (above 2200°C), the extreme hardness, the brittleness, and the chemical inactivity of silicon carbide are due to these strong covalent bonds.

27.14 Silicon Halides

All of the tetrahalides of silicon, SiX_4, and several mixed halides, such as $SiCl_2F_2$, have been prepared. **Silicon tetrachloride, $SiCl_4$,** can be prepared by direct chlorination at elevated temperatures or by heating silicon dioxide with chlorine and carbon. The equations are

$$Si(s) + 2Cl_2(g) \longrightarrow SiCl_4(g)$$
$$SiO_2(s) + 2C(s) + 2Cl_2(g) \longrightarrow SiCl_4(g) + 2CO(g)$$

Silicon tetrachloride is a covalent tetrahedral molecule containing four covalent Si—Cl bonds. It is a low-boiling (57°C), colorless liquid that fumes strongly in moist air, producing a dense smoke of finely divided silica as the Si—Cl bonds are replaced by Si—O bonds.

$$SiCl_4(g) + 2H_2O(g) \longrightarrow SiO_2(s) + 4HCl(g)$$

The Si—Cl bonds in silicon tetrachloride can be replaced by Si—C bonds in a stepwise reaction with a stoichiometric amount of $Na^+CH_3^-$ or another compound that contains a carbon anion.

$$SiCl_4 + Na^+CH_3^- \longrightarrow Cl_3SiCH_3 + NaCl$$

$$SiCl_4 + 2ClMgC_2H_5 \longrightarrow Cl_2Si(CH_2CH_3)_2 + 2MgCl_2$$

Elemental silicon ignites spontaneously in an atmosphere of fluorine, forming gaseous **silicon tetrafluoride, SiF_4.** The reaction of hydrofluoric acid with silica or a silicate also produces SiF_4.

$$SiO_2(s) + 4HF(g) \longrightarrow SiF_4(g) + 2H_2O(l) \qquad \Delta H° = -191.2 \text{ kJ}$$

$$CaSiO_3(s) + 6HF(g) \longrightarrow SiF_4(g) + CaF_2 + 3H_2O$$

Silicon tetrafluoride hydrolyzes in water and produces both **fluorosilicic acid** and **orthosilicic acid.**

$$3SiF_4 + 4H_2O \longrightarrow \underset{\substack{\text{Orthosilicic} \\ \text{acid}}}{H_4SiO_4(s)} + 4H^+ + \underset{\substack{\text{Fluorosilicic} \\ \text{acid}}}{2SiF_6{}^{2-}}$$

Fluorosilicic acid is a stronger acid than sulfuric acid. It is stable only in solution, however, and on evaporation of the solvent it decomposes according to the equation

$$H_2SiF_6(aq) \longrightarrow 2HF(g) + SiF_4(g)$$

27.15 Silicon Dioxide (Silica)

The usual crystalline form of silicon dioxide is **quartz,** a hard, brittle, clear, colorless solid. It is used in many ways—for architectural decorations, semiprecious jewels, optical instruments, and frequency control in radio transmitters. The contrasts in structure and physical properties between silica and its carbon analogue, carbon dioxide, are interesting. Solid carbon dioxide (dry ice) contains single CO_2 molecules held together with very weak intermolecular forces. The low melting point and volatility of dry ice reflect these weak forces between molecules. Each of the two oxygen atoms is attached to the central carbon atom by double bonds. In contrast, silicon does not form double bonds; hence quartz crystallizes as a hexagonal closest packed array of oxygen atoms, with silicon atoms occupying $\frac{1}{4}$ of the tetrahedral holes. A silicon atom in quartz is linked to four oxygen atoms by single bonds directed toward the corners of a regular tetrahedron; the SiO_4 tetrahedra share oxygen atoms. This gives a three-dimensional, continuous silicon-oxygen network. A quartz crystal is a macromolecule of silicon dioxide. The ratio of silicon to oxygen atoms is 1 to 2, and the simplest formula is SiO_2. The formula SiO_2 does not, however, represent a single molecule as does the formula CO_2.

At 1600°C quartz melts to give a viscous liquid with a random internal structure. When the liquid is cooled, it does not readily crystallize but usually supercools and forms **silica glass.** The SiO_4 tetrahedra in this glass have the random arrangement characteristic of supercooled liquids, and the glass has some very useful properties. Since silica glass is highly transparent to both visible and ultraviolet light, it is used for mercury vapor lamps, which give radiation rich in ultraviolet light. It is also used in certain optical instruments that operate in the ultraviolet region. The coefficient of expansion of silica glass is very low, so it is not easily fractured by sudden changes in temperature. It is insoluble in water and inert toward acids, except hydrofluoric acid.

$$SiO_2(s) + 4HF(aq) \longrightarrow SiF_4(g) + 2H_2O(l)$$

This reaction is used in the quantitative separation of silica from other oxides since both products are volatile. Hot solutions of metal hydroxides and fused metal carbonates both convert silica into soluble silicates (SiO_4^{4-} and SiO_3^{2-}). Typical reactions are

$$SiO_2 + 4OH^- \longrightarrow SiO_4^{4-} + 2H_2O$$
$$SiO_2 + Na_2CO_3 \longrightarrow Na_2SiO_3 + CO_2$$

The latter reaction is employed in the conversion of silicate rocks to soluble forms for analysis and is part of the glass-making process (Section 27.18).

27.16 Natural Silicates

Silicates are salts containing anions composed of silicon and oxygen. There are many types of silicates because the silicon-to-oxygen ratio can vary widely. In all silicates, however, sp^3 hybridized silicon atoms are found at the centers of tetrahedra with oxygen at the corners, and silicon is tetravalent. The variation in the silicon-to-oxygen ratio occurs because the silicon-oxygen tetrahedra may exist as discrete, independent units or may share oxygen atoms at corners, edges, or more rarely faces, in a variety of ways. The silicon-to-oxygen ratio varies according to the extent of sharing of oxygen atoms by silicon atoms in the linking together of the tetrahedra.

It is convenient to classify the silicates into groups, based on how the silicon-oxygen tetrahedra are linked.

1. Individual SiO_4^{4-} tetrahedra exist as independent groups in some minerals. Examples include **olivine**, Mg_2SiO_4, and **zircon**, $ZrSiO_4$. Positively charged metallic ions (Mg^{2+}, Zr^{4+}) bind together the negatively charged SiO_4^{4-} ions, which have the tetrahedral structure shown in Fig. 27-7. (Note that only the oxygen atoms are shown in the figure; a silicon atom is in the center of each tetrahedron but is not shown.)

2. Two SiO_4 tetrahedra share one corner oxygen and form discrete $Si_2O_7^{6-}$ ions (Fig. 27-8) in **hardystonite, $Ca_2ZnSi_2O_7$,** and **hemimorphite, $Zn_4(OH)_2Si_2O_7 \cdot H_2O$,** for example. The cations are between the $Si_2O_7^{6-}$ ions and bind them together.

3. Three SiO_4 tetrahedra share corners and form closed rings in **benitoite, $BaTiSi_3O_9$.** The $Si_3O_9^{6-}$ ions (Fig. 27-9) are held together by the metal cations. The rings are arranged in sheets with their planes parallel. Six SiO_4 tetrahedra share corners to form a closed ring in **beryl (emerald), $Be_3Al_2Si_6O_{18}$.**

4. Chains of SiO_4 tetrahedra in which each tetrahedron shares two oxygen atoms are present in **diopside, $CaMg(SiO_3)_2$,** and other minerals. Such a chain (Fig. 27-10) has an empirical formula of SiO_3^{2-}, but no SiO_3^{2-} ions are present as independent entities. Parallel chains extend the full length of the crystal and are held together by the positively charged metal ions lying between them.

Figure 27-7. The tetrahedral structure of the SiO_4^{4-} ion. (Only the oxygen atoms are shown; the silicon atom, not shown, is at the center.)

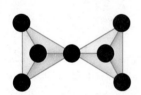

Figure 27-8. The structure of the $Si_2O_7^{6-}$ ion. (A silicon atom, not shown, is at the center of each tetrahedron.)

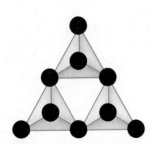

Figure 27-9. The structure of the $Si_3O_9^{6-}$ ion. (A silicon atom, not shown, is at the center of each tetrahedron.)

Figure 27-10. A portion of an SiO_3^{2-} chain. (A silicon atom, not shown, is at the center of each tetrahedron.)

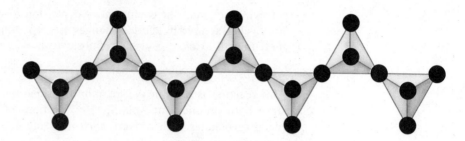

5. Double silicon-oxygen chains form when SiO_4 tetrahedra in single chains share oxygen atoms (Fig. 27-11). Metal cations link the parallel chains together. The fact that these ionic bonds are not as strong as the silicon-oxygen bonds within the chains accounts for the fibrous nature of many of these minerals. **Asbestos** is a chain-type silicate.

Figure 27-11. A portion of a double silicon-oxygen chain. (A silicon atom, not shown, is at the center of each tetrahedron.)

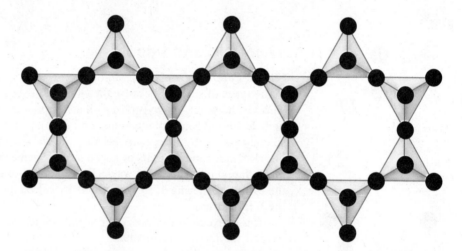

6. Silicon-oxygen sheets are formed by sharing of oxygen atoms between double chains. Metal ions form ionic bonds between the sheets. These ionic bonds are weaker than the silicon-oxygen bonds within the sheets; thus minerals with this structure cleave into thin layers. Examples include the micas **talc, $Mg_3Si_4O_{10}(OH)_2$**, and **muscovite, $KAl_2(AlSi_3O_{10})(OH,F)_2$**. Note that aluminum atoms can substitute for silicon atoms, as in muscovite. When this occurs, a larger charge on the cations is required to produce a neutral entity.

7. Three-dimensional silicon-oxygen networks with some of the tetravalent silicon replaced by trivalent aluminum are also observed. The negative charge that results is neutralized by a distribution of positive ions throughout the network. Examples are **feldspar, $K(AlSi_3O_8)$**, and the **zeolites**, such as **$Na_2(Al_2Si_3O_{10}) \cdot 2H_2O$.**

27.17 Silicones

Polymeric organosilicon compounds containing Si—O—Si linkages are known as **silicones.** They have the general formula $(R_2SiO)_n$ and may be linear, cyclic, or cross-linked polymers.

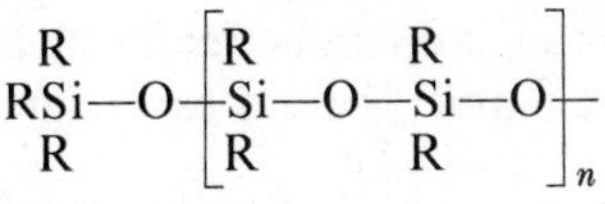

R   R
Si
R   O   O   R
Si   Si
R   O   R

R   ⌈ R   R ⌉
RSi—O—│Si—O—Si—O—├
R   │ R   R │n

Linear silicone 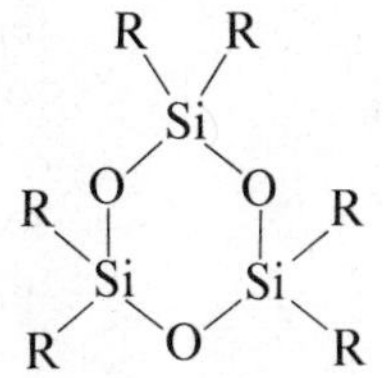   Cyclic silicone

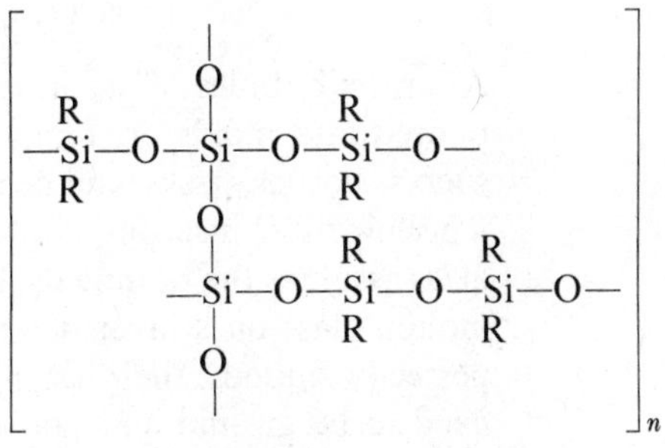

Cross-linked silicone

The R in the formulas represents methyl, —CH$_3$, ethyl, —C$_2$H$_5$, phenyl, —C$_6$H$_5$, or some other hydrocarbon group. The linear and cyclic silicones are produced by hydrolyzing organochlorosilanes of the type R$_2$SiCl$_2$, followed by polymerization through the elimination of a molecule of water from two hydroxyl groups of adjacent R$_2$Si(OH)$_2$ molecules.

$$R_2SiCl_2 + 2H_2O \longrightarrow R_2Si(OH)_2 + 2HCl$$

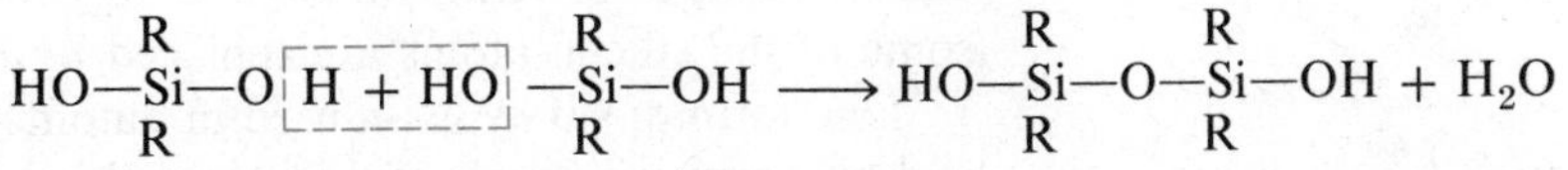

R   R   R   R
HO—Si—O⌈H + HO⌉—Si—OH ⟶ HO—Si—O—Si—OH + H$_2$O
R   ⌊_____⌋ R   R   R

Organosilicon polymers incorporate some of the properties of both hydrocarbons and oxysilicon compounds. They are remarkably stable toward heat and chemical reagents and are water-repellent. Depending on the extent of polymerization and the molecular complexity, silicones may be oils, greases, rubberlike substances, or resins. They are used as lubricants, hydraulic fluids, electrical insulators, and moisture-proofing agents. The viscosity of silicone oils changes so little as the temperature changes that they can be employed as lubricants where there are extreme variations in temperature.

Paper, wool, silk, and other fabrics can be coated with a water-repellent film by exposing them for a second or two to the vapor of **trimethylchlorosilane, (CH$_3$)$_3$SiCl**. The —OH groups on the surface of the material react, and the surface becomes coated with a thin film (monolayer) of (CH$_3$)$_3$Si—O— groups. This layer repels water because it is similar to a hydrocarbon film.

$$(Surface)—OH + Cl—Si(CH_3)_3 \longrightarrow (Surface)—O—Si(CH_3)_3 + HCl$$

27.18   Glass

Glass is a supercooled liquid consisting of a complex mixture of silicates. It is transparent, brittle, and entirely lacking in the ordered internal structure characteristic of crystals. When heated, it does not melt sharply. Instead it gradually softens until it becomes liquid.

The glass used for windowpanes, bottles, dishes, and so forth is a mixture of sodium silicates and calcium silicates with an excess of silica. It is made by heating together sand, sodium carbonate (or sodium sulfate), and calcium carbonate.

$$Na_2CO_3 + SiO_2 \longrightarrow Na_2SiO_3 + CO_2(g)$$
$$Na_2SO_4 + SiO_2 \longrightarrow Na_2SiO_3 + SO_3(g)$$
$$CaCO_3 + SiO_2 \longrightarrow CaSiO_3 + CO_2(g)$$

After the bubbles of gas have been expelled, a clear viscous melt results. This is poured into molds or stamped with dies to produce pressed glassware. Articles such as bottles, flasks, and beakers are formed by taking a lump of molten glass on a hollow tube, inserting it into a mold and blowing with compressed air until the glass assumes the outline of the mold. Plate glass (float glass) is made by pouring molten glass on a layer of very pure molten tin. Since the molten tin surface is perfectly smooth, the glass floating on it is also perfectly smooth and does not need to be ground and polished after hardening.

Glassware is annealed by heating it for a time just below its softening temperature and then cooling it slowly. Annealing lessens internal strains and thereby reduces the chances of breakage from shock or temperature change.

If sodium is replaced by potassium in a glass melt, a higher-melting, harder, and less-soluble glass is obtained. If part of the calcium is replaced by lead, a glass of high density and high refractive index is formed. This variety of glass is called **flint glass** and is used in making lenses and cut-glass articles. **Pyrex glass,** which is used for test tubes, flasks, and other laboratory glassware because it is resistant to sudden changes in temperature and to chemical action, is a borosilicate glass (some of the silicon atoms are replaced by boron atoms).

One form of **safety glass** used in automobiles contains a thin layer of plastic held between two pieces of thin plate glass. Adhesion of the glass to the flexible plastic reduces the danger from flying glass and jagged edges if it is broken.

Enamels on kitchen utensils, sinks, and bathtubs and glazes on pottery are made of easily melted glass containing opacifiers (opaquing agents), such as titanium dioxide and tin(IV) oxide.

27.19 Cement

Portland cement is essentially powdered calcium aluminosilicate, which sets to a hard mass when treated with water. It is made by pulverizing a mixture of limestone ($CaCO_3$) and clay (an aluminosilicate) and then roasting the powder in a rotary kiln at about 1500°C. This treatment yields marble-sized lumps called clinker, which are ground with a little gypsum, $CaSO_4 \cdot 2H_2O$, to a very fine powder.

Concrete is made by adding water to a mixture of Portland cement, sand, and stone or gravel. The reactions taking place as the Portland cement mixture sets (starts to harden to form concrete) are complex and not completely understood. It is known that calcium aluminate hydrolyzes, forming calcium hydroxide and aluminum hydroxide. These compounds then react with calcium silicates, forming calcium aluminosilicate in the form of interlocking crystals. Portland cement sets rapidly (within 24 hours) and then hardens slowly over a period of years.

27.20 The Electrical Conductivity of Silicon

Silicon is semimetallic in its chemical properties (Section 27.11) and in its electrical conductivity. It conducts electricity better than nonmetals (which are insulators), but not as well as metals; it is classified as a semiconductor.

Band theory accounts for the variation in electrical conductivity among metals, semiconductors, and insulators. According to this theory, atomic orbitals of the atoms in a crystal combine to yield molecular orbitals that extend through the entire crystal. As the number of atoms (and thus the number of atomic orbitals) increases, the number of molecular orbitals increases (Fig. 27-12). As the number of molecular orbitals, or **energy levels** as they are sometimes called, increases, the difference in energy between them becomes smaller and smaller until there is very little difference in energy between adjacent molecular orbitals. The effective result is a continuous *band* of molecular orbitals, or energy levels, that extends through the entire crystal. There is one energy level in the band for each atomic orbital that participates in forming it. Each of the energy levels in the band can contain two electrons. However, only the electrons in the higher-energy portion of a band are sufficiently free, or mobile, to cause electrical conductivity.

To gain a better understanding of band theory, consider the case of lithium, whose electronic structure is $1s^2 2s^1$. A very small crystal of lithium contains about 10^{18} atoms. For this many atoms the energy difference between energy levels is so small that the levels are essentially continuous, and they constitute a band. In a lithium crystal one band arises from the $1s$ atomic orbitals and is fully occupied with electrons. However, the second band, arising from the $2s$ atomic orbitals, is only half-filled with electrons (Fig. 27-12). Between the two bands is a **gap**, a range of energy in which no energy levels, or molecular orbitals, are located. No electrons can be found with these energies since there are no energy levels for them to occupy.

Figure 27-12. Atomic orbitals in a Li atom; molecular orbitals in Li_2, Li_3, and Li_4; and bands in a lithium crystal (Li_n).

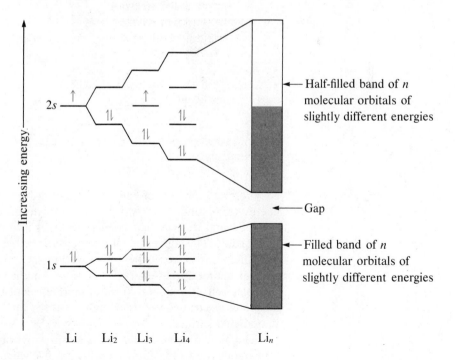

The spacing and the filling of the bands in a substance determine whether the substance is a **conductor,** a **nonconductor** (insulator), or a **semiconductor** (a poor conductor). A substance such as lithium, which contains a partially filled band [Fig. 27-13(a)], is a conductor. If the bands are completely filled or completely empty and the energy gap between them is large [Fig. 27-13(b)], the substance is an insulator. Diamond is an example of such an insulator. A substance that contains a completely filled band and a completely empty band can behave as a semiconductor if the energy gap between the bands is so small that electrons from the filled band can be promoted to energy levels in the empty band by thermal energy (heat). The previously empty band then contains a few electrons that can conduct an electric current [Fig. 27-13(c)]. Since the number of free electrons is much smaller than that in a metal, the material is a semiconductor, as is silicon. Alternatively, electrons can be provided to the empty band by addition of impurities with extra electrons (for example, arsenic with the valence shell $4s^2 4p^3$ is added as an impurity to silicon, $3s^3 3p^2$). The extra electrons occupy what would be an empty band in the pure material (in this case, pure silicon).

Figure 27-13. (a) Partially filled bands are found in metals. (b) Completely filled and completely empty bands separated by a large gap are found in insulators. (c) A completely filled band and a completely empty band separated by a small gap are found in semiconductors at 0 K. When warmed to room temperature, a few electrons are thermally promoted from the previously filled band to the previously empty band, as shown.

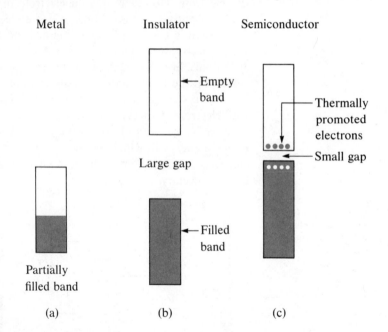

27.21 The Solar Cell

An enormous amount of energy is given off by the sun. The earth receives more energy from sunlight in two days than is stored in all known reserves of fossil fuels. Several devices that convert sunlight into electrical energy are in use or under study. The conventional photoelectric cell, used in electric-eye doors, transforms light into electrical energy (Section 4.1) but delivers as power only about 0.5% of the total light energy it absorbs. Another type of photoelectric device is the **solar cell.** It is about 20 times more efficient than the photoelectric cell and generates electric power from sunlight at the rate of 90 watts per square yard of illuminated surface.

The basic unit of a solar cell is a thin wafer of very pure silicon containing a tiny amount of arsenic. Since arsenic has five valence electrons, compared to four for silicon, replacement of silicon atoms with arsenic atoms creates **n-type silicon** (silicon that contains electrons in what would be an empty band in pure silicon). A thin layer of silicon containing a trace of boron is placed on the surface of the wafer. Since boron has only three valence electrons, replacement of silicon atoms with boron atoms forms **p-type silicon** (silicon that is missing electrons in what would be a filled band in pure silicon). Holes (a few vacant energy levels in an otherwise filled band) thus exist at the surface of the wafer, and a junction, called a p-n junction, exists between the body of the wafer and its thin surface layer. Electrons diffuse through the junction from the wafer to the holes; at the same time holes, or electron vacancies, move to the body of the wafer. This results in a net positive charge within the body of the wafer, which was neutral prior to the diffusion of electrons (negative charges) out of it, and a net negative charge on the surface layer. Thus an electrostatic potential develops between the two regions, until it is just large enough to prevent any further charge migration. When the diffusion of electrons is just balanced by the electrostatic potential, equilibrium is established between the two forces and hence also between electrons and holes. At equilibrium a net difference in potential exists between the two regions.

In a solar cell one electrical lead is attached to the body of the wafer, and one lead, to the surface. When the cell is exposed to sunlight, energy from the sunlight causes electrons to be released from their positions in the lattice near the p-n junction, thereby upsetting the equilibrium between electrons and holes and causing additional electrons to move across the junction into the body of the wafer. Thus a sufficient electromotive force is built up to cause electrons to move through the electrical leads, and a current flows through the wire. The device is thus an electric cell, with the positive terminal at the p-contact and the negative terminal at the n-contact (Fig. 27-14).

In actual practice a series of such wafers makes up a solar cell. The first practical application of the solar cell was as a power source for eight telephones on a rural line in Georgia in 1955. Such cells are used to power communication devices in satellites and spacecraft, especially those designed to remain in space for long periods.

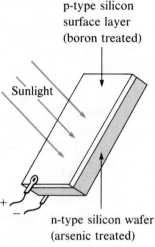

p-type silicon surface layer (boron treated)

Sunlight

+

−

n-type silicon wafer (arsenic treated)

Figure 27-14. A solar cell.

FOR REVIEW

SUMMARY

Although boron and silicon are members of different groups in the Periodic Table, they exhibit similar chemical behavior, an example of a diagonal relationship found in several places in the table. Both boron and silicon are metalloids.

Boron exhibits some metallic characteristics, but the large majority of its chemical behavior is that of a nonmetal that exhibits an oxidation number of $+3$ in its compounds (although other oxidation numbers are known). Its valence shell electron configuration is $2s^2 2p^1$, so it forms trigonal planar compounds with three single covalent bonds. The resulting compounds are Lewis acids since the unhybridized p orbital does not contain an electron pair. The most stable boron compounds are those that contain oxygen and fluorine. Diborane, the simplest stable boron hydride, contains **two-electron three-center bonds,** as does elemental boron.

Silicon is a metalloid with a valence shell electron configuration of $3s^2 3p^2 3d^0$. It commonly forms tetrahedral compounds in which silicon exhibits an oxidation number of $+4$. Although the d orbital is unfilled in four-coordinated silicon compounds, its presence makes these compounds much more reactive than the corresponding carbon compounds. Silicon forms strong single bonds with carbon, giving rise to the stability of silicon carbide and **silicones.** Silicon also forms strong bonds with oxygen and fluorine.

Silicates contain oxyanions of silicon and are important components of minerals and glass. Elemental silicon is a **semiconductor** that is used in **solar cells.** The electrical conductivity of silicon, as well as that of other semiconductors, conductors, and insulators, can be explained by the **band theory.**

KEY TERMS AND CONCEPTS

Band theory (27.20)
Conductor (27.20)
Energy levels (27.20)

Insulator (27.20)
Semiconductor (27.20)
Solar cell (27.21)

Three-center bonds (27.2)
Two-center bonds (27.2)

EXERCISES

Boron

1. What is the electron configuration of a boron atom?
2. Why is boron limited to a maximum coordination number of 4 in its compounds?
3. Describe the molecular structure and the hybridization of boron in each of the following molecules or ions:
   (a) BCl_3 (c) B_2H_6 (e) $B(OH)_4^-$
   (b) BH_4^- (d) $B(OH)_3$ (f) H_3BCO
4. Write a Lewis structure for each of the following molecules or ions:
   (a) BCl_3 (c) $B(OH)_3$ (e) H_3BCO
   (b) BH_4^- (d) $B(OH)_4^-$
5. Write two equations for reactions in which boron exhibits metallic behavior and two equations for reactions in which boron exhibits nonmetallic behavior.
6. Write balanced equations describing the reactions of boron with elemental fluorine, carbon, oxygen, and nitrogen, respectively.
7. Why is B_2H_6 said to be an electron-deficient compound?
8. Identify the oxidizing agent and the reducing agent in the chemical reaction for the preparation

of diborane from boron trifluoride and lithium aluminum hydride.
9. Write equations to show the formation of fluoroboric acid from boron trifluoride.
10. Show by equations the relationship of the boric acids to boric oxide.
11. Why does an aqueous solution of borax turn red litmus blue?
12. Explain the chemistry of the use of borax as a flux in soldering and welding.
13. Why is boron nitride sometimes referred to as inorganic graphite?
14. Why should the physical properties of borazine be similar to those of benzene?
15. Complete and balance the following equations. (In some instances, two or more reactions may be correct, depending on the stoichiometry.)
    (a) $B + Cl_2 \longrightarrow$
    (b) $B + S_8 \longrightarrow$
    (c) $B + SO_2 \longrightarrow$
    (d) $B_2H_6 + NaH \longrightarrow$
    (e) $BCl_3 + C_2H_5OH \longrightarrow$
    (f) $HBO_2 + H_2O \longrightarrow$
    (g) $B_2O_3 + CuO \longrightarrow$

(h) $BF_3 + NH_3 \longrightarrow$

16. A 0.7849-g sample of boron was oxidized in air, giving 2.5274 g of B_2O_3. From this information and the atomic weight of oxygen, calculate the atomic weight of boron.   *Ans. 10.81*

17. From the data given in Appendix J, determine the standard enthalpy change and standard free energy change for each of the following reactions:

(a) $BF_3(g) + 3H_2O(l) \longrightarrow B(OH)_3(s) + 3HF(g)$
    Ans. $\Delta H° = 87\,kJ$; $\Delta G° = 44\,kJ$

(b) $BCl_3(g) + 3H_2O(l) \longrightarrow$
    $B(OH)_3(s) + 3HCl(g)$
    Ans. $\Delta H° = -109.9\,kJ$; $\Delta G° = -154.7\,kJ$

(c) $B_2H_6(g) + 6H_2O(l) \longrightarrow$
    $2B(OH)_3(s) + 6H_2(g)$
    Ans. $\Delta H° = -510\,kJ$; $\Delta G° = -601.5\,kJ$

Silicon

18. What is the electron configuration of a silicon atom? What orbitals are present in the valence shell of silicon? Include the unfilled orbitals in the valence shell.

19. Describe the hybridization and the bonding of a silicon atom in elemental silicon.

20. Describe the molecular structure and the hybridization of silicon in each of the following molecules or ions:

(a) $SiCl_4$ (c) $(CH_3)_2SiH_2$ (e) Si_3H_8
(b) $SiO_4{}^{4-}$ (d) $SiF_6{}^{2-}$

21. Write a Lewis structure for each of the following molecules or ions:

(a) $SiCl_4$ (c) $(CH_3)_2SiH_2$ (e) Si_3H_8
(b) $SiO_4{}^{4-}$ (d) $SiF_6{}^{2-}$

22. Write two equations for reactions in which silicon exhibits metallic behavior and two equations for reactions in which silicon exhibits nonmetallic behavior.

23. Heating a mixture of sand and coke in an electric furnace produces either silicon or silicon carbide. What determines which will be formed?

24. In terms of molecular and crystal structure, account for the low melting point of dry ice (solid CO_2) and the high melting point of SiO_2.

25. How does the following reaction show the acidic character of SiO_2? Which acid-base theory applies?

$$CaO + SiO_2 \xrightarrow{\Delta} CaSiO_3$$

26. Describe the conversion of silicate rocks to soluble forms for the purpose of analysis.

27. Account for the existence of the great variety of silicates in terms of how SiO_4 tetrahedra are linked.

28. Give equations for the reactions involved in the production of common window glass.

29. How does the internal structure of silica glass differ from that of quartz?

30. How does the structure of acetone, $(CH_3)_2CO$, differ from that of the silicone, $[(CH_3)_2SiO]_3$?

31. Compare and contrast the chemistry of carbon and that of silicon.

32. Write an equation for the reaction of silicon with each of the following (heated if necessary for reaction):

(a) F_2 (c) O_2
(b) NaOH   (d) N_2

33. Write equations to contrast the hydrolysis of SiF_4, $SiCl_4$, and CCl_4.

34. Write an equation for the combustion of SiH_4; of Si_3H_8.

Additional Exercises

35. The experimentally determined ratio of hydrogen atoms to boron atoms in a gas is 3 to 1. How could it be proved that the gas actually consists of B_2H_6 molecules?

36. A hydride of boron contains 88.5% boron and 11.5% hydrogen by mass. The density of the hydride at 125°C and 450 torr is 2.21 g/L. A 5.0-mL gaseous sample of this hydride at 125°C and 450 torr decomposes to the elements, giving 35 mL of H_2 under the same conditions. Determine the molecular formula of the hydride and show that all data are consistent with that formula.   *Ans. $B_{10}H_{14}$*

37. Write balanced equations describing at least two different reactions in which solid silica exhibits acidic character.

38. Which would you expect to be a better Lewis base: the carbide ion, C^{4-}, or the silicide ion, Si^{4-}?

39. In what ways does the chemistry of silicates resemble that of phosphates?

40. Carbon forms carbonic acid, H_2CO_3, while silicon forms silicic acid, H_4SiO_4. Explain. (*Hint:* Draw the Lewis structures for the two compounds.)

41. (a) From the table of bond energies given in Section 18.7, calculate the approximate enthalpy change for the following reaction:

$$SiH_4(g) + 4HF(g) \longrightarrow SiF_4(g) + 4H_2(g)$$

Ans. −452 kJ

(b) Using bond energy data calculate the approximate enthalpy change for the corresponding reaction with methane:

$$CH_4(g) + 4HF(g) \longrightarrow CF_4(g) + 4H_2(g)$$

Ans. +436 kJ

(c) Why do the two enthalpy changes differ so greatly?

(d) Which reaction is more likely to proceed spontaneously?

42. Silicon reacts with sulfur at elevated temperatures. If 0.0923 g of silicon reacts with sulfur to give 0.3030 g of silicon sulfide, determine the empirical formula of silicon sulfide.

Ans. SiS$_2$

43. A hydride of silicon prepared by the reaction of Mg_2Si with acid exerted a pressure of 306 torr at 26°C in a bulb with a volume of 57.0 mL. If the mass of the hydride was 0.0861 g, what is its molecular weight? What is the molecular formula of the hydride?   *Ans. 92.0; Si$_3$H$_8$*

28

NUCLEAR CHEMISTRY

The chemical changes (transformations of one form of matter into another) we have studied thus far have involved only the electronic structures of atoms. Their nuclear structures did not change. However, changes in nuclei do occur and these changes are the basis for a branch of science called **nuclear chemistry,** which is on the borderline between physics and chemistry. It began with the discovery of radioactivity and has become increasingly important during the past 70 years.

THE STABILITY OF NUCLEI

28.1 The Nucleus

The nucleus of an atom is composed of protons and, except in ^1_1H, neutrons (Sections 4.5 and 4.6). The number of protons in the nucleus is called the atomic number, Z, of the element; the sum of the number of protons and number of neutrons is the mass number, A. Atoms with the same atomic number but different mass numbers are isotopes of the same element (Section 4.6). When talking about a single type of nucleus, the term **nuclide** is often used.

Protons and neutrons, collectively called **nucleons,** are packed together tightly in a nucleus. A nucleus is quite small (about 10^{-13} cm) compared to the entire atom (10^{-8} cm). Nuclei are extremely dense; they average 1.8×10^{14} grams per cubic centimeter. This density is extremely large compared to the densities of familiar materials. Water, for example, has a density of 1 gram per cubic centimeter. Osmium, the densest element known, has a density of 22.6 grams per cubic centimeter. If the earth's density were equal to the average nuclear density, the earth's radius would be only about 200 meters (650 feet).

To hold positively charged protons together in the very small volume of a nucleus requires a very strong force since the positive charges repel one another strongly at very short distances. The **nuclear force** is the force of attraction between nucleons that holds the nucleus together. This force acts between protons, between neutrons, and between protons and neutrons. The nuclear force is very different from and much stronger than the electrostatic force that holds negatively charged electrons around a positively charged nucleus. In fact, the nuclear force is the strongest force known and is about 30–40 times stronger than electrostatic repulsions between protons in a nucleus. Although the exact nature of the nuclear force is unknown, it is known that it is a short-range force, effective only within distances of about 10^{-13} centimeter.

28.2 Nuclear Binding Energy

Although the nature of the nuclear force is unknown, the magnitude of the energy changes associated with its action can be determined. As an example, let us consider the nucleus of the helium atom, $^4_2$He, which consists of two protons and two neutrons. If a helium nucleus were to be formed by the combination of two protons and two neutrons without change of mass, the mass of the helium atom (including the mass of the two electrons outside the nucleus) would be

$$(2 \times 1.0073) + (2 \times 1.0087) + (2 \times 0.00055) = 4.0331 \text{ amu}$$

(See Section 4.3 or Appendix D for masses of the proton, neutron, and electron.) However, the mass is only 4.0026 atomic mass units. This difference between the calculated and experimental masses, the **mass defect** of the atom, indicates a loss in mass of 0.0305 atomic mass unit. The loss in mass accompanying the formation of an atom from protons, neutrons, and electrons is due to conversion of that much mass to **nuclear binding energy.** Such conversions are extremely exothermic; thus the reverse reaction, decomposition of an atom to protons, neutrons, and electrons is extremely endothermic, and very difficult to accomplish.

The nuclear binding energy can be calculated from the mass defect by the **Einstein equation**

$$E = mc^2$$

where E represents energy in joules, m stands for mass in kilograms, and c is the speed of light in meters per second. A tremendous quantity of energy results from the conversion of a small quantity of matter to energy.

EXAMPLE 28.1   Calculate the binding energy for the nuclide $^4_2$He in joules per mole and electron-volts per nucleus.

The difference in mass between a $^4_2$He nucleus and two protons plus two neutrons is 0.0305 amu. In order to use the Einstein equation to convert this mass defect into the equivalent energy in joules, the mass defect must be expressed in kilograms (1 amu = 1.6605×10^{-27} kg).

$$0.0305 \text{ amu} \times \frac{1.6605 \times 10^{-27} \text{ kg}}{1 \text{ amu}} = 5.06 \times 10^{-29} \text{ kg}$$

The nuclear binding energy in *one* nucleus is found from the Einstein equation ($E = mc^2$).

$$E = 5.06 \times 10^{-29} \text{ kg} \times (3.00 \times 10^8 \text{ m s}^{-1})^2$$
$$= 4.54 \times 10^{-12} \text{ kg m}^2 \text{ s}^{-2}$$
$$= 4.54 \times 10^{-12} \text{ J}$$

Units of kg m^2 s^{-2} are equivalent to J (Section 1.15). The binding energy in 1 mole of nuclei is

$$4.54 \times 10^{-12} \text{ J/nucleus} \times 6.022 \times 10^{23} \text{ nuclei/mol} = 2.73 \times 10^{12} \text{ J mol}^{-1}$$
$$= 2.73 \times 10^9 \text{ kJ mol}^{-1}$$

Binding energies are often expressed in millions of electron-volts (MeV) per nucleus rather than in joules. The conversion factor is

$$1 \text{ MeV} = 1.602189 \times 10^{-13} \text{ J}$$

giving

$$4.54 \times 10^{-12} \text{ J/nucleus} \times \frac{1 \text{ MeV}}{1.602189 \times 10^{-13} \text{ J}} = 28.3 \text{ MeV/nucleus}$$

The changes in mass in all ordinary chemical reactions are negligible because only chemical bonds change, that is, they form or break. On the other hand, the changes in mass in nuclear reactions are very significant. Bringing together the proper number of protons and neutrons to form a nucleus releases very large amounts of energy. If the nuclear reaction of 2 moles of neutrons with 2 moles of hydrogen atoms to give 1 mole of helium atoms could be made to occur, 2.73×10^9 kilojoules of energy would be released (Example 28.1). For comparison, burning 1 mole of methane releases only 8.9×10^2 kilojoules, about three million times less energy.

The binding energy per nucleon (that is, the total binding energy for the nucleus divided by the sum of the numbers of protons and neutrons present in the nucleus) is greatest for the nuclei of elements with mass numbers between 40 and 100 and decreases with mass numbers less than 40 or greater than 100 (Fig. 28-1). The most stable nuclei are those with the largest binding energy per nucleon, those in the vicinity of iron, cobalt, and nickel in the Periodic Table. The binding energy per nucleon in helium is

$$\frac{28.3}{4} = 7.08 \text{ MeV/nucleon}$$

The binding energy per nucleon in $^{56}_{26}$Fe is almost 25% larger than in $^{4}_{2}$He.

Figure 28-1. Binding energy curve for the elements.

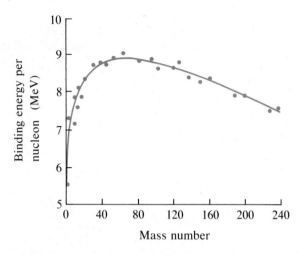

EXAMPLE 28.2   Calculate the binding energy per nucleon for the iron isotope with an atomic number of 26 and a mass number of 56, $^{56}_{26}$Fe, which has an atomic mass of 55.9349 amu.

First we determine the mass defect, the difference between the calculated mass of 26 protons, 30 neutrons, and 26 electrons and the observed mass of a $^{56}_{26}$Fe atom.

$$\text{Mass defect} = [(26 \times 1.0073) + (30 \times 1.0087) + (26 \times 0.00055)] - 55.9349$$
$$= 56.4651 - 55.9349$$
$$= 0.5302 \text{ amu}$$

Now we calculate the nuclear binding energy in millions of electron-volts.

$$E = mc^2$$
$$= 0.5302 \text{ amu} \times \frac{1.6605 \times 10^{-27} \text{ kg}}{1 \text{ amu}} \times (3.00 \times 10^8 \text{ m s}^{-1})^2$$
$$= 7.92 \times 10^{-11} \text{ kg m}^2 \text{ s}^{-2}$$
$$= 7.92 \times 10^{-11} \text{ J}$$
$$7.92 \times 10^{-11} \text{ J} \times \frac{1 \text{ MeV}}{1.602189 \times 10^{-13} \text{ J}} = 494 \text{ MeV}$$

The binding energy per nucleon is found by dividing the total nuclear binding energy by 56, the number of nucleons in the atom.

$$\text{Binding energy per nucleon} = \frac{494 \text{ MeV}}{56}$$
$$= 8.82 \text{ MeV}$$

The iron-56 nucleus is one of the more stable nuclei.

28.3   Nuclear Stability

A nucleus is stable if it cannot be transformed into another configuration without adding energy from the outside. A plot of the number of neutrons versus the number of protons for stable nuclei (the curve shown in Fig. 28-2) shows that the stable isotopes fall into a narrow band, which is called the **band of stability.** The straight line in Fig. 28-2 represents equal numbers of protons and neutrons for comparison. The figure indicates that the lighter, stable nuclei, in general, have equal numbers of protons and neutrons. For example, nitrogen-14 has 7 protons and 7 neutrons. Heavier stable nuclei, however, have slightly more neutrons than protons, as the curve shows. For example, iron-56 has 30 neutrons and 26 protons. Lead-207 has 125 neutrons and 82 protons, a neutron-to-proton ratio approximately equal to 1.5.

Figure 28-2. A plot of number of neutrons versus number of protons for naturally occurring nuclei. Each dot shows the number of protons and neutrons in a known stable nucleus. All isotopes of elements with atomic numbers greater than 83 are unstable.

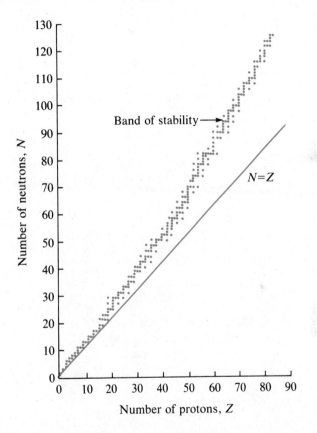

The isotopes that fall to the left or the right of the band of stability have unstable nuclei and are **radioactive.** Note that all isotopes with atomic numbers above 83 are radioactive. Radioactive nuclei change spontaneously (decay) to other nuclei that fall either in or closer to the band of stability. The exact nature of these changes toward more stable nuclei will be discussed in subsequent sections.

28.4   The Half-Life

The number of atoms of a radioactive element that disintegrate per unit of time is a constant fraction of the total number of atoms present. The time required for $\frac{1}{2}$

of the atoms in a sample to decay is called its **half-life.** The half-lives of radioactive nuclides vary widely. For example, the half-life of $^{142}_{58}Ce$ is 5×10^{15} years, that of $^{226}_{88}Ra$ is 1590 years, that of $^{222}_{86}Rn$ is 3.82 days, and that of $^{216}_{84}Po$ is 0.16 second. Only $\frac{1}{2}$ of a sample of radium-226 will remain unchanged after 1590 years, and at the end of another 1590 years the sample will have decayed to $\frac{1}{4}$ of its initial mass, and so on. The half-lives of a number of nuclides are listed in Appendix L.

Radioactive decay is a first-order process. The rate constant k for a radioactive disintegration can be expressed in terms of the half-life, $t_{1/2}$, by the following equation (which applies to all first-order reactions; see Section 15.7):

$$k = \frac{0.693}{t_{1/2}}$$

The relationship between the initial amount of a radioactive isotope and the amount remaining after a given period of time is expressed by the logarithmic relationship that applies to all other first-order reactions (Section 15.7).

$$\log \frac{c_0}{c_t} = \frac{kt}{2.303}$$

where c_0 is the initial amount and c_t is the amount remaining at time t.

EXAMPLE 28.3   Calculate the rate constant for the radioactive disintegration of $^{60}_{27}Co$, an isotope used in cancer therapy. It decays with a half-life of 5.2 years to produce $^{60}_{28}Ni$.

$$k = \frac{0.693}{t_{1/2}}$$

$$= \frac{0.693}{5.2 \text{ yr}} = 0.13 \text{ yr}^{-1}$$

EXAMPLE 28.4   Calculate the fraction and the percentage of a sample of the $^{60}_{27}Co$ isotope that will remain after 15 years.

$$\log \frac{c_0}{c_t} = \frac{kt}{2.303}$$

$$= \frac{(0.13 \text{ yr}^{-1})(15 \text{ yr})}{2.303} = 0.847$$

$$\frac{c_0}{c_t} = \text{antilog } 0.847 = 7.031$$

The fraction remaining will be the amount at time t divided by the initial amount, or c_t/c_0.

$$\frac{c_t}{c_0} = \frac{1}{7.031} = 0.14$$

Hence, 0.14, or 14%, of the $^{60}_{27}Co$ originally present will remain after 15 years.

EXAMPLE 28.5   How long would it take for a sample of $^{60}_{27}Co$ to disintegrate to the extent that only 2.0% of the original amount remains?

$$\frac{c_t}{c_0} = 0.020$$

Therefore,

$$\frac{c_0}{c_t} = \frac{1}{0.020} = 50$$

$$\log 50 = \frac{(0.13 \text{ yr}^{-1})t}{2.303}$$

$$t = \frac{2.303 \log 50}{0.13 \text{ yr}^{-1}} = 30 \text{ yr}$$

EXAMPLE 28.6   The half-life of $^{216}_{84}Po$ is 0.16 s. How long would it take to reduce a sample of it to the negligible amount of 0.000010% of the original amount (1.0×10^{-7} times the original amount)?

$$k = \frac{0.693}{0.16 \text{ s}} = 4.33 \text{ s}^{-1}$$

$$\frac{c_t}{c_0} = 1.0 \times 10^{-7}$$

Hence

$$\frac{c_0}{c_t} = 1.0 \times 10^7$$

$$\log(1.0 \times 10^7) = \frac{(4.33 \text{ s}^{-1})t}{2.303}$$

$$t = \frac{2.303 \log(1.0 \times 10^7)}{4.33 \text{ s}^{-1}} = 3.7 \text{ s}$$

These examples indicate part of the problem of disposal of radioactive isotopes. In 3.7 seconds any reasonable quantity of $^{216}_{84}Po$ would be reduced to a negligible quantity and would itself no longer be a problem. However, the burst of radiation during the first second or so would be extremely intense and very hazardous. Furthermore, $^{216}_{84}Po$ is a part of a series (see Table 28-1, Section 28.8), decaying into $^{212}_{82}Pb$, which is itself radioactive with a half-life of 10.6 seconds, producing $^{212}_{83}Bi$, which has a half-life of 60.5 minutes, and so on, until finally the stable isotope $^{208}_{82}Pb$ is reached. Hence, in coping with radiation effects and isotope disposal, all steps in the decay of the material must be taken into account. In contrast to the few seconds required for the decay of $^{216}_{84}Po$, it takes 30 years for 98% of a sample of $^{60}_{27}Co$ to disintegrate. Even after 200 years, 91.6% of the original quantity of $^{226}_{88}Ra$, an isotope with a half-life of 1590 years that is sometimes used in cancer treatment, would still be present. Obviously, we cannot simply leave $^{226}_{88}Ra$ or other radioactive isotopes lying around on the assumption that the radiation danger will disappear within a short period.

28.5 The Age of the Earth

One of the most interesting applications of the study of radioactivity has been in determining the age of the earth. To estimate the lower limit for the earth's age, scientists determine the age of various rocks and minerals, assuming that the earth must be at least as old as the rocks and minerals in its crust. One method is to measure the relative amounts of $_{37}^{87}Rb$ and $_{38}^{87}Sr$ in rock. $_{37}^{87}Rb$ decays to $_{38}^{87}Sr$ with a half-life of 4.7×10^{10} years. Thus 1 gram of $_{37}^{87}Rb$ would produce 0.5000 gram of $_{38}^{87}Sr$ and leave 0.5000 gram of $_{37}^{87}Rb$ after decaying for 47 billion years. The $_{37}^{87}Rb$-to-$_{38}^{87}Sr$ ratio in the Amitsog gneiss, a rock formation found in southwestern Greenland, has shown this formation to have an age of 3.75×10^9 years. This is the oldest known rock on earth.

NUCLEAR REACTIONS

The following sections describe reactions of nuclei that result in changes in their atomic numbers, mass numbers, and energy states. Such reactions are called **nuclear reactions.** Nuclear reactions result: (1) from the spontaneous decay of naturally occurring or artificially produced radioactive nuclei, (2) from the fission of unstable heavy nuclei, (3) from the fusion of light nuclei, and (4) from the bombardment of nuclei with other nuclei or with other fast-moving particles.

28.6 Equations for Nuclear Reactions

An equation that describes a nuclear reaction identifies the nuclides involved in the reaction, their mass numbers and atomic numbers, and the other particles involved in the reaction. These particles include:

1. **alpha particles ($_2^4He$, or α),** helium nuclei consisting of two neutrons and two protons
2. **beta particles ($_{-1}^0e$, or β),** high-speed electrons
3. **positrons ($_{+1}^0e$, or β^+),** particles with the same mass as an electron but with 1 unit of *positive* charge
4. **protons ($_1^1H$, or p),** nuclei of hydrogen atoms
5. **neutrons ($_0^1n$, or n),** particles with a mass approximately equal to that of a proton but with no charge

Nuclear reactions also often involve **gamma rays (γ).** Gamma rays are electromagnetic radiation, somewhat like X rays in character but with higher energies and shorter wavelengths.

Examples of equations for some types of nuclear reactions of historical interest follow.

1. The first radioactive element, naturally occurring polonium, was discovered by the Polish scientist Marie Curie and her husband, Pierre, in 1898. It decays by the reaction

$$_{84}^{212}Po \longrightarrow _{82}^{208}Pb + _2^4He \tag{1}$$

2. Technetium, a radioactive element that does not occur naturally on the earth and was first prepared in 1937, decays by the reaction

$$\ce{^{98}_{43}Tc} \longrightarrow \ce{^{98}_{44}Ru} + \ce{^{0}_{-1}e} \qquad (2)$$

3. One of the many fission reactions of uranium in the first nuclear reactor (1942) is

$$\ce{^{235}_{92}U} + \ce{^{1}_{0}n} \longrightarrow \ce{^{87}_{35}Br} + \ce{^{146}_{57}La} + 3\,\ce{^{1}_{0}n} \qquad (3)$$

4. One of the fusion reactions present during the detonation of a hydrogen bomb is

$$\ce{^{3}_{1}H} + \ce{^{2}_{1}H} \longrightarrow \ce{^{4}_{2}He} + \ce{^{1}_{0}n} \qquad (4)$$

5. The first element prepared by artificial means was prepared in 1919 by bombarding nitrogen atoms with α particles.

$$\ce{^{14}_{7}N} + \ce{^{4}_{2}He} \longrightarrow \ce{^{17}_{8}O} + \ce{^{1}_{1}H} \qquad (5)$$

A nuclear reaction is not correctly written unless it is balanced, that is, unless the sum of the mass numbers of the reactants equals the sum of the mass numbers of the products, and the sum of the atomic numbers of the reactants equals the sum of the atomic numbers of the products. If the atomic number and the mass number of all but one of the particles in a nuclear reaction are known, the particle can be identified by balancing the reaction.

EXAMPLE 28.7 The reaction of an α particle with magnesium-25 ($\ce{^{25}_{12}Mg}$) produces a proton ($\ce{^{1}_{1}H}$) and a nuclide of another element. Identify the new nuclide produced.

The nuclear reaction may be written as

$$\ce{^{25}_{12}Mg} + \ce{^{4}_{2}He} \longrightarrow \ce{^{A}_{Z}X} + \ce{^{1}_{1}H}$$

where A is the mass number and Z is the atomic number of the new nuclide, X. Since the sum of the mass numbers of the reactants must equal the sum of the mass numbers of the products, $25 + 4 = A + 1$, or $A = 28$. Similarly, the atomic numbers must balance, so $12 + 2 = Z + 1$ and $Z = 13$. The element with atomic number 13 is aluminum. Thus the product is $\ce{^{28}_{13}Al}$ (an unstable nuclide with a half-life of 2.3 min that decays by β decay to $\ce{^{28}_{14}Si}$).

28.7 Radioactive Decay

The spontaneous change of an unstable nuclide into another is called **radioactive decay.** The unstable nuclide is often called the **parent nuclide;** the nuclide that results from the decay, the **daughter nuclide.** The daughter nuclide may be stable, or it may further decay. Different types of radioactive decay can be classified by the particles or electromagnetic radiation produced:

1. The loss of an α particle by a radioactive nuclide, $\boldsymbol{\alpha}$ **decay,** occurs primarily in heavy nuclei ($A \geq 200$, $Z > 83$). Loss of an α particle gives a daughter nuclide with a mass number that is 4 units smaller and an atomic number that is 2 units smaller than those of the parent nuclide. Consequently, the daughter nuclide has a larger neutron-to-proton ratio than the parent nuclide, and, if the parent nuclide lies below the band of stability (Fig. 28-2), the daughter nuclide will lie closer to the band.

2. A nuclide with a large neutron-to-proton ratio (one that lies above the band of stability in Fig. 28-2) can reduce this ratio by electron emission, or **β decay,** in which a neutron in the nucleus decays into a proton that remains in the nucleus and an electron that is emitted.

$$\ _{0}^{1}n \longrightarrow\ _{1}^{1}H +\ _{-1}^{0}e$$

Electron emission does not change the mass number of the nuclide but does increase the number of its protons and decrease the number of its neutrons. Consequently, the neutron-to-proton ratio is decreased, and the daughter nuclide lies closer to the band of stability than did the parent nuclide.

3. Certain artificially produced nuclides in which the neutron-to-proton ratio is low (nuclides that lie below the band of stability in Fig. 28-2) undergo **β⁺ decay,** the emission of a positron. During the course of β^{+} decay, a proton is converted into a neutron with the emission of a positron. The neutron-to-proton ratio increases, and the daughter nuclide lies closer to the band of stability than did the parent nuclide.

When a positron and an electron interact, they annihilate each other. All of their mass is converted into energy—two 0.511-MeV γ rays are produced.

$$\ _{-1}^{0}e +\ _{+1}^{0}e \longrightarrow 2\gamma \qquad (0.511 \text{ MeV each})$$

4. A proton is converted to a neutron when one of the electrons in an atom is captured by the nucleus (this capture is called orbital **electron capture**).

$$\ _{1}^{1}H +\ _{-1}^{0}e \longrightarrow\ _{0}^{1}n$$

Like β^{+} decay, electron capture occurs when the neutron-to-proton ratio is low (the nuclide lies below the band of stability in Fig. 28-2). Electron capture has the same effect on the nucleus as that of positron emission; the atomic number is decreased by 1 as a proton is converted into a neutron. This increases the neutron-to-proton ratio, and the daughter nuclide lies closer to the band of stability curve than did the parent nuclide.

5. A daughter nuclide is not always produced in its ground state. It may be formed in an excited state and then decay to its ground state with the emission of a γ ray, a quantum of high-energy electromagnetic radiation. Figure 28-3 illustrates the relationships for the decay of uranium-233. Note that there is no change of mass number or atomic number during emission of a γ ray. **γ-ray emission** is observed only when a nuclear reaction produces a daughter nuclide that is in an excited state.

Figure 28-3. The decay of $_{92}^{233}U$ to $_{90}^{229}Th$. The decay can occur by three paths. Two paths give excited states of the $_{90}^{229}Th$ nuclide, which decay to the ground state by emission of γ rays.

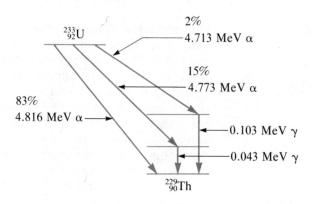

28.8 Radioactive Series

All isotopes of elements with atomic numbers larger than that of bismuth (83) are radioactive. A few elements with lower atomic numbers, such as potassium and rubidium, have some naturally occurring isotopes that are radioactive.

The naturally occurring radioactive isotopes of the heavier elements fall into chains of successive disintegrations, or decays, and all the species in one chain constitute a radioactive family, or series. Three of these series include most of the naturally radioactive elements of the Periodic Table. They are the **uranium series,** the **actinium series,** and the **thorium series.** Each series is characterized by a parent (first member) with a long half-life and a series of decay processes that ultimately lead to a stable end-product, that is, an isotope on the band of stability (Fig. 28-2). In all three series the end-product is an isotope of lead: $^{206}_{82}Pb$ in the uranium series, $^{207}_{82}Pb$ in the actinium series, and $^{208}_{82}Pb$ in the thorium series.

Successive transformations in the natural disintegrations in a radioactive series take place as described by the **displacement laws** formulated by Rutherford, Soddy, and Fajans:

1. When an atom emits an α particle, the product is an isotope of an element two places to the left of the parent in the Periodic Table.
2. When a β particle is emitted, the product is an isotope of an element one place to the right of the parent in the Periodic Table.

The steps in the thorium decay series are given in Table 28-1 (next page) as an illustration of one natural chain of successive decays.

28.9 Synthesis of Nuclides

Atoms with stable nuclei can be converted to other atoms **(transmuted)** by bombardment with other nuclei or with high-speed particles. The first artificial nucleus was made in Lord Rutherford's laboratory in 1919. He bombarded nitrogen atoms with high-speed α particles emanating from a naturally radioactive isotope of radium and observed the nuclear reaction

$$^{14}_{7}N + ^{4}_{2}He \longrightarrow ^{17}_{8}O + ^{1}_{1}H \tag{1}$$

The $^{17}_{8}O$ nucleus is stable, so this reaction does not lead to further changes.

Many artificially induced nuclear reactions produce unstable nuclei, which then undergo radioactive disintegration. The disintegration in these cases is referred to as **artificial radioactivity.** For example, boron-10 reacts with an α particle to form nitrogen-15 and a neutron.

$$^{10}_{5}B + ^{4}_{2}He \longrightarrow ^{13}_{7}N + ^{1}_{0}n \tag{2}$$

Nitrogen-13 is radioactive and decays by β^+ decay.

$$^{13}_{7}N \longrightarrow ^{13}_{6}C + \beta^+$$

In an abbreviated notation for **transmutation reactions,** the bombarding particle and the product particle are written in parentheses between the symbols for the reactant and product nuclides. Thus Equations (1) and (2) can be written as

$$^{14}_{7}N(\alpha, p) \, ^{17}_{8}O \quad \text{and} \quad ^{10}_{5}B(\alpha, n) \, ^{13}_{7}N$$

respectively, where α, p, and n are symbols for an α particle, a proton, and a

Table 28-1   The Thorium Decay Series

Mass number

232 ^{232}Th

230 α

228 ^{228}Ra $\xrightarrow{\beta}$ ^{228}Ac $\xrightarrow{\beta}$ ^{228}Th

226 α

224 ^{224}Ra

222 α

220 ^{220}Rn

218   α

216   ^{216}Po

214   α

212   ^{212}Pb $\xrightarrow{\beta}$ ^{212}Bi $\xrightarrow{\beta}$ ^{212}Po

210   α α

208   ^{208}Tl $\xrightarrow{\beta}$ ^{208}Pb

       81   82   83   84   85   86   87   88   89   90

                                          Atomic number

neutron; $^{14}_{7}$N and $^{10}_{5}$B are symbols for the reactants; and $^{17}_{8}$O and $^{13}_{7}$N are symbols for the products.

In addition to α particles and neutrons, β particles, protons, and other light nuclei can be used to cause transmutation reactions. Generally, the charged particles are accelerated to the kinetic energies that will produce reactions by machines (accelerators) that use magnetic and electric fields to accelerate the particles. In all accelerators the particles move in a vacuum to avoid collisions with gas molecules.

Since neutrons are not charged, they cannot be accelerated to the velocities necessary for nuclear transmutations. Neutrons used in transmutation reactions are therefore obtained from radioactive decay and other nuclear reactions occurring in a nuclear reactor (Section 28.12).

Many known elements have been transmuted into other known elements. Also, a considerable number of new elements have been synthesized. These include element 43, technetium (Tc); element 85, astatine (At); element 87, francium (Fr); and element 61, promethium (Pm).

Prior to 1940 the heaviest known element was uranium, whose atomic number is 92. In 1940 McMillan and Abelson were able to make element 93, neptunium (Np), by bombarding uranium with high-velocity neutrons. The nuclear reactions are

$$\mathrm{^{238}_{92}U} + \mathrm{^{1}_{0}n} \longrightarrow \mathrm{^{239}_{92}U}$$

$$\mathrm{^{239}_{92}U} \xrightarrow{\text{23 min}} \mathrm{^{239}_{93}Np} + \mathrm{^{0}_{-1}e}$$

Neptunium-239 is also radioactive, with a half-life period of 2.3 days, and converts to plutonium (Pu), whose atomic number is 94.

$$\mathrm{^{239}_{93}Np} \longrightarrow \mathrm{^{239}_{94}Pu} + \mathrm{^{0}_{-1}e}$$

Elements 95 through 107 and element 109 have also been prepared artificially. The elements beyond element 92 (uranium) are called **transuranium elements.** Elements 89 through 103 make up the **actinide series.** Element 109 was prepared in the summer of 1982 by a West German group who bombarded a target of bismuth-209 with accelerated nuclei of iron-58.

$$\mathrm{^{209}_{83}Bi} + \mathrm{^{58}_{26}Fe} \longrightarrow \mathrm{^{266}_{109}X} + \mathrm{^{1}_{0}n}$$

The researchers identified the element even though they made only one atom of it. The decay products of the atom were characteristic of those expected for element 109. It began to decay about 5×10^{-3} second after its formation, with the emission of an α particle and the formation of the same isotope of element 107 that the research team had first produced the preceding year.

$$\mathrm{^{266}_{109}X} \longrightarrow \mathrm{^{262}_{107}X} + \mathrm{^{4}_{2}He}$$
$$\longrightarrow \mathrm{^{258}_{105}Ha} + \mathrm{^{4}_{2}He}$$
$$\downarrow \text{Electron capture}$$
$$\mathrm{^{258}_{104}Rf} \longrightarrow \text{Smaller nuclei}$$

Equations describing the preparation of some other isotopes of the transuranium elements are given in Table 28-2.

The name *rutherfordium,* with the symbol Rf, has been suggested for element 104 by American scientists who synthesized the element. Soviet scientists, who also worked out the synthesis of the element, have suggested the name *kurchatovium.* It has been proposed that element 105 be named *hahnium,* with the symbol Ha, to honor Otto Hahn, the late German scientist who won the Nobel Prize for the discovery of nuclear fission.

Both a Soviet and an American group have announced the synthesis of element 106, by quite different methods. The two groups have agreed not to propose any name for the element until a decision is reached as to who will get credit for being the first to synthesize it. The failure of these two groups to reach agreement is responsible, in part, for the systematic names, suggested by the International Union of Pure and Applied Chemistry (Section 4.13), for the elements beyond element 103.

Table 28-2 Preparation of the Transuranium Elements

Name	Symbol	Atomic Number	Reaction
Neptunium	Np	93	$^{238}_{92}U + ^{1}_{0}n \longrightarrow ^{239}_{93}Np + ^{0}_{-1}e$
Plutonium	Pu	94	$^{238}_{92}U + ^{2}_{1}H \longrightarrow ^{238}_{93}Np + 2\,^{1}_{0}n$
			$^{238}_{93}Np \longrightarrow ^{238}_{94}Pu + ^{0}_{-1}e$
Americium	Am	95	$^{239}_{94}Pu + ^{1}_{0}n \longrightarrow ^{240}_{95}Am + ^{0}_{-1}e$
Curium	Cm	96	$^{239}_{94}Pu + ^{4}_{2}He \longrightarrow ^{242}_{96}Cm + ^{1}_{0}n$
Berkelium	Bk	97	$^{241}_{95}Am + ^{4}_{2}He \longrightarrow ^{243}_{97}Bk + 2\,^{1}_{0}n$
Californium	Cf	98	$^{242}_{96}Cm + ^{4}_{2}He \longrightarrow ^{245}_{98}Cf + ^{1}_{0}n$
Einsteinium	Es	99	$^{238}_{92}U + 15\,^{1}_{0}n \longrightarrow ^{253}_{99}Es + 7\,^{0}_{-1}e$
Fermium	Fm	100	$^{239}_{94}Pu + 15\,^{1}_{0}n \longrightarrow ^{254}_{100}Fm + 6\,^{0}_{-1}e$
Mendelevium	Md	101	$^{253}_{99}Es + ^{4}_{2}He \longrightarrow ^{256}_{101}Md + ^{1}_{0}n$
Nobelium	No	102	$^{246}_{96}Cm + ^{12}_{6}C \longrightarrow ^{254}_{102}No + 4\,^{1}_{0}n$
Lawrencium	Lr	103	$^{250}_{98}Cf + ^{11}_{5}B \longrightarrow ^{257}_{103}Lr + 4\,^{1}_{0}n$
Rutherfordium	Rf	104	$^{249}_{98}Cf + ^{12}_{6}C \longrightarrow ^{257}_{104}Rf + 4\,^{1}_{0}n$
Hahnium	Ha	105	$^{249}_{98}Cf + ^{15}_{7}N \longrightarrow ^{260}_{105}Ha + 4\,^{1}_{0}n$
—	—	106	$^{206}_{82}Pb + ^{54}_{24}Cr \longrightarrow ^{257}_{106}X + 3\,^{1}_{0}n$
			$^{249}_{98}Cf + ^{18}_{8}O \longrightarrow ^{263}_{106}X + 4\,^{1}_{0}n$

The unexpected stability of elements 104 and 105, together with extrapolation of known facts, convinced scientists that additional elements could be synthesized. Since then, elements 106, 107, and 109 have been reported, although all three are very unstable. Whether or not additional elements can be made cannot be predicted at this time. However, their synthesis may prove difficult; isotopes containing more than 157 neutrons are exceedingly unstable (an observation for which there is as yet no explanation).

Glenn Seaborg has suggested extending the Periodic Table to include new elements whose synthesis might be possible (Table 28-3). (It may be helpful to review the material in Chapter 4 on filling shells and subshells by electrons.) Element 104 is the first element of the periodic grouping Seaborg calls the **transactinide series,** that is, the elements beyond the actinide series. The chemical properties of elements 104 through 121 can be predicted by the method of Mendeleev—comparison with their lighter analogues in the table. Element 104 should be an analogue of hafnium; 105, an analogue of tantalum; and so on to element 118, a noble gas analogous to radon.

The most striking feature of Seaborg's extension of the Periodic Table is the addition of another inner transition series of elements starting with atomic number 121 and extending through atomic number 153. He calls this grouping the **superactinide series.** This series, after element 121, would include elements in which the 5g orbitals fill (see Table 4-4). Element 121 would receive the first 7d electron (the $n = 7$ major shell being the second from the outside) and be analogous to scandium, yttrium, lanthanum, and actinium. The first of eighteen 5g electrons could enter at element 122, for which the $n = 5$ shell is the fourth from

Table 28-3 Predicted Locations of New Elements in the Periodic Table

																s	
																1 H	2 He

s														*p*			
1 H																	
3 Li	4 Be											5 B	6 C	7 N	8 O	9 F	10 Ne
11 Na	12 Mg					*d*						13 Al	14 Si	15 P	16 S	17 Cl	18 Ar
19 K	20 Ca	21 Sc	22 Ti	23 V	24 Cr	25 Mn	26 Fe	27 Co	28 Ni	29 Cu	30 Zn	31 Ga	32 Ge	33 As	34 Se	35 Br	36 Kr
37 Rb	38 Sr	39 Y	40 Zr	41 Nb	42 Mo	43 Tc	44 Ru	45 Rh	46 Pd	47 Ag	48 Cd	49 In	50 Sn	51 Sb	52 Te	53 I	54 Xe
55 Cs	56 Ba	[57–71] *	72 Hf	73 Ta	74 W	75 Re	76 Os	77 Ir	78 Pt	79 Au	80 Hg	81 Tl	82 Pb	83 Bi	84 Po	85 At	86 Rn
87 Fr	88 Ra	[89–103] †	104 Rf	105 Ha	106	107	108	109	110	111	112	113	114	115	116	117	118
119	120	[121–153] ‡	154	155	156	157	158	159	160	161	162	163	164	165	166	167	168

								f								
*LANTHANIDE SERIES	57 La	58 Ce	59 Pr	60 Nd	61 Pm	62 Sm	63 Eu	64 Gd	65 Tb	66 Dy	67 Ho	68 Er	69 Tm	70 Yb	71 Lu
†ACTINIDE SERIES	89 Ac	90 Th	91 Pa	92 U	93 Np	94 Pu	95 Am	96 Cm	97 Bk	98 Cf	99 Es	100 Fm	101 Md	102 No	103 Lr

			g							*f*						
‡SUPER ACTINIDES	121	122	123					139	140						153

the outside; the eighteenth 5g electron would enter at element 139. This would presumably be followed by addition of fourteen 6f electrons (third major shell from the outside) for elements 140 through 153, making this latter series an inner transition series analogous to the lanthanide and actinide series. Following the superactinide inner transition series, elements 154 through 168 would be analogous to elements 104 through 118, with the addition of the remaining ten 7d electrons in the second major shell from the outside and the 6p electrons in the outermost major shell. Element 168 would be another noble gas—or should we expect a noble liquid?

28.10   Nuclear Fission

The greater stability of the nuclei of elements with mass numbers of intermediate values suggested the possibility of spontaneous decomposition of the less stable nuclei of the heavy elements into more stable fragments of approximately half their sizes. Two German scientists, Hahn and Strassman, reported in 1939 that uranium-235 atoms bombarded with slow-moving neutrons split into smaller fragments, consisting of elements near the middle of the Periodic Table and several neutrons. The process is called **fission.** Among the fission products were barium, krypton, lanthanum, and cerium, all of which have more stable nuclei than uranium.

$$^{235}_{92}\text{U} + ^{1}_{0}\text{n} \longrightarrow \text{fission fragments} + \text{neutrons} + \text{energy}$$
$$\text{(Isotopes of Ba, Kr, etc.)}$$

The sum of the atomic numbers of the fission products is 92, the atomic number of the original uranium nucleus. A loss of mass of about 0.2 atomic mass unit per uranium atom occurs in these fission reactions. This mass is converted into a fantastic quantity of energy—fission of 1 pound of uranium-235 produces about 2.5 million times as much energy as is produced by burning 1 pound of coal.

Fission of a uranium-235 nucleus produces, on the average, 2.5 neutrons as well as fission fragments. These neutrons may cause the fission of other uranium-235 atoms, which in turn provide more neutrons, setting up a **chain reaction.** Nuclear fission becomes self-sustaining when the number of neutrons produced by fission equals or exceeds the number of neutrons absorbed by splitting nuclei plus the number lost to the surroundings. The amount of a fissionable material that will support a self-sustaining chain reaction is called a **critical mass.** The critical mass of a fissionable material depends on the shape of the sample as well as on the type of material.

An atomic bomb contains several pounds of fissionable material, $^{235}_{92}\text{U}$ or $^{239}_{94}\text{Pu}$, and an explosive device for compressing it quickly into a small volume. When fissionable material is in small pieces, the proportion of neutrons that escape at the relatively large surface area is great, and a chain reaction does not take place. When the small pieces of fissionable material are brought together quickly to form a body with a mass larger than the critical mass, the relative number of escaping neutrons decreases, and a chain reaction and explosion result. The explosion of an atomic bomb can release more energy than the explosion of thousands of tons of TNT.

Chain reactions of fissionable materials can be controlled in a nuclear reactor (Section 28.12).

28.11   Nuclear Fusion

The process of combining very light nuclei into heavier nuclei is also accompanied by the conversion of mass into large amounts of energy. The process is called **fusion** and is the focus of an intensive research effort to develop a practical thermonuclear reactor (Section 28.14). The principal source of energy in the sun is the fusion of four hydrogen nuclei into one helium nucleus. Four hydrogen nuclei have a mass that is 0.7% greater than that of a helium nucleus; this extra matter is converted into energy during fusion.

It has been found that a deuteron, ^2_1H, and a triton, ^3_1H, which are the nuclei of the heavy isotopes of hydrogen, will undergo fusion at extremely high temperatures (**thermonuclear fusion**) to form a helium nucleus and a neutron.

$$^2_1H + {}^3_1H \longrightarrow {}^4_2He + {}^1_0n$$

This change is accompanied by a conversion of a portion of the mass into energy and is the nuclear reaction of the hydrogen bomb. In a hydrogen bomb a fission bomb (uranium or plutonium) is exploded inside a charge of deuterium and tritium to provide the temperature of many millions of degrees required for the fusion of the deuterium and tritium.

If the fusion of heavy isotopes of hydrogen can be controlled, hydrogen from the water of the oceans could supply energy for future generations.

NUCLEAR ENERGY

28.12 Nuclear Power Reactors

Any **nuclear reactor** that produces power by the fission of uranium or plutonium by bombardment with slow neutrons must have at least five components (see Fig. 28-4).

Figure 28-4. A light-water nuclear reactor. This reactor uses pressurized liquid water at 280°C and 150 atm as both a coolant and a moderator.

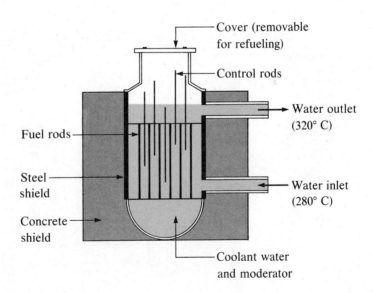

Cover (removable for refueling)

Control rods

Water outlet (320° C)

Fuel rods

Steel shield

Water inlet (280° C)

Concrete shield

Coolant water and moderator

1. A NUCLEAR FUEL. A fissionable isotope (commonly $^{235}_{92}U$, $^{233}_{92}U$, or $^{239}_{94}Pu$) must be present in sufficient quantity to provide a self-sustaining chain reaction. Most reactors used in the United States use pellets of the uranium oxide U_3O_8, in

which the concentration of uranium-235 has been increased (enriched) from its natural level to about 3%. The U_3O_8 pellets are contained in a tube (a fuel rod) of a protective material, usually a zirconium alloy. The core of a typical nuclear power reactor in the United States has about 40,000 kilograms of enriched U_3O_8, contained in several hundred fuel rods.

Naturally occurring uranium is a mixture of several isotopes; uranium-238 is the most abundant. About 1 in every 140 uranium atoms is the uranium-235 isotope. To obtain a higher concentration of uranium-235, it is necessary to separate uranium-235. The most successful method separates $^{235}_{92}UF_6$ from $^{238}_{92}UF_6$ by fractional diffusion of large volumes of gaseous UF_6 at low pressure through porous barriers. This method is based on the fact that the ligher $^{235}_{92}UF_6$ molecules diffuse through a porous barrier faster than do the heavier $^{238}_{92}UF_6$ molecules. The enriched UF_6 is chemically converted to U_3O_8.

2. A MODERATOR.

Neutrons produced by nuclear reactions move too fast to cause fission. They must be slowed down before they will be absorbed by the fuel and produce additional nuclear reactions. In a reactor neutrons are slowed by collision with the nuclei of a moderator such as heavy water (D_2O; Section 12.5), graphite, carbon dioxide, or light (ordinary) water. These materials are used because they do not react with or absorb neutrons. Most reactors in operation in the United States use light water as the moderator.

3. A COOLANT.

The coolant carries the heat from the fission reaction to an external boiler and turbine where it is transformed into electricity (Fig. 28-5). The coolant is a gas or liquid that is pumped through the reactor core. Some coolants also serve as moderators.

Figure 28-5 Power-generating plant employing a nuclear power reactor. In a coal-fired power plant the steam is generated in a boiler. Note that the reactor coolant is contained in a closed system and does not come in contact with outside cooling water. The reactor shielding has been omitted for clarity.

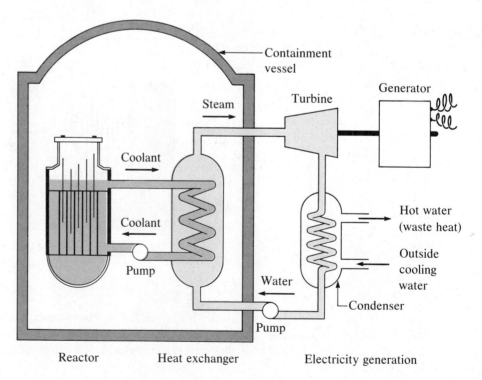

4. A CONTROL SYSTEM. A nuclear reactor is controlled by adjusting the number of slow neutrons present to keep the rate of the chain reaction at a safe level. Control is maintained by control rods that absorb neutrons. Rods containing cadmium or boron-10 are often used. Boron-10, for example, absorbs neutrons by the reaction $^{10}_{5}B$ (n, α) $^{7}_{3}Li$. The chain reaction can be completely stopped by inserting all of the control rods into the core between the fuel rods.

5. A SHIELD AND CONTAINMENT SYSTEM. A nuclear reactor produces neutrons and other particles by radioactive decay of the products resulting from fusion. In addition, a reactor is very hot, and high pressures result from the circulation of water or another coolant through it. A reactor must withstand high temperatures and pressures and protect operating personnel from the radiation. A reactor container often consists of three parts: (1) the reactor vessel, a steel shell that is 3–20 centimeters thick and absorbs much of the radiation produced by the reactor; (2) a main shield of 1–3 meters of high-density concrete; and (3) a personnel shield of lighter materials to absorb γ rays and X rays. In addition, reactors are often covered with a steel or concrete dome designed to contain any radioactive materials released by a reactor accident.

The energy produced by a reactor fueled with enriched uranium results from the fission of $^{235}_{92}U$ (Section 28.10) as well as from the fission of $^{239}_{94}Pu$. Plutonium-239 forms from $^{238}_{92}U$ present in the fuel (Section 28.9). In any nuclear reactor only about 0.1% of the mass of the fuel is converted into energy. The other 99.9% remains in the fuel rods as fission products and unused fuel. All of the fission products absorb neutrons, and, after a period of several months to a year, depending on the reactor, the fission products must be removed by changing the fuel rods. Otherwise, the concentration of these fission products will increase until the reactor can no longer operate because of their absorption of neutrons.

Spent fuel rods contain a variety of products, consisting of unstable nuclei ranging in atomic number from 25 to 60, some transuranium elements, including $^{239}_{94}Pu$ and $^{241}_{95}Am$, and unreacted $^{235}_{92}U$ and $^{238}_{92}U$. The unstable nuclei and the transuranium elements give the spent fuel a dangerously high level of radioactivity. The long-lived isotopes $^{90}_{38}Sr$ and $^{137}_{55}Cs$ and the shorter-lived isotope $^{131}_{53}I$ are particularly dangerous because they can be incorporated into human bodies if the radioactive material is dispersed in the environment. Consequently, it is absolutely essential that this material not be allowed to be released into the biosphere. It takes about 400 years for the radioactivity of $^{90}_{38}Sr$ and $^{137}_{55}Cs$ to decrease to a reasonably safe level. Other nuclides such as $^{241}_{95}Am$ and $^{239}_{94}Pu$ have much longer half-lives and require thousands of years to decay to a safe level. The ultimate fate of the nuclear reactor as a significant source of energy in the United States probably rests on whether or not a scientifically and politically satisfactory technique for processing and storing the components of spent fuel rods can be developed.

28.13 Breeder Reactors

A **breeder reactor** is a nuclear reactor that produces more fissionable material than it consumes. Because the supply of naturally occurring uranium is limited (some estimates suggest that the known reserves will only last for another 50 years of full-scale use), the conversion of nonfissionable material to nuclear fuel in a breeder reactor may provide a long-term energy supply. The breeder reactor is constructed with a blanket of fertile material such as $^{238}_{92}U$ or $^{232}_{90}Th$ surrounding the core of fissionable material. Extra neutrons produced by the fission in the core are captured by the blanket of fertile material to form (breed) more fissionable atoms. Hence the breeder reactor is a source of energy that at the same time produces more fuel than it consumes.

Breeder reactors are based on one of two sets of nuclear reactions. One set begins with $^{238}_{92}U$ and the other with $^{232}_{90}Th$. Both nuclides are the most abundant isotopes of their respective elements. $^{238}_{92}U$ produces the fissionable nuclide $^{239}_{94}Pu$, according to the reactions described in Section 28.9. $^{232}_{90}Th$ produces the fissionable nuclide $^{233}_{92}U$ by the following series of reactions:

$$^{232}_{90}Th + ^{1}_{0}n \longrightarrow ^{233}_{90}Th$$

$$^{233}_{90}Th \xrightarrow{\text{23 min}} ^{233}_{91}Pa + ^{0}_{-1}e$$

$$^{233}_{91}Pa \xrightarrow{\text{27 day}} ^{233}_{92}U + ^{0}_{-1}e$$

Experimental breeder reactors are in operation in the United States, Great Britain, Russia, and France.

Breeder reactors present challenges in addition to those outlined in Section 28.12. Not only must spent fuel be stored, but the blanket of fertile material must be processed (by remote control) to separate the radioactive fuel from other radioactive by-products, and the large amounts of plutonium produced must be safely stored. Plutonium-239 decays with a half-life of 24,000 years by emission of an α particle; thus it will remain in the biosphere for a very long time if it is dispersed. Plutonium is one of the most toxic substances known, estimates of a fatal dose being as low as 1 microgram (10^{-6} g). Moreover, plutonium can be used to make bombs (Section 28.10), and the existence of quantities of the pure material could be attractive to terrorist groups and to some countries not yet capable of producing their own atomic weapons.

28.14   Fusion Reactors

A fusion reactor is a nuclear reactor in which fusion reactions of light nuclei (Section 28.11) are controlled. At the time of this writing, there are no self-sustaining fusion reactors operating in the world, although small-scale fusion reactions have been run for very brief periods.

Fusion reactions require very high temperatures, about 10^8 K. At these temperatures, all molecules dissociate into atoms, and the atoms ionize, forming a new state of matter called a **plasma.** Since no solid materials are stable at 10^8 K, a plasma cannot be contained by mechanical devices. Two techniques to contain a plasma at the necessary density and temperature for a fusion reaction to occur are currently under study. The techniques involve containment by a magnetic field and by the use of focused laser beams.

After a plasma of hydrogen isotopes is generated and contained, it will undergo fusion reactions when the temperature exceeds about 10^8 K. One such reaction is

$$^{2}_{1}H + ^{3}_{1}H \longrightarrow ^{4}_{2}He + ^{1}_{0}n$$

which proceeds with a mass loss of 0.0188 atomic mass unit, corresponding to the release of 1.69×10^9 kilojoules per mole of ^4_2He formed. Deuterium (^2_1H) is available from heavy water. Tritium (^3_1H) can be prepared by reaction of the neutrons from the fusion reaction with lithium.

$$^1_0\text{n} + {^6_3}\text{Li} \longrightarrow {^3_1}\text{H} + {^4_2}\text{He}$$

FOR REVIEW

SUMMARY

Protons and neutrons, collectively called **nucleons,** are held together by a short-range, but very strong, force called the nuclear force. The **nuclear binding energy** may be calculated from the **mass defect** (the difference in mass between a nucleus and the nucleons of which it is composed) by the **Einstein equation,**

$$E = mc^2$$

The binding energy per nucleon is largest for the elements with mass numbers between 40 and 100; these are the most stable nuclei.

Stable nuclei have equal numbers of neutrons and protons or a few more neutrons than protons. Nuclei that deviate from stable neutron-to-proton ratios are radioactive and decay by losing one of several different kinds of particles. These include α particles, ^4_2He; β particles, $^0_{-1}\text{e}$; positrons, $^0_{+1}\text{e}$; and **neutrons,** ^1_0n. Some nuclei decay by **electron capture.** Each of these modes of decay leads to a new nucleus with a more stable neutron-to-proton ratio. The kinetics of the decay process are first-order. The half-life of a radioactive isotope is the time that is required for $\frac{1}{2}$ of the atoms in a sample to decay. Each radioactive nuclide has its own characteristic half-life.

New atoms can be produced by bombarding other atoms with nuclei or high-speed particles. The products of these **transmutation reactions** can be stable or radioactive. A number of artificial elements including technetium, astatine, and the transuranium elements have been produced in this way.

Nuclear power can be generated through **fission,** or reactions in which a heavy nucleus breaks up into two or more lighter nuclei and several neutrons. Since the neutrons may induce additional fission reactions when they combine with other heavy nuclei, a **chain reaction** can result. Useful power is obtained if the fission process is carried out in a **nuclear reactor.** The conversion of light nuclei into heavier nuclei **(fusion)** also produces energy. At present, this energy has not been contained and is used only in nuclear weapons.

KEY TERMS AND CONCEPTS

Alpha decay (28.7)
Alpha particles (28.6)
Artificial radioactivity (28.9)
Atomic number (28.1)

Band of stability (28.3)
Beta decay (28.7)
Beta particles (28.6)
Breeder reactor (28.13)

Chain reaction (28.10)
Daughter nuclide (28.7)
Einstein equation (28.2)
Electron capture (28.7)

Fission (28.10)

Fusion (28.11)

Fusion reactor (28.14)

Gamma rays (28.6)

Half-life (28.4)

Mass defect (28.2)

Mass number (28.1)

Neutron (28.6)

Nuclear binding energy (28.2)

Nuclear force (28.1)

Nuclear reactor (28.12)

Nucleons (28.1)

Nuclide (28.1)

Parent nuclide (28.7)

Positrons (28.6)

Protons (28.6)

Radioactive decay (28.7)

Transmutation reactions (28.9)

EXERCISES

1. Write a brief description or definition of each of the following:
   (a) nucleon   (e) γ ray
   (b) α particle   (f) mass number
   (c) β particle   (g) atomic number
   (d) positron

2. Indicate the number of protons and neutrons in each of the following nuclei:
   (a) $^{14}_{7}N$ (c) xenon-135
   (b) chlorine-37   (d) $^{250}_{100}Fm$

Nuclear Stability

3. Which of the following nuclei lie within the band of stability shown in Fig. 28-2?
   (a) nitrogen-17   (d) $^{102}_{47}Ag$
   (b) $^{12}_{6}C$ (e) radon-210
   (c) $^{35}_{16}S$

4. Which of the following nuclei would you expect to be unstable?
   (a) $^{1}_{1}H$ (b) $^{31}_{15}P$ (c) $^{39}_{20}Ca$ (d) $^{233}_{90}Th$

5. Define and illustrate the term *half-life*.

S 6. The mass of the isotope $^{27}_{13}Al$ is 26.98154. (a) Calculate its binding energy per atom in millions of electron-volts. (b) Calculate its binding energy per nucleon. (See Appendix D.)
   Ans. (a) 224.96 MeV; (b) 8.332 MeV

7. A $^{7}_{4}Be$ atom (mass = 7.0169 amu) decays to a $^{7}_{3}Li$ atom (mass = 7.0160 amu) by electron capture. How much energy (in millions of electron-volts) is produced by this reaction?   *Ans. 0.8 MeV*

S 8. The mass of a deuteron ($^{2}_{1}H$) is 2.01355 amu; that of an α particle, 4.00150 amu. How much energy per mole of $^{4}_{2}He$ produced is released by the following reaction?

$$^{2}_{1}H + ^{2}_{1}H \longrightarrow ^{4}_{2}He$$

   Ans. 2.301 × 10⁹ kJ

9. The half-life of ^{239}Pu is 24,000 yr. What fraction of the ^{239}Pu present in nuclear wastes generated today will be present in 1000 yr?
   Ans. 0.97 (97%)

S 10. What percentage of $^{212}_{82}Pb$ remains of a 1.00-g sample, 1.0 min after it is formed (half-life of 10.6 s)? 10 min after it is formed?
   Ans. 2.0%; 9.1 × 10⁻¹⁶%

S 11. The isotope ^{208}Tl undergoes β decay with a half-life of 3.1 min.
   (a) What isotope is the product of the decay?
   (b) Is ^{208}Tl more stable or less stable than an isotope with a half-life of 54.5 s?
   (c) How long will it take for 99.0% of a sample of pure ^{208}Tl to decay?   *Ans. 20.6 min*
   (d) What percentage of a sample of pure ^{208}Tl will remain undecayed after an hour?
   Ans. 1.5 × 10⁻⁴%

12. Calculate the time required for 99.999% of each of the following radioactive isotopes to decay:
   S (a) $^{228}_{88}Ra$ (half-life, 1590 yr)   *Ans. 26,400 yr*
   (b) $^{214}_{84}Po$ (half-life, 1.6 × 10⁻⁴ s)   *Ans. 0.0027 s*
   (c) $^{232}_{90}Th$ (half-life, 1.39 × 10¹⁰ yr)
   Ans. 2.31 × 10¹¹ yr

S 13. The isotope $^{90}_{38}Sr$ is an extremely hazardous species in the fallout from a nuclear fission explosion. A 0.500-g sample diminishes to 0.393 g in 10.0 yr. Calculate the half-life.   *Ans. 28.8 yr*

14. If 1 g of $^{226}_{88}Ra$ (atomic weight = 226) produces 0.0001 mL of the gas $^{222}_{86}Rn$ (atomic weight = 222) at STP in 24 h, what is the half-life of ^{226}Ra in years?   *Ans. 2 × 10³ yr*

Nuclear Decay

15. Describe the types of radiation emitted from nuclei of naturally radioactive elements.

16. What is the change in the nucleus that gives rise to a β particle?

17. The loss of an α particle by a nucleus causes what

change in the atomic number and the mass of the nucleus? What is the change in the atomic number and mass when a β particle is emitted?

18. How do nuclear reactions differ from ordinary chemical changes?

19. Many nuclides with atomic numbers greater than 83 decay by processes such as electron emission. Rationalize the observation that the emissions from these unstable nuclides normally include α particles also.

20. Identify the various particles that may be produced in a nuclear reaction.

21. Write a balanced equation for each of the following nuclear reactions.
    (a) Thorium-230 undergoes α decay.
    (b) Thorium-231 decays to protactinium-231.
    (c) Boron-11 and a positron are produced by the decay of an unstable nucleus.
    (d) Neptunium-239 forms from the reaction of uranium-238 with a neutron, then spontaneously converts to plutonium-239.
    (e) Strontium-90 decays to yttrium-90.

22. Complete the following equations:
    Ⓢ(a) $^{27}_{13}Al + ^{2}_{1}H \longrightarrow ? + ^{4}_{2}He$
    Ⓢ(b) $^{7}_{3}Li + ? \longrightarrow 2 ^{4}_{2}He$
    Ⓢ(c) $^{9}_{4}Be + ^{4}_{2}He \longrightarrow ^{12}_{6}C + ?$
    (d) $^{23}_{11}Na + ^{2}_{1}H \longrightarrow ^{24}_{11}Na + ?$

23. Fill in the atomic number of the initial nucleus and write out the complete nuclear symbol for the product of each of the following nuclear reactions:
    Ⓢ(a) ^{65}Cu (n, 2n)   (e) ^{106}Rh (n, p)
    Ⓢ(b) ^{54}Fe (α, 2p)   (f) $^{27}Al(\alpha$ n)
    Ⓢ(c) ^{33}S (n, p)   Ⓢ(g) ^{14}N (p, γ)
    (d) ^{33}S (p, n)

24. Complete the following notations by filling in the missing parts:
    Ⓢ(a) ^{2}H (d, n)   Ⓢ(e) $^{232}Th($, n) ^{235}U
    Ⓢ(b)   (α, n) ^{30}P (f) ^{2}H (^{12}C,   ) ^{10}B
    (c)   (d, n) ^{238}Np (g) ^{24}Mg (α, n)
    (d) ^{10}B (α,   ) ^{13}C (h) ^{238}U ($^{12}_{6}C$, 4n)

25. Use the abbreviated system of notation as in Exercise 24 to describe:
    (a) the production of ^{17}O from ^{14}N by α particle bombardment

(b) the production of ^{14}C from ^{14}N by neutron bombardment

(c) the neutron-induced fission of $^{235}_{92}U$

(d) the production of ^{233}Th from ^{232}Th by neutron bombardment

(e) the production of ^{239}U from ^{238}U by deuteron bombardment

26. For each of the following unstable isotopes, predict by what mode(s) spontaneous radioactive decay might proceed:
    (a) ^{34}P (n/p ratio too large)
    (b) ^{156}Eu (n/p ratio too large)
    (c) ^{235}Pa (too much mass, n/p ratio too large)
    (d) $^{3}_{1}H$
    (e) ^{18}F
    (f) ^{129}Ba
    (g) ^{237}Pu

27. Which of the following nuclei is most likely to decay by positron emission: chromium-53, manganese-51, or iron-59? Explain your choice.

28. Explain in terms of Fig. 28-2 how unstable heavy nuclides (atomic number greater than 83) may decompose to form nuclides of greater stability (a) if they are below the band of stability and (b) if they are above the band of stability.

Nuclear Power

29. Distinguish between nuclear fission and nuclear fusion. Why are both of these processes exothermic?

30. How are atomic bombs and hydrogen bombs detonated?

31. Describe the components of a nuclear reactor.

32. Describe how the potential energy of uranium is converted into electrical energy in a nuclear power plant.

33. What is a breeder reactor?

34. List advantages and disadvantages of nuclear energy as a source of electrical power as compared to coal, fuel oil, natural gas, and water.

35. Discuss and compare the problems of radioactive wastes for radioactive substances of short half-life and those of long half-life.

29

COORDINATION COMPOUNDS

The hemoglobin in your blood, the blue dye in the ink in your ballpoint pen and in your blue jeans, chlorophyll, vitamin B-12, and the catalyst used in the manufacture of polyethylene all contain **coordination compounds,** or **complexes**—compounds in which negative ions and/or neutral molecules are attached to metal ions. Examples of simple complexes include $[Ag(NH_3)_2]^+$, $[Fe(CN)_6]^{3-}$, $[Co(NH_3)_6]^{3+}$, $[Pt(NH_3)_2Cl_2]$, $Fe(CO)_5$, and species such as $[Al(H_2O)_6]^{3+}$ and $[Zn(OH)_4]^{2-}$ mentioned in previous chapters. In this chapter we will consider the structures and properties of such complexes.

Formation of a complex requires two kinds of species: (1) an ion or molecule that has at least one pair of electrons available for coordinate covalent bonding, and (2) a metal ion or atom that has a sufficient attraction for electrons to form a coordinate covalent bond with the attaching group. Ions of the transition metals, inner transition metals, and a few metals near these series in the Periodic Table are especially prone to combine in complexes.

The enthalpies of formation for coordination compounds vary greatly, but many are very stable. For example, the enthalpy of formation of $[Ni(NH_3)_6]I_3$ is −808 kilojoules per mole while that of $[Co(NH_3)_6](NO_3)_3$ is −1282 kilojoules per mole (Appendix J).

NOMENCLATURE, STRUCTURES, AND PROPERTIES OF COORDINATION COMPOUNDS

29.1 Definitions of Terms

The metal ion or atom in a complex is called the **central metal ion** or **atom;** the groups attached to it are called **ligands.** Ligands may be either ions or neutral molecules. Within a ligand, the atom that attaches directly to the metal by a coordinate covalent bond (Section 5.4) is called the **donor atom.**

The **coordination sphere,** usually the atoms within square brackets in a formula, includes the central metal ion plus the attached ligands. The **coordination number** of the central metal ion is the number of donor atoms bonded to it. The coordination number for the silver ion in $[Ag(NH_3)_2]^+$ is 2; that for the copper ion in $[Cu(NH_3)_4]^{2+}$ is 4; and that for the iron(III) ion in $[Fe(CN)_6]^{3-}$ is 6. In each of these examples the coordination number is equal to the number of ligands in the coordination sphere, but such is not always the case. Some ligands, such as ethylenediamine,

$$
\begin{array}{ccccc}
& H & H & H & H \\
& | & | & | & | \\
H- & N- & C- & C- & N-H \\
& \cdot\cdot & | & | & \cdot\cdot \\
& & H & H &
\end{array}
$$

contain two donor atoms. Thus the coordination number for cobalt in $[Co(H_2NCH_2CH_2NH_2)_3]^{3+}$ is 6. Although the coordination sphere of this complex contains only three ligands, all six nitrogen atoms are bonded to the cobalt. The most common coordination numbers are 2, 4, and 6. Coordination numbers of 3, 5, 7, and 8 do occur, as well as, though much less commonly, 9, 10, 11, and 12.

When a ligand attaches itself to a central metal ion by bonds from two or more donor atoms, it is referred to as a **polydentate ligand,** or a **chelating ligand.** The resulting complex is a **metal chelate;** examples include

$$
\left[Co\left(\begin{array}{c} NH_2-CH_2 \\ | \\ NH_2-CH_2 \end{array} \right)_3 \right]^{3+} \qquad \left[Co\left(\begin{array}{c} O-C=O \\ | \\ O-C=O \end{array} \right)_3 \right]^{3-} \qquad \left[Cu\left(\begin{array}{c} NH_2-CH_2 \\ | \\ O-\!\!-C=O \end{array} \right)_2 \right]^{0}
$$

The complex heme in hemoglobin (Fig. 29-1) contains a polydentate ligand with four donor atoms.

29.2 The Naming of Complexes

The nomenclature of complexes is patterned after a system suggested by Alfred Werner, a Swiss chemist and Nobel laureate, whose outstanding work more than 80 years ago laid the foundation for a clearer understanding of these compounds. The following rules are used for naming complexes.

1. If a complex is ionic, name the cation first and the anion second, in accord with usual nomenclature.

2. Name the ligands first, followed by the central metal.

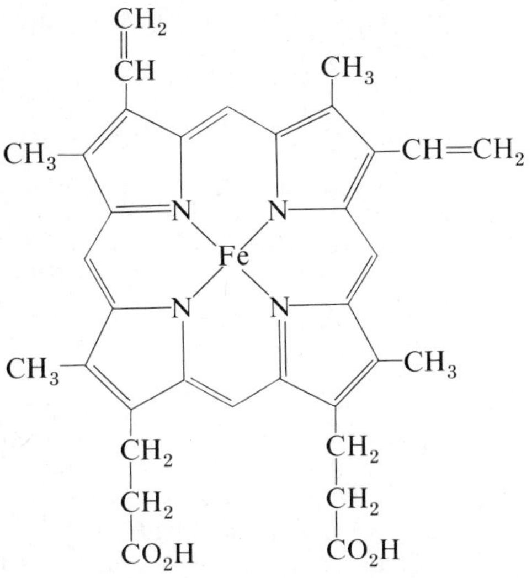

Figure 29-1. Heme, the square planar complex of iron found in hemoglobin.

3. Name the ligands alphabetically. (An older system names negative ligands alphabetically, then neutral ligands alphabetically, and finally positive ligands alphabetically.) Negative ligands (anions) have names formed by adding *-o* to the stem name of the group, for example,

F^-	*fluoro*	NO_3^-	*nitrato*
Cl^-	*chloro*	OH^-	*hydroxo*
Br^-	*bromo*	O^{2-}	*oxo*
I^-	*iodo*	NH_2^-	*amido*
CN^-	*cyano*	$C_2O_4^-$	*oxalato*
NO_2^-	*nitro*	CO_3^{2-}	*carbonato*
ONO^-	*nitrito*		

For most neutral ligands the name of the molecule is used. The four common exceptions are *aqua* (H_2O), *ammine* (NH_3), *carbonyl* (CO), and *nitrosyl* (NO).

4. If more than one ligand of a given type is present, the number is indicated by the prefixes *di-* (for two), *tri-* (for three), *tetra-* (for four), *penta-* (for five), and *hexa-* (for six). Sometimes the prefixes *bis-* (for two), *tris-* (for three), and *tetrakis-* (for four) are used when the name of the ligand contains numbers, begins with a vowel, is for a polydentate ligand, or includes *di-, tri-,* or *tetra-*.

5. When the complex is either a cation or a neutral molecule, the name of the central metal atom is spelled exactly as the name of element and is followed by a Roman numeral in parentheses to indicate its oxidation number. When the complex is an anion, the suffix *-ate* is added to the stem for the name of the metal (or sometimes to the stem for the Latin name of the metal), followed by the Roman numeral designation of its oxidation number. Examples in which the complex is a cation are as follows:

$[Co(NH_3)_6]Cl_3$	Hexaamminecobalt(III) chloride
$[Pt(NH_3)_4Cl_2]^{2+}$	Tetraamminedichloroplatinum(IV) ion
$[Ag(NH_3)_2]^+$	Diamminesilver(I) ion
$[Cr(H_2O)_4Cl_2]Cl$	Tetraaquadichlorochromium(III) chloride
$[Co(H_2NCH_2CH_2NH_2)_3]_2(SO_4)_3$	Tris(ethylenediamine)cobalt(III) sulfate

Examples in which the complex is neutral:

[Pt(NH$_3$)$_2$Cl$_4$]	Diamminetetrachloroplatinum(IV)
[Co(NH$_3$)$_3$(NO$_2$)$_3$]	Triamminetrinitrocobalt(III)
[Ni(H$_2$NCH$_2$CH$_2$NH$_2$)$_2$Cl$_2$]	Dichlorobis(ethylenediamine)nickel(II)

Examples in which the complex is an anion:

K$_3$[Co(NO$_2$)$_6$]	Potassium hexanitrocobaltate(III)
[PtCl$_6$]$^{2-}$	Hexachloroplatinate(IV) ion
Na$_2$[SnCl$_6$]	Sodium hexachlorostannate(IV)

29.3   The Structures of Complexes

The structures of many simple compounds and ions were discussed in Chapter 7. Spatial arrangements of atoms include geometries that are linear, trigonal planar, tetrahedral, square planar, square pyramidal, trigonal bipyramidal, and octahedral, among others. Now we can extend our study to the structures of coordination compounds. It may be helpful to reread Chapter 7 to refresh your memory of the structures discussed there. Many of these structures are found in coordination compounds as well as in simple compounds.

In 1893 Alfred Werner suggested that when ions or polar molecules are coordinated to a metal ion they are arranged in a definite geometrical pattern about it. This concept accounted for the properties of hydrates such as CoCl$_2 \cdot$6H$_2$O, ammoniates such as CoCl$_3 \cdot$6NH$_3$, and various so-called double salts such as PtCl$_4 \cdot$2KCl. Werner assigned the formulas [Co(H$_2$O)$_6$]Cl$_2$, [Co(NH$_3$)$_6$]Cl$_3$, and K$_2$[PtCl$_6$] to these compounds and pointed out that the properties of the complexes [Co(H$_2$O)$_6$]$^{2+}$, [Co(NH$_3$)$_6$]$^{3+}$, and [PtCl$_6$]$^{2-}$ could be explained by postulating that the six ligands are arranged about the central ion at the corners of a regular octahedron (Fig. 29-2).

Figure 29-2.   Octahedral structures of the [Co(H$_2$O)$_6$]$^{2+}$, the [Co(NH$_3$)$_6$]$^{3+}$, and the [PtCl$_6$]$^{2-}$ ions.

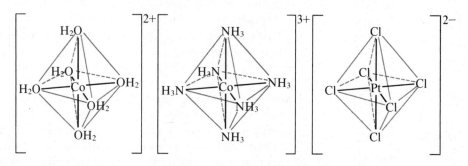

Many chemists use an abbreviated drawing of an octahedron [Fig. 29-3(a)] to show the geometry about a metal ion. However, a more realistic octahedron can be drawn by following the steps outlined in Fig. 29-3(b).

Complexes in which the metal shows a coordination number of 4 exist in either of two geometric arrangements, the square planar or the tetrahedral configurations. Examples of four-coordinated complex ions with the square planar configuration are [Ni(CN)$_4$]$^{2-}$ and [Cu(NH$_3$)$_4$]$^{2+}$ and with the tetrahedral configuration are [Zn(CN)$_4$]$^{2-}$ and [Zn(NH$_3$)$_4$]$^{2+}$ (Fig. 29-4).

Many other geometries are possible. Table 29-1 shows several of the known types.

Figure 29-3. Steps in drawing an octahedron.

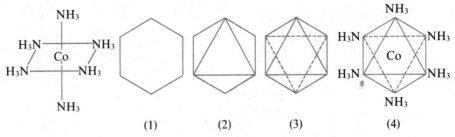

(1) (2) (3) (4)

(a) An abbreviated drawing of an octahedron.

(b) To construct a complete octahedron quickly and simply: (1) Draw a regular hexagon. (2) Draw a triangle inside the hexagon. (3) Draw another triangle with dashed lines upside-down inside the hexagon. (4) Add symbols for central metal and ligands.

Figure 29-4. (a) The square planar configuration of the $[Ni(CN)_4]^{2-}$ ion, and (b) the tetrahedral configuration of the $[Zn(CN)_4]^{2-}$ ion.

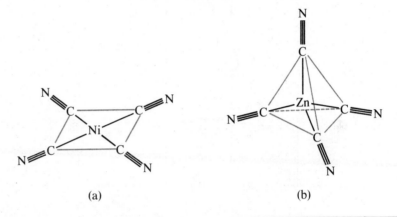

(a) (b)

Table 29-1 Geometric Shapes and Corresponding Hybridization of Orbitals for Several Complexes

Geometric Shape	Diagram of Shape	Coordination Number	Hybridization	Examples
Linear		2	sp	$[Ag(NH_3)_2]^+$, $Cu(CN)_2^-$
Trigonal planar		3	sp^2	$[HgCl_3]^-$
Tetrahedral		4	sp^3 (or sd^3)	$[Zn(CN)_4]^{2-}$, $[FeCl_4]^-$, $[CoBr_4]^{2-}$
Square planar		4	dsp^2	$[Ni(CN)_4]^{2-}$, $[Cu(NH_3)_4]^{2+}$

(continued)

Table 29-1 (continued)

Geometric Shape	Diagram of Shape	Coordination Number	Hybridization	Examples
Square pyramidal		5	d^2sp^2 (or d^4s)	$[VOCl_4]^{2-}$, $[Ni(Br)_3\{(C_2H_5)_3P\}_2]$
Trigonal bipyramidal		5	dsp^3 (or d^3sp)	$[Fe(CO)_5]$, $[Mn(CO)_4NO]$
Octahedral		6	d^2sp^3	$[Co(NH_3)_6]^{3+}$, $[PtCl_6]^{2-}$, $[MoF_6]^-$

29.4   Isomerism in Complexes

Certain complexes, such as $[Co(NH_3)_4Cl_2]^+$, the tetraamminedichlorocobalt(III) ion, have more than one form. These different forms with the same formula are **isomers** (Section 25.16). The $[Co(NH_3)_4Cl_2]^+$ ion has two isomers, one violet and the other green. Crystal structure analysis, using X-ray methods and other techniques, shows that the violet form has the ***cis* configuration** (chloride ions occupy adjacent corners of the octahedron) and the green form has the ***trans* configuration** (chloride ions occupy opposite corners), as shown in Fig. 29-5. Isomers such as these, which differ only in the way that the atoms are oriented in space relative to each other, are called **geometrical isomers,** or **stereoisomers.**

Figure 29-5. The *cis* and *trans* isomers of $[Co(NH_3)_4Cl_2]^+$.

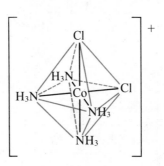

Violet, *cis* form

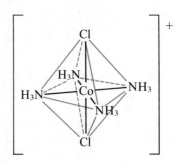

Green, *trans* form

A complex such as $[Cr(NH_3)_2(H_2O)_2(Br)_2]^+$ has a variety of different geometrical isomers. Each ligand could be *trans* to one like it (as shown in the figure on the left below) or the ammonia molecules could be *trans* to each other while the water molecules and bromine atoms are *cis* to their counterparts (figure on the right).

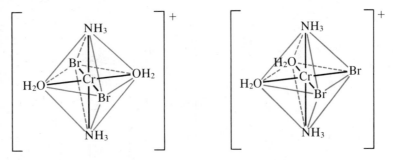

Similarly, the water molecules or the bromine atoms could be *trans* to each other while the other ligands are *cis* to ones like themselves:

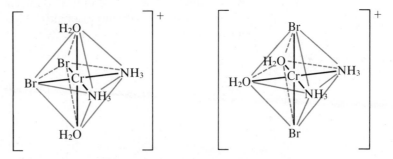

Finally, each group could be *cis* to one like itself; in this case two different arrangements are possible:

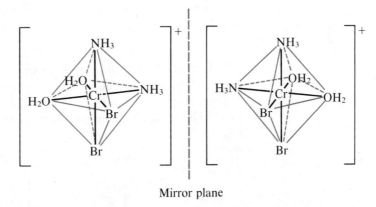

Mirror plane

These two geometrical isomers have the same arrangement of ligands in the coordination sphere (*cis*-H_2O, *cis*-Br, *cis*-NH_3), but they are mirror images of each other and are not superimposable. Therefore they are not geometrically identical. Isomers that are mirror images of each other but not identical are called **optical isomers** (Section 25.16). Mirror images of the other geometrical isomers of $[Cr(NH_3)_2(H_2O)_2(Br)_2]^+$ can be drawn, but the mirror images can be superimposed and so are identical. For example, the following mirror images are identical to each other; either may be turned 90° on an axis through the two corners occupied by the ammonia molecules and superimposed on the other:

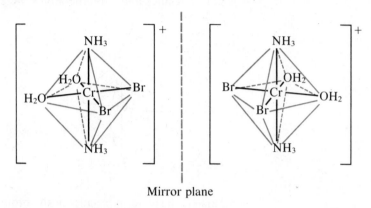

Mirror plane

The $[Cr(NH_3)_2(H_2O)_2(Br)_2]^+$ complex thus has a total of six geometrical isomers, two of which are a pair of optical isomers.

The tris(ethylenediamine)cobalt(III) ion, $[Co(H_2NCH_2CH_2NH_2)_3]^{3+}$, has two optical isomers, as shown in Fig. 29-6.

Figure 29-6. Optical isomers of $[Co(H_2NCH_2CH_2NH_2)_3]^{3+}$.

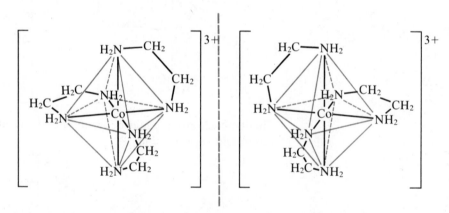

The $[Co(en)_2Cl_2]^+$ ion, where en stands for ethylenediamine, has two *cis* isomers, which are a pair of optical isomers, and one *trans* isomer. The *trans* configuration is symmetrical and has no optical isomerism. Its mirror image is superimposable on it and is therefore identical to it. The three isomers are shown in Fig. 29-7.

Figure 29-7. The three isomeric forms of $[Co(en)_2Cl_2]^+$. In these abbreviated formulas, N⌒N stands for $H_2NCH_2CH_2NH_2$.

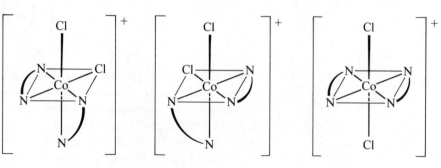

Cis forms (optical isomers)   *Trans* form

29.5 Uses of Complexes

Many uses of complexes are based on their colors, their solubilities, or the changes in chemical behavior of metal ions and ligands when they form complexes.

Chlorophyll (Fig. 29-8), the green pigment in plants, is a complex that contains magnesium. Plants appear green because chlorophyll absorbs yellow light; the reflected light consequently appears green (Section 29.10). The energy resulting from the absorption of this light is used in photosynthesis (Section 26.13). The square planar copper(II) complex phthalocyanine blue (Fig. 29-9) is one of many complexes used as pigments or dyes. This complex is used in blue ink, blue jeans, and certain blue paints.

The structure of heme (Fig. 29-1), the iron-containing complex in hemoglobin, is very similar to that of chlorophyll. In hemoglobin the red heme complex is bonded to a large protein molecule (globin) by coordination of the protein to a position above the plane of the heme molecule. Oxygen molecules are transported by hemoglobin in the blood by being bound to the coordination site opposite the binding site of the globin molecule.

Complexing agents are often used for water softening because they tie up such ions as Ca^{2+}, Mg^{2+}, and Fe^{2+}, which make water hard (Section 12.10). Com-

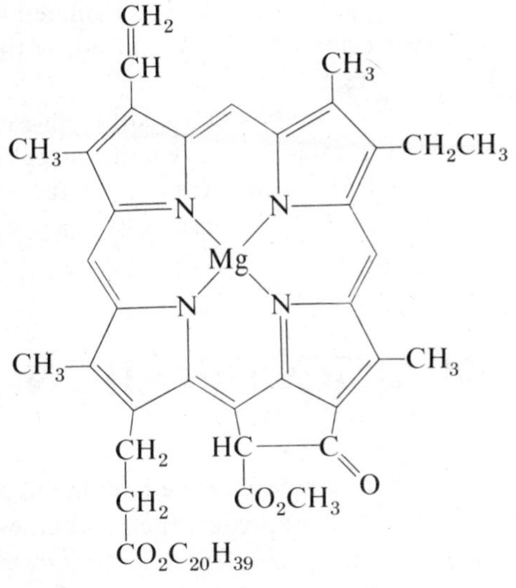

Figure 29-8. Chlorophyll, a square planar magnesium complex found in plants.

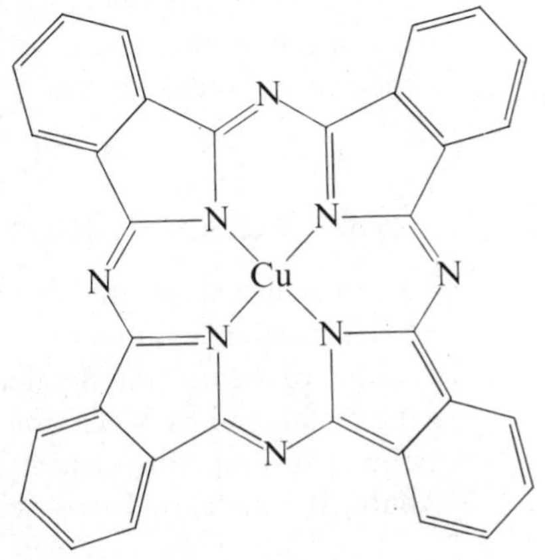

Figure 29-9. Copper phthalocyanine blue, a square planar copper complex found in some blue dyes.

plexing agents that tie up metal ions are also used as drugs. British Anti-Lewisite, $HSCH_2CH(SH)CH_2OH$, a drug developed during World War I as an antidote for the arsenic-based war gas Lewisite, is now used to treat poisoning by heavy metals such as arsenic, mercury, thallium, and chromium. The drug, abbreviated as BAL, is a ligand and functions by making a water-soluble chelate of the metal; this metal chelate is eliminated by the kidneys. Another polydentate ligand, enterobactin, which is isolated from certain bacteria, is used to form complexes of iron and thereby to control the severe iron build-up found in patients suffering from blood diseases such as Cooley's anemia. This disease prevents a patient's own blood from transporting oxygen adequately. Such patients need regular blood transfusions to survive, but as the new blood breaks down, the usual metabolic processes that remove excess iron are overloaded. This excess iron can build up to fatal levels in the heart, kidneys, and liver. Enterobactin forms a water-soluble complex with the excess iron, and this complex can be eliminated by the body.

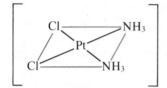

Figure 29-10. The square planar structure of the anticancer drug Platinol, *cis*-[Pt(NH$_3$)$_2$Cl$_2$].

The anticancer drug Platinol is *cis*-[Pt(NH$_3$)$_2$Cl$_2$], a square planar platinum(II) complex (Fig. 29-10). This substance, whose biological activity was first observed by scientists at Michigan State University in 1969, is believed to cross-link DNA strands (Section 26.11) and thereby interfere with mitosis (cell division).

Some complexing groups, when coordinated to certain metals, make the metal more easily assimilated by plants; in other cases, they keep the metal from being effectively utilized by the plant. Hence complexes can be effective in soil treatment.

In the electroplating industry it has been found that many metals plate out as a smoother, more uniform, better-looking, and more adherent surface when plated from a bath containing the metal as a complex ion. Thus complexes such as $[Ag(CN)_2]^-$ and $[Au(CN)_2]^-$ are used extensively in the electroplating industry.

BONDING IN COORDINATION COMPOUNDS

Any theory of the bonding in coordination compounds must explain four important properties: their stabilities, their structures, their colors, and their magnetic properties. The first modern attempt to explain the properties of complex compounds was made by applying the concepts of orbital hybridization (Sections 7.3 through 7.6) and valence bond theory. More recent models use crystal field theory, a model that examines the electrostatic attractions and repulsions between the central metal ion and the ligands, or they use molecular orbital theory.

29.6 Valence Bond Theory

The valence bond theory treats a metal-ligand bond as a coordinate covalent bond (Section 5.4) that forms when an orbital of the donor atom overlaps a hybrid orbital of the central metal atom. Electron pairs from the ligands are shared with the metal, and each electron pair occupies both an atomic orbital of a ligand and one of several equivalent hybridized orbitals of the metal. The ligand is a Lewis base; the metal, a Lewis acid.

In an octahedral complex such as $[Co(NH_3)_6]^{3+}$, a total of six electron pairs from the six ligands may be considered to have entered hybrid orbitals of the metal ion. As discussed in Section 7.6, six equivalent hybrid orbitals for the metal atom result from the hybridization of two d, one s, and three p orbitals; d^2sp^3 hybridization. All octahedral complexes, for example, $[Co(NH_3)_6]^{3+}$, $[SnCl_6]^{2-}$, $[Co(H_2O)_6]^{2+}$, $[Co(CN)_6]^{3-}$, $[Fe(CN)_6]^{3-}$, and $[Cr(en)_3]^{3+}$, can be described as being d^2sp^3 hybridized.

Let us look more closely at the octahedral $[Co(NH_3)_6]^{3+}$ ion. An isolated cobalt atom in its ground state has a $1s^2 2s^2 2p^6 3s^2 3p^6 3d^7 4s^2$ electron configuration (Section 4.12). Each atomic orbital can accommodate two electrons of opposing spin. The electrons enter the orbitals of a given type singly before any pairing of electrons occurs within those orbitals—Hund's Rule (Section 4.11). Each orbital in the cobalt atom can be represented by a circle as follows:

Co atom

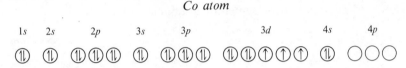

When the cobalt(III) ion, Co^{3+}, is formed, the atom loses three electrons, two from the $4s$ orbital and one from a completely filled $3d$ orbital, giving a structure possessing four unpaired electrons.

Co^{3+}

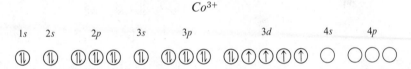

Six pairs of electrons are shared with the cobalt when six ammonia molecules bond to a cobalt(III) ion. It is postulated that the unpaired electrons in the $3d$ orbitals of a cobalt(III) ion pair as the electron pairs in the ligands approach, and the two empty $3d$ orbitals that result hybridize with the $4s$ and the three $4p$ orbitals to form six d^2sp^3 hybrid orbitals directed toward the corners of an octahedron. These hybrid orbitals accept the incoming electrons from the ammonia (indicated below in color). However, once the structure is formed, it is impossible to distinguish the source of the electrons, since all electrons are identical regardless of their origin. The distribution of electrons on cobalt in the $[Co(NH_3)_6]^{3+}$ ion is

$[Co(NH_3)_6]^{3+}$ (no unpaired electrons)

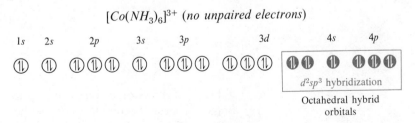

Octahedral complexes of cobalt(II), $[Co(CN)_6]^{4-}$, for example, contain one more electron than do those of cobalt(III). To permit d^2sp^3 bonding this extra electron must be promoted to a $4d$ (or perhaps a $5s$) orbital, where it is loosely held. This electron is readily removed, as indicated by the ease with which most Co(II) complexes are oxidized to Co(III) complexes.

$[Co(CN)_6]^{4-}$ *(one unpaired electron)*

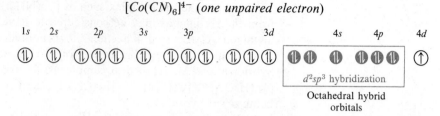

For four-coordinated complexes with a tetrahedral structure, the four bonding orbitals of the central atom come from the hybridization of one *s* and three *p* orbitals, or sp^3 hybridization. This bonding is analogous to that in methane (Section 7.3), except that in this case the coordinate covalent bonds are formed with both electrons of each shared pair coming from the ligand.

The distribution of electrons in the tetrahedral complex $[Zn(CN)_4]^{2-}$ is represented diagrammatically as follows:

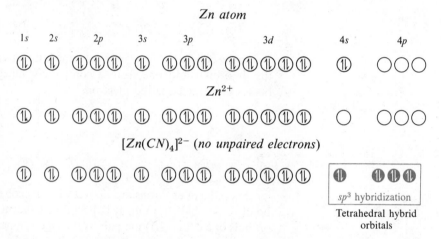

In the case of four-coordinated structures with a square planar structure, the four bonds to the central atom arise from dsp^2 hybridization, as illustrated for $[Ni(CN)_4]^{2-}$.

$[Ni(CN)_4]^{2-}$ *(no unpaired electrons)*

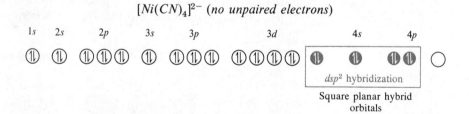

Table 29-1 shows other types of hybridization in coordination compounds.

Some properties of complexes cannot be explained satisfactorily by assuming hybridization and coordinate covalent bonding. For example, magnetic moments (Section 29.9) of a group of iron(II) complexes indicate that $[Fe(CN)_6]^{4-}$ has no unpaired electrons and $[Fe(H_2O)_6]^{2+}$ has four. Assuming covalent bonding, no unpaired electrons are expected for either complex, and $[Fe(CN)_6]^{4-}$ meets this expectation.

$$[Fe(CN)_6]^{4-}$$

$1s$	$2s$	$2p$	$3s$	$3p$	$3d$	$4s$	$4p$
⇅	⇅	⇅ ⇅ ⇅	⇅	⇅ ⇅ ⇅	⇅ ⇅ ⇅	⇅ ⇅ ⇅	⇅ ⇅ ⇅

However, the $[Fe(H_2O)_6]^{2+}$ ion with its four unpaired electrons does not meet the expectation.

If $[Fe(H_2O)_6]^{2+}$ is assumed to be "ionically" bonded by the electrostatic attraction of the Fe^{2+} ion for the negative end of the polar water molecule, the four unpaired electrons can be explained. Just as in the simple iron(II) ion, there are four unpaired electrons in the d orbitals of the iron atom since the ligand electrons do not enter these orbitals through sharing.

$$Fe^{2+} \ and \ [Fe(H_2O)_6]^{2+}$$

$1s$	$2s$	$2p$	$3s$	$3p$	$3d$	$4s$	$4p$
⇅	⇅	⇅ ⇅ ⇅	⇅	⇅ ⇅ ⇅	⇅ ↑ ↑ ↑ ↑	○	○ ○ ○

The difference in the bonding in $[Fe(CN)_6]^{4-}$ and in $[Fe(H_2O)_6]^{2+}$ is not surprising, since the cyanide ion makes an electron pair more readily available for sharing with a central metal ion than does the water molecule. Thus cyanide forms more stable complexes. Although the concept of two different types of bonding in complexes satisfactorily explains some facts, such as their magnetic behavior, it does not adequately explain certain other properties, such as their structural configurations and colors.

29.7 Crystal Field Theory

Even though coordination compounds have coordinate covalent bonding between the metal and the ligands, it is possible to understand, interpret, and predict the properties of many of these compounds, especially those of transition metals, on the assumption that there is no covalent bonding, but only simple electrostatic interactions. This totally ionic model of the bonding in complexes is called **crystal field theory.**

According to the assumptions of crystal field theory, the basic reason for the formation of a complex ion or molecule is the electrostatic attraction of a positively charged metal ion for negative ions or for the negative ends (the electron pairs) of polar molecules. The colors of complexes and their magnetic properties result from the electrostatic interactions of the electron pairs of the ligands with electrons in the d orbitals of the metal ion.

In Section 4.10 the shapes of the s, p, and d orbitals were given. An s orbital is spherical. A p orbital has a dumbbell shape, and the three p orbitals for a given level are oriented at right angles to one another along the x, y, and z axes (Fig. 4-16). The d orbitals, which occur in sets of five, consist of lobe-shaped regions and are arranged in space as shown in Fig. 4-17 and, reproduced within an octahedral structure, in Fig. 29-11.

The lobes in two of the five d orbitals, the d_{z^2} and $d_{x^2-y^2}$ orbitals, point toward the corners of the octahedron around the metal (Fig. 29-11). These two orbitals are called the e_g **orbitals** (the symbol actually refers to the symmetry of the orbitals, but we will use it as a convenient name for these two orbitals in an octahedral

Figure 29-11. Diagrams showing the directional characteristics of the five *d* orbitals. *L* indicates a ligand at each corner of the octahedron.

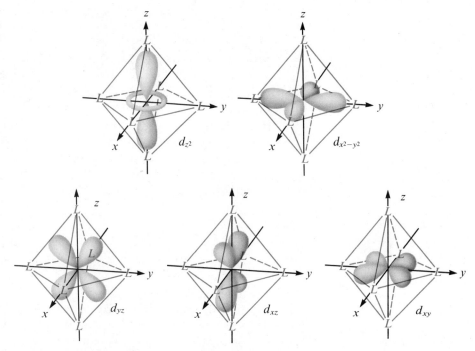

complex). The other three orbitals, the d_{xy}, d_{xz}, and d_{yz} orbitals, whose lobes point between the corners of the octahedron, are called the **t_{2g} orbitals** (again the symbol really refers to the symmetry of the orbitals). As six ligands approach the metal ion along the axes of the octahedron, the electron pairs of the ligands repel the electrons in the *d* orbitals of the metal ion. However, the repulsions between the electrons in the e_g orbitals (the d_{z^2} and $d_{x^2-y^2}$ orbitals) and the electron pairs of the ligands are greater than the repulsions between the electrons in the t_{2g} orbitals (the d_{xy}, d_{xz}, and d_{yz} orbitals) and the electron pairs of the ligands because the lobes of the e_g orbitals point directly at the ligand electron pairs while the lobes of the t_{2g} orbitals point between them. Thus electrons in e_g orbitals of a metal ion in an octahedral complex have higher potential energies than those of electrons in t_{2g} orbitals. The difference in energy may be represented as follows:

The difference in energy between the e_g and the t_{2g} orbitals is called the **crystal field splitting** and is symbolized by Δ or **10Dq.**

The size of the crystal field splitting (10Dq) depends on the nature of the six ligands located around the central metal ion. Different ligands have different concentrations of electrons in the orbital that points toward the central atom. Ligands with a low density of electrons produce a small crystal field splitting (small 10Dq value), while ligands with a high density of electrons produce a large crystal field splitting (large 10Dq value). The increasing crystal field splitting produced by ligands is expressed in the **spectrochemical series,** a short version of which is given here.

$$I^- < Br^- < Cl^- < F^- < H_2O < C_2O_4{}^{2-} < NH_3 < en < NO_2{}^- < CN^-$$

Increasing field strength of ligand electron pairs

In this series the ligands on the left project a low density of electrons toward the metal (low field strength), and those on the right have a large density of electrons (high field strength). Thus the crystal field splitting produced by an iodide ion (I^-) as a ligand is much smaller than that produced by a cyanide ion (CN^-) as a ligand.

In a simple metal ion in the gas phase, the electrons are distributed among the five $3d$ orbitals in accord with Hund's Rule, since the orbitals all have the same energy.

Ion	3d Orbitals	Ion	3d Orbitals
Ti^{3+}	⊕ ○ ○ ○ ○	Fe^{2+}	⊕ ⊕ ⊕ ⊕ ⊕
V^{3+}	⊕ ⊕ ○ ○ ○	Co^{3+}	⊕ ⊕ ⊕ ⊕ ⊕
Cr^{3+}	⊕ ⊕ ⊕ ○ ○	Co^{2+}	⊕ ⊕ ⊕ ⊕ ⊕
Cr^{2+}	⊕ ⊕ ⊕ ⊕ ○	Ni^{2+}	⊕ ⊕ ⊕ ⊕ ⊕
Mn^{2+}	⊕ ⊕ ⊕ ⊕ ⊕	Cu^{2+}	⊕ ⊕ ⊕ ⊕ ⊕
Fe^{3+}	⊕ ⊕ ⊕ ⊕ ⊕	Zn^{2+}	⊕ ⊕ ⊕ ⊕ ⊕

However, if the metal ion lies in an octahedron formed by six ligands, the energies of the d orbitals are no longer the same, and two opposing forces are set up. One force tends to keep the electrons of the metal ion distributed with unpaired spins within all of the d orbitals (it requires energy to pair up electrons in an orbital). The other force tends to reduce the average energy of the d electrons by placing as many of them as possible in the lower-energy t_{2g} orbitals. The d electrons will end up with the lowest possible total energy. If it requires less energy for the d electrons to be excited to the upper e_g orbitals than to pair in the lower t_{2g} orbitals, then they will remain unpaired. If it requires more energy to put d electrons in the upper e_g orbitals than to pair them in the lower t_{2g} orbitals, then they will pair.

In $[Fe(CN)_6]^{4-}$ the strong field of six cyanide ions produces a large crystal field splitting. Under these conditions the electrons require less energy to pair than to be excited to the e_g orbitals. Thus the six $3d$ electrons of the Fe^{2+} ion pair in the three t_{2g} orbitals. The result is in agreement with the experimentally measured magnetic moment of $[Fe(CN)_6]^{4-}$ (no unpaired electrons).

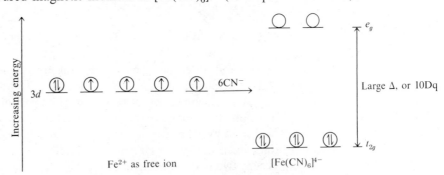

In $[Fe(H_2O)_6]^{2+}$, on the other hand, the weak field of the water molecules produces only a small crystal field splitting. Since it requires less energy to excite electrons to the e_g orbitals than to pair them, they remain distributed in all five $3d$ orbitals.

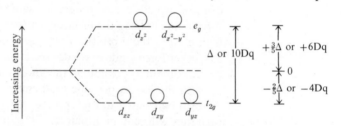

Thus four unpaired electrons should be present, as is verified by measuring the magnetic moment of $[Fe(H_2O)_6]^{2+}$. A similar line of reasoning can be followed for complexes of iron(III) to show why $[Fe(CN)_6]^{3-}$ has only one unpaired electron while $[Fe(H_2O)_6]^{3+}$ and $[FeF_6]^{3-}$ both have five unpaired electrons.

Actually, as was pointed out in Section 5.8, there is no sharp dividing line between covalent and ionic bonding. The Molecular Orbital Theory, discussed in Section 29.11, introduces a covalent component into the ionic viewpoint.

29.8 Crystal Field Stabilization Energy

As indicated earlier, the difference in energy between the e_g and t_{2g} orbitals in an octahedral complex is commonly symbolized by either Δ or 10Dq.

Since the splitting of the five d orbitals into two energy levels does not change the total energy, the zero line on the energy axis is the weighted average energy of these orbitals. Thus the e_g and t_{2g} energies are $+\frac{3}{5}\Delta$ and $-\frac{2}{5}\Delta$, respectively (or +6Dq and −4Dq, respectively). Hence

$$\left(\frac{3}{5}\Delta\right)2 + \left(-\frac{2}{5}\Delta\right)3 = 0$$

or
$$\left(\frac{3}{5} \times 10Dq\right)2 + \left(-\frac{2}{5} \times 10Dq\right)3 = (+6Dq)2 + (-4Dq)3 = 0$$

A value for each complex, called the **crystal field stabilization energy (CFSE),** can be calculated and is related to the stability of the complex. Actually the crystal field stabilization energy is a measure of how much more stable a complex is than a hypothetical complex that is identical in every way except showing no crystal field splitting. Consider, for example, an octahedral complex of the vanadium(III) ion, which has two d electrons (referred to as d^2). The two unpaired electrons occupy two of the lower-energy t_{2g} orbitals, whether the vanadium is coordinated to a ligand producing a weak field or to one producing a strong field. In calculating the crystal field stabilization energy, we take into account the fact that there are two electrons, each with an energy of −4Dq relative to the zero-energy line.

$$CFSE = 2(-4Dq) = -8Dq$$

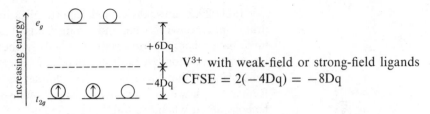

V^{3+} with weak-field or strong-field ligands
CFSE = $2(-4Dq) = -8Dq$

Now consider another metal ion, the manganese(II) ion, which has five d electrons (d^5). In the free ion these electrons are distributed evenly throughout the five d orbitals in accord with Hund's Rule. If the ion is coordinated to ligands that produce a weak field, the five electrons remain evenly distributed; that is, there are three unpaired electrons in the t_{2g} level and two in the e_g level.

Mn^{2+} with weak-field ligands
CFSE = $2(+6Dq) + 3(-4Dq) = 0$

The two electrons in the e_g orbitals are in the higher energy level, corresponding to $+6Dq$, and the three electrons in the t_{2g} orbitals are in the lower energy level, corresponding to $-4Dq$. Hence the crystal field stabilization energy is zero in this weak-field case.

$$CFSE = 2(+6Dq) + 3(-4Dq) = 0$$

If, however, the Mn^{2+} ion is coordinated to ligands that produce a strong field, all five electrons fall into the lower-energy nonbonding t_{2g} orbitals, and none are in the higher-energy e_g orbitals.

Mn^{2+} with strong-field ligands
CFSE = $5(-4Dq) = -20Dq$

The crystal field stabilization energy for Mn^{2+} in this strong-field case is $-20Dq$.

$$CFSE = 5(-4Dq) = -20Dq$$

In the weak-field case for Mn^{2+}, the splitting of the two energy levels (the 10Dq value) is much smaller than it is for the strong-field case. The possible gain in energy that would result if the fourth and fifth electrons occupied the lower-energy t_{2g} orbitals is not large enough to overcome the repulsion of the electrons for each other and to pair them up. Hence the fourth and fifth electrons go singly into the higher-energy e_g orbitals.

In general, the more negative the value of the crystal field stabilization energy, the more stable is the complex. A complex of Mn^{2+} with ligands of high field strength is more stable than one with ligands of low field strength. Sometimes the term **low-spin complex,** or **spin-paired complex,** is applied to complexes of a given metal with ligands of high field strength; the term **high-spin complex,** or **spin-free complex,** is sometimes used for complexes with ligands of low field strength.

Table 29-2 shows the crystal field stabilization energy for a variety of octahedral complexes with weak and strong fields. The calculated CFSE values in the table have been experimentally verified by spectroscopic methods. A relatively large negative number indicates that more energy is lost in the formation of the complex and that, in general, the complex is more stable for a given ligand. Remember, however, each ligand has its own specific 10Dq value, and this must be considered when the stabilities of complexes with different ligands are calculated. Weak-field ligands cause less splitting between the energy levels for a given metal ion and have smaller 10Dq values, in general, than do strong-field ligands attached to the same metal ion.

Table 29-2 Crystal Field Stabilization Energy (CFSE) of Octahedral Complexes[a]

Configuration	Weak Field (High Spin)				Strong Field (Low Spin)				Examples
	t_{2g}	e_g	Number of Unpaired Electrons	CFSE (Dq)	t_{2g}	e_g	Number of Unpaired Electrons	CFSE (Dq)	
d^0	0	0	0	0	0	0	0	0	Ca^{2+}, Sc^{3+}
d^1	1	0	1	-4	1	0	1	-4	Ti^{3+}
d^2	2	0	2	-8	2	0	2	-8	V^{3+}
d^3	3	0	3	-12	3	0	3	-12	Cr^{3+}, V^{2+}
d^4	3	1	4	-6	4	0	2	-16	Cr^{2+}, Mn^{3+}
d^5	3	2	5	0	5	0	1	-20	Mn^{2+}, Fe^{3+}
d^6	4	2	4	-4	6	0	0	-24	Fe^{2+}, Co^{3+}
d^7	5	2	3	-8	6	1	1	-18	Co^{2+}
d^8	6	2	2	-12	6	2	2	-12	Ni^{2+}
d^9	6	3	1	-6	6	3	1	-6	Cu^{2+}
d^{10}	6	4	0	0	6	4	0	0	Cu^+, Zn^{2+}

[a] Used by permission from *Concepts and Models of Inorganic Chemistry*, B. E. Douglas and D. H. McDaniel. Ginn, Blaisdell, New York, 1965. p. 351.

The largest CFSE value in the table for strong-field complexes is for the d^6 configuration, since in this case the lower-energy t_{2g} orbitals are completely occupied and the higher-energy e_g orbitals are empty. The largest CFSE values for weak-field complexes are for the d^3 and d^8 configurations. In the d^3 case all of the lower-energy t_{2g} orbitals are singly occupied, and the higher-energy e_g orbitals are empty. The d^8 configuration represents the lowest number of electrons for which the lower-energy t_{2g} orbitals are completely filled; however, since two electrons are present in the higher-energy e_g orbitals, the highest CFSE value for weak-field complexes (-12) is lower than that for strong-field complexes (-24). The CFSE is zero, indicating very low stability, at two places in the table for strong-field complexes: in the cases when no d electrons are present and when 10 d electrons are present. In the latter case all t_{2g} and e_g orbitals are filled, so that the stability gained in the formation of the complex by electrons going into lower-energy orbitals is exactly counterbalanced by the stability lost by electrons going into higher-energy orbitals. For weak-field complexes the d^0 and d^{10} configurations have CFSE values of zero, as does the d^5 configuration, in which all t_{2g} and e_g orbitals are singly occupied.

29.9   Magnetic Moments of Molecules and Ions

Molecules such as O_2 (Section 6.10) and NO (Section 6.13) and ions such as $[Co(CN)_6]^{4-}$ (Section 29.6) and $[Fe(H_2O)_6]^{2+}$ (Section 29.7) that contain unpaired electrons are **paramagnetic.** Paramagnetic substances tend to move into a magnetic field, such as that between the poles of a magnet. Many transition metal complexes have unpaired electrons and hence are paramagnetic. Molecules such as N_2 (Section 6.9) and C_2H_6 (Section 7.3) and ions such as Na^+, $[Co(NH_3)_6]^{3+}$ (Section 29.6), and $[Fe(CN)_6]^{4-}$ (Section 29.7) that contain no unpaired electrons are **diamagnetic.** Diamagnetic substances have a slight tendency to move out of a magnetic field.

An electron in an atom spins about its own axis. Because the electron is electrically charged, this spin gives it the properties of a small magnet, with north and south poles. Two electrons in the same orbital spin in opposite directions, and their magnetic moments cancel because their north and south poles are opposed. When an electron in an atom or ion is unpaired, the magnetic moment due to its spin makes the entire atom or ion paramagnetic. Thus a sample containing such atoms or ions is paramagnetic. The size of the magnetic moment of a system containing unpaired electrons is related directly to the number of such electrons; the greater the number of unpaired electrons, the larger is the magnetic moment. Therefore, the observed magnetic moment is used to determine the number of unpaired electrons present.

29.10   Colors of Transition Metal Complexes

When atoms absorb light of the proper frequency, their electrons are excited to higher energy levels (Chapter 4). The same thing can happen in coordination compounds. Electrons can be excited from the lower-energy t_{2g} to the higher-energy e_g orbitals, provided that the latter are not already filled with paired electrons.

The human eye perceives a mixture of all the colors, in the proportions present in sunlight, as white light and utilizes complementary colors in color vision. The eye perceives a mixture of two complementary colors, in the proper proportions, as white light. Likewise, when a color is missing from white light, the eye sees its complement. For example, as shown in Table 29-3, if red light is removed from white light, the eye sees the color blue-green; if violet is removed from white light, the eye sees lemon yellow; if green light is removed, the eye sees purple.

Consider $[Fe(CN)_6]^{4-}$ (Section 29-7). The electrons in the t_{2g} orbitals can absorb energy and be excited to the higher energy level. The necessary energy corresponds to photons of violet light. If white light impinges on $[Fe(CN)_6]^{4-}$, violet light is absorbed (to accomplish the excitation), and the eye sees the unabsorbed complement, lemon yellow. $K_4[Fe(CN)_6]$ is lemon yellow. In contrast, if white light strikes $[Fe(H_2O)_6]^{2+}$, red light (longer wavelength, lower energy) is absorbed. The eye sees its complement, blue-green. $[Fe(H_2O)_6]SO_4$, for example, is therefore blue-green.

As shown in Table 29-2, a coordination compound of the Cu^+ ion has a d^{10} configuration, and all the e_g orbitals are filled. In order to excite an electron to a higher level, such as the $4p$ orbital, photons of very high energy are needed. This energy corresponds to very short wavelengths in the ultraviolet region of the

Table 29-3   Complementary Colors

Wavelength, Å	Spectral Color	Complementary Color[a]
4100	Violet	Lemon yellow
4300	Indigo	Yellow
4800	Blue	Orange
5000	Blue-green	Red
5300	Green	Purple
5600	Lemon yellow	Violet
5800	Yellow	Indigo
6100	Orange	Blue
6800	Red	Blue-green

[a] The complementary color is seen when the spectral color is removed from white light.

spectrum. Since no visible light is absorbed, the eye sees no change, and the compound appears white or colorless. A solution containing $[Cu(CN)_2]^-$, for example, is colorless. On the other hand, Cu^{2+} complexes have a vacancy in the e_g orbitals, and electrons can be excited to this level. The wavelength (energy) of the light absorbed corresponds to the visible part of the spectrum, and Cu^{2+} complexes are almost always colored—blue, blue-green, violet, or yellow.

As we have noted earlier, strong-field ligands cause a large split in the energies of the d orbitals of the central metal atom (large 10Dq value). Transition metal coordination compounds with these ligands are yellow, orange, or red since they absorb higher-energy violet or blue light. On the other hand, coordination compounds of transition metals with weak-field ligands are blue-green, blue, or indigo since they absorb lower-energy yellow, orange, or red light.

29.11   The Molecular Orbital Theory

We have seen that a theory based primarily on covalent bonding is especially useful in explaining the structures of complexes and that a theory based primarily on ionic bonding is particularly useful in explaining the magnetic properties and the energies involved in the electron transitions that give rise to the colors of complexes. However, neither of these theories can explain certain other properties of complexes. Since most bonds have both ionic and covalent character, the theory of bonding in complexes can be improved by using molecular orbital concepts to introduce some contribution due to covalent bonding into the crystal field theory. In Chapter 6 we discussed the Molecular Orbital Theory, and in later chapters we applied the theory to a variety of substances. Now we can use it to understand the nature of the bonding in coordination compounds.

Molecular orbital energy diagrams for complexes are the same in principle as those for diatomic species (Chapter 6) but are, of course, more complicated since several atoms are involved, instead of just two, and a large number of electrons must be considered. The molecular orbital energy diagrams for $[CoF_6]^{3-}$, possessing weak-field ligands, and for $[Co(NH_3)_6]^{3+}$, possessing strong-field ligands, are shown in Fig. 29-12. Only the six d electrons of the Co^{3+} ion and the lone pair of electrons of each of the six ligands (a total of 18 electrons) are included.

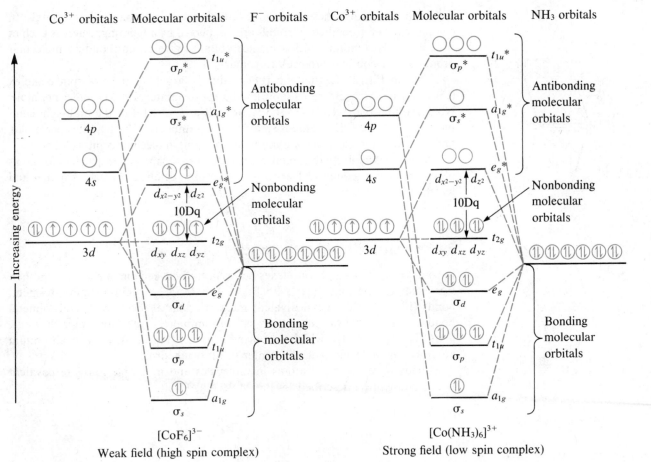

Figure 29-12. Molecular orbital energy diagrams for $[CoF_6]^{3-}$ and $[Co(NH_3)_6]^{3+}$.

The arrangement and symbols used in Fig. 29-12 are, in general, those described in Chapter 6. In addition, t_{1u}, a_{1g}, e_g, and t_{2g} are used beside the circles (orbitals) to label the various energy levels. In some cases the molecular orbital is also labeled $d_{x^2-y^2}$, d_{z^2}, or d_{xy}, etc., corresponding to the directional characteristics of the d orbital (Fig. 29-11). The antibonding and nonbonding energy levels use the $e_g{}^*$ and t_{2g} designations, respectively, in the molecular orbital energy level diagram. The energy difference between them is Δ or 10Dq (Sections 29.7 and 29.8).

Figure 29-12 shows that the ligand atomic orbitals are lower in energy than the corresponding metal atomic orbitals. This is indicative of some ionic character (Section 6.13); that is, some electronic charge has been transferred from the metal to the ligands.

Two of the d orbitals (the e_g orbitals: $d_{x^2-y^2}$ and d_{z^2}), are directed, as indicated in Section 29.7, toward the corners of the octahedron, where the ligands are located. In addition, the one $4s$ atomic orbital and the three $4p$ atomic orbitals of the metal are also so oriented with respect to the octahedron corners. Hence these metal atomic orbitals overlap with ligand orbitals, and six antibonding and six bonding molecular orbitals are formed. The other three d orbitals (the t_{2g} orbitals: d_{xy}, d_{xz}, and d_{yz}) do not point toward the ligand orbitals and hence are not in-

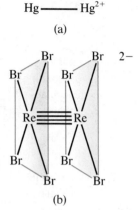

(a)

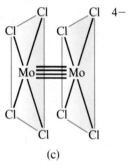

(b)

(c)

Figure 29-13. The structures of (a) Hg_2^{2+}, (b) $Re_2Br_8^{2-}$, and (c) $Mo_2Cl_8^{4-}$. In both (b) and (c) each metal atom is surrounded by a square plane of halide ions and occupies the apex of a square pyramid with respect to the other metal atom and its halide ions. The two planes are parallel to each other and are perpendicular to the rhenium-rhenium and molybdenum-molybdenum bonds.

volved in σ bonding; these molecular orbitals are called nonbonding orbitals. In complexes, therefore, the t_{2g} orbitals are unaffected by σ bonding, whereas each of the e_g orbitals combines with a ligand orbital to give an antibonding molecular orbital and a bonding molecular orbital.

Note in Fig. 29-12 that the 10Dq value is larger for the low-spin complex $[Co(NH_3)_6]^{3+}$. Measurements show it to be approximately 290 kilojoules per mole. Note also that the electrons in the t_{2g} orbitals are paired. For the high-spin complex $[CoF_6]^{3-}$ the 10Dq value is smaller (approximately 150 kilojoules per mole), and two unpaired electrons are in the antibonding molecular orbitals labeled e_g^*.

Analogous molecular orbital energy diagrams can be constructed for complexes that have geometrical structures other than octahedral, such as planar and tetrahedral.

29.12   Metal-to-Metal Bonds

In some compounds and complexes bonds occur between one metal and another. For example, the ions Hg_2^{2+}, $Re_2Br_8^{2-}$, and $Mo_2Cl_8^{4-}$ contain metal-to-metal bonds (Fig. 29-13). Several divalent metal acetates also contain metal-to-metal bonds. One typical example is chromium(II) acetate monohydrate, $[Cr(CH_3CO_2)_2 \cdot H_2O]_2$, which occurs as a dimer (Fig. 29-14). Each chromium atom in this molecule is in the center of an octahedron.

Bonds between metal atoms in molecules and in simple ions are covalent bonds and can be described by the Molecular Orbital Theory.

Figure 29-14. The structure of the chromium(II) acetate monohydrate dimer, which has a metal-to-metal bond. Each chromium atom is in the center of an octahedron. The octahedron is shown in color for one of the two chromium atoms.

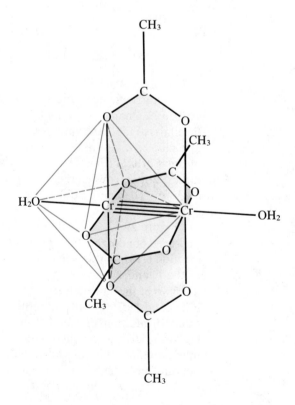

FOR REVIEW

SUMMARY

Metal ions can form **coordination compounds, or complexes,** in which a **central metal atom or ion** is bonded to two or more **ligands** by coordinate covalent bonds. Ligands with more than one donor atom are called **polydentate ligands** and form **chelates.** The common geometries found in complexes are tetrahedral and square planar (both with a **coordination number** of 4) and octahedral (with a coordination number of 6). *Cis* and *trans* **configurations** are possible in some octahedral and square planar complexes. In addition to these geometrical isomers, **optical isomers** (molecules or ions that are mirror images but not identical) are possible in certain octahedral complexes.

The bonding in coordination compounds can be described in terms of valence bonds, although this model of bonding does not explain the magnetic properties and colors very well. In the valence bond model the bonds are regarded as normal coordinate covalent bonds between a ligand, which acts as a Lewis base, and an empty hybrid metal atomic orbital.

The **crystal field theory** treats bonding between the metal and the ligands as a simple electrostatic attraction. However, the presence of the ligands near the metal ion changes the energies of the metal d orbitals relative to their energies in the free ion. Both the color and the magnetic properties of a complex can be attributed to this **crystal field splitting. Crystal field stabilization energy,** which also results from the crystal field splitting, can enhance the stability of a complex. The magnitude of the splitting (**10Dq** or **Δ**) depends on the nature of the ligands bonded to the metal. Strong-field ligands produce large splittings and favor formation of **low-spin complexes** in which the t_{2g} **orbitals** are completely filled before any electrons occupy the e_g **orbitals.** Weak-field ligands favor formation of **high-spin complexes.** Both the t_{2g} and e_g orbitals are singly occupied before any are doubly occupied.

The Molecular Orbital Theory can describe the bonding in coordination compounds, and it can also explain the colors and magnetic properties of these complexes.

KEY TERMS AND CONCEPTS

Central metal atom or ion
 (29.1)
Chelating ligand (29.1)
cis configuration (29.4)
Coordination number (29.1)
Coordination sphere (29.1)
Crystal field splitting (29.7)
Crystal field stabilization
 energy, CFSE (29.8)

Crystal field theory (29.7)
Diamagnetic (29.9)
Donor atom (29.1)
e_g orbitals (29.7)
Geometric isomers (29.4)
High-spin complex (29.8)
Ligands (29.1)
Low-spin complex (29.8)
Optical isomers (29.4)

Paramagnetic (29.9)
Polydentate ligand (29.1)
t_{2g} orbitals (29.7)
trans configuration (29.4)

EXERCISES

Structure and Nomenclature of Complexes

1. Explain what is meant by: (a) coordination sphere, (b) ligand, (c) donor atom, (d) coordination number, (e) coordination compound, and (f) chelating ligand.

2. Indicate the coordination number for the central metal atom in each of the following coordination compounds:
   (a) $[Pt(H_2O)_2Br_2]$
   (b) $[Pt(NH_3)(py)(Cl)(Br)]$ (py = pyridine, C_5H_5N)
   (c) $[Zn(NH_3)_2Cl_2]$
   (d) $[Zn(NH_3)(py)(Cl)(Br)]$
   (e) $[Ni(H_2O)_4Cl_2]$
   (f) $[Co(C_2O_4)_2Cl_2]^{3-}$

3. Using the metal ions Co^{3+}, Pt^{2+}, and Zn^{2+} and the ligands NH_3 and Cl^-, construct two examples each of square planar complexes, tetrahedral complexes, and octahedral complexes. Do not use isomers.

4. Give the coordination number for each coordinated metal ion in the following compounds:
   (a) $[Co(CO_3)_3]^{3-}$ (c) $[Co(NH_3)_4Br_2]_2SO_4$
   (b) $[Cu(NH_3)_4]^{2+}$ (d) $[Pt(NH_3)_4][PtCl_4]$
   (e) $[Cr(en)_3](NO_3)_3$ (en = ethylenediamine)
   (f) $[Pd(NH_3)_2Br_2]$ (square planar)
   (g) $K_3[Fe(CN)_6]$
   (h) $[Zn(NH_3)_2Cl_2]$ (tetrahedral)

5. Give the coordination numbers and write the formulas for each of the following, including all isomers where appropriate:
   (a) tetrahydroxozincate(II) ion (tetrahedral)
   (b) hexacyanopalladate(IV) ion
   (c) dichloroaurate(I) ion (*aurum* is Latin for gold)
   (d) diamminedichloroplatinum(II)
   (e) potassium diamminetetrachlorochromate(III)
   (f) hexaamminecobalt(III) hexacyanochromate(III)
   (g) dibromobis(ethylenediamine)cobalt(III) nitrate

6. Name each of the compounds given in Exercise 4 (including any isomers).

7. Sketch the structures of the following complexes. Indicate any *cis, trans,* and optical isomers.

(a) $[Pt(H_2O)_2Br_2]$ (square planar)
(b) $[Pt(NH_3)(py)(Cl)(Br)]$ (square planar, py = pyridine, C_5H_5N)
(c) $[Zn(NH_3)_2Cl_2]$ (tetrahedral)
(d) $[Zn(NH_3)(py)(Cl)(Br)]$ (tetrahedral)
(e) $[Ni(H_2O)_4Cl_2]$ (f) $[Co(C_2O_4)_2Cl_2]^{3-}$

8. Draw diagrams for any *cis, trans,* and optical isomers that could exist for each of the following (en is ethylenediamine):
   (a) $[Co(en)_2(NO_2)Cl]^+$ (d) $[Pt(NH_3)_2Cl_4]$
   (b) $[Co(en)_2Cl_2]^+$ (e) $[Cr(en)_3]^{3+}$
   (c) $[Cr(NH_3)_2(H_2O)_2Br_2]^+$ (f) $[Pt(NH_3)_2Cl_2]$

9. Name each of the compounds given in Exercise 8 (including any isomers).

10. Using NH_3 and Br^- as ligands, (a) give the formula of a six-coordinated chromium(III) complex that would not give ions when dissolved in water; (b) give the formula of a complex containing one six-coordinated chromium(III) ion that would give two ions when dissolved in water; and (c) sketch the isomers of the compounds formulated in (a) and (b).

11. Sketch each of the isomers for the ion $[Pt(en)(NH_3)_2Br_2]^{2+}$.

12. In this and other chapters we have discussed substances whose structures encompass a considerable variety of geometric shapes. (a) Name each shape; (b) draw a diagram of each; (c) give a specific example of each; (d) indicate the relationship of atomic orbitals to each of the geometric shapes; (e) discuss the relationship of the geometric shape to the number of groups attached to the central element. (*Hint:* You will find it useful to make use of coordination compounds in your answer, but do not confine your answer to coordination compounds nor to the material in this chapter alone.)

Bonding in Complexes

13. Draw orbital diagrams and indicate the type of hybridization you would expect for each of the following:
   (a) $[Cu(NH_3)_4]^{2+}$ (d) $[Co(NH_3)_6]^{3+}$
   (b) $[Zn(NH_3)_4]^{2+}$ (e) $[Fe(CN)_6]^{4-}$
   (c) $[Cd(CN)_4]^{2-}$

14. Show by means of orbital diagrams the hybridization for each of the examples given in Table 29-1.

15. Determine the number of unpaired electrons

expected for $[Fe(CN)_6]^{3-}$ and for $[Fe(H_2O)_6]^{3+}$ in terms of crystal field theory.

16. (a) Verify, by calculation, the crystal field stabilization energies listed in Table 29-2; (b) discuss the relationship of crystal field stabilization energies to the stabilities of complexes, and the limitations to the concept.

17. How many unpaired electrons will be present in each of the following:
    (a) $[CoF_6]^{3-}$ (high-spin)
    (b) $[Co(en)_3]^{3+}$ (low-spin)
    (c) $[Mn(CN)_6]^{3-}$ (low-spin)
    (d) $[Mn(CN)_6]^{4-}$ (low-spin)
    (e) $[MnCl_6]^{4-}$ (high-spin)
    (f) $[RhCl_6]^{3-}$ (low-spin)

18. Is it possible for a complex of a metal of the first transition series to have six or seven unpaired electrons? Explain.

19. The crystal field splitting in a tetrahedral complex is

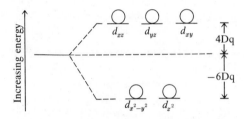

    (a) Show that the crystal field stabilization energy for a tetrahedral complex such as VCl_4 with a d^1 configuration is $-6Dq$.
    (b) All known tetrahedral complexes are high-spin. For what configuration will the value of the crystal field stabilization energy be greatest?
    (c) Calculate the CFSE value for each of the following tetrahedral complexes: $[NiCl_4]^{2-}$, $[FeCl_4]^{-}$, $[CoBr_4]^{2-}$, $[Zn(OH)_4]^{2-}$, $[MnCl_4]^{2-}$.

20. What is the significance of the relative energy values of ligand atomic orbitals, metal atomic orbitals, and the corresponding molecular orbitals for $[CoF_6]^{3-}$ and $[Co(NH_3)_6]^{3+}$?

21. For complexes of the same metal ion with no change in oxidation number, the stability increases as the CFSE of the complex increases. Which complex in each of the following pairs of complexes is the more stable? $[Fe(H_2O)_6]^{2+}$ or $[Fe(CN)_6]^{4-}$; $[Co(NH_3)_6]^{3+}$ or $[CoF_6]^{3-}$; $[Mn(H_2O)_6]^{2+}$ or $[MnCl_6]^{4-}$; $[Co(NH_3)_6]^{3+}$ or $[Co(en)_3]^{3+}$

22. Explain how the diphosphate ion, $P_2O_7^{4-}$ (Section 24.11), can function as a water softener by complexing Fe^{2+}.

23. Assume that you have a complex of a transitional metal ion with a d^6 configuration. Can you tell whether the complex is octahedral or tetrahedral if a measurement of the magnetic moment establishes that it has no unpaired electrons? (See Exercise 19 for the crystal field splitting in a tetrahedral complex.)

Additional Exercises

24. Determine the crystal field splitting of the d orbitals in a two-coordinated complex as the two ligands approach the metal along the z axis.

25. What is the crystal field splitting of the nickel ion in NiO? of the Fe^{3+} ion in $FeCl_3$? (See Section 11.16.)

26. The formation constant for $[Fe(CN)_6]^{3-}$ is 1×10^{44}; that for $[Fe(CN)_6]^{4-}$ is 1×10^{37}. Is this consistent with the difference in CFSE between the two complexes? CN^- is a strong-field ligand in both complexes. What other feature of the crystal field model can explain the difference in formation constants?

27. Calculate the concentration of free copper ion that is present in equilibrium with $1.0 \times 10^{-3}\,M$ $[Cu(NH_3)_4]^{2+}$ and $1.0 \times 10^{-1}\,M\,NH_3$.
    Ans. $8.3 \times 10^{-12}\,M$

28. Trimethylphosphine, $: P(CH_3)_3$, can act as a ligand by donating the lone pair of electrons on the phosphorus atom. If trimethylphosphine is added to a solution of nickel(II) chloride in acetone, a blue compound with a molecular weight of approximately 270 and containing 21.5% Ni, 26.0% Cl, and 52.5% $P(CH_3)_3$ can be isolated. This blue compound does not have any isomeric forms. What is the geometry and molecular formula of the blue compound?
    Ans. Tetrahedral; $NiCl_2[P(CH_3)_3]_2$

29. The standard reduction potential for the reaction

$$[Co(H_2O)_6]^{3+} + e^- \longrightarrow [Co(H_2O)_6]^{2+}$$

is about 1.8 V. The reduction potential for the reaction

$$[Co(NH_3)_6]^{3+} + e^- \longrightarrow [Co(NH_3)_6]^{2+}$$

is +0.1 V. Show by calculation of the cell potentials which of the complex ions, $[Co(H_2O)_6]^{2+}$ or

$[Co(NH_3)_6]^{2+}$, can be oxidized to the corresponding cobalt(III) complex by oxygen.

Ans. $[Co(NH_3)_6]^{2+}$

30. Using the radius ratio rule and the ionic radii given inside the back cover, predict whether complexes of Ni^{2+}, Mn^{2+}, and Sc^{3+} with Cl^- will be tetrahedral or octahedral. Write the formulas for these complexes.   *Ans.* $[NiCl_4]^{2-}$, *tetrahedral;* $[MnCl_6]^{4-}$, *octahedral;* $[ScCl_6]^{3-}$, *octahedral*

31. The complex ion $[Co(en)_3]^{3+}$ is diamagnetic. Would you expect the $[Co(en)_3]^{2+}$ ion to be diamagnetic or paramagnetic? the $[Co(CN)_6]^{3-}$ ion? Explain your reasoning in each case.

32. What type of tetrahedral complexes can form optical isomers?

30

METALS OF ANALYTICAL GROUP I

MERCURY, SILVER, AND LEAD

Mercury(I), silver, and lead(II) ions are precipitated as chlorides by a slight excess of hydrochloric acid; the chlorides of all other common metal ions are soluble in dilute acid. Thus Hg_2^{2+}, Ag^+, and Pb^{2+} make up Group I of the analytical scheme (Chapter 36). If the less familiar ions were considered, Analytical Group I would include Cu^+, Au^+, and Tl^+ because the chlorides of these ions are also insoluble.

The solubilities of mercury(I) chloride, Hg_2Cl_2, and silver chloride, $AgCl$, are low enough so that the mercury(I) ion and the silver ion are completely precipitated by the chloride ion in acid solution. However, lead chloride is somewhat soluble ($K_{sp} = 1.7 \times 10^{-5}$) and may even fail to precipitate if the concentration of the lead ion is too low or the temperature is too high (lead chloride is readily soluble in hot water). Because of the relatively high solubility of the chloride, lead is found in Analytical Group II as well as in Group I. Some properties of the metals of Analytical Group I are listed in Table 30-1.

Table 30-1   Some Properties of the Metals of Analytical Group I

Property	Mercury	Silver	Lead
Atomic number	80	47	82
Atomic weight	200.59	107.868	207.2
Oxidation numbers	+1, +2	+1, +2, +3	+2, +4
Reduction potential, V	(Hg_2^{2+}/Hg) +0.789 (Hg^{2+}/Hg) +0.854	(Ag^+/Ag) +0.80	(Pb^{2+}/Pb) −0.126
Density, g/cm^3, at 20°C	13.546	10.50	11.34
Melting point, °C	−38.87	960.8	327.4
Boiling point, °C	356.57	2193	1750

MERCURY

30.1   Periodic Relationships of Mercury

The three elements of Periodic Group IIB are zinc, cadmium, and mercury. Cadmium is in Analytical Group II (Chapters 31 and 37), and zinc is in Analytical Group III (Chapters 32 and 38). Each of these elements has 2 electrons in its outer shell and 18 in the underlying shell [$(n − 1)s^2$, $(n − 1)p^6$, $(n − 1)d^{10}$, ns^2]. They are representative metals and the +2 oxidation number prevails, although the +1 oxidation number of mercury (Hg_2^{2+}) is important. The melting points, boiling points, and heats of vaporization for the Group IIB members are lower than those for any other group of metals except the alkali metals (Group 1A). Of the three elements, zinc is the most reactive and mercury the least. The ions of all three are small in relation to their charges, so they show strong tendencies toward the formation of complex ions. The most frequently encountered complex ions with these elements are those with halides, cyanide, and ammonia. Coordination numbers of 4 and 6 are quite common in the complexes of these elements. The four-coordinated complexes are tetrahedral. The six-coordinated complexes have octahedral structures.

30.2   History, Occurrence, and Metallurgy of Mercury

Mercury was one of the few metals known in ancient times. The Greek philosopher Aristotle refers to the metal as liquid silver, or quick silver, because of its appearance and liquid state. The alchemists named the element after *Mercury,* the swift messenger of the gods of Roman mythology. One of the early uses of mercury was to extract gold from its ores, as gold is quite soluble in mercury. The alchemists used mercury in their attempts to transmute common metals into gold, assuming that mercury had unusual powers due to its liquid state and its ability to dissolve other metals, forming amalgams.

Elemental mercury is sometimes found in rocks and as an amalgam of silver and gold. However, the most important commercial source of the metal is the dark red sulfide, HgS, known as **cinnabar.** When cinnabar is roasted, metallic mercury distills from the furnace and is condensed to the liquid.

$$HgS(s) + O_2(g) \xrightarrow{\Delta} Hg(l) + SO_2(g) \qquad \Delta H° = −238 \text{ kJ}$$

Ordinarily, the roasting of a mineral in air results in the oxidation of the metal to an oxide. However, the mercury oxides are thermally unstable and decompose to the elements at the temperature of the furnace. Mercury may be purified by filtering it through chamois or a gold seal (the mercury wets the gold seal and hence passes through it; the impurities do not). It may also be purified by washing it with nitric acid to oxidize the metallic impurities or by distilling it in an atmosphere of oxygen to remove the more active metals.

30.3 Properties and Uses of Mercury

Mercury is a silvery white metal and is the only one that is liquid at room temperature. Its low freezing point ($-38.87°C$), its high boiling point ($356.6°C$), its uniform coefficient of expansion, and the fact that it does not wet glass make mercury an excellent thermometric substance. Because of its chemical inactivity, mobility, high density, and electrical conductivity, it is used extensively in barometers, vacuum pumps, and liquid seals, and for electrical contacts. Except for iron and platinum, all metals readily dissolve in, or are wet by, mercury to form **amalgams.** Sodium amalgam is used as a reducing agent, because it is less active than sodium metal alone. Amalgams of tin, silver, and gold are used in dentistry.

Mercury does not change chemically in air at ordinary temperatures but is slowly oxidized when heated, forming **mercury(II) oxide, HgO,** which decomposes to the elements at still higher temperatures. Being below hydrogen in the activity series, mercury is not attacked by hydrochloric acid but does dissolve in an excess of oxidizing acids such as cold nitric or hot concentrated sulfuric acid. With excess acid, mercury(II) salts are formed.

$$3Hg + 8H^+ + 2NO_3^- \longrightarrow 3Hg^{2+} + 2NO(g) + 4H_2O$$

$$Hg + 2H_2SO_4 \xrightarrow{\triangle} Hg^{2+} + SO_4^{2-} + SO_2(g) + 2H_2O$$

With excess mercury these oxidizing acids convert the metal to the corresponding mercury(I), Hg_2^{2+}, salts rather than to the mercury(II), Hg^{2+}, salts.

The halogens attack mercury, forming halides, and sulfur combines with it, forming mercury(II) sulfide. Mercury is displaced from solutions of its ions by all metals except silver, gold, and the platinum metals (Ru, Rh, Pd, Os, Ir, and Pt). A copper wire immersed in a solution of a mercury compound soon becomes amalgamated.

Mercury forms two series of compounds; in these series the metal has oxidation numbers of $+1$ and $+2$. The mercury atom uses both of its valence electrons in bonding in all of its compounds. For example, in mercury(II) chloride, $HgCl_2$, the chlorine atoms are covalently bonded to the mercury atom in a linear molecule, as shown by the Lewis structure

$$: \overset{..}{\underset{..}{Cl}} : Hg : \overset{..}{\underset{..}{Cl}} :$$

With the strongly electronegative sulfate and nitrate groups, mercury(II) forms ionic compounds in which dipositive mercury(II) ions, Hg^{2+}, are present. In the series of compounds in which mercury exhibits a $+1$ oxidation number, one valence electron of each of the mercury atoms serves to form an electron-pair bond between the two mercury atoms. The Hg_2Cl_2 molecule has a linear covalent structure.

$$: \overset{..}{\underset{..}{Cl}} : Hg : Hg : \overset{..}{\underset{..}{Cl}} :$$

When the second electron of each mercury atom is lost completely by electron transfer, as in the formation of $Hg_2(NO_3)_2$, the dimercury(I) ion $(Hg:Hg)^{2+}$ is formed.

The Hg_2^{2+} ion is the most common ionic species with a metal-metal bond (Section 29.12). Evidence from various sources points to the dimeric nature of this ion. X-ray analyses of crystalline mercury(I) chloride show that mercury atoms are bonded in pairs, and magnetic measurements of solutions of mercury(I) compounds show that they are not paramagnetic, meaning that there are no unpaired electrons as there would be if such solutions contained Hg^+ ions with an odd number of electrons per ion. On the contrary, $Hg:Hg^{2+}$ contains an even number of electrons, so that all electrons can be present as pairs. The molecular formulas of mercury(I) compounds are Hg_2X_2, not HgX (where X is an acid anion).

30.4 Mercury(I) Compounds

A number of mercury(I) compounds are known. One of these, **mercury(I) chloride, Hg_2Cl_2** (commonly called mercurous chloride), is a white, insoluble crystalline compound ($K_{sp} = 1.1 \times 10^{-18}$), which is of analytical importance because its insolubility serves to precipitate the mercury(I) ion.

$$Hg_2^{2+} + 2Cl^- \longrightarrow Hg_2Cl_2(s)$$

Mercury(I) chloride is manufactured by heating a mixture of mercury(II) sulfate, mercury, and sodium chloride.

$$HgSO_4 + Hg + 2NaCl \xrightarrow{\Delta} Hg_2Cl_2 + Na_2SO_4$$

Mercury(I) chloride is volatile and sublimes from the reaction mixture. It is used in the field of medicine under the name **calomel** as a cathartic and a diuretic (a stimulant for the organs of secretion).

When exposed to light, mercury(I) chloride slowly decomposes.

$$Hg_2Cl_2 \longrightarrow HgCl_2 + Hg$$

Hence it is usually stored in amber-colored bottles.

When mercury dissolves in cold, dilute nitric acid (with the mercury in excess), **mercury(I) nitrate, $Hg_2(NO_3)_2$,** is formed. This compound is important because it is the only readily soluble mercury(I) salt. This salt hydrolyzes in aqueous solution to form mercury(I) hydroxynitrate $Hg_2(OH)NO_3$.

$$Hg_2(NO_3)_2 + 2H_2O \rightleftharpoons Hg_2(OH)NO_3(s) + H_3O^+ + NO_3^-$$

The addition of nitric acid reverses the hydrolysis. On standing in contact with air, solutions of mercury(I) nitrate readily oxidize to mercury(II) nitrate. The accumulation of Hg^{2+} can be prevented by keeping a small amount of metallic mercury in the bottle in which the solution is stored.

$$Hg^{2+} + Hg \longrightarrow Hg_2^{2+}$$

Mercury(I) sulfide, Hg_2S, is unstable and immediately decomposes into mercury and mercury(II) sulfide when it is formed by passing hydrogen sulfide into a solution containing mercury(I) ions.

$$Hg_2^{2+} + H_2S + 2H_2O \longrightarrow Hg_2S(s) + 2H_3O^+$$
$$\phantom{Hg_2^{2+} + H_2S + 2H_2O \longrightarrow} \,\llcorner\!\!\longrightarrow Hg(l) + HgS(s)$$

Aqueous ammonia reacts with mercury(I) chloride in an oxidation-reduction reaction to form metallic mercury, as a finely divided black powder, and a complex white salt, **mercury(II) amidochloride.**

$$Hg_2Cl_2 + 2NH_3 \longrightarrow Hg + HgNH_2Cl + NH_4^+ + Cl^-$$

The formation of a black material (finely divided black metallic mercury) when white mercury(I) chloride is treated with aqueous ammonia is used as the test for the mercury(I) ion in qualitative analysis.

Alkali metal hydroxides precipitate the mercury(I) ion as black Hg_2O, which is unstable and decomposes into Hg and HgO.

Reducing agents, such as tetrachlorostannate(II) ion, readily reduce mercury(I) compounds to metallic mercury. This is used in qualitative tests for the ion.

$$Hg_2Cl_2(s) + [SnCl_4]^{2-} \longrightarrow 2Hg(l) + [SnCl_6]^{2-}$$

30.5 Mercury(II) Compounds

When a strong base is added to a solution of a mercury(II) compound, the oxide, HgO, precipitates. When precipitated from cold solutions, **mercury(II) oxide** is yellow, but when precipitated from hot solutions, it is red. Since the crystal structure is the same in both cases, this difference in color has been attributed to a difference in the size of the HgO particles. **Mercury(II) hydroxide** is unstable, losing the elements of a molecule of water to form the oxide.

Mercury(II) chloride, HgCl₂, may be formed by heating the metal with an excess of chlorine, by dissolving mercury(II) oxide in hydrochloric acid, or by the action of aqua regia (a solution of HNO_3 and HCl) on mercury. It is prepared commercially by heating mercury(II) sulfate with sodium chloride; the $HgCl_2$ sublimes from the reaction mixture.

$$2NaCl + HgSO_4 \xrightarrow{\Delta} Na_2SO_4 + HgCl_2$$

The compound $HgCl_2$ is also called bichloride of mercury and corrosive sublimate. In dilute solutions it is used as an antiseptic. It is moderately soluble in water but only slightly ionized, as indicated by the low electrical conductivity of its aqueous solutions. Its solubility can be increased by adding an excess of chloride ions, which cause the formation of the complex **tetrachloromercurate(II) ion** according to the equation

$$HgCl_2 + 2Cl^- \rightleftharpoons [HgCl_4]^{2-}$$

Iodide ions precipitate mercury(II) ions from solution as HgI_2, which is red. It is soluble in a solution containing an excess of iodide ions; tetraiodomercurate(II) ions are formed according to the equation

$$HgI_2 + 2I^- \longrightarrow [HgI_4]^{2-}$$

Hydrogen sulfide precipitates black **mercury(II) sulfide, HgS** ($K_{sp} = 1 \times 10^{-45}$), from solutions of mercury(II) salts, even from strongly acidic solutions. Because of this, mercury(II) is a member of Analytical Group II. The precipitate formed by the interaction of $HgCl_2$ in solution with H_2S is first white, then yellow, then red, and finally black. The white compound has the formula $HgCl_2 \cdot 3HgS$. When heated, HgS becomes bright red; the red sulfide is isomeric with the black and has, in the past, been used as the pigment **vermilion.**

Mercury(II) sulfide is dissolved by aqueous solutions of sodium sulfide, Na_2S, in the presence of an excess of hydroxide ions and forms the **thiomercurate ion.**

$$HgS + S^{2-} \rightleftharpoons [HgS_2]^{2-}$$

This reaction is used in the Group II procedures of the analytical scheme (Section 37.2).

30.6 Physiological Action of Mercury

Metallic mercury is an accumulative poison; the breathing of air containing mercury vapor or contact of the liquid metal with the skin should be avoided. All soluble mercury compounds are poisonous, but in small doses they may be medicinal. The fatal dose of mercury(II) chloride is 0.2–0.4 gram. The mercury(II) ion combines with the protein tissue of the kidneys and destroys their ability to remove waste products from the blood. Antidotes for mercury poisoning are egg white and milk; their proteins precipitate the mercury in the stomach. Mercury poisoning is treated with chelates like British Anti-Lewisite (Section 29.5).

SILVER

30.7 Periodic Relationships of Silver

Copper, silver, and gold are known as the **coinage metals** and have been used from very early times for the manufacture of ornamental objects and coins. They compose Group IB of the Periodic Table and may be classified as transition elements since they have valence electrons in two shells. Each has 1 electron in its outermost shell and 18 in the underlying shell $[(n - 1)s^2, (n - 1)p^6, (n - 1)d^{10}, ns^1]$. Since all of the alkali metals of Group IA and the coinage metals of Group IB have 1 electron in their outermost shell, the members of these two families might be expected to behave similarly. The elements of both families are good conductors of electricity and form a series of compounds with analogous formulas (Na_2O, Ag_2O; NaCl, AgCl; and Na_2SO_4, Ag_2SO_4). Other than these, similarities between the two families are almost entirely lacking. For example, the alkali metals are highly reactive, while the coinage metals are quite unreactive. This results, in part, from the fact that the atoms of the IB family are smaller than those of the IA family, and thus the valence electrons of the former are held more closely to the nucleus. Copper, silver, and gold, unlike the alkali metals, can use one or two electrons from the underlying shell in bond formation. Thus, copper, silver, and gold each exhibit oxidation numbers of $+1$, $+2$, and $+3$. However, the most common oxidation number for copper is $+2$; for silver, $+1$; and for gold, $+3$. Many of the ions and compounds of the coinage metals are colored, while the ions of the alkali metals are colorless and their only colored compounds are those in which the anion is colored, such as $K_2Cr_2O_7$ and $KMnO_4$. The hydroxides of the alkali metals are soluble and strongly basic, while those of copper and gold are insoluble and weakly basic. (Silver hydroxide is slightly soluble and strongly basic.) Most compounds of the alkali metals are soluble in water, while the majority of those of silver and gold are insoluble. Finally, the alkali metals show little tendency to form complex ions, while many stable complexes, such as

$[Cu(CN)_2]^-$, $[Ag(CN)_2]^-$, $[Au(CN)_4]^-$, $[Cu(NH_3)_2]^+$, $[Cu(NH_3)_4]^{2+}$, and $[Ag(NH_3)_2]^+$, are known for the coinage metals.

Among the coinage metals trends in properties are neither very regular nor readily interpreted. The chemical activity, elasticity, tensile strength, and specific heat decrease in the order copper, silver, gold; the density increases in this order. With their oxidation number of $+1$, all three metals form insoluble chlorides, but only silver is included in Analytical Group I. Compounds of copper(I) are not common and are not as important as those of copper(II), which is a member of Analytical Group II. Gold is considered a less common metal, and its chemical properties are relatively unimportant. Some physical properties of the coinage metals are given in Table 30-2.

Table 30-2 Some Physical Properties of the Coinage Metals

	Atomic Number	Atomic Weight	Atomic Radius, Å	Ionic (M^+) Radius, Å	Density, g/cm³, at 20°C	Melting point, °C	Boiling Point, °C
Copper	29	63.546	1.28	0.96	8.92	1083	2582
Silver	47	107.868	1.44	1.26	10.50	960.8	2193
Gold	79	196.9665	1.44	1.37	19.3	1063	2660

30.8 History of the Coinage Metals

These metals were the first to be used by primitive races because they are found in the free state, they tarnish slowly, and their appearance is pleasing to the eye. Ornaments of gold have been found in Egyptian tombs constructed in the prehistoric Stone Age. Gold and silver coins were used as a medium of exchange in India and Egypt thousands of years ago.

It is believed that copper was the first metal to be fashioned into utensils, instruments, and weapons; its use for such purposes began sometime during the Stone Age. The practice of alloying copper with tin and the use of the resulting bronze in place of stone introduced the Bronze Age.

The chemical symbol for gold (Au) is derived from the Latin name *aurum,* that for silver (Ag) from *argentum,* and that for copper (Cu) from *cuprum.*

30.9 Occurrence and Metallurgy of Silver

Silver is sometimes found in large nuggets but more frequently in veins and related deposits. It is frequently found alloyed with gold, copper, or mercury. It occurs combined as the chloride, AgCl (horn silver), and as the sulfide, Ag_2S, which is usually mixed with the sulfides of lead, copper, nickel, arsenic, and antimony. Most silver produced is a by-product of the mining of other metals such as lead and copper.

When lead is obtained from lead sulfide ore, it is usually mixed with some silver. The silver is extracted from the lead by the use of zinc, in which silver is about 3000 times more soluble than it is in lead. The lead is melted and thoroughly mixed with a small quantity of zinc. Lead and zinc are immiscible. Most of the silver leaves the lead and dissolves in the zinc. When mixing is stopped, the zinc rises to the surface of the lead and solidifies. The zinc-silver alloy is removed from the lead and the more volatile zinc is separated from the silver by distillation. This extraction procedure is known as the **Parkes process.**

The extraction of silver from its ores is dependent on the formation of the complex dicyanoargentate ion, $[Ag(CN)_2]^-$. Silver metal and all of its compounds are readily dissolved by alkali metal cyanides in the presence of air. Representative equations for the extraction of silver are

$$4Ag + 8CN^- + O_2 + 2H_2O \longrightarrow 4[Ag(CN)_2]^- + 4OH^-$$

$$2Ag_2S + 8CN^- + O_2 + 2H_2O \longrightarrow 4[Ag(CN)_2]^- + 2S + 4OH^-$$

$$AgCl + 2CN^- \rightleftharpoons [Ag(CN)_2]^- + Cl^-$$

The silver is precipitated from the cyanide solution by addition of either zinc or aluminum.

$$2[Ag(CN)_2]^- + Zn \longrightarrow 2Ag + [Zn(CN)_4]^{2-}$$

Silver is also obtained from the anode mud formed during the electrolytic refining of copper (Section 31.7).

30.10 Properties of Silver

Silver is a white, lustrous metal whose polished surface is an excellent reflector of light. As a conductor of heat and electricity it is second only to gold. Silver is noted for its ductility and malleability.

Silver is not attacked by oxygen in the air under ordinary conditions but tarnishes quickly in the presence of hydrogen sulfide or on contact with sulfur-containing foods such as eggs and mustard. Its reaction with hydrogen sulfide in the presence of air is

$$4Ag(s) + 2H_2S(g) + O_2(g) \longrightarrow 2Ag_2S(s) + 2H_2O(l) \qquad \Delta H° = -595.59 \text{ kJ}$$

Silver tarnish is a thin film of silver sulfide.

The halogens react with silver, forming halides, and the metal is soluble in oxidizing acids such as nitric acid and hot sulfuric acid.

$$3Ag + 4H^+ + NO_3^- \longrightarrow 3Ag^+ + NO(g) + 2H_2O$$

Since silver lies below hydrogen in the activity series, it is not soluble in nonoxidizing acids such as hydrochloric acid. Most silver salts are sparingly soluble in water, but silver nitrate is quite soluble. The common oxidation number exhibited by silver is $+1$, although the oxides AgO and Ag_2O_3 and some complexes containing Ag(II) and Ag(III) are known.

30.11 Uses of Silver

A large amount of silver is used for the making of coins, silverware, and ornaments. For most of its uses silver is too soft to wear well, and it is therefore hardened by alloying it with other metals, particularly copper. **Sterling silver** contains 7.5% copper, and jewelry silver contains 20% copper. Large amounts of silver are used in the preparation of dental alloys, photographic films, and mirrors.

The electroplating industry uses a large percentage of the silver produced. The object to be plated with silver is made the cathode in an electrolytic cell containing a solution of sodium dicyanoargentate, $Na[Ag(CN)_2]$, as the electrolyte. The anode is a bar of pure silver, which dissolves to replace the silver ions removed from the solution as plating at the cathode proceeds. The electrode reactions are

Cathodic reduction: $\quad [Ag(CN)_2]^- + e^- \longrightarrow Ag + 2CN^-$

Anodic oxidation: $\quad Ag + 2CN^- \longrightarrow [Ag(CN)_2]^- + e^-$

The film of deposited silver has a flat white appearance but can be given a brilliant luster by burnishing.

The characteristic luster can be restored to tarnished silverware by the following application of the activity series. The tarnished silver object is immersed in a solution of baking soda and salt, containing a teaspoonful of each to a liter of water, held in an aluminum vessel. (A piece of aluminum metal in a glass vessel will serve equally well.) The aluminum and silver must be in contact. After the tarnish is removed, the silver object should be thoroughly washed. The chemistry involved may be explained as follows: The silver sulfide of the film of tarnish dissolves somewhat in the solution, forming silver and sulfide ions. The more active aluminum displaces the silver from solution, causing it to plate out on the silver object, according to the equation

$$3Ag^+ + Al \longrightarrow 3Ag + Al^{3+}$$

The tarnish is removed with little loss of silver; however, the "antiquing" (actually silver sulfide tarnish), which often gives a desirable appearance by defining the crevices of the design, will be removed also. Silver polish can be used to remove tarnish, but the silver in the tarnish is lost.

Silver mirrors are formed by depositing a thin layer of silver on glass. This is accomplished by reducing a solution of silver nitrate in ammonia with some mild reducing agent, such as glucose or formaldehyde. The equation for the reaction may be written as

$$[Ag(NH_3)_2]^+ + e^- \text{ (from reducing agent)} \longrightarrow Ag + 2NH_3$$

30.12 Compounds of Silver

Silver oxide, Ag₂O, is formed when silver is exposed to ozone or when finely divided silver is heated in oxygen under pressure. Hydroxide bases act on silver nitrate to give a dark brown amorphous precipitate of silver oxide. Although the oxide is only slightly soluble in water, it dissolves enough to give a distinctly alkaline solution. The equilibrium is

$$Ag_2O(s) + H_2O \rightleftharpoons 2Ag^+ + 2OH^-$$

Silver oxide is a convenient reagent, both in inorganic and organic chemistry, for preparing soluble hydroxides from the corresponding halides because the silver halide formed at the same time may be removed conveniently by filtration. Cesium hydroxide, for example, may be prepared by the reaction

$$2Cs^+ + 2Cl^- + Ag_2O + H_2O \longrightarrow 2Cs^+ + 2OH^- + 2AgCl(s)$$

This reaction proceeds to the right because AgCl is much less soluble than Ag_2O. Silver oxide dissolves readily in aqueous ammonia to form the strong base **diamminesilver hydroxide.**

$$Ag_2O + 4NH_3 + H_2O \longrightarrow 2[Ag(NH_3)_2]^+ + 2OH^-$$

When the oxide is heated at atmospheric pressure in air, it readily decomposes according to the equation

$$2Ag_2O(s) \xrightarrow{\Delta} 4Ag(s) + O_2(g) \qquad \Delta H° = 62.09 \text{ kJ}$$

Silver fluoride, AgF, is very soluble in water, but the chloride, bromide, and iodide are relatively insoluble—the higher the atomic weight of the halogen, the lower is the solubility of the silver halide. The insoluble silver halides are formed as curdy precipitates when halide ions are added to solutions of silver salts. Silver chloride ($K_{sp} = 1.8 \times 10^{-10}$) is white; silver bromide ($K_{sp} = 3.3 \times 10^{-13}$) is pale yellow; and silver iodide ($K_{sp} = 1.5 \times 10^{-16}$) is yellow. When exposed to light, the halides of silver turn violet at first and finally black; during this process they are decomposed into the elements.

$$2AgX + \text{light} \longrightarrow 2Ag + X_2 \qquad (X = \text{halogen})$$

Silver chloride, AgCl, is readily soluble in an excess of dilute aqueous ammonia to form the diamminesilver complex.

$$AgCl(s) + 2NH_3 \longrightarrow [Ag(NH_3)_2]^+ + Cl^-$$

This reaction is used in separating silver chloride from mercury(I) chloride in the qualitative analysis scheme. Silver bromide will dissolve only if the ammonia is concentrated, and silver iodide is only scarcely soluble in this reagent. This behavior is used in the separation and identification of the halide ions in the anion analytical scheme (Section 41.11).

Silver nitrate, AgNO₃, is the only simple silver salt that is significantly soluble. It is obtained by dissolving silver in nitric acid and evaporating the solution. Organic materials, such as the skin, readily reduce silver nitrate to give free silver, which forms a black stain. Silver nitrate is heavily used in the production of photographic materials, as a laboratory reagent, and in the manufacture of other silver compounds.

LEAD

30.13   Periodic Relationships of Lead

The first two elements of Periodic Group IVA, carbon and silicon, have already been discussed in Chapters 25 and 27, respectively. The remaining three members of this family are germanium, tin, and lead.

Carbon and silicon are primarily nonmetallic in character, while germanium, tin, and lead, as is expected, become increasingly metallic in their properties with increasing atomic weight. In ionic compounds carbon and silicon are always found in the anion. On the other hand, germanium, tin, and lead form dipositive cations, Ge^{2+}, Sn^{2+}, and Pb^{2+}. The fact that the hydroxides of these ions are amphoteric is an indication of some nonmetallic character. All members of this periodic group form covalent compounds or anions in which the +4 oxidation number is exhibited; for example, CCl_4, $SiCl_4$, $GeCl_4$, $SnCl_4$, and $PbCl_4$ are low-boiling covalent liquids; Na_2CO_3, Na_2SiO_3, Na_2GeO_3, Na_2SnO_3, and Na_2PbO_3 are ionic compounds.

Germanium and the members of Periodic Group IVB (Ti, Zr, and Hf) are among the less familiar elements. The members of Group IVB usually show an oxidation number of +4 in their compounds; they are transition metals with two electrons in their outer shell and ten electrons in the underlying shell. Some physical properties of the Group IVA metals are given in Table 30-3.

Table 30-3   Some Physical Properties of the Metals of Group IVA

	Atomic Number	Atomic Weight	Atomic Radius, Å	Ionic (M^{4+}) Radius, Å	Density, g/cm^3, at 20°C	Melting Point, °C	Boiling Point, °C
Germanium	32	72.59	1.22	0.53	5.36	960	—
Tin	50	118.69	1.4	0.71	7.31	231.9	2337
Lead	82	207.2	1.75	0.84	11.34	327.4	1750

30.14   History, Occurrence, and Metallurgy of Lead

Because lead is easily extracted from its ores, it was known to the early Egyptians and Babylonians. It is mentioned in the Old Testament of the Bible. Lead pipes were commonly used by the Romans for conveying water, and in the Middle Ages lead was used as a roofing material. The stained glass windows of the great cathedrals of this age were set in lead.

The principal lead ore is **lead sulfide, PbS,** commonly called galena. Other common ores are **lead carbonate, PbCO₃,** called cerrusite, and **lead sulfate, PbSO₄,** called anglesite. Both of the latter may have been formed by the weathering of sulfide ores.

Lead ores are first concentrated by a series of selective flotation processes to remove the unwanted materials and a large part of the zinc sulfide that is usually associated with lead ores. The concentrated ore is then roasted in air to convert most of the sulfide to the oxide.

$$2PbS(s) + 3O_2(g) \xrightarrow{\Delta} 2PbO(s) + 2SO_2(g) \qquad \Delta H° = -710.4 \text{ kJ}$$

The roasted product is reduced in a blast furnace with coke and scrap iron.

$$PbO(s) + C(s) \xrightarrow{\Delta} Pb(s) + CO(g) \qquad \Delta H° = 106.8 \text{ kJ}$$

$$PbO(s) + CO(g) \xrightarrow{\Delta} Pb(s) + CO_2(g) \qquad \Delta H° = -65.7 \text{ kJ}$$

$$PbS(s) + Fe(s) \xrightarrow{\Delta} Pb(s) + FeS(s) \qquad \Delta H° = 0.4 \text{ kJ}$$

The lead obtained from the blast furnace contains copper, antimony, arsenic, bismuth, gold, and silver. The crude lead is melted and stirred to bring about the oxidation of antimony, arsenic, and bismuth. The oxides of these metals rise to the surface, and the molten lead is drained off for further refining. Gold and silver may be extracted from the lead by the Parkes process (Section 30.9) or by the electrolytic **Betts process.** In the Betts process thin sheets of pure lead are the cathodes, and plates of impure lead are the anodes. The electrolyte is a solution containing lead hexafluorosilicate, PbSiF₆, and hexafluorosilicic acid, H₂SiF₆.

30.15   Properties and Uses of Lead

Lead is a soft metal with little tensile strength and is the densest of the common metals except for gold and mercury. Lead has a metallic luster when freshly cut but quickly acquires a dull gray color when exposed to moist air. In air that contains moisture and carbon dioxide, lead becomes oxidized on its surface, forming a protective layer that is both compact and adherent; this film is probably lead hydroxycarbonate.

Unlike silver and mercury, lead lies above hydrogen in the activity series and therefore dissolves (though slowly) in dilute nonoxidizing acids. Concentrated nitric acid, an oxidizing acid, attacks it readily. It is not dissolved by pure water in the absence of air but does react with water in the presence of air to form the hydroxide.

$$2Pb + 2H_2O + O_2 \longrightarrow 2Pb(OH)_2$$

When heated in a stream of air, lead burns. It also reacts upon heating with sulfur, fluorine, and chlorine.

The major uses of lead depend on the ease with which it is worked, its low melting point, its great density, and its resistance to corrosion.

30.16   Compounds of Lead

Lead forms two well-defined series of compounds in which its oxidation numbers are $+2$ and $+4$. An atom of lead (like those of carbon, silicon, germanium, and tin) has four valence electrons; two of these are s electrons and the other two are p electrons. All four of the valence electrons are seldom, if ever, completely removed from the atom but are very often shared with electronegative elements. This fact accounts for the $+4$ oxidation number, which is typical of the Group IVA elements. Lead (and also tin) forms many compounds in which its two s valence electrons do not participate in the bonding but remain associated with the lead atom as a stable electron pair. When this happens, the element assumes the $+2$ oxidation number. Because of the high reduction potential of tetravalent lead,

$$PbO_2 + 4H^+ + 2e^- \longrightarrow Pb^{2+} + 2H_2O \qquad E° = +1.46 \text{ V}$$

$+2$ is the characteristic oxidation number.

The yellow, powdery **lead monoxide, PbO,** is obtained when lead is heated in air. **Lead dioxide, PbO$_2$,** is a chocolate-brown powder formed by oxidizing lead(II) compounds in alkaline solution. With sodium hypochlorite as the oxidizing agent, the equation for the oxidation is

$$Pb(OH)_3^- + ClO^- \longrightarrow PbO_2(s) + Cl^- + OH^- + H_2O$$

Lead dioxide is the principal constituent of the cathode of the charged lead storage battery (Section 20.17). Since lead(IV) tends to revert to the more stable lead(II) ion by gaining two electrons, lead dioxide is a powerful oxidizing agent.

Trilead tetraoxide, Pb$_3$O$_4$, called red lead, is prepared by carefully heating the monoxide in air at temperatures of 400–500°C. When red lead is treated with nitric acid, two-thirds of the lead dissolves as lead nitrate (oxidation number = +2), and the remaining third remains as lead dioxide (oxidation number = +4). The equation for the reaction is

$$Pb_3O_4 + 4H^+ + [4NO_3^-] \longrightarrow PbO_2(s) + 2Pb^{2+} + [4NO_3^-] + 2H_2O$$

When alkali metal hydroxides are added to solutions of lead(II) compounds, white **lead hydroxide, Pb(OH)$_2$** ($K_{sp} = 2.8 \times 10^{-16}$), precipitates. An excess of hydroxide ion causes lead hydroxide to dissolve according to the equation

$$Pb(OH)_2(s) + OH^- \longrightarrow Pb(OH)_3^-$$

which shows the amphoteric character of the Pb(OH)$_2$. Its reaction with hydrogen ions is given by the equation

$$Pb(OH)_2(s) + 2H^+ \longrightarrow Pb^{2+} + 2H_2O$$

Lead chloride, PbCl$_2$, may be formed by the direct union of the elements, by the action of hydrochloric acid on lead monoxide, or by precipitation from solutions containing lead(II) ions and chloride ions. It is soluble in hot water and in solutions of high chloride ion concentration, **tetrachloroplumbate(II) ions** being formed in the latter case.

$$PbCl_2(s) + 2Cl^- \rightleftharpoons PbCl_4^{2-}$$

These properties are important in the qualitative analysis of lead.

When either metallic lead or lead monoxide, PbO, is dissolved in nitric acid, **lead nitrate, Pb(NO$_3$)$_2$,** is formed. This salt is readily soluble in water, but unless the solution is made slightly acidic with nitric acid, hydrolysis occurs and hydroxynitrates precipitate.

$$Pb^{2+} + NO_3^- + 2H_2O \rightleftharpoons Pb(OH)NO_3(s) + H_3O^+$$

Lead nitrate is unstable at moderately high temperatures and decomposes in the same manner as the nitrates of other heavy metals.

$$2Pb(NO_3)_2 \xrightarrow{\triangle} 2PbO(s) + 4NO_2 + O_2$$

The *normal* **lead carbonate, PbCO$_3$,** may be prepared by the action of sodium hydrogen carbonate on lead chloride. **Lead hydroxycarbonate, Pb$_3$(OH)$_2$(CO$_3$)$_2$,** is formed when alkali metal carbonates are added to solutions containing the lead ion. This compound used to be important commercially as the paint pigment **white lead,** but because of the danger of lead poisoning its use has been largely curtailed, especially on objects used by children.

Lead sulfate, PbSO$_4$, forms when solutions of a soluble lead salt and a soluble sulfate are mixed. It is insoluble in water ($K_{sp} = 1.8 \times 10^{-8}$) but readily dissolves in solutions containing an excess of alkali metal halide or acetate ions. **Lead acetate, Pb(CH$_3$CO$_2$)$_2$,** is one of the few soluble compounds of lead; it is a weak electrolyte, indicating that it dissolves mainly as a covalent compound rather than as an ionic one. It is extremely toxic. **Lead chromate, PbCrO$_4$,** is insoluble in water but dissolves readily in acids and alkali metal hydroxides. Hydrogen sulfide precipitates black **lead sulfide, PbS** ($K_{sp} = 8.4 \times 10^{-28}$), which is insoluble in solutions of dilute acids and alkali metal sulfides.

FOR REVIEW

SUMMARY

Mercury(I), silver, and lead(II), the ions comprising Analytical Group I, are all precipitated as chlorides by a slight excess of hydrochloric acid. The chlorides of all other common metal ions are soluble in dilute acid. Because of the relatively high solubility of lead chloride, lead is found in Analytical Group II as well.

Mercury, silver, and lead form a variety of compounds, including oxides, hydroxides, chlorides, nitrates, and sulfates.

Mercury exhibits oxidation numbers of +1 and +2. The mercury(I) ion is dimeric, Hg_2^{2+}, and is the most common ionic species with a metal-metal bond. Silver can exhibit an oxidation number of +1, +2, or +3, but +1 is by far the most common in silver compounds. Lead exhibits an oxidation number of +2, as in PbO, or +4, as in PbO_2. The +2 oxidation number is more common.

Mercury, silver, and lead can all form coordination compounds, such as $[HgCl_4]^{2-}$, $[Ag(NH_3)_2]^+$, and $[Pb(OH)_3]^-$. The ability of silver to form the dicyano-argentate ion, $[Ag(CN)_2]^-$, is a key property used in the metallurgy of silver.

Lead is amphoteric; its hydroxide is soluble in nonoxidizing acids (producing the Pb^{2+} ion in solution) and in excess hydroxide ion (producing the $[Pb(OH)_3]^-$ ion in solution).

EXERCISES

Mercury

1. The roasting of an ore of a metal usually results in the conversion of the metal to the oxide. Why does the roasting of cinnabar produce metallic mercury rather than an oxide of mercury?
2. Why is mercury not attacked by hydrochloric acid even though it dissolves readily in nitric acid or hot sulfuric acid?
3. Why are mercury(II) halides spoken of as weak electrolytes?
4. What properties make mercury valuable as a thermometric substance?
5. Write Lewis structures for $HgCl_2$ and Hg_2Cl_2.
6. Hydrogen sulfide gas is bubbled through a solution marked mercury(I) nitrate. A precipitate forms. Of what substance(s) is the precipitate composed?
7. How many moles of ionic species would be present in a solution marked 1.0 M mercury(I) nitrate? How would you demonstrate the accuracy of your prediction?

8. What is the minimum number of grams of fish that one would have to consume to obtain a fatal dose of mercury, if the fish contains 5.0 parts per million of mercury by weight? [Assume all the mercury from the fish ends up as mercury(II) chloride in the body and that a fatal dose is 0.20 g of $HgCl_2$.] How many pounds of fish would this be?   *Ans. 30 kg; 66 lb*
9. How many pounds of mercury can be obtained from 10 ton of ore that is 7.4% cinnabar?
              Ans. 1.3 × 10³ lb
10. Write balanced chemical equations for the following reactions:
    (a) Mercury(II) oxide is added to a solution of nitric acid.
    (b) A solution of mercury(II) nitrate is stirred with liquid mercury.
    (c) A solution of mercury(I) nitrate is added to a solution of potassium sulfide.
    (d) An excess of zinc is added to a solution of

mercury(II) nitrate.

(e) Zinc is added to a solution containing an excess of mercury(II) nitrate.

Silver

11. What properties of silver have made it valuable as a coinage metal?

12. Explain the tarnishing of silver in the presence of sulfur-containing substances.

13. Compare the solubilities of silver halides in water and in aqueous ammonia.

14. Describe the Parkes process for extracting silver from lead.

15. Why does silver dissolve in nitric acid but not in hydrochloric acid?

16. Write equations for the electrode reactions when $Na[Ag(CN)_2]$ is used as the electrolyte in silver plating.

17. Dilute sodium cyanide solution is slowly dripped into a slowly stirred silver nitrate solution. A white precipitate forms temporarily but dissolves as the addition of sodium cyanide continues. Use chemical equations to explain this observation.

18. Show several ways in which coordination compounds play an important part in the chemistry of silver.

19. The formation constant of $[Ag(CN)_2]^-$ is 1.0×10^{20}. What will be the equilibrium concentration of Ag^+ if 1.0 g of Ag is oxidized and put into 1.0 L of 0.10 M CN^- solution?

Ans. 1.4×10^{-20} M

20. A 1.4820-g sample of a pure solid alkali metal chloride is dissolved in water and treated with excess silver nitrate. The resulting precipitate, filtered and dried, weighs 2.849 g. What percent by mass of chloride ion is in the original compound? What is the identity of the salt?

Ans. 47.55%; KCl

Lead

21. Describe the Betts process for the refining of lead.

22. Compare the nature of the bonds in $PbCl_2$ to that of the bonds in $PbCl_4$. Would you expect the existence of Pb^{4+} ions? Explain.

23. Show by suitable equations that lead(II) hydroxide is amphoteric.

24. What is the composition of the tarnish on lead?

25. Why should water to be used for human consumption not be conveyed in lead pipes?

26. Account for the solubility of $PbCl_2$ in solutions of high chloride ion concentration.

27. When elements in the first transition metal series form dipositive ions, they usually give up their s electrons. Does this apply to lead when it forms the Pb^{2+} ion?

28. A 100-mL volume of a saturated solution of lead(II) iodide contains 0.41 g of solute at 100°C. Calculate the solubility product at this temperature.   *Ans. 2.8×10^{-6}*

29. Calculate the molar solubilities of AgBr, Hg_2Cl_2, and $PbSO_4$.   *Ans. AgBr, 5.7×10^{-7} M; Hg_2Cl_2, 6.5×10^{-7} M; $PbSO_4$, 1.3×10^{-4} M*

31

METALS OF ANALYTICAL GROUP II

DIVISION A—LEAD, BISMUTH, COPPER, AND CADMIUM

Analytical Group II (see also Chapter 37) consists of the common metals whose ions form chlorides that are soluble in dilute acid but whose sulfides are precipitated by hydrogen sulfide in 0.3 M hydrochloric acid. Lead appears in Group II as well as in Group I because its chloride is of intermediate solubility in cold water. Mercury occurs in both Groups I and II because mercury(I) chloride is insoluble while mercury(II) chloride is soluble in dilute hydrochloric acid.

The ions Pb^{2+}, Bi^{3+}, Cu^{2+}, and Cd^{2+}, which comprise **Division A,** form sulfides that are insoluble in sodium sulfide solutions. This property permits the separation of these ions from those of **Division B,** Hg^{2+}, As^{3+}, Sb^{3+}, and Sn^{4+}, whose sulfides are soluble in water containing an excess of sodium sulfide.

If the ions of the less common metals were to be included in our scheme of analysis, Group II would contain ions of such metals as Au, Pt, Mo, Ru, Rh, Pd, Ir, Os, W, Re, and Te as well as the ones listed above.

The chemistry of lead was considered in Chapter 30 with that of the metals of Analytical Group I. The other metals of Division A will be considered in the first part of this chapter, and those of Division B will be discussed in the second part. Some important properties of the Division A metals are listed in Table 31-1.

Table 31-1   Some Properties of the Division A Metals of Analytical Group II

Property	Lead	Bismuth	Copper	Cadmium
Atomic number	82	83	29	48
Atomic weight	207.2	208.9804	63.546	112.41
Oxidation numbers	+2, +4	−3, +3, +5	+1, +2, +3	+2
Reduction potential, V	(Pb^{2+}/Pb)	(Bi^{3+}/Bi)	(Cu^{2+}/Cu)	(Cd^{2+}/Cd)
	−0.126	+0.32	+0.337	−0.403
Density, g/cm^3, at 20°C	11.34	9.78	8.92	8.65
Melting point, °C	327.4	271.0	1083	320.9
Boiling point, °C	1750	1420.0	2582	767

BISMUTH

31.1 Periodic Relationships of Bismuth

Bismuth is a member of Periodic Group VA, which also includes nitrogen, phosphorus, arsenic, and antimony. The elements of this group furnish an excellent illustration of the gradation in properties characteristic of groups of the Periodic Table. Nitrogen, the lightest member of the group, is a typical nonmetal that accepts electrons from active metals and forms the nitride ion N^{3-}, such as in magnesium nitride, Mg_3N_2, and lithium nitride, Li_3N. Bismuth is the heaviest member of the group and is almost entirely metallic in character. It gives up three of its five valence electrons to active nonmetals to form the tripositive ion, Bi^{3+}. The nonmetallic character of nitrogen and the metallic character of bismuth, each of which has five valence electrons, can be attributed, for the most part, to the difference in the distance between the valence shell and the nucleus for the two atoms. Because this distance is greater for bismuth, the attraction of the positive nucleus for the valence electrons is much smaller in an atom of bismuth than in an atom of nitrogen.

The properties of phosphorus, arsenic, and antimony are intermediate between those of nitrogen and bismuth, with nonmetallic character becoming less pronounced with increasing atomic weight. Atoms of each member of this family have five valence electrons and exhibit oxidation numbers of +3 and +5. (Chapters 23 and 24 cover the relationships among the members of this periodic group more completely.)

31.2 Occurrence and Metallurgy of Bismuth

Bismuth is most often found in the free state in nature. It sometimes occurs in compounds such as bismuth(III) oxide, Bi_2O_3 (called bismuth ocher), or bismuth(III) sulfide, Bi_2S_3 (called bismuth glance). It is produced in the United States as a by-product of the refining of other metals, particularly lead. Ores containing the free metal are treated by heating them in inclined iron pipes. The metal melts and flows away from the remaining materials. The oxide and sulfide ores are roasted and then heated with charcoal. As the bismuth is set free, it melts and collects beneath the less dense material.

31.3   Properties and Uses of Bismuth

Bismuth is a lustrous, hard, and brittle metal with a reddish tint. The metal burns when heated in air, giving the oxide Bi_2O_3. It oxidizes only superficially in moist air at ordinary temperatures to form a coating of oxide, which protects it against further oxidation. Bismuth combines directly with the halogens and sulfur. Oxidizing acids, such as hot sulfuric and nitric, dissolve it, forming the corresponding salts. Bismuth is dissolved slowly by hydrochloric acid in the presence of air, forming **bismuth(III) chloride, BiCl₃.** An unstable hydride of bismuth, BiH_3, called **bismuthine,** exists.

Like antimony, melted bismuth expands when it solidifies—a most unusual property. Because of this, it is used in the formation of alloys to prevent them from shrinking on solidification. Alloys of bismuth, tin, and lead have low melting points, which makes them useful for electrical fuses, safety plugs for boilers, and automatic sprinkler systems. Some of these alloys melt even in hot water. For example, Rose's metal (Bi, 50%; Pb, 25%; Sn, 25%) melts at 94°C; and Wood's metal (Bi, 50%; Pb, 25%; Sn, 12.5%; Cd, 12.5%) melts at 65.5°C.

31.4   Compounds of Bismuth

When bismuth is burned in air or when bismuth nitrate is strongly heated, yellow **bismuth(III) oxide, Bi₂O₃,** is formed. This oxide is basic as contrasted with the more acidic character of the lower oxides of the lighter members of Periodic Group VA. Thus Bi_2O_3 dissolves in acids to form salts. It does not exhibit acidic properties such as the ability to react with bases. Alkali metal hydroxides or aqueous ammonia precipitate white **bismuth hydroxide, Bi(OH)₃,** from solutions of bismuth(III) salts. When a suspension of bismuth hydroxide is boiled, it loses the elements of a molecule of water, forming **bismuth oxyhydroxide, BiO(OH)** ($K_{sp} = 1 \times 10^{-12}$).

Bismuth(V) oxide, Bi₂O₅, is produced by the action of very strong oxidizing agents on the trioxide. As expected due to the higher oxidation number of bismuth, this oxide is more acidic than Bi_2O_3 and shows its acidic character by dissolving in concentrated sodium hydroxide to form **sodium bismuthate, NaBiO₃.** This salt reacts with nitric acid to give **bismuthic acid, HBiO₃,** a very strong oxidizing agent used in the detection and estimation of manganese, which it oxidizes from Mn^{2+} to MnO_4^-.

When hydrogen sulfide is passed into a solution of a bismuth salt, brown **bismuth(III) sulfide, Bi₂S₃** ($K_{sp} = 1.6 \times 10^{-72}$), is precipitated. Bi_2S_3 does not dissolve in concentrated sulfide solutions to form thiosalts. This property is utilized in separating bismuth from arsenic and antimony, which form soluble thiosalts.

Bismuth(III) oxide dissolves in acids, forming the corresponding salts, such as the chloride $BiCl_3 \cdot 2H_2O$; the nitrate $Bi(NO_3)_3 \cdot 5H_2O$; and the sulfate $Bi_2(SO_4)_3$. Bismuth salts hydrolyze readily, forming hydroxysalts.

$$BiCl_3 + 2H_2O \rightleftharpoons Bi(OH)_2Cl(s) + 2H^+ + 2Cl^-$$
$$Bi(NO_3)_3 + 2H_2O \rightleftharpoons Bi(OH)_2NO_3(s) + 2H^+ + 2NO_3^-$$

Bismuth dihydroxychloride loses the elements of a molecule of water to form **bismuth oxychloride.**

$$Bi(OH)_2Cl \longrightarrow BiOCl(s) + H_2O$$

Bismuth dihydroxynitrate, when repeatedly washed with water, is converted to **bismuth hydroxide.**

$$Bi(OH)_2NO_3 + H_2O \longrightarrow Bi(OH)_3(s) + H^+ + NO_3^-$$

When dried, bismuth dihydroxynitrate decomposes to the oxynitrate $BiONO_3$.

31.5   Uses of Bismuth Compounds

Bismuth oxycarbonate, $(BiO)_2CO_3$, and **bismuth oxynitrate, $BiONO_3$** (under the names bismuth subcarbonate and bismuth subnitrate), are used in medicine for the treatment of stomach disorders, such as gastritis and ulcers, and skin diseases, such as eczema.

COPPER

The history and periodic relationships of copper were considered in Sections 30.8 and 30.7, respectively.

31.6   Occurrence of Copper

Copper occurs in both free and combined forms. The most important deposit of elemental copper in the world is in the Upper Peninsula of Michigan near Houghton. The largest single mass of elemental copper yet found, weighing 420 tons, is now on display in the Smithsonian Institution in Washington, D.C. The most important copper ores are the sulfides, such as chalcocite, Cu_2S, and chalcopyrite, $CuFeS_2$. Other ores are cuprite, Cu_2O melaconite, CuO, and malachite, $Cu_2(OH)_2CO_3$.

Copper is found in trace amounts in some plants. It is also found in the brightly colored feathers of certain birds and in the blood of certain marine animals such as lobsters, oysters, and cuttlefish, where it serves the same oxygen-carrying function as iron does in the blood of higher animals (Section 29.5).

31.7   Metallurgy of Copper

Ores that contain elemental (free) copper are pulverized, the unwanted rock material washed away, and the copper melted and poured into molds to cool. Oxide and carbonate ores are often leached with sulfuric acid to produce copper(II) sulfate solutions, from which copper metal may be obtained by electrodeposition. High-grade oxide or carbonate ores are reduced by heating them with coke mixed with a suitable flux (a substance that removes impurities by combining with them).

Sulfide ores are usually low-grade, containing less than 10% copper. These ores are first concentrated by the **flotation process.** In this process the pulverized ore is mixed with water to which a carefully selected oil has been added. A froth is produced by blowing air into the mixture. The metal-bearing particles have little or no attraction for water (such substances are **hydrophobic**). The particles do have an attraction for the oil and are therefore preferentially coated by it and adhere to the air bubbles in the froth, which floats on the surface. The particles of sand, rock, clay, etc., are wetted by water (such substances are **hydrophilic**) more readily than by the oil and sink to the bottom of the flotation vat. The froth containing the metal-bearing particles is removed, and the ore is recovered in a highly concentrated form.

The concentrate (or a high-grade sulfide ore if concentration was unnecessary) is then roasted in a furnace, at a temperature below its melting point, to drive off the moisture and remove part of the sulfur as sulfur dioxide. The remaining mixture, which is called calcine and consists of Cu_2S, FeS, FeO, and SiO_2, is **smelted** (extracted in the molten state) by mixing it with limestone (to serve as a flux) and heating the mixture above its melting point. The reactions taking place during the formation of the slag are

$$CaCO_3 + SiO_2 \longrightarrow CaSiO_3 + CO_2$$
$$FeO + SiO_2 \longrightarrow FeSiO_3$$

The Cu_2S, with some FeS, that remains after smelting is called matte. Reduction of the matte is accomplished in a **converter** by blowing air through the molten material. The air first oxidizes the iron(II) sulfide to iron(II) oxide and sulfur dioxide. Sand is added to form a slag of iron(II) silicate from the iron(II) oxide. After the iron has been removed, the air blast converts part of the Cu_2S to Cu_2O. As soon as copper(I) oxide is formed, it is reduced by the remaining copper(I) sulfide to metallic copper. The equations are

$$2Cu_2S + 3O_2 \longrightarrow 2Cu_2O + 2SO_2$$
$$2Cu_2O + Cu_2S \longrightarrow 6Cu + SO_2$$

The last traces of copper(II) oxide produced by the air blast are removed by reduction with H_2 and CO produced from methane or natural gas. The copper obtained in this way is called **blister copper** because of its characteristic appearance due to the air blisters it contains.

The impure copper is cast into large plates, which are used as anodes in the electrolytic purification of the metal (Section 20.6). Thin sheets of pure copper serve as the cathodes; copper(II) sulfate, in a solution of sulfuric acid, serves as the electrolyte. The impure copper passes into solution from the anodes, and pure copper plates out on the cathodes as electrolysis proceeds. Gold and silver in the anodes do not oxidize but fall to the bottom of the electrolytic cell as anode mud, along with bits of slag and Cu_2O. Silver, gold, and the platinum metals are recovered from the anode mud as valuable by-products. Metals more active than copper, such as zinc and iron, are oxidized at the anode, and their cations pass into solution, where they remain. The copper deposited on the cathode has an average purity of 99.955%. The value of the precious metals recovered from the anode mud is often sufficient to pay the cost of the electrolytic refining.

31.8 Properties and Uses of Copper

Copper is a reddish-yellow metal. It is ductile and malleable, so that it is readily fashioned into wire, tubing, and sheets. Copper is the best electrical conductor among the cheaper metals, but when used for this purpose it must be quite pure, since small amounts of impurities greatly reduce its electrical conductivity.

Copper is relatively inactive chemically. In moist air it first turns brown, due to the formation of a very thin, adherent film of either copper oxide or copper sulfide. Prolonged weathering causes copper to become coated with a green film of the hydroxycarbonate $Cu_2(OH)_2CO_3$, which is similar to the mineral malachite. This compound, or the hydroxysulfate, is responsible for the green color of well-weathered copper roofs, gutters, and downspouts. When heated in air, copper oxidizes forming copper(II) oxide along with some copper(I) oxide. Oxidizing acids, and nonoxidizing acids in the presence of air, convert copper to the corresponding copper(II) salts. Aqueous ammonia in the presence of air dissolves copper, producing a blue solution containing $[Cu(NH_3)_4]^{2+}$. Alkali metal hydroxides have no effect on the metal. Sulfur vapor reacts with hot copper to give both Cu_2S and CuS. Hot copper burns in excess chlorine to give $CuCl_2$.

The extensive and varied uses of copper make it second to iron in commercial importance. The chief use of copper is in the production of all types of electrical wiring. Copper is also used in the production of a great many alloys. Among the more important ones are brass (Cu, 60–82%; Zn, 18–40%), bronze (Cu, 70–95%; Zn, 1–25%; Sn, 1–18%), aluminum bronze (Cu, 90–98%; Al, 2–10%), and German silver (Cu, 50–60%; Zn, 20%; Ni, 20–25%). Other alloys containing copper are bell metal, gun metal, silver coin, sterling silver, gold coin, jewelry gold, and jewelry silver.

31.9 Oxidation Numbers of Copper

Copper forms two principal series of compounds, in which it has oxidation numbers of $+1$ and $+2$. When the single electron in the outermost $4s$ subshell of the copper atom is involved in bond formation, the oxidation number of $+1$ is exhibited. When an additional electron from the underlying $3d$ subshell is involved, the copper(II) ion results. When two electrons from the underlying shell are used in bonding, copper exhibits an oxidation number of $+3$; copper(III) compounds are rare, however. Copper(I) forms a complete series of binary compounds, such as the halides and the oxide, but copper(I) oxysalts are few in number and readily decompose in water. In contrast, copper(II) forms a complete series of oxysalts, while some of its binary compounds are unstable and decompose spontaneously.

31.10 Copper(I) Compounds

Solid salts of copper(I) show a variety of colors but in solution are generally colorless.

Copper(I) oxide, Cu_2O, may be prepared as a reddish-brown precipitate by boiling copper(I) chloride with sodium hydroxide.

$$2CuCl + 2OH^- \xrightarrow{\Delta} Cu_2O(s) + 2Cl^- + H_2O$$

When basic solutions of copper(II) salts are heated with reducing agents, copper(I) oxide precipitates.

$$2Cu^{2+} + 2OH^- + 2e^- \text{ (reducing agent)} \xrightarrow{\Delta} Cu_2O(s) + H_2O$$

This reaction is the basis for the test for the presence of reducing sugars in the urine in the diagnosis of diabetes. The reagent most often used for this purpose is Benedict's solution, which is a solution of copper(II) sulfate, sodium carbonate, and sodium citrate. The addition of a reducing sugar, such as glucose, causes the precipitation of the reddish-brown copper(I) oxide.

Copper(I) hydroxide, CuOH, is formed as a yellow precipitate when a cold solution of copper(I) chloride in hydrochloric acid is treated with sodium hydroxide. The hydroxide is unstable and decomposes into the oxide and water when heated.

When copper metal is heated with a solution of copper(II) chloride acidified with hydrochloric acid, an oxidation-reduction reaction takes place with the precipitation of **copper(I) chloride.**

$$Cu^{2+} + 2Cl^- + Cu \longrightarrow 2CuCl(s)$$

This white crystalline compound is readily soluble in concentrated hydrochloric acid, forming the complex ion $[CuCl_2]^-$.

$$CuCl + Cl^- \rightleftharpoons [CuCl_2]^-$$

The two chlorides of the $[CuCl_2]^-$ ion are bonded to the copper by coordinate-covalent bonds. Dilution with water brings about reprecipitation of the copper(I) chloride ($K_{sp} = 1.85 \times 10^{-7}$).

Among other stable and insoluble copper(I) binary salts are Cu_2S (black), $CuCN$ (white), $CuBr$ (white), and CuI (white).

31.11 Copper(II) Compounds

The more common compounds of copper are those in which the metal is bivalent. Solid copper(II) compounds may be blue, black, green, yellow, or white. However, dilute solutions of all soluble copper(II) salts have the blue color of the hydrated copper(II) ion, $[Cu(H_2O)_4]^{2+}$.

Copper(II) oxide, CuO, is a black insoluble compound that may be obtained by heating copper(II) carbonate or copper(II) nitrate or finely divided copper in oxygen.

$$CuCO_3 \xrightarrow{\Delta} CuO + CO_2$$

$$2Cu(NO_3)_2(s) \xrightarrow{\Delta} 2CuO(s) + 4NO_2(g) + O_2(g) \qquad \Delta H° = 420.5 \text{ kJ}$$

$$2Cu(s) + O_2(g) \xrightarrow{\Delta} 2CuO(s) \qquad\qquad\qquad \Delta H° = -157 \text{ kJ}$$

Because of the ease with which copper can be reduced either to an oxidation number of +1 or to the metal, hot copper(II) oxide is a good oxidizing agent. It is used as one in the analysis of organic compounds to determine the percentage of carbon they contain. Copper(II) oxide oxidizes the carbon to carbon dioxide, which is absorbed by sodium hydroxide. The gain in weight of the sodium hydroxide is a measure of the quantity of carbon dioxide produced and absorbed. The copper oxide is continuously regenerated by passing a stream of air through the tube in which the organic matter and the copper oxide are reacting.

When hydroxide ions are added to cold solutions of copper(II) salts, a bluish-green gelatinous precipitate of **copper(II) hydroxide, Cu(OH)$_2$** ($K_{sp} = 5.6 \times 10^{-20}$), is formed. With hot solutions CuO is obtained. Copper(II) hydroxide is somewhat amphoteric and dissolves slightly in solutions of concentrated bases, producing the **tetrahydroxocuprate ion, $[Cu(OH)_4]^{2-}$.** It is also soluble in aqueous ammonia, giving **tetraamminecopper(II) hydroxide, $[Cu(NH_3)_4](OH)_2$,** which is a strong base with a deep blue color due to the color of the $[Cu(NH_3)_4]^{2+}$ ion.

$$Cu(OH)_2(s) + 4NH_3 \longrightarrow [Cu(NH_3)_4]^{2+} + 2OH^-$$

The formation of this color is used as a qualitative test for the presence of the copper(II) ion in solution.

Copper(II) sulfate, CuSO₄, is the most important of the copper salts. The anhydrous salt is colorless, but when it is crystallized from aqueous solution the blue pentahydrate, $[Cu(H_2O)_4]SO_4 \cdot H_2O$, known as **blue vitriol,** is formed. As indicated by the formula, four of the water molecules are coordinated to the copper(II) ion. The fifth is attached to the sulfate ion and two of the coordinated water molecules by hydrogen bonds. The structure of the pentahydrate may be represented as shown below.

The dehydration of this salt proceeds in steps. First, loss of the two nonhydrogen-bonded water molecules yields $CuSO_4 \cdot 3H_2O$. Second, loss of the two other water molecules coordinated to the Cu(II) ion produces $CuSO_4 \cdot H_2O$. Finally, $CuSO_4$ is formed by loss of the water molecule attached to the sulfate ion.

Copper(II) sulfate is produced commercially by oxidizing copper(II) sulfide either directly to the sulfate or to the oxide, which is then converted to the sulfate by reaction with sulfuric acid. In the laboratory the anhydrous sulfate may be prepared by oxidizing copper with hot concentrated sulfuric acid.

$$Cu(s) + 2H_2SO_4(l) \longrightarrow CuSO_4(s) + SO_2(g) + 2H_2O(l) \qquad \Delta H° = -11.9\,kJ$$

Aqueous solutions of copper(II) sulfate are slightly acidic due to hydrolysis.

$$[Cu(H_2O)_4]^{2+} + H_2O \rightleftharpoons [Cu(H_2O)_3OH]^+ + H_3O^+$$

Copper(II) sulfate is used in the electrolytic refining of copper, in electroplating, in the Daniell cell, in the manufacture of pigments, as a color-fixing agent in the textile industry, and in the prevention of algae in reservoirs and swimming pools. Because anhydrous copper(II) sulfate is insoluble in alcohol and ether and readily takes up water, turning blue, this salt is used for detecting water in and removing it from liquids such as alcohol and ether. Of course, copper(II) sulfate cannot be used to remove or detect water in liquids with which it reacts.

Anhydrous **copper(II) chloride, CuCl₂,** may be prepared as a yellow crystalline salt by direct union of the elements. The blue-green hydrated salt $Cu(H_2O)_2Cl_2$ may be obtained by treating either the carbonate or the hydroxide of copper(II) with hydrochloric acid and evaporating the solution. Concentrated solutions of copper(II) chloride are green, because they contain both the hydrated copper(II) ion $[Cu(H_2O)_4]^{2+}$, which is blue, and the **tetrachlorocuprate(II) ion $[CuCl_4]^{2-}$,** which is yellow. (The combination of blue and yellow looks green to the eye.) When such a solution is diluted, water molecules replace the chloride ions of the $[CuCl_4]^{2-}$ ion, and the solution turns blue.

Copper(II) bromide, CuBr₂, is a black solid, which may be obtained by direct union of the elements or by the action of hydrobromic acid on copper(II) oxide or copper(II) carbonate.

It is interesting that copper(II) iodide, CuI_2, does not exist. When a solution containing the copper(II) ion is treated with an excess of iodide, an oxidation-reduction reaction occurs, with precipitation of **copper(I) iodide** and the formation of elemental iodine.

$$2Cu^{2+} + 4I^- \longrightarrow 2CuI(s) + I_2$$

This reaction is used in the quantitative determination of copper by titration of the liberated iodine with a standard solution of sodium thiosulfate. Once the amount of elemental iodine is known, the amount of copper can be calculated.

Copper(II) sulfide, CuS, black, is precipitated ($K_{sp} = 1.6 \times 10^{-48}$) by passing hydrogen sulfide into acid, basic, or neutral solutions of copper(II) salts. The sulfide readily dissolves in warm dilute nitric acid with the formation of elemental sulfur and nitric oxide.

$$3CuS + 8H^+ + 2NO_3^- \rightleftharpoons 3Cu^{2+} + 3S(s) + 2NO(g) + 4H_2O$$

Copper(II) sulfide is only slightly soluble in solutions of alkali metal sulfides.

31.12 Coordination Compounds of Copper

Copper possesses the property of forming both complex cations and anions that are quite stable. Copper(II) ions coordinate with four neutral ammonia molecules to form the deep blue **tetraamminecopper(II) ion.**

$$[Cu(H_2O)_4]^{2+} + 4NH_3 \rightleftharpoons [Cu(NH_3)_4]^{2+} + 4H_2O$$

The fact that the ammonia molecules displace the coordinated water molecules indicates the greater stability of the ammine complex. In fact, the ammine complex is so stable that most slightly soluble copper(II) compounds are dissolved by aqueous ammonia. The copper ion is located at the center of a square formed by the four attached groups, whether they are water, $[Cu(H_2O)_4]^{2+}$, or ammonia, $[Cu(NH_3)_4]^{2+}$. An example of a complex anion of copper(II) is $[CuCl_4]^{2-}$.

Copper(I) forms linear complex ions with ammonia, $[Cu(NH_3)_2]^+$, with the halides, $[CuX_2]^-$, and with cyanide ions, $[Cu(CN)_2]^-$. Unlike copper(II) complexes, most copper(I) complexes are colorless.

CADMIUM

The periodic relationships of cadmium were discussed in Section 30.1 along with those of zinc and mercury.

31.13 Occurrence, Metallurgy, and Uses of Cadmium

Cadmium is found in the rare mineral greenockite, CdS, and in small amounts (less than 1%) in several zinc ores. Most cadmium comes from zinc smelters and from the sludge obtained from the electrolytic refining of zinc.

In the smelting of cadmium-containing zinc ores, the two metals are reduced together. Because cadmium is more volatile (b.p. = 767°C) than zinc (b.p. = 907°C), their separation can be accomplished by fractional distillation.

Separation of the two metals is also possible by selective electrolytic deposition. Cadmium, being less active than zinc, is deposited at a lower voltage.

Cadmium is a silvery, crystalline metal resembling zinc. It is only slightly tarnished by air or water at ordinary temperatures. A large part of the cadmium produced is used in electroplating metals, such as iron and steel, to protect them from corrosion. The electrolyte solution contains tetracyanocadmate ions, $[Cd(CN)_4]^{2-}$, made by mixing cadmium cyanide and sodium cyanide. The equation for the cathodic reduction is

$$[Cd(CN)_4]^{2-} + 2e^- \longrightarrow Cd + 4CN^-$$

Cadmium-plated metals are more resistant to corrosion, more easily soldered, and more attractive in appearance than galvanized (zinc-coated) metals.

Several alloys of cadmium are easily fusible; these include Wood's metal (12.5% Cd), melting at 65.5°C, and Lipowitz alloy (10% cadmium), melting at 70°C. Some antifriction alloys used for bearings contain cadmium. These alloys melt at a higher temperature and have a lower friction coefficient than Babbitt metal (Section 31.20). Amalgamated cadmium, along with cadmium sulfate, is used in the Weston standard cell for measuring electrical potentials. Rods of cadmium are used in nuclear reactors (Section 28.12) to absorb neutrons and thus control the chain reaction.

31.14   Properties of Cadmium and Its Compounds

Cadmium dissolves slowly in either hot, moderately dilute hydrochloric acid or sulfuric acid, with the evolution of hydrogen. It is also dissolved by nitric acid, the oxides of nitrogen being evolved.

Cadmium oxide, CdO, is formed as a brown solid when cadmium burns in air. Alkali metal hydroxides react with cadmium salts to give white cadmium hydroxide, $Cd(OH)_2$, which is soluble in aqueous ammonia as a result of the formation of the soluble tetraammine complex $[Cd(NH_3)_4]^{2+}$. As is expected from the electron configuration of the cadmium ion, the four-coordinated complexes of cadmium, unlike those of copper, are tetrahedral (Section 29.3).

Cadmium chloride, CdCl$_2$, is converted almost completely into $[CdCl_4]^{2-}$ by high concentrations of the chloride ion. **Cadmium sulfide, CdS,** is formed as a bright yellow precipitate ($K_{sp} = 3.6 \times 10^{-29}$) by the action of hydrogen sulfide on solutions of cadmium salts. It is used as a paint pigment called **cadmium yellow.** Cadmium carbonate, phosphate, cyanide, and ferrocyanide are all insoluble in water. All cadmium compounds are soluble in an excess of sodium iodide, due to the formation of the complex ion $[CdI_4]^{2-}$.

DIVISION B—MERCURY, ARSENIC, ANTIMONY, AND TIN

The chemistry of mercury was considered in Chapter 30 with the metals of Analytical Group I. Some properties of the Division B metals are listed in Table 31-2.

Table 31-2 Some Properties of the Division B Metals of Analytical Group II

Property	Mercury	Arsenic	Antimony	Tin
Atomic number	80	33	51	50
Atomic weight	200.59	74.9216	121.75	118.69
Oxidation numbers	+1, +2	+3, +5	+3, +5	+2, +4
Reduction potential, V	(Hg^{2+}/Hg)	(As^{3+}/As)	(Sb^{3+}/Sb)	(Sn^{2+}/Sn)
	+0.854	+0.25	+0.21	−0.136
Density, g/cm³, at 20°C	13.546	5.73	6.68	7.31
Melting point, °C	−38.87	817 (36 atm)	630.5	231.9
Boiling point, °C	356.57	610 (sublimes)	1440	2337

ARSENIC

The periodic relationships of arsenic, a member of Periodic Group VA, have been discussed in Section 31.1.

31.15 Occurrence and Preparation of Arsenic

Arsenic is found both as the free element and combined. Its principal naturally occurring compounds are arsenopyrite, FeAsS, realgar, As_2S_2, and orpiment, As_2S_3. Arsenic is widely distributed in trace amounts in the sulfide ores of many metals; consequently, the metals and the sulfuric acid obtained from these sulfides frequently contain arsenic as an impurity. A large part of the arsenic used in the United States is a by-product from the gases of copper furnaces, from which arsenic(III) oxide is collected by Cottrell precipitators (Section 13.32).

Pure arsenic is obtained by heating arsenic-containing minerals in an inert atmosphere. When arsenopyrite is heated, it decomposes according to the equation

$$4FeAsS(s) \xrightarrow{\Delta} 4FeS(s) + As_4(g) \qquad \Delta H° = -88.7 \text{ kJ}$$

When arsenic ores are roasted in air, As_4O_6 is formed. Reduction of this oxide by heating with carbon produces the element.

$$As_4O_6(s) + 6C(s) \xrightarrow{\Delta} As_4(g) + 6CO(g) \qquad \Delta H° = 795 \text{ kJ}$$

31.16 Properties and Uses of Arsenic

Elemental arsenic exists in several allotropic forms. Ordinary **gray arsenic** is monoatomic (As). It is metallic in appearance and sublimes at 615°C, forming As_4 molecules, which are tetrahedral in structure like P_4 (Section 24.3). When arsenic vapor is cooled rapidly, an unstable yellow crystalline allotrope, which also consists of As_4 molecules, is produced. It is soluble in carbon disulfide. Arsenic vapor is yellow in color, has the odor of garlic, and is very poisonous. Arsenic is a member of an intermediate class of elements that possess some properties of both metals and nonmetals; that is, it is a metalloid (Section 8.3). Arsenic ignites when heated, producing white clouds of As_4O_6. It is slowly attacked by hot hydrochloric acid in the presence of air, forming arsenic(III) chloride, $AsCl_3$. Hot nitric acid readily oxidizes arsenic to arsenic acid, H_3AsO_4. Arsenic combines directly with sulfur, the halogens, and many metals. The trihalides (AsX_3) are either liquids or low-melting solids with properties suggestive of covalent rather than ionic bonding (for example, the absence of saltlike character).

Certain alloys contain arsenic as a hardening agent. Examples are bronze and lead shot.

31.17 Compounds of Arsenic

Arsine, AsH_3, a compound with a formula analogous to that of ammonia (NH_3), is formed by reducing arsenic compounds with zinc in acidic solution or by the reaction of a metallic arsenide such as zinc arsenide with acid.

$$As_4O_6 + 12Zn + 24H^+ \longrightarrow 4AsH_3(g) + 12Zn^{2+} + 6H_2O$$
$$Zn_3As_2 + 6H^+ \longrightarrow 3Zn^{2+} + 2AsH_3(g)$$

Arsine is not basic. It does not react with or dissolve in water or acids to form compounds analogous to ammonium compounds. It is unstable and decomposes to its elements when heated.

$$4AsH_3 \xrightarrow{\Delta} As_4 + 6H_2$$

In the Marsh test for arsenic (Fig. 31-1), the element is deposited as a black mirror on a cold glazed porcelain dish held in a flame in which arsine is being burned. This test is very sensitive and has been employed in the detection of arsenic poisoning.

Figure 31-1. Laboratory apparatus used in the Marsh test for arsenic.

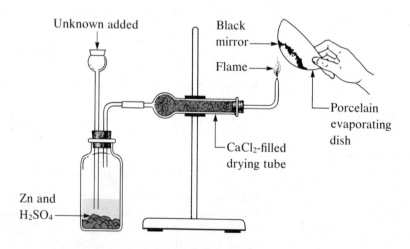

Unknown added

Black mirror

Flame

Porcelain evaporating dish

CaCl$_2$-filled drying tube

Zn and H$_2$SO$_4$

The substance commonly referred to as arsenic, or white arsenic, is actually **arsenic(III) oxide, As_4O_6.** It is the product of the combustion of the element or its compounds. It has a sweet taste and is a deadly poison. Arsenic(III) oxide is slowly dissolved by water with the formation of **arsenous acid, H_3AsO_3,** whose structure is not clearly known.

$$As_4O_6 + 6H_2O \rightleftharpoons 4H_3AsO_3 \quad [\text{or As(OH)}_3]$$

Arsenous acid is amphoteric as shown by the equations

$$As(OH)_3 \rightleftharpoons As(OH)_2{}^+ + OH^- \rightleftharpoons AsOH^{2+} + 2OH^- \rightleftharpoons As^{3+} + 3OH^-$$
$$H_3AsO_3 \rightleftharpoons H^+ + H_2AsO_3{}^-$$

Hence it acts as a base in the presence of strong acids and as an acid in the presence of strong bases.

$$As(OH)_3 + 3H^+ + 3Cl^- \rightleftharpoons AsCl_3 + 3H_2O$$
$$H_3AsO_3 + [Na^+] + OH^- \longrightarrow [Na^+] + H_2AsO_3{}^- + H_2O$$

Arsenous acid is known only in solution; when attempts are made to isolate it as a solid, it loses water and forms arsenic(III) oxide. All arsenites (salts of arsenous acid) except those of the alkali metals are insoluble in water.

Arsenic(V) oxide, which has the empirical formula As_2O_5 (the structure is not known), can be produced by heating **arsenic acid, H_3AsO_4.**

$$2H_3AsO_4 \xrightarrow{\Delta} As_2O_5 + 3H_2O$$

Arsenic acid is formed when arsenic(III) oxide is oxidized by concentrated nitric acid.

$$As_4O_6 + 8HNO_3 + 2H_2O \longrightarrow 4H_3AsO_4 + 8NO_2(g)$$

By careful heating, H_3AsO_4 may be converted by steps into **pyroarsenic acid, $H_4As_2O_7$,** then **meta-arsenic acid, $HAsO_3$,** and finally **arsenic(V) oxide, As_2O_5.**

Arsenic acid is a much stronger acid than arsenous acid, as is expected from the higher oxidation number of arsenic (see Section 14.4). It does not react as a base with strong acids as does arsenous acid. Arsenic acid is rather easily reduced, so that it can be used as an oxidizing agent. The arsenates (salts of arsenic acid) of calcium and lead are used as insecticides.

When hydrogen sulfide is passed into a hydrochloric acid solution of arsenic(III) chloride, yellow **arsenic(III) sulfide, As_2S_3,** is precipitated.

$$2AsCl_3 + 3H_2S \longrightarrow As_2S_3(s) + 6H^+ + 6Cl^-$$

Arsenic(III) sulfide dissolves readily in sodium sulfide, forming **sodium thioarsenite, Na_3AsS_3.**

$$As_2S_3 + 3S^{2-} \longrightarrow 2AsS_3{}^{3-}$$

Acids reprecipitate arsenic(III) sulfide from thioarsenite solutions.

$$2AsS_3{}^{3-} + 6H^+ \longrightarrow As_2S_3(s) + 3H_2S(g)$$

Arsenic(V) sulfide, As_2S_5, is a yellow compound that precipitates very slowly when hydrogen sulfide is passed through a solution of arsenic acid in hydrochloric acid.

$$2H_3AsO_4 + 5H_2S \longrightarrow As_2S_5(s) + 8H_2O$$

The arsenic acid oxidizes hydrogen sulfide in cold, weakly acidic solutions.

$$2H_3AsO_4 + 5H_2S \longrightarrow As_2S_3(s) + 2S(s) + 8H_2O$$

Iodide ions are a catalyst in the precipitation of arsenic(III) sulfide from solutions of the arsenate ion.

$$2I^- + AsO_4^{3-} + 8H^+ \longrightarrow As^{3+} + 4H_2O + I_2$$

$$I_2 + H_2S \longrightarrow 2H^+ + 2I^- + S(s)$$

$$2As^{3+} + 3H_2S \longrightarrow As_2S_3(s) + 6H^+$$

Arsenic(V) sulfide dissolves in sodium sulfide with the formation of **sodium thioarsenate, Na$_3$AsS$_4$.** The addition of an acid reprecipitates the As$_2$S$_5$.

31.18 Uses of Arsenic Compounds

The most important uses of arsenic compounds depend on their poisonous character. They are used as weed killers, cattle and sheep dips, and insecticides. The compounds most widely used for these purposes are Paris green [Cu$_3$(AsO$_3$)$_2 \cdot$ Cu(C$_2$H$_3$O$_2$)$_2$], lead arsenate, calcium arsenate, sodium arsenite, and arsenous acid.

ANTIMONY

31.19 History, Occurrence, and Metallurgy of Antimony

Antimony has been used, both as a metal and in compounds, for at least 5000 years. The Chaldeans used the metal in making vases and ornaments. The sulfide Sb$_2$S$_3$ was used by the Egyptians as a pigment and for painting eyebrows at least as early as 3000 B.C. Compounds of antimony were also used in medicine.

The principal ore of antimony is stibnite, Sb$_2$S$_3$, most of which is mined in China, Mexico, Argentina, Bolivia, and Chile. The sulfide is separated from earthy material by melting the ore; Sb$_2$S$_3$ is then reduced by heating with iron.

$$Sb_2S_3(s) + 3Fe(s) \xrightarrow{\Delta} 2Sb(s) + 3FeS(s) \qquad \Delta H° = -125 \text{ kJ}$$

The ore may also be roasted to the oxide, from which antimony may be obtained by reduction with carbon.

$$Sb_4O_6(s) + 6C(s) \xrightarrow{\Delta} 4Sb(s) + 6CO(g) \qquad \Delta H° = 777.4 \text{ kJ}$$

31.20 Properties and Uses of Antimony

Antimony is a lustrous, silver-white, brittle metal that is readily pulverized. It tarnishes only slightly in dry air but does oxidize slowly in moist air. The metal reacts readily with the halogens, phosphorus, and sulfur. It is oxidized, but not dissolved, by hot nitric acid, forming Sb$_4$O$_6$. It is slowly dissolved by hot concentrated sulfuric acid, forming Sb$_2$(SO$_4$)$_3$ and evolving SO$_2$. The fact that the metal and its oxides do not react with nitric acid to give the nitrate indicates that antimony is not very metallic. However, the metallic nature of the element is more pronounced than that of arsenic, which lies immediately above it in Group VA.

Lead hardened by the addition of 10–20% of antimony is used for shrapnel, bullets, and bearings. Because of its resistance to corrosion by acids, this alloy is used in making storage battery plates. **Babbitt metal** is an antifriction alloy of antimony with tin and copper, extensively used in machine bearings.

31.21   Compounds of Antimony

Stibine, SbH₃, is prepared by reducing other antimony compounds with zinc in acidic solution. Stibine is poisonous and unstable toward heat. When heated in the absence of air, it decomposes, The antimony, if in contact with a cold surface, will be deposited as a black mirror, as is arsenic when arsine is burned.

Antimony(III) oxide, Sb₄O₆, and **antimony(V) oxide, Sb₂O₅,** resemble the oxides of arsenic in their preparation and properties. Both antimony oxides form acids with water, and antimony(III) oxide exhibits some basic properties in forming salts with acids. The antimony in **antimonic acid, HSb(OH)₆,** has a coordination number of 6. **Sodium antimonate, NaSb(OH)₆,** is one of the very few relatively insoluble salts of sodium. Sb_2O_4 is also known; it is formed when either Sb_4O_6 or Sb_2O_5 is heated in air.

Antimony forms all of the **trihalides** and **pentahalides** except the pentabromide and the pentaiodide. The trihalides and pentahalides are prepared either by direct union of the elements or by the action of the hydrohalic acids on antimony oxides. Hydrolysis of antimony(III) chloride forms **antimony oxychloride,** which is white and insoluble.

$$SbCl_3 + H_2O \rightleftharpoons SbOCl(s) + 2H^+ + 2Cl^-$$

An excess of hydrochloric acid changes the oxychloride to **tetrachloroantimonous acid, HSbCl₄.**

Antimony(III) sulfide, Sb₂S₃, occurs in two strikingly different allotropic forms (see Section 9.10). A black form, found in nature as the mineral stibnite, forms when antimony and sulfur are heated together. A brilliant orange-red form is obtained by the reaction of hydrogen sulfide with slightly acidified solutions of antimony(III) compounds. The orange-red modification changes to the black one when heated or after standing in a dilute solution of an acid.

Antimony(III) sulfide dissolves in sodium sulfide solutions, forming **thioantimonite ions.**

$$Sb_2S_3(s) + 3S^{2-} \longrightarrow 2SbS_3^{3-}$$

Acidification of the solution causes Sb_2S_3 to reprecipitate.

$$2SbS_3^{3-} + 6H^+ \longrightarrow Sb_2S_3(s) + 3H_2S(g)$$

Excess hydrochloric acid dissolves antimony(III) sulfide with the formation of the **tetrachloroantimonate(III) ion, SbCl₄⁻,** and hydrogen sulfide.

$$Sb_2S_3(s) + 6H^+ + 8Cl^- \longrightarrow 2SbCl_4^- + 3H_2S(g)$$

Antimony(V) sulfide, Sb₂S₅, is orange and is similar to Sb_2S_3 in many of its reactions. Treatment of antimony(V) sulfide with an excess of hydrochloric acid causes reduction of the antimony to antimony(III).

$$Sb_2S_5(s) + 6H^+ + 8Cl^- \longrightarrow 2SbCl_4^- + 3H_2S(g) + 2S(s)$$

TIN

31.22   History, Occurrence, and Metallurgy of Tin

Tin found in ancient Egyptian tombs is evidence that the metal was known during the very early periods of history. Tin was obtained from ore deposits in England by the Romans and Phoenicians.

The most important and abundant ore of tin is called cassiterite, or tinstone, SnO_2. No important deposits have been found in the United States; but more than half of the tin produced in the world is used in this country.

Tin ore is crushed and washed with water to separate the lighter rocky material from the heavier ore. Roasting of the ore removes arsenic and sulfur as volatile oxides. Oxides of other metals are extracted by means of hydrochloric acid. The purified ore is reduced by carbon.

$$SnO_2(s) + 2C(s) \longrightarrow Sn(s) + 2CO(g) \qquad \Delta H^\circ = 360 \, kJ$$

The molten tin, which collects on the bottom of the furnace, is drawn off and cast into blocks (block tin). The crude tin is remelted and permitted to flow away from the higher-melting impurities, which are chiefly compounds of iron and arsenic. Further purification of the tin is accomplished by electrolysis, using impure tin anodes, pure tin cathodes, and an electrolyte of fluorosilicic acid (H_2SiF_6) and sulfuric acid.

31.23   Properties and Uses of Tin

Tin exists as three allotropic solids (see Section 9.10). They are **gray tin** (cubic crystals), **malleable tin** (tetragonal crystals), and **brittle tin** (rhombic crystals). The malleable form is silvery white with a bluish tinge. When a rod of it is bent, the crystals slip over one another, producing a sound described as tin cry. When malleable tin, or white tin, is heated, it changes to the brittle form. White tin changes slowly at low temperatures (below 13.2°C) into gray tin, which is powdery. Consequently, articles made of tin are likely to disintegrate in cold weather, particularly if the cold spell is lengthy. The change progresses slowly from a spot of origin; the gray tin that is formed catalyzes further change. This effect is, in a way, similar to the spread of an infection in a plant or animal body. For this reason, it is called tin disease or tin pest.

The metal dissolves slowly in cold dilute hydrochloric acid and rapidly in hot concentrated acid, $SnCl_2$ and H_2 being produced. Very dilute cold nitric acid converts tin to tin(II) nitrate. Concentrated nitric acid rapidly converts tin into insoluble, hydrated **metastannic acid, H_2SnO_3.** Alkali metal hydroxides dissolve tin, forming stannites and hydrogen.

$$Sn + 2OH^- + 2H_2O \longrightarrow [Sn(OH)_4]^{2-} + H_2(g)$$

The reactions with concentrated nitric acid and sodium hydroxide show that tin is not entirely metallic in character. Dry chlorine oxidizes tin to **tin(IV) chloride, $SnCl_4$,** and oxygen converts it to **tin(IV) oxide, SnO_2.** Tin burns with a white flame.

The principal use of tin is in the electrolytic production of **tin plate,** sheet iron coated with tin to protect it from corrosion. However, when a scratch through the tin exposes the iron, corrosion sets in rapidly. Iron is more active than tin and corrodes more rapidly when in contact with tin than otherwise. This is because iron and tin in contact with each other, and both in contact with moist air, comprise an electrolytic cell that promotes corrosion (Section 32.18). Tin is also used in making alloys, such as bronze (Cu, Zn, and Sn), solder (Sn and Pb), and type metal (Sn, Pb, Sb, and Cu).

31.24 Compounds of Tin

Depending on the method of preparation, **tin(II) oxide (stannous oxide), SnO,** is a black or green powder. It may be prepared by treating a hot solution of a tin(II) compound with an alkali metal carbonate or by heating tin(II) oxalate in the absence of air. The equations are

$$Sn^{2+} + CO_3^{2-} \longrightarrow SnO(s) + CO_2(g)$$

$$SnC_2O_4(s) \longrightarrow SnO(s) + CO_2(g) + CO(g)$$

When tin is burned in air, **tin(IV) oxide (stannic oxide), SnO$_2$,** is formed. The product is white when cold and yellow when hot.

Soluble bases precipitate tin(II) ions from solution as **tin(II) hydroxide, Sn(OH)$_2$** ($K_{sp} = 5 \times 10^{-26}$), which is white and gelatinous. An excess of base causes the hydroxide to dissolve, with the formation of the soluble **stannite ion.**

$$Sn(OH)_2(s) + OH^- \longrightarrow Sn(OH)_3^-$$

The stannite ion is an active reducing agent. It is used in qualitative analysis to test for bismuth because it reduces white bismuth hydroxide to black metallic bismuth.

$$2Bi(OH)_3 + 3Sn(OH)_3^- + 3OH^- \longrightarrow 2Bi + 3Sn(OH)_6^{2-}$$

Tin(II) chloride can be obtained as the dihydrate (SnCl$_2 \cdot$ 2H$_2$O) by evaporating a solution formed by the reaction of tin or tin(II) oxide with hydrochloric acid. The salt will hydrolyze in water, forming the hydroxychloride Sn(OH)Cl.

$$SnCl_2 + H_2O \rightleftharpoons Sn(OH)Cl(s) + H^+ + Cl^-$$

Tin(II) chloride is widely used as a reducing agent, because of the ease with which it is oxidized to tin(IV) compounds. It follows that aqueous solutions of tin(II) compounds must be protected against air oxidation.

Tin(IV) chloride is formed when an excess of chlorine reacts with metallic tin. It is a colorless liquid that is soluble in organic solvents such as carbon tetrachloride and is a nonconductor of electricity. These properties are typical of covalent compounds. When dissolved in water, the tetrachloride hydrolyzes (Section 31.25), but in hydrochloric acid, **hexachlorostannic acid** is produced.

$$SnCl_4 + 2HCl \longrightarrow H_2SnCl_6$$

Dark brown **tin(II) sulfide, SnS,** is precipitated ($K_{sp} = 8 \times 10^{-29}$) when hydrogen sulfide is passed into a solution of a tin(II) salt. Tin(II) sulfide is not dissolved by alkali metal sulfides or by alkaline polysulfides.

Tin(IV) sulfide is a yellow precipitate prepared by the reaction of hydrogen sulfide with tin(IV) ions in a moderately acid solution. Unlike tin(II) sulfide it dissolves in alkali metal sulfides with the formation of **thiostannate ions, SnS$_3^{2-}$.** The addition of acid reprecipitates SnS$_2$.

$$SnS_3^{2-} + 2H^+ \longrightarrow SnS_2(s) + H_2S(g)$$

Concentrated hydrochloric acid dissolves tin(IV) sulfide.

$$SnS_2(s) + 4H^+ + 6Cl^- \longrightarrow SnCl_6^{2-} + 2H_2S(g)$$

When an alkali metal hydroxide is added to a solution of a tin(IV) compound, a white precipitate forms. We would expect this compound to have the formula $Sn(OH)_4$ and to be acidic in character. However, neither this acid nor its salts have been isolated. On the other hand, alkali metal stannates are known that appear to be derived from an **orthostannic acid,** whose formula may be written as $H_2[Sn(OH)_6]$. The sodium salt, for example, may be written as $Na_2[Sn(OH)_6]$. Orthostannic acid readily loses water, yielding a **metastannic acid, H_2SnO_3,** and finally the anhydride SnO_2. When this anhydride is melted with sodium hydroxide, **sodium metastannate, Na_2SnO_3,** is formed. The metastannic acid is sometimes called **alpha-stannic acid** to distinguish it from **beta-stannic acid,** which is formed when hot concentrated nitric acid reacts with tin. Beta-stannic acid has the same composition as the alpha form, but it is insoluble in acids and only very slightly soluble in strongly alkaline solutions.

31.25 Hydrolysis of Covalent Halides

Except for the tetrahalides of carbon, the tetrahalides of the Group IVA elements hydrolyze when dissolved in water, for the most part as indicated by the general equation

$$MX_4 + 2H_2O \longrightarrow MO_2(s) \text{ (or a hydrate)} + 4HX$$

The mechanism by which the tetrahalides of silicon, germanium, tin, and lead hydrolyze may be as follows. Silicon tetrachloride is used as an example. The silicon atom in the $SiCl_4$ molecule carries a partial net positive charge because it is less electronegative than chlorine. During hydrolysis the positively charged silicon atom may attract hydroxide ions of water, forming a pentacovalent intermediate, according to the equation

One of the highly electronegative chlorine atoms is then easily lost from the intermediate as a chloride ion.

This process is repeated until all four chlorine atoms are replaced by hydroxide ions, $Si(OH)_4$ being the final product. The conversion of the hydroxide ions of water to coordinated hydroxy groups leaves an excess of hydrogen ions in solution, and hydrochloric acid is formed. Resistance of the carbon tetrahalides to hydrolysis is probably due to the fact that carbon is **coordinately saturated** (all the orbitals in its valence shell are filled with electrons) in these compounds, and it cannot form intermediates of the type suggested for silicon tetrachloride. Other halides such as sulfur hexafluoride, SF_6, do not hydrolyze when the central element is already coordinately saturated. On the other hand, tungsten hexachloride,

WCl$_6$, hydrolyzes, because the maximum coordination number for tungsten is 8 and it is therefore not coordinately saturated.

The silicon halides undergo complete hydrolysis; halides of more ionic character undergo less complete hydrolysis that may be suppressed by the addition of acids. Thus aqueous solutions of tin(IV) chloride can be prepared in the presence of hydrochloric acid.

FOR REVIEW

SUMMARY

Those metals whose chlorides are soluble in hydrochloric acid but whose sulfides are precipitated by hydrogen sulfide in 0.3 M hydrochloric acid comprise Group II of the analytical scheme. They are lead, mercury(II), bismuth, copper, cadmium, arsenic, antimony, and tin. Lead is found in Analytical Group II as well as in Analytical Group I because of the relatively high solubility of lead chloride. Mercury occurs in both groups because mercury(I) chloride is insoluble in dilute hydrochloric acid while mercury(II) chloride is soluble. The chemistry of lead and that of mercury were discussed in Chapter 30.

The ions Pb^{2+}, Bi^{3+}, Cu^{2+}, and Cd^{2+} comprise Division A of Analytical Group II and form sulfides that are insoluble in sodium sulfide solutions. Hence ions of Division A can be separated from those of Division B, Hg^{2+}, As^{3+}, Sb^{3+}, and Sn^{4+}, which are soluble in water containing an excess of sodium sulfide.

The metals of Analytical Group II form an interesting variety of compounds. The oxides and hydroxides of Bi^{3+}, Cu^{2+}, Hg^{2+}, and Cd^{2+} are basic, while those of Pb^{2+}, As^{3+}, Sb^{3+}, and Sn^{2+} can exhibit either acidic or basic properties depending on conditions.

Tin exists in three solid allotropic forms, gray tin (cubic crystals), malleable tin (tetragonal crystals), and brittle tin (rhombic crystals).

The elements of Analytical Group II all readily form alloys with a number of other metals. These alloys have a variety of useful properties.

The tetrahalides of elements of Periodic Group IVA, except those of carbon, hydrolyze when dissolved in water. A mechanism based on the occurrence of hydrolysis when the central element is not coordinately saturated can be formulated.

EXERCISES

Bismuth

1. What are the properties of metallic bismuth that make it commercially useful?
2. Compare the basicity and acidity of Bi_2O_3 and Bi_2O_5.
3. Write equations for the hydrolysis of $BiCl_3$.
4. Write balanced equations for the following reactions:
   (a) Bismuth is heated in air.

(b) Bismuth is added to hot concentrated sulfuric acid.

(c) BiH_3 is burned in air.

(d) Bi_2O_3 is dissolved in a concentrated solution of lithium hydroxide.

(e) Bi_2S_3 is roasted in air.

Copper

5. Describe the electrolytic refining process for metallic copper.

6. Copper plate makes up the surface of the Statue of Liberty in New York harbor. Why does it appear green?

7. In terms of electronic structure, explain the three oxidation numbers for copper.

8. Why must copper be quite free of impurities when used as an electrical conductor?

9. Explain the use of copper(II) sulfate in urine analysis.

10. Compare the stability of copper(I) halides to that of copper(II) halides.

11. A white precipitate forms when copper metal is added to a solution of copper(II) chloride and hydrochloric acid. The white precipitate dissolves in excess concentrated hydrochloric acid. Dilution with water causes the white precipitate to reappear. Write equations for the reactions involved.

12. How many grams of $CuCl_2$ contain the same mass of copper as in 100 g of $CuCl$?

Ans. 136 g

13. Why are most slightly soluble copper salts readily dissolved by aqueous ammonia?

14. The formation constant of $[Cu(NH_3)_4]^{2+}$ is 1.2×10^{12}. What will be the equilibrium concentration of Cu^{2+} if 1.0 g of Cu is oxidized and put into 1.0 L of 0.25 M NH_3 solution?

Ans. 1.1×10^{-11} M

15. Write an equation to explain the acidity of copper(II) sulfate solutions.

16. Sketch the structure of $Cu(H_2O)_4SO_4 \cdot H_2O$.

17. A solution made by dissolving anhydrous copper sulfate in pure water has a pH less than 7.0. Explain.

18. Would you expect CuS to dissolve in a 1 M NH_3 solution?

19. A 1.008-g sample of a silver-copper alloy is dissolved and treated with excess iodide ion. The liberated I_2 is titrated with a 0.1052 M $S_2O_3^{2-}$ solution. If 29.84 mL of this solution is required, what is the percentage of Cu in the alloy?

Ans. 19.79%

Cadmium

20. Why is cadmium chloride considered to be a weak electrolyte?

21. What property makes cadmium effective as a protective coating for other metals?

22. How many pounds of cadmium can be obtained from 10.0 ton of ore that is 5.0% greenockite?

Ans. 7.8×10^2 lb

23. Predict the favored direction of the reaction shown by the equation

$$[Cd(NH_3)_4]^{2+} + Cu^{2+} \Longrightarrow$$
$$Cd^{2+} + [Cu(NH_3)_4]^{2+}$$

24. Calculate the emf of the following electrochemical cell:

$$Cd \,|\, Cd^{2+}, M = 0.10 \,\|\, Ni^{2+}, M = 0.50 \,|\, Ni$$

Ans. 0.17 V

25. How many grams of CdS will precipitate when a solution obtained by dissolving 2.5 g of Wood's metal is treated with excess sulfide ion?

Ans. 0.40 g

Arsenic

26. Arsenic is classed as a metalloid. Explain.

27. Compare the metallic character of arsenic with that of antimony.

28. Account for the oxidation numbers of +3 and +5 for arsenic in terms of electronic structure.

29. Compare and contrast arsine with ammonia.

30. Which is more acidic, As_4O_6 or As_2O_5?

31. Draw molecular structures for pyroarsenic acid and meta-arsenic acid.

32. Write balanced equations to show the reaction of Na_2S with As_2S_3 and with As_2S_5.

33. What are the composition and use of Paris green?

34. Write balanced chemical equations describing the reaction of hot nitric acid and elemental arse-

nic; the reaction of hot concentrated sulfuric acid and elemental antimony; the oxidation of manganous ion to permanganate ion by a solution made by acidifying sodium bismuthate with nitric acid.

35. Why can it be said that iodide ions serve as a catalyst in the precipitation of arsenic(III) sulfide from arsenate ion?

36. Write balanced chemical equations for the following reactions:
    (a) Arsenic is burned in air.
    (b) Arsenic is added to a hot solution of hydrochloric acid.
    (c) As_2S_3 is roasted in air.
    (d) $AsCl_3$ is added to a solution of sodium sulfide.

Antimony

37. Can the term *amphoteric* be applied properly to the sulfides of antimony and arsenic? Explain why or why not.

38. Why is antimony used as a constituent of type metal?

39. Why is antimony insoluble in hot nitric acid?

40. What is the coordination number of antimony in the antimonate ion?

41. Write balanced chemical equations for the following reactions:
    (a) Sb_2S_3 is heated in air.
    (b) Sb_2S_3 is added to an excess of hot sulfuric acid.

Tin

42. Mercury(I) solutions can be protected from air oxidation to mercury(II) by keeping them in contact with metallic mercury. Can tin(II) solutions be similarly stabilized against oxidation to tin(IV) by keeping them in contact with metallic tin? [The standard reduction potential for tin(IV) to tin(II) is $+0.15$ V.]

43. How and why is tin plate applied to the surface of iron?

44. What is *tin pest,* also known as *tin disease?*

45. Does metallic tin react with HCl? with HNO_3? with NaOH?

46. Write balanced chemical equations describing:
    (a) the preparation of sodium metastannate
    (b) the purification of tin by electrolysis
    (c) the reaction of dry chlorine with tin
    (d) the formation of thiostannate ion
    (e) the dissolution of tin by basic solution
    (f) the reactions involved when an excess of aqueous NaOH is slowly added to a solution of tin(II) chloride
    (g) the thermal decomposition of tin(II) oxalate

47. Why is $SnCl_4$ not classified as a salt?

48. Why must aqueous solutions of $SnCl_2$ be protected from the air?

49. A 1.497-g sample of type metal is dissolved in nitric acid, and metastannic acid precipitates. This is dehydrated by heating to stannic oxide, which is found to weigh 0.4909 g. What percent tin was the original type metal sample?
    Ans. 25.83%

32

METALS OF ANALYTICAL GROUP III

NICKEL, COBALT, MANGANESE, IRON, ALUMINUM, CHROMIUM, ZINC

Analytical Group III (see also Chapter 38) consists of those common cations that are precipitated as sulfides or hydroxides in solutions of aqueous ammonia saturated with hydrogen sulfide, but which are not precipitated by hydrochloric acid or by hydrogen sulfide in acidic solutions (0.3 M in hydrogen ions). The sulfides of cobalt(II), nickel(II), manganese(II), iron(II), and zinc and the hydroxides of chromium(III) and aluminum precipitate in ammonium sulfide solutions.

Five of the metals of Analytical Group III (Cr, Mn, Fe, Co, and Ni) are members of the first transition series (Period 4) of the Periodic Table. This series begins with scandium and ends with copper; in it the number of electrons in the shell underlying the valence shell is being increased from 9 to 18. These elements have valence electrons in two shells; associated with this property are such characteristics as variation in oxidation number, color of the ions, and tendency toward the formation of stable complex ions. Successive members of each of the transition series show striking resemblances to each other. Atomic and ionic sizes are important in determining the properties of elements with similar electronic structures. Note that the following elements of Period 4 (and Analytical Group III) have atoms of very nearly the same radius (in Å): Cr, 1.26; Mn, 1.29; Fe, 1.26; Co, 1.26; Ni, 1.24. Zinc and aluminum are representative metals (valence electrons in one shell) with atomic radii of 1.33 Å and 1.43 Å, respectively. Some properties of the Analytical Group III metals are given in Table 32-1.

Table 32-1   Some Properties of the Metals of Analytical Group III

Property	Aluminum	Chromium	Manganese	Iron	Cobalt	Nickel	Zinc
Atomic number	13	24	25	26	27	28	30
Atomic weight	26.98154	51.996	54.9380	55.847	58.9332	58.70	65.38
Oxidation numbers	+3	+2	+2	+2	+2	+2	+2
		+3	+3	+3	+3	+3	
		+6	+4	+6	+4	+4	
			+6				
			+7				
Reduction potential, V	$\left(\dfrac{Al^{3+}}{Al}\right)$ -1.66	$\left(\dfrac{Cr^{3+}}{Cr}\right)$ -0.74	$\left(\dfrac{Mn^{2+}}{Mn}\right)$ -1.18	$\left(\dfrac{Fe^{2+}}{Fe}\right)$ -0.440	$\left(\dfrac{Co^{2+}}{Co}\right)$ -0.277	$\left(\dfrac{Ni^{2+}}{Ni}\right)$ -0.250	$\left(\dfrac{Zn^{2+}}{Zn}\right)$ -0.763
Density, g/cm³, at 20°C	2.70	7.1	7.20	7.86	8.71	8.9	7.14
Melting point, °C	660	1900	1244	1535	1493	1455	419.5
Boiling point, °C	2327	2642	2087	2800	3100	2800	907

NICKEL

Iron, cobalt, and nickel comprise the first horizontal triad of Group VIII of the Periodic Table. The second triad consists of ruthenium, rhodium, and palladium, and the third, of osmium, iridium, and platinum. In each of these horizontal series, the tendency to lose electrons decreases as the nuclear charge increases. Thus iron exhibits oxidation numbers of +2, +3, and +6, and cobalt, +2, +3, and +4. For nickel, oxidation numbers higher than +2 do exist (NiO_2 in the Edison cell, described in Section 20.17, for example) but are rare.

32.1 History, Occurrence, and Metallurgy of Nickel

Because nickel ores are difficult to reduce, early attempts to produce the metal from its ores were unsuccessful. Seventeenth-century metallurgists thought that nickel ores were copper ores because of the similarity in appearance; but when the ores failed to yield copper, they were called *Kupfernickel* (German). *Kupfer* means copper, and *nickel* means a devil, which was thought to prevent the extraction of copper from the ores. The metal was first obtained by Baron Axel Fredric Cronstedt, a Swedish chemist, in 1751.

Nickel occurs as an alloy with iron in meteorites. Metallic nickel and iron probably constitute most of the core of the earth. The most important ores of nickel are pentlandite, $(Ni, Cu, Fe)_9S_8$, and garnierite, which is a hydrated magnesium-nickel silicate of variable composition. Pentlandite is roasted and then reduced with carbon. This process results in the production of an alloy containing nickel, iron, and copper and known as Monel metal. Because of its resistance to corrosion, it has many important industrial uses. The process now used most extensively in the production of pure metallic nickel involves the separation of the sulfides of nickel, copper, and iron by selective flotation. After separation the nickel sulfide is converted to the oxide by roasting, and the oxide is reduced by carbon. The metal obtained in this way is approximately 96% pure. It is purified electrolytically to 99.98% purity. Gold, silver, and especially platinum are recovered from the anode mud.

32.2 Properties and Uses of Nickel

Nickel is a silvery white metal that is hard, malleable, and ductile. Like iron and cobalt, it is highly magnetic. It is not oxidized by air under ordinary conditions, and it is resistant to the action of bases. Dilute acids slowly dissolve nickel; hydrogen is evolved. Nickel can be made unreactive by treatment with concentrated nitric acid; nickel so treated will no longer displace hydrogen from dilute acids.

Because of its hardness, resistance to corrosion, and high reflectivity when polished, nickel is widely used in the plating of iron, steel, and copper. It is also a constituent of many important alloys, such as **Monel metal** (Ni, Cu, and a little Fe), and **Permalloy** (Ni and Fe), used in instruments for the electrical transmission and reproduction of sound. **German silver** is a nickel-zinc-copper alloy. **Nichrome** and **chromel,** alloys containing nickel, iron, and chromium, are resistant to oxidation at high temperatures and show high electrical resistance. They are used in electrical heating units such as electric stoves, pressing irons, and toasters. **Alnico** contains aluminum, nickel, iron, and cobalt; it is highly magnetic, able to lift 4000 times its own weight of iron. **Platinite** and **invar** are nickel alloys that have the same coefficient of expansion as that of glass and are therefore used for "seal-in" wires through glass, such as in electric light bulbs. Finely divided nickel is used as a catalyst in the hydrogenation of oils.

32.3 Compounds of Nickel

In its compounds nickel exhibits an oxidation number of +2 almost exclusively. One notable exception is nickel(IV) in the dioxide NiO_2, a black hydrous substance formed by the oxidation of nickel(II) salts in alkaline solution. This oxide forms one of the electrodes in the Edison storage cell (Section 20.17).

Nickel(II) oxide, NiO, is prepared by heating nickel(II) hydroxide, carbonate, or nitrate. When alkali metal hydroxides are added to solutions of nickel(II) salts, pale green **nickel(II) hydroxide, $Ni(OH)_2$,** precipitates ($K_{sp} = 1.6 \times 10^{-14}$). Both the oxide and the hydroxide are dissolved by aqueous ammonia with the formation of deep blue **hexaamminenickel(II) hydroxide,** $[Ni(NH_3)_6](OH)_2$.

$$Ni(OH)_2(s) + 6NH_3 \longrightarrow [Ni(NH_3)_6]^{2+} + 2OH^-$$

The soluble salts of nickel are prepared from nickel(II) oxide, hydroxide, or carbonate by treatment with the proper acid. The important soluble salts of the metal are the acetate, chloride, nitrate, and sulfate, and also the hexaamminenickel(II) sulfate. Ammoniacal solutions of the latter compound are used in nickel-plating baths. The hydrated nickel(II) ion, $[Ni(H_2O)_6]^{2+}$, imparts a pale green color to the solutions and crystallized salts of the nickel(II) ion.

Nickel(II) sulfide is produced as a black precipitate by the action of ammoniacal sulfide solutions on nickel(II) salts.

Nickel carbonyl, $Ni(CO)_4$, is formed as a colorless volatile liquid (boiling at 43°C) when carbon monoxide is passed over finely divided nickel. Nickel carbonyl readily decomposes at higher temperatures and deposits pure nickel metal. The **Mond process** for separating nickel from other metals is based on the formation and decomposition of nickel carbonyl. Carbon monoxide is passed over a mixture of nickel and other metals; nickel carbonyl is formed and carried along with the excess carbon monoxide, leaving other metals behind. When heated to 200°C, the nickel carbonyl decomposes, and the metal deposits as a fine dust.

COBALT

32.4   History, Occurrence, and Metallurgy of Cobalt

The name *cobalt* is derived from the German word *Kobold*, which means goblin. The reason for this was that early metallurgists believed that goblins carefully guarded the cobalt ores and prevented human beings from liberating the metal. Georg Brandt finally succeeded in isolating the metal in 1742. Even today, the metallurgy of cobalt (and of nickel) is difficult and complicated.

The common cobalt minerals are cobaltite, CoAsS, and smaltite, $CoAs_2$. The richest deposits of these minerals are found in Ontario, Canada. However, nearly all the cobalt produced is obtained as a by-product of the metallurgy of nickel, copper, iron, silver, and other metals. The metallurgy of cobalt involves its separation from nickel, copper, and iron, its conversion to Co_3O_4, and the reduction of this oxide with aluminum or hydrogen.

$$3Co_3O_4(s) + 8Al(s) \longrightarrow 9Co(s) + 4Al_2O_3(s) \qquad \Delta H° = -4029 \text{ kJ}$$

Purification of the metal is done by electrolytic deposition on a rotating stainless steel cathode.

32.5   Properties and Uses of Cobalt

Cobalt is similar to iron in appearance but has a faint tinge of pink. Like iron and nickel, it is magnetic. It dissolves slowly in warm dilute hydrochloric or sulfuric acid and more rapidly in dilute nitric acid. Like iron and nickel, cobalt is rendered passive (unreactive) by contact with concentrated nitric acid. It is not oxidized on exposure to air, but at red heat cobalt reduces steam, with evolution of hydrogen. The halogens, except fluorine, convert the metal to cobalt(II) halides. When cobalt is heated with fluorine, **cobalt(III) fluoride, CoF_3,** is formed.

Cobalt is alloyed with iron and small percentages of other metals for use in high-speed cutting tools and surgical instruments. Permanent magnets are made from the alloys **Alnico** (Al, Ni, Co, and Fe), **Hiperco** (Co, Fe, and Cr), and **Vicalloy** (Co, Fe, and V).

Finely divided cobalt metal is used as a catalyst in the hydrogenation of carbon monoxide and carbon dioxide (with the formation of hydrocarbons) and for the oxidation of ammonia.

32.6   Compounds of Cobalt

Cobalt(II) hydroxide, $Co(OH)_2$, is formed as a blue flocculent precipitate ($K_{sp} = 2 \times 10^{-16}$) when an alkali metal hydroxide is added to a solution of a cobalt(II) salt. The blue color of the precipitate changes to violet and then to pink, probably as a result of hydration. Cobalt(II) hydroxide is readily soluble in aqueous ammonia to form **hexaamminecobalt(II) hydroxide, [Co(NH$_3$)$_6$](OH)$_2$.** Solutions of the latter compound are oxidized by oxygen in air to various cobalt(III) compounds; the oxidation is accompanied by a color change in the solution.

When cobalt(II) hydroxide is heated in the absence of air, **cobalt(II) oxide, CoO,** results. This oxide is a black substance, but when dissolved in molten glass, it gives the glass a blue color. Such glass is called cobalt glass and contains **cobalt(II) silicate.** Ignition of cobalt(II) hydroxide or oxide in air gives rise to **cobalt(II, III) oxide, Co$_3$O$_4$.** Cobalt(III) oxide may be produced by gently heating cobalt(II) nitrate.

$$4Co(NO_3)_2 \xrightarrow{\Delta} 2Co_2O_3 + 8NO_2 + O_2$$

Note that three elements in this reaction change their oxidation numbers. (You will find balancing this equation by an oxidation-reduction method a challenging review.)

Cobalt(II) oxide and hydroxide readily dissolve in hydrochloric acid. Concentration of the resulting solution causes crystallization of **cobalt(II) chloride 6-hydrate, CoCl$_2$ · 6H$_2$O,** which is red. The hydrated cobalt(II) ion, [Co(H$_2$O)$_6$]$^{2+}$, exhibits a pink color in solution. When partially dehydrated, cobalt(II) chloride changes to a deep blue color. This is believed to result from a change in the coordination number of the cobalt(II) ion from 6 to 4.

$$[Co(H_2O)_6]Cl_2 \longrightarrow [Co(H_2O)_4]Cl_2 + 2H_2O$$
$$\text{Pink} \qquad\qquad\qquad \text{Blue}$$

The same change in color is effected by dissolving CoCl$_2$ · 6H$_2$O in alcohol. Writing made on paper with a dilute solution of the hydrated salt is almost invisible, but when the paper is warmed, dehydration of the salt occurs and the writing becomes blue. It fades again as hydration takes place from moisture of the air. This is the chemistry of one kind of "invisible" ink.

Black **cobalt(II) sulfide, CoS,** can be completely precipitated only from basic solutions. However, the precipitated sulfide is but slightly soluble in hydrochloric acid. Aqua regia readily dissolves it.

In addition to simple salts, oxides, and hydroxides, cobalt forms a large number of coordination compounds. Cobalt(II) simple salts are more stable than are those of cobalt(III); cobalt(III) coordination compounds, in contrast, are much more stable than the corresponding cobalt(II) ones. Thus the majority of the more stable complex salts of cobalt contain the metal with the +3 oxidation number. Some of the more important cobalt(III) coordination complexes are **hexaamminecobalt(III) chloride, [Co(NH$_3$)$_6$]Cl$_3$, potassium hexacyanocobaltate(III), K$_3$[Co(CN)$_6$],** and **sodium hexanitrocobaltate(III), Na$_3$[Co(NO$_2$)$_6$].** In these compounds the six ligands, NH$_3$, CN$^-$, and NO$_2^-$, occupy the corners of a regular octahedron with the cobalt at the center (see Section 29.3).

MANGANESE

32.7 Periodic Relationships of Manganese

Manganese, technetium, and rhenium comprise Periodic Group VIIB. Although rhenium was not discovered until 1925, its chemistry has since been widely studied. No stable isotope of technetium is known; however, the chemistry of the element has been studied by means of synthesized radioactive isotopes. Each of the three elements in Group VIIB has electrons in two valence shells, the outermost of which contains two electrons. Contrasted with these elements are the halogens of Periodic Group VIIA, which have seven electrons in their outer and only valence shell. The halogens are active nonmetals, and the Group VIIB elements are metals. However, compounds in which elements of these two groups have their higher oxidation numbers are strikingly similar. Thus Mn_2O_7 and Cl_2O_7 are both volatile, explosively unstable liquids; $KMnO_4$ and $KClO_4$ are both strong oxidizing agents that form isomorphous crystals of nearly the same solubility. In the following sections, only manganese will be discussed in detail.

32.8 Occurrence, Metallurgy, and Properties of Manganese

The most important ore of manganese is pyrolusite, MnO_2. Other manganese ores include braunite, Mn_2O_3, manganite, $MnO(OH)$ or $Mn_2O_3 \cdot H_2O$, hausmannite, Mn_3O_4, franklinite, $(Fe, Mn, Zn)O$, and psilomelane, $BaMn_9O_{16}(OH)_4$.

Nearly pure manganese may be produced by the high-temperature reduction of the dioxide by aluminum.

$$3MnO_2(s) + 4Al(s) \xrightarrow{\Delta} 3Mn(s) + 2Al_2O_3(s) \qquad \Delta H^\circ = -1791 \text{ kJ}$$

Since alloys of manganese and iron are used extensively in the production of steel, such alloys are usually produced instead of pure manganese. They are prepared by reducing the mixed oxides of manganese and iron with coke in a blast furnace. Alloys with a high percentage of manganese are called **ferromanganese,** and those with a low percentage of manganese are called **spiegeleisen** (German, meaning *mirror iron*).

Manganese is a gray-white metal with a slightly reddish tinge. It is brittle and looks like cast iron. It is readily oxidized by moist air and reacts slowly with water, forming manganese(II) hydroxide and hydrogen. It dissolves readily in dilute acids, forming manganese(II) salts. Manganese forms five oxides and five corresponding series of salts (Table 32-2). These compounds are considered in the following sections.

Table 32-2   Classes of Manganese Compounds

Oxidation Number	Oxide	Hydroxide	Character	Derivative	Name	Color
+2	MnO	$Mn(OH)_2$	Moderately basic	$MnCl_2$	Manganese(II) chloride	Pink
+3	Mn_2O_3	$Mn(OH)_3$	Weakly basic	$MnCl_3$	Manganese(III) chloride	Violet
+4	MnO_2	H_2MnO_3	Weakly acidic	$CaMnO_3$	Calcium manganite	Brown
+6	MnO_3	H_2MnO_4	Acidic	K_2MnO_4	Potassium manganate	Green
+7	Mn_2O_7	$HMnO_4$	Strongly acidic	$KMnO_4$	Potassium permanganate	Purple

32.9 Manganese(II) Compounds

Although manganese forms compounds in which it exhibits oxidation numbers of +2, +3, +4, +6, and +7, the only stable cation is the dipositive manganese ion, Mn^{2+}. The common soluble manganese(II) salts are the chloride, sulfate, and nitrate. Each imparts a faint pink color to its solutions.

Alkali metal hydroxides precipitate pale pink **manganese(II) hydroxide, Mn(OH)$_2$,** which is oxidized when exposed to air to the dark brown **manganese(III) oxyhydroxide, MnO(OH).** Manganese(II) hydroxide is only partially precipitated by aqueous ammonia and dissolves in solutions of ammonium salts of strong acids, according to the equation

$$Mn(OH)_2 + 2NH_4^+ \rightleftharpoons Mn^{2+} + 2NH_3 + 2H_2O$$

Manganese(II) hydroxide is entirely basic in character; it dissolves in acids but not in bases such as sodium hydroxide. **Manganese(II) oxide, MnO,** may be produced by heating the corresponding hydroxide in the absence of air.

From solutions of manganese(II) compounds alkali metal sulfides precipitate pink **manganese(II) sulfide, MnS** ($K_{sp} = 5.6 \times 10^{-16}$), which is readily soluble in dilute acids.

32.10 Manganese(III) Compounds

Dimanganese trioxide, Mn$_2$O$_3$, and MnO(OH) occur naturally, but the manganese(III) ion is unstable in aqueous solution and is readily reduced to the manganese(II) ion. **Manganese(III) chloride** is formed in solution by the reaction of hydrochloric acid with manganese dioxide at low temperature. When the solution is warmed, the manganese(III) chloride decomposes with the formation of manganese(II) chloride and chlorine.

$$MnO_2 + 4H^+ + 4Cl^- \longrightarrow MnCl_4 + 2H_2O$$
$$2MnCl_4 \longrightarrow 2MnCl_3 + Cl_2(g)$$
$$2MnCl_3 \longrightarrow 2Mn^{2+} + 4Cl^- + Cl_2(g)$$

32.11 Manganese(IV) Compounds

The most important manganese(IV) compound is the dioxide, MnO$_2$, which is amphoteric but relatively inert toward acids and bases. Cold, concentrated hydrochloric acid reacts with the dioxide, giving a green solution of the unstable **manganese(IV) chloride.** The sulfate Mn(SO$_4$)$_2$ is also unstable, but the complex salt K$_2$MnF$_6$ is not readily decomposed. When the dioxide is fused with calcium oxide, **calcium manganite, CaMnO$_3$,** is formed. The corresponding acid, manganous acid, H$_2$MnO$_3$, is unknown.

When manganese dioxide is heated to 535°C, it is transformed to **trimanganese tetraoxide, Mn$_3$O$_4$,** and oxygen, according to

$$3MnO_2(s) \xrightarrow{\Delta} Mn_3O_4(s) + O_2(g) \qquad \Delta H° = 172 \text{ kJ}$$

A historical note is that Scheele, independently of Priestley and Lavoisier, discovered oxygen by the reaction of concentrated sulfuric acid with manganese dioxide.

$$MnO_2 + 2H_2SO_4 \longrightarrow Mn(SO_4)_2 + 2H_2O$$
$$2Mn(SO_4)_2 + 2H_2O \longrightarrow 2MnSO_4 + 2H_2SO_4 + O_2$$

32.12 Manganates, Manganese(VI) Compounds

When an oxide of manganese is melted with an alkali metal hydroxide or carbonate in the presence of air or some other oxidizing agent (such as potassium chlorate or potassium nitrate), a **manganate** is formed.

$$2MnO_2 + 4KOH + O_2 \longrightarrow 2K_2MnO_4 + 2H_2O$$

Manganates are green and stable only in alkaline solution. The addition of water to solutions of manganates may bring about disproportionation, with the precipitation of manganese dioxide and the formation of the purple **permanganate ion, MnO_4^-.**

$$3MnO_4^{2-} + 2H_2O \longrightarrow 2MnO_4^- + 4OH^- + MnO_2(s)$$

The hypothetical manganic acid, H_2MnO_4, is too unstable to be prepared and isolated. Solutions containing the manganate ion become active oxidizing agents when acidified. The half-reaction is

$$MnO_4^{2-} + 8H^+ + 4e^- \longrightarrow Mn^{2+} + 4H_2O$$

32.13   Permanganates, Manganese(VII) Compounds

An important compound of manganese is **potassium permanganate, $KMnO_4$.** It is prepared commercially by oxidizing potassium manganate in alkaline solution with chlorine.

$$2MnO_4^{2-} + Cl_2 \longrightarrow 2MnO_4^- + 2Cl^-$$

The resultant purple solution deposits crystals when sufficiently concentrated. Potassium permanganate is a valuable laboratory reagent because it acts as a strong oxidizing agent. A solution of the free **permanganic acid, $HMnO_4$,** may be prepared by the reaction of dilute sulfuric acid and barium permanganate.

$$Ba^{2+} + 2MnO_4^- + [2H^+] + SO_4^{2-} \longrightarrow BaSO_4(s) + [2H^+] + 2MnO_4^-$$

Mn_2O_7, a dark brown, highly explosive liquid, is the anhydride of permanganic acid.

32.14   Uses of Manganese and Its Compounds

Manganese metal forms a number of important alloys. Two of these were mentioned in Section 32.8: spiegeleisen, which is a bright and lustrous iron alloy containing 5–20% manganese, and ferromanganese, which contains 70–80% manganese. Both of these alloys are used in making very hard steels for the manufacture of rails, safes, and heavy machinery. About 12.5 pounds of manganese metal are used in the production of every ton of steel to remove oxygen, nitrogen, and sulfur. An alloy called **manganin** (Cu, 84%; Mn, 12%; Ni, 4%) is used in instruments for making electrical measurements, because its electrical resistance does not change significantly with changes in temperature.

Manganese dioxide is used to correct the green color produced in glass by iron(II) compounds, to color glass and enamels black, as an oxidizing agent or dryer in black paints, and as a depolarizer in dry cells. Potassium permanganate is used as a disinfectant, a deodorant, and a germicide. Potassium permanganate is also used extensively as an oxidizing agent in analytical determinations.

IRON

32.15 History and Occurrence of Iron

Iron was known at least as early as 4000 B.C. and was probably used to some extent during prehistoric times. Since free iron is not commonly found in nature, that used in prehistoric times may well have been of meteoric origin. The early discovery and application of iron to the manufacture of tools and weapons were possible because of the wide distribution of iron ores and the ease with which the iron compounds in the ores could be reduced by carbon. For a long time charcoal was the form of carbon used in the reduction process. The production and use of iron on a large scale did not begin until about 1620, when coal was introduced as the reducing agent.

Iron is the most widely used of all metals. The amount of iron used is 14 times the amount of all other metals combined. The reasons for this are that its ores are abundant and widely distributed, the metal is easily and cheaply produced from the ores, and its properties can be varied over a wide range by the addition of other substances and by different methods of treatment, such as tempering and annealing.

Iron ranks second in abundance among the metals (aluminum is first) and fourth among all elements in the earth's crust. The central core of the earth is largely iron; this fact was deduced from studies of the earth's density and the rate of transmission of earthquake shock. Metallic meteorites are usually about 90% iron, and the remainder is principally nickel. Practically all rocks, minerals, and soils, as well as plants, contain some iron. Iron is present in the hemoglobin of the blood, which acts as a carrier of oxygen (Section 29.5).

The important iron-bearing ores are hematite, Fe_2O_3, and magnetite, Fe_3O_4. The presence of Fe_2O_3 accounts largely for the red coloration of many rocks and soils. Hematite often occurs as the reddish-brown hydrate $2Fe_2O_3 \cdot 3H_2O$, which is called limonite. Magnetite is a black, crystalline mineral, which is strongly magnetic and sometimes possesses polarity, in which case it is called lodestone. Black crystalline magnetite interlocked with crystals of silica in the form of hard rock is called taconite. Siderite is a white or brownish carbonate, $FeCO_3$. Pyrite, FeS_2, occurs as pale yellow crystals with a metallic luster, which are easily mistaken for gold and are therefore called *fool's gold*. Pyrite is quite abundant and is useful as a source of sulfur for the manufacture of sulfuric acid. However, since roasting does not remove all of the sulfur, pyrite is usually considered to be unsatisfactory as a source of iron, although low-grade pig iron can be made from it.

Taconite is the extremely hard rock that makes up some 95% by volume of the famous iron-bearing deposits of the Mesabi Range in upper Minnesota. Once regarded as useless for steel production, the taconite ores are now used extensively as a result of the development of satisfactory techniques for mining and concentrating them.

32.16 Metallurgy of Iron

The first step in the metallurgy of iron is usually roasting the ore to remove water, decompose carbonates, and oxidize sulfides. The oxides of iron are then reduced with coke in a blast furnace 80–100 feet high and 25 feet in diameter. Near the bottom of the furnace are nozzles, called tuyères, through which preheated air is blown into the furnace under pressure. The charge of roasted ore, coke, and flux (limestone or sand) is introduced continuously into the top of the furnace through cone-shaped valves arranged so that the charge is deposited uniformly around the circumference of the furnace near the walls. The entire stock in the furnace weighs several hundred tons.

As soon as the preheated air (500°C) under pressure enters the furnace, the coke in the region of the tuyères is oxidized to carbon dioxide with the liberation of a great deal of heat, which raises the temperature to about 1500°C. As the carbon dioxide passes upward through the overlying layer of white-hot coke, it is reduced to carbon monoxide.

$$CO_2 + C \longrightarrow 2CO$$

The carbon monoxide thus produced serves as the reducing agent in the upper regions of the furnace. The individual reactions are indicated in Fig. 32-1 and can be summarized as a single reversible reaction.

$$Fe_2O_3(s) + 3CO(g) \rightleftharpoons 2Fe(s) + 3CO_2(g) \qquad \Delta H° = -25 \text{ kJ}$$

The presence of an excess of carbon monoxide forces the reaction to proceed to the right. The iron ore is completely reduced before the formation of the slag begins in the middle region of the furnace; otherwise, part of the iron would be lost due to the reaction

$$FeO + SiO_2 \longrightarrow FeSiO_3$$

Just below the middle of the furnace, the temperature is high enough to melt both the iron and the slag. The molten iron and slag collect in two layers at the bottom of the furnace; the less dense slag floats on the iron and protects it against oxidation. Every hour or two, the liquid slag is withdrawn from the furnace. Blast furnace slag is often used in the manufacture of Portland cement. Four or five

Figure 32-1. Reactions that occur in a blast furnace.

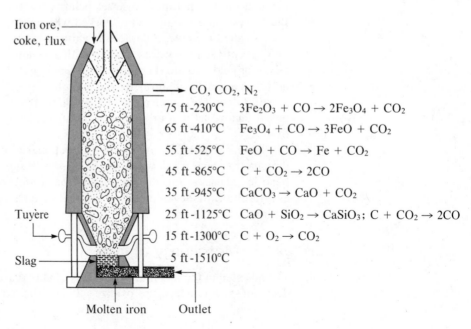

Iron ore, coke, flux

CO, CO₂, N₂

75 ft -230°C	$3Fe_2O_3 + CO \rightarrow 2Fe_3O_4 + CO_2$
65 ft -410°C	$Fe_3O_4 + CO \rightarrow 3FeO + CO_2$
55 ft -525°C	$FeO + CO \rightarrow Fe + CO_2$
45 ft -865°C	$C + CO_2 \rightarrow 2CO$
35 ft -945°C	$CaCO_3 \rightarrow CaO + CO_2$
25 ft -1125°C	$CaO + SiO_2 \rightarrow CaSiO_3$; $C + CO_2 \rightarrow 2CO$
15 ft -1300°C	$C + O_2 \rightarrow CO_2$
5 ft -1510°C	

Tuyère

Slag

Molten iron   Outlet

times a day, the molten iron is withdrawn through a tap hole into a large ladle that holds several hundred tons of the hot metal. The ladle containing the hot metal is then transferred to the pig-casting machine or to the steelmaking plant. A modern blast furnace can produce as much as 3000 tons of iron daily.

The hot exhaust gases, which issue from the top of the blast furnace and contain some unoxidized carbon monoxide, are mixed with air and burned to preheat the air used in the operation of the blast furnace.

The metal obtained from the blast furnace is called **pig iron.** It contains impurities of carbon (3–4%), silicon, phosphorus, manganese, and smaller amounts of sulfur. Iron reacts with carbon at the temperatures of the blast furnace to form **cementite, Fe$_3$C.**

$$3Fe(s) + C(s) \rightleftharpoons Fe_3C(s) \qquad \Delta H° = 25\ kJ$$

Pig iron is brittle and is usually converted to cast iron or steel. When pig iron is remelted and recooled, it is called **cast iron.** When pig iron is cooled rapidly, the carbon remains chemically combined with the iron in the form of the carbide Fe$_3$C. Because Fe$_3$C is white, the product has a light color and is therefore called **white cast iron.** It is brittle but hard and wear-resistant. When pig iron is allowed to cool slowly, the above equilibrium shifts so that some of the carbon separates from the iron as graphite. This gives the product a gray color, and it is called **gray cast iron.** In contrast to white cast iron, the gray variety is soft and tough. It is possible to control the cooling of cast iron in such a way that the surface cools rapidly, while the body of the casting cools slowly. This treatment results in a casting that has the toughness of gray cast iron and the wearing qualities of white cast iron.

Cast iron is especially suited to making molded products (castings), because it expands as it solidifies and consequently fills all details of the mold completely.

32.17   The Manufacture of Steel

The term **steel** is applied to many widely different alloys of iron. Steel is made from pig iron by removing the impurities and later adding substances such as manganese, chromium, nickel, tungsten, molybdenum, and vanadium to produce alloys with properties that make them suitable for specific uses. Most steels also contain small but definite percentages of carbon (0.04–2.5%). Thus a large part of the carbon contained in pig iron must be removed in the manufacture of steel.

Steels can be classified into three main categories: (1) **carbon steels,** which are primarily iron and carbon; (2) **stainless steels,** low-carbon steels containing 12% chromium; and (3) **alloy steels,** specialty steels that contain large amounts of other elements to impart special properties for specific uses. Of the 141 million tons of steel produced in the United States in a recent year, 124.6 million tons (88.4%) were carbon steels, 14.8 million tons (10.5%) were stainless steels, and 1.6 million tons (1.1%) were alloy steels.

The process used chiefly in the production of steel is the **basic oxygen process.** Steels for special purposes are made by the electric furnace method, which accounts for 14% of total steel production in this country. The basic oxygen process utilizes a cylindrical furnace with a basic lining, such as magnesium oxides and calcium oxides. A typical furnace is charged with 80 tons of scrap iron, 200 tons of molten pig iron, and 18 tons of limestone (to form slag) and operates under computer control. A jet of high-purity (99.5%) oxygen at 140–180 pounds of pressure per square inch is directed vertically through a water-cooled lance, 50 feet long and 10 inches in diameter, to a point 4–8 feet below the surface of the white-hot molten charge. The oxygen produces a vigorous reaction that oxidizes the impurities in the charge. In the central reaction zone, temperatures reach a level close to the boiling point of iron. The entire steel-making cycle is completed in one hour or less. Electrostatic precipitators (see Section 13.32) clean the gases resulting from the furnace reactions, making the furnaces virtually smokeless. The steel produced is of extremely high and uniform quality.

Various types of electric furnaces are used to obtain high-grade steel. Three distinct advantages of the electric furnace are (1) the temperature and composition of the metal bath can be controlled with high precision, (2) high temperatures are easily attainable, making possible the production of special steels with high melting points, and (3) the furnace can accept an all-scrap charge (basic oxygen furnaces are limited to a maximum of about 35% scrap, since excess scrap would have too great a cooling effect on the charge). However, the cost of power limits the use of the electric furnace. **Electric steel** is used as tool steel in the making of such articles as files, drills, knives, razor blades, watch springs, and axes.

Carbon is the most important alloying element used in the manufacture of steel. It may be present in three forms: in combination with iron as cementite, Fe_3C, as relatively large crystals of graphite, and as very small crystals of graphite (Sections 25.2 and 32.16).

Steel containing cementite in solid solution in iron contains as much as 1.5% carbon and is made by quenching the hot metal in water or oil. This steel is hard, brittle, and light-colored. If the metal is cooled slowly, the carbon is deposited largely as separate crystals of cementite or as particles of graphite; the product is softer and more pliable and has a much higher tensile strength than does steel that has been cooled rapidly.

Steels of low carbon content (up to 0.2%) are quite ductile and are known as **mild steels.** These are used in the manufacture of products such as sheet iron, wire, and pipe. **Medium steels** contain 0.2–0.6% carbon and are used in rails, structural steel, and boiler plate. Both mild and medium steels can be forged and welded. **High-carbon steels** contain 0.6–1.5% carbon. They are hard and brittle and are used in surgical instruments, razor blades, springs, and cutlery.

Steel that has been heated to redness and allowed to cool slowly is said to be **annealed.** Steel is **tempered** by rapid cooling in water or oil followed by controlled reheating to obtain the desired proportions of cementite and graphite. **Case-hardened steel** is made by heating low-carbon steel in closed containers with powdered carbon, followed by quenching in oil. The carbon reacts with the iron on the surface, forming cementite. Such a steel is very hard on its surface and tough in the interior. It is used in making axles and other articles that must resist wear and, at the same time, be tough and flexible enough to be safely subjected to sudden strains and shocks. Steel can also be case-hardened using nitrogen compounds. A very hard iron nitride is formed, and the process is therefore called **nitriding.**

Certain materials called **scavengers** are added to iron in the manufacture of steel to remove impurities, especially oxygen and nitrogen, and thus improve the quality of the product. The most important scavengers are aluminum, ferrosilicon, ferromanganese, and ferrotitanium. They react with dissolved oxygen and nitrogen forming oxides and nitrides, respectively. The oxides and nitrides are removed with the slag.

For every ton of steel produced, about 25–30 pounds of nonferrous metals are added or used as coatings. By the appropriate choice of the number and percentages of these elements, alloy steels of widely varying properties can be manufactured. Some of the important alloy steels and their features are given in Table 32-3.

Table 32-3   Alloy Steels

Name	Composition	Characteristic Properties	Uses
Manganese steel	10–18% Mn	Hard, tough, resistant to wear	Railroad rails, safes, armor plate, rock-crushing machinery
Silicon steel	1–5% Si	Hard, strong, highly magnetic	Magnets
Duriron	12–15% Si	Resistant to corrosion, acids	Pipes, kettles, condensers, etc.
Invar	36% Ni	Low coefficient of expansion	Meter scales, measuring tapes, pendulum rods
Chrome-vanadium	1–10% Cr, 0.15% V	Strong, resistant to strains	Axles
Stainless steel	14–18% Cr, 7–9% Ni	Resistant to corrosion	Cutlery, instruments
Permalloy	78% Ni	High magnetic susceptibility	Ocean cables
High-speed steels	14–20% W or 6–12% Mo	Retain temper at high temperatures	High-speed cutting tools
Nickel steel	2–4% Ni	Hard and elastic, resistant to corrosion	Drive shafts, gears, cables

32.18 Corrosion

Many metals, particularly iron, undergo corrosion when exposed to air and water. Losses caused by corrosion of metals total billions of dollars annually in the United States.

It has been shown that iron will not rust in dry air or in water that is free from dissolved oxygen. Both air and water are involved in the corrosion process. The presence of an electrolyte in the water accelerates corrosion, particularly when the solution is acidic. Strained metals corrode more rapidly than unstrained ones do. Heated portions of a metal corrode more rapidly than unheated ones do. Finally, iron in contact with a less active metal, such as tin, lead, or copper, corrodes more rapidly than iron that is either alone or in contact with a more active metal, such as zinc.

An electrochemical theory helps to explain why the presence of a less active metal than iron accelerates the corrosion of iron and the presence of a more active metal slows down the corrosion. It appears that very small electrochemical cells are set up when corrosion takes place. When iron is in contact with water containing an electrolyte, this half-reaction tends to occur:

Anodic oxidation: $Fe \longrightarrow Fe^{2+} + 2e^-$

Sometimes, one part of a piece of iron is more active than the rest and tends to go into solution more easily. This means that the more active part has a higher potential than the rest of the iron piece, and this part tends to dissolve in the electrolyte. On the less active parts of the iron (lower electrode potential), hydrogen tends to form according to the half-reaction

$$\text{Cathodic reduction:} \qquad 2H^+ + 2e^- \longrightarrow H_2$$

The accumulation of hydrogen on the surface of the less active part of the iron tends to polarize the electrode and stop the cell action. However, oxygen dissolved in any water present gradually removes the hydrogen in the same manner that MnO_2 depolarizes the dry cell (Section 20.16), and the electrochemical reaction proceeds; that is, corrosion occurs.

$$2H_2 + O_2 \longrightarrow 2H_2O$$

Iron(II) ions combine with hydroxide ions of the electrolytic solution, forming $Fe(OH)_2$. Iron(II) hydroxide is then readily oxidized by air to $Fe_2O_3 \cdot nH_2O$, which is **iron rust.**

Many methods and devices have been employed to prevent or retard corrosion. Iron can be protected against corrosion by coating it with: (1) an organic material such as paint, lacquer, grease, or asphalt; (2) another metal such as zinc, copper, nickel, chromium, tin, cadmium, or lead; (3) a ceramic enamel, like that used on sinks, bathtubs, stoves, refrigerators, and washers; or (4) an adherent oxide, which may be formed by exposing iron to superheated steam, thereby giving it an adherent coating of magnetic oxide, Fe_3O_4. A piece of iron can also be protected by treating it with a compound that provides a suitable ligand, such as phosphate, which ties the surface iron up in a complex so that it cannot react to form $Fe_2O_3 \cdot nH_2O$. Alloys of iron and certain other elements are often corrosion-resistant. Typical examples are stainless steel (Fe, Cr, and Ni) and duriron (Fe and Si).

Another method of preventing the corrosion of iron or steel involves an application of electrochemistry called **cathodic protection.** For example, corrosion of iron or steel water tanks can be retarded by suspending several stainless steel anodes in the tank; the tank serves as cathode. A small current is passed continuously through the system; the natural salts in the water make it conducting. A slight cathodic evolution of a protective coat of hydrogen takes place on the wall of the tank. Cathodic protection against corrosion is also used on iron and steel that is in contact with soil, such as underground pipe lines. Other applications of cathodic protection do not involve the use of a current from an external source. In these cases, the iron becomes a cathode when connected by a wire to a more active metal, such as zinc, aluminum, or magnesium. The difference in activity of the two metals causes a current to flow between them, producing corrosion on the more active metal and furnishing cathodic protection to the iron.

32.19 Properties of Iron

Pure iron is silvery white, capable of taking a high polish, ductile, relatively soft, and high in tensile strength. These properties are greatly modified by the amounts and nature of the impurities present, especially carbon. Pure iron is attracted by a magnet but does not retain magnetism. Iron dissolves in hydrochloric acid and dilute sulfuric acid, forming hydrogen and iron(II) ions, Fe^{2+}. Hot concentrated sulfuric acid forms Fe^{3+}, and SO_2 is evolved. Cold concentrated nitric acid induces **passivity** in iron (see also Section 32.26). When passive, a metal does not react with dilute acids that otherwise would readily dissolve it. The passivity is easily destroyed by a scratch or by shock. Hot dilute nitric acid oxidizes iron to Fe^{3+} with the evolution of nitric oxide. When heated to redness, iron reduces steam to produce hydrogen and iron(II, III) oxide, Fe_3O_4, which is the same oxide that is produced when iron burns in oxygen. When exposed to moist air at ordinary temperatures iron becomes oxidized, with a loose coating of partially hydrated iron(III) oxide, $Fe_2O_3 \cdot nH_2O$ (Section 32.18).

32.20 Compounds of Iron

Iron forms two principal series of compounds in which its oxidation numbers are $+2$ and $+3$; compounds in which iron is $+6$ are rare and less important. The $+2$ and $+3$ series are called iron(II), or ferrous, compounds and iron(III), or ferric, compounds, respectively. The oxides and hydroxides of iron(II) and (III) are basic in character with little or no acidic character.

Black **iron(II) oxide, FeO,** may be obtained by the thermal decomposition of iron(II) oxalate, FeC_2O_4.

$$FeC_2O_4 \xrightarrow{\Delta} FeO + CO + CO_2$$

Iron(II) hydroxide, Fe(OH)$_2$, is white when pure; it precipitates when alkali metal hydroxides are added to solutions of iron(II) salts. Ammonium salts and many organic acids greatly increase the solubility of iron(II) hydroxide. In the air the white hydroxide quickly turns green and then reddish-brown, due to oxidation to **iron(III) hydroxide.**

$$4Fe(OH)_2(s) + O_2(g) + 2H_2O(g) \longrightarrow 4Fe(OH)_3(s) \qquad \Delta H° = -532.2 \text{ kJ}$$

Iron(III) oxide, Fe$_2$O$_3$, occurs in nature as the mineral hematite. It may also be prepared by igniting $Fe(OH)_3$ or by roasting pyrite, FeS_2, in the air. It is a red powder, which is used as a paint pigment under the names rouge and Venetian red. When treated with either alkali metal hydroxides or aqueous ammonia, solutions of Fe^{3+} yield $Fe(OH)_3$, which is reddish-brown and insoluble in an excess of either reagent.

Iron(II, III) oxide, Fe$_3$O$_4$, is a mixed oxide, $FeO \cdot Fe_2O_3$, and is called magnetic oxide of iron, or lodestone. It is a valuable ore of iron.

Iron(II) chloride 4-hydrate, FeCl$_2$ $\cdot$ 4H$_2$O, is produced by the reaction of dilute hydrochloric acid with iron. It is pale green in color. Sulfuric acid reacts with iron or iron(II) oxide, forming green vitriol, or copperas, which crystallizes as the 7-hydrate $FeSO_4 \cdot 7H_2O$.

Iron(II) ammonium sulfate 6-hydrate, (NH$_4$)$_2$SO$_4$ $\cdot$ FeSO$_4$ $\cdot$ 6H$_2$O, Mohr's salt, is a double salt that crystallizes from an equimolar solution of ammonium sulfate and iron(II) sulfate. It is readily obtained in a pure state and is used extensively in laboratory work as a reducing agent; the iron(II) is readily oxidized to iron(III).

Iron(II) carbonate, FeCO$_3$, occurs in nature as the mineral siderite. It is also produced as a white precipitate ($K_{sp} = 2.11 \times 10^{-11}$) by the reaction of alkali metal carbonates with solutions of iron(II) salts in the absence of air. The carbonate dissolves in carbonic acid to form the hydrogen carbonate $Fe(HCO_3)_2$, a common constituent of hard waters.

Iron(II) sulfide, FeS ($K_{sp} = 1 \times 10^{-19}$), is precipitated when an alkali metal sulfide is added to a solution of an iron(II) salt. It is prepared commercially by heating a mixture of iron and sulfur. It reacts with acids, forming hydrogen sulfide.

An important iron(III) compound is **iron(III) chloride, FeCl₃,** which may be obtained by heating iron with chlorine. It is appreciably covalent, as is indicated by its volatility and its solubility in nonpolar solvents. When the anhydrous chloride is dissolved in water, the hydration of the ions is exothermic. Iron(III) chloride crystallizes from water as the brownish-yellow hydrate [Fe(H₂O)₆]Cl₃. This compound, like all soluble iron(III) salts, gives an acid solution by hydrolysis. The first step in the hydrolysis is

$$[Fe(H_2O)_6]^{3+} + H_2O \longrightarrow [Fe(H_2O)_5(OH)]^{2+} + H_3O^+$$

or

$$Fe^{3+} + H_2O \longrightarrow Fe(OH)^{2+} + H^+$$

Several additional steps also occur. Evidence for this is the appearance of the reddish-brown gelatinous [Fe(H₂O)₃(OH)₃], which is simply hydrated Fe(OH)₃.

Hydrated iron(III) chloride is used to coagulate blood, and thus stop bleeding, and in the treatment of anemia.

When a solution of an iron(III) salt is added to a solution containing thiocyanate ions, a blood-red color appears. An iron(II) salt yields a colorless solution. These facts are applied to the detection of iron(III) ions in the presence of iron(II) ions. It appears that the species responsible for the red color is the complex cation [Fe(H₂O)₅NCS]²⁺.

Other iron(III) salts include the sulfate Fe₂(SO₄)₃, and the iron **alums** formed from this salt, such as NH₄Fe(SO₄)₂ · 12H₂O.

When an excess of potassium cyanide is added to a solution of an iron(II) salt, **potassium hexacyanoferrate(II), K₄[Fe(CN)₆],** also called potassium ferrocyanide, is formed in solution.

$$Fe^{2+} + 6CN^- \longrightarrow [Fe(CN)_6]^{4-}$$

The complex hexacyanoferrate(II) ion is so stable that it does not give positive results in any of the common qualitative tests for iron(II) or cyanide ions. When the hexacyanoferrate(II) ion is oxidized by chlorine, **hexacyanoferrate(III) ion, [Fe(CN)₆]³⁻,** also called ferricyanide, is produced.

$$2[Fe(CN)_6]^{4-} + Cl_2 \longrightarrow 2[Fe(CN)_6]^{3-} + 2Cl^-$$

Solutions of potassium hexacyanoferrate(III) do not give positive results in the usual tests for either iron(III) ions or cyanide ions.

The addition of iron(II) ions to hexacyanoferrate(III) solutions produces a blue precipitate known as **Turnbull's blue.** Iron(III) ions react with hexacyanoferrate(II) ions to form **Prussian blue.** These precipitates have a deep blue color and are thought to be the same, with the approximate formula KFe[Fe(CN)₆] · H₂O. Three independent experimental methods have shown that the iron outside the coordination sphere has an oxidation number of +3 and the iron inside the coordination sphere has an oxidation number of +2. When solutions of iron(III) ions and hexacyanoferrate(III) ions are mixed, no precipitate forms but the mixture turns brown. Iron(II) ions and hexacyanoferrate(II) ions produce a white precipitate, K₂Fe[Fe(CN)₆]. These reactions serve as sensitive tests to distinguish between Fe²⁺ and Fe³⁺ in solution.

Blueprint paper is coated with a mixture of iron(III) ammonium citrate and potassium hexacyanoferrate(III); this gives the paper the bronze-green color of iron(III) hexacyanoferrate(III). When a drawing in black ink on tracing cloth is placed over the blueprint paper and the paper is exposed to light, the iron(III) ions are reduced by the citrate ions to iron(II) wherever light transmission occurs. The iron(II) ion immediately reacts with the hexacyanoferrate(III) ion to form an insoluble blue precipitate. After exposure, the paper is washed with water to remove the unchanged iron(III) hexacyanoferrate(III), leaving the body of the paper blue and the lines produced by the tracing white.

ALUMINUM

32.21　Periodic Relationships of Aluminum

Group IIIA of the Periodic Table consists of boron, aluminum, gallium, indium, and thallium. Boron is a nonmetallic element and has already been considered (Chapter 27). Aluminum is a common and abundant metal, while gallium, indium, and thallium are all scarce and almost without useful application. The Group IIIB elements (scandium, yttrium, the lanthanide elements, and the actinide elements) are less common elements and will not be considered here. All the elements of these two groups generally exhibit an oxidation number of $+3$. Oxidation numbers other than $+3$ are also exhibited by most of these elements.

32.22　Occurrence and Metallurgy of Aluminum

Aluminum is the most abundant metal and the third most abundant element in the earth's crust. It is too active chemically to occur as the element in nature and is usually found combined with oxygen. The most important ore of aluminum is the hydrated oxide $Al_2O_3 \cdot 2H_2O$, which is called bauxite. The mineral cryolite, Na_3AlF_6, is used in the metallurgy of aluminum, though it is not an ore of aluminum. Very large and widely distributed quantities of aluminum are found in complex aluminosilicate minerals, such as clays and feldspars. Anhydrous aluminum oxide occurs as corundum, ruby, and sapphire. Emery is a mixture of Al_2O_3 and Fe_3O_4.

Aluminum is produced by an electrolytic process invented in 1886 by Charles M. Hall, who began work on the problem while he was a student at Oberlin College. The process was discovered independently a month or two later by Paul L. T. Héroult in France.

The first step in the production of metallic aluminum from bauxite involves the purification of the mineral. The bauxite is dried and pulverized and then heated with sodium hydroxide under steam pressure. Aluminum oxide is converted to soluble sodium aluminate, while the iron oxide and silicates, which are present as impurities, remain undissolved.

$$Al_2O_3 \cdot 2H_2O + 2OH^- + H_2O \longrightarrow 2Al(OH)_4^- \quad \text{(aluminate ion)}$$

After the insoluble iron oxide and silicates are filtered from the solution, aluminum hydroxide is precipitated by hydrolysis of the aluminate. The precipitated aluminum hydroxide is ignited to the oxide, which is then dissolved in molten cryolite, Na_3AlF_6. This solution is electrolyzed. Aluminum metal is liberated at the cathode, and oxygen, carbon monoxide, and carbon dioxide are formed at the carbon anode. A mixture of fluorides often replaces cryolite as the electrolyte. A typical mixture is $2AlF_3 \cdot 6NaF \cdot 3CaF_2$.

Aluminum prepared by electrolysis is about 99% pure; the impurities are small amounts of copper, iron, silicon, and aluminum oxide. If aluminum of exceptionally high purity (99.9%) is needed, it may be obtained by the electrolytic **Hoopes process** (Fig. 32-2), which employs a molten bath consisting of three layers. The bottom layer is a molten alloy of copper and impure aluminum, the top layer is pure molten aluminum, and the middle layer (the electrolyte) consists of a molten mixture of the fluorides of barium, aluminum, and sodium with nearly enough aluminum oxide to saturate it. The densities of the layers are such that their separation is maintained during electrolysis. The bottom layer serves as the anode and the top layer as the cathode. As electrolysis proceeds, the aluminum in the bottom layer passes into solution in the electrolyte as Al^{3+}, leaving the copper, iron, and silicon behind in the anode, as they are not oxidized under these conditions. The aluminum ion is reduced at the cathode. During electrolysis, purified aluminum is drawn off from the upper layer, and the impure molten metal is added to the lower layer through a carbon-lined funnel.

Figure 32-2. Electrolytic cell for the purification of aluminum. (a) is the bottom portion of the cell in contact with the anode layer; (b) is an alloy of impure molten aluminum and copper, which acts as the anode and in which the anodic oxidation takes place; (c) is an electrolyte of a mixture of molten fluorides; (d) is the molten layer of pure aluminum at the cathodes; (e) is a funnel arrangement by which impure aluminum may be added to the bottom of the cell.

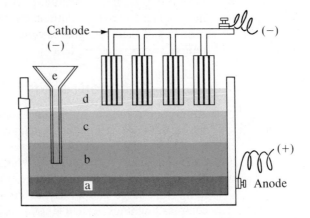

32.23 Properties and Uses of Aluminum

Freshly cut aluminum has a silvery appearance, but it soon becomes superficially oxidized and assumes a dull white luster. The metal is very light and possesses high tensile strength. Weight for weight it is twice as good a conductor of electricity as copper. Although aluminum is a relatively reactive metal, the tenacious coating of the oxide that forms on it prevents further corrosion. When heated in air, finely divided aluminum burns with a brilliant light. Because of the formation of a protective coat of oxide, aluminum does not decompose water. It is readily dissolved by hydrochloric acid and sulfuric acid, but is rendered passive by nitric acid. Aluminum dissolves in concentrated bases with the formation of aluminates and hydrogen.

$$2Al + 2OH^- + 6H_2O \longrightarrow 2Al(OH)_4^- + 3H_2(g)$$

This behavior is in accord with the amphoteric character of aluminum hydroxide.

When heated, aluminum combines directly with the halogens, nitrogen, carbon, and sulfur.

The facts that aluminum is an excellent conductor of heat and is lightweight and corrosion-resistant account for its use in the manufacture of cooking utensils. The most important uses of aluminum are in the aircraft and other transportation industries, which depend on the lightness, toughness, and tensile strength of the metal. Aluminum is also used in the manufacture of electrical transmission wire, as a paint pigment, and, in the form of foil, as a wrapping material.

Aluminum is one of the best reflectors of heat and light, including ultraviolet light. For this reason it is used as an insulating material and as a mirror in reflecting telescopes.

About half of the aluminum produced in this country is converted to alloys for special uses. **Duralumin** is an alloy containing aluminum, copper, manganese, and magnesium. Lightweight and nearly as strong as steel, it is useful in the construction of aircraft. **Aluminum bronzes** contain copper, and occasionally some silicon, manganese, iron, nickel, and zinc. These lightweight alloys have high tensile strength and great resistance to corrosion and are used extensively in the manufacture of crankcases and connecting rods for gasoline engines. **Alnico** is a magnetic alloy containing 50% iron, 20% aluminum, 20% nickel, and 10% cobalt that can lift more than 4000 times its own weight of iron.

When powdered aluminum and iron(III) oxide are mixed and ignited by means of a magnesium fuse, a vigorous and highly exothermic reaction occurs.

$$2Al(s) + Fe_2O_3(s) \longrightarrow 2Fe(s) + Al_2O_3(s) \qquad \Delta H° = -851.4 \text{ kJ}$$

This is known as the **thermite reaction** and is applied in the **Goldschmidt process.** The temperature of the reaction mixture rises to about 3000°C, so the iron and aluminum oxide become liquid. The process is frequently used in welding large pieces of iron or steel (Fig. 32-3). The thermite reaction is used in the reduction of metallic oxides that are not readily reduced by carbon, such as MoO_3 and WO_3, or that do not give pure metals when reduced by carbon, such as MnO_2 and Cr_2O_3. The reaction is also of particular value for the production of carbon-free alloys, such as ferrotitanium.

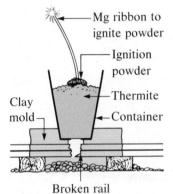

Figure 32-3. Thermite welding of a rail. Molten metal, mostly iron, flows out of the bottom of the container and welds the broken rail without having to remove it from the roadbed.

32.24 Compounds of Aluminum

Aluminum oxide, Al$_2$O$_3$, occurs in nature in pure form as the mineral corundum, a very hard substance that is used as an abrasive for grinding and polishing. Emery is aluminum oxide mixed with Fe_3O_4; it is also used as an abrasive. Several precious stones consist of aluminum oxide with certain impurities that impart color: ruby (red, by chromium compounds); sapphire (blue, by compounds of cobalt, chromium, or titanium); oriental amethyst (violet, by manganese compounds); and oriental topaz (yellow, by iron). Artificial rubies and sapphires are now manufactured by melting aluminum oxide (m.p. = 2050°C) with small amounts of oxides to produce the desired colors and then cooling the melt in such a way as to produce large crystals. These gems are indistinguishable from natural stones, except for microscopic, rounded air bubbles in the synthetic ones and flattened bubbles in the natural stones. The synthetic gems are used not only as jewelry but also as bearings ("jewels") in watches and other instruments and as dies through which wires are drawn. Very finely divided aluminum oxide, called **activated alumina,** is used as a dehydrating agent and a catalyst.

Various aluminum articles, such as drinking tumblers, are given a wear-resistant coating of oxide by anodic oxidation in an electrolyte of chromic, sulfuric, or oxalic acid. This coating readily absorbs dyes and pigments that give a pleasing decorative effect to the article.

When a base is added to a solution of an aluminum salt, a white, gelatinous precipitate of **aluminum hydroxide, Al(OH)₃** ($K_{sp} = 1.9 \times 10^{-33}$), is formed.

$$Al^{3+} + 3OH^- \longrightarrow Al(OH)_3(s)$$

or $\quad [Al(H_2O)_6]^{3+} + 3OH^- \longrightarrow [Al(OH)_3(H_2O)_3](s) + 3H_2O$

Aqueous solutions of sulfides or carbonates precipitate aluminum hydroxide, because such solutions contain hydroxide ions in considerable concentrations due to hydrolysis (see Chapter 16).

$$S^{2-} + H_2O \rightleftharpoons HS^- + OH^-$$

$$CO_3^{2-} + H_2O \rightleftharpoons HCO_3^- + OH^-$$

Aluminum hydroxide is amphoteric and is dissolved by both acids and bases.

$$[Al(OH)_4]^- \xrightarrow{OH^-} Al(OH)_3(s) \xrightarrow{3H^+} Al^{3+} + 3H_2O$$

or $\quad [Al(H_2O)_2(OH)_4]^- \xrightarrow{OH^-} Al(H_2O)_3(OH)_3(s) \xrightarrow{3H^+} [Al(H_2O)_6]^{3+}$

Aluminum oxide, obtained by heating aluminum hydroxide above 850°C, is insoluble in acids and bases. Products obtained below 600°C are soluble in these reagents.

Aluminum hydroxide is widely used to fix dyes to fabrics. Cloth soaked in a hot solution of aluminum acetate becomes impregnated with the aluminum hydroxide formed by hydrolysis of the salt. If aluminum hydroxide is precipitated from a solution containing a dye, the precipitate is colored. The aluminum hydroxide adsorbs the dye and holds it fast to the cloth. When used in this way, aluminum hydroxide is called a **mordant,** and the colored product is called a **lake.**

Aluminum hydroxide is also used in the purification of water because its gelatinous character enables it to carry down any suspended material in the water, including most of the bacteria. The aluminum hydroxide for water purification is ordinarily produced by the reaction of aluminum sulfate with lime.

$$2Al^{3+} + 3SO_4^{2-} + 3Ca^{2+} + 6OH^- \longrightarrow 2Al(OH)_3(s) + 3CaSO_4(s)$$

Anhydrous aluminum chloride, AlCl₃, is a white crystalline solid that sublimes at 180°C and is soluble in organic solvents. Measurements of its density show that it is dimeric in the vapor phase, and the formula should be written Al_2Cl_6. In these molecules each aluminum atom is tetrahedrally bonded to four chlorine atoms; two of these chlorine atoms are each bonded to two aluminum atoms.

Aluminum
chloride

During formation of the dimer one chlorine atom of each monomeric covalent molecule donates a pair of electrons to be shared with the aluminum atom of the other molecule, and thus two chlorine bridges are formed; each aluminum atom then possesses an octet of electrons. X-ray studies show that there are no discrete Al_2Cl_6 molecules in solid aluminum chloride. The formula for the solid compound is usually written $AlCl_3$, although the structure is actually a polymeric layered lattice that does not have individual $AlCl_3$ molecules.

Aluminum chloride can be prepared either by direct chlorination of aluminum or by heating bauxite with carbon and chlorine.

$$2Al(s) + 3Cl_2(g) \xrightarrow{\Delta} 2AlCl_3(s) \qquad\qquad \Delta H° = -1408 \text{ kJ}$$

$$Al_2O_3(s) + 3C(s) + 3Cl_2(g) \xrightarrow{\Delta} 2AlCl_3(s) + 3CO(g) \quad \Delta H° = -64.0 \text{ kJ}$$

If aluminum or aluminum hydroxide is treated with hydrochloric acid and the solution is evaporated, crystals of the hexahydrate $AlCl_3 \cdot 6H_2O$ are formed. When this salt is heated, aluminum oxide is produced.

$$2(AlCl_3 \cdot 6H_2O)(s) \longrightarrow Al_2O_3(s) + 6HCl(g) + 9H_2O(g) \quad \Delta H° = 977.4 \text{ kJ}$$

Aluminum sulfate 18-hydrate, $Al_2(SO_4)_3 \cdot 18H_2O$, is prepared by treating bauxite or clay with sulfuric acid.

$$Al_2Si_2O_5(OH)_4 + 6H^+ + [3SO_4^{2-}] \longrightarrow$$
Clay

$$2H_2SiO_3(s) + 2Al^{3+} + [3SO_4^{2-}] + 3H_2O$$

The metasilicic acid (H_2SiO_3) is removed by filtration. The sulfate is the cheapest soluble salt of aluminum, so it is used as a source of aluminum hydroxide for the purification of water, the dyeing and waterproofing of fabrics, and the sizing of paper.

When solutions of potassium sulfate and aluminum sulfate are mixed and concentrated by evaporation, crystals of a salt called **potassium alum, $KAl(SO_4)_2 \cdot 12H_2O$,** are formed. It is the commonest of a class of double salts known as **alums** and is frequently referred to simply as alum. The unipositive cation in an alum may be that of an alkali metal or silver ion, or the ammonium ion; the tripositive cation may be that of aluminum, chromium, iron, manganese, or some other ion. The selenate ion, SeO_4^{2-}, may replace the sulfate ion. Thus we have ammonium alum, $NH_4Al(SO_4)_2 \cdot 12H_2O$, sodium iron(III) alum, $NaFe(SO_4)_2 \cdot 12H_2O$, and potassium chrome alum, $KCr(SO_4)_2 \cdot 12H_2O$. The alums are all isomorphous; that is, they have the same crystal structure.

As was discovered in Section 27.16, many of the most important silicate rocks contain aluminum. Clay and sand are formed as disintegration products by the weathering of these rocks. Weathering involves the thawing and freezing of water in the rocks and the chemical action of water and carbon dioxide on them. For example, the chemical disintegration of feldspar may be represented by

$$2KAlSi_3O_8 + 2H_2O + CO_2 \longrightarrow 2K^+ + CO_3^{2-} + Al_2Si_2O_5(OH)_4 + 4SiO_2$$
Feldspar   Potassium carbonate   Clay   Sand

The soluble potassium carbonate formed is largely removed by water, and the residue of sand and clay remains as soil. One kind of pure clay has the formula shown above; it is white and is called **kaolin.** Ordinary clay is colored by compounds of iron and other metals. Porcelain and china are made of pure kaolin, while impure clays are used in the manufacture of earthenware products such as tile and brick. The buff or red color of such products is due to the presence of iron(III) silicate.

CHROMIUM

32.25 Periodic Relationships of Chromium

The elements that constitute Group VIB of the Periodic Table are chromium, molybdenum, and tungsten. These metals differ distinctly from the nonmetals of Periodic Group VIA, oxygen and sulfur, which we have already considered (Chapters 9 and 21). Chromium, molybdenum, and tungsten are each used in the production of commercially important alloy steels. These metals are typical transition metals; they show a variety of oxidation numbers, form highly colored compounds, and form stable complex ions. Like those of other transition metals, their lowest oxides are basic, the intermediate ones are amphoteric, and the highest ones are primarily acidic. The members of Groups VIA and VIB form some similar compounds in which they exhibit their higher oxidation numbers; thus sulfates (K_2SO_4) and chromates (K_2CrO_4) are analogous in formula and in some physical and chemical properties.

32.26 Occurrence, Metallurgy, and Properties of Chromium

Chromium does not occur as the element in nature. Its most important ore is chromite, $FeCr_2O_4$, sometimes called chrome iron ore. Practically all of the chromium used in the United States is imported. Because chromium is essential to the production of special alloy steels and the supply is often short, it is classed as a strategic material by the government.

Reduction of chromite by carbon in an electric furnace yields an alloy of iron and chromium, called ferrochrome, which is used in making chromium steels.

$$FeCr_2O_4 + 4C \longrightarrow Fe + 2Cr + 4CO$$

Pure chromium is prepared by a thermite reaction, reducing chromium(III) oxide with aluminum.

$$Cr_2O_3(s) + 2Al(s) \longrightarrow Al_2O_3(s) + 2Cr(s) \qquad \Delta H° = -536.0 \text{ kJ}$$

Chromium can also be produced by the electrolytic reduction of its compounds from aqueous solution, a process used in chromium plating.

Chromium is a very hard, silvery white, crystalline metal. It becomes passive when coated with a thin layer of oxide, which protects it against further corrosive attack. This property, along with its metallic luster, accounts for its extensive use in the plating of iron and copper objects, such as plumbing fixtures and automobile trim. Passivity is produced by treatment with concentrated nitric acid or chromic acid or by exposure of the metal to air.

Chromium is a transition metal that shows several oxidation numbers; the principal ones are +2, +3, and +6. The compounds of chromium are highly colored. The particular color exhibited is dependent to some extent on a specific oxidation number and on the structure of the compound.

32.27 Compounds of Chromium

Most compounds containing chromium(II) are oxidized by air, so they are prepared under nitrogen or some other inert atmosphere. **Chromium(II) chloride, CrCl$_2$,** can be prepared (under nitrogen) by dissolving chromium in hydrochloric acid or by reducing a solution of chromium(III) chloride with zinc in the presence of an acid. The anhydrous salt is colorless, but its solutions have the bright blue color of the hydrated ion $[Cr(H_2O)_6]^{2+}$. This ion is rapidly oxidized in air to the chromium(III) ion, $[Cr(H_2O)_6]^{3+}$; for this reason solutions of chromium(II) salts are frequently employed to remove the last traces of oxygen from gases. **Chromium(II) sulfate 7-hydrate, CrSO$_4 \cdot$ 7H$_2$O,** is prepared by dissolving the metal in sulfuric acid. **Chromium(II) acetate, Cr(C$_2$H$_3$O$_2$)$_2$,** forms as a red precipitate when a saturated solution of sodium acetate is added to a solution of chromium(II) chloride. Chromium(II) acetate is one of the few exceptions to the rule that acetates are soluble. **Chromium(II) hydroxide** is basic, and chromium(II) salts are only slightly hydrolyzed.

Green **chromium(III) oxide, Cr$_2$O$_3$,** is formed by heating the metal in air, by igniting ammonium dichromate, or by reducing a dichromate with sulfur or another similar reducing agent.

$$(NH_4)_2Cr_2O_7(s) \xrightarrow{\Delta} N_2(g) + 4H_2O(g) + Cr_2O_3(s) \qquad \Delta H^\circ = -300 \text{ kJ}$$

$$Na_2Cr_2O_7 + S \longrightarrow Na_2SO_4 + Cr_2O_3$$

This oxide is used as a pigment called **chrome green.** Bluish-green gelatinous **chromium(III) hydroxide** is precipitated ($K_{sp} = 6.7 \times 10^{-31}$) when a soluble hydroxide, a soluble sulfide, or a carbonate is added to a solution of a chromium(III) compound. Chromium(III) hydroxide is amphoteric, dissolving in both acids and bases.

$$Cr(OH)_3 + 3H^+ \longrightarrow Cr^{3+} + 3H_2O$$

$$Cr(OH)_3 + OH^- \longrightarrow [Cr(OH)_4]^- \quad \text{(chromite ion)}$$

Chromium(III) chloride 6-hydrate, CrCl$_3 \cdot$ 6H$_2$O, and the **sulfate 18-hydrate, Cr$_2$(SO$_4$)$_3 \cdot$ 18H$_2$O,** are the best known chromium(III) salts. Dilute solutions of the chloride are violet; the hexaaquachromium(III) ion, $[Cr(H_2O)_6]^{3+}$ is responsible for the color. In more concentrated solutions of the chloride, green $[Cr(H_2O)_4Cl_2]^+$ is formed. The hydrates of chromium(III) sulfate also exist in violet and green modifications. Chromium(III) forms alums of the type KCr(SO$_4$)$_2 \cdot$ 12H$_2$O, solutions of which are bluish-violet when cold and green when hot. Chrome alum is used in the tanning of leather, the printing of calico, and the waterproofing of fabrics, and as a mordant. It is the most important of the soluble chromium salts.

With its +6 oxidation number chromium is usually combined with oxygen in the form of the **chromium trioxide, CrO$_3$,** or the oxyanions chromate, CrO_4^{2-}, and dichromate, $Cr_2O_7^{2-}$. **Potassium chromate, K$_2$CrO$_4$,** is produced commercially by the oxidation of chromite ore that is mixed and heated in air with potassium carbonate.

$$4FeCr_2O_4 + 8K_2CO_3 + 7O_2 \xrightarrow{\Delta} 2Fe_2O_3 + 8K_2CrO_4 + 8CO_2$$

Lead chromate, PbCrO$_4$, and **barium chromate, BaCrO$_4$,** are both insoluble in water and are used as yellow pigments.

When an acid is added to a solution containing chromate ions, the solution changes from yellow to orange-red due to the formation of the **dichromate ion, $Cr_2O_7^{2-}$.**

$$2CrO_4^{2-} + 2H^+ \rightleftharpoons Cr_2O_7^{2-} + H_2O$$

The addition of a base or dilution with water reverses the reaction.

$$Cr_2O_7^{2-} + 2OH^- \rightleftharpoons 2CrO_4^{2-} + H_2O$$

The conversion of chromate to dichromate involves the formation of an oxygen bridge between two chromium atoms. The Lewis structure of the dichromate ion is

$$\left[\begin{array}{ccc} & \ddot{O} & \ddot{O} \\ \ddot{O} : Cr : \ddot{O} : Cr : \ddot{O} \\ & \ddot{O} & \ddot{O} \end{array} \right]^{2-}$$

A mixture of sodium dichromate and sulfuric acid is used as a cleaning solution for laboratory glassware. The cleansing action is due to the strong oxidizing power of the dichromate and the dehydrating action of the concentrated sulfuric acid.

The dichromates are used as oxidizing agents in many reactions. The dichromate ion is readily reduced to the chromium(III) ion by reducing agents such as sulfurous acid.

$$Cr_2O_7^{2-} + 3H_2SO_3 + 2H^+ \longrightarrow 2Cr^{3+} + 3SO_4^{2-} + 4H_2O$$

As the reduction progresses, a change from the orange color of the dichromate ion to the green color of the chromium(III) ion occurs.

The trioxide CrO_3 is produced as scarlet needle-shaped crystals when concentrated sulfuric acid is added to a concentrated solution of potassium dichromate.

$$Cr_2O_7^{2-} + 2H^+ \longrightarrow 2CrO_3(s) + H_2O$$

The anhydride CrO_3 reacts with water in different proportions forming **chromic acid, H_2CrO_4,** or **dichromic acid, $H_2Cr_2O_7$.**

$$CrO_3 + H_2O \longrightarrow 2H^+ + CrO_4^{2-}$$
$$2CrO_3 + H_2O \longrightarrow 2H^+ + Cr_2O_7^{2-}$$

These solutions are used as active oxidizing agents. When heated, the trioxide decomposes, liberating oxygen and forming green chromium(III) oxide.

$$4CrO_3(s) \longrightarrow 2Cr_2O_3(s) + 3O_2(g) \qquad \Delta H° = 497.1 \text{ kJ}$$

32.28   Uses of Chromium

A large portion of the chromium produced goes into steel alloys, which are very hard and strong. **Stainless steel,** which usually contains chromium and some nickel, is used in the manufacture of cutlery because of its corrosion resistance. Nonferrous chromium alloys include **nichrome** and **chromel** (Ni and Cr), which are used in various heating devices because of their electrical resistance. Chromium is widely used as a protective and decorative coating for other metals, in such products as plumbing fixtures and automobile trim.

The compounds of chromium have many uses, including uses as paint pigments and mordants and in the tanning of leather. Certain refractories used as linings for high-temperature furnaces are made by mixing pulverized chromite ore with clay or magnesia, MgO.

ZINC

32.29 Occurrence and Metallurgy of Zinc

Sphalerite, or zinc blende, ZnS, is the principal ore of zinc. Less important ores include zincite, ZnO, smithsonite, $ZnCO_3$, franklinite, a mixture of oxides of zinc, iron, and manganese, and willemite, Zn_2SiO_4.

In the metallurgy of zinc, sulfide ores are first concentrated by flotation and then roasted to convert zinc sulfide to zinc oxide. From carbonate ores zinc oxide is obtained by simply heating.

$$2ZnS(s) + 3O_2(g) \xrightarrow{\Delta} 2ZnO(s) + 2SO_2(g) \qquad \Delta H° = -878.3 \text{ kJ}$$

$$ZnCO_3(s) \xrightarrow{\Delta} ZnO(s) + CO_2(g) \qquad \Delta H° = 71.0 \text{ kJ}$$

The zinc oxide is reduced by heating it with coal in a fire-clay retort. As rapidly as zinc is produced, it distills out and is condensed. It contains impurities of cadmium, iron, lead, and arsenic, but it can be purified by careful redistillation.

An electrolytic process is also employed to produce zinc. The ore is roasted in special furnaces under conditions such that most of the sulfide is converted to the sulfate and the remainder is changed to the oxide. The roasted ore is leached with dilute sulfuric acid to dissolve the zinc sulfate and zinc oxide, as well as the sulfates and oxides of other metals. Silver sulfate and lead sulfate remain in the residue because they are insoluble. The resulting solution of zinc sulfate is treated with powdered zinc to reduce the less active metals such as copper, arsenic, and antimony to the elemental state. Any iron(II) present is oxidized by passing air through the solution. Manganese and iron are precipitated as hydroxides by adding a calculated amount of lime. After the solution is purified in this way, it is electrolyzed, and zinc is deposited on aluminum cathodes. Sulfuric acid is regenerated as electrolysis proceeds and is used in the leaching of additional roasted ore. At intervals the cathodes are removed, and the zinc is scraped off, melted, and cast into ingots. Electrolytic zinc has a purity of 99.95%.

32.30 Properties and Uses of Zinc

The periodic relationships of zinc were discussed in Section 30.1. Zinc is a silvery metal that quickly tarnishes to a blue-gray appearance. This color is due to an adherent coating of a basic carbonate, $Zn_2(OH)_2CO_3$, which protects the underlying metal from further corrosion. The metal is hard and brittle at ordinary temperatures but ductile and malleable at 100–150°C. When molten zinc is poured into cold water, it solidifies in irregular masses called **granulated,** or **mossy, zinc.** Zinc is a fairly active metal and will reduce steam at high temperatures.

$$Zn(s) + H_2O(g) \xrightarrow{\Delta} ZnO(s) + H_2(g) \qquad \Delta H° = -106.4 \text{ kJ}$$

Zinc dissolves in strong bases, liberating hydrogen and forming zincate ions.

$$Zn + 2OH^- + 2H_2O \longrightarrow [Zn(OH)_4]^{2-} + H_2(g)$$

A large amount of zinc is used in the manufacture of dry cells (Section 20.16) and in the production of alloys such as brass and bronze. About half of the zinc metal produced is used to protect iron and other metals from corrosion by air and water. Its protective action results from the basic carbonate film on its surface. Furthermore, it protects a less active metal such as iron because it forms an electrochemical cell with the iron, in which zinc serves as the anode and iron as the cathode (see Section 32.18). The zinc coating on iron may be applied in several ways, and the product is called **galvanized iron.**

32.31 Compounds of Zinc

Zinc oxide, ZnO, is produced when zinc vapor is burned and when zinc ores are roasted in an excess of air. The oxide is yellow when hot but pure white when cold. It is used as a paint pigment called zinc white or Chinese white; unlike white lead it does not blacken with hydrogen sulfide, because zinc sulfide is white. Zinc oxide is also used in the manufacture of automobile tires and other rubber goods and in the preparation of medicinal ointments.

Zinc hydroxide, Zn(OH)$_2$, is formed as a white gelatinous precipitate (K_{sp} = 4.5×10^{-17}) when a soluble hydroxide is added to a solution of a zinc salt.

$$Zn^{2+} + 2OH^- \longrightarrow Zn(OH)_2(s)$$

Zinc hydroxide is amphoteric and dissolves in an excess of base as well as in acids.

$$Zn(OH)_2 + 2OH^- \longrightarrow [Zn(OH)_4]^{2-} \quad \text{(zincate ion)}$$
$$Zn(OH)_2 + 2H^+ \longrightarrow Zn^{2+} + 2H_2O$$

Zinc hydroxide is soluble in aqueous ammonia, forming **tetraamminezinc hydroxide.**

$$Zn(OH)_2 + 4NH_3 \longrightarrow [Zn(NH_3)_4]^{2+} + 2OH^-$$

When zinc and chlorine react, **anhydrous zinc chloride, ZnCl$_2$,** is formed as a white deliquescent solid. It is also made by dissolving zinc metal, zinc oxide, or zinc carbonate in hydrochloric acid, evaporating the solution to dryness, and then fusing the residue to remove the last traces of moisture. The product, which melts at 262°C, is cast into sticks. The anhydrous salt is used as a caustic in surgery and as a dehydrating agent in certain organic reactions. Aqueous solutions of zinc chloride are acidic due to hydrolysis. Because of its acidic nature a solution of zinc chloride (with ammonium chloride) is used to dissolve the oxides on metal surfaces before soldering. Concentrated solutions of zinc chloride dissolve cellulose, forming a gelatinous mass, which may be molded into various shapes. Fiberboard is made of this material. Zinc oxide dissolves in concentrated solutions of zinc chloride, producing **zinc oxychloride, Zn$_2$OCl$_2$,** which sets to a hard mass and is used as a cement.

Zinc sulfate, ZnSO$_4$, is produced in large quantities by roasting zinc blende at low red heat.

$$ZnS(s) + 2O_2(g) \xrightarrow{\Delta} ZnSO_4(s) \quad \Delta H° = -777.0 \text{ kJ}$$

The product is extracted with water and recovered by recrystallization to form crystals of zinc sulfate 7-hydrate, $ZnSO_4 \cdot 7H_2O$, which is called white vitriol. It is used principally in the production of the white paint pigment lithopone, which is a mixture of barium sulfate and zinc sulfide formed by the reaction

$$Ba^{2+} + S^{2-} + Zn^{2+} + SO_4{}^{2-} \longrightarrow BaSO_4(s) + ZnS(s)$$

Zinc sulfide, ZnS, is found in nature as zinc blende and is formed in the laboratory when ammonium sulfide is added to a solution of a zinc salt ($K_{sp} = 1.1 \times 10^{-21}$). The sulfide is insoluble in acetic acid but dissolves readily in stronger acids such as hydrochloric or sulfuric.

$$ZnS(s) + 2H^+ \longrightarrow Zn^{2+} + H_2S(g)$$

Zinc sulfide is used as a white paint pigment, either alone or mixed with zinc oxide.

FOR REVIEW

SUMMARY

The metals of Analytical Group III are nickel, cobalt, manganese, iron, aluminum, chromium, and zinc. The ions of these metals are precipitated as sulfides or hydroxides in solutions of aqueous ammonia saturated with hydrogen sulfide, but they are not precipitated by hydrochloric acid or by hydrogen sulfide in solutions that are 0.3 M in hydrogen ions.

The chemistry of the metals is interesting, and many of their compounds are commercially important.

The metallurgy of iron leads to the manufacture of steel, which largely is accomplished by a process known as the basic oxygen process. The corrosion of iron, which occurs only in the presence of both water and air, takes place through an electrochemical process in which the presence of a less active metal than iron accelerates the corrosion of iron. The presence of a more active metal slows down the corrosion. The material known as iron rust is a hydrated form of iron(II) oxide, $Fe_2O_3 \cdot nH_2O$.

EXERCISES

Nickel

1. Why does cobalt precede nickel in the Periodic Table when cobalt has a higher atomic weight than nickel?
2. What is meant by *passivity?*
3. How is nickel rendered passive?
4. What is the composition of Alnico? What use is made of this alloy?

5. Write the equation for the dissolution of nickel(II) hydroxide in aqueous ammonia.
6. What is the physical nature of nickel carbonyl?

Cobalt

7. Sketch the geometrical structure of $[Co(CN)_6]^{3-}$.

8. Balance the following equation by oxidation-reduction methods:

$$Co(NO_3)_2 \longrightarrow Co_2O_3 + NO_2 + O_2$$

9. On the basis of the reactions of elemental fluorine, chlorine, and oxygen with metallic cobalt, predict which is the strongest oxidizing agent.

10. What is the percent by mass of cobalt in sodium hexanitrocobaltate(III)?   *Ans. 14.59%*

Manganese

11. How are the ferromanganese alloys produced?

12. Write a balanced equation for the reaction of manganese dioxide and hydrochloric acid.

13. Explain the dissolution of manganese(II) hydroxide in aqueous ammonium chloride in terms of an acid-base reaction.

14. Mention two uses of manganese dioxide.

15. Write the names and formulas for compounds in which manganese exhibits each of the following oxidation numbers: $+2$, $+3$, $+4$, $+6$, and $+7$.

16. How is potassium permanganate prepared commercially?

17. Would you expect a calcium manganite solution to have a pH greater or less than 7.0? Justify your answer.

18. How can the insolubility of barium sulfate be applied in the preparation of free permanganic acid?

Iron

19. Write the names and formulas of the principal ores of iron.

20. List the equations for the reactions that take place in the reduction of iron(III) oxide in a blast furnace.

21. What use is made of the hot exhaust gases from a blast furnace?

22. How many cubic feet of air are required per ton of Fe_2O_3 to convert it into iron in a blast furnace? Assume air is 19% oxygen.   *Ans. $3.5 \times 10^4 ft^3$*

23. Give the compositions and distinguishing properties of pig iron, cast iron, and steel.

24. Identify each of the following: hematite, magnetic oxide of iron, rouge, Mohr's salt, and Prussian blue.

25. What is the composition of iron rust?

26. Describe the chemistry of an invisible ink.

27. What compound of iron is a common constituent of hard water?

28. What is the gas produced when iron(II) sulfide is treated with a nonoxidizing acid?

29. Give the general equation for the hydrolysis of iron(III) salts. Write equations showing the steps in the hydrolysis of iron(III) chloride to produce a precipitate of hydrated $Fe(OH)_3$ in solution, $[Fe(H_2O)_3(OH)_3]$.

30. Iron(II) can be titrated to iron(III) by dichromate ion, which is reduced to chromic ion in acid solution. A 2.500-g sample of iron ore is dissolved and the iron converted to the iron(II) form. Exactly 19.17 mL of 0.0100 M $Na_2Cr_2O_7$ is required in the titration. What percentage of iron was the ore sample?   *Ans. 2.569%*

Aluminum

31. Describe the production of metallic aluminum by electrolytic reduction.

32. What is the reaction of sodium hydroxide with aluminum?

33. Illustrate the amphoteric nature of aluminum hydroxide by suitable equations.

34. How is the amphoteric nature of aluminum ion used in the processing of bauxite?

35. Why can aluminum, which is an active metal, be used so successfully as a structural metal?

36. What is the composition of alums as a class of double salts?

37. Explain the function of aluminum hydroxide in water purification.

38. Why is it impossible to prepare anhydrous aluminum chloride by heating the hexahydrate to drive off the water?

39. What is the Goldschmidt process?

40. Write balanced chemical equations for the following reactions:

(a) Gaseous hydrogen fluoride is bubbled through a suspension of bauxite in molten sodium fluoride.

(b) Metallic aluminum is burned in air.

(c) Metallic aluminum is added to a solution of sulfuric acid, ammonia is added, and the solution is evaporated to give crystals.

(d) Aluminum is heated in an atmosphere of chlorine.

(e) Aluminum sulfide is added to water.

(f) Aluminum hydroxide is added to a solution of nitric acid.

Chromium

41. Write a balanced chemical equation describing the oxidation by air of aqueous chromium(II) to chromium(III).
42. Write an equation for the decomposition of ammonium dichromate by ignition.
43. Account for the extensive use of chromium in the plating of iron and copper objects, such as the trim on automobiles.
44. Reduction of chromite ore by carbon yields an alloy of iron and chromium. How would you separate the two components of this alloy into solutions of their respective cations, if this were necessary?
45. What is the reaction between hydrogen ions and chromate ions? This is called a condensation reaction. Explain.
46. Write an equation to illustrate the action of the dichromate ion as an oxidizing agent.
47. What is the composition and chemical action of the cleaning solution used in the laboratory?

Zinc

48. How is zinc metal separated from the impurities cadmium, iron, lead, and arsenic during the refining process?
49. Why does pure zinc react only slowly with dilute hydrochloric or sulfuric acids?
50. What is galvanized iron, how is it produced, and what properties does it have that make it useful?
51. A dilute solution is dripped into a slowly stirred solution of sodium zincate, $Na_2[Zn(OH)_4]$. A white gelatinous precipitate is formed that, upon analysis, proves to be a hydroxide. Use a chemical equation to explain these observations.
52. What is lithopone, and how is it produced? In what way is lithopone superior to white lead as a paint pigment?
53. Write balanced chemical equations for the following reactions:
    (a) Zinc is burned in air.
    (b) An aqueous solution of ammonia is added dropwise to a solution of zinc chloride until the mixture becomes clear again.
    (c) Zinc is heated with sulfur.
    (d) Mossy zinc is added to a solution of hydrobromic acid.

Additional Exercises

54. From the electronic structure of thallium, what can you predict concerning the solubility of thallium(I) chloride? the basicity of thallium(I) hydroxide?
55. In what ways are the group VIB metals typical of transition metals in general?
56. Balance the following equations:
    (a) $Al + H^+ \rightleftharpoons Al^{3+} + H_2$
    (b) $NO_2^- + Al + OH^- + H_2O \rightleftharpoons$
        $$NH_3 + [Al(OH)_4]^-$$
    (c) $Cr_2O_7^{2-} + Fe^{2+} + H^+ \rightleftharpoons$
        $$Cr^{3+} + Fe^{3+} + H_2O$$
    (d) $SCN^- + MnO_4^- + H^+ \rightleftharpoons$
        $$SO_4^{2-} + CN^- + Mn^{2+} + H_2O$$
    (e) $Zn + OH^- \rightleftharpoons [Zn(OH)_4]^{2-} + H_2$

33

METALS OF ANALYTICAL GROUP IV

CALCIUM, STRONTIUM, AND BARIUM

Analytical Group IV (see also Chapter 39) contains those common cations that are precipitated as carbonates by ammonium carbonate from an aqueous ammonia solution containing ammonium chloride, but that are not precipitated by the reagents used in the precipitation of Analytical Groups I, II, and III. All the other cations, except magnesium, that form insoluble carbonates are separated in Groups I, II, and III. If ammonium chloride is not present, magnesium carbonate will precipitate with the ions of Group IV. However, precipitation of magnesium carbonate is not desirable because it is moderately soluble and of indefinite composition. The precipitation of Mg^{2+} is therefore postponed. Some properties of the Analytical Group IV metals are given in Table 33-1.

33.1 Periodic Relationships of Calcium, Strontium, and Barium

All elements of Periodic Group IIA (Be, Mg, Ca, Sr, Ba, and Ra), called the **alkaline earth metals,** have two electrons in their outer shells. They all show a single oxidation number, $+2$. They are all very reactive metals, the reactivity increasing with increasing atomic number. The members of Group IIB (Zn, Cd,

Table 33-1   Important Properties of the Metals of Analytical Group IV

Property	Calcium	Strontium	Barium
Atomic number	20	38	56
Atomic weight	40.08	87.62	137.33
Oxidation number	$+2$	$+2$	$+2$
Reduction potential, V	(Ca^{2+}/Ca) -2.87	(Sr^{2+}/Sr) -2.89	(Ba^{2+}/Ba) -2.90
Density, g/cm^3, at 20°C	1.54	2.60	3.5
Melting point, °C	850	770	704
Boiling point, °C	1490	1384	1638

and Hg), on the other hand, are heavier than the corresponding Group IIA elements of the same series, and the reactivities of Group IIB elements decrease with increasing atomic number. Beryllium, the first member of Group IIA, resembles aluminum, the second member of Group IIIA of the Periodic Table, in its chemical behavior. Beryllium and magnesium differ somewhat in their chemical properties from the other elements of Group IIA. For example, the hydroxides of beryllium and magnesium are nearly insoluble in water and are easily decomposed by heat into water and metal oxides. The hydroxide of beryllium, the smallest and most electronegative element of the group, is amphoteric, while the hydroxides of barium, strontium, and calcium are strong bases and their hydrated ions do not hydrolyze appreciably. Beryllium ion forms stable complex ions, but with a few exceptions the ions of other elements of Group IIA do not. Unlike the salts of the alkali metals (Chapter 34), many of the common salts of the alkaline earth metals are insoluble in water. Most common alkaline earth metal salts of weak or moderately strong acids are sparingly soluble in water (for example, the sulfates, phosphates, oxalates, and carbonates). Some are quite soluble in water, however, especially the cyanides, sulfides, and acetates. The solubility of the hydroxides of these metals in water increases with increasing atomic number, while the solubility of the carbonates and sulfates decreases with increasing atomic number.

33.2 Occurrence of Calcium, Strontium, and Barium

Because of their high chemical activity, these metals never occur as the elements in nature. Their most abundant naturally occurring compounds are the carbonates and the sulfates.

Calcium is the fifth most abundant element in the earth's crust and is widely distributed in nature. The common rock known as **limestone** is an impure calcium carbonate. Other natural forms of calcium carbonate include calcite, aragonite, chalk, marl, marble, marine shells, and pearls. Dolomitic limestone (dolomite) is $CaCO_3 \cdot MgCO_3$. Calcium sulfate occurs in anhydrite, $CaSO_4$, and gypsum, $CaSO_4 \cdot 2H_2O$. Other common minerals include fluorite, or fluorspar, which is CaF_2, and apatite, which may be formulated as $Ca_5(PO_4)_3(OH, Cl, F)$. Hydroxyapatite is the chief constituent of the bones and teeth of animals.

The minerals containing strontium are much rarer than those of calcium. The chief ones are celestite, $SrSO_4$, and strontianite, $SrCO_3$.

Barium occurs more abundantly than strontium but much less so than calcium. Its chief minerals are barite, or heavy spar, $BaSO_4$, and witherite, $BaCO_3$.

33.3 Preparation, Properties, and Uses of Calcium, Strontium, and Barium

The most important methods for producing these metals are those in which their molten salts (generally the chlorides) are electrolyzed. Because of the high chemical activity of these metals, their ions are not readily reduced by other methods. However, in the King process for producing barium, a mixture of barium oxide and barium peroxide is reduced with aluminum in a vacuum furnace. A temperature of 950–1100°C is required for the reaction to take place, and at the very low pressure (10^{-3} to 10^{-4} mm) that is maintained in the furnace the metallic barium distills off and is collected on a cold surface.

These metals are all silvery white, crystalline, malleable, and ductile. Calcium is harder than lead, strontium is about as hard as lead, and barium is quite soft. Calcium, strontium, and barium are all active metals, and their activities increase as their atomic numbers do. For example, calcium does not react with oxygen unless it is heated, while strontium oxidizes rapidly when exposed to air, and barium is spontaneously flammable in moist air. When heated, these metals combine with hydrogen, forming hydrides; with sulfur, forming sulfides; with halogens, forming halides; with nitrogen, forming nitrides; and with phosphorus, forming phosphides. Each of the metals displaces hydrogen from water, and the corresponding hydroxide is formed.

Calcium is used as a dehydrating agent for certain organic solvents, as a reducing agent in the production of certain metals, as a scavenger (to remove gases in molten metals) in metallurgy, as a hardening agent for lead used for covering cables and making storage battery grids and bearings, in steel-making when alloyed with silicon, and for many other purposes.

Elemental strontium is not abundant and has no commercial uses. Barium is used as a degasing agent in the manufacture of vacuum tubes, and alloys of barium and nickel are used in vacuum tubes and spark plugs because of their high thermionic electron emission. It is interesting that Mg^{2+} and Ca^{2+} are not poisonous but Be^{2+} and Ba^{2+} are very much so.

33.4 Oxides of Calcium, Strontium, and Barium

When the carbonates of these metals are heated sufficiently, they undergo a decomposition that yields the respective oxide and carbon dioxide.

$$MCO_3 \longrightarrow MO + CO_2$$

The temperature required for these decompositions increases with increasing basicity of the oxide. The temperature required to attain equilibrium pressure of carbon dioxide of 1 atm is about 900°C for $CaCO_3$, 1250°C for $SrCO_3$, and 1450°C for $BaCO_3$.

Pure calcium oxide is a white substance that emits an intense light, called limelight, when heated to a high temperature. Calcium oxide reacts vigorously and exothermally with water (Section 33.5) and is therefore often employed as a drying agent, for example, in the preparation of anhydrous alcohol and in the drying of ammonia.

In the production of barium oxide, barium carbonate is heated with finely divided carbon. The purpose of the carbon is to reduce the carbon dioxide that is produced and thus to shift the equilibrium

$$BaCO_3 \rightleftharpoons BaO + CO_2$$

to the right. This makes possible the conversion of the carbonate to the oxide at a lower temperature than the 1450°C otherwise necessary. Most of the barium oxide produced is used in the preparation of barium peroxide.

Calcium, strontium, and barium form peroxides of the general formula MO_2. Only barium peroxide is easily formed and important. It is produced when barium oxide is heated in air or oxygen to 500–600°C.

$$2BaO + O_2 \rightleftharpoons 2BaO_2$$

When barium peroxide is heated to 700–800°C, the reaction is reversed, and oxygen is liberated. The equilibrium can also be shifted by varying the partial pressure of the oxygen. Barium peroxide is used for bleaching various materials, of both plant and animal origin.

33.5 Hydroxides of Calcium, Strontium, and Barium

When the oxides of calcium, strontium, and barium react with water, the corresponding hydroxides are formed, and large amounts of heat are evolved.

$$CaO(s) + H_2O(l) \longrightarrow Ca(OH)_2(s) \qquad \Delta H° = -65.3 \text{ kJ}$$
$$SrO(s) + H_2O(l) \longrightarrow Sr(OH)_2(s) \qquad \Delta H° = -83.3 \text{ kJ}$$
$$BaO(s) + H_2O(l) \longrightarrow Ba(OH)_2(s) \qquad \Delta H° = -103 \text{ kJ}$$

Both the heat of hydration and the reactivity of the oxides toward water increase with increasing cation size. The solubility of the hydroxides in water increases in the same direction—the solubilities in moles per liter at 20°C are $Ca(OH)_2$, 0.025; $Sr(OH)_2 \cdot 8H_2O$, 0.043; and $Ba(OH)_2 \cdot 8H_2O$, 0.23.

Calcium oxide is known by various names, such as lime, quicklime, and unslaked lime, while **calcium hydroxide, Ca(OH)_2,** which is formed from reaction of CaO with water, is known commercially as slaked, or hydrated, lime. Calcium hydroxide is a dry white powder that forms a pasty mass with an excess of water. Because of the high heat of reaction of quicklime with water, it should always be stored where it cannot come in contact with water, since the slaking reaction (reaction of the oxide with water to form the hydroxide) generates enough heat to ignite paper or wood. Dry hydrated lime rather than quicklime is usually sold, because it can be shipped and stored in paper bags without risk of fire.

Calcium hydroxide is classed as a strong base and, because of its activity and cheapness, it is used more extensively in commercial processes than any other base. During 1982 over 14 million tons of lime and $Ca(OH)_2$ were produced in the United States, making them fifth in industrial production, after sulfuric acid, nitrogen, ammonia, and oxygen. Calcium hydroxide is sparingly soluble in water, only 1.7 grams dissolving in a liter of water at room temperature. A saturated solution of calcium hydroxide is often called **limewater.** A suspension of calcium hydroxide in water is known as **milk of lime.** Even though calcium hydroxide has a limited solubility, milk of lime furnishes a ready supply of hydroxide ions, because the solid continues to dissolve as hydroxide ions are removed from solution during chemical reaction; that is, the equilibrium shifts to the right.

$$Ca(OH)_2 \rightleftharpoons Ca^{2+} + 2OH^-$$
$$\text{(Solid)} \qquad\qquad \text{(Dissolved)}$$

Mortar, widely used in the construction industry, is prepared by mixing slaked lime with sand and water. One part of slaked lime is used with about three or four parts of sand, and enough water is added to give the mixture the consistency of paste. It dries and becomes hard (sets), when exposed to air, by adsorbing carbon dioxide and forming crystalline calcium carbonate.

$$Ca(OH)_2 + CO_2 \longrightarrow CaCO_3 + H_2O$$

The calcium carbonate cements together the particles of sand and unchanged calcium hydroxide. Many years are required for the complete conversion of the calcium hydroxide to carbonate, especially in heavy wall construction.

Ordinary **lime plaster** for coating walls and ceilings is similar to mortar, with some binding material such as hair or fiber to help hold it in place. Cement was discussed in Section 27.19.

The manufacture of at least 150 important industrial chemicals requires the use of lime. In fact, only five other raw materials are used more frequently than lime (or limestone, from which lime is made); these are salt, coal, sulfur, air, and water.

Strontium hydroxide is a rather active base, which is made by treating strontium carbonate with superheated steam.

$$SrCO_3 + H_2O \longrightarrow Sr(OH)_2 + CO_2$$

Strontium hydroxide is fairly soluble in hot water and crystallizes from solution as the octahydrate $Sr(OH)_2 \cdot 8H_2O$.

Barium hydroxide is made by slaking barium oxide and by treatment of barium carbonate with superheated steam. Barium hydroxide is the most soluble of the alkaline earth hydroxides, and it crystallizes from solution in the form of the hydrate $Ba(OH)_2 \cdot 8H_2O$. It is widely used in analytical chemistry because it has an advantage over hydroxides of sodium and potassium in that any carbonate that is formed due to contact with air precipitates out as barium carbonate, leaving the solution relatively pure.

33.6 Calcium Carbonate

Calcium carbonate occurs in two distinctly different crystal forms, which are known as calcite and aragonite. The first is common, while the second is comparatively rare.

Calcite is the low-temperature form of calcium carbonate and results when precipitation occurs below 30°C. Above this temperature calcium carbonate crystallizes in rhombic prisms of aragonite, the high-temperature form. Calcite is known in many varieties, one of which is onyx. Calcite possesses the property of **birefringence,** or double refraction; when a beam of light enters the crystal, it is broken into two beams (Fig. 33-1).

Limestone, by far the most abundant source of calcium carbonate, is second in abundance only to silicate rocks in the earth's crust. Limestone is used more extensively than any other building stone in the United States; more than 80% of it is quarried in southern Indiana. Coral reefs are composed largely of calcium carbonate of marine-animal origin. A pearl is built up of concentric layers of calcium carbonate deposited around a foreign particle, such as a small grain of sand, that has entered the shell of an oyster.

Calcium carbonate is only slightly soluble in pure water. However, it dissolves readily in water that contains dissolved carbon dioxide because it is converted to the more soluble **calcium hydrogen carbonate, $Ca(HCO_3)_2$.**

Figure 33-1. Calcite, or Iceland spar, has the unusual property of birefringence, or double refraction. Notice that the calcite produces two images of the word CALCITE, while the glass produces only one image. (*Courtesy of Bausch & Lomb Incorporated*)

$$CaCO_3 + H_2O + CO_2 \rightleftharpoons Ca^{2+} + 2HCO_3^-$$

This reaction is involved in the formation of limestone caves. When water containing carbon dioxide comes in contact with limestone rocks, it dissolves them. If water containing $Ca(HCO_3)_2$ finds its way into a cave where the $Ca(HCO_3)_2$ may liberate carbon dioxide, calcium carbonate may again be deposited, according to the reaction

$$Ca^{2+} + 2HCO_3^- \longrightarrow CaCO_3 + H_2O + CO_2$$

This reaction is the basis for the formation of **stalactites,** which hang from the ceiling of limestone caves, and of **stalagmites,** which form upward from the floor of the cave where the water has dripped (Fig. 33-2).

Calcium hydrogen carbonate is readily converted to calcium carbonate when its solutions are heated. This reaction is involved in the formation of scale in kettles and boilers and is the basis for one method of removing carbonate hardness from water (Section 12.11).

Figure 33-2. Stalactites and stalagmites are cave formations of calcium carbonate. These are in the Luray Caverns in Virginia. (*Courtesy of Caverns of Luray*)

33.7 Chlorides of Calcium, Strontium, and Barium

Calcium chloride, $CaCl_2$, is obtained commercially and may be produced in the laboratory by treating calcium oxide, hydroxide, or carbonate with hydrochloric acid. The salt crystallizes from water as the hexahydrate $CaCl_2 \cdot 6H_2O$, which is converted by heating to the monohydrate $CaCl_2 \cdot H_2O$. Complete dehydration occurs when the monohydrate is heated to a higher temperature, but some hydrolysis always takes place during the dehydration.

$$CaCl_2 + H_2O \xrightarrow{\Delta} CaO + 2HCl(g)$$

The anhydrous salt and the monohydrate are used extensively as drying agents for gases and liquids. Calcium chloride forms compounds with ammonia, $CaCl_2 \cdot 8NH_3$, and with alcohol, $CaCl_2 \cdot 4C_2H_5OH$, and so cannot be used for drying these substances. Calcium chloride 6-hydrate is very soluble in water and with ice makes an excellent freezing mixture, giving temperatures as low as $-55°C$. Solutions of calcium chloride are quite commonly used as cooling brines in refrigeration plants. Because it is a very deliquescent salt, $CaCl_2 \cdot 6H_2O$ is used to keep dust down on highways and in coal mines. Calcium chloride is frequently used to remove snow and ice from roads and walks.

Strontium chloride 6-hydrate, $SrCl_2 \cdot 6H_2O$, is made by treating strontium carbonate with hydrochloric acid and evaporating the resulting solution. It is used, as are other salts of strontium, to produce a red color in fireworks, railway fuses, and military flares.

Barium chloride, $BaCl_2$, is prepared by the high-temperature reduction of barium sulfate with carbon to form barium sulfide, which is then treated with hydrochloric acid.

$$BaSO_4 + 4C \xrightarrow{\Delta} BaS(s) + 4CO(g)$$
$$BaS + 2H^+(aq) + [2Cl^-(aq)] \longrightarrow Ba^{2+}(aq) + [2Cl^-(aq)] + H_2S(g)$$

When solutions of barium chloride are evaporated, the dihydrate $BaCl_2 \cdot 2H_2O$ separates. When a solution of the barium ion is required, this salt is usually employed. Like all other soluble barium salts, the chloride is very poisonous.

33.8 Sulfates of Calcium, Strontium, and Barium

The solubilities of the sulfates of calcium, strontium, and barium in water decrease with increasing atomic weight of the metal; in moles per liter at 20°C they are $CaSO_4 \cdot 2H_2O$, 4.9×10^{-3}; $SrSO_4$, 5.3×10^{-4}; and $BaSO_4$, 1×10^{-5}.

Anhydrous calcium sulfate occurs in nature as the mineral anhydrite, $CaSO_4$, and as the dihydrate, gypsum, $CaSO_4 \cdot 2H_2O$. The latter is found in enormous deposits. Even though it is low in solubility, calcium sulfate is largely responsible for the noncarbonate hardness of ground waters (Section 12.10).

When gypsum is heated, it loses water, forming the hemihydrate $(CaSO_4)_2 \cdot H_2O$.

$$2(CaSO_4 \cdot 2H_2O) \longrightarrow (CaSO_4)_2 \cdot H_2O + 3H_2O$$

The hemihydrate is known as **plaster of paris.** When it is ground to a fine powder and mixed with water, it sets by forming small interlocking crystals of gypsum. The setting results in an increase in volume and hence the plaster fits tightly any mold into which it is poured. Plaster of paris is used extensively as a component of plaster for the interiors of buildings. It is also used in making statuary, stucco, wallboard, and casts of various kinds.

Gypsum is also used in making Portland cement (Section 27.19), blackboard crayon (erroneously called chalk), plate glass, terra cotta, pottery, and orthopedic and dental plasters. Anhydrous calcium sulfate is sold under the name Drierite as a drying agent for gases and organic liquids.

Strontium sulfate occurs in nature in the mineral celestite, which is sometimes found in caves in the form of beautiful crystals whose faint tinge of blue suggested the name of the mineral.

The mineral barite, which is principally composed of **barium sulfate, $BaSO_4$,** is the usual source of barium in the production of other barium compounds. Since the sulfate is insoluble in acids, it is first converted to the sulfide by a high-temperature reduction with carbon. Barium sulfide is then dissolved in acids such as hydrochloric or nitric to produce the desired salts.

A mixture of $BaSO_4$ and ZnS is known as **lithopone** and is used as a white paint pigment. It is made by the reaction of aqueous solutions of barium sulfide and zinc sulfate.

$$Ba^{2+} + S^{2-} + Zn^{2+} + SO_4^{2-} \longrightarrow BaSO_4(s) + ZnS(s)$$

Barium sulfate is used in taking X-ray photographs of the intestinal tract because it is opaque to X rays. Even though Ba^{2+} is poisonous, $BaSO_4$ is so slightly soluble that it is nonpoisonous.

FOR REVIEW

SUMMARY

The ions in Analytical Group IV are those of calcium, strontium, and barium. These are the common ions that are precipitated as carbonates by ammonium carbonate from an aqueous solution containing ammonium chloride, but that are not precipitated by the reagents used in the precipitation of Analytical Groups I, II, and III. Ammonium chloride prevents the precipitation of magnesium carbonate, by hydrolysis that produces sufficient hydrogen ion to keep the relatively soluble magnesium carbonate dissolved.

Calcium, strontium, and barium, along with beryllium, magnesium, and radium, are the alkaline earth metals, Group IIA in the Periodic Table.

EXERCISES

Periodic Relationships

1. Account for the considerable differences in size between the alkaline earth metal atoms and their dipositive ions (see inside back cover of text).
2. How do the metals of Group IIA of the Periodic Table differ from those of Group IA in atomic structure and general properties?
3. List several specific differences in properties between elements in Groups IA and IB. Discuss the role of electron configuration with respect to these differences in properties.
4. The total of the first and second ionization energies for the beryllium atom is larger than the corresponding sum for helium. Why does beryllium exhibit extensive chemistry and helium very little?

Calcium

5. How is metallic calcium produced commercially?
6. Identify each of the following: quicklime, slaked lime, milk of lime, and soda lime.
7. Give the formula for each of the following naturally occurring calcium compounds: limestone, chalk, gypsum, fluorite, and hydroxyapatite.
8. Explain the formation of limestone caves, stalactites, and boiler scale.
9. Why is anhydrous calcium chloride not used for drying alcohol or ammonia?
10. Write a chemical equation describing the dehydrating action of calcium metal.
11. Give the chemistry of the preparation and setting of plaster of paris.
12. What is the essential reaction involved in the setting of mortar?

13. Crushed limestone is used in the treatment of acidic soils. Write equations for the reactions involved.
14. On the basis of atomic structure, explain why barium is more reactive than calcium.
15. Explain how calcium chloride 6-hydrate can be used to melt snow and ice on roads and walks and yet when mixed with ice can also be an excellent *freezing mixture,* much better than ice alone, for making ice cream. Are these two uses compatible?
16. What volume of calcium would have the same mass as 10.0 cm³ of copper?   *Ans. 57.9 cm³*

Strontium

17. Calculate the solubility of $SrSO_4$ in grams per 100 mL of water.   *Ans. 9.7×10^{-3} g/100 mL*
18. Calculate the pH of a saturated solution of strontium hydroxide.   *Ans. 12.93*

Barium

19. Give equations for the reaction of barium with each of the following: oxygen, carbon dioxide, hydrogen, nitrogen, sulfur, and water.
20. The barium ion is poisonous. Why, then, is it safe to take barium sulfate internally for making X-ray photographs of the intestinal tract?
21. Why is carbon added to barium carbonate when the latter is heated to make barium oxide?
22. How much water could be heated from 25°C to 100°C with the heat given off in hydrating 1.0 kg of BaO?   *Ans. 2.1 kg*

34

METALS OF ANALYTICAL GROUP V

SODIUM, POTASSIUM, MAGNESIUM, AND THE AMMONIUM ION

The metals comprising Analytical Group V (see also Chapter 40) are those whose cations fail to form precipitates with the reagents used in the precipitation of the first four groups. For this reason Group V is called the soluble group, and no single reagent will precipitate all four cations. Some properties of the metals of Analytical Group V are listed in Table 34-1.

Table 34-1   Some Properties of the Metals of Analytical Group V

Property	Sodium	Potassium	Magnesium
Atomic number	11	19	12
Atomic weight	22.98977	39.0983	24.305
Oxidation number	+1	+1	+2
Reduction potential, V	(Na^+/Na) -2.714	(K^+/K) -2.925	(Mg^{2+}/Mg) -2.37
Density, g/cm³, at 20°C	0.97	0.86	1.74
Melting point, °C	98	63.4	650
Boiling point, °C	889	757	1120

Magnesium is in the same group of the Periodic Table as calcium, strontium, and barium (discussed in Chapter 33) and is closely related chemically to these metals, but it is not precipitated by ammonium carbonate in the presence of other ammonium salts.

The ammonium ion is somewhat similar to ions of the alkali metals, particularly potassium. The ammonium group, NH_4, has not been prepared as a neutral molecule.

34.1   Periodic Relationships of the Alkali Metals

The metallic elements lithium, sodium, potassium, rubidium, cesium, and francium constitute Group IA of the Periodic Table and are known as the **alkali metals.** The heaviest of them, francium, occurs in nature in very small quantities as a short-lived radioactive isotope. The alkali metals and the metals of Group IIA (the alkaline earth metals) are often called the **light metals** because of their relatively low densities. The alkali metals have the largest atomic radii (Table 34-2) in their respective periods of the Periodic Table. There is a single electron in the outermost shell of each of the alkali metals. Since this electron is far removed from the nucleus of the atom, it is easily lost, and the alkali metals readily form stable positive ions. The difficulty with which such ions are reduced accounts for the fact that none of the alkali metals were isolated until Sir Humphry Davy, in 1807, produced sodium and potassium by the electrolytic reduction of their respective hydroxides. Predictably, the ease with which the valence electron is lost increases with increasing atomic radius. Of those members of the group that have been extensively studied, cesium is the most reactive; however, francium, with an atomic radius even larger than that of cesium, should be more reactive than cesium.

Table 34-2   Some Properties of the Alkali Metals

	Atomic Number	Atomic Weight	Atomic Radius, Å	Ionic Radius, Å	Density, g/cm³, at 20°C	Melting Point, °C	Boiling Point, °C
Lithium	3	6.941	1.52	0.60	0.534	180	1326
Sodium	11	22.98977	1.86	0.95	0.97	98	889
Potassium	19	39.0983	2.31	1.33	0.86	63.4	757
Rubidium	37	85.4678	2.44	1.48	1.53	38.8	679
Cesium	55	132.9054	2.62	1.69	1.87	28.7	690
Francium	87	223	2.7	—	—	—	—

The alkali metals all react vigorously with water, liberating hydrogen and giving solutions of the corresponding strong bases. They also react directly with oxygen, sulfur, nitrogen, hydrogen, and the halogens, giving ionic compounds. Most of the salts of the alkali metals are readily soluble in water; however, Li_2CO_3, Li_3PO_4, and LiF are relatively insoluble. In this respect, lithium resembles magnesium, the second member of Group IIA.

SODIUM AND POTASSIUM

34.2 Occurrence and Preparation of Sodium and Potassium

Sodium and potassium are not found as the free elements in nature, because of their high ease of oxidation. Their compounds, however, occur in abundance and are widely distributed. They appear in many complex silicate rocks such as feld-spars, of which the very abundant albite, $NaAlSi_3O_8$, and orthoclase, $KAlSi_3O_8$, are typical. As these minerals are disintegrated by weathering, the sodium and potassium are converted to soluble compounds, most of which are leached out of the soil by water and are eventually carried to the sea. As they pass through the soil, some of the potassium compounds are adsorbed by the colloidal particles of the soil and held there until used by plants for growth. Sodium compounds are much less retained by the soil than those of potassium, and little sodium is needed by plants. Changes in the earth's crust during past geological ages have, at various times and places, caused sections of the sea to become isolated and gradually to evaporate. Such geological changes have resulted in the formation of large depos-its of sodium and potassium salts. Sodium and potassium also occur in various parts of the world as chloride, sulfate, and borate minerals.

For many years metallic sodium was obtained by reducing sodium carbonate with carbon.

$$Na_2CO_3 + 2C \xrightarrow{\Delta} 2Na + 3CO$$

With the development of the dynamo for the production of electricity, Davy's original electrolytic method of obtaining sodium from molten sodium hydroxide became the industrial process for producing the metal. Most sodium is now pre-pared by the electrolysis of molten sodium chloride, mixed with either sodium carbonate or calcium chloride. The energy required to melt sodium chloride and the high vapor pressure of metallic sodium at this temperature prohibit the use of sodium chloride alone; the melting point of sodium chloride is 801°C, while that of the mixture is about 600°C. Sodium is shipped as 12-pound bricks in air-tight steel barrels or in tank cars that are loaded by pumping in the molten metal. To unload the cars, the metal is melted, and the liquid is forced out by nitrogen under pressure.

Potassium is produced on a relatively small scale, since this element offers few technical advantages over the less costly sodium. Potassium may be made by the electrolytic reduction of either molten potassium chloride or molten potassium hydroxide, but most of the production in the United States is by the reaction of sodium with molten potassium chloride.

$$Na + KCl \rightleftharpoons K + NaCl$$

The sodium is fed into the bottom of a column of melted potassium chloride, and potassium metal escapes at the top. It should be noted that the regular electromo-tive series does not apply exactly for molten salts. A variety of chemical factors determine the course of the reaction. Even though potassium is above sodium in the electromotive series, this particular reaction apparently proceeds because metallic potassium is more volatile than sodium and thus causes a shift in the equilibrium to the right.

34.3 Properties and Uses of Sodium and Potassium

Both sodium and potassium are soft enough to be cut by a knife; they are silvery white in color and are excellent conductors of heat and electricity. Both metals exhibit great reactivity. They tarnish immediately in moist air and react vigorously with water and acids, forming hydroxides or salts of the acids, respectively, and liberating hydrogen. The heat of reaction between these metals and water may cause the evolved hydrogen to ignite. One should never touch sodium or potassium with the hands, because the heat of reaction of the metal and the moisture on the skin may cause the metal to ignite. These metals are stored under kerosene or in sealed containers.

Sodium dissolves in mercury, forming an amalgam. This alloy is an active reducing agent more suitable than pure sodium for many uses because the mercury, being an inactive metal, retards the action of the sodium. Sodium forms alloys with many other metals, such as potassium, tin, and antimony, but not with iron. Potassium and sodium, if mixed in correct proportion, give an alloy that is liquid at room temperature.

Sodium and its compounds impart a yellow color to a Bunsen burner flame, and potassium and its compounds color the flame violet. The color of the potassium flame is best detected by observing it through cobalt glass (blue in color), which absorbs all wavelengths of light except blue and violet.

In general, potassium is more reactive than sodium; otherwise the chemical properties of the two metals are quite similar. Potassium and sodium differ primarily in the solubilities of their salts. Potassium salts are in general slightly less soluble than those of sodium; there are exceptions to this generalization, however.

Sodium is used as a reducing agent in the production of other metals, such as titanium and zirconium, from their chlorides or oxides. Its properties as a reducing agent make it useful in the manufacture of certain dyes, drugs, and perfumes. It is used in the preparation of sodium-lead alloys, which in turn are used in the production of tetraethyl lead from ethyl chloride. It is used in sodium lights for lighting highways because its yellow light penetrates fog well. The synthetic rubber industry consumes a considerable amount of sodium. The largest uses, however, are in the manufacture of compounds such as sodium peroxide and sodium cyanide that cannot be made directly from sodium chloride.

Potassium has no major uses for which sodium cannot be substituted. Sodium-potassium alloys, however, are growing in importance. Their low density, low viscosity, wide liquid range, and high thermal conductivity combine to make them particularly useful in industry as heat-transfer media although they must be handled carefully because they inflame spontaneously in air.

34.4 Compounds of Sodium and Potassium

1. SODIUM CHLORIDE. **Sodium chloride, NaCl,** is one of our most abundant minerals. Sea water contains 2.7% sodium chloride, and the waters of the Dead Sea and the Great Salt Lake contain 23%. There are vast deposits of rock salt in

the Stassfurt region of Germany, and large deposits in the United States are found in New York, Michigan, West Virginia, and California, and a very extensive bed (about 400–500 feet thick) underlies parts of Oklahoma, Texas, and Kansas. Most of the salt consumed comes from beds that lie below the surface of the earth, rather than from sea water. The salt is removed from these underground beds by mining or by forcing water down into the deposits to form saturated brines, which are then pumped to the surface. Natural sodium chloride contains other soluble salts that make it unsuitable for many uses. Salt is purified by dissolving it in water, concentrating the solution by evaporation (often under reduced pressure), and allowing the crystals to form again. Some of the impurities are more soluble than sodium chloride and for this reason remain in solution when it crystallizes. Some of the impurities that are less soluble than salt do not crystallize out because they are present in small amounts. Other less soluble impurities, such as calcium sulfate, do crystallize out with the salt, but are then removed by taking advantage of the decreasing solubility of calcium sulfate in sodium chloride solution as the temperature is raised above 80°C. Thus, if the brine is filtered hot, the resulting solution will not contain enough calcium sulfate to be saturated when cooled. Hence calcium sulfate cannot precipitate with sodium chloride, and sodium chloride crystallizes relatively free of the impurity.

Pure sodium chloride when the relative humidity is above 75%, or sodium chloride contaminated with chlorides of magnesium and calcium, absorbs water, causing the salt to cake. When in a shaker, the caking salt clogs the shaker. Basic magnesium carbonate or calcium aluminosilicate is usually added to serve as a drier, or anticaking agent.

Sodium chloride is the usual source of sodium or chlorine in almost all other compounds containing these elements. Its most extensive use is in the manufacture of chlorine, hydrochloric acid, sodium hydroxide, and sodium carbonate. Sodium chloride is an essential constituent of the foods of animals. It not only makes food more palatable, but it is the source of chlorine for the production of hydrochloric acid, a constituent of gastric juice. Sodium chloride is also a constituent of the blood and is essential for the life processes of the human body. However, excessive consumption of salt should be avoided, especially by those with potential heart problems or high blood pressure.

2. HYDRIDES. **Sodium hydride, NaH,** and **potassium hydride, KH,** are prepared by the direct union of the elements at slightly elevated temperatures and are decomposed at higher temperatures. They are saltlike compounds, with a crystal structure like that of sodium chloride (Section 11.18). When the molten hydrides are electrolyzed, the metal is formed at the cathode and hydrogen is evolved at the anode; thus the hydrides must be composed of metal ions and hydride ions, H^-. The hydrides are readily decomposed by water with the evolution of hydrogen. The hydride ions combine with hydrogen ions of the water,

$$H^+ + H^- \longrightarrow H_2$$

and leave an excess of hydroxide ions in solution.

$$NaH + H_2O \longrightarrow Na^+ + OH^- + H_2(g)$$

Sodium hydride is used widely in organic chemistry as a reducing agent and in the descaling of steel before it is plated or enameled.

3. OXIDES AND PEROXIDES. Oxides of most metals are produced by heating the corresponding metal hydroxides, nitrates, or carbonates. The oxides of the more active alkali metals cannot be prepared in this way, however, because their hydroxides, nitrates, and carbonates are stable, decomposing only at extremely high temperatures.

Sodium forms both **sodium oxide, Na_2O,** and **sodium peroxide, Na_2O_2.** Sodium oxide is prepared by heating sodium in a limited supply of air under carefully controlled conditions. The product is an extremely alkaline white powder. It is a very effective drying agent and is also used as a strong caustic. Sodium peroxide is the more important of the two oxides of sodium. It is manufactured by the reaction of sodium with an excess of air at 300–400°C. Sodium peroxide is a yellowish-white powder used extensively as an oxidizing and bleaching agent. Solutions of the compound are alkaline due to hydrolysis, as shown by the equations

$$O_2^{2-} + H_2O \rightleftharpoons HO_2^- + OH^-$$
$$HO_2^- + H_2O \rightleftharpoons H_2O_2 + OH^-$$

Potassium oxide, K_2O, may be obtained by heating potassium nitrate with elemental potassium in the absence of air. **Potassium peroxide, K_2O_2,** is formed by burning the metal in a calculated quantity of air. When exposed to an excess of air, potassium oxide combines readily with oxygen to give **potassium superoxide, KO_2,** in which the superoxide ion has a charge of −1. Potassium superoxide is most readily formed by burning the metal in an excess of oxygen and is used in gas masks for mine safety work. Moisture from the air reacts with the superoxide to give oxygen for the wearer of the mask to breathe.

$$4KO_2 + 2H_2O \longrightarrow 4KOH + 3O_2$$

When the breath is exhaled through the mask, carbon dioxide is absorbed by the potassium hydroxide.

$$CO_2 + KOH \longrightarrow KHCO_3$$

The wearer of the mask needs no air from outside the mask.

4. HYDROXIDES. **Sodium hydroxide** is frequently called **caustic soda** in commerce, and its solutions are sometimes referred to as lye, or soda lye. It is the seventh-ranked chemical in terms of production in the United States (over 9 million tons in 1982). Its commercial production is by two processes. The older process involves the action of slaked lime, $Ca(OH)_2$, on soda ash, Na_2CO_3, and is called the **lime-soda process.**

$$[2Na^+] + CO_3^{2-} + Ca^{2+} + [2OH^-] \longrightarrow CaCO_3(s) + [2Na^+] + [2OH^-]$$

The insoluble calcium carbonate is filtered out, and the filtrate is evaporated to yield solid sodium hydroxide. The residue is heated to remove the last traces of water and is then molded into sticks or pellets.

A second and more important industrial process for the production of sodium hydroxide is the electrolysis of aqueous sodium chloride (see Section 19.2 and 20.4).

$$[2Na^+] + 2Cl^- + 2H_2O \longrightarrow [2Na^+] + 2OH^- + H_2(g) + Cl_2(g)$$

Hydrogen is formed at the cathode, and chlorine is evolved at the anode. Sodium ions and hydroxide ions accumulate in the solution. The cell is designed so that the chlorine cannot react with the hydroxide ions to form hypochlorite ions.

Electrolytic sodium hydroxide contains some sodium chloride, most of which can be removed by a process called **fractional crystallization** because sodium chloride is less soluble than sodium hydroxide in water.

Pure sodium hydroxide can be obtained electrolytically from aqueous sodium chloride by using a mercury cathode cell. In this cell sodium, rather than hydrogen, is liberated at the cathode; the sodium, as it is reduced, forms an amalgam with the mercury. The amalgam is decomposed by water in a separate compartment to produce hydrogen and very pure sodium hydroxide, containing no chlorides.

$$2Na^+ + 2Cl^- + 2nHg \xrightarrow{\text{Electrolysis}} 2Na(Hg)_n + Cl_2(g)$$

$$2Na(Hg)_n + 2H_2O \longrightarrow 2Na^+ + 2OH^- + 2nHg + H_2(g)$$

Sodium hydroxide is an ionic compound that melts and boils without decomposition. When molten or in aqueous solution, sodium hydroxide reacts with silicate minerals and glass. The solid hydroxide and its solutions readily absorb carbon dioxide from the air, with the formation of sodium carbonate and water.

$$2OH^- + CO_2 \longrightarrow H_2O + CO_3^{2-}$$

Sodium hydroxide is highly soluble in water, giving off a great deal of heat and yielding strongly basic solutions. It is used extensively in the production of chemicals, rayon, lye, cleaners, textiles, soap, paper and pulp, and in petroleum refining.

Potassium hydroxide is manufactured in a manner similar to that described for the production of sodium hydroxide. The method involving the treatment of aqueous potassium carbonate with slaked lime has largely given way to the electrolytic method. Potassium hydroxide is very soluble in water and, being strongly deliquescent, is used as a dehydrating agent. It forms strongly alkaline solutions with water. Because it is more expensive than sodium hydroxide, which almost always serves equally well, its use is limited.

5. CARBONATES AND HYDROGEN CARBONATES. **Sodium carbonate** is, after sodium hydroxide, perhaps the most important manufactured compound of sodium. The total amount used in the United States in 1982 was 7.9 million tons, making it tenth among all industrially produced chemicals. It has the formula Na_2CO_3 and is commonly called **soda ash,** or simply **soda.**

The sodium carbonate produced in the United States is prepared from the mineral trona, $Na_2CO_3 \cdot NaHCO_3 \cdot 2H_2O$. Following recrystallization from water to remove clay and other impurities, trona is roasted to produce Na_2CO_3.

$$2(Na_2CO_3 \cdot NaHCO_3 \cdot 2H_2O) \xrightarrow{\Delta} 3Na_2CO_3 + 5H_2O(g) + CO_2(g)$$

Sodium carbonate is also manufactured by the **Solvay process.** This process is based on the reaction of ammonium hydrogen carbonate with a saturated solution of sodium chloride. Sodium hydrogen carbonate precipitates, since it is only slightly soluble in the reaction medium.

$$Na^+ + [Cl^-] + [NH_4^+] + HCO_3^- \longrightarrow NaHCO_3(s) + [NH_4^+] + [Cl^-]$$

The basic raw materials used in the process are limestone and common salt. The carbon dioxide is generated by heating limestone.

$$CaCO_3 \xrightarrow{\Delta} CaO + CO_2 \tag{1}$$

The ammonia is obtained by treating ammonium chloride, formed as a by-product of the process, with calcium hydroxide.

$$2NH_4^+ + [2Cl^-] + [Ca^{2+}] + 2OH^- \longrightarrow$$
$$2NH_3(g) + 2H_2O + [Ca^{2+}] + [2Cl^-] \tag{2}$$

Calcium hydroxide results from slaking the lime produced according to Equation (1).

$$CaO + H_2O \longrightarrow Ca^{2+} + 2OH^- \tag{3}$$

Aqueous ammonia reacts with an excess of carbon dioxide, yielding ammonium hydrogen carbonate.

$$NH_3 + CO_2 + H_2O \longrightarrow NH_4^+ + HCO_3^- \tag{4}$$

Sodium hydrogen carbonate is also a valued product of the Solvay process; after being freed of ammonium chloride by recrystallization, it is made commercially available. The major portion of the sodium hydrogen carbonate, however, is converted to sodium carbonate by heating.

$$2NaHCO_3 \xrightarrow{\Delta} Na_2CO_3 + H_2O + CO_2 \tag{5}$$

The carbon dioxide from this reaction is used to produce more sodium hydrogen carbonate. The only by-product of the entire process not reused in the process is calcium chloride.

Some sodium carbonate is produced from sodium hydroxide, as a by-product of the chlorine industry. We saw in Section 20.4 that the electrolysis of aqueous sodium chloride produces chlorine, hydrogen, and sodium hydroxide. By saturating the sodium hydroxide solution with carbon dioxide, sodium hydrogen carbonate is produced.

$$[Na^+] + OH^- + CO_2 \longrightarrow [Na^+] + HCO_3^-$$

The use of less carbon dioxide gives rise to sodium carbonate.

$$[2Na^+] + 2OH^- + CO_2 \longrightarrow [2Na^+] + CO_3^{2-} + H_2O$$

Sodium carbonate is used extensively in the manufacture of glass, other chemicals, soap, paper and pulp, cleansers, and water softeners and in the refining of petroleum.

When a solution of sodium carbonate is evaporated below 35.2°C, **sodium carbonate 10-hydrate, $Na_2CO_3 \cdot 10H_2O$**, known as **washing soda**, crystallizes out; above this temperature the monohydrate $Na_2CO_3 \cdot H_2O$ crystallizes out. Heating either hydrate produces the anhydrous compound. Solutions of sodium carbonate are basic, due to hydrolysis of the carbonate ions, as shown by the equation

$$CO_3^{2-} + H_2O \longrightarrow HCO_3^- + OH^-$$

For this reason sodium carbonate is generally used in commercial processes requiring a base that is not as strong as sodium hydroxide.

Sodium hydrogen carbonate, $NaHCO_3$, is commonly known as **bicarbonate of soda,** or **baking soda.** Its solutions are weakly basic due to hydrolysis of hydrogen carbonate.

$$HCO_3^- + H_2O \rightleftharpoons H_2CO_3 + OH^-$$

Acids react with sodium hydrogen carbonate with the formation of carbon dioxide.

$$HCO_3^- + H_3O^+ \longrightarrow 2H_2O + CO_2(g)$$

This reaction is involved in the leavening, or rising, process in baking. Baking powders contain a mixture of baking soda and an acidic substance, such as potassium hydrogen tartrate $KHC_4H_4O_6$ (cream of tartar), calcium dihydrogen phosphate, $Ca(H_2PO_4)_2$, or sodium aluminum sulfate, $NaAl(SO_4)_2 \cdot 12H_2O$. Starch or flour is also added to keep the powder dry. When the mixture is dry, no reaction takes place. As soon as water is added, carbon dioxide is given off. The acidic substance and water produce hydronium ions, which in turn react with the baking soda to form carbon dioxide.

$$HC_4H_4O_6^- + H_2O \rightleftharpoons H_3O^+ + C_4H_4O_6^{2-}$$
$$H_2PO_4^- + H_2O \rightleftharpoons H_3O^+ + HPO_4^{2-}$$
$$[Al(H_2O)_6]^{3+} + H_2O \rightleftharpoons H_3O^+ + [Al(OH)(H_2O)_5]^{2+}$$
$$HCO_3^- + H_3O^+ \longrightarrow CO_2(g) + 2H_2O$$

The carbon dioxide is trapped in the bread dough and expands when warmed. This produces spaces in the bread that give it a desirable lightness.

The lactic acid of sour milk and buttermilk or the acetic acid in vinegar will also furnish hydronium ions and thus serve the same purpose as the acidic constituent of baking powder. Thus it is possible to make bread rise by using either sour milk or buttermilk with ordinary sodium bicarbonate (baking soda).

Potassium carbonate cannot be prepared by the Solvay process because of the high solubility of potassium hydrogen carbonate in solutions of aqueous ammonia. Instead, it is obtained either by treating potassium hydroxide with carbon dioxide or from potassium chloride. The latter method involves heating potassium chloride under pressure with magnesium carbonate, water, and carbon dioxide, whereby the slightly soluble double salt $KHCO_3 \cdot Mg(HCO_3)_2 \cdot 4H_2O$ is formed.

$$2K^+ + [2Cl^-] + 3MgCO_3 + 3CO_2 + 11H_2O \xrightarrow{\Delta}$$
$$2\{KHCO_3 \cdot Mg(HCO_3)_2 \cdot 4H_2O\} + Mg^{2+} + [2Cl^-]$$

The double salt is removed from the solution by filtration and heated to 120°C, which causes the following reaction to take place:

$$2\{KHCO_3 \cdot Mg(HCO_3)_2 \cdot 4H_2O\} \xrightarrow{\Delta}$$
$$K_2CO_3 + 2MgCO_3 + 11H_2O + 3CO_2(g)$$

Finally, the potassium carbonate is leached from the solid mass, leaving a residue of insoluble magnesium carbonate, which is used again in the first step.

Potassium carbonate forms three hydrates, containing one, two, and three molecules of water, respectively, but it is usually sold in the anhydrous form. It is deliquescent and very soluble in water, forming a strongly alkaline solution. It is used in the manufacture of soft soap, glass, pottery, and various potassium compounds.

When a solution of potassium carbonate is saturated with carbon dioxide and concentrated by evaporation, crystals of **potassium hydrogen carbonate, KHCO$_3$,** are deposited. Solutions of this salt are weakly basic. When heated, potassium hydrogen carbonate is converted to potassium carbonate.

6. NITRATES. **Sodium nitrate,** or Chile saltpeter, is a very important compound that was discussed in connection with nitrogen and nitric acid (Chapter 23).

Potassium nitrate was one of the principal reagents used by the alchemists. It is formed in nature by the decay of organic matter and is prepared commercially from sodium nitrate and potassium chloride.

$$Na^+ + [NO_3^-] + [K^+] + Cl^- \longrightarrow NaCl(s) + [K^+] + [NO_3^-]$$

The sodium nitrate and potassium chloride are dissolved in hot water, and the solution is evaporated by boiling. Of the four compounds possible in a solution containing Na^+, NO_3^-, K^+, and Cl^-, sodium chloride is the least soluble in hot water, and hence it is the first to crystallize out when the concentration of the solution is increased by evaporation. On the other hand, potassium nitrate is very soluble in hot water, making it possible to separate the two salts by filtration. Sodium chloride has about the same solubility in both cold and hot water, so very little more of it crystallizes out when the filtrate is cooled. However, potassium nitrate is only slightly soluble in cold water, and thus it crystallizes out as the filtrate is cooled.

Potassium nitrate is a valuable fertilizer because it furnishes both potassium and nitrogen in forms that are readily utilized by growing plants. It is also used to some extent in preserving ham and corned beef, to which it imparts a red color, but studies are being conducted on the safety of using potassium nitrate as an additive to meat.

7. SULFATES. Extensive deposits of **sodium sulfate** occur in Canada, North Dakota, and the southwestern section of the United States, but most of the sodium sulfate used commercially is obtained as a by-product of the manufacture of hydrochloric acid from salt and sulfuric acid. The sodium sulfate thus formed is commonly called **salt cake** and is used in the manufacture of glass, paper, rayon, coal-tar dyes, and soap. Sodium sulfate ranked 39th in production among all chemicals in the United States in 1982, with a total production of 893,000 tons.

When sodium sulfate is crystallized from solution at temperatures below $32.28°C$, sodium sulfate 10-hydrate, $Na_2SO_4 \cdot 10H_2O$, is formed. This hydrate is called **Glauber's salt,** in honor of the alchemist Glauber who used it as a medicine in the seventeenth century. The anhydrous salt, Na_2SO_4, crystallizes from solutions at temperatures above $32.28°C$.

Sodium hydrogen sulfate is prepared by heating either sodium chloride or sodium nitrate to a moderate temperature with sulfuric acid.

$$NaCl + H_2SO_4 \xrightarrow{\Delta} NaHSO_4 + HCl(g)$$

$$NaNO_3 + H_2SO_4 \xrightarrow{\Delta} NaHSO_4 + HNO_3(g)$$

Potassium also forms two sulfates, **potassium sulfate, K_2SO_4,** and **potassium hydrogen sulfate, $KHSO_4$.** Potassium sulfate is obtained from natural salt deposits, and it is used as a fertilizer and in the preparation of alums. Potassium hydrogen sulfate is made by heating potassium sulfate with the proper quantity of sulfuric acid. It is used in analytical chemistry to convert metal oxides and silicates into sulfates. Heat converts potassium hydrogen sulfate to potassium pyrosulfate.

$$2KHSO_4 \xrightarrow{\Delta} K_2S_2O_7 + H_2O$$

When the pyrosulfate is strongly heated, it gives up sulfur trioxide.

$$K_2S_2O_7 \xrightarrow{\Delta} K_2SO_4 + SO_3$$

MAGNESIUM

The periodic relationships of magnesium were discussed with those of the other alkaline earth metals in Section 33.1.

34.5 Occurrence and Preparation of Magnesium

Elemental magnesium never occurs in nature because of its reactivity. However, compounds of magnesium are abundant and widely distributed. The chloride and sulfate are readily soluble in water and, consequently, are found in ground waters, to which they impart hardness (Sections 12.10 and 12.11). Sea water contains both the chloride and the sulfate, and compounds of magnesium concentrate in the mother liquor of sea water and in underground brines from which sodium chloride has been crystallized. Typical magnesium minerals are carnallite, $KCl \cdot MgCl_2 \cdot 6H_2O$, magnesite, $MgCO_3$, asbestos, $H_4Mg_3Si_2O_9$, dolomite, $MgCO_3 \cdot CaCO_3$, meerschaum, $Mg_2Si_3O_8 \cdot 2H_2O$, talc, or soapstone, $Mg_3Si_4O_{10}(OH)_2$, and brucite, $Mg(OH)_2$.

Magnesium metal is obtained from several different sources and prepared by several different methods.

Magnesium chloride is obtained from underground brines in Michigan, which contain about 3% magnesium chloride, 9% calcium chloride, 14% sodium chloride, and 0.1% bromine as bromide ion. The bromine is extracted first (Section 19.2), and the brine is then treated with a suspension of magnesium hydroxide to precipitate the hydroxides of iron and certain other metals. The filtrate is evaporated to crystallize out the sodium chloride. The magnesium and calcium chlorides are separated by fractional crystallization. Crystalline magnesium chloride 6-hydrate, $MgCl_2 \cdot 6H_2O$, is then completely dehydrated in an atmosphere of hydrogen chloride. The hydrogen chloride prevents hydrolysis, which would lead to the formation of basic magnesium salts (Section 34.7). Electrolysis of molten magnesium chloride produces magnesium metal and chlorine. Sodium chloride is usually added to lower the melting point of the electrolyte and increase the electrical conductivity. The product has a purity of 99.9%.

Sea water also serves as an important source of magnesium. Nearly 6 million tons of magnesium, as the chloride and the sulfate, are contained in each cubic mile of sea water. The raw materials for the process are sea water, oyster shells ($CaCO_3$) from shallow waters along the coast, salt, fresh water, and natural gas. Nearly 800 tons of sea water must be processed to obtain 1 ton of magnesium metal. Oyster shells are calcined to produce lime, which is slaked by adding water. In some operations the same sea water that was used for the recovery of bromine (Section 19.2) is treated with slaked lime to precipitate magnesium hydroxide. After filtration, the magnesium hydroxide can be treated with hydrochloric acid to produce magnesium chloride. The magnesium chloride is crystallized from solution by evaporation. The crystallized salt is then partially dehydrated by heating. The resulting magnesium chloride, which has a composition corresponding to

$MgCl_2 \cdot 1\frac{1}{2}H_2O$, is electrolyzed to produce magnesium and chlorine. The by-product chlorine is used in the burning of natural gas to produce the hydrogen chloride used in the process. In some operations magnesium carbonate and magnesium oxide are produced instead of magnesium metal.

34.6   Properties and Uses of Magnesium

Magnesium is a silvery white metal that is malleable and ductile at high temperatures. It is the lightest of the widely used structural metals. Although very active, magnesium does not undergo extensive reaction with air or water, due to the formation of a protective oxycarbonate film on its surface. Magnesium is soluble in acids, including carbonic acid, evolving hydrogen.

$$Mg + H_2CO_3 \longrightarrow MgCO_3(s) + H_2(g)$$

$$MgCO_3 + H_2CO_3 \longrightarrow Mg^{2+} + 2HCO_3^-$$

It also reacts with the alkali metal hydrogen carbonates and with various salts, which give an acid reaction by hydrolysis.

Magnesium decomposes boiling water very slowly, but it rapidly reduces the hydrogen in steam, forming magnesium oxide and hydrogen. The affinity of magnesium for oxygen is so great that it will react when heated in an atmosphere of carbon dioxide, reducing the carbon of the oxide to elemental carbon.

$$2Mg + CO_2 \xrightarrow{\Delta} 2MgO + C$$

The great reducing power of hot magnesium is utilized in preparing many metals and nonmetals, such as silicon and boron, from their oxides.

Magnesium will unite with most nonmetals. The brilliant white light emitted from burning magnesium makes it useful in flashlight powders, military flares, and incendiary bombs. When magnesium burns in air, both magnesium oxide, MgO, and magnesium nitride, Mg_3N_2, are formed.

Most of the magnesium produced commercially is used in making lightweight alloys, the most important of which are those with aluminum and zinc. Magnalium (1–15% Mg, 0–1.75% Cu, and the remainder Al) is lighter, harder, stronger, and more easily machined than pure aluminum. Other important magnesium alloys include duralumin (0.5% Mg, 0.5% Mn, 3.5–5.5% Cu, and the remainder Al) and Dowmetal (8.5% Al, 0.15% Mn, 2.0% Cu, 1.0% Cd, 0.5% Zn, and 87.85% Mg).

34.7   Compounds of Magnesium

Magnesium oxide, MgO, is sometimes called **magnesia.** It is formed commercially by heating magnesite, $MgCO_3$, to 600–800°C, which drives off carbon dioxide from the magnesium carbonate.

$$MgCO_3 \xrightarrow{\Delta} MgO + CO_2(g)$$

The product is a light, fluffy powder that still contains a small percentage of magnesium carbonate. It reacts slowly with water, forming magnesium hydroxide, and it is soluble in water containing carbon dioxide, forming magnesium hydrogen carbonate, a constituent of hard water. On the other hand, when magnesite is heated to above 1400°C, the product contains no magnesium carbonate and is a powder that is much more dense than the light form of the oxide. This form of magnesium oxide does not react with water, and it does not conduct heat well. Since it melts at 2800°C, it is used in making fire brick and crucibles, as a lining in furnaces, and in heat insulation.

Magnesium oxide does not slake readily, so magnesium hydroxide is prepared by treating a solution of a magnesium salt with an alkali metal hydroxide. The hydroxide is slightly soluble in water, but readily soluble in solutions of ammonium salts because of the acidity of the NH_4^+ ion.

$$Mg(OH)_2 + 2NH_4^+ \longrightarrow Mg^{2+} + 2NH_3 + 2H_2O$$

A suspension of magnesium hydroxide in water is called **milk of magnesia** and is used as a medicine to correct stomach hyperacidity and as a laxative.

Magnesium carbonate, $MgCO_3$, is found in nature as the mineral magnesite. A hydroxycarbonate whose formula can be written $3MgCO_3 \cdot Mg(OH)_2 \cdot 3H_2O$ is produced by precipitation. This compound is known commercially as **magnesia alba** and is used as a dental abrasive, a medicine, a cosmetic, and a silver polish. **Magnesium hydrogen carbonate, $Mg(HCO_3)_2$,** is a constituent of many hard waters.

Magnesium chloride crystallizes from aqueous solutions as the hydrated salt $MgCl_2 \cdot 6H_2O$. The hexahydrate is deliquescent, becoming moist in damp air. When heated, it undergoes hydrolysis and forms magnesium oxide, hydrogen chloride, and water.

$$MgCl_2 \cdot 6H_2O \xrightarrow{\Delta} MgO + 2HCl(g) + 5H_2O$$

Anhydrous magnesium chloride may be obtained by heating the hydrate in a current of hydrogen chloride, by burning magnesium in chlorine, or by carefully heating the double salt $NH_4Cl \cdot MgCl_2 \cdot 6H_2O$, in which case the water is driven off first and then the ammonium chloride.

Magnesium sulfate is found as the minerals kieserite, $MgSO_4 \cdot H_2O$, and epsomite, $MgSO_4 \cdot 7H_2O$. The heptahydrate in pure form is familiar as **Epsom salts.** It is used in medicine as a purgative, particularly in veterinary practice, in weighting cotton and silk, for polishing, and in insulation.

Magnesium ammonium phosphate, $MgNH_4PO_4$, is a slightly soluble crystalline salt, which is formed whenever a soluble phosphate is added to a solution containing magnesium, ammonium, and hydroxide ions. This white substance is used in the detection and estimation of either magnesium or phosphate ions in analytical chemistry.

Anhydrous magnesium perchlorate, $Mg(ClO_4)_2$, is a highly efficient drying agent, called Anhydrone. It rapidly absorbs up to 35% of its weight in water. The anhydrous salt is easily regenerated by heating the hydrate.

Several silicates of magnesium are of commercial importance. The mineral known as talc, or soapstone, is a hydrated magnesium silicate that feels greasy to the touch. It can be sawed and turned on a lathe, so is useful in the fabrication of tables, sinks, switchboards, and window sills. Asbestos is a calcium-magnesium silicate with a fibrous structure (Section 27.16), from which incombustible fabrics of considerable strength and durability can be made. Such materials have been widely used in making automobile brake linings, paper, drop curtains for theaters, cardboard, flooring, roofing, and covering for heating pipes and boilers, but their use has been curtailed due to the tendency of asbestos dust to induce lung cancers when inhaled over a period of time.

THE AMMONIUM ION AND ITS SALTS

34.8 The Ammonium Ion

In general, salts of the ammonium ion exhibit physical properties like those of salts of the potassium ion; the two ions are nearly the same size and have the same charge. The resemblance of ammonium and potassium salts is especially noticeable with respect to the formation of slightly soluble salts. There are two notable exceptions to the similarity between the salts of the ammonium and potassium ions: (1) The ammonium ion undergoes a slight hydrolysis, while the potassium ion does not,

$$NH_4^+ + H_2O \rightleftharpoons NH_3 + H_3O^+$$

and (2) ammonium salts decompose when heated, while the potassium salts melt.

$$NH_4Cl(s) \xrightarrow{\Delta} NH_3(g) + HCl(g) \qquad \Delta H° = 176.0 \text{ kJ}$$

$$NH_4NO_3(s) \xrightarrow{\Delta} N_2O(g) + 2H_2O(g) \qquad \Delta H° = -36.0 \text{ kJ}$$

$$NH_4NO_2(s) \xrightarrow{\Delta} N_2(g) + 2H_2O(g) \qquad \Delta H° = -227.2 \text{ kJ}$$

$$(NH_4)_2Cr_2O_7(s) \xrightarrow{\Delta} N_2(g) + 4H_2O(g) + Cr_2O_3(s) \qquad \Delta H° = -300 \text{ kJ}$$

The ammonium group, NH_4, thus far has not been isolated as a neutral species; it is always found as a positively charged ion, NH_4^+, in combination with a negative ion. When attempts have been made to isolate the neutral NH_4 unit, ammonia and hydrogen have always been obtained.

34.9 Ammonium Compounds

Ammonium chloride, NH_4Cl, is known commercially as **sal ammoniac.** It is produced by the reaction of ammonia with hydrochloric acid and is purified by sublimation. Ammonium chloride decomposes at 350°C into ammonia and hydrogen chloride.

$$NH_4Cl(s) \rightleftharpoons NH_3(g) + HCl(g) \qquad \Delta H° = 176.0 \text{ kJ}$$

For this reason it is used as a flux in soldering, because the hydrogen chloride that is formed when the salt is heated reacts with the films of metal oxides, converting them into chlorides that are either fusible or volatile, thus cleansing the metal surfaces to be joined by the solder. Ammonia and hydrogen chloride gases in the air of chemical laboratories combine to form the familiar white deposits of ammonium chloride so commonly seen on laboratory glassware and windowpanes.

Much of the ammonium chloride produced is used in the manufacture of dry cells (Section 20.16). It is also used in medicine, in dyeing, in calico printing, as a laboratory reagent, and as a fertilizer.

Ammonia reacts with nitric acid to form **ammonium nitrate, NH_4NO_3,** a white crystalline salt. Its production in the United States in 1982 was 7.3 million tons, ranking it twelfth among all chemicals. When heated to 166°C, ammonium nitrate melts and decomposes smoothly, with the formation of nitrous oxide and water. When detonated, sometimes simply by heating, it decomposes with explosive violence. The enormous explosion that destroyed a large part of Texas City and killed 576 people in April, 1947, was due to the decomposition of ammonium nitrate that was being loaded onto a ship in the harbor. Ammonium nitrate has been used for many years as an ingredient of explosives for warfare and mining. Its principal use, however, is as a fertilizer.

Ammonium hydrogen carbonate, NH_4HCO_3, is prepared by evaporating a solution made by treating an aqueous solution of ammonia with an excess of carbon dioxide. When heated, the white crystalline salt decomposes rapidly.

$$NH_4HCO_3 \rightleftharpoons NH_3 + H_2O + CO_2$$

Even at ordinary temperatures, ammonium hydrogen carbonate has a faint odor of ammonia. Treating a solution of ammonium hydrogen carbonate with an excess of ammonia produces **ammonium carbonate.**

$$HCO_3^- + NH_3 \rightleftharpoons NH_4^+ + CO_3^{2-}$$

When exposed to the air, $(NH_4)_2CO_3$ gives off ammonia more readily than does NH_4HCO_3 and thus is useful as smelling salts, since ammonia is an effective heart stimulant.

Large quantities of **ammonium sulfate, $(NH_4)_2SO_4$,** are produced when ammonia is absorbed in sulfuric acid. Its chief use is as a fertilizer to supply nitrogen to the soil. The production of ammonium sulfate in the United States in 1982 was 1.8 million tons, ranking it 29th among all chemicals.

When an aqueous solution of ammonia is saturated with hydrogen sulfide, **ammonium hydrogen sulfide, NH_4HS,** is formed. If the resulting solution is then treated with an equivalent amount of aqueous ammonia, **ammonium sulfide, $(NH_4)_2S$,** is formed.

FOR REVIEW

SUMMARY

The ions that comprise Analytical Group V, sodium, potassium, magnesium, and ammonium ions, are those that are not precipitated in Analytical Groups I, II, III, and IV. Sodium and potassium are alkali metals in Group IA of the Periodic Table. Many of the chemical and physical properties of the ammonium ion are similar to those of the alkali metal ions (particularly those of the potassium ion), as might be expected from the facts that the charge on the ammonium ion is $+1$ and that its size is about the same as that of the potassium ion (intermediate in size within the alkali metal group). Magnesium is an alkaline earth metal, in Group IIA of the Periodic Table.

EXERCISES

Alkali Metals

1. Why do the alkali metals not occur as the elements in nature?

2. Correlate the positive oxidation numbers of the alkali metals with their atomic structures.

Sodium

3. Why should metallic sodium never be handled with the fingers?
4. Why is sodium carbonate or calcium chloride added to the electrolyte in the electrolysis of molten sodium chloride?
5. By analogy with its reaction with water, suggest a chemical formula for the product of the reaction between sodium metal and ethyl alcohol.
6. Why must the chlorine and sodium hydroxide be kept separate in the manufacture of sodium hydroxide by the electrolysis of aqueous sodium chloride?
7. Outline the chemistry of the Solvay process for the production of sodium carbonate.
8. What volume of hydrogen measured at 25°C and 735 torr would be produced by the passage of a current of 0.674 A through an aqueous solution of sodium chloride for a period of 2.00 h?

   Ans. 0.636 L
9. Give two examples of slightly soluble sodium salts.
10. How much anhydrous sodium carbonate contains the same number of moles of sodium as in 100 g of $Na_2CO_3 \cdot 10H_2O$?   *Ans. 37.0 g*
11. Normal saline solution, a solution of sodium chloride in water used for washing blood cells, contains 0.85 g of NaCl in 0.100 L of solution. What is the sodium ion concentration in normal saline solution?   *Ans. 0.15 M*
12. What is the principal reaction involved when baking powder acts as a leavening agent?
13. What evidence is available to show that hydrogen is present as the negative hydride ion in sodium hydride?

Potassium

14. Why cannot potassium hydrogen carbonate be manufactured by the Solvay process?
15. Compare the solubilities of potassium and ammonium salts.
16. Cite evidence that the hydride and peroxide ions are strongly basic.
17. A 25.00-mL sample of KOH solution is exactly neutralized with 35.27 mL of 0.1062 *M* HCl.

What is the molar concentration of the KOH solution?   *Ans. 0.1498 M*

Magnesium

18. Outline the extraction of magnesium from sea water.
19. Magnesium is an active metal; it is burned in the form of ribbons and filaments to provide flashes of brilliant light. Why is it possible to use magnesium in construction and even for the fabrication of cooking grills?
20. Why cannot a magnesium fire be extinguished by either water or carbon dioxide? Suggest a method of extinguishing such a fire.
21. Write a chemical equation describing the dehydrating property of magnesium perchlorate.
22. When $MgNH_4PO_4$ is heated to 1000°C, it is converted to $Mg_2P_2O_7$. A 1.203-g sample containing magnesium yielded 0.5275 g of $Mg_2P_2O_7$ after precipitation of $MgNH_4PO_4$ and heating. What percent by mass of magnesium was present in the original sample?   *Ans. 9.577%*
23. Why cannot $MgCl_2 \cdot 6H_2O$ be dehydrated by heating in air? How may it be dehydrated?
24. Calculate the percentage of magnesium in asbestos.   *Ans. 26.31%*
25. Explain the dissolution of magnesium hydroxide in solutions of ammonium salts.

Ammonium

26. Why is a consideration of the chemistry of the ammonium ion and its salts introduced along with a discussion of the alkali metals and their ions?
27. When potassium ion is reduced, potassium metal results. What is (are) the product(s) when ammonium ion is reduced?
28. Write equations for the thermal decomposition of the following salts: NH_4Cl, NH_4NO_3, NH_4NO_2, and $(NH_4)_2Cr_2O_7$.
29. Write chemical equations describing the conversion of ammonium ion to nitride ion; of nitride ion to ammonium ion.
30. In terms of the oxidation numbers of nitrogen, explain why ammonium nitrate might explode.

SEMIMICRO
QUALITATIVE ANALYSIS

35

GENERAL LABORATORY DIRECTIONS

This course in qualitative analysis has two principal objectives. One of these is to give you the reasons for the analytical procedures and results in terms of the theory of ionic equilibria, especially that relating to weak electrolytes, solubility products, complex ions, and oxidation-reduction. The other objective is the practical one of introducing the sights and smells of chemical reactions and of teaching you careful laboratory manipulation, critical observation, and logical interpretation of observed results.

35.1 Classification of the Metals into Analytical Groups

Qualitative analysis pertains to the identification of the constituents present in a sample of a substance, a mixture of substances, or a solution. In the qualitative analysis of a solution that may contain any or all of the common metal ions, the first step is that of separating the ions into several groups, each of which contains ions exhibiting a common chemical property that is the basis of the separation. The separation of the common metal ions into groups is usually done as outlined briefly here:

THE METALS OF ANALYTICAL GROUP I. When dilute hydrochloric acid is added to a solution containing all of the common metal ions (and ammonium ion), mercury(I) chloride, silver chloride, and lead chloride precipitate. The chlorides of all the other common metal ions are soluble in this acid solution and can be separated from those of Group I by filtration or centrifugation.

THE METALS OF ANALYTICAL GROUP II.   After the Group I chlorides have been separated, the solution is made 0.3 M in hydrochloric acid, and the Group II metals are precipitated as sulfides by the addition of hydrogen sulfide to the solution. The precipitate formed consists of the sulfides of lead, bismuth, copper, cadmium, mercury(II), arsenic, antimony, and tin.

THE METALS OF ANALYTICAL GROUP III.   After the Group II sulfides have been separated, the solution is saturated with hydrogen sulfide, and then an excess of aqueous ammonia is added to it. Under these basic conditions, the sulfides of cobalt, nickel, manganese, iron, and zinc and the hydroxides of aluminum and chromium are precipitated.

THE METALS OF ANALYTICAL GROUP IV.   The Group IV metals, barium, strontium, and calcium, are precipitated as their carbonates from the filtrate or centrifugate of the Group III separation by ammonium carbonate in the presence of aqueous ammonia and ammonium chloride.

THE METALS OF ANALYTICAL GROUP V.   The filtrate from the Group IV separation contains the sodium, potassium, magnesium, and ammonium ions, which constitute Group V.

A flow sheet diagramming the separations of the metal ions into the various analytical groups is given in Table 35-1. Specific details for these separations are given in Chapters 36 through 40.

Even though rather definite directions are given for each analysis to be carried out, no two analyses will be exactly alike. For this reason, directions should never be followed blindly, to the letter, but with careful thought; procedures should be adapted to the particular problem at hand.

Note that formulas for precipitates are underlined and will be so indicated throughout Chapters 35 *through* 42. Solid materials are customarily indicated in one of three ways: (1) with the symbol (s) for the solid following the formula (this system has been used in the preceding chapters), for example, $AgCl(s)$; (2) with a descending arrow, $\downarrow$, following the formula, for example, $AgCl\downarrow$; or (3) by underlining the formula, for example, $\underline{AgCl}$. The latter method is especially well adapted to flow sheets and will be used in the following sections in flow sheets and in equations.

Chapter 30 through 34 are devoted to a consideration of the descriptive chemistry of the metals as they are grouped according to the analytical scheme. Chapters 36 through 40 cover the analysis procedures for the metals. You will find it helpful to study the descriptive and analytical chapters together for each analysis group.

35.2 Equipment

In semimicro qualitative analysis volumes of solutions from 1 drop to about 5 mL are employed, and small test tubes, centrifuge tubes, capillary syringes, and medicine droppers are used to carry out the separations and identifying tests.

Table 35-1   Flow Sheet of Group Separations (Formulas for Precipitates are Underlined)

Original solution	→ HCl	→ 0.3 M HCl, H₂S	→ NH₄Cl, NH₃+H₂O, H₂S	→ NH₃+H₂O, NH₄Cl, (NH₄)₂CO₃
Hg_2^{2+}	**Group I Chlorides:** $\underline{Hg_2Cl_2}$, $\underline{AgCl}$, $\underline{PbCl_2}$			
Ag^+				
Pb^{2+}	Pb^{2+}	**Group II Sulfides:** $\underline{PbS}$, $\underline{Bi_2S_3}$, $\underline{CuS}$, $\underline{CdS}$, $\underline{HgS}$, $\underline{As_2S_3}$, $\underline{Sb_2S_3}$, $\underline{SnS_2}$		
Bi^{3+}	Bi^{3+}			
Cu^{2+}	Cu^{2+}			
Cd^{2+}	Cd^{2+}			
Hg^{2+}	Hg^{2+}			
As^{3+}	As^{3+}			
Sb^{3+}	Sb^{3+}			
Sn^{4+}	Sn^{4+}			
Co^{2+}	Co^{2+}	Co^{2+}	**Group III Sulfides and Hydroxides:** $\underline{CoS}$, $\underline{NiS}$, $\underline{MnS}$, $\underline{FeS}$, $\underline{Al(OH)_3}$, $\underline{Cr(OH)_3}$, $\underline{ZnS}$	
Ni^{2+}	Ni^{2+}	Ni^{2+}		
Mn^{2+}	Mn^{2+}	Mn^{2+}		
Fe^{3+}	Fe^{3+}	Fe^{2+}		
Al^{3+}	Al^{3+}	Al^{3+}		
Cr^{3+}	Cr^{3+}	Cr^{3+}		
Zn^{2+}	Zn^{2+}	Zn^{2+}		
Ba^{2+}	Ba^{2+}	Ba^{2+}	Ba^{2+}	**Group IV Carbonates:** $\underline{BaCO_3}$, $\underline{SrCO_3}$, $\underline{CaCO_3}$
Sr^{2+}	Sr^{2+}	Sr^{2+}	Sr^{2+}	
Ca^{2+}	Ca^{2+}	Ca^{2+}	Ca^{2+}	
Mg^{2+}	Mg^{2+}	Mg^{2+}	Mg^{2+}	**Group V Soluble Ions:** Mg^{2+}, NH_4^+, Na^+, K^+
NH_4^+	NH_4^+	NH_4^+	NH_4^+	
Na^+	Na^+	Na^+	Na^+	
K^+	K^+	K^+	K^+	

Appendixes M and N are lists of apparatus and of solutions and solid reagents that each student will need in carrying out the laboratory work of this course. Some of the necessary equipment will be stocked in the laboratory desk or in the stockroom. Other equipment will have to be fabricated. Wash *all* of the apparatus before beginning your laboratory work.

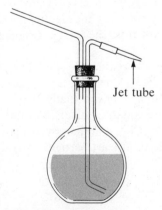

Figure 35-1. A laboratory-constructed wash bottle.

35.3 Wash Bottle and Stirring Rods

Wash bottles made of plastic are supplied to students in some laboratories. Alternatively, you may construct one using a 250-mL Florence flask and 6-mm glass tubing, as shown in Fig. 35-1. Make uniform bends with no constrictions, and fire-polish all ends of the glass tubing.

Keep the wash bottle filled with distilled water for use in the analytical procedures. Ordinary tap water contains such ions as Ca^{2+}, Mg^{2+}, Fe^{3+}, Al^{3+}, Cl^-, SO_4^{2-}, and HCO_3^-. Since these ions are among those being tested for in the unknown solutions, all water used in the procedures and for cleaning glassware must be distilled.

In addition, make at least five glass stirring rods, approximately 15 cm in length and 3 mm in diameter. Fire-polish both ends of each rod.

35.4 Capillary Syringes

Standard medicine droppers of approximately 1-mL capacity are used for measuring and transferring solutions of reagents. These droppers deliver approximately 20 drops per milliliter. A second type of dropper, called a capillary syringe, is used for the removal of liquids from precipitates held in small test tubes or centrifuge tubes. These may be supplied. If not, capillary syringes may be made from glass tubing by the following method. Heat the middle portion of a 7-in section of 8-mm glass tubing over a Bunsen burner flame, rotating until the glass softens. Remove the tube from the flame and slowly draw it out until the bore is about 1 mm. When the tube has cooled, cut the capillary at the midpoint and fire-polish the capillary ends. Flare the wide ends of the tubes by heating until soft and quickly pressing down against a flat metal surface. When the syringes are cold, attach medicine dropper bulbs to the flared ends. These syringes deliver approximately 40 drops per milliliter.

Figure 35-2. Reagent bottle equipped with dropper used in semimicro analytical work.

35.5 Reagents

The solids and solutions called for in the analytical procedures will be available in small (typically 10-mL) reagent bottles (Fig. 35-2). See Appendix N for a list of these reagents. Only a small quantity of the starred reagents will be needed during the course.

If you are to fill reagent bottles from stock bottles for your own use, be sure that your bottles are completely clean before filling them. To avoid mistakes, it is advisable to label each bottle before filling it.

35.6 Precipitation

Practically all of the precipitations are carried out in either 4-mL Pyrex test tubes or 2-mL conical test tubes. Check for completeness of precipitation by adding a drop of reagent to the solution (centrifugate) obtained from the separation of the precipitate. If the addition of more reagent to the solution shows that precipitation is incomplete, separate the mixture and test the second solution for completeness of precipitation.

The precipitating agent should be added slowly, preferably from a medicine dropper, and with vigorous shaking or stirring of the reaction mixture. The formation of larger crystals of the precipitate is favored by warming the solution, and separation of the precipitate should not be attempted before the crystals become large enough to settle.

A slight excess of the precipitating agent is added to reduce the solubility of the precipitate by the common ion effect (Section 16.7). However, a very large excess of the precipitating agent should be avoided, since it may actually increase the solubility of the precipitate. For example, in precipitating silver chloride a large excess of Cl^- will bring about the formation of $AgCl_2^-$, and thereby increase the solubility of AgCl. Many precipitates are dissolved, at least partially, by the formation of related complex ions.

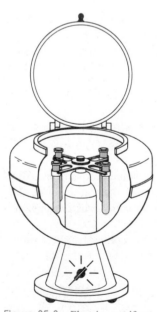

Figure 35-3. Electric centrifuge. Cutaway shows arrangement for holding test tubes used in centrifuging a precipitate.

35.7 Centrifugation of Precipitates

A precipitate may be separated from a liquid in a centrifuge (Fig. 35-3). Spinning a mixture of solid and liquid at high speed in a centrifuge forces the denser precipitate to the bottom of the containing tube by a centrifugal force that is many times the force of gravity. This accounts for the much shorter time required for settling a precipitate when centrifugation is employed. Colloidal precipitates require longer centrifugation than do crystalline precipitates because of the small size of colloidal particles.

Before centrifuging a test tube or centrifuge tube, prepare another tube to balance the first by filling an empty tube to the same level with water. Insert the tubes in opposite positions in the centrifuge, and turn the machine on. Allow the centrifugation to continue for at least 30 s. Turn the machine off and, after the rotation has stopped completely, remove the tubes.

35.8 Transfer of the Centrifugate

After centrifugation the precipitate should be found packed in the bottom of the tube. The supernatant liquid, or centrifugate, is separated from the precipitate by holding the tube at an angle of about 30° (Fig. 35-4) and removing the liquid by slowly drawing it into a capillary syringe. The tip of the syringe is held just below the surface of the liquid. As the pressure on the bulb is slowly released, causing the liquid to rise in the syringe, the capillary is lowered into the tube until all of the liquid is removed. As the capillary approaches the bottom of the tube, the tip must not be allowed to stir up the mixture by touching the precipitate.

Figure 35-4. Capillary syringe used to separate centrifugate from precipitate.

35.9 Washing of the Precipitate

The precipitate left in the tube after the removal of a supernatant liquid is still wet with a solution containing the ions of this liquid. The precipitate must be washed, usually with water, to dilute the solution adhering to the precipitate. The wash liquid is added to the precipitate, and the mixture is stirred thoroughly. The mixture is then centrifuged to cause the precipitate to settle again. After centrifugation, the washings are removed by a capillary syringe as described in Section 35.8. A precipitate is usually washed at least twice. The first wash liquid is ordinarily saved and added to the first centrifugate. If the precipitate must be transferred to another container, the reagent to be used is added, the mixture is well stirred, and then it is poured into the other container. After the precipitate has settled, the supernatant liquid may be employed to remove any precipitate remaining in the centrifuge tube.

Failure to wash precipitates thoroughly is one of the principal sources of error in qualitative analyses.

35.10 Dissolution and Extraction of Precipitates

When all or a part of a precipitate is to be dissolved by a reagent, the solvent is added to the precipitate that is in the centrifuge tube and the mixture is stirred. The mixture is then separated by centrifugation, and the operation is repeated using fresh solvent. Often the extraction of a precipitate is more efficient at an elevated temperature.

35.11 Heating of Mixtures or Solutions

Whenever it is necessary to heat a mixture in order to bring about a precipitation or to dissolve or extract a precipitate, the test tube or centrifuge tube is placed in

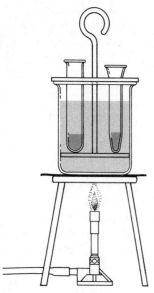

Figure 35-5. Individual water bath for heating reaction mixtures.

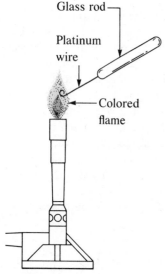

Figure 35-6. Platinum wire mounted in glass rod used in making flame tests for unknown ions.

a water bath (Fig. 35-5) maintained at a suitable temperature. The water in the water bath should be kept hot throughout the work period.

35.12 Evaporation

It is often necessary to heat a solution to boiling and hold it at the boiling temperature in order to concentrate it or to remove a volatile acid or base, or even to evaporate it to dryness. Evaporation should be carried out in a small casserole or porcelain evaporating dish. The contents of the container should be agitated constantly while the heating continues. The evaporation of solutions contained in small test tubes should be avoided because the contents of the tube may be lost due to overheating.

35.13 Cleaning Glassware

Because small amounts of contaminants may give rise to erroneous results, all glassware used in the analytical procedures should be thoroughly cleaned before it is used. The cleaning should be done with a brush and some cleansing powder such as a synthetic detergent. The apparatus should then be rinsed first with tap water and finally with distilled water. Test tube brushes and centrifuge tube brushes are available. Medicine droppers, capillary syringes, and stirring rods should be cleaned, rinsed, and stored in a beaker of distilled water.

35.14 Flame Tests

Flame tests are made using a platinum wire sealed in the end of a piece of glass rod. The wire is cleaned by first dipping the end of it in 6 M HCl and then heating it in the burner flame (Fig. 35-6). Rather than use the hottest portion of the flame, one should bring the looped end of the wire slowly up to the edge of the flame. The platinum wire should never come into the reducing part of the flame. The sequence of dipping in acid and heating in the flame should be repeated until the wire no longer imparts a color to the flame. The wire loop is dipped in the solution to be tested and then heated in the flame. Do not rely on memory to judge the color imparted to the flame; a known solution of the ion should be tested and its color compared to that given by the unknown.

35.15 Known and Unknown Solutions

You should know the details of the analytical procedures before attempting to analyze a solution of unknown composition. Therefore, known solutions containing all the ions of a given group are provided. You should become familiar with the separations and confirmatory tests for the ions of each group by practicing on a known solution before trying to determine the ions present in an unknown solution. You should especially note the quantities of precipitates obtained and the colors of precipitates and solutions as you proceed with the analysis of a known solution. These observations and the equations for the reactions involved should be recorded systematically in a notebook at the time the observations are made.

A typical set of laboratory assignments for knowns and unknowns is given in Appendix P.

35.16 Report Notebook

A notebook should be obtained (the type to be designated by your instructor) in which to record in a systematic way the results of the analyses that are performed. All observations that are made should be recorded *as soon as they have been completed.* An example of one method of keeping the systematic record is shown below.

Sample Unknown Report

Name: _____ Section: _____ Date: _____

Instructor's Approval: _____ Ions Found: _____ Grade: _____

Cation Unknown Report

No.	Substance	Reagent	Result	Inference or Conclusion	Precipitate or Residue	Centrifugate or Solution
1	Group I	HCl	White ppt	Group I present	One or more of $AgCl$, $PbCl_2$, Hg_2Cl_2	Possibly Pb^{2+} or Hg^{2+}
2	Ppt from 1	Hot water	No visible action	Hg_2^{2+} and/or Ag^+ present; Pb^{2+} uncertain	Hg_2Cl_2 and/or $AgCl$	Possibly Pb^{2+}
3	Filtrate from 2 or 1	K_2CrO_4	Yellow ppt	Pb^{2+} present	$PbCrO_4$	
4	Residue from 2	$NH_3 + H_2O$	Residue dissolves completely	Hg_2^{2+} absent; Ag^+ probable		$[Ag(NH_3)_2]^+$
5	Filtrate from 4	HNO_3	White ppt	Ag^+ present	$AgCl$	
6						

36

THE ANALYSIS OF GROUP I

36.1 Introduction

The procedures in this chapter are the first in the analysis of a solution of a known or unknown sample containing metal ions. Depending on the nature of the known or unknown, the solution may contain only the ions of Group I, or it may contain the ions of any or all of Groups I through V. If your sample is a general unknown, reserve a part of the original solution for the test for the ammonium ion in Group V. Review Section 35.1 for the classification of the metal ions into analytical groups before proceeding with the analysis of Group I. A schematic outline of the analysis for the Group I metals is given in the flow sheet in Table 36-1.

The flow sheet summarizes the chemistry of the separations and the identification tests in a form that is easily studied and remembered. Similar flow sheets are provided in this text to assist in the analysis of metals in Groups II through V.

A discussion of the chemistry involved in each step of the analysis of each group will be given just prior to the laboratory procedure for the step. Careful study of these discussions will help you learn the chemistry involved in the analysis. Additional details concerning the chemistry of the metals of the analytical groups are given in Chapters 30 through 34. Chapter 30 is concerned with the metals of Analytical Group I.

Table 36-1   Group I Flow Sheet

36.2 Precipitation of Group I

Hydrochloric acid is the reagent used to precipitate the metal ions of Group I. The chlorides of mercury(I), silver, and lead are the only ones of the cations under consideration in this qualitative scheme that are insoluble in acid solutions. The equations for the precipitation of the metal ions of Group I are

$$Hg_2^{2+} + 2Cl^- \rightleftharpoons Hg_2Cl_2$$
$$Ag^+ + Cl^- \rightleftharpoons AgCl$$
$$Pb^{2+} + 2Cl^- \rightleftharpoons PbCl_2$$

The solubilities and solubility products of the chlorides of this group are given in Table 36-2. These values show that the solubility of lead chloride in g/mL is about 2800 times greater than that of silver chloride and about 14,500 times greater than that of mercury(I) chloride. Thus the mercury(I) ion and the silver ion are in effect completely removed from a solution when hydrochloric acid is added to it. On the other hand, lead is not completely precipitated by chloride ion, and some of it is carried through to Group II.

Table 36-2   Solubilities of the Group I Chlorides

Salt	K_{sp}	Solubility, mol/L	Solubility, g/mL
Hg_2Cl_2	1.1×10^{-18}	6.5×10^{-7}	3.1×10^{-7}
AgCl	1.8×10^{-10}	1.3×10^{-5}	1.9×10^{-6}
$PbCl_2$	1.7×10^{-5}	1.6×10^{-2}	4.5×10^{-3}

A slight excess of hydrochloric acid is used in the precipitation of the Group I ions to prevent the precipitation of bismuth oxychloride and antimony oxychloride by hydrolysis of their chloride salts.

$$Bi^{3+} + Cl^- + H_2O \rightleftharpoons BiOCl + 2H^+$$
$$Sb^{3+} + Cl^- + H_2O \rightleftharpoons SbOCl + 2H^+$$

The hydrogen ions of the hydrochloric acid keep the above equilibria shifted to the left. In addition, a slight excess of chloride ions ensures more complete precipitation of Hg_2Cl_2, AgCl, and $PbCl_2$, due to the common ion effect (Section 16.7). For example, in a solution that is 0.30 M in Cl^-, the common ion effect will reduce the concentration of Ag^+ from 1.3×10^{-5} M in a saturated solution of AgCl to

6.0×10^{-10} *M*. On the other hand, a large excess of chloride ions must be avoided since silver chloride and lead chloride tend to react with chloride ions, forming soluble complex ions. The equations for these reactions are

$$AgCl + Cl^- \rightleftharpoons [AgCl_2]^-$$
$$PbCl_2 + 2Cl^- \rightleftharpoons [PbCl_4]^{2-}$$

PROCEDURE 1   *Precipitation of Group I: Ag^+, Hg_2^{2+}, and Pb^{2+}*. To 10 drops of the solution to be analyzed, add enough water to make a total volume of 1 mL. Add 2 drops of 6 *M* HCl to this solution (avoid using a large excess of HCl to eliminate the possibility of formation of $[AgCl_2]^-$ or $[PbCl_4]^{2-}$). Separate the precipitate by centrifugation. If your sample is a general unknown, reserve the solution (centrifugate) for the analysis of Groups II through V. If your sample is a Group I known or unknown, test the solution for lead as directed in Procedure 2 below. Treat the precipitate according to Procedure 2.

36.3   Separation and Identification of the Lead Ion

Lead chloride may fail to precipitate at all if the lead ion concentration is too low or if the temperature is too high. For this reason a test for lead should be made on the centrifugate from the original Group I separation.

    The fact that lead chloride is about three times as soluble in hot water as in cold water enables us to separate it from silver chloride and mercury(I) chloride by extraction of the mixed chlorides with hot water. The presence of the lead ion may be confirmed by precipitation of the slightly soluble yellow lead chromate.

$$Pb^{2+} + CrO_4^{2-} \rightleftharpoons \underline{PbCrO_4}$$

PROCEDURE 2   *Precipitate from Procedure 1: Hg_2Cl_2, AgCl, and $PbCl_2$*. Extract the precipitate twice with 10 drops of hot water to dissolve the $PbCl_2$. Reserve the residue for Procedure 3. Add 2 drops of 1 *M* K_2CrO_4 to the solution (hot water extract). A yellow precipitate of $PbCrO_4$ confirms the presence of lead.

36.4   Separation and Identification of the Mercury(I) Ion

When aqueous ammonia is added to a mixture of silver chloride and mercury(I) chloride, the silver chloride dissolves by forming soluble diamminesilver chloride.

$$\underline{AgCl} + 2NH_3 \rightleftharpoons [Ag(NH_3)_2]^+ + Cl^-$$

In the presence of ammonia, mercury(I) chloride undergoes auto-oxidation-reduction with the formation of finely divided metallic mercury, which is black, and mercury(II) amidochloride, $HgNH_2Cl$, which is white. The equation is

$$\underline{Hg_2Cl_2} + 2NH_3 \longrightarrow \underline{Hg} + HgNH_2Cl + NH_4^+ + Cl^-$$

The formation of this black (or gray) precipitate is sufficient proof of the presence of mercury(I) ions.

PROCEDURE 3   *Residue from Procedure 2: Hg_2Cl_2 and AgCl.* Extract the residue twice with 5 drops of 4 *M* aqueous ammonia. A black or gray residue confirms the presence of mercury(I). Reserve the aqueous ammonia extract for Procedure 4.

36.5   Identification of the Silver Ion

The presence of the silver ion may be confirmed by treating the aqueous ammonia extract of mercury(I) chloride and silver chloride with nitric acid. This causes the reprecipitation of white silver chloride. The hydrogen ion supplied by the nitric acid unites with the free ammonia in equilibrium with the diamminesilver ion and causes the equilibrium

$$[Ag(NH_3)_2]^+ \rightleftharpoons Ag^+ + 2NH_3$$

to shift to the right. The free silver ion is then precipitated by the chloride ion that is present in the solution. The net reaction is the sum of three equilibria.

$$2NH_3 + 2H^+ \rightleftharpoons 2NH_4^+$$
$$[Ag(NH_3)_2]^+ \rightleftharpoons Ag^+ + 2NH_3$$
$$\underline{Ag^+ + Cl^- \rightleftharpoons AgCl}$$
$$[Ag(NH_3)_2]^+ + Cl^- + 2H^+ \rightleftharpoons \underline{AgCl} + 2NH_4^+$$

PROCEDURE 4   *Solution from Procedure 3: $[Ag(NH_3)_2]Cl$.* Add 4 *M* HNO_3 until the solution is acid to litmus. Make sure the HNO_3 is stirred into the solution to neutralize all ammonia throughout before testing with litmus. A white precipitate (or cloudiness) of AgCl confirms the presence of silver.

FOR REVIEW

EXERCISES

1. What general statement can be made concerning the solubility of common chloride salts other than those of the cations of Analytical Group I?
2. Write out the flow sheet for the Group I analysis.
3. Give the color of each of the following: AgCl, $PbCl_2$, Hg_2Cl_2, $PbCrO_4$, Hg, $HgNH_2Cl$, and $[Ag(NH_3)_2]Cl$ (in solution).
4. Why must a large excess of chloride ions be avoided in the precipitation of the Group I chlorides?
5. In the case of an unknown containing only the cations of Group I, why is it advisable to make a confirmatory test for lead on the filtrate from the Group I precipitation?
6. Select a reagent used in the analysis of Group I that will in one step separate each of the following pairs:
   (a) Hg_2Cl_2, $PbCl_2$ (d) Hg_2^{2+}, Hg^{2+}
   (b) Hg_2Cl_2, AgCl   (e) AgCl, $PbCl_2$
   (c) AgCl, $CuCl_2$
7. In terms of ionic equilibria and solubility product theory, explain the dissolution of silver chloride in aqueous ammonia and its reprecipitation with nitric acid.

8. Show that the reaction of ammonia with mercury(I) chloride is an oxidation-reduction reaction.

9. What prevents the slightly soluble oxychlorides of bismuth and antimony from precipitating with the Group I chlorides?

10. A solution is 0.020 M in both Pb^{2+} and Ag^+. If Cl^- is added to this solution, what is the concentration of Ag^+ when $PbCl_2$ begins to precipitate? (K_{sp} for AgCl is 1.8×10^{-10}, and K_{sp} for $PbCl_2$ is 1.7×10^{-5}.)   *Ans. 6.2 $\times$ 10^{-9} M*

11. Calculate the concentration of Ag^+ in a 0.0010 M solution of $[Ag(NH_3)_2]Cl$ that is 0.25 M in aqueous ammonia. (K_f for $[Ag(NH_3)_2]^+$ is 1.6×10^7.)   *Ans. 1.0 $\times$ 10^{-9} M*

12. From the K_{sp} data given in Table 36-2, calculate the number of cations present in 1.0 mL of saturated solutions of Hg_2Cl_2, AgCl, and $PbCl_2$.
    Ans. 3.9 $\times$ 10^{14} Hg$_2$$^{2+}$ ions;
    8.1 $\times$ 10^{15} Ag$^+$ ions; 9.8 $\times$ 10^{18} Pb^{2+} ions

13. Calculate the solubility of Hg_2Cl_2 in 0.020 M NaCl. Compare this solubility with that of Hg_2Cl_2 in water. (K_{sp} for Hg_2Cl_2 is 1.1×10^{-18}.)
    Ans. 2.8 $\times$ 10^{-15} mol/L

14. Calculate the solubility of AgCl in 6.0 M HCl. (The formation constant for $Ag^+ + 2Cl^- \rightarrow$ $[AgCl_2]^-$ is 2.5×10^5, and K_{sp} of AgCl is 1.8×10^{-10}.)
    Ans. 2.7 $\times$ 10^{-4} mol/L

37

THE ANALYSIS OF GROUP II

The solution to be analyzed may be a Group II known or unknown, or it may be the solution from the Group I separation. In either case proceed according to Procedure 1 below. See Table 37-1 for the Group II flow sheet. You may also wish to refer to Chapter 31, which is devoted to the chemistry of the metals of Analytical Group II.

37.1 Precipitation of Group II Sulfides

1. SEPARATION OF GROUP II FROM GROUPS III–V. Ions of lead, bismuth, copper, cadmium, mercury(II), arsenic, antimony, and tin form sulfides that are insoluble in solutions that are 0.3 M in hydrogen ion. Cadmium sulfide is the most soluble of the sulfides of Group II, and zinc sulfide is the least soluble of the sulfides of Group III. Therefore, the separation of the sulfides of Group II from those of Group III is ensured if the sulfide ion concentration of the solution is controlled in such a way that the cadmium is completely precipitated as the sulfide without exceeding the solubility product of zinc sulfide. In a solution that is saturated with hydrogen sulfide and 0.3 M in hydrogen ion, the sulfide ion concentration is just right to effect the separation of cadmium from zinc and, hence, of Group II from Group III. The theory for these separations is discussed in Section 17.9.

Table 37-1 Group II Flow Sheet

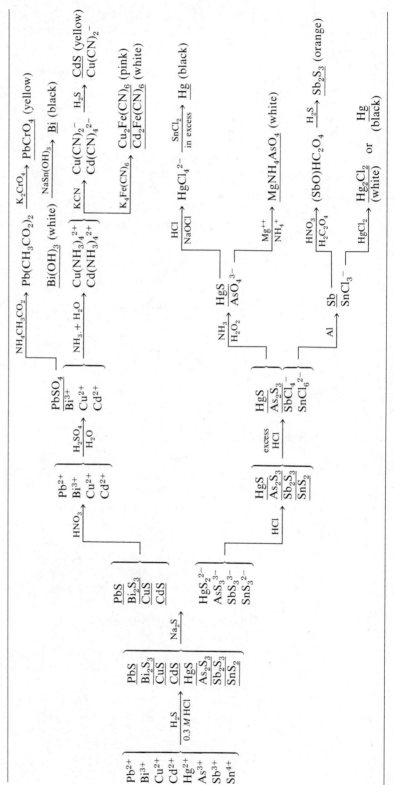

The hydrogen sulfide for the precipitation of the Group II cations is supplied by thioacetamide, CH_3CSNH_2, which hydrolyzes in hot acidic solutions with the formation of ammonium acetate and hydrogen sulfide.

$$CH_3CSNH_2 + 2H_2O \longrightarrow CH_3CO_2^- + NH_4^+ + H_2S$$

The hydrogen sulfide formed ionizes in aqueous solution according to the equation

$$H_2S \rightleftharpoons 2H^+ + S^{2-}$$

Hydrochloric acid furnishes hydrogen ions, which repress the ionization of hydrogen sulfide and thus cause a decrease in the sulfide ion concentration of the solution. (The ionization of hydrogen sulfide and its control are discussed in detail in Section 16.9.) The precipitation of the Group II sulfides may be represented by the following equations:

$$Pb^{2+} + H_2S \rightleftharpoons \underline{PbS} + 2H^+$$
$$2Bi^{3+} + 3H_2S \rightleftharpoons \underline{Bi_2S_3} + 6H^+$$
$$Cu^{2+} + H_2S \rightleftharpoons \underline{CuS} + 2H^+$$
$$Cd^{2+} + H_2S \rightleftharpoons \underline{CdS} + 2H^+$$
$$Hg^{2+} + H_2S \rightleftharpoons \underline{HgS} + 2H^+$$
$$2As^{3+} + 3H_2S \rightleftharpoons \underline{As_2S_3} + 6H^+$$
$$2Sb^{3+} + 3H_2S \rightleftharpoons \underline{Sb_2S_3} + 6H^+$$
$$Sn^{4+} + 2H_2S \rightleftharpoons \underline{SnS_2} + 4H^+$$

Note that the acidity of the solution is increased by the sulfide precipitation reactions since hydrogen ion is one of the products.

2. MECHANISM OF THE PRECIPITATION OF THE SULFIDES.   The concentration of the sulfide ion in a solution saturated with hydrogen sulfide and 0.3 M in hydrogen ions is 1.4×10^{-20} mol/L (Section 16.9). Because there are 6.022×10^{23} ions in a mole, only about 8 sulfide ions are present in each milliliter of solution. Yet practically all of the Pb^{2+} is precipitated immediately when hydrogen sulfide is passed into an acidic solution of a lead salt. The precipitation is so rapid that it is inconceivable that the reaction is one of simple union of lead and sulfide ions ($Pb^{2+} + S^{2-} \rightleftharpoons PbS$) when the sulfide ion is present in such a small concentration. However, the concentration of the hydrogen sulfide ion, HS^-, in a saturated solution of hydrogen sulfide that is 0.3 M in hydrogen ions is comparatively large relative to the low concentration of the sulfide ion, as seen from the calculations:

$$\frac{[H^+][HS^-]}{[H_2S]} = \frac{(0.3)[HS^-]}{0.1} = 1 \times 10^{-7}$$

$$[HS^-] = \frac{0.1}{0.3} \times 1 \times 10^{-7} = 3 \times 10^{-8}$$

Thus we find that the concentration of HS^- is 3×10^{-8} mol/L, or about 10^{12} times the concentration of the sulfide ion (1.4×10^{-20}). Therefore, it is reasonable to assume that an unstable intermediate hydrosulfide, $Pb(HS)_2$, is formed and immediately breaks down to form PbS and H_2S.

$$Pb^{2+} + 2HS^- \rightleftharpoons \underline{Pb(HS)_2} \rightleftharpoons \underline{PbS} + H_2S$$

A similar process is known to lead to the formation of oxides through the precipitation of unstable hydroxides. A notable example is the formation of silver oxide according to the equilibria

$$2Ag^+ + 2OH^- \rightleftharpoons \underline{2AgOH} \rightleftharpoons \underline{Ag_2O} + H_2O$$

Equilibrium is concerned only with the final result, not with the mechanism by which the result is obtained. This means that calculations involving the precipitation of metal sulfides are valid when concentrations of the sulfide ion are employed even though this ion may not be involved directly in the mechanism for the precipitation of the metal sulfide.

3. COLORS OF THE SULFIDES OF GROUP II.
The sulfides of lead(II), copper(II), and mercury(II) are all black. The sulfides of tin(IV), cadmium, and arsenic(III) are yellow, that of bismuth is dark brown or black, and that of antimony(III) is orange-red (sometimes black).

Tin(II) sulfide, SnS, forms a gelatinous precipitate that does not dissolve in sodium sulfide solution, while SnS_2 is readily soluble in this reagent. For this reason tin(II) is oxidized to tin(IV) by adding nitric acid and heating the original solution before the sulfides of the Group II ions are precipitated.

PROCEDURE 1   *Precipitation of Group II: Pb^{2+}, Bi^{3+}, Cu^{2+}, Cd^{2+}, Hg^{2+}, As^{3+}, Sb^{3+}, Sn^{2+}, Sn^{4+}.*
Add 5 drops of 4 M HNO_3 to 10 drops of the known or unknown (or to the centrifugate from the Group I separation) and evaporate the solution not quite to dryness, ending with a moist residue (Section 35.12). Cool and add 10 drops of water. Add 1 M aqueous ammonia until the solution (or mixture) is just basic to litmus. (Dip a stirring rod into the solution and then touch the moist end of the rod to a piece of litmus paper to test for acidity or basicity of solutions.) Add 1.0 M HCl until the solution is just acidic to litmus. Now add 2 drops of 6 M HCl and dilute the solution to 1.5 mL. Add 10 drops of 5% thioacetamide solution. The solution should now be 0.3 M in hydrogen ion if the directions have been followed carefully. Heat the solution contained in a test tube in a hot water bath for 10 min. Sufficient hydrolysis of the thioacetamide to get complete precipitation of the Group II sulfides is essential. Add 1 mL of water and heat the mixture for another 5 min. Separate the precipitate. If working on a general unknown, reserve the centrifugate for the analysis of Groups III through V. Wash the precipitate twice with 10 drops of 0.1 M HCl. Add the first 10 drops of wash solution to the original centrifugate and discard the rest.

37.2 Separation of the Group II Ions into Divisions A and B

The separation of the Group II ions into two divisions is based on the fact that HgS, As_2S_3, Sb_2S_3, and SnS_2 form soluble complex sulfides in alkaline sulfide solutions, while PbS, Bi_2S_3, CuS, and CdS do not form these complex ions.

The reagent used in this separation is sodium sulfide, which is produced by the reaction of hydrogen sulfide (formed by the hydrolysis of thioacetamide) with sodium hydroxide. The equation is

$$[2Na^+] + 2OH^- + H_2S \rightleftharpoons [2Na^+] + S^{2-} + 2H_2O$$

The presence of a high concentration of hydroxide ion, supplied by an excess of sodium hydroxide, reduces the extent of the hydrolysis of the sulfide ion.

$$S^{2-} + H_2O \rightleftharpoons HS^- + OH^-$$

This provides the relatively high concentration of sulfide ion required for the dissolution of the Division B sulfides, particularly HgS. The equations are

$$HgS + S^{2-} \rightleftharpoons [HgS_2]^{2-} \quad [\text{dithiomercurate(II)}]$$
$$As_2S_3 + 3S^{2-} \rightleftharpoons 2[AsS_3]^{3-} \quad [\text{trithioarsenite(III)}]$$
$$Sb_2S_3 + 3S^{2-} \rightleftharpoons 2[SbS_3]^{3-} \quad [\text{trithioantimonite(III)}]$$
$$SnS_2 + S^{2-} \rightleftharpoons [SnS_3]^{2-} \quad [\text{trithiostannate(IV)}]$$

The cations whose sulfides fail to dissolve in the presence of a high concentration of sulfide ion constitute Division A of Group II. They are Pb^{2+}, Bi^{3+}, Cu^{2+}, and Cd^{2+}.

PROCEDURE 2   *Precipitate from Procedure 1: PbS, Bi$_2$S$_3$, CuS, CdS, HgS, As$_2$S$_3$, Sb$_2$S$_3$, and SnS$_2$.*
Add 12 drops of 4 M NaOH and 4 drops of thioacetamide solution to the precipitate. Heat the mixture in a hot water bath for 5 min. Separate the mixture and reserve the solution. Treat the precipitate a second time with 12 drops of 4 M NaOH and 4 drops of thioacetamide solution. Heat the mixture in a hot water bath for 5 min. Separate the mixture and add the solution to that obtained from the first treatment. Save the combined solutions (thiosalts of the Division B ions) for Procedure 6. The residue consists of the sulfides of the ions of Division A.

37.3 Separation and Identification of Lead

The undissolved sulfides of Division A (PbS, Bi$_2$S$_3$, CuS, and CdS) are washed with a solution of ammonium nitrate. This is used rather than pure water to decrease the tendency of the sulfides to become colloidal. In a colloidal condition these sulfides would pass into the centrifugate and be partly lost.

The sulfides are then dissolved by treatment with hot dilute nitric acid. The acid oxidizes the sulfide ion to elemental sulfur and causes the metal sulfides to dissolve by holding the sulfide ion concentration at such a low level that the product of the concentrations of the ions is less than the solubility product. The two equilibria involved in the system, in the case of lead sulfide, are

$$\underline{PbS} \rightleftharpoons Pb^{2+} + S^{2-}$$
$$3S^{2-} + 8H^+ + 2NO_3^- \rightleftharpoons 3\underline{S} + 2NO(g) + 4H_2O$$

and the net reaction is

$$3\underline{PbS} + 8H^+ + 2NO_3^- \rightleftharpoons 3Pb^{2+} + 3\underline{S} + 2NO(g) + 4H_2O$$

Lead is separated from bismuth, copper, and cadmium by precipitating lead sulfate, PbSO$_4$, which is white. The sulfates of the other three metals are soluble. The nitric acid in the solution must first be removed because the sulfate ion is converted largely to hydrogen sulfate ion by hydrogen ions in high concentration.

$$SO_4^{2-} + H^+ \rightleftharpoons HSO_4^-$$

Under these conditions the concentration of sulfate ion is too low for the solubility product of lead sulfate to be exceeded. The removal of the nitric acid is accomplished by adding sulfuric acid and evaporating the solution to the point where sulfur trioxide fumes appear.

$$H_2SO_4 \longrightarrow H_2O + SO_3$$

The more volatile nitric acid will have been expelled before the SO$_3$ fumes appear. When the sulfuric acid solution is diluted with water, sulfate ions are formed and the lead precipitates as lead sulfate. The equations are

$$H_2SO_4 \rightleftharpoons H^+ + HSO_4^-$$
$$HSO_4^- \rightleftharpoons H^+ + SO_4^{2-}$$
$$Pb^{2+} + SO_4^{2-} \rightleftharpoons \underline{PbSO_4}$$

The lead sulfate is then dissolved in a solution of ammonium acetate with the formation of slightly ionized lead acetate. The equation is

$$\underline{PbSO_4} + 2CH_3CO_2^- \rightleftharpoons Pb(CH_3CO_2)_2 + SO_4^{2-}$$

When chromate ions are added to this acetate solution, yellow lead chromate, PbCrO$_4$, precipitates.

$$Pb(CH_3CO_2)_2 + CrO_4^{2-} \rightleftharpoons \underline{PbCrO_4} + 2CH_3CO_2^-$$

PROCEDURE 3   *Residue from Procedure 2: PbS, CuS, CdS, and Bi$_2$S$_3$.* Wash the residue with 1 mL of water to which 2 drops of 1 M NH$_4$NO$_3$ have been added. Discard the wash solution. Add 15 drops of 6 M HNO$_3$ to the residue and heat the mixture in a hot water bath for several minutes to dissolve the sulfides. Separate and discard any sulfur that is formed by the oxidation of sulfide ions, and transfer the solution to a casserole or evaporating dish. Add 5 drops of 4 M H$_2$SO$_4$ and evaporate the solution under the hood (very important) until white SO$_3$ fumes appear. (The SO$_3$ fumes are dense, not at all like steam. Do not prolong the heating after SO$_3$ fumes appear.) Cool and add 1 mL of water to the mixture. Warm the mixture and stir up the precipitate. (The precipitate, PbSO$_4$, may appear slowly, especially if the solution is too acidic.) Separate the mixture. Reserve the solution for Procedure 4. Wash the residue (PbSO$_4$) with a few drops of water and discard the wash solution. Extract the residue (Section 35.10) with a mixture of 5 drops of 1 M ammonium acetate (NH$_4$CH$_3$CO$_2$) and 1 drop of 1 M acetic acid (CH$_3$CO$_2$H). Add 1 drop of 1 M K$_2$CrO$_4$ to the extract. Scratch the inside wall of the test tube with a glass rod to initiate precipitation. The formation of a yellow precipitate confirms the presence of lead as the chromate, PbCrO$_4$.

37.4   Separation and Identification of Bismuth

When aqueous ammonia is added to the solution from the lead sulfate separation, the hydroxides of bismuth (white), copper (pale blue), and cadmium (white) are first precipitated. The equations are

$$Bi^{3+} + 3NH_3 + 3H_2O \longrightarrow \underline{Bi(OH)_3} + 3NH_4^+$$

$$Cu^{2+} + 2NH_3 + 2H_2O \longrightarrow \underline{Cu(OH)_2} + 2NH_4^+$$

$$Cd^{2+} + 2NH_3 + 2H_2O \longrightarrow \underline{Cd(OH)_2} + 2NH_4^+$$

When an excess of aqueous ammonia is added, the hydroxides of copper and cadmium dissolve by forming tetraammine complexes, but bismuth(III) hydroxide does not dissolve.

$$\underline{Bi(OH)_3} + NH_3 \longrightarrow \text{no reaction}$$

$$\underline{Cu(OH)_2} + 4NH_3 \longrightarrow [Cu(NH_3)_4]^{2+} + 2OH^-$$

$$\underline{Cd(OH)_2} + 4NH_3 \longrightarrow [Cd(NH_3)_4]^{2+} + 2OH^-$$

After it is separated from the complexes of copper and cadmium, bismuth(III) hydroxide is treated with sodium stannite, a strong reducing agent. The bismuth is reduced to the elemental state, in which condition it appears black. The equation is

$$2\underline{Bi(OH)_3} + 3[Sn(OH)_3]^- + 3OH^- \longrightarrow 2\underline{Bi} + 3[Sn(OH)_6]^{2-}$$
$$\text{Stannite ion} \qquad\qquad\qquad\qquad\qquad \text{Stannate ion}$$

Because sodium stannite is unstable and darkens on standing, it must be prepared just prior to being used, and the test for bismuth performed immediately. Sodium stannite is prepared by treating a solution of tin(II) chloride with an excess of sodium hydroxide. The equations are

$$Sn^{2+} + 2OH^- \rightleftharpoons \underline{Sn(OH)_2} \quad \text{(white)}$$

$$\underline{Sn(OH)_2} + \text{excess } OH^- \rightleftharpoons [Sn(OH)_3]^- \quad \text{(colorless)}$$

The darkening of the stannite solution on standing is due to the formation of black tin(II) oxide. The equation is

$$[Sn(OH)_3]^- \longrightarrow \underline{SnO} + H_2O + OH^-$$

PROCEDURE 4   *Solution from Procedure 3: Bi^{3+}, Cu^{2+}, and Cd^{2+}.* Add 15 M aqueous ammonia dropwise to the solution until it is distinctly basic to litmus (about 5 drops). The development of a deep blue color in the solution indicates the presence of the tetraamminecopper(II) ion, $[Cu(NH_3)_4]^{2+}$. The formation of a white precipitate indicates the presence of Bi^{3+} as $Bi(OH)_3$. Separate the mixture and reserve the solution for Procedure 5. Wash the precipitate with 15 drops of hot water, and then add several drops of fresh sodium stannite solution to the precipitate. The immediate formation of a black residue (finely divided metallic bismuth) confirms the presence of Bi^{3+}. (Preparation of sodium stannite: In a separate test tube, place 3 drops of 0.4 M $SnCl_2$ and add sufficient 4 M NaOH to dissolve the white precipitate that first forms. Be sure that you are not fooled by an initial precipitate that forms because of a temporary local excess of reagent and dissolves just with stirring.)

37.5 Identification of Copper and Cadmium

The presence of copper is apparent from the deep blue color exhibited by the complex ion $[Cu(NH_3)_4]^{2+}$. However, if the blue color is too faint to discern, as is the case with a trace of copper, the formation of pink copper(II) hexacyano-ferrate(II) serves as a more sensitive test. The equation for the reaction is

$$2[Cu(NH_3)_4]^{2+} + [Fe(CN)_6]^{4-} \longrightarrow \underline{Cu_2Fe(CN)_6} + 8NH_3$$

If cadmium is present, a white precipitate of cadmium hexacyanoferrate(II), $Cd_2Fe(CN)_6$, will form.

The presence of cadmium ions may be confirmed by forming yellow cadmium sulfide with hydrogen sulfide. However, if we attempt to precipitate CdS in the presence of copper, the black color of CuS obscures the yellow color of CdS. By first adding cyanide ions to a solution containing Cu^{2+} and Cd^{2+}, the very stable dicyanocopper(I) complex will be formed, which will not react with sulfide ions. The copper(II) ion is reduced to copper(I) by the cyanide ion, and the cyanide ion is oxidized to cyanogen, C_2N_2.

$$2[Cu(NH_3)_4]^{2+} + 6CN^- \longrightarrow 2[Cu(CN)_2]^- + C_2N_2(g) + 8NH_3$$
$$[Cu(CN)_2]^- + S^{2-} \longrightarrow \text{no reaction}$$

The corresponding tetracyanocadmium complex is much less stable than the dicyanocopper complex, and it will react with the sulfide ion to form yellow CdS.

$$[Cd(NH_3)_4]^{2+} + 4CN^- \rightleftharpoons [Cd(CN)_4]^{2-} + 4NH_3$$
$$[Cd(CN)_4]^{2-} + S^{2-} \longrightarrow \underline{CdS} + 4CN^-$$

PROCEDURE 5   *Solution from Procedure 4:* $[Cu(NH_3)_4]^{2+}$ *(blue) and* $[Cd(NH_3)_4]^{2+}$ *(colorless)*. If the solution is colorless, a trace of copper may be present. Place 10 drops of the solution in a test tube, acidify with acetic acid, and add several drops of 0.1 M $K_4Fe(CN)_6$. The formation of a pink precipitate, $Cu_2Fe(CN)_6$, indicates the presence of a small concentration of Cu^{2+}; the formation of a white precipitate indicates the absence of Cu^{2+} and the probable presence of Cd^{2+} as $Cd_2Fe(CN)_6$.

If copper(II) ions are present, add 2 drops of 1 M KCN to another 10 drops of the solution from Procedure 4, and then continue to add 1 M KCN dropwise until the solution becomes colorless. Add 2 drops of thioacetamide solution and heat the mixture in a hot water bath. The formation of either a yellow or an olive-green precipitate (CdS) confirms the presence of cadmium. If Cu^{2+} is found to be absent, test the second sample for Cd^{2+} as described above but leave out the KCN.

37.6 Reprecipitation of the Sulfides of the Division B Ions

To reprecipitate the sulfides of mercury, arsenic, antimony, and tin from the solution containing their complex thiosalts, it is necessary to increase the concentration of each cation to a value such that the ion product exceeds the solubility product. For example,

$$[Hg^{2+}][S^{2-}] > K_{sp}$$

This is accomplished by increasing the extent of dissociation of the complex ions through the addition of hydrochloric acid.

$$[HgS_2]^{2-} \rightleftharpoons Hg^{2+} + 2S^{2-}$$
$$2S^{2-} + 4H^+ \rightleftharpoons 2H_2S(g)$$

Net reaction: $[HgS_2]^{2-} + 4H^+ \rightleftharpoons Hg^{2+} + 2H_2S(g)$

As seen from the above equations, the added acid displaces the equilibria in the direction of an increased concentration of Hg^{2+} as a result of the formation and escape of the weak electrolyte H_2S as a gas. This causes an increase in the concentration of the cation to such an extent that the product of the concentration of the mercury(II) ion and the sulfide ion becomes larger than the solubility product, and reprecipitation of the sulfide occurs. The net reactions for these reprecipitations are

$$[HgS_2]^{2-} + 2H^+ \rightleftharpoons HgS + H_2S(g)$$
$$2[AsS_3]^{3-} + 6H^+ \rightleftharpoons As_2S_3 + 3H_2S(g)$$
$$2[SbS_3]^{3-} + 6H^+ \rightleftharpoons Sb_2S_3 + 3H_2S(g)$$
$$[SnS_3]^{2-} + 2H^+ \rightleftharpoons SnS_2 + H_2S(g)$$

PROCEDURE 6   *Solution from Procedure 2: Thiosalts of the Division B Ions, $[HgS_2]^{2-}$, $[AsS_3]^{3-}$, $[SbS_3]^{3-}$, and $[SnS_3]^{2-}$. Add 1 M HCl until the solution is just acidic to litmus. Heat the mixture in a hot water bath for several minutes. Separate the mixture and discard the solution. The precipitate consists of HgS, As_2S_3, Sb_2S_3, and SnS_2.*

37.7 Separation of Mercury and Arsenic from Antimony and Tin

To dissolve Sb_2S_3 and SnS_2, and leave HgS and As_2S_3 undissolved, the concentrations of the cations and sulfide ion must be made such that

$$[Sb^{3+}]^2[S^{2-}]^3 < K_{Sb_2S_3} \quad \text{and} \quad [Hg^{2+}][S^{2-}] > K_{HgS}$$

The addition of 6 M HCl reduces the sulfide ion concentration

$$S^{2-} + 2H^+ \rightleftharpoons 2H_2S$$

and the concentrations of antimony ion and tin ion

$$Sb^{3+} + 4Cl^- \rightleftharpoons [SbCl_4]^-$$
$$Sn^{4+} + 6Cl^- \rightleftharpoons [SnCl_6]^{2-}$$

to values such that these sulfides dissolve. The net reactions are

$$Sb_2S_3 + 6H^+ + 8Cl^- \longrightarrow 2[SbCl_4]^- + 3H_2S(g)$$
$$SnS_2 + 4H^+ + 6Cl^- \longrightarrow [SnCl_6]^{2-} + 2H_2S(g)$$

The sulfides of mercury and arsenic remain undissolved in the presence of 6 M HCl due to the very small values of their solubility products.

PROCEDURE 7   *Precipitate from Procedure 6: HgS, As$_2$S$_3$, Sb$_2$S$_3$, and SnS$_2$. Add 1 mL of 6 M HCl to the precipitate and stir the mixture. Heat the mixture in a hot water bath and then separate the residue. Add 15 drops of 6 M HCl to the residue and heat the mixture. Separate the residue and reserve the combined centrifugates for Procedure 11. Treat the residue according to Procedure 8.*

37.8   Separation of Arsenic from Mercury

The concentration of the sulfide ion in saturated solutions of the sulfides of mercury and arsenic is so small that it is not possible to dissolve these sulfides by addition of hydrogen ions. In these cases the product of $[H^+]^2$ and $[S^{2-}]$ does not exceed the ion product for H$_2$S, and equilibrium is established before these sulfides are dissolved.

In order to dissolve HgS and As$_2$S$_3$, it is necessary to reduce either the concentration of the sulfide ion by oxidation or the concentration of the cation by the formation of a complex ion. A mixture of aqueous ammonia and hydrogen peroxide is used to separate arsenic from mercury. In basic solution, hydrogen peroxide oxidizes As$_2$S$_3$.

$$As_2S_3 + 12OH^- + 14H_2O_2 \longrightarrow 2[AsO_4]^{3-} + 3SO_4^{2-} + 20H_2O$$

The arsenic(III) ion is oxidized to arsenate, $[AsO_4]^{3-}$, and the sulfide ion, to sulfate. Consequently, we have $[As^{3+}]^2[S^{2-}]^3 < K_{sp}$ for As$_2$S$_3$ and the As$_2$S$_3$ dissolves. Mercury(II) sulfide is not dissolved by this treatment.

PROCEDURE 8   *Residue from Procedure 7: HgS and As$_2$S$_3$. Add 12 drops of 4 M aqueous ammonia and 6 drops of 3% hydrogen peroxide to the residue. Stir the mixture and heat it in a hot water bath for 5–6 min. Separate the residue (HgS) and reserve the solution (AsO$_4$$^{3-}$) for Procedure 10.*

37.9   Identification of Mercury

In order to dissolve the extremely insoluble mercury(II) sulfide, it is necessary to reduce the mercury(II) and sulfide ion concentrations to the extent that $[Hg^{2+}][S^{2-}]$ will be less than the solubility product of HgS. This may be done by forming the complex ion $[HgCl_4]^{2-}$ and oxidizing the sulfide ion to sulfur. A mixture of hydrochloric acid and sodium hypochlorite is used to dissolve HgS. The chloride ions from hydrochloric acid combine with mercury(II) ions and form $[HgCl_4]^{2-}$; hypochlorite ions in acid solution oxidize the sulfide ions to sulfur. The equation is

$$\underline{HgS} + 2H^+ + 3Cl^- + ClO^- \longrightarrow [HgCl_4]^{2-} + \underline{S} + H_2O$$

The solution is then boiled to decompose excess hypochlorite ions, which would otherwise interfere with the confirmatory test for mercury by oxidizing tin(II) to tin(IV), thus destroying its reducing power. The decomposition of the hypochlorite ion is according to the equation

$$2H^+ + Cl^- + ClO^- \xrightarrow{\Delta} H_2O + Cl_2(g)$$

The presence of mercury(II) is confirmed by reducing $[HgCl_4]^{2-}$ to Hg_2Cl_2 (white) or Hg (black) by means of $[SnCl_3]^-$ ions.

$$2[HgCl_4]^{2-} + [SnCl_3]^- \longrightarrow \underline{Hg_2Cl_2} + [SnCl_6]^{2-} + 3Cl^-$$
$$\underline{Hg_2Cl_2} + [SnCl_3]^- + Cl^- \longrightarrow 2\underline{Hg} + [SnCl_6]^{2-}$$

The trichlorostannate(II) ion, $[SnCl_3]^-$, is formed when tin(II) chloride is added to the hydrochloric acid solution.

$$SnCl_2 + Cl^- \longrightarrow [SnCl_3]^-$$

PROCEDURE 9   *Residue from Procedure 8: HgS*. To the black residue add 6 drops of 5% NaClO and 2 drops of 6 M HCl. Stir the mixture, add 1 mL of water, and separate the sulfur from the solution. Heat the solution to boiling. Add 2 drops of 1 M SnCl$_2$ to the solution. The formation of a white, gray, or black precipitate confirms the presence of mercury.

37.10 Identification of Arsenic

The presence of arsenic (in the form of arsenate ions) is confirmed by the formation of white crystalline magnesium ammonium arsenate when magnesia mixture ($MgCl_2$, NH_4Cl, and aqueous ammonia) is added to a solution containing arsenate ions.

$$Mg^{2+} + NH_4^+ + AsO_4^{3-} \longrightarrow \underline{MgNH_4AsO_4}$$

PROCEDURE 10   *Solution from Procedure 8: AsO_4^{3-}*. Add 2 drops of 15 M aqueous ammonia and 5 drops of magnesia mixture to the solution. The formation of a white precipitate ($MgNH_4AsO_4$), frequently slow in forming, indicates the presence of arsenate ions.

37.11 Identification of Antimony and Tin

Antimony and tin in hydrochloric acid solutions are in the form of the complex ions $[SbCl_4]^-$ and $[SnCl_6]^{2-}$. Aluminum metal will reduce antimony to the metallic state and tin(IV) to tin(II) in hydrochloric acid solution, thus effecting a separation of the two metals.

$$[SbCl_4]^- + Al \longrightarrow \underline{Sb} + Al^{3+} + 4Cl^-$$
$$3[SnCl_6]^{2-} + 2Al \longrightarrow 3[SnCl_3]^- + 2Al^{3+} + 9Cl^-$$

To be exact, aluminum actually reduces tin to the metallic state. When all of the aluminum is gone, the metallic tin dissolves in the hydrochloric acid, forming $[SnCl_3]^-$ ions and liberating hydrogen.

The fact that the tin(II) ion is a strong reducing agent is used in the confirmatory test for the ion. $[SnCl_3]^-$ will reduce mercury(II) to mercury(I), or metallic mercury, depending on the amount of the tin(II) present in solution.

$$[SnCl_3]^- + 2HgCl_4{}^{2-} \longrightarrow [SnCl_6]^{2-} + \underline{Hg_2Cl_2} + 3Cl^-$$
$$[SnCl_3]^- + Hg_2Cl_2 + Cl^- \longrightarrow [SnCl_6]^{2-} + \underline{2Hg}$$

Antimony metal reacts with nitric acid to form the insoluble oxide Sb_4O_6. This oxide is soluble in oxalic acid, $H_2C_2O_4$, forming oxyantimony(III) hydrogen oxalate, $(SbO)HC_2O_4$.

$$4Sb + 4H^+ + 4NO_3{}^- \longrightarrow \underline{Sb_4O_6} + 4NO + 2H_2O$$
$$Sb_4O_6 + 4H_2C_2O_4 \longrightarrow 4(SbO)HC_2O_4 + 2H_2O$$

The presence of antimony is then confirmed by precipitating it as the orange-red sulfide, Sb_2S_3.

$$2(SbO)HC_2O_4 + 3H_2S \longrightarrow \underline{Sb_2S_3} + 2H_2C_2O_4 + 2H_2O$$

PROCEDURE 11   *Solution from Procedure 7: $[SnCl_6]^{2-}$ and $[SbCl_4]^-$. Boil the solution until all of the H_2S has been expelled. Add a volume of water equal to that of the solution and add 2 drops of 6 M HCl. Place a piece of aluminum wire about $\frac{1}{8}$ inch long in the solution, and heat the mixture until the aluminum has completely dissolved. Add 1 drop of 6 M HCl to the mixture and heat it for a few minutes. If antimony is present, black flakes of the metal will appear. Separate the mixture. Treat the black flakes with 3 drops of 4 M HNO_3 and several drops of 1 M oxalic acid. Add 2 drops of thioacetamide to the solution and place the test tube in a hot water bath. The formation of an orange-red precipitate of Sb_2S_3 confirms the presence of antimony. To the solution obtained from the separation of metallic antimony, add a few drops of 0.2 M HgCl_2. The formation of a white, gray, or black precipitate confirms the presence of tin.*

FOR REVIEW

EXERCISES

1. Give the color of each of the following: CuS, HgS, $[Cu(H_2O)_4]^{2+}$, PbSO_4, $[Cu(NH_3)_4]^{2+}$, $[Cu(CN)_2]^-$, MgNH_4AsO_4, PbS, Bi(OH)_3, Sb_2S_3, As_2S_3, SnS_2, $[Cd(NH_3)_4]^{2+}$, Bi (finely divided), Sb, and Bi_2S_3.

2. Why are the ions of lead and mercury found in both Groups I and II?

3. Explain in terms of ionic equilibria and solubility product theory why Group II can be separated from Group III by hydrogen sulfide in the pres-

ence of hydrogen ions at a concentration of 0.3 M.

4. In terms of ionic equilibria theory, discuss the effect of added hydrochloric acid on the concentration of sulfide ion in a solution of hydrogen sulfide. How would adding ammonia or hydroxide ions influence the concentration of sulfide ions?

5. Explain in terms of ionic equilibria and solubility product theory the dissolution of the sulfides of

copper, bismuth, cadmium, and lead in nitric acid.

6. Write the equation for the hydrogen-ion catalyzed hydrolysis of thioacetamide. Hydrogen sulfide, as one of the products of the hydrolysis of thioacetamide, is a diprotic acid; illustrate this property of hydrogen sulfide by suitable equations.

7. Explain the dissolution of lead sulfate in ammonium acetate and the reprecipitation of the lead as lead chromate in terms of ionic equilibria and solubility product theory.

8. Why is it necessary to remove the nitric acid present before attempting to precipitate lead as lead sulfate?

9. If a yellow precipitate is obtained when an unknown solution for Group II is treated with hydrogen sulfide, what ions are probably absent?

10. The Division A sulfides are washed with water containing ammonium nitrate. What is the function of the ammonium nitrate?

11. Why must sodium stannite, which is used in the identification of bismuth, be prepared just prior to its use?

12. How will the separation of Groups II and III be affected if the concentration of hydronium ion in the solution saturated with hydrogen sulfide is 0.1 M? 1 M?

13. If a Group II unknown contains copper in an appreciable concentration, this fact should be evident from an inspection of the unknown solution. Why?

14. Outline the separation of the following groups of ions, leaving out all unnecessary steps:
    (a) Bi^{3+}, As^{3+}, Sb^{3+} (c) Pb^{2+}, Cu^{2+}, Cd^{2+}
    (b) Ag^+, Hg^{2+}, Co^{2+}

15. Select a reagent used in the analysis of Group I or Group II that will separate each of the following pairs in one step:
    (a) As_2S_3, SnS_2 (e) Ag^+, Bi^{3+}
    (b) $[SbCl_4]^-$, $[SnCl_6]^{2-}$ (f) Ag^+, Fe^{3+}
    (c) CdS, HgS   (g) Bi^{3+}, Cd^{2+}
    (d) CuS, CdS

16. How many drops of 1 M aqueous ammonia will be required to neutralize the HCl in 2 mL of a Group II solution that is 0.5 M in HCl? (1 drop is 0.05 mL.) How many drops of 6 M HCl would be required to make the resulting solution 0.3 M in HCl?   *Ans. 20 drops; 3 drops*

17. Calculate the concentration of S^{2-} in an acid solution that is saturated with H_2S and has a pH of 0.52. (Saturated H_2S is 0.1 M.)
              Ans. 1.4 × 10⁻²⁰ M

18. How many grams of thioacetamide are required to precipitate quantitatively the bismuth and copper from 10 mL of a solution that is 0.010 M in each ion?   *Ans. 0.019 g*

19. Calculate the solubility of ZnS in a solution saturated with H_2S and 0.30 M in HCl.
              Ans. 0.076 M

20. Calculate the concentration of H^+, H_2S, HS^-, and S^{2-} in a 0.052 M solution of hydrogen sulfide.
    Ans. $[H^+] = [HS^-] = 7.2 × 10^{-5}$ M;
    $[S^{2-}] = 1.3 × 10^{-13}$ M;
    $[H_2S] = 0.052$ M

21. Calculate the concentration of H^+ required to prevent the precipitation of CdS from 0.010 M Cd^{2+} that is saturated with H_2S. (K_{sp} for CdS is 3.6 × 10⁻²⁹.) Is this an attainable concentration?
              Ans. 6.0 × 10² M

38

THE ANALYSIS OF
GROUP III

The solution to be analyzed may be a Group III known or unknown, or it may be the solution from the Group II separation. Table 38-1 is the Group III flow sheet. You will probably also find it helpful to refer to Chapter 32, which discusses the chemistry of the metals of Analytical Group III.

38.1 Precipitation of the Group III Ions

Analytical Group III contains the metallic ions Ni^{2+}, Co^{2+}, Mn^{2+}, Fe^{3+}, Al^{3+}, Cr^{3+}, and Zn^{2+}. These ions are not precipitated by hydrochloric acid (the Group I precipitant) or by sulfide ions in solutions that are $0.3\ M$ in hydrogen ion (the Group II precipitant). However, an ammonium sulfide solution precipitates Ni^{2+}, Co^{2+}, Mn^{2+}, Fe^{2+}, and Zn^{2+} as sulfides and Al^{3+} and Cr^{3+} as hydroxides.

The concentration of sulfide ions in the $0.3\ M$ hydrogen ion solution of the Group II precipitation is so small that the solubility products of the Group III sulfides is not exceeded. Hydrogen sulfide in an aqueous solution of ammonia has a much higher sulfide ion concentration, due to the formation of ammonium sulfide, according to the equation

$$2NH_3 + H_2S \rightleftharpoons 2NH_4^+ + S^{2-}$$

The resultant sulfide ion concentration is sufficiently large so that the solubility products of the sulfides of cobalt, nickel, manganese, iron, and zinc are ex-

Table 38-1 Group III Flow Sheet

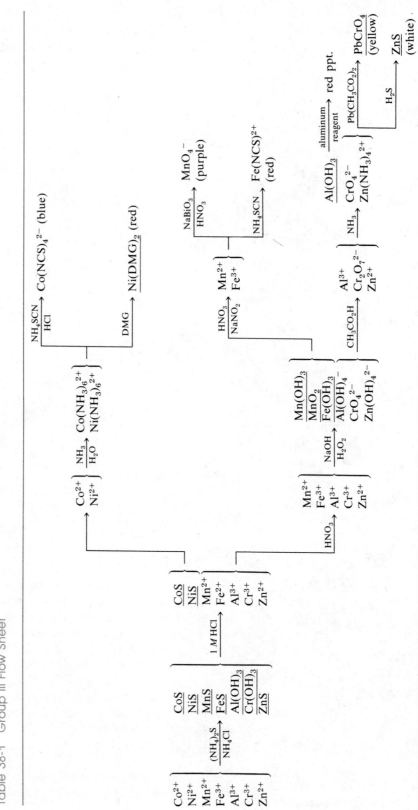

ceeded and precipitation occurs. Similarly, the hydroxide ion concentration of the ammonium sulfide solution is great enough to precipitate the hydroxides of aluminum and chromium. The solubility products of aluminum hydroxide and chromium hydroxide are lower than those of the corresponding sulfides; thus the hydroxides form, not the sulfides.

Hydrogen sulfide is produced for the precipitation of this group by the acidic hydrolysis of thioacetamide. When aqueous ammonia is added, the following precipitations occur:

$$Co^{2+} + S^{2-} \rightleftharpoons \underline{CoS} \text{ (brown or black)}$$
$$Ni^{2+} + S^{2-} \rightleftharpoons \underline{NiS} \text{ (brown or black)}$$
$$Mn^{2+} + S^{2-} \rightleftharpoons \underline{MnS} \text{ (light pink)}$$
$$Fe^{2+} + S^{2-} \rightleftharpoons \underline{FeS} \text{ (black)}$$
$$Zn^{2+} + S^{2-} \rightleftharpoons \underline{ZnS} \text{ (white)}$$
$$Al^{3+} + 3OH^- \rightleftharpoons \underline{Al(OH)_3} \text{ (white)}$$
$$Cr^{3+} + 3OH^- \rightleftharpoons \underline{Cr(OH)_3} \text{ (blue-green)}$$

In the event that iron is present as Fe^{3+} in the original solution, it will be reduced by hydrogen sulfide in the acidic solution according to the equation

$$2Fe^{3+} + H_2S \longrightarrow 2Fe^{2+} + S + 2H^+$$

Of the cations remaining in solution, only the magnesium ion, of Group V, forms an insoluble hydroxide. The hydroxide ion concentration is kept just low enough in the aqueous ammonia to prevent the precipitation of magnesium hydroxide by having ammonium chloride present. The ammonium ions from ammonium chloride exert the common ion effect with the ammonium ions from the ionization of ammonia, thus repressing the ionization of ammonia and reducing the hydroxide ion concentration of the solution. This point is considered quantitatively in Example 17.8 of Section 17.9.

Some evidence regarding the cations present in a solution containing the Group III ions can be obtained by noting the colors of the ions: Co^{2+}, pale red; Ni^{2+}, pale green; Zn^{2+}, colorless; Mn^{2+}, colorless in low concentrations; Fe^{2+}, pale green; Fe^{3+}, reddish brown; Al^{3+}, colorless; Cr^{3+}, dark green or blue. Due to the phenomenon of complementary colors, a solution containing certain combinations of colored ions may appear colorless.

PROCEDURE 1   *Precipitation of Group III: Co^{2+}, Ni^{2+}, Mn^{2+}, Fe^{3+}, Al^{3+}, Cr^{3+}, and Zn^{2+}.* (a) If the solution to be analyzed is a known or unknown for Group III only, take 10 drops of the solution, add 1 drop of 6 M HCl, dilute to 1 mL, and add 5 drops of 5% thioacetamide solution. Heat the solution in a hot water bath for at least 5 min. (b) If the solution to be analyzed is that from the Group II separation, add 5 drops of 5% thioacetamide solution and heat the mixture in a hot water bath for at least 5 min.

To the solution resulting from either procedure (a) or (b) above, add 5 drops of 15 M aqueous ammonia and stir up the precipitate. Heat the mixture for 5 min in the hot water bath. Separate the precipitate and wash it with a few drops of water. Reserve the solution for the analysis of Groups IV and V.

38.2   Separation of Cobalt and Nickel

Although the sulfides of cobalt and nickel will completely precipitate only from basic solutions, once formed, these sulfides are only slightly soluble in dilute HCl. It appears that these sulfides precipitate in a form (the α form) that is soluble in acid but change rapidly into other crystalline modifications (the β forms) that have much lower solubility products. The K_{sp} of α CoS is 5.9×10^{-21}, of β CoS is 8.7×10^{-23}, of α NiS is 3×10^{-21}, and of β NiS is 1×10^{-26}. These facts are used in separating CoS and NiS from MnS, FeS, Al(OH)$_3$, Cr(OH)$_3$, and ZnS, the latter five precipitates being soluble in 1 M HCl.

$$\underline{MnS} + 2H^+ \rightleftharpoons Mn^{2+} + H_2S(g) \qquad \underline{Al(OH)_3} + 3H^+ \rightleftharpoons Al^{3+} + 3H_2O$$
$$\underline{FeS} + 2H^+ \rightleftharpoons Fe^{2+} + H_2S(g) \qquad \underline{Cr(OH)_3} + 3H^+ \rightleftharpoons Cr^{3+} + 3H_2O$$
$$\underline{ZnS} + 2H^+ \rightleftharpoons Zn^{2+} + H_2S(g)$$

PROCEDURE 2   *Precipitate from Procedure 1: CoS, NiS, FeS, MnS, Al(OH)$_3$, Cr(OH)$_3$, and ZnS. Add 10 drops of 1 M HCl to the precipitate and stir the mixture. Separate the mixture immediately since prolonged contact with the acid causes some dissolution of CoS and NiS. Wash the sulfides that remain (CoS and NiS) with 4 drops of 1 M HCl. Reserve the combined centrifugates for Procedure 4.*

38.3   Identification of Cobalt and Nickel

The sulfides of cobalt and nickel readily dissolve in a mixture of nitric and hydrochloric acids due to the oxidation of the sulfide ion to sulfur.

$$3CoS + 8H^+ + 2NO_3^- \rightleftharpoons 3Co^{2+} + 2NO(g) + 3\underline{S} + 4H_2O$$
$$3NiS + 8H^+ + 2NO_3^- \rightleftharpoons 3Ni^{2+} + 2NO(g) + 3\underline{S} + 4H_2O$$

After boiling the solution containing the cobalt and nickel ions to remove oxides of nitrogen, which would destroy the reagents used in the confirmation of these ions, an excess of aqueous ammonia is added. Ammonia in excess reacts with cobalt ions and nickel ions to form complex ions.

$$Co^{2+} + 6NH_3 \rightleftharpoons [Co(NH_3)_6]^{2+} \text{ (pink)}$$
$$Ni^{2+} + 6NH_3 \rightleftharpoons [Ni(NH_3)_6]^{2+} \text{ (blue)}$$

In the ammoniacal solution, Ni^{2+} will react with dimethylglyoxime to form a very insoluble, bright red coordination compound. Co^{2+} forms a brown-colored soluble complex with dimethylglyoxime, which does not interfere with the test.

A concentrated solution of ammonium thiocyanate is used in the identification of the cobalt ion, with which the thiocyanate ion forms a complex ion, $[Co(NCS)_4]^{2-}$, that has a characteristic blue color.

$$[Co(NH_3)_6]^{2+} + 4NCS^- \rightleftharpoons [Co(NCS)_4]^{2-} + 6NH_3$$

When iron(III) ions are present as a contaminant, they interfere and form a bright red complex ion of the formula $[Fe(NCS)]^{2+}$. This interference may be avoided by converting the iron to the colorless and very stable hexafluoroferrate(III) ion, $[FeF_6]^{3-}$. The equation is

$$Fe^{3+} + 6F^- \rightleftharpoons [FeF_6]^{3-}$$

A second confirmatory test for cobalt involves its oxidation by the nitrite ion from Co^{2+} to Co^{3+} and the precipitation of the complex yellow salt $K_3[Co(NO_2)_6]$, tripotassium hexanitrocobaltate(III).

$$Co^{2+} + 6NO_2^- \longrightarrow [Co(NO_2)_6]^{4-}$$
$$[Co(NO_2)_6]^{4-} + NO_2^- + 2H^+ \longrightarrow [Co(NO_2)_6]^{3-} + NO(g) + H_2O$$
$$3K^+ + [Co(NO_2)_6]^{3-} \longrightarrow K_3[Co(NO_2)_6]$$

PROCEDURE 3   *Residue from Procedure 2: CoS and NiS.* Add 3 drops of 12 M HCl and 1 drop of 14 M HNO_3 to the residue and heat the mixture in a hot water bath. Separate any sulfur that forms and boil the solution to remove any excess nitric acid or oxides of nitrogen. Add sufficient 4 M aqueous ammonia to the solution to make it *slightly* basic to litmus. (Excess alkalinity favors the formation of the brown cobalt complex when testing for nickel with dimethylglyoxime.) Dilute the solution to 1 mL and divide it into three parts.

(a) *Test for Nickel.* Add 1 drop of dimethylglyoxime to one portion of the solution. The formation of a pink or red precipitate confirms the presence of nickel.

(b) *Test for Cobalt.* Acidify a second portion of the solution with 1 M HCl and add several crystals of NH_4NCS to it. Now add an equal volume of acetone and agitate the mixture. The development of a blue color proves the presence of cobalt. If the solution turns red when the NH_4NCS is added, iron(III) ions are present. Add 1 drop of 1 M NaF to the solution. Now, if the solution is bluish-green to green, the presence of cobalt is confirmed.

(c) *Test for Cobalt.* Acidify the third portion of the solution with 4 M CH_3CO_2H, and add several large crystals of KNO_2. Warm the mixture. The formation of a yellow precipitate confirms the presence of cobalt.

38.4 Separation of Iron and Manganese from Aluminum, Chromium, and Zinc

The solution containing Mn^{2+}, Fe^{2+}, Al^{3+}, Cr^{3+}, and Zn^{2+} is treated with nitric acid to remove the sulfide ion by oxidizing it to sulfur and to oxidize Fe^{2+} to Fe^{3+}. The equation for the oxidation of the iron is

$$3Fe^{2+} + 4H^+ + NO_3^- \longrightarrow 3Fe^{3+} + NO(g) + 2H_2O$$

It is desirable to have the iron as Fe^{3+} because $Fe(OH)_3$ is less soluble and less gelatinous in character than is $Fe(OH)_2$.

In order to separate iron and manganese from aluminum, chromium, and zinc, the solution is first treated with sodium hydroxide and then with hydrogen peroxide. The precipitations that first occur, as sodium hydroxide is added to the solution, are described by the following:

$$Mn^{2+} + 2OH^- \rightleftharpoons Mn(OH)_2 \text{ (white)}$$
$$Fe^{3+} + 3OH^- \rightleftharpoons Fe(OH)_3 \text{ (reddish-brown)}$$
$$Al^{3+} + 3OH^- \rightleftharpoons Al(OH)_3 \text{ (white)}$$
$$Cr^{3+} + 3OH^- \rightleftharpoons Cr(OH)_3 \text{ (dark green or blue)}$$
$$Zn^{2+} + 2OH^- \rightleftharpoons Zn(OH)_2 \text{ (white)}$$

The addition of an excess of NaOH dissolves the amphoteric hydroxides of aluminum, chromium, and zinc, but not the hydroxides of iron and manganese.

$$Al(OH)_3 + OH^- \rightleftharpoons [Al(OH)_4]^- \text{ (aluminate) (colorless)}$$
$$Cr(OH)_3 + OH^- \rightleftharpoons [Cr(OH)_4]^- \text{ (chromite) (green)}$$
$$Zn(OH)_2 + 2OH^- \rightleftharpoons [Zn(OH)_4]^{2-} \text{ (zincate) (colorless)}$$

Hydrogen peroxide is added to the resulting mixture to oxidize manganese(II) hydroxide to a mixture of manganese dioxide and manganese(III) hydroxide, both of which are much less soluble than $Mn(OH)_2$.

$$Mn(OH)_2 + H_2O_2 \longrightarrow MnO_2 + 2H_2O$$
$$2Mn(OH)_2 + H_2O_2 \longrightarrow 2Mn(OH)_3$$

Hydrogen peroxide oxidizes the chromite ion, $[Cr(OH)_4]^-$, to the chromate ion, CrO_4^{2-}. This is desirable because the reactions of the chromate ion permit better identification of chromium than do those of either Cr^{3+} or $[Cr(OH)_4]^-$.

$$2[Cr(OH)_4]^- + 3H_2O_2 + 2OH^- \longrightarrow 2CrO_4^{2-} + 8H_2O$$

The mixture containing MnO_2 and $Mn(OH)_3$ is brown, and solutions of CrO_4^{2-} are yellow.

PROCEDURE 4   *Solution from Procedure 2: Mn^{2+}, Fe^{2+}, Al^{3+}, Cr^{3+}, and Zn^{2+}.* Transfer the solution to a casserole, add 1 mL of 4 M HNO_3, and evaporate the solution to a moist residue. Take up the residue in 1 mL of water and transfer the solution to a test tube. Add 10 drops of 4 M NaOH beyond the amount of this reagent that is required to initiate precipitation. Now add 6 drops of 3% hydrogen peroxide to the mixture and heat it in a hot water bath for 5 min. Separate the mixture and wash the residue with 10 drops of water to which has been added 1 drop of 4 M NaOH. Save the combined centrifugates for Procedure 6.

38.5 · Separation and Identification of Manganese and Iron

The residue containing MnO_2, $Mn(OH)_3$, and $Fe(OH)_3$ is dissolved by a mixture of HNO_3 and $NaNO_2$. The nitrite ion in acidic solution reduces manganese to the Mn(II) ion. The iron(III) hydroxide is dissolved by nitric acid. The equations for these reactions are

$$2Mn(OH)_3 + 4H^+ + NO_2^- \longrightarrow 2Mn^{2+} + NO_3^- + 5H_2O$$

$$MnO_2 + 2H^+ + NO_2^- \longrightarrow Mn^{2+} + NO_3^- + H_2O$$

$$Fe(OH)_3 + 3H^+ \longrightarrow Fe^{3+} + 3H_2O$$

The presence of Fe^{3+} is confirmed by its reaction with the thiocyanate ion, SCN^-, to produce the blood-red complex ion $[Fe(NCS)]^{2+}$ or one of the possible complexes containing from one to six SCN^- groups.

$$Fe^{3+} + SCN^- \rightleftharpoons [Fe(NCS)]^{2+}$$

This test for iron may be conducted on a solution containing Mn^{2+} without interference by this ion. It should be noted that there are traces of Fe^{3+} existing in many reagents. Therefore a light pink color in this test may be due to iron as an impurity. Blank tests can be run on the reagents used in the test to establish this point.

Manganese may be identified in the presence of iron by oxidizing it to the permanganate ion by means of sodium bismuthate in nitric acid. The MnO_4^- ion is purple, but in dilute solutions it may appear pink.

$$2Mn^{2+} + 5BiO_3^- + 14H^+ \longrightarrow 2MnO_4^- + 5Bi^{3+} + 7H_2O$$

PROCEDURE 5   *Residue from Procedure 4: $Mn(OH)_3$, MnO_2, and $Fe(OH)_3$.* Treat the residue with 1 mL of 4 M HNO_3 and 2 drops of 1 M $NaNO_2$. Stir the mixture and heat it in a hot water bath. Separate any residue that remains. Heat the solution to boiling, then cool it, and divide it into two parts.

(a) *Test for Iron.* Dilute one portion of the solution to 1 mL and add 2 or 3 crystals of NH_4NCS. If iron is present, a dark red-red color will develop in the solution. (Faint or pale redness indicates   of iron due to impurities; a red color that does not p   indicates   nce of reducing agents but does not invalidate the tes

(b) *Test for M   anese.* To   ond portion of the solution, add a small quantity of soli   aBiO_3 and   w drops of 4 M HNO_3. The formation of a pink or purple   or that pers s confirms the presence of manganese. (Pink or purple that es not persist indicates the presence of reducing agents but does not invalidate the test.)   *real touchy!*

38.6 Separation and Identification of Aluminum

When acetic acid is added to the solution containing $[Al(OH)_4]^-$, CrO_4^{2-}, and $[Zn(OH)_4]^{2-}$, aluminum and zinc are converted to their cations.

$$[Al(OH)_4]^- + 4H^+ \rightleftharpoons Al^{3+} + 4H_2O$$

$$[Zn(OH)_4]^{2-} + 4H^+ \rightleftharpoons Zn^{2+} + 4H_2O$$

Acids react with yellow chromate, CrO_4^{2-}, and convert some of it to dichromate ion, $Cr_2O_7^{2-}$. The extent of the conversion depends on the concentration of the hydrogen ion.

$$2CrO_4^{2-} + 2H^+ \rightleftharpoons Cr_2O_7^{2-} + H_2O$$

The addition of excess aqueous ammonia precipitates aluminum as the hydroxide, converts $Cr_2O_7^{2-}$ to CrO_4^{2-}, and complexes the zinc as the soluble $[Zn(NH_3)_4]^{2+}$ ion.

$$Al^{3+} + 3NH_3 + 3H_2O \rightleftharpoons \underline{Al(OH)_3} + 3NH_4^+$$
$$Cr_2O_7^{2-} + 2OH^- \rightleftharpoons 2CrO_4^{2-} + H_2O$$
$$Zn^{2+} + 4NH_3 \rightleftharpoons [Zn(NH_3)_4]^{2+}$$

The presence of aluminum is confirmed by dissolving aluminum hydroxide in acetic acid and adding the aluminum reagent and $(NH_4)_2CO_3$. Aurin-tricarboxylic acid (aluminon) imparts a red color to the $Al(OH)_3$ formed.

PROCEDURE 6   *Solution from Procedure 4:* $[Al(OH)_4]^-$, CrO_4^{2-}, *and* $[Zn(OH)_4]^{2-}$. Add 4 M CH_3CO_2H to the solution until it is acid to litmus and then add 2 or 3 drops of the acid in excess. Now add 4 M aqueous ammonia until the solution is distinctly alkaline to litmus. If a white gelatinous precipitate forms, it is probably aluminum hydroxide. Separate the precipitate and reserve the solution for Procedure 7. Confirm the presence of aluminum by dissolving the precipitate in 4 M CH_3CO_2H and adding 2 drops of the aluminum reagent and enough 1 M $(NH_4)_2CO_3$ to make the solution basic. The formation of a reddish-colored precipitate confirms the presence of aluminum.

38.7 Identification of Chromium and Zinc

The presence of the chromate ion is confirmed by precipitating yellow lead chromate. Zinc ions do not interfere with the chromate test.

The zinc ions are precipitated by sulfide ions from a thioacetamide solution, white zinc sulfide being formed. The chromate ion may oxidize some of the sulfide ions to elemental sulfur, but ZnS is soluble in hydrochloric acid and sulfur is not. A second confirmatory test for zinc may be made by boiling all of the hydrogen sulfide out of the solution and precipitating the zinc as the hexacyanoferrate(II) ion, which is white.

$$2K^+ + Zn^{2+} + Fe(CN)_6^{4-} \longrightarrow \underline{K_2Zn[Fe(CN)_6]}$$

PROCEDURE 7   *Solution from Procedure 6:* CrO_4^{2-} *and* $[Zn(NH_3)_4]^{2+}$. Divide the solution into two parts.
(a) *Test for Chromium.* To one portion of the solution, add 1 M CH_3CO_2H until the solution is acid to litmus. Then add 2 drops of 0.1 M $Pb(CH_3CO_2)_2$. The formation of a yellow precipitate, $PbCrO_4$, confirms the presence of chromium.
(b) *Test for Zinc.* To the second portion of solution, add 5 drops of 5% thioacetamide solution and heat the mixture in a hot water bath. The formation of a white precipitate, ZnS, which is soluble in 4 M HCl, indicates the presence of zinc. Heat the solution to expel hydrogen sulfide and then neutralize it with 4 M aqueous ammonia. Now add 10 drops of 1 M HCl and 5 drops of 0.1 M $K_4[Fe(CN)_6]$. The formation of a white precipitate, $K_2Zn[Fe(CN)_6]$, proves the presence of zinc.

FOR REVIEW

EXERCISES

1. A Group III unknown is colorless. What cations are probably absent? Why should one not rely definitely on such an observation?
2. Why do aluminum and chromium precipitate as hydroxides rather than as sulfides in Group III?
3. What is the function of the ammonium chloride used in the Group III precipitant?
4. Why do the sulfides of Co^{2+}, Ni^{2+}, Mn^{2+}, Fe^{2+}, and Zn^{2+} precipitate in an ammonium sulfide solution but not in a 0.3 M HCl solution of hydrogen sulfide?
5. Why does iron precipitate as iron(II) sulfide rather than iron(III) sulfide in an ammonium sulfide solution?
6. Cite two examples of the use of complex ions in the analysis of Group III.
7. Account for the fact that CoS and NiS fail to precipitate in the 0.3 M HCl solution of Group II, yet they dissolve only very slowly in 1 M HCl.
8. Give the color of each of the following: $Fe(OH)_3$, Fe^{3+}, $Fe(OH)_2$, Fe^{2+}, $[Fe(NCS)]^{2+}$, $[FeF_6]^{3-}$.
9. Write equations showing the amphoteric nature of the hydroxides of zinc, chromium, and aluminum.
10. Outline the separation of the following groups of ions leaving out all unnecessary steps.
    (a) Ni^{2+}, Mn^{2+}, Zn^{2+}
    (b) Hg_2^{2+}, Hg^{2+}, Cu^{2+}, Fe^{2+}
    (c) Cd^{2+}, Co^{2+}, Ca^{2+}
11. Select a reagent used in Group III that will separate each of the following pairs.

    (a) Al^{3+}, Zn^{2+} (d) Mn^{2+}, Mg^{2+}
    (b) $[Zn(NH_3)_4]^{2+}$, CrO_4^{2-} (e) Fe^{3+}, Al^{3+}
    (c) CoS, ZnS
12. Why will $PbCrO_4$ precipitate when Pb^{2+} is added to a solution made up from $K_2Cr_2O_7$?
13. When and why is fluoride added in the test for cobalt using thiocyanate?
14. Show by the proper formulas that manganese can act as either a metal or a nonmetal, depending on its oxidation number.
15. What concentration of NH_4^+ must be present in 0.20 M aqueous ammonia to prevent the precipitation of $Mg(OH)_2$ if the solution contains 1.0×10^{-3} mol of Mg^{2+} per liter? (K_{sp} for $Mg(OH)_2$ is 1.5×10^{-11}.)   *Ans. 0.029 M*
16. A solution from the Group II separation is 0.3 M in HCl and has a volume of 5.0 mL. Calculate the number of drops of 15 M aqueous ammonia required to react with the HCl (assume that 1 drop is equal to 0.05 mL).   *Ans. 2 drops*
17. Calculate the concentration of Pb^{2+} in a saturated solution of $PbCrO_4$. (K_{sp} for $PbCrO_4$ is 1.8×10^{-14}.)   *Ans. 1.3×10^{-7} M*
18. Calculate the concentration of Pb^{2+} in a saturated solution of $PbCrO_4$ in the presence of 1.0×10^{-3} M Na_2CrO_4.   *Ans. 1.8×10^{-11} M*
19. Using 1×10^{-19} for the solubility product of FeS, show that 0.3 M HCl should dissolve or prevent the precipitation of FeS.
    Ans. These conditions permit an iron(II) ion concentration of 7 M

39

THE ANALYSIS OF GROUP IV

The solution to be analyzed may be a Group IV known or unknown, or it may be the solution from the Group III separation. The Group IV flow sheet appears in Table 39-1. Note that the chemistry of the metals of Analytical Group IV is also discussed in Chapter 33, to which you may find it useful to refer.

Table 39-1 Group IV Flow Sheet

$$
\begin{matrix}
\left.\begin{matrix} Ba^{2+} \\ Sr^{2+} \\ Ca^{2+} \end{matrix}\right\} & \xrightarrow[\substack{NH_4Cl \\ NH_3 + H_2O}]{(NH_4)_2CO_3} & \begin{matrix} \underline{BaCO_3} \\ \underline{SrCO_3} \\ \underline{CaCO_3} \end{matrix}
\end{matrix}
$$

$$
\xrightarrow{CH_3CO_2H} \left.\begin{matrix} Ba^{2+} \\ Sr^{2+} \\ Ca^{2+} \end{matrix}\right\} \xrightarrow[\substack{CH_3CO_2H \\ NH_4CH_3CO_2}]{K_2CrO_4} \begin{matrix} \underline{BaCrO_4} \\ Sr^{2+} \\ Ca^{2+} \end{matrix}\left.\right\}
$$

$$
\underline{BaCrO_4} \xrightarrow{HCl} Ba^{2+} \xrightarrow{H_2SO_4} \underline{BaSO_4}\ (white)
$$

$$
\left.\begin{matrix} Sr^{2+} \\ Ca^{2+} \end{matrix}\right\} \xrightarrow[\substack{K_2CrO_4 \\ alcohol}]{NH_3 + H_2O} \underline{SrCrO_4}\ (yellow)
$$

$$
Ca^{2+} \xrightarrow{(NH_4)_2C_2O_4} \underline{CaC_2O_4}\ (white)
$$

39.1 Precipitation of the Group IV Ions

Analytical Group IV contains the metallic ions Ba^{2+}, Sr^{2+}, and Ca^{2+}. These metals form chlorides, sulfides, and hydroxides that are soluble under the conditions that prevail in the precipitations of Groups I, II, and III. The carbonates of barium, strontium, and calcium precipitate in aqueous ammonia solutions containing ammonium carbonate.

The concentration of the carbonate ion in a solution of ammonium carbonate is too low to effect a complete precipitation of barium, strontium, and calcium due to the partial hydrolysis of the carbonate ion.

$$CO_3{}^{2-} + H_2O \rightleftharpoons HCO_3{}^- + OH^-$$

This hydrolysis is repressed by increasing the hydroxide ion concentration of the solution, the additional hydroxide ions being supplied by the buffer pair—aqueous ammonia and ammonium chloride. These conditions permit a carbonate concentration high enough to precipitate $BaCO_3$, $SrCO_3$, and $CaCO_3$, but not $MgCO_3$. At the same time the hydroxide ion concentration in this buffered solution is not high enough to precipitate $Mg(OH)_2$.

To bring the solution to the required ammonium ion concentration, the centrifugate from the Group III separation, which contains ammonium salts, is first evaporated to dryness and then heated strongly to expel the ammonium salts present. The equations are

$$NH_4Cl \xrightarrow{\Delta} NH_3(g) + HCl(g)$$

$$NH_4NO_3 \xrightarrow{\Delta} N_2O(g) + 2H_2O(g)$$

The necessary ammonium ion concentration is then obtained by adding the Group IV precipitant, which consists of ammonium carbonate, ammonium chloride, and aqueous ammonia in the required concentrations. The equations for the precipitations of the Group IV carbonates are

$$Ba^{2+} + CO_3{}^{2-} \rightleftharpoons \underline{BaCO_3} \text{ (white)}$$
$$Sr^{2+} + CO_3{}^{2-} \rightleftharpoons \underline{SrCO_3} \text{ (white)}$$
$$Ca^{2+} + CO_3{}^{2-} \rightleftharpoons \underline{CaCO_3} \text{ (white)}$$

PROCEDURE 1   *Precipitation of Group IV: Ba^{2+}, Sr^{2+}, and Ca^{2+}.* Evaporate the solution (10 drops of a Group IV known or unknown or the solution from the Group III separation) to dryness and ignite in a casserole to expel ammonium salts. Dissolve the residue in a mixture of 1 drop of 12 M HCl and 12 drops of water. Make the solution alkaline by adding 4 M aqueous ammonia. Add just 1 drop of aqueous ammonia in excess. Add 2 drops of 1 M $(NH_4)_2CO_3$, or more if necessary, to effect complete precipitation, and warm the mixture in a hot water bath. Allow the mixture to cool; separate the precipitate and reserve the solution for the Group V analysis.

39.2 Separation and Identification of Barium

1. DISSOLUTION OF THE CARBONATES. The carbonates of barium, strontium, and calcium are dissolved readily by strong acids such as hydrochloric acid or by weak acids such as acetic acid. The latter acid is used in these procedures, and the theory involved in these dissolutions is discussed in detail in Section 17.11. The equilibria involved, as applied to barium carbonate, are

$$\underline{BaCO_3} \rightleftharpoons Ba^{2+} + CO_3{}^{2-}$$
$$CH_3CO_2H \rightleftharpoons H^+ + CH_3CO_2{}^-$$

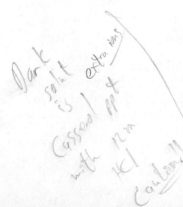

$$CO_3^{2-} + H^+ \rightleftharpoons HCO_3^-$$
$$HCO_3^- + H^+ \rightleftharpoons H_2CO_3$$
$$H_2CO_3 \rightleftharpoons H_2O + CO_2(g)$$

If the concentration of the hydrogen ion supplied by acetic acid is sufficiently high, the quantity of H_2CO_3 formed will exceed that in a saturated solution of CO_2 at atmospheric pressure and CO_2 will escape from solution. The loss of carbon dioxide from the system in this way causes all of the above equilibria to be shifted to the right until the precipitate dissolves completely. The net reaction is given by

$$BaCO_3 + 2CH_3CO_2H \rightleftharpoons Ba^{2+} + 2CH_3CO_2^- + H_2O + CO_2(g)$$

Similar equations may be written for the dissolution of the carbonates of strontium and calcium in acetic acid.

2. SEPARATION OF BARIUM FROM STRONTIUM AND CALCIUM.   By taking advantage of the fact that barium chromate is less soluble than strontium chromate and calcium chromate, barium can be separated from strontium and calcium. The solubility products are 2×10^{-10} for $BaCrO_4$ and 3.6×10^{-5} for $SrCrO_4$. Calcium chromate is relatively very soluble. The problem in this separation, then, is to control the concentration of the chromate ion so that the concentration of the barium ion will be reduced to at least 0.0001 M and strontium chromate will not be precipitated from a solution 0.1 M in the strontium ion. Let us first calculate the concentration of chromate ion required to reduce the barium ion concentration to 0.0001 M.

$$[Ba^{2+}][CrO_4^{2-}] = 2 \times 10^{-10}$$

$$[CrO_4^{2-}] = \frac{2 \times 10^{-10}}{[Ba^{2+}]} = \frac{2 \times 10^{-10}}{1 \times 10^{-4}} = 2 \times 10^{-6} \, M$$

Now let us determine the maximum concentration of chromate ion that can be present without precipitating any strontium chromate when the concentration of the strontium ion is 0.1 M.

$$[Sr^{2+}][CrO_4^{2-}] = 3.6 \times 10^{-5}$$

$$[CrO_4^{2-}] = \frac{3.6 \times 10^{-5}}{[Sr^{2+}]} = \frac{3.6 \times 10^{-5}}{10^{-1}} = 3.6 \times 10^{-4} \, M$$

Thus we find that the concentration of the chromate ion must be held between $2 \times 10^{-6} \, M$ and $3.6 \times 10^{-4} \, M$ to cause a nearly complete precipitation of $BaCrO_4$ without precipitation of any $SrCrO_4$.

In a solution containing chromate ions, we have the following equilibrium:

$$Cr_2O_7^{2-} + H_2O \rightleftharpoons 2CrO_4^{2-} + 2H^+$$

The concentration of chromate ions in a solution made by dissolving a dichromate salt in water can be controlled by adjusting the hydrogen ion concentration. The equation for the equilibrium constant is

$$\frac{[CrO_4^{2-}]^2[H^+]^2}{[Cr_2O_7^{2-}]} = 2.4 \times 10^{-15}$$

It is readily seen from this expression that the chromate ion concentration is a function of the hydrogen ion concentration. The more acidic the solution, the lower will be the chromate ion concentration; and the more basic the solution, the higher will be the chromate ion concentration. By substituting in the above expression, it can be shown that for a solution that is about 0.01 M in dichromate ions, the hydrogen ion concentration must be adjusted to about 1×10^{-5} M (pH = 5) to maintain the chromate ion concentration at about 5×10^{-4} M. This concentration of chromate ions is within the range (calculated above) necessary for the separation of Ba^{2+} from Sr^{2+}. A buffer composed of acetic acid and ammonium acetate is used to maintain the required hydrogen ion concentration.

As chromate ions are removed from the system through precipitation of barium chromate, the concentration of these ions does not change appreciably. The reason for this is that the excess of dichromate ions serves as a reservoir for chromate ions; and as CrO_4^{2-} is used in precipitating $BaCrO_4$, the equilibrium

$$Cr_2O_7^{2-} + H_2O \rightleftharpoons 2CrO_4^{2-} + 2H^+$$

shifts to the right with the generation of more CrO_4^{2-} from the relatively large supply of $Cr_2O_7^{2-}$. The hydrogen ions that are formed unite with the acetate ions of the buffer, and a constant pH is maintained.

3. DISSOLUTION OF BARIUM CHROMATE.

Even though barium chromate is precipitated from a weakly acidic solution, it is readily dissolved by strong acids. The hydrogen ion concentration of solutions of strong acids is sufficiently high to cause the chromate ion concentration to be reduced below the value required to maintain a saturated solution of barium chromate. Thus

$$[Ba^{2+}][CrO_4^{2-}] < K_{sp}$$

and the barium chromate dissolves. A solution of 12 M HCl is used in this analysis to bring about the dissolution of $BaCrO_4$.

4. IDENTIFICATION OF BARIUM.

The presence of the barium ion in the solution is confirmed by precipitating it as barium sulfate, using sulfuric acid as the precipitant.

Additional evidence for the presence of the barium ion is obtained by a flame test. When heated in the Bunsen burner flame, barium salts give it a yellow-green color.

PROCEDURE 2

Precipitate from Procedure 1: $BaCO_3$, $SrCO_3$, and $CaCO_3$. Dissolve the precipitate in a mixture of 2 drops of 4 M CH_3CO_2H and 4 drops of 1 M $NH_4CH_3CO_2$. Add 1 drop of 1 M K_2CrO_4 to the solution. The formation of a yellow precipitate indicates the presence of barium. Separate the mixture and reserve the solution for Procedure 3. Dissolve the precipitate ($BaCrO_4$) in 2 drops of 12 M HCl. Make a flame test (Section 35.14) with this solution. The barium ion imparts a weak and fleeting greenish-yellow color to the Bunsen burner flame. Add 1 drop of 4 M H_2SO_4 to the remainder of the solution. A white precipitate ($BaSO_4$) confirms the presence of barium.

39.3 Separation and Identification of Strontium

1. SEPARATION OF STRONTIUM FROM CALCIUM. Although the chromate ion concentration in an acetic acid-ammonium acetate buffered solution is too low to exceed the solubility product of strontium chromate, this compound will precipitate when the solution is made basic with aqueous ammonia. The equilibria are

$$Cr_2O_7^{2-} + H_2O \rightleftharpoons 2CrO_4^{2-} + 2H^+$$
$$2OH^- + 2H^+ \rightleftharpoons 2H_2O$$

and the net reaction is

$$Cr_2O_7^{2-} + 2OH^- \rightleftharpoons 2CrO_4^{2-} + H_2O$$

The resulting concentration of chromate ions is large enough so that the solubility product of strontium chromate is exceeded and precipitation occurs. The complete precipitation of strontium chromate is ensured by further reducing its solubility through the addition of enough ethyl alcohol to the solution to make it 50% alcohol by volume. Calcium chromate is soluble in this solution.

2. IDENTIFICATION OF STRONTIUM. The formation of a fine yellow crystalline precipitate of strontium chromate confirms the presence of strontium. A crimson color imparted to the Bunsen burner flame is characteristic of the strontium ion.

PROCEDURE 3 *Solution from Procedure 2: Sr²⁺ and Ca²⁺.* Add 4 *M* aqueous ammonia to the solution until the color changes from orange to yellow. Now add a volume of ethyl alcohol equal to the volume of the solution. The formation of a fine yellow precipitate indicates $SrCrO_4$. Separate the mixture and reserve the solution for Procedure 4. Dissolve the precipitate in 2 drops of 12 *M* HCl and make a flame test with this solution. A crimson color imparted to the flame is characteristic of the strontium ion.

39.4 Identification of Calcium

The solution from the strontium chromate separation contains the calcium ion. The addition of ammonium oxalate precipitates calcium oxalate, CaC_2O_4, a white crystalline salt.

$$Ca^{2+} + C_2O_4^{2-} \rightleftharpoons \underline{CaC_2O_4}$$

Calcium ions give a brick-red color to the Bunsen burner flame.

PROCEDURE 4 *Solution from Procedure 3: Ca²⁺.* Heat the solution to boiling and add 2 drops of 0.4 *M* $(NH_4)_2C_2O_4$. The formation of a white precipitate, which may form slowly, confirms the presence of calcium. Dissolve the precipitate in 12 *M* HCl and make a flame test with the solution. Calcium ions give a brick-red color to the flame.

PROCEDURE 5   *Original Solution: Ba^{2+}, Sr^{2+}, and Ca^{2+}.* Make flame tests with the original solution and compare with tests of known solutions that you make up yourself.

FOR REVIEW

EXERCISES

1. Why are ammonium salts expelled prior to the precipitation of Group IV?

2. What reagents compose the Group IV precipitant? What is the function of each of these chemicals in the precipitation?

3. In terms of ionic equilibria and solubility product theory, explain the dissolution of $BaCrO_4$ in HCl.

4. Barium chromate is precipitated from a solution of potassium chromate containing acetic acid. Why is the acetic acid present?

5. Write equations representing the various ionic equilibria involved in the dissolution of barium carbonate in acetic acid.

6. What are the characteristic colors imparted to the Bunsen burner flame by Ca^{2+}, Sr^{2+}, and Ba^{2+}?

7. Account for the color change from orange to yellow when aqueous ammonia is added to an acetic acid solution of potassium dichromate.

8. If the magnesium concentration in the unknown is quite high or if too large an amount of carbonate is added in precipitating Group IV, some magnesium carbonate will also precipitate. What effect would such an error have on the rest of the analytical scheme for Group IV?

9. Explain why ammonium carbonate rather than sodium carbonate is used in the precipitation of Group IV.

10. What characteristic properties cause Ca^{2+}, Sr^{2+}, and Ba^{2+} to be members of the same group of the analytical scheme?

11. Outline the separation of the following groups of ions leaving out all unnecessary steps:
    (a) Ag^+, Cu^{2+}, Zn^{2+}, Ca^{2+}, Na^+
    (b) Fe^{3+}, Ba^{2+}, Mg^{2+}
    (c) Pb^{2+}, Sn^{4+}, Al^{3+}, Sr^{2+}

12. What is the total weight of oxalates that can be obtained from a solution containing 3.00 mg each of Ba^{2+}, Sr^{2+}, and Ca^{2+}?   *Ans. 20.5 mg*

13. Calculate the hydroxide ion concentration of a solution that is 0.40 M in NH_3 and 0.30 M in NH_4^+.   *Ans. 2.4×10^{-5} M*

14. If sodium sulfate is added to a mixture originally 0.010 M in Sr^{2+} and 0.010 M in Ba^{2+}, what percentage of the barium remains unprecipitated before any $SrSO_4$ precipitates?   *Ans. 0.039%*

15. What concentration of carbonate ion is required to initiate precipitation of each of the following:
    (a) $CaCO_3$ from 0.010 M Ca^{2+} ($K_{CaCO_3} = 4.8 \times 10^{-9}$)
    (b) $SrCO_3$ from 0.010 M Sr^{2+} ($K_{SrCO_3} = 9.42 \times 10^{-10}$)
    Ans. (a) 4.8×10^{-7} M; (b) 9.42×10^{-8} M

16. How many drops (assume 1 drop = 0.050 mL) of 1.0 M acetic acid would be required to dissolve 5 mg of $SrCO_3$?   *Ans. 2 drops*

17. Will the concentration of OH^- in Exercise 13 be large enough to cause $Mg(OH)_2$ to precipitate in a 0.010 M solution of Mg^{2+}? (K_{sp} for $Mg(OH)_2$ is 1.5×10^{-11}.)   *Ans. No*

40

THE ANALYSIS OF GROUP V

The solution to be analyzed may be a Group V known or unknown, or it may be the solution from the Group IV separation. The Group V flow sheet is given in Table 40-1. The chemistry of the metals of Analytical Group V is also considered in Chapter 34, to which you may wish to refer.

Table 40-1   Group V Flow Sheet

$$
\begin{array}{c}
\left.\begin{array}{c} Na^+ \\ K^+ \\ Mg^{2+} \\ NH_4^+ \end{array}\right\}
\end{array}
$$

$\xrightarrow[\text{CH}_3\text{CO}_2^-, \text{H}_2\text{O}]{Zn^{2+}, UO_2^{2+}}$ $NaZn(UO_2)_3(CH_2CO_2)_9 \cdot 6H_2O$ (yellow)

$\xrightarrow[\text{NH}_3, \text{H}_2\text{O}]{NH_4^+, HPO_4^{2-}}$ $\begin{array}{c} MgNH_4PO_4 \text{ (white)} \\ \left.\begin{array}{c} K^+ \\ Na^+ \\ NH_4^+ \end{array}\right\} \end{array}$ $\xrightarrow[\text{heat}]{HNO_3}$ $\begin{array}{c} NH_3(g), N_2O(g) \\ \left.\begin{array}{c} K^+ \\ Na^+ \end{array}\right\} \end{array}$ $\xrightarrow{Na_3[Co(NO_2)_6]}$ $\begin{array}{c} K_2Na[Co(NO_2)_6] \\ \text{(yellow)} \end{array}$

$\xrightarrow[\text{heat}]{NaOH}$ $NH_3(g)$ $\xrightarrow{\text{moist red litmus}}$ blue litmus

40.1   Removal of Traces of Calcium and Barium

The cations in Group V (Na^+, K^+, Mg^{2+}, and NH_4^+) do not form precipitates with the reagents used to separate Groups I, II, III, and IV. In fact, the four ions of Analytical Group V have no precipitant in common.

If ammonium salts were present in excessive quantities during the precipitation of the carbonates of the Group IV ions, then trace amounts of barium and calcium ions may have come through into the Group V solution. Because these ions will interfere with the identification of the Group V ions, they must be removed. Ammonium oxalate is added to precipitate the calcium ions, and ammonium sulfate is added to precipitate the barium ions.

PROCEDURE 1   *Solution from the Group IV Separation: Na^+, K^+, Mg^{2+}, and NH_4^+ (Ca^{2+} and Ba^{2+}). Add to this solution 1 drop of 0.4 M $(NH_4)_2C_2O_4$ and 1 drop of 1 M $(NH_4)_2SO_4$. Separate and discard any precipitate that may form and use the solution for Procedure 2.*

40.2 Identification of Sodium

After the removal of the traces of barium and calcium, the sodium ion may be detected by the formation of the triple salt sodium zinc uranyl acetate 6-hydrate, which is a yellow crystalline compound.

$$Na^+ + Zn^{2+} + 3UO_2^{2+} + 9CH_2CO_2^- + 6H_2O \rightleftharpoons$$
$$NaZn(UO_2)_3(CH_2CO_2)_9 \cdot 6H_2O$$

The sodium reagent is a saturated solution of zinc acetate and uranyl acetate in acetic acid. Ammonium, magnesium, and potassium ions do not interfere with the test. The sodium ion imparts a yellow color to the Bunsen burner flame.

PROCEDURE 2   *Solution from Procedure 1: K^+, Na^+, Mg^{2+}, and NH_4^+. To 2 drops of the solution from Procedure 1 or a Group V known or unknown, add 1 M CH_3CO_2H until the solution is acid to litmus. Add 1 drop of this acidified solution to 5 drops of the sodium reagent. Shake the mixture and set it aside for an hour. The formation of a yellow crystalline precipitate indicates the presence of sodium.*

40.3 Separation and Identification of Magnesium

To a second portion of the solution for analysis is added an excess of aqueous ammonia and disodium hydrogen phosphate. The formation of a white crystalline precipitate, $MgNH_4PO_4$, is evidence of the presence of the magnesium ion. The equation is

$$Mg^{2+} + NH_4^+ + HPO_4^{2-} \rightleftharpoons MgNH_4PO_4 + H^+$$

Note that the hydrogen ion is one product of the reaction. Completeness of precipitation of the magnesium ammonium phosphate is ensured by making the solution basic with aqueous ammonia, which combines with the hydrogen ion and causes a shift of the equilibrium to the right.

After the magnesium ammonium phosphate is separated from the solution containing sodium ion and ammonium ion, it is dissolved in acetic acid. This treatment with acid causes the following equilibrium to be shifted to the right:

$$MgNH_4PO_4 + H^+ \rightleftharpoons Mg^{2+} + NH_4^+ + HPO_4^{2-}$$

The reason is that the phosphate ion in a saturated solution of $MgNH_4PO_4$ is removed from the solution by combining with hydrogen ions, making the product of the concentrations of the Mg^{2+}, NH_4^+, and PO_4^{3-} ions less than the solubility product. The magnesium reagent and sodium hydroxide are added to the solution containing magnesium ions. The magnesium hydroxide that is formed adsorbs the magnesium reagent [a dye, 4-(nitrophenylazo)-1-naphthol] and forms a blue precipitate, which confirms the presence of magnesium.

PROCEDURE 3   *Use 10 drops of the solution from Procedure 1 or 10 drops of a Group V known or unknown: K^+, Na^+, Mg^{2+}, and NH_4^+. Add 4 M aqueous ammonia to the solution until it is alkaline to litmus and then add 1 drop in excess. Now add 2 drops of 1 M Na_2HPO_4 to the solution. The formation of a white crystalline precipitate, often slow in forming, confirms the presence of magnesium. Separate the mixture and save the solution for Procedure 4. Dissolve the precipitate in a mixture of 2 drops of 1 M CH_3CO_2H and 3 drops of water. Add 1 drop of the magnesium reagent and an excess of 4 M NaOH to the solution. The formation of a blue precipitate confirms the presence of magnesium. (Do not confuse the sky-blue color of the precipitate with the purple color of the magnesium reagent. Be sure a precipitate exists.)*

40.4 Identification of Potassium

The identification of potassium is made using the solution obtained from the separation of magnesium ammonium phosphate. Ammonium ions must be removed prior to the test for the potassium ion because the precipitant for potassium ions will also precipitate ammonium ions. Concentrated nitric acid is added to the solution, it is evaporated to dryness, and then it is ignited to expel volatile ammonium salts.

$$NH_4Cl \xrightarrow{\Delta} NH_3(g) + HCl(g)$$

$$NH_4NO_3 \xrightarrow{\Delta} N_2O(g) + 2H_2O(g)$$

When a solution of $Na_3[Co(NO_2)_6]$ is added to an acidic solution of the residue, the potassium ion precipitates in the form of the yellow complex salt $K_2Na[Co(NO_2)_6]$.

$$2K^+ + Na^+ + [Co(NO_2)_6]^{3-} \longrightarrow \underline{K_2Na[Co(NO_2)_6]}$$

Potassium gives a characteristic violet flame test, which is masked by the intense yellow color produced by the sodium ion. By observing the flame through cobalt glass, which filters out the yellow color of the sodium ion, we may detect the potassium ion in the presence of the sodium ion.

PROCEDURE 4   *Solution from Procedure 3: K⁺, Na⁺, and NH₄⁺.* Add 4 drops of concentrated HNO_3 to the solution held in a casserole, evaporate it to dryness, and then heat the dry residue for several minutes. After the casserole is cool, add 1 drop of $1 M HCl$ and 5 drops of $4 M CH_3CO_2H$ to the residue and boil the resultant solution. Now add 2 drops of this solution to 5 drops of a saturated solution of $Na_3[Co(NO_2)_6]$. The formation of a yellow precipitate confirms the presence of potassium.

PROCEDURE 5   *Original Known or Unknown Solution for Group V or Solution from Procedure 1.* Make flame tests with this solution, and compare the results with known solutions of the ions in question. When testing for potassium, observe the flame through a cobalt glass. Sodium ions impart an intense yellow color to the flame. A trace of the sodium ion as contaminant will give a yellow coloration to the flame, so do not rely entirely on the flame test in reporting the presence of sodium ions. Potassium ions give a violet color to the flame.

40.5   Identification of Ammonium Ion

Since ammonium ions are added in the course of the analysis at various points, it is necessary to test the original solution of a known or unknown for this ion. The presence of the ammonium ion in a solution may be detected by the addition of a strong, nonvolatile base such as sodium hydroxide, with subsequent heating of the solution and liberation of gaseous ammonia.

$$NH_4^+ + OH^- \text{ (from NaOH)} \rightleftharpoons NH_3(g) + H_2O$$

The gaseous ammonia may be detected by moist red litmus paper and by its characteristic odor.

$$NH_3 + H_2O \rightleftharpoons NH_4^+ + OH^-$$

During the heating of the solution one must not permit the sodium hydroxide solution to come in contact with the litmus paper because this would void the test for ammonia.

PROCEDURE 6   *Original Solution of the Known or Unknown.* Add $4 M NaOH$ to the solution until it is basic. Moisten a piece of red litmus with distilled water and place it on the convex side of a watch glass. Place the watch glass on a beaker containing the solution and warm the solution gently. Avoid spattering of the solution by overheating and do not permit the litmus to come into contact with the solution. If the litmus turns blue within a short time, the presence of the ammonium ion is confirmed.

FOR REVIEW

EXERCISES

1. Why must the ammonium ion be removed before making the chemical test for the potassium ion?
2. Explain why the test for the ammonium ion must be made on a portion of the original sample during the analysis of either a Group V known or unknown or a general unknown.
3. Why is Group V often referred to as the soluble group?
4. Why are $(NH_4)_2SO_4$ and $(NH_4)_2C_2O_4$ added prior to the analysis of Group V?
5. If magnesium carbonate had, through an error in carrying out the procedure, precipitated with Group IV, how would you determine its presence? What error or errors might have caused its precipitation with Group IV?
6. When $MgNH_4PO_4$ is heated to a high temperature (1000°C) it is converted to magnesium pyrophosphate, $Mg_2P_2O_7$. Write a balanced equation for this reaction.
7. Write the equation for the dissolution of $MgNH_4PO_4$ in an acid. Explain this reaction in terms of solubility product and ionic equilibria theory.
8. Why may red litmus turn blue after prolonged exposure to the air of the laboratory?
9. What is the chemistry of the chemical test for potassium?
10. Explain the use of cobalt glass in the flame test for potassium.
11. What weight of $K_2Na[Co(NO_2)_6]$ can be formed from 2.5 mg of K^+?   *Ans. 14 mg*
12. How many mL of NH_3 gas measured at 20°C and 750 torr will be evolved when a solution containing 10 mg of NH_4Cl is treated with NaOH and heated?   *Ans. 4.6 mL*

41

THE ANALYSIS OF ANIONS

41.1 Introduction

The analytical scheme for anions includes 13 of the more common and important negative ions. The anions considered are carbonate, sulfide, sulfite, nitrite, sulfate, nitrate, phosphate, metaborate, oxalate, fluoride, chloride, bromide, and iodide. Among the anions that were covered by the analytical scheme for cations are arsenite, arsenate, stannite, stannate, permanganate, aluminate, chromate, dichromate, and zincate.

Although several schemes for anion analysis involving the separation of the ions into groups have been developed, the procedures are in general more complicated and unreliable than are those for cation analysis. Instead of using a systematic scheme of analysis involving the same solution throughout, the procedures outlined in this chapter involve a series of elimination tests that prove the absence of certain anions. Tests are then made on different samples of the unknown solution for the presence of the anions whose absence was not indicated in the preliminary elimination tests. (A flow sheet of the anion elimination tests is given in Table 41-1.)

The properties of the acids and their anions have been considered in the first part of the text in connection with the descriptive chemistry of the nonmetals. You will find it necessary to refer to earlier chapters in order to answer some of the questions that arise concerning the chemistry of the anions.

Table 41-1   Anion Elimination Flow Sheet

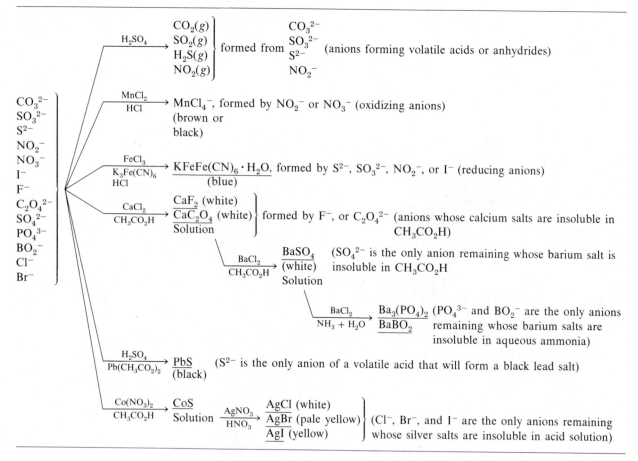

The anion knowns and unknowns are usually furnished in the form of mixtures of dry salts due to the fact that some of the anions react with one another in solution. Thus certain anions may be detected in freshly prepared solutions, while, after standing, these same anions may have undergone decomposition. Certain combinations of anions react almost immediately in alkaline solutions, and still other anions cannot exist together in acidic solutions. As an example of anion incompatibility, anions that are strong oxidizing agents usually react on acidification with anions that are strong reducing agents.

PRELIMINARY ELIMINATION TESTS FOR THE ANIONS

41.2   Anion Elimination Chart

Prepare an anion elimination chart like the one shown in Table 41-2. As you perform each elimination test, check off in the appropriate spaces those anions

Table 41-2   Anion Elimination Chart

Name: _____ Section: _____ Date: _____

Instructor's Approval: _____ Ions Found: _____

Anion Elimination Chart

Anion	Test I	Test II	Test III	Test IV	Test V	Test VI	Test VII	Test VIII	Make Confirmatory Tests for
CO_3^{2-}									
SO_3^{2-}									
S^{2-}									
NO_2^-									
NO_3^-									
I^-									
F^-									
$C_2O_4^{2-}$									
SO_4^{2-}									
PO_4^{3-}									
BO_2^-									
Cl^-									
Br^-									

that you have found to be absent. When all the elimination tests have been made, it will be apparent which anions may be present. Tests can then be made for the anions that have not been eliminated.

41.3 Elimination of Anions That Form Volatile Acid Anhydrides

The anions CO_3^{2-}, SO_3^{2-}, S^{2-}, and NO_2^- are derived from the weak acids H_2CO_3, H_2SO_3, H_2S, and HNO_2, respectively. As acids in solution, H_2CO_3, H_2SO_3, and HNO_2 are unstable and decompose, producing gases. H_2S does not decompose but has a limited solubility in water. When salts of these anions are treated with strong acids, the anions, being strong Brönsted bases, readily combine with the hydrogen ions from the strong acids. The equilibria are

$$CO_3^{2-} + H^+ \rightleftharpoons HCO_3^-$$
$$HCO_3^- + H^+ \rightleftharpoons H_2CO_3$$
$$H_2CO_3 \rightleftharpoons H_2O + CO_2(g)$$

$$SO_3^{2-} + H^+ \rightleftharpoons HSO_3^-$$
$$HSO_3^- + H^+ \rightleftharpoons H_2SO_3$$
$$H_2SO_3 \rightleftharpoons H_2O + SO_2(g)$$

$$S^{2-} + H^+ \rightleftharpoons HS^-$$
$$HS^- + H^+ \rightleftharpoons H_2S(g)$$
$$NO_2^- + H^+ \rightleftharpoons HNO_2$$
$$3HNO_2 \rightleftharpoons H^+ + NO_3^- + 2NO(g) + H_2O$$
$$2NO + O_2 \longrightarrow 2NO_2(g)$$

The high concentration of hydrogen ion furnished by the sulfuric acid used in this elimination test serves to shift the equilibrium to the right in each case. When the solubility of each gas is exceeded, it will escape from solution in the form of bubbles. Gentle heating decreases the solubilities of the gases and aids in the test. Boiling of the solution should be avoided, because bubbles of steam may be mistaken for bubbles of gaseous reaction products.

If no gas is evolved during acidification and gentle heating of an unknown sample, CO_3^{2-}, SO_3^{2-}, S^{2-}, and NO_2^- are absent. If a gas is evolved, then one or more of these anions is present, and it is necessary to make tests for the presence of each of these ions unless they are proved absent by other elimination tests.

TEST I   *Elimination of Anions That Form Volatile Acid Anhydrides:* CO_3^{2-}, SO_3^{2-}, S^{2-}, *and* NO_2^-. Put 25 mg of the solid unknown in a test tube and treat it with 2 drops of 1.5 M H_2SO_4. Examine the mixture for the formation of gas bubbles. If no gas is evolved, heat the mixture in the hot water bath. If no gas is evolved, then CO_3^{2-}, SO_3^{2-}, S^{2-}, and NO_2^- are proved to be absent. If these anions are absent, place a check mark after the formula for each of these anions in the Test I column of the anion elimination chart; leave these spaces blank if a gas is evolved.

If a gas is evolved, note its odor and color. Carbon dioxide is colorless and odorless. Sulfur dioxide is colorless and has the odor of burning sulfur. Hydrogen sulfide is colorless and has a characteristic odor. If the nitrite ion is present, nitrogen dioxide, with a reddish-brown color and characteristic odor, is evolved. The solution may be pale blue in color due to the presence of HNO_2 if the nitrite ion is present.

41.4   Preparation of the Solution for Analysis

The procedures outlined for the detection of the anions are applicable, in general, to salts containing only the alkali metals as cations. It is obvious that complications in the anion analysis will arise if certain metal ions are present to form precipitates, or possibly colored solutions, with the anions being detected.

Most of the metal ions may be removed from solution as insoluble carbonates, hydroxycarbonates, hydroxides, or oxides by treating a solution of the unknown with sodium carbonate. The ammonium ion, NH_4^+, is converted to NH_3, which is evolved from the hot solution. Since the carbonate ion is added in excess the identification of this ion must be made on the original sample. The CO_3^{2-} must be removed before certain other tests are made.

PROCEDURE 1   *Preparation of the Solution for Analysis: Removal of Heavy Metal Ions by* Na_2CO_3. Put 100 mg of the powdered unknown in a test tube and add 2 mL of 1.5 M

Na$_2$CO$_3$ solution. Heat the mixture for 10 min in the hot water bath. If ammonia is given off, continue the heating until no more gas is evolved. Separate any precipitate that forms. The remaining solution contains the anions in the form of sodium salts and will be referred to as the *prepared solution.*

41.5 Elimination of Oxidizing Anions

Note that the nitrite ion is included in the elimination tests for both oxidizing anions and reducing anions. The nitrite ion in acid solution is an oxidizing agent in the presence of strong reducing agents and a reducing agent in the presence of strong oxidizing agents.

The prepared solution is tested for the oxidizing anions NO_2^- and NO_3^- with a saturated solution of manganese(II) chloride in concentrated hydrochloric acid. The development of a dark-brown or black color is the result of the oxidation, in strongly acidic solution, of the manganese(II) ion, Mn^{2+}, to the manganese(III) ion, Mn^{3+}. The manganese(III) ion then unites with chloride ions and forms the dark-colored complex $MnCl_4^-$.

$$Mn^{2+} + HNO_2 + H^+ + 4Cl^- \longrightarrow MnCl_4^- + NO(g) + H_2O$$
$$3Mn^{2+} + 4H^+ + NO_3^- + 12Cl^- \longrightarrow 3MnCl_4^- + NO(g) + 2H_2O$$

TEST II *Elimination of Oxidizing Anions: NO_3^- and NO_2^-.* Add 6 drops of a saturated solution of $MnCl_2$ in 12 M HCl to 4 drops of the prepared solution and heat the mixture to boiling. The formation of a dark-brown or black color, due to $MnCl_4^-$, indicates the presence of NO_3^-, NO_2^-, or a mixture of these ions. Record the results of the test on the anion elimination chart.

41.6 Elimination of Reducing Anions

A sample of the prepared solution is acidified with hydrochloric acid and then iron(III) chloride and potassium hexacyanoferrate(III) are added. The prompt appearance of a blue to blue-green color or precipitate proves the presence of one or more of the reducing anions S^{2-}, SO_3^{2-}, NO_2^-, and I^-.

Iron(III), Fe^{3+}, forms a brown solution when fresh, and a green solution when old, with $Fe(CN)_6^{3-}$. In the presence of a strong reducing agent, the Fe^{3+} may be reduced to Fe^{2+}, or the $Fe(CN)_6^{3-}$ reduced to $Fe(CN)_6^{4-}$, with the formation of $KFeFe(CN)_6 \cdot H_2O$, a blue pigment. The equations are

$$2Fe^{3+} + S^{2-} \rightleftharpoons 2Fe^{2+} + S$$
$$2Fe^{3+} + SO_3^{2-} + H_2O \rightleftharpoons 2Fe^{2+} + SO_4^{2-} + 2H^+$$
$$2Fe^{3+} + 2I^- \rightleftharpoons 2Fe^{2+} + I_2$$
$$2Fe(CN)_6^{3-} + NO_2^- + H_2O \rightleftharpoons 2Fe(CN)_6^{4-} + 2H^+ + NO_3^-$$
$$K^+ + Fe^{2+} + Fe(CN)_6^{3-} + H_2O \rightleftharpoons KFeFe(CN)_6 \cdot H_2O$$
$$K^+ + Fe^{3+} + Fe(CN)_6^{4-} + H_2O \rightleftharpoons KFeFe(CN)_6 \cdot H_2O$$

TEST III   *Elimination of Reducing Anions: S^{2-}, SO_3^{2-}, I^-, and NO_2^-.* Mix 2 drops of a recently prepared saturated solution of $K_3Fe(CN)_6$ with 1 drop of 0.1 M $FeCl_3$, and add 2 drops of 6 M HCl. Add 2 drops of the prepared solution to this mixture and let the mixture stand for a few minutes. The development of a blue color or the formation of a blue precipitate indicates the presence of a reducing agent (S^{2-}, SO_3^{2-}, I^-, NO_2^-). Run a blank test with the same reagents leaving out the prepared solution, and compare the colors and intensities of the unknown and blank test solutions.

41.7 Elimination of Fluoride and Oxalate

Of the 13 anions that are included in this analytical scheme, only fluoride and oxalate form calcium salts that are insoluble in 4 M acetic acid.

$$Ca^{2+} + 2F^- \rightleftharpoons \underline{CaF_2} \text{ (white)}$$
$$Ca^{2+} + C_2O_4^{2-} \rightleftharpoons \underline{CaC_2O_4} \text{ (white)}$$

If no precipitate forms when calcium chloride is added to an acetic acid solution of the unknown, then fluoride and oxalate are absent.

TEST IV   *Elimination of Fluoride and Oxalate: F^- and $C_2O_4^{2-}$.* To 20 drops of the prepared solution, add 4 M CH_3CO_2H (count the drops) until the solution is just acid to litmus. Now add an equal number of drops of 4 M CH_3CO_2H in excess. Tap the test tube repeatedly until the excess of CO_2 from the Na_2CO_3 present is expelled. Add 8 drops of 0.1 M $CaCl_2$ and shake the contents of the tube. If a precipitate forms after a few minutes, it may be CaF_2, CaC_2O_4, or a mixture of these compounds. If no precipitate forms, F^- and $C_2O_4^{2-}$ are absent. Separate the precipitate and use the solution for Test V. Save the precipitate for the confirmatory test for $C_2O_4^{2-}$ and F^-.

41.8 Elimination of Sulfate

After the separation of fluoride and oxalate as insoluble calcium salts, barium chloride is added to precipitate the sulfate ion as white crystalline barium sulfate from the solution that is still acidic with acetic acid. If the sulfite ion is present in relatively large amounts, barium sulfite may precipitate. $BaSO_4$ is insoluble in dilute hydrochloric acid, while $BaSO_3$ will dissolve and evolve SO_2 when treated with hydrochloric acid.

Owing to the ease with which the sulfite ion is oxidized to the sulfate ion by oxygen in the air, if sulfite is present in the original solution, you may obtain a test for sulfate although none was present at the outset.

TEST V   *Elimination of Sulfate: SO_4^{2-}.* Test for completeness of precipitation in the solution from Test IV by adding 1 drop of 0.1 M $CaCl_2$. Now add 8 drops of 0.1 M

BaCl$_2$. Shake the mixture and let it stand for 5 min. If SO$_4^{2-}$ is present, a precipitate of BaSO$_4$ will form. Separate the mixture and reserve the solution for Test VI. If SO$_3^{2-}$ was present in the original solution, you may obtain a test for SO$_4^{2-}$ because sulfite is oxidized to sulfate by oxygen in the air.

41.9 Elimination of Phosphate and Metaborate

The solution from the barium sulfate separation is acidified with hydrochloric acid and heated to expel CO$_2$ from CO$_3^{2-}$ and SO$_2$ from SO$_3^{2-}$, if present. If they were not first removed, these ions would be precipitated as barium salts in the ammoniacal solution used to precipitate barium phosphate and barium metaborate.

$$3Ba^{2+} + 2PO_4^{3-} \rightleftharpoons \underline{Ba_3(PO_4)_2} \text{ (white)}$$
$$Ba^{2+} + 2BO_2^- \rightleftharpoons \underline{Ba(BO_2)_2} \text{ (white)}$$

Even though phosphate and metaborate are insoluble in water, they did not precipitate as their calcium or barium salts in the acetic acid solutions in which CaF$_2$, CaC$_2$O$_4$, and BaSO$_4$ were formed. Barium phosphate does not precipitate in an acetic acid solution due to the low concentration of free phosphate ions. The phosphate ion is a strong base

$$PO_4^{3-} + H^+ \rightleftharpoons HPO_4^{2-}$$

making its concentration so low that the solubility product of Ba$_3$(PO$_4$)$_2$ is not exceeded. Raising the pH of the solution by adding aqueous ammonia, on the other hand, increases the concentration of the phosphate ion

$$HPO_4^{2-} + OH^- \rightleftharpoons PO_4^{3-} + H_2O$$

and permits the solubility product of Ba$_3$(PO$_4$)$_2$ to be exceeded.

TEST VI   *Elimination of Phosphate and Metaborate: PO$_4^{3-}$ and BO$_2^-$.* Transfer the solution from Test V to a casserole, add 6 drops of 12 *M* HCl, and heat to expel SO$_2$ and CO$_2$. Make the solution alkaline with aqueous ammonia and then add 5 drops in excess. A white precipitate may be Ba$_3$(PO$_4$)$_2$, Ba(BO$_2$)$_2$, or a mixture of these. Because barium metaborate is prone to form supersaturated solutions, it is necessary to carry out the confirmatory test for BO$_2^-$ even though no precipitate forms in this elimination test.

41.10 Elimination of Sulfide

The addition of dilute sulfuric acid to a sample of the prepared solution causes hydrogen sulfide gas to be evolved if the sulfide ion is present. Hydrogen sulfide reacts with lead acetate and forms a black precipitate of lead sulfide.

$$Pb^{2+} + H_2S \rightleftharpoons \underline{PbS} + 2H^+$$

This elimination test also serves to confirm the sulfide ion if it is present.

41.11 Elimination of Chloride, Bromide, and Iodide

If the sulfide ion is present, it is first removed as the insoluble CoS before attempting to eliminate chloride, bromide, and iodide. Silver nitrate is then added to precipitate the halides as AgCl (white), AgBr (pale yellow), and AgI (yellow). Silver chloride is completely soluble in the mixture of aqueous ammonia and silver nitrate solutions used in Test VIII, silver bromide is partially soluble, and silver iodide is insoluble in this reagent.

$$AgCl + 2NH_3 \rightleftharpoons [Ag(NH_3)_2]^+ + Cl^-$$
$$AgBr + 2NH_3 \rightleftharpoons [Ag(NH_3)_2]^+ + Br^- \text{ (partially soluble)}$$
$$AgI + NH_3 \rightleftharpoons \text{no reaction}$$

These facts reflect the decreasing solubility of the silver halides with increasing atomic weight of the halogen. The solubility products are 1.8×10^{-10} for AgCl, 3.3×10^{-13} for AgBr, and 1.5×10^{-16} for AgI.

If a white precipitate forms when the aqueous ammonia extract of the silver halide precipitate is acidified, then the presence of the chloride ion is confirmed.

$$[Ag(NH_3)_2]^+ + Cl^- + 2H^+ \longrightarrow \underline{AgCl} + 2NH_4^+$$

A slight precipitate, yellow in color, may be silver bromide.

CONFIRMATORY TESTS FOR THE ANIONS

41.12 Identification of Carbonate

A sample of the original unknown is treated with hydrogen peroxide to oxidize any sulfite or nitrite that might be present. Then dilute sulfuric acid is added to decompose the carbonate and liberate carbon dioxide gas. The carbon dioxide gas evolved is brought in contact with a drop of barium hydroxide solution, which is held in a platinum wire loop suspended above the reaction mixture. The formation of a definite turbidity in the drop of barium hydroxide indicates the presence of carbonate.

$$CO_3^{2-} + 2H^+ \rightleftharpoons H_2O + CO_2(g)$$
$$CO_2 + Ba^{2+} + 2OH^- \longrightarrow \underline{BaCO_3} + H_2O$$

If the sulfite and nitrite were not removed by oxidation to sulfate and nitrate, respectively, sulfur dioxide and nitrogen dioxide would escape following acidification, enter the drop of barium hydroxide solution, and either prevent the formation of barium carbonate or mask its presence.

PROCEDURE 2   *Identification of CO_3^{2-}.* To 25 mg of the powdered solid unknown, add 10 drops of hydrogen peroxide and heat the mixture in the hot water bath. Place 1 drop of $Ba(OH)_2$ solution in a loop of platinum wire. Add 3 drops of 4 M H_2SO_4 to the unknown mixture and immediately hold the drop of $Ba(OH)_2$ over the reaction mixture. The formation of a definite turbidity in the $Ba(OH)_2$ solution indicates the presence of CO_3^{2-}.

41.13 Identification of Sulfate and Sulfite

The formation of a white precipitate of barium sulfate when barium chloride is added to a hydrochloric acid solution of the unknown serves to confirm the sulfate ion. Barium salts of the other anions are soluble in 6 M hydrochloric acid.

Bromine water is added to the solution obtained from the separation of barium sulfate for the purpose of oxidizing the sulfite ion to sulfate.

$$SO_3^{2-} + Br_2 + H_2O \rightleftharpoons SO_4^{2-} + 2Br^- + 2H^+$$

The appearance of a white precipitate (barium sulfate) indicates the presence of the sulfite ion in the original solution.

PROCEDURE 3   *Identification of SO_4^{2-}.* Acidify 5 drops of the prepared solution with 6 M HCl and add 2 drops in excess. Heat in the hot water bath to expel excess CO_2. Add 0.1 M $BaCl_2$ until precipitation is complete; then allow the mixture to stand. A white, finely divided precipitate of $BaSO_4$ confirms the presence of SO_4^{2-}. Separate the mixture and save the solution for Procedure 4.

PROCEDURE 4   *Identification of SO_3^{2-}.* Add 5 drops of bromine water and 2 drops of 0.1 *M* $BaCl_2$ to the solution from Procedure 2. Heat in the hot water bath and allow to stand for 5 min. A white precipitate of $BaSO_4$ shows the presence of sulfite in the original solution.

PROCEDURE 5   *Identification of S^{2-}.* The sulfide ion is confirmed in its elimination test (Test VII).

41.14   Identification of Nitrite

The oxidizing ability of the nitrite ion in acid solution is applied in the confirmatory test for this ion. The nitrite ion oxidizes iron(II) to iron(III), and nitric oxide is formed from the reduction of the nitrite ion.

$$NO_2^- + Fe^{2+} + 2H^+ \rightleftharpoons Fe^{3+} + NO + H_2O$$

The nitric oxide produced combines with some of the excess iron(II) ions in the reaction mixture, forming the brown complex $[Fe(NO)]^{2+}$.

$$Fe^{2+} + NO \longrightarrow [Fe(NO)]^{2+}$$

PROCEDURE 6   *Identification of NO_2^-.* To 5 drops of the prepared solution, add 4 *M* H_2SO_4 dropwise until the solution is acidic. Now add 5 drops of freshly prepared 0.1 *M* $FeSO_4$ solution. If NO_2^- is present, the solution will assume a dark brown color.

41.15   Identification of Nitrate

If nitrite, bromide, and iodide ions are absent, we may identify the nitrate ion by the brown ring test. Iron(II) sulfate is added to a sample of the prepared solution in a test tube and then acidified with dilute sulfuric acid. The test tube containing the solution is tilted slightly and concentrated H_2SO_4 is poured down the inside wall in such a way as to form a layer of acid in the bottom of the tube. If the nitrate ion is present, a brown ring will soon form at the interface of the two layers. The reduction of the nitrate ion to nitric oxide by iron(II) takes place rapidly only in a solution of very high hydrogen ion concentration and at relatively high temperatures, conditions that prevail at the interface of the two liquids.

$$3Fe^{2+} + NO_3^- + 4H^+ \rightleftharpoons 3Fe^{3+} + NO + 2H_2O$$
$$Fe^{2+} + NO \longrightarrow [Fe(NO)]^{2+}$$

If the nitrite ion is present, then it must be removed prior to the nitrate test because the entire solution will turn brown and the development of the brown ring will be obscured. The removal of the nitrite ion is accomplished by adding ammonium sulfate and evaporating nearly to dryness over a flame. The added ammonium ion reacts with the nitrite ion and forms nitrogen and water.

$$NH_4^+ + NO_2^- + heat \longrightarrow N_2(g) + 2H_2O$$

The residue is taken up in water, and the brown ring test for nitrate is conducted on the solution.

If bromide or iodide ions are present, they will be oxidized by the concentrated sulfuric acid used in the test for nitrate, forming bromine and iodine, respectively, both of which form colored layers. Therefore these ions must be removed before making the brown ring test for nitrate. This is accomplished by adding solid nitrate-free silver sulfate to an acidified sample of the prepared solution to precipitate the bromide and iodide ions as their silver salts.

$$Ag_2SO_4 + 2Br^- \longrightarrow \underline{2AgBr} + SO_4^{2-}$$
$$Ag_2SO_4 + 2I^- \longrightarrow \underline{2AgI} + SO_4^{2-}$$

PROCEDURE 7   *Identification of NO_3^- in the Absence of NO_2^-, Br^-, and I^-.* To 5 drops of the prepared solution in a test tube, add 4 M H_2SO_4 dropwise until the solution is acidic. Now add 5 drops of freshly prepared 0.1 M $FeSO_4$ solution. Add 5 drops of 18 M H_2SO_4, holding the test tube in an inclined position so that the sulfuric acid runs down the side of the test tube and forms a separate layer on the bottom. If NO_3^- is present, a brown ring will form at the interface of the two liquids, within a few minutes.

PROCEDURE 8   *Identification of NO_3^- in the Presence of NO_2^-. Removal of NO_2^-.* To 6 drops of the prepared solution add 4 M H_2SO_4 until the solution is acidic, and then add 4 drops of 1 M $(NH_4)_2SO_4$ solution. Place the mixture in a casserole and slowly evaporate the solution until only a moist residue remains (do not evaporate to dryness). Add 4 drops of water and evaporate to a moist residue a second time. Dissolve the residue in 10 drops of water and transfer the mixture to a small test tube. Perform the brown ring test for NO_3^- as described in Procedure 7 on the mixture resulting from the removal of NO_2^-.

PROCEDURE 9   *Identification of NO_3^- in the Presence of Br^- and I^-. Removal of Br^- and I^-.* To 6 drops of the prepared solution in a test tube, add 10 drops of water. Acidify the solution with 4 M CH_3CO_2H and then add 80 mg of powdered Ag_2SO_4 (nitrate-free). Stir and grind the mixture in the test tube for 2–3 min. Separate the precipitate and transfer the solution to a test tube. Perform the brown ring test for NO_3^- as described in Procedure 7 on the solution obtained from the removal of Br^- and I^-.

41.16   Identification of Oxalate

The precipitate formed during the elimination test for oxalate and fluoride ions (Test IV) is used for the identification of the oxalate ion. It consists of CaC_2O_4 and CaF_2 if both ions were present in the original solution. Treatment of the precipitate with sulfuric acid dissolves the calcium oxalate and forms oxalic acid.

$$CaC_2O_4 + 2H^+ \rightleftharpoons Ca^{2+} + H_2C_2O_4$$

Oxalic acid is a weak reducing agent that is readily oxidized to carbon dioxide and water by potassium permanganate in sulfuric acid solution. A dilute solution of the pink permanganate is employed, and the permanganate ion is reduced to the colorless manganese(II) ion by the oxalic acid. This bleaching of the colored permanganate solution shows the presence of the weak reducing agent oxalic acid.

$$5H_2C_2O_4 + 2MnO_4^- + 6H^+ \longrightarrow 10CO_2(g) + 2Mn^{2+} + 8H_2O$$

PROCEDURE 10	*Identification of $C_2O_4^{2-}$.* Wash the precipitate obtained in Test IV twice. Use 10 drops of water each time and discard the washings. To the residue add 10 drops of water and 10 drops of 4 *M* H_2SO_4, and shake the mixture. Add 0.002 *M* $KMnO_4$ dropwise until 1 drop imparts a permanent pink color to the solution. If $C_2O_4^{2-}$ is present, several drops of $KMnO_4$ should be required to give a permanent pink coloration to the solution. Run a blank by repeating the test but leaving out the precipitate from Test IV.

41.17   Identification of Fluoride

In the confirmatory test for the fluoride ion, a sample of the solid unknown is treated with concentrated sulfuric acid in the presence of silica, SiO_2.

$$2NaF + H_2SO_4 \longrightarrow Na_2SO_4 + 2HF(g)$$

The hydrofluoric acid produced reacts with the silica present and forms gaseous silicon tetrafluoride, SiF_4.

$$SiO_2 + 4HF \longrightarrow SiF_4(g) + 2H_2O$$

The gaseous SiF_4 is brought in contact with a drop of water contained in a loop of platinum wire suspended over the reaction mixture. The development of a white precipitate, H_4SiO_4, in the water confirms the presence of the fluoride ion.

$$3SiF_4 + 4H_2O \longrightarrow H_4SiO_4 + 4H^+ + 2SiF_6^{2-}$$

PROCEDURE 11	*Identification of F^-.* Place 25 mg of the powdered solid unknown in a test tube and add an equal volume of powdered silica. Select a cork to fit the test tube. Cut a notch on the side of the cork and bore a hole through its center. Through the hole, insert a glass rod to the end of which is sealed a platinum wire with a looped end. Add 2 drops of 18 *M* H_2SO_4 to the mixture in the tube. Now place the platinum wire loop with a drop of water in it in the test tube directly above the reaction mixture. Warm the test tube in the hot water bath and then set it aside to cool. The formation of a white precipitate or cloudiness in the drop of water confirms the presence of F^-.

41.18   Identification of Phosphate

Magnesia mixture ($MgCl_2$, NH_4Cl, and aqueous ammonia), when added to a slightly acidic solution of the unknown, will precipitate the phosphate ion as white crystalline magnesium ammonium phosphate. Since this compound tends to form supersaturated solutions, the precipitate may be slow in appearing.

$$Mg^{2+} + NH_4^+ + HPO_4^{2-} \rightleftharpoons MgNH_4PO_4 + H^+$$

PROCEDURE 12   *Identification of PO_4^{3-}*. Dilute 4 drops of the prepared solution with 10 drops of water. Make the solution just acidic with 4 M HNO_3 and add 10 drops of magnesia mixture. A white precipitate, often slow in forming, confirms the presence of PO_4^{3-}.

41.19   Identification of Metaborate

A sample of the prepared solution is evaporated to a small volume and then treated with methyl alcohol and sulfuric acid. The sulfuric acid converts the metaborate ion to orthoboric acid.

$$BO_2^- + H^+ + H_2O \longrightarrow H_3BO_3$$

In the presence of the dehydrating agent, concentrated sulfuric acid, orthoboric acid reacts with methyl alcohol and forms the volatile methyl orthoborate.

$$H_3BO_3 + 3CH_3OH \rightleftharpoons B(OCH_3)_3(g) + 3H_2O$$

The sulfuric acid present takes up the water produced in the reaction and shifts the equilibrium to the right. Methyl orthoborate burns with a characteristic green flame when ignited.

$$2B(OCH_3)_3 + 9O_2 \longrightarrow B_2O_3 + 6CO_2(g) + 9H_2O$$

PROCEDURE 13   *Identification of BO_2^-*. Evaporate 6 drops of the prepared solution to a small volume in a casserole. Add 1 mL of methyl alcohol and 5 drops of 18 M H_2SO_4. Transfer the mixture to a test tube and place the tube in the hot water bath. When the alcohol begins to boil, ignite the vapors. A green tinge to the flame confirms the presence of the metaborate ion.

41.20   Identification of Iodide, Bromide, and Chloride

The confirmatory test for chloride was outlined in the elimination test for chloride, bromide, and iodide (Test VIII); no further test for this ion is necessary.

In general the methods used for the detection of the halide ions in the presence of one another depend on the differences in the reduction potentials of the ions. Of the three ions chloride, bromide, and iodide, the iodide ion is the most easily oxidized, the bromide ion is second, and the chloride ion is the most difficult to oxidize.

The iodide ion may be oxidized to iodine by iron(III) in acid solution without affecting bromide or chloride ions.

$$2I^- + 2Fe^{3+} \rightleftharpoons I_2 + 2Fe^{2+}$$

The elemental iodine is then extracted from the aqueous solution by carbon tetrachloride or the less toxic dichloromethane. In either it exhibits a characteristic violet color.

After the complete removal of the iodide ion, a stronger oxidizing agent than iron(III) is used to oxidize the bromide to bromine. A dilute solution of potassium permanganate in the presence of nitric acid will oxidize bromide to bromine without interference from chloride.

$$10Br^- + 2MnO_4^- + 16H^+ \rightleftharpoons 5Br_2 + 2Mn^{2+} + 8H_2O$$

The free bromine is extracted with carbon tetrachloride or dichloromethane, to which it imparts a yellow or orange color, depending on its concentration.

PROCEDURE 14   *Identification of I⁻.* Dilute 6 drops of the prepared solution with 12 drops of water. Add 4 M HNO_3 until the solution is just acid to litmus, and then add 2 drops of the acid in excess. Treat the solution with 1 mL of 0.1 M $Fe(NO_3)_3$ and 10 drops of CCl_4 or CH_2Cl_2 and shake the test tube. The development of a violet color in the organic layer (the lower layer) proves the presence of I⁻. Remove and discard the organic layer by means of a capillary syringe. Add 10 more drops of CCl_4 or CH_2Cl_2, shake, and remove the organic layer. Repeat the extraction until the organic layer remains colorless, and use the aqueous solution in Procedure 15.

PROCEDURE 15   *Identification of Br⁻.* To the aqueous solution from Procedure 14, add 2 drops of 4 M HNO_3, then 0.1 M $KMnO_4$ solution dropwise until the solution remains pink. Extract the solution with CCl_4 or CH_2Cl_2. A yellow or orange color in the organic layer indicates the presence of Br⁻.

FOR REVIEW

EXERCISES

1. What is meant by incompatibility of anions? Give an example.
2. What principle is involved in the tests for the halide ions in the presence of one another?
3. Explain why the nitrite ion responds to both the test for oxidizing anions and that for reducing anions.
4. In terms of solubility product and ionic equilibria theory, explain the following:
   (a) $Ca_3(PO_4)_2$ is soluble in HCl, while $BaSO_4$ is not.
   (b) $CaCO_3$ is soluble in CH_3CO_2H, while CaC_2O_4 is not.

5. Make a list of anions that are reducing agents and list their oxidation products.
6. Make a list of anions that are oxidizing agents and list their reduction products.
7. If an anion unknown contains the sulfite ion, it may give a positive test for the sulfate ion even though sulfate was not used in making up the unknown. Explain.
8. An acidic solution contains silver ions. What anions are probably absent?
9. A neutral solution contains calcium ions. What anions are probably absent?
10. Explain how the nitrite ion interferes with the test

for nitrate. How is the nitrite ion removed prior to the test for nitrate?

11. A solution is slightly acidic and contains sulfide ions. What cations are probably absent?

12. A solution known to contain either Na_2CO_3 or Na_2SO_4 turns red litmus blue. Which salt is present?

13. Why is it necessary to extract the iodine in the confirmatory test for the iodide ion?

14. Outline a simple test to distinguish between the ions of each pair in the following:
    (a) CO_3^{2-} and SO_4^{2-} (d) PO_4^{3-} and Br^-
    (b) Cl^- and I^- (e) S^{2-} and SO_3^{2-}
    (c) F^- and BO_2^- (f) $C_2O_4^{2-}$ and Cl^-

15. If a solution is strongly acidic, what anions cannot be present in appreciable concentrations?

16. Would you expect the test for F^- to work if no powdered silica were added to the reaction mixture?

17. How many mL of a solution that is 0.010 M in Fe^{3+} ion would be needed to oxidize 1.0 mg of the iodide ion?   *Ans. 0.79 mL*

18. How many drops of 0.020 M $KMnO_4$ solution (assume 0.05 mL per drop) would be needed to react with 1.00 mg of oxalic acid and impart a pink color to the resulting solution?
                                          Ans. 5 drops

42

THE ANALYSIS OF
SOLID MATERIALS

The analytical schemes that have been outlined for cations and anions in the preceding chapters are usually applicable to solutions of the substances being analyzed. However, many substances are not readily soluble in water or even in acids. Thus special procedures for effecting the dissolution of certain solid substances must be used before proceeding with the analyses.

42.1 Dissolution of Nonmetallic Solids

The procedures outlined in this section are applicable to solid inorganic materials other than metals and alloys. The dissolution of metals and alloys is described in Section 42.2.

PROCEDURE 1   *For Samples Soluble in Water.* Grind the solid sample to a fine powder. Treat 100 mg of the powder with 1 mL of water. If no dissolution is apparent, heat the mixture in the hot water bath for a few minutes. If none of the sample appears to dissolve, evaporate a few drops of the supernatant liquid on a watch glass to determine whether partial dissolution has occurred. If a solid residue remains on the watch glass, heat the sample with fresh portions of water until complete dissolution has been effected. Concentrate the combined extracts by evaporation, and proceed with the analyses for cations and anions as outlined in the preceding chapters. Certain solid substances, such as $BiCl_3$ and $SbCl_3$, hydrolyze when treated with water to form new solid substances. In such cases Procedure 2 should be followed.

PROCEDURE 2   *For Samples Not Soluble in Water.* Warm a small sample of the solid substance with 6 *M* HCl or, if this has no effect, with 12 *M* HCl. If dissolution does not occur, or if it is incomplete, try 6 *M* HNO_3, 15 *M* HNO_3, and aqua regia (1 part 15 *M* HNO_3 to 3 parts 12 *M* HCl) in succession on small samples of the solid until a suitable solvent is found. If dissolution has been effected by any of these reagents, dissolve 100 mg of the sample in the effective solvent. Evaporate the solution nearly to dryness, and then take up the residue in 2 mL of water. Use this solution for the analytical schemes for cations and anions.

PROCEDURE 3   *Treatment of the Residue with Na_2CO_3 Solution.* If HCl, HNO_3, or aqua regia does not completely dissolve the solid, the residue remaining after the acid treatment may be treated with a Na_2CO_3 solution. Treat the residue with 2 mL of 1.5 *M* Na_2CO_3 in a casserole and boil the mixture for 10 min, replacing the water that is lost during evaporation. This procedure will convert many insoluble salts, such as $BaSO_4$, $CaSO_4$, and $PbSO_4$, and many oxides into acid-soluble carbonates. Separate the mixture and discard the solution unless it is to be used for anion analysis. Wash the precipitate with water and then dissolve it in a few drops of 6 *M* HNO_3. Dilute the solution with 1 mL of water and add it to the original acid solution that is to be evaporated. If dissolution of the residue was not complete, repeat the Na_2CO_3 treatment. The halides of silver, silicate salts, some oxides, and calcined salts are not dissolved by this procedure.

PROCEDURE 4   *Reduction of Silver Halides with Zinc.* Suspend the residue containing the insoluble silver halides in 5 drops of water, add 1 mL of 1 *M* H_2SO_4 and a few granules of zinc metal. Warm and stir the mixture for several minutes. Add more zinc if the evolution of hydrogen ceases. Separate the residue of precipitated silver. Wash the residue with water, dissolve it in 6 *M* HNO_3, and test the solution for Ag^+. Conduct tests for the halide ions on the solution from the silver separation.

PROCEDURE 5   *Fusion with Sodium Carbonate.* Silicates and certain oxides and calcined salts may not be taken into solution by treatment with acid or Na_2CO_3 solution or by reduction with zinc. Fusion with Na_2CO_3 is effective with many of these substances. Transfer the residue remaining after the zinc reduction of silver halides to a small nickel crucible. Add to the residue 100 mg of anhydrous Na_2CO_3, about half as much K_2CO_3, and a few milligrams of $NaNO_3$. Place the crucible in a small clay triangle and heat it in the hot flame of a Meker burner until the mixture fuses. Cool the crucible, add 1 mL of water, and warm the mixture until the solid mass has disintegrated. Separate the mixture and reserve the solution. Treat the residue with 8 drops of 6 M HNO_3 and warm in the hot water bath for a few minutes. Separate and add this nitric acid solution to the original solution. Analyze the solution for its cations.

42.2 Dissolution of Metals and Alloys

All of the common metals except Al, Cr, Mn, Fe, Sb, and Sn may be taken into solution with dilute nitric acid. Al, Cr, Mn, and Fe react superficially with nitric acid, forming an oxide film, which makes dissolution of these metals too slow to be practical. Metastannic acid, H_2SnO_3, and the insoluble oxides of antimony, Sb_4O_6, Sb_2O_4, and Sb_2O_5, are formed when nitric acid reacts directly with tin and antimony, respectively. Although these compounds are insoluble in nitric acid, they may be readily converted to the corresponding sulfides, which are soluble in nitric acid. Hydrochloric acid is in general not suitable as a solvent for alloys of unknown composition since the volatile hydrides of sulfur, arsenic, phosphorus, and antimony may be formed; consequently, these elements could escape detection. On the other hand, nitric acid oxidizes the first three of these elements to sulfate, arsenate, and phosphate, respectively, and antimony to oxides, all of which are nonvolatile. Aqua regia will generally dissolve alloys that resist the action of nitric acid alone.

PROCEDURE 1   *Preparation of the Sample for Dissolution.* Convert the metal sample into a finely divided state with a large surface for interaction with the solvent. This may be accomplished with a steel file, a mortar and pestle for brittle metals, a hammer for malleable metals, or a knife for soft metals.

PROCEDURE 2   *Selection of a Suitable Solvent.* Treat a small sample of the metal with 6 M HNO_3 in a test tube and warm the mixture if necessary. If the sample reacts completely, with or without the formation of a white precipitate, carry out Procedure 3. If the sample does not dissolve readily, try aqua regia (1 part of 15 M HNO_3 and 3 parts of 12 M HCl). If aqua regia fails to dissolve the sample, it may be treated with a mixture of concentrated HCl and Br_2 or fused with solid NaOH in a silver crucible.

PROCEDURE 3   *Dissolution by HNO_3.* Place 20 mg of the sample in a test tube, add 1 mL of 6 M HNO_3, and heat the mixture in the hot water bath. If a white residue forms, stir the mixture to remove the coating from the surface of the undissolved metallic particles. Add more HNO_3 if necessary to complete the reaction. After the metal has dissolved, transfer the solution or mixture to a casserole and evaporate it nearly to dryness. Add 5 drops of 15 M HNO_3 and again evaporate the solution nearly to dryness. Now add 5 drops of 6 M HNO_3 and 1 mL of water and transfer the solution to a test tube. If a clear solution is obtained, analyze it according to the procedures for the cations. If there is a residue (H_2SnO_3, Sb_2O_5, or a mixture of these), separate and analyze the solution according to the procedures for the cations, omitting the tests for tin and antimony. Add 10 drops of 4 M NaOH and 5 drops of 5% thioacetamide solution to the residue. Heat the mixture in the water bath for 5 min. Separate any residue and analyze the solution for tin and antimony.

PROCEDURE 4   *Dissolution by Aqua Regia.* Place 20 mg of the sample in a test tube, and add 10 drops of 15 M HNO_3 and 30 drops of 12 M HCl. Heat the mixture in the hot water bath until the reaction is complete. Transfer the reaction mixture to a casserole and evaporate it to a small volume (not to dryness). When it is cool, add 2 mL of water and 4 drops of 6 M HCl. Separate any precipitate that forms. Analyze the centrifugate, or the clear solution if no precipitate forms, by the procedures for Groups II, III, IV, and V. The precipitate may consist of AgCl, $PbCl_2$, or SiO_2. Analyze the precipitate for Group I.

APPENDIXES

APPENDIX A CHEMICAL ARITHMETIC

In the study of general chemistry, elementary mathematics is frequently used. Of particular importance and wide application are exponential arithmetic, significant figures, and logarithms.

A.1 Exponential Arithmetic

We use the exponential method of expressing very large and very small numbers. These numbers are expressed as a product of two numbers. The first number of the product, the *digit term*, is usually a number not less than 1 and not greater than 10. The second number of the product, the *exponential term*, is written as 10 with an exponent. Some examples of the exponential method of expressing numbers are

$$1000 = 1 \times 10^3 \qquad 0.01 = 1 \times 10^{-2}$$
$$100 = 1 \times 10^2 \qquad 0.001 = 1 \times 10^{-3}$$
$$10 = 1 \times 10^1 \qquad 2386 = 2.386 \times 1000 = 2.386 \times 10^3$$
$$1 = 1 \times 10^0 \qquad 0.123 = 1.23 \times 0.1 = 1.23 \times 10^{-1}$$
$$0.1 = 1 \times 10^{-1}$$

The power (exponent) of 10 is equal to the number of places the decimal is shifted to give the digit number. The exponential method is a particularly useful shorthand notation for very large and very small numbers. For example, $1,230,000,000 = 1.23 \times 10^9$; and $0.00000000036 = 3.6 \times 10^{-10}$.

1. ADDITION OF EXPONENTIALS. Convert all numbers to the same power of 10 and add the digit terms of the numbers.

EXAMPLE A.1   Add 5.00×10^{-5} and 3.00×10^{-3}.

$$3.00 \times 10^{-3} = 300 \times 10^{-5}$$
$$(5.00 \times 10^{-5}) + (300 \times 10^{-5}) = 305 \times 10^{-5} = 3.05 \times 10^{-3}$$

2. SUBTRACTION OF EXPONENTIALS. Convert all numbers to the same power of 10 and take the difference of the digit terms.

EXAMPLE A.2   Subtract 4.0×10^{-7} from 5.0×10^{-6}.

$$4.0 \times 10^{-7} = 0.40 \times 10^{-6}$$
$$(5.0 \times 10^{-6}) - (0.40 \times 10^{-6}) = 4.6 \times 10^{-6}$$

3. MULTIPLICATION OF EXPONENTIALS. Multiply the digit terms in the usual way and add algebraically the exponents of the exponential terms.

EXAMPLE A.3   Multiply 4.2×10^{-8} by 2×10^3.

$$(4.2 \times 10^{-8}) \times (2 \times 10^3) = (4.2 \times 2) \times 10^{(-8)+(+3)} = 8.4 \times 10^{-5}$$

4. DIVISION OF EXPONENTIALS. Divide the digit term of the numerator by the digit term of the denominator and subtract algebraically the exponents of the exponential terms.

EXAMPLE A.4   Divide 3.6×10^{-5} by 6×10^{-4}.

$$\frac{3.6 \times 10^{-5}}{6 \times 10^{-4}} = \left(\frac{3.6}{6}\right) \times 10^{(-5)-(-4)} = 0.6 \times 10^{-1} = 6 \times 10^{-2}$$

5. SQUARING OF EXPONENTIALS. Square the digit term in the usual way and multiply the exponent of the exponential term by 2.

EXAMPLE A.5   Square the number 4.0×10^{-6}.

$$(4.0 \times 10^{-6})^2 = 4 \times 4 \times 10^{2 \times (-6)} = 16 \times 10^{-12} = 1.6 \times 10^{-11}$$

6. CUBING OF EXPONENTIALS. Cube the digit term in the usual way and multiply the exponent of the exponential term by 3.

EXAMPLE A.6   Cube the number 2×10^3.

$$(2 \times 10^4)^3 = 2 \times 2 \times 2 \times 10^{3 \times 4} = 8 \times 10^{12}$$

7. TAKING SQUARE ROOTS OF EXPONENTIALS. Decrease or increase the exponential term so that the power of 10 is evenly divisible by 2. Extract the square root of the digit term and divide the exponential term by 2.

EXAMPLE A.7   Find the square root of 1.6×10^{-7}.

$$1.6 \times 10^{-7} = 16 \times 10^{-8}$$
$$\sqrt{16 \times 10^{-8}} = \sqrt{16} \times \sqrt{10^{-8}} = 4.0 \times 10^{-4}$$

A.2 Significant Figures

A beekeeper reports that he has 525,341 bees. The last three figures of the number are obviously inaccurate, for during the time the keeper was counting the bees, some of them would have died and others would have hatched; this would have made the exact number of bees quite difficult to determine. It would have been more accurate if he had reported the number 525,000. In other words, the last three figures are not significant, except to set the position of the decimal point. Their exact values have no meaning.

In reporting any information as numbers, only as many significant figures should be used as are warranted by the accuracy of the measurement. The accuracy of measurements is dependent on the sensitivity of the measuring instruments used. For example, if the mass of an object has been reported as 2.13 grams, it is assumed that the last figure (3) has been estimated and that the mass lies between 2.125 grams and 2.135 grams. The quantity 2.13 grams represents three significant figures. The mass of this same object as determined by a more sensitive balance might be reported as 2.134 grams. In this case we would assume the correct mass to be between 2.1335 grams and 2.1345 grams, and the quantity 2.134 grams represents four significant figures. Note that the last figure is estimated and is also considered to be a significant figure.

A zero in a number may or may not be significant, depending on the manner in which it is used. When one or more zeros are used in locating a decimal point, they are not significant. For example, the numbers 0.063, 0.0063, and 0.00063 have two significant figures. When zeros appear between digits in a number they are significant. For example, 1.008 grams has four significant figures. Also, the zero in 12.50 is significant. However, the quantity 1370 centimeters has four significant figures only if the accuracy of the measurement includes the zero as a significant digit; if the digit 7 is estimated, then the number has only three significant figures.

The importance of significant figures lies in their application to fundamental computation. When adding or subtracting, the last digit that is retained in the sum or difference should correspond to the first doubtful decimal place (indicated by underscoring in the following example).

EXAMPLE A.8   Add 4.383 g and 0.0023 g.

$$4.38\underline{3} \text{ g}$$
$$0.002\underline{3}$$
$$4.38\underline{5} \text{ g}$$

When multiplying or dividing, the product or quotient should contain no more digits than the least number of significant figures in the numbers involved in the computation.

EXAMPLE A.9   Multiply 0.6238 by 6.6.

$$0.623\underline{8} \times 6.\underline{6} = 4.\underline{1}$$

In rounding off numbers, increase the last digit retained by 1 if it is followed by a number larger than 5 or by a 5 followed by other nonzero digits. Do not change the last digit retained if the following digit is less than or equal to 4. If the last digit retained is followed by 5, increase the last digit retained by 1 if it is odd or leave the last digit retained unchanged if it is even.

A.3 The Use of Logarithms and Exponential Numbers

The common logarithm of a number is the power to which 10 must be raised to equal that number. For example, the logarithm of 100 is 2 because 10 must be raised to the second power to equal 100. Additional examples follow.

Number	Number Expressed Exponentially	Logarithm
10,000	10^4	4
1000	10^3	3
10	10^1	1
1	10^0	0
0.1	10^{-1}	-1
0.01	10^{-2}	-2
0.001	10^{-3}	-3
0.0001	10^{-4}	-4

What is the logarithm of 60? Because 60 lies between 10 and 100, which have logarithms of 1 and 2, respectively, the logarithm of 60 must lie between 1 and 2. The logarithm of 60 is 1.7782; or

$$60 = 10^{1.7782}$$

Every logarithm is made up of two parts, called the *characteristic* and the *mantissa*. The characteristic is that part of the logarithm that lies to the left of the decimal point; thus the characteristic of the logarithm of 60 is 1. The mantissa is that part of the logarithm that lies to the right of the decimal point; thus the mantissa of the logarithm of 60 is .7782. The characteristic of the logarithm of a number greater than 1 is 1 less than the number of digits to the left of the decimal point in the number.

Number	Characteristic	Number	Characteristic
60	1	2.340	0
600	2	23.40	1
6000	3	234.0	2
52,840	4	2340.0	3

The mantissa of the logarithm of a number is found in a logarithm table (Appendix B), and its value is independent of the position of the decimal point. Thus, 2.340, 23.40, 234.0, and 2340.0 all have the same mantissa. The logarithm of 2.340 is 0.3692; that of 23.40 is 1.3692; that of 234.0 is 2.3692; and that of 2340.0 is 3.3692.

The meaning of the mantissa and the characteristic can be better understood from a consideration of their relationship to exponential numbers. For example, 2340 may be written 2.34×10^3. The logarithm of (2.34×10^3) = the logarithm of 2.34 + the logarithm of 10^3. The logarithm of 2.34 is 0.3692 and the logarithm of 10^3 is 3. Thus, the logarithm of 2340 = 3 + 0.3692, or 3.3692.

The logarithm of a number less than 1 has a negative value, and a convenient method of obtaining the logarithm of such a number may be used. For example, we may obtain the logarithm of 0.00234 as follows. When expressed exponentially, $0.00234 = 2.34 \times 10^{-3}$. The logarithm of 2.34×10^{-3} = the logarithm of 2.34 + the logarithm of 10^{-3}. The logarithm of 2.34 is 0.3692 and the logarithm of 10^{-3} is -3. Thus the logarithm of $0.00234 = 0.3692 + (-3) = 0.3692 - 3 = -2.6208$. The abbreviated form for the expression $(0.3692 - 3)$ is $\bar{3}.3692$. Note that only the characteristic has a negative value in the logarithm $\bar{3}.3692$, and that the mantissa is positive. The logarithm $\bar{3}.3692$ may also be written as $7.3692 - 10$.

To multiply two numbers we add the logarithms of the numbers. For example, suppose we multiply 412 by 353.

Logarithm of 412	= 2.6149
Logarithm of 353	= 2.5478
Logarithm of product	= 5.1627

The number that corresponds to the logarithm 5.1627 is 145,400 or 1.454×10^5. Thus 1.45×10^5 is the product of 412 and 353.

To divide two numbers we subtract the logarithms of the numbers. Suppose we divide 412 by 353.

Logarithm of 412	= 2.6149
Logarithm of 353	= 2.5478
Logarithm of quotient	= 0.0671

The number that corresponds to the logarithm 0.0671 is 1.17. Thus, 412 divided by 353 is 1.17.

Suppose we wish to multiply 5432 by 0.3124. Add the logarithm of 0.3124 to that of 5432.

$$\begin{aligned} \text{Logarithm of } 5432 &= 3.7350 \\ \text{Logarithm of } 0.3124 &= \overline{1}.4948 \\ \hline \text{Logarithm of the product} &= 3.2298 \end{aligned}$$

The number that corresponds to the logarithm 3.2298 is 1697 or 1.697×10^3.

Let us divide 5432 by 0.3124. Subtract the logarithm of 0.3124 from that of 5432.

$$\begin{aligned} \text{Logarithm of } 5432 &= 3.7350 \\ \text{Logarithm of } 0.3124 &= \overline{1}.4948 \\ \hline \text{Logarithm of the quotient} &= 4.2402 \end{aligned}$$

The number that corresponds to the logarithm 4.2402 is 17,390 or 1.739×10^4.

Finding roots of numbers by means of logarithms is a simple procedure. For example, suppose we take the cube root of 7235. The logarithm of $\sqrt[3]{7235}$ or $(7235)^{1/3}$ is equal to $\frac{1}{3}$ of the logarithm of 7235.

$$\text{Logarithm of } 7235 = 3.8594$$
$$\tfrac{1}{3} \times 3.8594 = 1.2865$$

The number that corresponds to the logarithm 1.2865 is 19.34. Thus 19.34 is the cube root of 7235.

A.4 The Solution of Quadratic Equations

Any quadratic equation can be expressed in the following form:

$$ax^2 + bx + c = 0$$

In order to solve a quadratic equation, the following formula is used.

$$x = \frac{-b \pm \sqrt{b^2 - 4ac}}{2a}$$

EXAMPLE A.10   Solve the quadratic equation $3x^2 + 13x - 10 = 0$.

Substituting the values $a = 3$, $b = 13$, and $c = -10$ in the formula, we obtain

$$x = \frac{-13 \pm \sqrt{(13)^2 - 4 \times 3 \times (-10)}}{2 \times 3}$$

$$x = \frac{-13 \pm \sqrt{169 + 120}}{6} = \frac{-13 \pm \sqrt{289}}{6} = \frac{-13 \pm 17}{6}$$

The two roots are therefore

$$x = \frac{-13 + 17}{6} = 0.67 \qquad \text{and} \qquad x = \frac{-13 - 17}{6} = -5$$

Equations constructed on physical data always have real roots, and of these real roots only those having positive values are usually of any significance.

EXERCISES

In all exercises observe the principle of significant figures.

1. Carry out the following calculations:
   (a) $(4.8 \times 10^{-2}) \times (1.92 \times 10^{-4})$
   (b) $(6.1 \times 10^{-3}) \times (5.81 \times 10^{5})$
   (c) $(4.8 \times 10^{-4}) \div (2.4 \times 10^{-8})$
   (d) $(9.6 \times 10^{-4}) \div (2.00 \times 10^{3})$

2. Perform the following operations:
   (a) $\sqrt{90 \times 10^{-3}}$ (d) $(4.1 \times 10^{-2})^2$
   (b) $\sqrt[3]{2.7 \times 10^{-2}}$ (e) $(64 \times 10^5)^{1/2}$
   (c) $(1 \times 10^6)^4$ (f) $(5.2 \times 10^{-2})^{1/2}$

3. Add the following numbers: 2.863, 42.580, 0.02316, and 41.33.

4. Multiply 7.850 by 3.2; divide 4.6 by 2.075.

5. Carry out the following calculation:

   $$(5 \times 1.008) + (2 \times 12.011) + (126.9045)$$

6. Find the logarithms of the following numbers:
   (a) 51.5   (e) 4.80×10^3
   (b) 0.0515   (f) 4.80×10^{-3}
   (c) 512   (g) 1000
   (d) 51.2   (h) 34,200

7. Write the characteristic and the mantissa of each of the following logarithms. Check the values given to verify that they are correct.
   (a) $\log 8 = 0.9031$
   (b) $\log 800 = 2.9031$
   (c) $\log 0.08 = \bar{2}.9031$

   (d) $\log 0.8 = \bar{1}.9031$
   (e) $\log (8 \times 10^{-2}) = -1.0969$

8. Make the following calculations using logarithms:
   (a) 35.4×8.07 (d) $\dfrac{47.2 \times 8.50}{12.1 \times 306}$

   (b) $79.2 \div 0.082$ (e) $\dfrac{1.50 \times 0.082 \times 293}{740/760}$

   (c) $\sqrt[5]{4.270}$ (f) $(9.54 \times 1.20 \times 10^2)^{1/2}$

9. Find the value of x in the following:

   (a) $x = \dfrac{-5 \pm \sqrt{5^2 - 4(0.2)(1.7)}}{2(2)}$

   (b) $x =$

   $$\dfrac{11 \pm \sqrt{(-11)^2 - 4(3.5 \times 10^2)(-8.0 \times 10^{-3})}}{2(3.5 \times 10^2)}$$

10. Write the quadratic equations whose solutions are shown in Exercise 9.

11. Solve for x if $k = 1.8 \times 10^{-5}$ and $c = 3.8 \times 10^{-3}$:

    $$x^2 + kx - kc = 0$$

12. Solve for x if $k = 5.0 \times 10^{-4}$ and $c = 2.7 \times 10^{-2}$:

    $$k = \dfrac{x^2}{c - x}$$

APPENDIX B

Four-Place Table of Logarithms

No.	0	1	2	3	4	5	6	7	8	9	1	2	3	4	5	6	7	8	9
10	0000	0043	0086	0128	0170	0212	0253	0294	0334	0374	4	8	12	17	21	25	29	33	37
11	0414	0453	0492	0531	0569	0607	0645	0682	0719	0755	4	8	11	15	19	23	26	30	34
12	0792	0828	0864	0899	0934	0969	1004	1038	1072	1106	3	7	10	14	17	21	24	28	31
13	1139	1173	1206	1239	1271	1303	1335	1367	1399	1430	3	6	10	13	16	19	23	26	29
14	1461	1492	1523	1553	1584	1614	1644	1673	1703	1732	3	6	9	12	15	18	21	24	27
15	1761	1790	1818	1847	1875	1903	1931	1959	1987	2014	3	6	8	11	14	17	20	22	25
16	2041	2068	2095	2122	2148	2175	2201	2227	2253	2279	3	5	8	11	13	16	18	21	24
17	2304	2330	2355	2380	2405	2430	2455	2480	2504	2529	2	5	7	10	12	15	17	20	22
18	2553	2577	2601	2625	2648	2672	2695	2718	2742	2765	2	5	7	9	12	14	16	19	21
19	2788	2810	2833	2856	2878	2900	2923	2945	2967	2989	2	4	7	9	11	13	16	18	20
20	3010	3032	3054	3075	3096	3118	3139	3160	3181	3201	2	4	6	8	11	13	15	17	19
21	3222	3243	3263	3284	3304	3324	3345	3365	3385	3404	2	4	6	8	10	12	14	16	18
22	3424	3444	3464	3483	3502	3522	3541	3560	3579	3598	2	4	6	8	10	12	14	15	17
23	3617	3636	3655	3674	3692	3711	3729	3747	3766	3784	2	4	6	7	9	11	13	15	17
24	3802	3820	3838	3856	3874	3892	3909	3927	3945	3962	2	4	5	7	9	11	12	14	16
25	3979	3997	4014	4031	4048	4065	4082	4099	4116	4133	2	3	5	7	9	10	12	14	15
26	4150	4166	4183	4200	4216	4232	4249	4265	4281	4298	2	3	5	7	8	10	11	13	15
27	4314	4330	4346	4362	4378	4393	4409	4425	4440	4456	2	3	5	6	8	9	11	13	14
28	4472	4487	4502	4518	4533	4548	4564	4579	4594	4609	2	3	5	6	8	9	11	12	14
29	4624	4639	4654	4669	4683	4698	4713	4728	4742	4757	1	3	4	6	7	9	10	12	13
30	4771	4786	4800	4814	4829	4843	4857	4871	4886	4900	1	3	4	6	7	9	10	11	13
31	4914	4928	4942	4955	4969	4983	4997	5011	5024	5038	1	3	4	6	7	8	10	11	12
32	5051	5065	5079	5092	5105	5119	5132	5145	5159	5172	1	3	4	5	7	8	9	11	12
33	5185	5198	5211	5224	5237	5250	5263	5276	5289	5302	1	3	4	5	6	8	9	10	12
34	5313	5328	5340	5353	5366	5378	5391	5403	5416	5428	1	3	4	5	6	8	9	10	11
35	5441	5453	5465	5478	5490	5502	5514	5527	5539	5551	1	2	4	5	6	7	9	10	11
36	5563	5575	5587	5599	5611	5623	5635	5647	5658	5670	1	2	4	5	6	7	8	10	11
37	5682	5694	5705	5717	5729	5740	5752	5763	5775	5786	1	2	3	5	6	7	8	9	10
38	5798	5809	5821	5832	5843	5855	5866	5877	5888	5899	1	2	3	5	6	7	8	9	10
39	5911	5922	5933	5944	5955	5966	5977	5988	5999	6010	1	2	3	4	5	7	8	9	10
40	6021	6031	6042	6053	6064	6075	6085	6096	6107	6117	1	2	3	4	5	6	8	9	10
41	6128	6138	6149	6160	6170	6180	6191	6201	6212	6222	1	2	3	4	5	6	7	8	9
42	6232	6243	6253	6263	6274	6284	6294	6304	6314	6325	1	2	3	4	5	6	7	8	9
43	6335	6345	6355	6365	6375	6385	6395	6405	6415	6425	1	2	3	4	5	6	7	8	9
44	6435	6444	6454	6464	6474	6484	6493	6503	6513	6522	1	2	3	4	5	6	7	8	9
45	6532	6542	6551	6561	6571	6580	6590	6599	6609	6618	1	2	3	4	5	6	7	8	9
46	6628	6637	6646	6656	6665	6675	6684	6693	6702	6712	1	2	3	4	5	6	7	7	8
47	6721	6730	6739	6749	6758	6767	6776	6785	6794	6803	1	2	3	4	5	5	6	7	8
48	6812	6821	6830	6839	6848	6857	6866	6875	6884	6893	1	2	3	4	4	5	6	7	8
49	6902	6911	6920	6928	6937	6946	6955	6964	6972	6981	1	2	3	4	4	5	6	7	8
50	6990	6998	7007	7016	7024	7033	7042	7050	7059	7067	1	2	3	3	4	5	6	7	8
51	7076	7084	7093	7101	7110	7118	7126	7135	7143	7152	1	2	3	3	4	5	6	7	8
52	7160	7168	7177	7185	7193	7202	7210	7218	7226	7235	1	2	2	3	4	5	6	7	7
53	7243	7251	7259	7267	7275	7284	7292	7300	7308	7316	1	2	2	3	4	5	6	6	7
54	7324	7332	7340	7348	7356	7364	7372	7380	7388	7396	1	2	2	3	4	5	6	6	7
	0	1	2	3	4	5	6	7	8	9	1	2	3	4	5	6	7	8	9

Four-Place Table of Logarithms (continued)

No.	0	1	2	3	4	5	6	7	8	9	1	2	3	4	5	6	7	8	9
55	7404	7412	7419	7427	7435	7443	7451	7459	7466	7474	1	2	2	3	4	5	5	6	7
56	7482	7490	7497	7505	7513	7520	7528	7536	7543	7551	1	2	2	3	4	5	5	6	7
57	7559	7566	7574	7582	7589	7597	7604	7612	7619	7627	1	2	2	3	4	5	5	6	7
58	7634	7642	7649	7657	7664	7672	7679	7686	7694	7701	1	1	2	3	4	4	5	6	7
59	7709	7716	7723	7731	7738	7745	7752	7760	7767	7774	1	1	2	3	4	4	5	6	7
60	7782	7789	7796	7803	7810	7818	7825	7832	7839	7846	1	1	2	3	4	4	5	6	6
61	7853	7860	7868	7875	7882	7889	7896	7903	7910	7917	1	1	2	3	4	4	5	6	6
62	7924	7931	7938	7945	7952	7959	7966	7973	7980	7987	1	1	2	3	3	4	5	6	6
63	7992	8000	8007	8014	8021	8028	8035	8041	8048	8055	1	1	2	3	3	4	5	5	6
64	8062	8069	8075	8082	8089	8096	8102	8109	8116	8122	1	1	2	3	3	4	5	5	6
65	8129	8136	8142	8149	8156	8162	8169	8176	8182	8189	1	1	2	3	3	4	5	5	6
66	8195	8202	8209	8215	8222	8228	8235	8241	8248	8254	1	1	2	3	3	4	5	5	6
67	8261	8267	8274	8280	8287	8293	8299	8306	8312	8319	1	1	2	3	3	4	5	5	6
68	8325	8331	8338	8344	8351	8357	8363	8370	8376	8382	1	1	2	3	3	4	4	5	6
69	8388	8395	8401	8407	8414	8420	8426	8432	8439	8445	1	1	2	2	3	4	4	5	6
70	8451	8457	8463	8470	8476	8482	8488	8494	8500	8506	1	1	2	2	3	4	4	5	6
71	8513	8519	8525	8531	8537	8543	8549	8555	8561	8567	1	1	2	2	3	4	4	5	5
72	8573	8579	8585	8591	8597	8603	8609	8615	8621	8627	1	1	2	2	3	4	4	5	5
73	8633	8639	8645	8651	8657	8663	8669	8675	8681	8686	1	1	2	2	3	4	4	5	5
74	8692	8698	8704	8710	8716	8722	8727	8733	8739	8745	1	1	2	2	3	4	4	5	5
75	8751	8756	8762	8768	8774	8779	8785	8791	8797	8802	1	1	2	2	3	3	4	5	5
76	8808	8814	8820	8825	8831	8837	8842	8848	8854	8859	1	1	2	2	3	3	4	5	5
77	8865	8871	8876	8882	8887	8893	8899	8904	8910	8915	1	1	2	2	3	3	4	4	5
78	8921	8927	8932	8938	8943	8949	8954	8960	8965	8971	1	1	2	2	3	3	4	4	5
79	8976	8982	8987	8993	8998	9004	9009	9015	9020	9025	1	1	2	2	3	3	4	4	5
80	9031	9036	9042	9047	9053	9058	9063	9069	9074	9079	1	1	2	2	3	3	4	4	5
81	9085	9090	9096	9101	9106	9112	9117	9122	9128	9133	1	1	2	2	3	3	4	4	5
82	9138	9143	9149	9154	9159	9165	9170	9175	9180	9186	1	1	2	2	3	3	4	4	5
83	9191	9196	9201	9206	9212	9217	9222	9227	9232	9238	1	1	2	2	3	3	4	4	5
84	9243	9248	9253	9258	9263	9269	9274	9279	9284	9289	1	1	2	2	3	3	4	4	5
85	9294	9299	9304	9309	9315	9320	9325	9330	9335	9340	1	1	2	2	3	3	4	4	5
86	9345	9350	9355	9360	9365	9370	9375	9380	9385	9390	1	1	2	2	3	3	4	4	5
87	9395	9400	9405	9410	9415	9420	9425	9430	9435	9440	0	1	1	2	2	3	3	4	4
88	9445	9450	9455	9460	9465	9469	9474	9479	9484	9489	0	1	1	2	2	3	3	4	4
89	9494	9499	9504	9509	9513	9518	9523	9528	9533	9538	0	1	1	2	2	3	3	4	4
90	9542	9547	9552	9557	9562	9566	9571	9576	9581	9586	0	1	1	2	2	3	3	4	4
91	9590	9595	9600	9605	9609	9614	9619	9624	9628	9633	0	1	1	2	2	3	3	4	4
92	9638	9643	9647	9652	9657	9661	9666	9671	9675	9680	0	1	1	2	2	3	3	4	4
93	9685	9689	9694	9699	9703	9708	9713	9717	9722	9727	0	1	1	2	2	3	3	4	4
94	9731	9736	9741	9745	9750	9754	9759	9763	9768	9773	0	1	1	2	2	3	3	4	4
95	9777	9782	9786	9791	9795	9800	9805	9809	9814	9818	0	1	1	2	2	3	3	4	4
96	9823	9827	9832	9836	9841	9845	9850	9854	9859	9863	0	1	1	2	2	3	3	4	4
97	9868	9872	9877	9881	9886	9890	9894	9899	9903	9908	0	1	1	2	2	3	3	4	4
98	9912	9917	9921	9926	9930	9934	9939	9943	9948	9952	0	1	1	2	2	3	3	4	4
99	9956	9961	9965	9969	9974	9978	9983	9987	9991	9996	0	1	1	2	2	3	3	3	4
	0	1	2	3	4	5	6	7	8	9	1	2	3	4	5	6	7	8	9

APPENDIX C

Units and Conversion Factors

Base Units of International System of Units (SI)

Physical Property	Name of Unit	Symbol
Length	Meter	m
Mass	Kilogram	kg
Time	Second	s
Electric current	Ampere	A
Thermodynamic temperature	Kelvin	K
Luminous intensity	Candela	cd
Quantity of substance	Mole	mol

Units of Length

Meter (m) = 39.37 inches (in) = 1.094 yards (yd)
Centimeter (cm) = 0.01 m
Millimeter (mm) = 0.001 m
Kilometer (km) = 1000 m
Angstrom unit (Å) = 10^{-8} cm = 10^{-10} m

Yard = 0.9144 m (exact)
Inch = 2.54 cm (exact)

Mile (U.S.) = 1.60934 km

Units of Volume

Liter (L) = 0.001 m^3 = 1000 cm^3
Milliliter (mL) = 0.001 L = 1 cm^3

Liquid quart (U.S.) = 0.9463 L
Dry quart = 1.1012 L
Cubic foot (U.S.) = 28.316 L

Units of Weight

Gram (g) = 0.001 kg
Milligram (mg) = 0.001 g
Kilogram (kg) = 1000 g
Ton (metric) = 1000 kg = 2204.62 lb

Ounce (oz) (avoirdupois) = 28.35 g
Pound (lb) (avoirdupois) = 0.45359237 kg
Ton (short) = 2000 lb = 907.185 kg
Ton (long) = 2240 lb = 1.016 metric ton

Units of Energy

4.184 joule (J) = 1 thermochemical calorie (cal) = 4.184×10^7 erg
Erg = 10^{-7} J
Electron-volt (eV) = 1.602189×10^{-12} erg = 23.061 kcal mol^{-1}
Liter atmosphere = 24.217 cal = 101.32 J

Unit of Force

Newton (N) = 1 kg m s^{-2} (force that when applied for 1 second will give to a 1-kilogram mass a speed of 1 meter per second)

Units of Pressure

Torr = 1 mmHg
Atmosphere (atm) = 760 mm Hg = 760 torr = 101,325 N m^{-2} = 101,325 Pa
Pascal (Pa) = kg m^{-1} s^{-2} = N m^{-2}

APPENDIX D

General Physical Constants

Avogadro's number	6.022045×10^{23} mol^{-1}
Electron charge, e	$1.6021892 \times 10^{-19}$ coulomb (C)
Electron rest mass, m_e	9.109534×10^{-31} kg
Proton rest mass, m_p	$1.6726485 \times 10^{-27}$ kg
Neutron rest mass, m_n	$1.6749543 \times 10^{-27}$ kg
Charge-to-mass ratio for electron, e/m_e	1.7588047×10^{11} coulomb kg^{-1}
Faraday constant, F	9.648670×10^{4} coulomb/equivalent
Planck constant, h	6.626176×10^{-34} J s
Boltzmann constant, k	1.380662×10^{-23} J K^{-1}
Gas constant, R	8.20562×10^{-2} L atm mol^{-1} K^{-1}
	$= 8.31441$ J mol^{-1} K^{-1}
Speed of light (in vacuum), c	2.99792458×10^{8} m s^{-1}
Atomic mass unit ($= \frac{1}{12}$ the mass of an atom of the ^{12}C nuclide), amu	$1.6605655 \times 10^{-27}$ kg
Rydberg constant, R_{∞}	$1.097373177 \times 10^{7}$ m^{-1}

APPENDIX E

Solubility Products

Substance	K_{sp} at 25°C	Substance	K_{sp} at 25°C
Aluminum		$CoS(\alpha)$	5.9×10^{-21}
$Al(OH)_3$	1.9×10^{-33}	$CoS(\beta)$	8.7×10^{-23}
Barium		$CoCO_3$	1.0×10^{-12}
$BaCO_3$	8.1×10^{-9}	$Co(OH)_3$	2.5×10^{-43}
$BaC_2O_4 \cdot 2H_2O$	1.1×10^{-7}	**Copper**	
$BaSO_4$	1.08×10^{-10}	$CuCl$	1.85×10^{-7}
$BaCrO_4$	2×10^{-10}	$CuBr$	5.3×10^{-9}
BaF_2	1.7×10^{-6}	CuI	5.1×10^{-12}
$Ba(OH)_2 \cdot 8H_2O$	5.0×10^{-3}	$CuSCN$	4×10^{-14}
$Ba_3(PO_4)_2$	1.3×10^{-29}	Cu_2S	1.6×10^{-48}
$Ba_3(AsO_4)_2$	1.1×10^{-13}	$Cu(OH)_2$	5.6×10^{-20}
Bismuth		CuS	8.7×10^{-36}
$BiO(OH)$	1×10^{-12}	$CuCO_3$	1.37×10^{-10}
$BiOCl$	7×10^{-9}	**Iron**	
Bi_2S_3	1.6×10^{-72}	$Fe(OH)_2$	7.9×10^{-15}
Cadmium		$FeCO_3$	2.11×10^{-11}
$Cd(OH)_2$	1.2×10^{-14}	FeS	1×10^{-19}
CdS	3.6×10^{-29}	$Fe(OH)_3$	1.1×10^{-36}
$CdCO_3$	2.5×10^{-14}	**Lead**	
Calcium		$Pb(OH)_2$	2.8×10^{-16}
$Ca(OH)_2$	7.9×10^{-6}	PbF_2	3.7×10^{-8}
$CaCO_3$	4.8×10^{-9}	$PbCl_2$	1.7×10^{-5}
$CaSO_4 \cdot 2H_2O$	2.4×10^{-5}	$PbBr_2$	6.3×10^{-6}
$CaC_2O_4 \cdot H_2O$	2.27×10^{-9}	PbI_2	8.7×10^{-9}
$Ca_3(PO_4)_2$	1×10^{-25}	$PbCO_3$	1.5×10^{-13}
$CaHPO_4$	5×10^{-6}	PbS	8.4×10^{-28}
CaF_2	3.9×10^{-11}	$PbCrO_4$	1.8×10^{-14}
Chromium		$PbSO_4$	1.8×10^{-8}
$Cr(OH)_3$	6.7×10^{-31}	$Pb_3(PO_4)_2$	3×10^{-44}
Cobalt		**Magnesium**	
$Co(OH)_2$	2×10^{-16}	$Mg(OH)_2$	1.5×10^{-11}

Solubility Products (continued)

Substance	K_{sp} at 25°C	Substance	K_{sp} at 25°C
$MgCO_3 \cdot 3H_2O$	$ca\ 1 \times 10^{-5}$	$AgCl$	1.8×10^{-10}
$MgNH_4PO_4$	2.5×10^{-13}	$AgBr$	3.3×10^{-13}
MgF_2	6.4×10^{-9}	AgI	1.5×10^{-16}
MgC_2O_4	8.6×10^{-5}	$AgCN$	1.2×10^{-16}
Manganese		$AgSCN$	1.0×10^{-12}
$Mn(OH)_2$	4.5×10^{-14}	Ag_2S	1.0×10^{-51}
$MnCO_3$	8.8×10^{-11}	Ag_2CO_3	8.2×10^{-12}
MnS	5.6×10^{-16}	Ag_2CrO_4	9×10^{-12}
Mercury		$Ag_4Fe(CN)_6$	1.55×10^{-41}
$Hg_2O \cdot H_2O$	1.6×10^{-23}	Ag_2SO_4	1.18×10^{-5}
Hg_2Cl_2	1.1×10^{-18}	Ag_3PO_4	1.8×10^{-18}
Hg_2Br_2	1.26×10^{-22}	Strontium	
Hg_2I_2	4.5×10^{-29}	$Sr(OH)_2 \cdot 8H_2O$	3.2×10^{-4}
Hg_2CO_3	9×10^{-17}	$SrCO_3$	9.42×10^{-10}
Hg_2SO_4	6.2×10^{-7}	$SrCrO_4$	3.6×10^{-5}
Hg_2S	1×10^{-45}	$SrSO_4$	2.8×10^{-7}
Hg_2CrO_4	2×10^{-9}	$SrC_2O_4 \cdot H_2O$	5.61×10^{-8}
HgS	3×10^{-53}	Thallium	
Nickel		$TlCl$	1.9×10^{-4}
$Ni(OH)_2$	1.6×10^{-14}	$TlSCN$	5.8×10^{-4}
$NiCO_3$	1.36×10^{-7}	Tl_2S	1.2×10^{-24}
$NiS(\alpha)$	3×10^{-21}	$Tl(OH)_3$	1.5×10^{-44}
$NiS(\beta)$	1×10^{-26}	Tin	
$NiS(\gamma)$	2×10^{-28}	$Sn(OH)_2$	5×10^{-26}
Potassium		SnS	8×10^{-29}
$KClO_4$	1.07×10^{-2}	$Sn(OH)_4$	$ca\ 1 \times 10^{-56}$
K_2PtCl_6	1.1×10^{-5}	Zinc	
$KHC_4H_4O_6$	3×10^{-4}	$ZnCO_3$	6×10^{-11}
Silver		$Zn(OH)_2$	4.5×10^{-17}
$\frac{1}{2}Ag_2O\ (Ag^+ + OH^-)$	2×10^{-8}	ZnS	1.1×10^{-21}

APPENDIX F

Formation Constants for Complex Ions

Equilibrium	K_f
$Al^{3+} + 6F^- \rightleftharpoons [AlF_6]^{3-}$	5×10^{23}
$Cd^{2+} + 4NH_3 \rightleftharpoons [Cd(NH_3)_4]^{2+}$	4.0×10^6
$Cd^{2+} + 4CN^- \rightleftharpoons [Cd(CN)_4]^{2-}$	1.3×10^{17}
$Co^{2+} + 6NH_3 \rightleftharpoons [Co(NH_3)_6]^{2+}$	8.3×10^4
$Co^{3+} + 6NH_3 \rightleftharpoons [Co(NH_3)_6]^{3+}$	4.5×10^{33}
$Cu^+ + 2CN^- \rightleftharpoons [Cu(CN)_2]^-$	1×10^{16}
$Cu^{2+} + 4NH_3 \rightleftharpoons [Cu(NH_3)_4]^{2+}$	1.2×10^{12}
$Fe^{2+} + 6CN^- \rightleftharpoons [Fe(CN)_6]^{4-}$	1×10^{37}
$Fe^{3+} + 6CN^- \rightleftharpoons [Fe(CN)_6]^{3-}$	1×10^{44}
$Fe^{3+} + 6SCN^- \rightleftharpoons [Fe(CNS)_6]^{3-}$	3.2×10^3
$Hg^{2+} + 4Cl^- \rightleftharpoons [HgCl_4]^{2-}$	1.2×10^{15}
$Ni^{2+} + 6NH_3 \rightleftharpoons [Ni(NH_3)_6]^{2+}$	1.8×10^8
$Ag^+ + 2Cl^- \rightleftharpoons [AgCl_2]^-$	2.5×10^5
$Ag^+ + 2CN^- \rightleftharpoons [Ag(CN)_2]^-$	1×10^{20}
$Ag^+ + 2NH_3 \rightleftharpoons [Ag(NH_3)_2]^+$	1.6×10^7
$Zn^{2+} + 4CN^- \rightleftharpoons [Zn(CN)_4]^{2-}$	1×10^{19}
$Zn^{2+} + 4OH^- \rightleftharpoons [Zn(OH)_4]^{2-}$	2.9×10^{15}

APPENDIX G

Ionization Constants of Weak Acids

Acid	Formula	K_a at 25°C
Acetic	CH_3CO_2H	1.8×10^{-5}
Arsenic	H_3AsO_4	4.8×10^{-3}
	$H_2AsO_4^-$	1×10^{-7}
	$HAsO_4^{2-}$	1×10^{-13}
Arsenous	H_3AsO_3	5.8×10^{-10}
Boric	H_3BO_3	5.8×10^{-10}
Carbonic	H_2CO_3	4.3×10^{-7}
	HCO_3^-	7×10^{-11}
Cyanic	$HCNO$	3.46×10^{-4}
Formic	HCO_2H	1.8×10^{-4}
Hydrazoic	HN_3	1×10^{-4}
Hydrocyanic	HCN	4×10^{-10}
Hydrofluoric	HF	7.2×10^{-4}
Hydrogen peroxide	H_2O_2	2.4×10^{-12}
Hydrogen selenide	H_2Se	1.7×10^{-4}
	HSe^-	1×10^{-10}
Hydrogen sulfate ion	HSO_4^-	1.2×10^{-2}
Hydrogen sulfide	H_2S	1.0×10^{-7}
	HS^-	1.3×10^{-13}
Hydrogen telluride	H_2Te	2.3×10^{-3}
	HTe^-	1×10^{-5}
Hypobromous	$HBrO$	2×10^{-9}
Hypochlorous	$HClO$	3.5×10^{-8}
Nitrous	HNO_2	4.5×10^{-4}
Oxalic	$H_2C_2O_4$	5.9×10^{-2}
	$HC_2O_4^-$	6.4×10^{-5}
Phosphoric	H_3PO_4	7.5×10^{-3}
	$H_2PO_4^-$	6.3×10^{-8}
	HPO_4^{2-}	3.6×10^{-13}
Phosphorous	H_3PO_3	1.6×10^{-2}
	$H_2PO_3^-$	7×10^{-7}
Sulfurous	H_2SO_3	1.2×10^{-2}
	HSO_3^-	6.2×10^{-8}

APPENDIX H

Ionization Constants of Weak Bases

Base	Ionization Equation	K_b at 25°C
Ammonia	$NH_3 + H_2O \rightleftharpoons NH_4^+ + OH^-$	1.8×10^{-5}
Dimethylamine	$(CH_3)_2NH + H_2O \rightleftharpoons (CH_3)_2NH_2^+ + OH^-$	7.4×10^{-4}
Methylamine	$CH_3NH_2 + H_2O \rightleftharpoons CH_3NH_3^+ + OH^-$	4.4×10^{-4}
Phenylamine (aniline)	$C_6H_5NH_2 + H_2O \rightleftharpoons C_6H_5NH_3^+ + OH^-$	4.6×10^{-10}
Trimethylamine	$(CH_3)_3N + H_2O \rightleftharpoons (CH_3)_3NH^+ + OH^-$	7.4×10^{-5}

APPENDIX I

Standard Electrode (Reduction) Potentials

Half-Reaction	$E°$, V	Half-Reaction	$E°$, V
$Li^+ + e^- \longrightarrow Li$	-3.09	$[Cd(NH_3)_4]^{2+} + 2e^- \longrightarrow Cd + 4NH_3$	-0.597
$K^+ + e^- \longrightarrow K$	-2.925	$Ga^{3+} + 3e^- \longrightarrow Ga$	-0.53
$Rb^+ + e^- \longrightarrow Rb$	-2.925	$S + 2e^- \longrightarrow S^{2-}$	-0.48
$Ra^{2+} + 2e^- \longrightarrow Ra$	-2.92	$[Ni(NH_3)_6]^{2+} + 2e^- \longrightarrow Ni + 6NH_3$	-0.47
$Ba^{2+} + 2e^- \longrightarrow Ba$	-2.90	$Fe^{2+} + 2e^- \longrightarrow Fe$	-0.440
$Sr^{2+} + 2e^- \longrightarrow Sr$	-2.89	$[Cu(CN)_2]^- + e^- \longrightarrow Cu + 2CN^-$	-0.43
$Ca^{2+} + 2e^- \longrightarrow Ca$	-2.87	$Cr^{3+} + e^- \longrightarrow Cr^{2+}$	-0.41
$Na^+ + e^- \longrightarrow Na$	-2.714	✳ $Cd^{2+} + 2e^- \longrightarrow Cd$	-0.403
$La^{3+} + 3e^- \longrightarrow La$	-2.52	$Se + 2H^+ + 2e^- \longrightarrow H_2Se$	-0.40
$Ce^{3+} + 3e^- \longrightarrow Ce$	-2.48	$[Hg(CN)_4]^{2-} + 2e^- \longrightarrow Hg + 4CN^-$	-0.37
$Nd^{3+} + 3e^- \longrightarrow Nd$	-2.44	$ClO_4^- + H_2O + 2e^- \longrightarrow ClO_3^- + 2OH^-$	-0.36
$Sm^{3+} + 3e^- \longrightarrow Sm$	-2.41	$PbSO_4 + 2e^- \longrightarrow Pb + SO_4^{2-}$	-0.356
$Gd^{3+} + 3e^- \longrightarrow Gd$	-2.40	$In^{3+} + 3e^- \longrightarrow In$	-0.342
$Mg^{2+} + 2e^- \longrightarrow Mg$	-2.37	$[Ag(CN)_2]^- + e^- \longrightarrow Ag + 2CN^-$	-0.31
$Y^{3+} + 3e^- \longrightarrow Y$	-2.37	$Co^{2+} + 2e^- \longrightarrow Co$	-0.277
$Am^{3+} + 3e^- \longrightarrow Am$	-2.32	$[SnF_6]^{2-} + 4e^- \longrightarrow Sn + 6F^-$	-0.25
$Lu^{3+} + 3e^- \longrightarrow Lu$	-2.25	$Ni^{2+} + 2e^- \longrightarrow Ni$	-0.250
$\frac{1}{2}H_2 + e^- \longrightarrow H^-$	-2.25	$Sn^{2+} + 2e^- \longrightarrow Sn$	-0.136
$Sc^{3+} + 3e^- \longrightarrow Sc$	-2.08	$CrO_4^{2-} + 4H_2O + 3e^- \longrightarrow Cr(OH)_3 + 5OH^-$	-0.13
$[AlF_6]^{3-} + 3e^- \longrightarrow Al + 6F^-$	-2.07	✗ $Pb^{2+} + 2e^- \longrightarrow Pb$	-0.126
$Pu^{3+} + 3e^- \longrightarrow Pu$	-2.07	$MnO_2 + 2H_2O + 2e^- \longrightarrow Mn(OH)_2 + 2OH^-$	-0.05
$Th^{4+} + 4e^- \longrightarrow Th$	-1.90	$[HgI_4]^{2-} + 2e^- \longrightarrow Hg + 4I^-$	-0.04
$Np^{3+} + 3e^- \longrightarrow Np$	-1.86	$2H^+ + 2e^- \longrightarrow H_2$	0.00
$Be^{2+} + 2e^- \longrightarrow Be$	-1.85	$NO_3^- + H_2O + 2e^- \longrightarrow NO_2^- + 2OH^-$	$+0.01$
$U^{3+} + 3e^- \longrightarrow U$	-1.80	$[Ag(S_2O_3)_2]^{3-} + e^- \longrightarrow Ag^+ + 2S_2O_3^{2-}$	$+0.01$
$Hf^{4+} + 4e^- \longrightarrow Hf$	-1.70	$[Co(NH_3)_6]^{3+} + e^- \longrightarrow [Co(NH_3)_6]^{2+}$	$+0.1$
$SiO_3^{2-} + 3H_2O + 4e^- \longrightarrow Si + 6OH^-$	-1.70	$S + 2H^+ + 2e^- \longrightarrow H_2S$	$+0.141$
$Al^{3+} + 3e^- \longrightarrow Al$	-1.66	$Sn^{4+} + 2e^- \longrightarrow Sn^{2+}$	$+0.15$
$Ti^{2+} + 2e^- \longrightarrow Ti$	-1.63	$Cu^{2+} + e^- \longrightarrow Cu^+$	$+0.153$
$Zr^{4+} + 4e^- \longrightarrow Zr$	-1.53	$Co(OH)_3 + e^- \longrightarrow Co(OH)_2 + OH^-$	$+0.17$
$ZnS + 2e^- \longrightarrow Zn + S^{2-}$	-1.44	$[HgBr_4]^{2-} + 2e^- \longrightarrow Hg + 4Br^-$	$+0.21$
$Cr(OH)_3 + 3e^- \longrightarrow Cr + 3OH^-$	-1.3	$AgCl + e^- \longrightarrow Ag + Cl^-$	$+0.222$
$[Zn(CN)_4]^{2-} + 2e^- \longrightarrow Zn + 4CN^-$	-1.26	$Hg_2Cl_2 + 2e^- \longrightarrow 2Hg + 2Cl^-$	$+0.27$
$Zn(OH)_2 + 2e^- \longrightarrow Zn + 2OH^-$	-1.245	$ClO_3^- + H_2O + 2e^- \longrightarrow ClO_2^- + 2OH^-$	$+0.33$
$[Zn(OH)_4]^{2-} + 2e^- \longrightarrow Zn + 4OH^-$	-1.216	✳ $Cu^{2+} + 2e^- \longrightarrow Cu$	$+0.337$
$CdS + 2e^- \longrightarrow Cd + S^{2-}$	-1.21	$[Fe(CN)_6]^{3-} + e^- \longrightarrow [Fe(CN)_6]^{4-}$	$+0.36$
$[Cr(OH)_4]^- + 3e^- \longrightarrow Cr + 4OH^-$	-1.2	$[Ag(NH_3)_2]^+ + e^- \longrightarrow Ag + 2NH_3$	$+0.373$
$[SiF_6]^{2-} + 4e^- \longrightarrow Si + 6F^-$	-1.2	$O_2 + 2H_2O + 4e^- \longrightarrow 4OH^-$	$+0.401$
$V^{2+} + 2e^- \longrightarrow V$	$ca\ -1.18$	$[RhCl_6]^{3-} + 3e^- \longrightarrow Rh + 6Cl^-$	$+0.44$
✗ $Mn^{2+} + 2e^- \longrightarrow Mn$	-1.18	$Ag_2CrO_4 + 2e^- \longrightarrow 2Ag + CrO_4^{2-}$	$+0.446$
$[Cd(CN)_4]^{2-} + 2e^- \longrightarrow Cd + 4CN^-$	-1.03	$NiO_2 + 2H_2O + 2e^- \longrightarrow Ni(OH)_2 + 2OH^-$	$+0.49$
$[Zn(NH_3)_4]^{2+} + 2e^- \longrightarrow Zn + 4NH_3$	-1.03	$Cu^+ + e^- \longrightarrow Cu$	$+0.521$
$FeS + 2e^- \longrightarrow Fe + S^{2-}$	-1.01	$TeO_2 + 4H^+ + 4e^- \longrightarrow Te + 2H_2O$	$+0.529$
$PbS + 2e^- \longrightarrow Pb + S^{2-}$	-0.95	✗ $I_2 + 2e^- \longrightarrow 2I^-$	$+0.5355$
$SnS + 2e^- \longrightarrow Sn + S^{2-}$	-0.94	$[PtBr_4]^{2-} + 2e^- \longrightarrow Pt + 4Br^-$	$+0.58$
$Cr^{2+} + 2e^- \longrightarrow Cr$	-0.91	$MnO_4^- + 2H_2O + 3e^- \longrightarrow MnO_2 + 4OH^-$	$+0.588$
$Fe(OH)_2 + 2e^- \longrightarrow Fe + 2OH^-$	-0.877	$[PdCl_4]^{2-} + 2e^- \longrightarrow Pd + 4Cl^-$	$+0.62$
$SiO_2 + 4H^+ + 4e^- \longrightarrow Si + 2H_2O$	-0.86	$ClO_2^- + H_2O + 2e^- \longrightarrow ClO^- + 2OH^-$	$+0.66$
$NiS + 2e^- \longrightarrow Ni + S^{2-}$	-0.83	$[PtCl_6]^{2-} + 2e^- \longrightarrow [PtCl_4]^{2-} + 2Cl^-$	$+0.68$
$2H_2O + 2e^- \longrightarrow H_2 + 2OH^-$	-0.828	$O_2 + 2H^+ + 2e^- \longrightarrow H_2O_2$	$+0.682$
$Zn^{2+} + 2e^- \longrightarrow Zn$	-0.763	$[PtCl_4]^{2-} + 2e^- \longrightarrow Pt + 4Cl^-$	$+0.73$
$Cr^{3+} + 3e^- \longrightarrow Cr$	-0.74	$Fe^{3+} + e^- \longrightarrow Fe^{2+}$	$+0.771$
$HgS + 2e^- \longrightarrow Hg + S^{2-}$	-0.72	$Hg_2^{2+} + 2e^- \longrightarrow 2Hg$	$+0.789$

Standard Electrode (Reduction) Potentials (continued)

Half-Reaction	$E°$, V	Half-Reaction	$E°$, V
$Ag^+ + e^- \longrightarrow Ag$	$+0.7991$	$Cr_2O_7^{2-} + 14H^+ + 6e^- \longrightarrow 2Cr^{3+} + 7H_2O$	$+1.33$
$Hg^{2+} + 2e^- \longrightarrow Hg$	$+0.854$	$Cl_2 + 2e^- \longrightarrow 2Cl^-$	$+1.3595$
$HO_2^- + H_2O + 2e^- \longrightarrow 3OH^-$	$+0.88$	$HClO + H^+ + 2e^- \longrightarrow Cl^- + H_2O$	$+1.49$
$ClO^- + H_2O + 2e^- \longrightarrow Cl^- + 2OH^-$	$+0.89$	$Au^{3+} + 3e^- \longrightarrow Au$	$+1.50$
$2Hg^{2+} + 2e^- \longrightarrow Hg_2^{2+}$	$+0.920$	$MnO_4^- + 8H^+ + 5e^- \longrightarrow Mn^{2+} + 4H_2O$	$+1.51$
$NO_3^- + 3H^+ + 2e^- \longrightarrow HNO_2 + H_2O$	$+0.94$	$Ce^{4+} + e^- \longrightarrow Ce^{3+}$	$+1.61$
$NO_3^- + 4H^+ + 3e^- \longrightarrow NO + H_2O$	$+0.96$	$HClO + H^+ + e^- \longrightarrow \frac{1}{2}Cl_2 + H_2O$	$+1.63$
$Pd^{2+} + 2e^- \longrightarrow Pd$	$+0.987$	$HClO_2 + 2H^+ + 2e^- \longrightarrow HClO + H_2O$	$+1.64$
$Br_2(l) + 2e^- \longrightarrow 2Br^-$	$+1.0652$	$Au^+ + e^- \longrightarrow Au$	$ca +1.68$
$ClO_4^- + 2H^+ + 2e^- \longrightarrow ClO_3^- + H_2O$	$+1.19$	$NiO_2 + 4H^+ + 2e^- \longrightarrow Ni^{2+} + 2H_2O$	$+1.68$
$Pt^{2+} + 2e^- \longrightarrow Pt$	$ca +1.2$	$PbO_2 + SO_4^{2-} + 4H^+ + 2e^- \longrightarrow PbSO_4 + 2H_2O$	$+1.685$
$ClO_3^- + 3H^+ + 2e^- \longrightarrow HClO_2 + H_2O$	$+1.21$	$H_2O_2 + 2H^+ + 2e^- \longrightarrow 2H_2O$	$+1.77$
$O_2 + 4H^+ + 4e^- \longrightarrow 2H_2O$	$+1.23$	$Co^{3+} + e^- \longrightarrow Co^{2+}$	$+1.82$
$MnO_2 + 4H^+ + 2e^- \longrightarrow Mn^{2+} + 2H_2O$	$+1.23$	$F_2 + 2e^- \longrightarrow 2F^-$	$+2.87$

APPENDIX J

Standard Molar Enthalpies of Formation, Standard Molar Free Energies of Formation, and Absolute Standard Entropies [298.15 K (25°C), 1 atm]

Substance	$\Delta H°_{f298.15}$, kJ mol^{-1}	$\Delta G°_{f298.15}$, kJ mol^{-1}	$S°_{298.15}$, J K^{-1} mol^{-1}
Aluminum			
Al(s)	0	0	28.3
Al(g)	326	286	164.4
Al$_2$O$_3$(s)	−1676	−1582	50.92
AlF$_3$(s)	−1504	−1425	66.44
AlCl$_3$(s)	−704.2	−628.9	110.7
AlCl$_3 \cdot$6H$_2$O(s)	−2692	—	—
Al$_2$S$_3$(s)	−724	−492.4	—
Al$_2$(SO$_4$)$_3$(s)	−3440.8	−3100.1	239
Antimony			
Sb(s)	0	0	45.69
Sb(g)	262	222	180.2
Sb$_4$O$_6$(s)	−1441	−1268	221
SbCl$_3$(g)	−314	−301	337.7
SbCl$_5$(g)	−394.3	−334.3	401.8
Sb$_2$S$_3$(s)	−175	−174	182
SbCl$_3$(s)	−382.2	−323.7	184
SbOCl(s)	−374	—	—
Arsenic			
As(s)	0	0	35
As(g)	303	261	174.1
As$_4$(g)	144	92.5	314
As$_4$O$_6$(s)	−1313.9	−1152.5	214
As$_2$O$_5$(s)	−924.87	−782.4	105
AsCl$_3$(g)	−258.6	−245.9	327.1
As$_2$S$_3$(s)	−169	−169	164
AsH$_3$(g)	66.44	68.91	222.7
H$_3$AsO$_4$(s)	−906.3	—	—

Standard Molar Enthalpies of Formation, Standard Molar Free Energies of Formation, and Absolute Standard Entropies [298.15 K (25°C), 1 atm] (continued)

Substance	$\Delta H^\circ_{f298.15}$, kJ mol^{-1}	$\Delta G^\circ_{f298.15}$, kJ mol^{-1}	$S^\circ_{298.15}$, J K^{-1} mol^{-1}
Barium			
Ba(s)	0	0	66.9
Ba(g)	175.6	144.8	170.3
BaO(s)	−558.1	−528.4	70.3
BaCl$_2$(s)	−860.06	−810.9	126
BaSO$_4$(s)	−1465	−1353	132
Beryllium			
Be(s)	0	0	9.54
Be(g)	320.6	282.8	136.17
BeO(s)	−610.9	−581.6	14.1
Bismuth			
Bi(s)	0	0	56.74
Bi(g)	207	168	186.90
Bi$_2$O$_3$(s)	−573.88	−493.7	151
BiCl$_3$(s)	−379	−315	177
Bi$_2$S$_3$(s)	−143	−141	200
Boron			
B(s)	0	0	5.86
B(g)	562.7	518.8	153.3
B$_2$O$_3$(s)	−1272.8	−1193.7	53.97
B$_2$H$_6$(g)	36	86.6	232.0
B(OH)$_3$(s)	−1094.3	−969.01	88.83
BF$_3$(g)	−1137.3	−1120.3	254.0
BCl$_3$(g)	−403.8	−388.7	290.0
B$_3$N$_3$H$_6$(l)	−541.0	−392.8	200
HBO$_2$(s)	−794.25	−723.4	40
Bromine			
Br$_2$(l)	0	0	152.23
Br$_2$(g)	30.91	3.142	245.35
Br(g)	111.88	82.429	174.91
BrF$_3$(g)	−255.6	−229.5	292.4
HBr(g)	−36.4	−53.43	198.59
Cadmium			
Cd(s)	0	0	51.76
Cd(g)	112.0	77.45	167.64
CdO(s)	−258	−228	54.8
CdCl$_2$(s)	−391.5	−344.0	115.3
CdSO$_4$(s)	−933.28	−822.78	123.04
CdS(s)	−162	−156	64.9
Calcium			
Ca(s)	0	0	41.6
Ca(g)	192.6	158.9	154.78
CaO(s)	−635.5	−604.2	40
Ca(OH)$_2$(s)	−986.59	−896.76	76.1
CaSO$_4$(s)	−1432.7	−1320.3	107
CaSO$_4 \cdot$ 2H$_2$O(s)	−2021.1	−1795.7	194.0
CaCO$_3$(s) (calcite)	−1206.9	−1128.8	92.9
CaSO$_3 \cdot$ 2H$_2$O(s)	−1762	−1565	184
Carbon			
C(s) (graphite)	0	0	5.740
C(s) (diamond)	1.897	2.900	2.38
C(g)	716.681	671.289	157.987
CO(g)	−110.52	−137.15	197.56
CO$_2$(g)	−393.51	−394.36	213.6

Standard Molar Enthalpies of Formation, Standard Molar Free Energies of Formation, and Absolute Standard Entropies [298.15 K (25°C), 1 atm] (continued)

Substance	$\Delta H^\circ_{f_{298.15}}$, kJ mol^{-1}	$\Delta G^\circ_{f_{298.15}}$, kJ mol^{-1}	$S^\circ_{298.15}$, J K^{-1} mol^{-1}
$CH_4(g)$	−74.81	−50.75	186.15
$CH_3OH(l)$	−238.7	−166.4	127
$CH_3OH(g)$	−200.7	−162.0	239.7
$CCl_4(l)$	−135.4	−65.27	216.4
$CCl_4(g)$	−102.9	−60.63	309.7
$CHCl_3(l)$	−134.5	−73.72	202
$CHCl_3(g)$	−103.1	−70.37	295.6
$CS_2(l)$	89.70	65.27	151.3
$CS_2(g)$	117.4	67.15	237.7
$C_2H_2(g)$	226.7	209.2	200.8
$C_2H_4(g)$	52.26	68.12	219.5
$C_2H_6(g)$	−84.68	−32.9	229.5
$CH_3COOH(l)$	−484.5	−390	160
$CH_3COOH(g)$	−432.25	−374	282
$C_2H_5OH(l)$	−277.7	−174.9	161
$C_2H_5OH(g)$	−235.1	−168.6	282.6
$C_3H_8(g)$	−103.85	−23.49	269.9
$C_6H_6(g)$	82.927	129.66	269.2
$C_6H_6(l)$	49.028	124.50	172.8
$CH_2Cl_2(l)$	−121.5	−67.32	178
$CH_2Cl_2(g)$	−92.47	−65.90	270.1
$CH_3Cl(g)$	−80.83	−57.40	234.5
$C_2H_5Cl(l)$	−136.5	−59.41	190.8
$C_2H_5Cl(g)$	−112.2	−60.46	275.9
$C_2N_2(g)$	308.9	297.4	241.8
$HCN(l)$	108.9	124.9	112.8
$HCN(g)$	135	124.7	201.7
Chlorine			
$Cl_2(g)$	0	0	222.96
$Cl(g)$	121.68	105.70	165.09
$ClF(g)$	−54.48	−55.94	217.8
$ClF_3(g)$	−163	−123	281.5
$Cl_2O(g)$	80.3	97.9	266.1
$Cl_2O_7(l)$	238	—	—
$Cl_2O_7(g)$	272	—	—
$HCl(g)$	−92.307	−95.299	186.80
$HClO_4(l)$	−40.6	—	—
Chromium			
$Cr(s)$	0	0	23.8
$Cr(g)$	397	352	174.4
$Cr_2O_3(s)$	−1140	−1058	81.2
$CrO_3(s)$	−589.5	—	—
$(NH_4)_2Cr_2O_7(s)$	−1807	—	—
Cobalt			
$Co(s)$	0	0	30.0
$CoO(s)$	−237.9	−214.2	52.97
$Co_3O_4(s)$	−891.2	−774.0	103
$Co(NO_3)_2(s)$	−420.5	—	—
Copper			
$Cu(s)$	0	0	33.15
$Cu(g)$	338.3	298.5	166.3
$CuO(s)$	−157	−130	42.63
$Cu_2O(s)$	−169	−146	93.14
$CuS(s)$	−53.1	−53.6	66.5

Standard Molar Enthalpies of Formation, Standard Molar Free Energies of Formation, and Absolute Standard Entropies [298.15 K (25°C), 1 atm] (continued)

Substance	$\Delta H^\circ_{f298.15}$, kJ mol^{-1}	$\Delta G^\circ_{f298.15}$, kJ mol^{-1}	$S^\circ_{298.15}$, J K^{-1} mol^{-1}
$Cu_2S(s)$	−79.5	−86.2	121
$CuSO_4(s)$	−771.36	−661.9	109
$Cu(NO_3)_2(s)$	−303	—	—
Fluorine			
$F_2(g)$	0	0	202.7
$F(g)$	78.99	61.92	158.64
$F_2O(g)$	−22	−4.6	247.3
$HF(g)$	−271	−273	173.67
Hydrogen			
$H_2(g)$	0	0	130.57
$H(g)$	217.97	203.26	114.60
$H_2O(l)$	−285.83	−237.18	69.91
$H_2O(g)$	−241.82	−228.59	188.71
$H_2O_2(l)$	−187.8	−120.4	110
$H_2O_2(g)$	−136.3	−105.6	233
$HF(g)$	−271	−273	173.67
$HCl(g)$	−92.307	−95.299	186.80
$HBr(g)$	−36.4	−53.43	198.59
$HI(g)$	26.5	1.7	206.48
$H_2S(g)$	−20.6	−33.6	205.7
$H_2Se(g)$	30	16	218.9
Iodine			
$I_2(s)$	0	0	116.14
$I_2(g)$	62.438	19.36	260.6
$I(g)$	106.84	70.283	180.68
$IF(g)$	95.65	−118.5	236.1
$ICl(g)$	17.8	−5.44	247.44
$IBr(g)$	40.8	3.7	258.66
$IF_7(g)$	−943.9	−818.4	346
$HI(g)$	26.5	1.7	206.48
Iron			
$Fe(s)$	0	0	27.3
$Fe(g)$	416	371	180.38
$Fe_2O_3(s)$	−824.2	−742.2	87.40
$Fe_3O_4(s)$	−1118	−1015	146
$Fe(CO)_5(l)$	−774.0	−705.4	338
$Fe(CO)_5(g)$	−733.9	−697.26	445.2
$FeSeO_3(s)$	−1200	—	—
$FeO(s)$	−272	—	—
$FeAsS(s)$	−42	−50	120
$Fe(OH)_2(s)$	−569.0	−486.6	88
$Fe(OH)_3(s)$	−823.0	−696.6	107
$FeS(s)$	−100	−100	60.29
$Fe_3C(s)$	25	20	105
Lead			
$Pb(s)$	0	0	64.81
$Pb(g)$	195	162	175.26
$PbO(s)$ (yellow)	−217.3	−187.9	68.70
$PbO(s)$ (red)	−219.0	−188.9	66.5
$Pb(OH)_2(s)$	−515.9	—	—
$PbS(s)$	−100	−98.7	91.2
$Pb(NO_3)_2(s)$	−451.9	—	—
$PbO_2(s)$	−277	−217.4	68.6
$PbCl_2(s)$	−359.4	−314.1	136

Standard Molar Enthalpies of Formation, Standard Molar Free Energies of Formation, and Absolute Standard Entropies [298.15 K (25°C), 1 atm] (continued)

Substance	$\Delta H^\circ_{f298.15}$, kJ mol^{-1}	$\Delta G^\circ_{f298.15}$, kJ mol^{-1}	$S^\circ_{298.15}$, J K^{-1} mol^{-1}
Lithium			
Li(s)	0	0	28.0
Li(g)	155.1	122.1	138.67
LiH(s)	−90.42	−69.96	25
Li(OH)(s)	−487.23	−443.9	50.2
LiF(s)	−612.1	−584.1	35.9
Li$_2$CO$_3$(s)	−1215.6	−1132.4	90.4
Manganese			
Mn(s)	0	0	32.0
Mn(g)	281	238	173.6
MnO(s)	−385.2	−362.9	59.71
MnO$_2$(s)	−520.03	−465.18	53.05
Mn$_2$O$_3$(s)	−959.0	−881.2	110
Mn$_3$O$_4$(s)	−1388	−1283	156
Mercury			
Hg(l)	0	0	76.02
Hg(g)	61.317	31.85	174.8
HgO(s) (red)	−90.83	−58.555	70.29
HgO(s) (yellow)	−90.46	−57.296	71.1
HgCl$_2$(s)	−224	−179	146
Hg$_2$Cl$_2$(s)	−265.2	−210.78	192
HgS(s) (red)	−58.16	−50.6	82.4
HgS(s) (black)	−53.6	−47.7	88.3
HgSO$_4$(s)	−707.5	—	—
Nitrogen			
N$_2$(g)	0	0	191.5
N(g)	472.704	455.579	153.19
NO(g)	90.25	86.57	210.65
NO$_2$(g)	33.2	51.30	239.9
N$_2$O(g)	82.05	104.2	219.7
N$_2$O$_3$(g)	83.72	139.4	312.2
N$_2$O$_4$(g)	9.16	97.82	304.2
N$_2$O$_5$(g)	11	115	356
NH$_3$(g)	−46.11	−16.5	192.3
N$_2$H$_4$(l)	50.63	149.2	121.2
N$_2$H$_4$(g)	95.4	159.3	238.4
NH$_4$NO$_3$(s)	−365.6	−184.0	151.1
NH$_4$Cl(s)	−314.4	−201.5	94.6
NH$_4$Br(s)	−270.8	−175	113
NH$_4$I(s)	−201.4	−113	117
NH$_4$NO$_2$(s)	−256	—	—
HNO$_3$(l)	−174.1	−80.79	155.6
HNO$_3$(g)	−135.1	−74.77	266.2
Oxygen			
O$_2$(g)	0	0	205.03
O(g)	249.17	231.75	160.95
O$_3$(g)	143	163	238.8
Phosphorus			
P(s)	0	0	41.1
P(g)	58.91	24.5	280.0
P$_4$(g)	314.6	278.3	163.08
PH$_3$(g)	5.4	13	210.1
PCl$_3$(g)	−287	−268	311.7
PCl$_5$(g)	−375	−305	364.5

Standard Molar Enthalpies of Formation, Standard Molar Free Energies of Formation, and Absolute Standard Entropies [298.15 K (25°C), 1 atm] (continued)

Substance	$\Delta H^\circ_{f298.15}$, kJ mol^{-1}	$\Delta G^\circ_{f298.15}$, kJ mol^{-1}	$S^\circ_{298.15}$, J K^{-1} mol^{-1}
$P_4O_6(s)$	-1640	—	—
$P_4O_{10}(s)$	-2984	-2698	228.9
$HPO_3(s)$	-948.5	—	—
$H_3PO_2(s)$	-604.6	—	—
$H_3PO_3(s)$	-964.4	—	—
$H_3PO_4(s)$	-1279	-1119	110.5
$H_3PO_4(l)$	-1267	—	—
$H_4P_2O_7(s)$	-2241	—	—
$POCl_3(l)$	-597.1	-520.9	222.5
$POCl_3(g)$	-558.48	-512.96	325.3
Potassium			
$K(s)$	0	0	63.6
$K(g)$	90.00	61.17	160.23
$KF(s)$	-562.58	-533.12	66.57
$KCl(s)$	-435.868	-408.32	82.68
Silicon			
$Si(s)$	0	0	18.8
$Si(g)$	455.6	411	167.9
$SiO_2(s)$	-910.94	-856.67	41.84
$SiH_4(g)$	34	56.9	204.5
$H_2SiO_3(s)$	-1189	-1092	130
$H_4SiO_4(s)$	-1481	-1333	190
$SiF_4(g)$	-1614.9	-1572.7	282.4
$SiCl_4(l)$	-687.0	-619.90	240
$SiCl_4(g)$	-657.01	-617.01	330.6
$SiC(s)$	-65.3	-62.8	16.6
Silver			
$Ag(s)$	0	0	42.55
$Ag(g)$	284.6	245.7	172.89
$Ag_2O(s)$	-31.0	-11.2	121
$AgCl(s)$	-127.1	-109.8	96.2
$Ag_2S(s)$	-32.6	-40.7	144.0
Sodium			
$Na(s)$	0	0	51.0
$Na(g)$	108.7	78.11	153.62
$Na_2O(s)$	-415.9	-377	72.8
$NaCl(s)$	-411.00	-384.03	72.38
Sulfur			
$S(s)$ (rhombic)	0	0	31.8
$S(g)$	278.80	238.27	167.75
$SO_2(g)$	-296.83	-300.19	248.1
$SO_3(g)$	-395.7	-371.1	256.6
$H_2S(g)$	-20.6	-33.6	205.7
$H_2SO_4(l)$	-813.989	690.101	156.90
$H_2S_2O_7(s)$	-1274	—	—
$SF_4(g)$	-774.9	-731.4	291.9
$SF_6(g)$	-1210	-1105	291.7
$SCl_2(l)$	-50	—	—
$SCl_2(g)$	-20	—	—
$S_2Cl_2(l)$	-59.4	—	—
$S_2Cl_2(g)$	-18	-32	331.4
$SOCl_2(l)$	-246	—	—
$SOCl_2(g)$	-213	-198	309.7

Standard Molar Enthalpies of Formation, Standard Molar Free Energies of Formation, and Absolute Standard Entropies [298.15 K (25°C), 1 atm] (continued)

Substance	$\Delta H^{\circ}_{f298.15}$, kJ mol^{-1}	$\Delta G^{\circ}_{f298.15}$, kJ mol^{-1}	$S^{\circ}_{298.15}$, J K^{-1} mol^{-1}
SO$_2$Cl$_2$(l)	−394	—	—
SO$_2$Cl$_2$(g)	−364	−320	311.8
Tin			
Sn(s)	0	0	51.55
Sn(g)	302	267	168.38
SnO(s)	−286	−257	56.5
SnO$_2$(s)	−580.7	−519.7	52.3
SnCl$_4$(l)	−511.2	−440.2	259
SnCl$_4$(g)	−471.5	−432.2	366
Titanium			
Ti(s)	0	0	30.6
Ti(g)	469.9	425.1	180.19
TiO$_2$(s)	−944.7	−889.5	50.33
TiCl$_4$(l)	−804.2	−737.2	252.3
TiCl$_4$(g)	−763.2	−726.8	354.8
Tungsten			
W(s)	0	0	32.6
W(g)	849.4	807.1	173.84
WO$_3$(s)	−842.87	−764.08	75.90
Zinc			
Zn(s)	0	0	41.6
Zn(g)	130.73	95.178	160.87
ZnO(s)	−348.3	−318.3	43.64
ZnCl$_2$(s)	−415.1	−369.43	111.5
ZnS(s)	−206.0	−201.3	57.7
ZnSO$_4$(s)	−982.8	−874.5	120
ZnCO$_3$(s)	−812.78	−731.57	82.4

Complexes

Substance	$\Delta H^{\circ}_{f298.15}$, kJ mol^{-1}	$\Delta G^{\circ}_{f298.15}$, kJ mol^{-1}	$S^{\circ}_{298.15}$, J K^{-1} mol^{-1}
[Co(NH$_3$)$_4$(NO$_2$)$_2$]NO$_3$, *cis*	−898.7	—	—
[Co(NH$_3$)$_4$(NO$_2$)$_2$]NO$_3$, *trans*	−896.2	—	—
NH$_4$[Co(NH$_3$)$_2$(NO$_2$)$_4$]	−837.6	—	—
[Co(NH$_3$)$_6$][Co(NH$_3$)$_2$(NO$_2$)$_4$]$_3$	−2733	—	—
[Co(NH$_3$)$_4$Cl$_2$]Cl, *cis*	−997.0	—	—
[Co(NH$_3$)$_4$Cl$_2$]Cl, *trans*	−999.6	—	—
[Co(en)$_2$(NO$_2$)$_2$]NO$_3$, *cis*	−689.5	—	—
[Co(en)$_2$Cl$_2$]Cl, *cis*	−681.1	—	—
[Co(en)$_2$Cl$_2$]Cl, *trans*	−677.4	—	—
[Co(en)$_3$](ClO$_4$)$_3$	−762.7	—	—
[Co(en)$_3$]Br$_2$	−595.8	—	—
[Co(en)$_3$]I$_2$	−475.3	—	—
[Co(en)$_3$]I$_3$	−519.2	—	—
[Co(NH$_3$)$_6$](ClO$_4$)$_3$	−1035	−227	636
[Co(NH$_3$)$_5$NO$_2$](NO$_3$)$_2$	−1089	−418.4	350
[Co(NH$_3$)$_6$](NO$_3$)$_3$	−1282	−530.5	469
[Co(NH$_3$)$_5$Cl]Cl$_2$	−1017	−582.8	366
[Pt(NH$_3$)$_4$]Cl$_2$	−728.0	—	—
[Ni(NH$_3$)$_6$]Cl$_2$	−994.1	—	—
[Ni(NH$_3$)$_6$]Br$_2$	−923.8	—	—
[Ni(NH$_3$)$_6$]I$_2$	−808.3	—	—

APPENDIX K

Composition of Commercial Acids and Bases

Acid or Base	Specific Gravity	Percentage by Mass	Molarity	Normality
Hydrochloric acid	1.19	38%	12.4	12.4
Nitric acid	1.42	70%	15.8	15.8
Sulfuric acid	1.84	95%	17.8	35.6
Acetic acid	1.05	99%	17.3	17.3
Aqueous ammonia	0.90	28%	14.8	14.8

APPENDIX L

Half-Life Times for Several Radioactive Isotopes

(Symbol in parentheses indicates type of emission; $E.C.$ = K-electron capture, $S.F.$ = spontaneous fission; y = years, d = days, h = hours, m = minutes, s = seconds.)

$^{14}_{6}C$	5770 y	(β^-)	$^{226}_{88}Ra$	1590 y	(α)
$^{13}_{7}N$	10.0 m	(β^+)	$^{228}_{88}Ra$	6.7 y	(β^-)
$^{24}_{11}Na$	15.0 h	(β^-)	$^{228}_{89}Ac$	6.13 h	(β^-)
$^{32}_{15}P$	14.3 d	(β^-)	$^{228}_{90}Th$	1.90 y	(α)
$^{40}_{19}K$	1.3×10^9 y	$(\beta^-$ or $E.C.)$	$^{232}_{90}Th$	1.39×10^{10} y	$(\alpha, \beta^-,$ or $S.F.)$
$^{60}_{27}Co$	5.2 y	(β^-)	$^{233}_{90}Th$	23 m	(β^-)
$^{87}_{37}Rb$	4.7×10^{10} y	(β^-)	$^{234}_{90}Th$	24.1 d	(β^-)
$^{90}_{38}Sr$	28 y	(β^-)	$^{223}_{91}Pa$	27 d	(β^-)
$^{115}_{49}In$	6×10^{14} y	(β^-)	$^{233}_{92}U$	1.62×10^5 y	(α)
$^{131}_{53}I$	8.05 d	(β^-)	$^{234}_{92}U$	2.4×10^5 y	$(\alpha$ or $S.F.)$
$^{142}_{58}Ce$	5×10^{15} y	(α)	$^{235}_{92}U$	7.3×10^8 y	$(\alpha$ or $S.F.)$
$^{198}_{79}Au$	64.8 h	(β^-)	$^{238}_{92}U$	4.5×10^9 y	$(\alpha$ or $S.F.)$
$^{208}_{81}Tl$	3.1 m	(β^-)	$^{239}_{92}U$	23 m	(β^-)
$^{210}_{82}Pb$	21 y	(β^-)	$^{239}_{93}Np$	2.3 d	(β^-)
$^{212}_{82}Pb$	10.6 h	(β^-)	$^{239}_{94}Pu$	24,360 y	$(\alpha$ or $S.F.)$
$^{214}_{82}Pb$	26.8 m	(β^-)	$^{240}_{94}Pu$	6.58×10^3 y	$(\alpha$ or $S.F.)$
$^{206}_{83}Bi$	6.3 d	$(\beta^+$ or $E.C.)$	$^{241}_{94}Pu$	13 y	$(\alpha$ or $\beta^-)$
$^{210}_{83}Bi$	5.0 d	(β^-)	$^{241}_{95}Am$	458 y	(α)
$^{212}_{83}Bi$	60.5 m	$(\alpha$ or $\beta^-)$	$^{242}_{96}Cm$	163 d	$(\alpha$ or $S.F.)$
$^{207}_{84}Po$	5.7 h	$(\alpha, \beta^+,$ or $E.C.)$	$^{243}_{97}Bk$	4.5 h	$(\alpha$ or $E.C.)$
$^{210}_{84}Po$	138.4 d	(α)	$^{245}_{98}Cf$	350 d	$(\alpha$ or $E.C.)$
$^{212}_{84}Po$	3×10^{-7} s	(α)	$^{253}_{99}Es$	20.0 d	$(\alpha$ or $S.F.)$
$^{216}_{84}Po$	0.16 s	(α)	$^{254}_{100}Fm$	3.24 h	$(S.F.)$
$^{218}_{84}Po$	3.0 m	$(\alpha$ or $\beta^-)$	$^{255}_{100}Fm$	22 h	(α)
$^{215}_{85}At$	10^{-4} s	(α)	$^{256}_{101}Md$	1.5 h	$(E.C.)$
$^{218}_{85}At$	1.3 s	(α)	$^{254}_{102}No$	3 s	(α)
$^{220}_{86}Rn$	54.5 s	(α)	$^{257}_{103}Lr$	8 s	(α)
$^{222}_{86}Rn$	3.82 d	(α)	$^{263}_{106}(106)$	0.9 s	(α)
$^{224}_{88}Ra$	3.64 d	(α)			

APPENDIX M

Apparatus for Qualitative Analysis (one student)

50 Reagent bottles, dropper type, 10-mL
1 Rack for 50 reagent bottles
1 Test tube block
1 Flask, Florence, 250-mL
2 Beakers, 250-mL
2 Beakers, 50-mL
2 Beakers, 20-mL
1 Graduate, 10-mL
1 Casserole, 15-mL
6 Centrifuge tubes
6 Test tubes, 65 × 10 mm
6 Test tubes, 75 × 10 mm
1 Metal rack for a water bath

1 Micro burner
1 Bunsen burner
2 Watch glasses, 2.5-cm
1 File
1 Box of labels
1 Bottle of litmus paper, red
1 Bottle of litmus paper, blue
1 Test tube holder, small
1 Test tube brush, small, tapered
1 Test tube brush, small, not tapered
1 Wire, platinum, 2-in
1 Wire gauze
1 Ring stand (with ring) small

1 Spatula, micro (Monel metal)
1 Forceps
1 Two-hole rubber stopper to fit 250-mL
   flask
1 Wing top
1 Box of matches
1 Towel
1 Box of detergent, small
6 Medicine droppers, 1-mL
6 Capillary syringes
100-cm Glass tubing, 6-mm
100-cm Glass rod, 3-mm
1 Cobalt glass

APPENDIX N

Reagents for Cation Analysis

Group I Reagents

Hydrochloric acid, HCl, 6 M
Nitric acid, HNO$_3$, 4 M
Aqueous ammonia, NH$_3$ + H$_2$O, 4 M
*Potassium chromate, K$_2$CrO$_4$, 1 M

Group II Reagents (in addition to those listed for Group I)

Aqueous ammonia, NH$_3$ + H$_2$O, 6 M
Hydrochloric acid, HCl, 1.0 M
Thioacetamide, CH$_3$CSNH$_2$, 5% solution
Sodium hydroxide, NaOH, 4 M
Ammonium nitrate, NH$_4$NO$_3$, 1 M
Sulfuric acid, H$_2$SO$_4$, 4 M
Acetic acid, CH$_3$CO$_2$H, 1 M
Oxalic acid, H$_2$C$_2$O$_4$, 1 M
Ammonium acetate, NH$_4$CH$_3$CO$_2$, 1 M
Aqueous ammonia, NH$_3$ + H$_2$O, 15 M
*Potassium hexacyanoferrate(II), K$_4$Fe(CN)$_6$, 0.1 M
Potassium cyanide, KCN, 1 M
Hydrogen peroxide, H$_2$O$_2$, 3% solution
*Magnesia mixture: Dissolve 50 g of MgCl$_2$ · 6H$_2$O and
   70 g of NH$_4$Cl in 400 mL of water. Add 100 mL of 15 M
   aqueous ammonia and dilute to 1 L. Filter.
*Aluminum wire, Al
Mercury(II) chloride, HgCl$_2$, 0.2 M
Sodium hypochlorite, NaOCl, 5% solution

Group III Reagents (in addition to those listed for Groups I and II)

Hydrochloric acid, HCl, 12 M
Hydrochloric acid, HCl, 1 M
Nitric acid, HNO$_3$, 14 M
*Dimethylglyoxime, 1% solution: Dissolve 10 g in
   1 L of alcohol
Acetic acid, CH$_3$CO$_2$H, 4 M
*Ammonium thiocyanate, NH$_4$SCN, solid
Acetone, (CH$_3$)$_2$CO
*Sodium fluoride, NaF, 1 M
*Potassium nitrite, KNO$_2$, solid
*Sodium nitrite, NaNO$_2$, 1 M
*Sodium bismuthate, NaBiO$_3$, solid
*Aluminum reagent: Dissolve 1 g of the ammonium salt
   of aurin-tricarboxylic acid in 1 L of water
Lead acetate, Pb(CH$_3$CO$_2$)$_2$, 0.1 M
Ammonium carbonate, (NH$_4$)$_2$CO$_3$, 1 M

Group IV Reagents (in addition to those listed for Groups I–III)

Ethyl alcohol, C$_2$H$_5$OH
*Ammonium oxalate, (NH$_4$)$_2$C$_2$O$_4$, 0.4 M

---

*Fill reagent bottles only about one-fourth full of starred reagents.

Reagents for Cation Analysis (continued)

Group V   Reagents (in addition to those listed for Groups I–IV)

*Ammonium sulfate, $(NH_4)_2SO_4$, 1 M

*Sodium reagent: Mix 30 g of $UO_2(CH_3CO_2)_2 \cdot 2H_2O$ with 80 g of $Zn(CH_3CO_2)_2 \cdot 2H_2O$ and 10 mL of glacial acetic acid. Dilute the solution to 250 mL, let it stand for several hours, and filter. Use the clear solution.

*Sodium hexanitrocobaltate(III), $Na_3Co(NO_2)_6$: Dissolve 30 g of $NaNO_2$ in 97 mL of water; add 3 mL of glacial CH_3CO_2H and 3.3 g of $Co(NO_3)_2 \cdot 6H_2O$. Filter and use the clear solution.

*Disodium hydrogen phosphate, Na_2HPO_4, 1 M

*Magnesium reagent, 4-(nitrophenylazo)-1-naphthol: Dissolve 0.25 g of this reagent and 2.5 g of NaOH in sufficient water to make 250 mL of solution.

Reagents for Anion Analysis (in addition to those listed for cation analysis)

*Sulfuric acid, H_2SO_4, 1.5 M

Sodium carbonate, Na_2CO_3, 1.5 M

*Manganese(II) chloride, $MnCl_2$: Saturate 12 M HCl with $MnCl_2$.

*Potassium hexacyanoferrate(III), $K_3Fe(CN)_6$, a freshly prepared saturated solution

*Iron(III) chloride, $FeCl_3$, 0.1 M

Calcium chloride, $CaCl_2$, 0.1 M

Barium chloride, $BaCl_2$, 0.1 M

*Cobalt(II) nitrate, $Co(NO_3)_2$, 1 M

*Silver nitrate, $AgNO_3$, 0.1 M

*Bromine water, $Br_2 + H_2O$, saturated solution

*Barium hydroxide, $Ba(OH)_2$, saturated solution

*Iron(II) sulfate, $FeSO_4$, 0.1 M

*Sulfuric acid, H_2SO_4, 18 M

*Silver sulfate, Ag_2SO_4, solid (nitrate-free)

*Potassium permanganate, $KMnO_4$, 0.002 M

*Silica, SiO_2, powdered

*Methyl alcohol, CH_3OH

*Iron(III) nitrate, $Fe(NO_3)_3$, 0.1 M

*Carbon tetrachloride, CCl_4, or dichloromethane, CH_2Cl_2

*Potassium permanganate, $KMnO_4$, 0.1 M

*Lead acetate paper, filter paper moist with 0.1 M $Pb(CH_3CO_2)_2$

*Fill reagent bottles only about one-fourth full of starred reagents.

APPENDIX O   PREPARATION OF SOLUTIONS OF CATIONS

Stock Solutions

Stock solutions should contain the cations in question at a concentration of 50 mg per milliliter. These stock solutions may be prepared by grinding to a powder the weight of salt given below and adding enough water (or acid if specified) to make the volume 100 mL. Dissolution of the salts may be hastened by heating.

Known and Unknown Solutions

To prepare known or unknown solutions of cations, mix 20 mL of the stock solutions (40 mL of $AsCl_3$) of the cations desired and dilute the solution to 100 mL with water. This solution will contain 10 mg of cations per milliliter. This procedure allows for a maximum of five cations at a concentration of 10 mg of cations per milliliter. Dilution to 200 mL will allow a maximum of ten cations at a concentration of 5 mg per milliliter. Give each student about 1 mL of solution.

Stock Solutions of Cations (50 mg of cations per milliliter)

Group	Ion	Formula of Salt	Grams per 100 mL of Solution
I	Ag^+	$AgNO_3$	8.0
	Pb^{2+}	$Pb(NO_3)_2$	8.0
	Hg_2^{2+}	$Hg_2(NO_3)_2$	7.0 (dissolve in 0.6 M HNO_3)

Stock Solutions of Cations (50 mg of cations per milliliter) (continued)

Group	Ion	Formula of Salt	Grams per 100 mL of Solution
II	Pb^{2+}	$Pb(NO_3)_2$	8.0
	Bi^{3+}	$Bi(NO_3)_3 \cdot 5H_2O$	11.5 (dissolve in 3 M HNO$_3$)
	Cu^{2+}	$Cu(NO_3)_2 \cdot 3H_2O$	19.0
	Cd^{2+}	$Cd(NO_3)_2 \cdot 4H_2O$	13.8
	Hg^{2+}	$HgCl_2$	6.8
	As^{3+}	As_4O_6	3.3 (heat in 50 ml of 12 M HCl, then add 50 mL of water)
	Sb^{3+}	$SbCl_3$	9.5 (dissolve in 6 M HCl and dilute with 2 M HCl)
	Sn^{2+}	$SnCl_2 \cdot 2H_2O$	9.5 (dissolve in 50 mL of 12 M HCl. Dilute to 100 mL with water. Add a piece of tin metal)
	Sn^{4+}	$SnCl_4 \cdot 3H_2O$	13.3 (dissolve in 6 M HCl)
III	Co^{2+}	$Co(NO_3)_2 \cdot 6H_2O$	24.7
	Ni^{2+}	$Ni(NO_3)_2 \cdot 6H_2O$	24.8
	Mn^{2+}	$Mn(NO_3)_2 \cdot 6H_2O$	26.2
	Fe^{3+}	$Fe(NO_3)_3 \cdot 9H_2O$	36.2
	Al^{3+}	$Al(NO_3)_3 \cdot 9H_2O$	69.5
	Cr^{3+}	$Cr(NO_3)_3$	23.0
	Zn^{2+}	$Zn(NO_3)_2$	14.5
IV	Ba^{2+}	$BaCl_2 \cdot 2H_2O$	8.9
	Sr^{2+}	$Sr(NO_3)_2$	12.0
	Ca^{2+}	$Ca(NO_3)_2 \cdot 4H_2O$	29.5
V	Mg^{2+}	$Mg(NO_3)_2 \cdot 6H_2O$	52.8
	NH_4^+	NH_4NO_3	22.2
	Na^+	$NaNO_3$	18.5
	K^+	KNO_3	13.0

APPENDIX P   LABORATORY ASSIGNMENTS

The following list suggests a set of laboratory assignments.

1. Construct a wash bottle (if necessary), stirring rods, and capillary syringes.
2. Analyze a known solution containing all the cations of Group I.
3. Analyze an unknown solution containing Group I cations.
4. Analyze a known solution containing all the cations of Group II.
5. Analyze an unknown solution containing Group II cations.
6. Analyze a known solution containing all the cations of Group III.
7. Analyze an unknown solution containing Group III cations.
8. Analyze a known solution containing all the cations of Group IV.
9. Analyze a known solution containing all the cations of Group V.
10. Analyze an unknown solution containing cations from Groups IV and V.
11. Analyze a general unknown solution containing cations of all the groups.
12. Analyze a known salt mixture containing no oxidizing anions.
13. Analyze an unknown salt mixture containing no oxidizing anions.
14. Analyze a known salt mixture containing no reducing anions.
15. Analyze an unknown salt mixture containing no reducing anions.
16. Analyze a salt mixture for both cations and anions.
17. Analyze an alloy.

INDEX

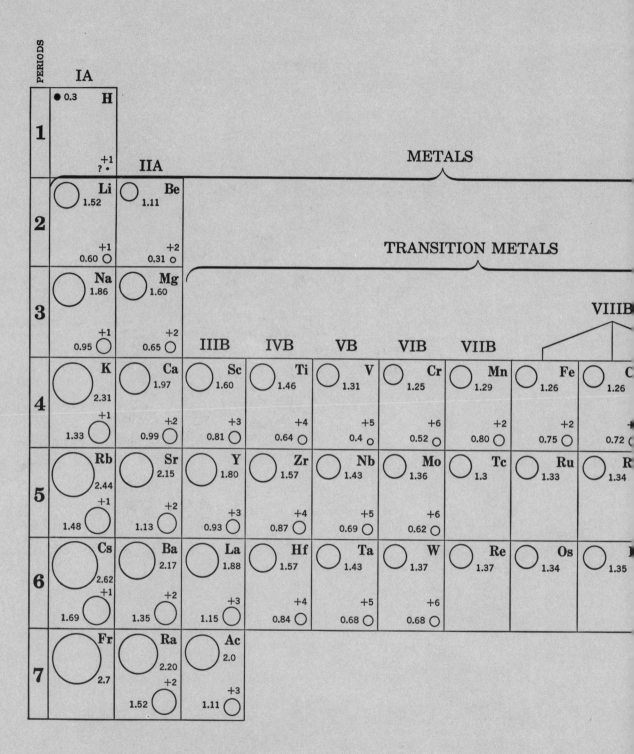